SAFETY AND HEALTH IN AGRICULTURE, FORESTRY, AND FISHERIES

EDITED BY

Ricky L. Langley, MD, MPH
Robert L. McLymore, Sr., EdD
William J. Meggs, MD, PhD
Gary T. Roberson, PhD, PE

**GOVERNMENT INSTITUTES
ROCKVILLE, MD**

Government Institutes, Inc., 4 Research Place, Rockville, Maryland 20850, USA.

Copyright © 1997 by Government Institutes, Inc.
All rights reserved.

Copyright does not extend to contributions prepared by George A. Conway, Richard D. Kennedy, Jennifer M. Lincoln, and Jan C. Manwaring. The United States Copyright Act provides that government employees cannot copyright material prepared as part of their official duties as government employees and have no right to assign or transfer copyright priviledges.

00 99 98 97   4 3 2 1

No part of this work may be reproduced or transmitted in any form or by any means, electronic or mechanical, including photocopying, recording, or any information storage and retrieval system, without permission in writing from the publisher. All requests for permission to reproduce material from this work should be directed to Government Institutes, Inc., 4 Research Place, Rockville, Maryland 20850, USA.

The information presented in this book is based on the best information available to the authors and publisher at this time. As science and technology improve, better solutions to workplace hazards and treatments for injuries and illnesses will become available. Readers should consult a broad range of sources of information before implementing any prevention or treatment plan. Neither the authors, editors, nor the publisher can be held responsible for specific or implied use of material contained herein.

Printed in the United States of America.

Library of Congress Cataloging-in-Publication Data
    Safety and health in agriculture, forestry, and fisheries
/ Ricky L. Langley . . . [et al.] editors.
    p.   cm.
Includes bibliographical references and index.
ISBN: 0-86587-552-9
    1. Agricultural laborers--Health and hygiene 2. Foresters--Health and hygiene
3. Fisheries--Health aspects 4. Agricultural laborers--Safety measures
5. Foresters--Safety measures 6. Fisheries--Safety measures I. Langley, Ricky L.
RC 965.A5S24 1997
363.11'963--dc21

96-46442
CIP

Dedication and Acknowledgments

This book is dedicated to those men and women who work or have worked in the environments of agriculture, forestry, and fisheries to feed, cloth, and house humanity.

The editors wish to acknowledge the efforts and time of the contributing authors. Without their efforts, a work of this magnitude would not have been a reality.

Summary of Contents

Foreword xix
   Dennis J. Murphy, Ph.D., C.S.P.

Foreword xxi
   Lewis Goldfrank, M.D.

Preface xxiii

Contributors xxv

About the Editors xxix

SECTION 1. General Issues Impacting Safety and Health in Agriculture, Forestry, and Fisheries

Chapter 1. Overview of Safety and Health in Agriculture, Forestry, and Fisheries 3
   William J. Meggs, M.D., Ph.D., F.A.C.E.P., Gary T. Roberson, Ph.D., Ricky L. Langley, M.D., M.P.H., and Robert L. McLymore, Sr., Ed.D.

Chapter 2. Ergonomics in Agriculture, Forestry, and Fisheries 9
   Tamara M. James, M.A., C.I.E., C.P.E.

Chapter 3. Occupational Vibration Exposure in Agriculture, Forestry, and Fisheries 53
   David Siebens, M.D., M.P.H.

Chapter 4. Noise and Hearing Loss in Agriculture, Forestry, and Fisheries 59
   David Siebens, M.D., M.P.H.

Chapter 5. Injuries Associated with Cold and Hot Environments 67
   Bruce A. Cohen, M.D., L.C.D.R., M.C., U.S.N.

Chapter 6. Electrical Safety in Agriculture, Forestry, and Fisheries 77
   Robert L. McLymore, Sr., Ed.D.

Chapter 7. Bite, Sting, and Envenomization Hazards in Agriculture and Forestry 89
   Charles S. Apperson, Ph.D., and Michael G. Waldvogel, Ph.D.

Chapter 8. Zoonotic Hazards in Humans in Agriculture, Forestry, and Fisheries 127
   Ricky L. Langley, M.D., M.P.H., and W.E. Morgan Morrow, B.V.Sc., M.S., Ph.D.

Chapter 9. Cancer in Farmers, Foresters, and Fishermen 143
   Paul A. James, M.D., and Stanley Schuman, M.D., Ph.D.

Chapter 10. Mental Health Issues in Agriculture, Forestry, and Fisheries 153
Patricia J. Gammon, Ph.D., and Laura Anibal, B.A.

Chapter 11. Disability in Agriculture, Forestry, Fisheries, and Migrant Workers 171
Barry L. Delks, M.S.

Chapter 12. Personal Protective Equipment in Agriculture, Forestry, and Fisheries 181
Timothy G. Prather, M.S.

Chapter 13. Safety and Health Programs in Agriculture, Forestry, and Fisheries 197
Gary M. Tencer, C.I.H.

SECTION 2. Safety and Health in Agriculture and Animal Production

Chapter 14. Epidemiology of Agricultural Injuries and Illnesses 215
Mark A. Purschwitz, Ph.D.

Chapter 15. Hazard Management and Safety with Agricultural Machines 233
Gary T. Roberson, Ph.D., P.E.

Chapter 16. Chemical Hazards of Farming 249
William J. Meggs, M.D., Ph.D., F.A.C.E.P., and Ricky L. Langley, M.D., M.P.H.

Chapter 17. Protecting Drinking Water Supplies 267
Gregory D. Jennings, Ph.D., Dorothy L. Miner, M.S., and Deanna L. Osmond, Ph.D.

Chapter 18. Grain Industry Health and Safety Issues 273
Amy Gitelman, M.P.H.

Chapter 19. Safety Around Beef Cattle 291
James B. Cowan, D.V.M., M.S.P.H.

Chapter 20. Safety Around Horses 315
W.E. Morgan Morrow, B.V.Sc., M.S., Ph.D., and Robert A. Mowrey, Ph.D.

Chapter 21. Skin Diseases in Farmers 321
William A. Burke, M.D.

Chapter 22. Respiratory Diseases Related to Work in Agriculture 353
Susanna G. Von Essen, M.D,. and Kelley J. Donham, D.V.M.

Chapter 23. Health and Safety of Migrant and Seasonal Farmworkers and Their Families ... 385
Judy Hayes Bernhardt, R.N., Ph.D., M.P.H.

Chapter 24. Safety on the Farmstead 397
Wilma S. Hammett, Ph.D., and Judieth E. Mock, Ed.D.

Chapter 25. Indoor Environmental Hazards in Animal Housing 419
Robert W. Bottcher, Ph.D., and Roberto D. Munilla, M.E.

Chapter 26. Industrial Hygiene Evaluations in Agriculture 439
William A. Popendorf, Ph.D., C.I.H., and Stephen I. Reynolds, Ph.D., C.I.H.

Chapter 27. Occupational Safety and Health Regulations in Agriculture 469
Regina C. Luginbuhl, M.S.

Chapter 28. EPA Worker Protection Standard for Agricultural Pesticides 483
Kay Glenn Harris, B.S.

SECTION 3. Safety and Health in Forestry

Chapter 29. Epidemiology of Forestry Injuries and Illnesses 495
R. Lane Tippens, M.D., M.P.H., and Ricky Langley, M.D., M.P.H.

Chapter 30. Timber Harvesting Safety 509
Thomas L. Bean, Ed.D,. and Linda K. Isaacs, M.S.

Chapter 31. Wildland Fires and Firefighting 521
Ricky L. Langley, M.D., M.P.H.

Chapter 32. Epidemiology and Prevention of Helicopter Logging Injuries 529
George A. Conway, M.D., M.P.H., and Jan C. Manwaring, B.S.

Chapter 33. Logging and the Law: OSHA Rules and Regulations for Forestry Workers 545
George E. Jeter, A.B.

SECTION 4. Safety and Health in Fisheries and Aquaculture

Chapter 34. Epidemiology of Fatal Injury in the U.S. Commercial Fishing Industry 557
Richard D. Kennedy, M.S., and Jennifer M. Lincoln, B.S.

Chapter 35. Safety at Sea 571
Dewayne Hollin, B.B.A., M.B.A.

Chapter 36. Drowning and Cold Water Survival 605
Thomas B. Faulkner, M.D., M.H.A.

Chapter 37. Diving Hazards 617
Edward D. Thalmann, M.D.

Chapter 38. Motion Sickness 643
David P. Thomson, M.S., M.D.

Chapter 39. Aquacultural Hazards 659
Robert M. Durborow, Ph.D.

Chapter 40. Skin Diseases in Fishermen 681
William A. Burke, M.D.

Chapter 41. Legal Concerns in the Fishing Industry 713
  Dennis W. Nixon, J.D., M.M.A., A.B.

Chapter 42. Federal Regulation of Fishing Vessels 721
  William J. Meggs, M.D., Ph.D., F.A.C.E.P.

APPENDICES

Appendix A. Agricultural Safety Checklist 727

Appendix B. Forestry Safety Checklist 733

Appendix C. Commercial Fishing Vessel Checklist 735

INDEX 737

Table of Contents

Foreword xix
Dennis J. Murphy, Ph.D., C.S.P.

Foreword xxi
Lewis Goldfrank, M.D.

Preface xxiii

Contributors xxv

About the Editors xxix

SECTION 1. General Issues Impacting Safety and Health in Agriculture, Forestry, and Fisheries

Chapter 1. Overview of Safety and Health in Agriculture, Forestry, and Fisheries 3

    Introduction, 3
    Epidemiology, 3
    Medical Illnesses in Farmers, Foresters, and Fishermen, 4
    Prevention and Regulation, 7

Chapter 2. Ergonomics in Agriculture, Forestry, and Fisheries 9

    Introduction, 9
    What Are CTDs?, 10
    CTD Risk Factors, 10
    Hand Tool Design, 16
    NIOSH Lifting Guidelines, 22
    Thermal Extremes, 26
    Developing an Ergonomics Program, 27
    Summary, 28
    References, 28
    Appendix A: Common Arm Cumulative Trauma Disorders, 30
    Appendix B: Sample Discomfort Survey, 31
    Appendix C: Ergonomic Checklist to Enhance Performance, Safety, and Comfort, 32
    Appendix D: Ergonomic Checklist for Tractors and Agricultural Machinery, 38
    Appendix E: Industrial and Agricultural Systems Standards, 41
    Appendix F: Guidelines for the Selection of Manual Material Handling Equipment, 42
    Appendix G: A Basic Program for Controlling Heat Stress, 44
    Appendix H: Draft OSHA Ergonomic Protection Risk Factor Checklists, 45

Chapter 3. Occupational Vibration Exposure in Agriculture, Forestry, and Fisheries 53

    Introduction, 53
    Physical Principles, 53
    Vibration Measurement and Standards, 54
    Whole Body Vibration, 54
    Hand-Arm Vibration, 55

Summary, 57
References, 58

Chapter 4. Noise and Hearing Loss in Agriculture, Forestry, and Fisheries 59

Introduction, 59
Physical Principles, 59
Normal Hearing, 61
Noise-Induced Hearing Loss, 62
Audiometry, 62
Regulatory Issues, 63
Prevention of Hearing Loss, 64
Summary, 65
References, 65

Chapter 5. Injuries Associated with Cold and Hot Environments 67

Introduction, 67
Cold Stress, 67
Heat Stress, 70
Prevention of Heat Stress, 73
Summary, 74
References, 75

Chapter 6. Electrical Safety in Agriculture, Forestry, and Fisheries 77

Introduction, 77
Electrical Systems' General Requirements: OSHA Standard 1910.303, 77
Electrocution Hazards on the Farm, 78
Removing Electrical Hazards, 80
Protecting Yourself and the Electrical System, 82
Summary, 86
References, 87

Chapter 7. Bite, Sting, and Envenomization Hazards in Agriculture and Forestry 89

Introduction, 89
What Are Arthropods?, 90
Epidemiology of Envenomization, 92
The Nature of Arthropod Venoms, 92
Blood Feeding, 94
Identifying the Arthropod(s) Based on the Injury, 94
Pest Arthropods, 96
Prevention of Stings and Bites, and Control of Arthropods of Public Health Significance, 120
Summary, 121
Glossary, 121
References, 122

Chapter 8. Zoonotic Hazards in Humans in Agriculture, Forestry, and Fisheries 127

Introduction, 127
Transmission and Infectivity of Disease, 127
Prevention and Control of Zoonotic Hazards, 128
Diagnosis and Treatment, 130

Summary, 132
References, 132
Appendix A: Tables of Occupational Infections, 134

Chapter 9. Cancer in Farmers, Foresters, and Fishermen 143

Introduction, 143
Causes of Cancer: An Overview, 143
Occupational Factors That May Be Linked to Cancer, 144
Regulatory Issues, 146
Approach to Hazard Education, 147
Summary, 147
References, 147

Chapter 10. Mental Health Issues in Agriculture, Forestry, and Fisheries 153

Introduction, 153
Stressors in Farming, Forestry Work, and the Fishing Industry, 153
Types of Mental Health Disorders, 154
The Nature of Mental Health Problems in Workers in Agriculture, Forestry, and Fisheries, 158
Interventions for Mental Health and Substance Abuse Problems for Workers in the Agriculture,
    Forestry, and Fisheries Industries, 167
Summary, 167
References, 169

Chapter 11. Disability in Agriculture, Forestry, Fisheries, and Migrant Workers 171

Introduction, 171
The Value of Agricultural Workers with Disabilities, 171
The Hazardous Nature of Agriculture, 172
The Needs of Agricultural Workers with Disabilities, 173
Availability of Resources, 176
Summary, 178
References, 178

Chapter 12. Personal Protective Equipment in Agriculture, Forestry, and Fisheries 181

Respiratory Protection Equipment, 181
Foot Protection, 189
Body Coverings, 190
Hand Protection, 192
Head Protection, 193
Summary, 195
References, 195

Chapter 13. Safety and Health Programs in Agriculture, Forestry, and Fisheries 197

Introduction, 197
Management Responsibilities, 197
Safety Policies and Procedures, 198
Regulations and Standards, 202
Functions of a Health and Safety Program, 205
Summary, 211
References, 211

SECTION 2. Safety and Health in Agriculture and Animal Production

Chapter 14. Epidemiology of Agricultural Injuries and Illnesses 215

    Introduction, 215
    Fatal Traumatic Injuries, 215
    Nonfatal Traumatic Injuries, 218
    Illnesses and Chronic Conditions, 220
    Summary, 227
    References, 227

Chapter 15. Hazard Management and Safety with Agricultural Machines 233

    Introduction, 233
    Machinery Safety Concepts, 233
    Persons at Risk, 234
    Hazard Recognition and Hazard Types, 234
    Management Strategies and Prevention Techniques, 243
    Summary, 246
    References, 247

Chapter 16. Chemical Hazards of Farming 249

    Introduction, 249
    Pesticides, 249
    Insecticides, 250
    Herbicides, 255
    Fungicides, 256
    Rodenticides, 257
    Federal Regulations of Pesticides, 257
    Fertilizers, 258
    Chemical Hazards Associated with Farm Equipment, 258
    Chemical Hazards of Animal Confinement, 259
    Chemical Hazards of Silos, 259
    Chemical Hazards of Animal Feed, Additives, and Disinfectants, 260
    Chemical Hazards of Mycotoxins, 260
    Chemical Hazards Associated with Crops, 261
    Prevention of Chemical Injuries in Agriculture, 261
    Summary, 261
    References, 261

Chapter 17. Protecting Drinking Water Supplies 267

    Introduction, 267
    The Condition of the Well, 268
    Summary, 272
    References, 272

Chapter 18. Grain Industry Health and Safety Issues 273

    Introduction, 273
    Grain Dust Characterization, 275
    Injury Statistics, 275
    Health Hazards, 276

Safety Hazards, 279
Safe Work Practices, 280
Rules and Regulations, 282
Summary, 283
References, 283
Appendix A: NIOSH Checklists Related to Grain Industry, 285

Chapter 19. Safety Around Beef Cattle 291

Introduction, 291
Principles of Injury Control, 291
Beef Cattle Psychology, 292
Handling Beef Cattle Safely, 294
Things to Keep in Mind When Handling Cattle, 295
Components of Beef Cattle Handling Facilities, 303
Summary, 312
References, 312

Chapter 20. Safety Around Horses 315

Introduction, 315
Accident Awareness on the Farm, 315
Stable Safety, 316
Fire Safety, 316
Choosing Your Horse, 316
Approaching a Horse, 317
Handling the Horse, 317
Leading the Horse, 317
Tying the Halter, 317
Saddling Up, 317
Mounting and Dismounting, 318
Clothing to Wear, 318
The Helmet, 318
Riding, Night Riding, and Trail Riding, 318
After the Fall, 318
Hauling Your Horse, Trailering, 319
Summary, 319
References and Resources, 319

Chapter 21. Skin Diseases in Farmers 321

Introduction, 321
Skin Diseases Due to Sun Exposure, 321
Exposure to Environmental Extremes, 325
Dermatoses Related to Exposure to Farm Chemicals, 326
Dermatoses Related to Plant Exposure on the Farm, 327
Arthropod Bites and Stings on the Farm, 333
Reptile and Animal Bites on the Farm, 339
Infections and Zoonoses in Farm Workers, 341
Hazard Prevention, 347
Summary, 348
References, 347

Chapter 22. Respiratory Diseases Related to Work in Agriculture ... 353

Introduction, 353
Organic Dust Exposures, 355
Airway Diseases, 363
Toxic Gas Exposures, 368
Summary, 371
References, 372

Chapter 23. Health and Safety of Migrant and Seasonal Farmworkers and Their Families ... 385

Introduction, 385
Work Environment, 386
Socioeconomic Conditions, 390
Health and Health Care Utilization, 391
Summary, 395
References, 395

Chapter 24. Safety on the Farmstead ... 397

Introduction, 397
Creating Safe Farmstead Environments for Children, 397
Safety for the Elderly, 403
Hazardous Household Products, 408
Pesticide Residues and the Farmstead, 412
Food Safety, 414
The Farm Pond, 415
Firearm Safety, 415
Summary, 416
References, 416

Chapter 25. Indoor Environmental Hazards in Animal Housing ... 419

Introduction, 419
Ventilation and Environmental Control, 421
Air Pollutants in Animal Housing, 429
Mechanical Hazards, 432
Electrical Hazards, 432
Noise, 433
Fire, 434
Children in Buildings, 435
Safety Signs, 435
Summary, 435
References, 436

Chapter 26. Industrial Hygiene Evaluations in Agriculture ... 439

Introduction, 439
Agricultural Occupational Disease, 440
Hazard Evaluation Methods, 445
Sample Collection and Analysis Methods, 452
Settings with Gaseous Respiratory Hazards, 456
Aerosol Respiratory Hazards, 460
Summary, 464

References, 465

Chapter 27. Occupational Safety and Health Regulations in Agriculture 469

Introduction, 469
From the Farm to the Factory, 469
State OSH Planning, 469
Current Safety and Health Risks, 470
What Is Being Done?, 470
Recordkeeping Requirements, 470
Regulations Protecting Agricultural Workers, 472
Chemical Hazards, 475
Dangerous Gases, 476
Farm Machinery, 477
Child Labor, 478
Summary, 480
References, 480

Chapter 28. EPA Worker Protection Standard for Agricultural Pesticides 483

Introduction, 483
The Agricultural Employer's Responsibilities, 483
Restrictions During Applications or Workers, 484
Handler Application Restrictions, 485
Specific Instructions for Handlers, 485
Summary of Provisions: General Scope and Applicability, 485
Protections for All Employees, 486
Protections for Workers, 487
Protections for Handlers, 488
Requests for Exception to REI, 489
Worker Protection Standard Questions and Answers, 489
Summary, 491
References, 491

SECTION 3. Safety and Health in Forestry

Chapter 29. Epidemiology of Forestry Injuries and Illnesses 495

Introduction, 495
Injury Analysis, 495
Acute Trauma, 496
Chronic Injury, 497
Physical Workload, 498
Noise and Exhaust, 499
Environmental Hazards, 499
Infectious Diseases, 499
Occupational Dermatitis, 501
Pesticide Exposure, 502
Occupational Respiratory Disease, 502
Psycho-Social Issues, 503
Protective Equipment, Safe Operations, and Preventive Health, 503
Summary, 505
References, 505

Chapter 30. Timber Harvesting Safety 509

Introduction, 509
Who Works in the Logging Industry?, 509
The Timber Harvesting System: Man and Machine Working Together, 509
How Are Loggers Injured or Killed?, 513
Hazards Involved in Logging, 513
Developing a Formal Safety Program, 517
Summary, 519
References, 519

Chapter 31. Wildland Fires and Firefighting 521

Introduction, 521
Epidemiology, 521
Factors Associated with Injuries and Illnesses Among Firefighters, 522
Fire Behavior, 524
Safety and Wildland Fire Suppression, 524
Fire Prevention, 525
Summary, 526
References, 526

Chapter 32. Epidemiology and Prevention of Helicopter Logging Injuries 529

Introduction, 529
Alaska Investigative Findings, 530
Overview of Hazards of Helicopters and Sling-Load Logging Operations, 534
U.S. Heli-Logging Experience, 537
Summary, 542
References, 543

Chapter 33. Logging and the Law: OSHA Rules and Regulations for Forestry Workers 545

Introduction, 545
Creation of OSHA Standards, 545
OSHA Enforcement, 546
General Requirements: From PPE to Explosives, 547
Summary, 554
Sources of Additional Information, 554

SECTION 4. Safety and Health in Fisheries and Aquaculture

Chapter 34. Epidemiology of Fatal Injury in the U.S. Commercial Fishing Industry 557

Introduction, 557
Recent Background, 557
Fishermen, Fishing Vessels, and Fisheries, 559
Population-At-Risk, 560
Capturing Fatal and Nonfatal Injuries, 561
Fatality Statistics, 562
Fatality Rates, 567
Fishery-Specific Fatality Rates, 568
Summary, 569

References, 569

Chapter 35. Safety at Sea ... 571

Introduction, 571
Save Lives, Save Losses, 571
Safety: A Common Problem, 572
Toward a Safer Vessel, 572
Reporting and Investigating Accidents, 587
Survival at Sea, 597
Survival Techniques and Equipment, 598
Other Safety Publications and Procedures, 603
Summary, 604
References, 604

Chapter 36. Drowning and Cold Water Survival ... 605

Drowning, 605
Cold Water Immersion, 606
Signs and Symptoms of Heat Loss (Hypothermia), 607
Drown-Proofing, 611
Treading Water, 612
Rescue Devices and Procedures, 613
Summary, 615
References, 615

Chapter 37. Diving Hazards ... 617

Introduction, 617
The Diving Environment, 618
Diving Practice, 622
Diving-Related Diseases, 628
Health Monitoring, 633
Regulations, 636
Medical Coverage of Diving Operations, 637
Available Resources, 638
Summary, 638
References, 639

Chapter 38. Motion Sickness ... 643

Introduction, 643
Anatomy and Physiology, 644
Etiology, 645
Treatment, 647
Summary, 654
References, 654

Chapter 39. Aquacultural Hazards ... 659

Introduction, 659
Health Issues, 659
Safety Issues, 668
Summary, 674
References, 674

Chapter 40. Skin Diseases in Fishermen 681

Introduction, 681
Skin Diseases in Fishermen, 681
Hazard Prevention, 707
Summary, 708
References, 708

Chapter 41. Legal Concerns in the Fishing Industry 713

Introduction, 713
Occupational Hazards, 713
Regulatory Framework, 716
Legal Remedies for Injured Fishermen, 717
Summary, 720
References, 720

Chapter 42. Federal Regulation of Fishing Vessels 721

Introduction, 721
Requirements for Commercial Fishing Industry Vessels, 721
Summary, 723
References, 724

APPENDICES

Appendix A. Agricultural Safety Checklist 727

Appendix B. Forestry Safety Checklist 733

Appendix C. Commercial Fishing Vessel Checklist 735

INDEX 737

Foreword

One cannot help but be impressed and inspired by the enormity of the task taken by Drs. Langley, McLymore, Meggs, and Roberson. The range of subjects that are addressed in *Safety and Health in Agriculture, Forestry, and Fisheries* and the depth in which many topics are discussed, make this book an enormously valuable contribution to the field of safety and health.

For the first time ever, professionals and students alike are able to pick up one book and find an in-depth treatment of hazards and injury prevention practices that heretofore have, for the most part, only appeared as independent articles in a wide range of research journals and proceedings of conferences and symposia.

This book takes a holistic approach to the many chemical, mechanical, biological, thermal, and environmental hazards encountered by those working in and around production agriculture, forestry, and fisheries. Common characteristics that bind these workers include: workers noted for their independence, ruggedness, and risk taking; work tasks that are often done in relative isolation from communities and emergency service providers; considerable exposure to adverse weather elements; and regular interaction with a wide variety of potentially hazardous tools, machines, and environments.

It is valuable, then, that the editors have chosen to organize Section 1 around several hazard and injury prevention topics that illustrate the commonality of safety and health problems faced by these workers. This approach should help to unify and support the many disparate professional groups (epidemiologists, medical and safety specialists, etc.) that are currently working in the agriculture, forestry, and fisheries industries.

Sections 2, 3, and 4 add detailed discussion on many topics specific to production agriculture, forestry, and fisheries, respectively. These chapters contain some of the most extensive compilations of research-based facts to support topic discussions as I have ever seen. It appears that chapter authors have approached their topics from as many different angles as possible. The result is a robust treatment of the subject matter.

Every professional who is concerned with occupational safety and health in agriculture, forestry, and fisheries should make this book one of the most frequently-consulted references in their personal libraries.

Dennis J. Murphy, Ph.D., C.S.P.
Professor
Department of Agricultural & Biological Engineering
Penn State University
Editor, *Journal of Agricultural Safety and Health*
Author, *Safety and Health for Production Agriculture*

Foreword

Remarkable progress has been made in the development of the diverse aspects of the discipline of occupational and preventive medicine. *Safety and Health in Agriculture, Forestry, and Fisheries* begins a critical analysis of the enormous subpopulation of workers in the agricultural, forestry, and fisheries industries. The recognition of the importance of the safety and health of these workers as well as an evaluation of their occupational and environmental health problems is timely.

The editors have developed analogous approaches for these three great industries. Although the analysis of the agricultural industry is broadest, the reviews of the logging and fishing industries are quite complete. The editors and individual authors define the industries in the broadest terms. Their review of epidemiology, occupational safety, disability, federal regulations and legal responsibility is extensive.

The authors review the agricultural industry from the point of view of: ergonomics, thermal stress, noise and vibration exposure, oncologic risk, zoonotic hazards, arthropod exposures, cutaneous disorders, respiratory diseases, and other diverse disorders relating to chemical exposure, machinery risk, and ultimately the principles of occupational safety.

The authors approach the epidemiology of agriculture, forestry, and fishery injuries and illnesses in great detail. There is special attention to the complex needs of immigrant workers and their health and safety. These sophisticated approaches to each industry set the standard for a review of occupational disease and injury—How, where, and why disorders occur and how they can be prevented.

All of us who care for workers in these industries will better be able to understand the environmental and occupational diseases and exposures these individuals face because of this text's depth and breadth. Those clinicians who devote their lives to the care of workers can use this text as a model in the study of the health and safety of their workers from any industry.

Lewis Goldfrank, M.D.
Bellevue Hospital Center
Author, *Goldfrank's Toxicological Emergencies*

Preface

It is our firm belief that knowledge is the key to safety and health in any field. Those who manage and harvest the earth's bounty, be it on farms, in the forests, or on the open waters, have not in the past received a fair share of efforts to increase knowledge about their occupational health and safety. This book was written to provide a comprehensive source of information on health and safety on farms, in forests, and on the waters. Agriculture, forestry, and fishing represent a group of related industries with similar environments and problems. Recognition of the hazards associated with these industries has been increasing in recent years, and the time for a comprehensive text is long overdue.

The work environment in these industries is constantly evolving. Small farms are vanishing in favor of large mechanized operations. Hazards change as new chemicals and machines are introduced. New knowledge about environmental problems in these industries can lead to operational changes. This work is intended to provide the reader with the current state of health principles and safety technology in these industries.

There are good reasons to combine agriculture, forestry, and fisheries in one book on health and safety. These fields are so interrelated that the standard industrial classification code of the United States government considers these fields as one industry, with subdivisions:

a) Agricultural production—crops
b) Agricultural production—livestock and animal specialties
c) Forestry
d) Fishing, hunting, and trapping.

The scope of this work is revealed by the wide variety of specialists who have contributed their expertise to make this book a reality. Agricultural engineers, medical toxicologists, occupational physicians, psychologists, emergency physicians, internists, attorneys, agricultural extension professionals, governmental regulators, ergonomists, agribusinessmen, veterinarians, entomologists, animal scientists, farm safety specialists, home economists, industrial hygienists, dermatologists, and marine fishery experts are all represented among the authors of various chapters.

This book was written for a broad audience. Farmers, forestry workers, fishermen, and those charged with safety and health in these fields should all find valuable and practical material in these pages. Agricultural extension experts and governmental regulators, as well as medical personnel, rescue workers, advanced students in agriculture, nurses, agricultural engineers, insurance adjusters, and attorneys with involvement in health and safety issues in these industries, should find this text useful.

This book is unique in its comprehensiveness and scope. It includes new paradigms for addressing health and safety. Hopefully this book will change the way safety and health are approached and help people to make better decisions in these three environments. It is the editors' fervent hope that in some way this book will lead to fewer illnesses, injuries, and deaths among those who provide us with earth's bounty.

Ricky L. Langley, M.D., M.P.H.
Robert L. McLymore, Sr., Ed.D.
William J. Meggs, M.D., Ph.D.
Gary T. Roberson, Ph.D., P.E.

Contributors

Laura K. Anibal, B.A.

Research Assistant
Carlow International Inc.
Falls Church, Virginia

Charles S. Apperson, Ph.D.

Professor and Extension Specialist
Department of Entomology
North Carolina State University
Raleigh, North Carolina

Thomas L. Bean, Ed.D.

Associate Professor and Safety Leader
Department of Food, Agriculture, and Biological Engineering
The Ohio State University
Columbus, Ohio

Judy Bernhardt, R.N., M.P.H., Ph.D.

Professor and Chair
Department of Community and Mental Health Nursing and Nursing Services Administration
East Carolina University
Greenville, North Carolina

Robert W. Bottcher, Ph.D.

Professor
Department of Biological and Agricultural Engineering
North Carolina State University
Raleigh, North Carolina

William A. Burke, M.D.

Associate Professor and Chief
Division of Dermatology
Department of Medicine
East Carolina University
Greenville, North Carolina

Bruce A. Cohen, M.D., L.C.D.R., M.C., U.S.N.

Fellow, Division of Occupational and Environmental Medicine
Department of Family and Community Medicine
Duke University
Durham, North Carolina

George A. Conway, M.D., M.P.H

Chief
Division of Safety Research, Alaska Field Station
National Institute for Occupational Safety and Health
Centers for Disease Control and Prevention
Anchorage, Alaska

James B. Cowan, D.V.M., M.S.P.H.

North Carolina Cooperative Extension Service Consultant
Public Health Officer, United States Air Force
Sheppard Air Force Base, Texas

Barry L. Delks, M.S.

Director, Breaking New Ground Resource Center
Department of Agricultural Engineering
Purdue University
West Lafayette, Indiana

Kelly J. Donham, D.V.M., M.S.

Professor and Associate Director for Agricultural Medicine
Department of Preventive Medicine and Environmental Health
University of Iowa
Iowa City, Iowa

Robert M. Duborow, Ph.D.

State Extension Specialist for Aquaculture
Cooperative Extension Program
Kentucky State University
Frankfort, Kentucky

Thomas B. Faulkner, M.D. M.H.A.

Clinical Instructor
Department of Family Medicine
Emory University School of Medicine
Clinical Faculty, Department of Occupational Health
Rollins School of Public Health, Emory University
Atlanta, Georgia

Patricia J. Gammon, Ph.D.

Clinical Psychologist
Division of Occupational and Environmental Medicine
Department of Community and Family Medicine
Duke University
Durham, North Carolina

Amy Gitelman, M.P.H.

Industrial Hygienist
Division of Occupational and Environmental Medicine
Department of Community and Family Medicine
Duke University
Durham, North Carolina

Wilma S. Hammett, Ph.D.

Associate Professor and Extension Home Furnishings Specialist
Extension Family and Consumer Sciences
North Carolina State University
Raleigh, North Carolina

Kay G. Harris, B.S.

Worker Protection Specialist
North Carolina Department of Agriculture
Raleigh, North Carolina

Dewayne Hollin, B.B.A., M.B.A.

Marine Business Management Specialist
Sea Grant College Program, Marine Advisory Service
Texas A&M University
College Station, Texas

Linda K. Isaacs, M.S.

Extension Associate
Department of Food, Agriculture, and Biological Engineering
The Ohio State University
Columbus, Ohio

Paul A. James, M.D.

Director
SUNY, Office of Rural Health
State University of New York at Buffalo
Buffalo, New York

Tamara M. James, M.A., C.I.E., C.P.E.

Ergonomist
Occupational and Environmental Safety Office
Duke University
Durham, North Carolina

Gregory D. Jennings, Ph.D.

Associate Professor
Department of Biological and Agricultural Engineering
North Carolina State University
Raleigh, North Carolina

George E. Jeter, A.B.

Education Specialist
North Carolina Division of Occupational Safety and Health
North Carolina Department of Labor
Raleigh, North Carolina

Richard D. Kennedy, M.S.

Statistician
Division of Safety Research
National Institute of Occupational Safety and Health
Centers for Disease Control and Prevention
Morgantown, West Virginia

Ricky L. Langley, M.D., M.P.H., F.A.C.P., F.A.C.O.E.M., F.A.C.P.M.

Assistant Clinical Professor
Division of Occupational and Environmental Medicine
Department of Family and Community Medicine
Duke University
Durham, North Carolina

Jennifer L. Lincoln, B.S.

Occupational Safety and Health Specialist
Division of Safety Research, Alaska Field Station
National Institute of Occupational Safety and Health
Center for Disease Control and Prevention
Anchorage, Alaska

Regina G. Luginbuhl, M.S.

Director
Agricultural Safety and Health Division
North Carolina Department of Labor
Raleigh, North Carolina

Jan C. Manwaring, B.S.

Environmental Health and Safety Specialist
Division of Safety Research
National Institute of Occupational Safety and Health
Centers for Disease Control and Prevention
Anchorage, Alaska

Robert L. McLymore, Sr., Ed.D.

Extension Farm Safety Specialist
Department of Biological and Agricultural Engineering
North Carolina State University
Raleigh, North Carolina

William J. Meggs, M.D., Ph.D., F.A.C.E.P.

Associate Professor
Division of Clinical Toxicology
Department of Emergency Medicine
East Carolina University
Greenville, North Carolina

Dorothy L. Miner, M.S.

Extension Associate
Department of Biological and Agricultural Engineering
North Carolina State University
Raleigh, North Carolina

Judieth E. Mock, Ed.D.

Associate Professor and Department of Extension Leader
Family and Consumer Sciences
North Carolina Cooperative Extension Service
North Carolina State University
Raleigh, North Carolina

W.E. Morgan Morrow, B.V.Sc., M.S., Ph.D.

Associate Professor
Department of Animal Sciences
North Carolina State University
Raleigh, North Carolina

Robert A. Mowrey, Ph.D.

Horse Commodity Coordinator
Department of Animal Sciences
North Carolina State University
Raleigh, North Carolina

Roberto D. Munilla, M.E.

Research Engineer
Department of Biological and Agricultural Engineering
North Carolina State University
Raleigh, North Carolina

Dennis W. Nixon, J.D., M.M.A., A.B.

Professor and Graduate Program Director
Department of Marine Affairs
University of Rhode Island
Kingston, Rhode Island

Deanna L. Osmond, Ph.D.

Extension Water Quality Specialist
Department of Biological and Agricultural Engineering
North Carolina State University
Raleigh, North Carolina

William A. Popendorf, Ph.D., C.I.H.

Professor of Industrial Hygiene
Department of Biology
Utah State University
Logan, Utah

Timothy G. Prather, M.S.

Assistant Extension Specialist
Department of Agricultural Engineering
University of Tennessee
Knoxville, Tennessee

Mark A. Purschwitz, Ph.D.

Assistant Professor and Extension Agricultural Safety and Health Specialist
Department of Biological Systems Engineering
University of Wisconsin-Madison
Madison, Wisconsin

Stephen J. Reynolds, Ph.D., C.I.H.

Assistant Professor
Department of Preventive Medicine and Environmental Health
University of Iowa
Iowa City, Iowa

Gary T. Roberson, Ph.D., P.E.

Associate Professor
Department of Biological and Agricultural Engineering
North Carolina State University
Raleigh, North Carolina

Stanley Schuman, M.D., Dr.P.H.

Professor
Department of Family Medicine
Medical University of South Carolina
Charleston, South Carolina

David P. Siebens, M.D., M.P.H.

Clinical Associate
Division of Occupational and Environmental Medicine
Department of Community and Family Medicine
Duke University
Durham, North Carolina

Gary M. Tencer, C.I.H.

Assistant Director
Occupational and Environmental Safety Office
Duke University
Durham, North Carolina

Edward D. Thalmann, M.D.

Captain, Medical Corps, U.S.N. (retired)
Assistant Medical Director, Divers Alert Network
Assistant Clinical Professor of Anesthesiology
Assistant Clinical Professor, Division of Occupational and Environmental Medicine
Duke University
Durham, North Carolina

David P. Thomson, M.S., M.D.

Department of Emergency Medicine
East Carolina University
Greenville, North Carolina

R. Lane Tippens, M.D., M.P.H.

Fellow
Division of Occupational and Environmental Medicine
Department of Community and Family Medicine
Duke University
Durham, North Carolina

Susanna G. Von Essen, M.D.

Associate Professor
Section of Pulmonary and Critical Care Medicine
Department of Internal Medicine
University of Nebraska Medical Center
Omaha, Nebraska

Michael G. Waldvogel, Ph.D.

Extension Specialist
Department of Entomology
North Carolina State University
Raleigh, North Carolina

About the Editors

Ricky L. Langley, M.D., M.P.H., F.A.C.P., F.A.C.O.E.M., F.A.C.P.M., is an assistant clinical professor in the Division of Occupational and Environmental Medicine, Department of Community and Family Medicine, at Duke University Medical Center. He received his M.D. and M.P.H. degrees from Bowman Gray School of Medicine and the University of North Carolina at Chapel Hill. Dr. Langley is board certified in internal medicine and preventive medicine, is a former president of the Carolinas Occupational Medical Association, and a fellow of the American College of Occupational and Environmental Medicine, American College of Preventive Medicine, and American College of Physicians. He is a member of the American Medical Association, American Conference of Governmental Industrial Hygienists, American Industrial Hygiene Association, North Carolina Medical Society, and the North Carolina Department of Labor Council on Agricultural Safety and Health.

Robert L. McLymore, Sr., Ed.D., is a farm safety specialist in the Department of Biological and Agricultural Engineering at North Carolina State University. He received his masters degree and doctorate in adult education from North Carolina A&T University and North Carolina State University, respectively. Dr. McLymore is a member of the American Society of Agricultural Engineers, National Institute for Farm Safety, North Carolina Safety Council, National Association of Farm Extension 4-H Agents, and North Carolina Association of Cooperative Extension Specialists.

William J. Meggs, M.D., Ph.D., F.A.C.E.P., is associate professor of emergency medicine in the Division of Clinical Toxicology, Department of Emergency Medicine at East Carolina University School of Medicine. He received his medical degree from the University of Miami and a doctoral degree in physics from Syracuse University. He is board certified in emergency medicine, allergy and immunology, internal medicine, and medical toxicology. Dr. Meggs is a fellow of the American College of Emergency Physicians and a member of the American Academy of Allergy and Immunology, American College of Physicians, Clinical Immunology Society, International Society of Bioelectricity, American Medical Association, North Carolina State Medical Society, and Society for Academic Emergency Medicine.

Gary T. Roberson, Ph.D., P.E., is a visiting associate professor in the Department of Biological and Agricultural Engineering at North Carolina State University. He received his doctoral and masters degrees in biological and agricultural engineering from North Carolina State University. A Professional Engineer, Dr. Roberson is a member of the American Society of Agricultural Engineers and the National Association of Colleges and Teachers of Agriculture.

SAFETY AND HEALTH IN AGRICULTURE, FORESTRY, AND FISHERIES

SECTION 1:

GENERAL ISSUES IMPACTING SAFETY AND HEALTH IN AGRICULTURE, FORESTRY, AND FISHERIES

1

OVERVIEW OF SAFETY AND HEALTH IN AGRICULTURE, FORESTRY, AND FISHERIES

William J. Meggs, M.D., Ph.D., F.A.C.E.P.
East Carolina University School of Medicine

Gary T. Roberson, Ph.D.
North Carolina State University

Ricky L. Langley, M.D., M.P.H.
Duke University Medical Center

Robert L. McLymore, Sr., Ed.D.
North Carolina State University

INTRODUCTION

Agriculture, forestry, and fisheries are widely recognized as being among the most hazardous industries in the United States and also in the world. To understand why high injury and high fatality rates occur in these industries, it is important to know the conditions in which individuals work and the hazards they face. By studying the epidemiology of injuries and illnesses in these industries, one can identify hazardous conditions or operations and hopefully modify these so as to reduce or prevent adverse effects from occurring.

Injuries and illnesses may be acute, subacute, or chronic in nature. Additionally, injuries and illnesses can be fatal or nonfatal. Nonfatal injuries and illnesses may be mild and the worker most often recovers without any sequelae, or the injury may be serious—in which case recovery may be prolonged and the worker left with a permanent impairment such as an amputation or chronic lung disease.

The epidemiology of fatal injuries is often more clear and data easier to obtain than for nonfatal injuries. Especially in these industries, minor injuries are often unreported and often viewed as a part of the job due to the frequent adverse circumstances in which these individuals work. Usually it is more difficult to accurately quantify the number of illnesses that occur in these industries. It may take years before debilitating symptoms of the illness appear. Therefore, failure to recognize the early manifestations of a work related illness and failure to recognize risk factors for development of an illness unnecessarily increase the burden of disease in these industries. In addition, the worker may have changed jobs before becoming ill, and work attribution of disease may fail to be correctly identified. This may not only have a negative impact on the affected worker (i.e. denial of workers compensation benefits) but failure to recognize these risk factors may endanger the health of currently employed workers.

Numerous studies have been done using various data sources to attempt to identify hazardous conditions on the farm, in the forest, and on the seas. A brief overview of safety and health hazards provided in this textbook follows.

EPIDEMIOLOGY

Injuries in agriculture, forestry, and fisheries account for over 1,000 fatalities per year, according to the National Safety Council. The primary agent responsible for farm mortality is the tractor. Other types of farm machinery, electrocution, animals, and falling objects account for most of the remaining farm fatalities.

Older farmers appear to be at great risk of fatal injuries. As one ages, physiological changes occur in the body, such as changes in reaction time and visual acuity, which may increase one's chance of an injury. Additionally, an injury to an elderly farmer with numerous chronic medical conditions likely increases one's probability of death from a serious injury compared to the same injury in a healthy, younger farmer. Additionally, scores of children die from farming mishaps yearly. Lack of training, horseplay, and riding on vehicles as extra passengers are all associated with unintentional childhood farm injuries. Injuries on farms tend to increase during planting and harvest seasons, as one would expect.

The agricultural industry leads most industries in chemical (especially pesticide) poisoning and respiratory diseases. Farmers are exposed to a variety of dusts and gases on the farm. Agricultural dust may cause hypersensitivity pneumonitis, organic dust toxic syndrome, bronchitis, and asthma. Gases such as nitrogen oxides and hydrogen sulfide may cause pulmonary damage and chemical asphyxiation.

Farmers are known to be at increased risk for certain cancers such and skin and lip cancers. The risk for other forms of cancer is controversial with some studies finding elevated risk while other do not. Sun exposure, farm chemicals, and possibly oncogenic animal viruses may be responsible for cancer in farmers. Other illnesses more common in farmers include hearing loss, zoonotic infections, back pain, and contact dermatitis.

Forestry workers have a high likelihood of injury at work. The annual fatality rate among loggers is nearly 23 times that of other U.S. workers. The primary causes of forestry injuries include being hit by falling trees or rolling logs, falls, chain saws, and machinery. Illnesses noted in the forestry industry include repetitive back strain, especially during planting of trees. Many chain saw operators develop hand and arm vibration syndrome, which causes hand pain and numbness. Inhalation of chain saw exhaust frequently causes upper respiratory and mucus membrane discomfort in operators.

Environmental hazards that may lead to illness in foresters include insect bites and stings and occasionally zoonotic and fungal infections. Lyme disease is clearly more likely in the forestry worker population. Seroprevalence studies in forestry workers have found a significant prevalence of antibodies to various tick borne and mosquito borne infectious agents. Sporotrichosis in foresters during tree planting has been traced to tree seedlings wrapped in sphagnum moss.

Occupational dermatitis is also common in the forestry industry. Skin and eye irritation, allergic contact dermatitis, and contact urticaria may occur from contact with a variety of plants and woods. Contact dermatitis from exposure to poison ivy, poison oak, and poison sumac represent the most frequent dermatologic complaints among forestry workers. Additionally, over 50 species of trees, especially tropical species, can cause dermatitis in workers. Respiratory complaints may occur after harvesting trees due to inhalation of wood dusts and various lichens and fungi found growing on the bark.

Forest and wild land fires cause an average of 16 deaths per year in U.S. firefighters. Additionally, repetitive muscle strain frequently occurs in responders and pulmonary illness and heat stress are occasionally reported.

New environmentally sensitive methods of timber harvesting using helicopters are increasingly being used in the Northwest. Several fatalities due to helicopter crashes have occurred. The number of crashes is likely to increase as this method of timber harvesting grows. Detailed analysis of the causes of the crashes should result in the design of safer equipment and safer operating procedures.

The commercial fishing industry has been identified as another high risk industry. During the 1980s, an estimated 60 to 100 deaths occurred annually. Workers at greatest risk are those that operate aboard easily capsized vessels and those who have insufficient training in ship board safety, especially regarding cold water survival techniques and the use of personal flotation devices. Most fatalities are due to drowning. Additional injuries that occur in the fishing industry include electrocutions, contact dermatitis from jellyfish and other marine plants and animals, zoonotic infections, and rarely bites and stings from marine life.

Commercial fishermen who may occasionally dive to repair their vessels or collect certain sea foods are at risk for numerous injuries and illnesses. A variety of injuries may occur from barotrauma, as well as from malfunctioning equipment. Divers may also develop dermatitis from their clothing as well as from the marine life previously noted. Diving equipment must be properly cleaned after each use to avoid possible contamination of the diver and other equipment as well as to prevent deterioration of the equipment. The diver must be thoroughly familiar with rescue procedures and the location of the nearest decompression facility must be known in advance of each dive.

MEDICAL ILLNESSES IN FARMERS, FORESTERS, AND FISHERMEN

Many forms of cancer are environmentally induced when the genetic machinery of tissue cells is changed by chemicals, radiation, or viruses. Damage that leads to uncontrolled proliferation of a tissue manifests itself clinically as cancer. Cancer risk may be reduced by avoiding those substances and exposures that induce cancer. Cancer is a heinous disease that can be difficult to treat and is often fatal. Hence, efforts to reduce cancer risk are more than justified. Since the discovery by Sir Percivall Pott in 1775 that exposure to polyaromatic hydrocarbons in soot lead to scrotal cancer in chimney sweeps, knowledge about occupational cancers has mushroomed.

Farm workers, lumber workers, and fishermen all work outdoors and are exposed to ultraviolet radiation in sun light, which has been firmly established to cause skin cancer in the laboratory setting. Epidemiological studies verify that workers chronically exposed to sunlight have an increased risk of melanoma, basal cell carcinoma, and squamous cell carcinoma of the skin. The phenoxy herbicides, of which 2,4-D and 2,4,5-T are examples, have been associated with non-Hodgkin's lymphoma and soft tissue sarcomas in humans. In farmers, there is an increasing risk of soft tissue sarcoma associated with increasing exposure to 2,4-D. Organic arsenical pesticides are associated with lung cancer, and this association may extend to chlorinated hydrocarbon pesticides.

A host of respiratory hazards exist in agricultural operations. The complex organic dusts found in animal confinement facilities can lead to chronic bronchitis and asthma. Farmers lung, an immunologically mediated hypersensitivity pneumonitis, occurs with sensitization to antigens produced by fungi and Thermophilic actinomycetes found on moldy hay. Oyster shell protein found in oyster shell dust can cause hypersensitivity pneumonitis. Bird breeders can get a hypersensitivity pneumonitis from exposure to dusts in areas in which birds are housed. For all etiologies, there is an acute febrile illness with cough, dyspnea, and pulmonary infiltrates that resolves within a few days of removal from the inciting exposure. Chronic exposure to the antigens can lead to permanent lung damage.

Silo filler's disease is caused by exposure to nitrogen oxides that form in the first few days when a silo is loaded with green silage. As the silage decomposes, the nitrous oxide appears as a yellow to red hazy gas over the silage. Acute exposure leads to pulmonary injury as the nitrogen dioxide combines with moisture in the airway to form acid. Symptoms of upper airway irritation, cough, shortness of breath, lightheadedness, choking, syncope, wheezing, chest pain, weakness, eye and throat irritation, pulmonary edema, and focal chest infiltrates may be delayed for several hours. Relapses after 3 weeks can occur from bronchiolitis obliterans, which is necrosis of small airways. Risk may persist for several weeks after a silo is loaded.

Silo unloader's syndrome occurs when a silo is opened and there is exposure to moldy elements at the top of a silo. Endotoxins from the cell wall of gram negative bacteria, as well as mycotoxins, proteinases, and histamine, may play a role in producing the flu-like symptoms of fever, malaise, muscle aches, headache, cough, chest tightness, wheezing, mild dyspnea, with nasal and throat irritation. Long term sequela usually do not occur.

Anhydrous ammonia fertilizer combines with water to form a caustic alkaline solution which can cause severe burns of the eyes, skin, and mucous membranes. Death can occur at high concentrations, while non-cardiac pulmonary edema can occur at lower concentrations. Eye exposures can lead to corneal abrasions with vision loss. Like the nitrogen dioxide exposures from silos, chronic sequela of bronchiolitis obliterans can occur if ammonia is inhaled. Immediate first aid, with removal from the exposure and sustained flushing of exposed areas with large amounts of water must be instituted on the scene. The most dangerous operation for anhydrous ammonia exposure is the transfer of ammonia between supply and applicator tanks.

Composting of animal wastes can produce asphyxiants such as methane and carbon dioxide, as well as metabolic poisons such as hydrogen sulfide. Closely related to cyanide in its toxicity, hydrogen sulfide blocks cellular respiration and can be rapidly fatal. Entry into manure pits is thus extremely dangerous and should only be undertaken with self contained breathing apparatus.

Pesticides are chemicals used to kill insects, weeds, and rodents. Organophosphate insecticides are immediately toxic to humans from ingestion, inhalation, and absorption through the skin. These pesticides block cholinesterase enzymes which down regulate the neurotransmitter acetylcholine at autonomic, central nervous system, and muscular sites of action. Secretions from the eyes, mouth, lungs, nose, and gut accompany muscle paralysis, seizures, and coma. In high doses, exposures can be rapidly fatal. Chronic sequela of acute poisoning include neuropsychiatric phenomena, permanent paralysis, and chemical sensitivity. Chronic exposure to organophosphate pesticides can lead to a syndrome with variable symptoms of headache, nausea, weakness or fatigue, chest tightness, abdominal pain, vertigo or incoordination, vomiting, perspiration, cough, vision disturbance, loss of appetite, dyspnea, nasal discharge, miosis, and wheezing.

Carbamate pesticides also block cholinesterase. Toxicity is similar to that of organophosphates and can be fatal, although, in general, carbamate poisoning is less severe and resolves more quickly. Pyrethrum and pyrethroid pesticides are respiratory sensitizers, and allergic asthma and rhinitis occurs when sensitized individuals are exposed. A syndrome of systemic pyrethroid toxicity has been described in China, with symptoms of facial burning and itching. Ingestion can cause gastrointestinal symptoms of nausea, vomiting, and epigastric pain. Fatigue, headache, and dizziness can occur, as well as weakness, fasciculations, seizures, coma, and death.

Many organochlorine pesticides such as DDT and chlordane have been banned in the United States due to environmental persistence and ecological damage. Acute toxicity is neurological, with tremors, muscle twitches, and paresthesias occurring in mild cases. Dizziness, fatigue, malaise, and headache can accompany these symptoms. At higher doses, seizures and coma can result. Nausea and vomiting are seen with ingestion. Respiratory failure can occur and is fatal without treatment. The cyclodiene compounds (aldrin, dieldrin, endrin, and toxaphene) have lower thresholds of toxicity and are more commonly associated with acute toxicity in humans.

In addition to the acute toxicity, there is toxicity associated with chronic exposure to organochlorine compounds. Chronic exposure to dieldrin has caused a symptom complex that includes headache, blurred vision, diplopia, tinnitus, dizziness, muscle twitches, diaphoresis, insomnia and nightmares, nystagmus, coordination difficulties, personality changes, and muscular fibrillation. Involuntary jerking of extremities has followed chronic exposures to aldrin, dieldrin, and thiodan, but is less likely to occur with DDT. Anorexia and weight loss, muscle weakness, tremor, anxiety, nervous tension, and fear have been reported with chronic DDT exposure. Dicofol, a derivative of DDT, has been associated with behavioral problems and decreased academic performance that persisted after an acute exposure.

A number of reports have linked chronic exposure to organochlorine compounds to chronic neurological symptoms. Chronic motor neuron disease has been linked to aldrin, lindane, and heptachlor. An increased risk of liver and other cancer was found in pesticide applicators in Rome who were exposed to organochlorine pesticides in the 1960s. Neurobehavioral defects have been found in children born to mothers with high tissue levels of organochlorines. These compounds are stored for long periods of time in fatty tissue and passed to the baby *in utero*. Based on animal data, it is thought that these deficits persist into adulthood. Chlordecone (Kepone) with chronic exposure causes tremor, nervousness, weight loss, rash, mental changes, weakness, ataxia, slurred speech, and loss of coordination.

Skin diseases are very common among agricultural workers. Sun exposure can cause sunburn, phototoxic and photoallergic reactions, "sun poisoning", the precancerous lesions of actinic keratoses, and increased risk of skin cancer. A variety of skin disorders are related to exposure to extremes of heat and cold. Contact with plants, fertilizers, animal feeds and additives such as antibiotics, pesticides, and chemicals used in maintenance of farm machinery can lead to contact dermatitis. The outdoor environment contains a variety of insects and animals which can inflict toxic envenomizations on farmers. Sun exposure causes increased prevalence rates and earlier onset of cataracts in certain fishermen.

Zoonotic infections are those transmitted from animals to man. Arthropods are often the vectors for transmitting zoonotic infections. Zoonotic infections such as Brucellosis can also be transmitted by direct contact with animal products. Some zoonotic infections—such as brucellosis, leptospirosis, Q fever, tularemia, and Lyme's disease—can cause prolonged and disabling illnesses.

HAZARDS

Hazards are omnipresent in the agriculture, forestry, and fisheries workplace. These hazards take on many forms: mechanical, biological, thermal, chemical, electrical, and environmental. There are hazards present in crop and livestock production, nursery and greenhouse operations, turfgrass production, logging, fishing, aquaculture and all other types of activity included under the umbrella of agriculture, forestry, and fisheries.

Mechanical hazards are common with machines used in agriculture, forestry, and fisheries. Machine components often have sharp edges, pinch points, entanglements, and crush points. Machines can overturn onto an operator or run over a worker if proper care is not taken. Regardless of how simple or complex the machine may be, some form of mechanical hazard may be present.

Chemical hazards are everywhere in the modern workplace. We easily recognize the danger of certain pesticides. However, equally hazardous in some cases are common chemicals such as fuels, lubricants, paints, solvents, and other chemicals used by machines and in buildings. In some cases, proper care is not taken in handling these compounds because the worker does not recognize the hazard.

Biological hazards such as insect bites and stings, exposure to poisonous plants, and snakes are also typical. Also included in the biological hazards are the hazards of dealing with livestock, wildlife, and marine animals.

Electrical hazards are present in buildings and workplaces where electricity is used for power tools, appliances, lighting, and convenience. Electrical failure may result in fire or severe electrical shock for the worker or victim. Lightening also poses a severe electrical hazard. Proper precautions should be take during electrical storms to minimize risk.

The presence of environmental hazards is due to the nature of the workplace common to many agricultural, forestry, and fisheries occupations. This workplace is outdoors and is often subject to environmental factors over which the worker has no control. In some cases, this environment can be improved by enclosures or cabs. In many cases it cannot. This workplace is unconventional by other industrial standards. Extremes of temperature, humidity, noise, vibration, light, and other factors can affect worker production.

Agricultural machines, such as tractors, combines, implements, and others, account for a large percentage of recorded injuries in agriculture. As machines continue to evolve and new designs are developed, additional hazards may be produced. Designers must recognize these hazards and develop effective engineering solutions to improve machine design. Equipment owners and users also share this responsibility to recognize hazards. Equipment owners and operators are responsible for their own safety as well as their employees and must take proper precautions to avoid injury.

Modern animal agriculture follows production practices different from those of the past. Often animals are grown to market size in enclosed, environmentally controlled production facilities. Maintaining a healthy environment in terms of air quality, temperature, humidity, and other factors is essential to efficient production as well as human health concerns for the workers involved. Workers must also understand how to properly handle and manage these animals, particularly the large ones such as swine, cattle, and horses. Injuries due to large animals rank second to machines as a major cause of farm injuries.

Injuries are common in forests as well. A large percentage of these injuries are related to machines used in logging and other forestry operations. In fisheries, a different set of hazards and concerns arise.

Fishing vessel operations and crews are concerned with many of the common hazards found in other occupations. In addition, fishermen are faced with hazards such as vessel collisions, grounding on reefs and bars, flooding, floundering, and capsizing.

Hazards are present in all types of occupations and in all types of workplaces. However, if hazards can be recognized, safety practices, engineering controls, and management actions can be applied to reduce risk. This text will attempt to expose the hazards discussed here and propose effective management and treatment options.

PREVENTION AND REGULATION

Strategies to prevent injuries and disability arise from a study of mortality and morbidity in the workplace. An analysis of accidents and injuries can lead to changes that will minimize those events in the future. Prevention of injuries associated with mechanical hazards may involve redesign of machinery, installation of safety shields, and institution of safe work practices. Chemical injuries may be prevented by replacement of hazardous chemicals with safer alternatives, wearing of protective clothing and respirators, and training in safe work practices. It is important to realize that the institution of a measure to reduce morbidity and mortality does not ensure success, and what seems to be an improvement may indeed backfire. Hence, once a safety measure has been instituted, it is essential that the outcome be monitored to ensure the intended result occurs. An example of a safety measure that backfired is the use of closed mixing systems for organophosphate pesticides. In a monitored situation these systems improved mixer efficiency so much that many more pounds of pesticide were handled by an individual worker, and exposure and toxicity actually increased.

Institution of safety practices can be voluntary or by government regulation. In the United States, regulation in farming, fisheries, and forestry is a complex legal framework involving multiple jurisdictions and agencies. The Federal Insecticide, Fungicide, and Rodenticide Act (FIFRA) empowers the Environmental Protection Agency (EPA) to register pesticides. The United States Coast Guard is authorized to enforce safety standards for commercial fishing vessels. The Occupational Safety and Health Administration (OSHA) regulates workplace safety. Regulations must be subjected to scrutiny to ensure the intended result without undo costs or compromise of job performance.

2

ERGONOMICS IN AGRICULTURE, FORESTRY, AND FISHERIES

Tamara M. James, M.A., C.I.E., C.P.E.
Duke University Medical Center

> Ergonomics has been defined as the science of matching the job to the worker and the product to the user. Occupational injuries and illnesses, such as cumulative trauma disorders, can often be traced to how the worker interacts with the task. In recent years, cumulative trauma disorders accounted for 50 percent of all occupational injuries. Ergonomic assessments of jobs and workplaces are instrumental in reducing these illnesses. In this chapter, risk factors for cumulative trauma disorders are identified and discussed. Use of anthropometric data in equipment design is discussed. Specific examples in agriculture are included. Physical work guidelines, including manual material handling and thermal exposures, are also presented. Development of an ergonomics program is identified as an important step in controlling injury and illness.

INTRODUCTION

Ergonomics is a word that comes from the Greek word *ergo*, which means work and *nomos* meaning natural law. The term first appeared in a Polish newspaper in 1857 (Karwowski 1991). The science of ergonomics dates back to World War II when a group of British Scientists, working for the armed services, looked at ways of improving soldier efficiency through the redesign of military systems and equipment. In the United States, *ergonomics* is often used interchangeably with the term *human factors engineering*. However, ergonomics is of European origin and tends to focus primarily on biomechanics, physiology, and engineering design. Human factors engineering originated from the field of psychology and tends to focus more on human performance and the design of systems.

Ergonomics has many different definitions. The simplest definition is: *Ergonomics is the scientific study of human work* (Pheasant 1991). Since ergonomics is also concerned with the interaction of human beings with tools, machines, and systems involved in performing work, the following phrase is more comprehensive and summarizes the general approach of ergonomics: *Ergonomics is the science of matching the job to the worker and the product to the user* (Pheasant 1991). The goal of ergonomics is to adapt the job or workplace to fit the person, rather than force the person to fit the job or workplace, with the ultimate goal of making the job or workplace safe, comfortable, and efficient, with no adverse health effects.

The idea that work can adversely affect the health of workers has actually been around for hundreds of years. In 1713, Bernardino Ramazzini, an Italian physician wrote:

> " For we must admit that the workers in certain arts and crafts sometimes derive from them grave injuries, so that where they hoped for a sustenance that prolonged their lives and fed their families, they are too often repaid with the most dangerous diseases and finally, uttering curses on the profession to which they had devoted themselves, they desert their post among the living."

Some of these adverse health effects are particularly prevalent in agriculture. It is widely known that agriculture ranks as one of the most hazardous occupations in this country. The U.S. Department of Labor's Bureau of Labor Statistics (BLS) reports that in 1994 the major industry group with one of the largest number of fatal work injuries was farming, forestry, and fishing (BLS 1995). For a number of years, researchers have worked to find ways to reduce or eliminate worksite hazards, using an ergonomic approach. This research is applied with a great deal of success in manufacturing and related industries and can benefit agricultural employers through the application of ergonomic principles as a means of controlling and reducing adverse health effects.

WHAT ARE CTDs?

Adverse health effects resulting from repeated exposure to trauma such as awkward postures and motions are known as cumulative trauma disorders or CTDs. CTDs are a problem throughout many different industries today including farming, forestry, and fishing. These CTDs are serious but often preventable. For example, in production agriculture there are many different activities such as operating tractors, milking cows, lifting and throwing bales of hay, picking cotton, lifting containers of produce, and bending over to pick crops. All have the potential of resulting in CTDs. The Bureau of Labor Statistics reports that CTD cases are increasing faster than any other occupational illness category. In 1980 CTDs comprised 18% of all occupational illnesses and by 1989 had increased to more than 50% of all work-related illness cases (BLS 1991).

Examining the separate meanings for each word of *cumulative trauma disorders* may help to provide a more practical understanding of what it is. The term *cumulative* means these injuries develop gradually over a period of weeks, months, or years as a result of repeated stress or strain on a specific body part such as muscles, tendons, or nerves. *Trauma* is an injury to some area of the body that is a result of mechanical stress. *Disorders* refer to physical ailments or abnormal conditions (Putz-Anderson 1988). In other words, CTDs are a family of muscle, tendon, nerve, or neurovascular disorders that are caused, precipitated, or aggravated by repeated exertions, stresses, or movements of the body over time. Low back pain is also considered a CTD. Examples of upper extremity CTDs are tendinitis, tenosynovitis, De Quervain's disease, epicondylitis (tennis elbow), trigger finger, hand-arm vibration syndrome, and carpal tunnel syndrome. Appendix A presents a list and descriptions of common upper extremity CTDs categorized by tendon disorders, nerve disorders, and neurovascular disorders.

Muscles, tendons, and nerves are the areas most affected by CTDs. Repeated and sustained exertions of a body part can lead to muscle fatigue. Muscle fatigue is generally characterized by discomfort and reduced strength or endurance. Recovery from muscle fatigue typically occurs within minutes or hours and as such is not hazardous. However, if a person continues to repeatedly exert muscles beyond normal fatigue levels, chronic fatigue sets in and muscle damage may result. This situation is common in individuals who work excessive overtime or have a similar second job or "moonlight." Individuals who have hobbies or activities outside the workplace that involve the use of the same muscles and tendons, often do not give the muscle set ample recovery time, resulting in chronic fatigue.

When chronic fatigue sets in, and muscles are not given adequate time to recover, muscles transfer the workload to other parts of the body, such as the tendons. Tendons are not designed to do the work of muscles, which means they are more at risk for being strained. This is the reason that chronic muscle fatigue is generally believed to be a risk factor for CTDs.

CTDs are widespread in many different occupations and industries, and are significant in aerospace, agriculture, automotive, clerical, data and word processing, electronics, fabric cutting, food processing, glassware, health care, manufacturing, postal services, metal forming, plastics molding, and the performing arts (Armstrong and Lackey 1994). The common denominator across these industries is they all rely heavily on hand-intensive work or manual materials handling. Table 2.1 lists sample jobs (from various industries) as well as common CTDs and occupational risk factors for those jobs.

CTD RISK FACTORS

The exact cause of a CTD is extremely difficult to determine. There are many different risk factors and combinations of risk factors that can cause a CTD. Generally the development of CTDs involves one or more of the risk factors listed in Table 2.2. This table presents occupational risk factors, or those that can cause, precipitate, or aggravate CTDs as well as non-occupational risk factors, or those that place certain individuals at greater risk of developing a CTD. Evidence supports the contribution of non-occupational risk factors in some situations and occupational risk factors in other situations (Armstrong et al. 1987).

Ergonomic Assessment and Hazard Prevention

To determine whether or not CTD risk factors are present in a particular work situation, an ergonomic assessment or survey should be conducted. A thorough ergonomic assessment must include a study of all aspects of the work environment. Many CTDs are often due to the faulty design and/or layout of items within the work environment or the design of the work methods. The work environment includes the entire work space within the direct reach of an individual and includes all relevant work objects such as tools, equipment, seating, and all associated controls and displays.

The first step in an ergonomic assessment is to document the job or tasks by gathering information that can be used to identify and quantify ergonomic stressors and to assist in the design of recommended solutions. The following information should be collected for each job (modified from Armstrong 1995):

- Job Title—the name of the job and a description of what the individual does.
- Production Standard—the amount and quality of work that is to be completed in a given period of time.
- Location—where the job takes place, the department, or building location.
- Tasks—the basic work elements or steps performed to accomplish an objective.
- Methods—the sequence of steps or the way a task is completed.
- Work Equipment—tools, machines, and personal protective equipment that are used to perform the task or job.
- Materials—products, materials, or other items that are handled or used by the worker.

Table 2.1. Cumulative trauma disorders by type of job performed (modified from Putz-Andersen 1988).

Type of Job	Disorder	Occupational Factors
1. Buffing/grinding	Tenosynovitis, thoracic outlet, carpal tunnel, De Quervain's	Repetitive wrist motions, prolonged flexed shoulders, vibration, forceful ulnar deviation, repetitive forearm pronation.
2. Punch press operator	Tendinitis of wrist and shoulder	Repetitive forceful wrist extension/flexion, repetitive shoulder abduction/flexion, forearm supination.
3. Overhead assembly (welders, painters, auto repair)	De Quervain's, thoracic outlet, shoulder tendinitis	Repetitive ulnar deviation in pushing controls, sustained hyperextension of arms, hands above shoulders.
4. Belt conveyor assembly	Tendinitis of shoulder and wrist, carpal tunnel, thoracic outlet	Arms extended, abducted, or flexed more than 60 repetitive forceful wrist motions.
5. Typing, keypunch, cashier	Tension neck, thoracic outlet, carpal tunnel	Static, restricted posture, arms abducted/flexed, high speed finger movement, palmar base pressure, ulnar deviation.
6. Small parts assembly (wiring, bandage wrap)	Tension neck, thoracic outlet, wrist tendinitis, epicondylitis	Prolonged restricted posture, forceful ulnar deviation and thumb pressure, repetitive wrist motion, forceful wrist extension and pronation.
7. Bench work (Glass cutters, phone operators)	Ulnar nerve entrapment	Sustained elbow flexion with pressure on ulnar groove.
8. Packing	Tendinitis of shoulder and wrist, tension neck, carpal tunnel, DeQuervain's	Prolonged load on shoulders, repetitive wrist motions, overexertion, forceful ulnar deviation.
9. Truck driver	Thoracic outlet	Prolonged shoulder abduction and flexion.
10. Core making	Tendinitis of the wrist	Prolonged shoulder abduction and flexion, repetitive wrist motions.
11. Stockroom, shipping	Thoracic outlet, shoulder tendinitis	Reaching overhead, prolonged load on shoulder in unnatural position.
12. Material handling	Thoracic outlet, shoulder tendinitis	Carrying heavy load on shoulders.

> **Table 2.2.** Occupational and non-occupational risk factors associated with CTDs (modified from Armstrong and Lackey 1994).
>
> **Occupational Risk Factors**
>
> Repeated and sustained exertions
> Awkward postures
> Forceful exertions
> Vibration
> Localized contact stresses
> Cold temperature
>
> **Non-Occupational Risk Factors**
>
> Systemic diseases such as diabetes and rheumatoid arthritis
> Acute trauma
> Congenital defects
> Body characteristics such as wrist size and shape
> Pregnancy
> Hormones
> Obesity
> Physical fitness, conditioning, and strength
> Smoking and nicotine ingestion
> Vitamin deficiency

The next step of an ergonomic assessment is to evaluate the job or tasks to identify work-related CTD risk factors using the list of Occupational Risk Factors presented in Table 2.2. To assist with the assessment, the following steps for analysis and control are provided (modified from Armstrong 1995).

Repeated and Sustained Actions—Performing the same activities or motions over and over again or maintaining the same position for prolonged periods of time. Following are steps for determining repetitiveness:

1. Determine the number of work items being produced or the tasks being performed each hour.
2. Define what the work items are.
3. Determine the number of steps or exertions required for each work object or task.
4. Repetitiveness is calculated as the number of exertions per hour.
5. Estimate the duration of each exertion.
6. Repetition rate can be stated as:

- **Very High**—Body parts are in constant motion and it is difficult to maintain the pace.
- **High**—Constant rapid motion, any wasted motions would result in falling behind.
- **Medium**—Steady motion with no difficulty keeping up and time allowed to pause.
- **Low**—There are long pauses in each work cycle and hands move only occasionally.
- **Very low**—All parts of the body are idle most of the time.

To reduce the number of repeated and sustained exertions, the following controls are suggested:

1. Adjust the work standard—modify the amount of work performed in a given amount of time. This may be difficult to implement.
2. Use mechanical aids—this includes jigs or fixtures for holding items, feeders for loading machines, or the use of power tools and power equipment in place of manual tools or equipment.
3. Use worker rotation—move workers through many different jobs during the day to help reduce repetition or excessive loading on any one body part. This approach may also be difficult to implement.
4. Use work enlargement—combining jobs with different motion patterns is effective however it may require redesign and changes to the entire job process.

Awkward Postures—Body postures that require substantial effort or result in compression or stretching of tissues in or around the joints. One tool used by ergonomists to collect symptomatic data from employees is a discomfort/comfort survey or questionnaire (see Appendix B). Surveys or questionnaires provide an accurate picture of the present work situation, and also help identify new or potential cases of CTDs. The following methods may also be used to assess awkward postures:

1. Observe the worker, using videotape to identify stressful postures.
2. Identify any factors that may affect posture such as work orientation, tool orientation, work layout, or space constraints.
3. Determine the frequency and duration of stressful postures.
4. Measure joint angles to help quantify postures.

Ideally, work should be performed in a way that minimizes prolonged or repeated exertions of the arms above shoulder height, the hands behind the back, extreme rotation of the forearm, bending of the wrist, or pinching with the fingers.

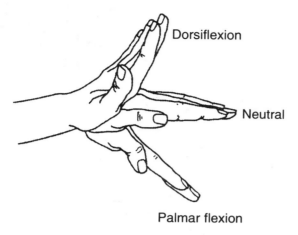
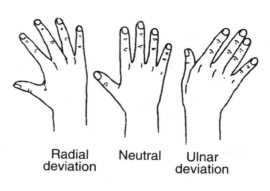

Figure 2.1. Examples of wrist postures (Sanders and McCormick 1993, reproduced with permission of McGraw-Hill.)

To reduce or eliminate these postures the following is suggested:

1. Adjust the work location in order to maintain comfortable or neutral body postures.
2. Adjust the work item so that neutral body postures can be maintained while working.
3. Select or design the tool size and shape so that a neutral or straight wrist and a comfortable grip can be maintained. (See Figure 2.1)

Forceful Exertions—An exertion to overcome weight, resistance, or inertia of the body or a work object. Forces are caused by gravity on work objects, tools, or the body; by friction; and by reaction forces. Bulky, poor-fitting, or stiff gloves and certain postures can also increase forces. Following are methods for analyzing the force requirements:

1. Examine the work methods and identify all exertions.
2. Identify and rate the factors affecting force such as weight, friction, gloves, posture, and maintenance of tools.
3. Determine the duration of the force.
4. Measure or estimate the force using some quantitative method such as a dynamometer or electromyography.

Obviously, the force requirement of a task should be minimized whenever possible. To reduce forces:

1. Use gloves or tape fingers to improve the grip on an object.
2. Avoid gloves that are too thick and interfere with gripping objects.
3. Pick up fewer objects at a time to reduce the total weight.
4. Select light tools to reduce the weight.
5. Use articulating arms to support tools.
6. Use gravity as an aid to easier material handling.
7. Use jigs and fixtures to hold work objects.
8. Use handles to make gripping objects easier.
9. Use hoists to support work objects.

Vibration—Exposure to vibrating objects. Operating power tools such as drills, drivers, chain saws, and grinders results in high levels of vibration in the hands and wrists. Vibration of the hands and wrists as well as vibration of the spine may occur while operating a motor vehicle such as a truck or tractor. The following approaches may minimize vibration exposure:

1. Identify vibration exposure by inspecting the work or by observing the worker.
2. Identify the factors affecting exposure to vibration.
3. Determine the frequency of exposure.
4. Quantify the frequency and acceleration using appropriate measurement instruments.
5. Compare these values with available vibration standards.

The source of vibration should be removed if possible. Some steps to minimize exposure to vibration are:

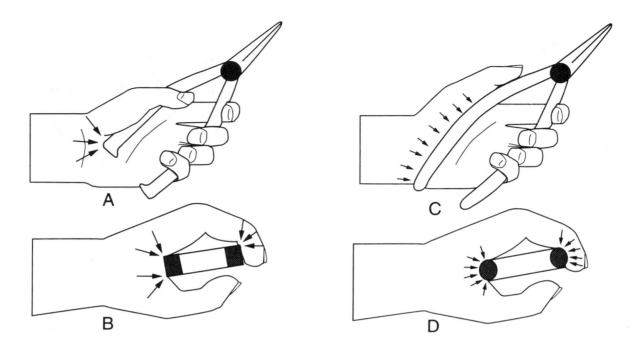

Figure 2.2. Examples of contact stresses caused by common tool use (Armstrong and Lackey 1994, reproduced with permission.)

1. Use anti-vibration gloves when using vibrating tools or equipment.
2. Dampen vibration by using flexible material on handles of tools or equipment.
3. Minimize exposure to vibration through job rotation or job enlargement.
4. Install special driving seats that dampen vibration.

Localized Contact Stresses—Contact stresses occur when soft tissue is compressed between the bone and hard or sharp objects such as tools or equipment. Compression of tendons in the fingers or nerves in the wrists are common CTDs. Figure 2.2 shows examples of contact stresses caused by common tool use. Following are steps to assess contact stresses:

1. Identify contact points between the body, tools or work objects, and work surfaces.
2. Measure or estimate the force of contact (such as the weight of the object) and the area and location of contact.
3. Determine the frequency and duration of the contact stress.

Methods for reducing or eliminating contact stresses include:

1. Use tools with longer handles to reduce stress concentrations in the palm of the hand.
2. Use tools with rounded edges on handles and ensure that work bench edges are also rounded or padded.
3. Use flexible-handle materials such as rubber on tools.
4. Use tools instead of the hand for pounding activities.
5. Provide padding for hands with gloves or some other type of material.

Cold Temperature—Prolonged exposure to temperatures between 32° F (0° C) and 68° F (20° C) results in reduced manual dexterity with accompanying nerve damage (Lockhart and Kiess 1971; Schiefer et al. 1984). Temperature affects the fingers through exposure to cold from the environment, exposure to cold exhaust from a tool, or handling cold materials. To prevent adverse effects on manual dexterity, finger temperatures should be kept above 77° F (25° C) (Armstrong and Lackey 1994). To assess the effects of cold temperature:

1. Check for exposure of the hand to cold less than 65° F (18°C).
2. Check the temperature of the environment.
3. Check for exhaust air.
4. Check the temperature of parts, materials, or tools.

The following controls will help prevent exposure to low temperatures:

1. Use gloves to increase finger temperature.
2. Construct tool handles using materials with low thermal conductivity.
3. Ensure exhaust air is directed away from workers.
4. Wear additional garments on the body.

The controls listed above to reduce or eliminate CTD risk factors are only a sample of the many available. As with any change to a job or a piece of equipment or tool, an evaluation should be conducted after implementation to assess the effectiveness of the change. Occasionally, through the elimination of one risk factor, another is created and the job or tool is made worse. A follow-up evaluation will prevent this from occurring.

Appendix C includes an ergonomic checklist for evaluating physical demands of various jobs or activities with the goal of enhancing performance, safety, and comfort. The information provided by the checklist may be used to identify problems or risk factors and to develop design improvements.

Anthropometry and Workplace Design

There is a great deal of variety between individuals in terms of their physical and mental characteristics, their abilities, and their limitations. These differences have an impact on the amount of work that can be performed and how well work can be performed. Ergonomics is concerned with measuring the variability of individuals and with matching the demands of tasks to the capabilities of these individuals.

Two approaches are used to improve the fit between task demands and individual variability: administrative controls, or fit the person to the job; and engineering controls, or fit the job to the person. Administrative controls address the capabilities of workers and include measures such as selection, and job skills such as training in proper lifting techniques. Engineering controls, on the other hand, improve work design through ergonomic improvements with the goal of eliminating CTD risk factors. Administrative controls are generally considered short-term, interim solutions while engineering controls are more desirable for long-term improvements.

A major factor to consider in evaluating a workplace is the worker's physical dimensions. Most workplaces do not fit the wide range of differences in the height and weight of individual workers. The workplace or equipment is often designed for the "average" person, or the designer uses himself as the "model." In order for the job to fit the person, the workplace or equipment must be suited to the body size of the operator.

The area of ergonomics that deals with measuring the human body for the purpose of optimal design of the workplace or products is called anthropometry. Using anthropometry to eliminate errors in design can have a tremendous impact on eliminating errors in operation in the workplace and in reducing accidents while using a piece of equipment. Some typical anthropometric measurements for a standing adult are shown in Figure 2.3.

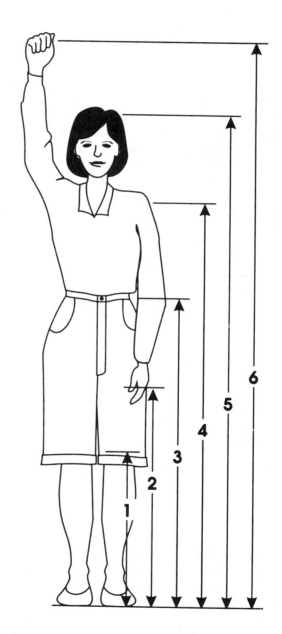

Figure 2.3. Typical anthropometric measurements (modified from McConville et al. 1981).

Table 2.3. U.S. civilian body dimensions of adults, all dimensions in inches (modified from McConville 1981).

	Men (percentile)				Women (percentile)		
	5	**50**	**95**		**5**	**50**	**95**
1. Knee Height	16.1	17.9	19.8	1. Knee Height	15.0	16.5	18.1
2. Knuckle Height	27.5	29.6	31.7	2. Knuckle Height	25.3	27.6	29.9
3. Elbow Height	39.4	43.3	46.9	3. Elbow Height	36.9	40.1	42.8
4. Shoulder Height	52.0	56.2	60.0	4. Shoulder Height	47.7	51.6	55.9
5. Standing Height	63.7	68.3	72.6	5. Standing Height	58.9	63.2	67.4
6. Overhead Reach	77.0	82.5	88.0	6. Overhead Reach	72.8	78.3	84.0

An example of anthropometric data is provided in Table 2.3. Measurements are generally taken for men and women, arms hanging relaxed at the sides, wearing light clothing, without shoes. Therefore, measurements should be increased accordingly to correct for the addition of shoes or additional clothing.

Anthropometric data is a tool for designers of workplaces, tools, and equipment. The goal of designers is to accommodate as many people as possible under extreme conditions. However, the design data that is used should represent the actual end users. For example, when designing for elderly people or certain ethnic groups, body measurements will vary greatly from those of the general population.

The first step in the design process is to define the user population. For adult populations this would include gender, ethnic origin, and age. The next step is to identify the limiting user. This is the largest or smallest individual who needs to be accommodated. Then the design approach should be based on how critical it is for a satisfactory match between the workplace or equipment and the users. The four design approaches using anthropometric data are as follows (Sanders and McCormick 1993):

Design for the Average Individual—There is no such thing as an average person. However, there are times when it is acceptable to design for the average as a compromise.

Design for Extreme Individuals—When designing certain items, it is best if designs accommodate all of a certain population. For instance, designing for the maximum population is appropriate if all people must be accommodated. The size of a door opening, the size of an escape hatch, and the strength of a supporting structure are all examples of items that should be designed for extremes. Likewise, designing for a minimum population is appropriate if some design feature must accommodate all people such as the distance of a control button or the force required to operate a control.

Design for a Range of Adjustment—This method of design is generally most preferable but is not always feasible. Equipment and facilities are often designed to be adjustable such as chairs, work benches, and automobile seats. Typically, designs for adjustability cover a range of 5th percentile for females to 95th percentile for males.

Design for the Entire Population—This approach is generally the most expensive, however it does accommodate the largest number of users.

Recommendations for workplace or equipment dimensions are not based solely on anthropometric data. Dimensions are also based on the behavior of individuals and the requirements of the work or the job itself. Working height is very important in the design of the workplace and equipment. If the work height is too low, the worker must lean forward, causing back discomfort. If the work height is too high, the shoulders must be raised, causing neck and shoulder discomfort. The appropriate work heights for various tasks are listed in Table 2.4.

HAND TOOL DESIGN

Another area where anthropometric data can be used to help fit the job to the person is in the design of hand tools. Hand tools are used to assist with the performance of various tasks such as cutting, pounding, turning, and scraping. Hand tools have been in existence since the Stone Age, and remarkably, interest in ergonomic design of tools has been in existence throughout history (Childe 1944).

Since the turn of the century, hand tools have gone from manual to powered. Since the forces while using tools are

Table 2.4. Recommended work heights for various activities (modified from Pheasant 1991 and Putz-Anderson 1988).

Activity	Height
Manipulative tasks involving moderate amount of force and precision.	2-4 inches (50-100mm) below elbow height.
Lighter and more delicate tasks (including writing).	2-4 inches (50-100mm) above elbow height.
Heavier tasks (particularly if they involve downward pressure on the work item).	4-10 inches (100-250 mm) below elbow height.
Lifting and handling tasks.	Knuckle height to a little above elbow height.
Two-handed pushing and pulling actions.	Slightly below elbow height.
Hand-operated controls.	Between elbow and shoulder height.
Computer input devices and keyboards.	Elbow height.

greater, the risk of injury is also greater. To help prevent or reduce the risk of injury, hand tools and handles should be selected or designed using the following guidelines (Putz-Anderson 1988; Helander 1995; Sanders & McCormick 1993):

- Handles should be provided—Handles are necessary to prevent exposure of the hand to vibration, exhaust air from motors, skin compression, and burns. Handles should be designed with a smooth surface without ridges, slightly compressible, and non-conductive. In general, triangular-shaped or oval-shaped handles are preferred. Avoid cylindrical handles. Grip size should be appropriate: .3 to .5 inches (8-13 mm) for tools requiring precision grips such as knives and 2 to 2.3 inches (50-60 mm) for tools requiring a power grip where the entire hand wraps around the handle. The minimum handle length for a tool is 4 inches (100 mm).

- Avoid contact stress—Considerable force is sometimes required to operate a hand tool. When this force is concentrated into the palm of the hand or pressure-sensitive areas of the hand, blood vessels and nerves become compressed. Select tools that are designed with large contact surfaces for even distribution of forces over a larger area of the hand.

- Design for minimum muscular effort—To minimize muscular effort, select tools that are balanced about the axis of the grip. In addition, use tools with the center of gravity close to the operator's body to minimize fatigue from holding the tool. Use of counterbalances for tools also help to minimize fatigue from prolonged use.

- Power with motors rather than with muscles—Use power tools whenever possible to reduce CTD risk factors such as forceful exertions and repeated exertions. For this to be effective, select the lightest tool possible.

- Bend the tool, not the wrist—This is a common recommendation for preventing CTDs in the use of hand tools. When the hand is in a neutral posture while using hand tools, the risk of developing CTDs such as carpal tunnel syndrome is minimized. This principle is illustrated in Figure 2.4, by a study conducted by Tischauer at Western Electric. For powered drivers, tools with pistol grips are most effective when the tool axis is horizontal. Tools with in-line handles are most effective when the tool axis is vertical. These principles are illustrated in Figure 2.5.

- Avoid vibration—Exposure to vibration from hand tools such as chain saws, jack hammers, and drills is a risk factor for vibration injuries such as Raynaud's disease, also known as vibration white finger. This disease is caused by vibration in the frequency range of 50 to 100 Hz which may permanently damage blood vessels and nerves.

- Avoid repetitive finger action—If the index finger is used repeatedly for operating triggers or buttons, a CTD known as "trigger finger" can result. This condition restricts movement of the index finger allowing a person to flex the finger, but not extend it. Thumb-activated triggers are better than index finger-activated triggers. Tools with triggers that utilize all four fingers are called "finger-strip" controlled. These tools are preferable as they allow the load to be spread across all fingers, minimizing the risk to any one finger.

- Remember Women and Lefthanders—Design tools for use by either hand with consideration for users of various cultures and genders. Hand size and grip strength vary considerably in these different populations.

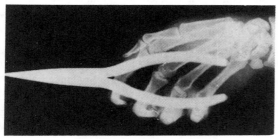 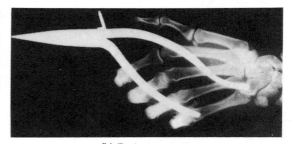

Figure 2.4. Pliers bent vs. bent wrists (Tischauer 1966, as presented by Sanders and McCormick 1993, reproduced with permission of McGraw-Hill).

Women make up approximately 50% of the world's population and lefthanders about 8 to 10%.

Design of Agricultural Machinery and Related Equipment

As more and more agricultural processes become mechanized, the demand for safety, productivity, and operator comfort increases. When purchasing or designing a piece of machinery, ergonomic factors are important. However, sometimes a balance between ergonomics and economics must be maintained.

Studies of farm tractors over the years have shown that even modern tractors have major ergonomic shortcomings. Ergonomic standards are often lower in many respects for tractors than for equipment used in other industries

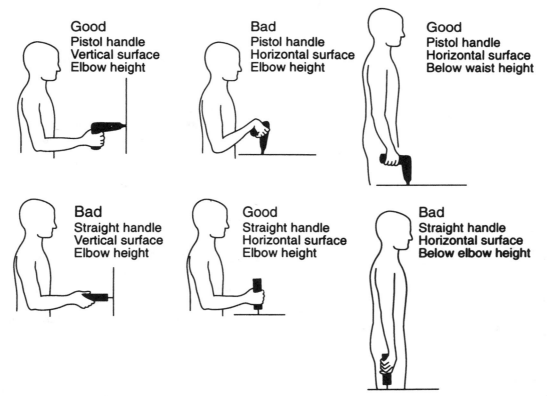

Figure 2.5. Examples of wrist postures with various tools and work surfaces (Helander, *A Guide to the Ergonomics of Manufacturing*, 1995, reproduced with permission.)

(Hansson 1990). Operators of agricultural machinery need to be able to sit comfortably, utilizing neutral body postures of the arms, legs, shoulders, head, and back. Exposure to dust, noise, and vibration must be minimized as well.

To aid in the process of designing or purchasing a tractor, Hansson (1991) has developed an "Ergonomic Checklist for Tractors and Agricultural Machinery." This checklist covers eleven different areas for individual assessment. The areas covered by the checklist include:

- The Cab
- The Seat
- Visibility from the Cab
- Entry/Exit
- Controls/Instrumentation
- Service and Maintenance
- Noise/Vibration
- Lights/Glare
- Cab Climate
- Dust and Exhaust
- Hitching of Implements

The checklist is helpful when comparing machines for purchase or when comparing design options for tractors. The checklist is provided in Appendix D. Appendix E lists standards for industrial and agricultural systems and equipment design standards.

Design of Controls and Displays

The design of controls and displays should always consider the expectations and the stereotypes of operators. For example, an operator expects to find the accelerator control on the right side of a vehicle and the brake on the left. Reversing the two controls would violate the operator's stereotype or expectation with disastrous consequences.

This idea of making sure that controls and displays are compatible with operators' expectations is particularly important in the design of agricultural equipment due to the unique aspects of the industry. For example, since the work is seasonal in nature, equipment may not be used frequently enough to assure familiarity, supervision is often minimal, and workers may be illiterate or unable to read English.

The purpose of a control and a display is to exchange information between the operator and the machine. A diagram of the man-machine interface is illustrated in Figure 2.6. The display provides information to the operator and the operator who perceives this information must understand it and make correct interpretations. Then the operator makes a decision and uses the controls to communicate the decision

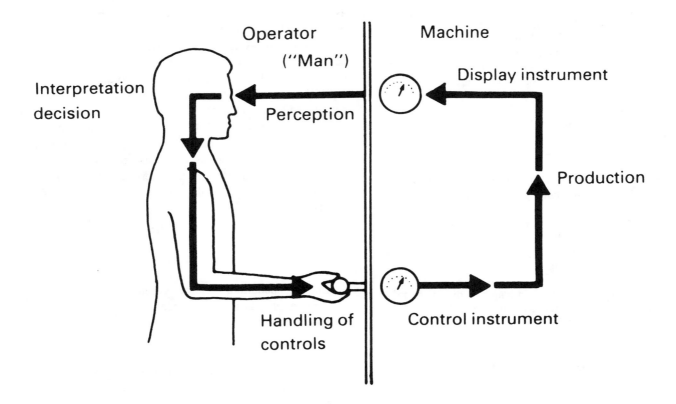

Figure 2.6. The Man-made machine system (Grandjean 1988, reproduced with permission)

to the machine. The display provides feedback to the operator about the results of his actions.

The information exchange from man to machine and from machine to man must be as clear and accurate as possible to prevent operator errors. This is achieved through proper coding for controls and displays. Some examples of coding are provided in Table 2.5.

When a number of controls or displays are involved in the operation of a piece of equipment, the following principles apply (Sanders and McCormick 1993).

1. The importance principle—the most important controls or displays should be the most prominent in the layout. Importance can be determined by frequency of use, amount of time used, or the need for quick and accurate response such as in emergency situations.
2. The frequency of use principle—the most frequently used controls should be within easy reach and the most frequently read displays in the direct line of sight.
3. The function principle—related controls and displays should be grouped together.
4. The sequence of use principle—controls or displays that have a sequential relationship to one another should be grouped together and placed in sequential order.

The location of controls and displays should be arranged according to how close they are to the operator in the workplace and the importance of the function controlled. Primary controls are used frequently and should be within easy reach of the operator. Secondary controls are used less frequently and can be located further away from the operator. Likewise, primary displays are read frequently and should require very little operator movement for viewing. Secondary displays are viewed less frequently and may be located outside the normal line of sight. Emergency displays may be located in either the primary or secondary area. However, consideration should be given to including an alarm or other signal to call attention to the emergency display.

Physical Work

In most industrial countries heavy physical work is not as common as it once was. Manual labor has been replaced by materials handling equipment, mechanization, and automation. However, within industries such as construction, agriculture, commercial fishing, and logging many occupations still require heavy physical work.

All human behavior and activity are controlled by muscle activity. Muscle constitutes 40 to 45% of total body weight for the average man and 25 to 35% for the average woman. When a person is at rest, 0.2 liters of oxygen are consumed per minute. During periods of intense physical work or activity that amount increases to 3 liters per minute. This is due almost entirely to the demands of muscles for metabolism.

Energy expenditure can sometimes be difficult to measure. Because of this, work is generally measured in terms of oxygen consumption:

Energy expenditure = 5 x oxygen consumption
(measured in kcal per minute)   (measured in liters per minute).

An individual's capacity to perform physically strenuous work is determined by his or her maximum aerobic power, which is measured by oxygen consumption. This is dependent on factors such as age, gender, and physical conditioning. Some examples of energy expenditure for various working and leisure activities are shown in Figure

Table 2.5. Coding for controls and displays (modified from Kaminaka 1991).

Labels	Labels should be located directly above the control and should be horizontal. Wording should be brief and should use familiar terms or words. Labels should be legible while the control is being operated.
Color	Colors should be easily distinguishable from one another. Different light levels can cause colors to be confused.
Shape	Shapes should be distinguishable through sight and feel.
Texture	The texture of surfaces should be easily distinguished but the use of gloves may interfere with this type of coding.
Size	Variations in length or diameter can be used as long as the differences can be detected.
Mode of Operation	For example, pushing a lever up typically indicates a movement forward.
Location	For example, the gas pedal in a vehicle is always to the right of the brake pedal.

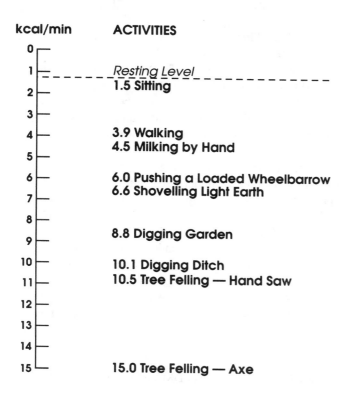

Figure 2.7. Typical levels of energy expenditure for various activities (modified from Pheasant 1991).

reason, work physiologists often study the physiological efficiency of various working methods, tools, and equipment. These measurements make it possible to design better jobs and equipment.

One such study was conducted in Germany and looked at various shoveling activities (Lehmann 1962). It was found that shoveling was most efficient when 18 to 22 pounds (8 to 10 kg) of load was shoveled 12 to 15 times per minute. When the material is light, a larger shovel is recommended. When the material is heavy, a smaller shovel is recommended. For fine-grained materials, the shovel should be slightly hollow and spoon shaped with a pointed tip. For course materials the cutting edge should be straight and the blade flat with a rim around the back and sides. For stiff materials such as clay, the cutting edge can be either straight or pointed, but the blade should be flat.

Another similar study evaluated the results of loosening the soil in a vegetable plot using two different hoes (Egli et al. 1943). In soft soil, a swivel hoe is more efficient than an ordinary chopping hoe. But if the soil is hard and dry either one work well as shown in Figure 2.8.

2.7. These are typical values for an average, 155 pound (70 kg) person performing activities under normal conditions.

A good indicator for assessing workload is heart rate or pulse rate. Simple measurements of heart rate are useful for determining if there are any problems with the current workload level. For example, about 100-140 beats per minute should be expected for workloads that are moderately intense.

The level of work intensity can also be expressed in terms of fatigue. Physiological fatigue or local fatigue refers to excessive use of a muscle set or body system. It may be experienced as a muscle cramp. At the worksite, general fatigue will lead to reduced performance, increased error rates, and the potential for accidents.

Local fatigue is classified as either acute or chronic. Acute fatigue is relieved by rest whereas chronic fatigue sets in when there is insufficient recovery time between work sessions. Fatigue generally depends on how hard as well as how long a person works. Proper work design can help control fatigue by reducing the workload while increasing the duration of work.

For all types of heavy physical work it is important to strive for maximum efficiency, or minimal energy use with minimal ergonomic stresses upon an individual. For this

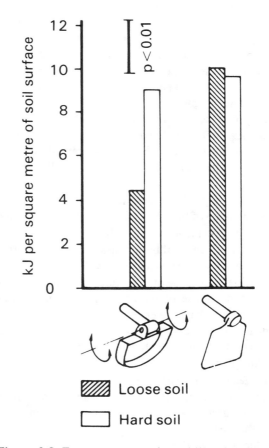

Figure 2.8. Energy consumption while using two types of hoes in two soil types (Egli et al. 1943, as presented by Grandjean, *Fitting the Task to the Man*, 1988, reproduced with permission.)

Manual Materials Handling

Manual materials handling (MMH) is a common problem for workers throughout the world. Individuals in jobs who must lift, lower, carry, push, or pull heavy materials have an increased rate of injuries, particularly of the low back. For the years 1985 through 1987 the Bureau of Labor Statistics reports that 71,098 compensation claims were filed by agricultural workers for all types of injuries. Of these, 26,450 or 37.2% were filed for a sprain or strain injury. Back injuries comprised 11,810 claims or 47% of all sprains/strains. Within that time period, sprains and strains were found to be highest in two agricultural industry groups: landscape/horticultural services and fruits/tree nuts (Bobick and Myers 1994).

Industrial Back Belts

Some employers approach the problem of lifting injuries through the use of "Lifting Belts" or "Back Belts." This approach should not be considered a "cure all" for back injury but instead indicates a need for revising lifting tasks and reducing the risks associated with lifting activities. Despite claims from vendors, there is very little scientific evidence to demonstrate that back belts prevent injury.

Over the last several years, the number of workers who use back belts has increased dramatically. These belts are currently worn by workers in a number of industries including airline baggage handling, warehouse operations, and grocery/produce handling. Because of the increased use, researchers at The National Institute of Occupational Safety and Health (NIOSH) became concerned with the question of whether or not back belts actually protect workers.

After reviewing the scientific data and literature addressing the issue of back belt usage, NIOSH determined that the results are inconclusive (NIOSH 1994). NIOSH believes the effectiveness of back belts cannot be supported or refuted based on the scientific evidence of studies performed to date. Because of these findings, NIOSH "does not recommend the use of back belts to prevent injuries among workers who have never been injured." Workers who rely on the use of back belts may actually increase their risk of injury by lifting more with the belt than they would normally lift without the belt.

Rather than relying on back belts to protect workers from lifting injuries, employers should establish a formal ergonomics program. Redesigning the work environment to make sure that tasks do not exceed a worker's capabilities is a more effective way of reducing the risks associated with lifting. Training in proper lifting methods and awareness of lifting hazards should also be included in this effort.

NIOSH LIFTING GUIDELINES

NIOSH addressed the concern for low back injuries by developing a set of guidelines that are used by industries throughout the United States. The NIOSH guidelines are not rules for safe lifting but rather indicate safe lifting limits. The guidelines were originally intended to be advisory, however the Occupational Safety and Health Administration (OSHA) has cited companies in the past for noncompliance with the guidelines.

The Revised NIOSH Lifting Equation (Putz-Anderson and Waters 1991) is a model that quantifies a lifting task and provides a means of determining and setting limits for safe lifting. The model is based on research from many different fields of study including physiology, biomechanics, and psychophysics. The Recommended Weight Limit (RWL) is the value calculated from the revised NIOSH lifting equation. The RWL is defined as the weight of the load that nearly all healthy workers could perform without an increased risk of developing low back pain over a prolonged period of time of up to eight hours (Waters et al. 1994). The RWL is defined by the following equation:

$$RWL = LC \times HM \times VM \times DM \times AM \times FM \times CM.$$

A description of the individual task multipliers and calculations are provided in Table 2.6. The equation assumes that in an "ideal" lifting situation, with all of the multipliers equal to one, the maximum weight that most workers can lift safely, or the RWL, is equal to the Load Constant (LC), or 51 pounds. Each multiplier is reduced (reducing the 51-pound weight) as the lift varies from an ideal lifting situation.

- The Horizontal Multiplier (HM) is based on the distance between the point where the hands hold the load and a point midway between the feet, measured horizontally with a tape measure.

- The Vertical Multiplier (VM) is based on the distance between the floor where the worker stands and the point where hands hold the load, measured vertically with a tape measure.

- The Distance Multiplier (DM) is based on the distance the load is raised or lowered by the worker, measured vertically.

- The Asymmetric Multiplier (AM) is based on the angle between the assymmetry line and the mid-sagittal line for a lift that begins or ends outside the mid-sagittal plane.

Table 2.6. Components of the RWL equation (Waters et al. 1994).

		U.S. Units	Metric Units
Load Constant	LC	51 lbs	23 kg
Horizontal Multiplier	HM	(10/H)	(25/H)
Vertical Multiplier	VM	$1 - (.0075 \times |V-30|)$	$1 - (.003 \times |V-75|)$
Distance Multiplier	DM	$.82 + (1.8/D)$	$.82 + (4.5/D)$
Asymmetric Multiplier	AM	$1 - (.0032 \times A)$	$1 - (.0032 \times A)$
Frequency Multiplier	FM	Table 3.7	Table 3.7
Coupling Multiplier	CM	Table 3.8	Table 3.8

Where H = distance from the mid-point of the line joining the ankle bones to a point on the floor directly below the mid-point of hand grasps; V = the vertical location of the hands from the floor; D = the vertical travel distance of the hands between the origin and destination of the lift; and A= the angle between the asymmetry line and the mid-sagittal line.

Table 2.7. Abridged frequency multiplier table (Waters et al. 1994).

Lifts Per Minute	<= 8 hours	<= 2 hours	<= 1 hour
.2	.85	.95	1.00
.5	.81	.92	.97
1.0	.75	.88	.94
2.0	.65	.84	.91
4.0	.45	.72	.84
6.0	.27	.50	.75
8.0	.18	.35	.60

Table 2.8. Coupling multiplier table (Waters et al. 1994).

Coupling	V < 30 inches	V >= 30 inches
Good - Comfortable Handles or Cut Outs	1.00	1.00
Fair - Uncomfortable Grips	.95	1.00
Poor - No Handles, Bulky Objects or Sharp Edges	.90	.90

- The Frequency Multiplier (FM) is a value that is determined from a table of values (Table 2.7) based on how often lifting is performed during a lifting session and the duration of the lifting session.

- The Coupling Multiplier (CM) is a value that is dependent on how easy or how difficult it is to grip the object and is also determined from a table of values (Table 2.8). Coupling refers to the nature of the hand-to-object gripping method which affects the force a worker must exert on the object during a lift.

The lifting equation, multipliers, and required measurements are demonstrated in the following example.

Figure 2.9 illustrates a task that is performed by a worker at an agricultural supply store twenty times throughout a shift. The worker is required to load 40-pound bags of seed from a hand truck to a shelf on a shelving unit. Measurements are taken and the data is recorded on a job analysis worksheet as indicated in Figure 3.10.

The vertical location of the hands is 15 inches at the origin and 36 inches at the destination. The horizontal location of hands is 18 inches at the origin and 10 inches at the destination. The lift is asymmetric in that the worker must turn approximately 45° in order to pick up the bag and unload the bag onto the shelf. (A = 45 for origin and destination). According to Table 2.8, the coupling is classified as fair because the bags do not have handles but they are semi-rigid and the worker can flex the fingers about 90°. The frequency is less than .2 lifts per minute for less than one hour.

As shown in Figure 2.10, the RWL for this activity is 18.9 pounds. The weight to be lifted is 40 pounds, which is greater than the RWL. This indicates that such a lift would be hazardous for the majority of healthy workers.

The Lifting Index (LI) is a term that provides a relative estimate of the level of physical stress associated with a particular manual lifting task. The estimate of the level of physical stress is defined by the ratio of the weight of the load lifted to the recommended weight limit. The LI is defined by the following equation:

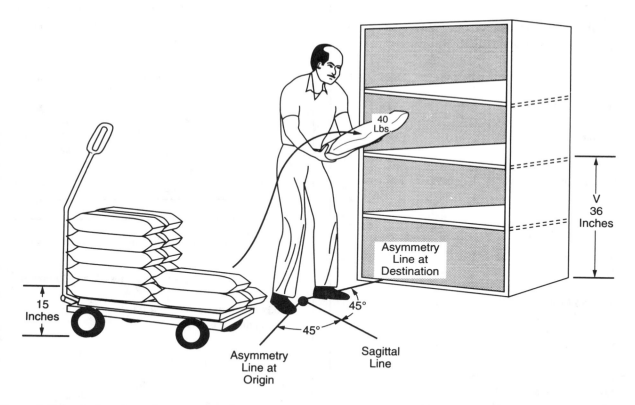

Figure 2.9. Agricultural supply store example.

$$LI = \frac{\text{Load Weight}}{\text{Recommended Weight Limit}} = \frac{L}{RWL}$$

The LI for the above example would be 2.1. Both the recommended weight limit (RWL) and the lifting index (LI) can be used to guide the ergonomic design of lifting activities in several ways (Waters et al. 1994):

1. The individual multipliers are used to identify specific job-related problems. The magnitude of each multiplier indicates the contribution of each task factor (horizontal, vertical, frequency, etc.)

2. The RWL is used to guide the redesign of existing manual lifting jobs or to design new manual lifting jobs. For example, if the task variables are fixed, then the maximum weight of the load is selected so as not to exceed the RWL; if the weight is fixed, then the task variables are optimized so as not to exceed the RWL.

3. The LI may be used to estimate the magnitude of physical stress for a task or job. From the NIOSH perspective, it is likely that lifting tasks with a LI > 1.0 pose an increased risk for low back pain for some percentage of the work force. The greater the LI, the smaller the number of workers capable of safely performing the lift. Two or more job designs can be evaluated by comparing the LI for each design.

4. The LI is used to prioritize ergonomic redesign. For example, a number of hazardous jobs could be rank ordered according to the LI and a strategy could be developed according to the rank ordering.

For the example illustrated in Figure 2.9, the following job modifications could be implemented to increase the RWL, reduce the LI, and decrease the risk of injury:

1. Bring the load closer to the worker and move the worker closer to the shelves to increase the HM value.
2. Reduce the angle of asymmetry to increase the AM. This could be done by moving the origin and destination points closer together.
3. Raise the height at the origin to increase the VM.

For a better understanding of the research behind the development of the RWL and the LI, consult the paper entitled Revised NIOSH Equation for the Design and

Figure 2.10. Single Task Job Analysis Worksheet (Waters 1994).

JOB ANALYSIS WORKSHEET

Department Agriculture Supply Store
Job Title Stock Clerk
Analyst's Name _____
Date _____

Job Description Loading bags onto shelves

STEP 1. Measure and record task variables

Object Weight (lbs)		Hand Location				Vertical Distance	Asymmetric Angle (Deg)		Frequency Rate lifts/min	Duration (hours)	Object Coupling
		Origin		Destination			Origin	Destination			
L (AVG)	L (max)	H	V	H	V	(D)	(A)	(A)	(F)		(C)
40	40	18	15	10	36	21	45	45	<.2	<1	Fair

STEP 2. Determine the multipliers and compute the RWL's

$$RWL = LC \times HM \times VM \times DM \times AM \times FM \times CM$$

ORIGIN   RWL = 51 x .56 x .89 x .91 x .86 x 1.0 x .95 = 18.9 Lbs

DESTINATION RWL = 51 x ___ x ___ x ___ x ___ x ___ x ___ = ___ Lbs

STEP 3. Compute the LIFTING INDEX

ORIGIN   LIFTING INDEX = $\dfrac{\text{OBJECT WEIGHT (L)}}{\text{RWL}} = \dfrac{40}{18.9} = 2.1$

DESTINATION LIFTING INDEX = $\dfrac{\text{OBJECT WEIGHT (L)}}{\text{RWL}} = $ ___ = ___

Evaluation of Manual Lifting Tasks (Waters et al. 1993) or consult the Applications Manual for the Revised NIOSH Lifting Equation (Waters et al. 1994).

Other Governmental organizations have addressed problems associated with MMH in different ways. In 1991, the Health and Safety Commission in the United Kingdom drafted a proposal for regulations and guidance for handling heavy loads at work. This document lists principles to follow when designing work for minimizing material handling injuries. The main points of the document are summarized in a checklist presented in Table 2.9.

Table 2.9. MMH checklist for employers (modified from Health and Safety Commission 1988).

1. Does the task involve:

Holding loads at a distance from the body?
Awkward postures, especially twisting or stooping?
Excessive movement of the load, especially
   -excessive lifting or lowering distances?
   -excessive carrying distances?
   -excessive pushing or pulling distances?
Risk of sudden movement or shifting of the load?
Frequent or prolonged physical effort?
Insufficient rest or recovery periods?

2. Is the load:

Heavy?
Bulky or unwieldy?
Difficult to grasp?
Unstable or are the contents likely to shift?
Sharp, hot or otherwise potentially damaging?

3. In the workplace are there:

Space constraints preventing good posture?
Uneven, slippery or unstable floors?
Variations in level of floors or work surfaces?
Extremes of temperature, humidity, or air movement?
Poor lighting conditions?

4. Does the job:

Require unusual strength, height, etc.?
Create a hazard to those who are pregnant or have a
   health problem?
Require special knowledge or training for safe
   performance?

When designing or purchasing manual material handling equipment, special attention should be given to features such as the type and size of the caster, the size of the equipment, and the location and type of handles or grips. Appendix F lists some guidelines for selecting or designing manual materials handling equipment.

THERMAL EXTREMES

Every year individuals in the agricultural industry suffer cold-related or heat-related injuries that result in temporary or even permanent disabilities or fatalities. Most of the health problems due to exposure of extreme temperatures may be avoided through proper precautions.

In general, the body maintains an internal temperature of 97° F to 99° F. However, when conditions are such that the amount of heat lost is greater than the amount of heat gain, the body temperature is lowered and hypothermia results. Hypothermia may even occur at temperatures above freezing. When the core body temperature drops below 95° F, vital organs are at risk of being damaged.

Older people and people with certain health problems such as diabetes, arthritis, and alcoholism tend to be more at risk for hypothermia. When the core body temperature drops below 95° F symptoms include shivering, confusion, loss of coordination, difficulty speaking, decreased breathing rate, and a decreased heart rate. When the body temperature reaches 80° F, loss of consciousness typically occurs. To prevent hypothermia, the following precautions are suggested for individuals (North Carolina OSHA 1990):

- Eat well and drink plenty of fluids, avoiding alcohol.

- Dress to hold in body heat, keeping in mind that half of the body's heat can be lost through the head, so a hat or head covering should be worn.

- Bedtime warmth is critical since air temperature and body temperatures generally fall while an individual sleeps.

- Exercise and other physical activities cause the body to produce an increase in heat. However, perspiration carries heat away from the body and decreases body temperature. Therefore, clothing with moisture control or layered clothing is important to protect against the cold.

- Keep the working and living space sufficiently warm, especially for those individuals with the health problems mentioned above.

When conditions are such that the amount of heat gain is greater than the amount of heat loss, a condition called hyperthermia results. Hyperthermia-related disorders include heat rash, fatigue, cramps, heat exhaustion, and heat stroke. Heat strokes occur when the core body temperature reaches 108° F and result in death about 50% of the time.

Symptoms of heat exhaustion typically include sweating, glassy eyes, a rapid heart rate, dizziness, near normal body temperature, headache, chills, and pale and clammy skin. Symptoms of heat stroke include hot, dry, flushed skin, elevated body temperature, confusion, disorientation, incoherent speech, convulsions, and unconsciousness.

Precautions to prevent heat disorders for individuals include (North Carolina OSHA 1990):

- Start the day well-rested. Fatigue and poor physical conditioning increase the risk for heat stress.

- Dress for the heat by wearing loose-fitting, lightweight clothing. Wearing a wide-brimmed hat will help shield against thermal radiation.

- Schedule heavy work during the cooler parts of the day or on alternate days. Whenever possible, reduce the workload or stay in the shade.

- Establish a work/rest schedule and a heavy work/light work schedule.

- Keep a balanced diet and frequently drink water or juice. When high temperature and humidity are present, the body requires even more fluids than are necessary to satisfy thirst.

- Special attention should be given to the elderly and others with health problems since they are more at risk during temperature extremes.

The U.S. Environmental Protection Agency (EPA), along with OSHA, has developed a *Guide to Heat Stress in Agriculture*. Appendix G lists eight steps of a basic program for controlling heat stress.

DEVELOPING AN ERGONOMICS PROGRAM

In terms of applicable standards or regulations, the draft ANSI Z-365 Standard on Cumulative Trauma Disorders provides guidance on developing ergonomics programs. Another document that provides guidance in this area is OSHA's "Ergonomics Program Management Guidelines for Meatpacking Plants," (OSHA 1990) published due to the high number of cumulative trauma disorders (CTDs) in that industry. However, injuries of this type are not confined to the meatpacking industry, and apply to all employers experiencing injuries caused by ergonomic hazards. In the guidelines, OSHA refers to ergonomic hazards as workplace conditions that pose a biomechanical stress to the worker. These hazardous workplace conditions include poor workstation layout, improper work methods, unsuitable tools or equipment, and job design problems that include work flow, posture, work/rest schedules, and repetition rate.

Also, in these guidelines, OSHA describes an effective ergonomics program as one that includes four major program elements: worksite analysis, hazard prevention and control, medical management, and training and education. The objectives of worksite analysis are to recognize, identify, and correct ergonomic hazards. Once hazards are identified, the next step is to design measures to prevent or control these hazards. The third element is implementation of a medical management system to eliminate or reduce the risk of injury though early detection and treatment. The final element is training and education to ensure that employees are sufficiently informed about ergonomic hazards to participate in their own protection.

Since the publication of the Meatpacking Guidelines, OSHA has released a Draft Proposed Ergonomic Protection Standard to several labor and management groups for discussion and suggestions only (OSHA 1995). It has not yet been released as an official document and it is not a substitute for the formal public comment and public hearing process. However, since it does provide a look at OSHA's strategy for preventing CTDs, it is discussed here as an aid to developing an ergonomics program.

The OSHA approach emphasizes the use of a risk-factor checklist to examine upper extremity, back, and lower extremity factors (see Appendix H) and to identify problem jobs. A score is assigned for tasks that involve lifting, pushing, pulling, kneeling, or squatting and the score is weighted by how long the activity is performed throughout the day. If a score exceeds a certain limit, the job is considered a "problem job" and the employer must take "further action" to reduce ergonomic hazards. Employers must continue efforts to control each problem job, using engineering controls and/or administrative controls, as long as the new risk-factor checklist score is above the limit.

According to the draft, employers would also have to provide ergonomic awareness and job-specific training for every employee working in a problem job as identified by the risk-factor checklist. Supervisors of these employees are also required to receive training. In addition to providing training for employees, employers will be required to evaluate the effectiveness of the training.

A medical management section, which is intended to help employers keep employees on the job and return them to work as quickly as possible after an illness or injury, is included in the draft. Once an individual has reported signs or symptoms of an illness or injury, the employer must obtain a medical assessment of the individual within five days. A "musculoskeletal disorder management plan" must also be obtained from a health care provider, including diagnosis, treatments, work restrictions, and follow-up examinations.

The last section of the draft deals with record keeping and requires each employer with ten or more employees to establish and maintain accurate records of the identification of problem jobs. These records should include the job titles and employee names working in problem jobs as well as copies of risk-factor checklists for these jobs. These records must be maintained for at least five years after the job is brought under control.

SUMMARY

The main objective of ergonomics is to make the job fit the person, not to make the person fit the job. This objective is generally accomplished by redesigning the job, tools, or workspace in order to reduce biomechanical stressors such as high force, repetition, and awkward postures. By improving the fit between the worker and the job, both the well-being and the productivity of the employee are positively impacted.

Applying ergonomics principles within the agricultural, forestry, and fisheries industries will have a positive impact on entire organizations. CTDs are a major cause of lost time in all of these industries. In addition to reducing CTDs and improving employee well-being, the implementation of a good ergonomics program will also lead to increased job satisfaction and morale, as well as reduced turnover and absenteeism. Past experience has shown that ergonomics programs have helped organizations reduce costs, while improving service and quality.

REFERENCES

American National Standards Institute (ANSI). 1995. *ANSI Z-365 Standard on Cumulative Trauma Disorders (draft)*. Chicago: National Safety Council.

Armstrong, T.J. 1995. *Analysis and Design of Jobs for Control of Upper Limb Musculoskeletal Disorders*. Ann Arbor, MI: University of Michigan, Center for Occupational Health and Safety Engineering.

Armstrong, T.J., L.J. Fine, S.A. Goldstein, V.R. Lifshitz, and B.A. Silverstein. 1987. Ergonomics considerations in hand and wrist tendonitis. *J. Hand Surgery* 12A(5) Part 2: 830-7.

Armstrong, T.J., and E.A. Lackey. 1994. *An Ergonomics Guide to Cumulative Trauma Disorders of the Hand and Wrist*. American Industrial Hygiene Association.

Bobick, T., and J.R. Myers. 1994. Agricultural related sprain and strain injuries, 1985-1987. *Intern. J. Ind. Ergo.* 14: 223-32.

Bureau of Labor Statistics, U.S. Department of Labor. 1991. *Occupational Injuries and Illnesses in the United States by Industry 1989*, Bulletin 2379. Washington, DC: U.S. Government Printing Office.

Bureau of Labor Statistics, U. S. Department of Labor. 1995. *National Census of Fatal Occupational Injuries, 1994*. Filename: BLS95.288. Washington, DC: U.S. Government Printing Office.

Childe, G. 1944. *The Story of Tools*. London: Cobbet.

Egli, R., E. Grandjean, and H. Turrian. 1943. Arbeitsphysiologische untersuchungen an hackgeraten. *Arbeitsphys.* 15: 231-34.

Grandjean, E. 1988. *Fitting the Task to the Man*. London: Taylor & Francis.

Hansson, J-E.. 1991. Ergonomic checklist for tractors and agricultural machinery. *Metodrapport 1*, National Swedish Institute of Occupational Health.

Hansson, J-E. 1990. Ergonomic design of large forestry machines. *Intern. J. Ind. Ergo.* 5: 255-66.

Hansson, J-E., K. Adolfsson, S. Kihlberg, L. Larsson. *The Agricultural Tractor as a Workplace. Research Report 15*. The Swedish National Institute of Occupational Health. (English Summary)

Health and Safety Commission. 1988. *Handling Loads at Work—Proposals for Regulations and Guidance*. London: Health and Safety Executive.

Helander, M. 1995. *A Guide to the Ergonomics of Manufacturing*. London: Taylor & Francis.

Hudock, S.D. 1994. *Ergonomics in Agriculture*. Cooperative Agricultural Surveillance Training (CAST).

Kaminaka, M.S. 1991. Equipment design: controls and displays. *Human Factors*. St. Joseph, MI: American Society of Agricultural Engineers.

Karwowski, W. 1991. Complexity, fuzziness, and ergonomic incompatibility issues in the control of dynamic work environments. *Ergonomics* 34: 671-86.

Lehmann, G. 1962. *Praktische Arbeitsphysiolgie.* 2nd edition. Stuttgart: Thieme Verlag.

Lockhart, J., and H. Kiess. 1971. Auxiliary heating of the hands during cold exposure and manual performance. *Human Factors* 13: 457-65.

McConville, J.T., K.M. Robinette, and T. Churchill. 1981. *An Anthropometric Data Base for Commercial Design Applications.* Yellow Springs, OH: Anthropology Research Project.

National Institute of Occupational Safety and Health (NIOSH). 1994. *Workplace Use of Back Belts: Review and Recommendations.* Publication No. 94-122. Washington, DC: Department of Health and Human Services.

North Carolina Division of Occupational Safety and Health. 1990. *A Guide to Farm Safety and Health.* Raleigh, NC: North Carolina Division of Occupational Safety and Health.

Occupational Safety and Health Administration. March 1995. *Draft Proposed Ergonomic Protection Standard, Unofficial Publication.* Washington, DC: U.S. Department of Labor.

Occupational Safety and Health Administration. 1990. *Ergonomics Program Management Guidelines for Meatpacking Plants,* OSHA Publication 3123. Washington, DC: U.S. Department of Labor.

Pheasant, S. 1991. *Ergonomics, Work and Health.* Gaithersburg, MD: Aspen Publishers.

Putz-Anderson, V. 1988. *Cumulative Trauma Disorders: A Manual for Musculoskeletal Diseases of the Upper Limbs.* London: Taylor & Francis.

Putz-Anderson, V., T. Waters. 1991. Revisions in *NIOSH Guide to Manual Lifting*, paper presented at national conference entitled, "A national strategy for occupational musculoskeletal injury prevention—implementation issues and research needs." Ann Arbor, MI: University of Michigan.

Ramazzini, B. 1713. De morbis, artificum. Trans. W.C. Wright (1940), *Diseases of Workers.* Chicago: University of Chicago.

Sanders, M., and E. McCormick. 1993. *Human Factors in Engineering and Design.* New York: McGraw-Hill, Inc.

Schiefer, R., R. Kok, M. Lewis, and G. Meese. 1984. Finger skin temperature and manual dexterity—some intergroup differences. *Appl. Ergo.* 15: 135.

Tischauer, E.R. 1966. Some aspects of stress on forearm and hand in industry. *J. Occup. Med.* 8: 63-71.

U.S. EPA and OSHA. May 1993. *A Guide to Heat Stress in Agriculture*, EPA-750-b-92-001. Washington, DC: U.S. Government Printing Office.

Waters, T., V. Putz-Anderson, A. Garg, and L. Fine. 1993. Revised NIOSH equation for the design and evaluation of manual lifting tasks. *Ergonomics* 36(7): 749-76.

Waters, T., V. Putz-Anderson., and A. Garg. 1994. *Applications Manual for the Revised NIOSH Lifting Equation.* U.S. Department of Health and Human Services, Publication # 94-110.

Woodson, W.E., B. Tillman, and P.L. Tillman. 1992. *Human Factors Design Handbook.* New York: McGraw-Hill, Inc.

Appendix A: Common Arm Cumulative Trauma Disorders (adapted from Hudock 1994).

Tendon Disorders:

Tendinitis - Inflammation, fraying or tearing of tendon fibers, resulting in pain and occasionally swelling.

Tenosynovitis - Tendon (synovial) sheath inflammation and swelling due to the production of excessive synovial fluid.

Stenosing tenosynovitis - Tendon surface becomes rough and irritated, causing synovial sheath to inflame, pressing on tendon. Sheath progressively constricts. An example is "De Quervain's disease" affecting the tendons that allow the thumb to move back and away from the hand.

Stenosing tenosynovitis crepitans - Swollen synovial sheath locks tendon in place resulting in jerking or snapping movement of joint. An example is "trigger finger" associated with flexing the index finger against resistance such as pulling a trigger on a tool.

Ganglionic cyst - Synovial sheath becomes swollen with excess fluid to the point of causing a bump under the skin, often on the wrist. This condition was also known as a "Bible bump" since large books, such as the Bible, were used to rupture the ganglion by pounding on it.

Lateral epicondylitis - Unsheathed tendons attached to the outer side of the elbow which control the finger extensor muscles become irritated. Associated with the use of the arm for impact or jerking motions. It is commonly known as "tennis elbow" or "pitcher's elbow."

Medial epicondylitis - Unsheathed tendons attached to the inner side of elbow, which control the finger flexor muscles, become irritated. Also known as "golfer's elbow."

Rotator cuff tendinitis - The most common shoulder tendon disorder. The four tendons that comprise the rotator cuff are the main source of stability and mobility for the shoulder. They rotate the arm inward and outward and away from the side. The tendons pass through a small opening between the humerus and the acromion. A bursa (a small fluid-filled sac) normally protects the tendons from irritation.

Nerve Disorders:

Carpal tunnel syndrome - Compression of the median nerve by swollen finger flexor tendons in the wrist, resulting in pain, numbness, and tingling in thumb and first three fingers of the hand.

Guyon canal syndrome - Compression of ulnar nerve as it passes through Guyon canal in wrist. Repetitive use of palm of hand as hammer cited as contributing factor. Symptoms occur in the ring and small finger.

Radial tunnel syndrome - Compression of radial nerve above the lateral epicondyle. Symptoms are similar to "tennis elbow."

Pronator teres syndrome - Compression of the median nerve as it passes through the pronator teres muscle in the forearm. Symptoms are similar to carpal tunnel syndrome and extend down forearm.

Cubital tunnel syndrome - Compression of the median nerve as it passes through the cubital tunnel behind the lateral epicondyle. Pain in elbow, forearm, and ring and small fingers. Associated with leaning elbows on hard surfaces and elbow flexion and wrist extension.

Neuromuscular Disorders:

Thoracic outlet syndrome - Compression of the nerves and blood vessels between the muscles of the neck and shoulder or between the first rib and clavicle. Burning, tingling, and numbness along arm, hand, and fingers. Associated with repeated reaching above shoulder level.

Hand-Arm Vibration syndrome - Numbness, tingling, and blanching of fingers due to closure of digital arteries often caused by extended use of vibrating hand tools and controls. Also called "Vibration White Finger Syndrome" or "Raynaud's Phenomenon."

Appendix B: Sample Discomfort Survey (Courtesy of Duke Ergonomics Program).

Work Station Survey

Name_____

Please complete the following survey in order to pinpoint specific areas of your work station that are in need of review. For each problem area below, please provide one or more symptoms, a frequency, and a severity rating.

Symptoms are:
- P Pain
- A Aching
- C Cramping
- Sw Swelling
- St Stiffness
- B Burning
- N Numbness
- T Tingling
- O Other

Frequency is rated as:
- 0 Never
- 1 Rarely
- 2 Occasionally
- 3 Frequently
- 4 Most of the time
- 5 All of the time

Severity is rated as:
- 0 None
- 1 Hardly Noticeable
- 2 Noticeable
- 3 Uncomfortable
- 4 Distracting
  (Interferes with work)
- 5 Painful
  (Prevents my working)

Problem Areas	Symptom(s)	Frequency	Severity
Headache	_____	_____	_____
Neck	_____	_____	_____
Shoulders	_____	_____	_____
Upper back	_____	_____	_____
Middle back	_____	_____	_____
Lower back	_____	_____	_____
Arms	_____	_____	_____
Elbow/Forearm	_____	_____	_____
Wrists	_____	_____	_____
Hands	_____	_____	_____
Fingers (all)	_____	_____	_____
Thigh	_____	_____	_____
Knee (while bending)	_____	_____	_____
Lower Leg	_____	_____	_____
Ankle/Foot	_____	_____	_____
Toes	_____	_____	_____

Please provide a frequency and severity rating for the following:

Visual Problems	Frequency	Severity
Blurred Vision	_____	_____
Burning Eyes	_____	_____
Watery Eyes	_____	_____
Dry Eyes	_____	_____
Itchy Eyes	_____	_____
Eyestrain	_____	_____
Redness	_____	_____
Ache: Front of eye	_____	_____
Back of eye	_____	_____
Entire eye	_____	_____

Appendix C: Ergonomic Checklist to Enhance Performance, Safety, and Comfort (Helander, *A Guide to the Ergonomics of Manufacturing*, 1995, reproduced with permission.)

A. Physical demands	Yes	If No, then Why? How to redesign?
Are hands at a convenient working height for the task?	____	_____
Are the joints mostly in a convenient neutral position?	____	_____
Are the wrists mostly in a straight, neutral posture?	____	_____
Can operator assume several different postures while working?	____	_____
Is this a dynamic rather than a static task?	____	_____
Can the task be performed with the torso and the head facing forward?	____	_____
Are primary items located within easy reach?	____	_____
Is frequent lifting below 20 kg (45 lbs)?	____	_____
Is occasional heavy lifting less than 25 kg (55 lb)?	____	_____
Are items to be lifted positioned between knuckle and shoulder height?	____	_____
Are there convenient aids for manual materials handling?	____	_____
Are there handles on items that are otherwise difficult to lift?	____	_____
Are hand tools appropriate for the task?	____	_____
Are hand tools comfortable and safe to use?	____	_____

For sitting tasks:

Are the feet firmly supported on the floor or by using a footrest?	____	_____

Appendix C *(continued)*

A. Physical demands
For sitting tasks *(continued)*:   Yes   If No, then Why? How to redesign?

Can the backrest be utilized while performing the task?   ____   _____

Are the elbow joints mostly at an intermediate angle?   ____   _____

Are primary items located within easy (5th percentile) reach—about 40 cm?   ____   _____

Is head bent slightly forward, rather than backward?   ____   _____

B. Task visibility   Yes   If No, then Why? How to redesign?

Are displays and dials easy to see from normal work position?   ____   _____

Is printed or displayed text large enough for reading (about 18-25 min. of arc)?   ____   _____

Are eyeglasses appropriate for task viewing distance?   ____   _____

Is illumination level uniform throughout working area?   ____   _____

Are illumination levels appropriate?
About 500 lux for VDT work
About 1000 lux for coarse assembly
About 2000 lux for fine assembly   ____   _____

Is direct glare from illumination sources and windows avoided?   ____   _____

Is indirect (reflected) glare avoided?   ____   _____

Is luminance contrast ratio in immediate task area less than 20:1?   ____   _____

C. Mental demands   Yes   If No, then Why? How to redesign?

Does the task involve moderate short-term memory load, rather than high?   ____   _____

Appendix C *(continued)*

C. Mental Demands *(continued)*:	Yes	If No, then Why? How to redesign?
Does the task involve few simultaneous factors, rather than several?	_____	_____
Is operator performance unpaced, rather than paced by the task?	_____	_____
Is the task varying, rather than repetitive and monotonous?	_____	_____
Can operator errors and slips easily be corrected?	_____	_____
Are special memory aids used?	_____	_____
Do display and controls follow population stereotypes?	_____	_____
Is the task easy to learn, rather than difficult?	_____	_____

D. Machine Design	Yes	If No, then Why? How to redesign?
Are tasks appropriately allocated between operators and machines?	_____	_____
Are manual controls easy to reach?	_____	_____
Are manual controls easy to distinguish from each other?	_____	_____
Are all machine functions and displays visible to the operator?	_____	_____
Can machine functions be handled through one command/control?	_____	_____
Are all controls on the machine necessary for the job?	_____	_____
Are location of controls and tools the same for similar machines?	_____	_____
Are memory aids used as a reminder of difficult task information?	_____	_____

Appendix C *(continued)*

D. Machine Design *(continued):*	Yes	If No, then Why? How to redesign?
Is it possible to operate machine without bending, twisting and far reaching?	_____	_____
Is there adequate body clearance for handling and maintenance tasks?	_____	_____
Are machine symbols and icons readily understood?	_____	_____
Are labels used to inform and remind operators of task information?	_____	_____
Are labels/symbols used to designate locations for frequently used items?	_____	_____

E. VDT Tasks	Yes	If No, then Why? How to redesign?
Are monitor screens positioned perpendicular to windows?	_____	_____
Can reflected glare on the screen be avoided?	_____	_____
Is the display located below a horizontal plane through the eyes?	_____	_____
Do the locations of display, documents and keyboard make it possible to sit straight without twisting the body?	_____	_____
Is a QWERTY keyboard used?	_____	_____
Are software functions understood and easy to use?	_____	_____
Are software functions and computer tasks routines easy to access?	_____	_____

F. Safety	Yes	If No, then Why? How to redesign?
Are there appropriate warning signs as a reminder of task hazards?	_____	_____
Is wording on warning signs relevant and informative?	_____	_____

Appendix C *(continued)*

F. Safety *(continued)*:	Yes	If No, then Why? How to redesign?
Are warning signs positioned where operators look?	____	_____
Is the workplace organized and clean with excellent house keeping?	____	_____
Are the floors even without drains or pit marks?	____	_____
Is it possible to perform the task without safety glasses or protective clothing?	____	_____
Has company established safety procedures and rules?	____	_____
Are safety rules and procedures prioritized by management and enforced?	____	_____
Does company analyze each reported accident or injury to improve safety?	____	_____
Do newly hired workers receive safety training?	____	_____
Do safety training programs present relevant task specific information?	____	_____
Are potential hazards clearly visible from the operators position?	____	_____
Have machine safety devices been installed, e.g., lockouts, and guards?	____	_____

G. Ambient Environment	Yes	If No, Then Why? How to redesign?
Is ambient noise below 85 dBA t protect against hearing damage?	____	_____
Is ambient noise level below 55 dBA to facilitate verbal communication?	____	_____
Is there a program to reduce noise pollution by redesign of machines and the work environment?	____	_____
Are vibration levels and frequencies so low as to not affect job performance?	____	_____

Appendix C *(continued)*

G. Ambient Environment *(continued):*	Yes	If No, then Why? How to redesign?

Is temperature and humidity within a comfortable range? _____ _____

Is it possible to perform work tasks without protective equipment? _____ _____

Can all work tasks be performed without risk of electric shock? _____ _____

H. Product and Process Design	Yes	If No, then Why? How to redesign?

Has product design been modified to improve productivity? _____ _____

Has product design been modified to create better jobs? _____ _____

Have the best machines been selected that maximize productivity? _____ _____

Have the best machines been selected that maximize operator convenience? _____ _____

Have processes been located so as to improve productivity? _____ _____

Have processes been located to improve operator convenience? _____ _____

Have machines and processes been selected to optimize task allocation between operators and machines? _____ _____

Appendix D: Ergonomic Checklist for Tractors and Agricultural Machinery (modified from Hansson 1991).

The Cab

1. Is the cab designed and oriented so that there is little risk of accidents during driving, loading, and unloading?
2. Is the cab free from protruding parts that can cause injury?
3. Is the cab size adequate?

Overall Assessment

Excellent_____
Good_____
Fair_____
Poor_____

The Seat

1. Is the seat well located in the cab?
2. Are the seat and armrests strong and securely mounted?
3. Is there enough movement in the seat?
4. Are the seat and back cushions a suitable size, shape, and angle?
5. Will the cushion and backrest provide adequate insulation and friction?
6. Is the seat equipped with vibration dampers?
7. Is there a warmer for the seat and backrest?

Overall Assessment

Excellent_____
Good_____
Fair_____
Poor_____

Visibility from the Cab

1. Is it easy to see the ground?
2. Is visibility good for attachment of implements?
3. Is upwards visibility adequate?
4. Is the machine equipped with effective windshield wipers and washers, defrosters, and rear-view mirrors?

Overall Assessment

Excellent_____
Good_____
Fair_____
Poor_____

Entry/Exit

1. Does the machine have the necessary hand grips and are they well placed?
2. Is it possible to enter and leave the cab without risk of accident?
3. Are foot and hand supports designed and located so that they are protected from damage and dirt?

Overall Assessment

Excellent_____
Good_____
Fair_____
Poor_____

Controls and Instrumentation

1. Are the controls of suitable types (pedals, levers, switches, etc)?
2. Is their function logical?
3. Are they conveniently located?
4. Do the controls move in a logical direction?
5. Do the controls offer the right resistance?
6. Is there little or no risk that a control will be moved unintentionally?
7. Does the machine have all the necessary signals and instruments?
8. Are the instruments of a suitable type?
9. Are they well placed?
10. Are they easy to read?

Overall Assessment

Excellent_____
Good_____
Fair_____
Poor_____

Appendix D *(continued)*

Noise and Vibration **Overall Assessment**

1. Is the sound level in the driver's seat below 85 dB(A)?   Excellent_____
2. Is the cab free from obviously disturbing noises below 70 dB(A)?   Good_____
3. Is the sound level outside the tractor low enough to allow working in the area without being disturbed by it?   Fair_____ Poor_____
4. Is the machine constructed so that the operator's exposure to vibration during normal work is at an acceptable level (wheel size, tire type, seat vibration dampers, etc.)?
5. Is the shock-absorbing equipment (seat, etc.) durable and firmly mounted?
6. Are the steering wheel and other hand controls free from vibration?

Lights/Glare **Overall Assessment**

1. Are the lights acceptable for driving on roads?   Excellent_____
2. Are they acceptable for driving on fields?   Good_____
3. Is the lighting good enough for stationary work?   Fair_____
4. Is the lighting good enough for connecting implements?   Poor_____
5. Is the machine painted so that there are no disturbing reflections or glare?
6. Are windshields tinted to reduce glare?

Cab Climate **Overall Assessment**

1. Is the heating system adequate and well designed?   Excellent_____
2. Can the driver protect himself from the sun's rays?   Good_____
3. Is the driver protected against drafts?   Fair_____
4. Do the fresh-air vents and cooling system work effectively and well?   Poor_____

Dust and Exhaust **Overall Assessment**

1. Is the cab equipped with cooling equipment?   Excellent_____
2. Is there an air filter?   Good_____
3. Is there sufficient over-pressure in the cab?   Fair_____
4. Is the cab free from the smell of exhaust?   Poor_____
5. Is it free from oil smells?

Hitching of Implements **Overall Assessment**

1. Does the tractor have some type of quick coupler?   Excellent_____
2. Does the tractor have a hydraulic upper-link?   Good_____
3. Does the tractor have a hydraulic lift cylinder?   Fair_____
4. Is the power take-off sufficiently shielded?   Poor_____
5. Is there an emergency shut-down?
6. Can the hydraulic lift be safely maneuvered from outside the cab?
7. Can an operator sitting in the cab easily see what must be seen to attach and detach implements?
8. Can implements be attached and detached from a good working position?
9. Can implements be attached without too much physical effort?
10. Does a change of the power take-off speed (540 or 1000 RPM) also involve a change of the power take-off shaft end?

Appendix D *(continued)*

Service and Maintenance

1. Can parts requiring service be reached easily and safely?
2. Can the work be done without risk of accident (slipping, being burned or cut, or being caught in machinery)?
3. Can the work be done in a comfortable position?
4. Can the work be done without heavy lifts or other undue physical exertion?
5. Can the work be done without getting dirty or being exposed to skin irritating substances such as oil?
6. Does the maintenance manual provide a clear and complete description of the work to be done?
7. Does the manual contain full descriptions of the operation of the components that have a direct bearing on ergonomic conditions and safety?
8. Is there a place to store tools?

Overall Assessment

Excellent_____
Good_____
Fair_____
Poor_____

Appendix E: Industrial and Agricultural Systems Standards (Woodson et al. 1992, reproduced with permission).

SAE J38   Lift arm safety device for loaders
SAE J67   Shovel dipper, clam bucket, and dragline bucket rating
SAE J94   Combination tail lamp and flood-lamp for industrial equipment
SAE J95   Headlamps for industrial equipment
SAE J96   Flash warning lamp for industrial equipment
SAE J98   Safety for industrial wheeled equipment
SAE J99   Lighting and marking of industrial equipment on highways
SAE J115   Safety signs for construction and industrial equipment
SAE J137c Lighting and marking of agricultural equipment on highways
SAE J153   Safety considerations for the operator
SAE J154   Operator enclosures (cabs)—human factors design considerations
SAE J167a Overhead protection for agricultural tractors—test procedures and performance requirements
SAE J168a Protective enclosures for agricultural tractors—test procedures and performance requirements.
SAE J169   Design guidelines for air-conditioning systems for construction and industrial equipment cabs
SAE J185   Access systems for construction and industrial equipment
SAE J208c Safety for agricultural equipment
SAE J209   Instrument face design and location for construction and industrial equipment
SAE J220. Crane boom stop
SAE J232   Industrial rotary mowers
SAE J284a Safety alert symbol for agricultural construction and industrial equipment
SAE J297   Operator controls on industrial equipment
SAE J298   Universal symbols for operator controls on industrial equipment
SAE J389a Universal symbols for operator controlsß on agricultural equipment
SAE J742b Front-end loader bucket rating
SAE J765a Crane load stability test code
SAE J774c Emergency warning device
SAE J833a U.S. male and female physical dimensions for construction and industrial equipment design
SAE J841e Operator controls on agricultural equipment
SAE J898a Control locations for construction and industrial equipment design
SAE J899   Operator's seat dimensions - construction and industrial equipment design
SAEJ909b   Attachment of implements to agricultural wheeled tractors equipped with quick-attaching coupler for three-point free link hitch
SAE J919b Operator sound-level measurement procedure for powered mobile construction machines—singular-type test
SAE J925   Minimum access dimensions for construction and industrial machinery
SAE J943a Slow-moving vehicle identification emblem
SAE J956   Remote and automatic control systems for construction and industrial machinery
SAE J974   Flashing warning lamp for agricultural equipment
SAE J975   Headlamps for agricultural equipment
SAE J976   Combination tail lamp and flood lamp for agricultural equipment
SAE J983   Crane and cable excavator basic operating control arrangement
SAE J994b Performance, test, and application criteria for electrically operated backup alarm devices
SAE J1001 Safety criteria for industrial flail mowers
SAE J1006 Performance test for air-conditioned agricultural equipment
SAE J1012 Agricultural equipment enclosure pressurization system test procedure
SAE J1013. Measurement of whole body vibration of the seated operator of agricultural equipment
SAE J1029 Lighting and marking of construction and industrial machinery
SAE J1040 Performance criteria for roll-over protective structures (ROPS) for construction, earth-moving, forestry, and mining machines
SAE J1051 Force-deflection measurements of seat and back cushions for agricultural, construction, and industrial equipment
SAE J1071 Operator controls for motor graders
SAE J1084 Operator protective structure performance criteria for certain forestry equipment
SAE J1105 Performance, test, and application criteria of electrically operated forward warning horn for mobile construction machinery
SAE J1129 Operator cab environment for heated, ventilated, and air-conditionedß construction and industrial equipment
SAE J1163 Method for determining operator seat location on agricultural and construction machines
SAE J1166 Operator station sound-level measurement procedure for powered mobile earth-moving machinery - work cycle test
SAE J1194 Roll-over protective structures (ROPS) for wheeled agricultural tractors

Appendix F: Guidelines for the Selection of Manual Material Handling Equipment
(Courtesy of Duke Ergonomics Program).

Guidelines for Caster Selection:

First and foremost consider specifying the caster that will make it easiest for the handler to start, control, and stop the handling device. Generally, the following are "rules-of-thumb":

1. The larger the diameter of the caster:
   - the easier the object will be to begin rolling from a dead stop;
   - the easier the handling device will roll;
   - the easier it will be to roll over floor imperfections (cracks, rough surfaces, sills, expansion joints, etc.).
   - the more weight a caster can support.

2. Typically, casters equipped with a ball-bearing raceway will be easier to start, stop, and control and will last longer than casters of a lesser design.

3. Where steering or control of the handling device is important, specify fixed casters on the end opposite the operator.

4. Where a combination of steering and a high degree of maneuverability is desired, a combination fixed/swivel caster should be specified.

5. If the handling device will be unattended or if there is the potential for its rolling out of control, the device should be equipped with one or more locking or braking casters.

Floor surface materials to be traveled is an important consideration in the caster selection process. Caster type definition should be based in part on the type(s) of surface(s) the handling device will travel. A caster that is ideally suited for one flooring type (i.e., carpet) may not be suited at all for another flooring type (i.e. concrete).

From an ergonomic standpoint:

- It is easier to push than it is to pull.

- Typically a person can push farther/longer than they can pull.

- Typically a person is more comfortable pushing rather than pulling.

- Install push bars, handles, and cushioned grips slightly below elbow height on equipment with casters to facilitate the pushing/controlling task.

Other Manual Materials Handling Equipment Specification Guidelines:

Make sure it is easy to grasp

1. Grips or handles should be provided on materials handling devices intended to be pushed or pulled and should be located between waist and shoulder height.

2. Handles on a materials handling device should feature a round profile rather than a square or rectangular shape (thereby reducing stress concentrators to the hands).

Appendix F *(continued)*

3. Use U-shaped handles approximately 4" X 1.5".

4. Handles should be padded where possible (cushioned interface with operator).

Make sure the device is of a size that is easy to manage:

1. Materials handling devices should allow good visibility of the travel path without having to assume an awkward posture (easy sighting over and around the materials handling device).

2. The materials handling device should allow the operator's body to be positioned close to and squarely in front of the item being handled (common center of gravity and minimized spine torsion).

3. Where possible have materials handling device minimize stressful movements; favor pushing over pulling, pulling rather than carrying, carrying rather than lowering, and lowering over lifting.

Make sure the device is stable:

1. Position heavy, large, or bulky materials low on the device in order to lower its center of gravity and thereby improve its stability. Conversely, light or small items should be positioned higher on the device.

2. Items being carried on the materials handling device should be easily accessible and in a comfortable work range (knees to chest).

Appendix G: A Basic Program for Controlling Heat Stress (U.S. EPA and OSHA 1993).

A Basic Program for Controlling Heat Stress

Step 1: Assign responsibility for heat stress problems.

Step 2: Train workers and supervisors.
- Train workers and supervisors in the control of heat stress and the recognition prevention and treatment of heat illnesses.
- Conduct safety meetings during heat spells.

Step 3: Acclimatize workers when they begin to work under hot conditions.
- Assign a lighter workload for 5-7 days.
- Allow longer res periods for 5-7 days.
- Assign work in the heat for less than 100 minutes per day.
- Gradually increase the time of work in the heat each day.
- Watch workers' response to working in the heat closely for 5-7 days.

Step 4: Account for the conditions of work and of the workers.
- Check weather conditions.
- Consider how heavy the work is.
- Consider whether the worker is to wear protective garments and equipment.
- Check if the worker is or has recently been sick or has had a sharp loss in weight.
- Check whether the worker is rested, is taking any medications, or appears to have consumed alcohol that day.

Step 5: Manage work activities.
- Set up rest breaks.
- Rotate tasks among workers.
- Schedule heavy work for cooler hours.
- Postpone non-essential tasks during heat spells.
- Monitor environmental conditions and workers.

Step 6: Establish a drinking water program.

Step 7: Take additional measures, as appropriate.
- Provide special cooling garments.
- Select lightest weight or "breathable" protective garments and cooler respirators that give adequate protection.
- Provide shade.
- Use air-conditioned mobile equipment.
- Modify pesticide usage and handling to reduce need for protective garments and equipment.

Step 8: Give first aid when workers become ill.
- Set up a first aid program.
- Take heat stroke victims to the nearest medical treatment facility.
- Follow up on incidents of heat illness.

Appendix H: Draft OSHA Ergonomic Protection Standard Risk Factor Checklists (OSHA 1995).

Draft — DO NOT CITE OR QUOTE   Draft — DO NOT CITE OR QUOTE

Instructions for Completing the Risk Factor Checklists

These general risk factor checklists offer a quick method for identifying important workplace risk factors that contribute to work-related musculoskeletal disorders (MSDs). The checklists are used to identify jobs that can be controlled with a quick fix or more thorough job analysis:

- CHECKLIST A is used to assess risk factors for upper extremities (hands, wrists, arms, shoulders, neck).
- CHECKLIST B is used to assess risk factors for the back and lower extremities.
- CHECKLIST C is used to assess manual handling tasks. The score from this checklist is used in Checklist B.

To calculate workplace risk factor scores, complete the following steps:

STEP 1. Read the description of each workplace risk factor carefully.

STEP 2. If the employee performs more than one major task, list each task in the space provided.

STEP 3. Estimate the amount of time the employee spends performing the job described. If the job consists of more than one task, you will need to estimate the hours the employee spends performing *each task*, then estimate the total hours for each workplace risk factor associated with the task. List the task in the space provided (see example below). For force and awkward postures, estimate the time spent in both static and repetitive activities.

STEP 4. Circle the risk factor score in Column C or D on Checklists A and B. If the employee performs tasks that involve the risk factor for more than 8 hours a day, circle the score in column D and add 0.5 point for each additional hour (over 8 hours) the employee is exposed to that risk factor; record the total in Column E.

STEP 5. Enter the score circled in Column C or D (also add the values in Column E) in the space provided in Column F.

STEP 6. Complete this process for all of the workplace risk factors in the job being evaluated. (Complete both Checklist A and Checklist B.)

STEP 7. Complete Checklist C and record the total score in the box provided for the manual handling score at the bottom of Checklist B.

STEP 8. Add the risk factor scores to generate a total score for each checklist. Record each total in the boxes provided.

STEP 9. If either Checklist A or Checklist B has a score higher than 5, the job is a problem job. **Do not add the scores from Checklists A and B.**

Task (Est. Time, Hours)	Risk Factor	Est. Time/Risk Factor (Hours)
Assembly (6 hours)	Repetition	5 hours
	Awkward shoulder posture	5 hours
	Pinch force	2 hours
	Neck bend	2 hours
Microscope work (1 hour)	Neck bend	1 hour

Draft — DO NOT CITE OR QUOTE   Draft — DO NOT CITE OR QUOTE

A Upper Extremity Risk Factors

Date: _____

Job: _____

Department: _____

Employee: _____

Analyst: _____

Comments: _____

Task (Est. Time, Hours)	Risk Factor	Est. Time/Risk Factor (Hours)

UPPER EXTREMITY RISK FACTOR SCORES   Page 1

A	B	C	D	E	F
RISK FACTOR CATEGORY	RISK FACTORS	TIME			SCORE
		2 to 4 Hours	4+ to 8 Hours	8+ Hours Add 0.5 per hour	
		Circle the score			
Repetition (Finger, Wrist, Elbow, Shoulder, or Neck Motions)	1. **Identical or Similar Motions Performed Every Few Seconds** *Motions or motion patterns that are repeated every 15 seconds or less. (Keyboard use is scored below as a separate risk factor.)*	1	3		
	2. **Intensive Keying** *Scored separately from other repetitive tasks in the repetition category; includes steady pace, as in data entry.*	1	3		
	3. **Intermittent Keying** *Scored separately from other repetitive tasks. Keying activity is less than 50 percent of the work*	0	1		

COMMENTS

(Continued)

Draft — DO NOT CITE OR QUOTE   Draft — DO NOT CITE OR QUOTE

ERGONOMICS IN AGRICULTURE, FORESTRY, AND FISHERIES / 47

A UPPER EXTREMITY RISK FACTOR SCORES — Page 2

A RISK FACTOR CATEGORY	B RISK FACTORS	C 2 to 4 Hours	D 4+ to 8 Hours	E 8+ Hours Add 0.5 per hour	F SCORE	COMMENTS
		TIME — Circle the score				
Hand Force (Repetitive or Static)	1. Grip More Than 10-Pound Load — Holding an object weighing more than 10 pounds or squeezing hard with hand in a power grip. (Power Grip)	1	3			
	2. Pinch More Than 2 Pounds — Pinch force of 2+ pounds as in the pinch used to open a small binder clip with the tips of fingers. (Pinch Grip)	2	3			
Awkward Postures	1. Neck: Twist/Bend — Twisting neck to either side more than 20°, bending neck forward more than 20° as in viewing items on a desk or bending neck backward more than 5°.	1	2			
	2. Shoulder: Unsupported Arm or Elbow Above Mid-Torso Height — Arm is unsupported if there is not an armrest when doing precision finger work, or when the elbow is above mid-torso height.	2	3			
	3. Rapid Forearm Rotation, Extremely Flexed Elbow — Rotating the forearm or resisting rotation from a tool. An example of forearm rotation is using a manual screwdriver.	1	2			
	4. Wrist: Bend/Deviate — Consider wrist bends that involve more than 20° of flexion (bending the wrist palm down) or more than 30° of extension (bending the wrist back). Consider extreme deviation toward the thumb or little finger.	2	3			
	5. Fingers — Forceful gripping to control or hold an object, such as click-and-drag operations with a computer mouse or deboning with a knife.	1	2			

(Continued)

48 / SAFETY AND HEALTH IN AGRICULTURE, FORESTRY, AND FISHERIES

Draft — DO NOT CITE OR QUOTE   Draft — DO NOT CITE OR QUOTE

A   **UPPER EXTREMITY RISK FACTOR SCORES**   Page 3

A	B	C	D	E	F	COMMENTS
RISK FACTOR CATEGORY	RISK FACTORS	\<-- TIME --\>			SCORE	
		2 to 4 Hours	4+ to 8 Hours	8+ Hours Add 0.5 per hour		
		Circle the score				
Contact Stress	1. Hard/Sharp Objects Press Into Skin *Includes contact of the palm, fingers, wrist, elbow, or armpit.*		2			
	2. Using the Palm of the Hand as a Hammer	2	3			
Vibration	1. Localized Vibration (Without Vibration Dampening) *Vibration from contact between the hands and a vibrating object, such as a power tool.*	1	2			
	2. Sitting on Vibrating Surface (Without Vibration Dampening)	1	2			
Environment	1. Lighting (Poor Illumination/Glare) *Inability to see clearly obscures characters or defects.*	0	1			
	2. Cold Temperature *Hands exposed to air temperature of less than 60°F for sedentary work, 40°F for light work, 20°F for moderate/heavy work; cold exhaust blowing on hands.*	0	1			
Work Organization	1. No Control Over Work Pace, Insufficient Breaks *Machine-paced, piece rate, constant monitoring, or daily deadlines. Workers with fixed or awkward postures (static or precision work) do not get frequent short breaks (micropauses). Enter 1 if one control factor is present or 2 if two or more control factors are present.*					

TOTAL UPPER EXTREMITY SCORE FOR CHECKLIST A   (sum of pages 1, 2 and 3)   [ ]

Draft — DO NOT CITE OR QUOTE   Draft — DO NOT CITE OR QUOTE

B **Back and Lower Extremity Risk Factors**

Date: _____
Job: _____
Department: _____
Employee: _____
Analyst: _____
Comments: _____

Task (Est. Time, Hours)	Risk Factor	Est. Time/Risk Factor (Hours)

BACK AND LOWER EXTREMITY RISK FACTOR SCORES — Page 1

A	B	C	D	E	F
		\multicolumn{3}{c}{TIME FACTOR}	
RISK FACTOR CATEGORY	RISK FACTORS	2 to 4 Hours	4+ to 8 Hours	8+ Hours Add 0.5 per hour	SCORE
		\multicolumn{3}{c}{Circle the score}	
Awkward Postures (Repetitive or Static)	1. Side Bending. Mild Bending of Torso More Than 20° But Less Than 45°		1	2	
	2. Severe Forward Bending of Torso More Than 45°	2	3		

COMMENTS

(Continued)

BACK AND LOWER EXTREMITY RISK FACTOR SCORES — Page 2

A	B	C	D	E	F	
RISK FACTOR CATEGORY	RISK FACTORS	\multicolumn{3}{c}{TIME FACTOR}		SCORE	COMMENTS
		2 to 4 Hours	4+ to 8 Hours	8+ Hours Add 0.5 per hour		
		Circle the score				
Awkward Postures (Repetitive or Static) *(Continued)*	3. Backward Bending of Torso	1	2			
	4. Twisting Torso	2	3			
	5. Prolonged Sitting Without Adequate Back Support *Back is not firmly supported by a back rest for an extended period.*	1	2			
	6. Standing Stationary or Inadequate Foot Support While Seated *Stand in one place (an assembly line or check stand) without sit/stand option or walking, or feet are not firmly supported when sitting.*	0	1			
	7. Kneeling/Squatting	2	3			
	8. Repetitive Ankle Extension/Flexion *Using a foot pedal to start or stop a machine cycle (as in sewing machine operations).*	1	2			
Contact Stress	1. Hard/Sharp Objects Press Into Skin *Includes contact against the leg.*	1	2			
	2. Using the Knee as a Hammer or Kicker	2	3			

(Continued)

BACK AND LOWER EXTREMITY RISK FACTOR SCORES — Page 3 — B

A RISK FACTOR CATEGORY	B RISK FACTORS	C 2 to 4 Hours	D 4+ to 8 Hours	E 8+ Hours Add 0.5 per hour	F SCORE	COMMENTS
		TIME FACTOR Circle the score				
Vibration	1. Sitting on Vibrating Surface (Without Vibration Dampening)		2			
Push/Pull	1. Moderate Load Moderate load = 20 pounds of initial force needed to push/pull an object, such as a shopping cart loaded with five 40-pound bags of dog food (200 pounds).	1	2			
	2. Heavy Load Heavy load = 50 pounds of initial force needed to push/pull an object, such as a two-drawer, full file cabinet across a carpeted floor.	2	3			
Work Organization	1. No Control Over Work Pace, Insufficient Breaks Machine-paced, piece rate, constant monitoring, or daily deadlines. Workers with fixed or awkward postures (static or precision work) do not get frequent short breaks (micropauses). Enter 1 if one control factor is present or 2 if two or more control factors are present.					

MANUAL HANDLING SCORE (from CHECKLIST C)

TOTAL BACK AND LOWER EXTREMITY SCORE FOR CHECKLIST B (sum of both pages 1 and 2 of CHECKLIST B)

C — Manual Handling (10 Pounds or More)

STEP 1: Determine If the Lift Is Near, Middle, or Far (Body to Hands)

- Use an average horizontal distance if a lift is made every 10 minutes or less.
- Use the largest horizontal distance if more than 10 minutes pass between lifts.

NEAR LIFT	MIDDLE LIFT	FAR LIFT
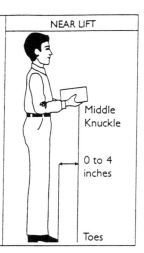 Middle Knuckle — 0 to 4 inches — Toes	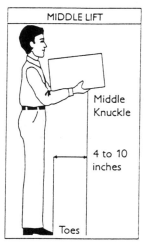 Middle Knuckle — 4 to 10 inches — Toes	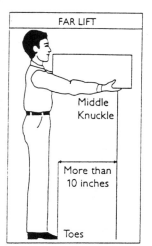 Middle Knuckle — More than 10 inches — Toes

STEP 2: Find the Lifting Zone and Estimate the Weight Lifted (Pounds)

- Use an average weight if a lift is made every 10 minutes or less.
- Use the heaviest weight if more than 10 minutes pass between lifts.
- Enter 0 in the total score if the weight is 10 lb or less.

	NEAR LIFT		MIDDLE LIFT		FAR LIFT	
DANGER ZONE	More Than 51 lb	5* Points	More Than 35 lb	6 Points	More Than 28 lb	6 Points
CAUTION ZONE	17 to 51 lb	3 Points	12 to 35 lb	3 Points	10 to 28 lb	3 Points
SAFE ZONE	Less Than 17 lb	0 Points	Less Than 12 lb	0 Points	Less Than 10 lb	0 Points

*If lifts are performed more than 15 times per shift, use 6 points.

STEP 2 SCORE: (enter 0, 3, 5, or 6) ☐

STEP 3: Determine the Points for Other Risk Factors

- Use occasional lifts if more than 10 minutes pass between lifts.
- Use the more than 1 hour points if the risk factor occurs with most lifts and lifting is performed for more than 1 hour.

FACTOR	OCCASIONAL LIFTS PERFORMED FOR 1 HOUR OR LESS IN TOTAL PER SHIFT	LIFTS PERFORMED FOR MORE THAN 1 HOUR IN TOTAL PER SHIFT
Twist torso during lift	1	1
Lift one-handed	1	2
Lift unstable loads (people, liquids, or loads that shift around or have unequal weight distribution)	1	2
Lift between 1 to 5 times per minute	1	1
Lift 5 or more times per minute	2	3
Lift above the shoulder	1	2
Lift below the knuckle	1	2
Carry objects 10 to 30 feet	1	2
Carry objects farther than 30 feet	2	3
Lift while seated or kneeling	1	2

STEP 3 SCORE: ☐

TOTAL SCORE — Add scores from Steps 2 and 3. Enter total score on Checklist B.

TOTAL: ☐

3

OCCUPATIONAL VIBRATION EXPOSURE IN AGRICULTURE, FORESTRY, AND FISHERIES

David Siebens, M.D., M.P.H.
Duke University Medical Center

> Vibration exposure is defined as either whole body or hand-arm. For each category, vibration exposure can be quantified in terms of frequency of vibration, acceleration, and duration of exposure. Standards exist which establish limits for vibration exposure for workers. Whole body vibration has been linked to spinal disorders, particularly lower back pain. Control measures such as seating design and exposure limits are discussed in this chapter. Hand-arm vibration has been linked to diseases such as vibration white finger and other phenomena. Vibrating tools such as chain saws and other power tools are identified as principle sources of this exposure. Control strategies are discussed including gloves and vibration reducing designs.

INTRODUCTION

For centuries, vibration has been known to cause a variety of health problems. Disease related to vibration has become increasingly common with the ubiquity of vibrating machinery and hand tools. Vibration is an important occupational health consideration, with millions of workers in the United States exposed. In 1973, it was estimated that 8 million Americans were exposed to significant occupational vibration, 6.8 million from motorized vehicles or vibrating structures such as trucks, or agricultural and construction equipment, and 1.2 million from the use of vibrating hand tools such as chain saws and other motorized tools, pneumatic tools, chippers, grinders, etc. (Wasserman 1973). Extensive use of vibrating machinery by workers in agriculture, forestry, and fisheries may put these workers at particular risk of vibration-related health problems. The purpose of this chapter is to give you a basic understanding of vibration, its adverse health effects, and strategies for the prevention of vibration-related illness.

Occupational vibration is divided into whole-body vibration (WBV), in which vibration is transmitted to the entire body through a vibrating structure such as a tractor seat, the deck of a ship, or the floor of a building; and hand-arm vibration (HAV), in which vibration impinges primarily on the upper extremities through use of a vibrating hand tool. Obviously, this distinction can become blurred when workers are subjected to both types of vibration simultaneously. However, because of the disparate nature of these two types of vibration exposure and the adverse health effects associated with each, this distinction is widely accepted and has been incorporated into health and safety standards for occupational vibration exposure, and strategies for prevention.

PHYSICAL PRINCIPLES

Familiarity with a few physical principles of vibration and some relevant terminology is essential to an understanding of occupational vibration and the related literature. Simply defined, vibration is back and forth, oscillating motion from a point of reference. *Frequency*, expressed in cycles per second (Hertz or Hz), describes the period or interval of oscillation. *Displacement* is the distance between the moving object and the point of reference and *velocity* is the time rate of change in displacement. The most widely used measure of vibration intensity is *acceleration*, which is the time rate of change in velocity expressed in meters/second^2. Acceleration is a convenient measure of vibration intensity because it is easy to measure, and from it, the other vibration parameters can be derived.

Closely related to frequency, is the concept of *resonance*. All mechanical systems have natural vibration frequencies at which externally applied vibration is amplified by the system itself. Familiar examples of resonance are loudspeakers, which are "boomy" in a certain frequency range, producing sound distortion; an old car may have a rattle or shake at 40 mph but not at 35 or 45 mph; a crystal goblet will shatter when subjected to sufficient sound intensity at its resonant frequency. As a mechanical system, the human body has resonant frequencies, and the body's natural amplification of external vibration can exacerbate the biologic effect of vibration at those frequencies. The resonant frequency of the torso of the seated subject is between 4 and 8 Hz, and that for the hand and fingers is around 100 to 200 Hz.

Finally, strategies for prevention of adverse health effects of vibration should recognize that the energy of vibration varies with the square of the vibration acceleration, and with the square of the exposure duration. This means that even small reductions in vibration acceleration or exposure time can have significant benefit in limiting the vibration energy transmitted to the worker (Wilder, et al. 1994).

VIBRATION MEASUREMENT AND STANDARDS

Vibration is typically measured with an accelerometer—a small, light weight apparatus containing piezo-electric crystals oriented in each of three mutually perpendicular axes. The axes along which human vibration is measured have been standardized. The reference point for WBV is the sternum, and for HAV it is the third metacarpal (one of the bones in the hand). Similarly, vibration can be measured at the source, such as the handle of a tool as illustrated in Figure 3.1. The electrical output from the accelerometer is analyzed by computer, and the averaged (root mean square) vibration acceleration at various frequency bands is derived for each axis and compared to the appropriate vibration standard.

There are separate standards for WBV and HAV, and the basic principles of these standards will be discussed here. A detailed discussion of these standards is beyond the scope of this chapter and anyone involved with occupational vibration, its measurement, control, or prevention should obtain and understand them thoroughly.

The standard for WBV is ISO 2631, which is identical to ANSI 3S.18. The standard establishes vibration exposure boundaries based on three factors—vibration frequency, acceleration, and exposure duration. The standard proposes a "fatigue decreased proficiency" boundary, beyond which workers would be expected to experience fatigue and difficulty with job performance. The standard also proposes a more conservative exposure boundary beyond which workers would be expected to begin to experience discomfort from vibration ("reduced comfort boundary") as well as a less conservative boundary ("exposure limit"). Vibration measurements along each axis are compared to the standard, and any acceleration along any axis above the standard exceeds the entire standard.

The standards for HAV include, ANSI S3.34, ACGIH threshold limit values for hand arm vibration, and NIOSH 89.106. The basic principles involved are similar to those for the WBV standard, but there are a few notable differences. Most importantly, the proposed vibration limits are established to prevent the occurrence of "vibration white finger" which is the most common health problem associated with HAV and will be discussed in a following section. Another difference is that there is only one family of exposure limits, with no provision for reduced comfort versus decreased proficiency, and so on.

It is important to recognize that there are few well-controlled epidemiologic studies establishing precisely how much vibration exposure is actually safe. Standards regarding vibration exposure have been, and will continue to be revised, as more and better information on vibration hazards becomes available.

WHOLE BODY VIBRATION

Exposure to WBV has been associated with numerous health effects including disorders of the spine, nervous system, gastrointestinal system, circulatory system, vestibular system, and the female reproductive system (Seidel and Heide 1986). The majority of epidemiologic evidence, however, suggests that occupational WBV exposure is a risk factor for low back pain and degenerative changes of the spine, the evidence for a relationship between WBV and other health problems being more equivocal (Hulshof and van Zanten 1987, Seidel and Heide 1986). The pathogenesis

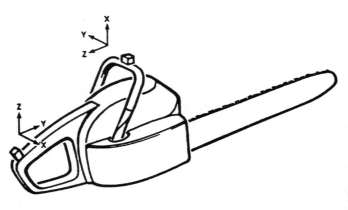

Figure 3.1. Coordinate system for vibration measurement in chain saws (NIOSH 1989).

of low back pain is complex, with physical, emotional, social, and economic factors all intertwined. Attribution of low back pain to a particular risk factor is often difficult because of the large number of confounding variables. However, one of the largest epidemiologic studies of the health effects of WBV, using data on 43,000 exposed workers concluded that WBV at or near the ISO Exposure Limit increases the risk of low back pain and degenerative changes of the spine (Seidel and Heide 1986). Agricultural tractor and heavy equipment driving has been associated with low back pain and degenerative disease of the spine in numerous studies (Hulshof and van Zanten 1987).

Low back pain is one of the most important occupational health problems. It is one of the leading causes of work-related disability payments, and among the most common causes of lost work time. Control of any factor causing low back pain, therefore, has enormous potential benefit.

The mechanism by which WBV causes low back pain is not clear. Current theories and the evidence for them have recently been reviewed (Wilder, et al. 1994). This review makes two points, which may have particular significance for agricultural and forestry workers. Vibration may increase the risk of low back pain in occupational settings in which heavy lifting is required in conjunction with vibration exposure. A truck driver, for example, may be well advised to recognize the increased risk of low back pain from vibration, and rest periodically during a long drive or before proceeding with any heavy lifting. Another point is that vibration can increase muscle fatigue in jobs which require awkward posture. The practice of driving a tractor while twisted around to look at the implement behind may put the operator at increased risk and should probably be minimized.

Fortunately, various types of cushions have been shown to have an impact on transmission of vibration from a vibrating seat to the worker (Pope, et al. 1989). Farmers in the United States appear to be well aware of the importance of good cushioning and seat suspension. It is estimated that they spend a total of more than 2 billion hours per year evaluating seats and other human factors features on their equipment.

Control of Whole Body Vibration

Several strategies for controlling WBV have been recommended (Wasserman and Taylor 1993).

- Workers should not be in contact with a vibrating surface for any longer than necessary.
- Machine controls should be located remotely from any vibrating surface whenever possible.
- Vehicles should use vibration isolated suspended or air-ride seats.
- Machinery generating vibration should be carefully maintained to prevent excessive vibration.

HAND-ARM VIBRATION

Vibration transmitted to the fingers, hands, and arms through the use of vibrating tools such as chain saws and other hand-held, gasoline-powered equipment, pneumatic tools, sanders, grinders, etc., cause a constellation of localized symptoms known as the "hand-arm vibration syndrome" (HAVS). The symptoms can include finger blanching, numbness, tingling, and loss of dexterity, which begin at the finger tips and move proximally with increasing exposure. In the past, this syndrome has been known as vibration white finger, dead finger, and dead hand. HAVS is divided into two components, vascular and sensorineural. The vascular component (finger blanching) results from constriction of digital blood vessels. The sensorineural component (tingling, numbness, loss of dexterity) results from digital nerve dysfunction (NIOSH 1989).

One of the early studies on HAV was performed in 1918 by Dr. Alice Hamilton on stone cutters using pneumatic hammers in Bedford, Indiana. Her work still affords an excellent description of the problem, its cause, and some preventive strategies. She wrote, "Among men who use the air hammer for cutting stone, there appears very commonly a disturbance in the circulation of the hands which consists of spasmatic contraction of the blood vessels of certain fingers, making them blanched, shrunken and numb. These attacks come on under the influence of cold, and are most marked, not while the man is at work with the hammer, but usually in the morning or after work....The fingers affected are numb and clumsy when the vascular spasm persists. As it passes over there may be decided discomfort and even pain, but the hands soon become normal in appearance and as a usual thing the men do not complain of discomfort between the attacks....The condition is undoubtedly caused by the use of the air hammer; it is most marked in those branches of stonework where the hammer is most continuously used and it is absent only where the air hammer is used little or not at all. Stonecutters who do not use the air hammer do not have this condition of the fingers. Apparently once the spastic anemia has been set up it is very slow in disappearing. Men who have given up the use of the air hammer for many years may still have their fingers turn white and numb in cold weather....The trouble seems to be caused by three factors—long-continued muscular contraction of the fingers in holding the tool, the vibrations of the tool, and cold. It is increased by too continuous use of the air hammer, by grasping the tool too tightly, by using a worn, loose air hammer and by cold in the working place. If these factors can be eliminated the trouble can probably be decidedly lessened" (Hamilton 1918).

The mechanism by which vibration causes HAVS is not clearly established. The mechanism is thought to involve specific anatomical changes to digital blood vessels and nerves as a consequence of mechanical damage (Pelmear and Taylor 1994). The etiology of HAVS is probably multifactorial with vibration, cold, strenuous muscular work and other factors all playing a role (Gemne 1982).

Prevention of Hand-Arm Vibration Exposure

Dr. Hamilton's description makes several important points about occupational HAV. Early recognition of HAVS is crucial, since prolonged HAV exposure can cause chronic, recurrent symptoms. There are data on pneumatic tool operators that suggest that the symptoms of HAVS, once severe, tend to be irreversible and persist even after the worker is no longer exposed to vibration (Bovenzi 1994). This makes worker and employer education about vibration hazards, and periodic medical surveillance exams to identify problems early, important preventive strategies.

Also clear from Dr. Hamilton's description is that there are several factors which increase the risk of disease from occupational vibration exposure. First, maintaining a tight grip on a vibrating tool increases the efficiency of vibration transmission from the tool to the hands and therefore raises the probability of vibration-related disease. Prolonged tight gripping may also be an independent risk factor for HAVS (Gemne 1982). Second, because tool wear and disrepair tend to worsen the intensity of vibration, regular maintenance of vibrating hand tools is important. Finally, HAV is more likely to cause disease in combination with exposure to cold. It is important to keep not only the fingers warm, but the rest of the body must be warm as well. A low body temperature tends to reduce blood flow to the extremities and may increase the risk of digital vascular constriction.

A syndrome of blanching, pain, and numbness of the fingers precipitated by cold and *unrelated to vibration* was first described in 1862 by a French physician named Maurice Raynaud. It affects as much as 10% of the population to some degree, and is more common in women. Raynaud's phenomenon is separated into primary Raynaud's *disease*, the cause of which is unknown, and the secondary Raynaud's *phenomenon*, which is caused by, or associated with vibration, trauma, certain drugs, or one of several diseases (rheumatoid arthritis, systemic lupus erythematosis, scleroderma, atherosclerosis, and several others) (Creager and Dzau 1994). The physician considering the diagnosis of HAVS must therefore take the other causes of secondary Raynaud's phenomenon into account.

Further details about establishing the diagnosis and treatment of HAVS are discussed in more detail elsewhere (Pelmear and Taylor 1991, Pelmear and Taylor 1994), but it is important to note that the severity of HAVS is currently graded by a detailed classification system. The Stockholm classification system is based on the worker's history, physical examination, and specific diagnostic tests (Gemne 1987; Brammer 1987) and is summarized in Table 3.1. This system separates the vascular effects (finger blanching), from the sensorineural effects (tingling, numbness, loss of dexterity). This distinction is made because even though these two types of symptoms often occur together, they can occur and progress independently of one another. The Stockholm classification system is a revision of the earlier Taylor Pelmear classification system, which is often mentioned in the HAV literature. The Taylor Pelmear system did not make the same distinction between vascular and neurologic effects.

Table 3.1. The Stockholm Workshop HAVS classification system.

Vascular Component

Stage	Grade	Description
0		No attacks
1	Mild	Occasional attacks affecting only the tips of one or more fingers
2	Moderate	Occasional attacks to distal and middle phalanges of one or more fingers
3	Severe	Frequent attacks affecting all phalanges of most fingers
4	Very severe	As in stage 3, with trophic skin changes in the finger tips

Sensorineural Component (SN)

Stage	Symptoms
OSN	Vibration exposed, no symptoms
1SN	Intermittent or persistent numbness with or without tingling
2SN	As in stage 1SN, with reduced sensory perception
3SN	As in stage 2SN, with reduced tactile discrimination and manipulative dexterity

Note: Staging is done for each hand separately.
(adapted from Gemne et al. 1987; Brammer et al. 1987).

Several studies have shown an increased risk of vibration-related disease associated with chain saw use among forestry workers (McKenna and Allen 1994, Koskimies et al. 1992, Farkkila et al. 1986). One study on forestry workers in Quebec found a prevalence of 30% among all chain saw users, and over 50% among those with 20 or more years of exposure (Theriault et al. 1982). The use of chain saws for prolonged periods, frequently in the cold, and in association with other strenuous use of the hands puts forestry workers at particular risk. Fortunately, chain saw manufacturers have reduced vibration with design improvements in the past two decades. The vibration of chain saws was shown to decrease from 14 to 2 meters/second/second between 1972 and 1990 with a resultant decrease in the prevalence of vibration induced symptoms (Koskimies 1992).

Control of Hand-Arm Vibration

The harmful effects of HAV can be minimized by several strategies including engineering controls, medical surveillance, work practices, and personal protective equipment (NIOSH 1983, Wasserman 1989).

Engineering Controls. Job tasks should be structured so as to minimize the need for use of vibrating equipment. Manufacturers have begun to redesign tools by balancing engines, mechanically isolating vibrating components through the use of damping materials.

Medical Surveillance and Worker Education. Pre-placement and periodic surveillance examinations should be performed on workers who are or will be exposed to HAV to minimize exacerbation of previously existing conditions, such as Raynaud's phenomenon, and ensure early detection of HAVS. Workers and their employers should be educated about the risks, signs, and symptoms of HAVS. Workers experiencing symptoms of HAVS should see a physician with appropriate training promptly.

Work Practices

- Vibrating hand tools should be maintained properly according to manufacturer's recommendations. Worn, poorly maintained tools tend to generate more vibration and increase the risk of HAVS.
- Work schedules should incorporate a 10-minute break for each continuous hour of vibration exposure.
- Workers should wear adequate clothing to keep their entire body warm (not only the hands) and keep their hands dry. If the hands become wet, the worker should stop, dry them and put on warm, dry gloves or mitts.
- Workers should let the tool do the work, gripping it as lightly as possible while still working safely and maintaining tool control.
- There is evidence that smoking increases the risk of HAVS, and workers exposed to HAV should be made aware of this and discouraged from smoking (Ekenvall and Lindblad 1989).
- Workers should run tools at reduced speeds where possible, and substitute a manual method of doing the work when practical.

Personal Protective Equipment. Anti-vibration gloves made of viscolastic damping materials *may* reduce transmission of vibration from hand-held tools. Anti-vibration materials should be evaluated carefully, because vibration transmission can actually be increased by the vibration characteristics of the material (Wasserman 1989). Gloves or mits offer protection from lacerations and keep the hands warm. Since the vascular effects of HAVS begin at the finger tip and move proximally, the practice of cutting the tips off the fingers of gloves to improve sensation and control may be inadvisable.

Treatment. Because HAVS tends to be irreversible, the mainstay of treatment must be prevention, as described above. For relief of symptoms associated with finger blanching, swinging the arms, along with maintaining body warmth are probably the most effective immediate measures. Among the proposed drug therapies for chronic or recurrent symptoms, the calcium channel blockers may be useful, particularly nifedipine. Currently, there is no effective treatment for the sensorineural symptoms associated with HAVS, other than removal from further vibration exposure. For a more detailed discussion on drug treatment see Pelmear and Taylor 1991.

SUMMARY

Occupational vibration exposure is an important and prevalent cause of disease and disability. The health consequences of prolonged vibration exposure can be chronic and irreversible, making preventive strategies particularly important. With adequate worker and employer education, vibration hazards can be significantly reduced by simple, low-tech and inexpensive means.

REFERENCES

American National Standards Institute (ANSI). 1986. *Guide for the Measurement and Evaluation of Human Exposure to Vibration Transmitted to the Hand, ANSI S3.34*. New York: ANSI.

American National Standards Institute (ANSI). 1979. *Guide for the Evaluation of Human Exposure to Whole-Body Vibration, ANSI S3.18*. New York: ANSI.

American Conference of Government Industrial Hygienists (ACGIH). 1984-1993. *Threshold Limit Values for Hand-Arm Vibration*. Cincinnati: ACGIH.

Boshuizen, H., C. Hulshof. and P. Bongers. 1990. Long-term sick leave and disability pensioning due to back pain disorders of tractor drivers exposed to whole-body vibration. *Int. Arch. Occup. Envir. Health* 62: 117-22.

Bovenzi, M., A. Franzinelli., L. Scattoni. and L. Vannuccini. 1994. Hand-arm vibration syndrome among travertine workers: a follow-up study. *Occup. Envir. Med.* 51: 361-65.

Brammer, A., W. Taylor, and G. Lundborg. Sensorineural stages of the hand-arm vibration syndrome. *Scand. J. Work Envir. Health* 13: 279-83.

Creager, M. and V. Dzau. 1994. *Raynaud's Phenomenon in Harrison's Principles of Internal Medicine*. New York: McGraw-Hill.

Ekenvall, L., and L. Lindblad L. Effect of tobacco use on vibration white finger disease. *J. Occup. Med.* 31: 13-16.

Farkkila, M., S. Aatola, J. Starck, et al. Hand grip force in lumberjacks: two-year follow-up. *Int. Arch. Occup. Envir. Health* 58: 203-08.

Gemne, G., I. Pyykko, W. Taylor, and P. Pelmear. 1987. The Stockholm Workshop Scale for the classification of cold-induced Raynaud's phenomenon in the hand-arm vibration syndrome (revision of the Taylor-Pelmear scale). *Scand. J. Work Envir. Health* 13: 275-78

Gemne, G. 1982. Pathophysiology and multifactorial etiology of acquired vasospastic disease (Raynaud syndrome) in vibration-exposed workers. *Scand. J. Work Envir. Health* 8: 243-49.

Hamilton, A. 1918. *A Study of Spastic Anemia in the Hands of Sstonecutter*. Industrial Accident and Hygiene Series, Bureau of Labor Statistics/Department of Labor Report 19, Bulletin 236. Washington, DC: U.S. Government Printing Office.

Hand-arm vibration syndrome. 1994. *J. Fam. Pract.* 38: 180-85.

Hulshof, C., and B.V. Van Zanten. 1987. Whole body vibration and low back pain: a review of epidemiologic studies. *Int. Arch. Occup. Envir. Health* 50: 205-20.

International Standards Organization (ISO). 1985. *Evaluation of Human Exposure to Whole-Body Vibration, ISO 2631*. Geneva: International Standards Organization.

Koskimies, I. Pyykko, J. Starck and R. Inaba. 1992. Vibration syndrome among Finnish forest workers between 1972 and 1990. *Arch. Occup. Envir. Health* 64: 251-56.

McKenna, K., A. Blann, and J. Allen. 1994. Vascular responses in chain saw operators. *Occup. Envir. Med.* 51: 366-70.

Mirbod, S., H. Yoshida, C. Nagata, et al. 1992. Hand-arm vibration syndrome and its prevalence in the present status of private forestry enterprises in Japan. *Int. Arch. Occup. Envir. Health* 64: 93-99.

National Institute for Occupational Safety and Health (NIOSH). 1989. *Criteria for a Recommended Standard: Occupational Exposure the Hand-Arm Vibration*. DHHS/NIOSH publication number 89-106.

Pelmear, P., and W. Taylor. Hand-arm vibration syndrome: clinical evaluation and prevention. *J. Occup. Med.* 33: 1144-49.

Pope, M., H. Broman, and T. Hansson. 1989. The dynamic response of a subject seated on cushions. *Ergonomics* 32: 1155-66.

Seidel, H., and R. Heide. 1986. Long term effects of whole-body vibration: a critical survey of the literature. *Int. Arch. Occup. Envir. Health* 58: 1-26.

Theriault, G., L. DeGuire, S. Gingras, et al. 1982. Raynaud's phenomenon in forestry workers in Quebec. *Can. Med. Assoc. J.* 126: 1404-08.

Wasserman, D. 1989. The control aspects of occupational hand-arm vibration. *App. Ind. Hyg.* 4: F22-26.

Wasserman, D., D. Badger, T. Doyle and L. Margolies. 1974. Industrial vibration: an overview. *J. Am. Soc. Safety Eng.* 19: 38-40.

Wilder, D., D. Wasserman, M. Pope, M. Pelmear and W. Taylor. 1994. Vibration. In: *Physical and Biological Hazards of the Workplace*. Wald and Stave., eds. New York, Van Nostrand Reinhold.

4

NOISE AND HEARING LOSS IN AGRICULTURE, FORESTRY, AND FISHERIES

David Siebens, M.D., M.P.H.
Duke University Medical Center

> Noise induced hearing loss results from mechanical damage to the ear and is one of the most common workplace hazards. Physical principles of noise—frequency, intensity, and time of exposure—are discussed along with the characteristics of the human ear and how it responds to sound. Audiometry is discussed as an effective tool to assess the ability to hear and how hearing may be affected by noise exposure. Occupational safety and health regulations designed to limit noise exposure are presented. Engineering controls are often the most effective methods of preventing noise induced hearing loss. Education programs about the harmful effects of noise exposure and the benefits of hearing protection are also discussed in this chapter.

INTRODUCTION

It is estimated that 28 million Americans have some degree of hearing loss, and approximately 10 million have some degree of hearing loss directly attributable to noise exposure. More workers are exposed to noise than to any other physical or chemical hazard, making noise-induced hearing loss one of the most prevalent occupational illnesses. Noise exposure is of particular concern in agriculture, forestry, and fisheries because of the extensive use of noisy machinery such as tractors, grain dryers and other farm implements, chain saws, chippers, trucks, power boats, etc. It has long been recognized that noise induced hearing loss is more prevalent among farmers, with about 50% of farmers over the age of 50 having significant hearing loss (Lierle and Reger 1958; Plakke and Dare 1992). Particularly worrisome is that rural children have a 2 1/2 times greater hearing loss than their urban counterparts (Emanuel 1990), suggesting that noise-induced hearing loss among agricultural workers is prevalent at an early age.

While noise-induced hearing loss is one of the most common occupational illnesses, it is also one of the most preventable and least treatable. One of the most constructive services the health and safety professional can offer the agricultural, forestry, or fishery worker is prevention of noise-induced hearing loss. The objective of this chapter is to give you sufficient understanding of noise hazards so that you can make a contribution to preventing a common and important occupational health problem.

PHYSICAL PRINCIPLES

A basic familiarity with several physical principles is essential to understanding noise hazards, their evaluation, and the associated regulatory issues. The physical properties of sound relevant to this discussion are frequency, intensity, and time.

Frequency

The frequency of sound is the rate at which pressure oscillations are produced by the sound source and is expressed in cycles per second, or Hertz (Hz). Young healthy adults can generally hear sound between 20 and 20,000 Hz, with the best hearing acuity in the 1000 to 5000 Hz range. The frequencies most important for hearing and understanding speech are between 500 and 3000 Hz.

Most sounds are made up of a broad spectrum of different frequencies, and it is the relative mixture and intensity of different frequencies that allows us to distinguish one sound from another. Pure tones, made up of only one frequency, are seldom encountered in the natural environment. An example of a pure tone is the hum of a tuning fork. Pure tones are important to this discussion because measurement of hearing acuity (discussed later) is based on the ability to hear pure tones at certain frequencies.

Intensity

Sound intensity is measured by establishing the magnitude of pressure oscillations above and below atmospheric pressure. The units of sound pressure (the pascal, or micro pascal) are cumbersome, however, and almost never used. The decibel is the most practical measure of sound intensity, but use of the decibel scale is often poorly understood and a brief review should make use of the decibel scale straight forward.

The sound pressure of the softest, audible sound is about 20 micropascals, and that of the loudest tolerable sound is about 20,000,000 micropascals. Since the use of numerous zeros makes the pascal scale cumbersome, a logarithmic, or decibel, scale was selected for simplification. The mathematics that follow are presented simply to give you an understanding of the decibel scale. It will not be necessary to use, or commit the following formula to memory.

The definition of the decibel (dB) is:

$$dB = 20 \log 10 (P/P0)$$

Where P is the sound pressure being measured, and P0 is a constant of 20 micropascals. Twenty micropascals is used because this is the approximate threshold for normal human hearing (the softest sound that can be heard) in the best heard frequencies. A sound pressure measurement of 20 micropascals would therefore translate to 0 dB.

$$dB = 20 \log 10 (20/20) = 0$$

Zero dB, then, is the sound intensity representing the approximate threshold of normal human hearing, but does *not* represent zero sound.

Since the dB is a logarithmic function, working with the dB scale can be less than intuitive at first. Take, for example, the fact that 0 dB is defined as the threshold of normal human hearing. This means that an individual with better than average hearing will be able to hear sounds of less than 0 dB. It is not uncommon for young adult to have hearing thresholds of -5 or -10 dB at certain frequencies. Another important example of how a logarithmic scale can lead to unexpected results is that a doubling of sound intensity results in a change of only 3 dB. The seemingly insignificant difference, then, between 90 and 93 dB in fact represents a doubling of sound intensity, and this will become important in the next section. The approximate intensity of a few familiar sounds is presented in Table 4.1.

Table 4.1. Approximate intensity of some familiar sounds.

Source	Sound intensity
Gunfire	130-170
Jet engine	130
Jackhammer	120
Chain saw	110-116
Personal cassette players (max)	110-115
Leaf blower	110
Two row corn picker	106
Tractors	73-102
Hog confinement building during feeding	100
Pneumatic conveyor	100
Irrigation pump	98
Fishing trawler engine room	97-104
One row beet puller	94
Fishing trawler fish processing plant	84-97
Grain drying	88-98
Cows during milking	94
Combine	81-86
Hog confinement building (non-feeding)	75
Vacuum cleaner	75
Conversational speech	60
Whisper	30

(McMahon 1988; Szczepanski 1991; Brookhouser 1994; NIOSH 1972).

Measurement of sound intensity with modern electronic meters is deceptively simple, and should be left to trained professionals. Sound intensity measurements are typically reported as dB on the A-scale, or dBA. The A-scale is simply an electronic means of enabling a sound level meter to "hear" like the human ear through the use of electronic filters. Sound at frequencies above the limit of human hearing (i.e., above 20,000 Hz) are filtered out completely, sound in poorly heard frequencies are given relatively less weight, and full weight is given to sound in the best heard frequencies (1000 to 5000 Hz). For example, a sound of 140 dB at 4000 Hz would be dangerously loud, but the same sound intensity at 40,000 Hz would be inaudible, harmless, and measured at 0 dB by a sound level meter operating in the A-scale mode. Sounds of equivalent intensity, then, can have different A-scale readings because of relative differences in frequency composition.

Time

Noise becomes increasingly hazardous with the duration of exposure. The relationship between noise intensity and duration of exposure has been well studied, and OSHA (1971, 1983) has established a permissible exposure limit (PEL) of 90 dBA for an 8-hour period. In theory, noise exposure below this level is safe, and would not be expected to harm the human ear. It is important to note, however, that susceptibility to noise-induced hearing loss can vary from one individual to another, and some experts advocate a lower noise exposure limit, believing that 90 dBA for 8 hours may in fact be harmful to some individuals.

In the occupational setting, noise levels are typically presented as a time weighted average (TWA) which combines noise intensity and exposure duration into a single number. The TWA simply provides a "common denominator" by which noise exposures of variable intensity and duration can be compared. The 8-hour work day has been selected as the referent exposure duration. The TWA is the sound intensity, which if present for 8 hours, would represent a comparable hazard to the sound being measured.

Recall from the previous section that a 3 dB change in intensity represents a doubling of sound energy. One might expect, then, that a doubling of exposure time would be equivalent to a 3 dB increase in intensity. However, intermittent noise exposure is typically less hazardous than continuous exposure of the same total duration. In the industrial setting, noise exposure is usually not continuous, and OSHA has therefore adopted a 5 dB rule in which a doubling of exposure time is considered equivalent to a 5 dB increase in intensity. Using the permissible exposure limit of 90 dBA for 8 hours as an example, this would be equivalent to 95 dBA for 4 hours, or 100 dBA for 2 hours, etc. All of these time-intensity combinations represent 90 dBA-TWA. Fortunately, modern noise dosimeters perform the TWA calculations for a particular exposure situation, sparing the safety professional the need to do so.

NORMAL HEARING

Noise-induced hearing loss (NIHL) results from mechanical damage to the ear from the sound energy itself. The mechanism by which noise damages the ear is easily understood with some familiarity of the basic anatomy and physiology of normal hearing.

The ear is divided into three sections: the outer ear, the middle ear, and the inner ear (Figure 4.1.) The outer ear

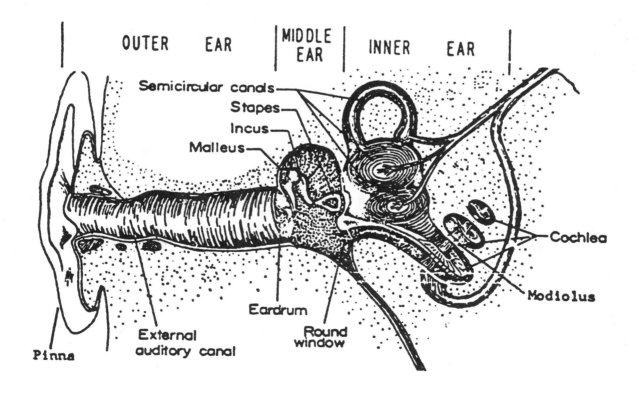

Figure 4.1. Cross-section of outer, middle, and inner ear structures (from NIOSH 1972).

includes the pinna (the visible part) and the canal leading to the tympanic membrane. The outer ear does not contribute much to hearing in humans, though reflection and resonance of sound off the pinna makes a minimal contribution to hearing at some frequencies.

The tympanic membrane separates the outer ear from the middle ear. Sound impinging on the ear drum causes it to vibrate, and this vibration is transmitted to the inner ear by a chain of three tiny bones, or ossicles, which are attached to the ear drum at one end, and the cochlea at the other. The anatomical relationship between the tympanic membrane, ossicles and cochlea is such that there is significant, mechanical amplification of the vibration impinging on the tympanic membrane. The middle ear is an air-filled space, connected to the nasopharynx by the eustachian tube. The eustachian tube equalizes the pressure of the middle ear with the atmospheric pressure, and its periodic opening accounts for the familiar sensation of the ears popping with changes in atmospheric pressure.

The inner ear contains the cochlea, which is the organ of hearing. The cochlea is a fluid-filled, snail-shaped structure with about 2.5 turns. Vibration of the ossicles is transmitted to fluid within the cochlea, causing the vibration of delicate hair cells. Movement of these hair cells converts the vibration into nerve impulses which are transmitted to the brain. The shape of the cochlea is such that high frequency sounds resonate at its base and low frequency sounds resonate at its apex. The brain then interprets nerve impulses coming from different regions of the cochlea as different pitches.

NOISE-INDUCED HEARING LOSS

Significant, long term noise exposure causes irreversible damage to the microscopic anatomy of the hair cells and supporting structures. Noise-induced damage initially affects the region of the cochlea responsible for hearing higher frequencies, in the 3000 to 6000 Hz range. If excessive noise exposure continues, the hearing loss extends to both higher and lower frequencies. When hearing loss extends into the 500 to 2000 Hz range, the subject begins to have difficulty understanding speech. Consonant sounds are in this frequency range, and when they are lost, the inability to differentiate consonant sounds makes speech sound garbled and difficult to understand ("dish", "fish" and "wish" all sound the same). Difficulty understanding speech is an indication that *significant* hearing loss has already occurred, but regrettably, it is often the first clue that a hearing problem exists.

This discussion is limited to noise-induced hearing loss, but there are many other causes of hearing loss. Hearing loss is divided into two main types, conductive, and sensorineural.

Conductive loss results from impaired conduction of vibration through the outer and middle ear to the inner ear. Examples of conductive loss are obstruction of the auditory canal by wax, damage to the tympanic membrane by a spark or sharp object, or damage to the ossicular chain from trauma or infection. Sensorineural loss results from damage to the inner ear, the auditory nerve, or brain. Noise induced hearing loss, then, is a type of sensorineural loss.

The American College of Occupational Medicine (ACOM 1989) has defined occupational, noise induced hearing loss as:

1. Sensorineural loss affecting the hair cells
2. Bilateral and symmetrical, affecting both ears equally
3. Never profound
4. Not progressive once noise exposure ceases
5. Decelerating with continued exposure
6. Greatest around 4000 Hz
7. Stable after 10 to 15 years
8. More severe with continuous noise exposure than with interrupted exposure

Hazardous noise exposures initially cause a temporary, reversible elevation in the hearing threshold (the softest sound the subject can hear) which lasts hours to days. This is called a temporary threshold shift. If hazardous noise exposure continues unabated, a permanent threshold shift will gradually develop. One exception to this pattern of gradual, permanent hearing loss is acoustic trauma, in which an intense, brief, exposure such as an explosion can cause an immediate and permanent loss.

AUDIOMETRY

Hearing acuity is measured by determining the hearing threshold at pure tone frequencies for each ear. The threshold is determined at standardized frequencies between 500 and 8000 Hz in a silent environment, usually a soundproof booth, using special earphones. The hearing threshold for each frequency is presented in either graphic or tabular form. Graphically, the hearing threshold is plotted on the vertical axis, against frequency along the horizontal axis. Hearing thresholds of up to 25 dB are considered normal in an adult, and a normal audiogram is presented in Figure 4.1. As mentioned previously, noise-induced hearing loss usually begins in the 3000 to 6000 Hz range, and a characteristic notch is typically seen in this range with noise-induced hearing loss. Figure 4.2 illustrates the audiometric pattern typical of early, noise induced hearing loss.

In the occupational setting, a simple, pure tone audiogram is used as a screening tool to identify and monitor workers with hearing loss so that further loss can be

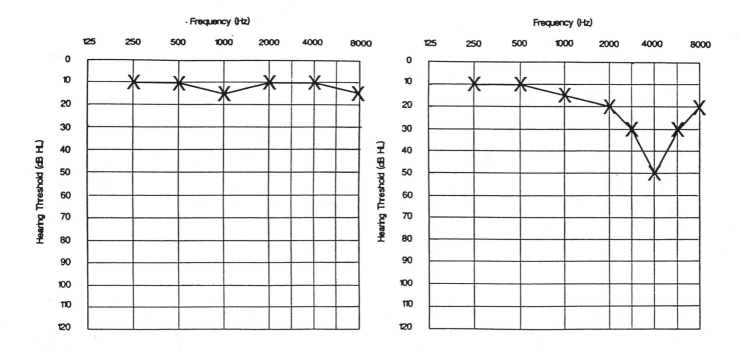

Figure 4.2. Audiogram showing: a) Normal hearing b) A pattern typical of noise-induced hearing loss. Hearing loss is most pronounced between 3000 and 6000 Hz. (NIOSH 1993).

prevented. More elaborate audiometric testing is performed by audiologists and otolaryngologists to clarify the precise nature and location of a hearing problem. All audiometry must be performed by specially trained personnel and interpreted by an audiologist, otolaryngologist, or other qualified physician.

REGULATORY ISSUES

The Occupational Health and Safety Act (1971) and subsequent Hearing Conservation Amendment (1983) establish guidelines and regulations for industrial noise exposure. In addition, several government agencies have separate (though similar) regulations on noise exposure in the industries they regulate. The Federal Aviation Administration, the Federal Railroad Administration, and the Mine Safety and Health Administration all regulate noise exposure in their respective industries. Among the most important things to realize about noise regulation is that the agricultural environment and small, independent businesses have not received much attention regarding noise exposure regulation. This gives the health and safety professional a special opportunity to assist and educate farmers, fishermen, and forestry workers about noise hazards and their prevention (Holt, et al. 1993).

The purpose of the OSHA regulations is to protect workers from occupational noise-induced hearing loss. As mentioned previously, the OSHA regulations establish a 90 dBA-TWA as the permissible exposure limit (PEL). Noise exposures above this level must be reduced by engineering controls (reducing noise at its source—e.g., improving the muffler on a tractor), administrative controls (altering work schedules to reduce individual exposure duration), or by the use of personal protective equipment (ear plugs or muffs).

The Hearing Conservation Amendment recognizes the potential hazard of borderline noise exposure below 90 dB, and requires workers exposed to more than 85 dBA-TWA to be enrolled in a hearing conservation program. The essential elements of a hearing conservation program include:

1. Noise hazard assessment by noise measurements
2. Baseline and annual audiometry
3. Noise exposure reduction by engineering controls, administrative controls, or, as a last resort, by personal hearing protection.
4. Worker education, training, and motivation

5. Recordkeeping

Further details on hearing conservation programs and noise regulation are beyond the scope of this text, but anyone responsible for noise hazards or a hearing conservation program should obtain and be familiar with the relevant regulations. In addition to the OSHA regulations, there are several useful resources available from the National Institute of Occupational Safety and Health (Suter 1990), the American Academy of Otolaryngology (1982), the American Industrial Hygiene Association (Berger et al. 1986) and others (Royster 1990).

PREVENTION OF HEARING LOSS

The most effective approach to reducing noise exposure is through engineering controls that reduce the potential for exposure at its source. An example of effective engineering control is the addition of cabs to tractors. Holt et al. (1993) found that 75% of tractors without cabs subjected the drivers to noise levels in excess of 90 dB, compared to only 18% of tractors with cabs. Newer tractors with cabs as original equipment had significantly lower noise levels than older tractors with retrofit cabs. All tractors with cabs built since 1980 had noise levels below 85 dB.

Educating workers about noise hazards is equally important. In order for workers to comply with noise reduction strategies, they must understand that their hearing is vitally important, is at risk, and that hearing loss can be prevented by following steps to limit noise exposure. Workers need to have a practical understanding of how much noise exposure is hazardous. Accurate sound level measurements are often unavailable in the real world. One clue that noise levels are approaching hazardous levels is the need to shout to converse with someone standing an arm's length away.

The benefit of almost any engineering control can be negated by inadequate worker education and motivation. Using the example of tractors again, the noise reduction achieved by the addition of a cab can be reduced by leaving the windows open, which is often done to enable the driver to hear any implement malfunction. In addition, radios are often used in the cab for entertainment, communication, or monitoring the weather, and are often turned up loud enough to overcome the ambient noise. Operating a tractor with the windows open and radio on typically adds about 5 dB to the noise in the cab, doubling the exposure intensity.

Education of the young is particularly important. Noise-induced hearing loss usually results from cumulative exposure over many years, and there is good evidence to suggest that noise-induced hearing loss among farm families begins in childhood. The need to educate rural children about noise hazards is illustrated by that fact that in one study, only 9% of farm children used any hearing protection in noisy environments (Broste 1989).

Administrative controls, which limit noise exposure by altering work schedules and practices, are frequently too disruptive to be effective. One work practice that can be effective, however, is maintaining as much distance from noise sources as possible. In an open space, noise intensity varies with the inverse square of the distance from the source. Doubling the distance from the source, then, yields a 4-fold decrease in noise exposure.

The cost and impracticality of many engineering and administrative controls make the use of personal hearing protection the most practical and common way to reduce noise exposure. Ear plugs and muffs are effective, but only if properly used. The most important factor in determining compliance with the use of hearing protection is comfort and ease of use. OSHA requires that workers be offered a selection of hearing protection alternatives, recognizing that hearing protection that is uncomfortable or interferes with job performance is unlikely to be used properly, if at all.

Ear plugs, caps, and muffs have a noise reduction rating, which indicates the number of decibels by which noise is reduced with proper use. For example, ear plugs with a noise reduction rating of 25 dB would be expected to result in an effective exposure of 75 dB when used properly in a 100 dB environment. The noise reduction rating is determined under ideal conditions, however, and many experts believe that using half the noise reduction rating is a more realistic estimation of the noise reduction actually achieved. The correct amount of attenuation is important, because too much noise reduction may subject the worker to other hazards resulting from inability to hear machinery, alarms, instructions from co-workers, etc.

Finally, everyone needs to be aware of noise hazards, regardless of their occupational exposure potential. Noisy leisure activities such as hunting, motorcycle or snowmobile riding, gardening, or use of power tools are likely to involve hazardous noise exposure. The use of personal cassette players at loud volume levels is an increasingly common noise exposure. Surveys have found that 80% of children in the middle and elementary school own or use a personal cassette player, with 5 to 10% doing so at potentially hazardous noise levels.

Educating workers in agriculture, forestry, and fisheries presents a significant challenge because these workers tend to be spread out over a large geographic area, and are not employed by a single organization. Migrant farm workers may be particularly difficult to reach. Disseminating information to farm workers may be best achieved through organizations with which the workers frequently have contact. Agricultural extension services, public health departments, and local health providers are all potential resources the health and safety professional can use to educate workers about minimizing noise hazards.

SUMMARY

Noise-induced hearing loss is a common, serious, and completely preventable occupational health problem. Workers in agriculture, forestry, and fisheries are at particular risk because of the extensive use of noisy machinery in these industries. Educating workers and employers about noise induced hearing loss and its prevention gives the health and safety professional an excellent opportunity to prevent an important, and all too common, health problem.

REFERENCES

American Academy of Otolaryngology. 1982. *Guide for Conservation of Hearing in Noise*. Ear and Neck Surgery Foundation, Inc. Rochester, MN: Custom Printing.

American College of Occupational Medicine. 1989. Occupational noise-induced hearing loss. Noise and Hearing Conservation Committee. *J. Occup. Med.*

Berger, E., et al., eds. 1986. *Noise and Hearing Conservation Manual*. 4th edition. Akron, OH: American Industrial Hygiene Association.

Brookhouser, P.E. 1994. Prevention of noise induced hearing loss. *Prev. Med.* 23: 665-69.

Broste, S.K., D. A. Hansen, et al. 1989. Hearing loss among high school farm students. *Am. J. Public Health* 79: 619-22.

Emanuel, D.A. 1990. A case for medical, environmental and safety screening. *Am. J. Ind. Med.* 18: 179-92.

Holt, J.J., S.K. Broste, and D.A. Hansen. 1993. Noise exposure in the rural setting. *Laryngoscope* 103: 258-62.

Lierle, D.M., and S.N. Reger. 1958. The effect of tractor noise on the auditory sensitivity of tractor operators. *Ann. Otology. Rhinology Laryngology* 67: 372-89.

McMahon, K.B. and C.D. Urbain. 1988. Farming in silence. *Farm J.* (November), pp. 13-15.

National Institute for Occupational Safety and Health (NIOSH). 1972. *Occupational Exposure to Noise, Criteria for a Recommended Standard*. Washington, DC: U.S. Department of Health Education and Welfare.

National Institute for Occupational Safety and Health (NIOSH). 1993. *Agricultural Noise Exposure and Control*. Cincinnati, OH: NIOSH, Cooperative Agricultural Surveillance Training.

Occupational Safety and Health Administration (OSHA). U.S. Department of Labor. 1983. Occupational Noise Exposure: Hearing Conservation Amendment; Final Rule. *Federal Register* 48: 9736-84.

Occupational Safety and Health Administration (OSHA). U.S. Department of Labor. 1971. Occupational Safety and Health Standards, National Consensus Standards and Established Federal Standards. *Federal Register* 36: 10518.

Plakke, B.L., and E. Dare. 1992. Occupational hearing loss in farmers. *Public Health Reports*. 107: 188-92.

Royster, J.D. and L.H. Royster. 1990. *Hearing Conservation Programs: Practical Guidelines for Success*. Chelsea, MI: Lewis Press.

Suter, A.H. and J. R. Franks, eds. 1990. *A Practical Guide to Effective Hearing Conservation Programs in the Workplace*. Cincinnati, OH: U.S. Department of Health and Human Services, NIOSH Report No. 90-120.

Szczepanski, C. and Z. Weclawik. 1991. Exposure of the crew of a fishing trawler-factory ship to noise. *Bull. Inst. Marit. Tropical Med.* 42: 67-70.

5

INJURIES ASSOCIATED WITH COLD AND HOT ENVIRONMENTS

Bruce A. Cohen, M.D., L.C.D.R., M.C., U.S.N.
Duke University Medical Center

Many factors influence adverse health effects of exposure to cold or hot environments. Type of exposure, type of work, duration of exposure, and age or physical condition of the worker are among the factors presented in this chapter. Potentially hazardous situations exist both indoors and outdoors. Workers from a wide range of occupations are affected. Cold stress may be identified by type of exposure: whole body, extremity, wind chill, contact, or airway. As the effects of exposure are manifested, tissue damage or detriments to body function may result. Heat stress occurs due to an imbalance in body temperature regulation: heat accumulation versus heat dissipation. Illnesses due to heat stress are discussed. Prevention and control measures such as acclimatization and personal protection are presented.

INTRODUCTION

Workers exposed to cold and hot environments manifest many different health effects. These are influenced by the following factors:

- Duration of exposure.
- Type of environment.
- Type of work being performed.
- Level of physical requirements.
- Protective clothing worn.
- Age.
- Current health status of the worker.
- Alcohol use.
- Smoking.
- Circulatory impairment.

These cold and hot environments may occur outdoors (i.e., farming, maintenance operations, construction, etc.), or indoors (i.e., meatpacking, bakeries, firefighting, etc.)

Problems arise when certain conditions exist:

- Long exposures to cold and heat in an "acceptable" environment.
- Long exposures to cold and heat in extreme environments.
- Excessive workloads.
- Inappropriate protective clothing.
- Inadequate rest and breaks.
- Poor nutrition.
- Low aerobic fitness.
- Lack of proper training to meet the requirements of the environment.

The actual incidence of thermal injuries and illnesses among industry workers is unknown with only the most severe situations attracting attention, usually in case history form. To date, there is no national surveillance system for monitoring these disorders.

COLD STRESS

Thermal stress generated by a cold environment can be defined by the action of its cooling power. It may be simple or complex. The nature of cooling includes whole-body cooling, extremity cooling, convective cooling (wind chill), conductive cooling (contact), and airway cooling. Additional factors include the work rate, inhibition of function (weight, reduction in reach, hearing and vision) caused by bulky outer garments, and additional stress factors, such as remoteness and darkness.

The temperature level (air temperature) is a poor predictor of protection requirements. Unfortunately, the general public uses temperature alone to gauge the level of protective clothing they wear.

Wind chill (the combination of air temperature and wind speed) is widely accepted for assessment of cold stress and the risk of cold injury to unprotected skin (Table 5.1).

The prevention of heat loss and the maintenance of efficient functioning is the primary goal in the cold environment. Effective clothing, shelter, exercise, and external heat sources are efficient means for combating cold stress. Changes in climactic conditions, accidents, poor preparation, and ignorance lead to cold casualties.

For the body to maintain thermal homeostasis in a cold environment it must limit heat loss and increase heat production.

A direct effect of cold stress is cooling of tissues. The most serious situation arises with whole-body cooling leading to hypothermia. Core temperatures of 33 to 35° C (91 to 95° F) are associated with violent shivering and reduced mental and physical work capacity. This gradually progresses to exhaustion, fatigue, neuromuscular incapacity, mental confusion, and unconsciousness.

Cold causes peripheral vasoconstriction to occur, which restricts blood flow to the extremities. This enables the core heat balance to be kept stable longer at the expense of superficial tissues (hands and feet). As the tissue temperature continues to decrease, discomfort develops and function and performance are impaired, eventually leading to numbness and loss of function. Cold extremities reduce work ability. When vasoconstriction cannot maintain body heat balance, muscular hypertonus and shivering become important mechanisms for increasing body temperature by causing metabolic heat production to increase several times the resting rate.

Heat loss from the head can be as high as 50% of the body's total heat loss. Suitable, protective headgear is always essential to prevent cold stress.

Heat loss by conduction is usually minimal, since there is a reflex withdrawal from cold objects. Metallic items and liquids have a high heat conductivity and can remove heat from a warm surface at high rates.

Airway cooling results by the inhalation of cold, dry air causing local cooling of the nasal mucosa and the upper respiratory tract.

Potential Occupational Exposures

- Out-of-door workers during cold weather (e.g., farmers, loggers).
- Refrigerated warehouse workers.
- Fishermen.
- Firemen.
- Divers.
- Refrigeration workers.
- Packing house workers.

Table 5.1. Wind chill index (U.S. Army Research Institute of Environmental Medicine).

Cooling Power of Wind on Exposed Flesh Expressed as Equivalent Temperature (under calm conditions)*

Estimated Wind Speed (in mph)	Actual Temperature Reading (°F)											
	50	40	30	20	10	0	-10	-20	-30	-40	-50	-60
	Equivalent Chill Temperature (°F)											
calm	50	40	30	20	10	0	-10	-20	-30	-40	-50	-60
5	48	37	27	16	6	-5	-15	-26	-36	-47	-57	-68
10	40	28	16	4	-9	-24	-33	-46	-58	-70	-83	-95
15	36	22	9	-5	-18	-32	-45	-58	-72	-85	-99	-112
20	32	18	4	-10	-25	-39	-53	-67	-82	-96	-110	-121
25	30	16	0	-15	-29	-44	-59	-74	-88	-104	-118	-133
30	28	13	-2	-18	-33	-48	-63	-79	-94	-109	-125	-140
35	27	11	-4	-20	-35	-51	-67	-82	-98	-113	-129	-145
40	26	10	-6	-21	-37	-53	-69	-85	-100	-116	-132	-148

(Wind speeds greater than 40 mph have little additional effect.)	**LITTLE DANGER** In < hr with dry skin. Maximum danger of false sense of security	**INCREASING DANGER** Danger from freezing of exposed flesh within one minute.	**GREAT DANGER** Flesh may freeze within 30 seconds.
	Trenchfoot and immersion foot may occur at any point on this chart.		

Cold Injuries

Cold injuries may be nonfreezing or freezing and involve the extremities and/or the body's core. The areas most likely involved are the cheeks, nose, ear, fingers, toes, hands, and feet. Table 5.2 lists some effects of cold exposure.

Frost nip: least severe cold injury; reversible ice crystal formation in the skin surface; develops slowly and painlessly and usually resolves spontaneously.

Chilblains: repeated exposure of bare skin to cold water or from wet cooling of an extremity over hours or days at temperatures slightly above freezing; usually involves hands and feet; trench foot (immersion foot) is a severe form; paresthesia (tingling) and anesthesia (decreased sensation) occur; 2 to 3 days following rewarming, hyperemia develops with intense pain, swelling, redness, heat, blistering, hemorrhage, and sometimes cellulitis and gangrene.

Table 5.2. Effects of cold exposure.

Effect	Body Temperature
1. Confusion	95° F (35° C)
2. Maximal shivering	95° F (35° C)
3. Amnesia for recent events	93° F (34° C)
4. Decreased time awareness	93° F (34° C)
5. Decreased attention span	93° F (34° C)
6. Decreased tactile sensation /grip strength	93° F (34° C)
7. Increased task times	93° F (34° C)
8. Decreased coordination (incl. ataxia)	91° F (33° C)
9. Hallucinations/delusions	91° F (33° C)
10. Shivering STOPS	90° F (32° C)
11. Bradycardia	90° F (32° C)
12. Decreased recognition	88° F (31° C)
13. Atrial and ventricular fibrillation	86° F (30° C)
14. Loss of consciousness	85° F (29° C)
15. Pupils fixed/dilated	85° F (29° C)
16. Respiration STOPS	75° F (24° C)
17. Flat EEG (? Reversible)	66° F (19° C)

Frostbite: the actual freezing of soft tissue; exposed areas of skin are most commonly involved. The freezing point of the skin is -1° C; however, with increased wind velocity, heat loss is greater and frostbite will occur more rapidly. For example, with a wind velocity of 20 mph, exposed skin will freeze within 1 minute at -10° C. The cold itself produces numbness and anesthesia, which may permit serious freezing to develop without the warning of acute discomfort. Frostbite injury may range from simple superficial injury with redness of the skin, transient anesthesia, and superficial bullae, to deep tissue freezing with persisting ischemia, thrombosis, deep cyanosis, and gangrene.

Hypothermia: systemic cold injury with a core temperature of less than 35° C is an extreme acute problem resulting from prolonged cold exposure and heat loss. When the vasoconstrictor mechanism is overpowered, sudden vasodilatation occurs with resultant rapid heat loss, and critical cooling ensues.

Alcohol and drugs (barbiturates, phenothiazides, antidepressants) may both increase heat loss and impair temperature regulation. Alcohol produces a decreased awareness of cold, causing a delayed response to being cold and, therefore, the likelihood of a deeper chill. Hypothermia can occur at air temperatures up to 18.3° C (65° F) or in water up to 22.2° C (72° F). In immersion, thermal conductance of water is 25 times that of air (up to 200 times in moving water).

Vascular Abnormalities: may be precipitated or aggravated by cold exposures, i.e., Raynaud's disease and phenomenon, acrocyanosis, thromboangiitis obliterans, and urticaria.

The final solution of a cold work problem depends on the balance between relief of cold stress and the individual effort involved in physiological and/or behavioral adjustments to maintain heat balance.

Protection from the Cold

- Clothing that resists wind and rain but "breathes", allowing perspiration to escape (e.g., head covering).
- Layering of garments to prevent overheating during exertion.
- Dry replacement clothes.
- Wind-protected, heated shelters for stationary work outdoors.
- Heated rest facilities and hot food and drinks should be available.
- Daily work and break schedules must take into account the current weather.
- Physically fit workers are less prone to develop cold related problems.
- Proper instruction and initiation of new cold environment workers.

HEAT STRESS

Body temperature is determined by the balance between heat accumulation (heat generated by physical activity—metabolic heat, and heat gained from the environment—environmental heat) and heat dissipation. Thermoregulatory responses act to maintain the body's thermal equilibrium. These regulatory mechanisms include vasodilation, increased heart rate, sweating, and increased body temperature. Factors that affect the efficiency of heat dissipation are temperature, relative humidity (moisture content), the speed of the surrounding air, and radiant temperature.

The industrial heat problem is one in which a combination of the above mentioned factors produces a working environment which may be uncomfortable or even hazardous because of imbalance of metabolic heat production and heat loss. Table 5.3 describes metabolic heat rates for estimating energy cost of work by task.

The main factors that influence the storage of heat within the body are expressed in the heat equation:

$$M + R + C + K - E = S$$

where:

M is the heat generated by metabolism, both basal and from any physical work being performed. These chemical reactions consume fuel, generate energy, and produce by-products that must be eliminated. Strenuous activity can produce an elevation in the amount of heat produced due to metabolism, on the order of 10 to 20 fold.

R is the radiant heat exchange, positive or negative, resulting from exposure of the body to remote surfaces at a greater or lower temperature than the body surface. Radiant heat transfer can contribute significantly to the heat load of workers; heat gain from the sun can be up to 300 kcal/hour.

C is the convective heat exchange, positive or negative, resulting from the differences in temperature between the body surface and air, or water, passing over it. When the air temperature exceeds the skin temperature heat is gained by the body.

K is conductive heat exchange to or from objects with which the body is in contact.

E is evaporative heat loss, significantly the evaporation of sweat or of water from wet clothing. The environmental limitation on evaporative cooling is determined by clothing insulation, air velocity, and humidity.

S is body heat storage, positive or negative. In the neutral state it is zero.

When heat loss fails to keep pace with heat gain, the core temperature rises. Certain physiological mechanisms initiate an attempt to increase heat loss from the body. The large blood vessels of the skin and subcutaneous tissues dilate, diverting much of the cardiac output to these superficial areas. Total cardiac output is also increased and sweat glands spread fluid over the skin's surface, increasing heat loss via evaporation.

Table 5.3. Metabolic heat rates (NIOSH 1986).

A. Body position and movement	kcal/min*	
Sitting	0.3	
Standing	0.6	
Walking	2.0–3.0	
Walking uphill	add 0.8 per meter rise	

B. Type of work	Average kcal/min	Range kcal/min
Hand work		
light	0.4	0.2–1.2
heavy	0.9	
Work one arm		
light	1.0	0.7–2.5
heavy	1.8	
Work both arms		
light	1.5	1.0–3.5
heavy	2.5	
Work whole body		
light	3.5	2.5–9.0
moderate	5.0	
heavy	7.0	
very heavy	9.0	

C. Basal metabolism	1.0

D. Sample calculation**	Average kcal/min
Assembling work with heavy hand tools	
1. Standing	0.6
2. Two-arm work	3.5
3. Basal metabolism	1.0
Total	5.1 kcal/min

* For standard worker of 70 kg body weight (154 lbs.) and 1.8 m² body surface (19.4 ft²).
**Example of measuring metabolic heat production of a worker when performing initial screening.

One of the most important adaptive processes affecting the ability of humans to function in heat stress environments is acclimatization. This is a process by which humans are able to adapt to hot environmental conditions through a series of physiological and psychological adjustments. It is characterized by the worker's ability to perform with less increase in core temperature and heart rate and less salt loss due to a lower concentration of sodium chloride in the sweat. Once acclimatized, an individual may work safely in conditions that previously would have been harmful. Acclimatization occurs in stages over a 1 to 2 week period and is retained only as long as the individual is continually exposed to the hot environment.

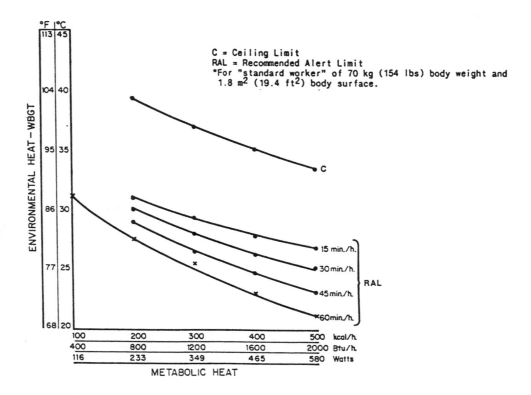

Figure 5.1. Recommended heat-stress alert limits for heat-unacclimatized workers (NIOSH 1986).

NIOSH has recommended limits for exposure to heat in both acclimatized and unacclimatized workers. For unacclimatized workers, the total heat exposure shall be controlled so that unprotected healthy workers are not exposed to combinations of metabolic and environmental heat greater than the applicable Recommended Alert Limits (RAL) given in Figure 5.1. For acclimatized workers the total heat exposure shall be controlled so that unprotected healthy workers in hot environments are not exposed to combinations of metabolic and environmental heat greater than the applicable Recommended Exposure Limits (REL) shown in Figure 5.2.

Environmental heat exposures are assessed by the Wet Bulb Globe Thermometer method described below. Environmental heat measurements shall be made at or as close as possible to the work area where the worker is exposed. Metabolic heat rates shall be measured or estimated from Table 5.4 to determine whether the total heat exposure exceeds the applicable RAL or REL.

Climate Evaluation Methods

Monitoring a workplace environment for heat exposure involves measuring heat-modifying factors and assessing work loads. The most commonly employed measurement is the Wet Bulb Globe Temperature (WBGT) index, which takes into account convective and radiant heat transfer, humidity, and wind velocity. It is calculated by integrating the readings of three separate instruments that indicate the "dry" bulb temperature (DBT), the natural "wet" bulb temperature (which takes into account convective and evaporative cooling, WBT), and the black globe temperature (which,

Table 5.4. Metabolic heat production during various agricultural activities.

Activity	Kcal/hour
Mowing	360-600
Weeding	180-480
Brush cleaning	360
Milking by hand	200-370
Tree felling with a saw	510-750
Stacking firewood	330-400
Tractor plowing	250
Thrashing	300
Log carrying	200

Modified from Astrand and Rodahl 1986.

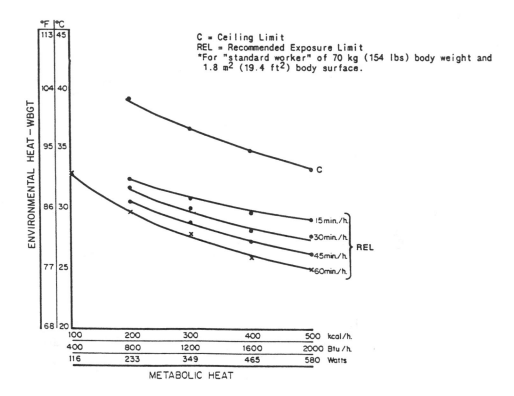

Figure 5.2. Recommended heat-stress exposure limits for heat-acclimatized workers (NIOSH 1986).

through the use of a black copper sphere, registers heat transfer by radiation, BGT).

WBGT = WBT x 0.7 + DBT x 0.1 + BGT x 0.2 (Outdoors)
WBGT = WBT x 0.7 + BGT x 0.3 (Indoors)

The WBGT predicts the rate of rise of the rectal temperature in healthy acclimatized persons during exercise.

International Organization for Standardization (ISO) Index

For the assessment of hot environments, a method based upon the WBGT index is described in document ISO 7243. If the WBGT reference value is exceeded a more detailed analysis can be made (ISO 7933) involving calculation, from the heat balance equation, of sweating required in a hot environment.

ISO 7730 provides an analytical method for assessing moderate environments and ISO TR 11079 (Technical Report) provides an analytical method for assessing cold environments involving the calculation of the clothing insulation required (IREQ) from a heat balance equation.

Potential Occupational Exposures

- Athletes.
- Farmers,
- Laundries.
- Confinement animal workers.
- Kitchen workers.
- Workers in boiler rooms, engine rooms, etc.
- Out-of-door workers during hot weather.
- Foundry workers.
- Glass products workers.
- Food cannery workers.
- Fire fighters.
- Mine and steam tunnel workers.

Heat Illnesses

Prolonged exposure to excessive heat may cause increased irritability, lassitude (leading to slower mental and physical job reactions), decrease in morale, increased anxiety, and inability to concentrate. These are indirect causes of accidents on the job. Direct causes of accidents are fogged glasses, sweat in the eyes, slippery hands, and dizziness or fainting.

Prickly Heat: also known as miliaria, is an acute, minor inflammatory disease of the skin; sweat gland ducts become clogged and the accumulating sweat is forced through the wall of the ducts into the surrounding tissues, causing an inflammatory reaction. There are three forms (in order of increasing severity): miliaria crystallina, miliaria rubra, and miliaria profunda.

Intertrigo: skin in the body folds, i.e., groin and axillae, becomes erythematous and macerated. This is commonly seen in obese workers.

Heat Cramps: spasms of the muscles usually occurring at rest following prolonged exertion; inadequate NaCl replacement with plain water will cause this.

Heat Edema: dependent edema of the ankles and feet in unacclimatized individuals; usually accompanied by prolonged standing or sitting; self-limited; elevate feet.

Heat Syncope: fainting in a hot environment, probably induced by blood pooling in dilated vessels of the skin and lower parts of the body, resulting in cerebral ischemia; self-limited.

Heat Exhaustion: Excessive sweating due to exertion in a hot and humid environment can cause extreme depletion of plasma volume and precipitate heat exhaustion; body core temperature rises to over 38° C (100.4° F) but remains below 40° C (104° F)—the cutoff for heatstroke; most common heat related illness; presenting symptoms include malaise, weakness, headache, anorexia, nausea, vomiting, tachycardia, and hypotension; minor cases are easily treated with fluid and salt repletion, along with rest in a cool environment.

Heatstroke: medical emergency; two types encountered in practice:

Classic: occurs without exertion in the elderly, in intoxicated individuals, and in persons debilitated by chronic illness; occurs during periods of sustained high temperatures and humidity; active sweating continues constantly but eventually fails.

Exertional: most common among young, otherwise healthy individuals unaccustomed to working in hot and humid environments; pathophysiology is uncertain; additional symptoms include elevated levels of serum enzymes, hypotension, rhabdomyolysis, disseminated intravascular coagulation, and acute renal failure.

A simple way to determine what hot weather "feels like" is to look at the heat index chart of the National Weather Service. (See Table 5.6). The combination of temperature and humidity gives the apparent temperature. The heat index is a measure of the contribution that humidity makes with high temperatures in reducing the body's ability to cool itself. It is not intended for the assessment of occupational heat stress, which should be determined as previously described, and it does not include a contribution from metabolism. However, it is readily available and can be utilized to categorize the potential for heat-related illness. For example, if the actual environmental temperature is 100° F and the humidity is 50%, the apparent temperature is 120° F.

PREVENTION OF HEAT STRESS

The three main preventive approaches to reducing heat stress are:

1. Shortening of the exposure.
2. Reduction of heat stress: may involve technical modifications or protective clothing.
3. Worker's adaptation capacity: based on acclimatization.

Table 5.5. Categorization of heat stress using the apparent temperature.

Category	Apparent Temp. (°F)	Heat Syndrome
Caution	80 to 90	Fatigue possible with prolonged exposure and physical activity
Extreme Caution	90 to 106	Heat cramps and heat exhaustion possible with prolonged exposure and physical activity
Danger	106 to 130	Heat cramps and heat exhaustion likely. Heat stroke possible with prolonged exposure and activity
Extreme Danger	>130	Heat stroke very likely

Source: U.S. Weather Service.

Table 5.6. The heat index (National Weather Service).

	\multicolumn{11}{c}{Enviromental temperature (F)}										
	70	75	80	85	90	95	100	105	110	115	120
0%	64	69	73	78	83	87	91	95	99	103	107
10%	65	70	75	80	85	90	95	100	105	111	116
20%	66	72	77	82	87	93	99	105	112	120	130
30%	67	73	78	84	90	96	104	113	123	135	148
40%	68	74	79	86	93	101	110	123	137	151	
50%	69	75	81	88	96	107	120	135	150		
60%	70	76	82	90	100	114	132	149			
70%	70	77	85	93	106	124	144				
80%	71	78	86	97	113	136					
90%	71	79	88	102	122						
100%	72	80	91	108							

(Relative Humidity (%))

Table 5.7 recommends methods to reduce heat exposure. Heat stress in industry is an important issue. Technical advances have lowered the incidence of heat stress in some industrial areas but are not always cost effective or efficient. Trained and knowledgeable workers and their physical and mental monitoring are the keys to reducing heat stress problems in the workplace.

SUMMARY

Injuries associated with cold and hot environments are generally underreported with only the most severe situations attracting attention. Most workers exposed to these extreme environments are able to deal with the hazards which confront them. Personal protective equipment (PPE) and a knowledge of both the task and environment are essential to the worker.

The prevention of heat loss and the maintenance of efficient functioning is the primary goal in the cold environment, while thermoregulatory response to elevated temperatures is essential in the hot workplace. Important factors to consider are: duration of exposure, type of work, physical requirements, PPE, age of the worker, and current health status.

Table 5.7. Checklist for controlling heat stress and strain.

Item	Actions to Consider
1. Controls	
M, Body heat production of tasks	Reduce physical demands of the work; powered assistance.
R, Radiative load	Interpose line-of-sight barrier, furnace wall insulation, metallic reflecting screen; heat reflective clothing; cover exposed body parts.
C, Convective load	If air temperature >35°C then reduce air temperature or reduce air speed across skin or wear clothing.
	If air temperature <35°C then increase air speed across skin or reduce clothing.
Emax, Maximum evaporative cooling by sweating	Increase by decreasing humidity or increasing air speed.
2. Work Practices	
Exposure Limit	Shorten duration of exposure; schedule very hot jobs in cooler part of day when possible; educate workers and supervisors on signs of overstrain.
Recovery nearby.	Provide air-conditioned space.
3. Personal Protection	
R, C, Emax	Provide cooled air, cooled fluid, or ice cooled conditioned clothing, reflective clothing or aprons.
4. Other Considerations	Break-in unacclimatized workers; provide water breaks at frequent intervals; evaluate cardiovascular status; fatigue or illness may temporarily contra-indicate exposure.
5. Heat Wave	Introduce heat alert program.

From NIOSH 1986.

REFERENCES

Astrand, P.O., and K. Rodahl. 1986. *Textbook of Work Physiology: Physiological Basis of Exercise.* 2nd edition. New York: McGraw-Hill.

Cordes, D., and D. Rea, eds. 1991. *Health Hazards of Farming. Occupational Medicine: State of the Art Reviews* (6)3: July-September. Philadelphia: Hanley and Belfus, Inc.

Eissing, G. 1995. Climate assessment indices. *Ergonomics* 38(1): 47-57.

Epstein, Y., et al. 1995. Exertional heatstroke: a preventable condition. *Isr. J. Med. Sci.* 31: 454-62.

Holmer, I. 1993. Work in the cold: review of methods for assessment of cold exposure. *Int. Arch. Occup. Envir. Health* 65: 147-55.

LaDou, J., ed. 1990. *Occupational Medicine.* Norwalk, CT: Appleton and Lange Publishers.

Lee-Chiong, T., et al. 1995. Heatstroke and other heat-related illnesses. *Postgraduate Medicine* 98(1).

Levy, B., and D. Wegman, eds. 1988. *Occupational Health: Recognizing and Preventing Work-Related Disease.* 2nd edition. Boston: Little, Brown and Company.

Meyer, J., and R. Rapp. 1995. Survey of heat stress in industry. *Ergonomics* 38(1): 36-46.

Morris, L. 1995. Practical issues in the assessment of heat stress. *Ergonomics* 38(1): 183-92.

National Institute for Occupational and Health (NIOSH). 1986. *Criteria for a Recommended Standard... Occupational Exposure to Hot Environments.* Revised criteria. DHHS (NIOSH) Publication No. 86-113. Washington, DC: U.S. Government Printing Office.

Parsons, K. 1995. International heat stress standards: A Review. *Ergonomics* 38(1): 6-22.

Raffle, P., et al., eds. 1988. *Hunter's Diseases of Occupations.* 8th edition. Boston: Little, Brown and Company.

Ramsey, J. 1995. Task performance in heat: a review. *Ergonomics* 38(1): 154-65.

Rom, W., ed. 1992. *Environmental and Occupational Medicine.* 2nd edition. Boston: Little, Brown and Company.

Simon, H. 1994. Hyperthermia and heatstroke. *Hospital Practice,* August 15.

Wald, P., and G. Stave, eds. 1994. *Physical and Biological Hazards of the Workplace.* New York: Van Nostrand Reinhold.

Williams J. 1994. *The USA TODAY 1995 Weather Almanac.* New York: Vintage Books.

6

ELECTRICAL SAFETY IN AGRICULTURE, FORESTRY, AND FISHERIES

Robert L. McLymore, Sr., Ed.D.
Department of Biological and Agricultural Engineering
North Carolina State University

> The use of electricity has proven to be invaluable to agriculture, forestry, and fisheries industries but some workers have found that if improperly used it can be the source of fires, injuries, and even death. Electrical hazards on the farm can result in electrical shock to humans or livestock and possibly result in a fire within structures or in operating equipment. These hazards can be eliminated when a better understanding of the principles, uses, and hazards associated with electricity are known and safety procedures are incorporated into the day-to-day routine. The overall goal of this chapter is to instruct persons when using equipment around sources of electricity on safe procedures to prevent incidents of electric shock, trauma, fires, and deaths.

INTRODUCTION

With the widespread use of electricity on the farm, more emphasis needs to be placed on using electricity and electrical equipment safely (Chamberlain and Hallman 1995). Each year in the United States an estimated 1,000 people die as a result of electrical injuries, and many more suffer severe injuries and disabilities. Those at highest risk of electrical injury are many young children in the home and people who work around sources of high-voltage electricity. More than 90% of the victims are male, most frequently between 20 and 34 years of age (Bailey 1989). In dozens of farms in the United States, electrocution is quick and deadly, killing an estimated 62 agricultural workers every year (Schwab and Miller 1992).

ELECTRICAL SYSTEMS' GENERAL REQUIREMENTS: OSHA STANDARD 1910.303

This condensed document contains information and requirements that owners and managers of agricultural businesses (agribusinesses) and other industries should understand about electrical use on their establishments. It is not intended to be totally inclusive but rather to highlight important information and legal requirements. A complete copy of the Standard 1910.303 of the Occupational Safety and Health Act is available by contacting your nearest OSHA office.

Requirements for All Systems

Conductors and electrical equipment are acceptable only if approved (Becker and Stephenson 1993). Listed or labeled equipment must be used or installed in accordance with any instructions included in the listing or labeling.

Electrical equipment must be free from recognized hazards that are likely to cause death or serious physical harm to employees. Safety of equipment must be determined using the following guidelines:

- Suitability of equipment for an identified purpose as may be evidenced by listing or labeling for that identified purpose,
- Mechanical strength and durability, including, for parts designed to enclose and
- Protect other equipment, the adequacy of that protection,
- Electrical insulation,
- Heating effects under conditions of use,
- Arcing effects,
- Classification by type, size, voltage, current capacity, and specific use,
- Other factors which contribute to the practical safeguarding of employees using or likely to come in contact with the equipment.

Conductors must be spliced or joined with suitable splicing devices or by brazing, welding or soldering with a fusible metal or alloy. Soldered splices must first be so joined or spliced as to be mechanically and electrically secure without solder, and then soldered. All splices and joints and the free ends of conductors must be covered with an insulation equivalent to that of the conductors, or with an insulating device suitable for the purpose.

Parts of electric equipment which in ordinary operation produce arcs, sparks, flames or molten metal must be enclosed or separated and isolated from all combustible material.

Electrical equipment may not be used without a manufacturer's name, trademark or other descriptive marking by which the organization responsible for the product may be identified. Other markings must be provided giving voltage, current, wattage, or other ratings as necessary. The markings must be of sufficient durability to withstand the environment involved.

Each required disconnecting method for motors and appliances must be legibly marked to indicate its purpose, unless located and arranged so the purpose is evident. Each service, feeder and branch circuit must be marked at its disconnecting means or over current device. These markings must also be of sufficient durability to withstand the environment involved.

Requirements of Systems of 600 Volts or Less

Sufficient access and working space must be provided and maintained about all electric equipment to permit ready and safe operation and maintenance of the equipment.

Except as required or permitted elsewhere, the dimension of the working space in the direction of access to live parts operating at 600 volts or less and likely to require examination, adjustment, servicing or maintenance while alive may not be less than indicated in Table 6.1. In addition, work space may not be less than 30 inches wide in front of the electric equipment. Distances must be measured from the live parts if they are exposed, or from the enclosure front or opening if enclosed. Concrete, brick or tile walls are considered to be grounded. Working space is not required in back of assemblies such as dead-front switchboards or motor control centers where there are no renewable or adjustable parts such as fuses or switches on the back, and where all connections are accessible from locations other than the back.

Required working space may not be used for storage. When normally enclosed live parts are exposed for inspection or servicing, the working space, if in a passageway or general open space, must be suitably guarded. Where there are live parts normally exposed on the front of switchboards or motor control centers, the working space in front of the equipment may not be less than three feet wide.

At least one entrance of sufficient area must be provided to give access to the working space about electric equipment. Entrances to rooms and other guarded locations containing exposed live parts must be marked with conspicuous warning signs forbidding unqualified persons to enter.

Illumination must be provided for all working spaces about service equipment, switchboards, panelboards and motor control centers installed indoors. Minimum headroom must be six feet three inches.

Live parts of electric equipment operating at 50 volts or more must be guarded against accidental contact by approved cabinets or other forms of approved enclosures.

ELECTROCUTION HAZARDS ON THE FARM

The most common risk of electrocution comes from over head power lines (Schawb and Miller 1992). Utility lines typically are not insulated, meaning that the lines are bare.

Table 6.1. Required working space for employees working near electricity.

Nominal Voltage to Ground	Minimum Clear Distance for Condition		
	a	b	c
0 - 150 V	3'	3'	3'
151 - 600 V	3'	3½'	4'

a. Exposed live parts on one side and no live or grounded parts on the other side of the working space, or exposed live parts on both sides effectively guarded by suitable wood or other insulating material. Insulated wire or insulated busbars operation at not over 300 volts are not considered live parts.
b. Exposed live parts on one side and grounded parts on the other side.
c. Exposed live parts on both sides of the workspace [not guarded as per condition (a)] with the operator between.

The lines may have been installed without insulation, or the insulation may have been removed by exposure to the elements (Schawb and Miller 1992).

To understand the hazards associated with electricity, it is important to know the basic principles of electricity and how electrical shocks occur. Electricity's basic principles can be explained with the terms voltage, current (amperage), and resistance. Voltage is the force that initiates the flow of electric charge (electrons). The actual flow of electric charge is called current. The rate of flow is measured in amperes. Resistance is based on how much a material impedes the current and regulates the rate of flow. Electricity usually takes the path of least resistance. When the human body becomes part of this path, the result is electric shock (Chamberlain and Hallman 1995).

Two wires are needed to complete an electric circuit—one wire to carry the current to an electric device, the other to return the electricity to the power source and finally to the ground. If the protective insulation on any of the wires or inside a piece of electrical equipment is defective, the current can follow a different path to ground. By coming into contact with a faulty electrical object, a person may act as a conductor to ground and experience a shock.

Several factors determine the effect electricity will have on the human body: the duration of contact, the amperage, the path the current takes through the body, and the electrical resistance of the body. A person standing in water is a better conductor than a person on dry ground. Taken together, these factors can produce some surprising results. For example, the current from a 7-watt Christmas tree bulb (60/1000 of an ampere) can kill a person if the current passes through the heart. Figure 6.1 shows the physiological effects of different current levels (Chamberlain and Hallman 1995).

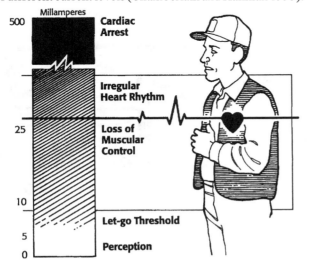

Figure 6.1. Increasing levels of current above the "let-go" threshold causes loss of muscular control, irregular heart rhythm, and finally, cardiac arrest (Chamberlain and Hallman 1995).

Overhead Power Lines

Many types of farm equipment can come in contact with overhead power lines. Tractors with front-end loaders, portable grain augers, fold-up cultivators, and equipment with antennas easily become an electric hazard and must be handled with care. Performance of certain tasks also pose electrocution hazards. Always use ladders with caution so that they do not come in contact with overhead power lines.

Numerous electrocutions on farms have been caused by contact with overhead wires. Transporting equipment requiring high clearance underneath low overhead wires can be especially dangerous. Fatalities have been attributed to portable augers coming into contact with overhead wires. These lives could have been saved if a few minutes had been taken to lower equipment. Other activities such as pruning trees or moving irrigation pipe or metal ladders near overhead wires can also be hazardous. Low overhead wires are frequently found in farm yards and along farm lanes between fields and roads. Anyone working in these situations should be made aware of the dangers overhead wires present. Hazards are also associated with buried service conductors. A person operating a trencher or excavating equipment may get electrocuted if the machine comes in contact with the underground wires. Keep a detailed map of all buried power lines to avoid such accidents (Chamberlain and Hallman 1995).

Forest workers must be aware of any electrical wires in the area they are harvesting. Additionally, fishermen must watch out for electrical cables around harbors and docks.

Grain Bin Regulations

Many electrocutions occur near grain bins and new requirements in the 1990 National Electric Safety Code (NESC) address those dangers. The Code requires raised power lines in locations where tall equipment is used. Existing buildings are exempt, but owners should upgrade these facilities.

According to the NESC, power lines must be as least 18 feet above the highest point on any grain bin with which portable augers or other portable filling equipment are used. The clearance must be maintained a specified distance (Use NESC code book to determine this value) around the bin, and sloped to meet the lower line clearance. If you are installing new grain bins, contact a licensed electrician or power company representative to help place electric service lines. Lines can also be buried to reduce risk of electrocution. Installation costs for overhead and underground power lines can vary; contact local representatives for these costs.

Determine Equipment Clearance

If you have overhead lines, ask your local utility company officials to help determine line height in each area of the farm. Never measure line heights yourself!!! Once you know the height of all power lines, you can determine the clearance needed for equipment that must travel underneath the line. Try to maintain at least a 10-ft clearance between the power line and the top of all equipment that must travel underneath it. Consider the possibility that some equipment during transport is actually taller than when in use. For example, a large, 16-row planter that folds up would be taller during transport than during field use (Schwab and Miller 1992).

Train Seasonal Employees

Busy planting and harvest seasons may require temporary employees, persons who may not be aware of potential dangers associated with overhead power lines. Always remind additional workers about overhead power lines. Give them special instructions such as,

"Never transport this cultivator through this gate because of the overhead line." Despite operating precautions, equipment can come into contact with electrical lines. It is important to know how to handle these situations.

If your tractor comes into contact with overhead power lines, *stay on the tractor*. Ask someone to contact the local utility company immediately to remove the danger. If there's an emergency, such as an electrical fire, and you need to leave the equipment, jump as far away from the equipment as possible. *Do not allow any part of your body to touch the equipment and the ground at the same time.* Once you get away from the equipment, never attempt to get back on or even touch the equipment. Many electrocutions occur when the operator dismounts and, realizing nothing has happened, tries to get back on the equipment. The best way to handle emergencies is by prevention. Respect electricity and avoid contact with overhead lines.

REMOVING ELECTRICAL HAZARDS

Electrical hazards on the farm can result in electrical shock to humans or to livestock and possibly result in a fire within structures or in operating equipment. Risks associated with electrical hazards on the farm are increased by the presence of moisture, especially by the dampness that is common in confined livestock areas. Animals are naturally grounded, making them more sensitive to low intensity electrical currents than humans. Humans have drier skin than animals and normally wear shoes or boots which provide greater resistance to electrical shock. Humans usually will not feel an electrical shock from stray current that a well-grounded animal does when standing on a damp concrete slab or damp ground. Animals experiencing even a minor electrical shock may be reluctant to drink from a waterer. On a dairy farm, for example, this reduced water intake may result in less milk production and a financial loss to the farmer (Tilma and Doss 1992).

There are several ways to reduce the electrical hazards in farm buildings and around work areas. You can protect yourself and your livestock from electrical shock by following these recommendations:

1. Select a shock protection system. If you have two-conductor circuits and a variety of tools, some with two-wire cords and plugs and some with three-wire cords and plugs, you have four alternatives:

   A. Have an electrician install ground-fault circuit interrupters (GFCI) permanently in each electrical circuit in the shop and other farm buildings.
   B. Plug in a portable GFCI when individual power tools are used.
   C. Convert your two-conductor circuits to three-wire, grounding-type circuits.
   D. Replace your present tools with new double insulated tools (see Figure 6.2).

Get the advice of a competent electrician who is familiar with agricultural wiring to help you decide the safest and most economical alternative for your operation. A combination of methods may be your best choice.

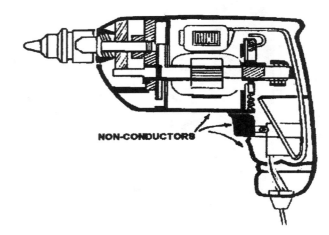

Figure 6.2. Double-insulated tools can reduce the risk of electrical shock because the tool does not conduct electrical leaks to the operator (Tilma and Doss 1992).

2. Purchase tools and equipment designed to prevent shock. Look for tools and equipment that carry the approval label of a recognized inspection and approval agency. The label "UL Listed" indicates that the item has been evaluated for electrical safety considerations by the Underwriters Laboratory. The PTI "Safety Seal" indicates approval by the Power Tool Institute. Approved tools are equipped with three-wire, grounding-type cords and plugs or a two-prong plug with double insulation on the tool. Buy either type if your shop has grounding-type circuits. Double insulated tools can reduce the risk of electrical shock if you have two-conductor circuits, with or without GFCI protection.
3. Avoid using grounding adapters. Grounding adapters are risky because two-conductor circuits do not have a grounding conductor to connect to the "pigtail" of the adapter.
4. If you find it necessary to use extension cords, refer to the guidelines in this chapter on Electrical Cords.
5. Do not abuse electrical power tools. The insulation on the conductors inside the tool may be damaged from rough handling. Dropping power tools, throwing them around, or picking them up by the power cord can destroy the insulation and connections inside the equipment. Avoid overheating by allowing tools to cool when they become hot from continuous use or from temporary overloads.

Ground-Fault Circuit Interrupters (GFCI)

A ground-fault circuit interrupter (GFCI) is a circuit breaker designed to prevent serious shock to people or animals under certain conditions. It can reduce the risk of shock when using electrical tools or appliances in damp or wet areas.

The GFCI works on the principle that the two wires supplying a single-phase electrical load must carry the same number of amperes (current) when the circuit is operating properly. If a ground-fault (path for the current to flow outside of the wire to the ground) occurs either to the grounding wire, or through a person or animal, some of the current will take an alternate route back to the system's grounding electrode. One of the wires will then carry less current than the other wire. When this occurs the GFCI will break the circuit, stopping the flow of electricity in a fraction of a second to reduce the electric shock hazard.

Ground-fault circuit interrupters come in several styles. They are commonly used as a receptacle outlet, part of an extension cord, or can be installed in the main electrical panel to replace an existing circuit breaker. When installed as a circuit breaker, the GFCI offers shock protection to an entire electrical branch. Ground-fault circuit interrupters are available for 120-V circuits with one hot wire and a neutral. A GFCI will work on the older two-wire electrical systems that have no ground wire. A 120-V, single-pole GFCI fits into the same size space as a standard single-pole breaker. There are also GFCIs for 240-V circuits using two hot wires.

All equipment plugged into a GFCI protected receptacle, including any two prong (two-wire) electrical plug, will have ground-fault protection. A portable GFCI is recommended for persons using power tools in damp or wet locations. The portable GFCI is plugged into an outlet and the power tool is plugged into the GFCI.

Certain conditions can result in "nuisance tripping" of a GFCI protected circuit or receptacle. Nuisance tripping can be reduced by avoiding:

- Circuits longer than 100 ft.
- Older non-double insulated power tools that contain faulty electrical insulation.
- Fluorescent or other types of electric-discharge lighting fixtures.
- Extension cords with cuts or splices where moisture has entered the cut or splice creating a "leak" (ground fault) or path for the current to flow outside of the wire to the ground.
- Permanently installed electric motors.

Installing a GFCI to prevent electrical shock from farm equipment seems like a good idea, but nuisance tripping may become a serious problem. The loss of a ventilation system in certain livestock facilities can be fatal to animals. Stock waterers may freeze in northern climates if the GFCI trips. Carefully consider the effects of loss of power to an agricultural circuit before installing GFCI protection. The most effective shock prevention system for agricultural equipment and circuits is proper equipment grounding conductor run with the circuit wires and connected to all metal agricultural equipment.

Extension Cords

Damaged or improperly used cords can result in electrical shocks or may start fires. If possible, avoid using extension cords in farming operations. Follow these precautions if you must use an extension cord:

- Do not use in wet areas.
- Do not try to repair a damaged extension cord or splice two wires together. Replace the cord.
- Keep cords away from sharp objects, heat, oil, and solvents that can damage insulation.

- Check an extension cord before each use for nicks and cuts. Replace or repair the cord if the insulation is damaged or worn.
- Use an extension cord with the correct size wiring (gauge) for the intended use. Do not overload an extension cord or use a "household" type extension cord to operate heavy-duty machinery. Overloading may cause excessive heating that may result in a fire.
- Use a grounded wire (three-prong with a "safety grounding" wire) for tools and machines having a grounded plug or use a portable GFCI (ground-fault circuit interrupter).
- Buy extension cords carrying a listing mark or certification of a recognized independent testing laboratory.
- Be sure the package for the cord indicates the maximum current and/or wattage rating of the cord. Two extension cords plugged together for additional length will reduce the amperage rating and increase the risk of an electric hazard.
- Route the cord to protect it from machinery and animals. Also, people should not be able to trip over or accidentally damage the cord.
- Extension cords deteriorate; do not use them in place of permanent electrical installations.

PROTECTING YOURSELF AND THE ELECTRICAL SYSTEM

Electric systems have built in features to safeguard equipment and wiring from excessive current. The most common circuit protection is a fuse. When excess current flows to the fuse, a portion of it breaks, cutting power to the circuit. To restore power, the fuse must be replaced. A circuit breaker offers the same protection, but is more convenient. A flip of a switch restores power after a circuit breaker has been tripped by an overload.

When a fuse blows or a circuit breaker is tripped, remember:

- The circuit breaker or the fuse should never be bypassed because this can damage equipment or start a fire if the circuit becomes overheated.
- Never replace a fuse with one that is larger than that specified for the circuit. A fuse that is too large will not protect against an overload, which can cause a fire.
- Do not replace fuses with pennies, nails, bar stock, or other objects. Many electrical fires have been caused by such substitutions.

Grounding

Grounding is a necessary safety feature of every electrical system (Chamberlain and Hallman 1995). It protects against electrical shock, fire, and damage to equipment and reduces hazards associated with lightning. Proper grounding for both the system and electrical equipment is particularly relevant for farmsteads.

Electricity is transported from the power company's supply lines to a main service entrance (Figure 6.3). Normally all power is metered at this point. The main disconnect should be located here so that all power can be manually turned off at one point. From the main service entrance, wires lead to each building or area service entrance through buried or overhead wires. These service entrances should also be equipped with disconnects so that power can be shut off to one site without affecting other areas. The main distribution system on a farmstead should always be large enough to accommodate present demand and future expansion. Proper installation of the electrical system is essential for safety. Local codes should always be followed because their main purpose is to provide users with safe systems. If no electric code exists for your area, the National Electric Code (NEC) is the minimum standard to follow. Only qualified electricians should install electrical systems (Chamberlain and Hallman 1995).

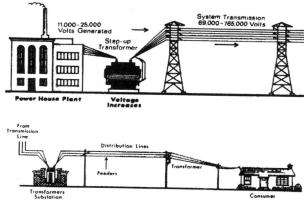

Figure 6.3. System grounding (National Research Council).

System grounding starts at the main service entrance (see Figure 6.4). The neutral of the main service is connected either to an all-metal water pipe or to a small metal rod that has been driven into the ground. These act as ground electrodes. If a ground rod is employed, it must be driven deep enough to always be in contact with moist soil (at least 10 feet). To ensure continuity of the ground throughout the system, each branch service entrance should be grounded with its neutral wire connected to the grounded main service neutral.

specific applications or recommendations for your farm wiring concerns.

Lockouts

The purpose of a lockout is to prevent equipment from being accidentally started and injuring people when it is being serviced or repaired. When servicing electrically powered equipment, a lock should always be placed on its switch (see Figure 6.5).

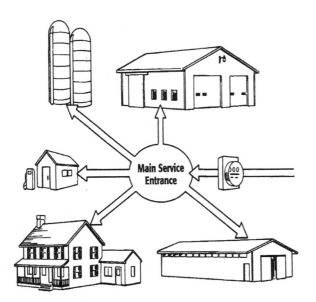

Figure 6.4. Central metering and distribution are common components of a farmsteads electrical system (Chamberlain and Hallman 1995).

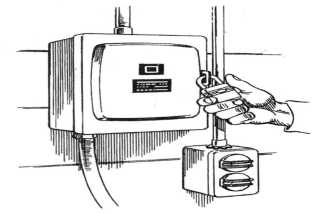

Figure. 6.5. Lock out the switch. Do not risk having a switch turned on while you are working on the electrical system. Note waterproof outlets (Chamberlain and Hallman 1995).

Equipment grounding ensures that a person who comes into contact with electrical equipment does not get a shock because of a ground fault. A ground fault can occur when wiring connections become loose or protective insulation is broken or defective. When current leaks and electrifies a metal box or fixture, it allows a conductive path to complete its circuit. An equipment ground serves as a conductor or path of low resistance (Chamberlain and Hallman 1995).

Problems can occur when using or moving equipment inside buildings if wiring is not securely fastened or protected. Interior wiring should also be protected from animal damage by encasing it in metal or plastic conduit or by elevating the wiring above the animal's reach (Tilma and Doss 1992).

These problems can be greatly minimized by the selection of appropriate cable wiring. Type "UF" cable wiring is rated for dry, damp, wet or corrosive locations. This type cable is permitted for surface and concealed wiring in buildings, and for direct burial in the earth. It may not be embedded in poured concrete.

Exercise care if electric conduit is used. Moist and extremely corrosive conditions may quickly destroy metal conduit. Animals may chew nonmetallic conduit or it may become damaged by impact from animals or machines. Check installations frequently and be prepared to replace damaged conduit or cable.

Type "NM" wire should not be buried because of the potential damage from moisture. Type "NM" electrical wire is not generally recommended for agricultural environments with animals or damp areas. Consult an electrician for specific applications or recommendations for your farm wiring concerns.

For example, if a silo unloader needs a repair, a lock on the control box will prevent the machine from starting while you are working inside the silo (Chamberlain and Hallman 1995).

Harsh Farm Environments

Many agricultural facilities have severe environments that require special attention when installing and maintaining electrical systems. Threats to the system come from a number of sources. The vapors from animal waste in confinement housing can corrode electrical components. High humidity in milking facilities can rapidly deteriorate conventional metal electric boxes. Physical damage to wiring, boxes and light fixtures can be caused by livestock, equipment, and people. To protect your assets from fire or the cost of rewiring, follow these recommendations when installing your electrical system.

In corrosive and damp environments:

- Use underground feeder (UF) electric cable.
- Make sure all control boxes, light fixtures, switches, and receptacles are made of corrosion-resistant materials.

- Install watertight covers on receptacles and switches and over light bulbs.
- Locate the distribution panel away from severe environments. If a clean, dry area, such as an office, is not available, mount the distribution panel outside.
- Make sure that every electrical system component or piece of equipment located outside is watertight.
- Run conductors through horizontal conduit and seal the conduit ends so moisture cannot enter the distribution panel. When conductors run from a warm, moist environment to a cold location, condensation can form and enter the distribution panel.
- Inside the farm building, mount wiring outside of walls to allow continuous inspection.

In dusty environments (such as grain or feed handling areas):

- Place protective enclosures over all light bulbs to protect them from dust and lessen the fire hazard (see Figure 6.6).
- Use explosion-proof switches. Fire from an explosion can occur in areas where fine dusts or harmful, highly flammable vapors come in contact with sparks from an electric switch.

Figure 6.6. When light bulbs are close to a work area, they can be bumped. Glass enclosures help, but a substantial guard further reduces risk of fire (Chamberlain and Hallman 1995).

In an areas where physical damage to the electrical system by livestock, equipment, or people is likely:

- Protect circuit boxes by thoughtfully choosing their location. Placing them around a corner or away from animals makes the boxes less vulnerable to abuse.
- Run conductors in conduit to protect them from physical damage by livestock.
- Use nonmetallic conduit in corrosive environments.
- Place guards over light bulbs located where they may get struck by equipment, and use enclosures to keep moisture and dust out (see Figure 6.6).
- Use only qualified electricians to repair damage to electrical installations.

Portable Elevators

Moving tall equipment such as a grain auger in the elevated position may result in an electrocution if it comes into contact with overhead power lines. Anyone moving an auger should lower it first. Check the height of all equipment to ensure that it will not touch overhead power lines during the move. Recently manufactured augers should have a warning label attached that indicates DANGER, Electrocution Hazard. Follow safety warnings listed on each piece of equipment (see Figure 6.7).

Figure 6.7. Check that elevators and augers don't come into contact with overhead power lines (Tilma and Doss 1992).

Standby Power

Many farm owners have standby generators on hand to use during power outages. When a standby generator is installed on single phase systems, it must be connected to the farm's wiring system through a double-throw transfer switch (see Figure 6.8). When the generator is in use, this switch disconnects the farm's electrical system from the normal power supply. There are two reasons why this is important. First, it prevents the generator from feeding power to power supply lines where repair persons may be working.

Second, the normal power supply cannot feed back to the generator and damage it when power from the electric utility is restored.

The transfer switch must be installed so that the generator is no more than 25 feet from the switch. Installation of the transfer switch should always be reviewed with the local electric supply company.

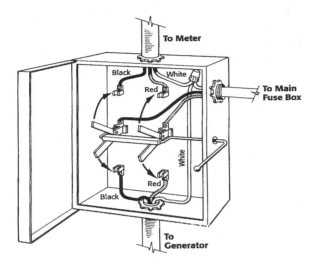

Figure 6.8. Double-pole, double-throw transfer switch (Chamberlain and Hallman 1995).

First Aid

Electric shock accounts for approximately 500 to 1,000 deaths annually in the United States and causes an additionally 5,000 patients to seek emergency treatment (Brown and Gaasch 1992; Cooper 1984; Kobernick 1982). Victims of electric shock experience a wide spectrum of injury, ranging from a transient unpleasant sensation from low-intensity current to instantaneous cardiac arrest from accidental electrocution (Budnick 1984).

Electric shock injuries result from the direct effects of current on cell membranes and vascular smooth muscle and from conversion of electric energy into heat energy as current passes through body tissues. Factors that determine the nature and severity of electric trauma include the magnitude of energy delivered, voltage, resistance to current flow, type of current, duration of contact with the current source, and current pathway. High-tension current generally causes the most serious injuries, although fatal electrocutions may occur with low-voltage (110V) household current (Budnick 1984). Skin resistance, the most important factor impeding current flow, can be reduced substantially by moisture, thereby converting what ordinarily might be a minor low-voltage injury into a life-threating shock (Wallace 1991). Contact with alternating current at 60-cycles per second (the frequency used in most household and commercial sources of electricity) may cause tetanic skeletal muscle contractions and prevent the victim from releasing the source of electricity, thereby leading to a prolonged duration of exposure, which allows skin resistance to be overcome. The repetitive frequency of alternating current also increases the likelihood of current exposure to the myocardium during the vulnerable recovery period of the cardiac cycle and can precipitate VF (ventricular fibrillation). (Geddes et al 1986).

Transthoracic current flow (e.g., a hand-to-hand pathway) is more likely to be fatal than a vertical path (Thompson and Ashwal 1983). However, the vertical pathway often causes myocardial injury, which has been attributed to the direct effects of current and coronary artery spasm (Chandra et al 1990; Ku et al 1989; Xenopoulos et al 1991).

Cardiopulmonary arrest is the primary cause of immediate death due to electrical injury (Homma et al 1990). Ventricular fibrillation or ventricular asystole may occur as a direct result of electric shock. Other serious cardiac arrhythmias, including ventricular tachycardia (VT) that may progress to VF, may result from exposure to low- or high voltage current (Jensen et al 1987).

Respiratory arrest may occur secondary to (1) electric current passing through the brain and causing inhibition of medullary respiratory center function; (2) tetanic contraction of the diaphragm and chest wall musculature during current exposure; (3) prolonged paralysis of respiratory muscles, which may continue for minutes after the shock has terminated. If respiratory arrest persists, hypoxic cardiac damage and arrest may occur.

Lightning

Lightning strike causes approximately 50 to 300 fatalities per year in the United States, with about twice that number of persons sustaining serious injury (Epperly and Stewart 1989; Duclos and Sanderson 1990). Lightning injuries have a 30% mortality rate, and up to 70% of survivors sustain significant morbidity (Cooper 1980).

The primary cause of death in lightning-strike victims is cardiac arrest, which may be due to primary VF or ventricular asystole (Kleiner and Wilkin 1978). Lightning acts as an instantaneous, massive direct current countershock, depolarizing the entire myocardium at once and producing asystole. In many cases cardiac automaticity may return spontaneously. However, concomitant respiratory arrest due to thoracic muscle spasm and suppression of the respiratory center may continue after return of spontaneous circulation. Unless ventilatory assistance is provided, hypoxic cardiac arrest may occur (Kleiner and Wilkin 1978).

Patients most likely to die of lightning injury if no treatment is forthcoming are those who suffer immediate cardiac arrest. Patients who do suffer cardiac arrest have an excellent chance of recovery if they receive immediate attention. Therefore, when multiple victims are struck simultaneously by lightning, usual triage priorities should be reversed. Rescuers should give highest priority to patients in respiratory or cardiac arrest (Kleiner and Wilkin 1978).

For victims in cardiopulmonary arrest, BLS (Basic Life Support) and ACLS (Advanced Cardiac Life Support) should be instituted immediately. The goal is to oxygenate the heart and brain adequately until cardiac activity is restored. Victims with respiratory arrest may require only ventilation and oxygenation to avoid secondary hypoxic cardiac arrest. Resuscitative attempts may have higher success rates in lightning victims than in patients in whom cardiac arrest was due to other causes, and efforts may be effective even when the interval before the resuscitative attempt is prolonged (Kleiner and Wilkin 1978).

In the industrial environment, direct current, in which the electrons flow in one direction with one positive lead and one negative lead, is the most common of injuries. Direct current produces a small entrance wound and a much larger exist wound (American College of Emergency Physicians 1992). These exit and entrance wounds are local lesions with a central area appearing charred, a middle zone of whitish to gray coagulation necrosis, and an outer area of brighter-red, edematous damaged tissue. Electricity flows from the point of contact to the ground producing heat within the body directly proportional to the distance between these two points and the resistance of the tissue in between. The effects of electric passage are generally worse with alternating current than with direct current. The current follows the lines of least resistance in the body. Skin has a high resistance in the dry state. Nerve, blood vessel, muscle, and bone exhibit greater resistance and, therefore incur greater damage in the reverse order of listing (American College of Emergency Physicians 1992).

It is very important to respond quickly to an electric shock emergency. The victim must first be separated from the source of the shock. The best method is to cut off the power source. Never touch a person until he or she has been separated from the power source, or you also risk being electrocuted (Chamberlain and Hallman 1995). Do not use a rope, wooden pole, or any other object to try to dislodge the victim from the current source. Do not try to cut the wire. Do not go anywhere near a high-tension line.

There is only one safe way to deal with a live high-tension wire and that is:

CALL THE ELECTRIC COMPANY

Wait until a qualified person has shut off the power before you approach the victim (Caroline 1991).

Send for professional medical help immediately. If the victim is not breathing, artificial respiration must be administered quickly. When there is no pulse, cardiopulmonary resuscitation (CPR) should be started (Chamberlain and Hallman 1995). Immediately after electrocution, respiration or circulation, or both, may fail. The patient may be apneic, mottled, unconscious, and in circulatory collapse from VF or asystole. The prognosis for recovery from electric shock is not readily predictable because the amplitude and duration of the charge usually are unknown. However, because many victims are young and without preexisting cardiopulmonary disease, they have a reasonable chance for survival, and vigorous resuscitative measures are indicated, even for those who appear dead on initial evaluation (Jensen et al. 1989).

SUMMARY

Proper installation of electrical systems is essential to electrical safety. This includes grounding the entire system and all electrical equipment. Protect the electrical system and your life by using only the appropriate fuses and circuit breakers. Follow local/state electric codes or the National Electric Code and hire a qualified electrician when installing or modifying your electrical system.

In the event of an electric emergency, do not use a rope, wooden pole, or any other object to try to dislodge the victim from the current source. Do not try to cut the wire. Do not go anywhere near a high-tension line. CALL THE ELECTRIC COMPANY. Wait until a qualified person has shut off the power before you approach the victim. If the victim is not breathing, artificial respiration must be administered quickly. When there is no pulse, cardiopulmonary resuscitation (CPR) should be started.

Finally, maintain your system by promptly repairing any damage or deterioration. Electrical safety is a critical component of any productive and safe business.

REFERENCES

American Burn Association. 1990. Hospital and prehospital resources for optimal care of patients with burn injury: Guidelines for developmental and operation of burn centers. *J. Burn Care Rehab.* 11: 98.

American College of Emergency Physicians Study Guide. 1992. New York: McGraw-Hill, Inc, pp. 693-94.

Amey, B.W., et al. 1985. Lightning injury with survival in five patients. *JAMA* 253: 243.

Arturson, M.G. 1985. The pathophysiology of severe thermal injury. *J. Burn Care Rehab.* 6: 129.

Beswick, D.R., S.D. Morse, and A.U. Barnes. 1982. Bilateral scapular fractures from low-voltage electrical injury. *Ann. Emerg. Med.* 11: 676.

Bingham, H. 1986. Electrical burns. *Clin. Plast. Surg.* 13: 75.

Browne, B.J., and W.R. Gaasch. 1992. Electrical injuries and lightning. *Emerg. Med. Clin. N. Amer.* 10: 211-29.

Budnick, L.D. 1984. Bathtub-related electrocutions in the United States, 1979 to 1982. *JAMA* 252: 918-20.

Caroline, N.L. 1991a. *Emergency Care In the Streets.* Boston: Little, Brown and Company, pp. 290-95.

Caroline, N.L. 1991b. *Emergency Medical Treatment: A Text for EMT- As and EMT-Intermediates.* 3rd edition. Boston: Little, Brown, 1991, Chapter 13.

Chamberlain, D., and E.M. Hallman. 1995. Electrical Safety on the Farm. Cornell Cooperative Extension Service. *Rural Safety and Health* 123: FSFS3.

Chandra, N.C., C.O. Siu, and A.M. Munster. 1990. Clinical predictors of myocardial damage after high voltage electrical injury. *Crit. Care Med.* 18: 293-97.

Cooper, M.A. 1980. Lightning injuries: prognostic signs for death. *Ann. Emerg. Med.* 9: 134-38.

Cooper, M.A. 1983. Of volts and bolts. *Emerg. Med.* 15(8): 99.

Cooper, M.A. 1984. Electrical and lightning injuries. *Emerg Med Clin North Am.* 2: 489-501.

Craig, S.R. 1986. When lightning strikes: Pathophysiology and treatment of lightning injuries. *Postgrad. Med.* 79: 109.

Davies, J.W.L. 1986. Toxic chemicals versus lung tissue. *J. Burn Care Rehab.* 7: 213.

Demling, R.H. 1985. Burns. *N. Engl. J. Med.* 313: 1389.

Dimick, A.R., L.H. Potts, S.E. Shaw, et al. 1985. Ten year profile of 1,271 burn patients. *J. Burn Care Rehab.* 6: 431.

Dixon, G.F. 1983. The evaluation and management of electrical injuries. *Crit. Care Med.* 11: 384.

Doyle, J.M. 1975. An Introduction to Electrical Wiring. Reston VA: Reston Publishing.

Duclos, P.J, and L.M. Sanderson LM. 1990. An epidemiological description of lightning-related deaths in the United States. *Int. J. Epidem.* 19: 673-79.

Epperly, T.D., and J.R. Stewart. 1989. The physical effects of lightning injury. *J. Fam. Pract.* 29: 267-72.

Frank, H.A., and T.L. Watchel, eds. 1981. Thermal injuries. *Top. Emerg. Med.* 3: 3.

Haponik, E.F., and A.M. Munster, eds. 1990. *Respiratory Injury: Smoke Inhalation and Burns.* New York: McGraw-Hill.

Geddes, L.A., J.D. Bourland, and G. Ford. 1986. The mechanism underlying sudden death from electric shock. *Med. Instrum.* 20: 303-15.

Hammond, J.S., and G. Ward. 1988. High-voltage electrical injuries: management and outcome of 60 cases. *South. Med. J.* 81: 1351.

Haynes, B.W. 1981. Emergency department management of minor burns. *Top. Emerg. Med.* 3: 3.

Herrick CN. 1975. Electrical Wiring: Principles and Practices. Prentice-Hall, Englewood Cliffs, N.J.

Homma, S., L.D. Gillam, and A.E. Weyman. 1990. Echocardiographic observations in survivors of acute electrical injury. *Chest* 97: 103-05.

Housinger, T.A., et al. 1985. A prospective study of myocardial damage in electrical injuries. *J. Trauma* 25: 122.

Hunt, J.L., R.M. Sato, and C.R. Baxter. 1987. Acute electric burns. *Arch. Surg.* 115: 434.

Jensen, P.J., P.E. Thomsen, J.P. Bagger, A. Norgaard, and U. Baandrup. 1987. Electrical injury causing ventricular arrhythmias. *Br. Heart J.* 57: 279-83.

Jones, R.A., and H.R. Spies. 1964. *Electric Wiring. Small Homes Council.* Circular Series G4.2. Urbana, IL: University of Illinois.

Journal of the American Medical Association. 1992. October 28.

Kinney, T.J. 1982. Myocardial infarction following electrical injury. *Ann. Emerg. Med.* 11: 622.

Kirstenson, S., et al. 1985. Lightning-induced acoustic rupture of the tympanic membrane. *J. Laryngol. Otol.* 99: 711.

Kleiner, J.P., and J.H. Wilkin. 1978. Cardiac effects of lightning strike. *JAMA* 240: 2757-59.

Kobernick, M. 1982. Electrical injuries: pathophysiology and emergency management. *Ann. Emerg. Med.* 11: 633-38.

Kotagal, S., et al. 1982. Neurologic, psychiatric, and cardiovascular complications in children struck by lightning. *Pediatrics* 70: 190.

Ku, C.S., S.L. Lin, T.L. Hsu, S.P. Wang, and M.S. Chang. 1989. Myocardial damage associated with electrical injury. *Amer. Heart J.* 118: 621-24.

Matthews, J.B., and C. Jelenko, III. 1979. Psychosocial support of the burn patient, his family and the burn team. *Life Sup. Nurs.* 2: 13.

Moran, K.T., et al. 1982. Lightning injury: Physics, pathophysiology, and clinical features. *Irish Med. J.* 79: 120.

Mortenson, M.L. 1983. Electricity sparks multi-system assessment. *Emerg. Med. Serv.* 12(1): 15.

National Fire Protection Association (NFPA). 1978. *National Electric Code 1978*. Boston: National Fire Protection Association.

Purdue, G.F., et al. 1986. Electrocardiographic monitoring after electrical injury: Necessity or luxury. *J. Trauma* 26: 166.

Ravitch, M.M., et al. 1961. Lightning stroke. *N. Engl. J. Med.* 264: 36.

Reichl, M., et al. 1985. Electrical injuries due to railway high tension cables. *Burns* 11: 423.

Rouse, R.F., and A.R. Dimick. 1978. The treatment of electrical injury compared to burn injury: a review of pathophysiology and comparison of patient management protocols. *J Trauma* 18: 43.

Salisbury, R. 1989. High-voltage electrical injuries. *Emerg. Med.* 21(13): 86.

Seward, P.N. 1987. Electrical injuries: Trauma with a difference. *Emerg. Med.* 19(9): 66.

Shires, G.T., and E.A. Black, eds. 1981. Second conference on supportive therapy in burn care. *J. Trauma* 21: 665.

Shires, G.T., and E.A. Black, eds. Proceedings of the NIH Consensus Development Conference in supportive therapy in burn care. *J. Trauma* 19: 855.

Solem, L., R.P. Fischer, and R.G. Strate. 1977. The natural history of electric injury. *J. Trauma* 17: 487.

Taussig HB. "Death" from Lightning- and the possibility of living again. Ann Intern Med 68:1345, 1968.

Taylor, C.O., J.C. Carr, and J. Rich. 1981. EMS system in action: Survival of two girls after direct lightning strikes. *EMT J.* 5(6): 419.

Thompson, J.C., and S. Ashwal. 1983. Electrical injuries in children. *AJDC* 137: 231-35.

Thygerson, A.L. 1980. Electric burns. *Emergency* 12(4): 35.

Tilma, C., and H.J. Doss. 1992. *Electrical Hazards on the Farm*. Center for Michigan Agricultural Safety and Health.

Wallace, J.F. 1991. Electrical injuries. In: *Harrison's Principles of Internal Medicine*. 12th edition. J.D. Wilson, E. Braunwald, K.J. Isselbacher, et al., eds. New York, NY: McGraw-Hill Book Co, Health Professions Division, pp. 2202-04.

Wilkinson, C., and M. Wood M. 1978. High-voltage electric injury. *Amer. J. Surg.* 136: 693.

Xenopoulos, N., A. Movahed, P. Hudson, and W.C. Reeves. 1991. Myocardial injury in electrocution. *Am Heart J.* 122: 481-84.

Yang, J.Y., et al. 1985. Electrical burn with visceral injury. *Burns* 11: 207.

7

BITE, STING, AND ENVENOMIZATION HAZARDS IN AGRICULTURE AND FORESTRY

Charles S. Apperson, Ph.D., and Michael G. Waldvogel, Ph.D.
Department of Entomology, Agromedicine Program
North Carolina State University

> Arthropods have both beneficial and detrimental effects on agriculture and humans. Everyone is familiar with the beneficial role of honeybees in pollinating plants. However some arthropods are also pests that bite or sting or serve as vectors of infectious diseases for both humans and animals. Thousands of people die each year from diseases transmitted by arthropods. Because of their usual outdoor activities, farmers, forestry workers, and fishermen are at high risk of arthropod bites and stings. This chapter discusses various aspects of arthropods that are harmful to humans including their venoms, blood feeding activities, and the epidemiology of bites and stings. The major pest arthropods are described and information on their distribution, life history, injury produced by biting or stinging, and first aid treatment is provided. Methods to prevent bites and stings and methods to control arthropods of public health significance are discussed.

INTRODUCTION

The biting and stinging activities of some animals in the phylum Arthropoda have a deleterious impact on human health and well-being. Hematophagous arthropods, such as mosquitoes and ticks potentially transmit pathogens during bloodfeeding that cause illness and death. For example, the mosquito-transmitted illness malaria was a leading cause of mortality early in the 20th century, accounting for 2.5 million deaths annually in Africa, Asia, and Europe (Bruce-Chwatt 1987). In the United States, where arthropod-borne illnesses are less common than in third world countries, the biting activity of hematophagous arthropods interferes with outdoor recreational activities. For example, the biting activity of nuisance mosquitoes is severe enough that in many geographic locales, mosquito control efforts are organized locally and supported through tax revenues. In the state of Florida, over $50 million are spent annually for mosquito abatement. The general public spends hundreds of millions of dollars each year, purchasing insecticides or repellents to use against noxious arthropods. The cost of medical treatments for stings inflicted by bees, wasps, and ants is unquantified but likely to be high.

There is no question that some arthropods have a detrimental effect on the quality of people's lives. Although rural residents are, perhaps, at greatest risk because they live and work in environments where more arthropod pests are produced, urban and suburban dwellers are also exposed to the noxious pests during outdoor activities in recreational and residential areas. Zoonoses, such as tick-borne Lyme disease and Rocky Mountain Spotted Fever, often result when suburbanization of rural landscapes brings people into closer association with wildlife that serve as reservoirs and hosts for these pathogens and their arthropod vectors.

In this chapter, we will review the major groups of arthropods that cause injury to people. Our objective is to provide sufficient detail to allow the reader to identify commonly occurring arthropods and the injuries that they cause. We will also provide some general information on first aid procedures for arthropod-inflicted injuries. Recognition and treatment of illnesses resulting from arthropod-borne pathogens is beyond the scope of this review.

The principal ways that arthropods cause injury can be grouped into the following broad categories. *Envenomization* results from stings or bites that are usually given in defense when the arthropod has been accidentally contacted or its nest or web disturbed. The *biting activity* of hematophagous arthropods, such as biting flies and ticks, can be annoying and usually causes localized swelling, redness, and itching. In rare instances, flies invade the muscle or skin of healthy individuals, causing a condition known as *myiasis*. Myiasis will not be covered in our review. Readers interested in this form of arthropod injury should consult Alexander (1984), Goddard (1993), and James (1947).

WHAT ARE ARTHROPODS?

The Phylum Arthropoda, which literally means "jointed legs", contains the vast majority of the biting/stinging animals of public health importance. All "arthropods" fit the following general description:

- A body made of segments that are grouped or fused together to form two or three distinct regions.
- Paired appendages, such as legs and antennae.
- An external covering or exoskeleton that is composed of chitin.

Over the years, a number of taxonomic arrangements have been proposed for the Arthropoda. Depending upon which classification scheme you follow, there are five to six major groups of arthropods. Figure 7.1 shows some of the more common groups of arthropods. For the purposes of this chapter, we are concerned with only two groups:

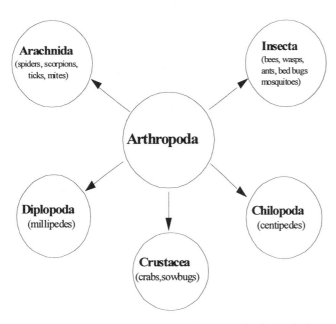

Figure 7.1. Representative classes of the phylum Artropoda.

Arachnida

This class includes spiders, scorpions, ticks, and mites. Arachnids have "simple eyes" and no antennae or wings. The mouthparts typically have two prominent structures, called *chelicerae*, which end in needle-like piercing tips. Arachnids have 4 pairs of legs and 2 body regions. The mouthparts and legs are attached to the anterior region, which is commonly called the *prosoma* or *cephalothorax*. The reproductive organs and the digestive and excretory systems are found in the posterior region, commonly called the *abdomen* or *opisthoma*.

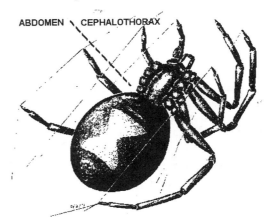

Figure 7.2. Arachnid body divisions (EPA 1992).

Insecta (Hexapoda)

Insects have 3 body regions: head, thorax, and abdomen. The mouthparts contain a pair of structures called *mandibles*, which may be modified for various modes of feeding, such as chewing, piercing-sucking, or sponging. The head bears the mouthparts, eyes (which may be simple, compound, or both), as well as a single pair of antennae. The thorax bears the appendages for locomotion: 3 pairs of legs, and 1 to 2 pairs of wings. Only adult insects have fully-developed and functional wings. Like arachnids, the insect abdomen contains most of the digestive and excretory systems, as well as the reproductive organs.

Our discussions of biting/stinging arthropods will be presented by taxonomic Order and, in some cases, by taxonomic Family. For example:

Insecta
  Diptera—mosquitoes, biting flies
  Hemiptera—bed bugs, assassin bugs
  Hymenoptera—bees, wasps, hornets and ants
  Lepidoptera—caterpillars

Arachnida
  Acarina—ticks and mites
  Araneae—spiders

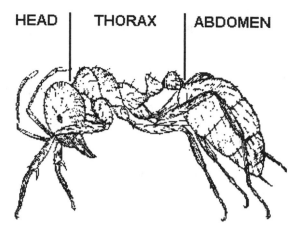

Figure 7.3. Divisions of the insect body (adapted from EPA 1992).

Arthropod Growth and Development

In order to grow and mature, arthropods must shed the outer layer of their exoskeleton during the process referred to as molting or ecdysis. With few exceptions, molting occurs only during the arthropod's juvenile stage. The number of molts varies among arthropods, but typically ranges from 3 to 15. Between molts, the form of the juvenile is referred to as an *instar*. Thus, a first instar immature is the form between egg hatch and the first molt. Maturation of the juvenile stage may occur over a period of several days to several months (or even several years). Arthropods mature and change form through the process of metamorphosis. Knowledge of the developmental stages of pestiferous arthropods is important to accurate identification, as well as a better understanding of their biology and behavior. Our discussions of insects, mites and ticks, involve 3 main types of metamorphosis:

Gradual Metamorphosis. Immatures in this group are called *nymphs*. They resemble the adult stage, except that they are smaller and lack wings. (Only adults have fully functional wings). Nymphs and adults are often found together and usually feed on the same host (or species of host). Common examples of insects with simple metamorphosis are bed bugs and cockroaches.

Complete Metamorphosis. Immatures in this group are called *larvae*. Typically, larvae have an appearance that is very different from that of the adult. They lack compound eyes and may not always have antennae. Larvae may be legless or, in the case of the Lepidoptera and some Hymenoptera, they may also have abdominal appendages called *prolegs*, which are used for locomotion. In agriculture, the larval stage is frequently the primary (or only) pest stage, whereas with biting/stinging insects such as bees, wasps and mosquitoes, the adult is the problem stage. Larvae continue to grow through a number of molts until they change into *pupae*. This non-feeding stage marks a time of great physiological activity, resulting in a complete change into the *adult* stage. Common examples of insects with complete metamorphosis are bees and moths.

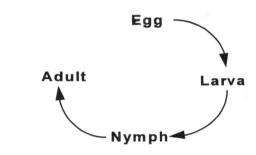

Figure 7.5. Complete metamorphosis.

Modified Metamorphosis. This is a type of metamorphosis is exhibited by mites and ticks. Eggs hatch to produce 6-legged larvae, which molt to produce 8-legged nymphs. The nymphs go through a series of molts before becoming adults. All 3 post-egg stages feed on hosts; however, they may not all feed on the same host or even the same species of host(s).

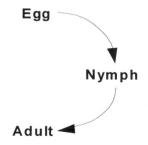

Figure 7.4. Gradual metamorphosis.

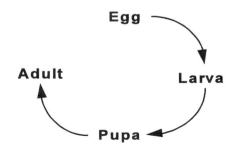

Figure 7.6. Modified metamorphosis.

EPIDEMIOLOGY OF ENVENOMIZATION

Few studies of the frequency and determinants of stinging or biting episodes have been conducted. While the number of deaths resulting from arthropod injury is likely to be accurately recorded, the annual incidence of stinging and biting episodes has likely been underestimated since physicians are usually not required to report non-fatal arthropod stings and bites to state health departments, and many (if not most) victims do not seek medical treatment. The importance of arthropod stings is placed in perspective when it is understood that far more deaths generally occur from wasp and bee stings than from most other animal-related injuries (Parrish 1963; Langley 1994). For example, in a recent retrospective study of coroners' records, Langley (1994) reported finding in North Carolina from 1972 to 1989 that 45.7 percent of animal-related fatalities (n=92) involved insects and/or spiders. In Utah from 1900 to 1990, Straight and Glenn (1993) reported that more deaths occurred from bees stings than from other venomous animals. Barnard (1973) estimated fatalities from bee and wasp stings at approximately 30 persons per year from 1963 through 1973.

Hymenoptera Stings

The numbers of Hymenoptera stings treated by physicians appears to be high. Keegan (1972) surveyed 135 physicians in Mississippi in 1971 and reported that 70.3 percent of 3,389 patients suffering from stings and bites were treated for wasp, bee, or ant stings, while 14.7 percent received medical treatment for spider bites, and only 3.6 percent were treated for snake bites. Golden et al. (1989) reported the frequency of systemic allergic reactions to bee and wasp stings among 269 test subjects to be 3.3 percent. However, 26.5 percent of the test subjects had IgE antibodies to Hymenoptera venom. Golden and coworkers concluded that: a) allergic reactions to the venom and asymptomatic sensitivity to stings are prevalent and, b) a majority of affected persons do not seek medical assistance. In the southeastern United States, envenomization from the red imported fire ant (RIFA) (*Solenopsis invicta*) appears to occur more frequently than from other arthropods. Adams and Lofgren (1982) studied the incidence of stings from the RIFA and other arthropods at Ft. Stewart, GA. Of 370 persons visiting the military base's infirmaries or hospital, 49 percent sought medical care for RIFA stings. From 1 to 3 percent of patients were treated for wasp or bee stings, or spider bites.

Estimates of the number of persons annually stung by RIFA's appear to be highly variable. From a survey of selected residents in Lowndes County, GA, Yeager (1978) reported that approximately 5 percent of 438 persons stung by the RIFA required medical attention. For the period July through September, projected attack rates per 100,000 persons exceeded 20,000 persons stung. A survey of 29,205 physicians from 13 southeastern states revealed that 2,022 (7%) reported treating approximately 20,700 patients on an annual basis for RIFA stings (Stafford et al. 1989). Based on the results of a mail survey of 1,012 physicians in South Carolina, Schuman et al. (1987) estimated the statewide RIFA sting rate for victims who sought medical treatment to be 221 cases per 100,000 in 1986. In a telephone survey of 240 households in Louisiana, Clemmer and Serfling (1975) found that 28.6 percent of family members were stung during summer months, but only 4.4 percent sought medical treatment. In another study, Adams and Lofgren (1981) reported that about 31 percent of 254 persons from 62 families were stung by fire ants at least once during the year. Lemke and Kissam (1989) surveyed 430 persons in several fire ant-infested counties in South Carolina and found that 89 percent had at least one family member that had been stung by RIFA's.

As would be expected, rural residents (as compared to urban residents) who frequently live adjacent to agricultural lands infested with RIFA are at greater risk of being stung. Both Adams and Lofgren (1981) and Yeager (1978) reported a greater frequency of RIFA stings for rural residents. RIFA attack rates are highest during warmer months from March to September or October (Yeager 1978; Adams and Lofgren 1981). However, as residential development increases, encroachment upon RIFA-infested land, as well as the potential movement of RIFA-infested nursery stock and sod will increase the likelihood of stings among suburban residents.

THE NATURE OF ARTHROPOD VENOMS

Arthropods use venoms in defense or to procure food. Most biting/stinging arthropods have evolved multicomponent venom systems. In-depth reviews of animal venoms are contained in Bücherl and Buckley (1971), Bettini (1978), and Tu (1984). Considerable effort has been expended recently in studying the molecular biology and genetics of various arthropod venom systems. Through technological advances in molecular genetics, the genes responsible for venom production have been cloned and inserted into baculoviruses (Carbonell et al. 1988; Stewart et al. 1991; Tomalski and Miller 1991, 1992; Tomalski et al. 1993). The resulting genetically-engineered viruses have expanded host ranges and faster kill times than conventional chemical insecticides (Duffey et al. 1995).

Venoms are generally delivered through a bite, sting, or by contact with the arthropod. The chemistry and mode of action of the venom components may vary considerably between groups of closely related arthropods. The severity of response to envenomization is highly variable depending on the amount of venom received, the site of the bite or sting, the mode of action of the venom, and the sensitivity of the victim (Ori 1984).

Spider Bites

Information on the frequency of spider bites is limited. Human contact with spiders probably occurs infrequently. Regardless, the mandibles (chelicerae) of most species are too short and fragile to penetrate the skin of humans. Spiders are predaceous and they possess poison glands and chelicerae to deliver their venom. The pharmacology and chemistry of spider venoms have been reviewed by Bücherl and Buckley (1971) and Tu (1984). Ori (1984) classified the mode of action of spider venoms as either neurotoxic, cytolytic (necrotic), or a combination of both. Discussions of spider envenomization usually focus on the black widow (*Latrodectus mactans*) and brown recluse (*Loxosceles reclusa*) spiders. The venoms of these public health arachnids are chemically similar in that the active components are primarily proteinaceous (Duchen and Gomez 1984); however, they are functionally dissimilar, having different modes of action. The proteins in black widow spider venom act at the neuromuscular junction level (Duchen and Gomez 1984). The syndrome resulting from the black widow spider bite, known as *latrodectism*, will be discussed in more detail later in this chapter. In contrast, brown recluse spider venom proteins are cytolytic and hemolytic, and commonly produce a local necrotic skin lesion (Ori 1984). Human envenomization by the brown recluse spider, referred to as *loxoscelism*, will be discussed later in this chapter. Case reports of envenomizations from spiders are reviewed by Ori (1984).

Hymenoptera Venoms

Venoms of social Hymenoptera including some ants, bees, wasps, and hornets have been studied extensively, and the allergens identified. These venoms are allomonal (used for defense) and they are immediately painful (Schmidt 1982), which confers a distinct advantage in protecting the insect's nest from predators. Schmidt (1982) divides venoms of social Hymenoptera into three principal groups: macromolecules (mainly enzymes), peptides and moderately-sized molecules, and micromolecules.

The allergenic proteins in honey bee venom include phospholipase A_2, hyaluronidase, acid phosphatase, allergen C, and the peptide melittin. Vespid wasp venoms contain hyaluronidase, a phospholipase A_1B (which is not related to the phospholipase of bee venom), a neurotoxic protein (antigen 5), and several minor proteins (Piek 1984). Venom components responsible for the hemolytic activity include melittin and phospholipase A. Some vasoactive biogenic amines, histamine and serotonin, are present in vespid venoms (Nakajima 1984). Ant venom, such as that of the RIFA, primarily contain water insoluble piperidine alkaloids and trace amounts of four allergenic proteins, Sol i *I-IV* (Hoffman et al. 1988a). Sol i *II* is a phospholipase that is structurally dissimilar to wasp and bee phospholipases. RIFA venom contains less than 0.1 percent protein by weight (Stableman and Lockey 1981). In contrast, 70 percent of Florida harvester ant (*Pogonomyrmex badius*) venom is protein, and contains six different enzymes, including phospholipase A_2, phospholipase B, hyaluronidase, lipase, acid phosphatase, and four esterases (Stablein and Lockey 1981). Unlike the venom of other arthropods, *P. badius* venom contains high amounts of lipase. The venom is very potent, and it is comparable in toxicity to the venom of most toxic snakes (Stablein and Lockey 1981).

Considerable antigenic cross-reactivity exists between venoms of various ant, wasp, and bee species. Hoffman et al. (1988b) found that persons who were allergic to honey bee, yellow jacket, and paper wasp venoms also exhibited an allergic reaction to fire ant venom allergen Sol i *I*. In the blood serum of persons allergic to vespid wasp stings, Hoffman (1985) found antibodies that cross-reacted with hyaluronidase, phospholipase, and antigen five components of yellow jacket, baldfaced (white-faced) hornet, and paper wasp venoms.

Urticatious Caterpillar Venom

Unlike stinging Hymenoptera, some insects in the order Lepidoptera use venoms solely to deter predators rather than to capture prey. Contact with larvae of these lepidopterans produces an immediate and painful dermatitis, as well as erythema and swelling of the skin, and pruritis. Additional systemic reactions may include nausea, headache, and lymphadenitis (El-Mallakh et al. 1986). The larvae are covered with spicules or hollow hairs that connect with poison cells. When the caterpillar is touched, the hairs break off in the skin releasing an allogenic toxin. The deterrent effects of allomonal venoms are obvious. Ori (1984) presents an extensive review of the classification and structure of urticating hairs. Irritative Lepidoptera possess two major types (spicule and spine hairs) and numerous subtypes of hairs. The venom associated with the urticatious hairs of the puss caterpillar, *Megalopyge urens*, contains hyaluronidase, and two proteinaceous substances that exhibit hemolytic and proteolytic activity (Ardao et al. 1966). Histamine has also been identified as a component of the venom of some irritative lepidoptera (Ori 1984). Caterpillar venoms are unusually rich in proteases and esterases compared to other arthropod venoms (Schmidt 1982). Protease enzymes are responsible for the intense pruritis that accompanies the dermatitis resulting from envenomization by irritative Lepidoptera.

BLOOD FEEDING

Hematophagous arthropods feed on vertebrates to obtain a meal of blood that is used as a protein source for egg development, and as a source of nutrients for the adult (O'Meara 1986). In the process of inserting their mouthparts into skin, these arthropods secrete saliva into the wound. The pharmacological properties of saliva, which facilitate the process of bloodfeeding, are correlated with the arthropods' mode of feeding (Ribeiro 1987). Arthropods that feed quickly, such as biting flies, have antihemastatic chemicals in their saliva to prevent blood platelet aggregation. Other arthropods, such as ticks, attach to their hosts for prolonged periods, allowing sufficient time for the hosts to develop inflammatory and immune reactions. In addition to anticoagulants, tick saliva contains antihemastatic chemicals, such as prostaglandin E2 that promote vasodilatation, as well as other anti-inflammatory and immunosuppressive agents (Ribeiro 1987). However, repeated exposure to the antigens in arthropod saliva usually stimulates the antibody production by the host. Antibodies are important to the development of immunity to ticks by wildlife and domestic animals. Humans also develop antibodies to the proteins in tick saliva; however, these antibodies do not provide humans with immunity to ticks, but instead they are involved in the inflammatory response to tick bites. In their studies of the epidemiology of Lyme disease, Schwartz and coworkers (1991, 1993) demonstrated the potential use of anti-tick saliva antibody in the blood serum of people residing in New Jersey-New York as a biologic marker of tick exposure, and a potential measure of the risk factor for exposure to the Lyme disease spirochete.

In rare instances, the attachment of female ticks results in the development of a neuromuscular syndrome known as tick paralysis. The salivary secretions of female ticks contain a toxin that produces an ascending flaccid paralysis in the victim (Gregson 1973). The paralysis rapidly regresses once the ticks are removed, but can be fatal if ignored (Gregson 1973). Tick paralysis has been reported to be caused worldwide by over 50 species (Gothe and Neitz 1991). The etiology and pathogenesis are reviewed in depth by Gothe and Neitz (1991).

Local reactions that occur around the site of insect bites may persist for a few hours, days or weeks. Hoffman (1995) characterized the various types of local reactions at bites sites as: whealing (forming a welt), erythema (redness), induration (callousness), pruritus (intense itching), local hyperthermia (warm to the touch), vesiculation (forming small blisters), hemorrhage, papular urticaria (a raised, smooth pimple-like elevation that itches) and pain.

IDENTIFYING THE ARTHROPOD(S) BASED ON THE INJURY

The physical appearance of the bite site, the number of bites and their location, and the time of day that the injury occurs can often be used to identify the arthropod responsible for the injury. Information presented in Tables 7.1 through 7.3 provide clues as to arthropods that are possibly responsible for bites and stings.

Table 7.1. Identification of biting or stinging arthropods based on the physical appearance of skin lesions.

Physical appearance of lesion	Arthropod(s) possibly involved
Swollen, erythmematic welt	mosquito, flea, biting gnat, horse or deer fly, stable fly, tick, conenose, or assassin bug, head, body, or pblic louse
Swollen, erythmetatic welt with two puncture wounds	spider, centimpede
Swollen, blanched welt with puncture in center	wasp, bee, hornet, or scorpion
Urticaria (dermatitis)	stinging caterpillar, chigger, northern fowl mite, larval tick
Vesicles filled with clear fluid	blister beetle, chigger
Vesicles filled with cloudy fluid	red imported fire ant
Nercrosis	spider

Table 7.2. Identification of biting or stinging arthropods based upon the location and pattern of lesions.

	Location of lesion					Number of lesions			Pattern of bite lesions		
	Legs/feet	Trunk	Genitals/groin	Arms/hands	Head/neck/face	Single	Few	Multiple	Scattered	Grouped	Linear
Bed bug		X					X	X		X	X
Hymenoptera[1]	X			X	X	X	X	X	X		
Biting midge					X		X	X	X		
Black fly				X				X	X		
Blister beetle				X		X			X		
Centipede	X			X		X					
Chigger	X	X		X	X			X		X	X
Conenose Bug					X	X					
Flea	X			X		X	X			X	
Horse/Deer fly	X			X			X	X	X		
Straw Itch mite		X						X	X		
Louse[2]		X	X				X	X	X		
Mosquito	X			X	X		X		X		
Northern fowl mite	X			X				X			
Scabies mite			X	X				X			
Scorpion	X			X		X					
Spider	X			X		X					
Stable fly	X			X			X				
Tick (all species)		X		X		X	X				
Tick (larva)	X						X				
Urticatious caterpillar				X			X			X	

[1] Paper wasp, yellow jacket, honeybee, and bald faced hornet.
[2] Head, body, and pubic lice.

Table 7.3. Identification of biting or stinging arthropods based on the time of day bites or stings are received.

Time of day	Arthropod(s) possibly involved
Daytime out-of-doors	container-breeding or floodwater mosquito, horse or deer fly, stable fly, Hymenoptera, conenose bug, wheel bug, assassin bug, wasp, bee, hornet, stinging caterpillar
Dusk/dawn out-of-doors	mosquito, biting gnat
Nightime indoors	bed bug, spider, flea
Nightime out-of-doors	mosquito

PEST ARTHROPODS

Centipedes (order Scolopendromorpha)

Centipedes are multi-segmented arthropods in the class Chilopoda. Most problematic species are in the family Scolopendridae. Centipedes are considered to be minor pests of public health importance. These predaceous arthropods are venomous, and occasionally people are bitten accidentally.

Description:
Adults—Centipedes have an extended, dorsoventrally flattened body, ranging from 3 to 250 mm (1/8 to 10 in.) in length. The head bears two sensory antennae, and the mandibles. The trunk is divided into 15 to 100 segments, depending on the species. Adults can vary from deep brown or blackish to yellow in color. The first trunk segment bears a pair of biting appendages, the forcipules, and the remaining segments each have a pair of legs. This distinguishes them from millipedes, which have 2 pairs of legs per body segment.
Nymphs—Immature centipedes resemble adults.

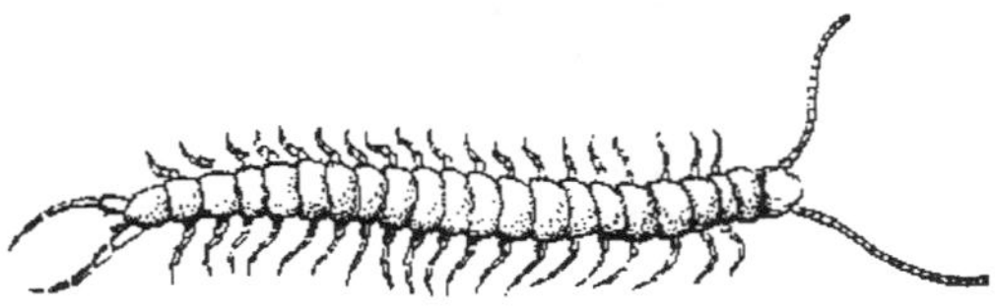

Figure 7.7. Centipede (EPA 1992).

Biology:
Distribution—The family Scolopendridae contains 5 genera and approximately 40 species. Many species are cosmopolitan in temperate, subtropical and tropical regions. The genus *Scolopendra* contains most nuisance species (Bücherl 1971).
Habitat—Centipedes live under stones, leaf litter, the bark of logs and trees, and other areas inhabited by insect and other small arthropods.
Life history—Eggs are laid in moist soil, and after hatching the nymphs pass through approximately 10 instars before reaching maturity. Adults live 3 to 5 years. Centipedes are nocturnal, and predaceous on small insects and other arthropods.

Injury:
Two puncture marks can usually be found at the bite site. A centipede bite usually results in a localized reaction, consisting of pain, swelling, erythema, and a superficial necrosis which resolves in 1 to 3 weeks (Minelli 1978).

First Aid:
Aspirin and an oral antihistamine should be effective in relieving pain and swelling at the bite site. If any necrosis occurs, the affected area should be thoroughly cleaned, an antibiotic ointment applied and the wound bandaged.

Scorpions (order Scorpionida)

Scorpions are placed in the class Arachnida along with spiders and mites. These arthropods are generally perceived to be public health menaces. Even though all species are venomous, most scorpions are harmless. Scorpions exhibit gradual metamorphosis.

Description:
Adult—All scorpions have 8 legs. Two enlarged appendages, pedipalps, are claw-like pincers for grasping prey. A large "tail", protruding from the abdomen contains a stinger on the last segment. Scorpions vary in color from blackish or brownish red to yellowish or straw colored. The body of the scorpion may contain a series of longitudinal stripes that are lighter in color. Adults vary in size from approximately 1 cm (0.4 in.) to as large as 12 cm (4.5 in.).
Nymph—Immature scorpions are similar in appearance to adults.

Biology:
Distribution—Scorpions occur worldwide in all temperate and tropical regions. The Buthidae is the largest family of scorpions, containing over 500 species and more than 40 genera (Shulov and Levy 1978).

Habitat—Some species prefer desert and semi-arid conditions while other species favor moist and milder climates. Scorpions are generally found in rock crevices or underneath objects such as stones, boards and, the bark of trees or logs.

Life History—Females give birth to living young, which crawl upon their mothers' backs and remain there until they molt in about 2 to 3 weeks. Most species molt 6 to 7 more times before reaching maturity in 6 months to several years. Scorpions are nocturnally active, preying on insects and arthropods, such as spiders. Some species of scorpions can survive in hibernation for several months without feeding.

Scorpions are attracted to peridomestic environments, such as piles of bricks, lumber or fire wood, because of the other arthropods that are found in these habitats. They occasionally wander indoors, especially in poorly constructed dwellings where cracks and crevices provide harborage for scorpions and their prey.

Injury:

Although all scorpions are venomous, most species produce a sting that is comparable to that of the honey bee. A scorpion sting results in a sharp burning sensation followed by swelling and the formation of a wheal. These local reactions usually resolve spontaneously in 1 to 2 days; however, fatal stings can be produced by some species, such as *Centroides sculpturatus* that occurs in the southwestern United States. Systemic symptoms of envenomization which are typical of neurotoxic poisons include: dizziness, profuse sweating, chills, abdominal cramping, nausea and vomiting, cardiovascular irregularities, and labored breathing. Death is usually the result of cardiac or respiratory failure. Other species of *Centroides* are involved in stinging episodes in Mexico and Central America, while in South America the genus *Tityus* contains some dangerously poisonous species. In Middle East and parts of Asia, the most poisonous scorpions are in the genera *Buthus*, *Palamneus*, *Leirus* and *Androctonus* (Goddard 1983). In the southeastern U. S., *Vejovis carolinianus*, the southern unstriped scorpion, commonly invades homes, and will occasionally accidentally sting people. Keegan (1980) presents an in-depth treatment of scorpions of medical importance.

First Aid:

The victim should be kept calm to slow the absorption of venom. A cold pack should be applied to sting site as soon as possible. For a severe envenomization, the victim should receive emergency medical care immediately.

Medical Treatment: Administration of antivenin within several hours of a severe envenomization is recommended (Keegan 1980; Hassan 1984).

Spiders (order Araneae)

Spiders are arthropods in the class Arachnida. They have a widespread, but largely undeserved reputation as being dangerous to the health and welfare of people and their pets. In truth, spiders are extremely beneficial because they prey on many insects that are considered to be pests. Encounters between people and spiders are usually accidental; the bites are a defensive response by the spider when its web or nest is disturbed. Only a few species, such as the black widow and brown recluse spiders, are of public health significance. Species causing public health problems are reviewed by Ori (1984) and Ebeling (1978).

Figure 7.8. Black widow spider (EPA 1992).

Black Widow (*Latrodectus mactans*)

Description:

Adult—Like all spiders, the black widow has 8 legs; however, males and females differ in appearance. Females are about 30-40 mm (1 1/5 to 1 1/2 in.) long with legs outstretched and have black, globular abdomens about 9 by 13 mm (1/3 to 1/2 in.) marked with a red or yellow hourglass shape underneath. Males are smaller, usually 16 to 20 mm (2/3 to 1 1/5 in.) long with legs outstretched. They are lighter in color and have a mid-dorsal red or pale brown stripe from which white or yellow streaks radiate. The male biting appendages (called pedipalps) at front of head are noticeably swollen.

Egg—Eggs are incorporated into grayish, silken balls, about 12 to 15 mm (ca. 1/2 to 2/3 in.) diameter, that are suspended in the spider's web. These egg masses contain 200 to 900 eggs.

Nymph—Entirely white at first, immature spiders called nymphs develop through 5 to 8 growth stages (instars). As they develop, nymphs become more similar in appearance to adult males, though smaller.

Biology:

Distribution—Though more abundant in the southern states, the black widow spider occurs throughout most of the Western Hemisphere. This species is usually found in association with human dwellings (McCrone and Stone 1965). The spider may hide in sheltered, dimly lit places such as barns, garages, basements, outdoor utility boxes, outdoor toilets, hollow stumps, trash, brush, and dense vegetation. Black widows usually seek dry sheltered sites such as buildings during periods of cold weather. Four other species of *Latrodectus* occur in the United States (Ebeling 1978). The northern black widow (*L. variolus*) is most common in the New England states but along its southern range it overlaps the distribution of *L. mactans*. The red widow (*L. bishopi*), and the brown widow (*L. geometricus*) are restricted to southern Florida (McCrone and Stone 1965). The widow spider, *L. hesperus*, occurs in the western United States and Canada.

Life History—The black widow spider overwinters as a young adult in buildings or in sheltered places outdoors. After a prolonged courtship, mating occurs in late spring. Soon afterward the female kills her mate and begins laying eggs. The grayish silken sac of eggs is attached to an irregular, tangled web with a funnel-shaped exit. Each female constructs 5 to 15 egg sacs, each containing 200 to 900 eggs.

Young spiders emerge from the sac in 10 to 30 days. They are cannibalistic at this stage, and only a few nymphs from each egg mass survive. They require 2 to 3 months to develop into adults. The parental adults die the same summer or autumn after laying eggs. The new generation of adults survives through the winter.

Feeding Habits—This spider feeds primarily on insects and other arthropods but, when disturbed, it may bite people or animals.

Injury:

The black widow usually bites people only when its web is disturbed, and the males do not bite. The female black widow bite is like a pin prick, but her neurotoxic venom causes localized pain within a few minutes after the attack. This pain spreads rapidly to arms, legs, chest, back and abdomen. Chills, vomiting, difficult respiration, profuse perspiration, delirium, partial paralysis, violent abdominal cramps and spasms may occur within a few hours of the bite.

The victim may go into shock and experience difficulties with vision. The victim usually recovers in 2 to 5 days; less than 5 percent of all black widow attacks are fatal (Ori 1984).

First Aid:

An ice pack should be applied immediately to the bite to slow absorption of the venom, and the victim should be transported to medical care. An antiseptic should be applied to the bite to prevent development of a secondary infection.

Medical Treatment:

Medical treatment should be administered immediately. Intravenous calcium gluconate (10% solution) may be administered to alleviate the intense pain that victims usually experience. An antivenom is commercially available (Geren and Odell 1984).

Brown Recluse (*Loxosceles reclusa*)

Description:

Adult—The brown recluse spider is usually grayish to reddish brown with a dark fiddle-like marking on the head and thorax. Sometimes, this spider may be light tan or cream colored. The body of the spider is 7 to 13 mm (1/4 to 1/2 in.) long with a leg span about the size of a half dollar. This species has 6 eyes arranged in 3 pairs. All specimens suspected to be brown recluse spiders should be examined by an expert for positive identification because this species can be confused with other spiders.

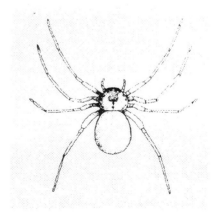

Figure 7.9. Brown recluse spider (Knecht 1986).

Egg Sac—When completed, this white, loosely spun, cocoon-like structure is about 4 mm (3/16 in.) high and 17 mm (3/4 in.) wide. The lower side is flat while the upper side is slightly convex. The sac is filled with tiny, yellow eggs.

Nymphs—Though white when newly hatched, the young spiders soon become grayish to reddish brown.

Biology:

Distribution—The brown recluse spider has been found throughout many of the southern and central states from Missouri east to North Carolina. A small number of specimens have been collected in Washington, D.C., Arizona, California, Florida, New Jersey, Pennsylvania, and Wyoming. In the western hemisphere, there are 50 species of *Loxosceles* (Ebeling 1978).

Feeding Habits—Like other venomous spiders, the brown recluse feeds primarily on insects and other arthropods. When disturbed, it may bite people and animals.

Life History—The brown recluse completes a single generation each year. Mating and oviposition usually occur during the summer months. Females produce 1 to 5 loosely spun egg sacs, each containing about 50 eggs. Egg sacs are incorporated into the large, irregular, sticky webs. Eggs incubate from 6 to 3 days, hatching sooner if temperatures are warm.

Immature spiders feed on paralyzed insects provided by their mother. They molt 5 to 8 times before becoming adults the next summer. From 9 to 200 days may elapse between molts. Though the average life span is 1.5 years, some brown recluse spiders live well over 2 years. They can survive several months without food.

Injury:

Reactions to the bite of this spider vary from mild to severe depending on the amount of venom and the victim's sensitivity to it. Mild to intense pain may be noticed as early as 2 to 6 hours after envenomization, but usually occurs the following day. The venom of the brown recluse spider is cytolytic. A reddening that develops at the site of the bite is quickly followed by the formation of a small, white blister which is inflamed and hard to the touch. The tissues of this blister eventually turn purple and later black. After about one week the central area is depressed and necrotic. The dead, black skin separates from the healthy tissue leaving an ulcer that may vary in size. The edges of the wound usually thicken while the central area fills with scar tissue. The wound heals slowly over a period of 3 months and usually leaves a permanent scar. Cosmetic surgery involving skin grafts may be required (Ori 1984).

First Aid:

An antiseptic and an ice pack should immediately be applied to the bite and the victim should be transported to medical care. If at all possible the spider should be collected so that it can be identified. *Loxosceles* antivenom is effective only if given within 36 hours after the bite (Ori 1984).

Medical Treatment:

For severely necrotic injuries, surgical excision of the of the wound followed by skin grafting may be needed. Rees et al. (1985) reported the use of dapsone, a leukocyte inhibitor, successfully reduced the need for early surgical excision.

Mites (order Acari)

Mites are not insects; they are more closely related to ticks and spiders, and other arachnids. On a world wide basis, mites are important nuisance pests and some are capable of transmitting disease agents, such as the bacteria that cause scrub typhus in Asia. Fortunately, the mites that people are commonly exposed to in the United States are not associated with any diseases.

Most mites are visible to the unaided eye and usually measure 3 mm (1/8 in.) or less in length. The majority of mites are free-living, but thousands are parasites of animals and plants. Although most of theses mites are external parasites, some inhabit the ear canals, lungs, intestine, and bladder of vertebrates. They can cause considerable discomfort to people because of their biting and bloodsucking behavior. Some mites can also cause serious allergic reactions, such as asthma, in some people.

Biology:

Most species of mites have the modified type of metamorphosis. Usually there is a single larval instar followed by 2 to 3 nymphal instars, then the adult. Mites, like ticks, have 4 pairs of legs as nymphs and adults but only 3 pairs of legs as larvae. The life histories of some common mites are described below. Additional general information on mites is provided in CDC (1975a) and Ebeling (1978).

Scabies Mite. *Sarcoptes scabiei*, commonly known as the scabies, mange, or itch mite, is a parasite of people. The female burrows into the outer layer of the skin (epidermis) where she feeds on tissue fluids and lays eggs. Eggs, laid at a rate of up to 3 per day for a period of 8 weeks, are cemented by the female to the floor of burrows in the epidermis of the skin. A fertilized female will lay about 200 eggs in her lifetime. Eggs hatch in 3 to 4 days and the larvae emerge from the burrows onto the surface of the skin and molt. A rash and intense itching will not occur until several weeks to a month after the initial infestation, when the nymphs burrow into the skin. The majority of mites occur in lesions between the fingers, on the sides of the feet, in folds of skin on the wrists, and in the bend of the knees and elbows. After feeding on tissue fluids, the nymphs molt to become adults. The life cycle, from egg to adult, can be completed in about 2 weeks. Holding the hand of or sleeping with an infected person appears to be a principal method by which the mites are spread. Transmission readily occurs within families, and within institutions such as nursing homes.

Scabies mites are host-specific. The varieties of scabies mites that infest domestic animals can penetrate the skin of humans but they cannot complete their life cycles. Treatment of scabies involves the application of a prescription insecticide-containing lotion to the body. Because of the time lag between the initial mite infestation and the appearance of symptoms, family members, or people coming in close contact with infested persons may require treatment.

Straw Itch Mite. *Peymotes ventricosus* normally parasitizes insects that bore into wood or stored grain. The straw itch mite commonly breeds in stored grain, straw, hay, and other dried grasses. People who come in contact with these mite-infested materials will usually be attacked. The bites of straw itch mites are characteristically found on the trunk of the body.

Control of the straw itch mite is usually accomplished through fumigation. Mite-infested materials should be tarped and fumigated by a person possessing the appropriate state pesticide applicator's license or certification.

Northern Fowl Mite. *Ornithonyssus sylviarum* is a common pest of domestic fowl, pigeons, starlings, house sparrows, and other wild birds that are commonly associated with people. The northern fowl mite overwinters in bird nests and in cracks and crevices of buildings. There may be several generations per year. Populations of this mite can build up at a very rapid rate. In the spring, nestling birds may be parasitized by thousands of mites. After the nestlings have matured and left their nest, large numbers of mites often invade buildings. In poultry houses, the northern fowl mite prefers to stay on birds but mites are often found on eggs, as well. While handling eggs or birds, poultry workers may be bitten.

Abandoned bird nests should be removed. Indoors, the mites are easily removed by vacuuming.

House Dust Mites. *Dermatophagoides pteronyssinus* and *D. farinae* are the most common species of house dust mites in the United States. These tiny mites are often extremely abundant in the dust of mattresses and rugs, and on carpets below beds. Inhalation of the dust containing mite feces and cast skins is a common cause of asthma in young children.

House dust mite populations can be greatly reduced by frequent and thorough vacuuming. Not only does vacuuming pick up mites, it also removes the organic debris on which the mites feed. Areas that should be especially targeted include mattresses, bed frames, rugs, and carpets (especially below beds), and on and under overstuffed furniture. Sanitization remains the key to reducing dust mite populations. Encasing mattresses and pillows in plastic covers and lowering the humidity indoors will provide some control of house dust mites as well.

Paper and Cable Mites. Complaints about biting problems for which a cause can not be found are often attributed to these fictitious pests. Cable and paper mites are often used as reasons to justify treatment of dwellings with pesticides. Application of pesticides inside or outside of dwellings when the pest arthropod has not been identified is a violation of the structural pest control laws in most states. Although mites are extremely small, they are usually detectable with the unaided eye. In the case of scabies and straw itch mites, the rash or bites that these mites leave allows them to be readily identified as the cause of the problem.

Chigger

Trombicula alfreddugesi is the species in the family Trombiculidae that is commonly referred to as the chigger or redbug.

Description:
Adult and Nymph—Both stages are about 1.5 mm (1/16 in.) long, hairy and brilliant red. Both the adult and nymph have 4 pairs of legs.
Egg—The minute, globular egg is light colored at first, changing to a darker color within 24 hours of being deposited.
Larva—The hairy larva is smaller than a pencil dot. It has 6 legs and is orange-yellow or orange-red in color.

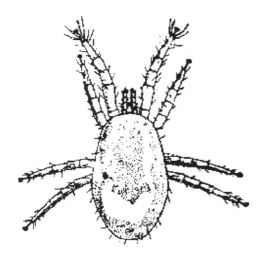

Figure 7.10. Bird mite (EPA 1992).

Biology:

Distribution—Redbugs are worldwide pests. *Trombicula alfreddugesi* is the most common chigger that attacks man in the United States (Azad 1986). Severe dermatitis is also caused by *Trombicula batatas* in the southern United States (Azad 1986).

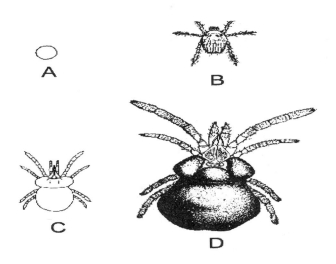

Figure 7.11. Chigger life stages: A - egg; B - larva; C - nymph; D - adult (Knecht 1986).

Habitat—Chiggers occur most frequently in berry patches, orchards, woodlands, and other areas where pine straw and leaf litter provide an environment of high humidity and organic matter.

Feeding Habits—Larval chiggers attach to the skin of people, domestic and wild animals, poultry and other birds. Nymphs and adults feed on insect eggs, small insects, and decaying organic matter (Sasa 1961).

Life History—Chiggers pass through four developmental stages: egg, larva, nymph, and adult. Eggs hatch in about a week after being deposited. The 6-legged larvae are parasitic and attach to vertebrate hosts until replete with tissue fluids. After feeding, larvae drop off their hosts and molt to nymphs which in turn molt to adults. Nymphs and adults have 8 legs and are not parasitic. An entire life cycle requires about 50 days; there may be as many as three generations per year in the southern United States (Wharton and Fuller 1952). Little is known about the biology of nymphs and adults or the overwintering stage.

Injury:

Several species of redbugs attack people causing intense pruritis and reddish welts on the skin. Chiggers prefer to feed on constricted areas of the body, such as around the waist just above the belt line and on the legs above the sock and shoe or boot tops. Chiggers do not burrow into the skin as commonly believed; they firmly attach to the skin with a cement-like salivary substance (Arlian and Vyszenski-Moher 1986). A feeding tube, called a *stylostome*, is created and used to transport digestive salivary secretions into the wound. The liquefied tissues are then imbibed by the mite via the stylostome. The replete larvae dislodge and drop to the ground after feeding for about 2 days on humans (Sasa 1961). However, most chiggers are dislodged by scratching within one hour of their attachment. An allergic reaction to the saliva injected by chiggers causes the swollen, reddish welts that itch intensely. No disease organisms are transmitted by chiggers in the United States.

First Aid:

Treatment is directed toward relieving inflammation and the intense itching that chigger bites cause. In this regard, use of aspirin and calamine lotion are usually effective. Topically applied corticosteroid and antihistamine lotions or creams may help in severe cases. To avoid mite bites, bathe in hot, soapy water and scrub down with a wash cloth immediately after walking through chigger-infested areas or being exposed to other mites.

Ticks (Order Acari)

Ticks are not insects. Of the arthropods that are encountered daily in home environments, they are most closely related to spiders. These ectoparasites are classified into two families: Ixodidae (the hard ticks) and Argasidae (the soft ticks). Most nuisance and public health species are hard ticks.

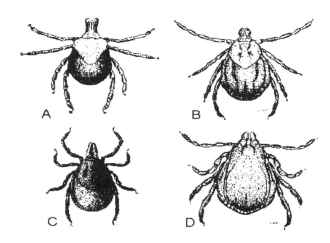

Figure 7.12. Common ticks: A - lonestar tick; B - American dog tick; C - Black-legged tick; D - Brown dog tick (EPA 1992).

Description:

Adult—All hard ticks have 4 pairs of legs as adults. The American dog tick, *Dermacentor variabilis*, has a reddish-brown body with a white, mottled pattern on the dorsal shield and slightly smaller than 6 mm (1/4 in.) in length. The brown dog tick, *Rhipicephalus sanguineus*, is uniformly reddish brown and is a bit smaller than the American dog tick. Engorged females can be 6 to 12 mm (1/4 to 1/2 in.) long. The lone star tick, *Amblyomma americanum*, is light brown and varies from 3 to 6 mm (1/8 to 1/4 in.) in length. The female has a conspicuous, silvery white spot at the end of its dorsal shield. The male, on the other hand, has several crescent-shaped, white markings on the posterior portion of its dorsal shield. The female black-legged tick, *Ixodes scapularis*, has a black dorsal shield and legs, and a brownish-orange abdomen. The male is uniformly black in color.

Egg—Eggs of American dog, brown dog, and lone star ticks are brown, shiny, oval and about 0.4 (1/64 in.) long and wide. The eggs are surrounded by a gelatinous material that prevents them from desiccating.

Larva—About 0.4 to 0.8 mm (1/64 to 1/32 in.) long, tick larvae have only 3 pairs of legs. Those of the American dog tick are pale yellow with reddish-brown margins along their backs. Brown dog tick and lone star tick larvae are light brown.

Nymph—Nymphs are about 0.8 to 3.2 mm (1/32 to 1/8 in.) long and have 4 pairs of legs. Lone star nymphs are reddish brown and the other two species are yellowish brown.

Biology:

Distribution—The American dog tick is especially abundant in eastern North America. The Brown dog tick is widely distributed throughout the world. The lone star tick is found from Texas, north to Oklahoma, eastward to the Atlantic Coast, and from Mexico to central and South America. In the southeast, the American dog tick is most common in Piedmont regions, whereas the lone star tick is most frequently encountered in the Coastal Plain areas of most states. The black-legged tick occurs in the midwest and along the east coast from Florida to Nova Scotia.

Life History—The lone star tick, brown, American dog, and black-legged ticks have similar life cycles. Adults of both sexes feed and mate on the host. A male may mate with several females before dying. After fertilization, a female drops off the host, deposits 3,000 to 5,000 eggs over a 3-week period and then dies. The eggs are deposited in protected areas such as in leaf litter in woodlands or cracks and crevices in dog houses. The brown dog tick is noted for laying its eggs indoors, in cracks and crevices. A gelatinous material surrounds the eggs and prevents them from desiccating. In about a month, the eggs hatch into larvae which are often called "seed" ticks. They climb low vegetation or the walls of pet quarters to await a host and they can live for several months without being found. Those successful in securing hosts feed for several days and then drop off to molt to the nymphal stage. Nymphs must also feed upon a host and may survive a year without a blood meal. Those nymphs that do find a host will feed for about a week, then drop off to molt to the adult stage. As with the nymphs, adult ticks await a host and can often survive for several years without one. Lone star, brown dog, and black-legged tick larvae rarely survive the winter; nymphs and adults are the overwintering stages. American dog ticks overwinter as adults or larvae. Black-legged tick adults are active in winter while adults of the other species are primarily active from spring to fall.

The lone star, brown, American dog, and black-legged ticks are referred to as 3-host ticks because each stage usually feeds on a separate host. American dog tick larvae and nymphs prefer to feed on wild rodents such as white-footed mice and meadow voles. Adult ticks attack people, dogs, livestock and wildlife such as raccoons and foxes. All stages of the brown dog tick prefer to feed on dogs; people and livestock are rarely attacked. All stages of the lone star tick attack people and other animals, such as cattle, sheep, horses, dogs, and deer. The black-legged tick attacks large to medium sized hosts such as people, deer, dogs, and raccoons, but in the southeast, lizards are preferred hosts of the immatures. The entire life cycle may require 3 months to several years depending on the availability of hosts.

Injury:

The American dog tick is responsible for transmitting disease organisms that cause Rocky Mountain spotted fever (RMSF) in man in the southeastern United States. The wood tick, *Dermacentor andersoni*, is the vector of RMSF in Rocky Mountain states. Infected ticks must remain attached for 6 hours before any pathogens are transmitted. Symptoms include headache, muscle and lower back pains, fever, chills, and a red macular rash that starts on the wrists and ankles and spreads to the trunk of the body. These symptoms occur within 2 to 12 days of a tick bite. The disease may be fatal if medication is not administered promptly. The rickettsial pathogens that cause RMSF are retained by ticks transtadially; i.e., as they grow from stage to stage. Disease organisms can even be transmitted to the next generation transovarioley (through the egg) by the female.

Human ehrlichiosis, another tick-borne illness, is caused by rickettsial pathogens *Ehrilichia chaffeensis*, *E. canis*, and *E. phagocytophila*. Ehrlichiosis, also called spotless RMSF, is becoming increasingly prevalent in the United States (CDC 1990, Yevich et al. 1995). Human ehrlichiosis is a febrile illness with symptoms similar to RMSF, except that usually no rash develops. It would appear that the lone star tick is the vector and white-tailed deer are reservoirs of the pathogen (Ewing et al. 1995).

In the United States, tick-borne Lyme disease is epidemic in areas of the northeast, midwest, and Pacific northwest (Lane et al. 1991). The disease is much less prevalent in southern states. The black-legged tick is the vector of the Lyme disease spirochete in the northeastern, midwestern, and southern states. The western black-legged tick, *Ixodes pacificus*, is responsible for spirochete transmission in the Pacific northwest. An infected tick must remain attached for at least 24 hours before any spirochetes are transmitted. Symptoms of Lyme disease usually begin with a characteristic skin rash that may be located anywhere on the body. The rash often starts as a small red pimple or spot at the site of the tick bite. Three to 32 days after the bite, there is a gradual expansion of redness. As the area of redness continues to grow, the center of the rash generally clears. The rash may have raised edges, is endurate, and hyperthermic. It may last up to 3 weeks. Fever, fatigue, headache, muscle and lower back pain may accompany the rash. Weeks to months after the rash, some patients develop arthritis, nerve and heart complications.

The feeding activity of the lone star tick often causes extreme nuisance. Larvae (seed ticks), which are abundant in late summer and early fall, are particularly pestiferous. It is not uncommon for people working out of doors to suffer several hundred bites. The swollen, reddish welts that result from seed tick bites are very similar in appearance to the welts caused by chigger bites. In rare cases, salivary secretions injected by lone star and American dog tick females (attached usually at the base of the skull) causes a condition known as "tick paralysis." Recovery is complete if the attached tick is removed before paralysis occurs. A severe infestation of brown dog ticks on pets may cause anemia.

First Aid:

To remove an attached tick, grasp it close to the point of attachment to the skin with forceps or fingers protected with tissue (Needham 1985). Pull directly away from the point of attachment with gradually increasing tension. If tick mouthparts break off in the skin, they should be removed with a sterilized needle. Broken mouth parts will not transmit disease organisms, but a secondary infection may result if the wound is not disinfected. Application of caustic, suffocating, or other materials to the attached tick will not induce a tick to detach (Needham 1985). Some of these materials may irritate the skin or kill the tick, making it more difficult to remove. After the tick has been removed, clean the bite with an antiseptic. The date of the tick bite should be marked on a calendar, as a reminder to seek medical assistance if any symptoms of tick-borne diseases occur within their respective incubation periods given above. The tick(s) should be saved for identification by preserving it in rubbing alcohol or by taping it to white typing paper or an index card with clear cellophane tape. It may be possible for ticks to be tested to determine if they are infected with pathogens. Inquiries concerning the availability of such diagnostic service should be directed to local or state health departments.

First aid treatment is directed toward relieving inflammation and the intense itching that tick bites cause. In this regard, use of aspirin and calamine lotion are usually effective. A corticosteroid and an antihistamine may be applied topically in severe cases.

Medical Treatment:

Rocky Mountain Spotted Fever, human erhlichiosis, and Lyme disease can be cured with antibiotics. Diagnosis of illness and antibiotic treatment should be made on the basis of clinical symptoms presented and the likelihood of contact with ticks.

Biting Flies (order Diptera)

The order Diptera contains some blood sucking flies that are major nuisance and public and veterinary health pests. The importance of blood-sucking insects, evolution of the blood-sucking habit, and other behavioral and physiological processes associated with bloodfeeding are reviewed by Lehane (1991).

Black Flies

Black flies are biting flies in the family Simuliidae.

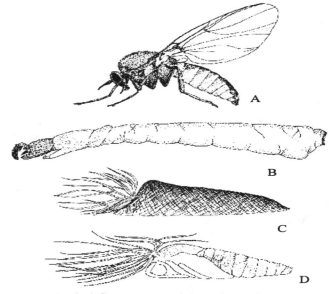

Figure 7.13. Black fly stages: A - adult; B - larva; C - cocoon; D - pupa (Knecht 1986).

Description:

Adult—Black flies are small (1 to 3 mm, 1/25 to 1/8 inches), dark flies with short legs. They have a distinct humpbacked shape (Figure 7.13), and are sometimes called buffalo gnats. Black flies have broad wings and their antennae are about as long as the head. Male flies have larger eyes, but generally are similar in appearance to females.

Egg—Black flies' eggs are oval but irregularly shaped. The eggs are white when first deposited but gradually darken as they mature.

Larva—Black fly larvae are slender, ranging from 5 to 15 mm (1/5 to 3/5 in.) long. Larvae are whitish-brown to blackish in color, and have a distinct head and anterior proleg. The head has slender antennae and two brush-like structures called head fans. They have a sucker-like structure on the posterior end.

Pupa—Black fly pupae are 2 to 3 mm (1/12 to 1/6 in.) long, and have a respiratory organ on the thorax which protrudes from the open end of the cocoon. The wing pads, legs, and other features of the adult fly can be identified, although these structures are closely appressed to the body of the pupae.

Biology:

Distribution—Black flies occur worldwide. The predominate species in most areas are in the genera *Simulium*, *Prosimulium*, and *Cnephia*.

Habitat—The immature stages of black flies occur in rivers, streams, and moving water that flows over spillways, or contains aquatic vegetation, rocks, or other objects just below the surface. Even a trickle of water can produce large numbers of flies.

Hosts—Female black flies feed on people, domestic and wild mammals, poultry and other birds. Males feed exclusively on plant nectar.

Life History—After taking a blood meal, females develop eggs, which are deposited in masses on objects near or under water. Some species deposit their eggs in flowing water. The eggs come to rest downstream on the bottom. First instar larvae remain near the eggs or spin a silken thread and drift on the current until the thread becomes entangled on a suitable substrate. Larvae anchor themselves by spinning a silken pad and clinging to it with a ring of hooks on the posterior end. They feed on organic material that is filtered from the flowing water by head fans. Densities of larvae can be extremely high. The sixth instar larvae pupate in a cone-like cocoon that faces downstream. New flies emerge through a slit in the pupal skin and are carried to the surface on a bubble. Males and females feed on plant nectars. Mating swarms of males can often be seen over bushes or other prominent landmarks. Depending upon the species, there may be 1 to 7 generations per year. Adults are active during the day time in the spring, summer and fall, and often migrate long distances (10 to 20 km) from breeding sites.

Injury:

The biting activity of black flies can be aggravating to people. Some people exhibit severe reactions, developing large erythematous welts after being bitten. Some species are a nuisance solely because they fly around the face and frequently land but do not bite. Black flies are vectors of *Leucocytozoon simondi*, which causes avian malaria in turkeys. In Central America and West Africa, black flies transmit a nematode, *Onchocerca volvulus*, that causes River Blindness in people.

First Aid:

Treatment is directed toward relieving inflammation and the intense pruritis that black fly bites cause. Aspirin and calamine lotion are usually effective remedies. Topical applications of corticosteroids may promote healing.

Stable Fly

Stomoxys calcitrans is a biting fly in the family Muscidae. It is commonly referred to as the biting house fly or dog fly.

Description:

Adult—The stable fly resembles the house fly, but is a bit larger. Adults are 6 to 8 mm (1/4 to 1/3 in.) long with four distinct dark longitudinal stripes on the back and several dark spots on the abdomen. A beak-like proboscis protrudes from the head.

Egg—The egg is about 1 mm (1/25 in.) long and curved on one side and straight and grooved on the opposite side. It is creamy white in color.

Larva—The first instar is about 1.25 mm (1/20 in.) long, and when fully mature, the third instar will measure 12 mm (1/2 in.) in length. Mature larvae are yellowish to creamy white in color.

Pupa—The skin of the mature larva hardens to form the pupal case (puparium), which is reddish brown in color. The puparium is 4 to 6 mm (1/6 to 1/4 in.) in length.

Biology:

Distribution—Stable flies occur throughout the New World.

Habitat—Immature stages are found in moist organic matter, such as straw, litter, manure mixed with straw, composting or rotting vegetable materials, waste silage or marine grasses that have washed on shore.

Hosts—The biting activity of stable flies is aggravating to people and domestic animals.

Life History—Females deposit 35 to 80 eggs at a time deep in organic matter. Approximately 10 egg layings will occur over a 4 to 6 week period. The eggs hatch in 1 to 3 days, and the first instars immediately begin to feed on the surrounding organic material. Larvae complete development in 14 to 26 days. Third instars migrate to dry areas where the puparium is formed from the skin of the fourth instar. Adults emerge in 5 to 10 days. Development time from egg to adult averages 20 to 25 days. There are several generations per year. In temperate regions, stable flies usually overwinter as larvae or pupae. Adults often migrate long distances (>100 km) in search of hosts. Both sexes blood feed. The bites of stable flies are painful, and the host usually disturbs the flies so that three to four feedings have to occur before a complete bloodmeal is obtained. At least three complete meals must be obtained before females begin to lay eggs.

Injury:

Stable flies are vicious and persistent biters. They commonly bite people around the ankles. Red, swollen welts result from the bite of stable flies. Their feeding activity can cause weight loss in cattle, and transmit pathogens that cause animal diseases such as hog cholera.

First Aid:

Treatment is directed toward relieving the pain and itching caused by stable fly bites. Aspirin and calamine lotion are usually effective remedies. Topical applications of corticosteroids may promote healing.

Biting Gnats

Biting gnats are biting flies in the family Ceratopogonide. These flies are also commonly referred to as no-see-ums, sandflies, and punkies.

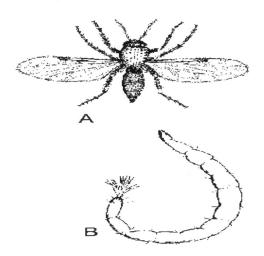

Figure 7.14. Biting gnat: A - adult; B - larva (Knecht 1986).

Description:

Adult—Some species are tiny, measuring 0.5 mm (1/50 in.) in length. Adults have grayish or yellowish bodies and clear or spotted wings.
Egg—The eggs are 0.25 mm (1/100 in.) in length and dark brown in color.
Larva—The larvae are worm-like, 3 to 4 mm (1/8 to 1/6 in.) long, and white with a brown head. Brush-like breathing structures protrude from the posterior ends of the larvae.
Pupa—This stage is dark-brown in color and may reach 4 mm (1/6 in.) in length.

Biology:

Distribution—Various species of sandflies occur worldwide in temperate and tropical regions.
Habitats—Ceratopogonid flies breed in coastal salt marshes, marine mud flats, the margins of freshwater swamps, tree holes and other semi-aquatic habitats that contain high amounts of organic matter.
Hosts—Only female sandflies bite, both males and females feed on plant nectars. People and large domestic and wild mammals are fed upon by sand flies.
Life History—Adults emerge in the spring after larvae or pupae have overwintered. Eggs hatch about one week after they have been deposited on sand or mud. There are 4 larval instars which take from 6 to 12 months to complete. Adults live about one month; females take several bloodmeals and lay several batches of eggs.

Injury:

Some ceratopogonids are economic and public health pests. The biting activity of *Culicoides* and *Leptoconops* spp. along the Atlantic and Gulf Coasts, and areas of the Caribbean, are a deterrent to tourism (Linley and Davis 1971). Ceratopogonids are important vectors of a variety of animal pathogens (Foote and Pratt 1954; Linley 1985).

First Aid:

As with other biting flies, treatment is directed toward relieving pain and itching caused by sand fly bites. Aspirin and calamine lotion are usually effective remedies. Topical applications of corticosteroids may promote healing.

Horse Flies and Deer Flies

Horse flies and deer flies, members of the family Tabanidae, are collectively known as tabanids. These blood-feeding pests are notoriously persistent in their pursuit of a blood meal.

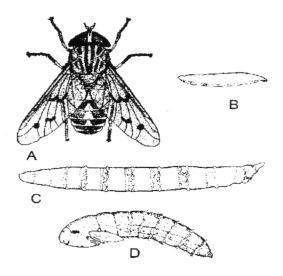

Figure 7.15. Adult horse fly: A - adult; B - egg; C - larva; D - pupa (Knecht 1986).

Description:

Adult—The majority of horse flies are in the genus *Tabanus*. These are moderate to large flies, 14 to 30 mm long (1 1/2 to 1 3/16 in.), with clear wings and grayish brown thoraxes. Most deer flies are in the genus *Chrysops*. These are small to medium sized flies, 10 to 13 mm (3/8 to 1/2 in.). Their wings are tinted smoky gray-brown or have dark patterns, and their bodies are greenish yellow in color with dark stripes.

Egg—The narrow eggs are cylindrical and 1 to 3 mm long. When laid, they are white but turn brown to black in a few hours. An egg mass consists of 200 to 500 eggs cemented together.

Larva—Mature larvae are approximately 10 to 1 9 mm (3/8 to 3/4 in.) long and up to 3 mm (1/8 in.) wide. Horse fly larvae are slightly larger than deer fly larvae. Horse fly and deer fly larvae are spindle-shaped and white, brownish or greenish in color.

Pupa—Pupae measure about 15 mm in length. Pupae turn dark brown within a few hours of their development. Pupae are rounded anteriorly and tapered posteriorly with a double row of spines on segments 2 through 7.

Biology:

Distribution—Deer and horse flies are found worldwide in temperate, subtropical, and tropical climates. There are over 4,100 species and 137 genera worldwide (Foil 1989). There are 26 genera and 22 species in America, north of Mexico (Goodwin et al. 1985).

Habitat—Biting flies are often encountered along hiking trails, roads, and lanes in or at the edge of woodlands, near floodplains and swamps, and on saltmarshes.

Life History—Egg masses are glued to the leaves and stems of plants growing at the edge of breeding sites, such as floodplains, swamps, marshes, ponds, or lakes (Axtell 1976). In about one week, eggs hatch and the larvae crawl down plant stems into the wet soil. Larvae are predaceous on nematodes and other semi-aquatic invertebrates. Larval development times vary among species, and may range from one month to a year. Mature larvae migrate to dry areas at the edge of breeding sites where they pupate for 1 to 4 weeks. Most species have one generation per year; however, some species may have a partial second generation produced within a single year. Adult emergence usually occurs from May through August.

Female horse and deer flies are day-active and persistent blood-feeding pests of cattle, horses, deer, and other warm-blooded animals, including people. Males do not feed on blood; both males and females feed on plant nectars.

Injury:

Female tabanids use their slashing mouthparts to inflict painful bites. The biting activity of deer flies and horse flies causes significant nuisance in coastal and inland recreational areas. The reaction of the host to the painful bites of tabanids usually interrupts their feeding activity. The interrupted feeding and consequent movement of tabanids from host-to-host, facilitates the mechanical transmission of pathogens via contaminated mouthparts (Foil 1989). These flies are actual or suspected vectors of important livestock diseases such as hog cholera virus, equine infectious anemia, anaplasmosis bacteria of cattle, tularemia, and many other disease agents of humans and animals (Foil 1989). Tabanid feeding can result in blood loss and consequent weight loss in cattle.

First Aid:

Treatment is directed toward relieving inflammation and the intense itching that tabanid bites cause. In this regard, use of aspirin and calamine lotion are usually effective. A corticosteroid and an antihistamine may be applied topically in severe cases.

Mosquitoes

These insects in the family Culicidae are generally regarded to be one of the most important pests of people and domestic animals. Mosquitoes are vectors of important human and animal diseases. The biting activity of these noxious arthropods is often a major deterrent to outdoor recreational pursuits.

Description:

Adult—Mosquitoes are long-legged insects with one pair of wings, and are 3 to 4 mm (1/8 to 3/16 in.) in length. The wings, legs, back (thorax), and abdomen are variously colored. Mosquitoes can be distinguished from all other 2-winged insects by the long proboscis ("beak"), and the scales on their wings. Male mosquitoes have bushy antennae.

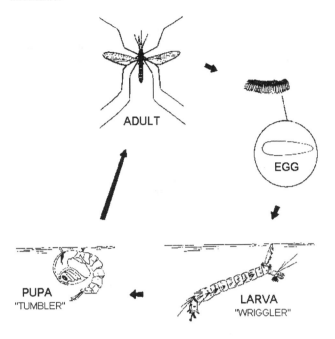

Figure 7.16. Culex mosquito life cycle (U.S. Public Health Service 1982 repr.)

Egg—The small, white, cylindrical eggs taper at one or both ends and darken within a few hours of when they are deposited. Eggs of *Anopheles* and *Aedes* mosquitoes are laid singly; those of *Culex* mosquitoes are deposited in raft-like masses.

Larva—Young larvae are 1 to 2 mm long which makes them difficult to see. Mature larvae are 6 to 8 mm (1/4 to 3/8 in.) long. Larvae vary in coloration from whitish to dark brown or black. The head is distinct and usually dark colored for all 4 instars. Upon close inspection, most instars are covered with bristles or hairs. Larvae often move through the water in characteristic wriggling motion; hence, the name "wriggler".

Pupa—Smaller than mature larvae, pupae are comma-shaped with a slender curved abdomen. They move through the water in a tumbling motion; hence, they are commonly called "tumblers".

Biology:

Distribution—Mosquitoes are worldwide in occurrence. In the United States, 167 species and 13 genera of mosquitoes are currently recognized (Darsie and Ward 1981).

Life History—Mosquitoes are active during twilight hours and at night; however, around the home, the mosquitoes that breed in discarded containers are active during the day.

All mosquitoes have one common requirement—they need water to complete their life cycle. Generally, there are two types of mosquito lifecyles. Floodwater mosquitoes, such as *Aedes* spp., lay individual eggs on the sides of tree holes or discarded containers, or in depressions in the ground that will hold water. The eggs can lay dormant for several years. Some eggs will hatch when they are flooded by rainfall. Several flooding and drying cycles are usually required for all of the eggs to hatch that are laid by a particular female mosquito. Permanent water mosquitoes lay eggs directly on the surface of water. The eggs are attached to one another to form a raft (*Culex* spp.) or the individual eggs (*Anopheles* spp.) float on the water. These eggs hatch in 24 to 48 hours and, the larvae feed on microorganisms and organic material in the water; however, some mosquitoes prey on the larvae of other mosquito species and are considered to be beneficial. In about 7 to 10 days after eggs hatch, larvae change to the pupal or "tumbler" stage in preparation for adult life. Female mosquitoes begin to seek an animal to feed on several days after emerging from water. Male mosquitoes mate with females one to 2 days after the females emerge. Males do not bite; they feed on plant juices.

Since mosquitoes need water to complete their life cycle, the source of a mosquito problem can be just about anywhere that water can collect. Bird baths, boats, canoes, discarded tires, plant pots and other such objects collect rainwater and become breeding sites for mosquitoes around the home. Most urban mosquito populations can be controlled by eliminating discarded containers from backyards, replacing the water in bird baths regularly, and storing boats, canoes, and other objects so that they do not collect rainwater.

Farm ponds and lakes usually do not breed mosquitoes if they contain fish and are free of weeds, algae, or floating debris that serve as harborage for mosquito larvae. Municipal and farm animal waste lagoons may also become breeding sites. Permanent natural bodies of water such as swamps usually contain a large array of predatory insects and fish that keep mosquitoes from reaching nuisance levels. Human activities may create mosquito breeding sites or increase the production of mosquitoes in natural bodies of water. Road building and maintenance often impedes the drainage of runoff from rainfall, creating a mosquito breeding site. Clogged drainage ditches along roads can become productive mosquito breeding sites. Logging and construction activities often leave tire ruts in the soil. These depressions are ideal breeding sites for "floodwater" mosquito species.

Injury:

The biting activity of mosquitoes can be extremely annoying. Saliva secreted by mosquitoes during feeding causes an inflammatory reaction, resulting in the formation of the characteristic welt. Mosquito bites can be very pruritic and if the skin is broken through scratching, a secondary infection and scarring may result. Mosquitoes will bite any area of exposed skin, although some species do exhibit distinct preferences for certain body regions (CDC 1982).

Mosquito feeding often results in the transmission of microorganisms that cause human and animal diseases (CDC 1982). Viruses causing illness such as dengue fever, eastern equine encephalitis, St. Louis encephalitis, and La Crosse encephalitis are mosquito-borne. Mosquitoes transmit filarial nematodes causing human filariasis and protozoans that cause malaria.

First Aid:

Treatment is directed toward relieving inflammation and the intense itching that mosquito bites cause. In this regard, use of aspirin and calamine lotion are usually effective. A corticosteroid and an antihistamine may be applied topically in severe cases.

Fleas (Order Siphonaptera)

Cat (*Ctenocephalides felis*) and dog (*Ctenocephalides canis*) fleas, ectoparasitic insects in the family Pulicidae (order Siphonaptera), are commonly encountered in and around the home. Their blood feeding activity can be irritating to people and pets.

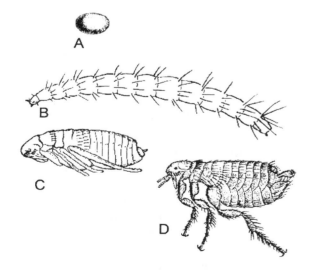

Figure 7.17. Flea life stages: A- egg; B - larva; C - pupa; D - adult (Knecht 1986).

Description:

Adult—The adult flea is brown, wingless, and compressed laterally. This insect is 0.75 to 1.5 mm (1/32 to 1/16 in.) in length and dark, reddish-brown. Its legs are well-developed for jumping.

Egg—A flea egg is oval and about 1/64 in. in diameter.

Larva—The legless, white thread-like larva varies in length from about 6 to 12 mm (1/4 to 1/2 in.). They have a pale brown head with no eyes.

Pupa—The pupa is shaped like the adult and develops inside a white, oval silken cocoon spun by the larva. The cocoon is about 4 mm (3/16 in.) long and is often covered with sand, lint, or dust.

Biology:

Distribution—Fleas, as a group, have a world wide distribution. The cat and dog fleas are the most common species in contact with people.

Feeding Habits—The laterally compressed shape of the adult allows easy movement through hairs on the host animal's body. Both sexes and all species of fleas feed on warm-blooded animals; however, each flea species usually feeds on a limited array of host animals.

Life History—Females deposit eggs on a host animal, such as a cat or dog, a few at a time. The female cat flea may lay up to 400 in her lifetime. Eggs usually fall off of the host and hatch in 2 to 14 days. Larvae generally feed on organic matter and dried blood excreted by adults. Indoors, fleas develop in carpeting and overstuffed furniture; outdoors, flea larvae usually develop where host animals rest, such as in pet quarters. After 1 to 5 weeks, larvae spin cocoons and pupate. Under favorable conditions, a new generation of adults emerges 1 to 3 weeks later. However, when conditions are not favorable, fully developed adults may remain in the cocoon for up to one year until they are disturbed by specific environmental cues. Cat and dog fleas, for example, jump from their cocoons when carbon dioxide and vibration signal that a host is nearby. Thus, people returning home from vacation may encounter a large number of hungry fleas. Indoors, in association with pets, an entire life cycle may be completed in as few as 2 to 3 weeks; outdoors, several months to 2 years may be required depending on the availability of host animals and environmental conditions (Ebeling 1978). Temperatures >95 °F and humidity <50 percent can be lethal to fleas.

Adult fleas cannot breed or survive for extended periods of time without feeding on blood. The dog flea, for example, may live about 60 days without feeding. If fed, adults may survive for 230 days without another meal. Temperatures of 65 to 70 °F and a humidity of 70 percent are optimal for flea development.

Injury:

Cat and dog fleas will readily bite people usually on the ankles and other areas of the legs. Pets and people may develop sensitivity to flea bites. Some people that are severely allergic suffer an intense pruritis. The swollen, red welts are caused by a local skin reaction to the salivary secretions that the flea releases while biting. Scratching flea bites may break the skin and lead to secondary infections and scarring.

Fleas may transmit plague and typhus pathogens, but these diseases occur infrequently in the United States.

First Aid:

Treatment is directed toward relieving inflammation and the intense pruritis that flea bites cause. In this regard, aspirin and calamine lotion are usually effective. Topical applications of corticosteroids may promote healing.

Human Lice (order Anoplura)

There are three species of human lice (family Pediculidae), whose common names indicate their preferred feeding site. Head lice (*Pediculus humanus capitis*), body lice (*Pediculus humanus humanus*) and pubic lice (*Pthirus pubis*) feed exclusively on humans.

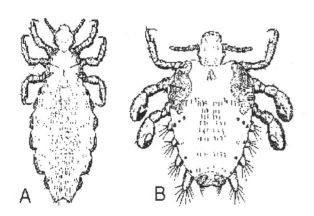

Figure 7.18. Human lice: A - body louse; B - pubic louse (Knecht 1986).

Description:

Adult—Head and body lice are 2 to 4 mm (1/12 to 1/6 in.) long and 1 to 2 mm (1/25 to 1/12 in.) wide, and grayish in color. The wingless adults have piercing mouthparts and stout legs, each with a single hooked claw. The body louse is slightly larger than the head louse. Both lice are races of the same species. The pubic louse is also wingless and has piercing mouthparts, and heavy legs that end in a claw. A public louse is brownish-gray in color. The abdomen is broader than long and has small protuberances along the sides. Females are 1.5 to 2 mm (1/32 to 1/12 in.) long, while the males are slightly smaller.

Egg—Each oval-elongate egg is white in color, and slightly less than 1 mm (1/25 in.) long. Each egg is glued to the base of a hair shaft with a cement-like substance. Head louse eggs are commonly referred to as "nits".

Nymph—Each nymph develops through 3 instars, which resemble the adult but is smaller and more slender.

Biology:

Distribution—Head, body, and pubic lice occur worldwide. The body louse tends to be less common in tropical regions, probably because of the lesser amounts of clothing worn due to the warmer climate.

Life History—Information on the biology of human lice and the epidemiology of louse infestations is presented in CDC (1975b) and Orkin et al. (1977). Body lice and head lice look identical, but their location on people is a reasonably accurate diagnostic clue. Head lice are found almost exclusively on the scalp, whereas body lice may be found either on the body or clothing. Unlike head lice, body lice attach their eggs to the fibers of undergarments, particularly along inside seams and other areas of close body contact. A female body louse usually deposits to 10 eggs per day, and a total of 270 to 300 eggs over her lifetime. The eggs are incubated by the person's body heat and hatch in about one week. Temperatures above 100°F or below 75°F reduce or completely stop egg hatch. Development time (egg to adult) is about 3 to 5 weeks.

Body lice are most common during the winter months, when people tend to wear layers of clothing, creating a warm, moist environment that is ideal for louse development. An important point to note is that body lice spend most of their life on clothing and crawl onto the host to feed for short periods. Although as many as 1,000 body lice have been removed from heavily infested clothing, ten lice per person is probably more typical. Body louse problems are more likely to occur in situations where the same clothing is worn constantly for several days or weeks. Since body lice can survive away from a person for 48 hours, they can survive in clothing that is removed nightly but worn again the following day.

Head lice live on the skin among the hairs on a person's head. They glue eggs ("nits") to the base of the hair shaft, frequently behind the ears or on the nape of the neck. Nits hatch in about 5 to 10 days and the light-colored nymphs begin feeding immediately. Development takes about 18 to 20 days; adults can live for about one month. Head lice can infest clothing and other items that come in contact with the head (e.g., hats, shirt collars, brushes, combs, etc.), but they cannot survive off of a person for more than 36 to 48 hours.

In order to survive, these lice require temperature and humidity conditions similar to what you would find associated with human bodies.

Pubic or crab lice are usually found on the hairs in the pubic areas; however, they can also be found among the coarser hairs of the chest, armpits, eyebrows, eyelashes, mustaches and beards. Unlike body lice, crab lice are dark gray-to-brown in color. The name "crab lice" comes from their flattened, oval crab-like shape. Their second and third pairs of legs have thumb-like projections that aid in grasping the host's hair. Eggs of pubic lice are dark brown and smaller than those of head and body lice. The female lays about 30 eggs during her 3 to 4 week lifespan, usually on the hairs of the pubic region. Crab lice are more sedentary than body and head lice, and usually pass their entire life cycle in the same area where the eggs were deposited.

Injury:

Head lice have little real medical impact. Their feeding activity irritates the scalp, causing intense itching. They are not known to transmit any disease organisms, but a secondary infection may result if the skin is broken by repeatedly scratching the area. The most notable impact of head lice is the personal embarrassment experienced with being identified publicly as having lice (e.g., in a classroom). Children, particularly those of elementary school age, are most likely to get head lice because of their close contact and social interactions with each other (e.g., sharing hats, combs, and brushes) creates opportunities for the lice to be spread among them. Children who become infested in school will carry lice home and may infest family members who unknowingly become a source for recurring louse problems in the home.

Although body lice can transmit certain disease organisms, this problem is generally confined to underdeveloped countries where poor sanitation and overcrowding are major contributing factors. Feeding activity by the lice causes significant skin irritation, swelling and the formation of red welt-like marks. Severe infestations can lead to allergic reactions and skin disorders, such as impetigo and eczema.

Crab louse bites produce discrete, round slate-gray to bluish colored swellings on the skin. Proteins in the louse's saliva may cause an allergic reaction and intense itching. As with head lice, secondary bacterial infections may result from constant scratching and breaking the skin. Lice infesting the eyelids can cause severe inflammation and swelling around the eyes. Crab lice are not known to transmit disease organisms.

Solving Head Lice Problems. Solving head lice problems is not difficult if everyone cooperates. Parents should notify the school principal or nurse immediately if their child is identified as having head lice. While they may be embarrassed by the problem, parents should not hesitate to contact the school. Two very important points should be remembered. First, just because a child has head lice does not necessarily mean that he/she is the source of the problem. Second, and most importantly, if school officials do not take any action because they are unaware of the problem, lice will likely spread to other children in the class. If the problem continues unchecked, eventually even those children who were treated previously will become reinfested and they must be subjected to treatment again.

Parents, school administrators, teachers, and other school staff should handle these situations discretely, being sensitive to the children's situation, which is difficult enough for them without the problem becoming a matter for public discussion. Children exhibiting symptoms of head lice (excessive head scratching and/or abrasions on the scalp) need to be checked for the presence of nits or lice by someone who can positively identify the eggs or the insects. The whitish-to-cream colored eggs are small (about 1 mm in length) and oval with a distinct cap. To the unaided eye, nits can easily be confused with dandruff, globules of hair oil or even dried flakes of hairspray. Upon closer examination with a magnifying glass, the eggs are more easily recognized. On average, the actual number of head lice found on a person is low: about 5 to 10. Therefore, careful inspection of the head, particularly along the back of the neck and around the ears is important. Another useful way to check for lice is to have the person hold their head over a light-colored towel or piece of cardboard, then brush their hair and scalp vigorously towards the towel. Dislodged lice should be visible on the towel. Once a louse infestation is confirmed, you should take the following steps:

Step 1. Treat all infested persons. Check all family members, classmates, and playmates. Anyone who is infested should receive proper treatment. There are several pediculicidal (lice-killing) shampoos/rinses that can be purchased at most drug stores. Another product, which contains the insecticide lindane, is available by prescription only. These products work well against nymph and adult lice, but not all of them will kill eggs. One or two additional treatments at 7 to 10 day intervals are sometimes needed in order to eliminate the infestation entirely. Read and follow the instructions on the product labels carefully. Particularly when children are involved, we strongly recommend that you seek advice from your doctor or child's pediatrician about the best way to proceed in solving the problem quickly and safely.

Step 2. Remove nits from hair. Mechanical removal of louse eggs helps reduce the number of lice that might hatch on the scalp. Since children who are declared "nit-free" by a doctor or school nurse can return to class sooner, removing nits has a positive impact on their morale and lessens disruption to their school and social activities. Special combs are provided with the pediculicides and the instructions for their use must be followed explicitly. Be persistent and thorough. Because the eggs are literally cemented to the hair shaft, they are not easily dislodged. Contrary to popular belief, using vinegar in a shampoo or as a rinse is not effective because it does not dissolve the glue that binds the nits to the hair shaft. There are also special non-insecticidal rinses that dissolve the glue binding the eggs to the hair. Nits are more easily removed after shampooing/rinsing, when the hair is still damp. Several combing sessions may be needed to remove all of the nits.

Step 3. Delouse personal items. Although head lice cannot survive for long periods off of people, sanitation helps end the problem more quickly. Items such as brushes and combs should be washed thoroughly in hot water (at least 130°F) for 5 to 10 minutes. Washable clothing, hats, head bands, bed linen, and other personal items should be washed in hot soapy water, then dried in a clothes drier for at least 20 to 30 minutes. Non-washable clothing can be dry-cleaned, but the expense is hard to justify when a simpler solution would be to isolate theses items for a few days. Vacuuming mattresses, upholstery, and carpeting will pick up stray lice. Although hair with attached lice eggs may fall out and become attached to stuffed animals or similar toys, such items do not serve as reservoirs for the lice. If there is some concern about these items carrying lice, the simplest solution is to vacuum them or stick them into a plastic garbage bag and then into a freezer overnight. Spraying furniture, carpeting, and bedding with an insecticide serves no real purpose other than providing some margin of psychological comfort. Although such sprays kill an occasional stray louse, family members who are already being treated with insecticidal shampoos would be exposed unnecessarily to additional pesticides. We do not recommend spraying insecticides except under extraordinary circumstances.

The extent of delousing activities in a school depends largely on the age of the students and the layout of the classroom. As in the home, vacuuming carpeting and/or sleeping mats can help. Mats with vinyl or other non-fabric coverings can be cleaned with hot, soapy water. Clothing or personal items that students leave in closets, storage areas, or desks should be removed and cleaned. Application of insecticides in the classroom is *not* necessary.

Treatment for Body Lice. Treatment for body lice is virtually identical to the procedures for head lice. First, the infested person must be treated with pediculicides (lotions or shampoos). As in the case of head lice, body lice move rapidly from person to person upon brief contact or when clothing is shared. Other family members (or classmates) with whom the person comes in contact should be checked and treated, if necessary. Clothing, bed linen, and other personal items that have been used recently must be deloused as outlined in Step #3 for head lice. Vacuuming mattresses, carpets, and upholstery will help remove stray lice.

Treatment of Pubic Louse Problems. Crab lice are spread primarily through sexual contact. It is possible, but extremely rare, that they could be acquired through contact with infested toilet seats, clothing, or bedding. As with other louse problems, successful treatment is based on a combination of sanitation and pediculicides. Family members who share a bed with the infested person must be examined and treated if necessary. Undergarments and bed linens should be washed in hot water for at least 20 minutes, then dried on a high setting. Louse control products are available either directly at a drugstore or by prescription; however, we strongly recommend that infested persons first seek advice from their personal physician. Because crab lice can only survive off of people for about 24 hours, insecticidal treatments in the home, workplace, school, or other areas are neither necessary nor recommended.

Bed Bugs (order Hemiptera)

Bed bugs (*Cimex lectularis*) and bat bugs (*Cimex adjunctus*) are insects in the family Cimicidae. These cimicids are pests of man and domestic animals, in addition to bats and wild birds.

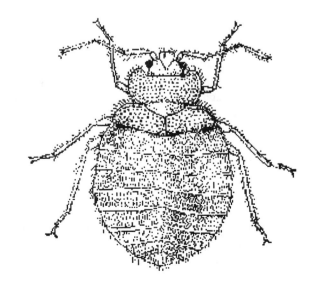

Figure 7.19. Bed bug (Knecht 1986; EPA 1992).

Description:

Adult—Bed bugs and bat bugs are reddish-brown, oval, flattened insects from 3.5 to 7.5 cm (1.4 to 3.8 in.) long and 1.5 to 3 mm (1/16 to 1/8 in.) 1.5 to 3 mm wide before feeding. Engorged adults are swollen and dull red in color. Adults are wingless. The eggs are deeply pigmented and a collar-like protrusion extends around the head. Bat bugs, unlike bed bugs have long body hairs.

Egg—The white oval eggs are each about 0.8 mm (1/32 in.) long.

Nymph—The first nymphal growth stages (instars) resemble the adult although they are smaller in size. A newly hatched nymph is almost colorless.

Biology:

Distribution—Bed bugs occur throughout the world. Bat bugs occur primarily in the eastern United States (Usinger 1966).

Hosts—Though bed bugs prefer human hosts, these parasitic insects will feed on many warm-blooded animals, including poultry, rats, dogs, rabbits, and cats. The bat bug prefers to feed on bats but it will feed on people.

Life History—Bed bugs feed and breed all year long when conditions permit. In homes, bed bugs are usually found in the bedroom where they hide during the day in mattresses, in the bed frame, and in cracks and crevices along wall baseboards. Both males and females come out at night to feed. Under favorable conditions, each female lays 200 to 500 eggs in cracks and crevices. When the insects feed regularly, eggs are laid in batches of 10 to 50 at 3 to 15 day intervals. Under laboratory conditions, the development time from egg to adult ranges from 37 to 128 days.

Poultry houses are often heavily infested with bed bugs. Side curtains are a favored hiding place. Heavy infestations can often be detected by the presence of black excrement on the poultry eggs.

Bed bugs can withstand long periods of starvation. For example, an adult male can live for 140 days under laboratory conditions without taking a blood meal. Immature bed and bat bugs must feed before proceeding to the next growth stage.

The life history of the bat bug is poorly known. When bats roost in homes, bat bugs may migrate into living areas in search of a host.

Injury:

The saliva of bat and bed bugs is highly allergenic to some people. Swollen, red welts that develop at the site of the bites are extremely pruritic. The welts may last for a week or more. Bed and bat bugs are not known to transmit any disease organisms through their bites.

First Aid:

Treatment is oriented toward relieving inflammation and the intense pruritis caused by the bites. In this regard, use of aspirin and calamine lotion are usually effective. Antihistamines and corticosteroid lotions may be used if the pruritis is severe.

Conenose Bugs, Assassin Bugs, and the Wheel Bug (order Hemiptera)

The family Reduviidae contains some aggressive predators. Some species, such as the conenose bugs, are exclusive blood suckers while other species prey on insects and occasionally bite people.

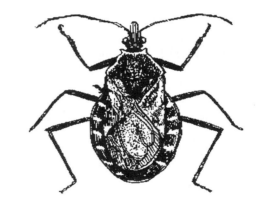

Figure 7.20. Conenose bug (EPA 1992).

Description:

Adult—Reduviids are large (1.5 to 2.5 cm; 3/4 to 1 in.), elongate insects with freely articulating heads, prominent eyes and 3-segmented beaks. The beaks attach at the front of the head and are folded back underneath. Colors vary between species, some being reddish brown to dark brown or black with red markings about the abdomen. The wheel bug, *Arilius critatus*, has a prominent rounded crest just behind the head.

Nymph—Most nymphs are similar in appearance to the adults but may differ in color. *Reduvius* spp. and *A. cristatus* have 5 nymphal instars.

Egg—*Reduvius* spp. lay individual eggs that are oval, smooth, slightly translucent and yellowish-gray in color (Ryckman and Ryckman 1967). In contrast, *Arilus critatus* lays masses of brown, bottle-shaped eggs. The egg masses, which contain about 125 eggs, are glued to twigs, the bark of trees, and on fence rails (Ebeling 1978).

Biology:

Distribution—Reduviids in the genus *Triatoma* are mostly confined to the western hemisphere. *Arilus critatus* occurs most frequently in the eastern and southern United States.

Habitat—The habitat of Reduviids varies between species. Conenose bugs (*Triatoma* spp.) are most commonly found in nests of its principal host the wood rat (*Neotoma*). *Reduvius* spp. and *Arilus critatus* (the wheel bug) occur out-of-doors on and around buildings and trees.

Life History—Eggs hatch within 3 to 5 weeks after they are laid. Each of the 5 instars of blood sucking Reduviids (such as *Triatoma* spp. and *Reduvius* spp.) must feed on a host. Reduviids require a long time to develop from egg to adult. For example, for *R. senilis*, Ryckman and Ryckman (1967) found the development period to average 653 days. Wheel and assassin (*Reduvius* spp.) bugs are predaceous on other insects. *Triatoma* spp. feed on the blood of vertebrates. In the southwestern United States, wood rats are favored hosts.

Injury:

The bite of conenose bugs is usually not painful. However, hours to days after being bitten, pruritic welts will usually develop at the site of the bites. In the western hemisphere, south of the United States border, conenose bugs are vectors of a protozoan (*Trypanosoma cruzi*) that causes Chagas' disease. The pathogens are not transmitted through a bite but they are excreted in the conenose bug's feces as it feeds. Infection occurs when the victim rubs the feces into the skin when scratching the bites.

In contrast to conenose bugs, the bites of assassin and wheel bugs are exceedingly painful. The bite area becomes inflamed and swollen, and pruritic.

First Aid:

Treatment is directed toward relieving inflammation and the intense pruritis that reduviid bites cause. An ice pack should be immediately applied to the bite area. Aspirin and calamine lotion are usually effective in relieving pain and swelling. Antihistamines and corticosteroid lotions may be used if the pruritis is severe.

Stinging Caterpillars (order Lepidoptera)

Larvae of some species of moths and butterflies, are covered with urticating spines or hairs that break off in the skin releasing an irritating toxin. There are many families of Lepidoptera that contain urticatious species, most notably Megalopygidae, Limacodidae, and Saturniidae.

Description:

Stinging caterpillars are the immature stages of several species of moths. The larva of the Io moth (*Automeris io*) is pea-green and has greenish spines tipped with black. A creddish stripe edged with white extends down the entire length of the caterpillar's abdomen. The Io moth larva is about 60 mm (2 3/4 in.) long when it is fully grown.

Slug caterpillar larvae are short fleshy and slug-like. A common species (*Euclea chloris*) is about 19 mm (3/4 in.) in length and is yellowish green in color. The many knob-like bumps on its body are covered with spine-like hairs.

The hackberry leaf slug (*Norape ovina*), also known as the flannel moth caterpillar, has a few long, plumose setae in its early instars. In older larvae, the long setae are more noticeable and there is a short, heavy spine adjacent to each spiracle. This spotted caterpillar is about 24 mm (1 in.) long when fully grown.

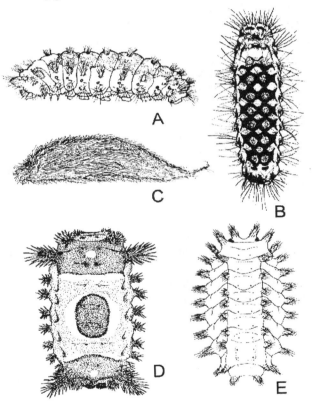

Figure 7.21. Urticatious caterpillars: A - Flannel moth caterpillar; B - Hackberry leafslug; C - Puss caterpillar; D - Saddleback caterpillar; E - Slug caterpillar (Knecht 1986).

The puss caterpillar (*Megalopyge opercularis*) is roughly pear-shaped and densely covered with hair-like setae. The long setae may be pale yellow, gray, reddish brown, or light brown. This larva is about 24 mm (1 in.) long when fully grown.

The saddleback (*Sibine stimulea*) is one of the more distinctive stinging caterpillars. Its short, stout green body has a white saddle-shaped spot with a bright, purplish-brown center. Its thoracic legs are very tiny and prolegs are absent. This caterpillar grows to about 24 mm (1 in.) long when fully grown.

Biology:
Distribution—Urticatious caterpillars are distributed world-wide in all tropical, subtropical, and temperate climates. Kawamoto and Kumada (1984) present an extensive list of more than 200 species together with their distributions. In the southeastern United States, all of the aforementioned species commonly occur.
Habitat—Stinging caterpillars are foliage-feeding insects but they are usually not numerous enough to cause substantial damage to plants. Most trees and ornamental shrubs found in the landscape of homegrounds will support infestations of stinging caterpillars.
Life History—In northern latitudes, only one generation of caterpillars is produced. In the southern areas, a second generation may develop. Stinging caterpillars are usually most abundant in the late summer and fall.

Injury:
Stinging episodes usually involve mature caterpillar larvae which occur in the late summer and early fall. Rotberg (1971) provides an informative description of the sequence of clinical events following contact with a venomous caterpillar. As mentioned in the section on envenomization, stinging caterpillars are covered with hair-like, fragile hollow spines filled with a venom. When the spines are broken by contact, venom passes through the spines into the skin. An intense burning pain immediately results which is followed within several minutes by reddening and swelling at the site of contact. Whitish circular papules may develop within 15 minutes. The affected area may coalesce into a single raised wheal or some distinct papules may be evident. Intense pruritis results and may persist for one or more hours. The dermatitis may persist for several days depending on the duration and site of contact. Often, raised vesicles develop at the site and these may be broken open by scratching, leading to a weeping sore which ultimately dries to form a scabby crust. Secondary infection of the lesion may occur.

First Aid:
No specific medical treatment is available. First aid is directed toward relieving pain and inflammation. In this regard, use of aspirin and calamine lotion are usually effective. Topically applied creams of antihistamines and corticosteroids are also effective in relieving swelling and pain.

Blister Beetles (Order Coleoptera)

Beetles in the family Meloidae contain a crystalline anhydride of cantharidic acid called cantharidin. When handled, the beetles exude the vesicating cantharidin, producing blisters.

Description:
Adult—Blister beetles are narrow, elongate insects, ranging from 0.5 to 17 mm (1/3 to 2/3 in.) long. The pronotum (body section between the wings and head) is narrower than the wings or the head. Adult *Epicuata* spp. vary in color from black covered with gray pubescence, to light brown with circular black spots, to blackish with a median yellow stripe and each wing cover bordered in yellow.
Larva—There are a variety of larval forms. There are 7 instars. A long-legged larva called a triungulin hatches from each egg. The next instar is similar to the first but has shorter legs. The third through fifth instars become increasingly grub-like. The sixth instar shortens and thickens, and lacks functional appendages. It is the overwintering stage, and is darker and has a thicker cuticle than the other instars. This stage molts to become a white, legless form.
Pupa—The seventh instar develops into the pupa, which resembles the adult.

Biology:
Distribution—*Epicauta vittata*, the stripped blister beetle, is most often collected in the eastern United States, but it ranges westward as far as Montana. The black blister beetle, *Epicauta pennsylvanica*, is common in the eastern states but ranges from Mexico through much of the United States. *Epicauta cinerea*, the clematis blister beetle, is most commonly collected in the southwestern United States. Several species of blister beetles are infrequent public health pests in Europe (Goddard 1993).
Habitat—Adults are usually found on plants. Some species are important pests of crops and garden plants, such as potatoes and tomatoes.
Life History—There is usually one generation of blister beetles produced each year. Blister beetle eggs are laid in holes in the ground. After hatching from eggs, triungulins search for and feed on grasshopper eggs. Larvae of other species feed on bees' eggs. Larvae overwinter in the sixth instar, and molt in the spring. The seventh instar molts to the pupal stage. Adults emerge in about 2 weeks.

Injury:
The vesicating fluid is released if the beetle is agitated. The beetles swell after imbibing air. Internal pressure increases until breaks occur in joints in the exoskeleton.

Fluid is released, relieving the pressure. If the beetles are handled, cantharidin contacts the skin, producing fluid-filled blisters (Lehman et al. 1955).

First Aid:

The blisters should be reabsorbed within a few days if they are not broken open. If the blisters rupture, then the affected areas should be thoroughly cleaned, treated with a topical antibiotic and bandaged.

Order Hymenoptera

Stinging Hymenoptera can be categorized behaviorally as either social or solitary. Social Hymenoptera form colonies that have a division of labor in the form of castes. Common examples of social Hymenoptera include honey bees, hornets, and ants. The workers, which are all females, are responsible for nest building and defense, foraging for food, as well as feeding and caring for the queen and her brood or offspring. New offspring are produced at regular intervals by the queen. At some point during the year, males (drones) are produced. Their sole purpose is to mate and fertilize the next generation of queens. Depending upon the species, the colony may die out annually or persist for several years.

With solitary wasps, the fertilized female constructs individual nests or clusters of brood chambers for her offspring. She is solely responsible for nest construction and for providing her offspring with food. Common examples of solitary Hymenoptera include cicada killers, *Specius speciosus*, and mud daubers (family Specidae).

Ants

Ants are insects in the family Formicidae. There are more than 10,000 species worldwide. Most ants are innocuous, but some species are beneficial, preying on pest insects in gardens or agricultural crops. A small number of species are considered to be pests, most often because they sting.

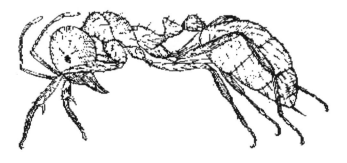

Figure 7.22. Red imported fire ant (Knecht 1986).

Red Imported Fire Ant, *Solenopsis invicta*

Description:

Adult—The worker (wingless female) is a reddish or dark brown and may be large (6 mm=1/4 in.) or small (3 mm=1/8 in.). The winged males are black and have colorless to pale brown wing veins. Winged females are mostly light brown.

Immature Stages—All stages in the life cycle of the fire ant are contained inside the ant's nest. The smooth, white, ellipsoidal egg is only about 0.4 mm (1/64 in.) long. The larva is translucent, soft-bodied, legless and pear-shaped. The worker larva grows from about 0.4 mm (1/64 in.) when newly hatched to about 3mm (1/8 in.) when fully grown. Some larvae grow into winged adults. The soft-bodied, white pupae resemble adults in size and shape.

Biology:

Distribution—Red imported fire ants (RIFAs) were introduced through the port of Mobile, Alabama in the early 1940s and subsequently spread into Mississippi, Louisiana, Texas, Florida, Georgia, Arkansas, Oklahoma, Tennessee, South Carolina, and North Carolina. The rapid spread of the RIFA has been largely accomplished through the movement of ant-infested nursery plants (Lofgren 1986). When it is established, the tremendous reproductive capacity of the RIFA allows it to populate an area rapidly.

Habitat—The RIFA prefers to establish colonies in open, sunny areas such as highway right-of-ways, pastures and other areas disturbed by human activities. The northward expansion of this species appears to be limited by low winter temperatures (Lofgren 1986).

Life History—In addition to the fertilized queen, the ant colony is comprised of winged, reproductive females and males, and three classes (sizes) of ants called minor, medium, and major worker ants. Like other ants, the RIFA queen lays and cares for the initial egg cluster. Later she only lays eggs and the workers take care of the other functions within the colony. A mature colony may contain 100,000 to 500,000 workers and a few thousand reproductive forms. The winged reproductive forms mate in flight and the males die shortly afterward. After shedding her wings, a fertilized female constructs a cell in the ground where she lays the initial cluster of eggs. As the colony grows in size, workers construct "honeycombed" galleries below ground to accommodate new ants. Workers also excavate tunnels in the mound of soil that protrudes above ground. Fire ant mounds can vary in size and shape depending on the type of soil. In clay soils, the dome-shaped mounds of mature colonies may measure 60 cm (2 feet) in diameter and height. Mounds constructed in sandy soils are usually flat and spread out over an area of a square meter (1 square yard).

Damage or Injury:

In urban areas, these ants typically invade lawns, parks, playgrounds, school yards, cemeteries, and golf courses, but may even move indoors. These ants have a painful sting and will attack anything that disturbs their nest. Each ant is capable of stinging several times. The sting causes a burning sensation followed within 24 hours by formation of a white pustule-like welt, which is characteristic of fire ant stings. Formation of a pustule does not indicate that the sting site is infected. The pustule contains a sterile, milky-colored fluid that results from the necrotic action of the fire ant's venom. The pustule may be extremely pruritic and, if scratched and broken open, infection and scarring may occur. Sometimes, the blisters may take as long as 10 days to heal. For some people that are highly allergic, fire ant stings may be life threatening. Symptoms that occur shortly after the sting may include chest pains, difficult breathing, severe swelling at the site of the sting, and nausea. Rhoades and coworkers (1977) after surveying allergists in Jacksonville, Florida reported that systemic allergic reactions occurred at the rate of 3.8 per 100,000 persons. From medical records at Fort Stewart, Georgia, Adams and Lofgren (1982) found that reactions to fire ant stings usually included swelling (81%) and pustule formation (51%); however, some people suffered respiratory distress (7.1%). People who are allergic to wasp and bee stings may also be allergic to fire ant stings.

First Aid:

No treatment is usually required for most mild reactions to fire ant stings. Local reactions can be treated with ice packs to reduce swelling and pain, and an antihistamine or calamine lotion to relieve itching. If hives or swelling develop in an area of the body away from the sting site, medical assistance should be obtained immediately. If the pustules are broken open, an antiseptic should be applied to prevent infection.

Medical Treatment:

Some people may exhibit a life threatening allergic reaction when stung by fire ants. If difficulty in breathing, swelling of the throat or face, unconsciousness, or hives occur within a few minutes following a fire ant sting, medical assistance should be obtained immediately. People who are severely allergic to fire ant stings should consider carrying an epinephrine syringe first aid kit and undertaking desensitization therapy to fire ant venom.

Harvester Ants (*Pogonomyrmex* spp.)

Description:

Adult—Harvester ants are large (10 mm; 3/8 in.) and reddish brown in color. Worker ants have large, wide heads. Winged females (virgin queens) are winged and reddish-brown in color.

Immature Stages—All stages in the life cycle of harvester ants are contained inside the ant's nest. The eggs are smooth, white, and ellipsoidal in shape. Larvae are translucent, soft-bodied, legless and pear-shaped. The worker larvae grow from about 0.5 mm (1/64 in.) when newly hatched to about 8 mm (5/16 in.) when fully grown. Some larvae grow into winged adults. The soft-bodied, white pupae resemble adults in size and shape.

Biology:

Distribution—Various species of harvester ants occur throughout North America. Problematic species include the California harvester ant, *P. califonricus*, which occurs primarily in California and Mexico. The red harvester ant, *P. barbatus*, is distributed from Louisiana and Kansas to Utah and California, and into Mexico. The western harvester ant, *P. occidentalis*, occurs from Arizona and Oklahoma northward into Canada. The Florida harvester ant, *P. badius*, occurs east of the Mississippi and is the only harvester ant in the southeastern United States.

Life History—Harvester ants construct low, flat or crater-shaped nests. One or more entry/exit holes in the center of the nest lead down into a honeycomb of galleries where the queen lays eggs and the brood is reared. There is a large vegetation-free area around each nest. Harvester ants gather and feed on seeds. Each mature colony consists of one egg-laying queen, a small number of winged males and females, and several thousand worker ants.

Injury:

The sting of worker harvester ants is painful. Harvester ant stings do not result in the formation of a pustule-like blister like that from RIFA stings.

First Aid:

No treatment is usually required for most mild reactions to harvester ant stings. Local reactions can be treated with ice packs to reduce swelling and pain, and an antihistamine or calamine lotion to relieve itching.

Paper Wasps, Yellowjackets, and Hornets

These stinging insects in the order Hymenoptera are commonly encountered on home grounds. Social wasps of the family Vespidae that are most problematic for people are placed in two subfamilies, the Vespinae (yellowjackets) and the Polistinae (paper wasps).

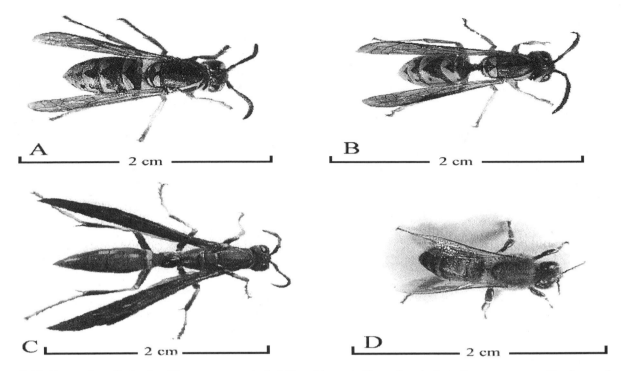

Figure 7.23. Examples of stinging Hymenoptera: A- baldfaced hornet; B - yellowjacket; C - paper wasp; D - honey bee (photographs courtesy of Miles Guralnick, Vespa Laboratories, Inc., Spring Mills, PA).

Description:

Adult—Paper wasps (*Polistes* spp.) and yellowjackets (*Vespula* spp.) are usually 13 to 25 mm (1/2 to 1 in.) long. Paper wasps have long legs, and are reddish brown to almost black in color. They have a slender, spindle-shaped abdomen. Yellowjackets are covered with black and yellow markings. The baldfaced "hornet" (*Dolichovespula maculata*) is black with a black and white abdomen and white face. The European or giant hornet (*Vespa crabro*), the only true hornet in the United States, is 20 mm (3/4 in.) long and covered with dark brown and deep yellow markings.

Egg—The eggs are small, white, and sausage-shaped.

Larva—The larvae are legless, white or cream colored, and grub-like with dark brown heads. Mature larvae of most species are 1.3 cm (1/2 in.) long.

Pupa—Pupae, which resemble adults, are slightly shorter than fully grown larvae. Pupae develop inside silken cocoons in cells within the nest.

Biology:

Distribution—Yellowjackets and paper wasps occur in most temperate areas of the world. The baldfaced hornet is not a true hornet. It is the most widespread yellowjacket wasp, occurring throughout the United States. In the United States, the European hornet is largely restricted to the eastern third of the country. A thorough treatment of the geographic distribution of yellowjackets, paper wasps, and the European hornet is presented in Akre et al. (1981) and Ebeling (1978).

Habitat—Yellowjackets, hornets, and paper wasps construct papery nests from plant fibers. Hornet nests are constructed above ground in trees or shrubs but yellowjacket nests are usually built below ground. Baldfaced hornet nests are usually free-hanging and visible, but the nests of the European hornet may be hidden. Paper wasps construct umbrella-shaped nests in dimly lit areas such as under the eaves of homes or storage sheds.

Life History—After constructing a small nest, the queen lays a single egg in each cell. The queen feeds masticated insects to the larvae when they are 2 to 3 days old. In 12 to 18 days, the larvae mature and spin silken caps over their cells. New adult workers, that emerge about 12 days later, attend and feed developing larvae, and enlarge the nest. After raising the first brood, the queen remains in the nest and deposits eggs. Within their life times, hornet and yellowjacket queens produce about 1,500 eggs while paper wasp queens lay several hundred eggs. The nests of paper wasps, yellowjackets, and hornets are abandoned each year. In the fall, a few virgin queens and males leave the nest and mate. Males and workers eventually die, but the queens overwinter in protected areas such as under the bark of trees or in rock crevices. In the spring, queens emerge and construct new nests.

Injury:

Sting reactions can be classified as systemic or local. Systemic reactions include symptoms that occur away from the sting site, and usually start minutes after the stinging episode. These symptoms, which vary in severity, include urticaria (hives), a feeling of anxiety, respiratory distress (shortness of breath or difficulty in breathing), cold sweats, and fainting. Generally, stings result in localized reactions, involving swelling, erythema, and pain. Deaths from anaphylactic reactions are immunologically mediated, and usually occur within 60 minutes of the sting (Hoffman 1985).

First Aid:

Local reactions can be treated with ice packs to reduce swelling and pain, and an antihistamine or calamine lotion to relieve itching. If urticaria or swelling develop in an area of the body away from the sting site, medical assistance should be obtained immediately.

Medical Treatment:

Some people may exhibit a life threatening allergic reaction when stung. If difficulty in breathing, swelling of the throat or face, unconsciousness, or hives occur within a few minutes following a sting, medical assistance should be obtained immediately. People who are severely allergic to Hymenoptera venom should consider carrying an epinephrine syringe first aid kit and undertaking desensitization therapy.

Honey Bee (*Apis mellifera*)

Of the 20,000 species of bees only about 5 percent are social. These species, including the honey bee, are in the family Apidae. Honey bees are not indigenous to North America, but were introduced by early colonists.

Description:

Adult—There are three castes of adults—worker, queen, and drone. The worker is 10 to 15 mm (2/5 to 3/5 in.) long, and brown or black in color. A short, dense pile of buff or pale colored branched hairs covers most of the body. The queen is slightly larger, ranging from 15 to 20 mm (3/5 to 4/5 in.) in length, and the drones are slightly shorter but more robust than the queen.

Larva—The larva is pear-shaped, pearly white in color and slightly translucent. Each larva passes through several instars before pupating.

Pupa—The pupa resembles the adult. The developing legs and wings are closely appressed to the body of the pupa.

Egg—Each egg is white and oval in shape.

Biology:

Distribution—*Apis mellifera* is found worldwide in all temperate, tropical, and subtropical areas. An aggressive subspecies, the Africanized bee *Apis mellifera scutella*, is now distributed throughout Central American and Mexico, and is now found in areas of southwestern Texas, Arizona, and California. The African bee was originally imported into Brazil in the 1950s to boost honey production. Swarms of the African bee escaped and adapted to the surrounding environment, and cross bred with local honey bees to produce a hybrid. Unfortunately this "Africanized bee" retained most of the aggressive behavioral traits of the African bee. The Africanized bee migrated northward at the rate of 50 miles per year, until arriving in southwest Texas in the early 1990s.

Life History—Unlike most other nonparasitic bees, honey bees lay their eggs in unprovisioned cells of the nest, and feed their larvae until they pupate. The queen can lay as many as 1,500 eggs a day, placing each egg in a cell of the brood comb. Males are produced from unfertilized eggs, while females are produced from fertilized eggs. The colony selects which eggs are to become queens by feeding the larvae special food. Worker bees are neuter females that contain incompletely developed reproductive tracts. Young workers care for the brood, build the comb, and protect the nest. Older workers forage for nectar and pollen.

Unlike wasps and hornets, honey bees occupy the same nest from year-to-year. Many workers and the queen survive the winter. In the spring and summer when the nest becomes crowded, a new queen is produced and the colony divides by swarming. The swarm settles on an object such as a tree branch for a few hours to several days before finding a new site for their nest. Swarms usually are not a problem unless they occur in a troublesome location or they are disturbed. The local Cooperative Extension Center can usually help locate a beekeeper who will remove the swarm.

Damage or Injury:

Swarms of honey bees often establish a nest in the walls of a building. These colonies should be removed immediately because the bees will build combs of wax in which to store honey and rear young bees. If the colony has been left in place for several weeks to a month before the bees are killed, the siding must be removed so that wax combs and dead bees can be removed; otherwise, odor problems will result from the decaying wax and bees.

Honey bees are extremely beneficial insects. They pollinate agricultural and horticultural crops and, produce honey and wax. Honey bees are often mistakenly blamed for stinging episodes involving wasps. However, honey bees will sting if their nest is disturbed. Unlike wasps, honey bees sting only once. Because the stinger is barbed at the end, it remains embedded in the victim. The worker bee subsequently dies when the stinger and accompanying poison gland are pulled loose from its body. Only worker bees sting, drones do not have a stinger. Envenomization usually produces localized pain, swelling, and redness, which resolve spontaneously in a few days. Some people exhibit systemic allergic reactions to honey bee stings. The symptoms can vary in severity from anxiety, constriction in the chest, and wheezing to respiratory failure, rapid drop in blood pressure, unconsciousness, and death. For severely allergic people, the onset of symptoms usually occurs within minutes of the sting.

First Aid:

The stinger should be removed immediately by scrapping it off. Do not use tweezers to remove the stinger as this may result in the injection of venom into the skin. A cold pack should be applied to the sting site as soon as possible to reduce swelling and pain, and an anihistamine or calamine lotion can be applied to relieve itching. If urticaria or swelling develop in an area of the body away from the sting site, medical assistance should be obtained immediately.

Medical Treatment:

An injection of epinephrine should immediately be administered to victims exhibiting an anaphylactic reaction after being stung. People who are severely allergic to a Hymenoptera sting should carry a syringe kit containing an injection pen of epinephrine, and consider undertaking desensitization therapy.

Miscellaneous Bees and Wasps

Bumble Bees

Bombus spp. are social insects in the family Apidae. Bumble bees usually nest in the ground or hollow logs, and feed their young pollen and honey. Nests usually only contain a few hundred adults. Adults are medium to large-sized and densely covered with black hairs, with areas of bright orange or yellow hairs. Colonies die out each year; fertilized queens overwinter and found new colonies in the spring. Bumble bees will sting if their nest is disturbed.

Sweat Bee

Halictus spp. are in the family Halictidae. These solitary bees are commonly called "sweat" bees because in their search for moisture they are attracted to human sweat. A mild sting may occur when the bees are crushed or swatted as they attempt to land. Sweat bees are small to medium in size, and black to metallic blue or green in color. Each female digs a short tunnel in the ground and provisions it with pollen and nectar before laying eggs.

Mud Daubers

Sceliphron caementarium and *Chalybion californicum* are solitary wasps in the family Sphecidae. These solitary wasps are medium-sized and have a thin, thread-like waists. Mud daubers build cylindrical mud tubes that encase cells containing their young. The mud tubes are constructed side-by-side in vertical rows in dimly lit sheltered areas, such as under bridges or the eaves of buildings. When building their nests, females sting spiders and insect larvae. Several paralyzed arthropods are placed in each cell of the mud nest before the female lays an egg in it. The developing larva feeds on the paralyzed arthropods and spins a cocoon in which it pupates the following spring. Mud daubers do not usually defend their nests, and they rarely sting unless handled.

Cicada Killers

Sphecius spp. are large solitary insects, reaching 40 mm (1.5 in.) in length as adults. The body is black with yellow markings similar to a yellowjacket wasp except larger. Several species of Cicada Killers occur in the United States. Females construct burrows in the ground, and in so doing, a small but conspicuous pile of soil is deposited around the opening. Cicada Killers provision their nests with insects, such as cicadas, which are paralyzed by stinging. An egg is laid on top of the paralyzed insect, which the developing larva uses as food. Cicada Killers commonly build their nests in loose soil or bare spots in yard areas, where they can often be seen in large numbers flying randomly at knee-high level. These wasps are usually not aggressive towards people, but they will occasionally sting people doing yardwork, or children walking barefoot on yards where wasps are constructing nests.

Velvet Ants

Velvet ants are actually wasps in the family Mutillidae. They are also called cowkillers because their sting is very painful.

Adult—Velvet ants are medium-sized 10 to 25 mm (7/16 to 1 in.) long insects that resemble ants, but they lack the node on the petiole between the abdomen and thorax that is characteristic of ants (Ebeling 1978). The black colored bodies of velvet ants have a velvety appearance because they are densely covered with white, yellow, golden, orange, or bright red hairs. Males have wings while females are wingless.

Biology:

Distribution—The Mutillidae, comprised of about 2,000 species, are worldwide in distribution but most common in Australia, Central and South America (Rathmayer 1978).

Habitat—Velvet ants predominantly occur in sandy, dry soils.

Life History—Mutillids are solitary wasps. Their larvae are parasitic on larvae and pupae of wasps, bees, beetles, and flies. The female velvet ant searches the ground and locates the nest of a suitable host. She lays an egg on the prepupal or pupal stage of the host. The velvet ant larva consumes the host in a few days and then spins a cocoon within the pupal case of the host. Mutillids possess a powerful sting which is apparently used in defense and not to paralyze prey. The chemistry and pharmacology of the venom are unknown (Rathmayer 1978). Additional information on the biology of North American mutillids is provided in Mickel (1928, 1974).

Injury:

Velvet ants inflict a painful sting. According to Rathmayer (1978), the venom causes a heavy burning sensation.

First Aid:

Local reactions can be treated with ice packs to reduce swelling and pain, and an antihistamine or calamine lotion to relieve itching.

PREVENTION OF STINGS AND BITES, AND CONTROL OF ARTHROPODS OF PUBLIC HEALTH SIGNIFICANCE

Avoidance of arthropod stings and bites is largely a matter of exercising common sense. A large number of envenomizations occur when people, out of curiosity, purposely handle arthropods. The following protective measures are offered as means of avoiding incidental contact with arthropods.

Hymenoptera. The following precautions are offered to reduce the likelihood of stings.

- Wear boots or protective footwear when hiking.
- When picnicking, keep all food covered (especially meats and sweets). Pay special attention to open soft drink containers and glasses to be sure that there are no yellowjackets or bees on or in the vessel before taking a drink.
- Do not eat or rest near trash or recycling bins. Food residue is a major attractant for some stinging insects especially in the summer.
- Wasps or bees that land on people or food should not cause alarm. The insects should be gently brushed away, not swatted or vigorously waved at. Movements should be slow to avoid provoking a defensive sting.
- Before choosing a spot for a picnic or to rest, the area should be thoroughly inspected for ant and wasp nests.
- If a nest is disturbed, leave the area quickly.
- People who are allergic to Hymenoptera venoms should carry an emergency sting kit that contains epinephrine.

Spiders. Follow these suggestions to reduce the chances of being bitten by a spider:

- Always check for spiders before sticking bare hand(s) into dark corners or areas.
- Always wear work gloves when handling boxes, firewood, lumber, and other items that have been stored or stacked undisturbed for some time.
- Stored clothing should be shaken vigorously to dislodge any spiders and inspected carefully before wearing.
- If bitten by either a black widow spider or brown recluse, immediately apply an ice or cold-pack to the bite and seek medical assistance. If the culprit spider can be caught, bring it along for positive identification.

Ticks and Chiggers. The following suggestions are offered to avoid tick bites.

- On homegrounds, keep weeds and tall grass mowed. This will discourage tick hosts, such as rodents, from becoming established. Mowing will also help to lower moisture levels, making the environment less suitable for ticks.
- When hiking in tick infested areas, walk on established paths or trails. Avoid hiking through brushy or shrubby areas.
- Layer clothing; i.e., tuck pant cuffs into sock or boot tops and shirt tails into pants. This minimizes the amount of bare skin that is exposed to ticks.
- Wear light-colored clothing so that ticks are more readily visible.
- Apply an insect repellent to exposed skin and clothing.

- Self-inspection for ticks on clothing or skin should be carried out at least twice daily. Attached ticks should be promptly removed. Note the date on a calendar.
- Check camp sites for ticks by dragging a light colored cloth over vegetation.

Mosquitoes. Personal protection from mosquitoes can be achieved using the following suggestions.

- Maintain window screens and screen door to prevent mosquitoes from flying indoors.
- Wear long sleeves shirts and pants when hiking, camping, or engaging in outdoor recreational activities in mosquito-infested areas.
- Wear light colored clothing because some mosquito species are attracted to dark colors.
- Apply an insect repellent to exposed skin and clothing.

SUMMARY

The bites and stings of arthropods delivered in defense or in pursuit of bloodmeal have a deleterious impact on the health and welfare of people and domestic animals. In this chapter, we have reviewed the major groups of arthropods of public and veterinary health significance with the intent of providing sufficient information so that common pests and their injuries could be identified. Basic biological information on each pest is given along with some general first aid procedures for arthropod-inflicted injuries.

GLOSSARY

Acari-An order in the phylum Arthropoda. Mites, ticks, and spiders are examples of arthropods classified in the order Acari.

allomone-A chemical primarily produced for defense by a living organism. Examples would include Hymenoptera venoms.

allergic-Exhibiting a hypersensitvity reaction to a foreign protein such as a wasp or bee sting. Mild allergic reactions may be localized at the site of the envenomization, and include swelling, pain, and redness and hyperthermia. See anaphylactic reaction.

anaphylactic reaction-Exhibiting an extreme hypersensitivity reaction to foreign proteins, such as a wasp or bee sting. Anaphylactic reactions are usually systemic; i.e., they occur away from the site of the envenomization. Hives, dizziness, nausea, and labored breathing are symptons of an anaphylactic reaction, which may be life-threatening.

Arthropoda-A phylum in the animal kingdom. Arthropods are characterized by having segmented body regions and jointed appendages.

cantharidin-The vesicating body fluid of blister beetles.

cytolytic-Causing lysis or breakdown of cells.

dengue fever-A febrile human illness caused by a virus that is transmitted through the bite of an infected *Aedes aegypti* mosquito.

dermatitis. A skin rash resulting from contact with or bites from an arthropod.

eastern equine encephalitis-A febrile illness of the central nervous system caused by a virus that is transmitted through the bite of a mosquito.

ectoparasitic-A mode of existance in which an arthropod feeds on a vertebrate host. Examples of ectoparasites include fleas, ticks, chiggers, and biting flies.

ehrlichiosis- Human ehrlichiosis is a tick-borne illness, caused by rickettsial pathogens *Ehrilichia chaffeensis*, *E. canis*, and *E. phagocytophila*.

envenomization-Delivery of a chemical secretion or venom on or below the skin through a bite or sting of an arthropod.

erythema-A redness of the skin.

forcipules-The biting appendages on the first body segment of a centipede.

hematophagous-Blood feeding; taking a blood meal.

host-An animal that is fed upon or parasitized by another animal. For example, humans are hosts for feeding mosquitoes, and infected mosquitoes are hosts for microorganisms.

hyperthermia-Increased temperature locally or systemically..

induration-Callousness or hardness.

instars-A growth stage of subadults in the life cycle of arthropods.

La Crosse encephalitis-A human illness, caused by a virus that is transmitted through the bite of a mosquito. The illness is characterized by swelling of the brain.

larva-A growth stage following the egg stage in the life cycle of an arthropod exhibiting complete development.

latrodectism-The medical syndrome caused by the bite of a black widow spider.

loxoscelism-The medical syndrome caused by the bite of a brown recluse spider.

Lyme Disease-A human illness caused by a spirochete, *Borrelia burgdorferi*, that is transmitted through the bite of the black-legged tick, *Ixodes scapularis*. The disease was named after the town of Old Lyme, CT where cases were first described.

lymphadenitis-Inflammation of the lymph glands.

metamorphosis-The process by which arthropods mature and change form.

molt-The physical process whereby the exoskelton of arthropods is shed.

myiasis-Injury caused by flies that invade the muscle or skin of otherwise healthy animals.

necrotic-Initiating necrosis or decay of tissue.

neurotoxic-A poisonous substance that acts on the nervous system.

nit-The egg of a head louse.

nymph-A growth stage following the egg stage in the life cycle of an arthropod exhibiting incomplete or gradual development.

papular-A smooth pimple-like elevation.

pediculosis-The condition of being infested with head or body lice.

pedipalps. Biting appendages on male spiders; grasping appendages on scorpions.

peridomestic. In or around human dwellings.

proboscis-The mouthparts of a biting insect, such as a mosquito.

proleg-The unjointed, leg-like fleshy appendage on the thorax or abdomen of insect larvae.

pruritis-An intense itching sensation.

pronotum-A body section, of insects, between the head and abdomen.

pupa-The growth stage between the larval and adult stages in the life cycle of an arthropod exhibiting complete metamorphosis.

puparium-The hardened skin of the third instar larva of flies, which encloses the developing pupa.

redbug-A common name for a chigger.

Rocky Mountain Spotted Fever-A febrile illness caused by *Rickettsia rickettsii*, a bacterium, that is transmitted through the bite of a tick.

spirochete-A specialized bacterium. *Borrelia burgdorferi* is the spirochete that causes Lyme disease.

St. Louis encephalitis-A febrile human illness of the central nervous system that is caused by a virus that is transmitted throught the bite of a mosquito.

stylostome-The feeding tube that forms around the mouthparts of a larval trombiculid mite. The stylostome is thought to be produced, in part, by host tissue reactions.

tick paralysis-An ascending, flaccid paralysis resulting from the attachment of a tick. Neurotoxins released by the tick are thought to be the cause of the paralysis. Ultimately death occurs, but the paralysis is rapidly resolved once the tick is removed.

triungulin-The first instar larva of a blister beetle. This agile and active larva feeds on the eggs of grasshoppers and bees.

urticaria-An itching skin rash, consisting of small, red welts.

vector-An animal that transmits a pathogenic organism to a susceptible host.

venom-A chemical secretion that is used for defensive purposes that is delivered through a sting.

vesiculation-Formation of small clear blisters.

wheal-A welt formed in reaction to an insect bite, such as a mosquito bite.

zoonosis-A disease of wildlife that affects people. Rocky Mountain spotted fever and Lyme disease are examples.

REFERENCES

Adams, C.T., and C.S. Lofgren. 1981. Red imported fire ants (hymenoptera: formicidae): frequency of sting attacks on residents of Sumter County, Georgia. *J. Med. Entomol.* 18: 378-82.

Adams, C.T., and C.S. Lofgren. 1982. Incidence of stings or bites of the red imported fire ant (hymenoptera: formicidae) and other arthropods among patients at Ft. Stewart, Georgia. *J. Med. Entomol.* 1: 366-70.

Akre, R.D., A. Greene, J.F. MacDonald, P.J. Landolt, and H.G. Davis. 1981. The yellowjackets of America north of Mexico. *USDA Agriculture Handbook.* No. 552. Washington, DC: U.S. Government Printing Office.

Alexander, J.O. 1984. *Arthropods and Human Skin.* New York: Springer-Verlag.

Ardao, M.I., C.S. Perdomo, and M.G. Pellaton. 1966. Venom of the *Megalopyge urens* (Berg) caterpillar. *Nature* (London) 20: 113-1140.

Arlian, L.G., and D.L. Vyszenski-Moher. 1986. Nutritional ecology of parasitic mites and ticks. In: *Nutritional Ecology of Insects, Mites and Spiders.* F. Slansky, Jr., and J.G. Rodriguez, eds. New York: John Wiley & Sons, Inc., pp. 765-70.

Azad, A.F. 1986. *Mites of Public Health Importance and Their Control.* World Health Organization, Vector Biology and Control Division, Vector Control Training and Information Series XIII. Genva: WHO/VBC/86.31.

Axtell, R.C. 1976. Coastal horse flies and deer flies (Diptera: Tabanidae). In: *Marine Insects.* L. Cheng, ed. New York: North-Holland Publishing Co., pp. 415-45.

Barnard, J.H. 1973. Studies of 400 hymenoptera sting deaths in the United States. *J. Allergy Clin. Immunol.* 52: 25-264.

Bettini, S., ed. 1978. Arthropod venoms. *Handbook of Experimental Pharmacology.* Vol 48. New York: Springer-Verlag.

Bruce-Chwatt, L.J. 1987. Malaria and its control: present situation and future prospects. *Ann. Rev. Public Health* 8: 75-110.

Bücherl, W. and E.B. Buckley, eds. 1971. *Venomous Animals and Their Venoms*. New York: Academic Press.

Bücherl, W. 1971. Venomous chilopods or centipedes. In: *Venomous Animals and Their Venoms*. W. Bücherl and E. B. Buckley, eds. New York: Academic Press.

Carbonell, L.R., M.R. Hodge, M.D. Tomalski, and L.K. Miller. 1988. Synthesis of a gene coding for an insect specific scorpion neurotoxin and attempts to express it in a baculovirus. *Gene* 73: 409-18.

Centers for Disease Control (CDC). 1975a. *Mites of Public Health Importance and Their Control*. Atlanta, GA: U.S. Dept. of Health and Human Services, Public Health Service, Centers for Disease Prevention and Control. DHEW Pub. No. (CDC) 75-8297.

Centers for Disease Control (CDC). 1975b. *Lice of Public Health Importance and Their Control*. Atlanta, GA: U.S. Dept. of Health and Human Services, Public Health Service, Centers for Disease Prevention and Control. DHEW Pub. No. (CDC) 75-8265.

Centers for Disease Control (CDC). 1982. *Mosquitoes of Public Health Importance*. Atlanta, GA: U.S. Dept. of Health and Human Services, Public Health Service, Centers for Disease Prevention and Control. HHS Pub. No. (CDC) 82-8140.

Centers for Disease Control (CDC). 1990. Rocky Mountain spotted fever and human ehrlichiosis--United States, 1990. *MMWR* 3: 281-84.

Clemmer, D.L., and R.E. Serfling. 1975. The imported fire ant: dimensions of the urban problem. *South. Med. J.* 68: 1133-38.

Darsie, R.F., Jr., and R.A. Ward. 1981. Identification and geographical distribution of the mosquitoes of North America, north of Mexico. *Mosq. Syst. Suppl.* 1: 1-313.

Duchen, L.W. and S. Gomez. 1984. Pharmacology of spider venom. In: *Handbook of Natural Toxins. Vol. 2. Insect Poisons, Allergens, and other Invertebrate Venoms*. A. Tu, ed. New York: Marcel Dekker, pp. 483-512.

Duffey, S.S., K. Hoover, B. Bonning, and B.D. Hammock. 1995. The impact of host plant on the efficacy of baculoviruses. In: *Rev. Pestic. Toxicol.* R. M. Roe and R. J. Kuhr, eds. Raleigh, NC: Toxicology Communications, Inc., pp. 137-275.

Ebeling, W. 1978. *Urban Entomology*. Los Angeles: University of California, Division of Agricultural Sciences.

El-Mallakh, R.S., D.L. Baumgartner, and N. Fares. 1986. "Sting" of puss caterpillar, *Megalopyge opercularis* (lepidoptera: megalopygidae): First report of cases from Florida and review of literature. *J. Florida Med. Assoc.* 73: 521-25.

Environmental Protection Agency (EPA). 1992. *Urban Integrated Pest Management: A Guide for Commercial Applicators*. J. Coxx, ed. Washington, DC: U.S. EPA, Certification and Training Branch, Office of Pesticide Programs, Document No. EPA 735-B-92-001.

Ewing, S.A., J.E. Dawson, A.A. Kocan, R.W. Baker, C.K. Warner, R.J. Panciera, J.C. Fox, K.M. Kocan, and E.F. Blouin. 1995. Experimental transmission of *Ehrlichia chaffeensis* (rickettsiales: ehrlichieae) among white-tailed deer by *Amblyomma americanum* (acari: ixodidae). *J. Med. Entomol.* 32: 368-74.

Foil, L.D. 1989. Tabanids as vectors of disease agents. *Parasitol. Today* 5: 88-96.

Foote, R.H. and H.D. Pratt. 1954. *The Culicoides of the Eastern United States (Diptera, Heleidae)*. U.S. Dept. HEW, PHS, Public Health Monograph No. 18. Washington, DC: U.S. Government Printing Office.

Geren, C.R., and G.V. Odell. 1984. The biochemistry of spider venoms. In: *Handbook of Natural Toxins. Vol. 2. Insect Poisons, Allergens and Other Invertebrate Venoms*. A.T. Tu, ed. New York: Marcel Dekker, pp. 441-81.

Golden, D.B., D.G. Marsh, A. Kagey-Sobotka, L. Freidhoff, M. Szklo, M.D. Valentine, and L.M. Lichtenstein. 1989. Epidemiology of insect venom sensitivity. *J. Am. Med. Assoc.* 262: 240-44.

Goddard, J. 1993. *Physician's Guide to Arthropods of Medical Importance*. Boca Raton, FL: CRC Press.

Goodwin, J.T., B.A. Mullens, and R.R. Gerhardt. 1985. *The Tabanidae of Tennessee*. Knoxville, TN: University of Tennessee Agricultural Experimental Station, Bulletin 642.

Gothe, R., and A.W.H. Neitz. 1991. Tick paralyses: pathogenesis and etiology. In: *Advances in Disease Vector Research*, Vol. 8. K.F. Harris, ed. New York: Springer-Verlag, pp. 177-204.

Gregson, J.D. 1973. *Tick Paralysis: An Appraisal of Natural and Experimental Data*. Canada Dept. Agriculture Monograph.

Hassan, F. 1984. Production of scorpion antivenin. In: *Handbook of Natural Toxins. Vol. 2. Insect Poisons, Allergens and Other Invertebrate Venoms*. A.T. Tu, ed. New York: Marcel Dekker, pp. 577-605.

Hite, J.M., W.J. Gladney, J.L. Lancaster, and W.H. Whitcomb. 1966. *Biology of the Brown Recluse Spider*. Fayetteville, AR:

University of Arkansas, Bulletin 711.

Hoffman, D.R. 1985. Allergens in hymenoptera venom. XV: the immunologic basis of vespid venom cross-reactivity. *J. Allergy Clin. Immunol.* 75: 611-13.

Hoffman, D.R. 1995. Allergic reactions to biting insects. In: *Monograph on Insect Allergy*. M.I. Levine and R.F. Lockey, eds. Milwaukee, WI: American Academy of Allergy and Immunology, Committee on Insect Allergy.

Hoffman, D.R., D.E. Dove, and R.S. Jacobson. 1988a. Allergens in hymenoptera. XX. isolation of four allergens from imported fire ant (*Solenopsis invicta*) venom. *J. Allergy Clin. Immunol.* 82: 818-27.

Hoffman, D.R., D.E. Dove, J.E. Moffitt, and C.T. Stafford. 1988b. Allergens in hymenoptera. XXI. Cross-reactivity and multiple reactivity between fire ant venom and bee and wasp venoms. *J. Allergy Clin. Immun.* 82: 828-834.

James, M.T. 1947. *The Flies that Cause Myiasis in Man*. Washington, DC: U.S. Dept. Agriculture Publication No. 631. U.S. Government Printing Office.

Kawamoto, F. and N. Kumada. 1984. Biology and venoms of *Lepidoptera*. In: *Handbook of Natural Toxins. Vol. 2. Insect Poisons, Allergens and Other Invertebrate Venoms*. A.T. Tu, ed. New York: Marcel Dekker, pp. 21-330.

Keegan, H.L. 1972. Venomous bites and stings in Mississippi. *J. Miss. State Med. Assoc.* 12: 459-99.

Keegan, H.L. 1980. *Scorpions of Medical Importance*. Jackson, MS: University of Mississippi Press.

Knecht, T.W. 1986. *Insect and Other Pests of Man and Animals*. 1986. Raleigh, NC: North Carolina Agricultural Extension Service.

Lane, R.S., J. Piesman, and W. Burgdorfer. 1991. Lyme borreliosis: relation of its causative agent to its vectors and hosts in North America and Europe. *Ann. Rev. Entomol.* 36: 587-609.

Langley, R.L. 1994. Fatal animal attacks in North Carolina over an 18-year period. *Am. J. Forensic Med. Pathol.* 15: 160-67.

Lehane, M.J. 1991. *Biology of Blood-Sucking Insects*. London: Harper Collins Academic.

Lehman, C.F., J.L. Pipkin, and A.C. Ressmann. 1955. Blister beetle dermatitis. *Arch. Dermatol.* 71: 36.

Lemke, L.A., and J.B. Kissam. 1989. Public attitudes on red imported fire ant (*hymenoptera: formicidae*) infestations in homes and recreational areas. *J. Entomol. Sci.* 24: 446-53.

Linley, J.R. 185. Biting midges (diptera: ceratopogonidae) as vectors of nonviral animal pathogens. *J. Med. Entomol.* 22:58-65.

Linley, J.R. and J.B. Davis. 1971. Sandflies and tourism in Florida and the Bahamas and Caribbean area. *J. Econ. Entomol.* 64: 264-78.

Lofgren, C.S. 1986. History of imported fire ants in the United States. In: *Fire Ants and Leaf-Cutting Ants: Biology and Management*. C.S. Lofgren and R.K. Vander Meer, eds. Boulder, CO: Westview Press, pp. 36-47.

McCrone, J.D. and K.J. Stone. 1965. The widow spiders of Florida. *Arthropods of Florida and Neighboring Land Areas*. Vol. 2. Gainesville, FL: Department of Agriculture.

Mickel, C.E. 1928. Biological and taxonomic investigations on the mutillid wasps. *Bull. U.S. National Mus.* 143: 1-351.

Mickel, C.E. 1974. Mutillidae miscellanea: taxonomy and distribution. *Ann. Entomol. Soc. Amer.* 67: 461-71.

Minelli, A. 1978. Secretions of centipedes. In: *Arthropod Venoms: Handbook of Experimental Pharmacology*, S. Bettini, eds. Vol 48. New York: Springer-Verlag, pp. 73-85.

Nakajima, T. 1984. Biochemistry of vespid venoms. In: *Handbook of Natural Toxins. Vol. 2. Insect Poisons, Allergens, and Other Invertebrate Venoms*. A. Tu, ed. New York: Marcel Dekker, pp. 100-133. .

Needham, G.R. 1985. Evaluation of five popular methods for tick removal. *Pediatrics* 75: 997-1002.

O'Meara, G.F. 1986. Nutritional ecology of blood-feeding diptera. In: *Nutritional Ecology of Insects, Mites and Spiders*. F. Slansky, Jr., and J.G. Rodriguez, eds. New York: John Wiley & Sons, Inc., pp. 741-764.

Ori, M. 1984. Biology of and poisonings by spiders. In: *Handbook of Natural Toxins. Vol. 2. Insect Poisons, Allergens, and Other Invertebrate Venoms*. A. Tu, eds. New York: Marcel Dekker, pp. 37-440.

Orkin, M., H.I. Maibach, L.C. Parish, and R.M. Schwartzman. 1977. *Scabies and Pediculosis*. Philadelphia, PA: J.B. Lippincott Co.

Parrish, H.M. 1963. Analysis of 460 fatalities from venomous animals in the United States. *Am. J. Med. Sci.* 245: 129-41.

Piek, T. 1984. Pharmacology of hymenoptera venoms. In: *Handbook of Natural Toxins. Vol. 2. Insect Poisons, Allergens, and Other Invertebrate Venoms*. A. Tu, ed. New York: Marcel Dekker, pp. 135-85.

Rathmayer, W. 1978. Venoms of sphecidae, pompilidae, mutillidae and bethylidae. In: *Arthropod Venoms. Handbook of Experimental Pharmacology.* Vol 48. S. Bettini, ed. New York: Springer-Verlag, pp. 661-90.

Rees, R.S., P. Altenberg, J.B. Lynch and L.E. King, Jr. 1985. Brown recluse spider bites: a comparison of early surgical excision versus dapsone and delayed surgical excision. *Ann. Surg.* 202: 659-63.

Rhoades, R.B., W.L. Schafer, M. Newman, R. Lockey, R.M. Dozier, P.F. Wubbena, A.W. Tower, W.H. Schmid, G. Neder, T. Brill, and W.J. Wittig. 1977. Hypersensitivity to the imported fire ant in Florida. Report of 104 cases. *J. Fla. Med. Assoc.* 64: 247-54.

Ribiero, J.M.C. 1987. Role of saliva in blood-feeding by arthropods. *Ann. Rev. Entom.* 32: 463-78.

Rotberg, A. 1971. Lepidopterism in Brazil. In: *Venomous Animals and Their Venoms.* W. Bücherl and E.B. Buckley, eds. New York: Academic Press, pp. 157-68.

Ryckman, R.E. and A.E. Ryckman. 1967. Epizootiology of *Trypanosoma cruzi* in southwestern North America. Part XI: Biology of the genus *Reduvius* in North America and the possible relationship of *Reduvius* to the epizootiology of *Trypanosoma cruzi.* (hemiptera: Reduviidae) (kinetoplastida: trypanosomidae). *J. Med. Entomol.* 3: 326-34.

Sasa, M. 1961. Biology of chiggers. *Ann. Rev. Entomol.* 6: 221-44.

Schmidt, J.O. 1982. Biochemistry of insect venoms. *Ann. Rev. Entomol.* 27: 339-68.

Stablein, J.J. and R.F. Lockey. 1981. Ants and human anaphlaxis. In: *Monograph on Insect Allergy.* M.I. Levine and R.F. Lockey, eds. Pittsburgh, PA: American Academy of Allergy, Committee on Insect Allergy, pp. 61-67.

Stahnke, H.L. 1978. The genus *Centroides* (Buthidae) and its venom. In: *Arthropod Venoms. Handbook of Experimental Pharmacology.* Vol. 48. S. Bettini, ed. New York: Springer-Verlag, pp. 279-307.

Schuman, S.H., C.H. Banov, S.T. Caldwell, P.M. Horton, and J.B. Kissam. 1987. Impact of fire ant sting morbidity in South Carolina. *J.S.C. Med. Assoc.* 83: 249-52.

Schwartz, B.S., D.P. Ford, J.E. Childs, N. Rothman, and R.J. Thomas. 1991. Anti-tick saliva antibody: a biologic marker of tick exposure that is a risk factor for lyme disease seropositivity. *Am. J. Epidem.* 134: 86-95.

Schwartz, B.S., R.B. Nadelman, D. Fish, J.E. Childs, G. Forester, and G.P. Wormser. 1993. Entomologic and demographic correlates of anti-tick saliva antibody in a prospective study of tick bite subjects in Westchester County, New York. *Am. J. Trop. Med. Hyg.* 48: 50-57.

Shulov, A., and G. Levy. 1978. Venoms of buthinae. In: *Arthropod Venoms. Handbook of Experimental Pharmacology.* Vol. 48. S. Bethni, ed. New York: Springer-Verlag, pp. 309-12.

Stafford, C.T., L.S. Hutto, R.B. Rhoades, W.O. Thompson, and L.K. Impson. 1989. Imported fire ant as a health hazard. *South. Med. J.* 82: 1515-19.

Stewart, L.D.M., M. Hirst, I.P. Ferber, A.T. Merryweather, P.J. Cayley, and R.D. Possee. 1991. Construction of an improved baculovirus insecticide containing a insect-specific toxin gene. *Nature* 352: 85-88.

Straight, R.C. and J.L. Glenn. 1993. Human fatalities caused by venomous animals in Utah, 1900-1990. *Great Basin Naturalist* 53: 30-34.

Tomalski M.D. and L.K. Miller. 1992. Expression of a paralytic neurotoxin gene to improve insect baculoviruses as biopesticides. *Biotechnology* 10: 545-54.

Tomalski, M.D. and L.K. Miller. 1991. Insect paralysis by baculovirus-mediated expression of a mite neurotoxin gene. *Nature* 352: 82-85.

Tomalski, M.D., T.P. King, and L.K. Miller. 1993. Expression of hornet genes encoding venom allergen antigen-5 in insects. *Arch. Insect Biochem. Physiol.* 22: 303-13.

Tu, A.T., ed. 1984. *Handbook of Natural Toxins. Vol. 2. Insect Poisons, Allergens and Other Invertebrate Venoms.* New York: Marcel Dekker.

Usinger, R.L. 1966. *Monograph of Cimicidae (Hemiptera: Heteroptera).* Thomas Say Foundation, Vol. 7. College Park, MD: Entomologic Society of America.

Wharton, G.W., and H.S. Fuller. 1952. *A Manual of Chiggers.* Washington, DC: Entomologic Society of Washington.

Yeager, W. 1978. Frequency of fire ant stinging in Lowndes County, Georgia. *J. Med. Assoc. Georgia.* 2: 101-02.

Yevich, S.J., J.L. Sánchez, R.F. DeFraites, C.C. Rives, J.E. Dawson, I.J. Uhaa, B.J.B. Johnson, and D.B. Fishbein. 1995. Seroepidemiology of infections due to spotted fever group *Nickettsiae* and *Ehrlichia* species in military personnel exposed in areas of the United States where such infections are endemic. *J. Infect. Dis.* 171: 1266-73.

8

ZOONOTIC HAZARDS IN HUMANS IN AGRICULTURE, FORESTRY, AND FISHERIES

Ricky L. Langley, M.D., M.P.H.
Duke University Medical Center

W.E. Morgan Morrow, B.V.Sc., M.S., Ph.D.
North Carolina State University

Although agricultural workers are usually more at risk from the machinery they operate (e.g., tractor accidents) they are also at risk from the microorganisms that inhabit their work environment. The range of pathogens is very broad including bacteria, viruses, rickettsia and chlamydia, fungi, parasites, and infestations with arthropods (e.g., chiggers or fleas). Many infections can be fatal including rabies, Lassa fever and Ebola virus. Fortunately, most infections are not fatal and are easily prevented. The preventive steps to take vary depending on the agent of concern but most require only very simple procedures such as wearing protective clothing (including gloves), wearing a NIOSH-approved respirator (face mask), applying insect repellents, and practicing simple hygiene to break the oral-fecal cycle (e.g., hand washing). For workers likely to be exposed to zoonotic pathogens, there are vaccines available for many diseases, although in the United States some are only available from the Centers for Disease Control and Prevention.

INTRODUCTION

A zoonotic illness is an infectious disease transmitted between vertebrate animals and humans. There are over two hundred known zoonotic diseases, and the number is increasing yearly. Zoonotic diseases are responsible for millions of infections and tens of thousands of deaths each year worldwide. In the United States, tens of thousands of infections occur yearly. Fortunately, few deaths occur in the United States (CDC 1989). Among the professions associated with zoonotic infections are livestock management, forestry work, fish and shellfish processing, waste management, and recreation management.

TRANSMISSION AND INFECTIVITY OF DISEASE

Zoonotic agents may be transmitted by any of several routes. Arthropods frequently act as vectors in the transmission of diseases between humans and animals. Other methods of transmission include animal bites and scratches, inhalation of infectious aerosols and ingestion of contaminated food and water. Additionally, transmission may occur from accidental puncture wounds and cuts from needles or scalpels used to treat infected animals.

Numerous factors increase the susceptibility of workers to infection. The resistance or susceptibility of the host may be affected by temperature extremes, stress, fatigue, poor general health, underlying disease, pregnancy, vaccination status, immunosuppression, age, race, and sex (Tulis 1994; Kligman 1991). Additionally, the exposure route and dose, as well as the virulence of the specific pathogen are important factors. Following exposure to an organism, the infective process may lead to clinical, subclinical, or asymptomatic disease.

Etiologic Agents

Zoonotic diseases may be caused by bacteria, viruses, rickettsia, chlamydia, parasites, and fungi (Joklit et al.. 1992). A brief discussion of each class of zoonotic agent follows.

Bacteria. Bacteria are single-celled organisms with cell walls and a nucleus containing DNA and RNA. Procaryotes such as bacteria do not have a membrane-bound nucleus as do eucaryotes. Bacteria are morphologically characterized by shape as cocci, bacilli, and spirilla. Bacteria may be differentiated based on the gram stain and other biochemical tests. Bacteria are usually considered to be gram positive or gram negative. Gram negative bacteria possess a lipopolysaccharide component of the cell wall that displays toxic properties and is referred to as endotoxin. There are thousands of known species of bacteria, and many zoonotic infections are due to bacteria.

Viruses. Viruses are responsible for the great majority of human infections. Viral classification depends on morphology, the presence of envelopes surrounding the viral capsid, and the type of genetic material (RNA or DNA). Viruses are obligate intracellular parasites and cannot multiply outside the host cell. Their survival in the environment outside the host cell is limited, ranging from a few hours to a few weeks.

Rickettsia. Rickettsia are intracellular parasites containing both RNA and DNA. They are usually coccobacillary in morphology and resemble gram negative bacteria. They are usually transmitted to humans through arthropod vectors. However, certain agents such as *Coxiella burnetii*, the agent responsible for Q fever, may be transmitted through contaminated aerosols.

Chlamydia. Chlamydia are obligate intracellular parasites that differ from viruses in that they are susceptible to antibiotics. They contain both DNA and RNA, possess a cell wall similar to that of gram negative bacteria, and synthesize their own proteins, nucleic acids, and lipids. The primary reservoir of zoonotic chlamydia are psittiacine birds and domestic chickens and turkeys. Psittacosis is readily transmitted by the aerosol route.

Fungi. Fungi are eucaryotic organisms and are comprised of molds and yeast. Fungi contain at least one nucleus, a nuclear membrane, endoplasmic reticulum, and mitochondria. Most fungi are obligate or facultative aerobes. Most zoonotic fungal infections are spread by direct contact or the aerosol route.

Parasites. Parasites are eucaryotic organisms, either unicellular or multicellular. Parasites often have complex life cycles that involve sexual and asexual reproductive states. A parasite that lives on the surface of the host is called an ectoparasite; if it lives in the tissues, it is an endoparasite. On a global basis, parasitic infection in association with malnutrition is a primary cause of morbidity and mortality.

See Appendix A for Tables of Occupational Infections.

PREVENTION AND CONTROL OF ZOONOTIC HAZARDS

For most societies the cost of totally eliminating the risk of contracting all the infectious diseases is too high. Minor infections are rarely a serious concern. We are willing to accept the occasional inconvenience of a cold or the flu for the convenience of working and being with other people. The more serious the consequences of the disease, however, the more cost we will incur and the more inconvenience we are prepared to suffer to combat the illness. Public health departments and organizations, such as the World Health Organization (WHO), help ensure that we are protected from infectious diseases by protecting the quality of our drinking water, managing the sewage system to break the fecal-oral cycle of disease transmission, and appropriately vaccinate our children. If taken at the societal level, these preventive measures can dramatically reduce the risk of infectious disease. The ultimate responsibility, however, lies with individuals. They must be aware of the health risks they are taking and take appropriate preventive action.

Understanding a pathogen's mode of transmission enables agriculture workers to adopt preventive practices that minimize the risk of disease transmission. Diseases can be transmitted five ways: though skin contact, penetration, inhalation, insect bites, and ingestion. Some disease can be transmitted a variety of ways. For example, tularemia can be transmitted by penetration, inhalation, or insect bites. Others can be transmitted only one way, such as Cat Scratch Fever which is only transmitted through a cat biting or scratching its victim.

Skin Contact

It is relatively easy to prevent skin contact with infectious disease—gloves, shoes, long pants, and long sleeved shirts can be worn, as needed. For example, to prevent the direct transfer of the fungi responsible for ringworm and sporotrichosis protective gloves and clothing should be worn when handling animals or working with known infected people. These fungi can also be spread indirectly so it is important not to handle contaminated clothing or other objects. If possible, work clothes should be washed at the worksite and due care should be exercised when handling contaminated articles. Schistosomiasis can be prevented by donning waders or full suits before entering contaminated water. Wearing shoes will prevent Cutaneous Larva Migrans.

Diseases and infestations that can be spread by direct skin contact with infectious agents or vectors include:

- Ringworm
- Scabies
- Fleas
- Anthrax
- Orf
- Stomatis
- Erysipeloid
- Monkeypox
- Leptospirosis
- Brucellosis
- Q-Fever
- Pasteurellosis
- Glanders
- Ornithosis
- Lymphocytic Choriomeningitis
- Strongyloidiasis
- Creeping Eruptions
- Tetanus

Penetration

Preventing skin penetration by infectious agents is similar to the procedures for preventing contact except the protective clothing usually needs to be more substantial. Herpes B is acquired from the bite of infective monkeys and can be prevented by wearing a heavy glove or gauntlet. Leather or mesh gloves can prevent the bites and scratches of feral cats, which can result in Cat Scratch Fever. Animal handlers, however, often find these gloves too restrictive and accept the risk. Rarely, rabies may be spread by inhalation but most of the time this infection is established through the bite of an infective animal. Because the bite of a rabid animal can be fatal you should never attempt to handle a known rabies carrier, even with protective gloves. A suspected rabid animal should be killed, taking care not to damage the head because the brain is needed to establish a diagnosis. Needlesticks, the most common injuries experienced by veterinarians and capable of introducing a variety of infectious agents, are best prevented by not recapping needles or using forceps or other mechanical devices to remove the needle. Again, the wearing of protective gloves is often regarded as too restrictive.

Diseases that can be spread by a penetrating injury include:

- Staphyloccus infection
- Streptococcus infection
- Pasteurellosis
- Rabies
- Cat Scratch Fever
- Rat Bite Fever
- Plague
- Tularemia
- Lymphocytic Choriomeningitis
- Herpes B

Inhalation

Wearing NIOSH-approved face masks will prevent users from inhaling infectious diseases when working in high risk situations where diseases such as tuberculosis, Q-fever, and brucellosis are prevalent. Unfortunately, major outbreaks of disease have occurred in situations where face masks were not used. In 1995, several people contracted tuberculosis from a fellow bar patron. This emphasizes the importance of other preventive measures such as quarantine, testing, and vaccination.

Diseases that can be spread by inhalation of the pathogens:

- Tuberculosis
- Q-Fever
- Pasteurellosis
- Tularemia
- Glanders
- Anthrax
- Lymphocytic Choriomeningitis

Insect Bites

Covering exposed skin and applying insect repellents to clothing and exposed skin will prevent transmission of many insect-borne diseases. Effective repellents include citronella, DEET, and neem oil. Permethrin can be applied to clothing and bed netting for further protection.

Diseases that are spread by insect bites include:

- Plague
- Tularemia
- Typhus
- Rocky Mountain Spotted Fever
- Lyme Disease
- Babesiosis
- Rickettsialpox
- Relapsing Fever
- Trypanosomiasis
- Leishmaniasis
- Malaria
- Encephalitis
- Ehrlichiosis

Ingestion

Breaking the oral-fecal cycle depends primarily on observing personal hygiene. At work you should always wash your hands before handling food or touching your mouth. Contaminated water can be decontaminated by adding iodine tablets to the water or by boiling or filtering the water.

Diseases that are spread by ingesting pathogens include:

- Visceral larvae migrans
- Strongyloidiasis
- Hydatid disease
- Giardiasis
- Amoebiasis
- Toxoplasmosis
- Cryptosporidiosis
- Campylobacteriosis
- Salmonellosis
- Shigellosis
- Hepatitis A
- Lymphocytic Choriomeningitis

Cleaning and Disinfection

Because most disinfectants are much less effective in the presence of organic matter, to decontaminate your work area you should first clean the area and then, only after it is clean, apply a disinfectant. After mechanically removing most of the dirt, clean the surface with a good detergent. Rinse off the detergent, then apply an appropriate disinfectant. To treat urine scale, first apply an acid solution so water soluble salts are formed that can be washed away. Some detergents have disinfectants already incorporated and they work well; however, never mix them yourself because some mixtures generate poisonous gasses (e.g., chlorine and ammonia) or are corrosive and may damage the surface you are trying to disinfect. Disinfectants are generally more effective the longer they are in contact with the contamination. Be sure the disinfectant chosen is effective against the pathogens of interest. In summary, always follow the directions on the label. Factors to consider when choosing the appropriate disinfectant include cost, efficacy, safety, corrosiveness, and the surfaces that need to be disinfected (Table 8.1).

Vaccines for the Zoonotic Diseases

Vaccines available include:

Available in United States:
- Anthrax
- Japanese Encephalitis
- Cholera
- Hepatitis A and B
- Plague
- Rabies
- Tetanus
- Yellow Fever
- Tuberculosis (BCG)

Not Available in United States:
- Junin
- Tick-borne Encephalitis
- Rift Valley Fever
- Congo-Crimean Hemorrhagic Fever
- Omsk Hemorrhagic Fever
- Kysanur Forest disease
- Leptospirosis

Vaccine Under Development
- Dengue
- Lyme disease
- Malaria
- Rocky Mountain Spotted Fever
- Paratyphoid
- African Trypanosomiasis

Contact CDC or U.S. Army
- Eastern Equnie Encephalitis
- Western Equine Encephalitis
- Venezuelan Equine Encephalitis
- Swine Influenza
- Monkey Pox
- Tularemia
- Q-Fever

DIAGNOSIS AND TREATMENT

The diagnosis of a possible zoonotic infection in humans depends on the likelihood of exposure to an infected animal or vector. The occupational history may suggest risk factors for exposure. Other lines of inquiry include a travel history, especially to third world countries. Questions on hobbies such as hunting or spelunking may provide clues to the cause of a person's illness. Does the patient have ill pets? Has he or she been recently bitten by ticks, fleas, or other arthropods? Has he or she been previously immunized against certain zoonotic agents? What symptoms does the patient have and when did they begin?

Once the history has been obtained a physical examination is performed to look for signs of infection. Cultures of appropriate body fluids, secretions, excretions, or skin lesions should be obtained. Occasionally, skin tests

Table 8.1. Characteristics of disinfectants.

	Chlorine compounds	Iodophors	Phenolics	Quaternary Ammonium Compounds	Alcohol	Aldehydes	Chlorhexidine	Sodium hydroxide
Common brands[1]:	Chlorox	Betadine, Pharmadine	Environ, Lysol	Zephiran	Rubbing alcohol	DC&R, Formalin	Nolvasan	Lye
Efficacy:								
Bacteria	Yes	Yes	Yes	Yes but only partial against gram (-)	Yes	Yes	Partial	Yes
Bacterial Spores	Partial	Partial	No	No	Partial	Yes	Partial	Yes
Viruses	Partial	Partial	Partial	Partial	Yes	Yes	Partial	Yes
Fungi	Yes	Yes	Partial	Partial	Yes	Yes	Partial	Yes
Activity in organic material:	Very poor	Very Poor	Excellent	Fair	Fair	Good	Good	Good
Residual activity:	Only chloramines have good residual activity	Up to 7 days when combined with detergents	Very good	No	No	None, unless using slow-release forms	Yes	Yes
Disadvantages:	Inactive in presence of organic material	Should not be used with alkali soaps.	Cresol has a strong odor	Can not use with soaps, limited spectrum of activity	Irritant	Noxious fumes	Narrow spectrum of activity	Very caustic
Notes:	Good for disinfecting clean equipment. Cheap and broadly effective	Good for disinfecting skin or clean equipment	Good for disinfecting soiled boots, facilities, or equipment	Effective at greater than 110°F	Good disinfectant for surface disinfection such as bench tops.	Formaldehyde gas works best at 80-140°F	Effective at high pH.	Heat is not detrimental. Lye is cheap and effective, but very caustic

[1] Reference to a product is not intended to be an endorsement to the exclusion of others which may be similar.

may help diagnose infections due to mycobacteria or fungi. Acute and convalescent serologic testing may confirm a suspected illness. Newer diagnostic techniques such as polymerase chain reaction (PCR) test may aid in the identification of an infectious agent.

Table 8.2. Safety checklist.

To Avoid Injuries in General:

- Take the time to assess the risks.
- Get assistance if needed.
- Don't rush (don't be overconfident).
- Don't exceed your physical limitations.
- Remain alert and aware of opportunities for injury.
- Wearing protective gear where appropriate, e.g., protection for eyes, ears, dust masks, shin guards, and skid resistant steel toe shoes.

To Avoid Injuries from Needle Sticks and Sharps:

- Don't recap needles.
- Use forceps or other mechanical devices for removing the needle..
- Remove blades with forceps or use disposable handles.
- Use leather gloves if you must handle broken glass, etc.

To Avoid Infectious Diseases and Animal Allergies:

- Ensure your rabies and tetanus shots are up-to-date.
- Wear gloves.
- Wash your hands after handling animals.
- Launder all you work clothes at work, not at home, because you may be unwittingly exposing your family to a serious disease.
- Start the day with clean work clothes and change them if they become soiled or after contact with a possible zoonotic agent.
- Don't eat, drink, or smoke around your cases.
- Use insect repellant when working in the field—check your body for ticks.
- Avoid self treatment—you can misdiagnose a serious disease with fatal consequences.
- If your allergy symptoms are not relieved by using a mask, see your physician.
- If you have life-threatening reactions to insect bites always carry an anaphylaxis kit.

The treatment of a zoonotic infection depends on the organism identified. Specific treatment recommendations are beyond the scope of this chapter. For many agents, effective therapy is available, although a prolonged duration of therapy may be needed. Unfortunately, few antiviral medications are available and supportive therapy is all that can be done for the majority of patients.

SUMMARY

Although agricultural workers are usually more at risk from the machinery they operate (e.g., tractor accidents), they are also at risk from less tangible microorganisms that inhabit their work environment. The range of pathogens is very broad including bacteria, viruses, ricksettia and chlamydia, fungal, viral, multicellular parasites, and infestations with insects (e.g., chiggers or fleas). Many infestations can be fatal including the infamous Lassa fever and Ebola virus. Fortunately, most are not fatal and are easily prevented.

While public health agencies can, and should, institute protection at the societal level, the primary responsibility still rests with the individual. It is the individual's responsibility to ensure they are aware of the zoonotic risks associated with moving into a new geographic area or coming into contact with insect or animal vectors. Once one is aware of the risk then it is important to take the appropriate preventive measures outlined in this chapter. The preventive steps will vary depending on the agent of concern but most require very simple procedures such as wearing protective clothing, including gloves, wearing a NIOSH-approved face mask, applying insect repellents, and practicing simple hygiene to break the oral-fecal cycle. For workers exposed to zoonotic diseases there are vaccines available for many diseases, although in the United States some are only available from the Centers for Disease Control and/or the U.S. Army.

REFERENCES

Benenson, A.S. 1995. *Control of Communicable Diseases Manual.* Washington, DC: American Public Health Association.

Centers for Disease Control (CDC). 1989. Summary of notifiable diseases in the U.S. *MMWR* 38:3.

Gilson, G., J. Maciulla, B. Nevils, L. Izquierdo, M. Chatterjee, and L. Curet. 1994. Hantaa virus pulmonary syndrome complicating pregnancy. *Amer. J. Obst. Gyn.* 171: 550-54.

Gleicher, N., S. Gall, B. Sibai, U. Elkayom, R. Galbraith, and G. Sarto, eds. 1992. *Principles and Practice of Medical Therapy in Pregnancy.* 2nd edition. Norwalk, CT: Appleton and Lange.

Heinsohn, P., R. Jacobs, and B. Concoby, eds. 1995. *Biosafety Reference Manual.* Fairfax, VA: American Industrial Hygiene Association.

Mead, P.B., and D. Hager, eds. 1992. *Infectious Diseases Society for Obstetrics and Gynecology: Infection Protocols for Obstetrics and Gynecology.* Montvale, NJ: Medical Economics Publishing.

Jasper, P., A. Peedicayil, S. Nair, and R. George. 1989. Hydatid cyst obstructing labour; a case report. *J. Tropical Med. Hyg.* 92: 393-95.

Joklit, W., H. Willett, D. Amos, C. Wilfert, eds. 1992. *Zinsser Microbiology.* 20th edition. Norwalk, CT: Appleton and Lange.

Kligman, E., W. Peate, and D. Cordes. 1991. Occupational infections in farming. In: *Health Hazards of Farming.* D. Cordes and D. Rea, eds. *Occup. Med. State of the Arts Reviews*, 6:3. Philadelphia: Hanley and Belfus, Inc.

Langley, R.L., W.H. Pryor, Jr, and K.F. O'Brian. 1995. Health hazards among veterinarians: a survey and review of the literature. *J. Agromed.* 2: 23-52.

Mandell, G., R. Douglas, and J. Bennet, eds. 1990. *Principles and Practice of Infectious Diseases.* 3rd edition. New York: Churchill Livingston, Inc.

Moore R.M., Jr, Y.M. Davis, and R.G. Kaczmarek. 1993. An overview of occupational hazards among veterinarians, with particular reference to pregnant women. *Am. Ind. Hyg. Assoc. J.* 54: 113-20.

Schnurrenberger, P.R., and W.T. Hubbert. 1981. *An Outline of the Zoonoses.* Ames, IA: The Iowa State University Press.

Silberberg L., P. Rollin, G. Kerouani, and D. Courdrier. 1993. Hemorrhagic fever with renal syndrome and pregnancy; a case report. *Trans. Royal Soc. Tropical Med. Hyg.* 87: 65.

Sweet R., R. Gibbs, eds. 1995. *Infectious Diseases of the Female Genital Tract.* Baltimore, MD: Williams and Wilkins.

Tulis, J., and W. Stopford. 1994. General principles of microbiology and infectious diseases. In: *Physical and Biological Hazards of the Workplace.* P. Wald and G. Stave, eds. New York: Van Nostrand.

Wormser, G.A. 1995. Vaccine against lyme disease? *Ann. Intern. Med.* 123: 627-28.

Appendix A: Tables of Occupational Infections.

Bacterial Infections:

Disease	Agent	Mode of Transmission	Typical Symptoms	Worst case	Most important prevention	Human vaccine available	Special risks for females or fetus?
Anthrax	*Bacillus anthracis*	Direct contact, ingestion of uncooked meat, inhalation of the spores	Cutaneous form: pruritic macules that form vesicles then necrotic skin lesions. Pulmonary form: fever, malaise, dypsnea, cyanosis with death following. Intestinal form: Anorexia, vomiting, diarrhea, fever, malaise.	Death if untreated	Personal hygiene. Wear a mask when working in high-risk situations.	Yes, from CDC	No
Brucellosis	*Brucella spp.*	Ingestion of contaminated food, esp. milk, and aerosol.	Chronic and recurrent flu-like symptoms	Rarely death	Personal hygiene especially when treating abortions. Wear a mask when working in high-risk situations. Don't drink unpasteurized milk.	No	Has been associated with miscarriages
Campylo-bacteriosis	*Campylobacter fetus* ssp. *intestinalis* and *C. fetus* ssp. *jejuni*	Injection of contaminated food or water. Humans may be carriers.	Chills, sweats, fever, cough, headache, weight loss, inappetence, vomiting, and diarrhea.	Death. Case fatality 50% in infants.	Avoid drinking untreated water, handle food carefully, don't drink unpasteurized milk.	No	Yes, fetal death
Cat Scratch Fever	*Bartonella henselae*	Cat scratch, or bite.	Fever, malaise, and lymphadenitis. Red papule usually develops at site.	Potentially fatal if immune suppressed.	Adequate restraint when handling cats, use tranquilizers.	No	No
Erysip-eloid	*Erysipelothrix rhusiopathiae*	Contact exposure through wounds	Dark red swollen lesion that grows outwards, usually on the hands.	Endocarditis and death if untreated	Personal hygiene. Wear a mask when working in high-risk situations.	No	No
Lepto-spirosis	*Leptospira interrogans*	Direct contact with urine through skin lesions, rarely inhaled as droplet or ingested.	Flu-like symptoms, occasionally jaundice.	Death. Case fatality rate increases with age and may be greater than 20% for patients not receiving renal dialysis.	Personal hygiene. Wear a mask when working in high-risk situations.	Not generally in USA. Is available in other countries	Yes, fetal death
Listeriosis	*Listeria monocytogenes*	Direct contact, ingestion. Present in manure, soil, silage, and sewage.	Meningoencephalitis, endocarditis, pneumonia, conjunctivitis, skin pustules.	Fetal death, death	Personal hygiene especially when treating animal abortions.	No	Yes, fetal death
Lyme Disease	A spirochete, *Borrelia burgdorferi*.	Tickborne, usually *Ixodes scapularis*	Flu-like symptoms. Distinctive, expanding "bull's eye" primary skin lesion.	Cardiac, neurologic, rheumatologic complications, and rarely death.	Spray tick repellant on clothing when working in the field especially in summer. Check for ticks after work. Prophylactic antibiotic treatment if pregnant.	Vaccine trials are in progress	Possibly, has been associated with premature birth, fetal death and congenital abnormalities.
Pasteur-elosis	Various including: *Pasteurella multocida*, *Pasteurella haemolytica*, and *Pasteurella pneumotropica*	Animal contact, typically an animal bite or scratch.	Pain, swelling and cellulitis at the infection site. Lymphadenopathy and sepsis can occur.	Very rarely death.	Avoid aggressive animals, clean bites and scratches immediatly with soap and water.	No	No
Plague, Bubonic plague	*Yersinia pestis*	Usually from the bites of an infected rodent fleas. May be from direct exposure (aerosols) or contact with infected tissues.	Bubonic form: fever and lymphadenitis in the nodes draining the site of the flea bite. Systemic form: Can travel via the blood to site throughout the body including the brain. Localizing in the lungs leads to the Pneumonic form. Aerosolized droplets of sputum can infect other people.	Death. Untreated bubonic plague has a case fatality rate of 50%.	Flea repellant. Local control programs include public education, personal hygiene, and aggressive rat control including rat poison. Wear a mask when working in high-risk situations.	Yes, usually 3 doses of a killed vaccine with boosters.	No
Rat Bite Fever	*Spirillum minor* and/or *Streptobacillus moniliformis*	Transmitted directly by the bite of a rat or other rodents including squirrels.	Extensive swelling at the bite site and flu-like symptoms.	Death. Untreated, they have a case fatality rate of 10%.	Rat control	No	No

Disease	Agent	Mode of Transmission	Typical Symptoms	Worst case	Most important prevention	Human vaccine available	Special risks for females or fetus?
Tetanus	*Clostridium tetani*	Organism spores directly introduced into the body usually by a puncture wound, abrasion or laceration and sometimes ignored because the wound seems so trivial.	Intermittent then continuous muscular spasms beginning with the jaw, followed by the neck, arms, trunk, and legs.	Death by asphyxiation. Case fatality rate 30-90%	Immediatly clean all wounds, periodic vaccination.	DTP, triple antigen (diptheria, tetanus, and pertussis) vaccine is usually given to children and protects to 10 years. Give booster doses every 10 years.	No
Tuber-culosis	*Mycobacterium spp.*	Ingestion, inhalation of aerosols, or direct contact.	Primary stage, which may last for months, is mild but the secondary can be severe with fever, cough, fatigue, night sweats, coughing blood. If the bacteria leaves the lung patients can suffer lymphadenitis, meningitis, osteomyelitis, or pericarditis.	Chronic debilitation and death if untreated.	Personal hygiene. Don't drink unpasteurized milk. Wear a face mask when working in high-risk situations.	BCG (Bacillus Calmette-Guerin) is used outside the USA.	Disseminated tuberculosis may be more severe for pregnant women. Babies may be infected in-utero.
Tularemia	*Francisella tularensis*	Direct contact with infected animal, aerosol, or ingestion, or arthopods bites.	Ulcer at site of wound and lymphadenopathy.	Death if untreated, case fatality rate is 4-6%	Personal hygiene	Yes, from CDC	No

Note: All bacterial diseases above occur in the USA.

Rickettsial and Chlamydial:

Disease	Agent	Mode of Transmission	Typical Symptoms	Worst case	Most important prevention	Human vaccine available?	Special risks for females or fetus?
Ehrlichiosis	Ehrlichia sennetsu and Ehrlichia chaffeenis	Tickborne	Flu-like symptoms, occasionally a rash.	Rarely death.	Avoid tick infested areas, wear light-colored long pants, long sleeves, and apply tick repellant to clothing. Check for ticks on your body after work.	No	No
Psittacosis	Chlamydia psittaci	Inhalation of the agent from the dried excretment of infected birds, especially psittacines.	Variable but can include fever, headache, rash, diarrhea, vomiting, and respiratory disease.	Rarely death in the very young or old.	Personal hygiene and wear a face mask or respirator particularly when dealing with an outbreak.	No	Fetal death, and increased likelihood of a severe illness if pregnant.
Rocky Mountain Spotted Fever	Rickettsia rickettsii	Tickborne	Flu-like symptoms leading to a rash. Most suffer a severe headache.	Death for 5-7% of cases.	Avoid tick infested areas, wear light-colored long pants, long sleeves. Apply tick repellents on clothing when working, especially in summer. Check for ticks particularly after working with dogs.	No	May cause a miscarriage or premature delivery.
Q fever	Coxiella burnetti	Aerosol	Flu-like symptoms with hepatitis and pneumonia.	Rarely death	Personal hygiene and wearing a face mask particularly when dealing with an outbreak.	Yes, contact CDC[1]	Small risk for miscarriage and stillbirth.
Scrub typhus	Rickettsia tsutsu-gamushi	By the bite of infective laval mites.	A skin ulcer and then headache with profuse sweating. Later a fever and dark red eruptions over the body, coughing and pneumonia.	Rarely death, but in untreated older people case fatality rate may reach 60%	Avoid mite infested areas. Wear clothing impregnated with miticidal chemicals (e.g., benzyl benzoate) and apply mite repellants (e.g., diethyltoluamide) to exposed skin and outside surface of clothes.	No	No
Typhus fever (epidemic)	Rickettsia prowazekii	Louse-borne	Flu-like symptoms followed in 5-6 days by macular eruptions, initially on the upper body then on the entire body.	Case fatality rate increase with age up to 40%	Personal hygiene and lice eradication.	Yes, CDC	No
Typhus fever (murine, endemic)	Rickettsia typhi	Infective rat fleas defecate Rickettsia typhi and contaminate their bite-site and other open wounds.	Similar to, but milder than, regular Typhus Fever	Overall case fatality rate is 2%. It increases with age.	Personal hygiene and flea and rat eradication.	No	No

[1] Vaccine is available from Centers for Disease Control, Atlanta, GA 30333. (404) 329-3145

Fungal:

Disease	Agent	Mode of Transmission	Typical Symptoms	Worst case	Most important prevention	Human vaccine available?	Special risk for females or fetus?
Blasto-mycosis	*Blastomyces dermatitidis*	Inhalation of spore-laden dust	Acute: fever and cough Chronic: cough, chest pain, weight loss, low fever, and weakness.	Death possible if untreated.	Unknown. It lives in soil, dogs, cats, and horses.	No	Lower prevalence in females.
Coccidio-mycosis	*Coccidioides immitis*	Inhalation of infective arthroconidia.	No symptoms or flu-like.	Disseminating form frequently fatal, with lung lesions and abscesses throughout the body.	Control dust and wear a face mask.	No	Lower prevalence in females. However, organism is more likely to spread and result in a more severe disease in pregnant females
Hist-plasmosis	*Histoplasma capsulatum* var. *capsulatum*	Inhalation of airborne conidia.	Usually no symptoms, 5 types are recognized: 1. Asymptomatic 2. Acute benign, flu-like symptoms 3. Acute disseminating, liver and spleen enlarged and prostration. 4. Chronic disseminating, fever, anaemia, hepatitis, endocarditis, meningitis, and mucosal ulcers. 5. Chronic pulmonary, signs similar to chronic pulmonary tuberculosis.	3. Acute disseminating form is usually found in the young and, without treatment, is usually fatal. 4. Chronic disseminating form is usually fatal unless treated.	Control dust, wear a face mask, and spray infectious areas with 3% formalin.	No	The chronic pulmonary form is more common in males over 40.
Hist-plasmosis	*Histoplasma capsulatum* var. *duboisii*	Unknown	Sub acute granuloma of the skin or bone.	Granulomas may spread throughout the body.	Unknown, but it is not transmissable person-to-person.	No	No
Ringworm	Most commonly: *Microsporum canis*, *Trichophyton -menta-grophytes*, and *Trichophyton verrucosum*	Contact with infective animals or people.	Scaly skin, hair loss, itching.	A kerion (large open wound) formation	Routine personal hygiene, wear gloves.	No	Some species may have higher prevalence among females, others may have a higher prevalence among males.
Sporo-trichosis	*Sporothrix schenckii*	Contact with infective animals	Raised, red, papules that enlarge and ulcerate. Lymphatics draining the area become firm and form a nodular chain.	May spread to involve joints and the brain.	Routine personal hygiene, wear gloves	No	No

Viral:

Disease	Agent	Mode of Transmission	Typical Symptoms	Worst case	Most important prevention	Human vaccine available?	Special risk for females or fetus?
Lympho-cytic Chorio-meningitis	Arenavirus	Inhaled and injested.	Flu-like symptoms for a few days then a complete recovery. Other people may suffer inflammation of the brain, testicles or salivary glands.	Very rarely fatal.	Personal hygiene and wear a face mask when working in high-risk situations	No	No
Newcastle disease	Paramyxo-virus	Contact with infected birds. Bird live- vaccines are pathogenic for humans.	Conjunctivitis, flu-like symptoms.	Flu-like symptoms	Personal hygiene	No	No
Orf, contagious ecthyma	Parapoxvirus	Direct contact with infected animal through skin lesions.	Local painless papular lesion usually on the hands or arms.	Swollen lymph nodes can become painful, occular damage	Personal hygiene	No	No
Rabies	Lyssavirus	Infective-animal bites	Personality change, aggression, fever, paralysis, and others.	Usually fatal	Vaccination. Use extreme caution when handling suspected rabid animals.	Yes	No
Ebola	Filovirus	Direct contact with infective blood, secretions, organs, semen, or by aerosols.	Malaise, fever, myalgia, headache, and pharyngitis, vomiting, diarrhea, maculopapular rash, and kidney and liver involvement.	Death. Case fatality rates up 90% are reported.	Reservoirs of infection are unknown. Avoid contact with infective people.	No	The virus can be transmitted through semen.
Lassa	Arenavirus	Contact with urine of infected rodents in dust or on food. Contact with the blood, urine or pharyngeal secretions of infective people.	Malaise, headache, fever, sore throat, cough, nausea, vomiting, diarrhea, myalgia, chest and abdominal pain. Commonly the face or neck swell.	Death	Deter rodents from colonizing the home and work.	No	No
Hantaan	Hantavirus	Inhalation of dust contaminated with infected rodent excreta	Malaise, headache, fever, sweats, prostration, hemorrhage, and renal abnormalities	Death	Rodent control	No	Pregnancy complications
Sin Nombre	Hantavirus	Inhalation of dust contaminated with infected rodent excreta	Malaise, headache, fever, sweats, prostration, and pulmonary edema.	Death	Rodent control	No	Fetal damage may occur from hypoxemia

Viral Diseases Transmitted by Mosquitoes and Ticks:

Disease	Agent	Mode of Transmission	Typical Symptoms	Worst case	Most important prevention	Human vaccine available?	Special risk for females or fetus?
California encephalitis	bunyavirus	Bite of infective mosquito	fever, encephalitis, meningitis	Death	Mosquito control. Spray mosquito repellant, DEET is very effective, on clothes and leave no exposed skin.	No	Insufficient data
Colorado tick fever	orbivirus	Bite of infective tick	fever, arthralgia	Encephilitis or death	Tick control, use DEET.	No	Can cause fetal abnormalities in laboratory animals
Dengue	flavivirus	Bite of infective mosquito, especially *Aedes aegypti*	fever, vomiting, headache, abdominal pain, arthralgia, diarrhea, hemorrhage	death rare unless hemorrhagic form: CFR 40-50% if untreated, less than 5% if treated.	Mosquito control. Spray mosquito repellant, DEET is very effective, on clothes and leave no exposed skin.	Vaccine trial underway	Congenital transmission may occur
Eastern equine encephalitis	alphavirus	Bite of infective mosquito	fever, encephalitis, meningitis, arthralgia	death	Mosquito control. Spray mosquito repellant, DEET is very effective, on clothes and leave no exposed skin.	No	Insufficient data
Everglades viral encephalitis	alphavirus	Bite of infective mosquito	fever, encephalitis, meningitis		Mosquito control. Spray mosquito repellant, DEET is very effective, on clothes and leave no exposed skin.	No	Insufficient data
La Crosse	bunyavirus	Bite of infective mosquito	fever, encephalitis, meningitis	death	Mosquito control. Spray mosquito repellant, DEET is very effective, on clothes and leave no exposed skin.	No	Insufficient data
Powassan	flavivirus	Bite of infective tick	fever, meningitis	Encephalitis	Tick control, use DEET.	No	Insufficient data
Rio Bravo	flavivirus	Unknown	fever, meningitis	Encephalitis		No	Insufficient data
St. Louis encephalitis	flavivirus	Bite of infective mosquito	fever, encephalitis, meningitis	death	Mosquito control. Spray mosquito repellant, DEET is very effective, on clothes and leave no exposed skin.	No	May infect fetus
Tacaiuma	bunyavirus like	Bite of infective mosquito	fever		Mosquito control. Spray mosquito repellant, DEET is very effective, on clothes and leave no exposed skin.	No	Insufficient data
Venezuelan equine encephalitis	alphavirus	Bite of infective mosquito	fever, encephalitis, meningitis, arthralgia	death	Mosquito control. Spray mosquito repellant, DEET is very effective, on clothes and leave no exposed skin.	Yes, CDC	Possibly teratogenic,
Western equine encephalitis	alphavirus	Bite of infective mosquito	fever, encephalitis, meningitis, arthralgia	death	Mosquito control. Spray mosquito repellant, DEET is very effective, on clothes and leave no exposed skin.	No	Possibly teratogenic, may infect fetus, produce stillborn, or abortions.
Yellow fever	flavivirus	Bite of infective mosquito, especially *Aedes aegypti*	Fever, headache, nausea and vomiting.	Death following prostration, jaundice, renal failure and generalized hemorrages.	Mosquito control. Spray mosquito repellant, DEET is very effective, on clothes and leave no exposed skin.	Yes	No

All viruses causing an encephalitis may produce long term neurologic problems in 1-20% of cases.

Parasites:

Disease	Agent	Mode of Transmission	Typical Symptoms	Worst case	Most important prevention	Special risk for females or fetus?
Amebiasis	A protozoan: Entamoeba histolytica	Fecal-oral, usually through contaminated water.	Most infections cause no symptoms. Fever, chills, diarrhea, alternating with constipation or remission.	Skin ulceration esp. in the perianal region. Abscesses in the liver, lung or brain	Personal hygiene. Safe water supply.	Pregnancy may increase the severity of the disease. May impair fertility by damage to reproductive organs.
Chagas Disease	Trypanosoma cruzi, a systemic protozoa.	Contamination of wounds by infective feces of blood-sucking cone-nosed bugs (triatomids).	Varies from no symptoms to swollen eyelids, conjunctivitis, fever, malaise, lymphadenopathy, and hepatosplenomegaly.	Myocarditis and meningoencephalitis. Death	Insect control. Use insecticides and bed nets.	Increased chance of premature labor, abortion, or congenital infection.
Crypto-sporidiosis	Cryptosporidium sp., a coccidian protozoan.	Fecal-oral, often through contaminated water.	Diarrhea, vomiting, wasting, abdominal pain. May be no symptoms.	Severe prolonged diarrhea with wasting and death.	Personal hygiene. Safe water supply.	Unknown
Cutaneous Larva Migrans	The larvae of the cat and dog hookworms: Ancyclostoma braziliense and Ancyclostoma caninum and Necator americanus	Skin penetration by parasite lavae.	Itchy red papular rash that develops into vesicular eruptions as the larvae move. The disease will spontaneously cure within weeks or a few months.	Secondary bacterial infection. Rarely systemic involvement.	Wear shoes in endemic areas. Treat infected cats and dogs or keep them at home and especially off beaches and wet-sandy areas like playgrounds.	Unknown
Hydatids	Echinococcus granulosus and Echinococcus multilo-cularis	Ingestion of eggs present in contaminated dog feces	Abdominal mass and eosinophilia	death	Proper disposal of the viscera of herbivores. Do not feed viscera to dogs.	Impaired fertility due to direct damage to reproductive organs. Cysts may complicate labor.
Filariasis	Lymphatic dwelling filariae: Wuchereria bancrofti, Brugia malayi, and Dirofilaria immitis	By the bite of an infective female mosquito.	Lung lesions, skin swellings, recurrent fever, lymphadenopathy, coughing.	Gross swelling of limbs, called elephantiasis.	Mosquito control. Spray mosquito repellant, e.g., DEET, on clothes and leave no exposed skin.	Possibly congenital infections. Females may be resistant to certain filaria.
Giardiasis	Giardia lamblia a flagellate protozoan	Ingestion of cysts in fecally contaminated water or food.	Usually no symptoms. Diarrhea with foul-smelling flatus, greasy stools, abdominal cramps, bloating, fatigue, and weight loss.	Imflammation of synovial membranes.	Personal hygiene, clean water supply.	No
American Leishmaniasis	Tissue flagellates: Leishmania mexicana and, Leishmania brasiliensis	By the bite of infective sand flies.	Nodular lesions on the skin and mucocutaneous regions that may be painful and ulcerate. Leishmania mexicana causes self-limiting cutaneous lesions.	Leishmania brasiliensis can cause painful mucocutaneous lesions that can reoccur years later and sometimes prove fatal.	Fly control. Spray fly repellant, e.g., DEET, on clothes and leave no exposed skin. Avoid sand-fly infested areas and forested areas particularly after sundown.	Pregnancy may enable the organism to spread throughout the body.
Cutaneous Leishmaniasis	Tissue flagellates: Leishmania tropica, Leishmania major, and Leishmania aethiopica	By the bite of infective sand flies.	Nodular lesions on the skin and mucocutaneous regions that may be painful and ulcerate.	Destruction of the naso-pharyngeal tissue.	Fly control. Spray fly repellant, e.g., DEET, on clothes and leave no exposed skin. Avoid sand-fly infested areas and forested areas particularly after sundown.	No
Visceral Leishmaniasis	Tissue flagellates: Leishmania donovani, Leishmania infantum, and Leishmania chagasi	By the bite of infective sand flies.	Fever, swollen liver and spleen, lymphadenopathy, anaemia, weakness and weight loss. Skin lesions may appear after other signs disappear.	Fatal if untreated	Fly control. Spray fly repellant, e.g., DEET, on clothes and leave no exposed skin. Avoid sand-fly infested areas and forested areas particularly after sundown.	Pregnancy may enable the organism to spread throughout the body. May infect fetus.

Disease	Agent	Mode of Transmission	Typical Symptoms	Worst case	Most important prevention	Special risk for females or fetus?
Malaria	A protozoan: *Plasmodium falciparum*, *Plasmodium vivax*, *Plasmodium malariae*, and *Plasmodium ovale*. Simian malaria is caused by other Plasmodium species that have non-human primates as their primary host.	By the bite of infective female anopheline mosquito.	*P. falciparum* is the most serious and causes various signs including fever, chills, sweats, and headache. These symptoms can reoccur every few years for the life of the patient. The other malarias cause similar though less severe symptoms and can reoccur.	*P. falciparum* infections can progress to include icterus, coagulation defects, shock, renal and liver failure, acute encephalopathy, and coma. Case fatality rate for the young and compromised is at least 10%.	Mosquito control. Spray mosquito repellant, e.g., DEET, on clothes and leave no exposed skin. In areas where malaria is known to exist, take appropriate suppressive drugs.	Females have more clinical attacks in pregnancy. Late malaria may reappear. Fetal growth retarded, decreased fetal body weight, miscarriage, stillbirth or congenital infection.
Sleeping sickness	The systemic protozoa: *Trypanosoma gambiense* and *Trypanosoma rhodesiense*	By the bite of infective *Glossina ssp.*, the tsetse flies.	A trypanosomal chancre, a hard, red, painful nodule develops at the fly-bite site. Flu-like symptoms with insomnia, irritability and enlarged lymph nodes develop ten days later.	Frequently fatal. CNS signs develop in chronic cases.	Fly control. Use insecticides and bed nets. Spray repellant, e.g., DEET, on clothes and leave no exposed skin.	Abortion, premature birth, congenital infection.
Strongyloidiasis	The nematodes: *Strongyloides stercoralis* and *Strongyloides fuelleborni*	Infective larvae penetrate the skin.	Abdominal pain, diarrhea, and urticaria. Can include nausea, weight loss, vomiting, weakness, constipation, and a very itchy dermatitis around the anus.	Massive hyperinfection leading to chronic wasting and death.	Dispose of human feces. Wear shoes.	Possible transmission through human milk
Schistosomiasis	The blood flukes: *Schistosoma japonicum*, *Schistosoma mansoni*, and *Schistosoma haematobium*	Infective larvae (cercariae which have developed in snails) penetrate the skin of people in fresh water. Swimmer's itch develops when the cercariae from the other blood flukes of birds and mammals penetrate peoples' skin.	*S. japonicum* and *S. mansoni* cause papular dermatitis, dry cough, diarrhea, abdominal pain and swollen liver and spleen. *S. haematobium* infects the urinary tract and causes a painful or difficult urination and blood in the urine.	Death. Symptoms vary according to where the worms live. Can include liver fibrosis, CNS disease, bladder or colorectal cancer.	Snail control. Do not expose skin to contaminated water.	Fetal death possible. Increased risk of ectopic pregnancy.
Toxoplasmosis	Protozoan: *Toxoplasma gondii*	Ingestion of oocysts (especially in cat feces) or eating cysts in meat or milk.	Lymphadenopathy, stiff neck, anorexia, arthralgia, rash, chorioretinitis, uveitis.	Fetal death if non-immune pregnant female is infected especially in the second trimester	Treat infected cats and use extreme caution when disposing of cat feces.	Congenital abnormalities and fetal death.
Visceral Lava Migrans	The Nematodes: *Toxocara canis*, *Toxocara cati*.	Ingestion of infective parasite eggs present in the feces of dogs and cats.	Symptoms depend on to where the larvae migrate. Can include: chronic eosinophilia, enlarged liver, fever, appetite loss, weight loss, abdominal discomfort, muscle and joint pain, skin rash, and coughing.	Rarely fatal. Loss of sight if larvae migrate to an eye.	Personal hygiene. Treat infected cats and dogs or keep them at home.	No

Note: Vaccines currently are not available for any of the parasitic diseases.

Infestations:

Disease	Agent	Mode of Transmission	Typical Symptoms	Worst case	Most important prevention	Human vaccine available?	Special risk for females or fetus?
Chiggers	Usually harvest mites	Mite bites	Itching	Secondary infection	Spray repellants on clothing when working in the field	No	No
Flea bite dermatitis	Fleas	Flea bites	Itching and red papules especially on the legs.	Bacterial superinfection	Flea control especially for pets.	No	No
Myiasis	Fly larvae	Flies laying eggs in wounds.	Pain, itching, larvae visible on wounds, furuncle may form.	Larvae may penetrate deeply resulting in destructive lesions.	Treat wounds.	No	No
Scabies	*Saroptes scabiei*	Direct skin-to-skin contact	Itchy lesions around around the base of fingers, wrist, and elbows.	Extensive body infestation	Personal hygiene, avoid skin contact with infected people.	No	No
Tick paralysis	Toxin from the saliva of ticks	Tick bite.	Anorexia, irritability, lethargy, weakness, and incoordination.	Death	Spray tick repellant on clothing when working in the field especially in summer. Check for ticks after work and remove them.	No	No

9

CANCER IN FARMERS, FORESTERS, AND FISHERMEN

Paul A. James, M.D.
SUNY at Buffalo School of Medicine and Biomedical Sciences

Stanley Schuman, M.D., Ph.D.
Medical University of South Carolina

> The overall cancer rate of farmers is less than that of the population at large due to reduced tobacco and alcohol use, increased physical activity levels, and a healthier diet. Nonetheless, certain types of cancer are more common in farmers. Sunlight exposures of farmers, foresters, and fishermen may predispose them to skin cancers. Sunlight exposure is the best documented occupational risk for cancer in these groups and can be minimized through protective clothing. Workers in these occupations are also exposed to a number of agricultural chemicals that are carcinogenic, but good work habits and preventive measures can minimize exposure to these chemicals.

INTRODUCTION

Cancer is the second leading cause of death in the United States and, although the rates for most cancers are remaining stable (NCHS 1994), there is continuing concern about causes of cancer relating to environmental factors and work habits. Farmers, foresters, and fishermen are perceived to be at increased cancer risk from these hazards, and much has been written in response to this concern. Though the evidence suggests that farmers, and to a lesser extent fishermen and foresters, actually have a reduced risk of cancer when compared to the general population, it is prudent to seek methods of reducing risks of potential exposures to cancer causing agents. This chapter will present guidelines for prevention of cancer based on current evidence in the medical literature relevant to farmers, fishermen, and foresters.

A coherent and unifying analysis reflecting all the risks of cancer for these populations is difficult. Cancer is not one disease but a class of diseases that affect every organ system of the body and that are caused by diverse etiologies, many as yet unknown. The work environments, exposures, and tasks of farmers, foresters and fishermen are equally diverse, providing little reproducibility for definitive studies to assess the contributing factors for cancer among these populations.

The populations of workers in agriculture, forestry, and fishing are diverse and difficult to define, especially for epidemiological studies. They are represented by racial and ethnic diversity with significant cultural and socioeconomic differences as well. Each of these occupations have significant numbers of workers who are part-time, or who participate in these endeavors as an avocation. Thus, tracking these individuals over time is difficult, especially when attempting to control for their exposures in other activities.

Despite these differences, farming, forestry, and fishing share qualities that could place their workers at increased risk of cancer. Specifically, these occupations have shifted increasingly to a reliance on mechanical and chemical technology in the last century, which has expanded their potential to shelter and feed an industrial world. It is important that the benefits of this technology be weighed against the risks of cancer to workers.

This chapter will review the causes of cancer, in general, and those types believed to have higher rates among some populations of farmers, foresters, and fishermen. It will assess the work practices of these groups as well as review the current literature to assess risk factors amenable to interventions. Finally, it will develop a set of guidelines for prevention through education and risk-factor reduction.

CAUSES OF CANCER: AN OVERVIEW

The term cancer implies a malignant, neoplastic process characterized by genetic changes in a single cell where growth is not regulated, cellular differentiation is uncoordinated, and metastases may develop. In general, cancer development is facilitated by either environmental hazards or host predispositions. Several classes of environmental carcinogenic exposures have been demonstrated to be carcinogenic. Chemicals, radiation, and infectious agents are the predominant environmental and occupational classes of carcinogens. Additionally, host factors that predispose to cancer reside in the genetic DNA code of the individual and are often represented by intergenerational occurrences of cancer in a family. Finally, social behaviors and nutritional factors play an important role in cancer causation and must be considered in any program to reduce cancer in these workers.

OCCUPATIONAL FACTORS THAT MAY BE LINKED TO CANCER

Chemicals

Chemical carcinogens have been studied the most and provide the most controversy in determining risks to individuals. In agriculture, these include pesticides, solvents, fuels, fertilizers and wood preservatives. Other chemical carcinogens may be natural and part of plants or degradation products from plants, microorganisms, or animals themselves. Prioritizing between these natural carcinogens and the man-made ones has added to the controversy of the role of pesticides in human cancers (Ames 1987). Because humans are exposed to a wide array of chemicals whose cancer causing effects may not be manifested for decades, the study of chemicals and cancer has relied upon generalizations from animal studies with high exposures. Unfortunately, the degree to which these generalizations are valid to humans with low exposure is not known.

Farming

Farmers are perhaps the most intensely studied occupational group with respect to cancer. Several reviews of chemical carcinogenesis have been published (Burmeister 1990; Blair 1991; Blair et al. 1992; Schenker et al. 1995; Zejda et al. 1993), with special attention focused on farm practices such as pesticide use, geographic locales (Wong and Foliart 1993) and special populations such as women, (McDuffie 1994; Wiklund and Dich 1994), or migrant and seasonal farmworkers (Zahm and Blair 1993). The uniqueness of each farming environment, with its diversity of geography, climate, crops, and livestock, provides immense variation of exposure to numerous potential carcinogens, both known and unknown. Yet, despite this, the overall incidence of cancer among farmers is lower than the general population (Blair et al.

1993; Brackbill 1994; Brownson 1993; Davis et al. 1993; Delzel and Grufferman 1985; Dubrow and Wegman 1984; Fincham et al. 1992; Gallagher et al. 1985; Gallagher et al. 1984; Gunnarsdottir et al. 1984; Keller and Howe 1994; Lynge 1990; Lynge and Thygesen 1988; Reif et al. 1989a; Saftlas et al. 1987; Spinelli et al. 1990; Stark et al. 1987).

Of the studies for chemical carcinogens, only two pesticides have been proven to be definite carcinogens in humans—vinyl chloride and arsenic (CSA 1988; Vainio et al. 1985). While virtually no chemical can be proven safe if exposed to large concentrations, the evidence of carcinogenicity continues to develop for certain chemicals. Table 9.1 lists representative chemicals and the evidence of carcinogenicity based on animal, human, and in vitro studies (CSA 1988). However, the reader is cautioned that the exposure dosage at which these chemicals may cause cancer in humans is unknown.

Table 9.1. Chemical compounds and evidence of human carcinogenesis.

Category 1: Sufficient Evidence for Casual Relationship Between Agent and Human Cancer

Arsenicals   Vinyl Chloride

Category 2: Probable Carcinogenic to Humans

Chlorophenols   Ethyl Thiourea
p-dicholorobenzene   Formaldehyde
DDT   Phenoxyacids &
Ethylene Dibromide   Byproducts
Ethylene Oxide   2,4,6 - Trichlorophenol

Category 3: Cannot Be Currently Classified as Carcinogenic to Humans

Aldrin/Dieldrin   (4-chloro- 2- methyl-
Naphthylthiouren   phenoxy) acetic acid
Chlordane   OP[1] - Methylparathion,
2, 4-D   Parathion
Heptachlor   Pentachlorophenol
Lindane   2,4,5 Trichlorophen-
Trichlofon   oxyacetic acid

Herbicides have been intensely studied regarding their potential for carcinogenesis. Agent orange, used in the Vietnam War, has been studied and remains controversial. The herbicide 2,4-D (a phenoxyacid) has been linked in a dose-response relationship to Non-Hodgkin's Lymphoma in 2 studies (Hoar et al 1986; Zahm et al 1990). Though the evidence remains insufficient to confirm causality, caution with these chemicals remains prudent not only for acute toxicity but for cancers.

Although total cancer mortality rates are lower for farmers, several epidemiological studies have demonstrated increased rates of certain types of cancer among specific farm populations over the years. Specific cancer sites studied have included lip, esophagus, stomach, colo-rectal, liver, pancreas, skin cancer (melanoma), prostate, testicular, bladder, kidney, brain, nasal, lung, soft tissue sarcoma, non-Hodgkin's lymphoma, Hodgkin's lymphoma, multiple myeloma and leukemia. Of these, those with reported excesses among farmers have included lip and skin cancer, leukemia, Hodgkin's lymphoma, Non-Hodgkin's lymphoma, multiple myeloma, brain cancer, soft-tissue sarcoma, stomach cancer, prostate and testicular cancer. (See Table 9.2).

However, with the exception of skin and lip cancer, these epidemiological studies in man fail to yield conclusive evidence of causality. First, many do not provide an adequate time sequence of exposures (Cooper 1984). Additionally, the study groups may be too small to detect an adequate statistical difference. Because of the diversity of the work and the multitude of other exposures, other carcinogens may be interacting with the compound in question, confounding the analysis.

Proper control groups are thus extremely difficult to identify, and many studies lack this necessary ingredient. Additionally, different studies have demonstrated conflicting results. This inconsistency on replication makes a coherent and informed policy of cancer prevention difficult (Susser 1973). Currently, large prospective studies are underway in several states to clarify these important concerns.

Foresters

No direct studies have been performed on loggers and foresters as a population. However, again, several epidemiological studies of associations have been performed. An increased risk of nasopharyngeal cancer was strongest for foresters and loggers in one study in New Zealand (Gallagher et al 1985), however, a case control study in France did not demonstrate this association (Luce et al 1992). Arsenic has been used as a wood preservative in pressure treated lumber and thus should be considered as a risk for those involved with this chemical processing.

Fishermen

The literature review of fishermen revealed no conclusive evidence of chemical exposures that may be carcinogenic. One report hypothesized that arsenic levels in surface waters, both as a natural element and as a pollutant from industrial use, may contaminate workers and increase the risk of cancer

---

Table 9.2. Sites of excess cancer reported among farmers in epidemiological studies.*

+**Lip** (Brackbill et al., 1994; Dardanoni et al., 1984; Fincham et al., 1992; Haguenoer et al., 1990; Keller, 1970; Pukkala et al., 1994)

Brain Cancer (Reif et al., 1989b)

+**Skin** (Autier et al., 1994; Dardanoni et al., 1984; Graham et al., 1985; Haguenoer et al., 1990; Keller, 1970; Lee and Strickland, 1980; Pukkala et al., 1994; Whitaker et al., 1979)

Soft Tissue Sarcoma (Balarajan and Acheson, 1984; Vineis et al., 1987; Wingren et al., 1990; Woods et al., 1987)

Leukemia (Brown et al., 1990; Brownson and Reif, 1988; Burmeister et al., 1982; Donham et al., 1980; Fasal et al., 1968; Milham, 1971; Oleske et al., 1985)

Stomach Cancer (Armijo et al., 1981; Burmeister et al., 1983; Chow et al., 1994; Higginson, 1966; Jeyaratnam et al., 1987; Kneller et al., 1990; Kraus et al., 1957; Sigurjonsson, 1967)

Hodgkin's Lymphoma (Balarajan, 1988; Bernard et al., 1987; Brownson and Reif, 1988; Dubrow et al, 1988; Fasal et al., 1968; Giles et al., 1984; Hoar et al., 1986; LaVecchia et al., 1989; Matthews et al., 1984; Pearce et al., 1985)

Prostate Cancer (Brownson et al., 1988; Fincham et al., 1992; Morrison et al., 1993; Pearce et al., 1987; Schuman et al, 1987)

Non-Hodgkins Lymphoma (Armijo et al., 1981; Balarajan, 1988; Brownson and Reif, 1988; Burmeister et al., 1983; Cantor, 1982; Cantor and Blair, 1984; Francheschi et al., 1989; Giles et al., 1984; Weisenburger, 1993; Woods et al., 1987)

Testicular Cancer (Balarajan, 1988; Haughey et al., 1989; Hayes et al., 1990; Jensen et al., 1984; McDowell and Balarajan, 1984; Mills et al., 1984; Van den Eeden et al., 1991; Wiklund et al., 1986)

Multiple Myeloma (Boffeta et al., 1989; Burmeister et al., Cantor and Blair, 1984; Cuzick et al., 1988; Gallagher et al., 1983; Giles et al., 1984; McLaughlin et al., 1988; Nandakumar et al., 1986; Pearce et al., 1985, 1986; Steineck and Wiklund, 1986; Tollerud et al., 1985)

*This reference list does not include negative studies and thus should be interpreted with caution.
+The literature consistently and reproducibly finds these cancers in excess.

(DeRenzi 1990). Another study hypothesized that the use of chrysoidine dyes used by fishermen to color bait may potentially increase the risk of urothelial cancers in fishermen (Cartwright et al. 1983). Yet, neither study demonstrated an association. A study by Ng suggests an association between fishing and nasal cavity and sinus cancers (Ng 1988). In British Columbia, fishermen had an elevated risk of stomach cancer. Standardized mortality ratios between farmers and fishermen in England and Japan suggest that ischemic heart disease and lung cancer are significantly higher for fishermen than for farmers in both countries (Naruse et al 1985).

REGULATORY ISSUES

In the United States, the Environmental Protection Agency (EPA) has been charged with primary responsibility for regulating pesticides since 1970. Other federal agencies, including the United States Department of Agriculture (USDA), the Food and Drug Administration (FDA), the Occupational Safety and Health Administration (OSHA), and the Consumer Product Safety Commission are also given the power to limit human exposure to specific chemicals. In general, these agencies have regulated a substance when it is expected to cause an increase of more than four cases of cancer per thousand persons (CSA 1988). The agencies have, in general, not regulated any chemical if its expected increase in cancer rate is less than one in a million. Between these two rates, regulation is on a case-by-case basis and is usually based upon the cost of the regulation and its relationship to the number of lives saved. Finally, it is difficult to regulate safe practices for farming and fishing because many businesses are family-owned and often have less than the eleven employees required for federal regulation by the Occupational Safety and Health Administration (OSHA).

Despite the lack of scientific evidence, there are approaches that may reduce the risk of cancer by limiting exposures to carcinogenic chemicals.

Guidelines:

1. Use chemicals only when indicated. For pesticide use, develop an Integrated Pest Management (I.P.M.) program.
2. Adhere to label instructions.
3. Wear protective clothing and clean these appropriately.
4. Use appropriate equipment for mixing and application.
5. Be knowledgeable of re-entry times and appropriate First-Aid for exposures to chemicals.

Radiation

A second form of environmental carcinogenic exposure is radiation. Because each of these populations work outside and are exposed to significant levels of sunlight, they may be exposed to increased levels of ultraviolet radiation, a proven risk factor for skin cancer, particularly melanoma (Lee and Strickland 1980). However, the role of sun exposure in populations from recreational exposure may be greater than that from occupational exposures (Autier et al 1994). Other exposures to radiation are possible from sources such as radon, but there is no evidence that farmers, fishermen, and foresters are at increased risk for this exposure. Electromagnetic radiation is a third type of exposure under study, but no studies have linked this exposure to these occupational activities.

Guidelines:

1. Wear protective clothing to prevent high dosage UV exposure.
2. Use UVB blocker (SPF 15 or higher) on exposed skin surfaces during high risk times.

Infectious Agents

A third group of environmental carcinogens are infectious agents; specifically, viral infections. Certain viruses such as Human Immunodeficiency Virus (HIV) and Epstein-Barr Virus (EBV) have been linked to human cancers. Yet, neither of these has been demonstrated to have a higher incidence among farmers, foresters, and fishermen than in the general population. One published report raised a concern about transfer to humans of a zoonotic infection from a bovine virus linked to leukemia, but no association was demonstrated to leukemia in humans (Donham et al 1980). The linkage of zoonotic diseases to human cancer is an area of continued study; and, if this is found to be an etiologic mechanism, farmers, foresters, and fishermen could be at increased risk.

Guideline:

1. Appropriate hygienic standards should be maintained in work practices. There is no other evidence to suggest other risks from this etiologic mechanism.

Genetic Factors

Genetic factors are an important determinant in the development of some cancers. Racial and ethnic characteristics may place the host at a higher risk for some forms of cancer. Distinguishing the cause of elevated rates of cancer between genetic predispositions and environmental factors is extremely difficult. However, there is little evidence that the genetic makeup of farmers, fishermen and foresters places them at higher risk for cancer.

Guideline:

1. Based on specific family, racial, and ethnic characteristics, certain forms of cancer screening may be indicated. However, there is no indication that this risk is higher for fishermen, farmers, and foresters and should be addressed on an individual basis.

Nutritional and Social Factors

Nutritional factors have also been implicated as causes for cancer. Alcohol is perhaps the most prevalent predisposing nutritional factor to cancer. Yet, the relative lower rate of cancers overall and the anecdotal evidence that these populations (especially farmers) may have healthier diets with less alcohol intake suggest that this population is relatively protected by their nutritional practices. Obesity itself has been linked to excess rates of colon, rectal, gallbladder, biliary tract, breast, endometrial, ovarian, cervical, and prostate cancers (Lee 1994). Thus, the importance of maintaining an active lifestyle cannot be overemphasized.

The linkage between smoking and cancer has been well documented. Yet, farmers in general have lower rates of lung cancer than the general population, and this is likely related to a lower incidence of smoking among this group. This fact is felt to be secondary to general factors of healthy lifestyle among farmers. Finally, an active lifestyle is postulated to be protective for certain forms of cancer.

Guidelines:

1. Reduce incidence of smoking.
2. Increase physical activity.
3. Reduce alcohol consumption.
4. Eat healthy diet, especially fiber.

APPROACH TO HAZARD EDUCATION

A systematic approach is beneficial for health care workers who are attempting to lower the risk of cancer in farmers, foresters, and fishermen. To develop this approach, one must first look at the worker or population of workers for genetic or hereditary predispositions to cancer. If present, these workers should be advised of the increased risk and appropriate surveillance and screening should be planned.

Secondly, individual risk assessment should be performed—not only is the family history important, but also a work history and exposure history are vital. Work habits as well as social and cultural habits are important risk factors that may be amenable to change. Education and strategies for risk factor reduction should be discussed and encouraged.

Thirdly, a work place assessment and evaluation of work practices is an important part of the strategy. Are safety precautions being adhered to and protective clothing and gear being used? If work practices incur risk of cancer, surveillance of exposure should be routinely monitored.

Health promotion for workers is very important as well as education on the possible etiologies of cancer and ways of reducing risks. Continued research into the work practices of each of these industries will be vital to appropriate education and preventive practices in the future.

SUMMARY

In summary, chemical carcinogenic exposures are certainly present for farmers, fishermen, and foresters, but as yet no significant and persistent hazards have been clearly proven. Yet, the use of potent chemicals, particularly pesticides, should be considered a risk to the development of cancer, and continued diligence at protection through education and safe work practices is likely to reduce this perceived risk.

A well-documented cause of cancer in these occupations is UV radiation. Yet, the exposure to this carcinogen can also be mitigated through education and protection. Additionally, the exposure to UV radiation from recreational activities must be considered when attempting to reduce this risk. Personal lifestyle factors are also important and the risks of cancer from smoking and alcohol should be considered for those who choose these significant exposures.

REFERENCES

Ames, B.N., R. Magaw, and L.S. Gold. 1987. Ranking possible carcinogenic hazards. *Science* 236: 271-90.

Armijo, R., M. Orellana, E. Medina, et al. 1981. Epidemiology of gastric cancer in Chile: I. Case-control study. *Int. J. Epidem.* 10: 53-56.

Autier, P., J.F. Dore, F. Lejeune, K.F. Koelmel, O. Geffeler, P. Hille, J.P. Cesarini, D. Lienard, A. Liabeuf, M. Joarlette, et al. 1994. Recreational exposure to sunlight and lack of information as risk factors for cutaneous malignant melanoma. *Melanoma Res.* 4(2): 79-85.

Axelson, O. 1987. Pesticides and cancer risks in agriculture. *Med. Oncol. Tumor Pharmac.* 4(3-4): 207-17.

Balarajan, R. 1988. Malignant lymphoma in agricultural and forestry workers in England and Wales. *Public Health* 102: 585-92.

Balarajan, R., E.D. Acheson. 1984. Soft tissue sarcomas in agriculture and forestry workers. *J. Epidem. Comm. Health* 38: 113-16.

Bernard, S.M., R.A. Cartwright, C.M. Darwin, et al. 1987. Hodgkin's disease: case control epidemiological study in Yorkshire. *Brit. J. Cancer* 55: 85-90.

Blair, A., M. Dosemeci, and E.F. Heineman. 1993. Cancer and other causes of death among male and female farmers from twenty-three states. *Am. J. Ind. Med.* 23(5): 729-42.

Blair, A., H. Malker, K.P. Cantor, L. Burmeister, K. Wiklund. 1985. Cancer among farmers. a review. *Scand. J. Work Environ. Health* 11(6): 397-407.

Blair, A., S.H. Zahm. Cancer among farmers. *Occup. Med.* 6(3): 335-54.

Blair, A., S.H. Zahm, N.E. Pearce, E.F. Heineman, and J.F. Fraumeni, Jr. 1992. Clues to cancer etiology from studies of farmers. *Scand. J. Work Envir. Health* 18(4): 209-15.

Boffetta, P., S.D. Stellman, and L Garfinkel. 1989. A case-control study of multiple myeloma nested in the American Cancer Society prospective study. *Int. J. Cancer* 43: 554-59.

Brackbill, R.M., L.L. Cameron, V. Behrens. 1994. Prevalence of chronic diseases and impairments among U.S. farmers, 1986-1990. *Am. J. Epidem.* 139(11): 1055-65.

Brown, L.M., A. Blair, R. Gibson, et al. 1990. Pesticide exposures and other agricultural risk factors for leukemia among men in Iowa and Minnesota. *Cancer Res.* 50: 6585-91.

Brown, L.M., and L.M. Pottern. 1984. Testicular cancer and farming. *Lancet* i: 1356.

Brownson, R.C., J.C. Chang, J.R. Davis, and J.R. Bagby, Jr. 1988. Occupational risk of prostate cancer: a cancer registry-based study. *J. Occup. Med.* 30(6): 523-26.

Brownson, R.C., J.S. Reif, J.C. Chang, and J.R. Davis, Jr. 1989. Cancer risks among Missouri farmers, 1977-1987. *Cancer* 64(11): 2381-86.

Brownson, R.C., and J.S. Reif. 1988. A cancer registry-based study of occupational risk for lymphoma, multiple myeloma and leukemia. *Int. J. Epidem.* 17: 27-32.

Brownson, R.C., J.S. Reif, J.C. Chang, and J.R. Davis. 1989. Cancer risks among Missouri farmers. *Cancer* 64(11): 2381-86.

Burmeister, L.F. 1990. Cancer in Iowa farmers: recent results. *Am. J. Ind. Med.* 18(3): 295-301.

Burmeister, L.F., G.D. Everett, S.F. Van Lier, and P. Isacson. 1983. Selected cancer mortality and farm practices in Iowa. *Am. J. Epidem.* 118(1): 72-77.

Burmeister, L.F., S.F. Van Lier, and P. Isacson. 1982. Leukemia and farm practices in Iowa. *Am. J. Epidem.* 115: 720-28.

Cantor, K.P. 1982. Farming and mortality from non-Hodgkin's lymphoma: a case-control study. *Int. J. Cancer* 29: 239-47.

Cantor, K.P., and A. Blair. 1984. Farming and mortality from multiple myeloma: a case-control study with the use of death certificates. *JNCI* 72: 251-55.

Cartwright, R.A., M.R. Robinson, R.W. Glashan, B.K. Gray, P. Hamilton-Stewart, S.C. Cartwright, and D. Barham-Hall. 1983. Does the use of stained maggots present a risk of bladder cancer to coarse fishermen? *Carcinogenesis* 4(1): 111-13.

Chow, W.H., H.S. Malker, A.W. Hsing, J.K. McLaughlin, J.A. Weiner, B.J. Stone, J.L. Ericsson, and W.J. Blot. 1994. Occupational risks for colon cancer in Sweden. *J. Occup. Med.* 36: 647-51.

Chow, W.H., J.K. McLaughlin, H.S. Malker, J.A. Weiner, J.L. Ericsson, B.J. Stone, and W. Blot. 1994. Occupation and stomach cancer in a cohort of Swedish men. *Am. J. Ind. Med.* 26(4): 511-20.

Colton, T. 1986. Herbicide exposure and cancer. *JAMA* 256: 1176-78.

Cooper, H.M. 1984. The analysis and interpretation stage. In: *The Integrative Research Review. A Systematic Approach.* Vol. 2. Sage Publications, Inc., pp. 79-133.

Council on Scientific Affairs. 1988. Cancer risk of pesticides in agricultural workers. *JAMA* 260 (7): 959-66.

Cuzick, J., and B. De Stavola. 1988. Multiple myeloma—a case control study. *Brit. J. Cancer* 57: 516-20.

Dardanoni, L., L. Gafa, R. Paterno, G. Pavone. 1984. A case-control study on lip cancer risk factors in Rajusa (Sicily). *Int. J. Cancer* 34: 335-37.

Davis, D.L., A. Blair, and D.G. Hoel. 1993. Agricultural exposures and cancer trends in developed countries. *Envir. Health Persp.* 100: 39-44.

Delzell, E., and S. Grufferman. 1985. Mortality among white and nonwhite farmers in North Carolina, 1976-1978. *Am. J. Epidem.* 121(3): 391-402.

De Renzi, G.P., G. Rallo, A. Capri, S. Agostino, and C. Angioni. 1990. Carcinogenic hazards from arsenic in seawater, seafood, and marine aerosols. *Adv. Appl. Biotech.Ser.* 1990: 191-98.

Donham, K.J, J.W. Berg, and R.S. Sawin. 1980. Epidemiologic relationships of the bovine population and human leukemia in Iowa. *Am. J. Epidem.* 112: 80-92.

Dubrow, R., J.O. Paulson, and R.W. Indian. 1988. Farming and malignant lymphoma in Hancock County, Ohio. *Brit. J. Ind. Med.* 45: 25-28.

Dubrow, R., and D.H. Wegman. 1984. Cancer and occupation in Massachusetts: a death certificate study. *Am. J. Ind. Med.* 6(3): 207-30.

Fasal, E., E.W. Jackson, and M.R. Klauber. 1968. Leukemia and lymphoma mortality and farm residence. *Am. J. Epidem.* 87: 267-74.

Fincham, S.M., J. Hanson, and Berkel. 1992. Patterns and risks of cancer in farmers in Alberta. *Cancer* 69(5): 1276-85.

Finkelstein, M.M., and G.M. Liss. 1987. Selection bias in occupational case-control studies that use death registries to select subjects: a discussion and demonstration. *Am. J. Ind. Med.* 12(1): 21-31.

Francheschi, S., D. Serraino, E. Bidoli, et al. 1989. The epidemiology of non-Hodgkin's lymphoma in the Northwest of Italy: A hospital-based case-control study. *Leuk. Res.* 13: 465-72.

Gallagher, R.P, J.J. Spinelli, and J.M. Elwood. 1983. Allergies and agricultural exposures as risk factors for multiple myeloma. *Brit. J. Cancer* 48: 853-57.

Gallagher, R.P., W.J. Threlfall, P.R. Band, and J.J. Spinelli. 1985. Cancer mortality experience of woodworkers, loggers, fishermen, farmers, and miners in British Columbia. *Nat. Cancer Inst. Monographs* 69: 163-67.

Gallagher, R.P., W.J. Threlfall, E. Jeffries, P.R. Band, J. Spinelli, A.J. Coldman. 1984. Cancer and aplastic anemia in British Columbia farmers. *JNCI* 72(6): 1311-15.

Giles, G.G., J.N. Lickiss, M.J. Baikle, et al. 1984. Myeloproliferative and lymphoproliferative disorders in Tasmania. *JNCI* 72: 1233-40.

Graham, S., J. Marshall, B. Haughey, et al. 1985. An inquiry into the epidemiology of melanoma. *Am. J. Epidem.* 122: 606-19.

Gunnarsdottir, H., V. Rafnsson. 1991. Cancer incidence among Icelandic farmers, 1977-1987. *Scand. J. Soc. Med.* 19(3): 170-73.

Haguenoer, J.M., S. Cordier, C. Morel, et al. 1990. Occupational risk factors for upper respiratory and upper digestive tract cancers. *Brit. J. Ind. Med.* 47: 380-83.

Haughey, B.P., S. Graham, J. Brasure, et al. 1989. The epidemiology of testicular cancer in upstate New York. *Am. J. Epidem.* 130: 25-36.

Hayes, R.B., L.M. Brown, L.M. Pttern, et al. 1990. Occupation and risk for testicular cancer: a case-control study. *Int. J. Epidem.* 19: 825-31.

Higginson, J. 1966. Etiologic factors in gastrointestinal cancer in men. *JNCI* 37: 537-49.

Hoar, S.K., A. Blair, F.F. Holmes, et al. 1986. Agricultural herbicide use and risk of lymphoma and soft-tissue sarcoma. *JAMA* 256: 1141-47.

Jensen, O.M., J.H. Olsen, and A. Osterlind. 1984. Testis cancer risk among farmers in Denmark. *Lancet* i: 794.

Jeyaratnam, J., J. Lee, H.P. Lee, and W.O. Phoon. 1987. Stomach cancer incidence in a cohort of fishermen in Singapore. *Scand. J. Work Envir. Health* 13(6): 524-26.

Kawachi, I., N. Pearce, and I. Fraser. 1989. A New Zealand cancer registry-based study of cancer in wood workers. *Cancer* 64(12): 2609-13.

Keller, A.Z. 1970. Cellular types, survival, race, nativity, occupations, habits and associated diseases in the pathogenesis of lip cancers. *Am. J. Epidem.* 91: 486-99.

Keller, J.E., and H.L. Howe. 1994. Case-control studies of cancer in Illinois farmers using data from the Illinois state cancer registry and the U.S. census of agriculture. *Eur. J. Cancer* 30A(4): 469-73.

Kneller, R.W., Y. Gao, J.K. McLaughlin, et al. 1990. Occupational risk ractors for gastric cancer in Shanghai, China. *Am. J. Ind. Med.* 18: 69-78.

Kraus, A.S., M.L. Levin, and P.R. Gerhardt. 1957. A study of occupational associations with gastric cancer. *Am. J. Public Health* 47: 961-70.

LaVecchia, C., E. Negri, B. D'Avanzo, and S. Franceschi. 1989. Occupation and lymphoid neoplasms. *Brit. J. Cancer* 60: 385-88.

Lee, J.A.H., and D. Strickland. 1980. Malignant melanoma: social status and outdoor work. *Brit. J. Cancer* 41: 757-63.

Lee, P.R. 1994. *The Clinician's Handbook of Preventive Services.* International Medical Publishing, Inc., Virginia.

Luce, D., A. Leclerc, et al. 1992. Occupational risk factors for sinonasal cancer: a case-control study in France. *Am. J. Ind. Med.* 21(2): 163-75.

Lynge, E. 1990. Occupational mortality and cancer analysis. *Public Health Rev.* 18(2): 99-116.

Lynge, E., and L. Thygesen. 1988. Use of surveillance systems for occupational cancer: data from the Danish national system. *Int. J. Epidem.* 17(3): 493-500.

Maroni, M., and A. Fait. 1993. Health effects in man from long-term exposure to pesticides. A review of the 1975-1991 literature. *Toxicology* 78(1-3): 1-180.

Matthews, M.L.V., L.E. Dougan, D.C. Thomas, and B.K. Armstrong. 1984. Interpersonal linkage among Hodgkin's disease patients and controls in Western Australia. *Cancer* 54: 2571-79.

McDowall, M., and R. Balarajan. 1984. *Testicular Cancer and Employment in Agriculture.* i: 510-11.

McDuffie, M.H. Women at work: agriculture and pesticides. 1994. *J. Occup. Med.* 36(11): 1240-46.

McLaughlin, J.K, H.S. Malker, M.S. Linet, J. Ericsson, B.J. Stone, J. Weiner, W.J. Blot, and J.R. Fraumeni, Jr. 1988. Multiple myeloma and occupation in Sweden. *Arch. Envir. Health* 43: 7-10.

Milham, S., Jr. 1971. Leukemia and multiple myeloma in farmers. *Am. J. Epidem.* 94: 307-10.

Mills, P.K., G.R. Newell, and D. Johnson. 1984. Testicular cancer associated with employment in agriculture and oil and natural gas extraction. *Lancet* i: 207-10.

Morrison, H., D. Savitz, R. Semenciw, B. Hulka, Y. Mao, D. Morison, D. Wigle. 1993. Farming and prostate cancer mortality. *Am. J. Epidem.* 137(3): 270-80.

Morrison, H.I., K. Wilkins, R. Semenciw, Y. Mao, D. Wigle. 1992. Herbicides and cancer. *JNCI* 84(24): 1866-74.

Nandakumar, A., B.K. Armstrong, and N.H. DeKlerk. 1986. Multiple myeloma in Western Australia: a case-control study in relation to occupation, father's occupation, socioeconomic status and county of birth. *Int. J. Cancer* 37: 223-26.

Naruse, Y., S. Kagamimori, M. Watanabe, M. Minowa, Y. Iibuchi. 1985. Mortality rates for farmers and fishermen in Japan compared with England and Wales. *Soc. Sci. Med.* 21(2): 139-43.

National Center for Health Statistics. 1994. *Vital Statistics of the United States, 1991.* Washington, DC: Public Health Service

Neutel, C.I. 1990. Mortality in fishermen: An unusual age distribution. *Brit. J. Ind. Med.* 47(8): 528-32.

Ng, T.P. 1988. Occupational mortality in Hong Kong. *Int. J. Epidem.* 17(1): 105-10.

Oleske, D., H.M. Golomb, M.D. Farber, and P.S. Levy. 1985. A case-control inquiry into the etiology of hairy cell leukemia. *Am. J. Epidem.* 121: 675-83.

Pearce, N.E., R.A. Sheppard, and J. Fraser. 1987. Case-control study of occupation and cancer of the prostate in New Zealand. *J. Epidem.Comm. Health* 41: 130-32.

Pearce, N.E., R.A. Sheppard, J.K. Howard, I. Fraser, B.M. Lilley. 1986. Leukemia among New Zealand agricultural workers. A cancer registry-based study. *Am. J. Epidem.* 124(3): 402-09.

Pearce, N.E., R.A. Sheppard, A.H. Smith, and C.A. Teague. 1987. Non-Hodgkin's lymphoma and farming: an expanded case-control study. *Int. J. Cancer* 39: 155-61.

Pearce, N.E., A.H. Smith, and D.O. Fisher. 1985. Malignant lymphoma and multiple myeloma linked with agricultural occupations in New Zealand. A cancer registry-based study. *Am. J. Epidem.* 121: 225-37.

Pearce, N.E., A.H. Smith, J.K. Howard, R.A. Sheppard, H.J. Giles, and C.A. Teague. 1986. Case-control study of multiple myeloma and farming. *Brit. J. Cancer* 54(3): 493-500.

Pukkala, E., A.L. Soderholm, and C. Lindquist. 1994. Cancers of the lip and oropharynx in different social and occupational groups in Finland. *Eur. J. Cancer B. Oral. Oncol.* 30B(3): 209-15.

Reif, J.S., N. Pearce, and J. Fraser. 1989a. Cancer risks in New Zealand farmers. *Int. J. Epidem.* 18(4): 768-74.

Reif, J.S., N. Pearce, and J. Fraser. 1989b. Occupational risks for brain cancer: a New Zealand cancer registry-based study. *J. Occup. Med.* 31(10): 863-67.

Saftlas, A.F., A. Blair, K.P. Cantor, L. Hanrahan, and H.A. Anderson. 1987. Cancer and other causes of death among Wisconsin farmers. *Am. J. Ind. Med.* 11(2): 119-29.

Schenker, M., and S. McCurdy. 1985. *Pesticides, Viruses, and Sunlight in the Etiology of Cancer Among Agricultural Workers. Cancer Prevention Strategies in the Workplace 1985.* C.E. Becker, and M.J. Coye, eds. Washington, DC: Hemisphere Publication Corporation.

Schumacher, M.C. 1985. Farming occupations and mortality from non-Hodgkin's lymphoma in Utah. A case-control study. *J. Occup. Med.* 27(8): 580-84.

Schuman, L.M., J.Mandel, C. Blackard, et al. 1977. Epidemiologic study of prostatic cancer: preliminary report. *Cancer Treat. Rep.* 61: 181-86.

Sharp, D.S., B. Eskenazi, R. Harrision, et al. 1986. Delayed health hazards of pesticide exposure. *Ann. Rev. Public Health* 7: 441-71.

Sigurjonsson, J. 1967. Occupational variations in mortality from gastric cancer in relation to dietary differences. *Brit. J. Cancer* 21: 651-56.

Smith, A.H., and M.N. Bates. 1989. Epidemiological studies of cancer and pesticide exposure. *ACS Symp. Ser.* (414): 207-22.

Spinelli, J.J, R.P. Gallagher, and P.R. Band, W.J. Threlfall, D. Raynor, and H. Schellekens. 1990. Occupational associations among British Columbia male cancer patients. *Can. J. Public Health.* 81(4): 254-58.

Stark, A.D., H.G. Chang, E.F. Fitzgerald, K. Riccardi, and R.R. Stone. 1987. A retrospective cohort study of mortality among New York State Farm Bureau members. *Archiv. Envir. Health* 42(4): 204-12.

Steineck, G., and K. Wiklund. 1986. Multiple myeloma in Swedish agricultural workers. *Int. J. Epidem.* 15: 321-25.

Sterling, T.D., and A.V. Arundel. 1986. Health effects of phenoxy herbicides. *Scand. J. Work Envir. Health* 12: 161-73.

Susser, M. 1973. Criteria of judgement. In: *Causal Thinking in the Health Sciences: Concepts and Strategies in Epidemiology.* The Oxford University Press, pp.140-63.

Tollerud, D.J., L.A. Brinton, B.J. Stone, et al. 1985. Mortality from multiple myeloma among North Carolina furniture workers. *JNCI* 74: 799-801.

Tsongas, T.A. n.d. *Occupational Factors in the Epidemiology of Chemically Induced Lymphoid and Hemopoietic Cancers. Toxicology of the Blood and Bone Marrow.* R.D. Irons, ed. Target Organ Series. New York: Raven Press, pp 149-77.

Vagero, D., and G. Persson. 1987. Cancer survival and social class in Sweden. 1987. *J. Epidem.Comm. Health* 41(3): 204-09.

Vainio, H., K. Hemminki, and J. Wilbourn. 1985. Data on the carcinogenicity of chemicals in the LARC Monographs Programme. *Carcinogenesis* 6: 1653-65.

Van den Eeden, S.K., N.S. Weiss, C.H. Strader, and J.R. Daling. 1991. Occupation and the occurrence of testicular cancer. *Am. J. Ind. Med.* 19(3): 327-37.

Vineis, P., B. Terracini, G. Ciccone, et al. 1987. Phenoxy herbicides and soft-tissue sarcomas in female rice weeders. *Scand. J. Work Envir. Health* 13: 9-17.

Weisenburger, D. 1993. Epidemiology of non-Hodgkin's lymphoma. Fifth International Conference on Malignant Lymphoma; Lugano, Switzerland, June 9-12.

Whitaker, C.J., W.R. Lee, and J.E. Downes. 1979. Squamous cell skin cancer in the Northwest of England, 1967-1969, and its relation to occupations. *Brit. J. Ind. Med.* 36: 43-51.

Wiklund, K., and J. Dich. 1994. Cancer risks among female farmers in Sweden. *Cancer Causes and Control* 5: 449-57.

Wiklund, K., J. Dich, L.E. Holm. 1986. Testicular cancer among agricultural workers and licensed pesticide applicators in Sweden. *Scand. J. Work Envir. Health* 12: 630-31.

Wiklund, K., and G. Steineck. 1988. Cancer in the respiratory organs of Swedish farmers. *Cancer* 61(5): 1055-58.

Wingren, G., M. Fredrikson, and H.N. Brage, B. Nordenskjold, O. Axelson. 1990. Soft tissue saracoma and occupational exposures. *Cancer* 66(4): 806-11.

Wong, O., and D.E. Foliart. 1993. Epidemiological factors of cancer in Louisiana. *J. Envir. Pathol. Toxic. Oncol.* 12(4): 171-83.

Woods, J.S., L. Polissar, R.K. Severson, L.S. Heuser, and B.G. Kulander. 1987. Soft tissue sarcoma and non-Hodgkin's lymphoma in relation to phenoxyherbicide and chlorinated phenol exposure in Western Washington. *JNCI* 78(5): 899-910.

Zahm, S.H, and A. Blair. 1993. Cancer among migrant and seasonal farmworkers: an epidemiologic review and research agenda. *Am. J. Ind. Med.* 24(6): 753-66.

Zahm, S.H., D.D. Weisenburger, P.A. Babbitt, et al. 1990. A case-control study of non-Hodgkin's lymphoma and the herbicide 2,4-Dichlorophenoxyacetic acid (2,4-D) in Eastern Nebraska epidemiology. *Epidemiology* 1: 349-56.

Zejda, J.E., H.H. McDuffie, and J.A. Dosman. 1993. Epidemiology of health and safety risks in agriculture and related industries. Practical applications for rural physicians. *West. J. Med.* 158(1): 56-63.

10

MENTAL HEALTH ISSUES IN AGRICULTURE, FORESTRY, AND FISHERIES

Patricia J. Gammon, Ph.D.
Licensed Clinical Psychologist
Duke University Medical Center

Laura Anibal, B.A.
Carlow International Inc.

> Stress is a major contributor to a wide range of mental and physical disorders. Sources of stress inducing situations in agriculture, forestry and fisheries are discussed in this chapter. Workers in these industries are subject to many uncontrollable factors and are often isolated in their work, which increases their stress level. In many cases, work related stress carries over to the home or family as well. Types of mental health disorders suffered by farmers, foresters, and fishermen are explained along with recommended methods of treatment. Strategies designed to serve this unique population are discussed.

INTRODUCTION

Prominent health concerns today, particularly as the effects of stress on physical health have become increasingly documented, include the effects of occupational stressors on the mental and physical well-being of workers. Substantial documentation shows that stress plays a role in a wide range of disorders, from coronary heart disease to gastrointestinal disturbances to the onset and exacerbation of depression and anxiety disorders. Each occupational field has its own set of unique stressors, as well as ones that all workers at one time or another share in common. Paradigms have been developed to look at occupational stress and how it impacts the welfare of workers.

A commonly used model for understanding the nature of occupational stress and its effects on workers has been described by Baker and Karasek (1995). They describe two dimensions of work as being critical. These dimensions include the amount of demand in the work environment, as well as the extent to which a worker has control over how he or she goes about meeting those demands. The model describes the most stressful work environment as one that involves high job demand for the worker, with low control over those factors influencing how a job is completed. The highest levels of stress are presumed to take the greatest toll on the welfare of the worker, impacting not only the worker, but his or her lifestyle, health behaviors, and family and social relationships.

STRESSORS IN FARMING, FORESTRY WORK, AND THE FISHING INDUSTRY

Few other fields of work are as effected by uncontrollable factors as those of farming, forestry work, and the fishing industry. Workers in these fields are subject to the changeable effects of weather conditions, natural disasters, and economic downswings. Their earnings are dependent on often unpredictable and uncontrollable economic factors, so estimating their earnings for any given year is often difficult. In addition, workers in these fields have little backup or support in the face of these uncontrollable events. Unlike workers in other occupational settings who at least have the added assistance of extensive networks of coworkers, those who work in farming, forestry, and fisheries often work in isolation, sometimes spending long days and nights in solitary work, without adequate emotional or social support. Their work at times keeps them removed from friends and family for extended periods of time. This distance can create much stress for their families as well, who sometimes must operate entirely in their absence, while other times they must provide support to their often very demanding work. And when uncontrollable events occur to entirely obliterate their prospects for continued work in their chosen profession, such workers are often isolated from practical resources, social support, and counseling services to both address their social/emotional distress and to provide supportive career redirection to lead them into new professional pathways. In addition, the rugged individualism and high level of independence that often characterizes workers in these fields may prevent them from seeking out whatever helpful resources or social support may be available, since they are so used to having to manage their work entirely on their own.

TYPES OF MENTAL HEALTH DISORDERS

Psychiatric and psychological disorders appear to be the result of an interaction between an individual's unique biological and genetic makeup, and a wide array of environmental and life experience influences. Some individuals may show a constitutional vulnerability toward certain disorders, such as depression or anxiety, while others may be more likely to react in different ways to stressful circumstances, such as with an exacerbation in physical health complaints and worry about chronic illnesses. Some may try to escape difficulties through excessive use of alcohol or other psychoactive substances, or in extreme frustration, turn to criminal activities to secure desired objects or to demonstrate power over others. Not only are constitutional factors involved in how an individual responds to environmental factors, but so too are cultural factors and social influences on what is acceptable for different groups. For example, women may admit to higher levels of depression than men because culturally they have greater permission to express these feelings, while men may try to escape depressed feelings through the use of alcohol and other substances because of the general lack of societal permission for males to respond emotionally to situations of concern.

Psychotic Disorders

Among psychiatric disorders which appear to be predominantly genetically-based are the psychotic disorders, including schizophrenia, which involve lack of contact with reality because of disordered thoughts, such as hearing voices or seeing things. These experiences are called hallucinations, and are typically accompanied by problems with social relationships and difficulty maintaining aspects of daily life functions. Additional psychotic disorders include delusional disorders, which involve fixed ideas (termed delusions) that other people are either out to harm one, spy on one, or that may involve preoccupation with one being someone of significant fame or notoriety. According to the current *Diagnostic and Statistical Manual* (DSM-IV 1994) published by the American Psychiatric Association, which describes all psychiatric disorders and lists diagnostic criterion for them, psychotic disorders are relatively rare in the population, with a prevalence of roughly 2%, but such disorders can significantly impair an individual's life functioning. These disorders are best treated with a combination of medication to curtail psychotic symptoms, and training and support in the development of improved skills for social interactions and daily life functioning. Medications are best provided by a psychiatrist, who is a medical doctor with special training in behavioral and psychological disorders. Additional assistance and psychosocial treatment can be provided by psychologists and/or social workers trained in behavioral management and the development and enhancement of life skills.

Mood Disorders

Another category of disorders that occur with greater frequency than psychotic disorders are mood disorders, which fall on a continuum of severity. The more severe mood disorders can have psychotic features, in which an individual experiences delusions or hallucinations. Mood disorders may involve primarily negative or depressed feelings, as in the case of the depressive and dysthymic disorders, or may involve cycling between periods of high energy or mania, and depressed moods, as in bipolar disorders. According to the DSM-IV, the lifetime prevalence of major depressive disorders in woman ranges from 10 to 25%, while for men it is from 5 to 12%. Depressive disorders may be triggered by stressful life events, while bipolar disorders appear to be rooted in the biology of the individual, and are less reactive to life stresses, per se. Treatment with medication is essential for individuals with bipolar disorder, making the services of a psychiatrist important. Depressive disorders can be treated with medication alone or psychotherapeutic techniques alone, but tend to be best treated with a combination of medication and psychotherapy. There are a wide array of antidepressant medications currently in use today, and a variety of effective psychosocial treatment approaches for mood disorders.

Anxiety Disorders

Another relatively common set of disorders are the anxiety disorders, which are further subdivided into several catagories. Panic attacks may mimic physical symptoms of a heart attack or other cardiovascular emergency and may occur in conjunction with a number of anxiety disorders, and therefore merit elaboration. An individual experiencing a panic attack feels intense fear or discomfort with at least four somatic symptoms, such as heart palpitations, shortness of breath or smothering sensations, chest pain, nausea, dizziness, sweating, chills or hot flushes, and fears of dying, losing control, or going crazy.

Individuals experiencing panic attacks may feel disconnected from themselves or from reality, and may experience sensations of numbness or tingling, trembling, or a feeling of choking. Many individuals experiencing anxiety attacks fear they are experiencing a catastrophic illness, and may be first diagnosed in a hospital emergency room or may seek assistance from their primary care physician. They may not identify anxiety as the root of their physical sensations.

Among the anxiety disorders are panic disorders, which involve recurrent panic attacks and may also involve agoraphobia. Agoraphobia technically means "fear of the marketplace," and is used to refer to states of anxiety about, or the consistent avoidance of, places in which one might experience difficulty or embarrassment in trying to escape if one were to experience a panic attack. Individuals with agoraphobia might try to avoid a diverse range of places, such as shopping malls, bridges or elevators, traveling in a plane or bus, and might become "homebound" due to fear of leaving the house at all. Or, they might become afraid of being at home alone. Agoraphobia can result in significant constraints on an individual's ability to function in life.

Other anxiety disorders include Specific Phobias, in which an individual is fearful of a specific object or situation; Social Phobia, in which an individual suffers an intense fear in social or performance situations in which they may be embarrassed; and Generalized Anxiety Disorder, in which one experiences high levels of anxiety and worry in response to a variety of situations or events. Generalized Anxiety Disorder has a lifetime prevalence of about 5%. According to DSM-IV, the anxiety may be associated with sleep difficulty, fatigue, problems with concentration, muscle tension, irritability, or restlessness, and may impair an individual's social and work functioning.

A somewhat less common anxiety disorder is Obsessive-Compulsive Disorder, which involves pervasive recurrent thoughts (termed obsessions), or ritualistic or repetitive actions (called compulsions), which are accompanied by noteworthy distress for the individual or cause significant impairment in functioning. There are also two anxiety disorders that occur in response to traumatic events, those being Post-Traumatic Stress Disorder (PTSD) and Acute Stress Disorder. Lifetime prevalence of PTSD has been estimated in community samples as ranging from 1 to 14%, depending on the populations examined. PTSD occurs in the wake of a traumatic stressor and involves the experience at the time of horror, intense fear, or helplessness, and the persistent reexperiencing of the trauma in distressing dreams, recollections, or flashbacks. Individuals try to avoid anything associated with the trauma, and experience persistent symptoms associated with arousal or anxiety, such as problems falling asleep, outbursts of anger, or hypervigilance. In PTSD, the disturbance is more than one month in duration after a trauma has occurred, while in Acute Stress Disorder, symptoms of distress similar to those in PTSD occur within one month of a severe stressor.

Somatoform Disorders

An additional class of disorders which manifest themselves largely in physical symptoms and health complaints are the Somatoform Disorders. While symptoms of these disorders are suggestive of medical illnesses, the symptoms cannot be fully explained medically. However, individuals most frequently present to medical settings with physical complaints, often unaware of the psychological contributing factors or causes. In fact, individuals with these disorders often prefer to view them in solely medical terms, and are resistant to psychological treatments. These disorders cause significant distress for the individual, and/or may involve impairment in social, occupational, or other areas of functioning.

Among these disorders is somatization disorder, which begins before age 30, extends over years, and involves combinations of pain and gastrointestinal, sexual, and pseudoneurological symptoms. A less severe disorder, that of undifferentiated somatoform disorder, involves a variety of physical complaints which have lasted at least 6 months. Conversion disorder involves symptoms that affect motor or sensory areas suggestive of a neurological disorder or other medical problem, and which appear to have developed in the wake of psychological conflicts or stressors. Classic examples of this syndrome are glove anesthesia, involving loss of feeling in a hand, "hysterical blindness," with no identifiable medical cause, or sensations of a "lump" in the throat, accompanied by difficulty swallowing. Prevalence rates of these disorders are not fully known, but are estimated to be in the 1 to 3% range of outpatient referrals to mental health centers (DSM-IV). They are seen with greater frequency in medical settings, are estimated as high as 5 to 15% of psychiatric hospitalizations in general hospitals, and from 25 to 30% of admissions to Veterans Administration hospitals (Kaplan et al. 1994).

Two additional somatoform disorders that occur at notably high frequency levels are hypochondriasis and pain disorder. Once again, individuals with these disorders primarily present for treatment at medical facilities rather than psychiatric treatment centers. Hypochrondriasis involves preoccupations and worries that one has a serious illness, often based on the misinterpretation of a number of symptoms or bodily sensations. Even when medical assistance rules out a serious medical condition, the individual still persists in their fear of having something medically serious. Such individuals are often resistant to psychological treatments, due to their insistence that physical illness is at the root of their difficulties. They often have accompanying anxiety and depression. Prevalence estimates of this disorder range from 4 to 9% of those presenting to a general medical practice.

Pain Disorder is another rather common somatoform disorder. It involves pain in one or more sites that cannot be fully accounted for by medical or neurological causes. Individuals with pain disorder show clinically significant distress, and some impairment in primary areas of functioning, such as in social or occupational functioning. Also, psychological factors are believed to be of significance in the onset, worsening, or maintenance of pain symptoms. The prevalence of this disorder is believed to be quite high, with as many as 10 to 15% of adults having some sort of work-related disability resulting from back pain alone (DSM-IV).

Additional psychiatric disorders with serious physical consequences are the Eating Disorders. This class of disorders is less prevalent overall, and is significantly more common in females than males. Anorexia Nervosa involves weight loss and refusal to maintain an appropriate weight for age and size (less than 85% of expected weight). There is accompanying distortion in one's perception of body weight and shape, and intense fear of "becoming fat" or gaining weight. Prevalence estimates are 0.5 to 1.0%, and this disorder is typically seen in adolescent girls, ages 14 to 18 years of age. Bulimia Nervosa involves episodes of binge eating, with alternating attempts at purging or preventing weight gain, such as self-induced vomiting, fasting, excessive exercise, or misuse of laxatives or diuretics. This disorder also occurs predominantly in adolescent and young adult females, with prevalence rates of about 1 to 3%.

Personality Disorders

The more longstanding, chronic, and pervasive disorders that characterize an individual's way of functioning in the world are the personality disorders. Personality disorders manifest themselves in the thoughts and perceptions of individuals, and in their behavior. Usually such patterns develop in adolescence or early adulthood, are quite persistent over time, and generate difficulties in a number of areas of an individual's functioning, impacting both relationships and functioning in occupational settings. Personality disorders are generally difficult to treat because individuals with these disorders often do not recognize problems in their own behavior, but more often generate distress in others around them.

The personality disorders are divided into three clusters in the DSM-IV. Cluster A involves paranoid, schizoid, and schizotypal personality disorders, all of which involve odd or eccentric behavior, and are characterized by some avoidance of interpersonal relationships. Research has suggested that this cluster of personality disorders are more commonly found in the relatives of individuals with schizophrenia than in control groups, with correlations being highest for schizotypal personality disorder.

Individuals with paranoid personality disorder have high levels of suspiciousness and distrust of others, and believe that others have negative intentions toward them. They anticipate exploitation or harm from others, when there is little or no evidence to support such a view. They have difficulty with close relationships and may be argumentative or show hostile aloofness (DSM-IV). Individuals with schizoid personality disorder are characterized by a striking detachment from others and an apparent lack of interest in close relationships. They appear to have a limited range of emotional expression, low interest in pleasurable activities, and appear indifferent to the reactions of others. The person with schizotypal personality disorder shows problems with close relationships as well as odd thinking and behavior, including cognitive and perceptual distortions. Their odd thinking may be characterized by paranoid ideation or preoccupation with paranormal phenomenon. They most likely have much discomfort in relating to others, show oddities in speech and appearance, and are not able to adjust to expected social norms.

Cluster B consists of the antisocial, borderline, histrionic, and narcissistic personality disorders. Commonalities of these disorders are an often dramatic, unpredictable, erratic, and emotional presentation, and their relationships tend to be unstable.

Individuals with antisocial personality disorder demonstrate a lack of concern for the rights of others and societal norms, and are often quite good at manipulation of others. They are typically involved in criminal behavior such as stealing, property damage, harassment of others, and pursuit of illegal occupations. Antisocial personality disorder is commonly associated with alcohol use disorders.

Individuals with borderline personality disorder demonstrate extreme instability in their relationships, self-perception, and emotions, and are highly impulsive. They are frequently in crisis, and go to extremes to avoid real or imagined abandonment. Recurrent suicidal threats or gestures are common, along with self-harming behaviors. Such individuals often have chronic feelings of boredom, an unstable sense of self, and often have high degrees of mood fluctuation, with frequent displays of anger and/or intense anxiety, depression, or irritability. Behavior can be quite impulsive, often in potentially harmful areas, such as substance abuse, reckless driving, or promiscuous sex. Biological relatives commonly have mood disorders, especially depression (Kaplan et al. 1994).

The individual with histrionic personality disorder often demonstrates much dramatic and emotional behavior. They are excitable and extroverted, but often have much difficulty sustaining committed relationships. The need to be the center of attention is strong, and seductive, provocative behavior is common, along with high suggestibility. There is a strong relationship between histrionic personality disorder and somatization disorder (Kaplan et al. 1994).

Individuals with narcissistic personality disorder demonstrate grandiosity in their sense of self-importance, are arrogant and haughty, envy others, and require excessive amounts of admiration. They feel they deserve special treatment, often have fantasies of unlimited success, power, beauty, or brilliance, and believe they should only associate with other high-status individuals. They are lacking in empathy, and tend to exploit others for their own advantage.

Cluster C personality disorders consist of the avoidant, dependent, and obsessive-compulsive personality disorders, and are characterized by high levels of anxiety or fearfulness. Individuals with avoidant personality disorder have a chronically negative self-perception, viewing themselves as inadequate or inferior to others. They are preoccupied with the possibility of being criticized or rejected in social situations, and are reluctant to take personal risks for fear of embarrassment. They are restrained in their relationships due to fear of shame or ridicule, and they avoid involvements with others unless they can ensure that they will be liked.

Table 10.1. Mental health disorders.

I. **Schizophrenia and Other Psychotic Disorders**
   Schizophrenia
      Paranoid Type
      Disorganized Type
      Catatonic Type
      Undifferentiated Type
      Residual Type

   Delusional Disorder
      Erotomatic
      Grandiose
      Persecutory
      Somatic

II. **Mood Disorders**
   Depressive Disorders
      Major Depression
      Dysthmic Disorder

   Bipolar Disorders
      Bipolar I Disorder
      Bipolar II Disorder
      Cyclothymic Disorder

III. **Anxiety Disorders**
   Panic Disorders
   Specific Phobias
   Social Phobias
   Post-traumatic Stress Disorder
   Generalized Anxiety Disorder

IV. **Adjustment Disorders**
   With Depressed Mood
   With Anxiety
   With Disturbance of Conduct
   With Mixed Disturbance of Emotion and Conduct

V. **Somatoform Disorders**
   Somatization Disorder
   Conversion Disorder
   Pain Disorder
   Hypochondriasis

VI. **Personality Disorders**
   Antisocial Personality Disorder
   Borderline Personality Disorder
   Histrionic Personality Disorder
   Narcissistic Personality Disorder
   Avoidant Personality Disorder
   Dependent Personality Disorder
   Obsessive-Compulsive Personality Disorder
   Paranoid Personality Disorder
   Schizoid Personality Disorder

VII. **Substance-Related Disorders**
   Alcohol-Related Disorder
   Amphetamine-Related Disorder
   Cannabis-Related Disorder
   Concaine Related Disorder
   Hallucinogen-Related Disorder
   Inhalant-Related Disorder
   Opiod-Related Disorder
   Sedative, Hypnotic or Anxiolytic-Related Disorder
   Phencylidine-Related Disorder

VIII. **Eating Disorders**
   Anorexia Nervosa
   Bulemia Nervosa

Persons with dependent personality disorder have an excessive need to be taken care of, demonstrate submissive and clinging behaviors, and are fearful of separation. They have difficulty making decisions, expressing disagreement, and initiating projects, and are fearful of being alone or left to take care of themselves. They go to extreme lengths to obtain caretaking and support from others. Individuals with obsessive-compulsive personality disorder are preoccupied with perfectionism, orderliness, and control. They tend to be stubborn and indecisive, lack flexibility, and have constricted emotions. They are preoccupied with details, rules, and organization, to the exclusion of the "big picture." They are overly involved in work and productivity, thereby excluding friendships and close relationships. They tend to be stingy with money, are hoarders, and can be inflexible about matters of rules, laws, and morality.

A further class of disorders involve the use of psychoactive substances, which may be legally or illegally obtained. Termed substance-related disorders, they range from incidents of excess use of alcohol or marijuana, to severe dependence, accompanied by often dangerous withdrawal symptoms, on these substances and a variety of others, including opioids and legally prescribed medications taken to excess, such as anti-anxiety medications or pain medications. The term "substance" can refer to a drug that is most commonly associated with abuse, a prescribed medication, or a toxin, according to DSM-IV. The Substance-Related Disorders are divided into two major groups—the Substance Use Disorders, which involve either dependence, which is probably what many people in common parlance refer to as "addiction", and Substance Abuse, which may involve discrete episodes of excess use of substances, without an individual necessarily having become physiologically or psychologically dependent on the drug or medication.

THE NATURE OF MENTAL HEALTH PROBLEMS IN WORKERS IN AGRICULTURE, FORESTRY, AND FISHING

The hazards of mental health and substance abuse issues for workers in agriculture, forestry, and fishing are well-documented in epidemiologic data. An examination of deaths resulting from alcohol-related causes, such as cirrhosis, digestive cancers, and suicide, suggested that the workers in these occupational areas are more likely to die from such causes, as are laborers, helpers, equipment cleaners and handlers, than are workers in other occupations (Brooks 1992). Suicide rates in the wake of economic stresses for farming were found to be higher in both farm and forestry workers than for workers in transportation (Ragland 1991). An additional study of suicides in Alabama from 1984 through 1989 found that people employed in farming, agriculture services, forestry and fisheries had intermediate levels of risk for suicide, unlike the low risks of those working in public administration (Liu and Waterbor 1994). The authors speculated that job stress, along with sociodemographic differences, occupational self-selection, and access to means of lethal self harm, might contribute to the different suicide rates.

Regarding other psychiatric disorders, researchers (Eisemann 1986) found that agricultural and forestry workers, along with blue collar and unskilled workers, were more highly represented in three groups of depressed patients, specifically those with unipolar depression, who suffered from only depressed feelings, and those with bipolar disorders, who suffered from periods of mania and depression, and those with unspecified depressive disorders. While depressed and bipolar groups of workers did not necessarily come from lower social classes, they tended to be subject to greater socioeconomic disadvantage, entering a downward spiral. Inevitably, economic stresses then became factors in the lives of these individuals, thereby further increasing depression and instability (Loboa and Neyer 1991).

Stressors in Agricultural Work

Rural farm families have been faced not only with the uncontrollable effects of nature, but also with economic changes that have increasingly spiraled their financial resources downward. The farm population, defined as those people living in rural settings with at least $1,000 of their yearly income resulting from agricultural sales, has shown a decline, with only about 2% of the U.S. population in farming, per se. However, about one-fourth of the population of the United States live in rural areas (Murray 1991) experiencing many of the stresses and disadvantages of workers in farming. Among farm families, poverty tends to be high, yet the availability of public resources tends to be quite low (Ellis and Gordon 1991). Mental health resources tend to be particularly sparse, and because of geographic isolation, farm families are often not willing or able to avail themselves of those resources that may be available.

In addition to the noteworthy stresses inherent in agricultural occupations, the Farm Crisis of the 1980s had a profound effect on the individual farmer, his family, and the nature of the farm industry itself. Two factors appeared to be operating simultaneously. Inflation was significantly increasing the value of land for farmers in the early 1970s, while increasing technological advances were obligating farmers to purchase and learn to use more complex farm equipment. Farmers therefore acquired high amounts of debt, expecting the financial successes of the early 1970s to continue. However, their debt-to-income ratio increased significantly as they were encouraged to purchase more advanced and costly equipment, so that when financial difficulties occurred, there was no financial cushion for farm families. The result was foreclosure for a number of farmers who had been in this profession for generations, and a change from the smaller farm passed on throughout generations of the family to the larger, more mechanized farming production operation. Issues involved in managing a larger staff of workers complicated the duties of the once self-reliant family farmer.

Concurrent with the economic losses was the beginning of changes in the structure and functioning of the farm family. Traditionally, farmers relied on family members, including wives and children, to be an integral part of work on the farm. Traditional family values were of much importance, and farms were passed on from generation to generation. With increasing influence of the media on children, and adolescents' disillusionment with the financial and physical strains of farming life, the close network of the farm family also began to dissolve. This further created strain for the family system, with a result of greater rebellion in adolescents, increased stress for farming parents, and higher rates of marital conflict and subsequent divorce (Olson 1986).

Hedlund and Berkowitz (1979) explored the nature of social-psychological stress for farm families by closely examining the stresses reported by twenty New York State farm families in intensive interviews conducted biennially. Citing a 1977 National Institute for Occupational Safety and Health (NIOSH) study that identified the job of farm manager as 12th of 130 occupations for high stress and high incidence of disease, the authors sought to identify precisely what made farming stressful for families. Noting that farm work required family members to work together, and often involved multigenerational involvement in the farm, they sought to address both family stresses and external stresses. They found that 30% of families reported marital stress or stress because of demands for intergenerational transfer of the farm; 20% reported stress related to sibling rivalry; and 35% of wives reported stress related to their farm role. At least one of these stresses was reported by 75% of families studied. As non-family stresses, the authors examined poor financial returns, over-involvement by state or federal agencies, feeling too "tied down" by farming, and not having any vocation. These non-family stresses were reported by 70% of farming families.

Additional studies of stresses for farmers were conducted by Walker et al. (1986, 1988). They conducted interviews from 140 Manitoba farmers, and found that the inability to control certain major factors influencing farm work were a significant source of stress for farmers. A variety of financial stresses were reported by 83% of respondents, including low prices for commodities, high debt load, increasing expenses, and irregular cash flow. Problematic weather conditions and the unpredictability of weather were reported as stressors by 75 percent of farmers. A third source of stress, listed as one of the top five stressors by 75% of male respondents, was government regulations and agricultural policies. Specific stressors included unrealistic quotas, farm subsidies, free trade proposals, and government "cheap food" policies.

The toll of stressors on farmers may be evidenced in the high suicide rates among farmers in 5 north central states. An examination of the 1,352 suicides of farm workers and ranchers in Wisconsin, Minnesota, North Dakota, South Dakota, and Montana, indicated that rates were highest among farmers, with suicides of farm laborers and farm women next highest in frequency (Gunderson et al. 1993). Lowest suicide rates were for child and adolescent farm residents. The authors concluded that prevention efforts to reduce suicides in these groups are essential. They felt there needed to be an expansion of local psychiatric inpatient hospitals, more intensive outpatient treatment programs which involve more extensive support than the traditional once a week therapy, and substantial education about depression, anxiety, and suicide for farmers and their family members. Figure 10.1 demonstrates the ways in which different factors and stresses can interact with one another to create substantial problems for farm families.

Mental Health Problems in Farm Families

In any environment in which workers and their families are subjected to stress and frustration, aggression may follow. As described above, farmers are likely to experience great amounts of stress. A specific group of farm workers subject to even greater levels of stress is migrant farm workers. The severe disruption to life that migrant workers experience through poor living facilities, having to be continually on the move, and difficult work environments, can take a profound toll on the family life of migrant workers as well. The toll may be most severe for the children in the family. A series of studies addressing the neglect and abuse of children in migrant farm families in five states—New York, New Jersey, Pennsylvania, Florida, and Texas—found that across all states, migrant children were significantly more likely to be abused than were other children, though there were differences in the rates of child abuse in each of the five states. The data for this study were taken from government censuses of school-age children conducted to discover which children qualified for educational services and were cross-referenced with Social Services records on confirmed cases of child abuse or neglect. The findings of higher abuse rates in the study were attributed in part to the extreme poverty of the migrant workers (Larson et al. 1990).

Rural children may be at risk in other areas as well. Spoth and Conroy (1993) noted that although preadolescent children from rural communities show a higher incidence of delinquent behavior, including substance abuse and conduct problems, their parents seem less aware of any problems as they are developing and worsening. Operating

Figure 10.1. Factors affecting farm workers.

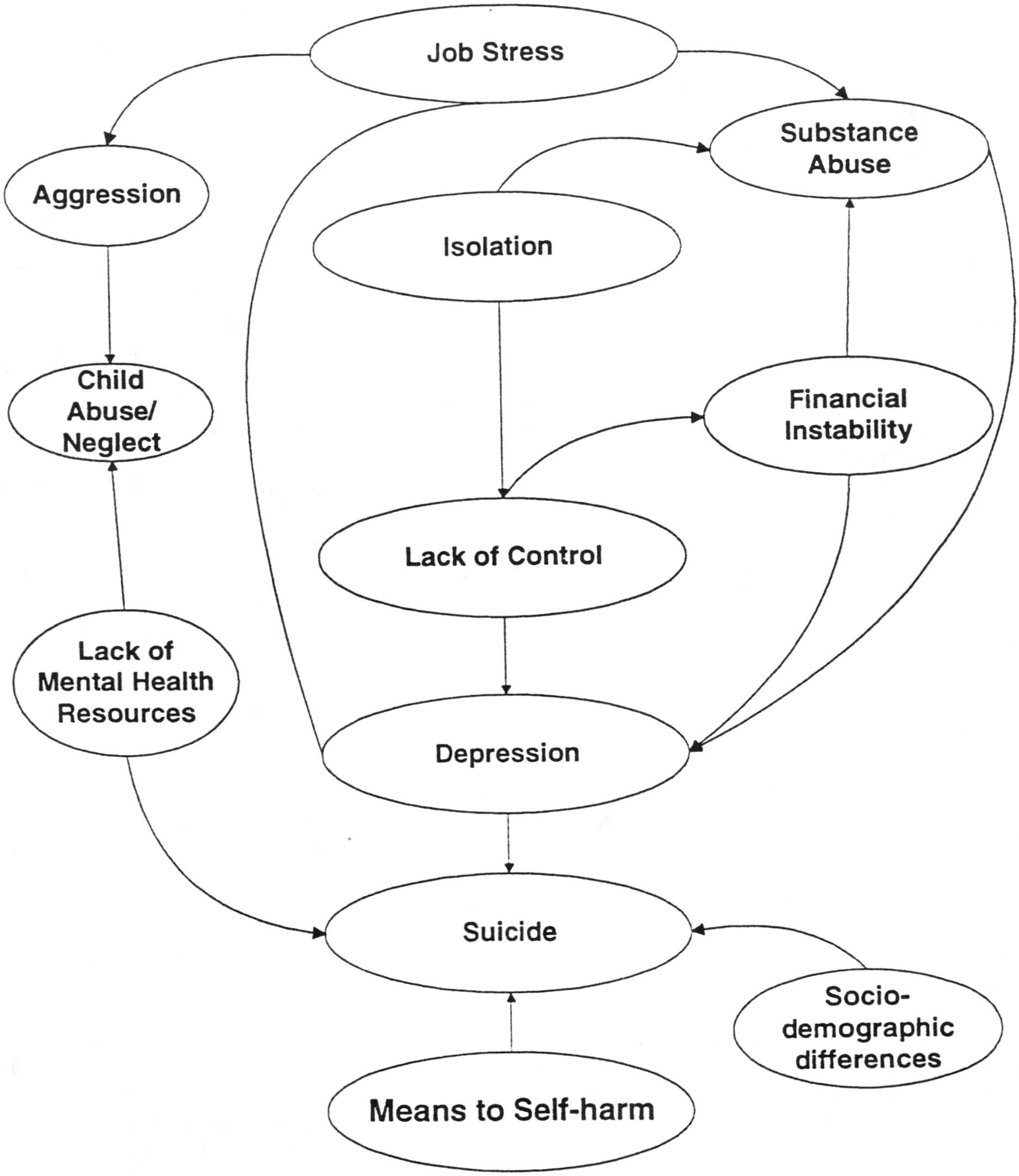

under the theory that when people believe themselves vulnerable to a certain situation (in this case to having a child with behavior problems) they will guard against or make efforts to correct the situation, these researchers came to the conclusion that the parents of these rural children were not taking actions to correct the problem behavior of their children because of not perceiving a potential problem. Education is therefore critical regarding risk for behavioral problems.

Attempts have also been made to discover if rural children experience more stressors and/or use different coping mechanisms. Atkins et al. (1993) administered the Feel Bad Scale and the School-Agers Coping Strategies Inventory to 157 rural and urban children in fourth, fifth, and sixth grade in Missouri. In spite of the fact that rural children tend to be underserved in their physical and mental health needs, this study found no significant difference between the two groups in either stressors or coping. Studies targeting those rural children who live and work on farms may be needed to clarify stressors unique to farm children.

A study by Bigbee (1987) found that, although levels of stress were similar in a study of urban and rural couples, the stressors which affected each group differed in type. A closer look at the stressors of farm women was carried out by Giesen et al. (1989). Women and their husbands from 169 farms in Holland were questioned on a number of areas to discern farm task load, home task load, husband support, and role conflict. A number of factors were found to contribute to role overload, including number of children, sex-role attitudes, a woman's intrinsic motivation, her perceived financial situation, tasks carried out, and number of hours spent on tasks. Husband support, number of tasks and hours spent at tasks, and self-esteem were found to reduce role overload. However, the role overload was not found to have a negative affect on farm women. A possible explanation for this finding was that the women took their difficult lifestyles for granted and therefore adapted to them. Also, it was found that the women were more affected by personal situations, such as self-esteem, rather than by external events.

Environmental circumstances can also have an effect on farm workers and their families. A study by Schiffman et al. (1995) discovered that people can be negatively affected by smells associated with farms. They administered the Profiles of Mood States to 44 people who lived near hog operations and 44 controls. The experimental group filled out the questionnaire when they could smell the odor of the farms, and the control group filled them out on two random days. There were significant differences in the scores between the two groups, with the "hog-smelling" group displaying more tension, depression, anger, fatigue, and confusion, and less vigor. Schiffman et al. offered several explanations for the findings, including a learned aversion to the odor; the subjects feeling less control over their environment; neural stimulation of immune responses, and direct physical effects from the molecules involved in the odor, such as increased adrenaline levels. The effects of these odors likely also influence those who are constantly in contact with them, again compounding the stressors which farm workers themselves and their families experience.

Prevention and Interventions

In the face of major stresses, particularly as a result of sudden changes in economic and environmental factors, the question remained as to what factors influenced which workers would remain in agriculture, and which would leave the field altogether. Schulman (1989) examined North Carolina farmers and speculated that the amount of perceived stress and the level of social support would be significant factors in which farmers survived. Interestingly, it was not the level of perceived stress which was of primary importance, but rather it was the perception of available social support that affected who thought they could survive in farming. These results suggest that developing supportive networks for workers in agriculture might provide the greatest buffer against the variety of stressors faced by agricultural workers. Use of an interview, such as that presented in Table 10.2, can be helpful for health care workers and counselors to identify areas of need which may benefit from social support or practical interventions.

Efforts have been made to address the unique mental health needs of farmers through model programs, such as one developed in Illinois called *Stress: Country Style* (Cecil 1988). This program was designed to respond to the unique concerns and difficulties experienced by farmers in response to the economic decline which triggered profound losses for farmers. The program included outreach counseling, community education about stress, and a crisis phone line. Mechanisms for linking individuals needing greater levels of intervention were developed in coordination with community mental health centers. What seemed to generally be of most benefit to stressed farm families, however, were programs which provided a combination of social support and practical advice and information on a variety of prominent concerns, such as financial issues, ways of securing loans, and ways of dealing with stress.

Case Illustration. A case example which illustrates the rugged individualism of farmers and the inclinations to want to care for themselves without trusting outside resources, is that of a farmer with a major depressive disorder who was seen in a North Carolina hospital. Having concerns about the distance to be traveled to a hospital for mental health care,

Table 10.2. Psychosocial stressors for agricultural workers: interview for clinicians

1. How much do you worry about having enough money to operate your farm effectively?
2. Are you concerned about having enough money to support your family?
3. How concerned are you about weather conditions affecting your crops?
4. Do you have concerns about the safety of your family members in operating farm equipment?
5. Do you have concerns about the safety of your staff in operating farm equipment?
6. Have you or anyone in your family been injured in farm work? If so, who was injured and with what type of injury?
7. How much do you worry about not knowing how much money you will be earning during a season because of varying market conditions?
8. Do you have any difficulty obtaining the medical services you feel your family needs?
9. How much debt have you acquired as a result of the need to purchase more advanced farm equipment?
10. Do you worry about your ability to repay those debts?
11. How many hours per day do you work?
12. Do you feel you have enough time left over from farm work to spend time with family and friends?
13. Do you think you have enough friends and acquaintances to enjoy leisure time with?
14. Do you believe you have enough time to enjoy recreational activities?
15. When was the last time your family had a vacation?
16. Have you ever been worried about members of your family having behavior or emotional problems or problems with their nerves?
17. Have you ever thought of obtaining counseling for yourself or your family members?
18. If you wanted to find a counselor for yourself or your family, how would you find one?

the costs of the treatment he needed, and wanting to be able to control his own treatment, a farmer decided to take his care into his own hands. He had previously been brought out of a severe major depression with electroconvulsive shock treatment (ECT). ECT is an effective treatment for intractable depression, manic episodes, and other serious disorders (Kaplan et al. 1994). While the patient is under anesthesia, electricity is administered to the brain through electrodes in order to induce seizures. This gentleman decided he could handle the treatment himself, however, so he set out to attach electrodes to the engine of his tractor, and see if he could give himself the needed electroshock treatment to eliminate his depression. Fortunately, he survived his attempt at self-care, but did end up on the neurological service of the hospital for injuries sustained from his self-imposed shock. The case illustrates some of the geographical, economic, and philosophical barriers to treatment that are commonly present for workers in agricultural areas.

Mental Health Issues in Forestry Workers

The forestry industry requires from its workers a number of different tasks, many of which involve heavy physical effort and significant risk of injury. Forestry workers are involved in all aspects of developing and transporting forest products, including establishing forests, maintaining them, and harvesting their products (Strehlke 1983). A primary operation in forestry work is "logging," which involves cutting down and transporting trees to obtain wood. Dangers not only involve much physical labor, but also exposure to dangerous tools, noise, and the risk of traumatic injury. Concerns about safety inevitably create stress for forestry workers, as do the sequellae of injuries which may occur on the job, in a working environment often quite distant from medical assistance.

Additional hazards for forestry workers are the demand for working under difficult weather conditions. Such conditions exacerbate the risk for accidents involving falling or slipping when work is conducted in rough terrain or during inclement weather, particularly in rain, snow, or icy conditions. Also, forestry workers often have significant distances to travel in order to get to and from work, particularly as the inclination to live in forestry camps has substantially decreased (Strehlke 1983). Often a worker must travel to a remote area on substandard roads. The risk for accidents in commuting is therefore enhanced.

Stressors in Forestry Work

Not only do the tasks of logging require high levels of physical labor and have high risks for physical injury, but there are physical and mental stresses in silvaculture work, which involves the cultivation of forest trees. Robinson et al. (1993) note that the work of tree planting results in symptoms that are consistent with musculoskeletal stress. They further confirmed this by taking blood samples during tree planting, and found a significant elevation of serum enzyme activity as the tree planting season began, which remained elevated throughout the season. While there have been studies emphasizing the physical demands of such work, the psychological demands have been less adequately studied, yet these demands appear to be high as well. Slappendel et al. (1993) emphasize the need for greater exploration of adjustments which could be made in the organization of work, i.e., adjustments in tasks demands, increased task variety, use of rest breaks, assessment of hours of work, supervision. They note the need for assessing advantages and disadvantages of mechanization so as to minimize the amount of psychological stress imposed by the work tasks in forestry.

In addition, forestry workers, like farmers, are at risk for pesticide exposure and the health consequences of such exposure. Further, in comparison with maintenance workers, Bovenzi et al. (1991) noted that forestry workers operating chain saws were at slightly greater risk for upper extremity cumulative trauma disorders (CTD'S). Forestry workers also showed higher rates of persistent upper limb pain, carpal tunnel syndrome, and muscle-tendon syndromes. Physiological stress, worry about pesticide poisoning, and persistent pain and cumulative trauma disorders can all have negative social/emotional and psychological consequences for forestry workers.

Apparent mental health concerns in forestry workers merit careful assessment by medical professionals and toxicologists, since apparent mental health problems may be caused by toxins in the environment. Green (1987) conducted a study of 1,222 male workers in forestry exposed to phenoxy acid herbicides and found higher numbers of deaths resulting from suicides in that group. They speculated that these suicides might result from neurological damage obtained from exposure to the herbicides.

Gardell (1973) examined the mental and physical health and job satisfaction of 370 rangers and employees in the Swedish Forest Service. They found a positive relationship between job satisfaction and mental health. Additionally, physical symptoms related to the head, back, and abdomen were related to higher levels of job stress and mental distress. The authors found that greater mechanization of forestry work, and increased needs for technical preplanning, actually caused higher levels of emotional strain accompanied by feelings of boredom. The result was lower scores on mental health tests, suggesting greater problems.

Psychosocial Stressors for Families of Forestry Workers

A further source of stress for forestry workers is the social and physical isolation their work places them in. Caldwell (1990) writes of the work of U.S. National Park Service Rangers, and comments on their frequent work in areas that are remote from other towns, subject to varieties of inclement weather, and characterized as having difficult terrain. Territories covered by the National Park Service include isolated areas in the Southwest, in which the nearest paved road may be 20 miles from a forestry site, and rugged areas of Alaska, in which there are park areas covering as many as 5 million acres, much of which is unsettled land.

Surveys of National Park Service rangers have identified problematic consequences of the remoteness of their work sites. Such isolation creates social stress, job stress, and family stress. Johnson et al. (1983) note that consequences of the stress of isolated areas include higher rates of alcoholism, drug abuse, and problems for families. Factors contributing to problems resulting from isolation included long distances to the nearest town, lack of opportunity for social relationships outside of the ranger group, and lack of opportunities to participate in a wider range of cultural and social activities.

Another common aspect of a National Park Service employee's life seen as an additional source of stress is the fact that many employees often find it necessary to live in government provided housing. Often these residences are located convenient to the site of the Ranger's job, but far from the nearest town, compounding the isolation problem. Nearly one-third (29%) of employees surveyed in a National Park Service study (Johnson et al. 1983) felt that using government quarters presented an inconvenience for their family. In terms of general satisfaction, although the majority of employees questioned in this same survey were reasonably satisfied with the government quarters they occupied, a substantial number were not (24%). This dissatisfaction was more common among families with more children: the satisfaction level decreased as the family surveyed grew larger, with those employees having no children reporting the greatest satisfaction, and those having two or more children reporting the least satisfaction. When questioned about how the government housing system could be improved, some of the suggestions involved occupancy or rent policies and quality of maintenance. Since Park Rangers are often required to live in these residences, it can be seen that such dissatisfaction can be a source of stress among the employees and their families.

In addition to this isolation, Park Rangers and their families are often subject to the stresses of frequent relocation. In order for employees to achieve higher levels of responsibility and greater pay, they must transfer to new positions, many times moving across the country. This sort of instability can bring additional stress to family members, and when the area the family is entering is a remote one, the adjustment is even further complicated. The difficulties that many children have in settling into a new environment, including recreating social networks, can be made even more complex by introduction into unfamiliar cultures. Such was the case for the children of a Park Ranger who received a high promotion by moving from the east coast, where his sons had just entered high school, to a small park in the west. Here, the boys not only had to readjust to a new high school society, never an easy process for an adolescent, but they also had to learn about and assimilate into the dominant local culture, in this case, that of the Navajo Native Americans. This learning process, though ultimately fulfilling, can bring additional stress and has the potential for a great deal of tension had the family not been as culturally tolerant (Anibal 1995).

A complex interaction of stressors can affect forestry workers and their families to compound the effects of stress experienced by the workers themselves. As in the case of dissatisfaction with government housing, some of the problems can be exacerbated with family size. Table 10.3 lists common stressors as they affect employees and their families. Particularly for families often isolated from social support, discomforts in the family can have significant mental health consequences, both for the stability of the marriage and the psychological welfare of the children.

There is therefore a need for educational programs to inform potential employees of the problems of living in isolated areas, along with counseling programs for the stressful consequences of isolation. The surveys taken by the National Park Service Cooperative Studies Unit, such as the one cited above, provide valuable information regarding employee attitudes which can help shape future policies. At present, National Park Rangers also have access to an Employee Assistance Program for referrals or to provide counseling for problems they or their family members may have which are able to be addressed in the forum of short term counseling. EAP's, because of their easy accessibility and prepaid costs, provide mental health assistance to many employees and family members who otherwise might not seek such help. These resources can help relieve some of the stressors on Park Rangers themselves and on their families.

Further stressors for workers in the National Park Service are the increasing role which they are having to take in law enforcement, and the inherent stresses and difficulties which come with law enforcement as well as dealing with disasters of various sorts, including forest fires. Both such activities often expose forestry workers to the acute traumas that emergency workers are regularly faced with, and often bring the worker into involvement with severe injury or death. Forestry workers may be the first on the site of airplane crashes, multi-casualty accidents, or may be involved in searches for missing persons feared injured or dead. A centrally important intervention in such situations is Critical Incident Stress Debriefing, which is now a part of the National Park Service Emergency Medical Service competencies (Anibal 1995).

Interventions

Critical Incident Stress Debriefings (CISD's) are a type of intervention which is optimally conducted within 24 to 48 hours after a traumatic incident has occurred. Growing out of traditions established in the military, CISD's are aimed at preventing emotional sequelae which can occur subsequent to exposure to a disaster or trauma. As far back as World War I, doctors became aware of a syndrome of symptoms, which often could be quite severe and incapacitating, that occurred in some soldiers after exposure to combat, and was termed "shell shock." More recently, this disorder has been labeled post-traumatic stress disorder (PTSD), and can occur in response to a variety of traumas, including exposure to natural disasters which result in serious injury or threatened loss of life, fires, and assaults. Critical Incident Stress Debriefings are aimed at providing immediate assistance for individuals and emergency workers who may be exposed to trauma, with the long range goal of decreasing the likelihood of the development of psychiatric sequelae or full blown PTSD (Kaplan et al. 1994).

Table 10.3. Stressors as they directly affect forestry workers and their families.

Stressor Members	Affects Employee	Affects Family
Risk of pesticide poisoning	X	
Risk of physical injuries	X	
Pressure to advance through relocation	X	
Relocation adjustments	X	X
Isolation	X	X
Job stability	X	
Cultural changes	X	X
Use of temporary housing units	X	X

Critical Incident Stress Debriefings are best conducted by an experienced mental health professional specifically trained in these interventions, and fully aware of the needs and experiences of their counselees. A facilitator conducting CISD's with forestry workers in the wake of a forest fire disaster, for example, in which forestry workers or firefighters were injured or killed, should be knowledgeable about the nature and concerns of firefighters and forestry workers, as well as about how to conduct debriefings and deal with distressing mental health issues which may arise.

Formal CISD's have 6 components, which are conducted over a 2 to 5-hour time period. Mitchell (1983) described the components beginning with an introductory phase, in which the process and rules of a CISD are described. Most essential is the need for absolute confidentiality, and the assurance that information revealed in the CISD will not be used against participants. The second phase involves a description by each participant of details about themselves, the incident itself, and their activities during the incident. In order to ensure disclosure of details and bring to memory all features of the disorder, individuals are asked to describe precisely what they saw, heard, smelled, and did during the incident. The third phase involves bringing feelings associated with the incident to the forefront. They are asked not only about how they felt during the incident, but also how they are feeling at present, and whether they ever felt similar things at other times in their lives. This phase may bring up previous traumatic incidents, and is one of the reasons why a trained mental health professional should conduct debriefings. It is also important that all participants express something of their feelings, so quieter or more reserved individuals are not left out.

While the first three phases set the stage and bring to life the incident and the feelings it evoked, the next three phases are geared toward identifying sequellae and establishing ways of coping. The fourth phase is the symptom phase, in which individuals discuss unusual things they may have experienced both at the time and in the present, and helps them address ways in which their life may have changed or ways in which things at home or work may have changed since the incident. Frequently, a variety of stress symptoms are brought up during this phase, which can then be addressed in the fifth phase. The fifth phase is a teaching phase, in which the common symptoms and features of stress response syndromes are explained and elaborated upon. This is where the facilitator can normalize the variety of symptoms and emotional experiences which are quite common in the wake of a critical incident. The final phase, that of re-entry, addresses any remaining questions or concerns, explains additional resources should symptoms persist or worsen, and helps the group develop a plan of action. It is not uncommon, for example, for a group dealing with loss of life in a forest fire, to develop a plan of action to provide support to the family of the worker who lost his or her life.

Mitchell (1983) estimates that the CISD sequence would take from 3 to 5 hours to complete. They also recommend a follow-up CISD a few weeks or months later to resolve any issues which may still remain.

Mental Health Issues for Fishermen

Risks Unique to the Industry. The perils of the sea have been well documented throughout literature, as well as their potential to trigger mental distress. Consider the figure of Captain Ahab in Melville's pivotal novel, *Moby Dick*. The one-legged captain is half crazed with rage over the injury that has befallen him from the creature that should have provided his livelihood, the great white whale. An examination of the factors that have contributed to Ahab's madness are not far from the experiences of many a fisherman. He suffered a severe injury threatening his survival; he spent many long days and nights in isolation, away from home and hearth while at sea. The perils of weather, including raging storms, unpredictable winds, extremes of heat and cold, and turbulent seas, were an ongoing threat. The uncertainties of the prospects for a good catch troubled himself and his crew. The potential conflicts, competitions and even boredom arising among a group of men in close quarters at sea caused considerable frustration. And one of the few solaces they had was the camaraderie shared around a jug of ale, providing an apparent easy escape from stress, but triggering greater problems for those inclined to consume too much. And the technical demands of operating a ship, including navigation and rigging, meant constant work demands and frequent sleepless nights for the ship's crew.

At least two million workers are employed in the fishing industry in the world, and it is by far one of the most dangerous professions. Because of the serious risks for injury and fatalities in this profession, the National Institute for Occupational Safety and Health (NIOSH) convened a workshop in 1992 to address issues of safety and health in the Fishing Industry (Myers and Klatt 1994). Causes of danger noted for fishermen, particularly those in the cold Alaskan waters, were the extremely cold weather, the likelihood of high winds, and the treacherousness of the seas. The authors noted that during peak fishing seasons, fishermen would often work round the clock, getting little or no sleep. In addition, workers would use dangerous machinery while on board a moving fishing vessel. They cited commercial fishing in Alaska as being thirty times more dangerous than the average U.S. job (Myers and Klatt 1992).

Fishing Industry Environments. The fishing industry is important in obtaining food for human use, and for securing by-products used in farming, medicine, and production of products such as soap and glue. (Puertas 1983).

The fishing industry includes both inland and sea fishing, with further subdivisions of sea fishing to include coastal fishing, ocean fishing, and high-sea fishing. Coastal fishing occurs relatively close to land (within 60 miles) and has a primary goal of supplying fresh fish and shellfish to coastal towns for human consumption. Workers are often in small vessels, have no fixed hours, and have work highly varied schedules according to season and weather. Often such workers also work in other fields, particularly farming. Ocean fishing is practiced primarily in Newfoundland, Iceland and Arctic waters, on large vessels, with the main catch being whale and cod (Puertas 1983).

The environments in which fishermen tend to work are likely to be quite dangerous. In Alaska and the Arctic, where a substantial number of fishermen work, traumatic injuries are the leading cause of death. Deaths by intentional injuries, in the form of suicides and homicides, and by unintentional injuries, through motor or air vehicle crashes, drownings, or resulting from alcohol or other drugs, cause 29% of all deaths for residents of Alaska, a significant portion of whom are fishermen, and account for over half of all years of potential life lost. For Alaska Natives, above other racial groups, injuries result in greater mortality and morbidity. Occupationally related fatalities are higher in Alaska than in any other state in the United States, with deaths among commercial fishermen, sailors, and loggers accounting for a sizable proportion of those occupational fatalities, along with airline pilots and truck drivers (Middaugh 1992). The dangerousness of fishing and logging in Alaska must take its toll psychologically for workers in these professions.

Riordan (1991) examined the experiences of stress in a sample of 252 commercial fishermen in Alabama, in comparison to a group of community businessmen. They found that for nearly all stressors they addressed, the fishermen encountered more than did their employees in community occupations. The types of issues that fishermen experienced as the most stressful, however, were safety concerns, long shifts of work, and carrying responsibility for the lives of others. It was also found that of personality variables related to stress, a greater sense of mastery was related to decreased stress, while optimism did not appear related to stress. This was in contrast to land-based employees, for whom higher optimism related to less stress, but greater mastery was related to increased stress. Perhaps it is important for the fisherman to be realistic in his/her assessment of risks and hazards, and to have both the ability and confidence to master the significant stressors in his/her environment. Stressors for Fishermen are depicted on Table 10.4.

Table 10.4. Stress among fishing workers.

Risk of Injuries Leading to Stress

Minor	*Major*
Puncture wounds	Damage to skull
Bruises	Damage to chest
Abrasions	Amputation
Frost bite	
Tenosynitis	

Occupational Circumstances Leading to Stress

Uncertainty about future availability of stock
New standards of safety requirements
New technologies available and in use
Increased competition
Adverse weather conditions
Long hours

Reactions to Stress

Somatic complaints
Anxiety and/or depression
Marital difficulties
Strain on family life
Alcohol and drug consumption

Substance Abuse. A further manner in which fishermen may attempt to cope with their occupational stresses is in the use of substances, with nicotine and alcohol being the most frequent substances of choice. Kline (1989) explored the use of smoking in an endeavor to alleviate job stress among shrimp fishermen. They found that in comparison to small business operators who were land-based, shrimp fishermen demonstrated higher rates of smoking. Also, hired captains smoked more than owner captains, and offshore captains more than inshore captains. The authors speculated that these rates of smoking may be explained through examination of smoking as an attempt to deal with job stress.

An additional substance used at extraordinarily high rates by seamen is alcohol. Granger (1993) noted that long trips away at sea might increase use of alcohol, tobacco, and illicit drugs in workers in the fishing industry. They reported a survey conducted in the northeast portion of Scotland which indicated that the frequency of alcohol treatment was higher in fishermen than in the rest of the population. Alcohol abuse contributes to health risks, as demonstrated in the high incidence of cirrhosis in fishermen, and also contributes to accidents resulting from drunkenness, including minor injuries and drownings.

A study of Polish seafarers (Nitka 1990) suggested that a number of stressors on the job cause seamen to use alcohol as an escape. In attempting to address whether specific personality traits, or the work at sea itself contributed to alcohol dependence, Nitka (1990) conducted a study examining medical and social correlates of alcohol abuse. Since none of the medical or social variables accounted for alcohol dependence, Nitka (1990) concluded that it was the nature of the work at sea rather than personality traits or medical issues that accounted for the high rates of alcohol use.

In an effort to look at psychological disorders in fishermen, Dolmierski et al. (1990) administered the MMPI, a measure of personality variables and psychopathology, to 588 seamen, including fishermen, employees of the fish processing plant, the engine crew, and workers on deck and in catering. They found the highest levels of psychological disturbance in the fishermen, with the lowest in those working on deck and in catering. It was noteworthy that they found what they termed "neuroses" to be most common, with 36% demonstrating psychosomatic neuroses. These would appear to fall into current diagnostic categories of somatoform disorders. Individuals with psychosomatic neuroses have numerous worries and anxieties about their physical health, and may be preoccupied with having a variety of illnesses for which there is no medical evidence. The authors attributed the high rates of this disorder as resulting from the factors which influenced how seamen lived and worked at sea.

The work of a fisherman not only takes a toll on himself, but also on the welfare and well-being of his family. One need only walk through a coastal town to see the waterfront dotted with homes with "widow's watches" lining the skyline, for women to look for the return of their absent spouses, often out at sea for extended periods of time. The anxieties in the wives are high as they try to run their homes without support of their husbands for times he is out at sea, and as they worry about whether he may succumb to the dangers of being at sea. Dixon et al. (1984) examined the effect of this role on fishermen's wives in a small U.S. island in the Atlantic Ocean. Comparing 67 wives of fishermen with 167 wives of nonfishermen, they found that fishermen's wives were not positive about their husband's chosen profession, and may actually try to dissuade their spouses and children away from of fishing as a way of life.

INTERVENTIONS FOR MENTAL HEALTH AND SUBSTANCE ABUSE PROBLEMS FOR WORKERS IN THE AGRICULTURE, FORESTRY, AND FISHERIES INDUSTRIES

For workers in these fields, who are selected for their independence, individualism, and desire to work in nature, traditional approaches to mental health problems might not be the most effective. Workers in these fields are less likely to contact a mental health professional, for example, than are workers in more traditional white collar employment. Therefore, family physicians need to address concerns in a wide number of areas, which may allow information on potential mental health problems to surface. It may be particularly helpful to administer a symptom checklist, for example, to workers in these fields when they come for routine visits. An example of such a symptom checklist is included in Table 10.5. There are also a number of checklists specifically focused on particular problem areas, such as measures to screen for depression, i.e., the Beck Depression Inventory or the Hamilton, or measures to determine risk for alcohol abuse or dependence, such as the SASSI.

Workers in all three of these fields are at particularly high risk for injury and illness as well. Therefore, programs geared toward prevention of disease and injury will be important. In addition, teaching coping skills in the aftermath of trauma is also needed. These might best be provided through local health care clinics. In addition, law enforcement officials need greater education about mental illness and serious psychiatric disorders, since in rural areas, particularly the seriously mentally ill are more likely to end up in the prison system or in offices of physical health care providers than with mental health case managers (Sullivan et al. 1996).

Educational programs in the community will be particularly important around topics which may effect workers in these fields. Overall programs on coping with stress, as well as those uniquely focused on the stressors encountered by workers and their families in these fields, will be particularly needed. These are the most likely routes such individuals may obtain access to helpful resources, and potentially obtain social support around the issues they struggle with.

SUMMARY

Workers in agriculture, forestry and the fishing industry, because of their rugged individualism, inclination to work independently and at times in isolation, and their trust in nature and the concrete as opposed to in philosophy and the abstract, are less likely to obtain mental health services in traditional ways. Therefore, creative strategies will need to

Table 10.5. Mental health symptom checklist.

Please indicate, by circling the number, how often you have been concerned or bothered by each of the following during the last three (3) months.

	None of the time	**Some** of the time	**Most** of the time	**All** of the time
Feeling anxious/stressed	0	1	2	3
Problems sleeping	0	1	2	3
Appetite problems	0	1	2	3
Work related concerns	0	1	2	3
Problems with supervisor	0	1	2	3
Feeling down or depressed	0	1	2	3
Feeling harrassed	0	1	2	3
Relationships concerns	0	1	2	3
Feeling unsafe	0	1	2	3
Concerns about sex	0	1	2	3
Physical health concerns	0	1	2	3
Feeling irritable	0	1	2	3
Feelings of anger	0	1	2	3
Losing your temper	0	1	2	3
Being treated unfairly	0	1	2	3
Concerns about your own alcohol/drug use	0	1	2	3
Concerns about another's alcohol/drug use	0	1	2	3
Bothered by your past	0	1	2	3
Feelings of hopelessness	0	1	2	3
Concerns about money	0	1	2	3
Legal problems	0	1	2	3
Current physical abuse	0	1	2	3
Past physical abuse	0	1	2	3
Current sexual abuse	0	1	2	3
Past sexual abuse	0	1	2	3
Marital problems	0	1	2	3
Career concerns	0	1	2	3
Feelings of loneliness	0	1	2	3
Uncomfortable with close relationships	0	1	2	3
Thoughts of suicide	0	1	2	3
Concerns about your children	0	1	2	3
Feeling fearful	0	1	2	3
Racial concerns	0	1	2	3
Other concerns: (please specifiy)___	0	1	2	3

be developed to serve these unique populations, possibly through agencies which provide practical assistance for areas of concern to these workers. Caregivers will likely need to meet workers in their home and work environments, and will need to understand their unique strengths and struggles. Only in an environment of trust and respect are these workers likely to admit to their deeper concerns and worries. Family involvement in mental health care will be important to the welfare of workers, as a stressful home environment results in a stressed worker.

REFERENCES

American Psychiatric Association. 1994. *Diagnostic and Statistical Manual of Mental Disorders.* Fourth Edition (DSM-IV). 1994. Washington, DC: American Psychiatric Association.

Anibal, C. 1995. National Park Service, personal communication.

Atkins, F., and S. Krnatz. 1993. Stress and coping among Missouri rural and urban children. *J. Rural Health* 9(2): 149-156.

Baker, D., and R. Darasek. 1995. Occupational stress. In: *Occupational Health: Recognizing and Preventing Work-Related Disease.* B. Levy and D. Wegman, eds. Boston: Little, Brown and Company, p. 381.

Bigbee, J.L. 1987. Stressful life events among women: a rural-urban comparison. *J. Rural Health* 3(1): 39-51.

Bovenzi, M., A. Zadini, A. Franzinelli, and F. Borgogni. 1991. Occupational musculoskeletal disorders in the neck and upper limbs of forestry workers exposed to hand-arm vibration. *Ergonomics* 34(5): 547-562.

Brooks, S. and T. Harford. 1992. Occupation and alcohol-related causes of death. *Drug Alc. Dep.* 29(3): 245-51.

Caldwell, B. 1990. Development of models for park rangers' perceived isolation of National Park Service areas. *Env. Behavior* 22: 5636-49.

Cecil, H.F. 1988. Stress, country style: Illinois response to farm stress. Special issue: Mental Health and the Crisis of Rural America. *J. Rural Comm. Psych.* 9(2): 51-60.

Dixon, R.D., R.C. Lowrey, J.C. Sabella, and M.J. Hepburn. 1984. Fisherman's wives: a case study of a Middle Atlantic coastal fishing community. *Sex Roles* 10(1-2): 33-52.

Dolmierski, R., M. Jezewska, I. Leszczynska, and J. Nitka. 1990. Evaluation of psychic parameters in seamen and fishermen with a long employment period. Part I. *Bull. Inst. Maritime Tropical Med. Gdynia* 41 (1-4): 115-21.

Eisemann, M. 1986. Social class and social mobility in depressed patients. *Acta Psychiatrica Scandinavica* 73(4): 399-402.

Ellis, J. and P. Gordon. 1991. Farm family mental health issues. *Occup. Med.: State of the Art Reviews* 6(3): 493-502.

Gardell, B. 1973. Job satisfaction among forestry workers. *Reports Dept. of Psych.* No. 385.

Giesen, C., A. Maas, and M. Vriens. 1989. Stress among farm women: a structural model approach. *Behav. Med.* 15(2): 53-62.

Grainger, C. 1993. Hazards of commercial fishing. *World Health Forum* 14(3): 313-15.

Green, L. 1987. Suicide and exposure to phenoxy acid herbicides. *Scand. J. Work Env. Health* 13(5): 460.

Gunderson, P., D. Donner, R. Nashold, L. Salkowicz, et al. 1993. The epidemiology of suicide among farm residents or workers in five north-central states, 1980-1988. *Am. J. Prev. Med.* 9(3 Suppl.): 26-32.

Hedlund, D. and A. Berkowitz. 1979. The incidence of social-psychological stress in farm families. *Int. J. Soc. Family* 9: 233-43.

Hsieh, H., et al. 1989. The relation of rural alcoholism to farm economy. *Comm. Ment. Health J.* 25(4): 341-47.

Johnson, D., D. Field, G. Machlis, and R. Converse. 1983. *The National Park Service Employee Survey of 1983: Statistical Abstract.* Seattle, WA: National Park Service, Cooperative Park Studies Unit, University of Washington.

Kaplan, H., B. Sadock, and J. Grebb. 1994. *J. Kaplan and Sadock's Synopsis of Psychiatry.* Baltimore: Williams and Wilkins.

Kline, A., M.C. Robbins, and J. Thomas, J. 1989. Smoking as an occupational adaptation among shrimp fisherman. *Human Org.* 48(4): 351-55.

Larson, W., J. Doris, and W. Alvarez. 1990. Migrants and maltreatment: comparative evidence from central register data. *Child Abuse Neglect* 14(3): 375-385.

Lempers J., and S. Clark-Lempers. 1989. Economic hardship, parenting, and distress in adolescence. *Child Dev.* 60: 25-39.

Liu, T. and J. Waterbor. 1994. Comparison of suicide rates among industrial groups. *Am. J. Ind. Med.* 25(2): 197-203.

Lobao, L., and K. Neyer. 1991. Consumption patterns, hardship, and stress among farm households. *Res. Rural Soc. and Dev.* 5: 191-209.

Mitchell, J. 1983. When disaster strikes. . . the critical incident stress debriefing process. *J. Emer. Med. Services* 8(1).

Murray, J. and P. Keller. 1991. Psychology and rural America: current status and future directions. *Am. Psychologist* 46(3): 220-31.

Myers, M.L., and M.L. Klatt, eds. 1994. *Proceedings of the National Fishing Industry Safety and Health Workshop.* U.S. Department of Health and Human Services.

Nitka, J. 1990. Selected medical and social factors and alcohol drinking in Polish seafarers. *Bull. Inst. Mar. Trop. Med. in Gdynia* 41(1-4): 53-57.

Olson, K., and R. Schellenberg. 1986. Farm stressors. *Am. J. Comm. Psych.* 14 (5): 555-69.

Puertas, N. 1983. Fishing industry. In: *Encyclopaedia of Occupational Health and Safety.* 3rd edition. Suigi Parmeggiani, ed. Geneva: International Labour Organisation, International Labour Office.

Ragland, J., and A.L. Berman. 1991. Farm crisis and suicide: dying on the vine? *Omega* 22(3): 173-85.

Riordan, C., G. Johnson, and J. Thomas. 1991. Personality and stress at sea. *J. Soc. Beh. Person.* 6(7): 391-409.

Robinson, D., D. Trites, and E. Banister. 1993. Physiological effects of work stress and pesticide exposure in tree planting by British Columbia silviculture workers. *Ergonomics* 36(8): 951-61.

Schiffman, E., A. Miller, M. Suggs, and B. Graham. 1995. The effect of environmental odors emanating from commercial swine operations on the mood of nearby residents. *Brain Res. Bull.* 37(4): 369-75.

Shulman, M., and P. Armstrong. 1989. The farm crisis: an analysis of social psychological distress among North Carolina farm operators. *Am. J. Comm. Psych.* 17(4): 423-41.

Slappendel, C., I. Laird, I. Kawachi, S. Marshall, and C. Cryer. 1993. Factors affecting work-related injury among forestry workers: a review. *J. Safety Res.* 21(1): 19-32.

Strehlke, B. 1983. Forestry industry. In: *Encyclopaedia of Occupational Health and Safety.* 3rd edition. Suigi Parmeggiani, ed. Geneva: International Labour Organisation, International Labour Office.

Spoth, R., and S. Conroy. 1993. Survey of prevention-relevant beliefs and efforts to enhance parenting skills among rural parents. *J. Rural Health* 9(3): 227-37.

Sullivan, G., C. Jackson, and K. Spritzer. 1996. Characteristics and service use of seriously mentally ill persons living in rural areas. *Psych. Services* 47(1): 57-61.

Walker, J., and L. Walker. 1988. Self-reported stress symptoms in farmers. *J. Clin. Psych.* 44: 10-16.

Walker, J., L. Walker, and P. MacLennan. 1986. An informal look at farm stress. *Psych. Reports* 59 (2): 427-30.

11

DISABILITY IN AGRICULTURE, FORESTRY, FISHERIES, AND MIGRANT WORKERS

Barry L. Delks, M.S.
USDA AgrAbility Project, Purdue University

> Farmers can find ways to continue farming after a disability. Assistive technology, modified equipment, mobility, farm management skills, farm safety, and accessibility to community activities often become key issues to the agricultural worker after a disabling injury or illness. This chapter identifies current approaches to assistive technology available to farmers and agricultural workers who desire to continue productive roles in the agricultural profession. Research on agricultural tools and equipment for farmers, ranchers, and migrant workers with disabilities, peer mentor programs, and resources and support for agricultural workers with disabilities are presented in this chapter as a beginning point for what is to come in this area.

INTRODUCTION

The overall goal of this chapter is to present the importance of persons with disabilities in the agricultural workplace. Within this goal are several specific objectives. These include reviewing the hazardous nature of agriculture; examining the needs of agriculture workers with disabilities; reviewing the importance of safety and prevention; and presenting resources and agencies that are available to assist agriculture, forestry, fisheries, and migrant workers with disabilities.

THE VALUE OF AGRICULTURAL WORKERS WITH DISABILITIES

Farming and agricultural production is a complex business with an honorable goal: to feed and cloth each of us. The 20 million jobs in the U.S. food and fiber systems account for 18% of all the nation's jobs. About 3.8 million jobs are in farm production, which includes farmers, hired farmworkers, and workers in forestry, fisheries, and agricultural services (Majchrowicz 1990).

Farmers and ranchers with disabilities have a very high rate of return to work after a disability injury. According to studies done by Sheldon (1992) and Allen (1993), the unemployment rate for farmers with disabilities is 9% or less. Stated another way, about 91% of farmers with disabilities continue to farm or work. Compared to the general population of persons with disabilities, the unemployment rate has been estimated at 50 to 70% (U.S. Census Bureau 1992). In part, this high rate of return to work demonstrates that farmers are fiercely independent and value their freedom and their work. Another reason so many individuals continue to farm after a disabling injury may be that when a farmer leaves the family farm great emotional and social upheaval occurs. Perhaps the departure is more significant for farmers than for other displaced workers because farmers are leaving not just an occupation, but a multigenerational way of life (Ferguson and Engles 1989).

The value of a person and his or her occupation is impossible to measure. Yet so many of us are quick to assume that because an individual is in a wheelchair, walks differently or needs some type of device to see or hear better, he has less value. The difference between an able bodied person and a disabled person may be one second in time. In an instant a finger, arm or leg can be amputated. A fall, exposure to chemicals, or the onset of an illness may cause a disability. We tend to look at an individual with disabilities and see what he/she cannot do. It is vital that we remember and utilize all the skills, talents, abilities and knowledge that an individual has, though some physical activities might be limited.

Many agricultural workers continue to farm, ranch, fish, or manage a forest after a disabling injury or illness. In fact, a project established in 1991 to assist agricultural workers with disabilities, the USDA AgrAbility Project, has helped more than 3,000 farmers and ranchers, using on-site technical assistance to enable them to continue to work after a disabling injury or illness (U.S. Dept. of Health and Human Services 1995). With this assistive technology most agricultural workers with disabilities are able to work successfully and continue to be an asset to their farm, family, community, and the food and fiber systems.

THE HAZARDOUS NATURE OF AGRICULTURE

Hazards in agriculture, forestry, fisheries, and migrant/seasonal work are numerous and are commonly faced every day. A partial list of the hazards are given in Table 11.1 (Murphy 1992). History has shown that the agricultural-related sector of the population is especially susceptible to disabling injuries. A recent report from the National Safety Council has classified agriculture as the most hazardous occupation in America (Ogilvie 1994).

Table 11.1. Hazards of agriculture.

Operation of tractor
PTO operations
Working with large animals
Pull-in points
Cutting points
Trucks and other vehicles
Power lines
All-terrain vehicles
Miscellaneous machinery and tools
Grain bins
Silos
Barns
Manure pits
Toxic gases and dust
Exposure to chemicals & loud noise
Exposure to extreme weather
Drivelines
Confinement buildings

The National Safety Council aggregrates fishing with agriculture and forestry. If accidents involving children in the agricultural workplace were included, agriculture's injury rate would be even higher. The National Safety Council estimates that farm injuries result in more than 1,100 deaths each year. More than 65,000 agricultural workers incur a permanent injury each year, with an additional 65,000 injured temporarily (Agricultural Safety and Health 1995). The Environmental Protection Agency (EPA) estimates that 300,000 farmworkers suffer acute illnesses and injuries from pesticide exposure each year (Wilk 1986).

In addition to the disabling injuries caused by accidents, farmers/ranchers and agricultural workers are also affected by other physical disabilities that restrict their ability to perform their jobs and participate actively in the rural community. A study of Indiana farm operators completed at Purdue University in 1981 revealed that 66% were affected by at least one physical impairment. More than 30% cited musculoskeletal impairments, 25% indicated hearing impairments, 24% cited cardiovascular impairments, and 22% indicated respiratory impairments. More than 17% indicated that they were no longer able to perform certain agricultural-related tasks on their farms and over 19% said that because of their physical impairments they were hindered or limited in their ability to perform essential farm-related tasks (Tormoehlen 1982). The severity of the disabling conditions impacting farmers and ranchers is reflected in Table 11.2.

Table 11.2. AgrAbility client survey.

Nature of Client's Disabilities	Percent	Number
Spinal cord injury, paraplegia	26.8	148
Spinal cord injury, quadriplegia	12.7	70
Amputation, upper limb	19.0	105
Amputation, lower limb	10.7	59
Arthritis	7.8	43
Hearing Impairment	6.2	34
Orthopedic Impairment	5.9	33
Visual Impairment	5.3	29
Polio or post polio	4.7	26
Head Injury	3.9	22
Heart Disease	3.9	22
Multiple Sclerosis	3.6	20
Hand Impairment	2.5	14
Muscular Dystrophy	2.4	13
Cancer	1.9	11
Diabetes	1.6	9
Respiratory impairment	1.5	8
Cerebral Palsy	.91	5
Burn injuries	.72	4
Amyotrophic lateral sclerosis	.54	3
Cognitive impairment	.54	3
Other	2.9	16
Total	*100%*	*707*

In addition to traditional farmers and ranchers, the U.S. Department of Health and Human Services estimates that there are more than 3.5 million migrant and seasonal farmworkers and dependents (DHHS 1990). Migrant labor is utilized to complete many agricultural tasks requiring exposure to high temperatures, sun, and pesticides. These workers often have impure water and lack of proper housing. Because of these conditions, migrant and seasonal workers have complex health problems, injuries, and illnesses. They are more likely to suffer from infectious diseases. Migrant and seasonal farmworkers have 20-fold greater risk of parasitic infection, 11-fold greater risk of gastroenteritis and infectious diarrhea, and 300-fold risk of infectious hepatitis than the general population (Ortiz 1974).

More than 130,000 injuries are reported each year, and many more injuries are never reported. A large number of disabled agricultural workers are not aware of assistance that could enable them to continue to farm or overcome physical barriers after a disabling injury or illness.

THE NEEDS OF AGRICULTURAL WORKERS WITH DISABILITIES

The Breaking New Ground Resource Center at Purdue University estimates that more than 500,000 farm/ranch family members and agricultural workers in the United States have physical disabilities that hinder them from completing essential work-related tasks or might eventually force them to prematurely discontinue farming. In its first four years, the USDA AgrAbility Project, established to serve agricultural workers with disabilities, responded to more than 67,000 individual requests for information regarding rehabilitation technology.

Agricultural workers are often isolated from rehabilitation services and resources which have the potential for reducing the impact of their disability. In addition to isolation, other factors make the needs of agricultural workers with disabilities unique. In a recent survey, Breaking New Ground assessed the needs of farmers and ranchers with severe disabilities (spinal cord injuries). This survey determined the distance from the source of mobility aids, difficulty of work-related tasks, and worksite accessibility. The needs of these farmers and ranchers with spinal cord injuries is summarized briefly in the following three sections.

Distance from Source of Mobility Aid

Distance was a major factor with respect to purchasing and servicing mobility aids. Over one-third of those surveyed purchased and obtained service for their primary mobility aid more than 100 miles away. More than 23% traveled 100 miles or more to purchase and service their manual wheelchairs.

Difficulty of Various Activities

Twenty-seven work-related activities were listed, and those surveyed were asked to rank the level of difficulty they had experienced when completing each task. A weighted average was used, with "No difficulty" = 1, "Some difficulty" = 2, "Difficult" = 3, "Very difficult" = 4, and "Need help" = 5. The 27 work activities were ranked by difficulty, with #1 being the most difficult (Table 11.3). Making heavy machinery adjustments or repairs (changing tires, wheel spacing, switching combine heads, etc.), loading or moving livestock, and harvesting logs or splitting firewood were listed as the most difficult activities requiring the most assistance. Maintaining farm buildings (painting and repair) was ranked as the second most difficult activity. Hitching implements to tractors, cleaning the milkhouse and working milking equipment, giving shots and attending to the medical needs of livestock, and castration/docking tails/clipping teeth all ranked third in level of difficulty and required assistance to complete. Work-related activities ranked the least difficult were 1) operating tractor or combine controls, 2) mowing the lawn and moving to and from fields, and 3) getting to and from farm buildings and handling common farm shop tools.

The responses tend to reflect the greater attention that has been given to assisting farmers with spinal cord injuries in operating agricultural equipment and completing

Figure 11.1. After a disabling accident, farmers continue to make significant contributions to the farm, family, and community.

Table 11.3. Difficulty of various farm/ranch-related activities after spinal cord injury (ranked in order of difficulty).

Weight average

1	Making heavy machinery adjustments or repairs	4.5
1	Loading or moving livestock	4.5
1	Harvesting logs or splitting wood	4.5
2	Maintaining farm buildings (painting and repair)	4.4
3	Cleaning milkhouse and washing milking equipment	4.3
3	Giving shots and attending to the medical needs of livestock	4.3
3	Castration/docking tails/clipping teeth	4.3
3	Hitching implements to tractor	4.3
4	Barn cleaning and handling manure	4.2
5	Maintaining the orchard (pruning, etc.)	4.1
5	Moving grain or concentrate to feed livestock	4.1
6	Milking	4.0
7	Routine machinery maintenance and repair	3.9
7	Moving hay and feeding hay to livestock	3.9
8	Fueling and routing maintenance of tractors, etc.	3.8
9	Making PTO connections	3.6
9	Feeding and watering young livestock	3.6
10	Welding	3.5
11	Gardening	3.4
11	Coupling hydraulic lines	3.4
12	Opening and closing barn doors, gates, etc.	3.2
13	Getting on or off tractor or other self-propelled machinery	3.1
14	Getting to and from farm buildings in most kinds of weather	2.5
14	Handling common farm shop tools	2.5
15	Moving to and from fields to check field work, crops, fences, etc.	2.4
15	Mowing the lawn	2.4
16	Operating tractor and combine controls	2.0

Table 11.4. Prioritized list of goals concerning worksite accessibility.

Weight average

1. Improve ability to effectively and safely use equipment and machinery, including accessing, operating, and maintaining equipment and hitching implements. 6.9

2. Improve overall mobility or accessibility around farmyard, buildings, and fields. 6.8

3. Improve ability to perform general maintenance activities around the farm, including effective use of hand tools, power tools, and maintenance materials. 6.0

4. Improve ability to manage farm/ranch operation successfully, including the maintenance of business records, sales and purchases, labor management activities. 5.8

5. Improve livestock-handling abilities related to feeding methods, healthcare needs, waste removal, and building sanitation. 5.3

6. Improve my ability to perform the following farm job _____. 5.0

7. Identify an alternative farm enterprise that would better suit my abilities and limitations. 4.4

8. Obtain part or full-time off-farm employment. 3.5

mechanical activities. Less attention has been given to the problems of completing livestock-related activities.

Prioritize Goals for Worksite Accessibility

Using a weighted average, with 1 = "Least important" and 8 = "Most important," worksite accessibility goals were prioritized and ranked. On this scale, a higher weighted average corresponds to a higher priority. Results are given in Table 11.4. The survey revealed farmers were most concerned with improving their ability to effectively and safely use machinery and equipment.

Improving overall mobility around the farmyard was listed second. Interviews with numerous farmers with spinal cord injuries confirmed the importance of this issue. Mobility in manual wheelchairs across gravel continues to be a common barrier for many farmers. The applicability of outdoor powered wheelchairs and all-terrain vehicles is a frequent question asked by this population.

The third and fourth priorities were improving their ability to perform general maintenance activities and improving farm management skills, respectively. Improving livestock handling abilities and abilities to perform miscellaneous farm chores were prioritized fifth and sixth, respectively.

Identifying alternative farm enterprises was listed seventh, and obtaining part- or full-time off-farm employment was listed eighth. This seems to confirm that farmers desire to stay on the farm and remain self-employed, and thus may often fail to consider off-farm employment (Field 1992).

Because the farm or ranch is not just a place of employment, but also a home, and because the operation may have been part of the family for generations, seeking off-farm employment may be the least desirable option for most farmers with disabilities. In addition to these factors, many farmers may have little or no off-farm employment experience. Thus, seeking off-farm employment may result in greater levels of anxiety and stress.

Distance from the source to purchase and service mobility aids, physically demanding tasks, and worksite accessibility are only a few of the needs faced by agricultural workers with disabilities. The needs vary from the type of disability, the type of operation and the geographic location. Numerous other factors influence the needs of agricultural workers with disabilities, but are hard to quantify. Family support, the attitude of the worker with a disability, and the funding available to modify or adapt equipment are such factors.

Prevention and Safety

There are many good resources that pertain to agricultural safety and prevention (Murphy 1992; Deere & Company 1994; Cockcroft 1989). To discuss disability in agriculture without discussing safety and prevention would be negligent. The topics within safety and prevention are very diverse including engineering for injury prevention; education and training; regulations; and human factors and ergonomics. Prevention and safety in agriculture can include very specific training on safely operating farm machinery, personal protective equipment, fire prevention, first aid, forest and woodlot safety, handling and storing chemicals, respiratory hazards, and numerous other topics. In occupations as diverse as agriculture, forestry, and fisheries, the need for prevention and safety training is great.

Agriculture has unique characteristics as compared to other careers. Agricultural workers are often self employed and generally work alone. A farmer may use dozens of different tools each day and work around large equipment. Farmers are exposed to extreme weather conditions and are responsible for the labor, management, and safety of their operation.

Other factors suggested for causing agricultural injuries are stress and the low value attached to safety decisions by farmers (Murphy 1981). A survey of Iowa grain farmers concluded: "The interactions among the farmer-equipment-agricultural environment are intimately interwoven, each influenced by the other. The combination of these factors along with physical stress, economic pressure and high seasonal workloads has introduced a wide variety of hazards to the people living or working within the farm environment. These hazards lead to an increased risk of accidents, injuries, illnesses and deaths to the farmer and family members" (Ogilvie 1990).

Farmers with disabilities need specific prevention and safety recommendations. A survey of 627 disabled farmers was designed to determine their perceived risk and educational needs. Data show that one of the most effective methods of delivering safety information was the Cooperative Extension Service, which was chosen by 30% of the respondents. Forty-two percent desired more safety information on machinery, while 40% and nearly 30%, respectively, thought more safety information on tractors and livestock handling was needed. Two recommendations that were developed as a result of this study are worthy of mentioning (Allen 1993).

1. Task-specific safety education material, especially on livestock and falls, needs to be developed for agricultural workers with disabilities.

Figure 11.2. Dust is just one hazard a farmer faces. This Airstream Dust Helmet provides a portable system to filter contaminants and allow some agricultural workers with respiratory impairments to continue to work.

Figure 11.3. A specially designed parlor makes the dairy farm more accessible to this farmer.

2. The success of agricultural workers returning to work following a disability appears to be high and the risk of injury similar to able-bodied farmers. Safety education materials need to be developed for agricultural workers with disabilities, since returning to the farm or ranch after a disabling injury is a viable option.

Prevention and safety may become more important to farmers with disabilities for two reasons. First, if the disability is the result of a farm accident, the farmer has a heightened awareness of the importance of safety. Secondly, many farmers with disabilities place a high value on the remaining physical abilities they have and implement safety strategies to prevent any future injuries.

The strategies of safety that apply to all agricultural workers apply to farmers with disabilities as well. The three most prevalent strategies are education, enforcement, and engineering. The most effective method of preventing injuries is removing hazards through proper equipment design and engineering methods (Aherin and Murphy 1992).

With the addition of assistive devices or modifications to equipment for workers with disabilities, proper engineering plays an important role in preventing future injuries or disabilities. Many agricultural workers who are recently injured may initially become despondent because they may be told they need to quit farming or it is impossible for them to farm. Fortunately, most farmers have a strong desire to continue working. Recent advancement in rural assistive technology promotes safe opportunities for agricultural workers with disabilities to continue their work.

AVAILABILITY OF RESOURCES

Availability of Rural Rehabilitation Technology Resources

Until recently, there have been few resources for the rehabilitation professional to turn to for solutions to rural rehabilitation problems. In the past few years AgrAbility Projects and other agencies are filling the void of rehabilitation needs of agricultural workers with disabilities.

The National AgrAbility Project is a cooperative project and is funded by the U.S. Department of Agriculture to help establish education and assistance programs for agricultural workers with disabilities. The project brings together rehabilitation professionals and Cooperative Extension Service Educators within each state to provide services directly to the agricultural worker or a family member who is disabled.

To document and provide a listing of agencies and programs, as well as resources that serve agricultural and rural families with disabilities, the Breaking New Ground

Figure 11.4. Although paralyzed, this farmer works more than 1,500 acres with the assistance of a vertical lift and a motorized wheelchair.

Resource Center at Purdue has produced a 12-page *Directory of Rural Assistive Technology Resources*. A few of the agencies from that directory are listed below.

AgrAbility Project
Breaking New Ground Resource Center
Purdue University
1146 Agricultural Biological Engineering Bldg.
West Lafayette, IN 47907-1146
(800) 825-4264 (Vc/TDD)
(317) 494-5088
(317) 496-1356 (FAX)

Association of Programs for Rural Independent Living (APRIL)
c/o Summit Independent Living Center
1280 South 3rd West
Missoula, MT 59801
Phone: (406) 728-1630
FAX: (406) 728-1632

Canadian Farmers with Disabilities
The Canadian Paraplegic Association
#3 - 3012 Louise Street
Saskatoon, Saskatchewan S7J 3L8
Canada
Phone: (306) 652-9644
FAX: (306) 652-2957

Handicapped Farmers Program
Saskatchewan Abilities Council
2310 Louise Avenue
Saskatoon, Saskatchewan
Canada S7J 2C7
Phone: (306) 374-4448

National Migrant Resource Program, Inc.
515 Capital of Texas Highway South
Suite 220
Austin, TX 78746
Phone: (512) 328-7682 or (800) 531-5120
FAX: (512) 328-8556

Rural Institute on Disabilities
52 Corbin Hall
Missoula, MT 59812
Phone: (406) 243-5467 or (800) 732-0323
TT: (406) 243-4200
FAX: (406) 243-2349

SUMMARY

Agricultural workers are as diverse as any population and have a vast array of abilities and skills. Like all of us, they desire the public to look upon their abilities, not their disabilities. Assistive technology, modified equipment, mobility, farm management skills, farm safety, and accessibility to community activities often become key issues to the agricultural worker after a disabling injury or illness. Self-determination, family support and assistive technology all play an important role in allowing the farmer with a disability to continue his work. Farmers who so desire will find a way to continue farming after a disability. Agricultural workers have acquired many talents, skills, and abilities and may need to be reminded of these skills immediately after the injury to help restore self-worth. Once self-worth is restored, encouragement from other farmers who have overcome similar disabilities is very helpful. Finally, providing the necessary resources, such as a list of agricultural tools, equipment, machinery, and buildings adapted to help the individual farmer keep farming, is vital in the rehabilitation process. Self-worth, peer and family support, and assistive technology give the farmer hope and the practical tools to continue farming.

Farming, agriculture production, forestry, and fisheries are complex industries with many hazards. The needs of agricultural workers with disabilities are diverse. However, with the intense desire of most agricultural workers to continue in their chosen occupation and with the new services and agencies for farmers with disabilities, farming remains a viable option for most. AgrAbility and the Breaking New Ground Resource Center will continue to research new areas, compile information on agricultural tools and equipment for farmers, ranchers and migrant workers with disabilities, coordinate peer mentor programs, and provide resources and support to offer the most positive options for agricultural workers with disabilities. Agricultural production is important to all of us. Addressing the diverse needs of agriculture, forestry, fisheries and migrant workers with disabilities will assist in keeping the food and fiber system productive.

REFERENCES

Agricultural Safety and Health: A Resource Guide. 1995. RICPS, No. 40.9.

Aherin, R.A., and D.J. Murphy. 1992. *Reducing Farm Injuries: Issues and Methods.* St. Joseph, MI: American Society of Agricultural Engineers.

Allen, P. 1993. An assessment of the risk and safety education training needs of farmers and ranchers with severe physical disabilities. 44:69-71.

Cockcroft. 1988. *Principles of Health and Safety in Agriculture.* Boca Raton, FL: CRC Press, Inc.

Ferguson, S.B., and D.W. Engles. 1989. American farmers: workers in transition. *Career Dev. Quarterly* 37: 240-48.

Field, W. 1992. *Assistive Technology Needs Assessments of Farmers and Ranchers with Spinal Cord Injuries.* Paralyzed Veterans Association, pp. 15-31.

Hathaway, L.R., ed. 1994. *Farm and Ranch Safety Management.* Moline, IL: Deere & Company Service Publications.

Majchrowicz, A.T., and M. Petrulis, 1990. Agriculture: a critical industry. *1990 Yearbook of Agriculture*, pp. 2-3.

Murphy, D. 1992. *Safety and Health for Production Agriculture.* St. Joesph, MI: American Society of Agricultural Engineers, pp. 15-42.

Murphy, D. 1981. Farm safety attitudes and accident involvement. *Accident Analy. Prev.* 13: 331-37.

Northeast Center for Agricultural and Occupational Health. 1994. Compendium of Northeastern Agriculture Health and Safety Resources, Northeast Center for Agricultural and Occupational Health.

Ogilvie, L. 1990. Agricultural Safety Equipment Usage: A Survey of Iowa Grain Farmers. Iowa City, IA: Thesis for the University of Iowa, pp. 1-159.

Ogilvie, L. 1994. *Accident Facts.* Chicago: National Safety Council.

Ortiz, J.S. 1974. Composite Summary and Analysis of Hearing Held by Department of Labor, Occupational Safety and Health Administration on Field Sanitation for Migrant Farmworkers (Docket #308, May 23-June 29).

Sheldon, E. 1992. The survey of employment experiences of farmers and ranchers with disabilities.

Tormoehlen, R. 1982. Nature and proportion of physical impairments among Indiana farm operators. Paper presented at American Society of Agricultural Engineers.

U.S. Department of Health and Human Services. 1995. *U.S.D.A. AgrAbility Summary*.

Wilk, Valerie. 1986. *Occupational Health of Migrant and Seasonal Farmworkers in the United States*. Second Edition. Washington, DC: Farmworker Justice Fund.

12

PERSONAL PROTECTIVE EQUIPMENT IN AGRICULTURE, FORESTRY, AND FISHERIES

Timothy G. Prather, M.S.
Department of Agricultural Engineering
University of Tennessee

When considering methods of preventing occupational injuries and illnesses, personal protective equipment is often necessary. There are numerous types of personal protective equipment, some of which are specialized, which are appropriate for use in agricultural settings. This chapter will provide an overview of the basic types of personal protective equipment which may benefit agricultural workers.

Due to the wide variety of personal protection devices available and the wide variety of agricultural work settings, this chapter is not intended to be all-inclusive. Furthermore, it must be emphasized that safety equipment should be selected on a case-by-case basis. The best choice will be based on the type and severity of hazards, proper fit for the worker, comfort and economic considerations. Using the wrong devices, or using the right devices improperly, can result in serious illness, injury or death. If not certain which types of glove, respirators, etc. to use, consult a reputable safety equipment dealer, industrial hygienist or other safety professional for assistance in selecting the best product for the job and training in its use.

RESPIRATORY PROTECTION EQUIPMENT

Agricultural workers may be exposed to many irritating and potentially harmful airborne contaminants. Examples of air contaminants found on farms include particles of dander, pollen, bacteria, mold spores, grain dust, hazardous chemicals, feed supplements, engine exhausts and welding fumes. These substances originate from the soil, animals and plants, decay or fermentation of crop materials and animal wastes, chemicals and fertilizers and operation of equipment.

Modern production practices may contribute to respiratory hazards, especially when systems are improperly designed or managed. For example, airborne dust, moisture and ammonia are of little concern when swine are kept outside because the contaminants cannot accumulate to harmful levels. However, confinement systems with controlled atmospheres can have serious problems with dusts, molds, moisture and toxic gases *if the total system is not engineered and managed properly.* The key word is *system,* because if any aspect of the facility is not operating as it should, the whole system suffers.

One problem with occupational illnesses, especially in agricultural operations, is that the victim may not associate the illness with job-related exposures. Symptoms of respiratory illness may be mistaken for the flu, common cold or simple exhaustion. Another problem is that people may perform a particular task for years without experiencing adverse health effects, but suddenly experience a severe reaction to even slight exposures to contaminants. The body can become "sensitized" and no longer tolerate even small exposures to a substance. Additionally, some exposures can lead to permanent lung injury and disability.

This section will attempt to explain the basic types of respiratory hazards and how to select the correct respirator for those hazards. This section is rather lengthy due to the need to explain all aspects of respirator selection and use. It is the author's experience that respirator selection and use is more confusing than any other type of personal protective equipment. Perhaps this is due to the fact that the hazards are often invisible, and certainly misunderstood and underestimated. Therefore, the following statement must be made clear before proceeding:

WARNING: Selecting and using a respirator involves more than simply purchasing one and putting it on. Using the wrong respirator, or using a respirator improperly, can result in serious illness or death. The respirator must be selected for the specific contaminant(s) in your workplace and it must fit properly. Use only respirators approved by NIOSH or MSHA for the hazards present in the workplace.

Types of Hazards

The basic types of respiratory hazards are:

Particulates. Particulates are particles of solid material and droplets of liquids which can be easily filtered from the air. There are several categories of particulates, as illustrated in Figure 12.1, and the respirators required for each type depend on the sizes of the particles or droplets. The particulates of interest in agriculture are:

- **Dusts and mists** are very small particles of solid material (dusts) or very small droplets of liquids (mists) that may remain in the air quite some time before settling, and may be carried a considerable distance by air currents. Dusts and mists may be small enough to be carried deep into the lungs. They may be toxic themselves and/or carry bacteria and fungi with them. Common dusts and mists on the farm are grain and feed dusts, mold, silica dust and pesticides. Dusts are considered to be particles 1 micron or more in diameter. Mechanical filter respirators *approved for toxic dusts and mists* provide adequate protection. Respirators with asbestos approval are easier to breathe through and may last longer than other models.

- **Asbestos-containing dusts** are particularly harmful because the sharp asbestos fibers can penetrate lung cells. Lung cancer can result from asbestos exposure. Mechanical filter respirators *approved for asbestos-containing dusts* are required for protection.

- **Fumes** are solid particles of burned or evaporated metal or other material. The very small particles condense and converge or clump together to form particles that may be much smaller than dusts and mists. Welding fumes are the most common fumes on the farm. Fumes are considered to be particles 0.001 to 1 micron in diameter. Mechanical filter respirators with *approval for fumes* must be used. These respirators may be used against dusts and mists as well.

Gases and Vapors. Gases and vapors are individual molecules in the atmosphere. Because the molecules are in the same size range as oxygen molecules, gases and vapors cannot be removed by the mechanical filter respirators used for dusts, mists and fumes. The respirator used will depend

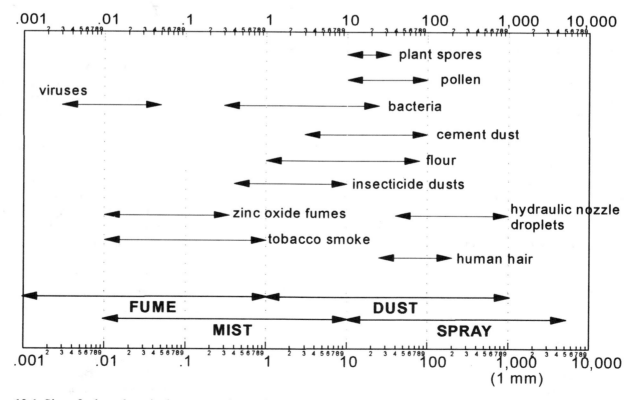

Figure 12.1. Size of selected particulate contaminants, in microns.

on the identity of the vapor or gas, its hazards and its concentration.

- **Vapors** result from of evaporation of liquids. Organic vapors from solvents, pesticides, lacquer thinner, spray painting, adhesives and gasoline are common on the farm. Be aware that some dusts and mists, such as pesticides and paints, give off organic vapors. A mechanical filter respirator will trap the particles, but the vapors will still enter the body. Chemical cartridge respirators must be used to remove vapors from the air. These usually contain activated carbon adsorbent material to trap the vapors. The cartridge must be approved for the type vapors to be encountered. *Cartridges approved for organic vapors are color coded with a black band or label.* Some conditions, such as pesticide application and spray painting, require an organic vapor respirator equipped with a pre-filter. This pre-filter removes the dusts and mists to prevent clogging of the chemical cartridge and to keep the dusts and mists out of the lungs.

- **Toxic gases** are in the gaseous state at normal temperature and atmospheric pressure, and did not result from evaporation. Some gases are so toxic that one breath can cause instant respiratory arrest and death. Examples include hydrogen sulfide (manure gas) and nitrogen dioxide (silo gas). Others, such as ammonia, are powerful irritants and can cause burns and irritation to body tissues. Carbon monoxide is particularly hazardous because it is absorbed by the blood over 200 times more readily than oxygen and causes death by asphyxiation. Low concentrations of certain toxic gases can be removed by chemical cartridge respirators. Others are so hazardous or so difficult to filter from the air that only respirators that supply fresh air to the user can be used. Use only respirators approved for the toxic gases and concentrations that will be encountered. Respirators that supply fresh air from a tank or compressor are required in some cases.

Oxygen Deficient Atmospheres. Oxygen deficient atmospheres cause death by asphyxiation (suffocation). The lack of oxygen may be caused by heavy gases or vapors displacing the oxygen or it could be caused by fuel-burning appliances using the available oxygen. Oxygen deficiency can be a problem in silos, grain bins, manure pits and other closed or confined spaces. The atmosphere normally contains about 21% oxygen. When the oxygen concentration in air drops to approximately 16%, you will become dizzy, experience a buzzing in the ears and have a rapid pulse. Oxygen-deficient atmospheres may cause inability to move and a semiconscious lack of concern about death. If you enter an area containing little or no oxygen, you may quickly lose consciousness with no warning symptoms. *The air must contain at least 19.5% oxygen* to ensure that workers can function properly. NEVER enter areas with less than 19.5% oxygen unless using air-supplying respirators. Air purifying respirators (mechanical filter, chemical cartridge respirators, gas masks and PAPRs) are not approved for use in atmospheres containing less than 19.5% oxygen. Only a supplied-air respirator or self-contained breathing apparatus can provide protection when there is inadequate oxygen.

Identify and Measure Air Contaminants

There are many respiratory hazards on farms, and some of these are listed in Table 12.1. The concentrations of the contaminants can vary significantly from one farm to the next, as well as from one location on your farm to the next. You must know the identity of the contaminants and their concentrations to determine the hazards. This information will help you select methods for controlling the hazards and protecting workers.

The senses of smell, taste and sight can warn of some problems, but are not always good indicators of potential hazards. Certain hazardous substances cannot be detected by smell until the concentrations are well above safe exposure limits. Others can be detected at low concentrations, but may "numb" the sense of smell as the concentrations increase. This leads a person to believe the danger is no longer present when, in fact, the danger has increased.

Several types of testing equipment are available to measure the concentration of air contaminants. Testing methods used will depend on the location, air contaminants and working conditions. Some instruments are simple, portable devices giving on-the-spot readings in minutes. Others require laboratory analysis of samples, but may offer more precise readings. Some instruments provide continuous readouts and can sound alarms when contaminants exceed preset levels. Some monitoring techniques report concentrations as time-weighted (averaged) levels over an entire work shift. For information about air testing in agricultural operations, contact your county Extension office, the State Department of Labor, Division of Occupational Safety and Health or an industrial hygienist.

Reducing Hazards

There are three ways to control worker exposure to respiratory hazards. These are listed below in order of preference:

Table 12.1. Selected respiratory hazards on farms.

Location:	Hazard:	Problem:	Protective Measures:
Anhydrous Ammonia	Ammonia gas	Irritation, burning May be overcome by gas; can cause permanent injury or death.	Keep ammonia equipment in good condition. Do not work in enclosed areas. Use ammonia respirator (color coded green).
Feed and grain centers	Grain dusts, including bacterial and mold spores and by-products	Asthma Bronchitis Flu-like illness (toxic organic dust syndrome (TODS))	Proper harvest and storage to prevent spoilage. Use toxic dust and mist respirator.
	Fumigants (toxic gas) and oxygen deficient atmosphere in recently fumigated structure	Irritation Swelling of vocal cords Constricted air passages Fluid in lungs Respiratory failure Overcome by vapors or suffocation due to insufficient oxygen May be fatal	Fumigate from outside structure. Observe reentry intervals. If entry cannot be avoided, use supplied-air respirator or self-contained breathing apparatus.
	Entrapment in grain	Suffocation	Shut off auger and lock out power before entering bin. Break crusted surface or vertical wall of spoiled grain with nonconducting pole from above.
Hay	Dusts, including bacterial and mold spores and by-products	Farmers lung TODS (both have flu-like symptoms)	Proper harvest and storage practices to prevent spoilage. Use toxic dust and mist respirator.
Livestock enclosures	Organic dusts (from animals, feed gases (ammonia, hydrogen sulfide)	Flu-like illness (TODS) Inflammation of nasal passages (allergic response) Bronchitis Asthma	Proper ventilation. Use feeds with low dust content. Use enclosed feeding systems. Use toxic dust and mist respirator.
Poultry houses and bird roosts	Fungus in old, dry bird and poultry droppings	Histoplasmosis	Wet area and perform tasks in a manner which minimizes dust formation. Use toxic dust and mist respirator.
	Administering live vaccine or working with infected animals	Newcastle disease	Use high efficiency particulate (HEPA) respirator and goggles.
Manure pits and tanks	Hydrogen sulfide gas and lack of oxygen	Fluid in lungs Sudden respiratory arrest and death	Do not enter pit or tank. If entry is necessary, ventilate and check for toxic gas and oxygen deficiency. Use supplied air (airline or SCBA) respirator.
Pesticide mixing and application	Toxic chemicals	Poisoning due to inhalation of toxic chemicals. Symptoms may vary according to the chemical and dose. Symptoms may be mistaken for fatigue or influenza.	Follow recommended work practices to minimize exposure. Wear recommended protective clothing. Use organic vapor respirator (color coded black) with dust/mist pre-filter approved for pesticides.

Table 12.1 (*continued*)

Location:	Hazard:	Problem:	Protective Measures:
Brake repair	Asbestos fibers	Lung cancer possible from inhalation of asbestos fibers.	**DO NOT** use compressed air to remove dust from parts, use wet cloth instead. Use non-asbestos containing replacement parts. Use toxic dust and mist respirator with asbestos approval.
Sandblasting, dry sanding and grinding painted surfaces	Silica dust Lead-based pigments	Irritation Possible lead poisoning from particles of old paint containing lead	Work outdoors or in well ventilated area. Use toxic dust and mist respirator.
Solvents, spray painting	Organic vapors	Headache, fatigue, dizziness, nausea, etc. from inhalation of organic vapors Possible long-term cancer risk	Use only with adequate ventilation. Use less volatile, less toxic or nonhazardous solvents. Use organic vapor respirator (color coded black). Respirator must have pre-filter when spray painting.
Hot metal work (welding, brazing, using cutting torch)	Zinc oxide fumes from heated galvanized metal. Other metal fumes could include lead, cadmium, nickel.	Metal fume fever (chills, headache, nausea, etc.)	Work only in well-ventilated area. Use toxic dust, mist and fume respirator.
Woodworking	Wood dusts	Irritation and allergic reactions	Work in well-ventilated area. Use toxic dust and mist respirator.
	Toxic wood preservatives	Irritation and possible poisoning from preservatives	Work in well ventilated area. Do not burn treated wood. Use toxic dust and mist respirator when cutting or sanding treated wood.
Silos, upright: Oxygen-limiting (airtight)	Lack of oxygen Toxic gases (primarily nitrogen dioxide)	Asphyxiation (suffocation) Possible sudden respiratory arrest and death from toxic gases	**DO NOT ENTER** an oxygen-limiting silo without a supplied-air respirator (airline or self-contained breathing apparatus).
Silos, upright: Conventional (non-airtight)	Toxic gases (nitrogen dioxide) produced during fermentation.	Sudden respiratory arrest and death Fluid in lungs Delayed reaction possible with permanent scarring of lung tissues. "Silo filler's disease"	**DO NOT ENTER** silo during filling and for at least 10 days after filling is completed without supplied-air respirator (airline or self-contained breathing apparatus). If silage surface is within 15-20 feet of blower spout, ventilate by running blower for 20 minutes before entry and during entry.
	Bacteria and mold spores and their products.	Flu-like illness (TODS)	Use toxic dust and mist respirator.

Engineering Controls. The best solution to any hazard is to eliminate or minimize the problem as much as possible. Eliminating or controlling hazards by selecting a safer process, better ventilation and maintaining a safer environment are engineering controls. A number of engineering controls are available to farmers. Proper management of harvest and storage conditions will minimize dust and mold problems. Good ventilation can keep levels of dusts, vapors and gases at safe levels in work areas. Nontoxic or less volatile products can be used to reduce hazards in some instances. Equipment controls can be relocated to keep people out of the hazardous areas in some cases. It might be impossible to eliminate some respiratory hazards from agricultural operations. Besides being very costly, *totally* eliminating all respiratory hazards may not be necessary.

Administrative Controls. If the hazard cannot be eliminated, keep the exposure time short enough to prevent health problems. People should be allowed to enter only if it is absolutely necessary to be where the hazards exist. Workers should be trained to understand the hazards and protective measures. Limiting duration of exposure to the hazards is called administrative control.

Use Respiratory Protection. If engineering and administrative controls are not feasible, appropriate *approved respirators and training in their proper use* must be provided for workers exposed to danger. The use of respirators is to be considered a last resort after other protective measures are found to be inadequate. However, respirators are often the only suitable option in agricultural settings.

Types of Respirators

Selecting and using a respirator involves more than simply purchasing one and putting it on. Using the wrong respirator, or using a respirator improperly, can result in serious illness or death. The respirator must be selected for the specific contaminant(s) in your workplace and it must fit properly. In addition, the respirator must be properly cleaned, inspected and stored after each use to prolong its life and help ensure protection for later uses.

Several types of respirators are available, each designed for a particular use. The best way to select a respirator is to visit a safety equipment dealer that carries several brands of respirators or consult an industrial hygienist. Either can assist with selection, fit testing and training you and your employees to use the respirator properly.

Respirators are divided into five basic categories for this publication. The categories are:

Mechanical Filter Respirators. These respirators are made of a fibrous filter and approved for use against particulates. Approvals may be for dusts and mists, welding fumes, asbestos dusts, etc. These respirators may be disposable paper masks or have a rubber or plastic facepiece and replaceable filter elements. Some may incorporate adsorbent materials for protection against some contaminants or nuisance (not harmful) levels of organic vapors. *A mechanical filter respirator does not provide oxygen.*

> **WARNING:** *Do not confuse a respirator with the nuisance dust masks commonly available at hardware and department stores. Dust masks are not approved for harmful contaminants and are unsuitable for agricultural work areas.*

Chemical Cartridge Respirators. These respirators have rubber or plastic facepieces and may have replaceable filter elements. Chemical cartridge respirators may be half-mask or full-facepiece, and are available with approvals for a variety of contaminants. Some companies market disposable chemical cartridge respirators. The main advantage of the disposable chemical cartridge respirator is reduced maintenance since there are no replaceable parts. The facepieces of disposable respirators may be softer and more comfortable since they are not intended to last as long as other types.

Chemical cartridges consist of canisters filled with adsorbent material (such as activated carbon) to capture contaminants. Many chemical cartridge respirators can be fitted with particulate pre-filters. By using the proper cartridges and pre-filters, it is possible to use chemical cartridge respirators in a variety of situations. *A chemical cartridge respirator does not provide oxygen.*

The cartridge needed depends on the particular contaminants present since different filter materials must be used for the various chemical hazards. Chemical cartridges are color coded so you can determine at a glance whether the correct respirator is in use. The color coding scheme specified by OSHA is shown in Table 12.2.

Gas Masks. Gas masks are similar to the chemical cartridge respirator, but have larger filter canisters and a full facepiece. Since the filter has greater capacity, a gas mask can be approved for protection against higher concentrations of contaminants than a chemical cartridge respirator. *A gas mask does not provide oxygen.*

Powered Air-Purifying Respirators (PAPRs). These respirators are similar to chemical cartridge respirators or gas masks, except they have a battery powered blower to force air through the filter elements. Using a PAPR can

Table 12.2. Color codes for respirators.

Contaminant from which protection is needed	Assigned color of respirator
Acid gasses	White
Hydrocyanic acid gas	White with 1/2-inch green stripe completely around cannister near the bottom
Chlorine gas	White with 1/2-inch yellow stripe completely completely cannister near the bottom
Organic vapors	Black
Ammonia gas	Green
Acid gases and ammonia gas	Green with 1/2-inch white stripe completely
Ammonia gases and organic vapors	Yellow
Hydrocyanic acid gas and chloropicrin	Yellow with 1/2-inch blue stripe completely around canister near the bottom
Acid gases, organic vapors and ammonia gases	Brown
Radioactive materials, except tritium and noble gases	Purple (magenta)
Particulates (dusts, fumes mists, fogs, or smokes) in combination with any of the above gases or vapors	Canister color for contaminant, as designated above, with 1/2-inch gray stripe completely around canister near the top
All the above atmospheric conditions	Red with 1/2-inch gray stripe completely around the canister near the top

Source: Repiratory Protection, 29 CFR 1910.134. OSHA Standards for General Industry 1986.

significantly reduce the physical stress of using a respirator. Some PAPRs provide protection from particulates only, but some models are approved for use with pesticides. *A powered air-purifying respirator does not provide oxygen.*

Air Supplying Respirators. Air supplying respirators come in two types - the air line respirator and the self-contained breathing apparatus (SCBA). *These are the only respirators that provide oxygen and can be used in oxygen deficient atmospheres or atmospheres that are immediately dangerous to life or health.* An air line respirator provides breathing quality air from an approved air pump located in a safe atmosphere or from a remote tank of breathing air.

Air-supplying respirators should be used only by specially trained individuals and they must be inspected and tested regularly. Use by untrained persons could lead to serious injury or death. *These are the only types of respirators that DO SUPPLY OXYGEN to the user.*

Selecting, Using, and Caring for a Respirator

IMPORTANT NOTICE: *Use only respirators approved by the National Institute for Occupational Safety and Health (NIOSH) or the Mine Safety and Health Administration (MSHA) for the particular hazards present. Use of a respirator not approved for those hazards could result in serious injury or death.*

All approved respirators will be permanently marked on the device itself and on the package with:

- The manufacturer's name
- The model number
- The hazard(s) it protects against
- The NIOSH approval number (example: TC-23C-1234)

Health Considerations of the User. Some people cannot use a respirator for medical reasons since using a respirator causes physical stress for the user. The effort required to suck the air through the filters and to blow the air back out through the filters or the exhalation port is tiring. In addition, the facepiece holds some moist, carbon dioxide laden air after the user exhales. This "used" air in effect reduces the amount of oxygen available to the respirator user with each breath. *People with known or suspected heart or lung disease must never be assigned to tasks requiring the use of a respirator unless approved by a physician.*

Some people cannot use respirators because it is difficult or even impossible to obtain a good fit. Facial hair, wrinkles, scars and missing teeth or dentures can prevent a good seal between the respirator and the face, allowing contaminated air to bypass the respirator and enter the lungs. For example, facial hair has a diameter of about 10 microns, which is over 10 times the size of dust and mist particles that pass deep into the lungs. The hair holds the respirator off the face, allowing contaminants to bypass the respirator.

Other conditions may exist that may make it difficult or impossible for some people to use respirators safely. A person who has any of the following conditions should not wear a respirator unless approved by a physician:

- Known or suspected heart disease, such as high blood pressure, artery disease or previous heart problems
- Known or suspected lung disease, such as asthma, emphysema or difficulty in breathing
- A beard or other facial hair (prevents a good facepiece-to-face seal)
- Scars or wrinkles (could prevent a good facepiece-to-face seal)
- Poor eyesight (may have difficulty reaching safety during an emergency)
- Eyeglasses cannot be worn with full facepiece respirators unless a special bracket is used to hold the spectacles entirely inside the facepiece (the temples of the frames prevent a good facepiece-to-face seal)
- Claustrophobia
- Missing or arthritic fingers (could cause difficulty putting on and adjusting the respirator)

Contact Lenses. Dusts, mists, vapors and gases can contaminate contact lenses and could damage the lenses or cause eye injury. For these reasons, *contact lenses should not be worn in contaminated atmospheres or where respirators are used*. In fact, OSHA does not permit wearing contact lenses when respirators are used. For more specific information regarding the use of contact lenses in contaminated environments, contact an eye specialist, the Department of Labor or an industrial hygienist.

Selecting the Respirator. After determining the identity and concentration of the air contaminants, select a respirator approved for those hazards. The brand is not as important as availability of replacement respirators, cartridges and parts. You cannot mix brands of cartridges and facepieces interchangeably. This results in an unapproved respirator and may cause serious injury or death. Safety equipment dealers or industrial hygienists can help you review manufacturers information and select a respirator suitable for your needs. They can also perform the needed training and fit testing so you and your employees can use the respirator properly.

Respirators are available in half-mask and full-face versions. Generally, the half-mask version is adequate in agricultural operations. Full-face respirators protect the eyes from irritation or injury, and are advisable when using anhydrous ammonia and other very hazardous materials. Powered air-purifying respirators (PAPRs) greatly reduce fatigue by using powered blowers to force air through the filters.

Fit Testing. To assure safety, your respirator must fit properly and you must learn to use it properly. Some brands of respirators fit different individuals better and more comfortably than others. The size and flexibility of the facepiece and the user's facial features will affect the fit. The Occupational Safety and Health Administration (OSHA) requires fit testing before workers begin tasks requiring the use of respirators. The fit testing process is actually a very good respirator selection and user training tool. Contact a safety specialist for complete fit testing procedures.

Qualitative Fit Testing. Qualitative fit testing should be done before the respirator is used the first time and at least annually afterward. The test involves wearing the respirator in the presence of a relatively harmless, but easily detected, air contaminant. The user will be asked to talk, move through a range of head motions and indicate if the contaminant can be smelled or tasted inside the respirator. Common substances used are saccharin aerosol, isoamyl acetate (banana oil) and irritant smoke.

Positive and Negative Pressure Tests. Positive and negative pressure tests should be done every time the respirator is used. To do the positive pressure test, block the exhaust port with the palm of your hand and exhale gently. You should not detect leakage around the facepiece. Block the inlets and inhale to do the negative pressure test. Again, you should not detect leakage around the facepiece. Adjusting the position of the respirator and tightening the straps will correct many fitting problems.

Using a Respirator. After selecting a respirator and completing the fit tests, work can begin. However, realize that breathing through a respirator is work and the user will tire more easily. The resulting stress and fatigue can reduce a person's ability to do some tasks. In hot weather, the respirator can be quite uncomfortable. You must allow frequent rest breaks when using respirators.

Some workers may grumble and complain about the stress and discomfort of using a respirator. Remember, employers must provide the safety equipment and ensure its use. Insist that the respirators be used as prescribed or the workers must be assigned to other tasks or dismissed. The consequences of not using the respirator can be serious illness, injury or death.

How Long Will the Respirator Last? The life of respirator filters or cartridges depend on the concentration of the contaminants, the rate of breathing, temperature and relative humidity and the capacity of the filtration system. Proper cleaning, care and storage will extend the life of the respirator by protecting the filters while not in use. The respirator or the filters should be replaced when signs of failure are evident.

Signs that the filters should be replaced are increased breathing resistance (due to clogging) or when the contaminants can be smelled or tasted inside the facepiece. Of course, detecting the contaminants inside the facepiece could mean the concentration of contaminants is dangerously high or that the respirator is not fitting properly.

Leave the area immediately when breathing resistance increases or contaminants are smelled or tasted, then find the problem and take corrective action. Special caution should be exercised when air contaminants do not have a distinct odor or taste at low, nonhazardous concentrations. In these cases follow the recommendations of the respirator manufacturer or an industrial hygienist concerning replacement frequency.

Workers cleaning grain bins might use several disposable dust respirators in a day since they are easily damaged and become wet with perspiration. On the other hand, a worker might use a PAPR several days in the same conditions before filters need to be replaced. Pesticide applicators may be able to use the same respirator cartridges an entire season or as little as a few days, depending on the respirator, how often it is used and the chemicals being handled.

Caring for a Respirator. After every use, the respirator should be carefully cleaned, dried and stored. Rubber, plastic and metal parts should be cleaned with a solution of mild detergent and a disinfectant. Examine all parts carefully for damage or deterioration. Pay special attention to flexible parts, such as the facepiece, gaskets, seals and valves. After, cleaning, carefully dry the respirator, replace the filters or cartridges if necessary, and place the respirator in a clean, airtight plastic bag. Store the respirator in a location away from excessive heat and safe from physical damage. Consult your respirator's instructions, your safety equipment supplier or one of the referenced publications for more information on respirator care and storage.

FOOT PROTECTION

Injuries to the feet are uncomfortable at best, and they can seriously interfere with the ability to accomplish farm and household chores. Although most foot injuries are not serious, they do result in pain and lost production. Injuries result from dropped objects, punctures and strains/sprains due to slips and falls are probably the most common, and most of these can be prevented through a combination of safe work practices and use of proper foot wear. While you may not always need to wear high-top steel-toe boots with steel shank and lug soles, you probably should wear them some of the time. However, sneakers and smooth soled shoes are not good choices for many farming activities because they do not provide protection from commonly encountered hazards.

Perhaps the most important aspect of good footwear is a non-slip sole. Good traction is needed to prevent slips and falls while walking on various surfaces, and especially when climbing ladders and stairs. Smooth leather soles common on western boots and dress shoes may be too slippery on many surfaces. Soles with lugs or texture provide better traction, especially when made of the soft, yet long-wearing materials found in quality boots.

A steel shank can provide improved puncture resistance as well as improved support in jobs which place concentrated loads on the feet, such as prolonged work on ladders or repeatedly pressing machinery brake pedals.

Steel Toe Safety Shoes

Perhaps the most controversial and misunderstood feature of work boots is the steel toe cap. The steel toe cap is intended to protect the toes from injuries caused by dropped objects or maybe a cow stepping on the toes. While steel toe caps are very effective in preventing injuries to the toes, they will not protect the rest of the foot. However, most of the misunderstanding concerns the toe cap itself. Two common misconceptions are:

They're not comfortable. Like any shoes or boots, safety shoes will not be comfortable unless they fit properly. Most reputable safety shoe dealers can fit all sizes and widths from AAAA to EEE. If they fit properly, the steel toe cap will only be noticed if it prevents an injury, when you are squatting or otherwise have your toes bent far back or if you kick something and push your toes against the toe cap.

A 1,500 pound cow might step on the toe cap and mash it down on my toes so that I can't get the boot off. This probably would not happen since the toe cap must support at least 1,500 pounds to meet the weakest approval for steel toe shoes, and the toe caps of better safety shoes support 2,500 pounds. Also, since a cow cannot stand on one foot, it is unlikely she could crush the toe cap unless she stomped really hard. But, if she did manage to mash the cap down on your toes, what would have been left of your toes without the toe cap? Should the unlikely happen, the sole can be cut off the boot to free the foot.

Safety shoes are available in a wide variety of men's and women's styles, including dress shoes, western boots as well as leather and rubber work boot styles. Some special purpose boots are available which incorporate instep guards to prevent injuries behind the toes, Kevlar pads to prevent cuts from chain saws and with caulks (screw-in spikes) to prevent slips when working on logs or ice.

Rubber Boots

Water proof boots are recommended for tasks involving prolonged contact with water or whenever there is a risk of exposure to hazardous chemicals, such as pesticides. Just as when selecting rubber gloves, make sure the material is suitable for the exposure. To prevent water and chemicals entering the top of the boot, place the coveralls *outside* the boot to shed water onto the ground.

An often overlooked consideration when using rubber boots is perspiration removal. Because rubber boots do not breath, perspiration cannot readily escape, so the socks and feet will eventually be soaked with perspiration. Keeping the feet dry is essential for good health, so proper measures must be taken. Socks of polypropylene and cotton blends can help wick the moisture away, and moisture absorbent liners are available as well. Remove the boots and allow the feet to "breath" and dry occasionally. To help prevent foot health problems always wash and dry the feet at the end of each day, allow footwear to dry and air out between uses, and begin each day with clean, dry socks.

BODY COVERINGS

The skin is your body's largest organ and performs a number of vital functions—keeping moisture in, sweating to cool your body and keeping harmful agents from entering the body. Because the skin is exposed to so many hazards, it is important that it be protected to keep you healthy. While the skin can adequately protect the body from many hazards, such as most bacteria, it can easily be penetrated or damaged by many chemicals. It has been estimated that as much as 80% of the pesticides entering the body enter through the skin. Another alarming fact is the increased incidence of skin cancers resulting from sun exposure. People who spend long hours in the sun are at increased risk of developing skin cancer unless protective measures are followed.

Agricultural hazards affecting the skin include pesticides, cleaning compounds, disinfectants, solvents, paints, wood preservatives, used motor oil, irritating dusts and prolonged contact with water. Friction, scrapes and cuts are also common sources of skin injuries and irritation. Properly selected clothing can prevent or minimize the chances of many skin disorders.

"Normal" clothing can prevent many types of skin problems. For example, a light colored long-sleeve shirt can protect the arms and upper body from sunburn and reduce the risks of skin cancers, and the long-sleeve shirt may actually be cooler than a short-sleeve shirt when working in the sun because it will reflect the sunlight. Other clothing that helps prevent sunburn and skin cancer includes a wide-brimmed hat to protect the ears, face and neck and long pants to protect the legs.

Additional protective clothing is needed for many farm activities. Some types of protective clothing follow.

Leather Aprons. Leather aprons and leather sleeves should be worn when welding to prevent burns from splattering molten metal and slag. Whether the leather safety gear is worn or not, wear clothing that has low flammability characteristics. Many synthetic fabrics melt easily and may burn rapidly.

Rubber Aprons. Rubber aprons are needed when handling liquids for prolonged periods of time, or whenever handling concentrated chemicals. Wear the rubber apron even if other protective clothing is being worn as well. Keeping the skin clean and dry is always important, but the groin area is especially sensitive. Chemicals can penetrate the skin of the groin area more than 10 times more readily than through the forearm (Figure 12.2). A splash of a concentrated pesticide here could result in a high dose of toxins entering the body very quickly.

Chain Saw Safety Chaps. Chain saw safety chaps or pants minimize the risk of cutting the legs when operating a chain saw. The chaps will not prevent all cuts, but have been proven effective in reducing the severity of leg injuries from chain saw cuts. Made from Kevlar® or other ballistic materials, the chaps prevent many cuts by covering the cutters with fibers to reduce their cutting efficiency, giving the operator a small amount of additional time to regain control of the saw. Some people use chain saw chaps when loading square bales of hay to reduce the number of scratches and scrapes on their legs from carrying and boosting the bales upward with their knees. The chaps also help protect the legs from thorns when walking through brushy areas. However, chain saw chaps are not intended to protect from machetes and knives used in Christmas tree pruning.

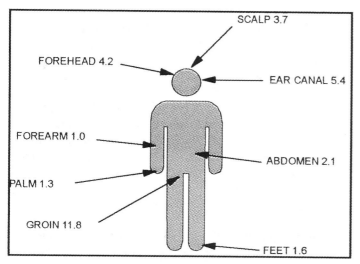

Figure 12.2. Absorbtion rates of chemicals through the skin of various parts of the body. Numbers are rates of absorption in comparison to the forearm.

Chemical Resistant Coveralls. Chemical resistant coveralls are available made from various materials that provide varying levels of protection and comfort. Some of the common materials include:

- **Tyvek**, a nonwoven polyethylene fabric used to make a variety of low-cost, disposable chemical resistant garments. There are several variations of Tyvek:

Standard Tyvek is essentially non-porous, so it provides excellent protection from pesticide dusts and mists. Standard Tyvek has been shown to provide a barrier to pesticide mists that is more than 20 times as effective as cotton chambray fabric (the study suggests that woven fabrics may wick the chemicals toward the body). Although standard Tyvek provides good protection from dusts and mists, splashes of liquids may penetrate the fabric, and can certainly seep through the seams. As mentioned above, wear a rubber apron when handling concentrated chemicals. However, since Tyvek is non-porous, it does not breathe, so you will get hot and sweaty after a while.

Polylaminated Tyvek features a plastic film bonded to the Tyvek fabric. This provides a tougher, more liquid resistant fabric. A hooded coverall of polylaminated Tyvek, used with the proper gloves, boots and respirator, is recommended for using orchard sprayers and provides excellent protection from pesticide mists. Polylaminated Tyvek garments also resist tearing better than standard Tyvek, and they also make rather good emergency rainsuits.

Perforated Tyvek is not generally recommended for pesticide application because the perforations allow chemicals to enter. However, the perforated Tyvek coveralls are excellent for dirty tasks such as equipment maintenance, painting, handling insulation, etc.

Some other synthetic, nonwoven fabrics similar to Tyvek are used for disposable garments, and may have names such as *Durafab*. These offer features and protection comparable to Tyvek. Durafab is unique among disposable fabrics because it features microscopic pores which are large enough to allow moisture vapor molecules to escape, yet small enough to block droplets of liquid water in most cases. These coveralls offer improved comfort when compared to standard Tyvek at comparable cost, but not as much protection as polylaminated Tyvek.

Regardless of the material used for disposable garments, the key word is *disposable*. These garments are not washable, and may provide reduced protection if they are laundered. These garments should be disposed of when damaged or contaminated, especially if contaminated with concentrated chemicals.

Gore-tex and similar fabrics are popular for outdoor garments because they allow water vapor molecules to escape through microscopic pores, yet the holes are too small to permit liquid water to enter. These fabrics provide good comfort and keep the wearer dry, even in rain. Garments incorporating Gore-tex and similar materials are excellent choices for equipment operators and for many outdoor recreation activities.

A variety of specialized garments are available for use in specialized tasks. For example, meat cutters might consider Kevlar® or chain link mesh aprons for protection from accidental cuts should a knife slip. Contact a safety equipment dealer for information concerning your specialized needs.

For protection from cold, *goose down, wool, Thinsulate, Hollofil* and other insulating materials are incorporated into a variety of garments for protection from cold temperatures. Each type of insulation has advantages and disadvantages. Generally, the advantages of synthetic materials over goose down are lower costs and better insulation when wet. Wool is also an excellent cold weather fabric which retains its insulating properties well when wet. Whatever your choice of cold weather garments, remember to dress in layers so you can add or remove layers to adjust to changing conditions and prevent soaking your clothes with perspiration.

HAND PROTECTION

While we often think of gloves when considering hand protection, there are other options. Of course, we should attempt to eliminate as many of the hazards as possible, such as filing down burrs and sharp edges, insulating or shielding hot surfaces and using safe work practices to minimize risks of injury. In this section we will discuss gloves and barrier creams. However, one of the best methods of protecting the hands (and preventing dermatitis elsewhere) is to thoroughly wash the hands with soap and water and dry them with single-use towels when finished with a task, and always before eating or going to the bathroom.

Barrier creams protect the hands by forming a protective film on the skin that reduces the amount of contact with irritating or harmful chemicals. The barrier cream is applied like a hand lotion, and can be washed off with soap and water. There are at least four types of barrier creams available, each intended for certain classes of hazards. One type may offer protection from oils and solvents, another protects from caustics, coolants and fertilizers, another protects from epoxies and still another protects from water based hazards. A barrier cream will not provide protection from cuts and scrapes, nor will it protect your hands from strong chemicals or from prolonged exposure to milder chemicals. However, barrier creams are an excellent choice for certain maintenance tasks because they permit normal use of hand tools and they do minimize exposure to chemicals such as used engine oil.

Types of Gloves

Gloves are available in a wide variety of styles and materials, each having its advantages and disadvantages. Since there is no single type of glove that is suitable for all tasks, it is important to understand a few basics of selecting gloves.

An evaluation of the hazards helps in determining the type gloves best suited for the job. Do you need protection from cuts and scrapes, protection from heat or protection from chemicals? Are the hazards low, moderate or high? Do you need to wear them all day, or intermittently? Do you need good manual dexterity for handling controls, tools or small objects? Glove selection becomes complicated when a situation involves several hazards or types of tasks.

How can you determine the best glove for the job? Pesticide labels are required to give specific recommendations for personal protective equipment, including gloves. You *must* use the PPE stated on the pesticide label to comply with federal and state regulations. For products other than pesticides, check the label for PPE recommendations or refer to a chemical resistance chart (featured in safety product catalogs) and select a material which provides good to excellent resistance from the chemicals you will be exposed to. Never use leather or fabric gloves when handling toxic chemicals since the chemicals readily penetrate these materials, and they can never be fully decontaminated of many pesticides.

Whatever material your "rubber" gloves are made of, remember these points:

- Gloves should be thick enough to provide adequate chemical resistance and prevent punctures or tears, yet thin enough to provide a good grip and manual dexterity. Latex examination gloves (like doctors use) are not adequate for protection from most chemicals, especially pesticides.
- Unlined gloves are normally recommended for pesticide use since they can be easily washed and decontaminated. Flock lined gloves are more comfortable, however, and often cost less than unlined gloves. Any lined gloves should be disposed of if contaminated inside by chemicals.
- The color of the gloves will affect comfort, especially outdoors. Black gloves will absorb heat from the sunlight, making them uncomfortable, so select a lighter color for outdoor work.

Glove cut, length and type and thickness of material affects the comfort and performance of any glove. Some gloves may be so heavily constructed that they will last forever—because they are too stiff or otherwise uncomfortable to be used on a routine basis. Gloves that are too heavy or uncomfortable, or which do not fit well, can significantly reduce a worker's productivity by making it difficult to handle tools and other objects. Those gloves may also be dangerous in certain situations. Other gloves may be very comfortable, but not durable enough to provide satisfactory performance and economy. The ideal glove fits well, provides the needed protection, permits good productivity and is economical. If there are several workers, it is likely that several styles or sizes of gloves will be needed. Try several styles of gloves to determine which provides the best comfort, grip and dexterity needed for the intended tasks.

Glove Materials

There are many materials used for gloves today, and although no single material can meet all your needs, you can probably meet all your needs with three or four types of gloves. Your choices will depend on the needs for protection from chemical and physical hazards.

Fabric. Fabric (jersey, cotton flannel, knit, etc.) gloves are inexpensive and suitable for many tasks where protection is needed from minor cuts and scrapes and where. Cloth gloves are also ideal for tasks where protection is needed from friction or when a glove might aid in gripping objects. Gloves with dimples or rubberized fabric on the palms and fingers are excellent when using hand tools and other similar tasks. Some fabrics are also used for high temperature applications, such as welding gloves. These may be more flexible than leather welding gloves, but can result in serious burns if they are wet when a hot object is picked up unless the gloves have a moisture barrier. A significant advantage of cloth gloves is that they breath well, minimizing perspiration buildup.

Leather. Leather is possibly the best all-around choice for protection from cuts, scrapes, friction and other physical hazards. It is tough, flexible, breathable and inexpensive and can be sewn into a wide variety of gloves. Leather may be used in conjunction with other materials for specialized purposes. For example, Kevlar® (the material used in body armor for police officers) is used in gloves for chain saw operators to provide extra protection from cuts should the left hand come in contact with the moving chain. Another example is the insulating materials sewn inside gloves used by welders to protect them from burns.

Rubber Gloves. Rubber gloves are needed for protection from the wide variety of chemicals used on the farm, around the shop and in the home. However, there are several "rubber" materials used for gloves today, and it is important that the correct material be used. Using the wrong material can allow chemicals to penetrate through the gloves or cause the gloves to deteriorate. The most common materials used for rubber gloves are:

- Butyl—highest resistance to penetration by most gases and water vapor.
- Neoprene—broad resistance to a variety of chemicals. Resists most oils, grease and caustics.
- Nitrile—superior protection against a wide variety of solvents, harsh chemicals, fats and petroleum products, plus excellent resistance to cuts, snags, punctures and abrasion.
- PVC—excellent abrasion resistance and protection from oils, acids and caustics.
- Natural rubber—excellent resistance to a variety of acids, solvents and alcohols, plus good cut resistance.

Note: Due to differences in glove manufacturers chemical resistance ratings of their products, no chemical resistance chart is presented here. Instead, always refer to the manufacturer's chemical resistance chart and glove selection guidelines when selecting rubber gloves.

Cut-Resistant Material. Cut-resistant materials such as Kevlar®, steel reinforced fabrics and chain link mesh are used in gloves designed for tasks where cuts are a major risk, such as handling glass and other sharp objects, and especially for meat cutting operations. Although these materials can be cut, they offer superior protection while permitting the user to grasp tools and other objects. Cut-resistant gloves often feature slip-resistant materials or textures on the fingers and palm to minimize the risk of sharp objects slipping through the hands, thus reducing the risks of a cut.

Caring for Gloves

The life of leather gloves can be extended by keeping the leather clean and keeping it from drying out excessively. Treating the leather with a leather care product such as mink oil or Snow Seal® will keep the leather soft and flexible (more comfortable to wear) and help it last longer. Fabric gloves can be laundered according to the label directions, and rubber gloves can be washed with detergent. However, any glove which has been contaminated with concentrated pesticides or other highly toxic or dangerous chemicals should be disposed of properly. Likewise, any frayed gloves should be disposed of to eliminate the risk of being caught on machinery.

HEAD PROTECTION

Any head injury has the potential to be serious, and any injury which results in brain damage can cause memory loss, affect the ability to reason, changes in personality and emotions, etc. Any of these changes can result in disability and interfere with the ability to earn a living.

Head injuries can result from work and leisure activities, so head protection must be considered at all times. Examples of agricultural activities with potential for head injuries include construction projects, machinery maintenance, horseback riding and operating ATVs. Fortunately, there is a wide variety of protective equipment suitable for the activities normally found in farming and recreational activities.

Bump Caps. Bump caps are lightweight plastic caps designed to protect the head from bumps and scrapes encountered in tasks such as building and machinery maintenance. However, a bump cap will not provide protection from impact, such as a dropped tool or other heavy object. Bump caps are also recommended for tasks where cleanliness and sanitation are high priorities, such as in food processing plants or when handling pesticides. A bump cap or hard hat can be decontaminated with soap and water, but a baseball cap or cowboy hat cannot be completely decontaminated of pesticides.

Safety Helmets. Safety helmets (commonly referred to as "hard hats") are primarily intended to protect the head from falling objects, although they may also protect from flying objects. A hard hat consists of a sturdy shell, usually made of plastic, and a suspension that holds the shell at least 1¼ inches from the head. When an object strikes the hat, the force is distributed through the suspension to a large area of the head and neck, preventing puncture wounds and concussion injuries in most cases. Hard hats are recommended for all construction and timber harvesting activities and any other tasks involving the risk of bumps or falling objects. While a hard hat may not be able to protect a person from a severe, direct blow, it can deflect many glancing blows which might otherwise result in serious injury or death.

Several accessories are available that can be mounted on a hard hat, including welding helmets, face shields, hearing protectors and communications devices. These can be useful because they keep together several PPE items needed for a particular task. For example, a hard hat with a face screen and hearing protectors is ideal for chain saw operation.

Although hard hats are heavier than baseball caps, they are cooler than baseball caps because there is an air space between the head and shell. This coolness is welcomed in summer, but a liner is often needed in winter to maintain comfort.

NOTE: Use safety helmets which are certified by the Safety Equipment Institute (SEI). Never use a metal hard hat when working around electrical systems.

Motorcycle Helmets. Motorcycle helmets are designed to protect the head from severe impacts that occur in motorcycle and ATV mishaps. It is estimated that over 60% of ATV related deaths result from head injuries, and most of the victims were not wearing a helmet. A helmet should be worn at all times by every motorcycle or ATV rider, regardless of age or skill level. A full face helmet is recommended to protect the face from injuries, and may also help protect the neck from excessive strain in a crash.

Never use any unapproved helmet, or one designed for other purposes, when riding a motorcycle or ATV. You can determine the approvals of a helmets by checking the labeling inside the helmet. At the minimum, the labels should indicate that the helmet meets DOT requirements. Most helmets will also meet the SNELL Foundation requirements, which are more strict than DOT standards.

Bicycle Helmets. Bicycle helmets are lightweight, ventilated helmets that are intended to prevent head injuries from bicycling mishaps. They are not as strong nor as heavy as motorcycle helmets, and they are not intended to be used on motorized vehicles because ATVs and motorcycles routinely travel at higher speeds and result in much harder blows to the head than normally experienced in bike crashes.

There are two basic types of bike helmets—soft shell and hard shell. Hard shell helmets are recommended because the hard plastic outer shell is more likely to slide on the ground or pavement than a soft shell helmet. A soft shell helmet can be gouged or caught on rough surfaces, causing the neck to be bent excessively and resulting in neck injuries.

Other specialized head gear may be appropriate for certain farm activities. For example, helmets are recommended for horseback riding, but they must be strong enough to protect from running into limbs and falling to the ground, as well as able to prevent a head injury should the horse fall onto the rider (yes, it has happened). A bicycle helmet might protect from the limbs and a fall, but could not support the weight of the horse. Look for a helmet designed for horseback riding.

Regardless of the type protective head gear, it can be damaged by accidents, abuse or improper care. Remember these points:

- Always replace any protective head gear that has received a hard blow, because it may have sustained damage that is not visible to the naked eye.
- Never wash any protective head gear with anything stronger than mild detergent and water—solvents can weaken or destroy the plastics used for protective head gear.
- Protect bicycle and motorcycle helmets from extreme heat. Do not leave them inside cars (trunk or passenger compartment) parked in the hot summer sun, because the temperature may be high enough to weaken materials and glues used in making the helmets. If they must be left in a car, put them on the floor and place a light colored towel or other object

over them. The floor is the coolest place, and the cover will reflect the suns heat.

SUMMARY

This review of personal protective equipment was intended to increase your awareness of the many hazards in agricultural work settings and methods of preventing injuries and illnesses from those hazards. Every situation is a unique combination of tasks, hazards, environmental conditions and human characteristics. For this reason, every situation must be evaluated and safety procedures implemented that consider these factors. Always consult a safety specialist if you need assistance.

Finally, the best personal protective equipment in the world is useless if not used properly at all times. Workers who fail to work safely or refuse to use safety equipment are a liability, just like a bull with good bloodlines but a bad habit of tearing down fences. They may do a good job, but they also cost a lot in down time and repairs. Their behavior must be changed or they must be replaced.

REFERENCES

American Lung Association and The Institute of Agricultural Medicine and Occupational Health, The University of Iowa. 1986. Agricultural Respiratory Hazards Education Series, Pm-1222-9. Ames, IA: Cooperative Extension Service, Iowa State University.

American National Standards Institute. 1983. Safety in Welding and Cutting. ANSI/ASC Z49.1. Cincinnati, OH: American National Standards Institute.

Cheremisoff, P.N. 1984. *Management of Hazardous Occupational Environments*. Lancaster, PA: TECHNOMIC Publishing Co., Inc., pp. 81-110.

Fundamentals of Machinery Operation. 1974. Moline, IL: John Deere Service Publications, pp. 43-44, 212.

National Institute for Occupational Safety and Health (NIOSH). 1978a. *Respiratory Protection: An Employer's Manual*, DHEW (NIOSH) Publication No. 78-193A. Cincinnati, OH: U.S. Department of Health, Education and Welfare, Public Health Service, Centers for Disease Control, NIOSH, Division of Technical Services.

National Institute for Occupational Safety and Health (NIOSH). 1978b. *Respiratory Protection: A Guide for the Employee*, DHEW (NIOSH) Publication No. 78-193B. Cincinnati, OH: U.S. Department of Health, Education and Welfare, Public Health Service, Centers for Disease Control, NIOSH, Division of Technical Services.

National Safety Council. 1977. *Accident Prevention Manual for Industrial Operations*. Seventh Edition. Chicago, IL: National Safety Council, pp. 1059; 1104-07; 1242-54; 1484-92.

National Safety Council. 1988. *Fundamentals of Industrial Hygiene*. Third Edition. Chicago, IL: National Safety Council, pp 77-92.

Occupational Safety and Health Administration (OSHA). 1993a. OSHA Standards for General Industry, 29 CFR 1910.252(b)(2)(i). Washington, DC: U.S. Department of Labor.

Occupational Safety and Health Administration (OSHA). 1993b. OSHA Standards for Welding, Cutting and Heating (draft revisions). 29 CFR 1926.351. Washington, DC: U.S. Department of Labor.

Occupational Safety and Health Administration (OSHA). 1986a. Respiratory Protection, 29 CFR 1910.134. OSHA Standards for General Industry. Washington, DC: U.S. Department of Labor, Occupational Safety and Health Administration.

Occupational Safety and Health Administration (OSHA). 1986b. Ventilation, 29 CFR 1910.94. OSHA Standards for General Industry. 1986. Washington, DC: U.S. Department of Labor, OSHA.

Occupational Safety and Health Administration (OSHA). 1986c. Lead, 29 CFR 1910.1025. OSHA Standards for General Industry. Washington, DC: U.S. Department of Labor, OSHA.

Olishifski J.B., and F.E. McElroy. 1977. *Fundamentals of Industrial Hygiene*. Chicago, IL: National Safety Council, pp. 165-83; 384-95; 634-54.

Respiratory Protection, Section 19, *Safety and Health Manual*. 1989. Knoxville, TN: Institute of Agriculture, University of Tennessee.

Vallen Safety Supply Company. 1995. *1995 Safety Catalog*. Houston, TX: Vallen Safety Supply Company.

Weeks J.L., B.S. Levy, G.R. Wagner. 1991. *Preventing Occupational Disease and Injury*. Washington, DC: American Public Health Association, pp 272-77.

13

SAFETY AND HEALTH PROGRAMS IN AGRICULTURE, FORESTRY, AND FISHERIES

Gary M. Tencer, C.I.H.
Duke University Medical Center

> A health and safety program within agriculture, forestries, and fisheries is important in enhancing the ability of the company to carry out its mission. There are five generally accepted functions of management: planning, organizing, staffing, directing, and controlling. These functions are equally appropriate to the management of safety and health programs as they are to running any type of business. All of these components are interconnected; each relies on the others. Planning is the most important function of management. By integrating the interactions between the other four functions into a documented strategy for anticipating and resolving program needs, good planning allows the delivery of safety and health services with optimal effectiveness and efficiency. This chapter allows the reader to compare various aspects of planning a comprehensive health and safety program that utilizes the major functions of management. Specific information is presented on hazard recognition, hazard assessment, and development of safety protocols. Administration controls are training are also covered.

INTRODUCTION

The prime objective of a health and safety program within an agriculture, forestry, and fishery industry is to enhance the ability to carry out the mission of the company. By reducing the injury or death of human resources as well as damage to facilities due to mishaps, the company is able to carry out its mission more effectively and efficiently. The health and safety program, therefore, is nothing more than an organized, well-managed approach to this problem. In this chapter, we will explore general management of health and safety, the regulatory environment, and the specific functions of a health and safety program.

MANAGEMENT RESPONSIBILITIES

Philosophy of the Organization

It is not difficult to see that in most organizations, the philosophies of management tend to affect not only the organization's product or service, but also the general flavor of the organization. There are very successful organizations with questionable management philosophies just as there are organizations which fail despite having the best of philosophies. For the most part, however, an organization that can develop and implement constructive and supportive philosophies will tend to succeed.

Of paramount importance are the following concepts:

- Responsibility—A sense of responsibility should be held by all levels of the organization. An attitude that management is looking after the employees, who in turn are looking after the organization's interests, is essential for a good safety and health program.

- Judgment and Conceptualization—The health and safety program must function in the real world. All policies and program components must be based on need which is truly characterized through verification of condition. A program that is not designed with this real-world view is not likely to be effective.

- Communication—Effective communication is a concept that receives almost universal support, yet is so difficult to do right. To be truly effective, communication must be bi-directional. Safety policies must be known and understood by the employee population; management must be aware of employee concerns about safety.

Functions of Safety Management

There are five generally accepted functions of management: planning, organizing, staffing, directing, and controlling. These functions are as appropriate to the management of safety and health programs as they are to running any sort of business. Although each of the functions summarized below can be analyzed individually, the reality is that, in actual practice, they are interconnected to the extent that each relies on the others.

Planning. *Planning* is the most important function of management. By integrating the interactions between the other four functions into a documented strategy for anticipating and resolving program needs, good planning allows the delivery of safety and health services with optimal effectiveness and efficiency.

Organizing. In general, *organizing* is nothing more than bringing order to the relative chaos that exists within the world. Organizing is essential in producing effective plans, communicating the goals and objectives of the health and safety program, and serving as the way to exploit available resources toward the accomplishment of the objectives established by the planning process. Resources include not only the operating budget, but also human resources which must be applied to the accomplishment of the plan.

Staffing. The *staffing* function has two critical parts within the safety and health program. On the micro scale is developing the team responsible for the management of the health and safety program. The bigger view of staffing requires consideration for the quality of the entire employee population; employees best prepared for their assigned tasks through adequate safety and health training will be better partners with management.

Directing. The function of *directing* attempts to promote an environment of participative action through effective communication of the objectives of the health and safety program. Each member of the organization must be made aware of all aspects of his or her job tasks and responsibilities as well as how their actions, as individuals, support the function of the entire organization.

Controlling. Controlling can be described simply as "sticking with the plan." Through established monitoring efforts, the manager of the health and safety program will be able to check on the actions set in motion by the health and safety program as well as the results obtained. Disparities which exist between these actions and results compared to what was expected are then corrected by appropriate means.

SAFETY POLICIES AND PROCEDURES

The above discussion of management issues was a general introduction to the functions and philosophies that are all rolled into the thing we call management. There are issues specific to health and safety management which need a bit more discussion.

Safety and the Chain of Supervision

The responsibility for the safety and health of the employees of an organization is a line function. Administrative responsibility and safety responsibility go hand in hand. Effective environmental, safety, and health performance can result only if all persons, from the head of the organization down to the individual worker, are responsible and accountable for safety conditions. Gone are the days when the health and safety program could be seen as a sort of token program with responsibilities only for the minimum required to allow the company to avoid regulatory scrutiny, while being available as a convenient scapegoat should something catastrophic happen. An organization should rely on the same chain of supervision that it employs for its primary mission to effectively ensure that the safety and health components are covered.

While accidents may be attributed to engineering flaws, alcohol/drug use, or illness, they almost always, upon evaluation, result from some sort of failure on the part of organization's management. Table 13.1 presents many examples of common causative factors which can be attributed to management deficiencies.

Staffing the Health and Safety Program

The concept of health and safety as a line management responsibility was discussed above. The trap that occasionally occurs behind this philosophy is the expression "Health and Safety are everybody's responsibility"; in reality this means that nobody has direct responsibility. In order to avoid this trap, all organizational efforts must pivot around some central focus—in this case, the Health and Safety Manager, Group, or Committee. Although a "staff" management function, this focus, with the authority and support of top line management, has direct responsibility to ensure that the program is effective.

Staffing strategies for the health and safety program are dependent on a variety of internal and external factors. The former include the organization's size and its geographical spread as well as the character of its operations; the latter are primarily reflections of the regulatory environment in which the organization operates. While many formulae have been proposed for health and safety program staffing, they tend

Table 13.1. Accident causes traced back to management responsibilities.

Immediate Cause	Example	Possible Underlying Causes	Possible Management Failures
			INADEQUATE:
1. Poor Housekeeping	An employee trips and falls over equipment left in an aisle.	Hazards not recognized	Supervisory training
			Supervisory safety indoctrination
	Material poorly piled on a high shelf falls off.	Facilities inadequate	Planning, layout
2. Improper use of tools, equipment, facilities	Using the side of a grinding wheel instead of the face and the wheel breaks.	Lack of skill, knowledge.	Employee training
		Lack of proper procedures	Established operational procedure
			Enforcement of proper procedure
	Someone using forklift truck to elevate people- person falls off.	Lack of motivation	Supervisory safety indoctrination
			Employee training
	Someone using compressed air to clean dust off clothes- eye injury.		Employee safety
3. Unsafe or defective equipment, facilities	Portable electric drill without ground wire	Not recognized as unsafe	Supervisory safety indoctrination
	Axe or hammer with loose head		Employee training
	Car with defective brakes, steering	Poor design or selection	Employee safety consciousness Planning, layout, design
			Supervisory safety documentation
		Poor maintenance	Equipment, materials, tools Maintenance, repair system
4. Lack of proper procedures	No requirement to check for gas fumes before starting engine - explosion	Omissions	Operational procedures
	No definite instructions requiring power to be locked out before maintenance is done	Errors by designer	Planning, layout, design
		Errors by supervisor	Supervisory proficiency

[1] Fine, William T.; *A Management Approach in Accident Prevention:* Naval Surface Weapons Center, White Oak TR 75-104 (AD A014562); Silver Spring, Maryland; July 1975

Table 13.1. *(continued)*

5. Improvising unsafe procedures.	"Rube Goldberg" haphazard temporary expedients, without proper planning	Inadequate training	Established operational procedure
			Enforcement of proper procedure
			Supervisory safety indoctrination
			Employee training
		Inadequate supervision	Employee safety consciousness
			Supervisory safety indoctrination
			Employee selection, placement
6. Failure to follow prescribed procedures	Shortcuts bypassing safety precautions	Need not emphasized	Enforcement of proper procedures
			Supervisory safety indoctrination
	Operation will only be done once - take a chance	Procedures unclear	Operational procedures
7. Job not understood	Employee uses wrong method, doesn't follow instructions	Instructions complex	Operational procedures
			Planning, layout, design
		Inadequate comprehension	Employee selection, placement
8. Lack of awareness of hazards involved	Not realizing rotation shaft was dangerous	Inadequate instructions	Supervisory safety indoctrination
	Not realizing fumes were dangerous		Employee training
			Employee safety consciousness
	Not realizing that hydrogen from battery charging operation could explode	Inadequate warnings	Planning, layout, design
			Safety rules, measures, equipment
			Operational procedures
9. Lack of proper tools, equipment, facilities	Cart too small for hauling large items	Need not recognized	Planning, layout, design
			Supervisory safety indoctrination
	Auto maintenance done without proper wrenches - cut knuckles	Inadequate supply	Equipment, materials, tools
		Deliberate	Morale, discipline
10. Lack of guards, safety devices	Machine has exposed belt and gear - severe cut	Need not recognized	Planning, layout, design
	No warning device on vehicle - pedestrian hit		Safety rules, measures, equipment
			Supervisory safety indoctrination
			Employee safety consciousness
		Inadequate availability	Equipment, materials, tools
	No guard rail on a scaffold 10 feet high	Deliberate	Moral, discipline, laziness
11. Lack of protective equipment, clothing	Eye protection not used in shop	Needs not recognized	Planning, layout, design
	Dermatitis because employee did not use protective lotion or gloves		Safety rules, measures, equipment
			Supervisory safety indoctrination
			Employee safety consciousness
	Not using respirator when spraying paint	Inadequate availability	Equipment, materials, tools
			Operational procedures

Table 13.1. *(continued)*

	Materials handler not wearing safety shoes	Deliberate	Morale, discipline, laziness
12. Exceeding prescribed limits, load, speed, strength, etc.	Driving vehicle too fast - accident	Warnings inadequate	Safety rules
			Proper procedures
	Overtaxing a crane or hoist or elevator, lifting beyond rated capacity - load drops	Instructions inadequate	Employee training
		Lack of comprehension	Employee training
			Employee selection, placement
13. Inattention; neglect of obvious safe practice	Welder picking up hot metal in bare hands	Lack of motivation	Enforcement of safety rules
			Enforcement of proper procedures
			Employee training
	Walking under suspended load		Employee safety consciousness
	Not cleaning up broken glass on floor	Inadequate comprehension	Employee selection, placement
14. Fatigue, reduced awareness, hypnosis	Putting excessive hours on hazardous machine operations	Excessive physical or mental requirements	Planning, layout, design
			Employee selection, placement
	Repetitive inspection of small parts		Operational procedures
15. Misconduct; deliberate failure to use protective clothing	Failure to install safety guards	Low morale	Supervisory training
		Poor attitude	Employee selection, placement
	Desire to speed (for thrills)	Malassignment	Planning, layout, design
			Employee selection, placement
			Employee training

to be valid only for very narrowly defined operations. If there were any universal rule of thumb, it would be this: an organization with an employee population approaching five hundred should strongly consider adding a full-time person to coordinate the health and safety program.

In deciding what qualifications are necessary to the program, the organization should be aware of the specialties available within the wide field of safety and health. These may include:

- Safety Engineers
- Industrial Hygienists
- Occupational Health Nurses
- Occupational Medicine Physicians
- Loss Control Specialists
- Workers' Compensation Managers
- Ergonomists

Committees

A Safety and Health Committee can be very useful for larger organizations. With cross section representation, this committee can serve a number of very important functions, but only when backed by a good organizational health and safety program as well as management support. Meeting on a periodic or *ad hoc* basis, an effective Health and Safety Committee can:

- Establish policies and procedures under which the organization's health and safety program will operate.
- Review proposed policies, procedures, and standards appropriate to provide a safe and healthful work environment. Recommend approval to top management.
- Review current accident, injury, and illness trends, discuss their prioritization and suggested resolution.

- Serve as a forum of communication between the Health and Safety Manager, top management, and the organization, through its representatives assigned to the committee.
- Appoint subcommittees as necessary to consider the impact of significant health and safety issues.

REGULATIONS AND STANDARDS

Safety and Health standards are based on procedures and precautions designed to minimize risk. Based on the source of the standard, they are monitored and validated against current scientific or empirical knowledge to assure their continued adequacy.

Regulations are the subset of standards promulgated as legal requirements by federal, state, or local authority. Non-regulatory standards developed by various organizations for specific purposes are usually called *consensus standards*. These may be considered to be guidelines which can be applied within an organization's health and safety program. Quite frequently a consensus standard will be incorporated by reference into a regulation, or adopted as an inspection guideline by a regulatory authority, giving it a mandatory compliance status.

A summary of current regulations and standards is presented below.

Occupational Safety and Health Administration

This group of standards, embodied in the Code of Federal Regulations, Title 29, is relatively comprehensive in its applicability in matters of workplace health and safety. They are regulatory in status and are enforced by either on the federal level, or on the state level in states with approved programs Part 1910—*General Industry*, and Part 1928—*Agriculture* are especially appropriate to the agricultural, forestry, and fisheries industries.

Environmental Protection Agency

These standards, embodied in the Code of Federal Regulations, Title 40, are primarily intended to ensure protection of the environment. They do, however, address workplace health and safety issues related to agricultural pesticide use in Part 170.

Other Federal Regulations

Other standards which can be pertinent to a health and safety program are embodied in the following sections of the Code of Federal Regulations:

- Title 7— Agriculture
- Title 9— Animals and Animal Product
- Title 33— Navigation and Navigable Waters
- Title 36— Parks, Forests, and Public Property
- Title 49— Transportation
- Title 50— Wildlife and Fisheries

Consensus Standards

Consensus standards are in some ways more valuable to a health and safety manager. They are, in most cases, more specific in their guidance and, because they do not have the burden of the rulemaking process that can confound the implementation of new or modified regulations, they are normally more current in their guidance. One caveat, though, is that because they do offer more flexibility in their application, they are not "cookbooks" and should be considered for use only by competent health and safety specialists. Consensus standards do undergo rigorous scientific and practical scrutiny as part of their approval process.

The *American Conference of Governmental Industrial Hygienists (ACGIH)* publishes, on an annual basis, Threshold Limit Values (TLVs™) for Chemical Substances and Physical Agents along with Biological Exposure Indices (BEIs™). The TLVs are upper limits of hazards, as quantified, which represent conditions which should not cause adverse effects to workers. BEIs are intended to be a tool for assessing workplace exposures by analyzing blood, urine, or expired air for the specific chemical or its metabolite.

The *National Fire Protection Association (NFPA)* publishes a number of general and specific standards which are intended to protect property and life from loss due to fire emergencies. These standards are frequently incorporated into state and local fire and building codes as well as the above-mentioned Federal Regulations.

The *American National Standards Institute (ANSI)* has as its mission to promote standardization of all possible conditions of modern life, an important endeavor in the current global economy. Among the very detailed health and safety standards are:

Standards Pertaining to Agriculture, Forestry, or Fisheries

ANSI A10.15-1995: Dredging, Safety Requirements for
ANSI O2.1-1983: Sawmills, Safety Requirements for
ANSI Z245.5-1990: Baling Equipment, Safety Requirements for
ANSI/HTI B173.4-1991: Axes, Safety Requirements
ANSI/ASAE S338.2 (R1993): Safety Chain for Towed Equipment

ANSI/HTI B209.3-1990: Wood Splitting Wedges, Safety Requirements
ANSI B175.1-1991: Gasoline Powered Chain Saws, Safety Requirements for
ANSI B157.1-1981: Scales, Weighing Devices, and Weighing Systems, Safety Requirements for
ANSI/OPEI B71.4-1990: Commercial Turf Care Equipment, Safety Specifications
ANSI/OPEI B71.8-1986: Walk-Behind Powered Rotary Tillers, Safety Specifications for
ANSI/SAE J684-MAY87: Trailer Couplings, Hitches, and Safety Chains—Automotive Type
ANSI Z359.1-1992: Personal Fall Arrest Systems, Subsystems, and Components, Safety Requirements for
ANSI/ASAE S395 (R1991): Self-Propelled, Hose-Drag Agricultural Irrigation Systems, Safety for

Other Standards

ANSI Z16.2-1995: Information Management for Occupational Safety and Health
ANSI A10.33-1992: Safety and Health Program Requirements for Multi-Employer Projects
ANSI/NFPA 1500-1992: Fire Department Occupational Safety and Health Program
ANSI A10.6-1990: Demolition, Safety Requirements for
ANSI Z535.1-1991: Safety Color Code
ANSI Z535.3-1991: Criteria for Safety Symbols
ANSI/SAE J115-JAN87: Safety Signs
ANSI A40-1993: Plumbing, Safety Requirements for
ANSI A10.13-1989: Steel Erection, Safety Requirements for
ANSI A14.3-1992: Ladders—Fixed, Safety Requirements
ANSI Z117.1-1995: Confined Spaces, Safety Requirements for
ANSI Z535.2-1991: Environmental and Facility Safety Signs
ANSI Z535.4-1991: Product Safety Signs and Labels
ANSI/HTI B173.7-1990: Hatchets, Safety Requirements
ANSI/NFPA 101-1994: Life Safety Code
ANSI/UL 30-1990: Metal Safety Cans
ANSI A10.5-1992: Material Hoists, Safety Requirements for
ANSI C2-1993: National Electrical Safety Code
ANSI O1.1-1992: Woodworking Machinery, Safety Requirements
ANSI A40 Table 1-1995: Plumbing, Safety Requirements for
ANSI A14.1-1994: Ladders—Portable Wood, Safety Requirements for
ANSI A14.2-1990: Portable Metal Ladders, Safety Requirements for
ANSI B24.1-1985 (R1991): Forging Machinery, Safety Requirements for
ANSI PH3.60-1979 (R1992): Multiflash Products, Safety Levels for
ANSI S2.60-1987: Balancing Machines—Enclosures and Other Safety Measures
ANSI Z49.1-1994: Welding, Cutting and Allied Processes, Safety in
ANSI/ASHRAE 15-1994: Mechanical Refrigeration, Safety Requirements for
ANSI/HTI B173.1-1990: Nail Hammers, Safety Requirements
ANSI/HTI B209.1-1991: Metal Chisels, Safety Requirements
ANSI/HTI B209.8-1991: Star Drills, Safety Requirements
ANSI/NFPA 170-1994: Public Fire Safety Symbols
ANSI/NFPA 1501-1987: Fire Department Safety Officer
ANSI/SAE AS 1043F: Safety Wire Holes, Standard Dimensions for
ANSI/SAE J98-NOV92: Industrial Wheeled Equipment, Safety for
ANSI A14.4-1992: Job-Made Wooden Ladders, Safety Requirements for
ANSI/SAE ARP 1384B: Passenger Safety Information Cards
ANSI/NFPA 101M-1995: Alternative Approaches to Life Safety
ANSI MH29.1-1994: Industrial Scissors Lifts, Safety Requirements for
ANSI/IEEE 1228-1994: Software Safety Plans, Standard for
ANSI/UL 1989-1994: Safety for Standby Batteries, Standard for
ANSI/UL 1996-1994: Safety for Duct Heaters, Standard for
ANSI A10.3-1995: Powder Actuated Fastening Systems, Safety Requirements for
ANSI A10.8-1988: Construction and Demolition Operations—Scaffolding, Safety Requirements
ANSI A10.11-1989: Safety Nets Used During Construction, Repair, and Demolition Operations
ANSI K61.1-1989: Storage and Handling of Anhydrous Ammonia, Safety Requirements for the
ANSI Z245.3-1977 (R1990): Stability of Refuse Bins, Safety Requirements for the
ANSI/API 527-1991: Seat Tightness of Safety Relief Valves
ANSI/IEEE 80-1986 (R1991): AC Substation Grounding, Safety in
ANSI/NFPA 70E-1995: Electrical Safety Requirements for Employee Workplaces
ANSI/NFPA 150-1995: Race Track Stables, Fire Safety in
ANSI/NFPA 1982-1993: Personal Alert Safety Systems (PASS)
ANSI/SAE AMS 2825A (R1993): Material Safety Data Sheets
ANSI/SAE ARP 594D: Fuel Pump Thermal Safety Design
ANSI/UL 1313-1993: Nonmetallic Safety Cans for Petroleum Products
ANSI/UL 1431-1992: Personal Hygiene and Health Care Appliances

ANSI A14.5-1992: Ladders—Portable Reinforced Plastic, Safety Requirements

ANSI A120.1-1992: Powered Platforms for Building Maintenance, Safety Requirements for

ANSI/API Spec 14A-1994: Subsurface Safety Valve Equipment

ANSI/NFPA 101A-1995: Guide On Alternative Approaches to Life Safety

ANSI/UL 1090-1995: Safety for Electric Snow Movers, Standard for

ANSI/UL 407-1995: Safety for Manifolds for Compressed Gases, Standard for

ANSI/UL 912-1995: Safety for Highways Emergency Signals, Standard for

ANSI/HTI B209.6-1990: Glaziers' Chisels and Wood Chisels, Safety Requirements

ANSI/IEEE 510-1992: Safety in High-Voltage and High-Power Testing

ANSI A10.22-1990: Rope—Guided and Nonguided Worker's Hoists, Safety Requirements

ANSI A10.28-1990: Work Platforms Suspended from Cranes or Derricks, Safety Requirements

ANSI B7.7-1990: Abrading Materials with Coated Abrasives Systems, Safety Requirements for

ANSI B11.6-1984 (R1994): Lathes, Safety Requirements for the Construction, Care, and Use of

ANSI B133.12-1981 (R1994): Gas Turbines—Procurement Standard, Maintenance and Safety

ANSI Z244.1-1982 (R1993): Lock Out/Tag Out of Energy Sources, Safety Requirements for the

ANSI/API 526-1984 (R1992): Flanged Steel Safety Relief Valves

ANSI/HTI B209.4-1991: Hand Tools—Nail Sets, Safety Requirements

ANSI/HTI B209.10-1990: Stud, Screw and Pipe Extractors, Safety Requirements

ANSI/NSF 53-1996: Drinking Water Treatment Units—Health Effects

ANSI/NSF 60-1988: Drinking Water Treatment Chemicals—Health Effects

ANSI/SAE AIR 1639 (R1989): Pneumatic Starting Systems, Safety Criteria for

ANSI/SAE ARP 4171: Food and Beverage Service Carts, Safety Considerations of

ANSI/SAE J697-MAY88: Safety Chain of Full Trailers or Converter Dollies

ANSI/UL 539-1990: Single and Multiple Station Heat Detectors, Safety for

ANSI/RIA R15.06-1992: Industrial Robots and Robot Systems, Safety Requirements

ANSI MH30.1-1993: Safety, Performance, and Testing of Dock Leveling Devices, Specifications

ANSI/UL 1492-1994: Safety for Audio-Video Products and Accessories, Standard for

ANSI/UL 2044-1994: Safety for Commercial Circuit Television Equipment, Standard for

ANSI/SAE AS 3509: Cable, Safety, Kit, Nickel Alloy, UNS N06600

ANSI/SAE AS 4536: Safety Cable Kit Procurement Specification and Requirements for Use

ANSI/UL 1419-1995: Safety for Professional Video and Audio Equipment, Standard for

ANSI/UL 1479-1995: Safety for Fire Tests of Through-Penetration Firestops, Standard for

ANSI/UL 1692-1995: Safety for Polymeric Materials—Coil Forms, Standard for

ANSI/UL 1971-1995: Safety for Signaling Devices for the Hearing Impaired, Standard for

ANSI A14.7-1991: Mobile Ladder Stands and Mobile Work Platforms, Safety Requirements for

ANSI Z229.1-1982: Shops Fabricating Structural Steel and Steel Plate, Safety Requirements for

ANSI/HTI B173.3-1991: Hand Tools—Heavy Striking Tools, Safety Requirements

ANSI/HTI B209.7-1990: Ripping Chisels and Flooring/Electricians' Chisels, Safety Requirements

ANSI/OPEI B71.6-1990: Powered Shredder/Grinders and Shredder/Baggers, Safety Specifications

ANSI A10.10-1990: Temporary and Portable Space Heating Devices and Equipment, Safety Requirements for

ANSI A10.31-1995: Construction and Demolition—Digger Derricks, Safety Requirements, Definition

ANSI B11.4-1993: Machine Tools—Shears, Safety Requirements for Construction, Care, and Use

ANSI B11.9-1975 (R1987): Grinding Machines, Safety Requirements for the Construction, Care, and Use

ANSI B11.10-1990: Metal Sawing Machines, Safety Requirements for Construction, Care, and Use of

ANSI B154.1-1995: Rivet Setting Equipment, Safety Requirements for Construction, Care, and Use of

ANSI B169.1-1990: Construction, Care, and Use of Machinery, Safety Requirements for the

ANSI MH16.2-1984: Industrial and Commercial Steel Storage Racks, Safety Practices for the Use of

ANSI/ALI B153.1-1990: Automotive Lifts, Safety Requirements for the Construction, Care, and Use of

ANSI/HTI B173.2-1991: Hand Tools—Ball Peen Hammers, Safety Requirements

ANSI/HTI B173.5-1991: Body Repair Hammers and Dolly Blocks, Safety Requirements

ANSI/HTI B173.8-1991: Riveting, Scaling, and Tinners' Setting Hammers, Safety Requirements

ANSI/HTI B209.9-1991: Hand Tools—Nail Puller Bars, Safety Requirements

ANSI/NFPA 1983-1995: Fire Service Life Safety Rope, Harnesses, and Hardware

ANSI/UL 372-1994: Primary Safety Controls for Gas- and Oil-Fired Appliances

ANSI/UL 495-1987: Power-Operated Dispensing Devices for LP-Safety Gas

ANSI/ASME A90.1-1992: Belt Manlifts (includes revision service), Safety Standard for

ANSI Z400.1-1993: Hazardous Industrial Chemicals—Material Safety Data Sheets, Preparation

ANSI Z390.1-1995: Accepted Practices for Hydrogen Sulfide (H2S) Safety Training Programs

ANSI/SAE AS 3511: Cable, Safety, Kit—Corrosion Resistant Steel, UNS S30400

ANSI/UL 1784-1995: Safety for Air Leakage Tests for Door Assemblies, Standard for

ANSI/UL 297-1995: Safety for Acetylene Generators Portable, Medium-Pressure, Standard for

ANSI B208.1-1982: Portable Pipe Threading Machines and Portable Power Drives, Safety Requirements

ANSI/HTI B209.5-1991: Hand Tools—Brick Chisels and Brick Sets, Safety Requirements

ANSI/SAE J284-JAN91: Safety-Alert Symbol for Agricultural Equipment (same as ANSI/ASAE S350)

ANSI/UL 248-8-1995: Safety for Low-Voltage Fuses—Part 8: Class J Fuses, Standard for

ANSI A10.4-1990: Personnel Hoists and Employee Elevators for Construction and Demolition Operations

ANSI A1264.1-1995: Workplace Floor and Wall Openings, Stairs, and Railings, Safety Requirements for

ANSI B11.3-1982 (R1994): Power Press Brakes, Safety Requirements

ANSI B11.14-1983: Coil Slitting Machines/Systems, Safety Requirements for the Construction, Care,

ANSI/AIHA Z9.3-1994: Spray Finishing Operations—Safety Code for Design, Construction, and Ventilation

ANSI Z97.1-1984 (R1994): Glazing Materials Used in Buildings, Safety Performance Specifications

ANSI Z268.1-1982: Design, Use and Maintenance of Metal Scrap Processing Equipment, Safety Requirements

FUNCTIONS OF A HEALTH AND SAFETY PROGRAM

The three primary functions of any effective health and safety program are:

- Identification of hazardous conditions.
- Evaluation of the hazard.
- Control of the hazard.

Hazard Identification

Hazard identification, unfortunately is usually retrospective. Following an injury or illness, it is very easy to realize that a hazardous condition existed. An effective health and safety program will rely more on prospective identification and anticipation. This is especially important for the agriculture, forestry, and fishery industries because of the relatively changeable nature of the work environment.

Safety and Health Surveys. The primary component of this strategy is a process to systematically survey work areas and operations on a regular basis for any factors which could affect health or safety. Conditions change constantly, workers develop habits that are contrary to good practice, and tools and equipment wear and develop faults or are modified in such a way that they circumvent safety controls. Personal or operation changes may cause training levels to be inadequate. Unless a programmed approach toward discovery of these changes is implemented, the organization could find the potential for loss to quickly reach a critical level.

In addition to their use in identifying hazardous conditions, surveys serve a number of other important purposes. Adequacy of the health and safety training program can be assessed by observing work techniques and use of personal protective equipment as well as by querying the worker. Worker and supervisor awareness of general or specific hazardous conditions by involving them interactively in the survey process. Surveys can promote cooperative attitudes toward correction.

Most organizations may wish to consider a multi-level survey strategy. Certain operations can be monitored continuously as part of daily routine. Operation supervisors or work area managers can conduct frequent walk-through surveys. Less frequent but more comprehensive surveys should be conducted by the health and safety program staff. Unannounced or unscheduled surveys should be done whenever warranted.

One specialized type of survey is the "Baseline Audit". This survey is done by the Health and Safety Manager for the purpose of establishing a initial reference snapshot of an operation. This baseline is then used to focus future identification and management of health and safety within this operation. For example, assume an inventory of chemicals is obtained during the survey. At some future time the regulations change requiring additional actions for use of a specific chemical; the chemical inventories can then be reviewed to identify those operations which then need assistance.

Surveys must be objective and consistent. The surveyor must have good knowledge of the operation and the job responsibilities of the worker. The surveyor must also have a good idea of what he or she expects to see—or not see. One of the best tools for this purpose is a checklist specifically designed for the operation. It is easy to allow familiarity to affect objectivity, the surveyor must guard against losing the ability to observe.

Hazard Reporting. In the agriculture, forestry, and fishery industries, communication of hazardous conditions upward from the organization's general worker population is the second critical need in striving for proactive hazard identifications. Workers, because they are more intimate with their work environment, will be the first to observe conditions which may cause injury or illness. This same closeness may result in a familiarity which may cause the worker to not connect a situation with a hazardous condition; however when other parts of the health and safety program are effective in providing adequate levels of awareness, the effects of this familiarity factor will be minimized. For a hazard reporting mechanism to be functional, the following are essential:

- It must be standardized and well-understood by all employees. The method should be explained in orientation and periodic training sessions.
- It must not be burdensome. An employee will be disinclined to use the reporting mechanism if he or she must spend an inordinate amount of time or effort to comply. Reporting forms should be readily available and easy to understand.
- It must support the organization's chain of supervision since many health and safety problems can be resolved in lower levels of management. Some employees, however, may be concerned about submitting notification of a potential problem to one's supervisor. The possibility of such a report "disappearing" before reaching the health and safety manager—or worse, reprisals against the employee—must be eliminated administratively and perceptually.
- It must ensure a response that is appropriate, adequate, and timely. The person submitting the report must be informed of the actions involved with assessing the reported problem and the corrective actions when necessary.

In any employee hazard reporting system, questions about anonymity may arise. Allowance for anonymous reporting should be made in the case that an employee does not want to call attention to him or herself; however, employees should be urged to view reporting as an integral part of the health and safety program and not as complaining. The causative factors in a hazardous condition are much easier to evaluate with the help of the identifier. Granted, some reports will be indicative of employee relations problems. These should be managed as such. When these are apparent or when the demand for anonymity exceeds perhaps 5 to 10% of the reports, there may be management problems at play which must be solved before the hazard reporting system can be effective.

Design Review. This third hazard identification activity is clearly anticipatory. Design reviews are basically evaluations of any plans for engineering or architectural modifications to ensure that any problems or omissions in the design which affect health or safety are brought to light for discussion and resolution. In this review, consideration must be given not only to how the end product of the design may affect safety and health, but the effects of the construction or fabrication process as well.

The organization must ensure that the input of the health and safety program is sought early in the design process. Often, specific knowledge about deficiencies in current operations or equipment can be used to eliminate or circumvent the same problem in the new design.

Hazard Assessment

Once a hazardous condition is identified, the next step is to evaluate it through both qualitative and quantitative means. The objective is to characterize the condition sufficiently to know whether it presents a significant safety or health hazard and, if so, what can be done to control it.

Various mechanisms can be used to measure the specific factors in any hazardous condition. The mechanism should be selected based on the qualities of the hazard under the conditions they are presented. Some possible assessment strategies are shown by Tables 13.2 and 13.3.

In making judgments about the significance of a health or safety problem, it is a good idea to keep in mind the difference between the terms *Hazardous Condition* and *Hazard*. A hazardous condition has a potential to cause injury or illness under certain circumstances; however a hazard exists only when this potential is realized.

One example of this concept involves the presence of an agricultural chemical as the hazardous condition. Chemicals, as most know, may be highly toxic materials, yet a 200-gallon bulk storage tank of the chemical does not in itself present a hazard. Conditions which allow the toxic nature of this material to be realized may include rupture of the tank or dispensing, mixing, or applying the chemical. Other conditions which may result in a safety hazard include improper storage or transport of dispensed quantities.

Table 13.2. Health hazards.

Hazard	Most Likely Exposure	How Assessed
Chemical	Skin exposure—may cause intoxication, skin damage, or chronic illness	Observation of use Inspection of PPE Medical surveillance
	Ingestion—may cause acute poisoning or chronic illness	Observation of opportunity
	Inhalation—may cause acute poisoning or chronic illness	Observation of use Air sampling of representative workers using validated methods Measurement of ventilation Medical surveillance
Noise	May cause temporary or permanent hearing loss	Direct Measurement Dosimetry Audiometry
Heat	May cause acute illness—heat stroke/heat exhaustion	Direct measurement Measurement of ventilation

Table 13.3. Safety hazards.

Hazard	Most Likely Exposure	How Assessed
General	Slips/trips/falls	Observation Direct measurement
	Cuts/punctures/crush injuries	Observation of guarding, fail-safes Direct measurement Inspection of tools
	Falls	Observation of work Inspection of PPE
Chemical	Fire and explosion	Inventory of quantities Observation of incompatibilities Observation of chemical use Observation of ignition sources Tests of suppression systems
Electrical	Fire and shock	Measurement of voltage, polarity, and ground quality Inspection of equipment and tools Observation of work practices Inspection of PPE
Ergonomic	Cumulative trauma	Observation of work practice Inspection of tools and equipment Measurement of vibration

Another good example of an operation with multiple hazardous conditions would be a sawmill. With remotely operated, automated log handling and sawing equipment, there is a low potential for injury unless the controls are circumvented for maintenance reasons.

Hazard assessment, therefore, is the process of inventorying hazardous conditions, evaluating their potential effects on safety and health through observation and measurement, and finally making a judgment about the probability for this potential to be realized. Any strategy employed to prioritize hazards must take into account:

- The type of injury, illness, or other loss most likely to be caused by the hazardous conditions.
- The likelihood that circumstances will come about which allow the mishap to happen.

Hazard Control

Control of hazards is, in some respects, the easiest step in the process of preventing injuries and illnesses, but only if the identification and assessment steps were adequate. Once you know that you have a problem, the factors that have resulted in the problem, and a good understanding of its significance, you are well on your way toward a resolution. On the other hand, health and safety problems, when misidentified or poorly characterized, may never be resolved, even after much time and expense.

Hazard control mechanisms usually are grouped into three categories, *Engineering Controls, Personal Protective Equipment (PPE),* and *Administrative Controls.* It is important to realize that control is not an on/off switch. You don't just chose a method and apply it. Control involves a gradient of actions from *effective/expedient* toward the objective of *most effective/most efficient.* These actions will include a combination of measures from the three control categories which will evolve as control is gained.

Engineering Controls. Engineering controls attempt to remove the hazard from the work environment, to disable or render it harmless, or to guard it in such a way as to provide a barrier between the worker and the hazard. Engineering controls are required, by regulation, to be the primary method of control when feasible.

Some good examples of engineering controls are:

- Ventilation of grain handling areas to prevent buildup of explosive atmospheres.
- Rollover protection and fault-actuated kill switches on powered agricultural equipment.
- Anti-kick back devices and spark arrestors on chain saws.
- Proximity sensors which stop an operation when a worker approaches too closely.

There are many examples of good controls, but probably many more examples of poorly designed or applied controls. Avoid problems when implementing any engineering control by considering the some of the well-known pitfalls.

A control should be simple. A complicated control is more expensive, difficult to install and maintain, more likely to misfunction, and will confuse workers and maintenance personnel.

A control must not significantly interfere with the operation. A poorly designed control can negatively affect production, or even create a new hazard. Workers will always have a bit of reluctance in accepting the constraints of a newly implemented control; however, they will rapidly accept it if it is functional and noninterfering. Otherwise, they will find a way of disabling or removing the control. On the other side of the organization, any control that measurably affects output may be unacceptable to management.

A control should be automatic and reliable. Controls should run in the background and not rely on frequent or complicated human intervention to actuate or verify. Their function must be relied on and workers must be warned of nonfunctional conditions. Again, controls may be circumvented if they are seen as a burden. Even worse, if a control is nonfunctional yet perceived to be working, a worker may develop a false sense of security which could lead to high potential for an incident.

A control should not cause additional hazards. Like the above recommendations, this is not always possible to achieve. In every case, however, the planning process must consider all hazards inherent in the control design in order to minimize their effect.

Personal Protective Equipment. The agricultural, forestry, or fisheries industries, unlike other industries, are less likely to have fixed workplaces. Because of this, these industries have less reliance on engineering controls and proportionately more on the use of Personal Protective Equipment (PPE). It is important, nevertheless, that engineering controls are considered and a rational determination is made that they are not feasible prior to considering PPE. If engineering controls are planned, PPE will certainly be important for worker protection during the design and installation. It will serve additionally to augment other control strategies for specific hazards.

Many types of PPE are available as shown by Table 13.4. PPE is seldom universal in its applicability. It must be selected with care to ensure that it is functional as protection against the specific hazard, it is sized correctly for the individual who must wear it, and it will be reasonably comfortable for the intended duration of wear.

When PPE is required for a particular operation, its use must be monitored to ensure that workers are wearing it correctly and that it continues to provide effective protection.

Administrative Controls. Administrative controls are those that attempt to limit or eliminate worker exposure to hazards by means other than PPE or engineering controls. This type of control relies on the scheduling or work tasks and events, developing detailed work practices or procedures which are known to reduce hazard, or in substitution of a hazardous material with a less hazardous replacement. Some good examples are:

- Scheduling crop spraying for times when no field work is being done.
- Prohibiting access to a fumigation area until the process cycle is complete.
- Rotating workers through certain operations to reduce their daily cumulative exposure to an agricultural chemical.
- Requiring the buddy system for certain hazardous operations.

Administrative measures are seldom the sole mechanism of control. They should be employed to augment or enhance the other PPE or engineering controls and not used to avoid compliance with regulatory requirements for training or medical surveillance.

Table 13.4. Personal protective equipment.

Hazard	Recommended PPE
Heavy Impact or Falling Object	Head: hard hat Knees, Legs, and Ankles: fiber metal leggings Feet and Toes: steel box toe shoes or foot guards
Moderate Impact	Head: hard hat Feet and Toes: steel box toe shoes
Large Flying Particles or Objects	Head: hard hat Eyes: goggles, spectacles with side shields Face: plastic face shields Fingers, Hands, and Arms: leather gloves or mittens, sleeves Trunk: leather or canvas fiber aprons, coats, or jackets Knees, Legs, and Ankles: leather, fiber metal, or flame-resistant duck pants, knee guards, shin guards, leggings or spats
Small Flying Particles	Head: abrasive blasting hood Eyes: goggles, spectacles with side shields, plastic eye shields Fingers, Hands, and Arms: leather or duck fabric gloves or mittens, sleeves Trunk: leather or canvas fiber aprons, coats, or jackets Knees, Legs, and Ankles: leather, fiber metal, or flame-resistant duck pants, knee guards, leggings, or spats
Dusts	Eyes: goggles, spectacles with side shields, plastic eye shields Face: plastic face shields Respiratory: approved dust, airline, or abrasive blasting respirator
Sparks and Metal Spatter	Head: cotton or wool cap (flame retardant treated) Eyes: goggles, spectacles with shields, plastic eye shields Face: plastic face shields Fingers, Hands, and Arms: leather flame resistant duck or aluminum fabric gloves or mittens, sleeves Trunk: leather aprons, coats, or jackets Knees, Legs, and Ankles: leather fiber metal, flame-resistant duck pants, knee guards, shin guards, leggings, or spats
Splashing Metal	Eyes: goggles, spectacles with side shields, plastic eye shields Face: wire screen shield Fingers, Hands, and Arms: leather flame-resistant duck or aluminum fabric gloves or mittens, sleeves Trunk: leather aprons, coats, or jackets Knees, Legs, and Ankles: leather fiber metal or flame-resistant duck pants, knee guards, shin guards, leggings, or spats Feet and Toes: leather shoes, foundry shoes
Splashing Liquids and Chemicals	Head: plastic-rubber hat Eyes: goggles, hoods Face: plastic face shields, hoods Respiratory: chemical-resistant suits with air supply Fingers, Hands, and Arms: rubber, natural rubber, plastics, synthetic fabrics, coated glass fiber, or other chemical-resistant gloves or mittens, sleeves Trunk: rubber, plastic, or other chemical-resistant material Knees, Legs, and Ankles: rubber, plastic, or other chemical resistant material Feet and Toes: nonskid shoes with rubber, neoprene, or wood soles, rubber or neoprene overshoes Whole Body: coveralls, overalls, or liquid hazard suite of impervious materials
Mists, Vapors, Gases, Fumes, and Smoke	Eyes: goggles Face: plastic face shields for mists Respiratory: Immediately Dangerous to Life: self-contained apparatus Respiratory: Not Immediately Dangerous to Life: air line respirator, chemical cartridge respirator with filter for specific contaminant Fingers, Hands, and Arms: rubber, natural rubber, plastic synthetic fiber, coated glass fiber, or other chemical-resistant gloves or mittens, sleeves. Trunk: rubber, plastic, or other chemical-resistant material Knees, Legs, and Ankles: rubber, plastic, or other chemical resistant material Feet and Toes: conductive shoes (for explosive gases, vapors, or other materials)
Hot Materials	Fingers, Hands, and Arms: leather gloves, mittens, hand pads, or finger cots; leather, or flame resistant duck sleeves Trunk: leather aprons, coats, or jackets Knees, Legs, and Ankles: leather, or flame resistant duck pants, knee guards, shin guards, leggings, or spats.

Table 13.4. *(continued)*

Hazard	Recommended PPE
Hot Materials *(continued)*	Feet and Toes: wood soles
Heat	Head: cotton or wool cap Fingers, Hands, and Arms: leather aluminized fabrics, glass fiber insulated gloves, mittens, or hand pads; flame-resistant duck or reflective fabric (for radiant heat) sleeves Trunk: flame-resistant fabrics, aluminized fabrics for radiant heat Knees, Legs, and Ankles: flame-resistant duck, aluminized fabrics for radiant heat Feet and Toes: leather or wood soles, thermal insulated shoes Whole Body: aluminized garments for radiant heat, vortex tube with air cooled suits, cool vests
Moisture and Water	Head: plastic-rubber hat Fingers, Hands, and Arms: rubber, oiled fabrics, plastic, coated glass fiber gloves, mittens or finger cots, rubber oiled fabrics, or plastic sleeves Trunk: rubber or plastic material Knees, Legs, and Ankles: rubber or plastic material Feet and Toes: nonskid shoes, leather or wood soles, rubber or neoprene overshoes Whole Body: garments of rubber, plastic, or other impervious material
Slips and Falls	Feet and Toes: nonskid shoes, wood soles, slip-resistant soles and heels (cord and cork) Whole Body: Climbing spikes and belts, safety lanyards
Cuts and abrasions	Head: hard hat Fingers, Hands and Arms: leather, metal or Kevlar® mesh, or cotton canvas gloves, mittens, hand pads, or finger cots, leather sleeves Trunk: leather or canvas fiber aprons, coats, or jackets Knees, Legs, and Ankles: leather or fiber metal pants, knee guards, shin guards, leggings, or spats Feet and Toes: steel box toe, wood soles
Dermatitis	Head: plastic-rubber hat, cotton or wool cap Face: plastic face shield, protective barrier creams Fingers, Hands, and Arms: rubber, synthetic rubber plastic or cotton gloves, protective barrier creams Trunk: rubber or plastic material Knees, Legs, and Ankles: rubber or plastic material Feet and Toes: rubber boots, wood soles, shower sandals (paper or wood)
Electricity and Electric Shock	Head: plastic-rubber or plastic hat (Class B) Fingers, Hands, and Arms: rubber gloves and sleeves resistant to 10,000 volts for three minutes Trunk: rubber material Knees, Legs, and Ankles: rubber material Feet and Toes: electrical hazard footwear
Explosives	Head: cap (flame retardant) Fingers, Hands, and Arms: gloves Trunk: powder uniform Feet and Toes: conductive shoes
Machinery	Head: cap (long hair), cotton, or wool caps Fingers, Hands, and Arms: flame-resistant duck sleeves Trunk: rubber, plastic, or canvas fiber aprons, coats, or jackets Knees, Legs, and Ankles: fiber material or flame-resistant duck pants, knee guards, shin guards, leggings, or spats Feet and Toes: steel box shoes
Reflected Light and Glare	Eyes: goggles, spectacles with side shields with filter lenses
Welding	Head: leather skull cap Eyes: goggles-welders' eyecup, helmets, or hand shields with filter lenses Face: helmets or hand shields with filter lenses, face shield Hands: flame proof gauntlet gloves
Radiant Energy (intense)	Eyes: helmets (filter lenses with metal or plastic spectacles, hand shields (filter lenses) with metal or plastic spectacles Face: helmets or hand shield with filter lenses
Laser Radiation	Eyes: protective eyewear
Noise	Ears: plug or insert, cup or muff ear protectors, helmet

Recordkeeping

Proper records are essential to any health and safety program. Although you may design your recordkeeping activities around regulatory requirements, there are other reasons for doing a good job in this function. Good records allow you to develop a history of health and safety concerns which can be used for a number of reasons:

- Analysis injury and illness trends.
- Establishment of representative baselines for operations.
- Documentation of safe conditions in claims against the organization.
- Tracking of the corrective action process.

Types of Records. *Injury and Illness Logs* must be kept for all incidents that are work related. These logs must be summarized at the end of each calendar year and communicated to the affected employees. Detailed procedures for gathering and maintaining these records is contained in *Code of Federal Regulations, Title 29, Part 1904*.

Employee Medical and Exposure Records must be kept if they are established in response to a regulatory requirement, an injury or illness, or the implementation of any hazard assessment process that quantifies exposures to a workplace hazard.

Training Records should be maintained for any type of training given to employees to enhance their awareness of health and safety hazards in their work areas. The records should include not only attendance records but a course outline or other description of the training.

Training

Each organization must provide a health and safety program designed to empower each employee to be a partner in the prevention of workplace injury and illnesses. Workers who are provided information to enhance their understanding of the workplace hazards they may encounter will develop an attitude of safety awareness that will carry through to their work practices. An effective training presentation may be general or specific, but will always cover the following:

- Recognition of hazards which may be found in the work area.
- Discussion of the factors of work tasks that affect safety or health
- Use of proper practices, procedures, or control strategies.
- Response to new or unexpected hazards.

Most work assignments will demand some safety training or orientation. The factors related to the specific hazards should dictate the training method and the depth of the coverage. Training should also be tailored to the level of employee; a pesticide applicator may need specific information on the chemical used, using PPE, etc., while his or her supervisor will need more training on control measures, administrative procedures, or how to conduct basic safety training.

SUMMARY

Health and Safety Programs are nothing more than the systematic application of effective management techniques to the problem of protecting the human and physical resources of an organization. Through the thoughtful planning, organizing, staffing, directing, and controlling, the organization can define its philosophy, appropriate safety policies and procedures, and support structure in a manner which integrates with its overall mission.

The functions of an effective Health and Safety Program include the identification and evaluation of hazards, design and implementation of control measures, recordkeeping, and training.

REFERENCES

Boylston, R., Jr. 1991. *Managing Health and Safety Programs.* New York: Van Nostrand Reinhold.

Clayton, G.D., and F.E. Clayton, eds. 1991. *Patty's Industrial Hygiene and Toxicology. Volume I: General Principles.* 4th edition. New York: John Wiley & Sons, Inc.

Garrett, J.T., L.J. Cralley, and L.V. Cralley, eds. 1988. *Industrial Hygiene Management.* New York: John Wiley & Sons, Inc.

George, C.S., Jr. 1972. *The History of Management Thought.* Englewood Cliffs, NJ: Prentice-Hall, Inc.

Harrison, L., ed. 1994. *Environmental, Health, and Safety Auditing Handbook.* 2nd edition. New York: McGraw-Hill, Inc.

Robey, D. 1982. *Designing Organizations: A Macro Perspective.* Homewood, IL: Richard D. Irwin, Inc.

Saccaro, J.A. 1994. *Developing Safety Training Programs.* New York: Van Nostrand Reinhold.

Terrell, M.J. 1995. *Safety and Health Management in the Nineties: Creating a Winning Program.* New York: Van Nostrand Reinhold.

SECTION 2:

SAFETY AND HEALTH IN AGRICULTURE AND ANIMAL PRODUCTION

14

EPIDEMIOLOGY OF AGRICULTURAL INJURIES AND ILLNESSES

Mark A. Purschwitz, Ph.D.
Department of Biological Systems Engineering
University of Wisconsin-Madison

Each year several hundred agricultural workers are fatally injured and thousands experience nonfatal injuries or develop an illness due to farm work. Machinery, primarily the tractor, is the leading agent of fatal farm injuries. Many studies have found that young children and elderly farm workers have a high incidence of fatal farm injuries. This chapter reviews the epidemiology of injuries that occur on the farm including the agents responsible for the injuries, the ages and sex of the victims, the months of occurrence, and the nature of the injuries. An overview of the various types of illnesses and chronic conditions attributed to farm work is discussed including zoonotic infections, cancer, respiratory disorders, dermatitis, hearing loss, and musculoskeletal disorders.

INTRODUCTION

The epidemiology of agricultural injuries and illnesses can be categorized into three main areas: fatal traumatic injuries; nonfatal traumatic injuries; and illnesses and chronic conditions.

FATAL TRAUMATIC INJURIES

The epidemiology of fatal traumatic agricultural injuries is more clear than the other epidemiologic areas, as fatality data has been easier to obtain, and thus more readily available, than morbidity data.

Totals and Rates

The National Safety Council (NSC), using data on unintentional work deaths from the Bureau of Labor Statistics' Census of Fatal Occupational Injuries (CFOI), reported 833 agricultural work-related fatalities in 1993 and 779 in 1992. This includes agriculture, forestry, and fishing, based on Standard Industrial Classification (SIC) codes 01, 02, 07, 08, and 09. Using this CFOI data and their own estimation procedures, NSC estimated 890 agricultural fatalities for 1994 and a death rate of 26 per 100,000 workers (NSC 1995). The denominator was based on BLS Current Employment Estimates data; only deaths to persons age 14 and older were included.

Myers and Hard (1995), reporting on 1980 to 1989 data from the National Traumatic Occupational Fatality (NTOF) database of the National Institute for Occupational Safety and Health (NIOSH), found 5,823 fatalities in agricultural production, for an annual average of 582 and a death rate of 22.9 per 100,000 workers. An additional 904 deaths were identified in the agricultural services sector. The denominator was based on BLS data on workers employed in agriculture; only deaths to persons age 16 and older were included.

In contrast, the number of unintentional work deaths in 1994 for all industries was estimated at 5000, with a death rate of 4 per 100,000. Other industries of note were mining and quarrying (160 deaths and a rate of 27 per 100,000); construction (910 deaths and 15 per 100,000); and transportation and public utilities (740 deaths and 12 per 100,000)(NSC 1995).

Agent of Injury

A variety of agents of injury result in farm mortality. These include tractors and other farm machines; animals; trees and woodcutting; electrocution; confined spaces such as manure pits, silos, and grain bins; other structures such as barns and sheds; falls; tools; motor vehicles used for farm use; drowning; fire; chemicals; and others. Farmers perform a variety of tasks, such as using an array of farm machinery; raising large animals; clearing trees; using electrical devices;

handling, storing and removing farm products and wastes; constructing buildings; handling chemicals; maintaining and repairing machines and structures; working near bodies of water; and other potentially dangerous tasks. This is all in addition to the management responsibility necessary for any business.

Myers and Hard (1995) reported the three leading agents of fatal injury in production agriculture, based on NTOF data, were:

machinery including tractors	41.7%
motor vehicles	17.4
electrocution	6.1

Bernhardt and Langley (1993) reported on 123 North Carolina farm fatalities during 1984 to 1988, based on medical examiner data, and found the agent of injury to be as follows:

tractors	61.8%
machines	13.8
electrocutions	7.3
falling objects	6.5
other	10.6

Purschwitz and Skjolaas (1995) reported 121 fatal farm injuries in Wisconsin for 1992 to 1994, based on newspaper clippings and death certificate reports from the state vital statistics agency. They listed agent of injury as:

tractors	34.7%
machines	25.6
animals	9.9
tree felling	5.0
confined spaces	4.1
trucks	4.1
falls	3.3
other	13.2

Sheldon et al. (1995) described the agents of fatal injury to children under age 18 in Wisconsin and Indiana. Of the 460 victims, the agents of injury were as follows:

tractors	43.7%
other machinery	24.8
confined spaces	7.2
farm trucks	4.3
falls	3.9
drownings	2.4
other	13.7

Tractors and other machines are the primary agents of fatal farm injury, resulting in more deaths than any other agents. Separating out the tractor from other farm machines shows that tractors are the single most common agent of fatal injury on U.S. farms.

Stallones (1990) conducted a hand-search of death certificates for fatal injuries occurring on farms in Kentucky. This included fatalities involving house fires, firearms and drownings. Farm equipment (including tractors) still accounted for almost 49% (237/486) of the fatalities.

Examples of other states and provinces reporting tractor-related fatalities as the leading proportion of their fatal farm injuries, for various numbers of years, are as follows (Purschwitz 1992):

Pennsylvania	54.6%	(106/194)
Ontario	49.2	(207/421)
Minnesota	43.7	(51/119)
Indiana	37.0	(112/303)
Nebraska	31.8	(238/749)

Tractors are involved in several incidents that commonly result in death or serious injury. The single most common fatal incident is the tractor rollover, when the tractor overturns on its operator (and passengers if any). Tractor rollovers cause more farm-related deaths than any other single incident. For example, in Kentucky in 1994, 28 tractor-related fatalities were reported; of these, 23 were tractor rollovers (CDC 1995a). Bernhardt and Langley (1993) cited unpublished data revealing that 55.8% (191/342) of North Carolina tractor-related fatalities were rollovers.

Another common incident is the tractor runover, where the victim is run over by the wheel of the tractor. These runovers often occur when a person falls from the tractor, sometimes as an extra rider. Another scenario is when the operator starts the tractor from the ground, typically standing in front of a rear wheel; if safety systems are bypassed or inoperable, the tractor can start up and run over the victim. A third runover scenario is for an unseen bystander to be run over; this is especially a problem for young children who may be nearby but do not perceive the danger.

Examples of states and provinces reporting tractor rollovers and runovers as a percentage of their tractor fatalities, for various number of years, are as follows (Purschwitz 1992):

	tractor fatalities	rollovers	runovers
Georgia	202	75.7%	13.9%
Indiana	112	67.8	18.8
Nebraska	238	64.7	35.3
Pennsylvania	106	60.4	27.4
Ontario	207	59.9	21.7
United States	2,439	47.7	26.9

Other farm machines are also involved in many fatal incidents. Farm machines are designed to grab, control, slice, chop, wrap, convey, compress, spread, crush, or otherwise process agricultural materials, and the functional components can cause great damage to the human body. Machines typically require significant power transmission, and these power transmission components can also inflict severe damage. Entanglement of the victim in a machine component may result in fatal injury from mechanical damage or blood loss.

Another fatal machinery incident is the runover, where a victim is run over by a machine. This might occur after falling from the tractor pulling the machine, falling from the machine itself, or being an unseen bystander. Persons working beneath the machine may be fatally crushed if it becomes unstable or the support mechanism fails.

Examples of states and provinces reporting machinery-related fatalities as a percentage of their farm fatalities, for various numbers of years, are as follows (Purschwitz 1992):

Indiana	22.4%	(77/303)
Minnesota	22.7	(27/119)
Ontario	16.7	(68/421)
Pennsylvania	14.4	(28/184)

Animals can be an agent of fatal injury on farms. Farm animals, while domesticated, are typically larger than humans, and can inflict damage either intentionally or unintentionally. Bulls cause several fatalities each year; dairy bulls in particular are often extremely vicious and territorial. Cows can be very protective of newborn calves. Horses, particularly stallions, can fatally attack people, or kick when frightened. Boars can inflict serious damage with their tusks, and a sow with new piglets can be very protective and bite.

Langley (1994) found that of 92 animal-related deaths in North Carolina during 1972 to 1989, 18 were work-related. Nine of those were cattle, 5 of which were bulls. Rabl and Auer (1992) reported on 10 cases of farm animal-related fatalities in Austria seen at one medical center since 1975, including death of an unborn child (the mother was kicked by a cow). Two of the 10 cases involved being hit by a bull; 7 involved being kicked or hit by a cow; and one involved being hit by a sheep.

Farmers die in confined spaces from toxic gases, lack of oxygen, or suffocation following physical entrapment. Kelley and Field (1995) identified 236 fatal entrapments in grain bins or grain transport vehicles in 24 states and provinces over a 31-year period. Examples of individual confined space incidents include two manure pit incidents reported in 1992 in Minnesota, just three days apart, which killed two farmers, a son, and an uncle (CDC 1993a). Three fatalities from silo gas were reported in a single incident in Ontario (Canada Plan Service 1982). Purschwitz and Skjolaas (1994) reported a fatality from entering an oxygen-limiting silo. An Iowa farm owner died of carbon monoxide poisoning while using a gasoline-powered high-pressure washer in an unventilated swine building (CDC 1993b).

Age and Sex of Victim

Farm injuries kill persons of all ages. Victims have been as young as 2 months. The author has personally seen the death certificate of a 9-month old who was run over by a tractor. At the upper end of the age spectrum, victims may be in their 90s.

Myers and Hard (1995) found that of the 6,711 agricultural production and service sector fatalities from the NTOF database in 1980 to 1989, 6,610 (98.5%) involved males, and only 101 (1.5%) involved females. They reported the age distribution and fatality rates (per 100,000 workers) for males as follows:

age 16-24	13.6%	15.3/100,000 workers
25-34	17.6	19.6
35-44	13.3	20.6
45-54	14.7	26.8
55-64	18.2	31.6
65+	22.6	60.5

Clearly, from these data, older farmers are at risk. Unlike many other occupations, which have a retirement age at or near 65, farmers do not have to retire. They can continue to farm as long as they are able, especially with a scaled-down operation. Furthermore, because older farmers often help children or neighbors, particularly at the busiest times of planting or harvest, they can continue to be exposed to farm hazards their entire lives.

Examples of states and provinces reporting fatalities to workers age 65+, as a percentage of their farm fatalities, for various numbers of years, are as follows (Purschwitz 1992):

Pennsylvania	33.0%	(63/191)
Indiana	28.1	(85/303)
Minnesota	27.7	(33/119)
Nebraska	15.6	(117/349)
Ontario	15.4	(65/421)

Many children die each year of farm-related injuries. Rivara (1985) reported 300 fatalities per year, but these included all fatal injuries occurring on farms, whether recreational or not. However, the mere fact that children die in farm work or worksite related injuries is evidence of the serious nature of farm injuries. Farms are homes as well as places of work and recreation. Because children live on farms, and often enter or are taken into the workplace, they are exposed to similar farm hazards as the adults.

Examples of states and provinces reporting fatalities to children age 15 or younger, as a percentage of their farm fatalities, for various numbers of years, are as follows (Purschwitz 1992):

Wisconsin	28.6%	(247/863)
Nebraska	20.6	(154/749)
Minnesota	19.3	(23/199)
Ontario	19.0	(80/421)
Pennsylvania	17.3	(33/191)

(age 14 and younger)

Month of Occurrence

Fatalities tend to increase during the months of exposure, e.g., the busy months of planting and harvest—or typically spring and fall. In states where the harvest season begins in summer for hay crops, the number of fatalities during summer months tends to be elevated as well. Fatalities among children tend to be higher in the summer months when they are off from school and assisting in farm work or present in the workplace.

In the following table, Murphy (1991) reported the month of occurrence for 194 Pennsylvania farm fatalities during 1985 to 1989. Steel (1994) reported on 1,479 farm fatalities to farmers age 55 and over, from 1980 to 1993, using consolidated National Safety Council data from a variety of states and sources. Tormoehlen (1986) reported on 247 farm fatalities to children age 15 and under in Wisconsin during 1970 to 1984.

	Murphy	Steel	Tormoehlen
January	5.2%	5.1%	2.0%
February	1.0	3.6	4.9
March	10.3	4.5	2.4
April	8.8	7.4	7.7
May	7.7	10.5	5.3
June	13.9	10.5	17.0
July	10.3	12.9	20.2
August	10.8	12.0	13.0
September	13.4	10.6	6.9
October	8.8	12.8	10.1
November	6.2	6.1	7.3
December	3.6	4.1	3.2

NONFATAL TRAUMATIC INJURIES

Nonfatal injuries occur much more often than fatal injuries. However, due to the lack of reporting systems and mandated reporting, data are typically available only through surveys or other special studies, or through data collection mandated by a particular state law.

Totals and Rates

Gerberich et al. (1993) conducted an intensive population-based survey of farm households in five north central states. They found one farming-related injury for every five farms (19.4 per 100 farms), and 58.3 injuries per 1,000 persons. The rate for males was 86.1 per 1,000; the rate for females was 24.8 per 1,000. The highest injury rate for any subgroup was 158.5 per 1,000 persons, for males age 30 to 39. The highest rate for females was for age 20 to 29, 53.7 per 1,000 persons.

Nordstrom et al. (1995) conducted a 2-year population-based prospective study of farm work-related injuries in a large clinic and hospital serving central Wisconsin, where the farms are predominantly dairy. A unique demographic and medical records linkage system was used, along with an agricultural census. There were 510 cases of injury, of which 62% occurred to farm residents, and 38% to nonfarm residents. The rate of injury for residents of dairy farms was 326 cases per 10,000 person-years, compared with 141 cases per 10,000 person-years for residents of farms other than dairy. The overall rate of injury for the entire population of the surveillance area was 51.8 cases per 10,000 person-years. Adult male farm residents had rates of 556.9 injuries per 10,000 person-years and 21.3 injuries per million hours of farm work.

Pratt et al. (1992) conducted a 2-year prospective study of 600 farmers and farm workers on 201 dairy farms, and found that 151 persons suffered 200 injuries, for a rate of 166 injuries per 1,000 worker-years. Owner-operators were injured the most often; they accounted for 62% of injuries despite making up only 33% of the workers. Relatives (e.g., spouses or children) of the owner/operators accounted for 24% of the injuries, and hired workers accounted for 14%. Owner/operators had an work-hour injury rate of 1.54 injuries per 1,000 hours of work, whereas relatives had a rate of 0.89 per 1,000 hours and hired workers a rate of 0.50 per 1,000 hours.

Zhou and Roseman (1994) surveyed Alabama farms for unintentional injuries involving agricultural activities, and found 71 injuries among 718 farmers, or a cumulative incidence of 9.9% per year. The 71 injuries were distributed among 56 people. The rate based on work hours was 4.15 per 100,000 farming hours.

The Minnesota Farming Health Survey reported 106 injuries in the farm households sampled, for a rate of 78 injuries per 1000 farm household members (CDC 1995b).

Schelp (1992) studied 1,500 Swedish farms within an area with health care facility surveillance of all injuries. There were 174 farmwork-related injuries in one year. Farmers had the highest annual injury rate, 88.5 work-related injuries per 1,000 workers. Workers in manufacturing were second, with a rate of 75 per 1,000. Of the farmers and farm workers, owners had a rate of 98.8 per 1,000, compared with hired workers of 37.7 per 1,000.

Agent of Injury

The types of agents of injury for nonfatal injury are similar to those of fatal injury, but different agents are at the top of the list with regard to the number of injuries caused. Animals are a major agent of nonfatal injury. Farm machines are also major agents; tractors are, too, but they are not the major agent of nonfatal injury as they are for fatal injury. Morbidity data will vary by type of agriculture; for example, an area that produces few livestock would not be expected to have many animal-inflicted injuries.

Pratt et al. (1992) found agents of injury as follows:

machinery	35%
animals	32
falls	8
respiratory	7
chemicals	2
other	16

Stueland et al. (1990) conducted close surveillance of farm-related injuries in a large clinic and hospital in central Wisconsin. The farming in this area is primarily dairy. In a 2-year period there were 913 patients with such injuries. Agents of injury for the 913 cases were as follows:

animals	22.5%
falls	17.4
tractors, implements	13.8
tools	11.7
fixed hazards	
(e.g., structures)	7.7
exposure	5.6
other vehicles	3.3
other	18.0

Lee et al. (1995) noted that for the 210 victims (of the 913 above) who were female, 40% of the injuries involved animals, and less than 10% involved tractors or implements.

Zhou and Roseman (1994) found machinery accounted for 28.6% of injuries; falls 23.2%; and animals 12.5%. There was a single case of acute chemical poisoning among the 71 injuries (1.8%).

Stueland et al. (1991) reported on the 246 farm-related injuries to children (age less than 19 years) which were part of the previously mentioned study of 913 cases. Agents of injury included tractors, implements, structures, animals, tools, falls, and exposure. For the children under age 6, falls accounted for nearly half the injuries, while for children aged 6 to 18, falls accounted for less than 15% of the injuries. Injuries to males aged 6 to 18 primarily involved tractors, implements, structures, and tools, while for females aged 6 to 18, injuries primarily involved animals.

Nature of Injury

Stueland et al. (1990) found the injuries suffered by the 913 cases (some had more than one injury) distributed as follows:

contusion/abrasions	25.3%
lacerations	24.1
fractures	18.4
foreign bodies	6.2
puncture wounds	6.2
corneal abrasions	6.0
head injuries	5.1
cellulitis	5.1
amputations	2.7
burns	1.8
back injuries	1.3
other	14.0

There can be somewhat unusual, yet serious, injuries on farms. Proust (1993) described hand injuries involving high pressure injection. Roerig (1993) reported on four cases involving women being scalped by entanglement of their hair in a rotating shaft. Neidich (1993) described fourteen cases of child ingestion of alkaline dairy pipeline cleaner.

Age of Victim

Gerberich et al. (1993) reported rates of injury per 100,000 hours worked by age as follows:

	males	females
5-9 years	11/100,000 hrs.	6/100,000 hrs.
10-14	8	7
15-19	6	2
20-24	3	2
25-29	5	8
30-39	6	6
40-49	4	4
50-59	4	4
60-69	5	5
70+	5	1
all ages	5	5

Pratt et al. (1992) found that injured workers tended to be older, worked more hours, and had heavier workloads. Peak ages were 31 to 40 and 51 to 60. The 463 men in the study accounted for 181 injuries, for an annual rate of 195/1000 workers, while the 137 women accounted for only 19 injuries, for an annual rate of 69/1000.

Stueland et al. (1990) reported the following age distribution of the 913 injury victims:

0-18:	27.0%
19-25:	13.5
26-45:	35.7
46-65:	19.2
>65:	4.6

Lee et al. (1995) reported that for the 210 female victims (of the 913 described above), the peak 10-year age category was 10 to 19, as opposed to the peak 10-year category for males of 20 to 29.

For the 246 children (age less than 19 years) who were among the 913 victims, Stueland et al. (1991) reported that 48 involved children less than 6 years old. Peak age groups for injury were 4 to 8 and 14 to 18.

Month of Occurrence

From the New York dairy study by Pratt et al. (1992), about 30% of the injuries occurred in the summer and about 33% in the fall. Zhou and Roseman (1994) noted that in Alabama 16.3% of injuries occurred in September, and 14.3% in the spring, the two peak times.

June and July correspond to busy hay making periods on Wisconsin dairy farms, and Stueland et al. (1990) found the following distribution of the 913 injuries by month:

January	4.4%
February	5.3
March	6.6
April	9.1
May	8.1
June	15.5
July	11.0
August	10.4
September	9.7
October	7.6
November	6.1
December	6.2

Acute Pesticide Poisoning

Acute agricultural work-related pesticide poisoning is not a common event, compared with other agricultural trauma. Nordstrom et al. (1995) noted that in the central Wisconsin dairy area the agent of injury was 100 times more likely to be mechanical than toxic. However, acute poisoning does occur, particularly in states where much hand labor is done, and because it may be perceived as being more common than it is, it deserves discussion.

Maddy et al. (1990) reviewed 40 years of California data on acute illness/injury and death associated with pesticide exposure from 1949 to 1988. They felt that California has some of the world's best data on this type of incident, although there are shortcomings in the reporting, and concluded that although pesticide usage is estimated to have increased four-fold during the 40 years, the number of pesticide-related occupational illnesses per year has increased very little. Cholinesterase inhibitors and methyl bromide were most often involved with serious occupational systemic poisoning during that time. From 1878 to 1987, there was an annual average of one occupational death, 15 suicides, and an estimated 5 nonsuicide, nonoccupational deaths, from acute pesticide poisoning. In 1987, 1,507 cases of potentially pesticide-related occupational illness were identified, with only 744 of these demonstrating some systemic toxic symptoms. Poison control centers handled approximately 17,000 pesticide exposures that year, almost all of which were nonoccupational.

According to the U.S. Centers for Disease Control (CDC 1994), there were 1154 reported cases of organophosphate poisoning in California from 1982 to 1990, 495 of which were from one insecticide (mevinphos). Between June and August 1993, 26 cases of mevinphos illness were reported in Washington orchards.

London et al. (1994) studied pesticide poisoning records in one area of South Africa for 1987 to 1991; such poisonings were reportable to the national health department. Of 225 cases reported, only 22 were occupational accidents on farms; there were 80 suicides. Underreporting was noted, however; of 135 cases of pesticide poisoning admitted to provincial hospitals, only 30 had notifications traceable to the health department.

ILLNESSES AND CHRONIC CONDITIONS

Occupational illnesses on farms include respiratory illnesses, cancer, dermatitis, and zoonoses. Chronic conditions include hearing loss and musculoskeletal disorders.

Respiratory Illnesses

There are numerous farm respiratory hazards. These include molds and dusts from grain, hay, and silage; animal dander; toxic gases and particulate matter from livestock wastes; toxic gases, dust, or oxygen deprivation in crop storage structures; vapors from fuels and solvents; pesticide vapors; and welding fumes. Some can result in acute toxic exposures from respiratory poisoning or asphyxiation, while others result in illnesses; this section discusses the latter.

Von Essen (1991) provides an excellent overview of illnesses due to airborne dusts on farms. These include hypersensitivity pneumonitis, organic dust toxic syndrome (ODTS), chronic bronchitis, asthma, and mucous membrane inflammation. Brackbill et al. (1994) combined 5 years of National Health Interview Survey Data and found, for chronic respiratory disease, farmers had an age-adjusted prevalence risk ratio (PRR) of 1.3 versus other currently employed workers.

Zejda et al. (1993a) summarized eight representative cross-sectional studies of chronic respiratory symptoms in farmers and found that in some studies 5 to 21% of farmers had chronic bronchitis, over 20% had chest wheeze, 0.4 to 1.5% had hypersensitivity pneumonitis, and 6 to 15% had ODTS.

Malmberg (1990) reported that in Scandinavian farm populations, 40 to 50% have symptoms of hypersensitivity of the eyes, nose, or lungs, 2 to 3% have asthma, and an additional 3 to 4% have symptoms of airway obstruction provoked by exposure to dust. Senthilselvan and Dosman (1995) found 8.1% of Saskatchewan farmers self-reported allergic rhinitis.

Swine Producers. Swine producers have respiratory hazards from their operations. Donham (1990) reviewed 14 studies from the U.S., Sweden, Canada, and the Netherlands, and found the prevalence of respiratory symptoms in swine workers to be 2 to 4 times that of comparison groups. Pulmonary function studies have showed decrements in flow rates over the workshift, with minor decrements in flow volume. Symptoms and prevalence rates from the studies as reviewed by Donham were:

Chronic Symptoms:

cough:	12-55%
phlegm:	12-55
wheezing and tightness of chest:	12-33
shortness of breath:	12-33
ODTS:	10-30

Acute Symptoms (directly associated with presence in the working environment):

cough:	38-75%
wheezing and tightness of chest:	20-55

Donham et al. (1990a) found that confinement swine workers had significantly higher rates of cough, phlegm, chest tightness, shortness of breath, and eye/nose/throat irritation associated with work than did nonconfinement swine workers.

Carvalheiro et al. (1995) found increased incidence of ODTS, mucous membrane irritation, and chronic bronchitis among farmers working with swine or other animals. Cormier et al. (1991) found a slight but increased prevalence of chronic bronchitis among swine building workers versus nonfarming referents in Quebec, and more evidence of airflow obstruction. Subjects working more than 3 hours per day had a higher prevalence of chronic bronchitis than workers with fewer hours (21.9% vs. 13.3%).

Zejda et al. (1993a) found that Canadian swine producers had significantly more symptoms of chronic bronchitis than grain farmers or nonfarming controls. Over one-fourth of the swine producers had "asthmatic tendencies", wheezing without having a cold. There were significant effects on lung function based on the daily duration of work; the longer the workday, the more symptoms and reduced lung function. Schwartz et al. (1992) noted that swine producers suffered airway injuries manifested by air trapping, enhanced airway response to methacholine challenge, and thickening of the basement membrane.

Iverson et al. (1990) reported 28.3% of 1175 Danish swine farmers had symptoms of shortness of breath, wheezing, and dry cough, and 10.9% had asthma. These figures were higher than for dairy farmers (7.4% shortness of breath, wheezing, dry cough; 5.5% asthma). Age was found to be an important factor; 3.6% of farmers aged 31 to 50 had asthma and 17.9% had chronic bronchitis, while 11.8% of farmers aged 51 to 70 had asthma and 33.0% had chronic bronchitis.

Dairy Producers. Dairy producers are also at risk. Rylander et al. (1990) found in a Swedish study that swine producers and dairy producers reported flu-like symptoms (reported as organic dust toxic syndrome), cough during work, and throat and nose irritation. Swine producers reported more chest tightness and eye irritation. Marx et al. (1993) reported a "significant increase in reactivity to certain inhaled allergens, among those dairy farmers reporting barn-associated respiratory symptoms, that is unrelated to past exposures to causative agents of farmer's lung disease." In a 1984 evaluation of over 1500 dairy farmers in the Marshfield Clinic study area of central Wisconsin, a prevalence rate of farmers lung was found of 4.2/1000/year in 1977, and a combined prevalence rate of farmers lung and organic toxic dust syndrome of 8.82/1000/year (Marx et al. 1990).

Dalphin et al. (1994) found French dairy farmers who used barn drying systems on their forage, resulting in fewer microorganisms in the barn air, clearly had better respiratory function than those who used traditional storage methods. Dalphin et al. (1993) found an excess of chronic bronchitis in dairy farmers, and indicated it occurs more frequently in farmers with previous episodes of acute lung reactions.

Husman et al. (1990) found 13.6% of 2866 Finnish farmers had one or more acute incidents of ODTS related to work, not including 23 farmers who had symptoms of farmers lung. Fifty-two percent acquired symptoms less than 4 hours after exposure. Cattle tending was significantly more associated with ODTS than swine production. Holness and Nethercott (1989) reported a prevalence rate of allergic rhinitis of 21% among workers on primarily dairy and beef farms in Ontario.

Poultry Producers. Poultry producers are at risk of respiratory problems comparable to swine producers. Donham et al. (1990b) found poultry workers had decrements of lung airway function significantly associated with the workshift, indicating that, like swine workers, the main health effect appears to be airway inflammation manifested by bronchitis and perhaps hyperactive airways. Reynolds et al. (1993) found symptoms in turkey producers to include cough, phlegm production, wheezing, breathlessness, and allergies; symptoms are more frequently reported in winter, when the concentration of airborne agents is higher. Persons raising turkeys for more than 10 years had a higher prevalence of symptoms, and the prevalence increased with increased years of work.

Grain Producers and Handlers. Grain dust is also a respiratory hazard. Bernhardt and Langley (1993) said grain dust can cause acute respiratory inflammatory response— nasal stuffiness, rhinorrhea, sore throat, acute bronchitis, occupational asthma—as well as ODTS, hypersensitivity pneumonitis, and eye and skin irritation.

Hurst et al. (1990) found Saskatchewan grain farmers had significantly more cough, phlegm, shortness of breath, chest tightness, and lower lung flow rates than control groups. According to Hurst and Dosman (1990), "Grain dust should be regarded as a dust with toxic properties, not just a nuisance dust." They found that grain handlers had a dose-dependent relationship between total dust exposure and decreased lung function, particularly lower flow rates and more chronic bronchitis, and that grain handlers had a annual decline in lung function greater than that of control groups. They also found that 6 to 30% of grain farmers get grain fever, which resembles influenza, during or shortly after exposure to grain dust.

Terho (1990) found that one Finnish farmer in ten had either asthma, chronic bronchitis, or farmers lung, and these were associated with farming types that involved handling grain or feed, or the combined effects of grain and animal dusts. In Canada, Senthilselvan et al. (1993) found a significant association of grain farming with asthma and other symptoms in men, and wheezing in women. There was an increasing trend with duration of farming.

Zejda et al. (1993b) summarized five respiratory cross-sectional studies of grain elevator workers and found, depending on the study, that 17 to 49% had chronic bronchitis, 28 to 42% had chest wheeze, and 16 to 32% had ODTS. Rublaitus et al. (1994) found that 16.5% of Ohio grain farmers experienced flu-like symptoms in connection with dusty work in the previous year; 10.5% experienced chest tightness in connection with work, and 63.6% had experiences some respiratory symptoms attributable to dust work environments.

Mushroom Workers. Hypersensitivity pneumonitis, or farmers lung, is associated with mold spores from a variety of crops. Sanderson et al. (1992) reported mushroom workers lung. Approximately 20% of the more heavily exposed workers on a Florida mushroom farm reported occasional symptoms; 7 workers had to either leave their jobs or transfer to an isolated area in order to be free of symptoms.

Cancer

General. Blair and Zahm (1991) reviewed the literature regarding cancer among farmers. Of 357 descriptive studies reported, 165 showed increased relative risks, but only 44 were statistically significant. There were 73 studies showing statistically significant decreased relative risks. The review indicates that, overall, farmers are at less risk for death from cancer than the general population. Risks of such common cancers as lung, bladder, colon, and esophagus are consistently lower. However, farmers seem to be at elevated risk of several cancers, such as leukemia, Hodgkins disease, non-Hodgkins lymphoma, multiple myeloma, and cancers of the lip, stomach, skin, prostate, brain, and connective tissue. It was noted that except for cancer of the lip, the elevated risk was relatively small, with considerable variability among studies.

Blair et al. (1992) performed meta analyses for relative risk (RR), and reported statistically significant excesses for the following with 95% confidence intervals:

lip	RR=2.08 (1.80-2.40)
Hodgkin's disease	1.16 (1.03-1.29)
skin melanoma	1.15 (1.04-1.21)
multiple myeloma	1.12 (1.04-1.21)
stomach	1.12 (1.09-1.14)
prostate	1.08 (1.06-1.11)
leukemia	1.07 (1.03-1.11)

The RR and 95% CI for those cancers reported with non-significant excesses were:

non-Hodgkins lymphoma	RR=1.05 (0.98-1.12)
cancer of connective tissue	1.06 (0.91-1.24)
brain	1.05 (0.99-1.12)

They found substantial and significant deficits for all cancers combined; for cancers of the lung, esophagus, bladder, colon, liver and kidney; for ischemic heart disease; and for total mortality. They concluded that several of the tumors excessive in farmers are rising in the general population and are also excessive in immunodeficient patients; the tumors vary in frequency, histology, and prognosis; and the tumors do not fall into obvious groups.

Blair et al. (1993) conducted a proportionate mortality ratio (PMR) study using 5 years' of death certificates in 23 states, for farmers in crop and livestock farming. They found deficits from all cancer combined, as well as cancers of the lung and liver. They also found a number of cancer excesses, which varied by race, sex, and region of the country. Proportionate cancer mortality ratios, with 95% confidence intervals, for those cancers reported as excessive among white male farmers were:

pancreas	1.14 (1.07-1.20)
bone	1.34 (0.99-1.77)
lip	2.31 (1.43-3.53)
prostate	1.18 (1.14-1.22)
kidney	1.10 (1.01-1.19)
eye	1.58 (0.92-2.52)
brain	1.15 (1.05-1.26)
thyroid	1.34 (0.95-1.83)
all lymphatic and hematopoietic tissue	1.22 (1.18-1.27)

Burmeister (1990) computed proportionate mortality ratios for white male Iowa farmers for 1979 to 1986 compared with nonfarmers, and found elevated PMR's for the following cancers (no confidence intervals were given):

lip	1.18
large intestine	1.03
pancreas	1.04
skin	1.33
brain	1.22
lymphoma	1.22
multiple myeloma	1.00
other blood-related	1.23
Hodgkins	1.45
prostate	1.24
overall cancer	1.03

Austin et al. (1995) calculated PMRs for Alabama farmers, and found lower PMRs for all malignant neoplasm when compared to both the Alabama and U.S. populations, including respiratory, digestive, and lymphopoietic. They found elevated PMRs for all external causes of death.

Dean (1994) found in Ireland that farmers, farm managers, and relatives assisting farmers do not have any significantly increased risk of dying from brain, lymphatic, or hematopoietic cancers compared with the general population. Farm laborers actually had fewer than expected deaths from brain or hematopoietic cancer. They concluded there is no significant increase in multiple myeloma deaths, that males have fewer brain, lymphatic, and hematopoietic cancers, and when both sexes are taken together, fewer leukemia deaths.

Franceschi et al. (1993) studied cancer in Italian farmers and concluded there was a significant relative risk for cancer of the oral cavity and pharynx (OR=1.6), and observed a nonsignificant elevated OR of 1.5 for pancreatic cancer and 1.6 for Hodgkins disease. They observed significantly decreased ORs of 0.6 for colon cancer and 0.6 for bladder cancer. Females had a significant elevated risk of multiple myeloma of 2.4.

Forastiere et al. (1993) conducted a case-referent study in Italy of cancer deaths, and found farmers had decreased risk of lung and bladder cancer and melanoma, and nonsignificant excess risk of stomach, rectal, kidney, and nonmelanoma skin cancer. There was an excess risk of stomach and kidney cancer for farmers with more than 10 years experience, and of stomach, rectal, and pancreatic cancers among licensed pesticide users with greater than 20 years experience.

Possible Associations of Cancer with Pesticides. Chronic exposure to pesticides, insecticides, herbicides, and fungicides has been linked in various studies with elevated risk of some cancers. Ongoing studies are necessary to link exposure to cases and test the significance and strength of the linkages. In a discussion of cancer and chlorophenoxy herbicides, Bond and Rossbacher (1993) noted the differences between ecological, case-referent, and retrospective cohort studies:

1. Ecological studies do not consider exposure and disease relationships at the individual level, and are useful only for generating hypotheses.
2. Case-referent studies are difficult to do well, and may be weakened by selection bias or exposure misclassification due to recall or interviewer bias. With regard to chlorophenoxy herbicides, there is a "remarkable" variability in risk estimates and between herbicide use and Hodgkins disease or soft-

tissue sarcoma, and some studies show an association with non-Hodgkins lymphoma, while others do not.
3. Retrospective cohort studies need large numbers of subjects to be able to detect or rule out modest increases in risk of uncommon diseases.

They concluded that there is some suggestive evidence linking increased risk of cancer with chlorophenoxy herbicides, but the evidence is inconsistent and far from conclusive. They said animal studies have given no proof of carcinogenicity, and recommended large prospective case-referent studies.

Cantor et al. (1992) found an association between pesticide use and non-Hodgkins lymphoma in Iowa and Minnesota, with odds ratios of 1.2 for men who had ever farmed, and/or of 1.5 or more for those who had personally handled, mixed, or applied any of several pesticide groups. Stronger associations were found for first use prior to 1965, or when personal protective equipment was not used.

However, Zahm et al. (1993a) reported three case-control studies in Iowa, Minnesota, Nebraska, and Kansas, and concluded that it was unlikely that exposure to atrazine a triazine herbicide explained any appreciable amount of observed increase in non-Hodgkins lymphoma. In fact, in a later publication, Zahm et al. (1995) noted that the risk among farmers who personally handled atrazine was lower than those who used it on their farms but did not handle it. In an ecologic study in Ontario, McLaughlin (1995) found no evidence of association between lymphoma and phenoxy herbicides like 2,4-D, but a general pattern of lower relative risks in counties with higher quantities of pesticide use, which suggested that pesticide use has not affected lymphoma rates in Ontario. The only statistically significant associations were with insecticides, "other herbicides" (other than triazine or phenoxy), and "all pesticides combined".

Fleming and Timmeny (1993) conducted a literature review of aplastic anemia cases which reported an association with pesticide exposure. They concluded that an etiologic association with organochlorine and organophosphate exposures is suggested, but pointed out that pesticide exposure can be missed if not purposely sought, and that selective questioning of persons with the disease can introduce overreporting bias because cases are more likely to report exposures than controls.

Morrison et al. (1994) provided updated results of a 1971 to 1987 study of 155,547 farmers in three western Canadian provinces. The mortality rate for all major causes of death was low (SMR=0.72). For non-Hodgkins lymphoma, the SMR was 0.79. Based on census data, estimated relative risks for individual farmers increased, but not significantly, as the number of acres sprayed increased; the increase was significant when more than 380 hectares (939 acres) were sprayed. There was no significant risk increase based on updated data on fuel/oil expenditures. They concluded that any exposure-disease relationship between 2,4-D (a phenoxy herbicide) and non-Hodgkins lymphoma is "at best tenuous".

In Iowa case-control studies, Burmeister (1990) found nonsignificant elevated odds ratios for multiple myeloma and pesticide exposure, which he said "hints at risk". He found no linkages between leukemia and dairy farming.

Figa-Talamanca et al. (1993) studied licensed Italian agricultural pesticide applicators and concluded there was a significantly elevated mortality ratio for brain cancer, but not a significant increase in haematolymphopoetic cancers.

Hansen at al (1992) conducted a 10-year prospective study of 4,015 employed gardeners in greenhouses, nurseries, and public parks and gardens, an occupation often involving pesticides. They found only a slightly elevated, nonsignificant standard mortality rate (SMR) of 104 (95% CI=91-120) for all cancers. The SMR for lymphatic and hematopoietic cancers was 1.43 (CI=83-229), and for soft-tissue sarcomas was 455 (CI=94-1328) but these were not statistically significant, except for a soft tissue sarcoma SMR for men of 526 (CI=109-1538), based on three cases. For lymphatic leukemia the SMR for both sexes combined was a nonsignificant 251 (CI=92-546), but for men was 275 (101-594), based on six cases. For non-Hodgkins lymphoma, there were six cases among men and two among women, for nonsignificant SMRs of 173 (63-376) for men, 364 (44-1314) for women, and 200 (86-393) for all.

Garry et al. (1994) surveyed Minnesota pesticide applicators and grouped them into users of a) herbicides only; b) herbicide/insecticides/fumigants; and c) fungicides. They compared prevalence of chronic diseases such as hypertension, heart disease, skin cancer, neurologic disorders, kidney disease, asthma and other lung disorders, and allergies, and found the same rate as reported for the general population in national health survey data. Between groups, the fumigant-user group had significantly higher prevalence of the diseases.

Special Populations. McDuffie (1994) reviewed a representative sample of studies of agricultural exposures and cancer among women. Reported in the 21 studies, which vary in size and method, were excesses of non-Hodgkins lymphoma, leukemia, multiple myeloma, soft-tissue sarcoma, and cancers of the breast, ovary, lung, bladder, cervix, and sinonasal cavities. She concluded that there may be a number of host factors that alter disease risk, but that sufficient evidence exists to mount campaigns aimed at greater use of protective equipment and preventive measures.

Zahm et al. (1993b) did a case-control study in eastern Nebraska and found no increased risk of non-Hodgkins lymphoma in women who ever lived or worked on a farm. Neither use of herbicides nor insecticides was associated with NHL. There were small nonsignificant associations for the relatively few women who personally handled insecticides (odds ratio 1.3) or herbicides (OR 1.2); for those women handling organophosphates, there was a statistically significant OR of 4.5. Pesticide-related risks appeared to be greater among women with family histories of cancer.

In a study of cancer in women in Sweden, Linet et al. (1994) found an elevated risk of 1.9 for chronic myeloid leukemia among female farmers, and of 5.3 for women employed in animal breeding.

Dosemeci et al. (1994) studied death certificates in three states of African-American farmers/farm managers/farm supervisors/farm workers, and found an elevated odds ratio for cancer of 1.4 compared with all other occupations. They said farming contributed one-third of the excess risk.

Zahm and Blair (1993), noting that most cancer risk studies are done on the farm owner/operator, summarized 11 descriptive and analytic studies on migrant workers. The studies found elevated risks for various cancers as follows (no numeric values given):

1. liver
2. lung
3. lung, buccal cavity and pharynx, bone, soft tissue, skin
4. liver, larynx
5. multiple myeloma, stomach, liver, cervix
6. prostate, lung
7. no cancer excess
8. overall cancer excess vs. farmers
9. prostate, liver
10. liver
11. testis

Gold and Severs (1994) examined a group of studies with respect to childhood cancers, and concluded that there is no overwhelming or strongly consistent evidence for an association of any tumor site with parental agricultural occupational exposure to pesticides. The studies varied with respect to risks found for different tumors and statistical significance.

Holly et al. (1992) studied paternal occupations of 43 cases of Ewings bone sarcoma, a rare tumor that is the second most common bone tumor in children. Paternal occupations involving agriculture were found in seven cases versus five controls, for an adjusted relative risk of 8.8 95% (CI=1.8-42.7). They reported another study with a RR of 3.0 for paternal agricultural occupation. They said they "did not address how paternal exposure could affect the child".

Pearce and Reif (1990) discussed overall cancer mortality, cancers with increased or decreased risk, chemicals, viruses, and various populations. They said, "Given the wide variety or exposures, it is perhaps not surprising that the reasons for the increased risks of certain cancers for agricultural workers remain unclear."

Other Illnesses

Parkinson's Disease. Hertzman et al. (1994) conducted a case-control study of Parkinson's disease and found a significant association with occupations in which handling or contacting horticultural (tree fruit) pesticides (odds ratio 2.3), but no specific chemicals were associated with the disease. They concluded that occupations involving agricultural chemical use may predispose people to development of Parkinson's, but it is likely that pathogenesis is multi-factorial rather than related to a specific agent. Hubble et al. (1993) concluded that pesticide use can be considered a risk factor, but not a cause of, development of Parkinson's.

Birth Defects. Lin et al. (1994) found no significant association between parental farming occupations or potentially pesticide-exposed occupations and overall limb-reduction defects; there was a weak positive association with "associated limb defects", but negative association with isolated limb defects. They found no association between limb reduction defects and residence in a farming or high-pesticide-use county.

Nurminen et al. (1995) used data from a complete Finnish birth defects registry and found a small excess of orofacial clefts related to maternal agricultural work in the first trimester, but was not able to associate these defects with the agricultural chemicals of concern. A slightly elevated risk of central nervous system defects was found, but none for skeletal defects. Rowland (1995), commenting on the study and others, said that evidence of pesticide causation of human birth defects is weak. He said that interpretation is difficult because of statistical adjustments made, differences between cases and controls, and said that "authors overstate their ability to estimate total exposure."

Dermatitis. Hand labor which requires hand contact with fruits or vegetables can result in dermatitis from the plant product itself, or from contact with chemicals used on the plants. Dermatitis can also result from contact with solvents or cleaners. In addition, in those cases where impervious gloves or boots must be worn, dermatitis can result from the warm, moist conditions inside the gloves or boots.

Schenker and McCurdy (1990) reported that skin disorders comprised over two-thirds of the agricultural occupational illnesses reported to the U.S. Bureau of Labor Statistics in 1984. The rate for migrant workers was 33.4 cases per 10,000 person-years, or almost three times the rate for manufacturing employees and over five times the rate for all private sector employees. In a case-control study of Italian agricultural workers authorized to use pesticides, Cellini and Offidani (1994) found 12% reported dermatitis, versus 6% of nonagricultural controls.

Susitaival et al. (1994) found that 16% of women and 7% of men on Finnish farms reported dermatoses; risk factors were daily handling of disinfectants, handling silage preservatives, milking cows, and serving machinery. Finkelstein at al (1994) reported an outbreak of phytodermatitis from hand harvest of celery with high psoralen contact, substances which cause a toxic dermal reaction upon exposure to ultraviolet radiation.

Zoonoses. Farmers are at risk for zoonoses if they come into contact with diseased animals or their body fluids, breathe certain aerosols from diseased animals, receive needle-stick wounds during treatment of livestock, or come into contact with droppings from diseased birds or rodents. These include such diseases as leptospirosis, a bacterial infection that may be life threatening in rare cases; lyme disease; brucellosis, a potentially serious bacterial infection from infected animals or raw milk, with a wide range of symptoms; ovine enzootic abortion, a strain of chlamydia in infected sheep which can lead to women aborting in their third trimester; and ornithosis/psittacosis, a flu-like disease inhaled from infected birds and which can lead to heart inflammation, hepatitis, or pneumonia (Williams 1994). Contact with mice or mouse droppings is a major risk factor for hantavirus in the southwest U.S. (Zeitz et al. 1995).

In a serological investigation of past zoonotic infection among randomly-selected farmers in Northern Ireland, Stanford et al. (1990) found evidence of antibodies to a number of agents, including brucella, leptospira, and hantavirus. Thomas et al. (1994) reported on a case of leptospirosis from an unvaccinated dairy herd in New Zealand. Leptospirosis results in headache, nausea, fatigue, vomiting, fever, and blurred vision, and is transmitted through blood, urine, tissue, or direct contact with mucous membranes or abrasions.

Lengerich et al. (1993) reported in Wisconsin that dairy farmers and other persons in contact with cattle are at a greater risk of Cryptosporidium infection than persons without such contact. Fishbein and Raoult (1992) reported in France a case of Q-fever transmitted from a vaccinated herd of goats to workers and those who drank unpasteurized milk from the herd. Peel et al. (1991) reported on actinobacillus bacteria isolated from wounds due to horse and sheep bite.

Not to be overlooked is tetanus, which can be contracted from many farm hazards. Luisto and Seppalainen (1992) reported on Finnish cases and said the farming population was especially at risk. For example, tetanus was contracted by farm family members from splinters, a cow stepping on a foot, animal bites, being hit by flying objects, and scratches or lacerations from machinery and stones.

Green Tobacco Sickness. Persons harvesting tobacco by hand have been known to suffer from "green tobacco sickness", which may cause nausea, vomiting, weakness, diarrhea, headache, dizziness, and sometimes fluctuation in heart rate (McKnight et al. 1994; Hipke 1993). The route of exposure is dermal; as the leaves are harvested, sap coats the workers hands, and dissolved nicotine is absorbed through the skin.

Tuberculosis. Tuberculosis has been shown to be a problem in some migrant labor camps. In a random sample of migrant farmworkers in North Carolina, Ciesielski et al. (1991) found active TB in 3.6% of U.S.-born black workers and 0.47% of Hispanic workers. Positivity tests occurred in 76% of Haitians, 54% of U.S.-born blacks, and 33% of Hispanics.

Chronic Conditions

Hearing Loss. The relationship of farm noise to hearing loss is well-established. Farm machines, along with some animal confinement structures, produce loud noise which can result in severe cumulative hearing loss. A number of investigators have reported increased hearing loss in farmers compared with other occupational groups or the general population (Donham and Horvath 1988).

May et al. (1990) randomly selected and tested 49 full-time dairy farmers and found high frequency average threshold hearing loss of greater than 20dB in 55% of right ears and 65% of left ears. The left ear is the ear facing the tractor exhaust when looking over the right shoulder at a trailing implement. Ejercito et al. (1989) reported hearing loss for male and female farmers at the 4000 Hz level of 54 and 32 dB, respectively, compared with losses of 30 and 22 dB for nonfarm men and women. Broste et al. (1989) have carefully documented increased prevalence of hearing loss among high school students actively engaged in farm work. Holt et al. (1993) documented high noise levels for operators of farm tractors.

Musculoskeletal Disorders. Holness and Nethercott (1995) found that 71% of a self-selected sample of Ontario swine producers reported chronic back problems. Dupuis (1989) indicated of 429 patients with lower back pain, of various occupations, 48% of the farmers in the group said the onset of their pain was prior to age 30. Bobick and Myers (1992) reported on 52,527 workers' compensation claims from agricultural workers in 1985 to 1986, and found that 36.4% were sprain/strain injuries. Meyers et al. (1995) discuss cumulative trauma disorders and ergonomics applications in agriculture.

Comparing farmers versus other currently employed workers, Brackbill et al. (1994) found a statistically significant age-adjusted prevalence risk ratio (PRR) of 1.4 for farmers and arthritis. Nordstrom (1994) reported the risk of osteoarthritis to be 4.3 times greater in farmers aged 20 to 60 than white-collar workers, and cited a Swedish study showing male farmers to have 3.78 times the risk as urban males. May (1994) said that osteoarthritis of both knee and hip are common problems for dairy farmers aged 55 to 65.

Migrant workers perform much manual labor and are at risk of musculoskeletal injuries. Mobed et al. (1992) reported that musculoskeletal complaints ranked second and third of all physical problems reported on two different surveys of migrant workers.

SUMMARY

Farm operators, their family members, farm employees, and others who may visit or help on the farm, are all at risk of traumatic injuries or work-related illnesses. There are many potential agents of injury and illness. Tractors and other machines are leading agents of fatal injury, and also are involved in many nonfatal injuries. Animals are a leading agent of nonfatal injury. Dusts and molds can lead to respiratory illnesses. Farmers have deficits of some types of cancers and increased prevalence of others, possibly related to chemical exposure. Chronic conditions such as hearing loss and back pain are prevalent among farmers and farm workers.

REFERENCES

Austin, S.G., N. Huang, and C.W. Woernie. 1995. PMR study of mortality among Alabama workers and farmers. *Amer. J. Ind. Med.* 27(1): 29-36.

Bernhardt, J, and R.L. Langley. 1993. Agricultural hazards in North Carolina. *N. C. Med. J.* 54(10): 512-15.

Blair, A., M. Dosemeci, and E.F. Heineman. 1993. Cancer and other causes of death among male and female farmers from twenty-three states. *Amer. J. Ind. Med.* 23(5): 729-42.

Blair, A., S.H. Zahm, N.E. Pearce, E.F. Heineman, and J.F. Fraumeni, Jr. 1992. Clues to cancer etiology from studies of farmers. *Scand. J. Work Envir. Health* 18(4): 209-15.

Blair, A., and S.H. Zahm. 1991. Cancer among farmers. *Occup. Med.* 6(3): 335-54.

Bobick, T.G., and J.R. Myers. 1992. *Back Injuries in Agriculture: Occupations Affected.* NIFS Paper 92-01. Columbia, MO: National Institute for Farm Safety.

Bond, G.G. and R. Rossbacher. 1993. A review of potential human carcinogenicity of the chlorophenoxy herbicides MCPA, MCPP, and 2,4-DP. *Brit. J. Ind. Med.* 50(4): 340-48.

Brakbill, R.M., L.L. Cameron, and V. Behrens. 1994. Prevalence of chronic diseases and impairments among U.S. farmers 1986-1990. *Amer. J. Epidem.* 139(11): 1055-65.

Broste, S.K., D.A. Hansen, R.L. Strand, and D.T. Stueland. 1989. Hearing loss among high school farm students. *Amer. J. Public Health* 79(5): 619-22.

Burmeister, L.F. 1990. Cancer in Iowa farmers: recent results. *Am. J. Ind. Med.* 18(3): 295-301.

Canada Plan Service. 1982. *Silo gas.* Plan M-7410. Ottawa, Canada: Canada Plan Service, Agriculture.

Cantor, K.P., A. Blair, G. Everett, R. Gibson, L.F. Burmeister, L.M. Brown, L. Schuman, and F.R. Dick. 1992. Pesticides and other agricultural risk factors for non-Hodgkin's lymphoma among men in Iowa and Minnesota. *Cancer Res.* 52(9): 2447-55.

Carvalheiro, M.F., Y. Paterson, E. Rubenowite, and R. Rylander. 1995. Bronchial reactivity and work-related symptoms in France. *Amer. J. Ind. Med.* 27: 65-74.

Cellini, A., and A. Offidani. 1994. An epidemiological study of cutaneous diseases of agricultural workers authorized to use pesticides. *Derm.* 189(2): 129-32.

Centers for Disease Control and Prevention . 1995a. Farm-tractor-related fatalities—Kentucky, 1994. *MMWR* 44(26): 481-84.

Centers for Disease Control and Prevention. 1995b. Eye injuries to agricultural workers—Minnesota, 1992-1993. *MMWR* 44(18): 364-66.

Centers for Disease Control and Prevention. 1994. Occupational pesticide poisoning in apple orchards—Washington, 1993. *MMWR.* 42(51-52): 993-95.

Centers for Disease Control and Prevention.1993a. Fatalities attributed to entering waste manure pits—Minnesota, 1992. *MMWR* 42(17): 325-29.

Centers for Disease Control and Prevention. 1993b. Unintentional carbon monoxide poisoning from indoor use of pressure washers—Iowa, January 1992-January 1993. *MMWR* 42(40): 777-79; 785.

Ciesielski, S.D., J.R. Seed, H.D. Espusito, and N. Hunter. 1991. The epidemiology of tuberculosis among North Carolina migrant farmworkers. *JAMA* 265(13): 1715-19.

Cormier, Y., L.P. Boulet, G. Bedard, and G. Tremblay. 1991. Respiratory health of workers exposed to swine confinement buildings only or to both swine confinement buildings and dairy barns. *Scand. J. Work Envir. Health* 17(4): 269-75.

Dalphin, J.C., J.C. Polio, D. Pernet, M.F. Maheu, B. Toson, A. Dubiez, E. Monnet, J.J. Laplante, and A. Depierre. 1994. Influence of barn drying of fodder on respiratory symptoms and function in dairy farmers of the Doubs region of France. *Thorax* 49(1): 50-3.

Dalphin, J.C., D. Pernet, A. Dubiez, D. Debieuvre, H. Allemand, and A. Depierre. 1993. Etiologic factors of chronic bronchitis in dairy farmers. Case control study in the Doubs region of France. *Chest* 103(2): 417-21.

Dean, G. 1994. Deaths from primary brain cancers, lymphatic and haematopoietic cancers in agricultural workers in the Republic of Ireland. *J. Epidem. Comm. Health* 49(4): 364-68.

Donham, K.J. 1990. Health effects from work in swine confinement buildings. *Amer. J. Ind. Med.* 17: 17-25.

Donham, K.J., J.A. Merchant, D. Lassise, W.J. Popendorf, and L.F. Burmeister. 1990a. Preventing respiratory disease in swine confinement workers: intervention through applied epidemiology, education, and consultation. *Amer. J. Ind. Med.* 18(3): 241-61.

Donham, K.J., B. Leistikow, J. Merchant, and S. Leonard. 1990b. Assessment of U.S. poultry worker respiratory risks. *Amer. J. Ind. Med.* 17: 73-74.

Donham, K.J., and E.P. Horvath. 1988. Agricultural occupational medicine. In: *Occupational Medicine.* 2nd edition. C. Zenz, ed. Chicago, IL: Year Book Medical Publishers, p. 935.

Dosemeci, M., R.N. Hoover, A. Blair, L.W. Figgs, S. Devesa, D. Grauman, and J.F. Fraumeni, Jr. 1994. Farming and prostate cancer among African-Americans in the southeastern United States. *J. Natl. Cancer Inst.* 86(22): 1718-19.

Dupuis, P.R. 1989. Low back pain in the farmer: an outline. In: *Principles of Health and Safety in Agriculture.* J.A. Dosman and D.W. Cockcroft, eds. Boca Raton, FL: CRC Press, pp. 370-72.

Ejercito, V.S., D.A. Hansen, and W.E. Pierce. 1989. Prevention of hearing loss among farmers. In: *Principles of Health and Safety in Agriculture,* J.A. Dosman and D.W. Cockcroft, eds. Boca Raton, FL: CRC Press, pp. 327-28.

Figa Talamanca, I., I. Mearelli, P. Valente, and S. Bascherini. 1993. Cancer mortality in a cohort of rural licensed pesticide users in the province of Rome. *Int. J. Epidem.* 22(4): 579-83.

Finkelstein, E., U. Afek, E. Gross, N. Aharoni, L. Rosenberg, and S. Halevy. 1994. An outbreak of phytophotodermatitis due to celery. *Int. J. Dermatol.* 33(2): 116-18.

Fishbein, D.B., and D. Raoult. 1992. A cluster of Coxiella burnetii infections associated with exposure to vaccinated goats and their unpasteurized dairy products. *Am. J. Trop. Med. Hyg.* 47(1): 35-40.

Fleming, L.E., and C.J. Timmeny. 1993. Aplastic anemia and pesticide: an etiologic association? *J. Occup. Med.* 35(11): 1106-16.

Forastiere, F., A. Quercia, M. Miceli, L. Settimi, B. Terenzoni, E. Rapiti, A. Faustini, P. Borgia, F. Cavarini, and C.A. Perucci. 1993. Cancer among farmers in central Italy. *Scand. J. Work. Env. Health* 19(6): 382-29.

Franceschi, S., F. Barbone, E. Bidoli, S. Guarneri, D. Serraino, R. Talamini, and C. La Vecchia. 1993. Cancer risk in farmers: results from a multi-site case-control study in northeastern Italy. *Int. J. Cancer* 53(5): 740-45.

Garry, V.F., J.T. Kelly, J.M. Sprafka, S. Edwards, and J. Griffith. 1994. Survey of health and use characterization of pesticide appliers in Minnesota. *Arch. Env. Health* 49(5): 337-43.

Gerberich, S.G., R.W. Gibson, L.R. French, P. Carr, C.M. Renier, P.D. Gunderson, F. Martin, J. A. True, J. A. Shutske, and K. Brademeyer. 1993. The regional rural injury study - I (RRIS-I): A population-based effort. Report to the Centers for Disease Control, Atlanta, GA.

Gold, E.B., and L. E. Sever. 1994. Childhood cancers associated with parental occupational exposures. *Occup Med.* 9(3): 495-539.

Hansen, E.S., H. Hasle, and F. Lander. 1993. A cohort study on cancer incidence among Danish gardeners. *Am J Ind Med.* 21(5): 651-60.

Hertzman, C., M. Weins, B. Snow, S. Kelly, and D. Calne. 1994. A case-control study of Parkinson's disease in a horticultural region of British Columbia. *Mov. Discord.* 9(1): 69-75.

Hipke, M.E. 1993. Green tobacco sickness. *South Med. J.* 86(9): 989-92.

Holly, E.A., D.A. Aston, D.K. Ahn, and J.J. Kristiansen. 1992. Ewing's bone sarcoma, paternal occupational exposure, and other factors. *Amer. J. Epidemiol.* 135(2): 122-29.

Holness, D.L., and J.R. Nethercott. 1995. What actually happens to the farmers? Clinical results of a follow-up study of hog confinement farmers. In: *Agricultural Health and Safety: Workplace, Environment, Sustainability.* H.H. McDuffie, J. A. Dosman, K. M. Semchuk, S. A. Olenchock, and A. Senthilselvan, eds. Boca Raton, FL: CRC Press, pp.269-74.

Holness, D.L., and J.R. Nethercott. 1989. Respiratory status and environmental exposure of hog confinement and control farmers in Ontario. In: *Principles of Health and Safety in Agriculture.* J. A. Dosman and D. W. Cockcroft, eds. Boca Raton, FL: CRC Press, pp. 69-71.

Holt, J.J., S.K. Broste, and D.A. Hansen. 1993. Noise exposure in the rural setting. *Laryngoscope* 103(3): 258-62.

Hubble, J.P., T. Cao, R.E. Hassanein, J.S. Neuberger, and W.C. Koeller. 1993. Risk factors for Parkinson's disease. *Neurology* 43(9): 1693-7.

Hurst, T.S., J.A. Dosman, B.L. Graham, D. Hall, P. Van Loon, P. Bhasin, and F. Froh. 1990. Respiratory symptoms and pulmonary function in Saskatchewan farmers. *Amer. J. Ind. Med.* 17: 59.

Hurst, T.S., and J.A. Dosman. 1990. Characterization of health effects of grain dust exposures. *Amer. J. Ind. Med.* 17: 27-32.

Husman, K., E.O. Terho, V. Notkola, and J. Nuutinen. 1990. Organic dust toxic syndrome among Finnish farmers. *Amer. J. Ind. Med.* 17(1): 79-80.

Iversen, M., R. Dahl, J. Korsgaard, E.J. Jensen, and T. Hallas. 1990. Cross-sectional study of respiratory symptoms in 1,175 Danish farmers. *Amer. J. Ind. Med.* 17(1): 60-1.

Kelley, K.W., and W.E. Field. 1995. *Characteristics of Flowing Grain-Related Entrapments and Suffocations in On-Farm Grain Storage Facilities and Grain Transport Vehicles.* NIFS Paper 95-2. Columbia, MO: National Institute for Farm Safety.

Langley, R.L. 1994. Fatal animal attacks in North Carolina over an 18-year period. *Amer. J. Foren. Med. Pathol.* 15(2): 160-67.

Lee, B.C., P.M. Layde, and D.T. Stueland. 1995. The injury experience of women on family farms. In: *Agricultural Health and Safety: Workplace, Environment, Sustainability.* H.H. McDuffie, J.A. Dosman, K.M. Semchuk, S.A. Olenchock, and A. Senthilselvan, eds. Boca Raton, FL: CRC Press, pp. 269-74.

Lengerich, E.J., D.G. Addiss, J.J. Marx, B.L. Ungar, and D.D. Juranek. 1993. Increased exposure to cryptosporidia among dairy farmers in Wisconsin. *J. Infect. Dis.* 167(5): 1252-55.

Lin, S., E.G. Marshall, and G.K. Davidson. 1994. Potential parental exposure to pesticides and limb reduction defects. *Scand. J. Work Envir. Health* 20(3): 166-79.

Linet, M.S., J.K. McLaughlin, H.S. Malker, W.H. Chow, J.A. Weiner, B.J. Stone, J.L. Ericsson, and J.F. Fraumeni, Jr. 1994. Occupation and hematopoietic and lymphoproliferative malignancies among women: a linked registry study. *J. Occup. Med.* 36(11): 1187-98.

London, L., R.I. Ehrlich, S. Rafudien, F. Krige, and P. Vurgarellis. 1994. Notification of pesticide posioning in the western Cape, 1987-1991. *South Afr. Med. J.* 84(5): 269-72.

Luisto, M., and A.M. Seppalainen. 1992. Tetanus caused by occupational accidents. *Scand. J. Work Envir. Health* 18(5): 323-6.

Maddy, K.T., S. Edmiston, and D. Richmond. 1990. Illness, injuries, and deaths from pesticide exposures in California 1949-1988. *Rev. Envir. Contam. Toxicol.* 114: 57-123.

Malmberg, P. 1990. Health effects of organic dust exposure in dairy farmers. *Amer. J. Ind. Med.* 17(1): 7-15.

Marx, J.J., J.T. Twiggs, B.J. Ault, J.A. Merchant, and E. Fernandez-Caldas. 1993. Inhaled aeroallergen and storage mite reactivity in a Wisconsin farmer nested case-control study. *Am. Rev. Respir. Dis.* 147(2): 354-58.

Marx, J.J., J. Guernsey, D.A. Emanuel, J.A. Merchant, D.P. Morgan, and M. Kryda. 1990. Cohort studies of immunologic lung disease among Wisconsin dairy farmers. *Amer. J. Ind. Med.* 18(3): 263-68.

May, J.J. 1994. A study of osteoarthritis of the knee and hip in dairy farmers. *Proceedings of the NIOSH Symposium on Efforts to Prevent Injury and Disease Among Agricultural Workers,* August 25-27, 1993, Lexington, KY. DHHS (NIOSH) Publication Number 94-119. 1994.

May, J.J., M. Marvel, M. Regan, L.H. Marvel, and D.S. Pratt. 1990. Noise-induced hearing loss in randomly selected New York dairy farmers. *Amer. J. Ind. Med.* 18(3): 339-43.

McDuffie, H.H.1994. Women at work: agriculture and pesticides. *J. Occup. Med.* 36(11): 1240-46.

McKnight, R.H., E.J. Levine and G.C. Rodgers Jr. 1994. Detection of green tobacco sickness in a regional poison center. *Vet. Hum. Toxicol.* 36(6): 505-10.

McLaughlin, J.R. 1995. Pesticide utilization and lymphoma in Ontario—an ecologic analysis. In *Agricultural Health and Safety: Workplace, Environment, Sustainability.* H.H. McDuffie, J.A. Dosman, K.M. Semchuk, S.A. Olenchock, and A. Senthilselvan, eds. Boca Raton, FL: CRC Press, pp. 157-64.

Meyers, J., L. Bloombert, J. Faucett, I. Janowitz, and J.A. Miles. 1995. Using ergonomics in the prevention of musculoskeletal cumulative trauma injuries in agriculture: learning from the mistakes of others. *J. Agromed.* 2(3): 11-24.

Mobed, K., E.B. Gold, and M.B. Schenker. 1992. Occupational health problems among migrant and seasonal farm workers. *West. J. Med.* 157(3): 367-73.

Morrison, H.I., R.M. Semenciw, K. Wilkins, Y. Mao, and D.T. Wigle. 1994. Non-Hodgkin's lymphoma and agricultural practices in the prairie provinces of Canada. *Scand. J. Work Envir. Health* 20(1): 42-47.

Morrison, H., D. Savitz, R. Semenciw, B. Hulka, Y. Mao, D. Morison, and D. Wigle. 1993. Farming and prostate cancer mortality. *Amer. J. Epidem.* 137(3): 270-80.

Murphy, D.J. 1991. Pennsylvania farm fatalities during 1985-1989. Extension Circular 390. University Park, PA: Penn State University, Cooperative Extension.

Myers, J.R., and D.L. Hard. 1995. Work-related fatalities in the agricultural production and services sectors, 1980-1989. *Am. J. Ind. Med.* 27(1): 51-63.

National Safety Council. 1995. *Accidents Facts.* 1995 Edition. Chicago: National Safety Council, pp.48-50.

Neidich, G. 1993. Ingestic of caustic alkali farm products. *J. Pediatr. Gastroenterol Nutr.* 16(1): 75-77.

Nordstrom, D.L, P.M. Layde, K.A. Olson, D. Stueland, L. Brand, and M.A. Follen. 1995. Incidence of farm-work-related acute injury in a defined population. *Amer. J. Ind. Med.* 28:551-564.

Nordstrom, D.L. 1994. Epidemiology of osteoarthritis. *Wisc. Med J.* 93(12): 636-68.

Nurminen, T., K. Rantala, K. Kurppa, and P.C. Holmberg. 1995. Agricultural work during pregnancy and selected structural malformations in Finland. *Epidem.* 6(1) : 23-30.

Pearce, N., and J.S. Reif. 1990. Epidemiologic studies of cancer in agricultural workers. *Am. J. Ind. Med.* 18(2): 133-48.

Peel, M.M., K.A. Hornidge, M. Luppino, A.M. Stacpoole, and R.E. Weaver. 1991. Actinobacillus spp. and related bacteria in infected wounds of humans bitten by horses and sheep. *J. Clin. Microbiol.* 29(11): 2535-8.

Pratt, D.S., L.H. Marvel, D. Darrow, L. Stallones, J.J. May, and P. Jenkins. 1992. The dangers of dairy farming: the injury experience of 600 workers followed for two years. *Amer. J. Ind. Med.* 21(5): 637-650.

Proust, A.P. 1993. Special injuries of the hand. *Emerg. Med. Clin. North Am.* 11(3): 767-79.

Purschwitz, M.A., and C. A. Skjolaas. 1995. *1994 WisconsinFarm-Related Fatalities.* Departmental bulletin. Department of Agricultural Engineering, University of Wisconsin—Extension/Madison.

Purschwitz, M.A., and C.A. Skjolaas. 1994. *1993 Wisconsin Farm-Related Fatalities.* Departmental bulletin. Department of Agricultural Engineering, University of Wisconsin—Extension/Madison.

Purschwitz, M.A. 1992. Farm and agricultural injury statistics.In *Safety and Health for Production Agriculture.* D.J. Murphy, ed. St. Joseph, MO: American Society of Agricultural Engineers, pp.43-70; 231-235.

Rabl, W., and M. Auer. 1992. Unusual death of a farmer. *Am. J. Forensic Med. Pathol.* 13(3): 238-42.

Reynolds, S.J., D. Parker, D. Vesley, D. Smith, and R. Woellner. 1993. Cross-sectional epidemiological study of respiratory disease in turkey farmers. *Am. J. Ind. Med.* 24(6): 713-22.

Rivara, F.P. 1985. Fatal and nonfatal farm injuries to children and adolescents in the United States. *Pediatrics* 76(4): 567-73.

Roerig, S. 1993. Scalping accidents with shielded PTO units: four case reports. *AAOHN J* 41(9): 437-39.

Rowland, A.S. 1995. Pesticides and birth defects. *Epidemiology* 6(1): 23-30.

Rublaitus, S.M., J.R. Wilkins, L.A. Jones, G.L. Mitchell, T.L. Bean, and J.M. Crawford. 1994. Symptoms of respiratory disease among Ohio farm operators. *Abstracts of Agricultural Safety and Health: A National Conference on Detection, Prevention, and Intervention, May 24-26, 1994.* Columbus, OH. Cincinnati, OH: NIOSH.

Rylander, R., N. Essle, and K.J. Donham. 1990. Bronchial hyperreactivity among pig and dairy farmers. *Amer. J. Ind. Med.* 17: 66-69.

Sanderson, W., G. Kullman, J. Sastre, S. Olenchock, A. O'Campo, K. Musgrave, and F. Green. 1992. Outbreak of hypersensitivity pneumonitis among mushroom farm workers. *Am. J. Ind. Med.* 22(6): 859-72.

Schelp, L. 1992. The occurrence of farm-environmental injuries in a Swedish municipality. *Accid. Anal. Prev.* 24(2): 162-66.

Schenker, M.B., and S.A. McCurdy. 1990. Occupational health among migrant and seasonal farmworkers; the specific case of dermatitis. *Am. J. Ind. Med.* 18(3): 345-51.

Schwartz, D.A., S.K. Landas, D.L. Lassise, L.F. Burmeister, G.W. Hunninghake, and J.A. Merchant. 1992. Airway injury in swine confinement workers. *Ann. Intern. Med.* 116(8): 630-5.

Senthilselvan, A., and J.A. Dosman. 1995. Risk factors for allergic rhinitis in farmers. In: *Agricultural Health and Safety: Workplace, Environment, Sustainability.* H.H. McDuffie, J.A. Dosman, K.M. Semchuk, S.A. Olenchock, and A. Senthilselvan, eds. Boca Raton, FL: CRC Press, pp.75-78.

Senthilselvan, A., Y. Chen, and J.A. Dosman. 1993. Predictors of asthma and wheezing in adults. Grain farming, sex, and smoking. *Am. Rev. Respir. Dis.* 148(3): 667-70.

Sheldon, E.J., W.E. Field, and R.L. Tormoehlen. 1995. Fatal farm work-related injuries involving children and adolescents in Wisconsin and Indiana. In: *Agricultural Health and Safety: Workplace, Environment, Sustainability.* H.H. McDuffie, J.A. Dosman, K.M. Semchuk, S.A. Olenchock, and A. Senthilselvan, eds. Boca Raton, FL: CRC Press, pp.355-362.

Stallones, L. 1990. Surveillance of fatal and non-fatal farm injuries in Kentucky. *Am. J. Ind. Med.* 18(2): 223-34.

Stanford, C.F., J.H. Connolly, W.A. Ellis, E.T. Smyth, P.V. Coyle, W.I. Montgomery, and D.I. Simpson. 1990. Zoonotic infections in Northern Ireland farmers. *Epidem. Infect.* 105(3): 565-70.

Steel, S. 1994. *Analyzing work accident fatality data for aged farmers and ranchers in the U.S.* NIFS Paper 94-1. Columbia, MO: National Institute for Farm Safety.

Stueland, D., P. Layde, and B.C. Lee. 1991. Agricultural injuries in children in central Wisconsin. *J. Trauma* 31(11): 1503-9.

Stueland, D., T. Zoch, P. Stamas Jr, G. Krieg, and W. Boulet. 1990. The spectrum of emergency care of agricultural trauma in central Wisconsin. *Amer. J. Emerg. Med.* 8(6): 528-30.

Susitaival, P., L. Husman, M. Horsmanheimo, V. Notkola, and K. Husman. 1994. Prevalence of hand dermatoses among Finnish farmers. *Scand. J. Work Envir. Health.* 20(3): 206-12.

Terho, E.O. 1990. Work-related respiratory disorders among Finnish farmers. *Am. J. Ind. Med.* 18(3): 269-72.

Thomas, M.C., A. Chereshsky, and K. Manning. 1994. An outbreak of leptospirosis on a single farm in east Otago. *N. Zeal. Med. J.* 107(982): 290-91.

Tormoehlen, R. 1986. *Fatal farm accidents occurring to Wisconsin children, 1970-1984.* ASAE Paper 86-5514. St. Joseph, MI: American Society of Agricultural Engineers.

Von Essen, S. 1992. Airborne dusts. *Papers and Proceedings of the Surgeon General's Conference on Agricultural Safety and Health,* April 30-May 3, 1991, Des Moines, IA: DHHS (NIOSH) Publication Number 92-105, pp. 204-15

Williams, N. 1994. Zoonotic infections: The animal connection. *Occup. Health Lond.* 46(8): 278-80.

Zahm, S.H., D.D. Weisenburger, K.P. Cantor, F.F. Holmes, and A. Blair. 1995. Non-Hodgkin's lymphoma and the use of atrazine: results from three case-control studies. In: *Agricultural Health and Safety: Workplace, Environment, Sustainability.* H.H. McDuffie, J.A. Dosman, K.M. Semchuk, S.A. Olenchock, and A. Senthilselvan, eds. Boca Raton, FL: CRC Press, pp.151-56.

Zahm, S.H., D.D. Weisenburger, K.P. Cantor, F.F. Holmes, and A. Blair. 1993a. Role of the herbicide atrazine in the development of non-Hodgkin's lymphoma. *Scand. J. Work Envir. Health.* 19(2): 108-14

Zahm, S.H., D.D. Weisenburger, R.C. Saal, J.B. Vaught, P.A. Babbitt, and A. Blair. 1993b. The role of agricultural pesticide use in the development of non-Hodgkin's lymphoma in women. *Arch. Envir. Health* 48(5): 353-58.

Zahm, S.H., and A. Blair. 1993. Cancer among migrant and seasonal farmworkers: an epidemiologic review and research agenda. *Am. J. Ind. Med.* 24(6): 753-66.

Zeitz, P.S., J.C. Butler, J.E. Cheek, M.C. Samuel, J.E. Childs, L.A. Shands, R.E. Turner, R.E. Voorhees, J. Sarisky, and P.E. Rollin, et al. 1995. A case-control study of hantavirus pulmonary syndrome during an outbreak in the southwestern United States. *J. Infect. Dis.* 171(4): 864-70.

Zejda, J.E., T.S. Hurst, C.S. Rhodes, E.M. Barber, H.H. McDuffie, and J.A. Dosman. 1993a. Respiratory health of swine producers. Focus on young workers. *Chest* 103(3): 702-9.

Zejda, J.E., H.H. McDuffie and J.A. Dosman. 1993b. Epidemiology of health and safety risks in agriculture and related industries. Practical applications for rural physicians. *West. J. Med.* 158(1): 56-63.

Zhou, C. and J.M. Roseman. 1994. Agricultural injuries among a population-based sample of farm operators in Alabama. *Am. J. Ind. Med.* 25(3): 385-402.

15

HAZARD MANAGEMENT AND SAFETY WITH AGRICULTURAL MACHINES

Gary T. Roberson, Ph.D., P.E.
North Carolina State University

> Agricultural machinery accounts for the largest share (25.5%) of recorded farm injuries. Increased mechanization of agriculture has brought many improvements in workload demands but has also created new hazards which must be faced by the workers. It is essential that machine operators and owners understand the basic concepts of machinery safety in order to recognize, manage, and control these hazards. This chapter discusses many types of hazards found on agricultural machines. Each hazard is defined and illustrated by examples and hazard pictorials. Types of injury associated with each hazard are discussed as well as factors which contribute to injury. Hazard recognition is stressed as a necessary skill for safe operation of equipment. Management and control strategies are presented to reduce injuries and enhance safety. Basic guidelines for safe machine operation are also presented.

INTRODUCTION

Modern agriculture is becoming increasingly diverse and often highly mechanized. While there is still a large demand and need for manual labor, mechanization is constantly increasing. Machines are available to perform a wide range of tasks in all facets of production such as tillage, planting, chemical application, harvesting, and material handling. Unfortunately, with all the advantages mechanization offers, there is a dark side to the picture due to the additional hazards generated by mechanization.

Agriculture consistently ranks among the nation's most hazardous occupations. Agricultural tractors and machines, added together, account for a large percentage of all recorded farm injuries as shown in Figure 15.1 (Roberts 1990). Understanding how agricultural machines function and how to recognize hazards around these machines are key skills for productive and safe operation. Growers, farm managers, farm workers, machinery dealers, machinery manufacturers, and others involved in agricultural production must recognize the problems associated with modern agriculture and be prepared to deal with them. While increased diversity and mechanization are important assets for agriculture, the problems they pose for safety must not be overlooked.

MACHINERY SAFETY CONCEPTS

Concerned persons, from machine operators to machine designers and health professionals, need to be familiar with some of the basic concepts of safety.

Safety means being relatively free from hazard. Although there is no perfectly safe machine, facility, or operation, safety can be improved through proper design, good management, and the use of safe work practices.

A *hazard* is any set of circumstances that may cause harm. Hazards may be mechanical, biological, chemical, thermal, or some other type. Hazards can be rated according to the severity of their consequences—that is, the degree of harm they can cause.

Risk is the *probability* of exposure combined with the hazard severity. Some hazards, although they may cause serious harm, are unlikely to occur and therefore represent a low risk. Other hazards, while not so serious, may represent a greater risk because they are more likely to occur.

Exposure occurs when a part of the body comes into contact with a hazard or a hazardous environment. Exposure may occur through the absorption, ingestion, or inhalation of some substance or through direct contact with an object or machine part. Exposure is *acute* if it occurs only briefly; it is *chronic* if it continues over an extended period of time.

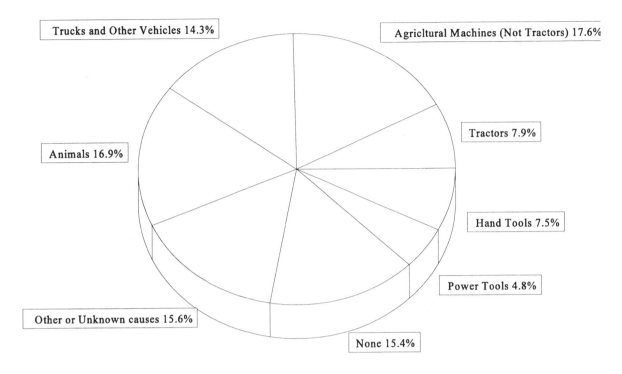

Figure 15.1. Cause of farm injuries.

Injury is a cut, fracture, sprain, or other type of bodily harm that results from exposure to a hazard. Injuries can be minor, requiring no more than first aid, or severe, requiring extensive medical treatment and involving lost workdays. Death can be the result of some of the more serious injuries.

Illness, as used in this chapter, means any abnormal condition or disorder, other than an injury, caused by exposure to materials and equipment in the work environment. Examples of illnesses are skin disorders, respiratory diseases, and poisonings caused by exposure to dust, toxic agents, cumulative trauma, or other agents.

PERSONS AT RISK

Many agricultural machines require only one operator. The operator controls the power and function of the machine from the designated station, often a platform or enclosed cab. In many cases, there is no one but the operator in the vicinity of the machine. In such cases, the operator is the only person at direct risk. However, because the operator is working alone, there may be no one around the area to help or summon help should an injury occur or illness suddenly develop. Equipment operators in these situations should always be aware of this problem and arrange for routine check-ins with their family, friends, or supervisors. Modern techniques, such as cellular phones or emergency message devices would be beneficial in these conditions as well.

In other cases, agricultural machines are being used in close proximity to other people. These people may be other equipment operators or they may be bystanders or children. Sometimes these onlookers are not familiar with the design or function of the machine and may not be aware of the hazards.

For example, an onlooker or bystander may not be aware of the hazard potential from pesticide spray drift, objects thrown by a mower, or an implement which rapidly lowers to the ground. An alert equipment operator should always be aware of these potentially hazardous situations and take the necessary action to avoid injuries. Above all, agricultural machines are not toys or recreational vehicles. People not required to operate a machine should be kept well clear at all times.

HAZARD RECOGNITION AND HAZARD TYPES

Hazards on machines can be identified by a number of descriptions or categories. Hathaway (1987) identified eleven common machinery hazards that operators should recognize and be prepared to deal with. Murphy (1992) grouped hazards by source: tractors, machinery, respiratory, animals, agricultural chemicals, and others. Roberson and McLymore (1994) identified thirteen hazard categories common to agricultural machines. Considering these sources, a composite list of hazard categories to be discussed in this chapter is presented in Table 15.1.

Table 15.1. Hazard categories.

Hazard Type	Description
1. Pinch Point	Areas where moving parts close together.
2. Entanglement	Components that might catch our clothing, arms, or legs, and wrap it around or pull you in.
3. Sharp Edges	Blades or other sharp edges that cut or shear.
4. Crush Point	Area where components close together trapping a hand, arm, foot, leg, or body.
5. Visibility	The operator or others may not be able to see properly due to obstructions on or around the machine.
6. Thrown Objects	Machine components that rotate at high speeds can pick up objects and throw them with great force for some distance.
7. Energy Release	Energy that can be released suddenly or not controlled.
8. Falling	May be caused by slippery or cluttered surfaces, riding on machines while not in a proper seat, or at other locations.
9. Speed	Travel speed on highways or in the field may cause hazards.
10. Work Environment	Extremes of heat and cold or environmental quality may tax the operator.
11. Cumulative Trauma	Noise, vibration, and repetitive motion causes nerve and muscle damage.
12. Chemicals	Liquids and gases used on machines can be harmful.
13. Run over	Tractors or other machines can run over and crush people beneath them.
14. Roll over	Tractor or other machines when pushed to their limits can overturn onto the operator.

The reader will note that some hazards may fall under more than one category. For example, a rotary mower blade has a sharp edge or shear hazard as well as an energy release hazard due to its high speed, plus a thrown object hazard if it picks up a rock while in use. A component that exhibits more than one hazard can be considered a compound hazard.

Most machines have multiple hazards, where more than one hazard or hazardous component can be found on a machine. For example, a round baler has an entanglement hazard in the header area, a crushing hazard at the bale chamber door, and an energy release hazard with the hydraulic hoses, to name a few. The presence of multiple hazards makes modern equipment dangerous for users who are not properly trained or prepared.

In the sections which follow, the hazard types identified in Table 15.1. will be explained in greater detail. Sample pictorials taken from the American Society of Agricultural Engineers (ASAE) standards will be used to illustrate many of the hazards. Common preventative strategies will also be discussed. It may be impossible to identify every hazard or provide an example on the pages that follow. Hopefully these pages will stimulate thought and discussion and the reader will be able to develop a basis for hazard identification.

Pinch Point Hazards

The pinch point may well be the most common of the machine hazards. A pinch point occurs where machine parts move together with at least one of the components traveling in a circle (Murphy 1992). Belts running onto pulleys, chains running onto sprockets, gears, and rollers are all examples of common pinch point hazards (Figure 15.2).

Pinch point hazards most often cause injury to the fingers or hands when an exposure occurs. In some cases, the toes or feet may be exposed, resulting in injury. Clothing

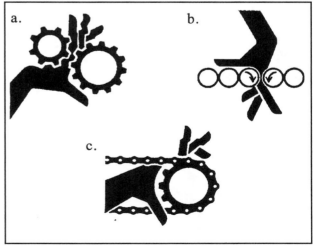

Figure 15.2. Pinch Point Hazards: a) gears, b) rollers, c) chain and sprocket (ASAE S441.2, 1995).

can also be captured by the pinch point hazard which in turn may lead to exposure of other parts of the body. The result is often cuts, bruises, fractures and even amputations, particularly of the fingers. While the injury is not often life threatening, it is serious since the worker may lose the use of his/her hands or feet either temporarily or permanently.

Exposure to pinch point hazards can be reduced by keeping all guards and shields in place on equipment. Equipment operators must understand that some components have to be exposed in order to function. The operator should know where these exposed components are and how to avoid the hazard while working with the machine. Always disengage and turn off the power to a machine when servicing or adjusting equipment. Take the key out if necessary to prevent someone from starting the machine while it is being worked on.

Entanglement Hazard

Entanglement hazards are similar to pinch points but are larger and in many cases more aggressive (Roberson and McLymore 1994). Any exposed rotating shaft can catch or snag the clothing and wrap it around. PTO drive lines and other rotating shafts on equipment are common entanglement hazards of this type. In addition, components which roll or pull together, such as feed rolls, spindles, or gathering chains found in harvesting machines, are particularly dangerous entanglement hazards. These may start as pinch point hazards but become more dangerous as the arm or leg, or whole body is entangled (Figure 15.3).

Entanglement hazards often lead to broken bones, mangled limbs, crushed arms and legs, or amputated limbs. Entanglement in larger machines such as balers or combines may involve the entire body being crushed or badly mangled.

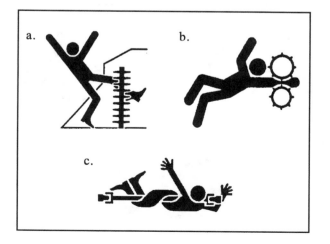

Figure 15.3. Entanglement hazard pictorals: a) picker spindles, b) feed rools, c) driveline (ASAE S441.2, 1995).

A common cause of exposure to an entanglement hazard is attempting to remove crop material from a machine that is plugged or stopped up. Workers should not attempt to clear an obstruction on any machine while the power is on. Another common entanglement occurs when someone reaches or steps across a rotating driveline. Equipment operators should avoid the area where these components operate and be sure to leave shields in place. It only takes a fraction of a second to become entangled in a machine. It may take hours for rescue workers to remove someone from such a machine.

Many of the same precautions outlined for pinch point hazards apply to entanglement hazards as well. PTO's and drive lines should always be shielded. Recognize that some hazards such as headers, tillers, and mowers cannot be fully shielded an still function. They must be somewhat exposed. Workers should always be aware of the safety zone around entanglement hazards. Once the hazard is recognized and the safety zone established, exposure is minimized.

Many entanglement hazards are also equipped with safety stops or emergency shut off switches. These devices can shut off power to the equipment if an operator is at risk. Some such devices are activated automatically whenever something out of the ordinary is detected, others must be engaged manually. In a panic situation, the person entangled may not be able to react and engage the safety stop or shut off. Further, like any other machine component, there is always the possibility that the safety device itself may fail. As valuable as these devices are, they are no substitute for knowledge of potential hazards and how to avoid them.

Sharp Edge Hazard

Sharp edges and other shear points are common on many machines. Cutterbars, mower blades, and knives are easy to recognize as shear points. Blades on mowers, chippers, harvesters, saws, and tillage tools must be kept sharp for good performance. Workers should appreciate this feature and take the necessary precautions while working with this type of equipment or component (Figure 15.4).

Not all sharp edges occur on blades or cutting tools. Sharp edges also occur on machines where components pass close enough together to create a shearing action. Augers running in tubes, paddle conveyors, lift arms, and telescoping tubes are examples of this type of shear point.

Mower blades, choppers, chain saw teeth, and other cutters typically travel at high speeds. They can slice through clothing and the human body with ease. Augers and other sharp edges can trap a hand or arm between a moving edge and a stationary edge and shear through it with equal ease.

Figure 15.4. Sharp edge hazard pictorials: rotary mower, b) auger, c) tiller, d) trimmer (ASAE S441.2 1995).

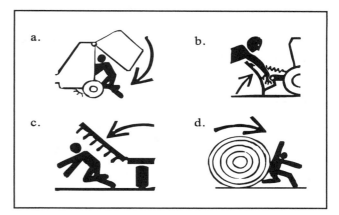

Figure 15.5. Crushing hazard pictorials: a) bale chamber, b) blade mount, c) falling boom, c) bale rolling. (ASAE S441.2 1995).

Workers should take care around sharp surfaces and edges. Augers and conveyors should have protective cages around the intake to prevent contact with the shear points. Keep these guards in place at all times. Mower blades are enclosed by the deck. However, not all blades can be fully enclosed and still function. Keep fingers and toes clear of the blade area. Also, operators of small push mowers should not walk backwards or pull a mower toward themselves, especially on slopes where they may slip and fall.

Crush Point Hazard

Areas where components may close together trapping a hand, arm, foot, leg, or the whole body are crush points. Hitching implements presents many crushing opportunities. Likewise, any heavy load or object may cause a crush hazard if you reach or work under it and the load falls. Many crushing hazards exist where the parts of the machine come together during operation. Examples include the steering joint of an articulated tractor, implements raising against the tractor frame or roll bar, or a bale or other heavy object falling from a loader bucket.

Crushing hazards result in severe bruising and fractures of the arms, legs, or upper body. The upper body can also be crushed or squeezed, causing suffocation.

Keep all guards and shields around crushing hazard components in place at all times. Operators should always stand clear of moving machines or components that can crush. Wait until the machine stops before you attempt a connection or adjustment. In situations where falling objects are a potential hazard, have the machine equipped with a canopy or other overhead protective structure. Pay attention to raised loads, be sure they are supported properly before attempting to work around the hazard area.

Visibility Hazard

Visibility is a critical issue on many machines. Visibility hazards occur where a machine operator may not be able to see properly (Roberson and McLymore 1994). In some cases, a hazard may be completely obscured from view by the machine or something attached to the machine. Obstructions on machines or the location of the operator's station may block his or her view of other hazards on the machine. For example, saddle tanks on tractors may prevent the operator from seeing objects in front or to the side of the tractor. A large trailer or implement may block the operator's view to the rear. Combine operators may also have a poor view to the rear of their machines. While not being able to see a hazard does not cause injury directly, the inability to see the hazard may lead to exposure before the operator or worker can recognize the problem.

Equipment operators should pay close attention to the way equipment is mounted. Check out other options for attachment mounting to see if another option will serve just as well and restore operator visibility. Use mirrors to enhance your view of the rear or other areas of the machine. Consider backup alarms for large equipment. These alarms may alert someone that the operator can't see to the approach of a hazardous machine.

Thrown Object Hazard

Machine components that rotate at high speeds can pick up objects and throw them with great force for a great distance. Rotary cutters or mowers often throw rocks or gravel. Fans may pick up grain or other debris. Compressed air can blow dirt or grit into the eye or through the skin. Chain saws and other power tools can throw chips toward the face and eyes. String trimmers can throw sand and gravel when the string hits the ground. These objects may also ricochet and travel in unpredictable directions.

Figure 15.6. Thrown object hazard pictorial: rotary mower (ASAE S441.2 1995).

Objects thrown at the body can easily cause puncture wounds, bruises, or lacerations. Objects thrown into the eye can permanently damage the eye and may result in blindness.

Moving components should be kept shielded as much as possible. Discharge chutes or chain deflectors on mowers are designed to absorb energy or deflect the object away. These safety devices should not be removed. Covers, screens, or diffusers are available for fans or air compressors. As discussed for other hazards, some of the components described above cannot be fully shielded and still function. The operator should be aware of this and be prepared to take proper precautions.

Workers should also wear safety glasses and perhaps a face shield while working with this type of equipment. Even with the shields and guards in place, objects can still be thrown and strike the operator or a bystander.

Energy Hazard

Energy can be stored in many ways. A compressed spring, hydraulic fluid under pressure, and fans and wheels turning at high speeds all possess energy. Properly applied, energy allows machines to accomplish many tasks. However, energy can pose a hazard if it is not properly controlled.

Energy can be released suddenly and unexpectedly when a failure occurs or when an operator uses a machine incorrectly. Hydraulic fluid leaking from a pressurized system is an example of energy release. A spring can release energy if it breaks or slips out of its mounting while under load. Contact with electrical wires can cause a dangerous release of electrical energy. Raised loads store potential energy which can be released if a latch or fastener fails, allowing the object to fall. Wheels, blades, and other rotating devices store kinetic energy which may prove hazardous. Likewise, infrared and ultraviolet radiation produced by welders can be a hazardous form of energy.

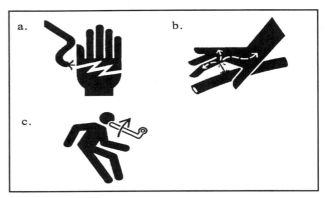

Figure 15.7. Electrical hazard pictorials: a) electrical shock, b) high pressure fluid, c) high speed moving components (ASAE S4441.2 1995).

Releasing spring energy can throw the object against an operator resulting in injury. Fluid under pressure squirting through a small hole can penetrate the skin and enter the bloodstream causing poisoning. Dangerous electrical voltages can cause electrocution and death. Grain dust can be explosive if certain conditions are present. Objects rotating at high speeds, such as blades, wheels, or free spinning winch handles, are energy releases which can cause injury if they strike someone. Excessive speed on chains, belts, or shafts may cause them to break and throw objects at the operator or severely damage the machine itself, making it unsafe.

Several precautions can be taken to avoid energy hazards. First, safely release any energy in the system and turn off power before working around objects that store energy. Lower raised loads to the ground or support them with safety bars or pins. Check valves, controls, or switches to be sure all energy is safely released. Replace belts, chains, shafts, and gears when they are worn. If these break under load, the energy can be released as a thrown object. Operators should also avoid excessive speed on all powered components since excessive speed can lead to failure and pose a hazard. Remember that energy, when controlled, is a useful tool. Uncontrolled, energy can be very hazardous.

Falling Hazard

Injuries resulting from falling hazards are also common. Slippery or cluttered surfaces on machines can cause a person to fall. Mud, oil, or water on a platform can prove quite hazardous. In most cases, steps and platforms on agricultural machines are difficult to maneuver on under good conditions. Any surface where someone must walk or stand must be kept clean and clear of obstructions.

Riding on a machine such as a tractor in a place other than an operator's seat may result in that person losing their balance and falling from the machine and into the path of the tractor or its implement. If the machine does not have a properly designed and installed seat for an additional rider, there should not be anyone other than the operator on the machine. This is a controversial subject. The argument for an extra rider may be made in the case of teaching someone to drive a tractor (Murphy and Steel 1995). A supervisor on board for the untrained operator may be advisable. Regardless, riding in any position other than the seat and properly using the seat belt can result in a fall.

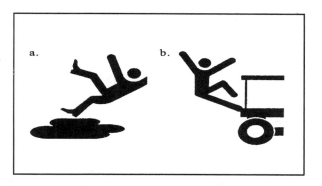

Figure 15.8. Falling hazard pictorials: a) slippery surfaces, b) falling from tractor (ASAE S441.2 1995).

Falling can result in injury if you fall onto a hard surface or if you fall onto a sharp tool. Falling from a machine into water such as a stream or pond can result in drowning. Falling from a tractor or other machine can lead to a number of other hazards as the machine passes over you.

The safest place to ride on a machine is in the operator's seat. If the machine has a roll over protective structure (ROPS), use your seat belt as well. The seat belt is designed to hold you safely in the seat as you operate the machine. Do not allow extra riders on the machine. Pick up all loose tools or objects on equipment decks or platforms. Keep mud and debris cleared from surfaces to avoid slipping.

Speed Hazard

Agricultural machines, such as tractors and combines, are not designed for high speed travel. While modest speeds can be attained on smooth highways, tractors and combines are out of place on public roadways. The lack of speed can pose a hazard when other vehicles—cars and trucks—are traveling two or three times as fast as the tractor or combine. If a tractor traveling at 15 mph is struck from the rear by a car at 55 mph, the net impact is still 40 mph. The author does not advocate designing tractors and combines for faster speeds. Instead, operators need to be aware of the hazard posed by the high speed vehicles they are likely to encounter.

On the other hand, the maximum speed a tractor can deliver is often too fast for the rough terrain found in a field. Excessive speed can result in tractor overturn, especially in sharp turns or on rough ground. When driving too fast, the operator has less time to react to events around him or her. Driving a machine too fast on rough ground may cause the operator to be thrown from the machine which can lead to other hazards. Drive slowly on rough ground and keep the tractor under control at all times.

If you must travel on highways with agricultural equipment, use slow moving vehicle (SMV) emblems and other appropriate lighting or marking. It may be advisable to have someone traveling ahead of the tractor or combine in a car or truck with warning lights or a sign to warn of the approaching slow moving or large machine.

Work Environment Hazards

The work environment on or around a machine poses many potential hazards. Included are surface temperatures, air temperatures, humidity, and lack of ventilation. Each of these can pose a hazard if the worker is exposed to conditions beyond the body's ability to accommodate.

Figure 15.9. Slow moving vehicle symbol. (ASAE S276.4, 1995).

Radiators and mufflers on machines are good examples of hot surfaces. Cold surfaces can be found around air conditioning equipment or around gas handling equipment such as anhydrous ammonia. Contact with extremely hot or cold surfaces can cause painful injuries to the body as illustrated in Table 15.2.

Table 15.2. Temperature effects for human contact.

Temp. °F	Effect
212	Second degree burn of 15 second contact
180	Second degree burn on 30 second contact
160	Second degree burn on 60 second contact
140	Pain, some tissue damage
120	Pain, burning heat
91	Warm, neutral feeling
54	Cool
37	"Cool Heat" sensation
32	Pain
Below 32	Pain, Tissue damage

Air temperature, humidity, and ventilation rates affect worker performance (Suggs 1991). Temperature and high humidity extremes force the body to work harder to maintain its optimum status. Lack of ventilation can increase operator discomfort and lead to a build up of toxic gases. Table 15.3 provides some desirable ranges to consider. Air temperature, humidity, and ventilation all affect a worker's productivity and alertness.

Table 15.3. Worker environment.

Ventilation	23 CFM per Adult Worker
Relative Humidity	30% to 65%
Temperature	68°F to 84°F

A well-designed and maintained operator environment, such as a cab or platform, and proper selection of clothing help to insure peak performance from machine operators and thus from machines.

Cumulative Trauma Hazards

Cumulative trauma is defined as chronic exposure to a hazard where the effects develop over time. Noise, vibration, and repetitive motions are examples of cumulative trauma. Each of these hazards can cause nerve, muscle, or joint damage. Cumulative trauma hazards are common in one form or another on most machines. A cumulative trauma develops due to repeated exposure. The effects may not be immediately noticeable.

Prolonged exposure to excessive noise levels can lead to permanent hearing loss. Table 15.4 outlines the allowable exposure limits. Engines, grinders, saws, tractors, combines, and many other machines can generate dangerously high noise levels. Exposure to vibration in excess of the body's tolerance limits can impair nervous and/or circulatory function or lead to motion sickness as shown in Table 15.5. Repetitive motions lead to injury of joints, Table 15.6, or at least fatigue, which reduces the operators ability to perform.

Table 15.4. Noise exposure levels.

Duration, Hours per Day	Sound Level, Decibels (A scale)
16	85
8	90
4	95
2	100
1	105
1/2	110
1/4 or less	115

Many engines and machines produce dangerous noise and vibration levels. The operator is often seated or standing close to the source or these noise levels. Vibration is transmitted to the body through the feet, the buttocks, and the hands and arms.

Table 15.5. Cumulative trauma disorders.

Body Area/Disorder	Symptoms and Effects
Lower Back	Pain in the Lumbar region
Carpal Tunnel Syndrome	Pain and numbness of the wrist
Tendinitis	Pain in the hands, wrist, elbow, knee, or ankle
Ganglionic Cyst	Swelling and pain, usually in the wrist

Table 15.6. Vibration effects.

Type of vibration	Vibration Frequency (cycles per second)	Vibration Source and Effects
Motion Sickness	0.01 to 1.0	Land, sea, or air vehicle. Nausea, vomiting, malaise
Whole Body	1 to 80	Transmitted form frame of machine through legs or buttocks. Causes discomfort and reduced proficiency due to operator fatigue.
Hand-Arm	5 to 1500	Transmitted from controls, handles, and power tools through the hands and arms. Causes numbness, and reduced circulation.

Tractors, combines, chain saws, and other powered tools and machines produce noise and vibration. Operators should pay close attention to the design of a piece of equipment, particularly how the operator relates or interfaces with the machine. If hazardous noise levels are present, use hearing protectors—ear plugs or ear muffs—which can reduce noise levels by up to 20 DB. Gloves are effective to some extent in reducing vibration to the hands. A well-designed seat is a must to minimize whole body vibration. Proper adjustment of the seat mechanism is important to getting full benefit from the design.

Chemical Hazards

Modern agricultural machines use a variety of chemicals to enhance performance. Some of these chemicals can prove harmful if an operator is exposed. Chemicals can be inhaled, ingested, splashed in the eye, or absorbed through the skin. Effects of this chemical exposure may be immediate or long term. Further, some chemicals lead to fire or explosion hazards.

Table 15.7 illustrates some of the chemical compounds found on agricultural machinery and the hazards associated with them. While commonplace on many machines, each may pose a serious threat given the wrong circumstances and should not be taken lightly.

Table 15.7. Chemicals used on machines

Chemical	Where Used
Gasoline	Engines
Diesel fuel	Engines
Antifreeze	Engines
Motor oil	Engines
Freon	Air conditioners
Calcium chloride	Liquid tire ballast
Grease	Chassis/bearings
Carbon monoxide	Engine exhaust
Carbon dioxide	Engine exhaust/ Closed cabs
Transmission/hydraulic oil	Transmission/ hydraulic systems/ differentials
Ether/starting fluid	Engines

Wear proper clothing when operating or working around machinery. Gloves, shoes, goggles, or other personal protective equipment (PPE) are very effective at reducing chemical exposure. Avoid fumes from chemicals, splashing, or accidental swallowing. Be careful of chemical spills; be sure they are cleaned up quickly and properly.

Run Over Hazard

The run over hazard is among the most severe hazards associated with agricultural machinery. Run over is defined

Figure 15.10. Run over hazard pictorial (ASAE S441.2, 1995).

as the machine running over a person, crushing them beneath the wheels or tracks of the machine. Run overs occur when someone stands or walks in the path of an oncoming machine.

Run overs can be caused by several circumstances. Some run overs have occurred due to an operator or extra rider falling from a machine into its path or into an implement. Another possible cause is someone standing or working in the path of a machine and not realizing the machine is approaching. This is often coupled with a visibility hazard on the machine where the operator also is unable to see the person in danger. This may be especially true when children are involved. A child may not recognize the approach of a large machine as a hazard. Finally, many run overs occur due to jump starting or bypass starting a tractor or machine. Jump starting is done when a machine fails to start on its own due to a failure of the battery or starting circuit. A fresh battery is attached to the machine and the starting circuitry is bypassed. Unfortunately, this also bypasses the safety interlocks in the starting system. As a result, the machine can start with the transmission engaged and lurch forward as the engine comes up to speed. The person attempting the jump start may be trapped in front of the wheel with no time to escape.

Run overs can be avoided by following a few simple guidelines. First, do not allow extra riders on a machine. If there is not a properly designed and installed seat for an extra person, do not allow them to ride. Second, the operator must be alert. Check the path to be sure no one is in the path of travel. Sound your horn or use backup alarms to help warn others of the approaching machine. When jump starting, use the proper jump starting procedure as outlined by the manufacturer. Be sure the transmission is disengaged and never take shortcuts. By taking proper precautions, the run over hazard can be avoided.

Rollover Hazard

The rollover hazard also ranks as one of the most severe machinery hazards. A rollover occurs when the machine overturns, often with the operator trapped underneath the machine. The overturn can be to the side, the rear, or the front. Tractors or other machines when pushed beyond their limits can rollover or overturn onto the operator or nearby persons.

Rollovers can result from a number of factors. Steeply sloping terrain or unstable ground can contribute to an overturn because the machine is constantly tilted to the side and its center of gravity has shifted closer to its outer support (Murphy 1992). Placing too much load on the hitch, hitching improperly, or carrying a load improperly can also contribute to a rollover. Improper loading often makes the

Figure 15.11. Rollover hazard pictorial (ASAE S441.2, 1995).

machine unstable or poorly balanced. Too little weight on the front when carrying a heavy implement can likewise make the tractor unstable. High speeds, rapid acceleration, or rapid stopping can make the machine unstable as well. High speeds and sharp turns can equally contribute to a rollover.

If a tractor or other machine rolls over, the operator can become trapped under the machine and crushed, resulting in severe injury or death.

Tractors manufactured since October 25, 1976 should be equipped with a rollover protective structure (ROPS) unless it was designed for orchard or feedlot use. Many older machines can be fitted with ROPS as well. Check with the manufacturer about availability. If your tractor has a ROPS, be sure seat belts are worn.

Hitch loads properly, maintain safe speeds, operate with caution at all times, and use the roll over protection provided with the tractor. It may not be possible to prevent every rollover, but the risk of injury can be reduced.

Evaluating Hazards

Equipment owners or operators can practice hazard recognition by evaluating their machines for hazards. Saran et al. (1991) describes a procedure based on the injury severity, probable frequency, and exposure to hazards. When implemented, the procedure reflects the overall relative potential of a machine or procedure to result in an injury or illness.

Roberson and McLymore (1994) suggest a simplified procedure for hazard evaluation. In addition to recognizing hazards, the procedure allows for a comparison between machines to see which one offers the best protection or to determine what improvements need to be made. It should be noted that the procedure provides a relative index, not an absolute assessment. Each operator may perceive a situation differently, leading to differing scores. Figure 15.12 offers an example of the evaluation form.

To use the procedure, carefully look over the machine and try to identify as many hazards as possible. Now study each hazard and assign a hazard rating (3=high, 2=moderate, 1=low) according to its severity or potential to cause injury or illness. For example, a hazard that may cause serious injury or death should be rated high (3). A hazard that could result in a minor injury could be rated low (1). Now rate the hazard according to the probability (3=high, 2=moderate, 1=low) that an operator is exposed during routine or foreseeable operation or use of the machine. Once each hazard has been evaluated, multiply the hazard rating and probability rating together to determine the risk rating for each hazard. Finally, risk ratings for each hazard are added together to determine the relative safety index for the machine or procedure. The lower the relative safety index, the "safer" the machine or procedure is perceived to be.

MANAGEMENT STRATEGIES AND PREVENTION TECHNIQUES

Prevention of injuries and illnesses is important to equipment manufacturers, dealers, farm managers, and farm workers. The effort to prevent these injuries and illnesses begins with the concept and design phase of a machine and continues through to its use in the workplace.

Safety Hierarchy

Manufacturers utilize a safety hierarchy (Sullivan & Matthews 1991), shown in Table 15.8, to aid in the evaluation of hazards and how to deal with those hazards.

Table 15.8. Safety hierarchy.

Priority	Action
First	Eliminate hazard
Second	Apply latest safety technology
Third	Use warning signs
Fourth	Train and instruct
Fifth	Prescribe personnel protective equipment

The first option should always be to eliminate the hazard. Manufacturers typically review the history of machines and incorporate those changes or modifications which have a proven potential for injury or illness reduction. By designing to eliminate hazards from the machine, the machine becomes safer to operate. However, these design changes do not affect machines already in the field unless the manufacturer offers a recall or upgrade program. Owners or operators should inquire about these changes and see if their machines

```
Type of Equipment: _____

Item        Hazard       Hazard           Hazard          Probability          Risk
Number      Type         Description      Rating    X     Rating       =       Rating

_____      _____       _____           _____    X     _____       =       _____
_____      _____       _____           _____    X     _____       =       _____
_____      _____       _____           _____    X     _____       =       _____
_____      _____       _____           _____    X     _____       =       _____
_____      _____       _____           _____    X     _____       =       _____
_____      _____       _____           _____    X     _____       =       _____
_____      _____       _____           _____    X     _____       =       _____
_____      _____       _____           _____    X     _____       =       _____

Relative Safety Index (Sum of all Risk Ratings)..................._____
```

Figure 15.12. Sample hazard evaluation form (Roberson and McLymore 1994).

can be improved. Unfortunately, it is not always possible to retrofit older machines with the latest safety designs.

Eliminating hazards is a noble goal for any designer and should be done whenever feasible. Unfortunately, it is impossible to achieve a complete elimination of hazards. Many machines—for example, headers of combines, baler pickup units, or mower blades—have to be somewhat exposed in order to perform their intended function. We may be able to improve a design and reduce the number of hazards or exposure to hazards, but machines cannot be made foolproof.

Many modern machines include protective systems designed to enhance an operator's safety. These include guards and shields, roll over protective structures (ROPS), interlocks, and controls. Safety technology of this type is quite effective in preventing injuries. However, it is not always fail safe. Guards are sometimes removed and not replaced, seat belts provided with ROPS equipped tractors are not always worn, and interlocks or controls are sometimes bypassed by persons who do not recognize the risk involved. The bottom line is that even a safety device can fail and is certainly no substitute for safety conscious behavior.

Safety messages are often placed on machines by using signs and decals. These safety signs point out some of the hazards present, what injuries or illnesses may be caused, and how to avoid the hazard. Safety signs, as illustrated in Figure 15.13. (ASAE S441.2 1995), use a signal word and safety alert symbol to attract attention, a message panel to describe the hazard and its consequences, and often a pictorial panel which illustrates the hazard. The signal words and their application are outlined in Table 15.9 (ASAE S441.2 1995).

Table 15.9. Warning signs.

Signal Words	Application
Caution	Potentially hazardous situation that may cause minor to moderate injury. Also used to warn against unsafe practices.
Warning	Potentially hazardous situation that may cause serious injury or death. Also used to warn against unsafe practices or removal of guards.
Danger	Hazardous situation that will result in serious injury or death if not avoided. Often used for machine components that cannot be guarded due to their intended function

Figure 15.13. Typical safety sign (ASAE S441.2 1995).

Another approach to enhance safety is effective education and training about machine hazards and how to avoid them. Manufacturer sponsored demonstrations, dealer training, and operator manuals are all parts of this program. Effective training and education carries over to the equipment operator. As the owner of a machine, you should provide proper training for its operation. As an operator, you should expect safety training. A very important element is the operator's manual for a piece of equipment. A well-written manual should contain important information on maintenance, operation, and safety (ASAE EP 363.1 1995). This manual should never be overlooked or taken lightly.

Education and training should also include learning to recognize or establish safety zones around equipment hazards, where appropriate. A safety zone is an area the worker should not enter while a machine is in operation because the probability of exposure is greatly increased. Workers should always be aware of the safety zones and the hazards within the zone. Once the hazard is recognized and the safety zone established, exposure is minimized.

Finally, several techniques for prevention of injuries and illnesses are available for the agricultural worker. For example, each worker should have the basic Personal Protective Equipment (PPE) needed for the job assigned. PPE will vary according to the job at hand but may include many of the items listed in Table 15.10.

Table 15.10. Personal protective equipment.

Safety Glasses	Protect the eyes from flying particles.
Gloves	Protect against heat, cold, cuts, vibration, and chemicals.
Clothing	Suited to environment and chemical protection.
Shoes	Protect against crushing, chemicals, and other hazards.
Hearing Protection	Reduce dangerous noise levels.
Headgear	Protect the head and/or face from hazard exposure.
Respirator	Guards against respiratory chemical hazards.

Standards and Regulations

The American Society of Agricultural Engineers (ASAE) lists 45 design and performance standards related to safety (see Table 15.11 (ASAE 1995)). Compliance with ASAE and other standards is voluntary. However, the equipment industry recognizes the value of these standards. Many manufacturers diligently observe and implement the provisions of these standards to enhance the safety of their products.

Table 15.11. ASAE safety standards for agricultural machines.

Standard Number	Standard Title
S474	Agricultural Rotary Mower Safety
S333.2	Agricultural Tractor Auxiliary Power Take-Off Drives
S350	Safety Alert Symbol for Agricultural Equipment
S365.2	Braking System Test Procedures and Braking Performance Criteria for Agricultural Field Equipment
S338.2	Safety Chain for Towed Equipment
EP443.1	Color Coding Hand Controls
S248.3	Construction and Rating of Equipment for Drying Farm Crops
S392	Cotton Module Builders
EP282.2	Design Values for Emergency Ventilation and Care of Livestock and Poultry
EP409.1	Safety Devices for Chemigation
S397.2	Electrical Service and Equipment for Irrigation
S318.11	Safety for Agricultural Field Equipment
S355.1	Safety for Agricultural Loaders
S354.3	Safety for Farmstead Equipment
S361.3	Safety for Portable Agricultural Auger Conveying Equipment
S440.2	Safety for Powered Lawn and Garden Equipment
S395	Safety for Self-Propelled, Hose Drag Agricultural Irrigation systems
S373.1	Safety for Self-Unloading Forage Boxes
S203.13	Front and Rear Power Take-Off for Agricultural Tractors
S304.6	Graphical Symbols for Operator Controls and Displays on Agricultural Equipment
S493	Guarding for Agricultural Equipment
EP367.2	Guide for Preparing Field Sprayer Calibration Procedures
S351	Hand Signals for Use in Agriculture
S279.9	Lighting and Marking of Agricultural Field Equipment on Highways
S317	Improving Safety on Enclosed Mobile Tank for Transporting and Spreading Agricultural Liquids and Slurry
S207.12	Operating Requirements for Tractors and Power Take-Off Driven Implements
S335.4	Operator Controls on Agricultural Equipment
S310.4	Overhead Protection for Agricultural Tractors-Test Procedures and Performance Requirements
S515	Pallet Load Transfer System for Vegetable Harvesters, Shuttle Vehicles, and Road Trucks
S525	Agricultural Cabs - Environmental Air Quality - Definitions, Test Methods, and Safety Procedures
EP371	Preparing Granular Applicator Calibration Procedures
S383.1	Roll over Protective Structures (ROPS) for Wheeled Agricultural Tractors
S519	Roll over Protective Structures (ROPS) for Wheeled Agricultural Tractors (ISO Compatible)
S216	Self Powered Electric Warning Lights
S355.1	Safety for Agricultural Loaders
S441.2	Safety Signs
S363.2	Wiring and Equipment for electrically driven or controlled Irrigation Machines
S276.4	Slow-Moving Vehicle Emblem
EP363.1	Technical Publications for Agricultural Equipment
S431	2000-RPM Front and Mid PTO for Lawn and Garden Ride-On Tractors
S370.3	2000-RPM Power Take-Off for Lawn and Garden Ride-On Tractors

OSHA regulations pertaining to agricultural machinery include sections on slow moving vehicles, roll over protection, and safety for field and farmstead machinery among other topics. The Act also requires that employees who operate tractors and machinery be informed of safe operating practices when initially assigned and at least annually thereafter. Table 15.12 is a compilation of employee operating instructions suggested by the OSHA standards (OSHA 1928 Subpart C App A 1996; OSHA, 1928.57 1996). The owner or manager of any farm should review the OSHA requirements and provide positive measures to prevent accidents.

Fair Labor Standards Act (FLSA) regulations establish minimum ages for children employed in jobs which are considered hazardous (Eickhof 1991). Hazardous occupations related to machinery are listed in Table 15.13.

Most occupations have a minimum age requirement of 16 unless certain exemptions apply. The farm operator should carefully review these requirements and the exemptions if minors are employed.

Table 15.12. Employee operating instructions.

1. Securely fasten your seat belt if the tractor has a ROPS.
2. Where possible, avoid operating the tractor near ditches, embankments, and holes.
3. Reduce speed when turning, crossing slopes, and on rough, slick, or muddy surfaces.
4. Stay off slopes too steep for safe operation.
5. Watch where you are going, especially at row ends, on roads, and around trees.
6. Do not permit others to ride.
7. Operate the tractor smoothly - no jerky turns, starts, or stops.
8. Hitch only to the drawbar and hitch points recommended by tractor manufacturers.
9. When tractor is stopped, set brakes securely and use park lock in available.
10. Keep all guards in place when the tractor or machine is in operation.
11. Stop engine, disconnect the power source, and wait for all machine movement to stop before servicing, adjusting, cleaning, or unclogging equipment. If the machine must be running to make adjustments, carefully follow the procedure provided by the manufacturer.
12. Make sure everyone is clear of the machine before staring or engaging power.
13. Lock out electrical power before performing maintenance or service on farmstead equipment.

SUMMARY

Agricultural machines contribute to a large percentage of injuries. However, as discussed in this chapter, this is not an uncontrollable situation. The effort to reduce injuries begins with the design of the equipment and continues to the machine operator. Any weak link in this chain can lead to injury or illness.

Hazard recognition is an essential safety skill. Whether practiced by a designer or a machine operator, the ability to recognize hazards, understand the consequences, and know how to avoid the hazards is critical. It is not possible to completely remove hazards from all machines. If it were, there would be no potential for injury or illness at all. A "safe" machine would then be possible.

Given that hazards exist and it is unlikely, if not impossible, to remove them completely, we must practice hazard management or control. Hazards can be evaluated to determine severity or risk. Once evaluated and analyzed, the safety hierarchy can be employed to determine the proper course of action.

Table 15.13. FLSA guidelines.

Operating a tractor of over 20 horsepower.

Operating or assisting in operating a corn picker, cotton picker, grain combine, hay mower, forage harvester, hay baler, potato digger, feed grinder, crop dryer, mobile pea viner, forage blower, auger conveyor, self-unloading wagon, power post hole digger, power post driver, or nonwalking rotary tiller;

Operating a trencher, earth-moving equipment, forklift, potato combine, or power-driven circular, band, or chain saw;

Harvesting or loading timber with a butt diameter of more than 6 inches;

Driving a bus, truck, or automobile when transporting passengers;

Working in a horizontal silo while operating a tractor for packing purposes;

Handling, transporting, or applying anhydrous ammonia and chemicals identified by the words "Danger" and "Poison".

Education and training are included as key elements of any hazard management or control strategy. Without proper training, needless injuries can occur despite the application of other safety technology. Agricultural machines are hazardous. As mechanization increases, new hazards will arise. However, by practicing hazard recognition, management and control, injuries and illness can be avoided.

REFERENCES

American Society of Agricultural Engineers (ASAE). 1995. *ASAE Standards*. St. Joseph, MI: ASAE.
S442.1. Safety Signs for Agricultural Equipment.
S276.4. Slow Moving Vehicle Identification Emblem.
EP363.1. Technical Publications.

Eickhoff, W.D. 1991. *North Carolina Farm Labor Rules and Regulations*, AG-444. Raleigh, NC: North Carolina Cooperative Extension Service.

Hathaway, L.R., ed. 1987. *Fundamentals of Machine Operation: Agricultural Safety*. Moline, IL: Deere & Company.

Murphy, D.J. 1992. *Safety and Health for Production Agriculture*. St. Joseph, MI: ASAE.

Murphy, D.J., and S. Steele. 1995. *Extra Riders on Farm Equipment*. University Park, PA: Pennsylvania Cooperative Extension Service, The Pennsylvania State University.

Occupational Safety and Health Administration (OSHA). 1996. 1928.51. Roll-over protective structures (ROPS) for tractors used in agricultural operations. http://www.osha-slc.gov/OshStd_data/1928.0051.html.

Occupational Safety and Health Administration (OSHA). 1996. 1928 Subpart C, App A. Employee operating instruction. http://www.osha-slc.gov/OshStd_data/1928_Subpart_C_App_ A.html.

Occupational Safety and Health Administration (OSHA). 1996. 1928.57. Guarding of farm field equipment, farmstead equipment, and cotton gins. http://www.osha-slc.gov/OshStd_data/1928.0057.html.

Roberson, G.T., and R.L. McLymore. 1994. *Agricultural Machinery and your Health and Safety*. AG-481-2. Raleigh, NC: North Carolina Cooperative Extension Service.

Roberts, D.L. 1990. *A Guide to Farm Safety and Health*. NC-OSHA Industry Guide #10, , Raleigh, NC: Division of Occupational Safety and Health, Department of Labor.

Roberts, D.L. and W.J. Becker, eds. 1991. *Human Factors: A Series of Quality Instructional Materials*. St. Joseph, MI: ASAE.

Saran, C., D. Wheless, and C.W. Suggs. 1991. SHEP Farm Machine Index. ASAE Paper No. 91-5006. St. Joseph, MI: ASAE.

Suggs, C.W. 1991. *Thermal Environment of Agricultural Workers: Environmental Stress. Human Fators: A Series of Quality Instructional Materials*. St. Joseph, MI: ASAE.

Sullivan, H.D. and E.J. Matthews. 1991. *Product Liability. Human Factors: A Series of Quality Instructional Materials*. St. Joseph, MI: ASAE.

16

CHEMICAL HAZARDS OF FARMING

William J. Meggs, M.D., Ph.D., F.A.C.E.P.
East Carolina University School of Medicine

Ricky L. Langley, M.D., M.P.H.
Duke University Medical Center

> The contemporary farm is a highly sophisticated operation in which dozens of chemical products are used. Like all chemical products, toxicity can result from exposures resulting from improper use and accidents. Insecticides and herbicides used to kill insects and weeds are toxic to humans. A variety of toxic rodenticides are used on the farm. Farm equipment requires hydrocarbon products for fuel, lubrication, and cleaning. Fertilizers such as anhydrous ammonia can be very toxic if mishandled. The air in animal confinement facilities can contain endotoxins and other hazardous substances. The decay of hay in silos can produce nitrogen oxides that can have devastating pulmonary toxicity. Asphyxiation can result from collections of gases such as methane in manure pits. The toxic chemicals found on farms are discussed. Mechanisms of toxicity, health hazards of exposures, and strategies to prevent chemical poisonings on the farm are discussed.

INTRODUCTION

A wide variety of toxic chemical products are available for use by farmers, as given in Table 16.1. Insecticides and herbicides used to kill insects and weeds are toxic to humans. A variety of toxic rodenticides are used on the farm. Farm equipment requires hydrocarbon products for fuel, lubrication, and cleaning. Fertilizers such as anhydrous ammonia can be very toxic if mishandled. The air in animal confinement facilities can contain endotoxins and other hazardous substances. The decay of hay in silos can produce nitrogen oxides that can have devastating toxicity.

In this chapter the toxic chemicals to which farmers are exposed will be discussed. The mechanisms of toxicity will be outlined, clinical consequences of exposures will be discussed, and regulations will be presented. Strategies to prevent chemical poisonings on the farm will be given.

PESTICIDES

Pesticides are substances used to kill or repel pests. In agriculture, insecticides, herbicides, fungicides, and rodenticides are widely used to enhance productivity by removing harmful pests. These toxins are often harmful or deadly to humans. In the United States, there were over 61,000 pesticide exposures reported to poison centers in 1994. Of these exposures, 2,413 were of moderate or major toxicity, and there were 8 deaths. Of these deaths, 4 involved organophosphate pesticides with or without other exposures, 2 were from organochlorine pesticides, and a single fatality each occurred with boric acid and an insect repellent (Litovitz 1995). Worldwide, current estimates are that 2 million cases of pesticide poisoning occur annually, with 40,000 fatalities (Foo 1985).

Table 16.1. Examples of chemical hazards on the farm.

Pesticides:
insecticides	rodenticides
fungicides	fumigants
herbicides	

Gases:
ammonia	hydrogen sulfide
carbon monoxide	oxides of nitrogen
carbon dioxide	welding fumes
methane	

Hydrocarbons and Related Chemicals:
gasoline	mineral spirits
kerosene	toluene
diesel fuel	grease
antifreeze	transmission fluid
motor oil	hydraulic oil
paint thinners	freon

> **Table 16.1** *(continued)*
>
> *Biological Agents:*
>   endotoxins   mycotoxins
>
> *Antimicrobial Agents:*
>   antibiotics   sanitizers
>   disinfectants
>
> *Fertilizers*
>
> *Detergents*

A number of pesticides have been canceled, suspended, or restricted due to toxicity to humans or the environment. Table 16.2 lists pesticides that have been banned or restricted in some way. In addition to toxicity, allergic reactions can occur to pesticides. Table 16.3 lists pesticides of known allergenicity.

> **Table 16.2.** Suspended, cancelled, and restricted pesticides.
>
> | | |
> |---|---|
> | Alar | Disinfectants labeled |
> | Aldrin | "germ proof" |
> | Amitraz | EBDCs |
> | Arsenic trioxide | EDB |
> | Benomyl | Endrin |
> | BHC | EPN |
> | Bithionol | Flouroacetamide |
> | Bromoxynil | Heptachlor |
> | Bromoxynil Butyrate | Kepone |
> | Cadmium | Lead arsenate |
> | Calcium arsenate | Lindane |
> | Captafol | Mercury |
> | Captan | Metaldehyde |
> | Carbon Tetrachloride | Mirex |
> | Chloranil | Monocrotophos |
> | Chlordane | OMPA |
> | Chlordimeform | 10,10'- Oxybi- |
> | Chlorobenzilate | sphenoxarsine |
> | Copper arsenate (basic) | Oxyfluorfen |
> | Creosote | Parathion (Ethyl) |
> | Cyanazine | PCNB |
> | Cyhexatin | Pentachlorophenol |
> | DBCP | Phenarsazine chloride |
> | Daminozile | Polychlorinated |
> | DDD (TDE) | biphenyls |
> | DDT | Polychlorinated |
> | 2,4-D | terphenyls |
> | Dinoseb | Pronamide |
> | Quaternary ammonium | Strobane |
> | compounds | Strychnine |
> | Safrole | 2,4,5-T |
> | Seed Treatments | 2,4,5-TCP |
> | Silvex/2,4,5-T | Thallium sulfate |
> | Sodium arsenate | TOK |
> | Sodium arsenite | Toxaphene |
> | Sodium cyanide | Ttributyltin (TBT) |
> | Sodium flouride | Trifluralin |
> | Sodium mono- | Vinyl chloride |
> | flouroacetate | |
>
> Source: Office of Compliance Monitoring, Office of Pesticides and Toxic Substances, U.S. EPA, February 1990.

> **Table 16.3.** Pesticides known to cause allergies.
>
> | | |
> |---|---|
> | Allidochlor | Dichloropropene |
> | Anilazine | Lindane |
> | ANTU | Maneb |
> | Barban | Nitrofen |
> | Benomyl | Propachlor |
> | Captafol | Pyrethrum/pryrethroids |
> | Captan | Rotenone |
> | Dazomet | Thiram |
> | Dichloropropane | Zineb |
>
> Source: Allergy and Pesticides, North Carolina Cooperative Extension Service, January 1994.

INSECTICIDES

Organophosphate Insecticides

Organophosphates were discovered during World War II, when they were developed as nerve gas agents. These pesticides are extremely toxic to humans and other animals. Organophosphate pesticides are extensively used, having to a large degree replaced chlorinated hydrocarbon pesticides such as DDT in the United States even though their acute toxicity to humans is greater. Their advantages over chlorinated hydrocarbon pesticides are less persistence in the environment, less chronic adverse effects on birds and wildlife, lack of long term deposition in human adipose tissues, and fewer chronic health effects in humans. The toxicity of organophosphate pesticides includes acute toxicity, chronic toxicity, and chronic sequela of acute poisonings. Poisonings can occur by inhalation, skin contact with systemic absorption, and accidental or intentional ingestion. Examples of organophosphates include acephate, chlorpyrifos, diazinon, dichlorvos, malathion, parathion, and trichlorfon. The nerve gas sarin that was released in the Tokyo subway (Suzuki 1995) is an extremely toxic cholinesterase inhibitor. Other nerve gases of this class are soman, GF, and GB.

Acute Poisoning. Organophosphates irreversibly block the enzyme acetylcholinesterase which degrades the neurotransmitter acetylcholine. Hence, acetylcholine accumulates at nerve synapses and the neuromuscular junction, and a poisoned person has increased stimulation at the sites of action of acetylcholine. The autonomic nervous system (muscarinic receptors), the neuromuscular junction of skeletal muscle (nicotinic receptors), and the central nervous system are all affected with increased stimulation. The parasympathetic branch of the autonomic nervous system has acetylcholine as a presynpatic and postsynaptic neurotransmitter. Excess parasympathetic stimulation leads to bradycardia, hypotension, salivation, lacrimation (tearing), urinary incontinence, diarrhea, vomiting, abdominal cramping, bronchospasm, and bronchorrhea. The bronchospasm and bronchorrhea simulate an asthma attack and can lead to hypoxia, which in turn can result in tachycardia. In one study of organophosphate poisoning, tachycardia was four times more common than bradycardia (25.3% vs. 6.3%) (Agarwal 1993).

For the sympathetic branch of the autonomic nervous system, acetylcholine is a presynpatic neurotransmitter. In most cases of poisoning, parasympathetic stimulation dominates sympathetic stimulation. Occasionally sympathetic domination is seen, with tachycardia and hypertension rather than bradycardia and hypotension.

Organophosphate poisoning leads to accumulation of acetylcholine at the neuromuscular junction. Clinical consequences are muscle fasciculations, muscle weakness, and paralysis. Paralysis of the respiratory muscles can lead to respiratory failure and death. Central nervous system effects of poisoning include confusion, hallucinations, seizures, and coma.

Opthamological changes associated with acute organophosphate poisoning include pinpoint or miotic pupils, decreased visual fields, myopia, and decreased visual acuity (Rengstorff 1994). Clinical effects that sometimes occur with poisoning include fever, pancreatitis (Zweiner and Ginsberg 1988; Yelamos 1993; Weizman 1992), cardiac arrhythmias (Ludomirski 1982), liver function abnormalities, hypoglycemia, and hyperglycemia (Namba et al. 1971).

Aggressive behavior was reported in four persons poisoned with organophosphate or carbamate pesticides, with two homicides. This behavior resolved after exposure ceased, occurred in individuals with no history of violence or antisocial behavior, who showed remorse, and had no subsequent history of violent behavior. Enhanced activation of cholinergic receptors in the hypothalamus was proposed as the mechanism of this violent behavior and is supported by animal studies (Devinsky 1992).

Treatment of Acute Organophosphate Poisoning. Atropine, an anticholinergic agent, is used to reverse the muscarinic effects of acute organophosphate pesticide poisoning. The dose of atropine is titrated to stop the copious secretions, including the pulmonary secretions which are the most dangerous and can lead to asphyxiation. Large doses as well as prolonged therapy with atropine may be necessary in order to dry the respiratory tract. Atropine infusion for as long as 5 weeks has been reported with severe organophosphate poisoning (Beards 1994) but lead to a paralytic ileus. Atropine is known to cause tachycardia, but atropine should not be withheld in an organophosphate poisoned patient with tachycardia. The tachycardia may be due to hypoxia from bronchorrhea. The indication for using atropine in acute organophosphate poisoning is respiratory tract secretions, and the dose necessary to dry these secretions, no matter how high, should be used. Doses many times those needed for other indications such as sinus bradycardia in the setting of ischemic cardiac disease may be needed.

A class of compounds called oximes displaces organophosphates from cholinesterases. These compounds can reverse muscle paralysis from organophosphate poisoning. One oxime, 2-PAM or pralidoxime, is available in the United States for the treatment of acute organophosphate poisoning. Aging is a term used to describe the phenomena of permanent and irreversible binding of the organophosphate to cholinesterases, which occurs after 1 to 2 days. Once aging occurs, the enzymatic activity of the cholinesterase is permanently destroyed, and new enzyme must be synthesized before clinical symptoms resolve. After aging has occurred, pralidoxime is no longer effective in removing the organophosphate from the enzyme. Hence, it is imperative that the enzyme be unblocked with pralidoxime before aging occurs.

Systemic poisoning from organophosphates can occur from inhalation, skin absorption, and ingestion. The severity and rapid onset of poisoning is such that Emergency Medical Services should be summoned immediately after poisoning. The victim should be decontaminated immediately. Any exposed clothing should be removed. Leather belts and shoes should be discarded because organophosphates can permanently bind to them. Rescuers should be careful to avoid becoming poisoned. Urine and phlegm from poisoned persons can be toxic to rescuers and pass through skin and latex gloves.

Intermediate Syndrome Following Acute OP Poisoning. An intermediate syndrome has been described, with neurological symptoms beginning 1 to 4 days after acute OP poisoning. Paralysis of neck flexor muscles, muscles innervated by cranial nerves, proximal limb muscles, and respiratory muscles occurs 48 to 96 hours after poisoning, with some patients requiring ventilatory support. Symptoms resolved by 18 days in most patients, though one patient developed a delayed polyneuropathy (Senanayake et al. 1987; Mani et al. 1992). Use of 2-PAM is thought to decrease or prevent the incidence of the intermediate syndrome.

Chronic Neurological Sequela of Acute Organophosphate Poisoning. Permanent neurological deficits described in persons poisoned with organophosphate pesticides include neuropsychiatric deficits and paralysis. Studies have found a broad spectrum of neuropsychiatric disability among subjects with a past history of acute poisoning relative to control groups (Savage 1988; Rosenstock et al. 1991; Steenland et al. 1994). Severity of the deficits was worst for those most severely poisoned, with severity measured by days of hospitalization or disability (Steenland et al. 1994). Persistent peripheral neuropathy has also been described, with decreased finger and toe vibrotactile sensitivity (Steenland et al. 1994). A study of farm workers who suffered a single acute poisoning found poisoned workers performed poorly relative to matched controls on all neuropsychological tests, including verbal and visual attention, visual memory, visuomotor speed, sequencing and problem solving, motor steadiness, and dexterity (Rosenstock et al. 1991). A review of the literature concluded that the existing literature supports in a consistent fashion the following sequela of OP poisoning: impaired vigilance and reduced concentration, reduced information processing and psychomotor speed, memory deficit, linguistic disturbances, depression, anxiety, and irritability (Ecobichon and Joy 1994).

Neuropsychological effects have been documented with chronic occupational exposure to organophosphate pesticides. A controlled study of sheep farmers exposed to organophosphate pesticides in sheep dip found significantly worse performance on tests to assess sustained attention and speed of information processing relative to controls, as well as "vulnerability to psychiatric disorder" as determined by a questionnaire. Short term memory and learning testing was not affected in this study (Stephens and Spurgeon 1995). A study in India of workers chronically exposed to organophosphates found a significantly higher prevalence of peripheral neuropathy in exposed workers (Ernest et al. 1995). Vibrotactile thresholds are elevated in workers exposed to some organophosphates relative to controls (McConnell et al. 1994).

Paralysis has been reported as a complication of OP poisoning. Cases of delayed bilateral recurrent laryngeal nerve paralysis have been reported 25 to 35 days after poisoning with chlorpyrifos, parathion, and methamidophos (de Silva et al. 1994). Chronic fatigue (Corrigan et al. 1994) and the multiple chemical sensitivity syndrome have been associated with organophosphate poisoning (Corrigan et al. 1994; Tabershaw and Cooper 1966; Meggs and Cleveland 1992) with as many as 17.5% of persons poisoned reporting a persistent intolerance to chemicals (Tabershaw and Cooper 1966).

Chronic Organophosphate Exposure. Chronic exposures to organophosphate pesticides occur among farm workers, pesticide manufacturing plant workers, and exterminators. Symptoms of chronic exposures are not as dramatic as seen in acute exposures. The most common symptom developing in exposed orchard sprayers was headache, followed by nausea, weakness or fatigue, and chest tightness. Other symptoms were abdominal pain, vertigo or incoordination, vomiting, perspiration, cough, vision disturbance, loss of appetite, dyspnea, nasal discharge, miosis, and wheezing (Sumerford et al. 1953).

Children exposed several times a year to 3% malathion by aerial spraying in Japan were reported to develop oculopathy, with optic neuropathy and retinal degeneration. Oculopathy was thought to resolve with treatment with prinfinium bromide and vitamins B12, C, and E (Ishikawa et al. 1993).

Biomarkers of Exposure to Organophosphates. Since organophosphates inhibit plasma and red blood cell cholinesterase, functional assays of these enzymes can be used as a marker of exposure. Decreases in the enzymes from baseline measurements are of greater value than a single level because of the variation in measured values over a population of people. An individual whose baseline enzyme levels are high in the normal range can become significantly poisoned, have a decrease in enzyme level, and still remain in the normal range. Studies of exposed greenhouse workers show decreases in cholinesterase levels during seasons in which pesticides are used (Lander and Hinke 1992). A study in Israel found symptoms associated with chronic exposure in individuals with normal cholinesterase levels (Richter et al. 1992).

In spite of this difficulty, decreases of cholinesterase activity from baseline is thought to correlate with toxicity. Mild symptoms of headache and parasympathetic stimulation develop with a 60% decrease in cholinesterase levels. Symptoms resolve in one day. With 60 to 90% decreases in levels, moderate symptoms develop. Muscle weakness, tremor, and neuropsychiatric symptoms can develop and persist for up to 2 weeks. With a 90 to 100% reduction in levels, severe symptoms, with seizures, cyanosis, pulmonary edema, respiratory failure due to muscle weakness, coma, and death can occur (Namba 1971).

Serum cholinesterase levels have poor prognostic value in cases of acute organophosphate poisoning. These levels do not correlate with amount of atropine required or need for mechanical ventilation (Nouira et al. 1994). In a study of workers chronically exposed to organophosphates, cholinesterase levels were not found to correlate with symptoms (Matchaba-Hove 1995).

Prevention of Organophosphate Poisoning. Closed systems for mixing and loading pesticides have been developed, but unfortunately these systems do not always minimize exposure. In a study conducted among workers at airports in Nicaragua, workers using closed systems were found to have lower cholinesterase levels than workers mixing and loading pesticides by hand (McConnell et al. 1992). A lack of training in the use of closed systems may have contributed to the increased exposure. Workers using closed systems applied 3,250 pounds/worker/14 days while workers using convention systems only applied 849 pounds/worker/14 days. This increase in amounts of pesticides handled may account for the greater decrease in cholinesterase activity with closed systems (McConnell et al. 1992).

100% cotton gloves were found effective in preventing organophosphate poisoning by skin absorption in an in vitro test system (Keeble et al. 1993). One study found wearing of protective clothing and frequency of application influenced cholinesterase levels, while use of gloves and face masks did not (Lander and Hinke 1992).

Aerial spraying leads to increased drift of pesticides away from the target and has been associated with illness among residents living near the sprayed fields. Subacute poisonings—with headache, fatigue, diarrhea, cramps, and respiratory poisoning—occur without dramatic reductions in cholinesterase levels. Measures recommended to minimize problems with aerial spraying are (Gordan and Richter 1991):

1. Using reduced amounts of organophosphate pesticides.
2. Using less toxic pesticides.
3. Reducing the length of the spray season.
4. Banning the use of flaggers.
5. Using tractor spraying instead of aerial spraying.
6. Spray planes should have enclosed, air conditioned cockpits.
7. Pilots should be properly trained and licensed.
8. Using protective clothing.

Avoidance of concentrated solutions, and using closed systems for mixing and loading, reduce toxicity if properly done. However, these systems greatly increase efficiency, and hence exposure, so one mixer or loader can mix and load several times as much pesticide as one not using closed systems. Proper training in the use of these systems is important. Ground sprayers must be careful not to spray into the wind, because poisonings have occurred from inhalation and skin contact under such circumstances.

In Israel, cotton yields increased though both organophosphate and total pesticide use decreased (Gordan 1991). This experience illustrates that efficient management of pests, with reduction of pesticide use, is possible. Integrated Pest Management (IPM) is a technique for reducing pesticide use by increasing surveillance, reacting to problems with spot rather than blanket treatments, while decreasing prophylactic or routine spraying in the absence of a problem. Decreased use of these toxins should reduce the number of poisonings while reducing environmental contamination.

Carbamate Insecticides

Carbamates are a class of structurally related compounds used as insecticides (aldicarb, carbaryl, carbofuran, isoprocarb, and xylycarb), herbicides (asulam, chlorbufam, phemedipham), fungicides (benomyl, carbendazim, thiophanates), and nematocides. Medicinal uses of carbamates include using physostigmine as a cholinergic agent to treat anticholinergic overdoses, and pyridostigmine as a treatment of myasthenia gravis. The mechanism of action of carbamates is the same as organophosphate pesticides, in that carbamates inhibit cholinesterases. Unlike organophosphates, the carbamate inhibition is reversible and the duration of action of poisoning is shortened. Chronic effects are thought less likely to occur with carbamates than organophosphates but have been reported (Garber 1987; Dickoff et al. 1987).

Acute Poisonings with Carbamates. Excessive parasympathetic stimulation, with pinpoint pupils (miosis), salivation, diaphoresis (sweating), lacrimation (tearing), bronchorrhea (excessive bronchial secretions), bronchospasm, vomiting, diarrhea, and abdominal cramps have been seen with acute carbamate poisoning. Muscle weakness with respiratory failure, seizures, and coma occur from action at neuromuscular junctions and central nervous systems receptors.

A carbamate poisoning of epidemic proportions occurred in California after watermelons were contaminated with aldicarb. Over 600 people were affected with cholinergic symptoms occurring within 2 hours of ingesting contaminated melon. Seventeen of these required hospitalization, 6 died, and 2 in utero deaths occurred (Jackson et al. 1986).

Like acute organophosphate poisonings, acute carbamate poisoning has been associated with pancreatitis (Weizman and Sofer 1992).

Treatment of Carbamate Poisoning. As in organophosphate poisoning, treatment is aimed at blocking the cholinergic crisis with atropine. Atropine should be titrated to a dose that reduces respiratory secretions. There are two reasons given for not using pralidoxime to treat carbamate poisoning. Since the binding half life of carbamates to cholinesterase is short, on the order of 30 minutes, and irreversible binding does not occur as in organophosphate poisoning, little need is seen for pralidoxime. Further, animal and human case reports have suggested that pralidoxime should not be used to treat carbaryl poisoning. A study of children poisoned with carbamates and treated with the oxime obidoxime chloride (Toxogonin) found no adverse reactions (Lifshitz et al. 1994). Toxogonin was found to be efficacious in a murine study of carbamate poisoning with physostigmine, pyridostigmine, and aldicarb but increased the toxicity of carbaryl (Sterri et al. 1979). Pralidoxime should be used to treat mixed poisonings with organophosphates and carbamates, and pesticide poisonings when the agent has not been identified but there are symptoms of cholinesterase inhibition.

Chronic Sequela of Carbamate Poisoning. Chronic toxicity has not been established to the extent documented for organophosphate poisoning, but case reports have appeared. A 27-year old man with carbaryl poisoning suffered acute cholinergic poisoning with coma and pinpoint pupils. Chronic sequela of severe ankle and toe weakness, loss of vibratory sensation and proprioception in the toes, and decreased sensation to pinprick and touch below the midcalf developed (Dickoff et al. 1987). A 55-year old woman developed a sensorimotor polyneuropathy with lower limb paresthesia and difficulty walking after an ingestion of m-tolyl methyl carbamate (Umehara et al. 1991).

Chlorinated Hydrocarbon Insecticides

Dichlordiphenyltrichloroethane (DDT) was the first and remains the prototype pesticide of the organochlorine, or chlorinated hydrocarbon class. Examples of other pesticides in this class are lindane, aldrin, dieldrin, chlordane, and heptachlor. These pesticides have been restricted or banned in the United States but are used throughout the world both to control agricultural pests and insects that transmit infectious diseases to humans. Persistence in the environment, accumulation in human tissues, concentration at the top of food chains, and toxicity to humans and wildlife resulted in curtailment of widespread use in the United States.

Acute Toxicity of Organochlorine Insecticides. The primary toxicity of organochlorine pesticides is neurological, with tremors, muscle twitches, and paresthesias occurring in mild cases. Dizziness, fatigue, malaise, and headache can accompany these symptoms. At higher doses, seizures and coma can result. Nausea and vomiting are seen with ingestion. Respiratory failure can occur and is fatal without treatment. The cyclodiene compounds (aldrin, dieldrin, endrin, and toxaphene) have lower thresholds of toxicity and are more commonly associated with acute toxicity in humans.

Chronic Toxicity of Organochlorine Compounds. Chronic exposure to dieldrin has caused a symptom complex that included headache, blurred vision, diplopia, tinnitus, dizziness, muscle twitches, diaphoresis, insomnia and nightmares, nystagmus, coordination difficulties, personality changes, and muscular fibrillation. Involuntary jerking of extremities has followed chronic exposures to aldrin, dieldrin, and thiodan (Jager 1970), but is less likely to occur with DDT. Anorexia and weight loss, muscle weakness, tremor, anxiety, nervous tension, and fear have been reported with chronic DDT exposure. Dicofol, a derivative of DDT, has been associated with behavioral problems and decreased academic performance that persisted after an acute exposure (Lessenger and Riley 1991).

A number of reports have linked chronic exposure to organochlorine compounds to chronic neurological symptoms. Chronic motor neuron disease has been linked to aldrin, lindane, and heptachlor (Fonseca et al. 1993). An increased risk of liver and other cancer was found in pesticide applicators in Rome who were exposed to organochlorine pesticides in the 1960s (Figa-Talamanca et al. 1993). Neurobehavioral tissue defects have been found in children born to mothers with high tissue levels of organochlorines. These compounds are stored for long periods of time in fatty tissue and passed to the baby in utero. Based on animal data, it is thought that these deficits persist into adulthood (Hall 1992).

Chlordecone (Kepone) with chronic exposure causes tremor, nervousness, weight loss, rash, mental changes, weakness, ataxia, slurred speech, and loss of coordination (Martinez et al. 1978).

Pyrethrins

Pyrethrins are a class of naturally occurring substances with potent insecticidal activity. These compounds are found in the chrysanthemum plant, *Pyrethrum cinerariae folium*, and other species. Pyrethroids are synthetic analogues of pyrethrins. These compounds are biodegradable and do not accumulate in human tissues. Pyrethrums and pyrethroids are allergens that have been associated with asthma, rhinitis, and hypersensitivity pneumonitis. In animal models, pyrethroids are neurotoxic. A human death associated with an inhalation exposure to a pyrethrin shampoo was attributed to severe bronchospasm (Wax and Hoffman 1994).

Table 16.4. Diagnostic criteria for acute occupational pyrethroid poisoning (He et al.).

1. Suspicious Cases
    having abnormal facial sensations
    miliary papules or contact dermatitis
    without significant signs and symptoms
2. Mild Acute Poisoning
    In additiion to the above skin symptoms or signs,
    having significant systemic symptoms, i.e.,
    dizziness,
    headache, nausea, anoxrexia, fatigue,
    with listlessness, vomiting or increased stomal secretion
    resulting in sick-leave for more than one day.
3. Moderate Acute Poisoning
    Having aggravation of the above systemic symptoms
    and signs, and occurrence of mild disturbance of consciousness, or
    muscular fasciculations in limbs.
4. Severe Acute Poisoning
    In addition to the above systemic symptoms or signs,
    having convulsive attacks, coma or pulmonary edema.

In humans, skin contact has lead to paresthesias with tingling and burning, with onset within 30 minutes of exposure and persisting up to 8 hours (LeQuesne et al. 1980). While symptomatic, a person exhibits no abnormalities on neurological examination or nerve conduction studies, and the symptoms probably arise from direct action of pyrethroids on sensory nerve endings. These symptoms were seen in 32 of 50 sprayers in a study of Chinese cotton field sprayers, but systemic poisoning was not seen (Zhang 1991). Burning of the mouth and disturbed taste can result from pyrethroid exposure, and these oral symptoms are reported to persist longer than dermatological symtoms (Grant 1993).

Systemic toxicity in humans can occur with pyrethroids. In China, pyrethroids are used extensively in agricultural practice, and a number of cases of systemic toxicity from the pyrethroids cypermethrin, deltamethrine, and fenvalerate have been reported. In a Chinese study, occupational exposures were associated with burning and itching of the face. Ingestion was associated with nausea, vomiting, and epigastric pain. Weakness, fatigue, headache, and dizziness can occur. Fasciculations, seizures, coma, and death have been reported in severe poisonings, which are more likely to occur with ingestion (He et al. 1989). Based on a review of 573 cases reported in the Chinese medical literature from 1983 to 1988, He and his collaborators presented a diagnostic criteria for acute pyrethroid poisoning that is given in Table 16.4. Treatment of severe pyrethroid poisoning is supportive, with no known antidotes.

HERBICIDES

Several classes of herbicides are in general use and can pose health hazards for farmers.

Chlorophenoxy Herbicides

2,4-dichlorophenoxyacetic acid (2,4-D) is effective against broad leaf plants and is commonly used as a weed killer on lawns and grain crops. Toxicity can result from ingestion, dermal contact, and inhalation, and multiple organ systems can be affected. Nausea, vomiting, diarrhea, pulmonary edema, cardiac arrhythmias, bradycardia, muscle twitches, and myotonia have been reported. Dermal exposure to concentrated solutions can cause chemical burns. Peripheral neuropathy has been associated with chronic exposure. Painful paresthesias and muscle stiffness have been described (Goldstein et al. 1959). Exposure of farmers has been shown to increase the risk of non-Hodgkin's lymphoma in a dose dependent fashion (Zahm et al. 1990). The defoliant Agent Orange that was used extensively in Vietnam by United States forces was a mixture of 2,4-D and 2,4,5-T.

2,4,5-trichlorophenoxyacetic acid (2,4,5-T) is an herbicide that was banned in the United States in 1979 due to contamination with the by-product, 2,3,7,8-tetrachlorodibenzo-p-dioxin (TCDD). Gastrointestinal symptoms, hypotension, tachycardia, and coma have been described with acute ingestion, and transient peripheral neuropathy has been described in the recovery phase (O'Reilly 1984; Hayes 1982).

Urea Substituted Herbicides

The urea substituted herbicides diuron, linuron, monolinuron, and monuron are inhibitors of photosynthesis. This class of herbicides is of low systemic toxicity, but methemoglobinemia can occur with ingestion. In cases of ingestion, the patient should be observed for cyanosis and elevated methemoglobin levels.

Endothall is a herbicide which is a mucous membrane and skin irritant. With ingestion, endothall has been reported to cause gastrointestinal hemorrhage, hypotension, disseminated intravascular coagulopathy, and death (Allender 1983; Day 1988).

Bipyridyl Herbicides

The bipyridyl herbicides paraquat and diquat are extensively used as weed killers. These quaternary nitrogen compounds inhibit photosynthesis, damage cell membranes, and have broad spectrum cidal activity against plants that is of short duration. Paraquat has low potential for systemic toxicity from incidental inhalation exposure with field use, but ingestion can be lethal. Skin exposure results in blistering and irritation. Ocular exposure can cause ulceration of the cornea. Cough and nose bleed have been reported following inhalation exposure. Ingestion of paraquat has resulted in nausea, vomiting, gastric ulcers, rhabdomyolysis, renal failure, and pulmonary fibrosis. Pulmonary fibrosis has a high mortality with death occurring within 2 weeks. With severe ingestion, multiple organ system failure, with cardiac, respiratory, renal, and hepatic dysfunction, leads to death within a day.

Paraquat ingestion demands immediate and early gut decontamination. The diatomaceous Fuller's earth and bentonite have been employed. Activated charcoal is equal to the diatomaceous earths in binding paraquat (Okonek 1982) and is more readily available. Oxygen supplementation should be avoided if possible because paraquat accumulates in the lung and causes tissue damage by lipid peroxidation and superoxide formation. Dialysis is thought to be ineffective in removing paraquat from blood, but hemoperfusion, though controversial, may be of benefit if prolonged (Okonek 1982).

Diquat is similar in structure to paraquat but does not cause pulmonary fibrosis. Like paraquat, it is caustic and can result in dermal burns from skin contact and gastrointestinal damage with ingestion. Hepatic, renal, and central nervous system toxicity occur, and the lethal dose is similar to that of paraquat.

FUNGICIDES

A number of substances are used as antifungal agents on highly perishable crops such as berries. Seeds are often treated with fungicides to prevent rot before germination. Packing paper and cartons for produce may be impregnated with fungicides. Fungicides are also found in paints, carpets, fabrics, and paper pulp. Organometallic compounds with cadmium, copper, mercury, and tin are in use, as well as carbamates and chlorinated aromatic compounds.

Allergic diseases have been associated with a number of fungicides. Occupational asthma has been associated with the fungicides captafol (Royce et al. 1993) and in a farmer exposed to tetrachlororisopathalonitrile (Honda et al. 1992). Contact dermatitis has been reported in associated with Kathon 893 (Oleaga et al. 1992), fluazinam (van Ginkel et al. 1995), dichlofluanid (Hansson et al. 1995), triforine (Ueda et al. 1994), bupirimate (McFadden et al. 1993), and Metasol D3T (Emslie 1993). The fungicide chlorothalonil has been associated with both contact urticaria and anaphylaxis (Dannaker et al. 1993), contact dermatitis (Bruynzeel et al. 1986) and has been reported to cause facial dermatitis from exposure to chlorothalonil containing paint (Eilrich and Chelsky 1991). The fungicide Dyrene was verified as the cause of an outbreak of contact dermatitis at a tomato-strawberry farm by patch testing (Schuman and Dobson 1985).

Acute circulatory failure has been associated with oral ingestion of the fungicide iminoctadine (Koyama et al. 1993).

Organometallic Compounds

Cadmium chloride, sulfate, and succinate are used as fungicides and are gastrointestinal and respiratory irritants. Acute inhalation can lead to pneumonitis and pulmonary edema, as well as dyspnea, cough, chest pain and fever.

Copper containing organic chemicals are available as fungicides. Numerous products and formulations are available. These compounds are irritants that effect the eyes, skin, and respiratory system. Systemic toxicity is rare and occurs from ingestion, with liver failure, renal failure, gastrointestinal irritation, and hemolysis being reported.

Organic mercurials are extremely toxic and are used as fungicides on seeds. Poisonings occur when seeds are ingested, and secondary poisonings of humans eating meat from animals fed contaminated seeds have occurred. Headache, paresthesia of the extremities, emotional lability, and muscle rigidity or spasticity can occur. Ingestion can be fatal, and survivors can have permanent neurological deficits.

Tin is available as triphenyl and tributylin salts for use as a fungicide. Occupational exposures are associated with irritant reactions of eyes, skin, and respiratory tract. Headache, nausea, emesis, seizure, and coma can arise in severe poisonings. Cerebral edema has been associated with ethyl tin exposure. Triphenyl tin acetate exposure through cutaneous absorption led to urticarial eruption, hepatic injury, slight glucose intolerance, and electroencephalogram abnormalities (Colosio et al. 1993). Exposure to bis (tributyltin) oxide in paint led to burning sensations of the nose and forehead, headache, nose bleed, cough, anorexia, nausea, and vomiting (Anonymous 1991).

Chlorinated Aromatic Fungicides

Organochlorine compounds based on the benzene ring are widely used as fungicides. These compounds include hexachlorobenzene, pentachloronitrobenzene, dicloran, chlorothalonil, and chloroneb. These compounds may be irritants, and human toxicity has been reported for hexachlorobenzene. Single dose exposures of hexachlorobenzene are tolerated, but chronic exposure from ingesting treated wheat seed has lead to impaired hemoglobin synthesis, accumulation of porphyrins, and hepatotoxicity (Schmid 1960). Chronic persistent hepatitis has been associated with diphenyl, which is used in packing paper for citrus fruit (Carella and Brett 1994).

Thiocarbamates

Thiocarbamate fungicides have very different toxicity from carbamates insecticides. They have weak or no cholinesterase inhibition. Thiram is an irritant and contact sensitizer, and has a disulfiram-like interaction with ethanol. Rarely occupational exposure to thiram followed by alcohol ingestion has lead to symptoms of headache, flushing, confusion, diaphoresis, and rash. The fungicide metam-sodium interacts with water to produce methyl isothiocyanate gas, which is an respiratory irritant that can lead to pulmonary edema. Ziram and ferbam are identical thiocarbamates except for a zinc or iron atom, respectively. Toxicity is primarily irritancy to eyes, respiratory system, and skin. Ziram has rarely been associated with neurotoxicity and hemolytic anemia. Ethylene bi dithiocarbamate (EBCD) compounds are maneb (manganese moiety), zineb (zinc moiety), nabam (sodium moiety), and mancozeb (zinc and maneb). These compounds are primarily irritants. A case of maneb exposure in a garden was associated with acute renal failure (Koizumi et al. 1979). Manganese exposure is known to produce a Parkinson's like syndrome, and cases of agricultural workers exposed to maneb have developed such a syndrome (Ferraz et al. 1988).

A number of other substances are used as fungicides, and except for skin irritancy, reports of human toxicity from these compounds have not been reported except for triadimefon. In addition to being an irritant, this compound can cause hyperactivity followed by sedation in humans (Morgan et al. 1989).

Thiazoles

Symptoms associated with 2-(thiocyanomethylthio) benzothiazole (TCMTB) in saw mills include dry skin around the eyes, blood-stained mucus from the nose, nose bleed, peeling skin, itching and burning skin, and skin redness or rash (Teschke et al. 1992), suggesting an irritant effect.

RODENTICIDES

A number of toxic rodenticides are used on the farm. In general, toxicity only arises from accidental or intentional ingestion. Yellow phosphorus is used as a rodenticide, and skin contact can cause severe burns. Warfarin ingestion seldom lead to coagulopathy in humans, but super warfarins such as brodifacoum can cause prolonged coagulopathy requiring treatment with vitamin K1. Red squill contains a cardioglycoside which causes severe nausea in humans, often leading to vomiting. However, cardiac toxicity requiring treatment with anti-digoxin antibody fragment has been reported. Vacor, a rodenticide that destroys beta islet cells in the pancreas, leading to diabetic ketoacidosis, and thallium, which causes painful paresthesias, gastrointestinal symptoms, cardiotoxicity, and alopecia, have both been banned in the United States.

FEDERAL REGULATION OF PESTICIDES

In the United States, the federal government regulates pesticides on the basis of the Federal Insecticide, Fungicide, and Rodenticide Act (FIFRA), originally passed in 1947, and the Federal Food, Drug, and Cosmetic Act (FFDCA). FIFRA gives the U.S. Environmental Protection Agency (EPA) the authority to register pesticides through the Office of Pesticide Programs. Levels of pesticide residues in food and feed crops are regulated through FFDCA. The EPA has the authority to ensure that adverse effects to humans and the environment do not result when pesticides are used according to the label, and has the authority to balance benefits against risks when a pesticide is registered. Over five thousand registration submissions per year are submitted to the EPA, including new agents and amendments of existing labels. About twenty new registrations for active ingredients are submitted yearly. A large body of data is required for registration of a new pesticide. The cost can be $10 million and accumulation of the data can take several years (Culleen 1994).

Certified private pesticide applicators are required to maintain records of applications of restricted use pesticides. These records must contain the product name, EPA registration number, amount applied, the size of area treated, and the precise date of application. In addition, the record must contain the crop, commodity, stored product, or site treated. Location must be specified in enough detail to allow the precise area treated to be identified 2 years later. The applicator's name and certification number must be recorded. Access to the information is restricted to health care professionals caring for exposed individuals, and state and federal representatives with identification. Fines of up to $500 can be levied for a first offense, but subsequent violations carry fines of not less than $1,000 per violation.

FERTILIZERS

Fertilizers are composed of nitrogen, phosphates, lime, and potash. A variety of forms and compositions exist, and chemicals can be applied singly or in combination. Granular fertilizer is not as hazardous as gases and liquids.

Anhydrous ammonia fertilizer is a gas at atmospheric pressure. Under pressure, it is a liquid. When dissolved in water, anhydrous ammonia forms an extremely alkaline solution which can cause severe burns of the eyes, skin, and mucous membranes (CDC 1993a). Ocular exposure can result in severe corneal burns. First and second degree burns of the skin can occur. Exposure time and concentration are determinants of the severity of the burns. The odor threshold for ammonia is 53 parts per million. At the irritant threshold of 400 ppm, irritation of the eyes, nose, and throat occur. Immediate eye injury occurs at exposures of 700 ppm. At 1700 ppm, laryngeal spasm and coughing can occur. Exposures to 2500 to 4500 ppm for 30 minutes can be lethal, and an exposure of 5000 ppm is rapidly fatal (Dalton 1978; Millec et al. 1989). Upper airway edema can develop rapidly and leads to cyanosis and asphyxiation. Non-cardiac pulmonary edema or adult respiratory distress syndrome has been reported after ammonia exposure. Chronic sequela to an acute exposure includes bronchiolitis obliterans and chronic cystic bronchiectasis (Dalton 1978; Millec et al. 1989).

After eye or skin exposure to ammonia, exposed areas should be decontaminated by flushing with water. Large amounts of water should be used, and irrigation should be prolonged. After eye exposures, several liters of water may be necessary, and irrigation should continue until after pH has normalized. After termination of the irrigation, the pH of tears should be checked after 10 minutes to allow for equilibration with irrigation repeated until pH normalizes at 10 minutes. The eye should be examined with fluorescein to check for corneal damage and treated appropriately. Chemical burns to the skin should be treated as any other burn of similar size, location, and thickness.

Inhalation exposures to ammonia can be devastating. At the scene, victims should be removed to fresh air while the emergency medical services system is activated, then transported to the nearest hospital emergency department. Patency of the airway and competency of respirations and oxygenation should be verified. Endotracheal intubation and mechanical ventilation should be instituted if indicated. Since the onset of pulmonary edema may be delayed, a victim without initial pulmonary symptoms should be observed for 12 hours. A follow up chest x-ray should be obtained before discharge from the emergency department.

Exposures to concentrated anhydrous ammonia most often occur during the transfer from supply tanks to applicator tanks. Damaged hoses, leaking valves, and disconnected hoses are the main sources of exposure (CDC 1993a).

Well water can become contaminated with nitrates from fertilizers, septic tanks, and animal manure. Infants of farm families using well water are particularly at risk for developing methemoglobinemia from contaminated water. This increased susceptibility of infants is due to their relative absence of methemoglobin reductase, lower levels of NADPH in red blood cells, and the sensitivity of fetal hemoglobin to oxidative stresses.

CHEMICAL HAZARDS ASSOCIATED WITH FARM EQUIPMENT

A variety of farm machinery is operated and repaired by farmers. The buildings in which the machinery is housed are also maintained and repaired by farmers. Solvents such as gasoline, diesel fuel, paint thinners and strippers, mineral spirits, kerosene, hydraulic fluids, and greases are used by farmers on the equipment (Shaver and Tong 1991). Many of these solvents are flammable and explosive. Farmers are exposed occupationally to solvents through skin contact, inhalation, and ingestion.

Solvents can be upper respiratory and mucus membrane irritants. Central nervous system symptoms can occur with systemic absorption, such as dizziness, dysphagia, slurred speech, and mental confusion. Seizure, coma, and death can occur in severe cases. Organic solvents can sensitize the myocardium to indigenous catecholamines, resulting in cardiac dysrhythmias and sudden death (Proctor et al. 1988). Defatting of the skin by solvents can cause dermatitis and at high concentrations can result in severe burns.

Gasoline siphoning with a hose can be started by sucking on the hose. This practice on the farm has lead to ingestion and aspiration, with severe lung injury from aspiration (Shaver and Tong 1991).

Farmers solder or weld broken equipment, so welding hazards such as inhalation of metal fumes, ozone, nitrogen dioxide, and carbon dioxide can exist on the farm. Metal fume fever, most commonly associated with zinc fumes and resulting from welding galvanized materials, can occur on the farm.

Use of gloves and protective clothing is essential when farmers work with chemical agents. Welding goggles and safety goggles can protect the eye from keratitis and chemical injuries. Chemicals on the skin should be removed immediately with water, and contaminated clothing should be removed. Children on farms present a special problem with regard to chemical exposures. Farm chemicals should be stored in sealed containers in secured areas, and children should be warned of chemical hazards.

CHEMICAL HAZARDS OF ANIMAL CONFINEMENT

For 25 years, confinement facilities to raise cattle, pigs, and poultry, with almost a million workers employed, have been developed. Large buildings frequently house thousands of animals and generate large amounts of waste products. The waste usually drains through floor grates into a containment area. In other systems, the wastes are washed into gullies in the concrete floor, and then into a storage pond or lagoon. Toxic gases are generated by organic wastes. Anaerobic bacterial digestion of manure produces over 150 different gases. A variety of alcohols, acids, amines, carbonyls, sulfides, nitrogen heterocycles, esters, mercaptans, disulfides, carbon monoxide, methane, ammonia, and hydrogen sulfide are produced. Oxygen depletion may occur near the surface of the manure. Many deaths have been recorded of farmers cleaning out manure pits, with NIOSH recording 104 fatalities in 68 incidents in animal confinement facilities from 1982 to 1992 (CDC 1993d).

Methane and carbon dioxide are simple asphyxiants. If levels of methane or carbon dioxide are elevated, this oxygen can be displaced and death may occur from oxygen deprivation. Methane can also accumulate to explosive levels. With ventilation system failure, carbon dioxide produced from animal expiration can rise to dangerous levels.

Hydrogen sulfide gas is a highly toxic metabolic poison similar to cyanide in its action. Cytochrome oxidase activity is blocked, so aerobic respiration ceases. The noxious rotten egg odor of hydrogen sulfide does not persist with continued exposure due to the effects of hydrogen sulfide on the olfactory apparatus. Above 150 ppm, the olfactory apparatus is paralyzed, and danger may persist but not be recognized. Any exposure to a rotten egg odor should lead to evacuation of the area until risk can be determined.

Entry to a manure pit should only be done with self contained breathing equipment and a partner (CDC 1993d).

Respiratory problems associated with confinement housing include the toxic organic dust syndrome, acute and chronic bronchitis, occupational asthma, chronic obstructive pulmonary disease, and hypersensitivity pneumonitis (do Pico 1992).

CHEMICAL HAZARDS OF SILOS

Silos are storage facilities for animal feed. Silage, which is a fermented form of animal feed, is formed when bacteria digest a variety of crops, including corn, oats, and alfalfa. There are three types of silo structures on farms: oxygen limited silos, conventional silos, and trench silos (Murphy 1992). Trench silos have earth or concrete retaining walls with no structural component over the top, and hence have not been associated with respiratory hazards. Oxygen limited silos have openings limited to the top of the silo, and at the bottom for unloading the silage. Hence oxygen levels are diminished, while carbon dioxide levels are high. Entry into an oxygen limited silo can be immediately life-threatening from asphyxiation. Toxic levels of oxides of nitrogen can occur in oxygen limited silos.

Conventional silos have a series of small doors along the entire height of the silo. These doors provide access from which silage is unloaded. Conventional silos neither limit oxygen nor capture carbon dioxide. The danger of asphyxiation is much less in a conventional silo than an oxygen limited silo. The primary hazard associated with conventional silos is from oxides of nitrogen that are produced during the fermentation of silage.

Hazards from silos include becoming trapped in the grain and drowning, death from asphyxiation due to oxygen depletion, and the high levels of oxides of nitrogen within the silo (CDC 1993b; Etherton 1991). Silo gases may seep out of the silo into nearby barns and affect both the animals and farmers in the barn.

As the silage ferments, concentrations of carbon dioxide and oxides of nitrogen increase. Levels begin to rise within a few hours of loading and reach peak concentrations in 48 to 60 hours. High concentrations may persist for several weeks. Nitrogen dioxide has an odor similar to bleach and appears as a slightly yellow or red-brown haze over silage. Inhaled nitrogen dioxide reacts with water in the airway to form acids which cause severe pulmonary injury, termed silo filler's disease. Duration of exposure is an essential determinant of clinical symptoms. Immediate upper airway irritation may occur, but symptoms may be delayed for several hours. Cough, shortness of breath, lightheadedness, choking, syncope, wheezing, chest pain, weakness, eye and throat irritation may all occur. With increasing exposure, pulmonary edema or focal bronchial pneumonia may occur. With severe exposures, symptoms may improve and then be followed by a relapse 2 to 3 weeks later. Symptoms of relapse are dyspnea, fever, and cough. The relapse is due to bronchiolitis obliterans, which represents a destruction of deep bronchial cells and subsequent sloughing of necrotic tissues. On occasion, an individual who develops bronchiolitis obliterans will develop persistent pulmonary dysfunction (do Pico 1992; Grover and Ellwood 1989)

Sudden death can occur upon entry to a silo due to high concentrations of nitrogen oxides. Concentrations of nitrogen dioxide greater than 200 ppm produces immediate loss of consciousness. Acute laryngospasm, respiratory arrest, and asphyxiation may be responsible for the sudden death. High levels of nitrogen dioxide are associated with oxygen depletion and high levels of carbon dioxide. Silo filler's disease must be differentiated from silo unloader's syndrome and farmer's lung disease.

Farmer's lung disease is a hypersensitivity pneumonitis related to inhalation of thermophilic actinomycetes and fungi. Acute farmer's lung presents with abrupt onset of high fever, chills, muscle aches, cough and shortness of breath within 4 to 8 hours after exposure to dust from decomposing feed material from the top layer of silage. Chest roentgogram can show patchy or diffuse ground glass infiltrates. Pulmonary function testing shows restrictive lung disease. Subacute farmer's lung disease has a more gradual onset, with fever, night sweats, and weight loss accompanying productive cough and dyspnea. Permanent impairment of pulmonary function occurs in a few cases. Bronchoalveolar lavage in farmer's lung disease demonstrates lymphocytosis in the airways. A mononuclear aveolitis with granuloma formation is seen on lung biopsy.

Silo unloader's syndrome, a form of organic dust toxic syndrome, is associated with a large exposure to moldy elements at the top of a silo generally occurring when a silo is opened or uncapped. During this activity, moldy silage is handled. This syndrome is thought to be due to endotoxins, which are products from the cell wall of gram negative bacteria. Mycotoxins, proteinases, and endogenous histamines may also play a role in this syndrome (CDC 1994; Dalton and Bricker 1978). Silo unloader's syndrome is characterized by flu-like symptoms, with fever, malaise, muscle aches, headache, cough, chest tightness, wheezing, mild dyspnea, nasal and throat irritation. The worker may appear toxic. Chest roentgenogram and pulmonary function tests are usually normal. There may be a mild neutrophilia on the blood cell count. Bronchoalveolar lavage and lung biopsy show neutrophilic alveolitis and bronchitis with no granulomas. Organic dust toxic syndrome is not associated with long term sequela. Most individuals are able to return to work in a few days with symptomatic treatment (do Pico 1992).

CHEMICAL HAZARDS OF ANIMAL FEED, ADDITIVES, AND DISINFECTANTS

Numerous chemicals, including antibiotics and trace metals, are added to animal feed as growth promoters and prophylaxis for disease. Contact dermatitis can occur from handling animal feeds. Ethylenediamenedihydroiodine, furazolidone, hydroquinone, and halquinol have all been implicated as causing contact dermatitis. Ethoxyquin, an antioxidant in animal feed, has been reported to cause contact dermatitis. The growth promoting factor quindoxin can cause photocontact dermatitis (Abrams et al. 1991).

Other antibiotics causing dermatitis include neomycin, ethylenediamine, thiobendazole, sulfacetamide, sulfmethazine, tetracyclines, and bacitracin. Allergic reactions to nitrofurazone and tylosin in animal feed can occur (Abrams et al. 1991). Veterinarians are also at risk for occupational diseases from additives to animal feed. Manganese oxide is used as an additive in animal and poultry feed and can potentially cause central nervous system damage (Tanaka 1994).

Disinfectants used on animal farms can cause allergic or irritant reactions. Hypochlorite, iodine, and phenols are known to cause contact dermatitis. Allergic contact dermatitis can develop to rubberized clothing worn while working with animals. Chromate is used to preserve milk to be tested for quality control purposes, and there are documented cases of chromate allergy in milk testers (Abrams et al. 1991).

CHEMICAL HAZARDS OF MYCOTOXINS

Mycotoxins are by-products of fungal metabolism which produce toxic effects when inhaled or ingested. Mycotoxins can contaminate foods and feeds such as nuts, corn, cotton seed, wheat, millet, sorghum, barley, peas, sesame, soy beans, cow peas, Brazil nuts, pistachio nuts, almonds, beans, and sweet potatoes. Drought, high temperatures, and insect infestation all favor the growth of fungi and the production of mycotoxins in the fields (Coulomb 1991).

Of the numerous mycotoxins, aflatoxins, ochratoxins, trichothecenes, and zearalenones are the most troublesome (Schneider and Dickert 1994). Mycotoxins are responsible for numerous deaths among animals each year. The production of mycotoxins are lessened by proper food and feed cultivation and storage practices.

The most widely known mycotoxins are the aflatoxins. These substituted coumarin compounds are produced by

Aspergillus flavus and *Aspergillus parasiticus*. Aflatoxins are known carcinogens and immune suppressors and have been associated with liver cancer. Reports of acute effects from aflatoxin ingestion are unusual in the United States. Reports of death due to hepatic failure and massive gastrointestinal hemorrhage have occurred among persons ingesting aflatoxin contaminated grains in other countries (Coulombe 1991). In the United States, the federal Food and Drug Administration has responsibility to regulate aflatoxin levels in foods.

CHEMICAL HAZARDS ASSOCIATED WITH CROPS

Green tobacco syndrome is caused by cutaneous absorption of nicotine from skin contact with wet tobacco plants. Symptoms of green tobacco syndrome are headache, nausea, and dizziness. Tobacco workers often seek emergency medical care during the tobacco season, and symptoms usually resolve with a few hours of hydration and observation. Up to 50% of tobacco handlers report symptoms. That nicotine is the causative agent is supported by the demonstration that symptoms correlate with urinary levels of nicotine and continine (Ghosh et al. 1986). This illness can be prevented by avoiding skin contact with tobacco plants.

PREVENTION OF CHEMICAL INJURIES IN AGRICULTURE

Pesticides

1. Adopt agricultural practices that minimize use of chemicals, such as Integrated Pest Management and organic farming, whenever possible.
2. Carefully read pesticide labels. Follow instructions for indications and safety precautions.
3. Store pesticides in proper containers in secured areas.
4. Only properly trained and certified personnel should handle and apply pesticides.
5. Be aware of the toxicity of pesticides. Have a plan for responding to a pesticide poisoning.
6. In the event of poisoning, responders should avoid becoming contaminated.
7. Wear appropriate protective clothing when handling and applying pesticides.

Fertilizers

Workers applying anhydrous ammonia should be educated to the potential dangers of exposure, and should know to clean immediately with water. They should use safety goggles, gloves, and protective clothing.

SUMMARY

A number of chemicals are used on the modern farm, many of which are dangerous to life and health. Particularly toxic are the pesticides designed to kill a variety of plant and animal pests. Fertilizers can be very caustic to skin and particularly eyes, and anhydrous ammonia spills can result in severe burns. Petroleum distillates are extensively used for the operation and maintenance of farm machinery and pose respiratory, central nervous system, and cardiac hazards through inhalation exposure. Animal confinement facilities house large numbers of animals in small spaces, leading to a complex indoor air environment which can be toxic. Decay of vegetation in silos can liberate nitrogen dioxide, a potent respiratory toxin. Farmers need to be fully informed of the dangers of farm chemicals, because with knowledge of dangers and practices to minimize exposures, the chemical hazards of farming can be minimized.

REFERENCES

Abrams, K., D. Hogan, and H. Maibach. 1991. Pesticide related dermatoses in agricultural workers. *Occup. Med. State of the Art Reviews* 6(3): 463-92.

Agarwal, S.B. 1993. A clinical, biochemical, neurobehavioral, and sociopsychological study of 190 patients admitted to hospital as a result of acute organophosphorus poisoning. *Envir. Res.* 62: 63-70.

Allender, W.J. 1983. Suicidal poisoning by endothall. *J. Anal. Tox.* 7: 79-82.

Anonymous. 1991. Acute effects of indoor exposure to paint containing bis(tributyltin) oxide—Wisconsin. *MMWR* 40: 280-81.

Beards, S.C., P. Kraus, and J. Lipman. 1994. Paralytic ileus as a complication of atropine therapy following severe organophosphate poisoning. *Anaesthesaia* 49: 791-93.

Bruynzeel D.P., and W.G. van Ketel. 1986. Contact dermatitis due to chlorothalonil in floriculture. *Contact Dermatitis* 14: 67-68

Carella, G., and P.M. Bettolo. 1994. Reversible hepatotoxic effects of diphenyl: report of a case and a review of the literature. *J. Occup. Med.* 36: 575-576.

Carlson, J.E., and J.W. Villaveces. 1977. Hypersensitivity pneumonitis due to pyrethrum. *JAMA* 237: 1718-19.

Centers for Disease Control (CDC). 1993a. *Anhydrous Ammonia Safety*. NIOSH Publ. 93-131.

Centers for Disease Control (CDC). 1993b. *NIOSH Warns Farmers of Deadly Risk of Grain Suffocation*. NIOSH Update Publ. 93-116, April 28.

Centers for Disease Control (CDC). 1993c. *NIOSH Warns of Carbon Monoxide Hazard Using Pressure Washers Indoors*. NIOSH Update Publ. 93-117, May 10.

Centers for Disease Control (CDC). 1993d. *NIOSH Warns: Manure Pits Continue to Claim Lives*. NIOSH Update Publ. 93-114, July 6.

Centers for Disease Control (CDC). 1994. *Preventing Organic Dust Toxic Syndrome. NIOSH Alert*, Publ. 94-102, April.

Colosio, C., M. Tomasinis, S. Cairoli, et al. 1993. Occupational triphenyltin acetate poisoning: a case report. *Brit. Jour. Ind. Med.* 48: 136-9

Corrigan P.M., S. MacDonald, A. Brown, et al. 1994. Neurasthenic fatigue, chemical sensitivity, and GABA receptor toxins. *Med.Hyp.* 43: 195-200.

Coulombe, R.A., Jr. 1991. Aflatoxins in mycotoxins and phytoalexins. In: R.P. Sharma, ed. Boca Raton: CRC Press, pp 103-64.

Culleen, L.E. 1994. Pesticide registration in the United States: overview and new directions. *Quality Assurance* 3: 291-99.

Dalton, M., and D. Bricker. 1978. Anhydrous burn of the respiratory tract. *Texas Med.* 74: 51-4.

Dannaker C.J., H.I. Maibach, and M. O'Malley. 1993. Contact urticaria and anaphylaxis to the fungicide chlorothalonil. *Cutis* 52: 312-15.

Day, L.C. Delayed death by endothall, a herbicide. 1988. *Vet. Human Tox.* 30: 366.

de Silva H.J., P.S. Sanmuganathan, and N. Senanyake. 1994. Isolated bilateral recurrent laryngeal nerve paralysis: a delayed complication of organophosphorus poisoning. *Hum. Exp. Tox.* 13: 171-3.

Devinsky O., J. Kernan, and D.M. Bear. 1992. Aggressive behavior following exposure to chilinesteruse inhibitors. *J. Neuropsych. Clin. Neuroscience* 4: 189-94.

Dickoff, D.J., O. Gerber, and Z. Turofsky. 1987. Delayed neurotoxicity after ingestion of carbamate pesticides. *Neurology* 37: 1229.

do Pico, G. 1992. Hazardous exposure in lung disease among farm workers. *Clinics in Chest Med.* 13: 311-28.

Ecobichon, D.J., and R.M. Joy. 1994. *Pesticides and Neurological Disease*. 2nd Edition. Boca Raton, FL: CRC Press.

Eilrich, G.L., and M. Chelsky. 1991. Facial dermatitis caused by chlorothalonil in a paint. *Contact Dermatitis* 25: 141-4.

Emslie, E.S. 1993. Contact dermatitsis due to Metasol D3T in a Jay Cushion. *Contact Dermatitis* 29: 4.

Ernest, K., M. Thomas, M. Paulose, et al. 1995. Delayed effect of exposure to organophosphates. *Indian J. Med. Res.* 101: 81-4.

Etherton, J., et al. 1991. Agricultural machine related deaths. *AJPH* 198: 766-8.

Ferez, H.B., P.H.F. Bertolucci, J.S. Pereira, et al. 1988. Chronic exposure to the fungicide *maneb* may produce symptoms and signs of CNS manganese intoxication. *Neurology* 38: 550-3.

Figa-Talamanca, I., I. Mearelli, P. Valente. 1993. Mortality in a cohort of pesticide applicators in an urban setting. *Int. J. Epidem.* 22: 674-6.

Fonseca, R.G., L.A. Resende, M.D. Silva, and A. Camargo. 1993. Chronic motor neuron disease possibly related to intoxication with organochlorine insecticides. *Acta Neur. Scand.* 88: 56-8.

Foo, G.S. 1985. *The Pesticide Poisoning Report: A Survey of Some Asian Countries*. Malaysia: Internation Organization of Consumers Unions.

Forget, G. 1991. Pesticides and the third world. *J. Tox. Env. Health* 32: 11-31.

Garber, M. 1987. Carbamate poisoning: the other insecticide. *Pediatrics* 79: 734.

Ghosh, S.K., H.N. Saiyed, V.N. Gokani, and M.U. Thakker. 1986. Occupational health problems among workers handling Virginia tobacco. *Int. Arch. Occ. Env. Health* 58: 47-52.

Ghosh, S.K., V.N. Gokani, J.R. Parikh, et al. 1987. Protection against "green symptoms" from tobacco in Indian harvesters: a preliminary intervention study. *Arch. Env. Health* 42: 121-4.

Goldstein, N.P., P.H. Jones, and J.R. Brown. 1959. Peripheral neuropathy after exposure to an ester of dichlorophenoxyacetic acid. *JAMA* 171: 1306-9.

Gordan, M., and E.D. Richter. 1991. Hazards associated with aerial spraying of organophosphates in Israel. *Rev.Env.Health* 9: 229-38.

Grant, S.M. 1993. An unusual case of burning mouth sensation. *Brit. Dental J.* 175: 378-80.

Grover, J., and P. Ellwood. 1989. Gases in forage tower silos. *Annals of Occ. Hyg.* 33: 519-35.

Hall, R.H. 1992. A new threat to public health: organochlorines and food. *Nutrition Health* 8: 33-43.

Hansson, C., and J. Wallengren. 1995. Allergic contact dermatitis from dichlofluanide. *Contact Derm.* 32: 116-17

Hayes, W.H., Jr. 1982. *Pesticides Studied in Man.* Baltimore: Williams and Wilkins.

He, F., S. Wang, L. Liu, et al. 1989. Clinical manifestations and diagnosis of acute pyrethroid poisoning. *Arch. Tox.* 63: 54.

Honda, K.I., H. Kohrogi, M. Ando, et al. 1992. Occupational asthma induced by the fungicides tetrachlorisophthalonitrile. *Thorax* 47: 760-61.

Ishikawa, S., M. Miyata, S. Aoki, and Y. Hanai. 1993. Chronic intoxication of organophosphorous pesticide and its treatment. *Folia Medica Cracoviensia* 34:139-151.

Jackson, R.J, S.W. Statton, L.K. Goldman, et al. 1986. Aldicarb food poisoning from contaminated melons, California. *MMWR* 35: 16.

Jager, K.W. 1970. *An Epidemiological and Toxicological Study of Long-Term Occupational Exposure.* New York: Aldrin, Dieldrin, Endrin, and Tlodrin, p. 234.

Keeble, V.B., L. Correll, and M. Ehrich. 1993. Evaluation of knit glove fabrics as barriers to dermal absorption of organophosphorus insecticides using an invitro test system. *Toxicology* 81: 195-203.

Koizumi, A., S. Shiojima, M. Omiya, et al. 1979. Acute renal failure and maneb exposure. *JAMA* 242: 2583-85.

Koyama, K, M. Yamashita, T. Miyauchi, and K. Goto. 1993. A fungicide containing iminoctadine causes circulatory failure in acute oral poisoning. *Vet. Hum. Tox.* 35: 512.

Lander, F., and K. Hinke. 1992. Indoor application of anti-cholinesterase agents and the influence of personal protection on uptake. *Arch. Env. Cont. Tox.* 22: 163-6.

LeQuesne, P.M., U.C. Maxwell, and S.T. Butterworth. 1980. Transient facial sensory symptoms following exposure to synthetic pyrethroids. a clinical and psychological assessment. *Neurotoxicology* 2: 1.

Lessenger, J.E., and N. Riley. 1991. Neurotoxicities and behavioral changes in a 12-year-old male exposed to dicofol, an organochlorine pesticide. *J. Ind. Health* 33: 255.

Lifshitz, M., M. Rotenberg, S. Sofer, et al. 1994. Carbamate poisoning and oxime treatment in children: a clinical and laboratory study. *Pediatrics* 93: 652-55.

Litovitz, T.L., L. Felberg, R.A. Soloway. 1995. 1994 Annual Report of the American Association of Poison Control Centers Toxic Exposure Surveillance System. *Am. J. Emer. Med.* 13: 551-597.

Ludomirski, A., H.O. Klein, P. Sarelli, et al. 1982. Q-T prolongation and polymorphous ("torsades des pointes") ventricular arrhythmias associated with organophosphorus insecticide poisoning. 49: 1654-58.

Mani, A., M.S. Thomas, and A.P. Abraham. 1992. Type II paralysis or intermediate syndrome following organophosphorous poisoning. *J. Assoc. Physicians of India* 40: 542-4.

Martinez, A.J., J.R. Taylor, S.A. Houff, and E.R. Isaacs. 1977. Kepone poisoning: clinic oneuropathological study. *Neurotoxicology.* Vol I. Roizin L et al., eds. New York: Raven Press.

Matchaba-Hove, R.B., S. Siziya. 1995. Organophosphate exposure in pesticide formulation and packaging factories in Harare, Zimbawe. *Central Afr. J. Med.* 41: 40-4.

McConnell, R., M. Keifer, L. Rosenstock. 1994. Elevated quantitative vibrotactile threshold among workers previously poisoned with methamidophos and other organophosphate pesticides. *Am. J. Ind. Med.* 25: 325-34.

McConnell, R., M. Cordon, D.L. Murray, and R. Magnotti. Hazards of closed pesticide mixing and loading systems: the paradox of protective technology in the Third World. *Brit. Jour. Ind. Med.* 49: 615-9.

McFadden, J.P., M. Kinoulty, and R.J. Rycroft. 1993. Allergic contact dermatitis from the fungicide bupirimate. *Contact Derm.* 28: 47.

Meggs, W.J., and C.H. Cleveland, Jr. 1992. Rhinolaryngoscopic examination of patients with the Multiple Chemical Sensitivity Syndrome. *Arch. Env. Health* 48: 14-18.

Millec, T., J. Kucan, and C.E. Smoot III. 1989. Anhydrous injuries. *J. Burn Care Rehab.* 10: 448-53.

Morgan, D.P. 1989. *Recognition and Management of Pesticide Poisonings,* 4th edition. U.S. Environmental Protection Agency EPA-5440/9-88-001.

Murphy, D.J. 1992. *Safety and Health for Production Agriculture.* St. Joseph, MI: American Society of Agricultural Engineers.

Namba, T. 1971. Cholinesterase inhibition by organophosphate compounds and its clinical effects. *Bull. OMS* 44: 289.

Namba, T., C.T. Nolte, J. Jackrel, and D. Grob. 1971. Poisoning due to organophosphate insecticides. 50: 475-90.

National Safety Council. 1993. *Accident Facts, 1993 edition.* Itasca, IL: National Safety Council.

Nouira, S., F. Abroug, S. Elatrous, et al. 1994. Prognostic value of serum cholinesterase in acute organophosphate poisoning. *Chest* 106: 1811-14.

Okonek, S., H. Setyadhama, A. Burchert, and E.G. Krienke. 1982a. Activated charcoal is as effective as Fuller's Earth Orbentonite in paraquat poisoning. *Klin. Wschr.* 60: 207.

Okonek, S., L.S. Weileman, J. Majdanzic, et al. 1982b. Successful treatment of paraquat poisoning. activated charcoal per OS and continuous "hemoperfusion." *J. Toxicol. Clin. Tox.* 19: 807.

Oleaga, J.M., A. Aguirre, N. Landa, M. Gonzalez, and J.L. Diaz-Perez. 1992. Allergic contact dermatitis from Kathon 893. *Contact Derm.* 27: 345-46.

O'Reilly, J.F. 1984. Prolonged coma and delayed peripheral neuropathy after ingestion of phenoxyacetic weed killers. *Postgrad. Med. J.* 60: 76-7.

Proctor N., J. Hughes, and M. Fischman. 1988. *Chemical Hazards of the Workplace,* 2nd edition. Philadelphia: Lippincott Co.

Purschwitz, M., and W. Field. 1990. Scope and magnitude of injuries in the agricultural workplace. *Am. J. of Ind. Med.* 18: 179-92.

Rengstorff, R.H. 1994. Vision and ocular changes following accidental exposure to organophosphates. *J. Appl. Tox.* 14: 115-18.

Richter, E.D, P. Chuwers, Y. Levy, et al. 1992. Health effects from exposure to organophosphate pesticides in workers and residents in Israel. *Israel J. Med. Science* 28: 584-98.

Rosenstock, L., M. Keifer, W. Daniell, et al. 1991. Chronic central nervous system effects of acute organophosphate pesticide intoxication. *Lancet* 338: 223-7.

Royce, S., P. Wald, D. Sheppard, and J. Balmes. 1993. Occupational asthma in a pesticides manufacturing worker. *Chest* 103: 295-6

Savage, E., T. Keefe, L. Mounce, et al. 1988. Chronic neurological sequela of acute organophosphate pesticide poisoning. *Arch. Env. Health* 43: 38-45.

Schneider, E., and K. Dickert. 1994. Health costs and benefits of fungicide use in agriculture: a literature review. *J. Agro.Med.* 1:19-39.

Schmid, R. 1960. Cutaneous porphyria in Turkey. *NEJM* 263: 397-8.

Schuman, S.H, and R.L. Dobson. 1985. An outbreak of contact dermatitis in farm workers. *J. Am. Acad. Derm.* 13: 220-223

Senanayake, N., and L. Karalliedde. 1987. Neurotoxic effects of organophosphorus insecticides: an intermediate syndrome. *NEJM* 316:761-3.

Shaver, C, and T. Tong. 1991. Chemical hazards to agricultural workers. *Occup. Med. State of the Arts Reviews* 6: 391-413.

Sfeukazza, S., and W. Beckett. 1991. The respiratory health of welders. *Am. Rev. of Res. Dis.* 143: 1134-48.

Smith, P.W. 1977. *Medical Problems in Aerial Applications.* Federal Aviation Administration, U.S. Department of Transportation.

Smith, J., D. Rogers, and R. Sykes. 1983. Farm tractor associated deaths--Georgia. *MMWR* 32: 481-82.

Steenland, K., B. Jenkins, and R.G. Ames, et al. 1994. Chronic neurological sequela to organophosphate pesticide poisoning. *Am. J. Pub. Health* 84: 731-6.

Stephens, R., A. Spurgeon, I.A. Calvert, et al. 1995. Neuropsychological effects of long-term exposure to organophosphates in sheep dip. *Lancet* 345: 1135-39.

Sterri, S.H., B. Rognerud, S.E. Fiskum, and S. Lyngaas. 1979. Effects of toxogonin and P2S on the toxicity of carbamates and organophosphorus compounds. *Acta Pharmacologica et Toxicologica* 45: 9-15.

Sumerford, W.T, W.J. Hayes, Jr, J.M. Johnson, K. Walker, and J. Spillane. 1953. Cholinesterase response and symptomatology from exposure to organic phosphorus insecticides. *Am. Med. Assoc. Arch. Ind. Hyg. Occup. Med.* 7: 383.

Suzuki, T., H. Morita, K. Oho, K. Maekawa, R. Nagai, and Y. Yazaki. 1995. Sarin poisoning in Tokyo subway. *Lancet* 345: 980.

Tabershaw, I.R, and W.C. Cooper. 1966. Sequelae of acute organic phosphate poisoning. *J. Occup. Med.* 8: 5-19.

Tanaka, S. 1994. Manganese and its compounds in occupational medicine.

Teschke, K., C. Hertzman, M. Wiens, H. Dimich-Ward, et al. 1992. Recognizing acute health effects of substitute fungicides: are first aid reports effective? *Am. J. Ind. Med.* 21: 375-82.

Ueda, A., K. Aoyama, F. Manda, et al. 1994. Delayed-type allergenicity of triforine (saprol). *Contact Derm.* 31: 140-45

Umehara, F., S. Izumo, K. Arimura, M. Osame. 1991. Polyneuropathy induced by M-tolyl methyl carbamate intoxication. *J. Neurology* 238: 47-8.

U.S. Department of Agriculture. 1992. *Agriculture Statistics, 1992.* Washington, DC: U.S. Government Printing Office.

van Ginkel, C.J, and N.N. Sabapathy. 1995. Allergic contact dermatitis from the newly introduced fungicide fluazinam. *Contact Dermatitis* 32: 160-2.

Yelamos, P., F. Diez, F. Laynez, and J.F. Pena. 1993. Acute pancreatitis in acute poisoning caused by organophosphate insecticides. Report of 3 Cases. *Medicina Clinica* 101: 154-55.

Wax, P.M, and R.S. Hoffman. 1994. Fatality associated with inhalation of a pyrethrin shampoo. *J. Tox. Clin. Tox.* 32: 457-60.

Weizman, Z., and S. Sofer. 1992. Acute pancreatitis in children with anticholinesterase insecticide intoxication. *Pediatrics* 90:204-6.

Zahm, S.H., D.D. Weisburger, P.A. Babitt, R.C. Saal, J.B. Vaught, K.P. Cantor, and A. Blair. 1990. A case-control study of non-Hodgkin's lymphoma and the herbicide 2,4-dichlorophenoxyacetic acid (2,4-D) in Eastern Nebraska. *Epidem.* 1: 349-56.

Zenz, C., O.B. Dickerson, E.P. Horvath. 1994. Occupational Medicine. 3rd edition. St. Louis: Mosby Yearbook, Inc., pp 542-48.

Zhang, Z.W. 1991. Exposure levels and biological monitoring of pyrethroids in spraymen. *Chinese J. Prev. Med.* 25: 85-8.

Zhang, Z.W., J.X. Sun, S.Y. Chen, et al. 1991. Levels of exposure and biological monitoring of pyrethroids in spraymen. *Brit. J. Ind. Med.* 48: 82-6.

Zweiner, R.J., and C.M. Ginsburg. 1988. Organophosphate and carbamate poisoning in infants. *Pediatrics* 88: 121.

17

PROTECTING DRINKING WATER SUPPLIES

Gregory D. Jennings, Ph.D.
Dorothy L. Miner, M. S.
Deanna L. Osmond, Ph.D.
North Carolina State University

If you drink water from a well or spring, the water comes from the ground. Most groundwater is safe to drink. If pollution gets into groundwater, your well or spring water may not be safe. Many things we do in our homes and on our farms can pollute the groundwater. If groundwater becomes polluted, it is nearly impossible to clean up. The only ways to get safe drinking water are to treat your existing water, drill a new well, or get water from another source. All of these options are expensive and inconvenient. This chapter can help you keep your drinking water safe.

INTRODUCTION

Many things can pollute our drinking water. Human waste from septic systems and animal manure can pollute the groundwater. Pesticides, fertilizers, fuels, and cleaning products can also pollute the water if they are not stored or handled properly and leak into the ground. Everyone is responsible for protecting drinking water. People using private wells or springs should take additional precautions to prevent pollution, since their wells are not routinely tested.

There are several types of contaminants that can enter your water:

Microbial Pathogens. Pathogens are disease-producing microorganisms, which include bacteria, viruses, and parasites. Pathogens in drinking water are serious health risks. They get into drinking water when the water source is contaminated by sewage and animal waste, or when wells are improperly sealed and constructed. They can cause gastroenteritis, salmonella infection, dysentery, shigellosis, hepatitis, giardiasis, and other gastrointestinal infections causing diarrhea, vomiting, abdominal cramps, and gas. The presence of coliform bacteria, which are generally harmless bacteria, may indicate that pathogens have also entered the drinking water system.

People worry the most about potentially toxic chemicals in water.

These include organic, inorganic, and radioactive elements:

Organics. This group of contaminants includes pesticides, including herbicides (weed killers), insecticides (bug killers), and fungicides. This group also includes volatile organic chemicals (VOCs), such as gasoline and fuel additives, solvents, degreasers, and adhesives. Some of the common VOCs are benzene, trichloroethylene (TCE), toluene, and vinyl chloride. Possible chronic health effects include cancer, central nervous system disorders, liver and kidney damage, reproductive disorders, and birth defects.

Inorganic Elements. These contaminants include toxic metals such as arsenic, barium, chromium, lead, mercury, and silver. These metals can get into your drinking water from natural sources, industrial processes, and the materials used in your plumbing system. Toxic metals are regulated in public water supplies because they can cause acute poisoning, cancer, and other health effects. Nitrate is another inorganic contaminant. The nitrate in mineral deposits, fertilizers, sewage, and animal wastes can contaminate water. Nitrate can cause "blue baby syndrome" in infants.

Radioactive Elements. Radon is a radioactive contaminant that results from the decay of uranium in soils and rocks. It is usually more of a health concern when it enters a home as a soil gas than when it occurs in water supplies. Radon in air is associated with lung cancer. Radon can enter the air from water supplies during showering.

THE CONDITION OF THE WELL

What Shape Is Your Well In?

One of the easiest ways to protect well water from pollution is to make sure that the well is in good shape and placed in the right location. A poorly built or maintained well can allow pollutants to enter water directly. The closer the well is to sources of pollution, the more likely the well will become polluted. For instance, if the well casing is cracked and pesticides that are being mixed near the well are spilled, then the pesticides can easily leak into the well and pollute your drinking water.

How Can You Get Help?

Information in this chapter will help you focus on potential problems with your drinking water that may be caused by a poorly placed, constructed, or maintained well. Read the information and answer the questions. Before you begin answering the questions, gather any records you have about your well. Go to the well. Walk around the area near it and look at it closely. If you have more than one well, focus on the well that provides drinking water for your family and then on the others. Your answers to the following questions will help you assess your well condition and possible health risks.

- If you answer a or b, then there are few, if any, problems.
- If you answer c or d, then you may have a problem and should check the condition of your well.

Where Is the Well Located?

A well's location is important in determining its pollution potential. Stormwater runoff (water that flows over the land during a storm) can carry pollutants such as bacteria, oil, and pesticides. Wells in the path of stormwater runoff can become polluted if stormwater runoff flows into a well that is not properly sealed. A well that is downhill from pollutants such as a livestock yard, a leaking gasoline tank, or a failing septic system runs a greater risk of becoming polluted than a well that is uphill from these sources of pollution.

Circle the answer that best describes the position of your well:

a. Uphill from all pollution sources. No surface water runoff reaches well. Surface water flows away from the well.
b. Uphill from most pollution sources. No surface water runoff reaches the well if drainage is working correctly.
c. Downhill from many pollution sources, or from any one especially hazardous source. Some surface runoff may reach well.
d. Settling or depression around casing. Surface water runoff from feedlot, pesticide or fertilizer mixing area, fuel storage, or farm dump collects near the well.

How Close Is the Well to Sources of Pollution?

Many states have laws that do not allow wells to be built near sources of pollution. Check your state laws to determine any minimum "separation distances." These minimum distances are set in order to make use of the natural protection soil provides. However, state well codes do not mention every farm activity and structure. When no distances are mentioned for the specific activity or structure you have in mind, provide as much separation as possible between your well and any potential source of pollution. If your farmstead is located on soils that soak up water very quickly (such as sandy soils) or on thin soil that lies over bedrock, maximum separation is needed. If the source or activity presents a high risk of pollution, keep it as far away from your well as possible. The law requires that existing wells meet only the distance requirements in effect at the time the well was built. For your own sake, you should meet current regulations and exceed them if you can.

If your state does not have minimum separation distances, use the following guidelines. In general, wells should be located at least 100 feet from the following pollution sources:

- Any source of sewage, such as septic tank and drainfield, cesspools and privies, sewer lines which are not water-tight, and any sludge-spreading or wastewater irrigation operations.
- Any source of animal waste, such as animal feedlots or manure piles, animal barns, or lagoons.
- Any source of chemical contamination, including fertilizer, pesticide, herbicide (insect and weed killers), or other chemical storage areas; buried gasoline and oil tanks.

Wells should be located at least 50 feet from the following pollution sources:

- Building foundations
- Streams, lakes, ponds

Circle the answer that best describes the position of your well:

a. All separation distances are greater than minimum requirements.
b. All separation distances are at least 3/4 the minimum requirements.
c. All separation distances are at least 1/2 the minimum requirements.
d. Some separation distances are less than 1/2 the minimum requirements.

How Well Does the Soil Filter Out Pollutants?

Soil can filter pollutants picked up by stormwater runoff as it travels down to groundwater. The ability of soil to filter your water depends on the type of soil around the well. Water passes quickly through sand, so sandy soil cannot filter out pollutants. Water and pollutants move more slowly through clay, so clay soils have more time to filter out pollutants. Soils high in organic matter also filter pollutants.

Circle the answer that best describes the soil near your well:

a. Fine-textured soils (clay loams, silty clay).
b. Medium textured soils (silt, loam) with high organic matter.
c. Medium or coarse textured soils with low organic matter.
d. Coarse textured soils (sands, sandy loam).

How Quickly Does Water Reach Your Well?

Another factor that influences groundwater pollution is the depth from the soil surface to the water table or to fractured bedrock. The water table is the top of the ground water. Groundwater can be stored in soil or rock. Groundwater reaching fractured bedrock can move quickly down to wells. The farther water and pollutants have to move through the soil to reach the top of the water table, the longer the soil will have to filter the groundwater.

Circle the answer that best describes the depth of the water table:

a. Water table or fractured bedrock deeper than 50 feet.
b. Water table or fractured bedrock deeper than 25 feet.
c. Water table or fractured bedrock deeper than 10 feet.
d. Water table or fractured bedrock shallower than 10 feet.

How Are Your Well Casing and Cap?

When wells are drilled, the driller installs a steel or plastic lining pipe called a "casing" to keep the borehole from collapsing. Wells cased below the water table offer greater protection from pollution since they help ensure that surface water is filtered through soil, sand, or rock before entering the well.

You can inspect your well casing for holes or cracks at the surface and, using a light, check the inside of the casing. If the well casing moves when you push on it, the casing might not keep out pollutants. In areas of shallow (less than 20 feet from the surface), fractured bedrock, listen for water running down the well when the pump is not on. If you hear water running, there could be a crack or hole in the casing, or the well is not cased down into the water table. Both conditions are bad for your water quality because a poor casing may not keep out contaminants.

To prevent pollutants from flowing into the well, the driller should install a tight-fitting cap. This cap prevents insects or surface water from entering the well. A screened vent in the cap allows air to enter the well. The cap should be installed firmly so that children cannot remove it easily. Check the well cap to see that it fits tightly. Electrical wiring should be enclosed in the conduit to protect you from being shocked. If the well has a vent, make sure it faces the ground, is tightly connected to the well cap or seal, and is properly screened. The well code requires that all private wells have a seal. Not all wells have caps. Some wells may have pumping equipment attached at the surface.

Circle the answer that best describes the construction of your well:

a. No holes or cracks. Cap tightly secured. Screened vent.
b. No defects visible. Well vented but not screened.
c. No holes or cracks visible. Cap easily removed.
d. Holes or cracks visible. Cap loose or missing. Can hear water running.

Are Casing and Grout Deep Enough?

The space between the casing and the sides of the well hole provides a direct channel for stormwater runoff to reach the groundwater. To seal off that channel, the driller fills the space with grout, such as cement, concrete, or a special type of clay called bentonite. Both the grout and the casing prevent pollution from getting into the well. In addition, grout must extend deep enough to seal off any layers of poorer quality water that make contact with the well casing.

Circle the answer that best describes the casing and grouting of your well:

a. Cased and grouted to required depth.
b. Cased and grouted below water table of your well.
c. Cased, but not grouted.
d. No casing. No grout.

Is the Well Protected at the Ground Surface?

The well casing extends above the ground to prevent stormwater runoff from entering the well directly. Some well codes require that at least 12 inches of casing pipe extend above the ground after the final grading of the surrounding land. The wellhead should be surrounded 2 feet in all directions by a concrete pad, which should slope away from the well. The concrete pad stabilizes the casing and the soil around it, and the slope of the pad keeps stormwater runoff from entering the well.

Circle the answer that best describes your wellhead:

a. Casing extends more than 12 inches above surface and concrete pad extends 2 feet in all directions.
b. Casing above ground level and concrete pad 1 to 2 feet in all directions.
c. Casing above ground level and no concrete pad.
d. Casing below ground level or in pit or basement and no concrete pad.

How Old Is Your Well?

The age of your well is an important factor in predicting whether your water might be polluted. A well constructed more than 60 years ago is likely to be located at the center of the farmstead, which means it is probably surrounded by many activities that can cause pollution. It may also be more shallow than a newer well and may have a thinner casing that can corrode more easily. (Even wells with modern casings that are 30 to 40 years old can be corroded.) Older well pumps are more likely to leak lubricating oils into the well. All of these characteristics of older wells can contribute to the pollution of your well water. If you have an older well, you might wish to have it examined by a county health department representative, state environmental regulatory agency groundwater specialist, or a qualified well driller.

Circle the answer that best describes the age of your well:

a. Less than 15 years old.
b. 15 to 30 years old.
c. 30 to 60 years old.
d. More than 60 years old.

Is Your Well Drilled or Dug?

Wells that have been dug rather than drilled pose the highest risk of pollution because they are shallow and often poorly protected from stormwater runoff. A dug well is a large-diameter hole (usually more than 2 feet wide), which often has been constructed by hand.

Driven wells, also known as sand point wells, pose a moderate to high risk of being polluted. They can only be installed in areas of relatively loose soils, such as sand, because they are constructed by driving a small-diameter pipe into the ground.

Other types of wells include jetted wells, in which water under high pressure washes away the soil, and bored wells, in which an earth auger removes the soil. Drilled wells are made either by rotary drilling or by percussion drilling. (Some people refer to drilled wells as "punched.") Drilled wells for farm use are commonly 4 to 8 inches in diameter. Bored wells are commonly 18-24 inches in diameter. Drilled, jetted, or bored wells are the safest types.

Circle the answer that best describes how your well was installed:

a. Drilled
b. Jetted or bored.
c. Driven (sand point).
d. Dug.

Are You Preventing Backflow?

Backflow occurs when water (and possibly pollution) flows backwards through the pipes from the house to the well. There should be anti-backflow devices, known as check valves, on all faucets with hose connections, or there should always be air gaps between hoses or faucets and the water level. Without anti-backflow devices, you risk having polluted water in laundry tubs, sinks, washing machines, pressure washers, outside hydrants, or swimming pools flow back through the plumbing into your well water.

If a vacuum forms in a water supply pipe, the backflow that can result is called backsiphoning. Backsiphoning from pesticide mixing tanks or pressure washers allows chemicals to flow back into the well through the hose. An anti-backflow device should be used when filling pesticide sprayer tanks, to prevent the chemical mixture from flowing back into the well and polluting groundwater. If you don't have an anti-backflow device, the hose must be kept out of the tank when filling the pesticide sprayer. Another option to use if you don't have an anti-backflow device is to use an inexpensive plastic container. The container is filled with water at the well and then used to fill the sprayer away from the farmstead and the well.

Water supplies that have cross connections between them (connections between two otherwise separate pipe systems) also put your water at risk because the water in one pipe system can become polluted by the other.

Circle the answer that best describes anti-flow devices attached to your well:

a. Anti-backflow devices (such as check valves) installed on all faucets with hose connections and no cross-connections between water supplies.
b. Anti-backflow devices installed on some faucets with hose connections.
c. No anti-backflow devices. Air gap maintained.
d. No anti-backflow devices. Air gap not maintained. Cross-connections between water supplies.

Do You Have any Unused Wells?

Many farms have old, unused wells on the property. If unused wells are not properly filled and sealed, they can provide a direct route into the groundwater for stormwater runoff carrying pollutants. These wells can also allow pollutants to move from one groundwater system to another. Wells should also be filled so that children and animals cannot fall into them.

You cannot always see unused wells. A depression in the ground may indicate an old well. Pipes sticking out of the ground around existing or past farmsteads are the most obvious signs of an unused well. Other places to check for unused wells include basements of houses, under front steps of houses, or near old cisterns.

A well that has been permanently closed by approved methods is considered an abandoned well. A license is not required to properly abandon a well, but you must meet the minimum well code requirements for abandonment. Use of unacceptable materials and methods can lead to well collapse and groundwater pollution.

Some states have regulations include the following requirements for well abandonment:

- The pump, piping, and any other obstructions must be removed from the well. Casings and screens should be removed if doing so will not cause or contribute to groundwater pollution.
- Any casing that is not properly grouted must either be removed or properly grouted.
- The well must be chlorinated to disinfect it before it is sealed.
- The entire depth of the well must be filled with cement, grout, or clay. Specific requirements vary according to well type and local geological characteristics.

Circle the answer that best describes any abandoned wells located on your property:

a. No unused, unsealed wells.
b. Unused wells sealed and filled.
c. Unused well on property more than 100 feet from supply well. Not capped or filled.
d. Unused well less than 100 feet from supply well. Not capped or filled.

Is Your Plumbing Adding Lead to Your Water?

Copper pipes installed before 1986 may have lead in the solder. Solder used in plumbing connections after 1986 was not allowed to contain lead. Brass, used in fixtures and well pumps, also contains lead. Very old houses may even have lead plumbing. Lead from these sources can leach into your drinking water and become a health hazard. Water from wells and springs may be corrosive or acidic, and this can increase the amount of lead that leaches into the water.

Circle the answer that best describes your plumbing:

a. Plastic piping, no brass in fixtures or well pump
b. Plastic piping, brass in fixtures or well pump
c. Copper piping, age unknown or installed after 1986
d. Copper piping installed before 1986, or lead piping.

Has Your Well Been Tested Recently?

Well water should be tested once a year. You can have your water tested by either a public or a private laboratory. A list of certified labs may be available from your health department or cooperative extension office. Although it would be expensive and difficult to test your water for every possible pollutant, some basic tests should be conducted. If you take the samples yourself, you must carefully follow the instructions that come with the collection bottle.

Water should be tested once a year for bacteria and nitrate, which can cause health problems. Yearly testing is necessary because groundwater travels and may pick up pollutants elsewhere. So even if you are doing everything you can to prevent your well from being contaminated, it may become polluted from other people's activities. If your water has high bacteria or nitrate levels, talk to a county health specialist. There may be problems with the location or construction of your well.

If your well draws from sandy soil or granite bedrock, testing once for corrosivity is also important. The tests to check for corrosivity include hardness, alkalinity, pH, conductivity, and chloride. Test once to find out how corrosive the water may be to your plumbing system.

- Test for pollutants that are most likely at your farmstead.
- Test for lead if you have lead pipes or soldered copper joints or brass fixtures.
- Test for volatile organic chemicals (VOCs) if you have an underground fuel storage tank, or if there has been a nearby use or spill of oil, petroleum, or solvent.
- Testing for pesticides can be expensive but it is important if the potential for pesticide pollution is high, such as after a spill or if your well is downhill from fields where pesticides have been applied. Testing for pesticides may also be justified if your well has high nitrate levels or if your well is shallow or not properly cased and grouted.

It is important to record test results and to note changes in water quality over time. In addition to water analysis results, keep records of your well construction and of maintenance done on the well and pump.

Circle the answer that best describes the tested water quality of your well:

a. Consistent satisfactory water quality. Bacteria, nitrate, and other tests meet standards.
b. Occasional deviation from standards with bacteria, nitrate, and other tests and no cross-connections.
c. Bacteria, nitrate, and other tests most often do not meet standards and no cross-connections.
d. No water tests done. Water discolored after rainstorms or during spring melt. Noticeable changes in color, clarity, odor, or taste.

SUMMARY

If your county has a well permit program, contact the county health department before beginning any well construction or repairs. Your county health department or local cooperative extension center can be a valuable source of information on the following topics:

- New well or spring construction and site selection
- Well inspection and maintenance
- Registered well drillers (or call your state environmental regulatory agency)
- Unused well abandonment
- Construction records for existing wells
- Well water testing: Advice on appropriate tests to run, list of certified testing laboratories, assistance interpreting test results, health risks
- Backflow prevention
- Water pollution and health risks
- Water treatment devices
- Groundwater information

REFERENCES

Murdock, B.S. 1991. *Environmental Issues in Primary Care*. Minneapolis, MN: Minnesota Department Health.

Midwest Plan Service (MWPS). 1979. *Private Water Systems Handbook*. MWPS-14. Ames, IA: Midwest Plan Service.

Wagenet, L., K. Mancl, and M. Sailus. 1995. *Home Water Treatment*. NRAES-48. Ithaca, NY: Cooperative Extension.

18

GRAIN INDUSTRY HEALTH AND SAFETY ISSUES

Amy Gitelman, M.P.H.
Duke University Medical Center

> Grain processing has many potential health and safety hazards. After harvest, grain is moved to farm storage bins, silos, or transferred to elevators. Hazards ranging from silo gases, fires, explosions, and a host of others are present. Grain dust is a serious hazard in the grain handling industry posing explosion, fire, and health hazards. Deaths due to suffocation under grain or from silo gases have also been documented. Other hazards, such as machine entanglements, falls, and electrocutions are also possible in grain handling facilities. Health hazards due to grain dust or other exposures range from asthma, rashes, grain fever, allergic reactions, and eye, nose, and sinus irritations. Guidelines for exposure limits and training requirements for grain handling facilities are presented.

INTRODUCTION

In 1992, the National Safety Council ranked agriculture as the second most deadly occupation in the United States, with 42 deaths per 100,000 workers (Wilk 1993). In 1995, the Bureau of the Census reported that there were 1,037,000 U.S. farmers excluding horticultural farmers. The 1994 agricultural statistics provided by the USDA reported 1,832,000 unpaid farm labor workers and 828,000 hired farm labor workers. In 1992, 2.7 million workers were employed in farming and 18.3 million workers were employed to store, transport, process, and merchandise products from the nation's farms (Jansson 1992). The agricultural production sector (i.e. farms) employed approximately 2.5 million workers annually from 1980 through 1989 (Myers and Hard 1995).

Grain processing (including harvesting, storing, handling, and shipping) has many potential health and safety hazards. During harvest and storage season, potential hazardous situations can develop due to increased storage capacities, larger and faster handling capacities, and automation. Among manufacturers, there were 108,000 grain mill products employees. In 1993, 82,393,000 acres of feed grains and 65,926,000 acres of food grains were harvested. Feed grains included corn for grain, oats, barley, and sorghum for grain. Food grains included wheat, rye, and rice. The total area planted for grains was 319,553,000 acres.

Grain includes the seed or fruit of food plants (e.g., wheat, rye, oats, barley, corn, legumes, oil seeds). After grain is harvested at a farm, it is transferred to a series of grain elevators. Grain elevators serve as collection, storage, and transfer points for grain. Grain sampling, weighing, blending, drying, cleaning, and fumigating may also occur during various stages of grain processing. A schematic of grain processing is provided in Figure 18.1.

At country grain elevators, grain is graded, weighed, cleaned, and stored until shipment to terminal elevators or local mills. Figure 18.2 depicts a typical grain elevator. At terminal elevators, grain is graded, cleaned, dried, and stored. At a basic terminal elevator, grain is unloaded from or loaded into railcars and trucks using a trackshed, processed in a workhouse, stored in an annex of concrete silos, and transferred to ships or nearby mills using transfer galleries (Farant and Moore 1978).

On a farm, forages and grains may be stored in a silo structure. Silos are designed for fermentation and storage of field crops (Engberg 1993). Oxygen limiting silos, conventional silos, and trench (bunker) silos are the three types of silo structures existing on farms (Murphy 1992). Oxygen limiting silos, common on dairy farms, are mainly used for forages or to store high moisture grains. They are designed to prevent oxygen from entering the silo and to allow carbon dioxide generated during fermentation to accumulate. During anaerobic fermentation of green silage, silo gases including nitrogen dioxide, nitric oxide, and carbon dioxide are generated (Engberg 1993).

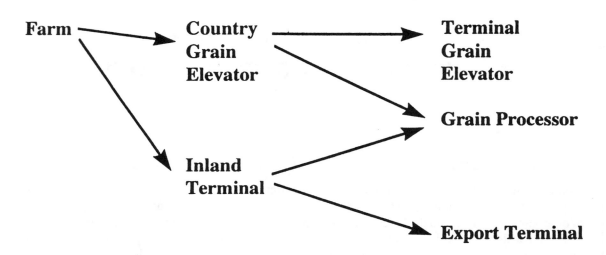

Figure 18.1. Grain processing schematic.

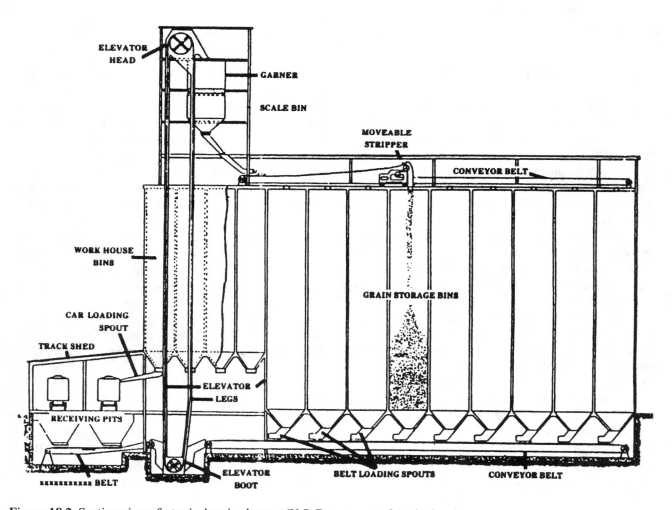

Figure 18.2. Section view of a typical grain elevator (U.S. Department of Agriculture)

GRAIN DUST CHARACTERIZATION

Grain dust is generated in each phase of grain processing, because each phase involves movement of the grain. The movement of grain from a railway boxcar to a grain elevator in to a storage bin generates grain dust. Processing grain at a mill results in grain dust. Grain dust is an explosive risk, a fire risk, and a health risk to exposed people (Warren 1992).

Grain dust exposure has been associated with several occupations in the grain industry. The highest airborne dust concentrations tend to occur during grain loading and unloading (Bardana et al. 1992). Farant and Moore (1978) measured total dust concentrations in terminal, transfer, and country grain elevators ranging from 0.18 to 781 mg/m^3. Their results indicated that elevator men performing housekeeping and maintenance chores or working in transfer galleries were more likely to be exposed to high grain concentrations. Table 18.1 lists occupations and individuals with potential for grain dust exposure.

Table 18.1. Occupations and individuals with grain dust exposure.

- Farmers during harvest, loading, and unloading of grain
- Mill workers
- Grain elevator/storage agents and workers
- Dock workers and longshoremen
- Grain processing workers
- Workers manufacturing cereals and feeds
- Grain seed workers
- Bakers
- Individuals living downwind from grain-unloading facilities

Adapted from Bardana et al. 1992.

Grain dust is a complex mixture of 60 to 75% organic and 25 to 40% inorganic material (Brown 1988). Grain dust composition is dependent on variables including contamination, humidity, and age (Van Fleet et al. 1992). Typical components of grain dust include cereal grain fragments and their decomposition products, plant matter fragments, soil and trace chemical elements, fertilizers, pesticides, or herbicides, microorganisms (mites, fungi, bacteria, endotoxin), insects or insect fragments, hair, feathers and excreta from birds or rodents, and other contaminants accumulated during grain harvesting and processing (metal fragments, lubricating oils, paint chips) (Brown 1988). Grain dust also contains approximately 5% silica (Dosman and Cockcroft 1989).

Approximately 40% of grain dust particles are respirable (Zejda et al. 1993) and can penetrate the terminal bronchioles. These respirable particles have been characterized as less than 5 microns in mean aerodynamic diameter. Grain dust particles range in size from respirable particles 10 microns or less in diameter to particles 100 microns or greater (Farant and Moore 1978).

Many of the components of grain dust are potential antigens (protein or carbohydrate substances capable of stimulating an immune response). Table 18.2 lists the potential antigens present in grain dust. Long term exposure to an antigen can lead to disabling conditions such as asthma or pulmonary fibrosis (Engberg, L 1993).

Table 18.2. Potential antigens present in grain dust.

- Plant matter from grains including durum and spring wheat, oats, barley, rye, and corn
- Plant matter from seeds including rapeseed, flax, and sunflower seeds
- Non-grain plant matter
- Bacteria
- Cellulose hairs and spikes
- Fungal matter
- Grain rusts, smuts
- Grain storage mites
- Insect parts and excreta
- Pollens
- Mineral particles
- Pesticides and fertilizers
- Rodent matter including hairs and excreta
- Silica dust
- Soil, minerals and trace elements
- Starch granules
- Wheat weevil debris

Adapted from Bardana et al. 1992.

INJURY STATISTICS

The Bureau of Labor Statistics (BLS), National Safety Council, and NIOSH show agriculture to be among the four most hazardous industrial divisions in the United States (Myers and Hard 1995). The top four leading causes of grain and silage handling deaths from 1985 to 1989 reported by NIOSH were:

1. Suffocation under grain or silage or by gases in bins or silos.
2. Entanglements in augers on all types of machinery.
3. Falls from machinery, bins, and buildings.
4. Electrocutions from machinery contacting an overhead power line (NIOSH 1995).

The average annual fatality rates for agricultural production sector by cause of death are provided in Table 18.3. Machinery accidents were the most common cause of death among agricultural service sector.

Table 18.3. Frequency and fatality rates per 100,000 workers for agricultural production sector by cause of death, 1980-1989.

Cause of Death	Agricultural Deaths	Production Rate
Air transport	121	0.5
Caught in/flying	102	0.4
Drowning	83	0.3
Electrocution	354	1.4
Environmental	318	1.2
Explosions	53	0.2
Falling Objects	316	1.2
Falls	258	1.0
Fires	74	0.3
Homicide	177	0.7
Machinery	2,427	9.5
Motor Vehicles	1,013	4.0
Poisoning	65	0.3
Suicide	136	0.5
Suffocation	131	0.5
Water transport	39	0.2
Other/unknown	156	0.6

Adapted from Myers and Hard 1995.

Table 18.4. Health effects associated with grain dust exposure.

- Grain dust asthma
- Extrinsic allergic alveolitis
- Grain fever
- Skin rashes
- Eye irritation
- Nasal and sinus irritation

Adapted from Bardana, et al. 1992.

HEALTH HAZARDS

The predominant health hazards during grain processing are categorized into pulmonary and nonpulmonary effects from grain dust exposure. Adverse health effects associated with grain dust exposure (irritation of throat, eyes, skin, and lungs) were first reported by Bernardino Ramazzini in 1713 (Brown 1988). Grain handlers have been shown to have lower mean spirometry values than a control population (Bardana et al. 1992). They also have experienced a higher prevalence of eye, nasal, and chest symptoms including cough, sputum production, wheezing, and shortness of breath (Bardana, et al. 1992). Table 18.4 lists the adverse health effects reported with grain dust exposure.

Studies examining the effects of grain dust exposure and smoking on respiratory symptoms and lung function have suggested that these exposures are additive except in the least exposed workers (5 years or less) where a synergistic effect was observed in peripheral airways dysfunction tests (Cotton et al. 1983). Herbert et al. (1981) found that there may be an interaction between the effects of inhaled dust and smoke that causes a deterioration of lung function. Reddy (1995) found that respiratory symptoms are significantly higher among grain workers who have smoked compared to grain workers who have never smoked or who quit smoking. A number of previous studies also have supported these findings (Reddy 1995).

Pulmonary Health Effects

Pulmonary health disorders associated with grain dust exposure include acute respiratory symptoms, acute and/or chronic airway obstruction, asthma, chronic bronchitis, pulmonary fibrosis, and allergic alveolitis (Van Fleet et al. 1992). Asthma is the predominant clinically reported lung disease among grain handlers (Warren 1992). Chronic bronchitis, characterized as chronic cough and phlegm production, is common in all workers regularly exposed to grain dust (Warren 1992).

Grain workers are exposed to many respiratory sensitizers (e.g., animal or vegetable matter, dusts containing molds, barley, wheat, or animal protein) that can cause hypersensitivity pneumonitis (Engberg 1993). Hypersensitivity pneumonitis is characterized by cough, shortness of breath, fever, malaise, muscle aches, joint pain, and weight loss (Engberg 1993). Once an individual develops hypersensitivity pneumonitis, increasingly smaller quantities of the respiratory sensitizer can precipitate reactions. In this situation, respiratory protection may not be adequate to prevent recurrent problems.

Silos may become respiratory hazards due to build up of hazardous gases, such as carbon dioxide and nitrous oxides. Conventional or stave silos are the most common type of silo on farms. These silos have a series of hatch doors used to unload silage. During forage fermentation, nitric oxide (NO), nitrogen dioxide (NO_2), and nitrogen tetroxide (N_2O_4) are generated. Both NO_2 and N_2O_4 are respiratory irritants. When nitrogen dioxide is inhaled, it combines with moisture in the respiratory tract and forms nitric acid. Symptoms of nitrogen dioxide exposure usually do not occur until several hours after exposure has occurred. Symptoms of nitrogen dioxide exposure include burning of respiratory tract tissue, fluid leakage in the lungs, and death.

Farmer's lung disease (a hypersensitivity pneumonitis), silo unloader's syndrome (organic dust toxic syndrome, ODTS), and grain fever are associated with exposure to grain dust in silos and grain bins. Farmer's lung disease (FLD) is described as a specific, delayed hypersensitivity reaction (sensitization) to certain thermophilic bacteria and fungal spores. As exposures reoccur, the reactions to FLD become more severe, causing both respiratory and systemic responses. FLD responses include fever, shortness of breath, fatigue, muscle ache, lung inflammation and death. The acute symptoms of Farmer's Lung Disease (FLD) are usually more severe than ODTS. Farmer's lung disease can progress to chronic lung scarring and interstitial fibrosis, whereas ODTS does not (Popendorf 1995). FLD usually occurs in the winter when hay and moldy crops are fed to animals by the farmer (Shaver and Tong 1991).

Organic dust toxic syndrome (ODTS) occurs as a direct effect of exposure to extremely high concentrations of endotoxin and/or other toxic-microbe related components of agricultural dusts (Popendorf 1995). Acute symptoms include flu-like ailments (fever, chills, fatigue, weakness), breathing difficulties, and cough (Ehlers et al. 1993). Chronic exposure to grain dust may cause an increased prevalence of respiratory ailments including cough, wheeze, shortness of breath, and phlegm, as well as an increased decrement in ventilatory capacity (Morgan and Seaton 1995). ODTS differs from FLD in that it does not cause permanent lung damage or death.

Organic Dust Toxicity Syndrome (ODTS) is referred to by several names including atypical farmer's lung, pulmonary mycotoxicosis, toxic organic dust syndrome (TODS), and silo unloader's syndrome (SUS) (Murphy 1992).

Silo filler's syndrome (SFS) is caused by exposure to nitrogen oxides including NO, NO_2, and N_2O_4. The nitrogen oxides are produced during the fermentation of nitrates contained in the silage (Shaver and Tong 1991). A warning sign of dangerous conditions is a reddish brown gas above the silage. Dead birds or flies under the silo chute may also indicate the presence of NO_2. Since NO, NO_2, and N_2O_4 are all gases that are heavier than air, the gas buildup may be reduced by keeping the silo chute doors open to the silage level (NIOSH 1995). Fatal concentrations of gas can still accumulate in low spots of the silage surface. Figure 18.3 depicts the accumulation of silo gas. The risk of silo filler's disease can be reduced by avoiding entry into freshly filled silos for at least 2 weeks (Engberg 1993). SFS usually occurs in the late summer and early fall harvesting months.

Asphyxiation is a hazard in the grain industry. Oxygen-depleted environments can occur in air tight silos and grain bins (Ehlers et al. 1993). During grain fermentation, carbon dioxide is produced. Since carbon dioxide is denser than air, it can displace oxygen and create an oxygen-deficient atmosphere. These areas should be restricted to personnel who have been trained to wear air supplied respirators when entering these environments. A warning sign should also be placed on suspect areas to remind people of the possibility of an oxygen-depleted environment.

Barn allergy, characterized by rhinitis and allergic asthma, is caused by hypersensitivity to the excreta of storage mites found in hay, straw, and grain (Parkes 1994).

The results of several representative cross-sectional studies assessing chronic respiratory symptoms in grain workers are provided in Table 18.5. The cross sectional studies show that chronic bronchitis may affect 17 to 49% of grain workers. These results were associated with the duration of grain dust exposure and were more frequent than in control subjects. Chest wheezing was found in 24 to 42% of grain elevator workers. The prevalence of hypersensitivity among grain workers was not known in these studies. Organic dust toxic syndrome occurred in 16 to 32% of grain workers. This is sometimes referred to as "grain fever."

Nonpulmonary Health Effects

Non-pulmonary health disorders associated with grain dust exposure include grain fever, dermatitis, conjunctivitis, rhinitis, and chemical poisoning. Grain fever is characterized by flu-like symptoms (malaise, fever, shivering) that develop during and/or after occupational exposure to high concentrations of grain dust (Dosman and Cockcroft 1989). Dermatitis or "grain itch" involves skin inflammation. Conjunctivitis is inflammation of the mucous membranes lining the eyelids and covering the front of the eyeball. Rhinitis is inflammation of the nose and nasal mucous membranes (Van Fleet et al. 1992).

Noise exposure is also a prominent health hazard associated with grain and feed mill industries. Conveyors, grain dryers, tractors, motors, and augers are typical noise sources (Van Fleet et al. 1992). The chapter on noise and vibration contains detailed information on the health effects of noise exposure.

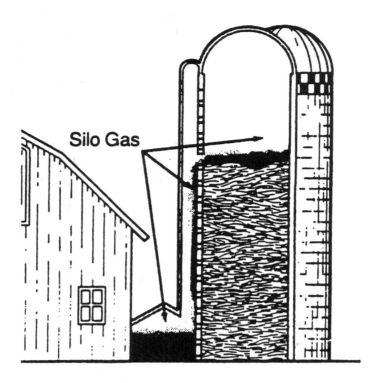

Figure 18.3. Silo gas is heavier than air and will displace oxygen. Wear a self-contained breathing apparatus if you must enter the silo during the first 4 to 6 weeks after filling stops (Tilma and Doss 1992).

Table 18.5. Chronic respiratory symptoms in grain workers (Zedja et al).

Number of Subjects	Chronic Disorders presented on a % basis				
	Chronic Bronchitis	Chest Wheeze	Hypersensitivity Pneumonitis	Organic Dust Toxic Syndrome	Reference
610	34	32	-----	----	Chan-Yeung et al, 1980
80	32	----	----	16	Cockroft et al, 1983
390	17	28	----	----	Cotton et al, 1983
310	49	42	----	32	doPico et al, 1984
582	35	24	----	----	Yach et al, 1985

Fumigants

Fumigation typically occurs in rail cars, grain elevators, or grain terminals. Fumigants, in smoke, vapor or gas form, are used to retard insect infestation and mold growth. Fumigants may also be used for bacterial control and rodent control. Fumigants can be single chemicals, such as chloroform, or combinations of multiple chemicals, such as "80/20", a mixture of 80% carbon tetrachloride and 20% carbon disulfide. From the 1950s through the 1970s, carbon disulfide was the fumigant of choice for insect control in the grain industry (Chapman et al. 1994).

Many fumigants have been banned or restricted in use, such as ethylene dibromide and dibromochloropropane (DBCP), due to their carcinogenic, mutagenic, and reproductive hazards. The U.S. Environmental Protection Agency prohibited the sale or distribution of carbon disulfide-carbon tetrachloride grain fumigants in 1986 (Chapman et al. 1994).

Fumigants can enter the body by inhalation, ingestion, or by absorption through the skin. Respiratory tract injury is the most common result of fumigant exposure. Dermal injury associated with fumigant exposure ranges from mild chemical burns to vesiculation and ulceration (Shaver and Tong 1991). If the eyes are exposed to liquid fumigants, corneal ulceration may develop.

"80/20," commonly used in the grain industry to control insect infestation, is associated with renal toxicity due to carbon tetrachloride exposure and neurotoxicity due to chronic carbon disulfide exposure (Chapman et al. 1991). An early cardinal sign of chronic low level carbon disulfide exposures is visible tremor (Chapman et al. 1991, 1994). Neurotoxic symptoms associated with carbon disulfide exposure include dizziness, fatigue, blurred vision, tremor, and gait problems (Peters et al. 1988). Carbon disulfide can produce irreversible, progressive changes, such as Parkinson-like movement disorders and peripheral neuropathy, that can worsen for 3 to 10 years after exposure has ceased (Chapman et al. 1994).

When applying fumigants, manufacturer's label instructions should be followed. Due to the dangers of fumigants, some producers have licensed professional fumigators apply the fumigants for them.

Mycotoxins

Mycotoxins are toxic substances produced by fungi, especially mold. Uncapping a silo containing moldy produce is an acute risk to farmers. Aflatoxins, one of the most serious mycotoxins, may be produced in moldy grain. Epidemiologic studies have associated exposures to aflatoxin B1 with an increased risk of cancer of the liver and biliary tract among grain mill workers. Respiratory diseases including chronic bronchitis, asthma, hypersensitivity pneumonitis, and acute febrile illness can occur due to exposure to bacteria, fungi, vegetable dusts, animal danders, and insects.

These respiratory conditions can be reduced by adopting preventive measures that reduce microbial growth in grain dusts. Microbial growth can be limited by drying grain to a safe water content and maintaining it during storage (Parkes 1994).

Treatment

The most efficacious treatment is avoidance of the environment known to produce symptoms. For those who are unable to change occupations, smoking cessation as well as proper respiratory protection are recommended (Bardana et al. 1992). Health problems should be discussed with your physician.

SAFETY HAZARDS

There are many safety hazards associated with the grain industry. Silo loading and unloading, machinery hazards, fires and explosions, and surface hazards are discussed in this section.

Silo Loading and Unloading Hazards

Silo loading and unloading are associated with safety hazards. Entering a grain bin can result in entrapment and suffocation, especially during unloading. When the grain bridges during unloading, the grain flow stops and a crusted layer of grain that appears solid on the surface covers a large hole underneath. Figure 18.4 depicts a worker attempting to walk on bridged grain. If grain bridges, the unloader should be stopped and a long wooden pole (not a metal one) should be used to break up the bridge and get the grain flowing again (Figure 18.5). Entrapment and burial can occur within seconds (NIOSH 1983). If a grain bin needs to be entered, stop the unloading auger, use a safety harness with rope, and secure it to a point outside of the bin. Have one or two people outside of the bin to provide help in the event of an emergency. As an additional precaution, permanent ladders could be installed on the insides of all grain bins.

Machinery Hazards

Crushing or amputation of an extremity are common and severe grain auger injuries. A grain auger operates on a screw-type principle. It is a finned rotating shaft inside a tube, used to propel grain (Sterner 1991). Grain elevators and feed mill employees frequently work with moving machinery

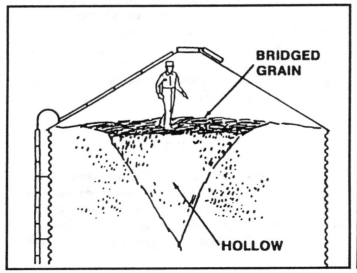 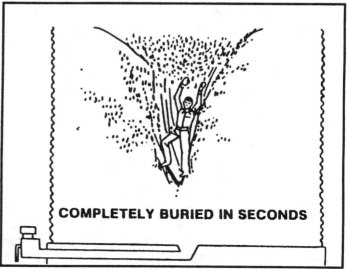

Figure 18.4. Crusted, spoiled grain can also result in a grain bin suffocation (Baker 1969).

including conveyors, drive motors, drive belts, gears, pulleys, and augers. Hazards associated with moving machinery include pinch points, nip points, and entanglement. Loose clothing (a shoelace, loose string, thread, flap of cloth, or jacket corner) can become entangled in moving machinery very easily.

Fire and Explosion Hazards

Grain dust explosions are the primary cause of injury, death, and property damage in grain and mill industries (Van Fleet EL et al. 1992). Explosions are caused when an ignition source, grain dusts, oxygen, and a confined space (e.g., elevator) occur together. Grain dust ignition sources that have a high probability of occurrence include hot bearings, welding and cutting, belt slippage and misalignment, and foreign objects caught in machinery (Van Fleet EL et al. 1992). Fire, rather than an explosion, may occur when an ignition source, grain dusts, and oxygen are present.

The explosive properties of common grain dusts are presented in Table 18.6. The table illustrates that grains are of varying volatility. Each type of grain dust has its own lower explosive limit (LEL). As grain is handled, grain dust is produced. The more grain is handled in a confined space, the greater the probability is that the grain dust's LEL will be exceeded (Van Fleet EL et al. 1992). Once the LEL is exceeded, there is a greater likelihood of an explosion.

Preventive measures include minimizing dust accumulation, reducing atmospheric oxygen to 12% and using spark proof electrical equipment. Overheating of machinery (e.g., grain-drying equipment) can be reduced by regularly checking and servicing bearings, belts, and conveyors (NIOSH 1995). Liquefied petroleum (LP) gas hazards, used for grain drying, can be reduced by using only LP gas specified pipe, tubing, hose, fittings, and valves. The National Fire Protection Association has recommendations for storing LP gas.

Surface Hazards

Loose grain and grain dust may interfere with floor traction and lead to slips and falls. Grain spills should be cleaned up as soon as possible to minimize loss of floor traction. Floors aisles and passageways should be kept clean and dry.

SAFE WORK PRACTICES

Due to the potential for grain dust explosions, OSHA requires employers to provide annual employee training. The training must address safety precautions pertinent to the facility, the recognition of hazards associated with dust accumulation and common ignition sources, the prevention of dust accumulation and common ignition sources, and specific procedures and practices relevant to the employee's job.

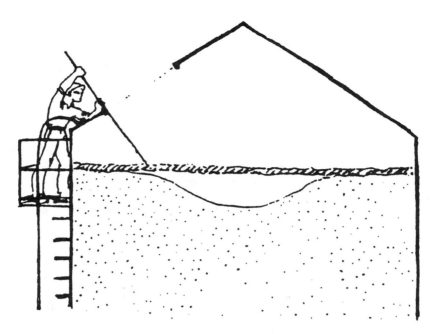

Figure 18.5. Surface crusts or material sticking to bin walls should be broken up from outside the bin with a wooden pole or weighted line. Never walk onto a surface crust or enter a storage structure below material sticking to the walls (NIOSH 1995).

Table 18.6. Explosive properties of common grain dusts (Van Fleet et al. 1992).

Grain Dust	Maximum Pressure, kPA	Maximum Rate of Pressure Rise, MPa/s	Ignition Temperature		Minimum Ignition Energy, J	Lower Explosive Limit, g/m^3
			Cloud °C	Layer °C		
Alfalfa	455	7.6	460	200	0.32	100
Cereal grass	360	3.5	550	220	0.80	200
Corn	655	41.0	400	250	0.04	55
Flax shive	560	5.5	430	230	0.08	80
Grain dust: winter wheat, corn, oats	790	38.0	430	230	0.03	55
Rice	640	18.0	440	220	0.05	50
Soy flour	540	5.5	540	190	0.10	60
Wheat flour	655	26.0	380	360	0.05	50
Wheat straw	680	41.0	470	220	0.05	55

Examples of specific safety procedures include:

1. Cleaning procedures for grinding equipment
2. Clearing procedures for choked legs
3. Housekeeping procedures
4. Hot work procedures
5. Preventive maintenance procedures
6. Lockout/tagout procedures (Van Fleet et al. 1992).

Respiratory hazards and noise exposures can be reduced by use of PPE. In the grain industry, air purifying respirators, self-contained breathing apparatus (SCBA), hearing protection, gloves, and safety glasses or goggles may be required. Selecting the appropriate PPE for the job is crucial to reducing hazards. A dust respirator will provide no protection against chemicals. An air purifying respirator will provide no protection in an oxygen deficient atmosphere.

The input end of grain augers should be protected by a screen so that only grain and nothing larger can be pulled up the auger tube (Sterner 1991). Close fitting clothing should be worn when working near power augers, because loose floppy clothes could be caught in the auger's moving parts.

Health and Safety Checklist for Working In or Near Conventional Silos

1. Avoid entering the silo for 2 weeks after filling.
2. If the silo must be entered: ventilate the silo thoroughly by leaving silo doors open and running the blower one hour before entry and during the time anyone is inside the silo.
3. Avoid low places in the silo, because toxic gas (NO_2) may accumulate.
4. Never jump down into the silage. If entry is necessary, use a self-contained breathing apparatus, wear a lifeline, and have rescuers nearby.
5. Avoid ascending the silo chute if 1 or 2 day old silage is present at any level in the silo.
6. Never open the silo door just above the silage without first checking for silo gas. Check for silo gas by using a good light and looking down the silo from above.
7. If the silo must be entered immediately after filling:
   a. Check the airflow direction in the silo chute.
   b. Check for silo gas.
   c. Leave the blower running.
   d. Avoid using 1 to 2 day old silage.
8. Only let trained workers near the silo. Do not allow children to play near the silo.
9. Warning signs of silo gas exposure include shortness of breath, rapid breathing, coughing, or chest pain. (Adapted from Shaver and Tong 1991).

NIOSH's health and safety checklist for the grain industry is provided at the end of this chapter. It covers many aspects of health and safety included in 29 CFR 1910. Checklists are provided for walking and working surfaces, electrical wiring, manlifts, occupational health and environmental control, occupational noise exposure, personal protective equipment, general environmental controls, medical and first aid, fire protection, automatic sprinklers, compressed air, materials handling and storage, machine and machine guarding, welding, cutting and brazing, and recordkeeping.

RULES AND REGULATIONS

The American Conference of Governmental Industrial Hygienists (ACGIH) has established an 8-hour time weighted average (TWA) level of exposure for grain dust as 4 mg/m^3 for inhalable (total) dust or particulate. OSHA has established an 8-hour permissible exposure limit (PEL) for grain dust as 10 mg/m^3.

OSHA has established training requirements to address grain dust explosion potential. Annual employee training must address:

1. Safety precautions associated with the facility
2. Hazard recognition and preventive measures related to dust accumulation and common ignition sources
3. Specific safety procedures and practices for:

   - Cleaning grinding equipment
   - Clearing choked legs
   - Housekeeping
   - Hot work (welding operations)
   - Preventive maintenance
   - Lockout/ tagout (Van Fleet et al. 1992)

OSHA has established a confined space entry permit program (29 CFR 1910.272) to minimize fatalities resulting from entry into confined spaces such as bins, silos, and tanks. According to 29 CFR 1910.272, the confined space entry permit program should include:

1. Written standard operating practices for entering confined spaces
2. Documented employee training
3. Use of permits prior to entry
4. Disconnecting, blocking off, locking out, and tagging energy sources including all mechanical, electrical, hydraulic, and pneumatic energy sources which may present a danger to employees in confined space

5. Atmospheric testing for oxygen content, combustible gas or vapor, and toxic agents
6. Precautions for fire and explosions
7. Use of body harnesses with a lifeline or boatswain's chair when entering a silo or confined space from the top
8. Appropriate respirator use as needed
9. Avoiding working alone
10. Rescue equipment
11. A rescue plan for each entry (Van Fleet et al. 1992).

SUMMARY

The grain industry has many health and safety issues associated with it. The predominant health issues associated with the gain industry are pulmonary and nonpulmonary effects from exposure to gain dust, fumigants, and mycotoxins. The predominant safety issues associated with the grain industry are silo loading and unloading hazards, machinery hazards, and fire and explosion hazards. Workshops and training seminars need to be targeted to personnel in the grain industry to raise awareness of the grain industry's health and safety issues.

REFERENCES

Awad el Karim, M.A., et al. 1986. Respiratory and allergic disorders in workers exposed to grain and flour dusts. *Arch. Env. Health* 41(5): 297-301.

Baker, D.E. 1969. Safe storage and handling of grain. *Agricultural MU Guide*. Columbia, MO: University Extension, University of Missouri-Columbia, pp. 1-4.

Bardana, E.J., A. Montanaro, and M.T. O'Hollaren. 1992. *Occupational Asthma*. Philadelphia: Hanley & Belfus, Inc.

Broder, I., et al. 1983. Variables of pulmonary allergy and inflammation in grain elevator workers. *J. Occup. Med.* 25(1): 43-7.

Broder, I., et al. 1979. Comparison of respiratory variables in grain elevator workers and civic outside workers of Thunder Bay, Canada. *Am. Rev. Resp. Disease* 119: 193-203.

Brown, M.A. 1988. Health hazards of storing, handling, and shiping grain. NIOH and NIOSH basis for an occupational health standard: Grain dust.

Chan-Yeung, M.R. Wong, and L. MacLean. 1979. Respiratory abnormalities among grain elevator workers. *Chest* 75(4): 461-67.

Chapman, L.J. et al. 1991. Finger tremor after carbon disulfide-based pesticide exposures. *Arch. Neurol.* 48: 866-70.

Chapman, L.J. et al. 1994. Neurotoxic illness in the grain industry: Index cases of fumigant Parkinsonism. In: *Supplement to Agricultural Health and Safety: Workplace, Environment, Sustainability*. H.H. McDuffie et al., eds. Saskatoon, Canada: University of Saskatchewan, pp. 111-22.

Cordes, D.H., and D.F. Rea. 1991. Health hazards of farming. In: *Occupational Medicine: State of the Art Reviews* 6(3), Hanley and Belfus, Inc.

Cotton, D.J., et al. 1982. Effects of smoking and occupational exposure on peripheral airway function in young cereal grain workers. *Am. Rev. Respir. Dis.* 126: 660-5.

Cotton, D.J., et al. 1983. Effects of grain dust exposure and smoking on respiratory symptoms and lung function. *J. Occup. Med.* 25(2): 131-41.

Cummings, P.H. 1991. Farm accidents and injuries among farm families and workers: A pilot study. *AAOHN Journal* 39(9): 409-15.

doPico, G.A., et al. 1977. Respiratory abnormalities among grain handlers. *Am. Rev. Resp. Dis.* 115: 915-27.

doPico, G.A., et al. 1981. Pulmonary reaction to grain dust and its constituents. *Chest* 80(1): 57S-60S.

Dosman, J.A., and D.W. Cockcroft. 1989. *Principles of Health and Safety in Agriculture*. Boca Raton, Florida: CRC Press, Inc.

Ehlers, J.K., et al. 1993. Health and safety hazards associated with farming. *AAOHN Journal* 41(9): 414-21.

Engberg, L. 1993. Women and agricultural work. *Occup. Med.: State of the Art Reviews* 8(4): 869-82.

Etherton, J.R., et al. 1991. Agricultural machine-related deaths. *Am. J. Public Health* 81(6): 766-8.

Farant, J.P., and C.J. Moore. 1978. Dust exposures in the Canadian grain industry. *AIHA Journal* 39(3): 177-94.

Hard, D.L., et al. 1992. A model agricultural health promotion systems program for building state-based agricultural safety and health infrastructures. *Scand. J. Work Env. Health* 18(2): 46-8.

Herbert, F.A., et al. 1981. Respiratory profiles of grain handlers and sedentary workers. *CMA Journal* 125: 46-50.

Jansson, B., 1992. On farm injuries and safety. *Am. J. Ind. Med.* 21: 619-22.

Morgan, W.K. and A. Seaton. 1995. *Occupational Lund Disease*, 3rd Edition. Philadelphia: W.B. Saunders Company.

Murphy, D.J. 1992. *Safety and Health for Production Agriculture*. St. Joseph, MI: American Society of Agricultural Engineers.

Myers, J.R., and K.L. Hard. 1995. Work-related fatalities in the agricultural production and service sectors, 1980-1989. *Am. J. Ind. Med.* 27: 51-63.

National Institute for Occupational Safety and Health (NIOSH). 1988. Basis for an occupational health standard: grain dust. In: *Health Hazards of Storing, Handling, and Shipping Grain*. U.S. Department of Health and Human Services.

National Institute for Occupational Safety and Health (NIOSH). 1975. *Health and Safety Guide for Grain Mills*. U.S. Department of Health, Education, and Welfare.

National Institute for Occupational Safety and Health (NIOSH). 1983. *Occupational Safety in Grain Elevators and Feed Mills*. U.S. Department of Health and Human Services.

National Institute for Occupational Safety and Health (NIOSH). 1995. *Safe Grain and Silage Handling*. 1995. U.S. Department of Health and Human Services, pp 1-5.

Parkes, W.R 1994. Extrinsic allergic bronchioloalveolitis (hypersensitivity pneumonia). In: *Occupational Lung Disorders*. Oxford, Boston: Butterworth-Heinemann, pp. 683-721.

Peters, H.A., et al. 1988. Extrapyramidal and other neurologic manifestations associated with carbon disulfide fumigant exposure. *Arch. Neurol.* 45: 537-40.

Popendorf, W., et al. 1985. A synopsis of agricultural respiratory hazards. *Am. Ind. Hygiene Assoc. J.* 46(3):154-61.

Reddy, M. 1995. Respiratory health surveillance and wellness promotion for grain workers. In: *Agricultural Health and Safety: Workplace, Environment, Sustainability*. H.H. McDuffie, ed., et al., pp. 31-5.

Shaver, C.S., and T. Tong. 1991. Chemical hazards to agricultural workers. *Occup. Med.: State of the Art Reviews* 6(3): 391-413.

Smith, R. 1993. Perils in the fields. *Occup. Health and Safety* 78-81.

Statistical Abstract of the U.S. 1995. U.S. Department of Commerce, Economics and statistical administration, Bureau of the Census, 155th edition, pp. 419, 749.

Sterner, S. 1991. Farm injuries: How can the family farm be made a safer place? *Postgrad. Med.* 90(2):141-50.

Tilma, C. and H. Doss. 1992. *Confined Space Hazards a Threat to Farmers*. Center for Michigan Agricultural Safety and Health.

USDA. 1994. *Agricultural Statistics*, pp. 333-36.

Van Fleet, E.L., O. Frank, and J. Rosenbeck. 1992. A guide to safety and health in feed and grain mills. NC-OSHA Industry Guide #29: 1-42.

Warren, C.P.W. 1992. Health and safety in the grain industry. In: *Environmental and Occupational Medicine*. W.N. Rom, ed. Boston: Little, Brown and Company, pp. 381-91.

Wilk, V.A. 1993. Health hazards to children in agriculture. *Am. J. Ind. Med.* 24: 283-90.

Zejda, J.E., H.H. McDuffie, and J.A. Dosman. 1993. Epidemiology of health and safety risks in agriculture and related industries: practical applications for rural physicians. *Western J. Med.* 158: 56-63.

Zhou, C., and J.M. Roseman. 1994. Agricultural injuries among a population-based sample of farm operators in Alabama. *Am. J. Ind. Med.* 25: 385-402.

Appendix A: NIOSH Checklists for the Grain Industry.

NIOSH CHECKLISTS

Since safe conditions depend on vigilance for possible hazards and immediate remedial action, periodic inspections are one of the most important aspects of a successful safety and health program.

Management will find a checklist, such as the one presented on the following pages, helpful in performing a self-inspection of its facility. Because businesses vary, it is best that each business develop a customized list from the information in this booklet and a walk-through inspection.

Using this checklist, the manager, supervisor, or employee representative makes periodic inspections (preferably at least once each month) to identify problem areas so that corrective action may be taken.

Reference made in the "Checklist" subtitles refers to appropriate sections of "general industry standards, Title 29 Code of Federal Regulations Part 1910."

NIOSH CHECKLISTS (Cont.)

WALKING AND WORKING SURFACES
AISLES AND FLOOR (29 CFR 1910.22)

	Yes	No
Are all places of employment kept clean and orderly?	☐	☐
Are floors, aisles and passageways kept clean and dry and all spills cleaned up immediately?	☐	☐
Are floor holes, such as drains, covered?	☐	☐
Are permanent aisles appropriately marked?	☐	☐
Are wet surface areas covered with non-slip materials?	☐	☐

STORAGE LOFTS, SECOND FLOORS, ETC. (29 CFR 1910.22, .23)

Are signs showing floor-load capacity present?	☐	☐
Are platforms, storage lofts, balconies, etc. that are more than 4 feet above the floor protected with standard guardrails?	☐	☐
Are all platforms, lofts, and balconies (where people or machinery could be exposed to falling objects) guarded with standard 4-inch toeboards?	☐	☐

STAIRS (29 CFR 1910.24)

| Are there standard stair rails or handrails on all stairways having 4 or more risers? | ☐ | ☐ |

NIOSH CHECKLISTS (cont.)

	Yes	No
Are all stairways at least 22 inches wide?	☐	☐
Do stairs have at least a 7-foot overhead clearance?	☐	☐
Do stairs angle no more than 50° and no less than 30°?	☐	☐

LADDERS (29 CFR 1910.25., .26, .27)

Have defective ladders (e.g. broken rungs, side rails, etc.) been tagged as "DANGEROUS, DO NOT USE" and removed from service for repair or destruction?	☐	☐
Is it prohibited to use the top of an ordinary step ladder as a step?	☐	☐
Do fixed ladders have at least 3½ feet of extension at the top of the landing?	☐	☐
Is the distance between the centerline of rungs on a fixed ladder and the nearest permanent object in back of the ladder at least 7 inches or more?	☐	☐
Do fixed ladders longer than 20 feet have cages?	☐	☐
Do fixed ladders longer than 30 feet have landing platforms?	☐	☐
Do all fixed ladders have a preferred pitch of 75°-90°?	☐	☐

NIOSH CHECKLISTS (cont.)

EGRESS (29 CFR 1910.36-.38)

	Yes	No
Are all exits marked with an exit sign and illuminated by a reliable light source?	☐	☐
Is the lettering at least 6 inches high with the principle letter strokes at least ¾ of an inch wide?	☐	☐
Is the direction to exits, when not immediately apparent, marked with visible signs?	☐	☐
Are doors or other passageways, that are neither exits nor access to an exit, and located where they may be mistaken for exits, appropriately marked "NOT AN EXIT", "TO BASEMENT", "STOREROOM", etc.?	☐	☐
Are exit doors side-hinged?	☐	☐
Are all doors that must be passed through to reach an exit or way to an exit, always free to access with no possibility of a person being locked inside?	☐	☐
Are all exit routes always kept free of obstructions?	☐	☐

NIOSH CHECKLISTS (cont.)

NATIONAL ELECTRICAL CODE
ELECTRICAL WIRING

	Yes	No
Have exposed wires, frayed cords and deteriorated insulation been repaired or replaced?	☐	☐
Are junction boxes, outlets, switches and fittings covered?	☐	☐
Is all metal fixed electrical equipment grounded?	☐	☐
Does all equipment connected by cord and plug have grounded connections?	☐	☐
Are electrical appliances such as vacuums, blowers, vending machines, etc. grounded?	☐	☐
Are all portable electrical hand tools grounded? (Double insulated tools are acceptable without grounding.)	☐	☐
Are breaker switches identified as to their use?	☐	☐
Do flexible cords and cables not run through holes in wall or ceiling or through doorways or windows?	☐	☐
Are flexible cords and cables free from splices or taps?	☐	☐

NIOSH CHECKLISTS (cont.)

	Yes	No
Are flexible cords and cables fastened so that there is no direct pull on joints or terminal screws?	☐	☐
Are flexible cords and cables never substituted for fixed wiring?	☐	☐
Are flexible cords and cables not attached to building surfaces?	☐	☐
Is electrical equipment accessible, in good repair and approved for the location?	☐	☐

NIOSH CHECKLISTS (cont.)

MANLIFTS (29 CFR 1910.68)

	Yes	No
Are all floor openings guarded?	☐	☐
Is the area at the bottom landing kept clear of stairs, ladders, and debris?	☐	☐
Is a fixed metal ladder provided for the entire travel of the manlift?	☐	☐
Are both runs of the manlift adequately illuminated?	☐	☐
Are brakes provided which are capable of stopping and holding 250 pounds on each step?	☐	☐
Are signs posted giving instructions for the use of the manlift?	☐	☐
Are the manlifts inspected at least every 30 days with a record made of the findings?	☐	☐

NIOSH CHECKLISTS (Cont.)

OCCUPATIONAL HEALTH AND ENVIRONMENTAL CONTROL (29 CFR 1910.93, .94, .95)

	Yes	No
Is management aware of the hazards caused by various chemicals used in the plant?	☐	☐
Is employee exposure to these chemicals kept within the acceptable levels?	☐	☐
Are eye wash fountains and safety showers provided in areas where chemicals, such as caustics, are used?	☐	☐
Are all containers, such as vats, storage tanks, etc. labeled as to their contents?	☐	☐
Are employees required to wear personal protective equipment (gloves, eye protection, respirators, etc.) when handling hazardous materials?	☐	☐
If internal combustion engines are used, is carbon monoxide kept within acceptable levels?	☐	☐
Is employee exposure to welding fumes controlled by ventilation, use of respirators, exposure time or other means?	☐	☐
Is vacuuming used wherever possible rather than blowing or sweeping dust?	☐	☐
Is adequate ventilation provided to minimize exposure to grain dust and fumigants?	☐	☐
Have administrative or engineering controls been established to prevent over-exposure to grain dust and fumigants?	☐	☐

NIOSH CHECKLISTS (cont.)

OCCUPATIONAL NOISE EXPOSURE
(29 CFR 1910.95)

	Yes	No
If a noise problem is suspected, have noise levels been accurately measured?	☐	☐
If a noise problem exists, have plans to reduce noise levels by engineering methods been formulated (e.g. enclosure, maintenance, different methods of processing)?	☐	☐
If engineering controls cannot reduce the noise to safe levels: have administrative controls, such as limiting worker exposure in a given area, been started?	☐	☐
Are affected employees given annual audiometric tests if necessary?	☐	☐
Do all employees in high-noise areas wear hearing protection?	☐	☐
Are annual noise surveys made to re-evaluate problem areas?	☐	☐

NIOSH CHECKLISTS (Cont.)

PERSONAL PROTECTIVE EQUIPMENT
(29 CFR 1910.132-137)

	Yes	No
Is personal protective equipment provided, used, and maintained wherever it is necessary?	☐	☐
Is employee-owned personal protective equipment, such as gloves, protective shoes, etc., adequate, and properly maintained?	☐	☐
Is eye protection available where debris or flying objects could be a hazard?	☐	☐
Are respirators provided and worn during dusty operations, paint spraying, etc.?	☐	☐
Is the proper respirator in use for the hazards present? (For example, dust masks do not protect against solvent vapors.)	☐	☐
Is hearing protection used when noise exceeds 90 dBA (e.g. when using hammer mills)?	☐	☐

RESPIRATORY PROTECTION DEVICES
(29 CFR 1910.134)

	Yes	No
Are respirators provided when necessary?	☐	☐
Are there written standard operating procedures for the selection and use of respirators?	☐	☐

NIOSH CHECKLISTS (Cont.)

	Yes	No
Is the user instructed and trained in the proper use of respirators?	☐	☐
Where practicable, are respirators assigned for use by employees individually?	☐	☐
Are respirators cleaned and disinfected after use?	☐	☐
Are respirators stored in a convenient, clean, and sanitary location?	☐	☐
Are routinely-used respirators inspected during cleaning?	☐	☐

NIOSH CHECKLISTS (cont.)

GENERAL ENVIRONMENTAL CONTROLS
SANITATION (29 CFR 1910.141-149)

	Yes	No
Are restrooms and washrooms kept in clean and sanitary condition?	☐	☐
Are covered receptacles for sanitary napkins provided in the women's restroom?	☐	☐
Are covered receptacles for waste food kept in clean and sanitary condition?	☐	☐
Is all water that is provided for drinking, washing and cooking, suitable for drinking?	☐	☐
Are all outlets for water that is not suitable for drinking, clearly posted as "UNSAFE FOR DRINKING, WASHING OR COOKING"?	☐	☐
Are employees prohibited from eating in areas where toxic materials are present?	☐	☐

NIOSH CHECKLISTS (cont.)

MEDICAL AND FIRST AID (29 CFR 1910-151)

	Yes	No
Is at least one employee on each shift currently qualified to render first aid in the absence of a nearby clinic or hospital? (Some states require first-aid trained persons regardless of nearby clinics or hospitals.)	☐	☐
Are first-aid supplies readily available, inspected and replenished?	☐	☐
Are first-aid supplies approved by a consulting physician, indicating that they are adequate?	☐	☐
Are medical personnel readily available for advice and consultation on matters of employee health?	☐	☐
Is there a first-aid kit easily accessible to the work area?	☐	☐
Are emergency phone numbers posted?	☐	☐
Where employees may be exposed to injurious corrosive materials, are they provided with quick-drenching and flushing facilities for immediate emergency use?	☐	☐

NIOSH CHECKLISTS (cont.)

FIRE PROTECTION (29 CFR 1910.157, .159, .160)

	Yes	No
Are extinguishers selected for the types of combustibles and flammables in the areas where they are to be used? Class A. Ordinary combustible material fires Class B. Flammable-liquid, or grease fires Class C. Energized-electrical-equipment fires	☐	☐
Are extinguishers fully charged and in designated places?	☐	☐
Are extinguishers located along normal paths of travel?	☐	☐
Are extinguisher locations free from obstruction or blockage?	☐	☐
Are extinguishers not mounted too high? If not exceeding 40 pounds, the top must not be higher than 5 feet above floor-greater than 40 pounds, the top must not be higher than 3½ feet above floor.	☐	☐
Have all extinguishers been serviced, maintained and tagged at intervals not to exceed 1 year?	☐	☐
Are all extinguishers checked (by management or designated employee) monthly to see if they are in place or if they have been discharged, etc.?	☐	☐
Have all extinguishers been hydrostatically tested according to schedules set for the type of extinguisher?	☐	☐

NIOSH CHECKLISTS (cont.)

AUTOMATIC SPRINKLER (if applicable)

	Yes	No
Is there at least one automatic water supply of adequate pressure, capacity and reliability?	☐	☐
Are water-flow alarms provided on all sprinklers?	☐	☐
Are the sprinkler systems periodically inspected and continuously maintained?	☐	☐
Is the clearance between sprinkler deflectors and the top of storage at least 18"?	☐	☐

DRY CHEMICAL SYSTEMS (if applicable)

	Yes	No
Does a competent inspector make annual inspections and perform tests on all dry chemical systems?	☐	☐
Are the inspector's reports kept on file?	☐	☐
Are visual inspections regularly made?	☐	☐
Are all dry chemical systems continuously maintained?	☐	☐
Is combustible material never piled within 36 inches of the sprinkler system for Class I storage?	☐	☐
Is combustible material never piled next to lights or within 18 inches of the sprinkler system for Class II and III storage?	☐	☐

NIOSH CHECKLISTS (cont.)

COMPRESSED AIR (29 CFR 1910.169)

	Yes	No
Are pulleys and belts on compressors and motors completely guarded?	☐	☐
Are flexible cords or plugs on electric motors periodically checked and replaced if in a deteriorated condition?	☐	☐
Do the relief valves operate properly?	☐	☐
Are air tanks drained regularly?	☐	☐
Is the pressure-relief device and gauge in good operating condition?	☐	☐
Is air pressure reduced to less than 30 psi when the nozzle, used for cleaning, is dead ended?	☐	☐
Have employees been instructed that the use of compressed air to blow debris from clothing or body is prohibited because it can enter the body and cause serious harm?	☐	☐

NIOSH CHECKLISTS (Cont.)

MATERIALS HANDLING AND STORAGE (29 CFR 1910.176-181)

	Yes	No
Is there safe clearance for equipment through aisles and doors?	☐	☐
Is stored material stable and secure?	☐	☐
Are storage areas free from tripping hazards?	☐	☐
Are only trained operators allowed to operate powered lift trucks?	☐	☐
Are appropriate overhead guards installed on powered lift trucks?	☐	☐
Is battery charging on electric units performed only in designated areas?	☐	☐
Are "NO SMOKING" signs posted near electric battery charging units?	☐	☐
On units using internal combustion engines, do the exhaust gases in the room not exceed allowable limits for carbon monoxide?	☐	☐
Are dock boards (bridge plates) used when loading or unloading from dock to truck or dock to rail car?	☐	☐

NIOSH CHECKLISTS (cont.)

	Yes	No
Are containers of combustibles or flammables, when stacked one upon the other, always separated by dunnage sufficient to provide stability?	☐	☐
Are racks and platforms loaded within the limits of their capacity?	☐	☐
Is all storage secured against sliding or collapsing?	☐	☐
Are all vehicles shut off prior to loading?	☐	☐
Have aisles been designated and kept clear to allow unhindered passage?	☐	☐
If motorized equipment, such as lift trucks, is used, are aisles permanently marked, providing sufficient clearance for passage of the equipment?	☐	☐
Are specifications posted for maximum loads which are approved for floors (except slabs with no basements), roof of a building, or some other structure?	☐	☐
Are lift trucks selected and approved for the atmosphere (e.g. grain dust) in which they operate?	☐	☐

NIOSH CHECKLISTS (cont.)

MACHINE AND MACHINE GUARDING (29 CFR 1910.212)

	Yes	No
Are belts, pulleys, and rotating shafts (air compressor, drill presses, etc.) properly guarded?	☐	☐
Are chains, sprockets and gears properly guarded?	☐	☐
Are all in-going nip points properly guarded?	☐	☐
Are rotating shafts that are not smooth properly guarded?	☐	☐
Are all rotating parts (lubrication, fittings, etc.) recessed or covered with collars?	☐	☐
Are all pieces of equipment with an electric motor or any electrical connection effectively grounded?	☐	☐
Are sprockets and V-belt drives within reach of platforms and passageways or less than 7 feet from the floor completely enclosed?	☐	☐
Are fans less than 7 feet above floor guarded, having openings ½ inch or less?	☐	☐

ABRASIVE WHEEL MACHINERY (Grinders) (29 CFR 1910.215)

	Yes	No
Is the work rest used and kept adjusted to within 1/8 inch of wheel?	☐	☐
Is the adjustable tongue on top side of grinder used and kept adjusted to within ¼ inch of wheel?	☐	☐

NIOSH CHECKLISTS (cont.)

	Yes	No
Do side guards cover the spindle, nut and flange and 75% of the wheel diameter?	☐	☐
Are bench and pedestal grinders permanently mounted?	☐	☐
Are goggles or face shields always worn when grinding?	☐	☐

NIOSH CHECKLISTS (Cont.)

HAND AND PORTABLE POWER TOOLS (29CFR 1910.242-244)

	Yes	No
Are tools and equipment (both company and employee-owned) in good condition?	☐	☐
Have mushroomed heads on chisels, punches, etc. been reconditioned or replaced if necessary?	☐	☐
Have broken hammer handles been replaced?	☐	☐
Have worn or bent wrenches been replaced?	☐	☐
Has compressed air used for cleaning been reduced to 30 psi when dead ended?	☐	☐
Have employees been instructed that the use of compressed air to blow debris from clothing or body is prohibited because it can enter the body and cause serious harm?	☐	☐
Have deteriorated air hoses been replaced?	☐	☐
Are portable abrasive wheels appropriately guarded?	☐	☐
Have employees been made aware of the hazards caused by faulty or improperly used hand tools?	☐	☐
Are only non-sparking tools used in areas contaminated by excessive grain dust?	☐	☐

NIOSH CHECKLISTS (cont.)

WELDING, CUTTING AND BRAZING (29 CFR 1910.252)

	Yes	No
Are fuel gas cylinders and oxygen cylinders separated by 20 feet or a barrier 5 feet high having a ½-hour fire resistance rating?	☐	☐
Are cylinders secured and stored where they cannot be knocked over?	☐	☐
Are cylinder protective caps in place except when the cylinder is in use?	☐	☐
Are compressed gas cylinders kept away from sources of heat, elevators, stairs, or gangways?	☐	☐
Are only instructed employees, who are judged competent by the employer, allowed to use oxygen or fuel gas equipment?	☐	☐
Do all cylinders (except those with fixed hand wheels) have non-adjustable wrenches, keys, or handles in place on valve stems while cylinders are in use?	☐	☐
Is welding always conducted at a safe distance from flammable liquids or dusty areas?	☐	☐
Are all compressed gas cylinders legibly marked for identifying the content?	☐	☐
Are the valves shut off when the cylinder is not in use?	☐	☐
Are flash shields provided to protect nearby workers from the welding flash?	☐	☐

NIOSH CHECKLISTS (cont.)

RECORDKEEPING (29 CFR 1904.2-.8)

	Yes	No
Is employee poster (OSHA or equivalent state poster) prominently displayed?	☐	☐
Have occupational injuries or illnesses, except minor injuries requiring only first aid, been recorded on OSHA Form Nos. 100 and 101, or equivalent?	☐	☐
Has a summary of all occupational injuries and illnesses been compiled at the conclusion of each calender year and been recorded on OSHA Form No. 102? Was it posted during the month of February?	☐	☐
Have all OSHA records been retained for a period of five years, excluding the current year?	☐	☐

NIOSH INFORMATION SOURCES

AMERICAN NATIONAL STANDARDS INSTITUTE (ANSI)
1430 Broadway, New York, N.Y. 10018

- ☐ A12.1   Floor and Wall Openings
- ☐ A14.1   Portable Wood Ladders
- ☐ A58.1   Minimum Design Load
- ☐ A64.1   Fixed Stairs
- ☐ B15.1   Mechanical Power Transmission
- ☐ C1   National Electric Code
- ☐ Z4.1   Sanitation in Places of Employment

NATIONAL FIRE PROTECTION ASSOCIATION (NFPA)
470 Atlantic Ave.
Boston, Mass. 02210

- ☐ NFPA-10-1970   Installation of Portable Fire Extinguishers
- ☐ NFPA-101-1970   Life Safety Code
- ☐ NFPA-13A-1971   Sprinkler Systems, Maintenance
- ☐ NFPA-17-1969   Dry Chemical Extinguishing Systems
- ☐ NFPA-70-1971   National Electric Code

NATIONAL SAFETY COUNCIL
425 North Michigan Avenue
Chicago, Illinois 60611

NIOSH AND OSHA REGIONAL DIRECTORS

Trade associations and insurance companies can also provide useful information. The Small Business Administration will provide information concerning procedures for securing economic assistance on compliance with the OSHA Standards (if needed).

NIOSH FILMS

The NIOSH health and safety film "Grain Elevator and Feed Mill Safety" is available for free loan from the NIOSH Regional Offices and the Office of Technical Publications.

19

SAFETY AROUND BEEF CATTLE

*James B. Cowan, D.V.M., M.S.P.H.,
Public Health Officer, United States Air Force
North Carolina Cooperative Extension Service Consultant*

> Handling large animals, such as beef cattle, ranks second only to machinery as a cause of agricultural injuries. Animal handlers can reduce their risk of injury by understanding cattle psychology, applying correct cattle handling principles, and using well-designed and constructed cattle handling facilities and equipment.

INTRODUCTION

Injuries caused by beef cattle are a serious problem and contribute to agriculture's ominous title as one of the most dangerous occupations in the United States. Although a thorough epidemiological assessment of beef cattle inflicted injuries is lacking, several studies indicate that large farm animals, including beef cattle, are responsible for a significant proportion of nonfatal farm work injuries.

Surveys conducted in 31 in 1981 and 35 states in 1986 by the National Safety Council found animals accounting for 17% of all nonfatal farmwork injuries, ranking second behind agricultural machinery (25%) (NSC 1982, 1987). In 1988, interviews with farmers from Iowa, New York, South Carolina, and Washington revealed that 21% had been injured by a farm animal sometime during the year (Kendall 1990). A population-based study of agricultural injuries found farm animals, mostly cattle, as causing 12.5% of agricultural injuries among a random sample of 1,000 farm operators in Alabama during 1991 (Zhou and Roseman 1994). In Wisconsin, 134 patients who sustained farm animal-related injuries were admitted to a large rural trauma center between 1977 and 1983. The mechanisms of injury among these patients were fall from horse, 45 (33%); kicked by cow, 28 (21%); bovine assault, 25 (19%); equine assault, 17 (13%); kicked by horse, 11 (8%); and animal-drawn vehicle, 8 (6%). Duration of hospitalization ranged from 1 to 115 days with an average stay of 7 days (Busch et al. 1986). Cogbill et al. (1991) reviewed the medical records of patients admitted to the same trauma center over a 12-year period ending in December 1989. Of the 739 patients admitted for an injury while farming, large farm animals were responsible for the majority of injuries; (30%, n=225). Although animals predominantly cause nonfatal injuries, they have killed people. During a 4-year period ending in 1980, farm animals were responsible for 25 deaths in Wisconsin (Busch 1986). From 1979 through 1985, farm animals caused 3.7% of fatal injuries on Kentucky farms (Stallones 1990). In 1983, livestock accounted for 18% of farm deaths in Alberta, Canada (Denis 1988).

Despite the results of these studies being, at best, an approximation of the true incidence of injuries caused by beef cattle, there is agreement among animal behaviorists, agricultural injury experts, and health professionals that measures to prevent and control farm animal inflicted injuries, in general, are warranted (Steele-Bodger 1969; Busch et al. 1986; Grandin 1989a; Stallones 1990; Cogbill et al. 1991; Stueland et al. 1991). Moreover, these groups often mention educational intervention as an appropriate strategy for controlling these injuries. The aim of this chapter is to prevent beef cattle-related injuries by educating injury control professionals and students and the agricultural community about the correct principles for safely handling beef cattle.

PRINCIPLES OF INJURY CONTROL

To fully appreciate and effectively develop and apply injury control strategies, one must first have a conceptual understanding of the factors leading to injury. The host-agent-environment model of classic Epidemiology provides a useful framework for understanding the mechanism of action of injury occurrence. Consider the situation that occurs when a beef cattle producer or ranch hand suffers a broken jaw after being kicked by a cow. Within the physical and social environment, this injury results from the transmission of mechanical energy (agent) from the cow (vector) against the animal handler (host) in sufficient force to exceed the individual's resistance to injury.

Based on this framework of injury causation, Haddon (1980) published ten general strategies of injury control. Each strategy is specified below along with an example illustrating its usefulness in developing ideas for controlling beef cattle inflicted injuries.

- Prevent the creation of the hazard in the first place. Example: Breeders should select animals with a calm temperament.

- Reduce the amount of hazard brought into being. Example: Castrate cattle while they are young rather than waiting until they are older, bigger and stronger.

- Prevent the release of the hazard that already exists. Example: To prevent cattle from turning on their handler, ensure handling facilities are free of obstacles.

- Modify the rate or spatial distribution of release of the hazard from its source. Example: Install working chutes that enable handlers to control an animal's speed.

- Separate, in time or space, the hazard and that which is to be protected. Example: Use proper techniques for moving cattle efficiently while maintaining a safe distance between handler and animal.

- Separate the hazard and that which is to be protected by interposition of a material barrier. Example: When deworming cattle restrain them in a headgate and squeeze chute.

- Modify relevant basic qualities of the hazard. Example: Administer a sedative to reduce an animal's excitement and pain when providing medical care. Note, there are few sedative medications approved for use in beef cattle, and none available over-the-counter. Also, be aware of withdrawal times, especially if cattle are being sent to market.

- Make what is to be protected more resistant to damage from the hazard. Example: Handlers who are healthy and strong are more resistant to injury than someone who is sick or out of shape.

- Begin to counter the damage already done by the environmental hazard. Example: Establish an emergency medical service that provides prompt and competent medical care to rural communities, including the treatment of cattle inflicted injuries.

- Stabilize, repair and rehabilitate the object of damage. Example: Following surgical repair of a torn ligament, provide rehabilitation to restore maximum function to the limb.

Haddon made two points regarding the application of these strategies. First, the principles serve as a tool for identifying and assessing the value of actual and possible control strategies. Second, the effectiveness of the strategy analysis does not depend on a thorough understanding of the factors influencing the injury event. For example, placing a cow in a squeeze chute protects the handler from being kicked even though the circumstances leading to a kick may not be fully understood.

From a more general perspective, injury control interventions can be separated into the areas of work practice and engineering controls. Work practice controls reduce the likelihood of exposure by modifying the procedures involved in performing the task. As will be presented later in the chapter, using the concepts of "flight zone" and "point of balance" to move cattle efficiently and safely, are examples. Engineering controls reduce exposure by either removing the hazard or isolating the worker from the hazard. This includes designing and constructing cattle handling facilities that enable handlers to work cattle from a safe distance. Central to the successful development and application of work practice and engineering controls, is an understanding of beef cattle psychology.

BEEF CATTLE PSYCHOLOGY

Beef cattle psychology is the study of how cattle sense and react to the world around them. The way cattle behave is influenced by their ability to see, smell, and hear and by their maternal and herd instincts (Grandin 1983, 1989a).

Sense of Sight

With their eyes positioned on the sides of their heads, cattle have a wide angle view of their world and a narrow blind spot directly behind them (Figure 19.1). Although their peripheral vision is enormous, cattle have problems seeing because their view of their environment is like looking through a fish-eye lens—everything appears bent and distorted, especially at the periphery. For example, a fence post that looks straight to us appears curved to cattle.

Because their view is distorted, cattle react strongly to contrast and movement. They will balk or turn when approaching bright sunlight or a shadow. To them, a shadow on the ground appears to be a deep hole and they will look at it closely before proceeding further. Quick movements, like a farm hand flapping his arms to get cattle moving or a tarp blowing in the wind, will spook these animals, causing them to balk or become excited.

Sense of Hearing

Noise is very stressful to cattle. Like any animal, cattle are disturbed by loud, abrupt noises or sounds that are new to them, especially high-pitched noises. Sudden, high pitched noises such as a gate slamming, a telephone ringing, gas bleeding off a hydraulic line, or a crack of a whip bother these animals.

Sense of Smell

Cattle have an excellent sense of smell and depending on the situation, scent will often be the dominant factor affecting a cow's behavior. For example, a cow's sense of smell can tell her if she is being separated from her calf and this will often cause her to become stressed and dangerous. Odors are also the primary means of sexual communication between cows and bulls. A bull fenced in a pen can smell a cow in heat from a significant distance. If he smells her he will begin to act aggressively.

Herd Instinct

Another characteristic of cattle that contributes to their behavior is their herd instinct. Cattle are social animals and are more comfortable and feel safer in a group. Their herd instinct evolved as a means of protecting themselves against predators and pests. Even though cattle were domesticated thousands of years ago, their herd instinct remains strong. When isolated from the rest of the herd, a single animal will become very stressed and easily upset.

Two primary characteristics of cattle herd instinct are following the leader and the herd social order. Among members of a herd, there are followers and leaders. The "leader" is the animal that is almost always the first member of the group to move in a particular direction. When this animal heads off to go somewhere, the rest of the group will follow. A herd's social order is similar to the pecking order among a flock of chickens. One animal asserts dominance over a weaker member but, in turn, is dominated by a stronger animal. Usually, dominant animals are not the herd leaders, nor are subordinate cattle necessarily followers. By observing a herd, you can identify the dominant and subordinate cattle. When they are grazing, dominant cattle

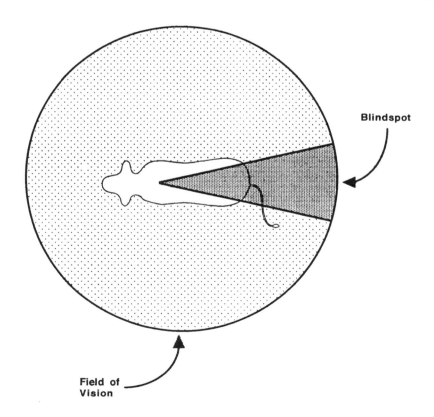

Figure 19.1. The animal's field of vision.

are usually in the middle of the group where they can get the most protection, and the subordinate cattle are on the periphery. Also, when they are at the feeder, the dominant cattle will get at the food by pushing subordinate cattle away.

Maternal Instinct

Maternal instinct in cattle is very strong. Normally, a cow will be wary of people, especially strangers, and will be protective of her young. Her protective nature is strongest during the first 2 weeks after her calf is born and can remain strong for as long as she is taking care of the calf. As a result, handlers risk injury whenever they try to do anything to a calf when the mother is nearby and not restrained. This includes assisting in delivery, examining the newborn, castrating, and ear tagging.

HANDLING BEEF CATTLE SAFELY

Safe and efficient handling of beef cattle requires proper cattle handling practices and correctly designed and installed cattle handling facilities. This section builds on information presented in the previous section "Beef Cattle Psychology."

Beef Cattle Handling Practices

Proper cattle handling practices are dependent on two important concepts, flight zone and point of balance (Grandin 1984, 1990c). To correctly handle cattle, a handler must fully understand and consistently apply these concepts. Furthermore, for beef cattle operations with hired workers, management must train employees on how to handle cattle properly and provide incentives for careful and caring animal handling. Rough handling must not be tolerated.

Flight Zone

Flight zone is a term used to describe an animal's personal space (Figure 19.2). It is an imaginary circle around the animal. When a handler enters its flight zone, the animal will move away. When a handler steps out of the flight zone, the animal will stop. If the flight zone is penetrated too deeply, the animal will often panic. The shaded area in Figure 19.2 labeled "blind spot" is the area where the handler cannot be seen as they approach the animal. Entering an animal's flight zone in its blindspot may agitate the animal and cause it to kick.

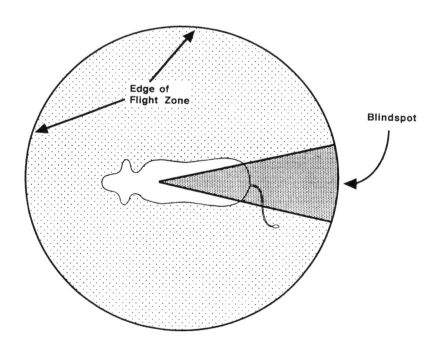

Figure 19.2. The animal's flight zone.

Having a flight zone is not unique to cattle; people have their own flight zones as well. For example, when someone enters the personal space of another individual, the normal response is for that person to back away. Cattle behave the same way. They do not want people to come too close.

The size of a flight zone varies from one animal to the next and is influenced by several factors, including how wild or tame an animal is. The tamer the animal, the smaller its flight zone. Cattle on the range have flight zones approaching 300 feet, while the flight zone of feedlot cattle may be as little as 5 to 25 feet. In some cases, if cattle are extremely tame, they may not have a flight zone at all. Animals without a flight zone or with very small flight zones are often difficult to drive. Usually, tame animals are moved efficiently with a feed bucket or using a halter.

Similarly, if cattle are treated gently and carefully by people with whom they are familiar, they will let these individuals get closer to them than with strangers. Some producers only let people work their cattle after the animals have become familiar with them. Very often, the movement of animals from pasture to the handling facility proceeds more smoothly when strangers are kept at a distance.

The breed of cattle or line within the breed affects the size of the flight zone. Breeds that are easily excited, such as those with Brahman blood, tend to have larger flight zones.

The size of the flight zone also varies according to the direction from which an animal is approached. In most instances, the flight zone will be larger when the handler approaches head on. When cattle are confined in a single-file chute their flight zones become smaller.

Point of Balance

The second concept, point of balance, is the place where there is a balance between moving an animal forward or moving it backward. In Figure 19.3, it is an imaginary line that starts from the animal's shoulder and goes out at a 90 degree angle until it intersects the edge of the flight zone. A handler uses an animal's point of balance and flight zone to move it in an orderly manner. Placing oneself to the left of the imaginary line at position A, moves the animal backward, and placing oneself to the right of the imaginary line at position B, moves the animal forward.

THINGS TO KEEP IN MIND WHEN HANDLING CATTLE

For a handler to successfully move cattle by applying the principles of flight zone and point of balance, he or she must

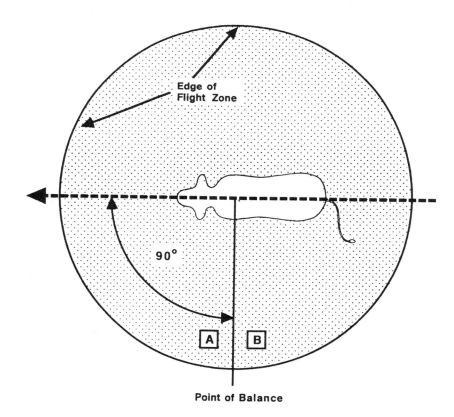

Figure 19.3. The animal's point of balance.

first keep some specific points in mind (Grandin 1989b, 1990a, 1990c, 1995a, 1995b):

- Because cattle cannot see directly behind themselves, handlers should avoid approaching them from the rear. If someone must come up from behind, the individual should let the animal know by making some gentle sounds. An animal that is startled from behind will kick. If cattle are approached in their blind spot after they are already moving, they will zigzag back and forth as they try to keep an eye on the person. Cattle proceed more smoothly when handlers remain in the animal's field of vision.

- Handlers should not use any quick movements. Cattle are very sensitive to movement. Handlers who flap their arms when working cattle will upset them. By keeping movements smooth and consistent, cattle will remain calm and more manageable.

- Handlers should not move cattle by whooping, hollering, or screaming at them. Cattle need to be handled deliberately, confidently, and calmly. Getting them excited makes the job more difficult.

- Only small amounts of noise are needed to move cattle. For example, rustling a stick (such as mop handle) with plastic strips attached is often enough to get an animal to proceed in the right direction.

- Because of their herd instinct, it is easier to work cattle in groups rather than managing them alone. If a handler is having difficulty moving a single animal out of the pen, he or she should release a couple of animals into the pen and move all three animals out together.

- Before handling a calf, it should be separated from the cow. If a handling facility is available, the handler should allow the cow to walk into a pen and then cut the calf off from behind with the gate. Under field conditions, the handler should get the calf into the bed of a pickup truck while someone else diverts the mother's attention. The calf should be handled in the truck bed or outside the pasture. The calf is returned to its mother after procedures are accomplished. Whenever possible, the cows are kept upwind when separating them from their offspring. This will reduce their ability to smell their calves and make the separation easier.

- Cattle have long memories, especially of painful events. Therefore, handlers should consistently handle cattle with a minimum amount of excitement. Cattle treated poorly in the past will be more difficult to handle in the future. Nose-tongs and electric cattle prods are rarely needed. Halters and long sticks or sticks with plastic streamers should be substituted for nose-tongs and cattle prods.

- Often, if a handler fails to restrain cattle on the first attempt, it will be more difficult to restrain them the next time around. Handlers will increase the probability of first time success, if they plan ahead by scouting the work area and making sure it is clear of obstacles, checking equipment and facilities to ensure they are in good working order, and familiarizing themselves with the tasks they will be performing. An inexperienced handler should be very patient; increased speed will come with practice.

- When working cattle in confined areas such as chutes, pens, and alleys, the handler must be especially careful because it is very easy to deeply invade the animal's flight zone or enter its blind spot. If either situation occurs, the animal may try to jump the fence or turn back on the handler. To prevent cattle from bolting in this situation, the handler should act quickly and back out of the animal's flight zone. If the animal turns on the handler, the person should just get out of the way. Once the handler is outside the flight zone, the animal will normally settle down. Occasionally, however, reluctant cattle may need to be dominated in order to proceed. If cattle get the idea that anytime they don't want to go somewhere that they don't have to, then they get quickly spoiled.

- As cattle proceed down a working chute, they may start moving too quickly and ride up on one another. When this occurs, the handler should not step up to the chute and try to push the cattle down with a stick or cattle prod. This will upset the animals more because the handler will have entered deeply into their flight zone. Rather, the person should step back and leave them alone. They will eventually settle down on their own.

Using Flight Zone and Point of Balance Concepts

Handlers should consider the concepts of flight zone and point of balance whenever they interact with cattle. Using these principles in different combinations allows handlers to successfully and safely move cattle forward and backward, change an animal's direction of forward movement, maintain forward movement of a herd of cattle, direct stragglers to regroup with a herd, and easily move cattle into and out of a pen (Grandin 1989b).

Moving Cattle Forward

Before moving an animal forward, the handler should be standing within the shaded triangle and just outside the flight zone (Figure 19.4). To determine the edge of the animal's flight zone, the handler approaches the cow confidently and deliberately from slightly behind the shoulder. As the individual approaches, the animal's first reaction is to look at the person. When the handler has entered the flight zone, the animal will begin to move away.

Moving from position A to position B represents the point of entering the animal's flight zone. The handler must remember not to penetrate this area too deeply or the animal will panic and flee or turn back and run over the individual. The handler must always be alert to how the animal is reacting to his or her presence. Once a cow begins to move away, the handler can keep it moving straight ahead by moving back and forth between positions A and B. When stationed at position B, the animal is pressed forward. By moving back to position A pressure is relieved before the animal panics. As the handler moves along, he or she will repeat the pattern of entering the flight zone and backing away.

To stop the animal's forward progress, the handler moves to position A and remains there. After taking a few more steps, the cow will realize that the person is no longer entering its personal space and will stop.

Moving Cattle Backward

Depending on the situation, there are times when an animal will need to be moved backward. If a handler has been moving a cow forward by entering and retreating from its flight zone, the individual can get her to back up by placing himself or herself in front of its point of balance (Figure 19.4).

The handler moves from position B to position C, while being careful not to cut across the flight zone. If the animal's personal space is invaded too deeply, it will be spooked and run or turn back. Proper movement from position B to C requires the handler to follow along the imaginary edge of

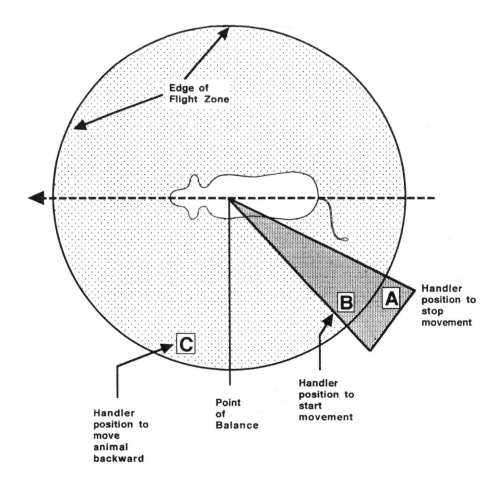

Figure 19.4. Moving the animal forward and backward.

the flight zone. The handler must be patient and allow the animal to move backward at its own pace.

Changing the Animal's Direction of Forward Movement

Steering cattle is like steering a car. First, the handler gets the animal moving forward by entering its flight zone. Once it is moving, the handler shifts from position B to position C (Figure 19.5). Although position C is not any closer to the animal than position B, it is closer to the animal's point of balance.

When the handler is stationed at position C, the animal will circle ahead in the direction shown. A cow circles around the person because it wants to keep the individual in sight at all times and does not want to be cut off. If the animal did not circle around the handler, it would not be able to see the person. Movements between position B and C take advantage of the animal's natural circling behavior.

Moving a Herd of Cattle with One Handler

The concepts of flight zone and point of balance apply to a herd just as they apply to individual animals. When combined with the natural tendency of cattle to follow one another, these concepts help handlers manage cattle more easily. As shown in Figure 19.6, the handler initiates cattle movement by approaching the group until the individual is stationed at position 1, just inside the group's flight zone. In response to the handler's movement, the herd will proceed in the direction shown. Because the handler is near the rear of the group, he or she will have to be careful to avoid the animals' blind spot. As mentioned before, if the handler enters their blind spot, the herd will begin zigzagging back and forth as they try to keep an eye on the person.

Once the cattle are moving, the handler walks at an angle until he/she reaches position 2. The distance between positions 1, 2, 3, and 4, will depend on what works best for each handler and the herd they are managing. Because the

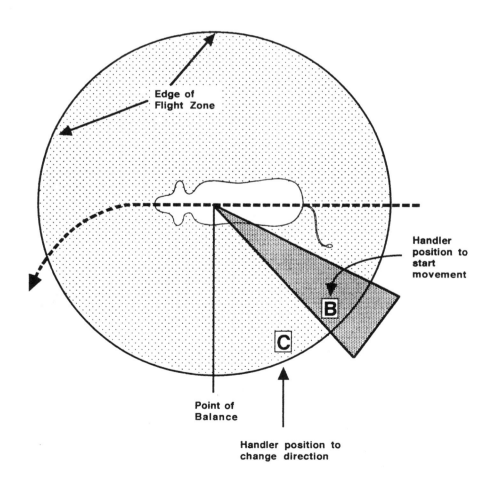

Figure 19.5. Changing the animal's direction of forward movement

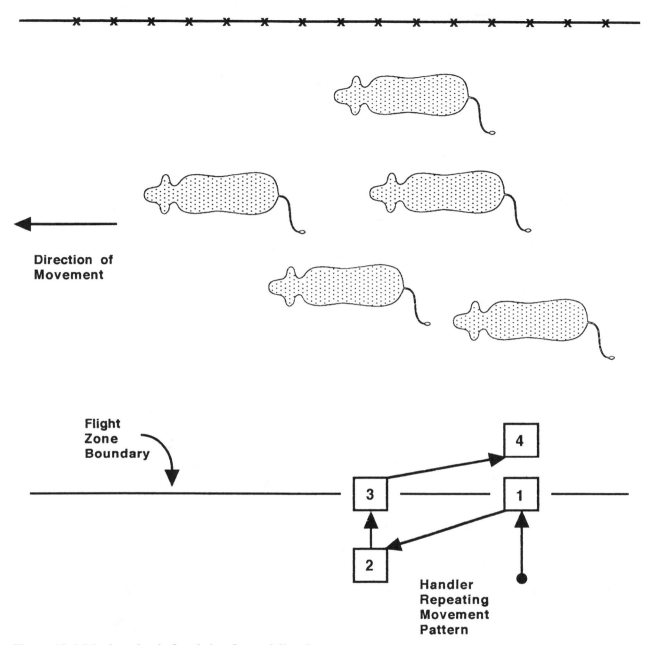

Figure 19.6. Moving a herd of cattle in a forward direction.

handler is walking at an angle between position 1 and 2, the amount of pressure on the group's collective flight zone will decrease as position 2 is approached. Consequently, as the cattle begin to slow down, the handler again increases pressure on the flight zone by walking straight toward the cattle until he or she reaches position 3. A handler's presence at position 3 puts pressure on the group's leaders to move forward and the rest of the group follows along.

As the group speeds up, the handler turns and heads up in the opposite direction of the herds' movement until reaching position 4. Concurrently, the handler approaches the group at a slight angle and increases pressure on the herd's flight zone. Although it may seem that this action will halt the group's progress, the handler is actually motivating them forward by crossing their point of balance.

To keep the group moving along, the handler will repeat this pattern. The length and angle of movements in the pattern will vary according to the characteristics of the herd. Only with practice will the handler discover the length and angle of each movement that works best for the herd being handled.

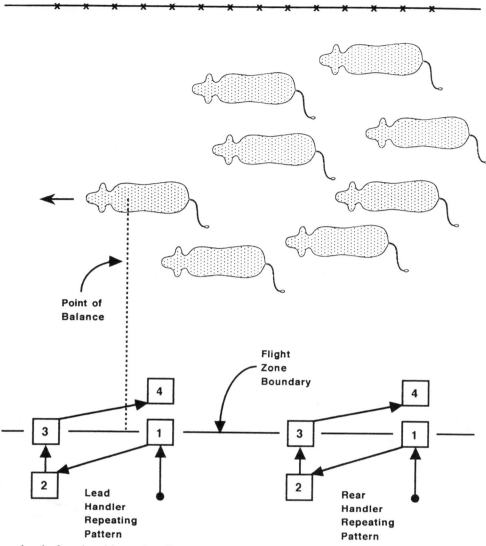

Figure 19.7. Moving a herd of cattle with two handlers.

Moving a Herd of Cattle with Two Handlers

If a second handler is available to directly control the group leader, controlling and moving the group is easier to manage. The rear handler moves in and out of the group's collective flight zone, following the same pattern used when one handler is moving the herd. The lead handler should be positioned just behind the lead animal's point of balance. Moving in and out of the leader's flight zone, the lead handler will adhere to the same pattern as the rear handler (Figure 19.7).

To prevent any animals from escaping between the lead and rear handlers, the two handlers should remain as close together as possible. Cattle in the rear are of little concern. They will have a strong desire to follow the rest of the herd even when the rear handler is in front of them.

Moving Stragglers Back into the Group

Occasionally, cattle break away from the herd and become stragglers. These animals can be managed fairly easily if the handler does not chase them from behind. This action will only upset and excite these animals. The follow-the-leader instinct is very strong and handlers should allow this characteristic of cattle to do most of the work. With just a little help, stragglers can be drawn back into the rest of the group (Figure 19.8).

From position 1, the handler approaches the stragglers at an angle until he/she crosses over both their flight zone and point of balance. As the handler walks toward these animals, pressure is gradually applied to their flight zone. After crossing the point of balance, the stragglers will speed up to regroup with the cattle in front of them. Just past the point of balance, at position 2, the handler turns around and heads

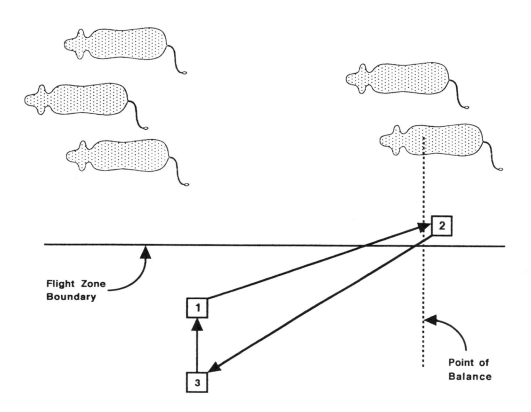

Figure 19.8. Moving stragglers back into the group.

back at an angle toward position 3. From position 3, the handler returns to position 1 and resumes the normal pattern of movement.

Moving Cattle into a Pen

Before the cattle arrive, the handler should open and secure the gate to the pen. Once cattle are moving, it's best not to interrupt their forward momentum. A handler should not hold the gate open while another handler drives cattle in from the rear. The cattle will be more hesitant to enter the pen because they do not want to get too close to the person at the pen's entrance.

On the other extreme, cattle should not be rushed into the pen. They will enter the pen in a controlled manner if they walk past the handler at a distance that does not upset them. For very gentle cattle, the correct position for the handler is position 1, just outside the cattle's flight zone (Figure 19.9). While cattle are entering the pen, the handler alternately increases and decreases pressure on the flight zone by moving back and forth between positions 1 and 2.

Handlers should not make any side-to-side movements and should be careful not to enter the cattle's flight zone too deeply. The objective is to place just enough pressure on the herd to keep them against the fence and under control as they enter the pen. Once the leader enters the pen, the rest of the group will follow. If the leader balks due to poor cattle handling technique, the herd will follow suit.

Emptying a Pen and Sorting at the Gate

Cattle should leave a pen in a controlled manner and not be allowed to run out wildly. The proper method of emptying a pen is for the handler to move back and forth between positions 1 and 2 (Figure 19.10). By moving between these two positions, the handler alternately increases and

302 / SAFETY AND HEALTH IN AGRICULTURE, FORESTRY, AND FISHERIES

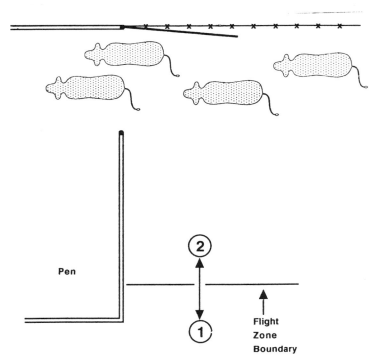

Figure 19.9. Emptying a pen and sorting at the gate.

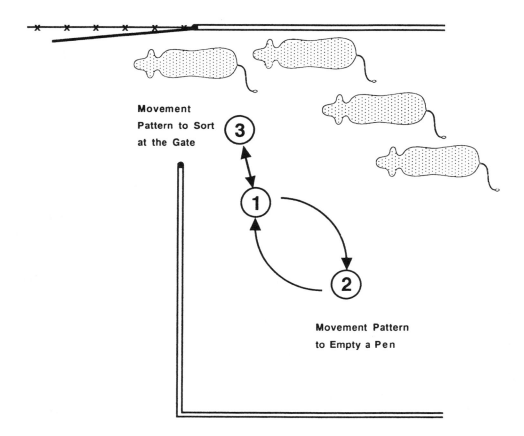

Figure 19.10. Moving cattle into a pen.

decreases pressure on the cattle as they leave the pen. This will ensure that they move steadily and in an orderly manner. If they race out, they can hurt themselves or the handlers who are working them.

Sometimes the handler may want to sort cattle as they are leaving the pen. In order to allow some cattle out of the pen while leaving others in, the handler moves in a deliberate and calm manner between position 1 and 3. The handler should not move from side-to-side or the cattle may slip past. When the handler is stationed at position 1, pressure is decreased on the animal's flight zone allowing it to leave the pen. Moving to position 3 increases pressure on the flight zone and the animal is held back. By paying attention to each animal's flight zone, the handler can increase pressure on those animals to be held back and decrease pressure on others to be let out. To facilitate sorting, the handler should plan ahead by knowing which animals will be sorted and using the herd following instinct to their advantage. Also, no one should be pushing cattle aggressively while a handler is sorting.

Designing and Installing Safe and Efficient Beef Cattle Handling Facilities

To safely and efficiently work beef cattle, properly planned, designed, and installed cattle handling facilities are needed in addition to correctly applied cattle handling practices (Grandin 1983, 1990b, 1990c). Furthermore, achieving success requires close integration of work practice and engineering controls.

Whether a completely new facility is being built or minor changes are being made to an existing facility, the designer and builder must take into account the management practices the producer wants to implement, how cattle sense and react to their surroundings, the environment, and the interaction between handler and animal (Baxter 1995; Fraser 1995). Ultimately, a superior beef cattle handling facility will be an integrated system founded upon proper selection, design, and construction of facility components, and with thoughtful consideration about site location and facility layout (Collins 1991; Curtis 1995; Jacobson 1995; Jungbluth 1995)

COMPONENTS OF BEEF CATTLE HANDLING FACILITIES

The size and complexity of a beef cattle handling facility will depend on the number of cattle that need to be managed, the financial resources available to the producer and the management practices that will be performed (Jungbluth 1995). Unfortunately, a majority of producers who raise beef cattle do not spend enough time or money planning, designing or installing a capable handling facility. Consequently, if producers elect to perform recommended herd health practices (vaccinating, identifying, castrating, dehorning, implanting, deworming, checking for pregnancy, etc.) with an inadequate facility, the risk of injury to themselves and others is increased and productivity is diminished (Lane 1986a; Cowan 1994).

A complete beef cattle handling facility will normally contain the following components: headgate, scale, holding or squeeze chute, working chute, crowding pen, holding pen(s), and loading chute (Grandin 1983; Lane 1986b, 1992). For smaller operations, the producer may elect to borrow some of these components, when needed, rather than permanently installing them.

Headgate

As a primary piece of equipment for restraining cattle, the headgate must be strong, safe, easy to operate, and work smoothly and quietly. Headgates come in four basic designs: self-catching, scissors-stanchion, positive-control, and full-opening stanchion. As the name implies, the self-catching headgate closes automatically as the animal's head enters through it. The scissors-stanchion design incorporates two biparting halves that pivot at the bottom while the positive-control type locks tightly around the animal's neck. The full opening stanchion has two biparting halves that open and close like a pair of sliding doors.

A producer has a choice of purchasing self-catching, scissor, and full-opening stanchions with either straight or curved stanchion bars. Although both types of bars are acceptable, the curved bar stanchion, as compared to the straight bar stanchion, provides more control of the head, but at the expense of increasing the likelihood of choking the animal. The positive-control headgate poses the greatest threat of choking cattle and putting pressure on their carotid arteries. To operate a headgate safely, the operator must be familiar with the piece of equipment and ensure it is mechanically sound.

Holding Chute

The holding chute is directly attached to the headgate. A well-designed holding chute allows the handler to safely perform various management practices on the restrained animal. Useful design parameters include squeeze action, removable side panels for safe and easy access to different parts of the animal, and a non-slip floor. Conventional squeeze chutes in combination with a curved bar stanchion headgate must be properly adjusted to prevent cattle from laying down and pressing the bottom part of their neck against the stanchion. If the holding chute does not have a squeeze mechanism, then a straight bar stanchion is required.

As with a headgate, a handler should engage a squeeze chute to obtain sufficient pressure to provide a feeling of being held. Excessive pressure will cause struggling while moderate pressure can have a calming effect. An animal will also be held more comfortably in a squeeze chute if the bars that contact the animal are of sufficient diameter or width to distribute pressure over a large area; this will minimize pressure points. When using hydraulic or pneumatically powered restraint devices, the handler must be cautious to ensure an appropriate amount of pressure is applied. Equipping these devices with pressure regulating devices will prevent a careless operator from applying too much pressure.

To prevent the animal from backing up before their head is caught, the chute should be equipped with a blocking gate or bar. The gate or bar will also prevent the next animal in line from entering the chute before the first animal is released. Installing a service-gate at the back of the chute provides ready access to the animal's rear and allows castration, pregnancy testing, and other herd health practices to be safely accomplished.

Working Chute

The working chute leads cattle from the crowding pen to the holding chute. At a minimum, the chute should be of sufficient length to hold four or five animals at one time. To prevent animals from moving backwards, "back up" bars should be placed at intervals within the chute. A chute with sloped sides has the advantage of restricting an animal's feet to a narrow path and prevents them from turning around. It also allows the chute to accommodate animals of different sizes.

Crowding Pen

A crowding pen is used to easily move cattle from the holding pen to the working chute. The pen should be about 150 square feet in area; enough space to hold five or six head of cattle. Inherent to its proper design and function is the

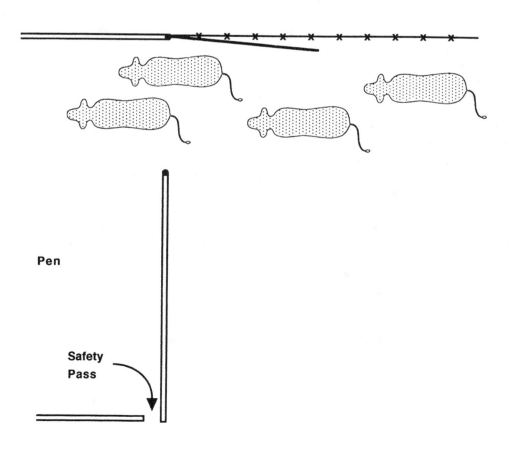

Figure 19.11. Holding pens.

pen's shape. The pen should form a gradual V as it approaches the working chute. A solid crowding gate of 10 to 12 feet is installed for handlers to use in pushing animals from the crowding pen into the working chute.

Holding Pens

The size of the beef cattle operation will influence the number and size of holding pens that are installed. Each pen should, at a minimum, provide 20 square feet per animal. The holding pen(s) should be conveniently located to allow for the smooth transfer of cattle from pasture to the crowding pen. When cattle will be moved by handlers on foot an escape route should be provided by installing safety posts and/or safety passes in the corners of the pen (Figure 19.11) and at 40 to 50 foot intervals along the sides of large pens. Step-overs are another alternative for providing a quick means of escape.

Scales

Scales should be located just off of the working chute. When positioned here, a gate can be opened to direct cattle over the scales only when the animals need to be weighed. If scales are placed within the working chute, cattle must cross over them every time they are worked. This reduces the scale's service life and increases repair costs.

Loading Chute

Chutes for loading and unloading cattle also need to be designed and constructed properly. The loading chute should be positioned in an area that enables the vehicle driver to view and approach the chute from the driver side of the vehicle. When a loading chute is in place, there should not be any gaps between the trailer and chute. Gaps can cause foot and leg injuries if an animal's leg slips into it. Injuries can also occur when the incline of the loading chute is too steep. Portable or adjustable chutes should be no steeper than 25 degrees, and the slope of a permanently installed chute should not exceed 20 degrees (Figure 19.12).

A chute that can be adjusted to different heights is more flexible and functional than a chute that is fixed at one height. Chutes should also have solid sides and a floor that is stair-stepped or cleated for sure footing. Lastly, the location of the chute is important. Because cattle are more likely to balk if they associate loading with the discomfort of being restrained, the loading chute should not be located near the squeeze chute or headgate.

Select Materials and Equipment for Durability

The components selected to comprise the beef cattle handling facility should be built of materials and equipment that are strong and of high quality. Inferior products save a few dollars in the beginning, but in the long run cost more as a result of repairs or injuries to handlers and cattle when the materials do not perform as intended. When building a new facility, it is important to use a flexible design that can be enlarged to accommodate more cattle or additional animal husbandry procedures. As a cattle producing operation grows, the facility should be able to grow along with it.

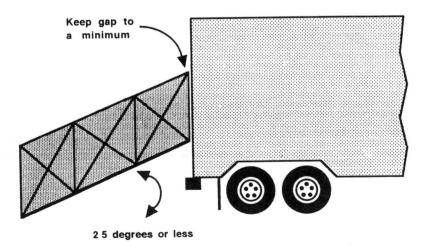

Figure 19.12. Loading chute.

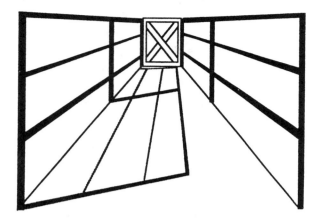

Figure 19.13. Shadows cause balking.

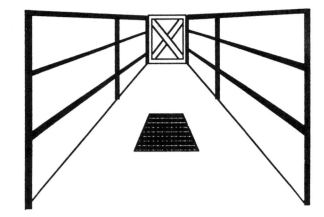

Figure 19.14. Drainage grates cause balking.

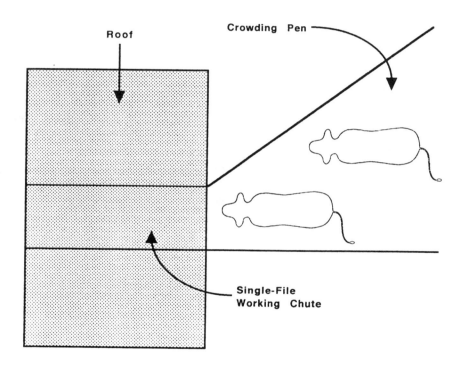

Figure 19.15. Crowding cattle as they enter a dark area causes balking.

Measures to Prevent Balking

During facility design and construction, planners and builders must collaborate to ensure the facility incorporates components whose characteristics minimize balking among cattle. Whenever cattle fail to proceed smoothly through a handling facility, operational efficiency is compromised and there is an increased probability that handlers will interact with these animals. Subsequent to this interaction is the possibility of injury to handlers or cattle. If balking is prevented, efficiency and safety are enhanced (Grandin 1983, 1984; Gill, 1992b).

Many environmental stimuli will cause cattle to balk. Cattle will hesitate to move forward when they see a shadow or anything else that creates strong contrasts between light and dark. A fence is one of many types of objects that can cast a shadow across an alley, loading chute or any other area where cattle are being handled (Figure 19.13). Solid fencing helps reduce shadows that cause balking. Similarly, drainage grates located in alleys and working chutes or reflections off of standing water produce contrasts that bother cattle. An animal walking down an alley with a grate (Figure 19.14) would hesitate or refuse to go forward when it came to the grate. Instead of using drainage grates, a facility should be built with sloped floors to direct water into a ditch outside the fence.

Cattle also balk when they experience the contrast of moving from a bright, sunny area into a dark area. This situation occurs when cattle proceed toward a handling facility that has a roof over a single-file working chute (Figure 19.15). In this situation, cattle are forced to move simultaneously from a crowding pen into a working chute and from bright sunlight into dark. Balking can be reduced if the first 10 to 15 feet of the working chute are in sunlight (Figure 19.16). As a result, cattle are lined up in the chute before they enter the roofed working area. Another alternative is to leave the facility completely unroofed; however, this makes it more difficult to work cattle in poor weather. Lamps can also be installed to illuminate the entrance. They must not be aimed at the eyes of approaching animals or they will not proceed. Indoor facilities should have diffuse lighting to reduce shadows and contrasts. For outdoor facilities, chutes and alleys should be laid out in a north-south rather than east-west orientation. Although cattle do not balk when walking directly into sunlight, it does make them uncomfortable.

With their wide-angle vision, cattle are easily distracted by people, tractors, and other things found on the farm. Distracted cattle are more difficult to handle. Distractions can be reduced by making the sides of the working chute, loading chute, and crowding pen solid (Figure 19.17). Solid chutes reduce shadows and distractions and force cattle to see only one means of escape—that which is in front of them. Cattle will also proceed more smoothly when they are forced to concentrate on those animals they are following.

In Figure 19.17, the gate at the end of the alley is solid while in Figure 19.18 a cow can see through the gate. One-way gates in alleys and chutes should not be solid. Being able to gaze through the gate contributes to the animal's perception that the view in front of them is a means of escape. Cattle are more likely to balk if the chute or alley appears to be a dead end.

A curved chute works better than a straight chute because it prevents the animal from seeing the trailer, squeeze chute, or headgate until the last moment (Figure 19.19). Balking is reduced because the animal does not realize it has approached a dead end until it is already there. However, the curved chute itself can appear as a dead end if it is not properly designed. As the cattle view the chute from the crowding pen, they should be able to see at least two body lengths up the chute. If they cannot see that far ahead, they will balk.

A second advantage of a curved chute is that it uses the natural tendency of cattle to circle around their handler. Cattle choose to do this because it keeps the handler at a comfortable distance away and within sight at all times. When working cattle in a curved chute, the handler should

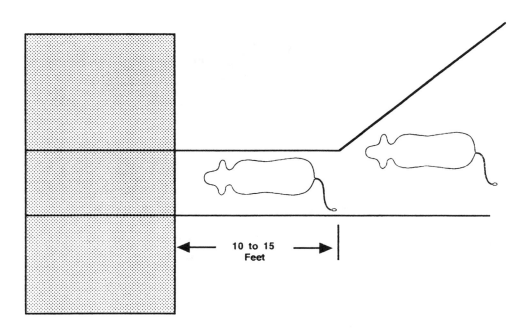

Figure 19.16. Crowding cattle before they enter a dark area reduces balking.

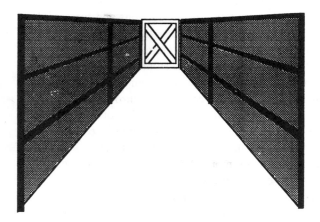

Figure 19.17. Solid gates cause balking.

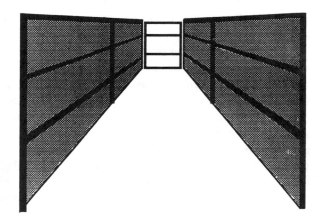

Figure 19.18. Gates that are not solid reduce balking.

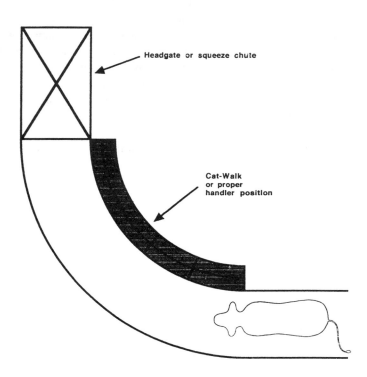

Figure 19.19. Proceeding a headgate or squeeze chute with a curved chute reduces balking.

be positioned along the inner radius of the chute. If an elevated catwalk is installed (Figure 19.19), it should be located along the chute's inner radius. Handlers should avoid working cattle from the outside radius. Finally, whether curved or straight, the chute should be oriented so that it directs cattle back to the area where they entered the facility. This takes advantage of the tendency of cattle to return to their starting point.

Other Points to Remember When Designing, Upgrading, or Using a Facility

- Facilities should be designed with a minimum of square corners and sharp turns. When it is laid out with smooth curves and fewer corners, cattle are less likely to get bunched up.

- Facilities should have nonslip flooring, especially in areas where floors are made of concrete. Concrete floors need to be deeply grooved with a diamond pattern to prevent slipping. Animals will panic when they walk on slick floors.

- The facilities need to be inspected periodically. This is especially important just prior to their use in handling cattle. The inspection should ensure the following:
  —Alleys, chutes, and pens are free of obstacles that may cause cattle to balk.
  —Headgates, chutes, and gates are functioning properly. Specifically, gates should operate quietly and close and latch properly. Rubber pads on metal gates will prevent banging and clanging.
  —Gates are equipped with tiebacks to prevent them from swinging closed inadvertently.
  —Posts and fences are not splintered and protruding bolts or nails are absent.

Layout of Beef Cattle Handling Facilities

Determining where the facility will be located is an important part of facility design. The design's potential for maximum production efficiency will not be achieved if the handling facility is situated poorly. Obviously, there is greater flexibility when installing a new facility as compared to improving an existing structure. To make loading and unloading cattle more convenient, the facility should be located in an area that is served by an all-weather road and is in close proximity to cattle pasture. Moving cattle into the facility is easier if it is located at the corner of a pasture or along a fence where two pastures join. The facility should also be near shade and water.

Cattle are more easily managed when the facility is built on level ground with good drainage. Note that with extended use, cattle will work the soil out of the facility, creating a dam effect around the perimeter. This causes rain water to "pool" in the facility. To minimize this problem, soil should be built up where the facility will be installed to raise it slightly above its surroundings. If the facility must be located on a slope, it should be laid out in a manner that requires cattle to move uphill. This is important for two reasons. First, cattle move downhill with great difficulty. And second, if the area where cattle will be managed is at the bottom of the hill, handlers and cattle will be working in an area that is wet and muddy.

The variety of cattle handling facility arrangements used by beef cattle producers in the United States is immense. They can range from virtually nothing to large, elaborate facilities capable of safely and functionally handling several hundred head of cattle. To gain an overall appreciation of the different types of designs, three general layouts are presented: the tobacco barn situation, an inexpensive adaptation of an existing structure, and two complete facilities that are capable of safely handling 1 to 15 and 25 to 50 head of cattle, respectively (Midwest Plan Service 1987; Gill 1992a; Neel 1992; Cowan et al. 1994).

Tobacco Barn Situation

A significant number of producers build facilities using existing structures that were designed for some other purpose. At one extreme, for example, is a facility layout which is based on the use of a tobacco barn (Figure 19.20). Although this design is better than a rope and tree, it is not very good for handling cattle. Often this type of facility is used when a producer decides to raise a few head of cattle as a hobby or as a small source of supplemental income. In this situation, the only time cattle are handled is when they are loaded up for market. To do this, the animals are enticed into the barn and then a borrowed portable headgate or trailer is placed against the doorway.

Typically, the producer and maybe one or two workers enter the barn and get behind the cattle. From here, they whoop and holler to force cattle toward the doorway with an often unreasonable expectation of getting one of the animals to voluntarily enter the headgate, squeeze chute, or trailer. What usually happens, however, is that the cattle turn back and circle around over and over, not leaving an easy means of escape for the handlers. Often handlers will end up getting kicked or run over.

Eventually, the job will get done, but not until both the cattle and handlers are completely stressed. What should have taken 15 to 30 minutes to accomplish ends up taking 2 to 3 hours. With a reasonable investment of time and money, this facility could be substantially improved to handle a small number of cattle easily and safely.

Adapting an Existing Structure

Using the before mentioned tobacco barn, Figure 19.21 shows how this existing structure can be modified to adequately handle a few head of cattle. The main components of this design are a crowding pen, a 12-foot gate, a short working chute, and a portable headgate, squeeze chute, or trailer. The most important feature of this facility layout is the 12 foot gate that has the freedom to swing over 180 degrees. A single handler can place cattle into the middle of the crowding pen and then slowly and methodically move them into the working chute. Once the gate is swung around, it is latched in a closed position. After the gate is closed, the handler can stand outside the chute and work cattle toward the headgate, squeeze chute, or trailer.

310 / SAFETY AND HEALTH IN AGRICULTURE, FORESTRY, AND FISHERIES

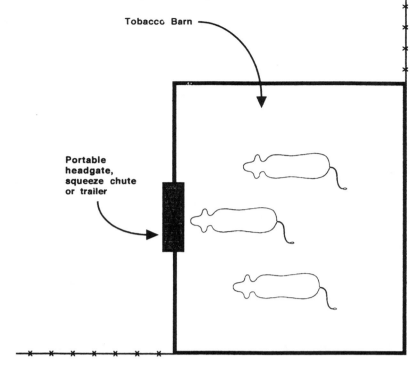

Figure 19.20. Tobacco barn facility layout.

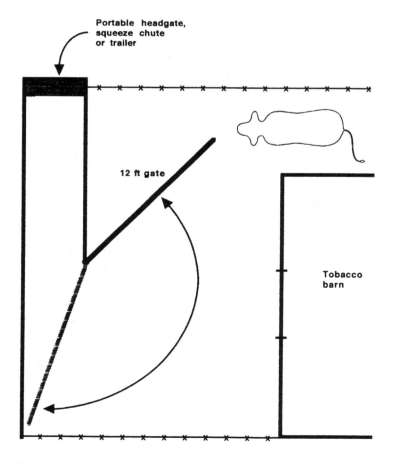

Figure 19.21. Improved tobacco barn facility layout.

Intermediate Design

While the previous layout can handle only three cows at a time, the intermediate facility design can handle as many as fifteen. It is appropriate for producers raising between 20 and 40 head of cattle. The basic facility (Figure 19.22) is a robust design that incorporates several improvements for simplifying cattle handling.

First, the handler can move a group of up to 25 cattle into the holding pen before moving them into the crowding pen. Second, the facility has fewer corners in which the cattle can get bunched. As a result, they can be handled more easily and more quickly. Third, the gate of the crowding pen can be latched in different positions as cattle are moved into the working chute.

Being able to latch the gate in different positions reduces the number of handlers because no one needs to hold the gate. In most cases, one individual can operate this facility. Possessing a fundamentally sound design, the facility can be modified according to the producer's needs and financial resources. For example, additional pens can be easily added to handle larger groups of cattle.

The first step in using the intermediate facility is to move a group of no more than 20 cattle into the holding pen. As the handler moves the group in, some of the cattle will proceed into the crowding pen. Next, the gate is swung around with no more than 15 head of cattle being moved toward the working chute. As cattle advance down the chute and are processed, the handler moves the gate forward in steps to apply pressure on the cattle and keep them moving forward. The use of "back-up" bars prevents animals from moving backwards.

Once the first group of cattle has been worked, the handler can process a second group and the cycle continues as before. Depending on the producer's needs, a portable headgate, squeeze chute, scales or trailer can be temporarily mounted at the end of the working chute.

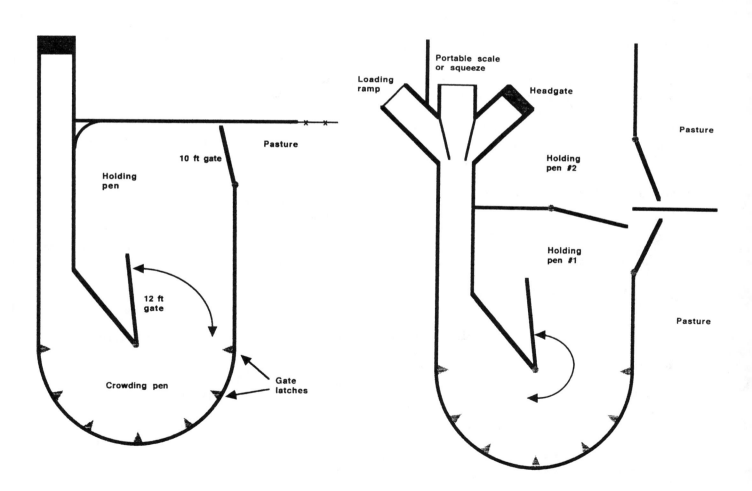

Figure 19.22. Intermediate facility layout.

Figure 19.23. Modification of failities to accomodate more animals.

Complete Facility for 25 to 50 Head of Cattle

As shown in Figure 19.23, a producer can modify an intermediate facility to accommodate more animals. The most notable addition is a second holding pen which doubles the facility's maximum capacity from 25 to 50 head of cattle. Other changes include a permanently mounted loading ramp and headgate. By leaving the middle exit open, the producer can borrow a squeeze chute or scale when necessary.

This facility can be further modified to handle even more animals. A well-designed facility provides a strong foundation for making additional improvements without having to rebuild the facility from scratch.

SUMMARY

Presently, our understanding of beef cattle inflicted injuries, including their distribution among members of the agricultural community and the factors influencing their occurrence, is limited. Nonetheless, beef cattle producers can prevent these injuries by applying correct cattle handling practices and installing properly designed and constructed handling facilities. Successful development, execution and integration of work practice and engineering controls depends on an understanding of beef cattle psychology, handler-cattle interactions, the physical and social environment, and the resources and production objectives of the beef cattle producer. Ultimately, a collaborative effort between animal behaviorists, facility designers, builders, producers, and handlers increases the likelihood that a beef cattle operation will be rewarded with increased productivity and enhanced safety.

REFERENCES

Baxter, M. 1995. There's more to design than behavior. In: *Animal Behavior and the Design of Livestock and Poultry Systems.* Ithaca, New York: Northeast Regional Agricultural Engineering Service.

Busch, H.M., T.H. Cogbill, J. Landercasper, and B.O. Landercasper. 1986. Blunt bovine and equine trauma. *J Trauma* 6: 559-61.

Cogbill, T.H., E.S. Steenlage, J. Landercasper, and P.J. Strutt. 1991. Death and disability from agricultural injuries in Wisconsin: a 12-year experience with 739 patients. *J Trauma* 12: 1632-37.

Collins, W.H. 1991. Facilities concepts for dairy beef production. *Holstein Beef Production.* Ithaca, New York: Northeast Regional Agricultural Engineering Service, pp. 223-34.

Cowan, J.B., R. Melton, and R. McLymore. 1994. *Improving Beef Cattle Handling for Increased Profitability and Safety.* Raleigh, NC: North Carolina Cooperative Extension Service, North Carolina State University.

Curtis, S.E. 1995. Animal systems are not jigsaw puzzles. In: *Animal Behavior and the Design of Livestock and Poultry Systems.* Ithaca, New York: Northeast Regional Agricultural Engineering Service, pp. 256-58.

Curtis, S.E. 1995. Ecological design: philosophy and practice for animal systems. In: *Animal Behavior and the Design of Livestock and Poultry Systems.* Ithaca, NY: Northeast Regional Agricultural Engineering Service, pp. 1-8.

Denis, W.B. 1988. Causes of health and safety hazards in Canadian agriculture. *Internt. J. Health Servs.* 3: 419-36.

Fraser, D. 1995. Behavioral concepts important to environmental design. In: *Animal Behavior and the Design of Livestock and Poultry Systems.* Ithaca, NY: Northeast Regional Agricultural Engineering Service, pp. 30-44.

Gill, W. 1992a. *Adapting Existing Structures for Beef Cattle Handling Facilities.* Knoxville, TN: Agricultural Extension Service, University of Tennessee.

Gill, W. 1992b. *Stress Management for Cattle Producers Starts with Handling Facilities.* Knoxville, TN: Agricultural Extension Service, University of Tennessee.

Grandin, T. 1983. Welfare requirements of handling facilities. In: *Farm Housing and Welfare.* S.H. Baxter, M. R. Baxter, and J.A.D. MacCormack. Martinus Nijhoff Publishers.

Grandin, T. 1984. Reduce stress of handling to improve productivity of livestock. *Vet. Med.* 6: 827-31.

Grandin, T. 1989a. Behavioral principles of livestock handling. *Prof. Anim. Sci.* 2: 1-28.

Grandin, T. 1989b. *Cattle Behavior During Handling and Corral Design for Ranches.* Grandin Livestock Handling Systems, Inc.

Grandin, T. 1990a. Bovines: rules of restraint. *Lge Animal Vet.* 4: 24-8.

Grandin, T. 1990b. Design of loading facilities and holding pens. *Appl. Animal Beh. Sci.* 28: 187-201.

Grandin, T. 1990c. *Recommended Animal Handling Guidelines for Meat Packers.* Washington, D. C.: American Meat Institute.

Grandin, T. 1995a. Little things do matter: debugging animal environments. In: *Animal Behavior and the Design of Livestock and Poultry Systems.* Ithaca, NY: Northeast Regional Agricultural Engineering Service.

Grandin, T. 1995b. Restraint of livestock. In: *Animal Behavior and the Design of Livestock and Poultry Systems*. Ithaca, NY: Northeast Regional Agricultural Engineering Service.

Haddon, W. 1980. Advances in the epidemiology of injuries as a basis for public policy. *Public Health Reports* 95: 411-21.

Jacobson, L.D. 1995. Combining equipment and facilities into an efficient system. In: *Animal Behavior and the Design of Livestock and Poultry Systems*. Ithaca, NY: Northeast Regional Agricultural Engineering Service, pp. 259-62.

Jungbluth, T. 1995. The design of systems. In: *Animal Behavior and the Design of Livestock and Poultry Systems*. Ithaca, NY: Northeast Regional Agricultural Engineering Service, pp. 9-19.

Kendall, T., K.J. Donham, D. Yoder, and L. Ogilvie. 1990. The farm family perception of occupational health: a multistate survey of knowledge, attitudes, behaviors, and ideas. *Am. J. Ind. Med.* 18: 427-31.

Lane, C.D. 1986a. *Evaluate Your Beef Cattle Handling System*. Knoxville, TN: Agricultural Extension Service, University of Tennessee.

Lane, C.D. 1986b. *Facilities for Backgrounding Cattle*. Knoxville, TN: Agricultural Extension Service, University of Tennessee.

Lane, C.D. 1992. *Beef Cattle Handling Facilities*. Knoxville, TN: Agricultural Extension Service, University of Tennessee.

Midwest Plan Service. 1987. *Beef Housing and Equipment Handbook*. Ames, IA: Iowa State University.

National Safety Council (NSC). 1982. *1982 Farm Accident Survey Report*. Chicago, IL: NSC.

National Safety Council (NSC). 1987. *Accident Facts, 1987 Edition*. Chicago, IL: NSC.

Neel, J.B. 1992. *The Need for Beef Cattle Handling Facilities in Tennessee*. Knoxville, TN: Agricultural Extension Service, University of Tennessee.

Stallones, L. 1990. Surveillance of fatal and non-fatal farm injuries in Kentucky. *Am. J. Ind. Med.* 18: 223-34.

Steele-Bodger, A. 1969. Hazards of animal handling. *Ann. Occup. Hyg.* 12: 79-85.

Stueland, M.D., P. Layde, B.C. Lee. 1991. Agricultural injuries in children in central Wisconsin. *J. Trauma* 11: 1503-09.

Wright, K.A. 1993. Management of agricultural injuries and illness. *Nursing Clin. North Am.* 28: 254-66.

Zhou, C., J.M. Roseman. 1994. Agricultural injuries among a population-based sample of farm operators in Alabama. *Am. J. Ind. Med.* 25: 385-402.

SAFETY AROUND HORSES

W.E. Morgan Morrow, B.V.Sc., M.S., Ph.D.
Robert A. Mowrey, Ph.D.
North Carolina State University

> Horseback riding is a popular but sometimes dangerous recreational activity. Most if not all of the injuries are avoidable if proper prevention measures are taken. Horse safety begins with accident awareness around the farm itself. Guidelines for safety include steps to ensure the safety of visitors and children as well as guidelines for workers. Safety practices are identified for all aspects of work and interaction with the horse. These guidelines include how the horse is selected and handled, placement and work inside the stall, fire safety around the farm or barn, and proper saddling and riding habits. Personal protective equipment and proper clothing is emphasized to ensure rider safety.

INTRODUCTION

Horseback riding is a popular recreational and professional sport around the world. Part of the attraction of the sport is the thrill a rider gets from controlling a large and powerful animal. Unfortunately, with the thrills come the spills. In the United States, about 145,000 emergency-room visits resulting from horseback injuries were recorded in the National Electronic Injury Surveillance System from 1991 to 1992. Because not all injuries are reported, these represent only the minimum number of injuries and probably only the most serious. In a recent study of patients with acute injuries treated at a major Swedish hospital, 31% of patients were admitted with horse-related injuries. About half of the patients were injured when they fell off their horse after they lost control, and the other half injured during other activities such as grooming or riding in a cart. Abrasions, contusions, and sprains constituted most (71%) of the injuries, however, 5% suffered cerebral concussion (Bjornstig et al. 1991).

Similar studies in Europe (Edixhoven et al. 1981; Gierup et al. 1976) confirm that falls are a major cause of injury. Although no estimates of fatalities exist for the United States, a review of medical examiners' reports from 1976-1987 identified 205 deaths from horse related injuries (CDC 1990). A study of fatal animal attacks in North Carolina (Langley 1994) identified 20 horse-related fatalities over an 18-year period. All these studies serve to indicate the hazardous nature of working with horses and the need to take preventative measures to avoid injury. Fortunately, most, if not all, horse-related injuries are preventable. We recommend that anyone associated with horses take a course in basic first aid and cardiopulmonary resuscitation and follow the guidelines hereunder.

ACCIDENT AWARENESS ON THE FARM

It takes time to prepare for, and hopefully prevent, accidents but it is time well spent; lives can be saved.

- Dissuade unwelcome visitors and trespassers. Post "no trespassing" signs around the perimeter of fenced-in horse areas.
- Keep the perimeter fence in good repair.
- Be especially careful to ensure that dangerous animals (e.g., stallions and bulls) are not accessible to the public.
- Don't leave visitors unsupervised.
- Children are especially "at risk" on the farm. Watch them closely.
- Any water container should be inaccessible to children.
- Electric boxes should be locked.
- Keep machinery sheds locked and keys in a locked box.
- Post hazardous areas.
- Keep chemical storage areas locked and posted.
- Take special care if you allow hunting or fishing on the farm.

For additional information, refer to the chapters on electrical, machinery, and farmstead safety.

STABLE SAFETY

- Keep a list of emergency phone numbers, in large type, on the wall, by the telephone.
- Keep a first aid manual and kit in a prominent area.
- Ensure the first aid kit is fully stocked; replace items as they are used.
- Ensure all areas have adequate lighting.
- Ground all water sources.
- Ensure all areas are free of sharp protruding objects, e.g., nails that can injure people and horses.
- Keep all building structures, especially the flooring, in good repair.
- Ensure flooring provides adequate traction.
- Stack hay and other products safely.
- Install railings in hay lofts.
- Install hand rails on all stairs.
- Store all equipment safely.
- Ensure all fans have guards.
- Ensure bridges can carry the required weight.
- Trim all low branches on trees.
- Ensure outside overhead wires are at least 14 feet high; underground wiring is preferable.
- Post speed signs, traffic and parking directions.
- Enclose all electrical wires in conduit.
- Clean up all debris.
- For additional information refer to the chapters on electrical, machinery, and farmstead safety.

FIRE SAFETY

Most barn fires occur in either summer or winter. In winter, they usually result from electrical appliances. In summer, hay will spontaneously combust, and lightning strikes are also more common. Rodents can chew through wires and cause fire-starting sparks at any time of the year. If horses are to survive, the fire must be extinguished within about 30 seconds.

- Buildings must have properly grounded lightning rods.
- Keep the hay in a building separate from other livestock.
- Enclose hay storage area lights in a safety shield.
- Maintain a strict rodent control program.
- Construct barns from nonflammable materials.
- Use fire-retardant paints.
- Ensure every third or fourth stall has partition walls that extend to the ceiling and delay the fire.
- Maintain a strict "No Smoking" policy; post "No Smoking" signs.
- Keep sand bucket "ash trays" at all entrances.
- Keep and maintain fire extinguishers at all exits, near the electrical breaker box, and in the middle of long aisles.
- Fire extinguishers should be at least 5 lb., "all class" (ABC), consisting of a dry-chemical, and maintained in fully operational condition.
- Keep a dedicated hose ready and install hydrants at either end of the barn.
- Ensure fire exits are clearly marked.
- Install smoke detectors and sprinkler systems.
- Keep manure piles at least 20 feet from all barns and 100 feet from open water courses.
- Keep a halter and lead shank at all stalls in case of emergency.
- Ensure all new employees practice a fire drill and it is repeated at least annually.
- Clean up all debris.
- Ensure your barn complies with the local authority's building code.
- Ensure your electrical system at least meets local codes and the National Electrical Code, NFPA 70.
- Racetrack stables should comply with NFPA 150.
- Barns and stables that include living quarters have additional electrical and fire codes.

Managing Flammables and Accelerants

- Flammables include hay, straw, dust, and bedding.
- Accelerants increase the rate that a fire spreads and includes gasoline, oil, and kerosene.
- Aerosol cans can explode if left in the sun or exposed to heat.
- Keep all flammables, accelerants, and aerosols in a barn separate from the horses.
- Keep a list of accelerants and all farm chemicals in a separate building.
- The list should include their name, date of purchase and expiration, quantity, and location. Firefighters can use this list to help them fight the fire and protect both human and animal life.
- Ask your local fire department to check out the barn.

For additional information refer to the chapters on electrical and farmstead safety.

CHOOSING YOUR HORSE

One of the most important things any rider can do to ensure a safe ride is to choose the horse matched to his or her level of experience.

- Don't overestimate your ability. Be sure you have the skills now to handle the horse.
- Novices should never ride stallions. Even experienced riders must be very careful working with stallions or young newly broken horses.

APPROACHING A HORSE

It is important to use common sense when approaching any horse regardless of how well you think you know it. All riders, regardless of their experience, must be careful when around horses.

- Make the horse aware of your presence, talk to it as you approach.
- Always approach a horse from the side, never from behind or directly in front.
- Don't make any sudden or loud noises.
- Recognize your horse's moods and be extra careful if he appears "grouchy".
- Swishing tails and laid back ears indicate a ill-natured horse.

HANDLING THE HORSE

Know your horse. Be calm and confident around a horse; nervousness is infectious.

- Never tease a horse.
- If you are in a field catching a horse, always be aware of your surroundings and know how the other horses, or other animals, are reacting.
- If you have to walk around a horse stay at least 8 ft. away, well out of range. If that is not possible, or appropriate, stay as close as possible to the horse. If you stay close, the force of impact from a kick is diminished.
- Never crawl under the belly of a horse.
- When approaching a horse, first contact it by placing your hand on its neck or shoulder.
- Maintain physical contact as you move to the area of interest, e.g., to inspect its feet.
- To inspect the horse's foot, place one hand on the horse and rub the top of the leg with the other hand. Gradually move your hand down the leg. If the horse struggles then reverse your moves and start all over again. Move the hand down to the back of the knee, cannon, and fetlock. Then gently squeeze the tendon, and grab the pastern and start to lift it as you gently push the body of the horse away so it moves its weight bearing to the other leg. Move your body as you place the foot between your knees then rest the horses leg on your legs. The final position should be comfortable for both you and the horse. Some horses are difficult to handle and should be left to a more experienced handler.
- To work with a horse's tail, stand to one side and face the rear. Grasp the tail and bring it to you. Do not stand behind the horse.
- Don't place your head either above or below a horse's head. Even a playful toss of the horse's head can leave you with major injuries.
- Never leave a halter on a loose horse unless it is a "break-away" type. The simple act of the horse scratching its face with its hind foot can result in the foot becoming entangled in the halter and resulting personal injury for anyone trying to save the horse.
- Always carry a sharp knife so you can cut a rope or lead should a horse become entangled.
- Don't hand feed. Horses can't distinguish between fingers and feed until it is too late! If you must, then be sure to hold the feed in your open palm and not in your finger tips.

LEADING THE HORSE

- Always lead from the left side and hold the lead close to the halter, about 1 to 2 ft.
- Hold the extra length in your left hand but never wrap the lead shank around your hand. Carry the lead wrapped in a "Figure-8" pattern in the middle so you can hold the horse or release it without trapping you hand.
- Never attach the lead rope to your body or clothing.

TYING THE HALTER

- Tie high, level with the horses mouth, and short, leaving about a 2 feet lead.
- Always tie a quick release knot.
- Tie the horse away from other horses so they can't fight.
- Never tie a horse by the reins or with a bit in its mouth; they can break and/or injure the horse's mouth.
- Never go over or under a tie rope.

SADDLING UP

- Ensure all the equipment is in good working condition.
- Groom the horse prior to saddling; that will remove troublesome burrs, etc., that could irritate the horse.

- Place an ample saddle pad over the withers then pull it back to the middle of the withers. This smoothes out the horse's hair and makes it more comfortable.
- Place the saddle smoothly onto the back of the horse.
- Pull the front legs forward prior to tightening the girth; this ensures the skin is not pinched.
- Tighten up the girth gently but firmly.
- With a Western Double-Rigged saddle, fasten the front cinch first. When unsaddling, release the rear cinch first.
- On an English Saddle, use the "peacock" stirrup. These have an elastic strap on the outside of the iron that disengages if the foot catches when the rider falls.
- Check the cinch with your fingers after saddling, before you mount, and after riding about a half mile.
- Keep control of the horse when bridling by fastening the halter around the horses neck.
- A horse's mouth is sensitive; be considerate when inserting the bit.

MOUNTING AND DISMOUNTING

- Always mount or dismount a horse in the open, never in confined spaces or near solid structures.
- Hold the reins firmly so you have control, but don't pull on them as you mount.
- Keep the near rein shorter so if the horse moves away it will circle away from you.
- Never strap yourself, or anyone else, to the saddle.

CLOTHING TO WEAR

- Always wear boots with high heels and hard toes. The high heels prevent your feet from slipping through the stirrup and trapping your foot should you fall. The hard toes protect your feet should a horse stand on you.
- Always wear long-sleeved shirts and long pants when riding; they protect you from sun damage, insect bites, and chaffing.
- Apply ultraviolet light protectives to exposed skin.

THE HELMET

The most common cause of death in riders is head injury which mainly result from falls (Bjornstig et al. 1991; Edixhoven et al. 1981; CDC 1990; Langley 1994; Ornehult et al. 1989). Research has shown that helmets are very effective at preventing head injuries (d'Abreu 1976).

Unfortunately, few nonprofessional riders wear them and few seem to recognize the hazard. Protective headgear should be impenetrable, padded, and stay on during a fall. It should be SEI certified and meet the appropriate standards.

- Always wear protective headgear; "broken brains" are hard to heal!
- Ensure it fits snugly and that the chin strap touches the riders jaw.
- Replace any helmet immediately should it be damaged.

RIDING, NIGHT RIDING, AND TRAIL RIDING

- Ride in confined areas until you are confident you know your horse and can control it.
- Keep your horse under control.
- Ensure your horse responds to leg and voice cues and reining.
- If your horse is startled try to remain calm; speak to it calmly, and try to keep control until it settles down.
- If your horse bolts, pull on one rein so he turns in a circle until he stops.
- Walk, don't gallop, up or down hills.
- Let your horse "find his own way" when traveling over difficult terrain.
- If you have to cross or travel a paved road, hold your horse to a walk.
- Only ride with a saddle and bridle; never bareback or with just a halter.
- Don't rush past slower riders. Pass at a walk or trot and allow at least 4 feet between horses.
- Either stay a full horse-length behind a horse or ride abreast.
- Be sure to walk the first and last half mile from and back home.
- Tie a red ribbon on the tail of "kickers" and stand back at least 8 feet.
- When night-riding wear white, reflectors, and carry a flashlight.

AFTER THE FALL

Despite our best efforts and adherence to safe riding practices, accidents still happen. Remember that falls can be fatal. Quick and appropriate action can save the life of a rider and prevent spinal injuries. If there are any head injuries at all be sure to call an ambulance, do not be dissuaded by the victim. Never give food or drink to any victim until they have been thoroughly checked.

Rider is conscious and seems OK:

- If the rider has taken a hard fall but tells you that they feel OK, do not assume that they are OK. Do not let them get up until a qualified medical professional has checked for spinal injuries.

Rider is unconscious:

- Determine if the fallen rider is breathing by placing your ear close to the rider's nose and mouth.
- If the fallen rider is not breathing or is turning blue or purple then he or she is not receiving oxygen and needs immediate assistance. People can survive only 5 to 6 minutes without oxygen.

A. If someone is available, send them for help immediately.
B. Remove the helmet, taking care not to twist the neck.
C. Gently position the fallen rider on his or her back. Try to not allow the spine to twist—this could worsen the injury.
D. Place your thumb in the victim's mouth, grasp the lower jaw, and gently pull forward to open the airway.
E. If he or she still isn't breathing, something else may be blocking the airway.
F. Look and feel for food or any obstruction. Remove any obstruction carefully to ensure it doesn't travel deeper into the airway.
G. If the victim is still not breathing then apply mouth-to-mouth or mouth-to-nose resuscitation.
H. Once you have restored breathing, cover the fallen rider to minimize heat loss, tell him or her not to move, and go for help.

Rider is hurt but still conscious:

- Don't move the fallen rider and don't allow him or her to move.
- Comfort the patient.
- Don't give him or her anything to drink or eat.
- Tell him or her not to move and then go for medical assistance.

HAULING YOUR HORSE, TRAILERING

- Ensure truck and trailer are in good working condition. Check the trailer floor boards, tires, and breaks.
- It is safer if two people load the horse.
- Chock the trailer wheels before starting and ensure that you are on level ground.
- Only load horses into trailers that are attached to the hauling vehicle.
- Remove the saddle, bridle, and any other equipment.
- Wrap the horse's legs with a pad and leg wrap.
- Apply the halter and use a 5 foot cotton, not nylon, lead shank.
- Never enter the trailer stall ahead of the horse unless there is an escape door.

Horses can be trained to enter a trailer by the use of butt ropes. One person should be on each side of the horse gently pulling the butt rope while a third person leads the horse into the trailer. Also, train your horse to respond to a forward-motion cue: make a clicking sound while gently pulling on the chain/rope under the chin. Release the pressure when the horse moves forward. The horse learns to respond to the clicking and will move forward. A trained horse will enter a trailer unaided particularly if, during the training period, it receives some grain to eat afterwards.

- As soon as the horse is in the trailer, secure the butt chain or bar.
- Tie the horse's head loosely but short enough to prevent the head from turning beyond 90 degrees.
- Double check all connections both electrical and mechanical.
- Check the connections, the trailer, and the horse after 5 miles and every time you stop.
- When unloading, first carefully drop the tailgate, then untie the horse, then undo the butt chain.
- Always load and unload the horse while the trailer is still hitched to the truck.

SUMMARY

Protecting yourself around horses requires a lot of common sense and attention to detail. Horse riding is a vigorous sport enjoyed by many but not without risk. The publicity surrounding Christopher Reeve's tragic fall has done a lot to sensitize the public to the inherent dangers of horse riding and increase their awareness of the need for safety. By following the precautions detailed in this chapter, riders and those who work around horses can dramatically decrease the risk of injury or death to themselves and their horses and increase their enjoyment of the popular sport of horseback riding.

REFERENCES AND RESOURCES

American National Red Cross. *American Red Cross Standard First Aid.* Washington, DC: American National Red Cross.

Basic Horse Safety Manual. Washington, DC: American Youth Horse Council, 1700K St. NW, #300, Washington, DC 20006. (202) 296-4031.

Bjornstig, U., A. Eriksson, and L. Ornehult. 1991. Injuries caused by animals. *Injury* 22:(4) 295-8.

Centers for Disease Control (CDC). 1990. Injuries associated with horseback riding: United States, 1987 and 1988. *MMWR* 39: 329-32.

Comerford, P.M., and J.T. Potter. *Safety Considerations for Equestrian Activities*. Lexington, KY: American Youth Horse Council, 4903 Iron Works Pike, 40511-8434.

CSN Rural Injury Prevention Resource Center, National Farm Medicine Center, 1000 North Oak Avenue, Marshfield, WI 54449-5790. (715) 389-4999 or fax (715) 389-3950.

d'Abreu, F. 1976. Brain damage in jockeys. *Lancet* 1: 1241.

Edixhoven, P., S.C. Sinha, D.J. Dandy. 1981. Horse injuries. *Injury* 12: 279-82.

Fire Safety in Agricultural Buildings. Reinsurance Association of Minnesota.

Gierup, J., M. Larsson, S. Lennquist. Incidence and nature of horse-riding injuries. *Acta Chir. Scand.* 142: 57-61.

Langley R.L. 1994. Fatal animal attacks in North Carolina over an 18-year period. *Am. J. Forensic Med. Pathol.* 15: 160-7.

Ornehult, L., A. Eriksson, and U. Bjornstig. 1989. Fatalities caused by nonvenomous animals: a ten-year summary from Sweden. *Accid. Anal. Prev.* 21: 377-98.

Rutgers Cooperative Extension, New Jersey Agricultural Experiment Station.
Horse Trailer Maintenance and Trailering Safety.
Handling Emergency Situations on the Farm
Safety Recommendations for the Stable, Barn Yard, and Horse/Livestock Structures.
Accident-Proofing Farms and Stables.
Fire Prevention and Safety Measures Around the Farm.
Farm Machinery and Equipment Safety, Parts 1 and 2.

21

SKIN DISEASES IN FARMERS

William A. Burke, M.D.
Associate Professor and Section Head, Dermatology
East Carolina University Department of Medicine

> The highest risk of skin disease of any occupation is found for farmers, who are exposed to harsh outdoor environmental conditions, high levels of solar radiation, and contact with irritant or allergenic plants. Working outside poses a risk of bites and stings from arthropods and other animals, with an additional risk of hypersensitivity reactions, zoonotic infections, and toxic envenomation. Farm chemicals including fertilizers, herbicides, pesticides, veterinary products, and animal feeds as well as their additives can lead to cutaneous irritant or allergic contact dermatitis. Current data on skin diseases of farmers is inadequate, but an expansion of our knowledge base on these disorders is being driven by the economic pressures of increasing medical insurance costs and the need for compliance with OSHA and worker's compensation laws.

INTRODUCTION

Agricultural workers carry the highest risk of skin disease of any occupation (Donham and Horvath 1988; Calvert 1989). Exposure to harsh outdoor environmental conditions including extremes of heat and cold as well as high levels of solar radiation can lead to a variety of skin disorders. Working outdoors can also lead to skin diseases related to contact with irritant or allergenic plants. Arthropod and other animal bites and stings are commonly seen in farm workers and some may lead to hypersensitivity reactions, zoonotic infections, or toxic envenomation. The skin of farmers frequently comes in contact with a variety of farm chemicals (fertilizers, herbicides, pesticides, etc.) as well as animal feeds/additives and veterinary products which can lead to cutaneous irritant or allergic contact dermatitis.

In spite of the high incidence of skin disease in farmers and other agricultural workers, there is a notable paucity of dermatologic medical data regarding this profession. Farms have traditionally been family-run operations in very rural, medically-isolated regions. The local physician is most frequently an overwhelmed generalist with little or no specialty support. In addition, except in the larger corporate farms, workman's compensation and Occupational Safety and Health Administration (OSHA) laws are not required or enforceable. Many farm laborers are uninsured part-time or migrant workers with economic or social pressures to remain outside of a physician's office. Farmers and their families, even when insured, tend to be independent, stoic individuals, who may not seek medical attention for perceived minor skin ailments and are often distrustful of institutions including the field of medicine. Finally, there is no central organized labor union for farmers or agricultural workers.

In the last few decades, with the tremendous increase of large corporate farms with their large numbers of employees, there has been an increased demand for information on related medical disorders including skin diseases. The economic pressures of increasing medical insurance costs as well as the need for compliance with OSHA and worker's compensation laws are driving this demand.

SKIN DISEASES DUE TO SUN EXPOSURE

Excessive Exposure

Farmers and agricultural workers are exposed to higher levels of ultraviolet solar radiation relative to the general population and this can cause both acute and chronic problems related to the skin. The most common acute reaction from excessive exposure to sunlight is sunburn which is primarily due to the ultraviolet B (290-320 mm) wavelengths (Warshauer and Steinbaugh 1983). Since the melanin pigment in melanocytes in the basal layer of the epidermis is protective, agricultural workers with lightly pigmented skin (skin types I or II—Table 21.1) are at the higher risk for developing not only sunburn and other acute sun-related reactions, but also problems related to chronic sun exposure such as skin cancer.

Table 21.1. Skin types, response to sunlight and sunscreen need.

Skin Type	Response to Sunlight	Skin Characterization	Recommended Sunscreen SPF* with Moderate Sun Exposure
I	Burns quickly; Doesn't tan	Very fair skinned white persons: blue eyes; red hair; freckles	15+
II	Burns easily; Pigments mildly	Fair skinned white persons: blue eyes; blonde hair	12-15
III	Burns moderately Pigments moderately	Average Caucasian: brown to dark hair; brown eyes	8-12
IV	Burns occasionally; Pigments easily	Hispanics; Asians; American Indians: Tan to light brown skin	6-8
V	Burns rarely; Pigments deeply	Brown skinned persons: Fair-skinned black persons; Asian Indians; some Middle Easterners	Rarely needed
VI	Almost never burns; Pigments intensely	Darkly pigmented black persons	Almost never needed

*SPF = sun protection factor

Acute reactions to sun exposure also include polymorphous light eruption, which is commonly called (in addition to severe sunburn) "sun poisoning". This reaction can arise from hours to several days following sun exposure and the clinical presentation of rash is polymorphous (erythematous papules and sometimes vesicles) which usually occur on the neck, trunk, and extremities (sometimes sparing the face) (Epstein 1980). The mechanism of this eruption is poorly understood, but it most commonly occurs in the spring of the year as outdoor activities increase. With increasing exposure to solar radiation, the intensity of the eruption generally decreases. Topical (or rarely systemic corticosteroids) may help, and hydroxychloroquine may be needed in chronic or more severe cases.

Many medications (including over-the-counter) as well as topical agents may cause acute phototoxic reactions (Table 21.2). This exaggerated sunburn is primarily due to the UVA (320 to 400 mm) wavelength of light, (Warshauer and Steinbaugh 1983) and thus the usual sunscreens may not be very protective since they are rated for use as UVB photoprotectants. The UVA phototoxic reaction may begin within several hours of sun exposure or, in some cases, may not peak until 48 hours afterwards. Like sunburn, little can be done therapeutically except for symptomatic care.

Photoallergic reactions are due to medications and chemicals (Table 21.3) where these agents are transformed by UVA light into a hapten which is allergenic (Type IV hypersensitivity/cell-mediated immunity) to the sensitized individual (Epstein 1972). Like other cell-mediated allergic reactions, this occurs approximately 7 to 10 days after the initial exposure, or 24 to 72 hours after re-exposure. Unlike the phototoxic eruption, this eruption itches rather than burns and, because sensitization is necessary, it occurs in only a small percentage of persons exposed to the medication/chemical and therefore is quite rare. Whereas the phototoxic eruption is most commonly seen in fair-skinned individuals and presents as a severe sunburn, the photoallergic eruption can be seen in all skin types and presents as a severe eczematous eruption. In addition to discontinuation of the offending agent, topical or systemic corticosteroid medications are helpful in controlling this reaction. In some cases, the dermatitis continues to worsen with repeated exposure to ultraviolet light even after the offending agent has been discontinued, and this problem is termed persistent light reaction.

Table 21.2. Some commonly used medications that may cause phototoxic reactions.

Antiarrhythmia medications
- amiodarone

Antibiotic/antifungal medications
- nalidixic acid
- quinolones (ciprofloxacin, lomefloxacin, norfloxacin)
- sulfonamides; sulfamethoxazole/trimethoprim
- sulfones (dapsone)
- tetracycline, doxycycline, oxytetracycline, demeclocycline
- griseofulvin

Antidepressant/antipsychotic medications
- imipramine
- protriptyline
- nortriptyline
- chlorpromazine
- promethazine
- triflupromazine

Antidiabetic medications
- sulfonylureas (chlorpropamide, tolbutamide)

Antihypertensive medications
- captopril
- thiazides

Cytotoxic/chemotherapeutic medications
- 5-fluorouracil
- dacarbazine
- vinblastin

Diuretic medications
- furosemide
- triamterine
- thiazides

Non-steroidal anti-inflammatory drugs (NSAID's)
- carprofen
- naproxen
- piroxicam
- sulindac
- tiaprofenic acid

Psoralens
- methoxsalen
- trioxsalen

Table 21.2 (continued)

Retinoids
- isotretinoin
- etretinate

Topical medications
- tretinoin
- coal tar

Table 21.3. Some commonly used medications that may cause photoallergic reactions.

Antiarrhythmia medications
- quinidine

Antibiotic/antifungal medications
- nalidixic acid
- sulfonamides
- griseofulvin

Antidepressant/antipsychotic medications
- chlorpromazine
- imipramine
- promethazine

Antidiabetic medications
- chlorpropamide
- tolbutamide

Antihypertensive medications
- thiazides

Non-steroidal anti-inflammatory drugs (NSAID's)
- indomethacin
- piroxicam

Topical agents
- PABA

There are numerous other medical diseases which may flare acutely with sun exposure including: lupus erythematosus, porphyria cutanea tarda and other porphyrias, reactivation of labial herpes simplex, solar urticaria, and pellagra.

Chronic Exposure

Chronic exposure to ultraviolet light (both UVA and UVB) is well known to be responsible for much of what we term "aging" of the skin. Histologically, there is both basophilic degeneration of the collagen fibers and loss of elastic fibers in the dermis (collectively known as solar elastosis). This elastosis is manifested clinically as wrinkles and furrows in the skin ("farmer's neck"). Increased telangiectasias, poikiloderma of Civatte, hypopigmented macules (idiopathic guttate hypomelanosis), and solar lentigos ("liver spots") are also due to chronic excessive ultraviolet radiation. Except for the hypopigmented macules, these changes are generally more pronounced in lightly pigmented individuals.

Chronic sun exposure also leads to an increased risk of skin and lip cancer in agricultural workers (Blair and Zahm 1991; Burmeister 1989). Actinic keratoses are commonly seen in farmers with lightly pigmented skin (skin types I to III) and are precancerous lesions which can sometimes develop into a squamous cell carcinoma.

These lesions histologically show epidermal dysplasia with atypical keratinocytes and disordered progression of maturation of epidermal cells from the lower portions of the epidermis to the flattened cells of the stratum granulosum. Actinic keratoses are most often treated by cryodestruction with liquid nitrogen, but topical agents such as lotions and creams containing 5-fluorouracil, masoprocol, and tretinoin may also be used. Regular use of sunscreens not only is important in prevention of actinic keratoses but also has been shown to help reverse these premalignant lesions.

Analogous to actinic keratosis, actinic cheilitis is lip damage due to chronic sun exposure and is more common on the lower lip than the upper. Treatment of this condition is similar to that described for actinic keratoses.

Of all the cutaneous malignancies, the basal cell epithelioma/carcinoma (BCE) is the most common (Kuflik and Janniger 1993). Farmers and agricultural workers with lightly pigmented skin are at highest risk for developing this tumor as well as the squamous cell carcinoma (SCC) and malignant melanoma (MM). A review of multiple studies (Blair and Zahm 1991) revealed a relative risk for development of non-melanoma skin cancers in farmers ranging from 0.8 to 1.8 (seven of eight studies showed increased risk). For lip cancer, all studies (nine of nine) showed an increased relative risk (range = 1.3 to 3.1). For melanoma, eight of twelve studies showed increased relative risk (0.5 to 6.3). In one study in North Carolina, (Delzell and Grufferman 1985) the proportional mortality ratio for malignant melanomas was higher for non-white as compared to white farmers indicating factors other than skin type and solar radiation may play a role here.

BCE's generally occur on sun-exposed skin and often present as small, pearly papules with overlying telangiectatic vessels. Without treatment, they may invade deeply and become ulcerative ("rodent ulcer"—Figure 21.1). These tumors can enlarge locally but only very rarely metastasize.

Treatment is most commonly with local excision, curettage/desiccation, cryosurgery, or radiation. In recurrent tumors or those in high risk areas such as periocular, nasal or auricular tumors, Mohs microscopic-directed surgery is indicated (Preston and Stern 1992).

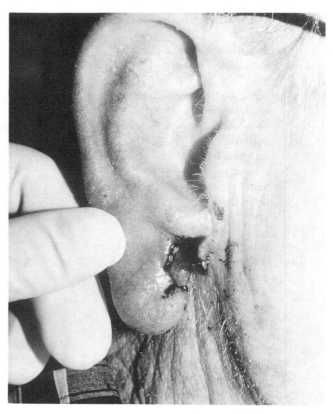

Figure 21.1. The basal cell carcinoma is the most common skin cancer and is due to chronic sun exposure. The ulcerative form of this tumor is commonly known as a "rodent ulcer."

The somewhat less common SCC (Figure 21.2) is also found primarily on sun-damaged skin and often arises from a pre-existing actinic keratosis. This hyperkeratotic lesion is generally pink in color and treatment is similar to a BCE. However, there is a somewhat higher concern here about metastasis and those SCC's occurring on the lip have an especially high propensity for metastasis. Bowen's disease is SCC-*in situ*. Treatment of SCC's is similar to that of a BCE as discussed above. Wedge resection or Mohs microscopic-directed surgery is generally recommended for SCC of the lip due to the higher chance of metastasis.

Although BCE's and SCC's occur predominantly after years of chronic sun exposure, MM's occur most commonly in light-skinned individuals (Skin Type I to II) with a history of repeated sunburns—especially at an early age (Armstrong

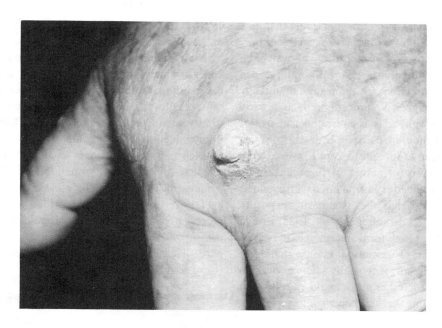

Figure 21.2. A squamous cell carcinoma on the dorsal hand. These cancers often arise from a pre-existing actinic keratosis.

1988; NIH 1992; Friedman et al. 1991). While the incidence of BCE's and SCC's is rising, the incidence of MM is of special concern since by the year 2000 an estimated 1 in 90 Caucasian Americans will develop this tumor during their life span. In Australia, the problem is even more epidemic with the Caucasian lifetime risk in males now at 1 out of 14 (Green 1982).

The treatment of MM is surgical excision. Those lesions less than 1 mm thick have less metastatic potential since these thin lesions are either *in situ* or have not yet penetrated to the depth of lymphatic or vascular channels. These lesions do not require the very wide excisions recommended in the past and currently a 1 cm margin is advised (NIH 1992). The dysplastic nevus in some cases may be a pre-melanoma and persons with large numbers of these (especially those with a family history of melanoma) have a high risk of developing MM.

Rare genetic syndromes at high risk for developing skin cancers include the basal cell nevus syndrome, xeroderma pigmentosum, and the dysplastic nevus syndrome. Persons with these disorders who work in agricultural occupations should be carefully counseled about their high risk of developing malignancies and need for sun avoidance.

EXPOSURE TO ENVIRONMENTAL EXTREMES

The farmer and other agricultural workers often have to work in extremes of heat and cold in addition to varying degrees of wetness. Heat leading to increased perspiration can produce miliaria ("heat rash"), which can be further divided into miliaria crystallina (small fragile vesicles), miliaria pustulosa (small pustules on an erythematous base), and miliaria rubra (the common red papular variety of "heat rash" or "prickly heat"). Cool compresses and a low-potency topical corticosteroid will improve this condition.

Just as a person ready to swing an ax will spit on his hands to increase the coefficient of friction, sweating in areas that rub (intertriginous areas) can lead to a frictional dermatitis known as intertrigo. Absorbent powders may help prevent this problem, but the acute rash is best handled with a low-potency topical corticosteroid (after excluding other intertriginous rashes such as dermatophyte infections, cutaneous candidiasis, or erythrasma).

Very low temperatures for extended periods of time without proper protection can lead to the obvious: frostbite. Caused by the actual freezing of soft tissue leading to vascular stasis, it usually occurs on exposed skin or the lower extremities. Frostbite often presents with numbness, cold pale tissue (even after rewarming), and swelling. Patients with frostbite are generally admitted to a hospital and a protocol followed (McCauley et al. 1983; Steele 1989). Emergency treatment of frostbite consists of rapid thawing using warm water at 40 to 42°C (104 to 108°F) usually for 15 to 30 minutes or until thawing is complete. It is important to realize that it is better to wait to thaw the injured area until evacuation occurs (even for hours) rather than have the area refreeze after thawing (Nelson 1991). In addition, it is important as the weeks pass to realize that tissue that may initially appear to have a poor prognosis for recovery may indeed be viable. A conservative approach of watching, waiting, protecting the involved area from injury, and preventing secondary infection are mainstays of therapy and are preferred over aggressive procedures such as early amputation. Tetanus prophylaxis is indicated and use of tobacco by the patient should be strictly banned.

Frostnip is the term used for "pre-frostbite" where there is no tissue lost. Chilblains (or pernio) is characterized by recurrent episodes of localized erythema often with cyanosis occurring primarily on the hands, feet, or ears after exposure to low but non-freezing temperatures (Nelson 1985; Brown 1991). Plaques or nodules may form and, in extreme cases, there may be vesiculation or ulceration. Intense itching or burning may occur. Chilblains is typically seen in young to middle-aged females and, other than avoidance of cold weather, there is no treatment other than supportive.

Trenchfoot (or immersion foot) is the term used to describe a syndrome that develops over a period of hours to days of continued exposure of wet feet to temperatures of 32 to 50° F (Smith et al. 1989). It may present similar to frostbite although there is no tissue freezing, and so the prognosis is better. The damage is neurovascular and continued hypersensitivity to cold is common.

Other diseases which flare with exposure to cold and can affect an agricultural workers' ability to work outdoors during cold weather include cryoglobulinemia and cold urticaria.

DERMATOSES RELATED TO EXPOSURE TO FARM CHEMICALS

The skin of today's farmers and agricultural workers invariably comes in contact with a wide variety of chemicals needed for running and maintaining a modern farm. Among others, these include fertilizers, pesticides, fungicides, herbicides, fumigants, rodenticides, wood preservatives, plant growth regulators, organic solvents, cleansers, veterinary products, and fuels. These groups of agents contain an extreme variety of chemical products and use of these substances varies substantially depending both on the products raised on the farm and the location of the farm itself. Skin reactions due to contact with farm chemicals can vary widely: irritant and allergic contact dermatitis, contact urticaria, photodermatitis, chloracne, and pigmentary disorders. "Acute contact dermatitis" (either due to a direct irritant or an allergic hypersensitivity reaction) is the most common agricultural related skin problem (Birmingham 1983). In agroworkers this skin reaction may be due to farm chemicals or plant exposure (covered in the next section).

Irritant contact dermatitis most commonly occurs on exposed areas such as hands and face and is due to various irritating chemical agents coming in contact with unprotected areas of skin. Even when gloves are worn to protect the hands, the offending chemical residue may build up in reusable gloves thereby occluding the chemical and leading to increased irritation (Figure 21.3). Clothing may become impregnated with these compounds, also leading to irritant problems in covered areas of skin. Severe acute irritant reactions are commonly termed "chemical burns". Common irritants on the farm include cleansers, soaps, germicidal agents, fertilizers, lime, pesticides, degreasers, solvents, preservatives, and ammonia. Liquid ammonia is sold in pressurized tanks and freezes as it vaporizes. This in particular can cause caustic burns with improper handling (Cordes and Rea 1988).

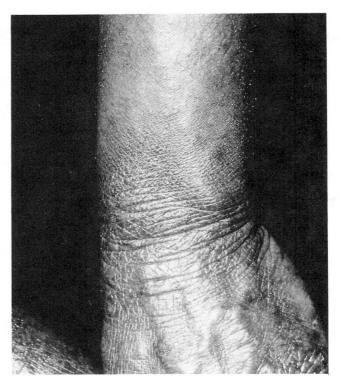

Figure 21.3. Irritant dermatitis of the hand and forearm in a dairy worker using an iodine containing solution to cleanse cow udders prior to attachment to a milking machine. Allergic patch testing provided assurance that this worker was not allergic to the solution.

Treatment of irritant dermatitis is symptomatic and may involve treating the active dermatitis with topical (and rarely systemic) corticosteroids. Education about protection of the skin from future exposure to the offending agent(s) through use of protective gloves and clothing is very important in prevention of continuing skin problems.

Allergic contact dermatitis due to farm chemicals can be disabling to a farmer and yet, has been poorly studied and is probably under-reported. Allergic contact dermatitis from fertilizers is usually due to cobalt or nickel (Pessegueiro 1990). Although many pesticides are irritants, positive patch tests to them are not common and are usually attributable to fungicides (especially captan, difolatan, and folpet) (Lisi et al. 1987). In a study of 216 agricultural workers from Japan, 29% were sensitive to captafol, 19% to sulfur compounds, 18% to organophosphates, and 10% to chlorinated hydrocarbons (Matsushita et al. 1980). Numerous pesticides have been implicated in allergic contact dermatitis (Table 21.4) (Lisi et al. 1987; Matsushita et al. 1980; Adams 1990; Fisher 1986; Abrams et al. 1991; Shaver and Tong 1991; Sharma and Kaur 1990; Bruynzeel 1991; Won et al. 1993; Schuman and Dobson 1985; Maibach 1986). Veterinary products (especially antibiotics such as neomycin, bacitracin, sulfonamides and penicillin) may also cause allergic contact dermatitis in agricultural/livestock workers. Allergic patch testing may be required for identifying the causative agent. There are a few studies on agricultural products for recommended concentrations for patch testing farm chemicals (Abrams et al. 1991; Fisher 1983; Garcia-Perez et al. 1984). Because many allergic contact dermatitis allergens are the same as those seen in nonagricultural workers (e.g., nickel, paraphenylenediamine, rubber additives, chromate, etc.) patch testing to standard dermatologic kits (e.g., TRUE® test kit) is also recommended.

Phototoxic reactions have rarely been reported with pesticide use, (DeEds et al. 1940) and contact urticaria/angioedema (Kentor 1986) and erythema multiform (Bhargava et al. 1977) have also been reported due to farm chemicals. Phenolic compounds can cause irreversible depigmentation and this can be seen with use of phenol cleansing agents.

Chloracne and porphyria cutanea tarda are well known side effects from use of Agent Orange as a herbicide in Vietnam. This is generally not seen on the farm, but chloracne has been reported in the manufacturing of the herbicide oxydiazol (Taylor et al. 1977).

The farmer and agricultural worker are indeed at risk for a wide variety of cutaneous eruptions due to exposure to pesticides and farm chemicals. Many of these products are complex and a variety of different agents are used for different regions of the country. Any farmer or agricultural worker with a skin problem that could potentially be related to farm chemicals needs careful interrogation and appropriate patch testing, not only to standard allergen kits but also to potentially allergenic farm chemicals.

Table 21.4. Pesticides implicated in allergic contact dermatitis.

Insecticides/Miticides (Acaricides)

Aldrin	Naled
Carbaryl	Oxydemton methyl
Carbofuran	Parathion
Dichlorvos (DDVP)	Promecarb
Fenvalerate	Pyrethrum
Lindane	Rodannitrobenzene
Malathion	Tetmosol
Methomyl	Thiometon

Herbicides

Alachlor	Glyphosate (Glyphosphate)
Amitrole	Nitralin
Atrazine	Nitrofen
Barban	Peridimethalin
Butachlor	Phenmedipham
Chloridazon	Proprazine
Dazomet	Trichlorobenzyl
Dichlofopmethyl	2,4D (2,4 dichlorephenoxy acetic acid)

Fungicides

Anilazine	Maneb
Barban	Mercury fungicide
Benomyl	Methyl-2-benzimidazole
Captafol	PCNB (pentachloro-nitrobenzene)
Captan	Phtahan
Chlorothalonil	Plondrel
Difoltan	Streptocycline
Dinocap	Thiophanate-methyl
Ditalimfos	Thiram
Dithianone	Zineb
Folpet	Ziram
Mancozeb (Manzeb)	

DERMATOSES RELATED TO PLANT EXPOSURE ON THE FARM

In the past, man rather than machines worked the fields and there was an obvious higher risk of exposure of skin to potentially harmful and sensitizing plants. In spite of this, agricultural workers and farmers continue to be at high risk for dermatoses related to plant exposure and it has been estimated that 50% of agricultural occupational skin disease is due to plants, trees, and natural vegetation. (O'Malley et al. 1988). Plants can harm human skin in several ways via contact: a) mechanical injury, b) pharmacologic injury, c) irritant contact phytodermatitis, d) allergic contact phytodermatitis, and e) phytophotodermatitis (Stoner and Rasmussen 1983).

Mechanical Injury

Mechanical injury by plants is common and can range from severe injuries such as a tree or coconut falling on a person to a more common and mundane injury such as the multiple thorn injuries one is likely to encounter when trying to clear blackberry vines from a fence row. Many plants (e.g., palms, palmettos) have sharp-edged leaves that can cause cutting injuries, while others (e.g., yucca, century plant) have spines that can easily penetrate human skin. Other plants (e.g., blackberry, rose, devil's walking stick, prickly ash, cacti, horse nettle, greenbrier, etc.) have thorns that can also cause penetrating injuries (Figure 21.4). Not only can thorns and spines lead to injury itself, but plant and other material may become imbedded leading to a foreign body granuloma reaction. They can also introduce infections such as sporotrichosis and other deep fungal infections, environmental mycobacterial infections, or the more common pathogenic bacteria such as *Staphylococci* or *Streptococci*.

Treatment of spine/thorn injuries involves removal of the remaining plant material. This may be difficult as plant parts are generally radiolucent and may be difficult to find. Large spines/thorns may be removed with splinter forceps, while smaller spines (e.g., prickly pear) may be removed with cellophane tape or school glue (allowed to harden and removed as a sheet) in non-hairy areas (Gelbard 1984).

Pharmacologic Injury

Some plants are capable of causing pharmacologic injury to the skin via small plant hairs containing histamine, acetylcholine, and serotonin (Figure 21.5). Contact with such plants in the families Urticaceae and Euphorbiaceae can cause an immediate painful "splash of grease" sensation in

Figure 21.4. Mechanical injuries by plants are often due to plant spines and thorns such as in this greenbrier (*Smilax* sp.).

Figure 21.5. Tiny plant hairs in the tread-softly (*Cnidoscolus stimulosa*) contain histamine, acetylcholine and serotonin, and contact with skin will lead to a painful urticarial eruption.

the affected skin followed by an urticarial eruption (Thurston 1974). Examples of plants causing this problem in the Urticaceae include the stinging nettle (*Urtica dioica*) and wood nettle (*Laportea canadensis*). Plants in the Euphorbiaceae that can cause this reaction are the spurge nettle or tread-softly (*Cnidoscolus stimulosis*) and the noseburn vine (*Tragia involucra*).

Irritant Contact Phytodermatitis

There are numerous families of plants which can cause an irritant contact phytodermatitis (Table 21.5). Unlike those plants which have saps that are sensitizers and can occasionally stimulate allergic reactions in some individuals, irritant reactions occur in anyone whose skin comes in contact with the irritating chemicals in these plants for a long enough period of time. While primary irritation from plants is common, it is generally mild enough to be ignored or be self-treated by agricultural workers. The most notable of the irritant plants are in the spurge (Euphorbiaceae) family and include such well-known plants as poinsettia, crown-of-thorns, and candelabra cactus. Although most spurges are found in ornamental gardens or as house plants, several are weeds including the petty spurge (*Euphorbia peplus*), (Calnan 1975) the sun spurge (*Euphorbia helioscopia*), and the mole plant (*Euphorbia lathyrus*). In ancient times, spurges were often planted on graves to deter grave robbers and used by beggars to produce dermatitis and blisters in order to promote pity. They were also used to blind people as a form of punishment, as sap from these plants produces a severe keratoconjunctivitis (Lovell 1993). Ingestion of these plants can cause gastroenteritis and ingestion by livestock may produce toxic symptoms (Schmidt and Evans 1980).

Another family, the Araceae, contains many irritant plants (such as dumb cane, *Dieffenbachia* sp.) which are a common horticultural/floral worker problem but are rarely a problem in agricultural workers in this country. These plants have calcium oxalate crystals which penetrate the skin allowing proteolytic enzymes found in the sap to penetrate the epidermal barrier. Jack-in-the-pulpit (*Arisaema triphyllum*) and skunk cabbage (*Symplocarpus foetidus*) are members of this family which may be found on agricultural land; however, they generally occur in moist forest land and are not generally a problem to agricultural workers. Certain bulbs such as daffodils (Amaryllidaceae) contain calcium oxalate as well as proteolytic enzymes and hand dermatitis is an occupational hazard of bulb handlers. The pineapple also contains calcium oxalate crystals as well as a proteolytic enzyme known as bromelin. Prior to mechanization of pineapple picking and packing in developed countries, agricultural workers in this industry often developed a severe dermatitis as well as abrasions and cuts from the spines on the pineapple leaf.

Table 21.5. Plant families commonly causing irritant phytodermatitis.

Euphorbiaceae (spurges)

Wild: cypress spurge, sun spurge, petty spurge, flowering spurge, mole plant, wild poinsettia, machineel tree
Ornamental: poinsettia, crown-of-thorns, candelabra cactus, pencil tree

Araceae

Wild: skunk cabbage, Jack-in-the-Pulpit, water arum
Ornamental: caladium, elephant ears, philodendron, dieffenbachia (dumb cane), anthurium, calla lily

Ranunculaceae

Wild: buttercups, pasque flower, larkspur, baneberry, marsh marigold
Ornamental: windflower (anemone), clematis

Brassicaceae

Mustards, radishes, true watercress, peppergrass, whitlow grass, shepherd's purse

The buttercup (family Ranunculaceae, Figure 21.6) also contains irritant plants, and these plants are a common cause of "meadow dermatitis". In addition, the family Brassicaceae (mustards and radishes) can cause an irritant dermatitis which, in severe cases, can blister. Irritating properties were the basis for previous use of "mustard plasters".

Irritant plant dermatitis should be treated with symptomatic care using antihistamines orally, and topical or rarely systemic corticosteroids. If blistering occurs, Burow's (aluminum acetate/1:20) solution may be helpful.

Allergic Contact Phytodermatitis

Allergic contact phytodermatitis in farmers and agricultural workers can be a major problem and source of disability. Unlike irritant contact phytodermatitis which can occur in anyone, allergic contact phytodermatitis requires a sensitized immune system (Type IV hypersensitivity/cell-mediated immunity), and therefore does not occur in all exposed individuals. In a study of occupational skin disease

Figure 21.6. The buttercup in the Ranunculaceae can cause an irritant plant dermatitis and is a common cause of "meadow dermatitis."

in California, (Mathias 1989) allergic contact phytodermatitis from poison oak was alone responsible for 49% of all skin problems combined. The Toxicodendron plants (poison ivy, poison oak and poison sumac) in the family Anacardiaceae are the most common offenders causing allergic contact phytodermatitis in farmers and agricultural workers as well as in the general population. Generally, poison ivy (*Toxicodendron radicans* or *T. rydbergii*) is the most common plant sensitizer east of the Rockies with eastern poison oak (*T. toxicarium*) or poison sumac (*T. vernix*) being less common. Poison ivy (Figure 21.7) grows as a vine or small shrub and is an extremely common plant found in the eastern United States. Eastern poison oak usually grows as a small shrub and poison sumac is usually seen as a small tree in low-lying swamp forests. Western poison oak (*T. diversilobum*) is one of the most common bushes along the coast and coastal mountain ranges of the western United States. It is a major cause of allergic contact dermatitis in agricultural workers and persons who are outdoors, but it can be an especially severe problem in firefighters (Vietmeyer 1985).

The sensitizing *Toxicodendron* plant oil, urushiol (from the Japanese word meaning "sap") is released when the plant is traumatized and turns black after oxidation (Guin 1980). Black spots on clothing similar to tar stains which occur after clearing brush may indicate contact with urushiol. These clothing spots remain allergenic even after multiple washings and can lead to occult re-exposure. Pets and farm animals are non-allergic but may get the urushiol on

Figure 21.7. Poison ivy (*Toxicodendron radicans*) is one of the most common plants in the Eastern United States. Although other plants which have a somewhat similar appearance, the old saying "leaves of three, leave them be" can be helpful in preventing episodes of re-exposure.

them potentially leading to secondary contact with farm workers.

When urushiol contacts the skin, it becomes "anchored" in place within minutes via a nonpolar aliphatic side chain that penetrates the lipid bilayer of epidermal cell membranes leaving the polar phenolic antigenic ring structure exposed. This "anchoring" property makes prompt washing (within 15 to 30 minutes) imperative in helping to prevent a reaction following exposure. Although a sensitized individual will generally develop the typical eruption in 24 to 72 hours, it may take up to 3 weeks after primary sensitization for the eruption to appear. Initially, the rash will be erythematous and edematous with rapid progression to vesicles and bullae. The classic presentation is a pruritic erythematous eruption with areas of linear vesiculation (Figure 21.8). The typical course of the rash without treatment and without re-exposure is about 3 weeks. Persons allergic to *Toxicodendron* are generally allergic to other members of the Anacardiaceae including the mango (urushiol in the peel of the fruit), cashew (peel), gingko (fruit), Brazilian pepper (entire plant), poisonwood (entire plant), Japanese lacquer tree (entire plant), and India marking nut tree (entire plant).

Another major cause of allergic contact dermatitis in farm workers is due to the Compositae family which is one of the largest group of plants (20,000 to 25,000 plant species). A partial listing of this group of plants is in Table 21.6. These plants contain various sesquiterpene lactones which are not only sensitizers but also irritants.

Table 21.6. Common plants in the Compositae.

Ragweed	Thistle
Daisy	Common fleabane
Marigold	Horseweed
Sunflower	Joe-Pye weed
Zinnia	Boneset
Dog fennel	Camphor weed
Daisy fleabane	Wild lettuce
Sneezeweed	Blazing star
Dandelion	Silverrod
Dahlia	Goatsbeard
Hawkweed	Ironweed
Parthenium	Cocklebur
Rabbit tobacco	Burdock
Black-eyed Susan	Chamomile
Indian blanket	Mayweed
Aster	Compass plant
Cornflower	Chicory
Tansy	Safflower
Goldenrod	Endive
Coreopsis	Chrysanthemum
Yarrow	Strawflower
Tickseed marigold	Feverfew
Ragwort	Ox-eye
Jerusalem artichoke	

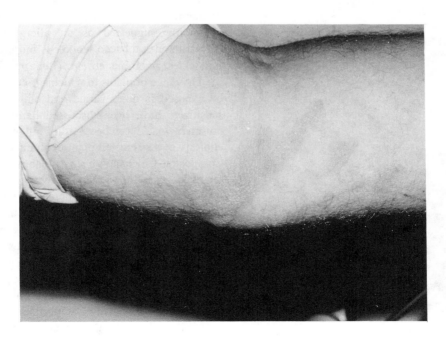

Figure 21.8. A typical linear vesicular eruption in a patient who had a history of exposure to poison ivy. Dermatitis due to the *Toxicodendron* group of plants is the most common cause of allergic contact dermatitis in the United States.

Ragweed (*Ambrosia* spp.) is one of the most common weeds on farms, and "ragweed dermatitis" provides the model for rashes from other members of the Compositae. Whereas the antigen which is responsible for IgE-mediated allergic rhinitis is water soluble and found in pollen, the antigen responsible for ragweed dermatitis is a sesquiterpene lactone (alpha-methylene-beta-butyrolactone) and is organic solvent soluble and found in and on the stems and leaves of the plant. Unlike urushiol in *Toxicodendron* plants, these sesquiterpene lactones are found on the surface of the plants and trauma to the plant does not have to occur for contact with the allergen. In addition, although the time course of the eruption is similar to *Toxicodendron* dermatitis, the eruption tends to be dry and eczematous rather than vesicular. Commonly, the pruritic eruption initially begins on the hands after weeding activities. Over time, repeated contact may lead to an airborne contact dermatitis involving exposed areas, and may eventuate in an entire-body erythroderma in extreme cases (Crounse 1980; Hjorth et al. 1976). Although mechanization of farming has likely led to a decrease in the number of sensitized individuals, this severe eruption can become a major source of disability to a farmer or agricultural worker. In highly sensitized individuals, the rash due to Compositae often flares in the fall/early winter as fields of weeds are mowed prior to planting with crops the next spring.

Dermatitis due to many other members of the Compositae family has been well documented, and many of these are common agricultural weeds (goldenrod—Figure 21.9, fleabane, dog fennel, dandelion, black-eyed susan, parthenium, tansy, sneezeweed, etc.). The most common cause of allergic contact dermatitis in the Compositae is chrysanthemum but this is seen more in the horticultural and floral industry than on the farm. Other plant families (Magnoliaceae—magnolia, tulip trees; Jubulaceae—liverworts; Lauraceae—sweet bay) may also contain sesquiterpene lactones which may cross-react with the Compositae.

Another cause of allergic dermatitis due to plants is seen in the tulip industry where agricultural workers commonly become sensitized and develop a rash known as "tulip fingers" or "tulip fire" (Verspyck 1969; Hausen 1982). This is due to the glucosides tuliposide A and tuliposide B which are hydrolysized into the sensitizers tulipalin A and B. These sensitizers are also found in the Peruvian lily (*Alstroemeria* spp.), now the most common sensitizing plant in floral workers.

The trumpet creeper vine (*Campsis radicans* or "cowitch") is often found growing on fencerows and abandoned farm buildings and can cause an allergic contact dermatitis (Lampe and McCann 1985). Although tobacco is commonly suggested as an allergenic plant, allergic dermatitis in a tobacco worker is more commonly due to fungicides (zineb, maneb, etc.) (Benezra et al 1985).

The range of plants to which agricultural workers could be exposed is quite varied and only the more common sensitizers have been discussed here. Except in typical poison ivy/oak, it can be very difficult to determine the exact etiology of a suspected plant dermatitis. Patch test kits for Compositae which have been available in the past have been removed from the market by the Food and Drug Administration. As many potential sensitizers are also potent irritants, interpretation of patch tests to plant parts is difficult to interpret, and controls are generally needed. In addition, there is always concern about sensitizing a patient during the process of testing. Because there is a lack of a standardized, approved patch test kit for plants, many adult-onset, idiopathic "chronic eczema" problems may actually be due to plants—especially Compositae.

In all types of allergic contact dermatitis due to plants, treatment is similar. First, re-exposure to the suspected plant must be prevented, and the allergic rash is treated with topical and/or systemic corticosteroids. It must be

Figure 21.9. Goldenrod (*Solidago altissima*) is one of the most common weeds on farmland. It is in the Compositae family of plants and may cause an allergic contact dermatitis in sensitized individuals.

remembered that a 3-week course of treatment is necessary as undertreatment often leads to a reflare of the dermatitis. Symptomatic relief with oral antihistamines and menthol-containing lotions may be also be beneficial. With *Toxicodendron* or other blistering plant rashes, Burow's solution compresses twice daily may be helpful. In severe refractory cases of extensive Compositae dermatitis, low dose chronic corticosteroids, azathioprine, or PUVA therapy may be beneficial.

Phytophotodermatitis

Phytophotodermatitis is due to plant-containing psoralens (furocoumarins) which are potent photosensitizers. There are several families that contain these photosensitizing chemicals: 1) the Umbelliferae (Apiaceae), 2) the Rutaceae, 3) the Moraceae, and 4) the Leguminaceae. The Umbelliferae contain many common weeds seen on farms including Queen Anne's lace (Figure 21.10), cow parsnip, bishop weed, and hogweed. In addition, important crops such as celery, carrots, parsnips, parsley and dill are also members of this family and can cause a photodermatitis in agricultural workers as well as others handling these plants (Berkeley et al. 1986; Smith 1985). Celery workers and handlers can also be exposed to psoralens in pink rot (*Sclerotina sclerotium*), a fungus occurring on celery (Wu et al. 1972).

The Rutaceae family contains the citrus fruit trees, and the best known photosensitizers in this group are limes and the Bergamot orange. Fig pickers (Moraceae) are also susceptible to phytophotodermatitis (Watemberg et al. 1991). The leguminous farm weed *Psoralea* can also lead to phytophotodermatitis in exposed farm workers.

Phytophotodermatitis is almost always of the phototoxic variety of photoreaction. This exaggerated sunburn is due to the UVA wavelength and can occur in any individual if the degree of plant exposure is coupled with sufficient sun exposure. Tanning beds also use UVA and can elicit the response and, as the UVA wavelength can pass through window glass, the eruption can also occur while driving in a car or after exposure near a sunny window indoors. After contact with to these plant saps followed by sun or UVA exposure, the skin becomes painful, inflamed and may often blister. Hyperpigmentation usually follows (Figure 21.11).

Treatment of phytophotodermatitis (like acute sunburn) is entirely symptomatic. Residual hyperpigmentation may take months to fade, and hydroquinone bleaching agents are generally not recommended.

ARTHROPOD BITES AND STINGS ON THE FARM

Arthropods are ubiquitous in nature and, because many agricultural work activities take place outdoors, farm workers often come in contact with them. Some types of arthropods are associated with exposure to livestock and still others are associated with wild animal feed or grain contact. Some

Figure 21.10. Queen Anne's lace or wild carrot (*Daucus carota*) in the Umbelliferae family can cause phytophotodermatitis when plant sap containing psoralens comes in contact with the skin and there is sufficient exposure to UVA light.

334 / Safety and Health in Agriculture, Forestry, and Fisheries

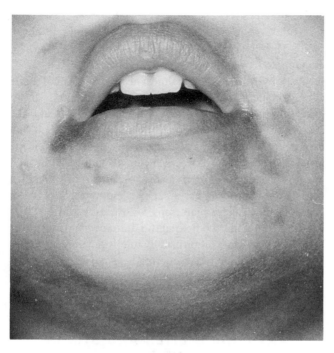

Figure 21.11. Phytophotodermatitis showing the residual hyperpigmentation which is characteristic. This person had chewed on limes while outdoors in sunlight.

Figure 21.12. A typical red imported fire ant (*Solinopsis invicta*) mound. The mound can easily be identified by molesting it with a stick as numerous ants will quickly pour forth from the molested area.

arthropods that cause problems in man are harmful to crops and livestock, whereas others are important pollinators of crops.

Bees, Wasps, and Ants

The order Hymenoptera is responsible for most fatalities due to bites and stings in humans (primarily due to anaphylaxis) (Parrish 1963). Honeybees are important crop pollinators and, because of this, some farms maintain beehives, thereby obtaining honey as a by-product. Honeybees are generally docile; stinging occurs only defensively. The stinger is barbed and will tear off the distal abdominal segments (containing the venom sacs) after a sting occurs. Venom continues to be injected and it is advisable to remove the stinger (without squeezing the abdominal remains) using a knife blade. After stinging, a pheromone is released attracting other bees. Although most honeybees are considered docile, the Africanized honeybees ("killer bees") are vicious stingers, often without provocation. These bees escaped into the wild in Brazil in 1957 after importation from Africa, and have since migrated up through Central America and Mexico and now are found in the Southwest in Texas and westward into California where there exists extensive rich farmland. Further migration of these aggressive bees is expected. In spite of many news reports, their sting is no more dangerous than that of the domestic honeybee.

Bumblebees (*Bombus* spp.) are larger bees that usually live underground. They have a mild disposition and generally only sting when molested. The carpenter bee (*Xylocopa* spp.) female chews a tunnel into dry wood of dead trees or wood of houses and outbuildings and will sting to protect its nest.

The wasp (*Polistes* spp.) is primarily aggressive around its nest site or when provoked. Wasps (including muddaubers, and potter wasps) often build their nests in protected areas in open farm outbuildings. The velvet ant (family Mutillidae) is actually a wingless wasp inaccurately but commonly known as a "cow killer ant". Hornets (*Vespa* and *Vespula* spp.) usually build large nests and are primarily aggressive when this nest is threatened. Yellow-jackets (*Vespula* spp.) are related and are known to be quite aggressive, even when away from its nest site which is usually underground. A common area today for a yellow jacket sting is on the lip as they are often attracted inside an opened soft drink can. Wasps, hornets, and yellow jackets all have smooth stingers and can sting multiple times.

The most notorious farm pest in the order Hymenoptera is the red imported fire ant (*Solenopsis invicta*). "Imported" from South America via ship ballast dumped on the shores of Mobile Bay, Alabama, earlier this century, this aggressive ant has since spread throughout most of the southern United States, now reaching as far north as North Carolina (Burke 1991). The fire ant mound (Figure 21.12) is usually quite noticeable on farmland, and can reduce valuable cropland and livestock pastures into a relatively worthless and unusable eyesore. Farming losses (especially in the soybean industry) with lower crop yields from infested fields have been studied (Apperson and Powell 1983; Adams et al. 1977). Loss occurs not only from decreased seed yield, but also from mounds interfering with and damaging harvesters. Other crops such as okra, potatoes, corn, eggplant, beans, and citrus are damaged by fire ants. It is common to find soybean and corn fields supporting 100 to 125 mounds/hectare and citrus groves in Florida are often very heavily infested (greater than 200 mounds/hectare) (Adams 1986).

The red imported fire ant attacks in large numbers when the nest is disturbed and livestock (especially young) as well as humans are at risk. Fire ant stings result in multiple sterile pustules as the ant bites and holds with mouthparts and then stings with its hindparts in an arc surrounding the bite. Agricultural workers who are sensitive to Hymenoptera venom are at high risk for anaphylaxis as the fire ant venom may cross-react with others in the order.

Treatment of mild Hymenoptera stings consists of removal of the stinger if present and symptomatic care. An ice cube can provide first aid relief and antihistamines may reduce edema. In the case of anaphylaxis, emergency treatment consists of epinephrine (0.3 to 0.5 ml 1:1000 for adult individuals) and rapid delivery to a medical care facility (Frazier 1968).

Flies, Gnats, and Mosquitos

The order Diptera is made up of flies, gnats, and mosquitos. The stable fly (family Muscidae) is common on farms where poultry or livestock are kept. Unlike the common housefly which looks similar, this fly delivers a bothersome bite. The deerfly or yellow fly (*Chrysops* spp.) is somewhat larger, and is an extremely aggressive fly which is especially common in lower lying moist areas on the farm. It can transmit tularemia and possibly (reported but not yet confirmed) Lyme disease. Horseflies (*Tabanus* spp.) are also common on farms around livestock, and can deliver painful bites (Figure 21.13). They have been reported to carry the same diseases as the deerfly. Myiasis is the term for parasitism by fly larvae. Although primarily a problem in tropical areas, botfly myiasis can occasionally occur in the southern United States as well as in migrant workers from more tropical areas. A pulsatile papule is diagnostic of furuncular myiasis (Davis and Shuman 1982).

Figure 21.13. The horsefly (*Tabanus* sp.) can deliver a painful bite and is considered a vector for tularemia. Secondary infection of the bite site is common.

As with all outdoor workers, mosquitoes are also a nuisance to farmers and other agricultural workers. In the United States, they may transmit viral encephalitis. Nocturnal biting midges and gnats/blackflys ("no see-ums") are merely a nuisance, but can be extremely aggravating.

Bugs and Beetles

In the Hemiptera ("stink bugs") order of insects, painful bites are commonly seen from assassin or kissing bugs (family Reduviidae, which have mouthparts adapted for sucking blood. Although most are nocturnal, a few such as the wheel bug (*Arilus cristatus*), are often seen in the daytime. The bloodsucking cone-nosed bug (*Triatoma sanguisuga*) (Figure 21.14) is representative of this group. Some aquatic hemiptera are commonly found in farm ponds and may bite. They include the giant waterbugs (family Belostomatidae), creeping waterbugs (family Naucoridae),

Figure 21.14. The cone-nose bug is in the order Hemiptera and can deliver a painful bite.

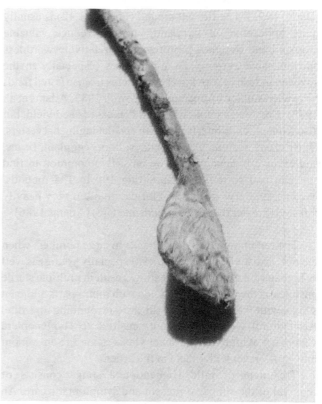

Figure 21.15. The puss caterpillar (larvae of the moth *Megalopyge opercularis*) has poisonous spines hidden within the longer hairs, and handling of this caterpillar can lead to a painful sting.

waterscorpions (family Nepidae) and backswimmers (family Notonectidae).

Hemiptera bites may be extremely painful and the wheelbug has been described as delivering one of the most painful arthropod bites known to man (Frazier 1969). When feeding, however, these insects usually bite painlessly. Local reactions such as blisters may occur and treatment is supportive (Burnett et al. 1987a).

In the order Coleoptera (the true beetles), the blister beetle (*Epicauta* and *Lytta* spp.) are most problematic to farmers. These beetles are commonly seen on garden plants, along fence rows (especially with goldenrod), and in flower beds or alfalfa crops (Burnett et al. 1987b). If a blister beetle is swatted against the skin, the vesicant cantharidin is released and blisters generally appear 2 to 5 hours later. Alfalfa contaminated with blister beetles has been reported to be fatal in horses (Beasley 1983). Rove beetles (family Staphylinidae) may also cause vesiculation and may also bite.

Caterpillars, Moths, and Butterflies

The order Lepidoptera (moths and butterflies) can also cause skin problems in agricultural workers. Dermatitis can arise either from contact with urticating hairs of various moths or with stings from certain species of caterpillars. In this country, moths containing urticating hairs (or seta) include the gypsy moth, the Douglas fir tussock moth, the browntail moth and the yellowtail or mulberry tussock moth (Dinehart et al. 1985; Berger 1986). Generally the eruption occurs within hours after exposure. Multiple small pruritic urticarial papules develop and can be associated with conjunctivitis, rhinitis, and dyspnea. Caterpillars that contain stinging hairs connected to venom sacs include the saddleback caterpillar, the puss or flannel moth caterpillar (Figure 21.15), and the Io moth caterpillar. Treatment of stings from these caterpillars is symptomatic.

Figure 21.16. The lone star tick (*Amblyomma americanum*) is shown above respectively in the nymph, adult male and adult female. This tick is a vector for ehrlichiosis and tuleremia.

Fleas

Fleas are in the order Siphonaptera and are generally named for the animal on which it parasitizes (e.g., rat fleas, dog fleas, cat fleas, hen fleas, etc.). Although fleas prefer to be species specific, they will often bite other hosts including man when the preferred host is unavailable. The number of flea bites therefore increases when animals are removed from their quarters. Except for secondary *Staphylococcus* or *Streptococcus* infection, other more serious infections secondary to flea bites are rare in this country (although endemic plague is occasionally seen in the western United States).

Mites, Ticks, Spiders, and Scorpions

The class Arachnida contains the mites, ticks, spiders, and scorpions. Mites belong to the order Acari and are extremely common tiny arthropods. *Sarcoptes scabiei* is the mite, well-known to cause the human infestation known as scabies. Zoonotic species-specific scabies infestations also occur in dogs, cats, swine, horses, mules, cattle, sheep, goats, and pigs (Chakrabarti 1990; Estes et al. 1983). Close contact with infested animals can lead to a self-limited pruritic eruption on humans.

Dogs, cats, and rabbits may be infested with *Cheyletiella* mites ("walking dandruff") and close contact may lead to multiple bites in man (Lee 1991; Shelley et al. 1984; Fox and Reed 1978). *Dermanyssus gallinae*, the poultry or chicken mite can infest a large number of birds including domesticated poultry. Although large scale poultry operations are usually well-versed in the effects of these mites and their control, farmers who have a small poultry business as a sideline are much more likely to have infestation of their poultry houses. These mites can bite and become a significant aggravation to the farmer and his workers (O'Donel 1984).

Another common problem in farmers is "grain itch" or "straw itch" which is due to *Pyemotes* mites that prey on the larvae of insects feeding on grain or straw. These bites may be trivial or form large bullae in sensitized grain or straw handlers. Itching can be quite severe and prevents the majority of affected persons from sleeping. Severely sensitized individuals may be unable to work (O'Donel 1984).

Probably the most common mite causing a pruritic rash in persons working outdoors is *Eutrombicula alfreddugesi* (the chigger, redbug, or harvest mite). The 6-legged larvae of this mite waits on vegetation or forest litter for a passing host. It attaches to human skin with its mouthparts and then secretes digestive enzymes which liquefy cells and provides the mite with a fluid diet. A feeding tube (stylostome) is formed and the mite continues to feed for up to 3 days. This generally coincides with the time the hypersensitivity reaction is beginning in sensitized individuals and the chigger has usually dropped off or been scratched off by the time the eruption is becoming clinical apparent (Jones 1987). For this reason, common home remedies such as application of nail polish "to kill the critter" are generally not worthwhile.

Infestation of barns and outbuildings with rats, bats, mice, or nesting birds can also lead to species-specific mites which can become a problem to humans if the host pest is eradicated (O'Donel 1984; Westrom and Milligan 1984; Fishman 1988; Aylesworth and Baldridge 1985). Although rickettsial-pox has a mite vector in the United States, mites are not otherwise considered transmitters of infectious agents in this country. Treatment of mite bite hypersensitivity reactions is generally symptomatic, although topical or occasionally systemic corticosteroids may be beneficial in severe reactions.

Ticks are also a common problem on farms and the hard ticks (Family Ixodidae) are most problematic for humans. The American dog tick (*Dermacentor variabilis*), wood tick (*Dermacentor andersoni*), lone star tick (*Amblyomma americanum*) (Figure 21.16), and black-legged or deer tick (*Ixodes scapularis* and *pacificus*) are of primary concern due to zoonoses such as Rocky Mountain spotted fever, Lyme disease, tick paralysis, ehrlichiosis, Colorado tick fever, relapsing fever, babesiosis, Q fever, tularemia, and tick-borne encephalitis (Petri 1988). When a tick attaches, it

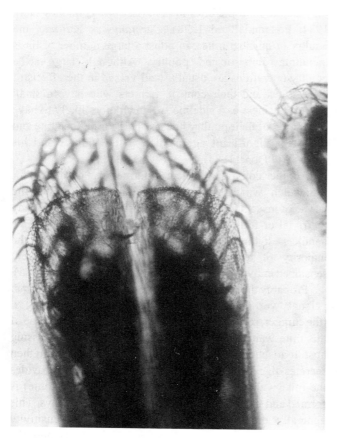

Figure 21.17. The proboscis of a tick contains numerous recurved teeth making it difficult to extract.

inserts its proboscis into the dermis. In addition to numerous recurved teeth on the proboscis (Figure 21.17), the tick secretes a glue-like cement making tick extraction difficult. Tick removal should be accomplished as soon as possible as the tendency for potential disease transmission increases during prolonged feeding as the tick becomes engorged. Although numerous anecdotal tick extraction methods have been advocated, removal of the tick is best accomplished by grasping the tick at the head with iris forceps (blunt curved forceps) and, without squeezing, gently pulling the tick out of the skin (Needham 1985). Tick larvae and nymphs are commonly called "seed ticks" and require a careful examination for identification and removal (Jones 1981). In southern California and Mexico, a soft tick, the pajaroello tick (*Ornithodoros* sp.) commonly causes local hypersensitivity and persons can develop a flu-like illness (Gentile 1989).

Spiders are also commonly found on farms and, although they can cause bites in humans, can also be beneficial in reducing populations of other insects. In the United States, spiders capable of producing serious bites on the farm include the black widow spider (*Latrodectus mactans*) and the brown recluse or fiddleback spider (*Loxosceles reclusa*).

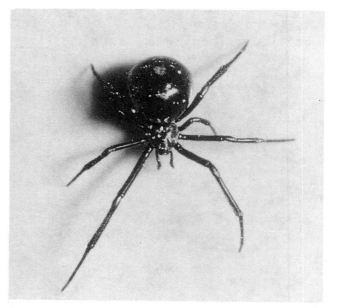

Figure 21.18. The black widow spider (*Latrodectus mactans*) has a venom which is a potent neurotoxin. The spider has a characteristic red hourglass shape on the ventral aspect of the abdomen.

The black widow spider is found in all states except Alaska (Rees and Campbell 1989). The spider is characterized by a shiny black body with rounded abdomen and a red hourglass shape on the ventral abdomen with either no or variable red markings dorsally (Figure 21.18). This spider prefers dark moist areas and on farm is commonly found in woodpiles, brick/cinderblock piles, barns, outbuildings, fields, and trash piles. In the past, outhouses were common sites for nesting and the anogenital area was a common bite site. In Europe and South Africa, black widow bites are associated with grain threshing and grape harvesting (Rees and Campbell 1989).

The black widow bite can be painless but more commonly feels like a pin prick, and usually occurs when the web is threatened. Generally there is no local tissue reaction at the bite site although two small fang marks may be seen with close inspection. The venom, a neurotoxin, causes depletion of acetylcholine at the neuromuscular junction and symptoms usually begin within one hour and peak at 1 to 6 hours (Cobb 1994). Pain and numbness spread from the envenomation site leading to generalized muscle spasms and cramps. Truncal and abdominal musculature are especially affected and the envenomation syndrome may mimic an acute surgical abdomen or heart attack. Pain, parasthesias, facial edema, hypertension, ptosis, headache, diaphoresis, nausea, vomiting, excessive salivation, dysphagia, and a low-grade fever may also occur (Mack 1994; Miller 1992).

Initial measures following a black widow spider bite should be directed at moving the victim to a medical facility. Ice may be used to reduce pain and swelling at the bite site. Calcium infusions, pain medications (morphine, meperidine), and muscle relaxants (methocarbamol, diazepam) are useful in treating the symptoms. An equine-derived antivenin is indicated for high risk individuals, particularly elderly persons (especially those with hypertension or cardiovascular disease) or young children. Generally one ampule of antivenin is sufficient in most cases (Miller 1992). As with all spider bites, tetanus prophylaxis is indicated.

The brown recluse spider (*L. reclusa*) can also be found on the farm and is most common in the South-Central states. Named "recluse" due to its shy habits, it is usually found outdoors in woodpiles or trash dumps or indoors in storage buildings, farm buildings, old vacant houses, or even in closets or behind furniture in occupied homes. The brown recluse is a medium-sized brown spider and has an inverted dark brown violin-shaped figure on the thorax just posterior to the head (hence the common name, "fiddleback spider"). The bite of the fiddleback is often painless but pain usually develops in the area within several hours. Initially, vasospasm and tissue ischemia occur at the site with a central whitish area of vasoconstriction surrounded by erythema ("bulls-eye" appearance). Blistering and tissue necrosis often follow centrally with an irregularly bordered erythematous to violaceous area surrounding the bite site and often extending inferiorly due to gravitational spread of the toxin. Tissue loss may be extensive and may take months to heal.

Systemic symptoms may also occur with rash, fever, chills, nausea, vomiting, malaise and arthralgias. Rarely hemolysis may occur and may be associated with hemoglobinuria, renal failure, and death (Foil and Norment 1979; Mack 1992). Treatment of brown recluse bites continues to be controversial but most authorities advocate the use of dapsone (King and Rees 1983). The use of prednisone is generally reserved for systemic and life-threatening loxoscelism (Hobbs and Harrell 1989; Anderson 1982).

Other spiders that can be found on farms and can cause milder bite reactions include the wolf spider (*Lycosa* spp.), jumping spider (*Phidippus* spp.), the common garden spider and house spider (*Chiracanthium* spp.), signature spider (*Argiope* spp.), fishing spider (*Dolomedes* spp.), running spider (*Herpyllus* spp.), and tarantulas (*Aphonopelma* and *Dugesiella* spp.) (O'Donel 1984; Campbell et al. 1987; Krinsky 1987). Symptomatic care, observation for secondary infection, and attention to tetanus prophylaxis are the necessary treatment for these milder bites.

Scorpions may be found across the southern United States but are especially prevalent in the Southwest. Scorpions in the genus *Centruroides* are the most important in causing severe envenomation. The sting of this scorpion causes pain, numbness, and local cramping in addition to more systemic manifestations such as hyperactivity, hyperesthesia, headache, hypertension, salivation, vomiting, abdominal cramping, and even death. A goat-derived antivenin is available at Arizona State University (for local use) (Burnett et al. 1985). Opioids should not be used as they may potentiate toxicity. Non-*Centruroides* scorpions (*Hadrurus*, *Vejovis*, *Diplocentrus* spp.) produce a painful local reaction similar to a bee sting where symptomatic care is all that is necessary.

Whipscorpions, although not toxic, are known as vinegarones because of their ability to secrete acetic and caprylic acid from glands at the base of the tail. Contact with the eyes may be irritating. Mastigoprotus (giant vinegarone or "Grampus") is found in the southern and southwestern United States among debris on soil and under logs.

Centipedes and Millipedes

Centipedes (*Chilopoda*) are elongate arthropods with one pair of legs per segment which live in rock crevices or under rocks or other objects in contact with the ground. Centipedes can bite and envenomate using a pair of hollow fangs. Although the large (up to 30 cm) centipedes found in Indopacific areas can cause more serious envenomation, those found in the United States are comparable to a bee sting and only symptomatic care and tetanus prophylaxis are indicated.

Millipedes (*Diplopoda*) look similar to centipedes but have two pairs of legs per body segment. They feed on decaying vegetative matter and are often found under objects on the ground or under forest leaf litter. Although they do not bite, they secrete irritating chemicals which can lead to skin or mucous membrane irritation after handling. Other than washing the area, there is no specific therapy.

REPTILE AND ANIMAL BITES ON THE FARM

One of the great fears of working outdoors in an agricultural setting is coming in contact with a venomous snake. In a previous study of animal bites in the United States, snake bites accounted for 30% of fatalities with rattlesnakes causing the majority of these (Parrish 1963). Of the 45,000 estimated snake bites annually, approximately 7,000 are caused by venomous snakes (Parrish 1966). The highest rates for venomous snake bites (10 to 20 per 100,000 population) are in the Southern states with North Carolina, Arkansas, Texas, Mississippi, and Louisiana leading the country (Sullivan and Wingert 1989). Most bites are on the extremities and occur during the warmer months of the year when snakes are more active. The percentage of venomous snake bites associated with agricultural activities ranges from 4 to 16% in various studies (Russell 1983).

The Crotalidae family are the pit vipers and, in this country, consist of the copperhead (*Agkistrodon contortrix*), the cottonmouth or water moccasin (*Agkistrodon piscivoris*), and a variety of rattlesnakes (*Crotalus* and *Sistrurus* spp.). Coral snakes (*Micrurus* and *Micruroides* spp.) are in the Elapidae family.

Pit vipers are named for the facial pits which are important in locating prey (Figure 21.19). They have vertically-oriented slitted pupils and a distinctive triangular head. The fangs are attached to venom glands located near the eye and angle of the upper jaw. These snakes account for most of the fatal poisonous snake bites in the United States: 68% by rattlesnakes, 6% by cottonmouth moccasins, and the remainder by other species (including non-indigenous and unidentified) (Parrish 1963). Less than 1% of envenomizations are due to coral snakes (Sullivan and Wingert 1989).

Following pit viper envenomation, the affected area develops pain and swelling with hemorrhagic bullae. Localized cyanosis, lymphangitis, and lymphaditis occur with spread of the venom. Hypotension, hemolysis, coagulopathies (low PT, PTT, platelet count, fibrinogen), pulmonary edema, and with some species neurotoxicity may also occur. First aid treatment in the field should be directed at resting and reassuring the victim, immobilizing/splinting an affected extremity, and rapid transportation to a medical facility. Old first aid "remedies" such as icepacks and incision with suction, are not generally advisable and, in many cases, may be harmful. Although lymphatic tourniquets may be beneficial (especially where there is a delay in getting medical attention), use of tourniquets by an inexperienced person can often lead to more severe problems such as limb loss.

Equine-obtained Crotalidae polyvalent antivenin is the treatment of choice in an envenomated victim. It is important to ensure that envenomation has occurred prior to administration of the antivenin as it has been estimated that in approximately 20% of poisonous snake bites, no venom is injected (Russell et al. 1975). Rattlesnake envenomation is generally the most severe, and the Eastern diamondback rattlesnake (*Crotalus adamanteus*) which can exceed 7 feet in length, is the most dangerous. Copperhead venom is the least toxic of the Crotalidae, and symptoms may be mild enough not to require antivenin therapy (Russell 1983). In severe envenomizations by Crotalidae, over 20 vials of antivenin may be needed to counteract the venom (Russell 1983; Sullivan and Wingert 1989).

Coral snakes are found in the southern United States and are small brightly colored (orange-red, yellow, and black

Figure 21.19. The Crotalidae family of snakes characteristically has a pit in front of the eye near the upper angle of the jaw. This specimen is a red-diamond rattlesnake.

alternating bands) snakes. Because there are many nonpoisonous snakes which mimic the colors of the coral snake, the old rhyme "red next to yellow, kill a fellow—red next to black, friend of Jack!" can be helpful in identification. Unlike the pit vipers, the coral snake is related to the cobras and sea snakes (Elapidae) and has a rounded head with round pupils and short fixed fangs. The venom is primarily neurotoxic and generally there is little or no reaction at the bite site. Nausea, vomiting, parasthesias, headache, muscle weakness, tremors, and eventual flaccid paralysis with death (often due to respiratory and cardiac failure) may occur (Kunkel 1985; Kitchens and Van Mierop 1987). First aid measures for a coral snake bite are similar to those for Crotalidae bites. An antivenin is available and generally requires four to six vials for treatment (Kitchens and Van Mierop 1987).

The Gila Monster (*Heloderma suspectum*) and beaded lizard (*H. horridum*) are poisonous lizards found only in the Great Sonoran Desert of the southwestern United States where some large ranches occur but where there is little agricultural activity. No antivenin is available as the bites are rarely fatal.

There is no data to document the numbers of animal bites from wild or domestic animals in agricultural workers. Many bites or injuries occurring in stables, barns, and livestock areas go unreported and victims may not even seek medical attention. Treatment is primarily directed at cleansing and irrigating the wound coupled with observation for secondary infection. Tetanus prophylaxis is critical as there have been twice as many cases of tetanus from animal bites each year in the United States as cases of rabies (Strassburg et al. 1981). In major "dirty" wounds, administration of human tetanus immune globulin is also important. Rabies prophylaxis may be indicted in carnivorous animal bites in certain areas. Rabies may occasionally also be found in infected non-carnivorous animals on the farm including cattle, swine, sheep, goats, horses, and mules.

Secondary infection following an animal bite is not uncommon and a wide variety of organisms have been isolated (Goldstein and Richwald 1987). *Pasteurella multocida* infections are common following domestic dog and cat bites but may also be seen with other animal bites as this bacterium has also been isolated from many farm animals including horses, cattle, sheep, swine, and poultry (Tindall and Harrison 1972). For animal bite related infections, a broad spectrum antibiotic such as amoxicillin/clavulanate is generally recommended.

INFECTIONS AND ZOONOSES IN FARM WORKERS

Animal-Related Infectation

Most skin infections in agricultural workers are due to *Staphylococci* and *Streptococci*, the most common cutaneous pathogens in nonfarm workers and the population as a whole. However, because of exposure to arthropod vectors as well as zoonotic infections in farm and wild animals, agricultural workers are at increased risk for more unusual diseases. In addition, traumatic injuries related to farm equipment use can lead to infections which are more commonly due to Gram-negative organisms (Agger et al. 1986). Finally, the influx of migrant workers from Central America and Mexico can "import" diseases not generally seen in the United States.

As has been previously discussed, farm workers are at high risk for bites from a variety of arthropods, some of which carry infectious agents. Probably the most serious of these infections is Rocky Mountain Spotted Fever. Caused by *Rickettsia rickettsii*, this disease generally presents from 2 to 14 days following a tick bite (usually the dog tick, *Dermacentor variabilis*, or wood tick, *D. andersoni*). Persons infected are generally quite ill with a flu-like illness with high fever, malaise, and post-orbital headache. There are no cutaneous findings early in the disease, but a nonspecific rash usually develops which becomes petechial and is characteristically located peripherally involving the palms and soles (Figure 21.20). Rarely, no rash occurs. Without treatment, the disease is often fatal, and persons with glucose-6-phosphate-dehydrogenase deficiency are especially at risk for serious disease. Diagnosis can best be confirmed by specific direct immunofluorescence staining of the rickettsial organisms in affected tissue. Serologic tests

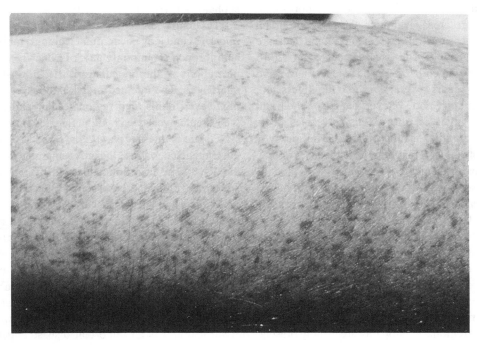

Figure 21.20. Rocky Mountain spotted fever due to *Rickettsia rickettsii* is commonly transmitted by a *Dermacentor* tick and characteristically produces a rash on the distal extremities which becomes petechial and commonly involves the palms and soles.

may also be useful but often require a convalescent titer in order to be helpful. Doxycycline, tetracycline, or chloramphenicol are useful in treatment (Spach et al. 1993; Woodward 1992).

Another rickettsial disease, ehrlichiosis, is due to *Ehrlichia chaffeensis*. The primary tick vector in transmitting disease to man is most likely the lone star tick (*Amblyomma americanum*) although other ticks can carry the organism. The incubation period ranges from 4 to 33 days (mean = 9d) (Eng et al. 1988). Infected persons develop a flu-like illness with elevations of liver enzymes and hematologic abnormalities. A rash may be seen, especially in children, but is generally nonspecific. Acute and convalescent indirect antibody titers can help in diagnosis, and tetracycline/ doxycycline is the drug of choice in adults (Jantausch 1994).

Q-fever or query fever is also due to a rickettsial organism, *Coxiella burnetii*. Reservoirs for the disease include cattle, sheep, as well as other farm and wild animals. The organism is shed in the animals' urine, feces, milk, and birth products (Angeloni 1994). Although Q-fever generally produces no rash, a nonspecific macular erythematous eruption has been reported in a small number of patients (Spelman 1982). Doxycycline or erythromycin are useful in treating the acute illness.

Endemic or murine typhus is seen in the United States primarily in the South and is caused by *Rickettsia typhi*. Although inhalation of dust from rat-infested buildings can lead to disease in man, most cases are due to direct contact of skin or mucosa with infected rat-flea feces often from around grain or feed storage areas. Although the rat is the main reservoir, mice, opossums, shrews, skunks, and cats are common hosts (Azad 1990). The disease occurs from 8 to 16 days after exposure with chills, fever, headache, nausea and vomiting. The rash occurs several days into the illness and, unlike Rocky Mountain Spotted Fever, begins on the trunk and only rarely involves the palms or soles. The erythematous rash may be macular or papular but rarely is petechial or hemorrhagic. Treatment is with doxycycline, tetracycline or chloramphenicol.

Anthrax is caused by a Gram-positive spore-forming rod (*Bacillus anthracis*) and is infrequent in this country (Kligman et al. 1991). Contact with infected animals or animal products is the usual source of infection and the disease may present with a cutaneous lesion ("malignant pustule"), pneumonia ("woolsorters disease"), or rarely with gastrointestinal disease or meningitis. When skin inoculation occurs, a pruritic papule begins which becomes vesicular, pustular, and necrotic (sometimes with satellite lesions). Fever, malaise, and regional adenopathy are usually present. Anthrax is rare in this country due to anthrax vaccination of susceptible animals, elimination of contaminated soil, and decontamination of hides (Ancona 1990). Penicillin, tetracycline, or erythromycin are generally curative.

Brucellosis is primarily caused by three Gram-negative bacteria: 1) *Brucella abortus* (from cattle), 2) *B. suis* (from pigs), and 3) *B. melitensis* (from sheep and goats). The disease is usually contracted via direct contact with secretions and excretions of infected animals. Chills, fever, headache, and malaise (i.e., a flu-like syndrome) are characteristic. In a 12-year prospective study (Ariza et al. 1989), skin lesions occurred in approximately 6% of patients and the rash usually occurred on the legs. Appearance of the rash varied from a violaceous papulonodular eruption to an erythematous papular eruption to an erythema nodosum-like eruption to rare cases of purpura. A contact dermatitis to *Brucella* has also been reported (Trunnel et al. 1985). Doxycycline and an aminoglycoside are generally used for treatment.

Erysipeloid is caused by *Erysipelothrix rhusiopathiae*, a Gram-positive pleomorphic rod. This organism causes swine erysipelas and it may be seen in pig farmers as well as abattoir workers. Poultry (especially turkeys) may also harbor the organism. The infection usually begins as a small inflamed papule on the hands within 1 to 3 days of inoculation, and then spreads into a bright-red area of cellulitis. Unlike with erysipelas, patients are usually afebrile and without regional adenopathy. The disease may rarely disseminate with endocarditis a frequent complication. Treatment is with penicillin, ampicillin, or a first generation cephalosporin (Barnett et al. 1983).

Tularemia or "rabbit fever" is due to *Francisella tulerensis* and may occur after cleaning a rabbit or following an arthropod (commonly a tick or deerfly) bite. Except for ulceration at the inoculation site, the skin is usually unaffected. The infected individual has an abrupt onset of high fever, chills, and regional adenopathy. Treatment is with streptomycin (Cerny 1994; Langley and Campbell 1995).

Plague is due to *Yersinia pestis* and is a rare disease in the United States where the only remaining endemic area is in the West. The infection is carried by prairie dogs, mice, voles, rats, rabbits, ground squirrels, and domestic cats, and infection in humans may occur by cleaning infected animals, by arthropod vector, or close contact with an infected cat (Craven et al. 1993; CDC 1994). The bubonic form is the most common and very enlarged tender lymph nodes ("buboes") occur proximal to the inoculation site. Treatment is with an aminoglycoside (gentamycin or streptomycin) (Butler 1994).

Glanders due to *Pseudomonas mallei* results from contact with infected mules, donkeys, and horses. Historically, it is an important disease and was especially important when man who intimately associated with equines as in cavalry units in wartime (Wilkinson 1981). Today it is not generally seen in the United States due to testing and eradication programs. Cutaneous involvement (known as farcy) may occur in the equine animals and also in man. Ceftazidime is the current drug of choice. Melliodosis due to *Pseudomonas pseudomallei* is acquired from soil and is almost exclusively seen in the tropics. In addition to humans, swine, sheep, and goats may be affected.

There are two spirochetal diseases which can involve skin and which are more common in outdoor/farm workers: leptospirosis and Lyme disease. *Leptospira interrogans* and other *Leptospira* species are the causative agents of leptospirosis and there are a variety of serotypes (Health and Johnson 1994). Domestic animals (cows, pigs, sheep, goats, horses, and dogs) as well as wildlife (squirrels, opossums, skunks, raccoons, foxes, rats, mice, rabbits, muskrats, nutria, deer, etc.) may harbor the disease and contaminate water or soil by shedding bacteria in urine. Direct contact through intact mucous membranes, abraded skin, or ingestion of contaminated water may lead to infection. Swimming in contaminated farm ponds or streams, or contact with infected urine during milking are activities that may lead to contact with the organism. Skin involvement is typically an erythematous macular eruption which usually occurs on the anterior lower extremities, and leptospirosis is often referred to as "pretibial fever". Jaundice may also develop and is associated with a poor prognosis. Treatment is usually with doxycycline or tetracycline (Kriesberg 1993).

Lyme disease has now been reported in most of the continental United States, but the majority of cases are seen in the Northeast and North-Central United States. The spirochete, *Borrelia burgdorferi*, can be found in a variety of wildlife (deer, mice, opossums, skunks, raccoons, foxes, squirrels, birds, lizards, etc.) and, at times, domestic animals (horses, cows, dogs, pigs, etc.). The primary reservoir is the white-tailed deer (*Odocoileus virginianus*) and the disease is rarely seen where there is not a significant deer population. The small *Ixodes scapularis* (also previously known as *I. dammini*) and *I. pacificus* ticks are the vectors in this country with the nymphal stage being the most important vector (Fish 1995).

The typical skin lesion, erythema migrans (EM) usually begins 3 days to 3 weeks after inoculation by an infected tick. A small papule begins at the site of the tick bite and a gradually expanding erythematous halo occurs with central clearing (Figure 21.21). The affected individual may have fever, malaise, and headache during the EM stage. Multiple annular secondary lesions may follow the primary lesion, and arthritis, carditis, and a meningoencephalitis may develop weeks to months later. Doxycycline or amoxicillin are most commonly used in treatment (Jantausch 1993).

There are two viral zoonoses that can affect the skin of farm workers: milker's nodules and orf. A paravaccinia virus causes milker's nodule (Figure 21.22). This disease is seen primarily in dairy farm workers especially those people who

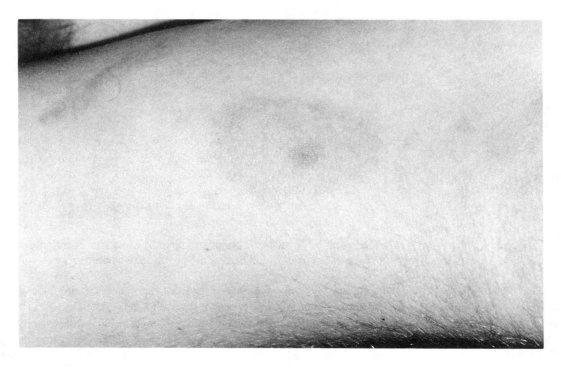

Figure 21.21. Lyme disease is commonly transmitted by an *Ixodes* tick and begins as a small erythematous papule which enlarges into an expanding erythematous halo.

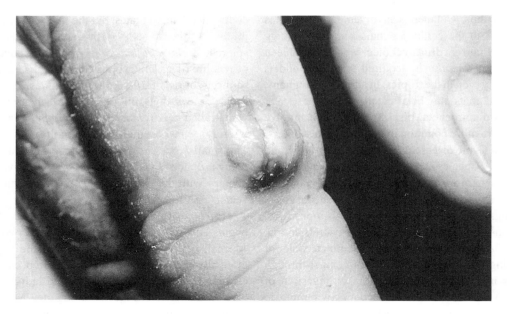

Figure 21.22. A milker's nodule is caused by a paravaccinia virus. This is a typical lesion in a dairy farmer.

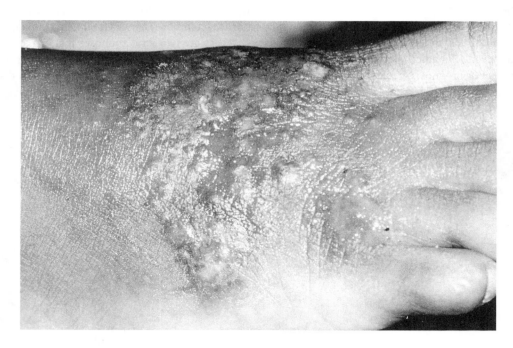

Figure 21.23. Dermatophyte infections due to geophilic or zoophilic organisms tend to be much more inflammatory and can mimic a pyoderma. A potassium hydroxide preparation is useful in distinguishing these two entities.

have not yet developed immunity (the young or newly hired workers). This is a benign and self-limited disease that is often ignored by farmers familiar with it. The virus occurs in infected udders and teats of cows (pseudocowpox), and hand milking provides for direct contact. A dime-sized vesicopustular nodule is characteristically seen on the hands or forearms, and at times there may be regional lymphadenopathy. Treatment is symptomatic and the lesion resolves over several weeks (Groves et al.).

Orf (or ecthyma contagiosum) is due to a poxvirus which infects the nostrils, lips, and buccal mucosa of sheep (especially lambs) and goats. The clinical presentation, course, and treatment are similar to milker's nodule (Zimmerman 1991).

Table 21.7. Zoophilic and geophilic dermatophyte fungi common on farms.

Organism	Source	Hair fluorescence
Microsporum canis	cats, dogs	positive
M. gallinae	poultry	negative
M. nanum	pigs, soil	negative
M. gypseum	soil	negative
Trichophyton verrucosum	cows	negative
T. equinum	horses	negative
T. mentagrophytes (var. mentagrophytes)	mice	negative

*occasionally weakly positive *in vivo*

Fungal-Related Infestation

Farm workers may also develop skin problems from a variety of fungi. Zoophilic and geophilic dermatophytes (Table 21.7) are well-known to cause inflammatory tinea infections (Figure 21.23) and at times, a kerion. Although these infections are generally easy to treat with topical or oral antifungal agents, they may be misdiagnosed as pyodermas due to the intensity of the inflammatory response (as compared to the more usual anthropophilic dermatophyte infections).

Mycetoma (or Madura foot) usually occurs following inoculation trauma and may be caused by either fungi or bacteria (McElroy et al. 1992). *Pseudoallescheria boydii* is the most common cause of fungal mycetoma in the United States (Figure 21.24), while *Madurella mycetomatis* is more common worldwide. Actinomycete bacteria (*Nocardia, Streptomyces, Actinomyces*) can also cause this problem. Following inoculation, a nodule develops over months at the primary inoculation site. Deeper invasion of tissue leads to sinus tracts, abscesses, scarring, and granuloma formation. The treatment is excision if the lesion is small. With more extensive involvement surgical debridement is helpful and intravenous miconazole is the best antifungal agent for *P. boydii* infection. With bacterial mycetoma, the prognosis using antimicrobial agents is good. With extensive eumycetoma, however, the prognosis is poor and, in refractory cases on extremities, amputation may be needed.

Deep fungal skin infections that can be seen more commonly in agricultural workers include sporotrichosis, chromomycosis, blastomycosis, and coccidioidomycosis. Sporotrichosis is due to *Sporothrix schenckii* and is primarily seen in persons such as farm workers who work with soil. Following inoculation, a nodule develops within several weeks and may spread via lymphatic channels into multiple nodules and regional lymphadenopathy. Occasionally, the disease presents as a granulomatous ulcer. Treatment is with potassium iodide, itraconazole, or ketoconazole, as well as topical applications of heat (Urabe 1986).

Chromomycosis (or chromoblastomycosis) is due to several species of dematiaceous fungi that form typical "copper penny" Medlar bodies noted in granulomas in routine histologic evaluation of lesions. The causative organisms are found in soil and on vegetation and inoculation generally occurs following a penetration injury. Typically, a granulomatous verrucous nodule grows over months. Treatment is generally with surgical excision, cryosurgery or heat. Itraconazole or fluconazole may be beneficial in more severe cases (Smith et al. 1993; Yu and Gao 1994).

Although both blastomycosis and coccidioidomycosis can form an infectious granuloma via primary inoculation of the skin, cutaneous lesions due to these organisms are usually due to disseminated disease commonly from a pulmonary source. Blastomycosis is due to *Blastomyces dermatitidis* and the organism is thought to reside in soil or rotting wood. Endemic areas include the Mid-Atlantic and Southeastern United States as well as the Mississippi valley and upper Midwest (Murphy 1989). The infection is acquired from inhaled spores presumably from contaminated dust or soil. Skin lesions may be verrucous nodules, plaques, ulcers, and may even mimic pyoderma gangrenosum. The current treatment of choice is itraconazole.

Coccidioidomycosis (San Joaquin or Valley Fever) also usually develops after inhalation of contaminated dust or soil. The organism, *Coccidioides immitis* is a soil saprophyte common in the Southwestern United States (Hedges and Miller 1990). Disseminated skin lesions may vary from papulopustules to granulomatous plaques to abscesses. Erythema nodosum, erythema multiforme, and a toxic erythema are sometimes seen with this infection. Treatment is usually with Amphotericin B or ketoconazole.

Although more of an infestation, one helminth problem that warrants mention when discussing potential zoonoses acquired on a farm is cutaneous larva migrans, or creeping eruption. Known colloquially as "catworms", the disease is most commonly caused by the larvae of the cat or dog hookworm, *Ancyclostoma braziliensis* or *A. canis*, although other hookworms including cattle hookworm have been

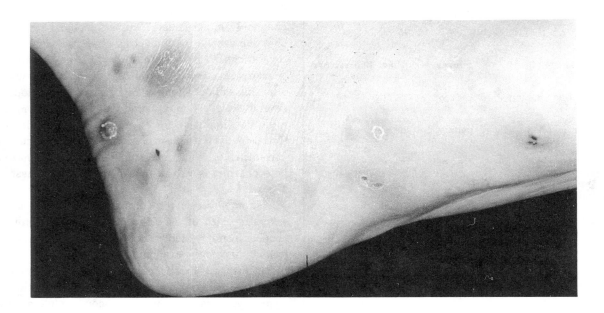

Figure 21.24. This patient stepped on a nail in a barnyard many years prior to this photograph. A biopsy confirmed the diagnosis of mycetoma and it subsequently grew *Pseudoallescheria boydii*, the most common cause of this disease in the United States. When due to a fungus, it generally does not respond well to antifungal medications and amputation may be required for a cure.

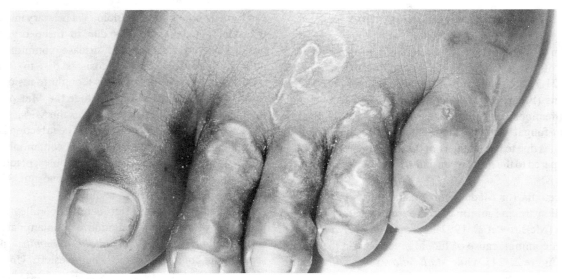

Figure 21.25. Localized creeping eruption in a migrant farm worker. The patient was treated with thiabendazole suspension applied locally.

implicated in this problem. The larvae penetrate the skin and migrate between the dermis and epidermis with each larva producing a superficial serpiginous tunnel (Figure 21.25). The feet, back, and buttock are most commonly involved and there is intense pruritus. In severe cases, there may be an associated hypersensitivity pneumonia and peripheral eosinophilia (Loeffler's syndrome) (Davies et al. 1993). Although the disease is self-limited, patients are generally treated with either oral thiabendazole or use of oral thiabendazole suspension applied topically to individual lesions (Rodilla et al. 1994). Prednisone may be helpful in controlling the inflammatory response in severe cases.

When the physician considers diseases in farm workers it must be remembered that many agroworkers are foreign immigrants and, indeed, many are in this country illegally. Not only does this present a problem in that diseased individuals will likely not seek medical attention until late in the course of their disease, but also these patients from other countries may import unusual diseases such as leishmaniasis, furuncular myiasis and even leprosy (Fleischer et al. 1990). Health care workers who see foreign migrant workers in their clinical settings need to be aware of the clinical manifestations of these as well as other unusual tropical diseases imported into this country.

HAZARD PREVENTION

In agricultural workers, one of the more severe yet easily preventable skin problem is that of skin cancer. Since a great deal of farm work occurs outdoors, exposure to excessive solar radiation can be a daily occurrence. Wearing of protective clothing is the mainstay of prevention with a wide-brimmed hat recommended. Many farmers prefer a cap and this does not adequately protect the ears or neck.

During extensive outdoor activities, regular use of a number 15 (or higher) sunscreen is advisable in skin types I to III, and the sunscreen should be applied at least 30 minutes before sun exposure. Reapplication at intervals may be necessary if perspiration is heavy or if the sunscreen is otherwise washed off. A lip balm containing a sunscreen will help to prevent actinic cheilitis and lip cancer. During periods of extreme exposure, zinc oxide or titanium dioxide can be used in high risk areas such as nose and lips. As chronic sun damage is cumulative, use of sunscreens in farm children is recommended and may start a worthwhile lifetime habit. When able, outdoor farm activities should be scheduled in the early morning and evening hours when sunlight is less intense. In addition, farm equipment such as tractors and harvesters should be fitted with cabs containing sunscreen glass, or at least a sun protective canvas canopy.

Doctors can help prevent phototoxic and photoallergic drug reactions in agroworkers by choosing medications that do not cause these reactions when practicable. While doctors should educate all farmers and agricultural workers about the need for sun protection, individuals diagnosed as having skin disorders which are exacerbated by sunlight may need additional counseling.

Prevention of intertrigo and heat rash can be difficult when a worker must work in a hot environment. Scheduling heavy activities during the cooler part of the day as well as use of absorbent powders or low potency topical corticosteroids in intertriginous areas may help. Wearing of warm protective clothing and staying indoors during times of extremely cold temperatures (especially with high winds) will obviously help prevent frostbite and other problems related to the cold.

Wearing of protective clothing including impermeable gloves and boots is important when working with farm chemicals. Washable cotton gloves may be worn inside heavy vinyl or rubber gloves for additional protection as well as absorbance of sweat. Facial shields or at least safety glasses may be necessary for eye and facial protection in certain situations where farm chemicals are utilized.

Dermatoses related to plant exposure are best avoided by remaining away from the offending plants. Protective clothing should be worn during mowing, brush clearing, "bush-hogging", "weed-eating", or "clear-burning" of vegetation. Highly sensitized individuals may be disabled from certain activities. Although topical "barrier" creams have been developed (Grevelink et al. 1992) to help prevent allergic contact dermatitis in sensitized individuals, some are non-efficacious (Smith et al. 1993), and others are unpleasant to use and compliance is oftentimes low. When clearing or harvesting plants mechanically or by hand that can potentially lead to phytophotodermatitis, protective clothing, and gloves as well as sun avoidance are advisable. Sunscreens are only marginally effective in protection in the UVA wavelength which stimulates these reactions.

Arthropod bites and stings are extremely common in agricultural occupations. *Hymenoptera*-sensitized individuals should be educated about their problem, and an epinephrine-injecting device prescribed which should be carried with them at all times. Insect repellents containing diethyltoluamide are very effective against non-Hymenoptera insects, ticks, and mites. As many zoonoses transmitted by ticks occur only after prolonged attachment by the tick, close inspection, and rapid removal of ticks can help prevent transmission of disease. Permethrin-containing clothing sprays are currently available in many sporting goods stores, but they have not yet been approved by the FDA since long-term safety studies are currently lacking.

When working where there is a possibility of a venomous snakebite, (high grass and weeds, swampy areas, rubble/dump areas, or when clearing old farm buildings), it is advisable to wear protective boots, gloves, long sleeves, and full-length pants (with leggings in very high risk areas). As many snake bite injuries are due to well-intentioned but improper first aid treatment, snake bite kits should be screened and persons at risk should be accurately trained in proper first aid techniques.

Bites from domestic and wild animals are difficult to predict and thereby difficult to prevent. Rabies vaccination of dogs and cats on the farm is extremely important, but vaccination of livestock is not considered economically feasible. Updated tetanus vaccinations and boosters are important in farm workers not only for "dirty" animal wound bite protection against tetanus, but also due to the "dirty" nature of other wounds and injuries on the farm. Testing programs for certain zoonoses as well as high-risk animal vaccination programs are important as recommended in certain locales. Workers at high risk can also be vaccinated against such illnesses as Q-fever, anthrax, and others.

Wearing of protective clothing such as safety glasses/face shields, aprons, gloves, and boots during milking activities or where there is potential for contact with animal secretions such as urine, feces, or placental products is also important. Pasteurization of milk has also significantly reduced disease transmission to humans including farm workers. Respiratory protective masks may be advisable where heavy dust or silage products are in the air.

Farm workers should not be allowed to eat, drink, or smoke in possibly contaminated areas or near livestock quarters/pens. Strict attention to handwashing here is necessary, even after wearing protective gear. Since many zoonoses are obtained from ill animals, prompt veterinary attention to animal abnormalities such as an obvious zoophilic dermatophyte infection can reduce the probability of contagion not only amongst livestock, but also transmission to farm workers.

SUMMARY

The agricultural worker has the highest risk of skin disease of any occupation and, yet, with proper preventive measures, many potentially serious skin diseases are preventable. Due to many factors presented earlier, there has been a paucity of research looking at skin problems as well as other health problems in agricultural occupations.

Education of agricultural field workers, farmers, agricultural extension agents, agricultural inspectors, and physicians who care for affected workers will likely have the greatest impact on prevention. A person sensitized to poison ivy who can not identify the plant is soon back in the physician's office. If the physician also cannot identify the plant, the patient continues to be uneducated about preventive measures. Because many agricultural workers do not speak English, educational programs will need multilingual development.

While such educational programs will cost money to implement, they are likely to be far less costly than increased governmental regulation, which will likely lead to concomitant regulatory agencies and inspection teams. While some regulation of this sort is necessary in agriculture, money spent on education about illnesses related to the skin as well as prevention of these diseases are key to reducing skin problems in the agroworker and in making the farm a safer place to work and visit.

REFERENCES

Abrams, K., D.J. Hogan, and H.I. Maibach. 1991. Pesticide-related dermatoses in agricultural workers. *Occup. Med: State of the Art Rev.* 6: 463-92.

Adams, R.M. 1990. *Occupational Skin Disease*. Philadelphia: W.B. Saunders.

Adams, C.T., J.K. Plumley, W.A. Banks, and C.S. Lofgren. 1977. Impact of the red imported fire ant, *Solenopsis invicta* Buren (Hymenoptera: Formicidae) on harvest of soybeans in North Carolina. *J. Elisha Mitchell Scient. Soc.* 93: 150-2.

Adams, C.T. 1986. Agriculture and medical impact of the imported red fire ants. In: *Fire Ants and Leaf Cutting Ants*. C.S. Lofgren, R.K. VanderMeer, eds. Boulder, CO: Westview Press, pp. 48-57.

Agger, W.A., T.H. Cogbill, H. Busch, Jr., J. Landercasper, and S.M. Callister. 1986. Wounds caused by corn-harvesting machines: an unusual source of infection due to gram-negative bacilli. *Rev. Infect. Dis.* 8: 927-31.

Ancona, A.A. 1990. Biologic causes. In: *Occupational Skin Disease*. R.M. Adams, ed. Philadelphia: W.B. Saunders, pp. 89-112.

Anderson, P.C. 1982. Necrotizing spider bites. *Am. Fam. Physician* 26: 198-203.

Angeloni, V. 1994. Rickettsial diseases. In: *Military Dermatology*. W.D. James, ed. Washington, DC: Office of the Surgeon General at TMM Publications, pp. 213-254.

Apperson, C.S., and E.E. Powell. 1983. Correlation of the red imported fire ant (Hymenoptera: Formicidae) with reduced soybean yields in North Carolina. *J. Econ. Entomol.* 76: 259-63.

Ariza, J., O. Servitje, R. Pallares, P.F. Viladrich, G. Rufi, J. Peyri, and F. Gudiol. 1989. Characteristic cutaneous lesions in patients with brucellosis. *Arch. Dermat.* 125: 380-83.

Armstrong, B.K. 1988. Epidemiology of malignant melanoma: intermittant or total accumulated exposure to the sun? *J. Derm. Surg. Oncol.* 14: 835-49.

Aylesworth, R., and D. Baldridge. 1985. Feather pillow dermatitis caused by an unusual mite, *Dermatophagoides scheremetewskyi*. *J. Am. Acad. Dermat.* 13: 680-81.

Azad, A.F. 1990. Epidemiology of murine typhus. *Ann. Rev. Entom.* 35: 553-69.

Barnett, J.H., S.A. Estes, J.A. Wirman, R.E. Morris, and J.L. Staneck. 1983. Erysipeloid. *J. Am. Acad. Dermat.* 9: 116-23.

Beasley, V.E. 1983. Cantharidin toxicosis in horses. *J. Am. Vet. Med. Assoc.* 182: 283-84.

Benezra, C., G. Ducombs, Y. Sell, and J. Foussereau. 1985. *Plant Contact Dermatitis*. Toronto: B.C. Decker, Inc.

Berger, T.G. 1986. Korean yellow moth dermatitis: report of an epidemic. *J. Assoc. Milit. Dermat.* 12: 32-5.

Berkley, S.F., A.W. Hightower, R.C. Beier, D.W. Fleming, C.D. Brokopp, G.W. Ivie, and C.V. Broome. 1986. Dermatitis in grocery workers associated with high natural concentrations of furocoumarins in celery. *Ann. Int. Med.* 105: 351-55.

Bhargava, R.K., V. Singh, and V. Soni. 1977. Erythema multiforme resulting from insecticide spray. *Arch. Dermat.* 113: 686-87.

Birmingham, D.J. 1983. *The Diagnosis of Occupationally-Induced Skin Diseases.* Kalamazoo, MI: The Upjohn Co.

Blair, A., and S.H. Zahm. 1991. Cancer among farmers. *Occup. Med.: State of the Art Rev.* 6: 335-54.

Brown, W.D. 1991. Heat and cold in farm workers. *Occup. Med.: State of the Art Rev.* 6: 371-89.

Bruynzeel, D.P. 1991. Contact sensitivity to Lannate®. *Contact Derm.* 25: 60-61.

Burke, W.A. 1991. The red imported fire ant (*Solenopsis invicta*): a problem in North Carolina. *N.C. Med. J.* 52: 153-58.

Burmeister, L.F. 1989. Cancer mortality in Iowa farmers. In: *Principles of Health and Safety in Agriculture.* J.A. Dosman and D.W. Cockcroft. Boca Raton, FL: CRC Press, pp. 268-73.

Burnett, J.W., G.J. Calton, and R.J. Morgan. 1987a. Triatoma: the "Kissing bug." *Cutis* 39: 399.

Burnett, J.W., G.J. Calton, and R.J. Morgan. 1987b. Blister beetles: "Spanish fly." *Cutis* 40: 22.

Burnett, J.W., G.J. Calton, and R.J. Morgan. 1985. Scorpions. *Cutis* 36: 393.

Butler, T. 1994. Yersinia infections. Centennial of the discovery of the plague bacillus. *Clin. Inf. Dis.* 19: 655-61.

Calnan, C.D. 1975. Petty spurge (*Euphorbia peplus*). *Contact Derm.* 1:128-30.

Calvert, J.F. 1989. Agricultural medicine. *Am. Acad. Fam. Phys. Monograph* 122: 1-2.

Campbell, D.S, R.S. Rees, and L.E. King. 1987. Wolf spider bites. *Cutis* 39: 113-14.

Center for Disease Control (CDC). 1994. Human plague-United States, 1993-1994. *JAMA* 271: 1312.

Cerny, Z. 1994. Skin manifestations of tularemia. *Int. J. Dermatol.* 33: 468-70.

Chakrabarti, A. 1990. Pighandler's itch. *Int. J. Dermat.* 29: 205-06.

Cobb, M.W. 1994. Arthropod and other animal bites. In: *Military Medicine. Part III. Disease and the Environment.* W.D. James, ed. Washington DC: Office of the Surgeon General at TMM Publications.

Cordes, D.H., and D.F. Rea. 1988. Health hazards of farming. *Am. Fam. Phys.* 38: 233-34.

Craven, R.B., G.O. Maupin, M.L. Beard, T.J. Quan, and A.M. Barnes. 1993. Reported cases of human plague infections in the United States, 1970-1991. *J. Med. Entomol.* 30: 758-61.

Crounse, R.G. 1980. Plant dermatitis due to the Compositae (Asteraceae) family. *J. Am. Acad. Dermatol.* 2: 417-24.

Davies, H.D., P. Sakuls, and J.S. Keystone. 1993. Creeping eruption. *Arch. Dermatol.* 129: 588-91.

Davis, E., C. Shuman. 1982. Cutaneous myiasis: devils in the flesh. *Hosp. Pract.* 17: 115-23.

De Eds, F., R.H. Wilson, and J.O. Thomas. 1940. Photosensitization by phenothiazine. *JAMA* 114: 2095-97.

Delzell, E., and S. Grufferman. 1985. Mortality among white and non-white farmers in North Carolina, 1976-1978. *Am. J. Epidem.* 121: 391-402.

Dinehart, S.M., M.E. Archer, J.E. Wolf, M.H. McGavran, C. Reitz, E.B. Smith. 1985. Caprito itch: dermatitis from contact with *Hylesia* moths. *J. Am. Acad. Dermatol* 13: 743-47.

Donham, K.J., and E.P. Horvath. 1988. Agricultural occupational medicine. In: *Occupational Medicine: Principles and Practical Applications.* C. Zenz, ed. Chicago: Yearbook Medical Publishers, Inc., pp. 933-957.

Eng, T.R., D.B. Fishbein, J.E. Dawson, C.R. Green, and M. Redus. 1990. Survey of human ehrlichiosis in the United States: 1988. *Ann. N.Y. Acad. Sci.* 590: 306-07.

Epstein, J.H. 1980. Polymorphous light eruption. *J. Am. Acad. Dermatol.* 3: 329-42.

Epstein, J.H. 1972. Photoallergy: a review. *Arch. Dermatol.* 106: 741-48.

Estes, S.A, B. Kummel, and L. Arlian. 1983. Experimental canine scabies in humans. *J. Am. Acad. Dermatol.* 9: 397-401.

Fish, D. 1995. Environmental risk and prevention of Lyme disease. *Am. J. Med.* 98 (Suppl 4A): 4A/2S-4A9S.

Fisher, A.A. 1983. Occupational dermatitis from pesticides: patch testing procedures. *Cutis* 31: 483-508.

Fisher, A.A. 1986. *Contact Dermatitis*. Philadelphia: Lea & Febiger.

Fishman, H.C. 1988. Rat mite dermatitis. *Cutis* 42: 414-16.

Fleischer, A.B, B.A. Maxwell, D.B. Baird, and J.T. Woosley. 1990. Hansen's disease (leprosy): the North Carolina experience. *Cutis* 45: 427-34.

Foil, L.D., and B.R. Norment. 1979. Envenomation by *Loxosceles reclusa*. *J. Med. Entomol.* 1: 18-25.

Fox, J.G., C. Reed. 1978. *Cheyletiella* infestation of cats and their owners. *Arch. Dermatol.* 114: 1233-34.

Frazier, C.A. 1969. *Insect Allergy*. St. Louis: Warren H. Green, Inc.

Frazier, C.A. 1968. *Diagnosis and Treatment of Insect Bites*. Clin. Symposia. Summit, NJ: CIBA Pharmaceutical Co.

Friedman, R.J., D.S. Rigel, M.K. Silverman, A.W. Kopf, K.A. Vossaert. 1991. Malignant melanoma in the 1990s: the continued importance of early detection and the role of the physician examination and self-examination of the skin. *CA Cancer J. Clin.* 41: 201-26.

Garcia-Perez A., B. Garcia-Bravo, and J.V. Beneit. 1984. Standard patch tests in agricultural workers. *Contact Derm.* 10:151-53.

Gelbard, M.K. 1984. Removal of small cactus spines from the skin. *JAMA* 252: 3368.

Gentile, D.A. 1989. Tick-borne diseases. In: *Management of Wilderness and Environmental Emergencies*. P.S. Auerbach and E.C. Geehr, eds. St. Louis: C.V. Mosby Co. pp.563-87.

Goldstein, E.J.C., G.A. Richwald. 1987. Human and animal bite wounds. *Am. Fam. Physician* 36: 101-09.

Green, A. 1982. Incidence and reporting of cutaneous melanoma in Queensland. *Austr J. Dermatol.* 23: 105-09.

Grevelink, S.A., D.F. Murrell, and E.A. Olsen. 1992. Effectiveness of various barrier preparations in preventing and/or ameliorating experimentally produced *Toxicodendron* dermatitis. *J. Am. Acad. Dermatol.* 27: 182-88.

Groves, R.W., E. Wilson-Jones, and D.M. McDonald. 1991. Human orf and milkers' nodule: a clinicopathologic study. *J. Am. Acad. Dermatol.* 25: 706-11.

Guin, J.D. 1980. The black spot test for recognizing poison ivy and related species. *J. Am. Acad. Dermatol.* 2: 332-33.

Hausen, B.M. 1982. Airborne contact dermatitis caused by tulip bulbs. *J. Am. Acad. Dermatol.* 7: 500-03.

Heath, S.E, R. Johnson. 1994. Leptospirosis. *J. Am. Vet. Med. Assoc.* 205: 1518-23.

Hedges, E., and Miller S. 1990. Coccidiomycosis: office diagnosis and treatment. *Am. Fam. Phys.* 41: 1499-1506.

Hjorth, N., J. Roed-Petersen, and K. Thompsen. 1976. Airborne contact dermatitis from Compositae oleoresins simulating photodermatitis. *Brit. J. Dermatol.* 95: 613-20.

Hobbs, G.D., and R.E. Harrell, Jr. 1989. Brown recluse spider bites: a common cause of necrotic arachnidism. *Am. J. Emer. Med.* 7: 309-12.

Jantausch, B.A. 1994. Lyme disease, Rocky Mountain spotted fever, ehrlichiosis: emerging and established challenges for the clinician. *Ann. Allergy* 73: 4-11.

Jones, J. 1987. Chiggers. *Am. Fam. Physician* 36: 149-52.

Jones, B.E. 1981. Human "seed tick" infestation: *Amblyomma americanum* larvae. *Arch. Dermatol.* 117: 812-14.

Kentor, P.M. 1986. Urticaria from contact with pentachlorophenate. *JAMA* 256: 3350.

King, L.E., and R.S. Rees. 1983. Dapsone treatment of a brown recluse bite. *JAMA* 250: 648.

Kitchens, C.S., and L.A.S. Van Mierop. 1987. Envenomation by the Eastern Coral Snake (*Micrurus fulvius fulvius*). *JAMA* 258: 1615-18.

Kligman, E.W, W.F. Peate, and D.H. Cordes. 1991. Occupational infections in farm workers. *Occup. Med.: State of the Art Rev.* 6: 429-46.

Kreisberg, R.A. 1993. An abundance of options. *N. Engl. J. Med.* 329: 413-16.

Krinsky, W.L. 1987. Envenomation by the sac spider *Chiracanthium mildei*. *Cutis* 40: 127-29.

Kuflik, A.S., and C.K. Janniger. 1993. Basal cell carcinoma. *Am. Fam. Physician* 48: 1273-76.

Kunkel, D. 1985. Venomous reptile bites. In: *Environmental Emergencies*. R.N. Nelson, D.A. Rund, M.D. Keller, eds. Philadelphia: W.B. Saunders. pp.199-218.

Lampe, K.F, and M.A. McCann. 1985. *AMA Handbook of Poisonous and Injurious Plants*. Chicago: American Medical Association.

Langley, R., and Campbell, R. 1995. Tularemia in North Carolina, 1965-1990. *N.C. Med. J.* 56: 314-17.

Lee, B.W. 1991.. *Cheyletiella* dermatitis: a report of fourteen cases. *Cutis* 47: 111-14.

Lisi, P., S. Caraffini, and D. Assalve. 1987. Irritation and sensitization potential of pesticides. *Contact Derm.* 17: 212-18.

Lovell, C.R. 1993. *Plants and the Skin*. Oxford: Blackwell Scientific Publications.

Mack, R.B. 1992. The bite of the spider woman: *Loxosceles reclusa* (the brown recluse). *N.C. Med. J.* 53: 200-03.

Mack, R.B. 1994. Will the defendant please rise: black widow spider poisoning. *N.C. Med. J.* 55: 86-8.

Maibach, H.I. 1986. Irritation, sensitization, photoirritation and photosensitization assays with a glyphosate herbicide. *Contact Derm.* 15: 152-56.

Mathias, C.G.T. 1989. Epidemiology of occupational skin disease in agriculture. In: *Principles of Health and Safety in Agriculture*. J.A. Dosman, D.W. Cockcroft. Boca Raton FL: CRC Press, pp. 285-87.

Matsushita, T., S. Nomura, and T. Wakatsuki. 1980. Epidemiology of contact dermatitis from pesticides in Japan. *Contact Derm.* 6: 255-59.

McCauley, R., D. Hing, M. Robson, and J. Heggers. 1983. Frostbite injuries: a rational approach based on the pathophysiology. *J. Trauma* 23: 143-47.

McElroy, J.A., C. Prestes, and W.P.D. Su. 1992. Mycetoma: infection with tumefaction, draining sinuses and "grains." *Cutis* 49: 107-10.

Miller, T.A. 1992. Latrodectism: bite of the black widow spider. *Am. Fam. Physician* 45: 181-87.

Murphy, P.A. 1989. Blastomycosis. *JAMA* 261: 3159-62.

National Institutes of Health (NIH). 1992. Diagnosis and treatment of early melanoma. NIH Consensus Statement 10(1): 1-26.

Needham, G.R. 1985. Evaluation of five popular methods for tick removal. *Pediatrics* 75: 997-1002.

Nelson, R.N. 1985. Peripheral cold injury. In: *Environmental Emergencies*. R.N. Nelson, D.A. Rund, and M.D. Keller, eds. Philadelphia: W.B. Saunders Co., pp. 25-40.

O'Donel, A.J. 1984. *Anthropods and Human Skin*. Berlin: Springer-Verlag. 422pp.

O'Malley, M., M. Thun, J. Morrison, C.G.T. Mathias, and W.E. Halperin. 1988. Surveillance of occupational skin disease using the supplementary data system. *Am. J. Indust. Med.* 13: 291-99.

Parrish, H.M. 1963. Analysis of 460 fatalities from venomous animals in the United States. *Am. J. Med. Sci.* 245: 129-41.

Parrish, H.M. 1966. Incidence of treated snakebites. *U.S. Public Health Rep.* 81: 269-76

Pessegueiro, M. 1990. Contact dermatitis due to nickel in fertilizers. *Contact Derm.* 22: 114-15.

Petri, W.A. 1988. Tick-borne diseases. *Am. Fam. Physician* 37: 95-104.

Preston, D.S, and R.S. Stern. 1992. Non-melanoma cancers of the skin. *N. Engl. J. Med.* 327: 1649-62.

Rees, R.S, and D.S. Campbell. 1989. Spider bites. In: *Management of Wilderness and Environmental Emergencies*. P.S. Auerbach and E.C. Geehr. St. Louis: C.V. Mosby Co., pp. 543-561.

Rodilla, F., J. Colomina, and J. Magraner. 1994. Current treatment recommendations for cutaneous larva migrans. *Ann. Pharmacother.* 28: 672-73.

Russell, F.E., R.W. Carlson, J. Wainschel, and A.H. Osborne. 1975. Snake venom poisoning in the United States: experiences with 550 cases. *JAMA* 233: 341-44.

Russell, F.E. 1983. *Snake Venom Poisoning*. Great Neck, NY: Scholium International, Inc.

Schmidt, R.J., and F.J. Evans. 1980. Skin irritants of the sun spurge (*Euphorbia helioscopia*). *Contact Derm.* 6: 204-10.

Schuman, S.H., and R.L. Dobson. 1985. An outbreak of contact dermatitis in farm workers. *J. Am. Acad. Dermatol.* 13: 220-23.

Sharma, V.K, S. Kaur. 1990. Contact sensitization by pesticides in farmers. *Contact Derm.* 23: 77-80.

Shaver, C.S., and T. Tong. 1991. Chemical hazards to agricultural workers. *Occup. Med.: State of the Art Rev.* 6: 391-413.

Shelley, E.D., W.B. Shelley, J.F. Pula, and S.G. McDonald. 1984. The diagnostic challenge of non-burrowing mite bites: *Cheyletiella yasguri*. *JAMA* 251: 2690-91.

Smith, C.H., J. Barker, and R. Hay. 1993. A case of chromoblastomycosis responding to treatment with itraconazole. *Brit. J. Dermatol.* 128: 436-39.

Smith, D.J., Jr, M.C. Robson, and J.P. Heggers. 1989. Frostbite and other cold-induced injuries. In: *Management of Wilderness and Environmental Emergencies*. P.S. Auerbach, E.C. Geehr. St Louis: C.V. Mosby Co., pp. 101-118.

Smith, D.M. 1985. Occupational photodermatitis from parsley. *The Practitioner* 229: 673-75.

Smith, W.B., J.M. Baunchalk, and W.J. Grabski. 1993. Lack of efficacy of barrier cream in preventing *Rhus* dermatitis. *Arch. Dermatol.* 129: 787-88.

Spach, D.H., W.C. Liles, G.L. Campbell, R.E. Quick, D.E. Anderson, Jr., and T.R. Fritsche. 1993. Tick-borne diseases in the United States. *N. Engl. J. Med.* 329: 936-47.

Spelman, D.W. 1982. Q-fever: a study of 111 consecutive cases. *Med. J. Austral.* 1: 547-53.

Steele, P. 1989. Management of frostbite. *The Physician and Sportsmedicine* 17: 135-44.

Stoner, J.G., and J.E. Rasmussen. 1983. Plant dermatitis. *J. Am. Acad. Dermatol.* 9: 1-15.

Strassburg, M.A., S. Greenland, J.A. Marron, and L.E. Mahoney. 1981. Animal bites: patterns of treatment. *Ann. Emer. Med.* 10: 193-97.

Sullivan, J.B., Jr., and W.A. Wingert. Reptile bites. 1989. In: *Management of Wilderness and Environmental Emergencies*. P.S. Auerbach and E.C. Geehr, pp.479-511.

Taylor, J.S., R.C. Wuthrich, K.M. Lloyd, and A. Poland. 1977. Chloracne from manufacture of a new herbicide. *Arch. Dermatol.* 113: 616-19.

Thurston, E.L. 1974. Morphology, fine structure and ontogeny of the stinging emergence of *Urtica dioica*. *Am. J. Bot.* 61: 809-17.

Tindall, J.P., and C.M. Harrison. 1972. *Pasteurella multocida* infections following animal injuries, especially cat bites. *Arch. Dermatol.* 105: 412-16.

Trunnel, T.N., M. Waisman, and T.L. Trunnell. 1985. Contact dermatitis caused by *Brucella*. *Cutis* 35: 379-81.

Urabe, H., and S. Honbo. 1986. Sporotrichiosis. *Int. J. Dermatol.* 25: 255-57.

Verspyck Mijnssen, G.A.W. 1969. Pathogenesis and causative agent of "tulip finger." *Brit. J. Dermatol.* 81: 737-45.

Vietmeyer, N. 1985. Science has got its hands on poison ivy, oak and sumac. *Smithsonian* 16: 88-95.

Warshauer, D.M., and J.R. Steinbaugh. 1983. Sunlight and protection of the skin. *Am. Fam. Phys.* 27: 109-15.

Watemberg, N., Y. Urkin, and A. Witztum. 1991. Phytophotodermatitis due to figs. *Cutis* 48: 151-52.

Westrom, D.R., M.P. Milligan. 1984. Rodent mite dermatitis. *J. Assoc. Milit. Dermatol.* 10: 19-20.

Wilkinson, L. 1981. Glanders: medicine and veterinary medicine in a common pursuit of a contagious disease. *Med. Hist.* 25: 363-84.

Won, J.H., S.K. Ahn, and S-C. Kim. 1993. Allergic contact dermatitis from the herbicide Alachlor®. *Contact Derm.* 28: 38-50.

Woodward, T.E. 1992. Rocky Mountain spotted fever: a present-day perspective. *Medicine* 71: 255-59.

Wu, C.M., P.E. Koehler, and J.C. Ayers. 1972. Isolation and identification of xanthotoxin (8-MOP) and bergapten (5-MOP) from celery infected with *Sclerotina sclerotium*. *Appl. Microbiol.* 23: 852-56.

Yu, R.Y., and L. Gao. 1994. Chromoblastomycosis successfully treated with fluconazole. *Int. J. Dermatol.* 33: 716-719.

Zimmerman, J.L. 1991. Orf. *JAMA* 266: 476.

22

RESPIRATORY DISEASES RELATED TO WORK IN AGRICULTURE

Susanna G. Von Essen, M.D.
University of Nebraska Medical Center

Kelley J. Donham, D.V.M.
University of Iowa

> Farmers are exposed to a number of environmental agents which can cause respiratory diseases such as organic toxic dust syndrome, acute and chronic bronchitis, occupational asthma, chronic obstructive pulmonary disease, and hypersensitivity pneumonitis. Hydrogen sulfide gas found in manure pits may be rapidly fatal. Nitrogen dioxide from the decay of silage can cause acute respiratory illness and permanent lung damage. The complex environment of animal confinement facilities can lead to respiratory illness, with overlapping patterns of asthma and bronchitis.

INTRODUCTION

Respiratory illness associated with work in agriculture has been recognized for several centuries. Ramazzini, a 17th century Italian physician, was one of the first to describe breathing disorders in those who handled grain (Ramazzini 1705). Subsequently, an eighteenth century Icelandic physician named Sveinn Palsson described "heysott", a disease of those who work with badly harvested and moldy hay in winter (Schullian 1982). Thackrah described "morning cough and expectoration" in corn millers in 1832.

In 1932, Campbell described a febrile illness associated with cough and dyspnea experienced after handling hay which he gave the name "farmer's lung". At this time it was recognized that farmer's lung is a form of hypersensitivity pneumonitis which can lead to pulmonary fibrosis. This work was furthered in the 1960s by Pepys (Pepys et al. 1962) who described the presence of serum allergic precipitins in farmer's lung, one of the main immunologic features of this disorder.

Beginning in 1985, a series of consensus conferences was held in Skokloster, Sweden (the Skokloster Conference Series, held under the auspices of the International Commission of Occupational Health, the Organic Dust Committee) which resulted in a much broader understanding of respiratory disorders in farmers and the mechanisms by which they occur (*AJIM* 10:3 1986; 17:1 1990; 25:1 1994). A new concept evolved, that of viewing agricultural exposures as producing a complex, overlapping group of clinical conditions. It was at this time that the term "organic dust toxic syndrome" (ODTS) was coined to describe the acute febrile flu-like illness seen after heavy organic dust exposure (Donham 1986d). It was recognized that this syndrome has features very similar to features with acute farmer's lung but also that it does not carry the risk of pulmonary impairment that can be seen with farmer's lung.

It was at this time that bronchitis affecting agricultural workers was defined as presenting both as an acute, self-limited illness that has cough as a main symptom as well as in a chronic form with cough and sputum production (doPico 1986; Rylander et al. 1994). It was recognized that asthma, as defined by reversible airway obstruction associated with airway inflammation, can be caused by agricultural exposures. Also, an asthma-like syndrome was defined in swine confinement workers, a group of workers that is growing in number. The asthma-like syndrome is associated with cough and wheezing, as is asthma itself. Lung function test values are generally within the normal range in healthy workers of this group although this condition is associated with decreases in baseline air flow after a shift at work in some individuals. It became apparent that there is a great deal of overlap in symptoms and pathophysiology between these entities in swine confinement workers and other groups (Figure 22.1).

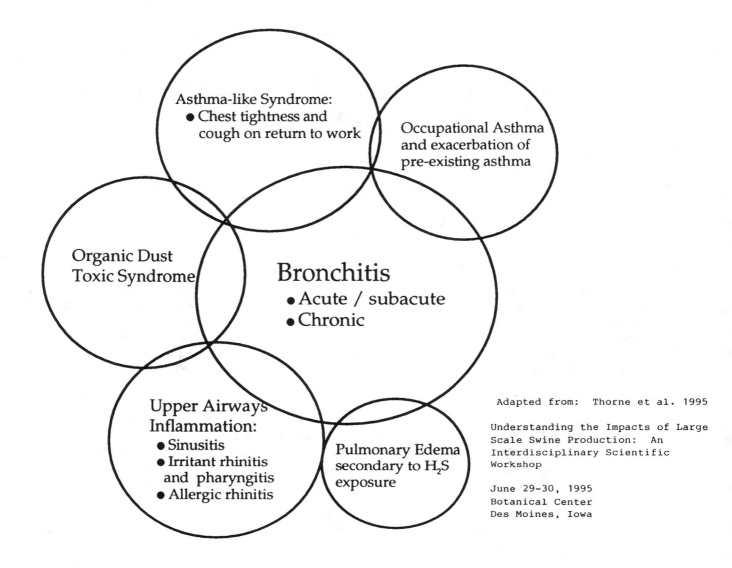

Figure 22.1. The spectrum of respiratory disease in swine confinement workers.

There is also evidence that respiratory disorders related to work in agriculture may lead to increased mortality from respiratory disease. A recent study indicated that mortality was increased in agricultural workers who filed workers' compensation claims for respiratory disease (Beaumont 1995).

Disorders of the upper and lower respiratory tract have been reported after exposure to a variety of work environments in the farm setting. The types of agricultural environments most commonly associated with respiratory complaints include grain farming and handling, working in animal confinement units, and dairy farming. Growing fruits and vegetables has also been associated with respiratory illness (Gamsky 1992). Persons affected include owner-operators, farm laborers, spouses, family members and veterinarians. It is estimated that 12 million people are exposed to the agricultural working environment in the United States (doPico 1992). Individuals who are involved in processing of agricultural products, such as grain handlers, are also at risk for work-related respiratory illness.

Agricultural respiratory diseases may result from exposures to a variety of dusts, gases, agricultural chemicals, and infectious agents. Dusts are the most ubiquitous exposures in agriculture, occurring in almost every type of agriculture enterprise. Dust exposures may be divided into those primarily consisting of organic components and those consisting mainly of inorganic components.

Field dust is the primary source of inorganic dust exposures. It may contain asbestos or silica. Lung changes consistent with silica exposures have been described in farm workers in the Central Valley of California (Shenker 1991). Fortunately, the radiographic findings have not been associated with respiratory impairment in farm workers.

Organic dust exposures to agricultural workers are the major respiratory exposure of agricultural workers. They are very common, affecting a high percentage of workers in many sectors of production agriculture. Disease caused by agricultural organic dust exposures may result from periodic, short-term exposures to dust containing large numbers of microbes. Examples of this type of exposure include uncapping a non-air tight silo and handling moldy grain or spoiled hay in a confined space. Dust levels are very high in these settings (e.g., 10 to 20 mg/m^3) and have very high microbial content (e.g., 10^6 to 10^9 microbes/m^3) and have high endotoxin content (e.g., 1,000 to 10,000 EU/m^3). Disorders caused by this type of exposure include organic dust toxic syndrome and hypersensitivity pneumonitis.

The second common exposure pattern is daily exposure to a lower level of organic dust. Typically, total dust levels are 2 to 9 mg/m^3, microbe counts are at 10^3 to 10^5 organisms/m^3 and endotoxin concentration is 50 to 900 EU/m^3. Examples of such exposures include work in a swine confinement unit, a dairy barn or a poultry growing facility. Usual symptoms seen with these exposures include those of acute and chronic bronchitis, an asthma-like syndrome and symptoms of mucous membrane irritation such as rhinitis.

Gases also play an important role in causing lung disorders in the agricultural setting. In the swine confinement buildings and in the poultry facilities ammonia levels are often in the range where chronic exposure contributes to the respiratory effects seen in workers (Figure 22.2). Exposure to the fertilizer anhydrous ammonia has both acute and long-term effects on the respiratory tract. Acute poisoning from hydrogen sulfide gas released from manure storage facilities in dairy barns and swine confinement units can cause fatalities. Inhalation of insecticidal fumigants can also lead to death.

Given the variety and extent of exposures in the agricultural setting to agents that cause lung disease, it is not surprising that many researchers believe that respiratory diseases are the major health problem of agricultural workers. One positive factor in this picture is that agricultural workers have a lower prevalence of smoking than does the general population. Approximately 15% of the farm population report that they smoke cigarettes. Still, studies in some sectors of agricultural production reveal a 25% prevalence of respiratory conditions in the workers.

The remainder of this review will report on a variety of agricultural respiratory conditions. Although this report will discuss individual conditions, the reader should understand that agricultural exposures are extremely varied. Exposures may occur to multiple agents at one time. More than one of the conditions described may occur simultaneously in the same patient and they may result from several different exposures or agents.

Figure 22.2. Swine confinement building.

ORGANIC DUST EXPOSURES

Organic Dust Toxic Syndrome

Definition. Organic dust toxic syndrome is a febrile illness that is common after heavy organic dust exposure (Von Essen 1990b; Rask-Andersen 1995; doPico 1992). This exposure is usually periodic and short-term in nature, lasting for several hours. This condition previously was called "precipitin negative farmer's lung", "pulmonary mycotoxicosis", "grain fever" and "silo-unloaders syndrome" (Edwards et al. 1974; doPico 1982a; Emanuel et al. 1975; Pratt and May 1984; Shelley 1979; doPico et al. 1977; Tse et al. 1973). The term "organic dust toxic syndrome" (ODTS) was chosen at a consensus conference and is now accepted internationally by many investigators (Donham 1986d). "Toxic alveolitis" (Malmberg 1988) has also been proposed as a term for this condition because of evidence of alveolar involvement and to differentiate it from hypersensitivity pneumonitis, which is felt to occur by different immunologic mechanisms. "Inhalation fever" has also been proposed as a unifying term for this illness (Rask-Andersen 1992).

Table 22.1. Epidemiology of organic dust toxic syndrome *vs.* acute farmer's lung.

	Organic Dust Toxic Syndrome	Acute Farmer's Lung
Incidence	1-33/100	2-30/10,000
Clustering of Cases	Yes	Uncommon
Season	Most common in summer, fall	Most common in winter, early spring
Sex	Most common in males (may be related to division of labor)	Male or female, depending on division of labor
Age	Median = 30's, 40's	Median = 30-50's
Nonsmokers vs Smokers, Former Smokers	May be more common in non-smokers	More common in non-smokers
Exposure History	Heavy exposure to organic dust, commonly from "moldy" grain, silage, "Moldy" hay or wood chips and others. May occur with first exposure	Repeated exposure to causative agent, including spoiled hay, grain, poultry and others
Causative agent	Endotoxin, possibly other agents	Antigens in the thermophilic actiomycetes Aspergillus spp. avian proteins and others

Similar symptoms are reported after exposure to a wide variety of organic materials (Weber et al. 1993; Holness et al. 1982; Rylander et al. 1976; Lundholm and Rylander 1980; Tse et al. 1980). Because it is not always possible to distinguish acute episodes of ODTS from acute episodes of hypersensitivity pneumonitis, some authors believe that these disorders represent parts of a spectrum of responses to complex organic dusts rather than completely distinct clinical entities (Weber et al. 1993; Cormier 1993a). Metal fume fever, seen after welding galvanized metal, has similar symptoms and pathophysiology (Blanc 1993a, 1993b).

Epidemiology. ODTS is a common illness seen after heavy exposure to organic dust and is much more common than farmer's lung (Table 22.1). The exposure that causes ODTS is usually short-term and episodic. OTDTS is commonly misdiagnosed as farmer's lung. The yearly incidence of ODTS in farmers was estimated to be 1/100 in a Scandinavian study (Malmberg et al. 1988). However, other researchers estimate that one-third of farmers experience a bout of ODTS in their lifetime. Many workers have multiple bouts of this disorder. ODTS symptoms are also seen in approximately 20 to 33% of those who work in hog confinement buildings (Holness 1989; Zhou 1991; Donham 1977; Larsson 1994; Merchant 1989). Determining the incidence of this disorder is complicated by the fact that the symptoms and exposure history resemble those of acute farmer's lung (Table 22.2). Also, laboratory testing needed to differentiate the two disorders from one another is not always performed. Many farmers do not seek medical care for this problem (Rask-Andersen 1989a).

Etiologic Agents. ODTS is seen after a variety of exposures to organic materials on the farm, such as straw, silage, grain dust, and wood chips (Rask-Andersen 1989b). It is more likely to occur if the material handled appears to be "moldy" and if the number of airborne microorganisms is high (Malmberg et al. 1985).

Exposure to endotoxin can cause symptoms similar to those of ODTS (Sandström 1994). Endotoxin levels can be high in the settings where ODTS is seen, with levels ranging from 100 to 5,000 EU/m^3 (Rask-Andersen 1989; Olenchock 1990; Thedell 1980). Endotoxin challenge has been shown to induce symptoms similar to those of ODTS (Rylander 1989). Three consensus conferences held on organic dusts and endotoxins have lead general agreement that endotoxin is an important agent in the pathogenesis of ODTS (*AJIM* 10:3 1986; 17:1 1990; 25:1 1994).

However, endotoxin levels measured in the environment do not always correlate well with the risk for ODTS (Rask-Andersen 1989a). Mycotoxins have also been considered in the effort to identify agents that can cause ODTS. Mycotoxins have been measured in dust collected at the location where farm worker became ill with ODTS and these have not been detectable or have been present in low amounts (May 1986b).

Table 22.2. Clinical features of organic dust syndrome vs. acute farmer's lung.

	Organic Dust Toxic Syndrome	Acute Farmer's Lung
Latency	4-12 Hours	4-8 Hours
Symptoms	Dry cough, chills, fever, malaise, dyspnea, myalgia, chest tightness, headache	Fever, chills, malaise, dry cough, dyspnea
Duration of Acute Symptoms	Usually less than 24 hours, may last for several days	12-36 hours
Physical Examination of the Chest	Normal or scattered rales	End inspiratory bibasilar rales
White Blood Count	Leukocytosis to 25,000 WBC/cμmm, (neutrophilia)	Leukocytosis to 25,000 WBC/mm (neutrophilia), also mild to moderate eosinophilia
Arterial Blood Gases	Normal or mild hypoxemia	Hypoxemia, which may be severe, A-a gradient
Serum Allergic Precipitins to Farmer's Lung	Usually negative	Usually positive
Chest X-ray	Normal or minimal interstitial infiltration	Patchy, ill defined, parenchymal densities, Interstitial strands
Pulmonary Function Tests	Normal or mild restriction and/or decreased DLCO	Moderate to severe restriction, decreased DLCO. May also see evidence of obstruction
Bronchoalveolar Lavage Findings	Elevated neutrophils	Elevated neutrophils, lymphocytes,
Open Lung Biopsy Results	Multifocal acute inflammation of terminal bronchioles, alveolar, interstitial area. Exudate consists of neutrophils, macrophages. Large numbers of fungal spores present	Acute granulomatous interstitial pneumonitis, with macrophages, foreign body giant cells, neutrophils, eosinophils

Clinical Features. Unlike farmer's lung, prior sensitization is not required in order to develop ODTS (Brinton 1987). Some workers have multiple episodes of ODTS. As outlined in Table 22.2, symptoms include fever, chills, muscle aches, shortness of breath, cough, headache, nausea, and chest tightness (Rask-Andersen 1989). These symptoms develop 4 to 8 hours after the causative exposure and in the majority of patients they last from less than 24 hours to 2 to 3 days (Rask-Andersen 1989). Often several persons working together in the same environment become ill (Rask-Andersen 1989).

A leukocytosis, consisting of increased peripheral blood neutrophils, is seen. In contrast to farmer's lung, chest X-rays are normal or are reported as showing minimal interstitial infiltrates, hypoxia is usually not seen and the majority of patients do not have abnormal pulmonary function tests (Emanuel et al. 1975; Marx et al. 1990; Malmberg 1985; Lecours et al. 1986). Pathological findings in the lung consist of an acute interstitial pneumonitis and bronchitis with the presence of many fungal elements (Emanuel et al. 1975).

Bronchoalveolar lavage is remarkable for the presence of increased numbers of neutrophils (Emanuel 1986, 1989; Lecours et al. 1986).

Serum precipitating antibodies to antigens that cause farmer's lung may be positive or negative (Rask-Andersen 1989b). The frequency of positive skin prick tests is no greater in those who have had ODTS than in a control population and mean total IgE levels are not greater (Rask-Andersen 1989b). Pulmonary function tests in those who have had ODTS are in the normal range (Malmberg 1985; May et al. 1990).

Findings seen in ODTS have been reproduced in the laboratory setting in animals and humans using both particulate matter and soluble extracts of silage or grain dust (Marx et al. 1981; Von Essen et al. 1988; Von Essen 1995). Soluble extracts of grain dusts have been shown to attract neutrophils in guinea pigs and to cause the release of neutrophil chemotactic activity from alveolar macrophages and bronchial epithelial cells in vitro (Von Essen et al. 1988; Von Essen 1995). Release of neutrophil chemotactic activity does not correlate with endotoxin levels in the dust extracts (Von Essen 1995; Von Essen et al. 1988). There is evidence that complement activation may play a role in causing the inflammation seen in ODTS (Edwards et al.1974; Von Essen et al. 1988; Olenchock et al. 1980).

Treatment and Prognosis. Treatment of ODTS consists of bed rest and antipyretics (Rask-Andersen 1989). Increased sensitivity to organic dust is sometimes reported after an episode of ODTS and many farmers reported symptoms of airway disease after an attack of ODTS (Rask-Andersen 1989; Marx et al. 1990). Farmer's lung can occur in persons who have had ODTS (Marx et al. 1990).

Prevention of ODTS includes minimizing risk by changing farming practices when possible in order to improve ventilation and reduce organic dust exposure by other means (Watson 1986; Donham 1995). It is also recommended that respirators approved by NIOSH and the Mine Safety and Health Administration be worn when exposure to organic dust cannot be avoided (NIOSH 1994).

Hypersensitivity Pneumonitis

Hypersensitivity pneumonitis is a generic term used to describe an acute, subacute, or chronic pulmonary condition with delayed febrile systemic symptoms, manifested by an influx of inflammatory cells to the lung parenchyma and the formation of granulomas there (Curtis and Schuyler 1994). Many different exposures have been reported that initiate this disorder, eliciting colorful names for specific conditions such as maple bark strippers disease and bird fancier's disease.

Hypersensitivity pneumonitis or extrinsic allergic alveolitis seen in farmers is often called farmer's lung (Chmelik et al. 1974; Sharma 1991; Hammar 1988; Fink 1991; doPico 1992). Agricultural workers are more likely to experience hypersensitivity pneumonitis than are other American workers (U.S. Health and Human Services 1994). The modern day description of this disorder was written by Campbell, who studied farmers working with moldy hay in the United Kingdom (Campbell 1932). It was not until 1963 that Pepys and colleagues made the association between symptoms experienced by the patients with farmer's lung, exposure to thermophilic actinomycetes and the presence of precipitating antibodies against moldy hay antigens (Pepys and Jenkins 1963).

Epidemiology. The incidence of farmer's lung has been estimated to be 2 to 30 per 10,000 farmers, (Malmberg et al. 1988; Grant et al. 1972; Terho 1986). Thus, it is thought to be much less common than ODTS (Table 22.1). Farmer's lung is most common in areas with relatively high rainfall and where dairy farming is practiced. For example, Finland appears to have a much higher rate of farmer's lung than other countries. The largest number of cases occurs during the indoor feeding period of cattle and on farms with least mechanization (Terho 1986, 1980; Grant et al. 1972; Rautalahti et al. 1987). Farmer's lung has also been reported in poultry farmers and those who raise mushrooms (Sastre 1990; Warren 1974; Boyer et al. 1974). The cumulative prevalence of farmer's lung has been estimated at 0.1 to 4.3% (Belin and Malberg 1986; Madsen et al. 1976; Grant et al. 1972). It has been reported in children as well as in adults (Kristiansen and Lahoz 1991; O'Connell et al. 1989; Bureau et al. 1979). It is suspected that this diagnosis is often overlooked (Madsen et al. 1976).

Workers vary in their susceptibility to farmer's lung. This phenomenon is not yet well understood. There is evidence that genetic factors may be important for susceptibility to farmer's lung although no connection has been found between HLA antigens and this disorder (Terho 1985, Terho 1984; Flaherty 1980). There is evidence from animal studies that viral infection enhances the lung response to inhalation of *Saccharopolyspora rectivirgula*, suggesting that viral infection may play an initiating role in farmer's lung (Cormier 1994a, 1994b). Farmer's lung is less likely to occur in cigarette smokers (Sharma 1991; Warren 1977).

Etiologic Agents. Etiologic agents of farmer's lung include the thermophilic actinomycetes with *Saccharopolyspora rectivirgula* (formerly *Micropolyspora faeni*) being the most common species. *Thermoactinomyces candidus*, *Thermoactinomyces vulgaris* and others from this genus, as well as *Aspergillus fumigatus* and *Aspergillus umbrosus* are important causes of farmer's lung (Kotimaa et al. 1991; Wenzel et al. 1974; Hollingdale 1974; Husman 1987a). Other mold antigens, including Pullaria, Penicillium, Botrytis and Cladeosporium can also cause farmer's lung (Belin and Malberg 1986). These organisms are found on moldy grain, hay and other vegetable matter. Farmers who handle moldy wood chips may develop the disorder from exposure to *Aspergillus fumigatus Paecillomyces*, and *Rhizopus* (Kolmodin-Hedman 1986). Less commonly, other fungi are associated with hypersensitivity pneumonitis in the farm setting, as are avian proteins (Cockcroft et al. 1983; Warren 1972; Boyer and Mosier 1974). Prior sensitization to the causative agent is required. Once sensitization has occurred, limited exposure to the causative agent may be sufficient to trigger an attack of farmer's lung (Williams 1963).

Immunopathogenesis of Farmer's Lung. The immunopathogenesis of farmer's lung is complex.

Understanding of the pathogenesis of farmer's lung has evolved over the years into two related lines of study. The line of research initiated by Pepys explored the immune process whereby specific inhaled antigens complex with specific IgG antibodies in the lung parenchyma (Pepys et al. 1962; Chmelik et al. 1974). Sensitization to the antigens that cause farmer's lung can be determined by using serologic tests to measure circulating serum precipitins. Identifying the presence of this sensitization can be helpful for diagnosing farmer's lung. Serum precipitins are an indicator of exposure to some of the common antigens incriminated in farmer's lung. They are usually detectable in the serum of patients with farmer's lung, although not always (Reynolds et al. 1977). Reasons for not detecting serum precipitins in a patient with the clinical features of farmer's lung include that there are many antigens that may cause farmer's lung in addition to those present in a battery of tests.

Precipitins can be measured using an enzyme-linked immunosorbent assay (ELISA) or by double diffusion or immunoelectrophoresis (Marx 1982; Dalphin 1994). The immunoelectrophoresis and ELISA assays appear to be a more effective method for diagnosis of farmer's lung (Marx 1982; Dalphin et al. 1994).

Precipitins are present during the first year in nearly all cases and fall to approximately 50% by the end of the second year if exposure ends (Chmelik et al. 1974; Kobayashi et al. 1963; Cormier and Belanger 1989; Rankin et al. 1967; Barbee et al. 1968). They may be absent in those with chronic farmer's lung (Barbee et al. 1968). IgG and IgA antibodies to farmer's lung antigens most commonly are present in persons with this disorder (Ojanen et al. 1990; Ojanen 1992). IgM antibodies to farmer's lung antigens may be present in serum as well. IgE receptors can also be present on the surface of alveolar macrophages of patients with farmer's lung, suggesting that IgE may play a role in the airway obstruction observed in some patients with this problem (Pforte et al. 1992). Pure antigens have only recently become available for testing for farmer's lung (Kumar et al. 1993). Avidity of the IgG antibodies increases with re-exposure and recurrence (Kaukonen et al. 1994).

Serum precipitins are positive in 10 to 83% of the farming population, including persons who do not have a history of farmer's lung (Cormier and Belanger 1989; Gangwar et al. 1991; Dalphin et al. 1994; Stanford et al. 1990). Thus, serum allergic precipitins serve only to support the diagnosis of farmer's lung in the presence of a compatible clinical picture and also as markers of exposure in asymptomatic farmers. Having serum precipitins to antigens that cause farmer's lung without having symptoms of this disorder does not correlate with increased risk for respiratory disease.

Formation of serum precipitating antibodies is associated with both immune complex (Type III) as well as cellular hypersensitivity (Type IV) reactions in the lung parenchyma (Pepys et al. 1962; Sharma 1991; Reynolds 1982a, 1982b). The Type III reaction can be initiated by antigen-antibody complexes which then activate complement and directly stimulate macrophages. Immunofluorescent studies in patients with farmer's lung have demonstrated the presence of IgG, IgA and IgM immune complexes and C3 complement in the lung (Wenzel et al. 1971; Ghose et al. 1974; Dall'Aglio et al. 1988; Wick 1992). Analysis of bronchoalveolar lavage fluid reveals both IgG and IgA immune complexes (Reynolds 1988; Solal-Céligny et al. 1982). Circulating immune complexes are found in the blood as well (Solal-Céligny et al. 1982). Although serum complement levels remain normal in humans in acute hypersensitivity pneumonitis, C1q and C3 is present in bronchoalveolar lavage fluid (Moore et al. 1974; Pesci et al. 1990).

Animal studies have been useful in understanding hypersensitivity pneumonitis and have demonstrated that immune complexes contribute to neutrophilic inflammation in experimental disease (Yoshizawa 1991). Animal studies also suggest that complement activation plays a role in farmer's lung (Olenchock and Burrell 1976).

Delayed hypersensitivity (Type IV) also contributes to the pathologic changes seen in farmer's lung. The importance of delayed hypersensitivity in farmer's lung is known from studies showing the presence of circulating sensitized lymphocytes that undergo blast transformation and produce migration inhibition factor from antigen exposure (Marx et al. 1973; Caldwell et al. 1973).

The initial response, seen within 24 hours of exposure, is an increase in neutrophils recoverable from the lung by bronchoalveolar lavage (Fournier 1985). This response cannot be distinguished from the neutrophil influx seen in organic dust toxic syndrome. The neutrophils recovered in early farmer's lung are activated (Vogelmeier 1993). A rapid increase in T-lymphocytes follows.

Studies of the T helper/t suppressor (CD4/CD8) ratio in farmer's lung are contradictory. An inverted T helper/T suppressor (CD4/CD8) ratio has been identified in farmer's lung by a number of investigators (Costabel et al. 1985; Orefice et al. 1991; Semenzato 1991; Trentin et al. 1988; Leatherman 1984; Cormier 1986). Others have found the T helper/T suppressor ratio to be normal in farmer's lung or that there is little change after recovery from the acute illness (Sibille et al. 1988; Cormier et al. 1987). The CD4/CD8 ratio has also been described as being increased in farmer's lung (Ando 1991). Patients who stop being exposed to the causative antigens can have normalization of their CD4/CD8 ratio 6 months after diagnosis, suggesting that dose and duration of exposure influence immunologic mechanisms of patients with farmer's lung (Trentin 1988).

Enhanced expression of Ia antigen, IL-2R markers, VLA-1, HLA-DR Ag and CD2 antigen has been noted on lung T cells recovered from animals and human subjects with farmer's lung, indicating that these T cells are activated (Yamaguchi 1991; Costabel et al. 1985; Trentin et al. 1990). While these cells proliferate in vitro in the presence of IL-2, they do not release this cytokine (Trentin et al. 1990). The pattern of T cell growth is usually polyclonal.

Increased numbers of cytotoxic cells have been identified in BAL fluid of patients with farmer's lung (Semenzato et al. 1988). These include LAK cells, lymphokine activated killer cells and non-HLA locus restricted T-cytotoxic cells. NK cells are increased in farmer's lung (Trentin et al. 1988; Denis 1993b)

Recently, research concerning farmer's lung has also focussed on the aspects of this disorder that appear to be direct toxic effects of the organic dust on the lung (Schuyler 1995). These aspects include release of cytokines, the presence of increased numbers of inflammatory cells as well as the presence of markers of fibrosis.

It has become apparent that cytokines play an important role in this disorder. The role of cytokines in farmer's lung has been elucidated in a mouse model. Release of large amounts of TNF-alpha and interleukin-1 from lung macrophages has been verified (Denis 1991a). Administration of anti-TNF-alpha blocks the influx of inflammatory cells, the rise of TNF-alpha and interleukin-1 in lung lavage fluid and the development of lung fibrosis. Interleukin-1, TNF-alpha, macrophage inflammatory protein-1 alpha and interleukin-8 release from alveolar macrophages are increased in human subjects with farmer's lung (Denis 1993b, 1995, 1991a). The release of interleukin-8, which is a potent chemoattractant for neutrophils, may account for the presence of increased numbers of these cells early in acute farmer's lung. Interleukin-4, which can down regulate macrophages, decreases the inflammation in a mouse model of farmer's lung (Ghadriian 1992).

Study of cytokines may also aid in understanding of the spontaneous regression of lung inflammation in farmer's lung (Denis 1993a). In a mouse model, macrophage and lymphocyte activity did not change but NK cell activity was higher after longer periods of challenge. At this time, the cytokine profile changed from the presence of high levels of tumor necrosis factor-alpha and granulocyte macrophage colony-stimulating factor to large amounts of interferon-gamma and interleukin-2 (Denis 1993a).

Mast cells may also be increased in number in BAL fluid of patients with farmer's lung and in lung tissue (Miadonna et al. 1994; Laviolette et al. 1991; Pesci 1991a, 1991b; Bjermer et al. 1988; Takizawa 1989). Histamine levels are elevated in the supernatants, indicating that the mast cells are activated (Miadonna et al. 1994). The role of the mast cell in causing farmer's lung symptoms has not yet been completely defined.

Abnormalities in other cell groups include an elevation in the number of eosinophils recovered on BAL (Calhoun 1991; Reynolds et al. 1977). The macrophage also plays a role in this disorder (Fink 1991). Production of reactive oxygen species by macrophages from patients with farmer's lung is increased, indicating that this cell type contributes to the airspace inflammation noted (Calhoun 1991). In animal studies, lymphocytes secrete alveolar macrophage migration inhibition factor in farmer's lung, indicating that there is a lymphocyte-macrophage interaction (Kawai et al. 1973). Macrophages from patients with farmer's lung express HLA DR, supporting the concept that these macrophages may have enhanced capacity to present antigen to T lymphocytes, leading to local proliferation of lymphocytes (Haslam 1987).

Markers of pulmonary fibrosis are elevated in bronchoalveolar lavage fluid of patients with farmer's lung (Lalancette et al. 1993; Bjermer et al. 1987; Cormier 1993b; Teschler et al. 1993; Larsson et al. 1992). These include hyaluronic acid, Type II procollagen, fibronectin, vitronectin and fibroblast growth factors. Albumin levels have also been shown to be elevated in lavage fluids after challenge, suggesting an increase in lung permeability.

Clinical Features. The symptoms of farmer's lung may be acute, subacute or chronic at presentation. Determinants of the clinical course include the frequency and intensity of the inhalation exposure and possibly host factors (Richerson 1989; Chmelik et al. 1974). Symptoms of acute farmer's lung resemble those of organic dust toxic syndrome (Table 22.2).

Persons with acute farmer's lung commonly complain of fever, chills, malaise, dyspnea, chest tightness, cough, and anorexia which begin 4 to 6 hours after inhalation of the offending substance (Hammar 1988; Pepys and Jenkins 1965; Dickie 1958; Hapke 1968; Richerson et al. 1989; Chmelik et al. 1974). Some patients with farmer's lung complain mainly of respiratory symptoms and others state that respiratory and constitutional symptoms are equal in severity. Symptoms usually last for a day to several days (Sharma 1991; Richerson et al. 1989).

Subacute farmer's lung may appear gradually over several days to weeks. Fatigue and weight loss are often prominent symptoms in this form of the disorder, in addition to cough and dyspnea. Subacute farmer's lung can progress to severe respiratory impairment (Richerson et al. 1989, Barbee et al. 1968).

In all forms of farmer's lung, rales are often heard on auscultation of the chest (Chmelik et al. 1974). There may also be cyanosis and signs of respiratory distress, but clubbing is not commonly seen (Richerson et al. 1989; Hapke 1968).

Acute and subacute farmer's lung symptoms are associated with a decreased forced vital capacity (FVC) on spirometry, which indicates that restriction is present (Hapke et al. 1968; Barbee et al. 1968). Compliance can be decreased (Hapke et al. 1968). A low diffusing capacity for carbon monoxide can also be seen, and the A-a gradient may be increased (Fink 1992; Sharma 1991). There may also be evidence of airways obstruction on spirometry, demonstrated by a decreased FEV_1/FVC and/or FEF_{25-75} (Sharma 1991; Barbee et al. 1968). Airway resistance may increase (Müller-Wening 1989). Many of these patients have bronchial hyperreactivity evident on methacholine challenge (Freedman 1981; Mönkäre et al. 1981). Evidence of hypoxemia can be present, demonstrated by a low paO_2 on arterial blood gases as well as a low oxygen saturation. Exercise stress testing may reveal signs of functional pulmonary impairment (Chmelik et al. 1974).

Leukocytosis with a predominance of neutrophils and a shift to immature cells is common in acute farmer's lung (Fournier et al. 1985). A mild eosinophilia may be seen. IgG, IgM, and IgA levels may be elevated in serum (Fink 1992).

It is important to remember that symptoms of acute farmer's lung can resemble those of ODTS (Table 22.2). Diagnostic criteria have been proposed to distinguish farmer's lung from ODTS and other disorders (Table 22.3).

Chronic farmer's lung may develop insidiously or may follow a series of well-documented acute attacks of farmer's lung (Richerson 1989; Chmelik 1974; Seal 1968). Prominent symptoms include dyspnea and fatigue. It may be difficult to distinguish chronic farmer's lung from idiopathic pulmonary fibrosis and other interstitial lung disorders (Chmelik et al. 1974).

Radiographic Findings. Chest X-ray findings may include a reticulonodular pattern, small round opacities or irregular opacities as seen in Figure 22.2 (Gurney 1992; Barbee 1968; Hapke 1968). The chest X-ray can also be normal (Gurney 1992; Epler 1978; Hodgson 1989; Arshad 1987). Computed tomography (CT) and high-resolution CT may be superior to conventional X-ray for identifying abnormalities in farmer's lung, although they are much more expensive (Figure 22.3). Findings reported have included widespread airspace consolidation, small rounded opacities, patchy air-space consolidation and a reticular pattern (Gurney 1992; Silver 1989; Lynch 1992, Bessis 1992). CT scans reveal evidence of emphysema in some patients with a history of farmer's lung (Lalancette 1993).

Table 22.3. The Terho criteria for the diagnosis of farmer's lung (Terho 1986a).

Main Criteria

- Exposure to offending antigens (history, measurements of microorganisms, serum allergic precipitins)
- Symptoms compatible with hypersensitivity pneumonitis
- Chest radiographic changes compatible with hypersensitivity pneumonitis

Additional Criteria

- Basal crepitating rales
- Impairment of pulmonary diffusion capacity
- pO_2 decrease at rest or during exercise
- Restrictive lung function impairment
- Lung biopsy changes compatible with hypersensitivity pneumonitis
- A positive provocation test (work exposure, inhalation challenge)

All main criteria and at least two additional criteria should be fulfilled. Other diseases with similar symptoms and clinical findings should be excluded. If the chest X-ray is normal, but if the criteria were otherwise fulfilled, the diagnosis is considered confirmed if a lung biopsy shows changes compatible with hypersensitivity pneumonitis.

Histopathology. Transbronchial lung biopsy or open lung biopsy is helpful for establishing the diagnosis of farmer's lung (Chmelik et al. 1974). Open lung biopsy, while more invasive, offers a better sample of lung tissue. Histologic findings in patients with acute and subacute farmer's lung include the presence of a pneumonitis with noncaseating granulomas containing foreign body giant cells and foamy macrophages or histiocytes. Clumps of epithelioid cells and lymphocytes may also be present. Birefringent particles can be seen within macrophages (Seal 1968). The findings can resemble those of sarcoidosis (Seal 1968). Neutrophils may be seen (Emanuel 1964; Dickie and Rankin 1958; Seal 1968). The airways may also show signs of involvement, with changes including loss of epithelial cells and the presence of inflammatory infiltrate in the terminal bronchioles or bronchiolitis obliterans (Heino et al. 1982, Chmelik et al. 1974; Seal et al. 1968). Pulmonary vasculitis is not seen in farmer's lung.

In chronic farmer's lung, nonspecific diffuse interstitial and focal peribronchiolar fibrosis is found (Seal et al. 1968; Chmelik et al. 1974). A lymphocytic infiltrate may be present. Emphysematous changes have been described (Seal et al. 1968). Vascular changes consistent with the presence of

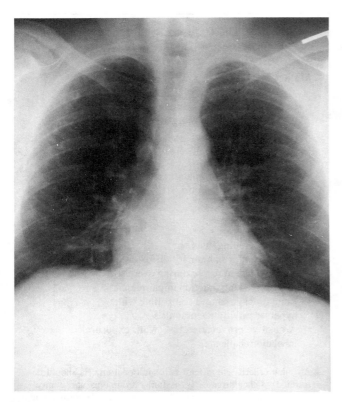

Figure 22.3. A. Chest X-ray of a patient with acute Farmer's Lung.

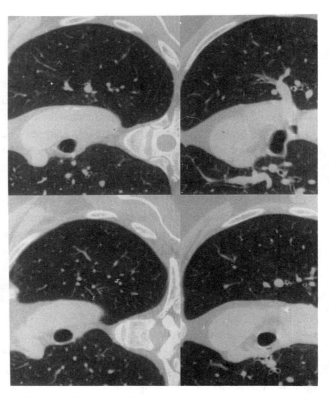

B. Thin cut CT scan of the thorax of a patient with acute Farmer's Lung, showing multiple small nodules.

pulmonary hypertension have been noted (Seal et al. 1968). Because these findings are not specific, this diagnosis depends on clinical history.

Findings from Bronchoscopy with Bronchoalveolar Lavage. Bronchoscopy with bronchoalveolar lavage (BAL) is a newer technique that can be used to analyze cells present in the lower respiratory tract and measure a variety of markers of inflammation in farmer's lung, thereby furthering understanding of the pathophysiology of this disorder. BAL is often done at the same time as transbronchial biopsy. The results of BAL often help establish the diagnosis of farmer's lung. BAL findings reflect the presence of inflammation seen on histologic sections, with the exception of underestimate the number of macrophages (Semenzato et al. 1985). BAL in acute farmer's lung is notable for the presence of increased numbers of neutrophils (Solal-Céligny 1982; Bernardo 1979; Costabel 1988; Fournier et al. 1985; Haslam 1987).

After several days, lymphocytes predominate in the lower respiratory tract while peripheral blood lymphocyte counts remain normal (Solal-Céligny et al. 1982; Bernardo et al. 1979; Costabel 1988; Reynolds 1977). The majority of these lymphocytes are T cells (Godard et al. 1981; Bernardo et al. 1979; Trentin et al. 1988; Semenzato 1991, 1988). The presence of activated lymphocytes indicates that pulmonary inflammation is present but it does not correlate well with clinical parameters or prognosis in farmer's lung (Leblanc et al. 1986). Lymphocytes in the lung may remain elevated for months to years (Leblanc et al. 1986; Cormier et al. 1987). The intensity of the alveolitis does not correlate with functional alterations.

It is of note that lymphocytes are elevated in BAL fluid in asymptomatic farmers who do not have a history of farmer's lung (Cormier et al. 1984; Solal-Céligny et al. 1982; Von Essen 1990a). These lymphocytes have a normal T-helper to T-suppressor ratio and normal lymphocytotoxicity (Cormier et al. 1984; Semenzato 1988). The presence of lectins and lymphocyte mitogens has been noted in grain dusts (Olenchock et al. 1986) This could contribute to the presence of increased numbers of lymphocytes in the lower respiratory tract.

Treatment and Prognosis. Preventing further episodes of farmer's lung is an important part of treatment for this problem. Primary prevention is also desirable, although farmers are not as likely to use respiratory protection if they do not have respiratory complaints (Schlenker and Parry 1990). Causative exposures can sometimes be reduced by changing work practices (Watson 1986). More efficient drying of hay and grain and use of mechanical feeding systems can decrease exposure to causative agents (de Haller 1986; Grant et al. 1972). Exposure can also be reduced through the use of appropriate particle filtering half-mask respirators, which has been shown to prevent or attenuate the acute reaction to challenge (Müller-Wening and Repp 1989; Hendrick et al. 1981; Kusaka 1993). Powered air purifying respirators (e.g., the Airstream helmet) have a higher assigned protection factor for protection against particulate exposures than do half-mask respirators (Lenhart and Reed 1989). Powered air purifying respirators have also been shown to attenuate but do not prevent the response (Müller-Wening and Schmidt 1990). Respirator use to prevent farmer's lung is influenced by a prior history of the disorder as well as participation in an occupational health intervention program (Virolainen et al. 1987).

Medical therapy of acute and subacute farmer's lung consists of oral corticosteroids (Kokkarinen et al. 1992; Cormier and Belanger 1989; Mönkäre and Haatela 1987). Use of oral corticosteroids shortens the episodes of illness and hastens the improvement in the DLCO, but does not decrease the chance of recurrence (Kokkarinen 1992, ARRD). Pulmonary function testing has shown that the paO_2 reaches maximum improvement within one month of the initial exam while the FVC improves for up to a year and the DLCO may improve for 2 years (Kokkarinen and Tykianen 1993).

Treatment of farmer's lung with corticosteroids or contact avoidance has been shown to be associated with a decrease in tumor necrosis factor-alpha, interleukin-1, macrophage inflammatory protein-1 alpha and interleukin-8 release from alveolar macrophages (Denis 1993b 1995). A decrease in NK cell activity was demonstrated as well. Corticosteroid therapy does not alter the long-term prognosis (Mönkäre and Haahtela 1987).

The prognosis in farmer's lung is variable. Rarely, acute episodes can lead to death (Barrowcliff and Arblaster 1968). The high intensity lymphocytic alveolitis is likely to persist and does not predict the outcome of the disease (Cormier 1987). Measuring markers or substrates of fibrosis in bronchoalveolar lavage fluid has not been a useful tool for predicting outcome (Lalancette et al. 1993). Patients may recover without impairment. Or, they may be left with evidence of obstruction or restriction on spirometry and with a reduced diffusing capacity for carbon monoxide (Lalancette et al. 1993; Kokkarinen et al. 1992; Kokkarinen and Tykianen 1993; Barbee et al. 1968). Those with five or more symptomatic recurrences are most likely to have restriction and a low DLCO (Braun et al. 1979). Farmer's lung frequently leads to disability (Hapke et al. 1968).

Patients may develop persistent airway obstruction after an episode of farmer's lung (Kokkarinen and Tykianen 1993; Cuthbert 1983; Lalancette et al. 1993; Barbee et al. 1968; Braun et al. 1979). This may have the features of asthma or emphysema. Chronic bronchitis has also been reported (de Haller 1986; Braun et al. 1979; Cormier et al. 1987). An isolated decrease in DLCO may be seen (Lalancette et al. 1993).

If a large amount of pulmonary fibrosis is present secondary to farmer's lung then chronic hypoxia and right heart failure can ensue, which can be fatal (Barbee et al. 1968). This outcome has been reported in 3 to 17% of patients with farmer's lung (Barbee et al. 1968; Frank 1958; Emanuel et al. 1964).

AIRWAY DISEASES

Asthma and Rhinitis

There are two clinical pictures associated with agricultural asthma and rhinitis, based on the causative agents and the response to these exposures. The minority of cases are those of an IgE-based specific antigen-antibody reaction in atopic individuals. The majority of cases are non IgE-mediated, and are related to chronic exposure to environmental irritants. Exposure to these irritants causes respiratory tract inflammation and symptoms similar to those of IgE-mediated diseases.

Epidemiology. Asthma affects 5 to 10% of people around the world. The prevalence of asthma and allergic rhinitis has been determined in several populations of farm workers. Estimates of asthma prevalence in a farm population have ranged from 6.5% to 13% (Parry and Schlenker 1994; Cuthbert et al. 1980). Reversible airways constriction may be seen in up to 25% of confinement livestock or grain workers.

Both rural and urban populations commonly suffer from allergic rhinitis, with the prevalence being 10% or more in the farming population (Heinonen et al. 1987; Parry and Schlenker 1994; Cuthbert et al. 1980). Allergic rhinitis is common in persons who suffer from asthma. People who suffer from allergic rhinitis may have asthma as well.

Atopy has been seen in over half of farming populations to substances in the agricultural environment. These substances include grasses, molds, hog and cattle dander and grain dust, (Parry and Schlenker 1994; Cuthbert 1980). One group of investigators found that respiratory symptoms and airways obstruction on spirometry are more common in farmers with a history of allergic rhinitis (Senthilselvan and Dosman 1995). In contrast, a recent study done in Wisconsin revealed that atopy is more common in farmers than in the nonfarming population but this did not correlate with respiratory symptoms (Marx et al. 1993). A significant correlation has been found in grain workers between cutaneous reactivity to grain dust and wheezing with grain dust exposure (doPico et al. 1977).

The work-related non-Ig-E mediated asthma is perhaps better defined as an asthma-like condition because bronchospasm demonstrable on spirometry is not consistently present in this group. This asthma-like syndrome is seen in workers chronically exposed to irritant agricultural dusts (Zejda 1994; Thorne 1996). Typically these are non-atopic individuals who have worked 6 or more years in swine or poultry confinement, dairy production or grain handling. Swine confinement workers commonly develop chronic bronchitis as well as the asthma-like syndrome (Figure 22.1). The asthma-like syndrome has been identified in 2 to 40% of workers as a daily complaint and in 38 to 63% as an occasional symptom (Donham et al. 1977; Donham 1984, 1986; Zhou 1991; Zuskin et al. 1991; Dosman et al. 1988; Holness 1989). Cross-shift decrements in lung function however are commonly found in swine, poultry and other livestock workers (Zuskin et al. 1991; Matson et al. 1983; Schwartz et al. 1995a). Symptoms of the asthma-like syndrome are more common in farmers as they age (Dosman et al. 1988). The condition is more prevalent in individuals after least 6 years of working in this setting for at least 2 hours per day. However, the condition is seen after shorter periods of exposure (Donham et al. 1995).

An irritant rhinitis is also seen in swine confinement workers (Donham et al. 1977). Rhinitis been reported in 47 to 74% of those who work in swine confinement.

Causative Agents. Storage mites are present in both grain and hay and have been associated with the presence of Ig E-mediated asthma (Ingram et al. 1979). Mites species identified include *Acarus siro*, *Tydeus spp.*, *Lepidoglyphus destructor*, *Tyrophagus longior*, *Glycyphagus domesticus*, *Chortoglyphus arcuatus* and *Mesostigmata* (Leskinen and Klen 1987; Terho 1987d; Marx et al. 1993). Reactivity to storage mite antigens has also been described in an asthmatic grain worker (Warren et al. 1983). In a group of Canadian asthmatics, those with occupational exposure to grain dust were more likely to have allergy to grain storage mites than were control asthmatics (Warren et al. 1983). In contrast, skin tests performed on a group of symptomatic and asymptomatic Finnish farmers revealed that only reactivity to the fodder yeast *Candida utilis* correlated with symptoms. Reactivity to storage mites, cereal grains and fungi did not correlate with complaints in a European study (Terho et al. 1987d). Raising horses appears to be more likely to be associated with allergic rhinitis symptoms than exposure to cattle, sheep, poultry, or swine.

Soybean sensitivity has also been described. Asthma epidemics have been associated with exposures to soybean dust in the nonagricultural setting (Antó et al. 1989, 1993; Weill et al. 1964). The majority of grainworkers complain of nasal and/or conjunctival irritation after exposure to grain dust (Tse et al. 1980). For some individuals who work with grain, these symptoms clearly are those of asthma while in others the clinical picture is that of bronchitis (Chan-Yeung et al. 1979; Davies et al. 1976).

There is some evidence that inhalation of pesticides can induce asthma (Senthilselvan et al. 1995; Senthilselvan 1992). Others have looked for pulmonary function test changes in agricultural workers heavily exposed to pesticides and did not find any differences from a control group (Rastogi 1989).

The swine confinement unit is a complex environment with high levels of ammonia and other gases, endotoxin and microbes (Figure 22.2). Air quality can vary greatly in quantity from one building to another and also with weather conditions (Donham 1986a, 1986b; Cormier 1990; Curtis 1974; Wathes 1987; Heber and Stroik 1987; Louhelainen et al. 1987; Butera et al. 1991; Thorne 1992; Meyer 1991). Hazardous substances found in the air of the swine confinement unit environment have been associated with work-exposure decrements in FEV_1 which is a component of the asthma-like syndrome in these workers. The hazardous substance and the observed recommended threshold concentrations include the following: ammonia levels ≥ 7.0 ppm, total dust ≥ 2.5 mg/m^3, respirable dust ≥ 0.35 mg/m^3 and endotoxin at 10 EU/m^3 (Donham 1995, 1989; Schwartz 1995a). The exposure threshold for respiratory illness is lower in farmers who smoke cigarettes. Asthma-like symptoms are nearly twice as prevalent in smoking swine workers.

Clinical Findings. In the agricultural worker with asthma, symptoms may begin during the exposure if the individual is having an immediate-type (IgE-based) reaction or 8 to 12 hours later if the reaction is delayed. In more typical non-allergic asthma, about a 2-hour exposure is necessary to induce symptoms. Rarely, fatal asthma attacks can occur during or after heavy organic dust exposure.

An accurate work history is critical to understanding the patient's problem. This will help the patient and the physician determine which exposures need to be avoided or minimized in the future.

Findings on physical examination in asthma include cough, wheezing, tachypnea, and decreased breath sounds to auscultation. However, the examination may be quite normal between exacerbations of asthma. Having the patient come to the clinic during an attack precipitated by a workplace exposure can be very helpful for establishing the diagnosis and assessing the severity of the problem.

Persons with the chronic or intermittent asthma-like syndrome associated with swine confinement usually have normal physical examinations of the chest. Bronchoscopy with bronchoalveolar lavage and endobronchial biopsy in these patients have not revealed the presence of increased numbers of inflammatory cells. However, endobronchial biopsies have revealed thickening of the basement membrane of airway epithelia, a finding also seen in asthma (Schwartz et al. 1992).

Physical findings in allergic rhinitis include a pale, edematous nasal mucosa. Because individuals with this problem are more prone to having sinusitis, there may also be sinus tenderness and purulent nasal secretions. In animal confinement workers, the differential diagnosis includes the irritant rhinitis for which these individuals are at risk.

Pulmonary Function Testing. Pulmonary function abnormalities in asthma include signs of airway obstruction that include a reduced FEV_1, FVC, FEV_1/FVC ratio and a decreased FEF_{25-75}. There may also be an increase in residual volume and total lung capacity, indicating the presence of air trapping. In asthma these changes are completely or partially reversible after inhalation of a beta-agonist bronchodilator medication. It must be remembered that pulmonary function can be normal in asthmatics when they are asymptomatic.

Farmers without a history of respiratory complaints almost always have unremarkable pulmonary function tests (Iversen et al. 1989). However, if there is a history of wheezing and nonproductive cough and dyspnea or known asthma, then obstruction is likely to be seen on spirometry (Iversen et al. 1989).

Swine confinement workers with respiratory symptoms have been found to have an increased residual volume and total lung capacity but do not usually have abnormal spirometric measures of airflow (Schwartz et al. 1992; Donham 1984b; Zuskin 1991b). Some persons who work in swine confinement units do develop significant airways obstruction (Holness 1995). Cross-shift decreases in lung function have been noted and have been found to be predictive of longitudinal change in spirometric function (Schwartz et al. 1995a). Measuring peak expiratory flow rate can also be useful in quantifying cross-shift decreases in airflow experienced by symptomatic swine confinement workers (Quinn et al. 1995).

Methacholine or histamine challenge testing helps make the diagnosis of asthma by provoking airways obstruction on spirometry. Positive methacholine challenge tests are often seen in symptomatic swine confinement workers, a finding that is unrelated to atopy (Schwartz et al. 1992; Zhou et al. 1991; Bessette et al. 1993). This is also seen after acute exposure to swine dust (Malmberg and Larsson 1993). It is important to remember that asymptomatic grainworkers and swine confinement workers with normal baseline spirometry may have a small amount of bronchial hyperreactivity (Gerrard et al. 1979; Schwartz et al. 1992; Mink et al. 1980).

Allergy Testing. In a group of grainworkers, skin reactivity was noted to both fungal extracts and mixed grain dust extract (Tse et al. 1980, Warren 1974). However, the presence of atopy does not correlate with the presence of asthma symptoms in grain workers (Chan-Yeung et al. 1979; Grzybowski et al. 1980). Radioallergosorbent (RAST) tests can also be done to detect IgE antibodies to grain and grain dust (Lewis et al. 1986). It must be noted that mild airway obstruction can be seen after acute grain dust exposure in nonallergic persons with no prior history of grain handling (Tse et al. 1980; doPico et al. 1981).

Animal allergy can be a cause of allergic symptoms in some farmers. Skin tests may be positive to cattle dander and hog dander in a third of farmers or more. In at least one study, this finding correlated with the presence of respiratory symptoms (Terho et al. 1987b).

Immunologic studies of swine confinement workers have shown that IgG and IgE antibodies are present in some of these workers to feed antigens and pig-derived antigens (Brouwer et al. 1986; Zuskin et al. 1991; Matson et al. 1983; Donham 1994). IgG and IgE antibodies do not correlate with pulmonary function abnormalities in most published studies. In one report, positive skin reactions correlated with cross-shift pulmonary function decreases (Zuskin et al. 1991). As noted above, airway hyperresponsiveness to methacholine is common in swine confinement workers. However, there is no relationship between this hyperreactivity and an antibody response or positive dermal prick test (Donham 1995).

Treatment. Eliminating or minimizing exposures that cause symptoms is a key component of managing asthma, the asthma-like syndrome and rhinitis in agricultural workers. Methods used to control exposures include environmental control of hazardous substances supplemented with NIOSH approved respirators that have been fit-tested and properly selected for the type of task to be completed.

Treatment of asthma in agricultural settings also includes standard medical management. Medications commonly used to treat asthma include inhaled beta-agonists, anti-inflammatory medications such as nedocromil sodium and inhaled or oral corticosteroids and theophylline. Intensification of medication use during times of unavoidable peak exposures can be useful for the management of the disorder. Some workers may need to change jobs in order to control their asthma symptoms, particularly the atopic asthmatics.

Respirators have been shown to reduce swine dust levels from 77 to 96% (Pickrell et al. 1992). While a relatively low percentage of swine confinement workers regularly wears respirators, education programs have been shown to increase respirator use (Gjerde et al. 1991).

Allergic rhinitis in the agricultural setting may also be controlled by wearing respirators and limiting exposures to allergens and irritants when possible. Medical treatment for allergic rhinitis includes corticosteroid nasal sprays and immunotherapy. Use of respirators is recommended as treatment for irritant rhinitis. It has not been determined if medical therapy is useful in the management of this problem.

Bronchitis

Epidemiology. Chronic bronchitis is defined as cough productive of sputum for at least 3 months of the year for 2 years (Meneely 1962). Airway obstruction may or may not be present on pulmonary function testing in chronic bronchitis. The term acute bronchitis refers to cough with or without sputum production that occurs episodically. Both acute and chronic bronchitis are seen after exposures in the agricultural setting. Chronic bronchitis can be related to work in a variety of aspects of agriculture but has been reported most frequently in workers exposed to grain dust, the swine confinement environment, and intensive poultry housing. Acute bronchitis has been described in grain workers (Warren et al. 1989) and both swine veterinarians and swine producers (Donham et al. 1977 and Donham et al. 1984a).

Chronic bronchitis is more common in non-smoking farmers than in the general population (Hodgkin et al. 1984; Von Essen 1993; Kiviloog et al. 1974; Miller et al. 1988). The prevalence of the disorder ranges from 3 to 30% in non-smoking general farming populations. (Babbott et al. 1980; Vergnenegre et al. 1995; Terho et al. 1987; Husman et al. 1987; Lefcoe and Wonnacott 1974; Vohlonen et al. 1987a; Terho 1987a, 1987c). One study revealed that a randomly selected group of non-confinement farmers had a chronic bronchitis prevalence of 12% compared to matched swine confinement farmers (25%) and to matched non-farming workers (3%) (Donham 1989). Risk for this disorder increases with age and years of exposure. The prevalence of chronic bronchitis is higher if the farmers also smoke cigarettes and if they are exposed to large amounts of organic dust (Babbott et al. 1980; Warren and Manfreda 1980; Depierre et al. 1988; Schlenker et al. 1989; Vohlonen 1987a). Having a history of farmer's lung appears to be a risk factor for this problem (de Haller 1986; Brain 1979; Cormier et al. 1987; Dalphin 1993b). A history of acute respiratory illness after heavy organic dust exposure is also a risk factor for chronic bronchitis (Dalphin 1993a). Acute bronchitis symptoms after grain dust exposure are seen in approximately 20% of farmers and are not related to atopy (Warren et al. 1989).

Chronic bronchitis is also increased in grain workers and in those who work in animal feed mills (Dosman 1980; Jorna et al.,1994; Tse 1973; doPico et al. 1984, 1977; Broder et al. 1979; Chan-Yeung et al. 1980, 1979). The prevalence of chronic bronchitis in nonsmoking grain workers has been found to be 25 to 30% (Tse 1973; doPico et al. 1977). No correlation has been found between precipitins to fungi, bacterial, grain, and the presence of chronic bronchitis (doPico et al. 1977).

As mentioned above, chronic bronchitis is common in persons who work in animal confinement. It has been reported in 5 to 58% of swine confinement workers (Cormier 1991; Wilhelmsson et al. 1989; Bongers 1987; Donham 1994). The prevalence of this disorder is greater in those who spend at least 2 hours each day in the swine buildings versus those with shorter daily exposures (Cormier 1991). Acute bronchitis has also been described in persons exposed to large amounts of dust in swine confinement buildings (Wang et al. 1996; Larsson 1992; Donham 1994).

Etiologic Agents. Risk for chronic bronchitis is associated with cumulative agricultural dust exposure (Jorna et al. 1994; Schwartz 1995a). Agricultural organic dust is a heterogenous material (Figure 22.4). Grain dust, for example, contains plant components, soil particles, bacteria, endotoxin, beta 1,3 glucans, fungi, and mycotoxins (Dashek et al. 1986; Lacey 1980; Goynes et al. 1986; Schaller and Nicholson 1980; Yoshida and Maybank 1980; Parnell et al. 1986; Von Essen 1995). The particles in grain dust vary in size, with about 40% in the respirable range (Brain and Mosier 1980). Some types of grain dust are more likely to cause symptoms than others, including the dusts from wheat, barley, rye, oats, and grain sorghum.

The risk for bronchitis from grain dust exposure is more strongly associated with endotoxin levels in the organic dust than with total dust levels (Schwartz 1995b). Acute bronchitis symptoms are also related to the dose of grain dust to which exposure occurs (doPico et al. 1983). While endotoxin is clearly very important, there is evidence from in vitro studies that the ability of grain dust to cause inflammation does not come from endotoxin alone (Von Essen et al. 1988; Von Essen 1995).

Features of the swine confinement unit environment that correlate with the presence of bronchitis symptoms in workers are the same as for the asthma-like syndrome seen in this population. Threshold concentrations associated with airway disease in workers are as follows: total dust levels $\geq$ 2.5 mg/m^3, respirable dust $\geq$ 0.35 mg/m^3, ammonia levels $\geq$ 7.0 and endotoxin $\geq$ 10.0 EU g/m^3 (Donham 1995, 1989). The exposure threshold for respiratory illness is lower in farmers who smoke cigarettes. The asthma-like symptoms are nearly twice as prevalent in smoking swine confinement workers. Pressure washing (removing dust using water under high pressure) is also related to dust concentrations in the air and to the presence of chronic respiratory symptoms (Preller et al. 1995).

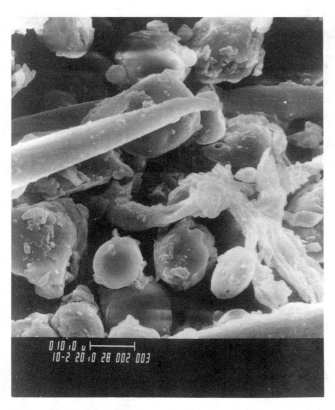

Figure 22.4. Scanning electron micrograph of grain dust.

Clinical Features. Workers with chronic bronchitis in the agricultural setting complain of cough and phlegm production which is often worsened by exposures in their work. They may also have dyspnea, chest tightness, and wheezing (Schwartz 1995a; doPico et al. 1977). They may complain of frequent "chest colds" and state that this type of illness lasts longer than expected. Eye, nasal and throat irritation symptoms may be present in these workers as well. Chest X-rays are unremarkable or show the presence of bronchial wall thickening.

Persons with acute bronchitis related to work in agriculture have episodic complaints similar to those with chronic bronchitis (Von Essen 1990a). A group of symptomatic grain farmers studied at harvest was noted to have increased neutrophils in bronchoalveolar lavage fluid (Von Essen et al. 1990a). Cough, phlegm, wheezing and/or chest tightness and dyspnea are also associated with acute grain dust exposure in grain elevators (doPico 1983). A peripheral blood leukocytosis has been noted after grain dust exposure (doPico 1983; Von Essen 1995).

Bronchoscopy performed in persons who are experiencing acute bronchitis symptoms after grain dust exposure after harvest revealed the presence of erythema and other visible signs of airways inflammation as well as an increased number of neutrophils (Von Essen et al. 1990a). Similar signs of airways inflammation were reported in swine farmers studied using bronchoscopy.

Symptoms of acute bronchitis can be induced by grain dust inhalation in previously unexposed persons and in grain workers using saline extracts of grain dusts (doPico 1982; Clapp et al. 1994; Von Essen 1995b). An acute neutrophilic lower respiratory inflammation was reproduced using this technique. The presence of IL-1ß, TNF-alpha, IL-6, IL-8, as well as IL-1 receptor antagonist in bronchoalveolar lavage fluid after challenge suggests that these substances mediate the inflammatory process induced by grain dust exposure (Turner and Nicolls 1995). These challenges have not provided evidence that complement activation plays an important role causing the in vivo findings even though grain dust has been shown to activate complement in vitro (doPico et al. 1982; Von Essen et al. 1988; Olenchock et al. 1980; Skea et al. 1988). Leukotrienes are released from lung tissue exposed to grain dust and therefore likely play a role in the clinical findings of exposed workers (Chan-Yeung et al. 1987).

Acute bronchitis symptoms have been described in swine confinement workers and in persons without prior exposure to this environment (Wang et al. 1996; Larsson et al. 1992; Donham 1994). Fever and malaise often accompany the respiratory symptoms. This syndrome is characterized by the presence of increased numbers of neutrophils in the lower respiratory tract, as demonstrated using bronchoscopy with bronchoalveolar lavage. A rise in serum TNF-alpha and IL-6 have be been demonstrated after acute exposure to the swine confinement unit environment, which likely contribute to neutrophil recruitment to the airways.

Respiratory illness in swine confinement workers is often difficult to precisely categorize as bronchitis or the asthma-like syndrome. There is often considerable overlap in symptoms. Also, one individual may experience more than one type of effect from exposure to this environment (Figure 22.2).

Pulmonary Function Testing. In bronchitis, spirometry may show fixed airway obstruction or minimal reversibility after beta-agonist inhalation. Some individuals have chronic bronchitis symptoms without airways obstruction. Severe airways obstruction in agricultural workers is usually associated with cigarette smoking.

Grain workers have reduced pulmonary function values compared to workers without organic dust exposure. The values than are diminished include FEV_1, FEV_1/FVC and FEF_{25-75} (Schwartz et al. 1995; Dosman 1980; Chan-Yeung et al. 1979). Reversibility with beta agonist bronchodilators is often present. Not all grain workers with symptoms have airflow obstruction. Severe airflow obstruction is rare in the grain worker population with bronchitis but the functional abnormalities can lead to disability and cause workers to leave the industry.

Acute decrements in pulmonary function can be demonstrated in the workplace where there is grain dust exposure (doPico et al. 1983). This has been reproduced in the experimental setting using saline extracts of grain dusts (doPico 1982b; Clapp et al. 1994; Von Essen 1995b).

As a group, farmers have higher baseline FVC and FEV_1 values than a normal control population. However, their FEF_{25-75} values are typically lower than expected (Donham et al. 1990). Farmers who have complaints of chronic bronchitis have small decreases in lung function compared to controls (Chen et al. 1991; Babbott et al. 1980; Iversen et al. 1989). It is typical that agricultural workers exposed to organic dusts have 5 to 20% declines in volumes and flow rates over a work period. When reviewing pulmonary function testing results in groups of farmers it must be considered that standard comparison populations may not be appropriate for use in analysis of the data.

When large groups of swine confinement workers are studied, their spirometry findings are usually in the normal range and comparable to that of non-confinement workers without disease (Donham et al. 1984; Zhou et al. 1991; Preller et al. 1995). Swine confinement workers with respiratory symptoms have been shown to have evidence of air trapping on pulmonary function tests (Schwartz et al. 1992).

Methacholine and histamine challenge testing often is positive in both grain workers and livestock farmers even in absence of respiratory symptoms and a history of atopy (Chan-Yeung et al. 1979; Mink et al. 1980; Cockcroft 1980; Hargreave 1980; Chatham et al. 1982; Enarson et al. 1989; Iversen et al. 1989). Results vary from the great sensitivity to methacholine seen in asthma to a lesser degree of bronchial hyperreactivity seen in chronic bronchitis (Hahn 1980; Rijcken et al. 1987; Ramsdell et al. 1982).

Treatment and Prognosis. Correct identification of work related respiratory hazards is essential for treatment and prevention of bronchitis. (Vohlonen 1987a). Worker protection must focus first on environmental control. Appropriate use of respirators should be considered additional protection. Use of respirators can reduce the amount of dust and ammonia inhaled by swine confinement workers (University of Iowa 1989). However, studies have shown that only approximately a third of producers usually wear a respirator when working in a hog barn (Zejda 1995). There is some evidence that wearing a respirator helps reduce pulmonary symptoms (Zejda 1995; Bongers 1987).

Building design features, such fully slatted floors, play a role in reducing levels of ammonia, dust and endotoxin and decreasing respiratory symptoms (Aarnick 1995; PIGS-Misset 1994; Bongers 1987). Increasing fat in the diet of the pigs has been shown to reduce dust and ammonia levels in the air as well as bacterial colony forming particle concentrations in swine confinement buildings (Chiba et al. 1987, 1985). Use of frequent cleaning (power washing) of the buildings significantly reduces aerosolized dust levels.

Smoking cessation is an important part of treatment for the agricultural worker with chronic bronchitis, as mentioned above. There is evidence that smoking and grain dust exposure are additive or synergistic in their negative effects on the lung (Chen et al. 1991; doPico et al. 1984). The same appears to be true for exposure to the swine confinement unit environment. Other aspects of medical therapy for chronic bronchitis in this setting include use of medications used to treat this problem when it is caused by cigarette smoking alone. These include inhaled or oral corticosteroids and various types of bronchodilator drugs.

Factors that predicted longitudinal decline in lung function in a group of Canadian grain workers included acute changes in lung function over the course of one work week as well as the presence of bronchial hyperreactivity (Tabona et al. 1984). The number of years worked in the industry is also important (Pahwa et al. 1994). The decline in lung function seen in nonsmoking as well as smoking workers but is greatest in the latter group. Interestingly, respiratory symptoms, initial lung function and skin reactivity to common allergens do predict worsening of lung function. Employment in the grain elevator industry is associated with respiratory impairment that continues into the retirement years and limits daily activities for some workers (Kennedy et al. 1994).

The prognosis of those with chronic respiratory symptoms related to work in swine confinement is not yet well defined. In one recently published study, 6 hog farmers with both respiratory symptoms and pulmonary function abnormalities were followed for 6 years (Holness 1995). Symptoms improved in those who stopped raising hogs but the FEV_1 declined in all of the individuals.

TOXIC GAS EXPOSURES

Effects of Silo Gas Exposure

Silage is made by chopping a green forage crop (usually corn but possibly alfalfa, oats or other forage) and placing it into an enclosed space that limits exposure of this material to the open atmosphere. Structures used for this purpose often are vertical cylinders made of concrete. Horizontal structures are also used but are less likely to be associated with human health hazards. Farmers often enter the vertical silos soon after filling to level the contents and to cover them with a plastic sheet. Bacteria and other microbes present in this setting create a process called "ensilage" that results in preserving the material for animal feed. Carbon dioxide and oxides of nitrogen accumulate in the space above the forage (Figure 22.5).

The gas that causes acute lung injury is nitrogen dioxide (NO_2), which begins to accumulate immediately, reaches its maximal concentration within 1 to 2 days and persists for up to 2 weeks (Douglas et al. 1989; May 1996). Levels of NO_2 can vary greatly from year to year, depending on the degree of nitrate stored in the plant materials. There may be dangerously high levels of nitrogen dioxide in a silo that had been entered without incident in the past. Plant stress from drought or an early freeze as well as excess use of nitrogen fertilizer can increase NO_2 levels.

When present in high concentration (>200 ppm), NO_2 causes the air to appear orange-brown in color and have an acrid odor. Under these circumstances it can induce syncope and cause death in minutes. Survivors may develop noncardiogenic pulmonary edema (Grayson 1956). The pulmonary edema may develop up to 48 hours following an acute exposure, making it necessary to monitor exposed patients over this time period.

At lower concentrations (50 to 100 ppm), NO_2 has a bleach-like odor and respiratory symptoms immediately after exposure may be mild. This may be followed by presentation weeks later with fever, cough, and dyspnea associated with a diffuse reticulonodular pattern on chest X-ray and the presence of bronchiolitis obliterans on lung biopsy (Lowry 1956; Ramirez and Dowell 1971; Horvath et al. 1978; Fleetham 1978; Cornelius 1960; Leib 1958; *MMWR* 1982). Corticosteroid therapy can be used to treat the bronchiolitis obliterans and prevent it from progressing to respiratory failure (Jones and Proudfoot 1973; Horvath et al. 1978).

Measures used to prevent NO_2 exposure include optimizing ventilation before entering the silo. Silos should not be entered for at least 10 days after filling (Douglas et al. 1989; Epler 1989). If the silo must be entered before this time, the silage blower (the machine that loads silage into the silo) should be operated for at least 30 minutes before entering.

Some silos are built of enameled metal plates with rubber seals around each plate for the near total exclusion of oxygen (May 1996). These are oxygen limiting silos. Some concrete silos are made air tight by coating the inside with epoxy resins. Attempts to enter these silos are likely to end with fatal asphyxia. Silage stored in these facilities does not release oxides of nitrogen.

Acute Poisoning from Livestock Confinement Building Pit Gases

Livestock confinement unit wastes are often stored in a liquid form in a space beneath the building housing animals prior to land application. If stored in a deep system (greater than 3 feet) for more than several weeks at a time, the system may become an anaerobic environment and produce hydrogen sulfide (H_2S). This method of waste handling is most commonly seen on pig farms (Figure 22.2) but is also used in dairy operations. This system reduces labor but places humans and animals at risk of poisoning by pit gases (Crapo 1989; Donham 1982; *MMWR* 1992; NIOSH 1993). The hazards of entering any confined space where this manure is stored is greatest in the summer months when weather conditions can accelerate the production and accumulation of hazardous gases (NIOSH 1993). Concentrations of toxic gases may only be hazardous when the pit is agitated prior to

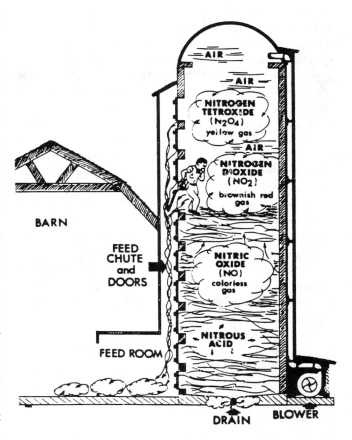

Figure 22.5. Nitrogen ozxides in a silo.

pumping and can be a risk to persons and animals in the buildings at this time.

Although as many as 168 gaseous compounds have been identified in the air of livestock confinement buildings, most are present in trace amounts (O'Neill and Phillips 1992; Hartung and Phillip 1994). Several gases can be present in concentrations that pose immediate danger to human life. Of these, hydrogen sulfide is directly toxic. Carbon dioxide and methane pose a risk to humans largely by displacing oxygen and creating a oxygen deficiency environment where asphyxiation can occur (NIOSH 1993). Ammonia levels can rise to acutely irritating levels in swine confinement buildings on rare occasions. The human health effects of these gases are discussed in more detail below.

Hydrogen sulfide is a byproduct of bacterial fermentation and is the most toxic of the components of manure gases. At high levels it acts through direct toxic effects on the lung and by uncoupling the oxidative phosphorylation system at the cellular level. Hydrogen sulfide has a rotten egg odor when present at usual low levels, near 1 ppm, but which overwhelms the olfactory sense when present at high levels, above 50 ppm. Exposure to 100 ppm produces pneumonitis, 250 ppm cause pulmonary edema and central nervous system depression, 500 ppm can be fatal in 30 minutes and concentrations over 700 ppm are rapidly fatal. Attempts at rescue without use of an oxygen supply can lead to multiple fatalities (MMWR report 1992; NIOSH 1993).

Carbon dioxide in swine confinement buildings comes largely from air exhaled by the animals but is also produced by microbial digestion of manure. Carbon dioxide levels in swine confinement buildings typically range from 1500 ppm to 3500 ppm but may exceed 5000 ppm. Carbon dioxide can reach lethal proportions in 6 to 10 hours if ventilation equipment fails (Fulhage 1980). Carbon dioxide, rather than a common health hazard, is more a surrogate measure for the general environmental air quality within buildings.

Methane has been associated with asphyxiation in manure storage areas. At concentrations >5% it is dangerously explosive. Explosion of manure gas has been reported.

Ammonia is released into the air from the waste of the animals. It has a characteristic pungent odor detectable at concentrations >10 ppm. Concentrations in swine confinement units usually remain below 10 ppm. Increased risk for chronic airways symptoms of workers is associated with ammonia ≥ 7 ppm. Levels in livestock confinement may exceed 100 ppm on rare occasions. At these levels, ammonia causes intense inflammation of mucous membranes of the eyes and the respiratory system.

Survivors of acute manure pit gas poisoning may suffer an altered level of consciousness during exposure, leading to aspiration of pit contents. This may be followed by a severe polymicrobial pneumonia (Osbern 1981).

Steps to prevent death from manure pit gases include the following: 1) agitating the slurry minimally when emptying the pit because hydrogen sulfide is rapidly released during agitation, 2) avoiding entry of a manure pit, 3) posting hazard signs on all manure pits, 4) providing access to all serviceable parts from outside the manure pit, 5) fitting all openings to manure pits with metal grill covers, 6) educating workers about risks and 7) entering pits only with appropriate equipment such as breathing apparatus and safety lines in the presence of another worker (NIOSH 1993).

Other Toxic Inhalation Exposures

Carbon Monoxide Poisoning. Carbon monoxide is an odorless, tasteless, colorless, and nonirritant gas. It rapidly diffuses across the alveolar-capillary membrane and binds to hemoglobin with an affinity more than 200-fold greater than that of oxygen. Complaints of patients with carbon monoxide poisoning include nausea, vomiting, headaches, malaise, weakness, fatigue, chest pain, palpitations, and dyspnea. The patient may be found comatose. Severity of carbon monoxide poisoning can be assessed by measuring pH and carboxyhemoglobin levels in arterial blood. Treatment consists of supplemental oxygen. Outcomes may be improved if the oxygen is given in a hyperbaric chamber. (Hardy and Thom 1994). When treating a pregnant patient who has been exposed to carbon monoxide it must be recalled that the unborn fetus is much more susceptible to the effects of carbon monoxide poisoning.

Carbon monoxide poisoning has been reported in swine confinement workers using gasoline-powered devices for pressure washing the buildings (NIOSH update 1993). This problem can be avoided by keeping machines themselves outside the building and bringing only the hoses inside. Carbon monoxide poisoning can also be caused in the agricultural setting by use of faulty heaters or ventilation malfunction. There are numerous additional sources of carbon monoxide exposure in farm operations such as operating tractors or other equipment inside buildings.

Fumigant Exposure. Fumigants are highly volatile, highly penetrating gaseous substances used to kill insects and other pests in grain or in empty storage facilities. The two fumigants most commonly used today in agriculture are phosphine and methyl bromide. Fatalities have been described after phosphine exposure from stored grain fumigated with aluminum phosphide pellets in order to reduce insect infestation (Garry et al. 1993; Heyndrickx 1976; Wilson 1980). The autopsy findings have included pulmonary edema and increased aluminum levels in blood (Garry et al. 1993).

Methyl bromide also acts as a toxin to the lung. It results in symptoms similar to phosphine exposure.

Anhydrous Ammonia Exposure. Anhydrous ammonia is a commonly used nitrogen fertilizer. It is transported, stored and applied to soil in its liquid form at -60°F. During application it passes from a tank pulled behind a tractor through tubing and is injected into the soil. It is highly soluble in water and penetrates human tissues readily. When it is in contact with human tissues, it forms ammonium hydroxide and dissociates into free hydroxyl ions. Acute injuries to farmers and commercial fertilizer applicators from this exposure include freezing of tissue, dermal burns, severe corneal burns, laryngeal edema necessitating tracheal intubation, and adult respiratory distress syndrome. Late effects of inhalation of large amounts of anhydrous ammonia include bronchiolitis obliterans and reactive airways dysfunction syndrome (Helmers et al. 1971; Leduc et al. 1992).

Mycotoxin Exposure. Fungi are ubiquitous in agricultural environments. Numerous mycotoxins produced by fungi have been described. Many of these mycotoxins have known health effects in animals when feed contaminated with these materials is ingested. However, when animals inhale mycotoxins, adverse health effects are not known to occur. Human health effects of mycotoxin exposure in the agricultural setting are less well characterized.

It has been demonstrated that certain mycotoxins are toxic to alveolar macrophages (Sorenson 1990). It has been proposed that mycotoxins may play a role in the high mortality from lung cancer in Louisiana sugarcane workers (Mulvey et al. 1983). Also, acute renal failure and respiratory distress has been described after inhaling ochratoxin from *Aspergillus ochraceus* after exposure in a granary (Di Paolo 1993).

Paraquat Poisoning. Paraquat is well-known to cause irreversible pulmonary fibrosis after accidental or intentional ingestion (Copland NEJM 1974). Rarely, pulmonary fibrosis occurs after heavy skin contamination with paraquat (Levin et al., Thorax 1979).

Grain Suffocation

Suffocation can occur in stored grain in several different circumstances. Moisture can cause the grain to cake or crust at the surface, a phenomenon that is often called bridging (NIOSH 1993a). Persons walking on the crusted surface or attempting to break up the bridged material have fallen through and been engulfed by grain. Workers may also be buried by the stored grain when the storage bin is being emptied from the bottom, creating a rapidly descending column of grain inside the bin which engulfs the victim. Steps to be taken to prevent grain suffocation in a storage bin include the following: 1) breaking up surface crusts from outside the bin, 2) not standing on top of stored material if possible, 3) wearing safety belts or harnesses equipped with properly fastened life lines, 4) turning off the unloading equipment before entering the bin and 5) turning on ventilating equipment prior to entry. It must be remembered that suffocation can also occur in grain filled gravity feed wagons.

Suffocation may occur by a different mechanism in bins or silos filled with high moisture corn (>13%) as this gives off carbon dioxide, thereby displacing oxygen. Most commonly suffocation would occur in tight silos, but may occur in conventional silos or in grain bins for approximately 2 weeks following filling.

SUMMARY

Respiratory disorders are common in a variety of agricultural settings. Those who handle grain are at risk for developing organic dust toxic syndrome, bronchitis, and exacerbations of asthma. Farmer's lung is an important problem in dairy farmers but is of less importance in those who handle grain and in other types of agricultural workers. Swine confinement workers often develop asthma-like syndrome or bronchitis. Organic dust toxic syndrome is seen in this population as well. Fatal toxic gas inhalations occur in farm settings. Important examples of this problem include nitrogen dioxide exposure in silos, hydrogen sulfide exposure in animal confinement manure pits and carbon monoxide in any setting where an internal combustion engine is used indoors.

REFERENCES

Aarnink, A.J., and A. Swierstra. 1995. The influence of slatted floor type on ammonia emission. *PIGS-Misset* 11: 36-39.

Ando, M., K. Konishi, R. Yoneda, and M. Tamura. 1991. Difference in the phenotypes of bronchoalveolar lavage lymphocytes in patients with summer-type hypersensitivity pneumonitis, farmer's lung, ventilation pneumonitis, and bird fancier's lung: Report of a nationwide epidemiologic study in Japan. *J. Allergy Clin. Immunol.* 87: 1002-09.

Anto, J.M., J. Sunyer, R. Rodriquez-Roisin, M. Suarez-Cervera, L. Vazquez, et al. 1989. Community outbreaks of asthma associated with inhalation of soybean dust. *N. Engl. J. Med.* 320: 1097-1102.

Anto, J.M., J. Sunyer, C.E. Reed, J. Sabria, F. Martinez, F. Morell, R. Codina, R. Rodriguez-Roisin, M.J. Rodrigo, J. Roca, and M. Saez. 1993. Preventing asthma epidemics due to soybeans by dust-control measures. *N. Engl. J. Med.* 329: 1760-63.

Arshad, M., S.R. Braun, and E.V. Sunderrajan. 1987. Severe hypoxemia in farmer's lung disease with normal findings on chest roentgenogram. *Chest* 91: 274-75.

Ault, B.J., J.J.Marx, J.A. Campbell, and J.A. Merchant. 1995. A ten-year follow-up survey of central Wisconsin farmers regarding health and farming status. *Agricultural Health and Safety: Workplace, Environment, Sustainability*. Chapter 3.8: 295-99.

Babbott FL, Gump DW, Sylvester DL, et al. 1980. Respiratory symptoms and lung function in a sample of Vermont dairymen and industrial workers. *Amer. J. Public Health* 70:241-245.

Barbee, R.A., Q. Callies, H.A. Dickie, and J. Rankin. 1968. The long-term prognosis in farmer's lung. *Amer. Rev. Respir. Dis.* 97: 223-31.

Barrowcliff, D.F., and P.G. Arblaster. 1968. Farmer's lung: a study of an early acute fatal case. *Thorax* 23: 490-500.

Beaumont, J.J., D.F. Goldsmith, L.A. Morrin, and M.B. Schenker. 1995. Mortality in agricultural workers after compensation claims for respiratory disease, pesticide illness, and injury. *J. Occup. Envir. Med.* 37: 160-69.

Belin, L., and P. Malmberg. 1986. Antibodies to microbial antigens in various farmer populations. *Amer. J. Ind. Med.* 10: 277-80.

Bernardo, J., G.W. Hunninghake, J.E. Gadek, V.J. Ferrans, and R.G. Crystal. 1979. Acute hypersensitivity pneumonitis: serial changes in lung lymphocyte subpopulations after exposure to antigen. *Amer. Rev. Respir. Dis.* 120: 985-94.

Bessis, L., P. Callard, C. Gotheil, A. Biagg, P. Grenier. 1992. High-resolution CT of parenchymal lung disease: precise correlation with histologic findings. *RadioGraphics* 12: 45-58.

Bessette, L., L.P. Boulet, G. Tremblay, and Y. Cormier. 1993. Bronchial responsiveness to methacholine in swine confinement building workers. *Arch. Env. Health* 48: 73-77.

Bjermer, L., A. Englstrom-Laurent, R. Lundgren, L. Rosenhall, R. Hallgren. 1988. Bronchoalveolar mastocytosis in farmer's lung is related to the disease activity. *Arch. Intern. Med.* 148:1362-65.

Bjermer, L., A. Englstrom-Laurent, R. Lundgren, L. Rosenhall, and R. Hallgren. 1987. Hyaluronate and type III procollagen peptide concentrations in bronchoalveolar lavage fluid as markers of disease activity in farmer's lung. *Brit. Med. J.* 295: 803-06.

Blanc, P.D., H.A. Boushey, H. Wong, S.F. Wintermeyer, and M.S. Bernstein. 1993a. Cytokines in metal fume fever. *Am Rev Respir Dis* 147:134-138.

Bongers, P., D. Houthuijs, B. Remijn, R. Brouwer, and K. Biersteker. 1987. Lung function and respiratory symptoms in pig farmers. *Br. J. Ind. Med.* 44: 819-23.

Blanc, P.D., H.A. Boushey, H. Wong, S.F. Wintermeyer, and M.S. Bernstein. 1993. The lung in metal fume fever. *Seminars in Resp. Med.* 14: 212-25.

Boyer, R.S, L.E. Klock, C.D. Schmidt, L. Hyland, K. Maxwell, R.M. Gardner, and A.D. Renzetti. 1974. Hypersensitivity lung disease in the turkey raising industry. *Am. Rev. Respir. Dis.* 109: 630-35.

Brain, J.D., and M.J. Mosier. 1980. Deposition and clearance of grain dusts in the human lung. In: *Occupational pulmonary disease: Focus on grain dust and health*. J.A. Dosman, D.J. Cotton, eds. New York: Academic Press, pp. 77-94.

Braun, S.R., G.A. doPico, A. Tsiatis, E. Horvath, H.A. Dickie, and J. Rankin. 1979. Farmer's lung disease: long-term clinical and physiologic outcome. *Am. Rev. Respir. Dis.* 119: 185-191.

Brinton, W.T., E.E. Vastbinder, Greene J.W., J.J. Marx, R.H. Hutcheson, W. Schaffner. 1987. An outbreak of organic dust toxic syndrome in a college fraternity. *JAMA* 258: 1210-12.

Broder, I., S. Mintz, M. Hutcheon, et al. 1979. Comparison of respiratory variables in grain elevator workers and civic outside workers of Thunder Bay, Canada. *Am. Rev. Respir. Dis.* 119: 193-203.

Brouwer, R., K. Biersteker, P. Bongers, B. Remijn, and D. Houthuijs. 1986. Respiratory symptoms, lung function, and IgG4 levels against pig antigens in a sample of Dutch pig farmers. *Am. J. Ind. Med.* 10: 283-85.

Bureau, M.A., C. Fecteau, H. Patriquin, M. Rolla-Pleszczynski, S. Masse, and R. Begin. 1979. Case Report-Farmer's lung in early childhood. *Am. Rev. Resp. Dis.* 119: 671-75.

Butera, M., J.H. Smith, W.D. Morrison, R.R. Hacker, F.A. Kains, and J.R. Ogilvie. 1991. Concentration of respirable dust and bioaerosols and identification of certain microbial types in a hog-growing facility. *Can. J. Anim. Sci.* 71: 271-77.

Caldwell, J.R., C.E. Pearce, C. Spencer, et al. 1973. Immunologic mechanisms in hypersensitivity pneumonitis. *J. Allergy Clin. Immunol.* 52: 225-32.

Calhoun, W.J. 1991. Enhanced reactive oxygen species metabolism of air space cells in hypersensitivity pneumonitis. *J. Lab. Clin. Med.* 117: 443-52.

Campbell, J.M. 1932. 1932. Acute symptoms following work with hay. *Brit. Med. J.* 1143-44.

Causative Agents for Organic Dust Related Disease. 1994. R. Rylander and Y. Peterson, eds. Special Issue of Proceedings of an International Workshop held in Skokloster, Sweden, April 6-9, 1992. *Amer. J. Indust. Med.* 25: 1-148.

Chan-Yeung, M., H. Chan, H. Salari, R. Wall, and K.S. Tse. 1987. Grain-dust extract induced direct release of mediators from human lung tissue. *J. Allergy Clin. Immun.* 80: 279-84.

Chan-Yeung, M., M. Schulzer, L. MacLean, et al. 1980. Epidemiologic health survey of grain elevator workers in British Columbia. *Amer. Rev. Respir. Dis.* 121: 329-38.

Chan-Yeung, M., R. Wong, and L. MacLean. 1979. Respiratory abnormalities among grain elevator workers. *Chest* 75: 461-67.

Chen, Y., S.L. Horne, H.H. McDuffie, and J.A. Dosman. 1991. Combined effect of grain farming and smoking on lung function and the prevalence of chronic bronchitis. *Intl. J. Epidem.* 20: 416-23.

Chiba, L.I., E.R. Peo, A.J. Lewis, M.C. Brumm, R.D. Fritschen, J.D. Crenshaw. 1985. Effect of dietary fat on pig performance and dust levels in modified-open-front and environmentally regulated confinement buildings. *J. Anim. Sci.* 61: 763-81.

Chiba, L.I., E.R. Peo, A.J. Lewis. 1987. Use of dietary fat to reduce dust, aerial ammonia and bacterial colony forming particle concentrations in swine confinement buildings. *Trans. ASAE* 30: 464-68.

Chmelik, F., D. Guillermo, C. Reed, H. Dickie. 1974. Allergy grand rounds. *J. Allergy Clin. Immun.* 54: 180-88.

Chronic bronchitis, asthma, and pulmonary emphysema. 1962. A statement by the Committee on Diagnostic Standards for Nontuberculous Respiratory Disease. *ARRD* 85: 762-68.

Chronic obstructive pulmonary disease and smoking in grain workers. 1977. *Ann. Int. Med.* 87: 784-86.

Clapp, W.D., S. Becker, J. Quay, J.L. Watt, P.S. Thorne, K.L. Frees, X. Zhang, H.S. Koren, C.R. Lux, and D.A. Schwartz. 1994. Grain dust-induced airflow obstruction and inflammation of the lower respiratory tract. *Am. J. Respir. Crit. Care Med.* 150: 611-17.

Cockcroft, D.W. 1980. Clinical assessment of nonspecific bronchial reactivity using a standardized histamine inhalation test: Relevance in early diagnosis. In: *Occupational Pulmonary Disease: Focus on Grain Dust and Health*. J.A. Dosman, and D.J. Cotton, eds. New York: Academic Press, pp. 161-66.

Cockcroft, D.W., B.A. Berscheid, and I.A. Ramshaw. 1983. Sporobolomyces: a possible cause of extrinsic allergic alveolitis. *J. Allergy Clin. Immun.* 72: 305-09.

Copland, G.M., A. Kolin, and H.S. Schulman. 1974. Fatal pulmonary intra-alveolar fibrosis after paraquat ingestion. *N. Engl. J. Med.* 291: 290-92.

Cormier, Y., E. Assayag, G. Tremblay. 1994a. Viral infection enhances response to micropolyspora faeni. *Amer. J. Ind. Med.* 25: 79-80.

Cormier, Y., J. Belanger. 1989. The fluctuant nature of precipitating antibodies in dairy farmers. *Thorax* 44: 469-73.

Cormier, Y., J. Belanger, and M. Laviolette. 1987. Prognostic significance of bronchoalveolar lymphocytosis in farmer's lung. *Amer. Rev. Respir. Dis.* 135: 692-95.

Cormier, Y., J. Belanger, J. Beaudoin, M. Laviolette, R. Beaudoin, and J. Hebert. 1984. Abnormal bronchoalveolar lavage in asymptomatic dairy farmers. *Amer. Rev. Respir. Dis.* 130: 1046-49

Cormier, Y., L-P Boulet, G. Bedard, and G. Tremblay. 1991. Respiratory health of workers exposed to swine confinement buildings only or to both swine confinement buildings and dairy barns. *Scand. J. Work Envir. Health* 17: 269-75.

Cormier, Y., M. Fournier, M. Laviolette. 1993a. Farmer's fever. Systemic manifestations of farmer's lung without lung involvement. *Chest* 103: 632-34.

Cormier, Y., M. Laviolette, A. Cantin, G.M. Tremblay, R. Begin. 1993b. Fibrogenic activities in bronchoalveolar lavage fluid of farmer's lung. *Chest* 104: 1038-42.

Cormier, Y., G.M. Tremblay, M. Fournier, E. Israel-Assayag. 1994b. Long-term viral enhancement of lung response to saccharopolyspora rectivirgula. *Am. J. Respir. Crit. Care Med.* 149: 490-94.

Cormier, Y., G. Tremblay, A. Meriaux, G. Brochu, and J. Lavoie. 1990. Airborne microbial contents in two types of swine confinement buildings in Quebec. *Am. Ind. Hyg. Assoc. J.* 51: 304-09.

Cornelius, E.A., E.H. Betlach. 1960. Silo-filler's disease. *Radiology* 74: 232-38.

Costabel, U., K.J. Bross, K.H. Ruhle, G.W. Lohr, and H. Matthys. 1985. Ia-like antigens on T-cells and their subpopulations in pulmonary sarcoidosis and in hypersensitivity pneumonitis. *Am. Rev. Respir. Dis.* 131: 337-42.

Costabel, U. 1988. The alveolitis of hypersensitivity pneumonitis. *Eur. Respir. J.* 1: 5-9.

Cotton, D.J., B.L. Graham, K.Y.R. Li, F. Froh, G.D. Barnett, and J.A. Dosman. 1983. Effects of grain dust exposure and smoking on respiratory symptoms and lung function. *J. Occup. Med.* 25: 131-41.

Curtis, J.L., and M. Schuyler. 1994. Immunologically mediated lung disease. *Textbook of Pulmonary Diseases*, Chapter 28, pp. 689-744.

Curtis, S.E. 1974. Seasonal effects on bacteria and dust in swine-house air. The George A. Young Conference on Advances in Swine Repopulation, pp. 66-68.

Cuthbert, O.D, and M.F. Gordon. 1983. Ten year follow-up of farmers with farmer's lung. *Brit. J. Ind. Med.* 40: 173-76.

Cuthbert, O.D, D.G. Wraith, J. Brostoff, and W.D. Brighton. 1980. The role of mites in hay and grain dust allergy. *Occupational Pulmonary Disease: Focus on Grain Dust and Health*. New York: Academic Press, pp. 469-75.

Dall'Aglio, P.P, A. Pesci, G. Bertorelli, E. Brianti, and S. Scarpa. 1988. Study of immune complexes in bronchoalveolar lavage fluids. *Respiration* 54: 36-41.

Dalphin, J.C.H, D. Pernet, A. Dubiez, D. Debieuvre, H. Allemand, and A. Depierre. 1993a. Etiologic factors of chronic bronchitis in dairy farmers. *Chest* 103: 417-21.

Dalphin, J.C., B. Toson, E. Monnet. D. Dubiez, A. Dubiez, J.J. Laplante, J.M. Aiache, and A. Depierre. 1994. Farmer's lung precipitins in Doubs (a department of France): prevalence and diagnostic value. *Allergy* 49: 744-50.

Dalphin, J.C., D. Debieuvre, D. Pernt, M.F. Maheu, J.C. Polio, B. Toson, A. Dubiez, E. Monnet, J. Laplante, and A. Depierre. 1993b. Prevalence and risk factors for chronic bronchitis and farmer's lung in French dairy farmers. *Brit. J. Ind. Med.* 50: 941-44.

Dashek, W.V., S.A. Olenchock, J.E. Mayfield, G.H. Wirtz, D.E. Wolz, and C.A. Young. 1986. Carbohydrate and protein contents of grain dusts in relation to dust morphology. *Env. Health Persp.* 66: 135-43.

Davies, R.J, M. Green, and N.M. Schofield. 1976. Recurrent nocturnal asthma after exposure to grain dust. *Am. Rev. Respir. Dis.* 114: 1011-19.

Definition and classification of chronic bronchitis for clinical and epidemiological purposes. 1965. *The Lancet* 1: 775-79.

deHaller, R. 1986. Respiratory symptoms and preventive aspects in farmers chronically exposed to moldy hay. *Am. J. Ind. Med.* 10: 288.

Denis, M., D. Bisson, and E. Ghadirian. 1993a. Cellular and cytokine profiles in spontaneous regression phase of hypersensitivity pneumonitis. *Exp. Lung Res.* 19: 257-71.

Denis, M., Y. Cormier, I. Tardif, E. Ghadirian, and M. Laviolette. 1991a. Hypersensitivity pneumonitis: whole micropolyspora faeni or antigens thereof stimulate the release of proinflammatgory cytokines from macrophages. *Am. J. Respir. Cell Mol. Biol.* 5: 198-203.

Denis, M. 1995. Proinflammatory cytokines in hypersensitivity pneumonitis. *Am. J. Respir. Crit. Care Med.* 151: 164-69.

Denis, M., M. Bedard, M. Laviolette, and Y. Cormier. 1993b. A study of monokine release and natural killer activity in the bronchoalveolar lavage of subjects with farmer's lung. *Am. Rev. Respir. Dis.* 147: 934-39.

Denis, M., Y. Cormier, M. Fournier, J. Tardif, and M. Laviolette. 1991b. Tumor necrosis factor plays an essential role in determining hypersensitivity pneumonitis in a mouse model. *Am. J. Respir. Cell Mol. Biol.* 5: 477-83.

Depierre, A., J.C. Dalphin, D. Pernet, et al. 1988. Epidemiological study of farmer's lung in five disticts of the French Doubs province. *Thorax* 43: 429-35.

Dickie, H.A., and J. Rankin. Farmer's Lung. 1958. *JAMA* 167: 1069-76.

DiPaolo, N., A. Guarnieri, F. Loi, G. Sacchi, A.M. Mangiarotti, and M. DiPaolo. 1993. Acute renal failure from inhalation of mycotoxins. *Nephron* 64: 621-25.

Donham, K.J., D. Zavala, and J. Merchant. 1984a. Acute effects of the work environment on pulmonary functions of swine confinement workers. *Am. J. Indust. Med.* 5: 367-75.

Donham, K.J, L.J. Scallon, W. Popendorf, M.W. Treuhaft, and R.C. Roberts. 1986a. Characterization of dusts collected from swine confinement buildings. *Am. Ind. Hyg. Assoc. J.* 47: 404-10.

Donham, K.J., W. Popendorf, U. Palmgren, and L. Larsson. 1986b. Characterization of dusts collected from swine confinement buildings. *Am. J. Ind. Med.* 10: 294-97.

Donham, K.J., P. Haglind, Y. Peterson, R. Rylander. 1986c. Environmental and health studies in swine confinement buildings. *Am. J. Ind. Med.*; 10: 289-93.

Donham, K.J., R. Rylander. 1986d. Epilogue: Health effects of organic dusts in the farm environment. *Am. J. Ind. Med.* 10: 339-40.

Donham, K.J. 1995. Health hazards of pork producers in livestock confinement buildings: from recognition to control. *Agricultural Health and Safety: Workplace, Environment, Sustainability*. Chapter 1. 9:43.

Donham, K.J., M. Rubino, T.D. Thedell, and J. Kammermeyer. 1977. Potential health hazards to agricultural workers in swine confinement buildings. *J. Occup. Med.* 19: 383-87.

Donham, K.J., J.A. Merchant, D. Lassise, and W.J. Popendorf. 1990. Preventing respiratory disease in swine confinement workers: Intervention through applied epidemiology, education, and consultation. *Am. J. Ind. Med.* 18: 241-61.

Donham, K.J., S.J. Reynolds, P. Whitten, J. Merchant, L. Burmeister, and W.J. Popendorf. 1995. Respiratory dysfunction in swine production facility workers: Dose-response relationships of environmental exposures and pulmonary function. *Am. J. Ind. Med.* 27: 405-18.

Donham, K.J., D.C. Zavala, J.A. Merchant. 1984. Respiratory symptoms and lung function amont workers in swine confinement buildings: A cross-sectional epidemiological study. *Arch. Env. Health* 39: 96-100.

Donham, K.J. 1994. Swine confinement buildings. In: *Handbook of Organic Dusts*. R. Rylander and R.R. Jacobs, eds. Boca Raton, FL: Lewis Publisher, CRC Press, pp. 219-232.

doPico, G.A., W.G. Reddan, S. Anderson, D. Flaherty, and E. Smalley. 1983. Acute effects of grain dust exposure during a work shift. *Am. Rev. Respir. Dis.* 128: 399-404.

doPico, G.A., W. Reddan, A. Tsiatis, et al. 1984. Epidemiologic study of clinical and physiologic parameters in grain handlers of northern United States. *Am. Rev. Respir. Dis.* 130: 759-65.

doPico, G.A., D. Flaherty, P. Bhansali, N. Chavaje. 1982a. Grain fever syndrome induced by inhalation of airborne grain dust. *J. Allergy Clin. Immun.* 69: 435-43.

doPico, G.A. 1992. Hazardous exposure and lung disease among farm workers. *Clinics in Chest Med.* 13: 311-28.

doPico, G.A. 1986. Health effects of organic dusts in the farm environment. Report on diseases. *Am. J. Ind. Med.* 10: 261-65.

doPico, G.A., S. Jacobs, D. Flaherty, J. Rankin. 1982b. Pulmonary reaction to durum wheat. *Chest* 1: 55-61.

doPico, G.A., W.G. Reddan, A. Tsiatis, D. Flaherty, S. Jacobs, J. Rankin. 1981. Pulmonary reaction to grain dust and its constituents. *Chest* 1: 57S-60S.

doPico, G.A., W. Reddan, D. Flaherty, et al. 1977. Respiratory abnormalities among grain handlers. *Am. Rev. Respir. Dis.* 115: 915-26.

Dosman, J.A., D.J. Cotton, and B.L. Graham. 1980. Chronic bronchitis and decreased forced expiratory flow rates in lifetime nonsmoking grain workers. *Am. Rev. Respir. Dis.* 121: 11-16.

Dosman, J.A., B.L. Graham, D. Hall, P. Pahwa, H.H. McDuffie, and T. Lucewicz. 1988. Respiratory symptoms and alterations in pulmonary function tests in swine producers in Saskatchewan: Results of a survey of farmers. *J. Occup. Med.* 30: 715-720.

Douglas, W.W., N.G.G. Hepper, T.V. Colby. 1989. Silo-filler's disease. *Mayo Clin. Proc.* 64: 291-304.

Edwards, J.H., J.T. Baker, B.H. Davies. 1974. Precipitin test negative farmer's lung - activation of the alternative pathway of complement by mouldy hay dusts. *Clin. Allergy* 4: 379-88.

Emanuel, D.A., F.J. Wenzel, C.I. Bowerman, and B.R. Lawton. 1964. Farmer's lung. *Am. J. Med.* 37: 392-401.

Emanuel, D.A., J.J. Marx, B. Ault, R.C. Roberts, M.J. Kryda, and M.W. Treuhaft. 1989. Organic dust toxic syndrome (pulmonary mycotoxicosis) - a review of the experience in central Wisconsin. *Respiratory Risks in Agriculture*, pp. 72-75.

Emanuel, D.A., F.J. Wenzel, and B.R. Lawton. 1975. Pulmonary mycotoxicosis. *Chest* 67: 293-97.

Emanuel, D.A., J. Marx, B. Ault, M. Treuhaft, R. Roberts, and M. Kryda. 1986. Pulmonary mycotoxicosis revisited. *Am. J. Ind. Med.* 10: 305-06.

Enarson, D.A., S. Vedal, M. Chan-Yeung. 1989. Assessment of airway responsiveness in working populations. In: *Principles of Health and Safety in Agriculture*. J.A. Dosman and D.W. Cockcroft, eds. Boca Raton, FL: CRC Press, pp. 19-22.

Epler, G.R., T.C. McLoud, E.A. Gaensler, et al. 1978. Normal chest roentgenograms in chronic diffuse infiltrative lung disease. *N. Engl. J. Med.* 298: 934-39.

Epler, G.R. 1989. Editorial: Silo-filler's disease: A new perspective. *Mayo Clin. Proc.* 64: 368-70.

Farant, J-P. 1989. Assessment of dust nature and levels in the grain industry. In: *Principles of Health and Safety in Agriculture*. J.A. Dosman and D.W. Cockcroft, eds. Boca Raton, FL: CRC Press, pp. 178-181.

Fink, J.F. 1991. The alveolar macrophage and hypersensivity pneumonitis. *J. Lab. Clin. Med.* 435-37.

Fink, J.N. 1992. Hypersensitivity pneumonitis. *Clin. Chest Med.* 13: 303-09

Flaherty, D.K., S.R. Braun, J.J. Marx, J.L. Blank, D.A. Emanuel, J. Rankin. 1980. Serologically detectable HLA-A, B, and C Loci antigens in farmer's lung disease. *Am. Rev. Respir. Dis.* 122: 437-43.

Fleetham, J.A., P.W. Munt, and B.W. Tunnicliffe. 1978. Silo-filler's disease. *Can. Med. Assoc. J.* 119: 482-84.

Fournier, E., A.B. Tonnel, P.H. Gosset, B. Wallaert, J.C. Amerisen, and C. Voisin. 1985. Early neutrophil alveolitis after antigen inhalation in hypersensitivity pneumonitis. *Chest* 88: 563-66.

Frank, R.C. 1958. Farmer's lung- a form of pneumoconiosis due to organic dusts. *Am. J. Roentgenology* 79: 189-215.

Franzese, A.C., G. Scala, S. Sproviero, G. Raucci. 1994. Farm workers' occupational allergy to tetranychus urticae: clinical and immunologic aspects. *Allergy* 49: 466-71.

Freedman, P.M., and B. Ault. 1981. Bronchial hyperreactivity to methacholine in farmer's lung disease. *J. Allergy Clin. Immun.* 67: 59-63.

Fulhage, C. 1980. Gases and odors from swine wastes. University of Missouri. 1880-1881.

Gamsky, T.E., S.A. McCurdy, S.J. Samuels, and M.B. Schenker. 1992. Reduced FVC among California grape workers. *Am. Rev. Respir. Dis.* 145: 255-56.

Gangwar, M., Z.U. Khan, S.N. Gaur, and H.S. Randhawa. 1991. Occurrence and significance of precipitating antibodies against thermophilic actinomycetes in the sera of dairy herd workers, Nangali, Delhi. *Antonie van Leeuwenhoek* 59: 167-75.

Garry, V.F., P.F. Good, J.C. Manivel, and D.P. Perl. 1993. Investigation of a fatality from nonoccupational aluminum phosphide exposure: measurement of aluminum in tissue and body fluids as a marker of exposure. *J. Lab. Clin. Med.* 122: 739-47.

Gerrard, J.W., J. Mink, S.C. Cheung, L.K.-T Tan, and J.A. Dosman. 1979. Nonsmoking grain handlers in Saskatchewan: Airways reactivity and allergic status. *J. Occup. Med.* 21: 342-46.

Ghadirian, E., and Denis. 1992. Murine hypersensitivity pneumonitis: interleukin-4 administration partially abrogates the disease process. *Microbial Patho.* 12: 377-82.

Ghose, T., P. Landrigan, R. Killeen, and J. Dill. 1974. Immunopathological studies. *Clin. Allergy* 4: 119-29.

Gjerde, C., K. Ferguson, C. Mutel, K. Donham, and J. Merchant. 1991. Results of an educational intervention to improve the health knowledge, attitudes and self-reported behaviors of swine confinement workers. *J. Rural Health* 7: 278-86.

Godard, P., J. Clot, O. Jonquet, J. Bousquet, and F.B. Michel. 1981. Lymphocyte subpopulations in bronchoalveolar lavages of patients with sarcoidosis and hypersensitivity pneumonitis. *Chest* 80: 447-52.

Goynes, W.R., B.F. Ingber, and M.S. Palmgren. 1986. Microscopical comparison of cotton, corn, and soybean dusts. *Env. Health Persp.* 66: 125-33.

Grant, I.W.B., W. Blyth, V.E. Wardrop, R.M. Gordon, J.C.G. Pearson, and A. Mair. 1972. Prevalence of farmer's lung in Scotland: A pilot survey. *Brit. Med. J.* 1: 530-34.

Grayson, R.R. 1956. Silage gas poisoning: nitrogen dioxide pneumonia, a new disease in agricultural workers. 45: 393-408.

Grzybowski, S., M. Chan-Yeung, and M.J. Ashley. 1980. Atopy and grain dust exposure. In: *Occupational Pulmonary Disease: Focus on Grain Dust and Health*. J.A. Dosman and D.J. Cotton, eds. New York: Academic Press, pp. 257-62.

Gurney, J.W. Hypersensitivity pneumonitis. 1992. *Radiologic Clin. N. Am.* 30: 1219-30.

Hahn, H.L. 1980. Nonspecific bronchial challenge and bronchial hyperreactivity. *Occupational Pulmonary Disease Focus on Grain Dust and Health*: pp. 153-60.

Hammar, S. 1988. Hypersensitivity Pneumonitis. *Pathol-Annu* 23: 195-215.

Hapke, E.J., R.M.E. Seal, G.O. Thomas, M. Hayes, and J.C. Meek. 1968. Farmer's lung. *Thorax* 23: 451-68.

Hardy, K.R., and S.R. Thom. 1994. Pathophysiology and treatment of carbon monoxide poisoning. *Clin. Toxic.* 32: 613-29.

Hargreave, F.E. 1980. Nonspecific airway reactivity in occupational pulmonary disease. In: *Occupational Pulmonary Disease: Focus on Grain Dust and Health*. J.A. Dosman and D.J. Cotton, eds. New York: Academic Press, pp51-64.

Hartung, J., V.R. Phillips. 1994. Control of gaseous emissions from livestock buildings and manure stores. *J. Agr. Engnr. Res.* 57: 173-89.

Haslam, P.L. 1987. Bronchoalveolar lavage in extrinsic allergic alveolitis. *Eur. J. Respir. Dis.*, Supplement 154, 71: 120-35.

Health Effects of Organic Dusts in the Farm Environment. 1986. Special Issue from the Proceedings of an International Workshop held in Skokloster, Sweden, April 23-25, 1985. R. Rylander, K. J. Donham, and Y. Peterson, eds. *Am. J. Ind. Med.* 10: 193-340.

Heber, A.J., and M. Stroik. 1987. Factors influencing dust characteristics in a survey of swine finishing units. *ASAE*, pp. 1-23.

Heino, M., S. Monkare, T. Haahtela, L.A. Laitinen. 1982. An electron-microscopic study of the airways in patients with farmer's lung. *Eur. J. Respir. Dis.* 63: 52-61.

Heinonen, O.P., M. Horsmanheimo, I. Vohlonen, E.O. Terho. 1987. Prevalence of allergic symptoms in rural and urban populations. *Eur. J. Respir. Dis.* 71: 64-69.

Helmers, S., F.H. Top, and L.W. Knapp. 1971. Ammonia injuries in agriculture. *J. Iowa Med. Soc.* LXI: 280.

Hendrick, D.J., R. Marshall, J.A. Faux, J.M. Krall. 1981. Protective value of dust respirators in extrinsic allergic alveolitis: clinical assessment using inhalation provocation tests. *Thorax* 36: 917-21.

Hendrick, A., C. Van Peteghem, M. Van Den Heede, and R. Lauwrert. 1976. A double fatality with children due to fumigated wheat. *Eur. J. Toxicol.* 9: 113-18.

Hodgkin, J.E., D.E. Abbey, G.L. Euler, and A.R. Magie. 1984. COPD prevalence in nonsmokers in high and low photochemical air pollution areas. *Chest* 86: 830-38.

Hodgson, M.J., D.K. Parkinson, and M. Karpf. 1989. Chest x-rays in hypersensitivity pneumonitis: A metaanalysis of secular trend. *Am. J. Ind. Med.* 16: 45-53.

Hollingdale, M.R. 1974. Antibody responses in patients with farmer's lung disease to antigens from micropolyspora faeni. *J. Hyg. Camb.* 72: 79-89.

Holness, D.L., I.G. Taraschuk, and R.S. Goldstein. 1982. Acute exposure to cotton dust. *JAMA* 247: 1602-03.

Holness, D.L., and J.R. Nethercott. 1989. Respiratory status environmental exposure of hog confinement and control farmers in Ontario. In: *Principles of Health and Safety in Agriculture.* J.A. Dosman, and D.W. Cockcroft, eds. Boca Raton, FL: CRC Press, pp. 69-71.

Holness, D.L. 1995. What actually happens to the farmers? Clinical results of a follow-up study of hog confinement farmers. *Agricultural Health and Safety: Workplace, Environment, Sustainability.* Chapter 1:49-52.

Horvath, E.P., G.A. doPico, R.A. Barbee, and H.A. Dickie. 1978. Nitrogen dioside-induced pulmonary disease. *Ann. Occup. Med.* 20: 103-10.

Hudgel, D.W. Bronchial Asthma. 1994. *Textbook of Pulmonary Disease* 1: 647-87.

Husman, K., I. Vohlonen, E.O. Terho, and R.A. Mantyjarvi. 1987a. Precipitins against microbes in mouldy hay in the sera of farmers with farmer's lung or chronic bronchitis and of healthy farmers. *Eur. J. Respir. Dis.* 71: 122-27.

Husman, K., M. Koskenvuo, J. Kaprio, E.O. Terho, and I. Vohlonen. 1987b. Role of environment in the development of chronic bronchitis. *Eur. J. Respir. Dis.* 71: 57-63.

Ingram, C.G., I.S. Symington, I.G. Jeffrey, and O.D. Cuthbert. 1979. Bronchial provocation studies in farmers allergic to storage mites. *The Lancet* 1330-32.

Iverson, M., R. Dahl, E.J. Jensen, J. Korsgaard, T. Hallas. 1989. Lung function and bronchial reactivity in farmers. *Thorax* 44: 645-49.

Jones, G.R., A.T. Proudfoot, and J.I. Hall. 1973. Pulmonary effects of acute exposure to nitrous fumes. *Thorax* 28: 61-65.

Jorna, T.H.J.M., P.J.A. Borm, J. Valks, R. Houba, and E.F.M. Wouters. 1994. Respiratory symptoms and lung function in animal feed workers. *Chest* 106: 1050-55.

Katila, M., T. Ojanen, and R. Mantyjarvi. 1986. A six-year follow-up of antibody levels against microbes present in the farming environment in a group of dairy farmers in finland. *Am. J. Ind. Med.* 10: 307-09.

Kaukonen, K., J. Savolainen, M. Viander, and E.O. Terho. 1994. Avidity of Aspergillus embrosus IgG antibodies in farmer's lung disease. *Clin. Exp. Immun.* 95: 162-65.

Kawai, T., J. Salvaggio, J.O. Harris, and P. Arquembourg. 1973. Alveolar macrophage migration inhibition in animals immunized with thermophilic actinomycete antigen. *Clin. Exp. Immun.* 15: 123-30.

Kennedy, S.M., H. Dimich-Ward, A. Desjardins, A. Kassam, S. Vedal, and M. Chan-Yeung. 1994. Respiratory health among retired grain elevator workers. *Am. J. Respir. Crit. Care Med.* 150: 59-65.

Kiviloog, J., L. Irnell, G. Eklund. 1974. The prevalence of bronchial asthma and chronic bronchitis in smokers and non-smokers in a representative local Swedish population. *J. Resp. Dis.* 55: 262-76.

Kobayashi, M., M.A. Stahmann, J. Rankin, and H.A. Dickie. 1963. Antigens in moldy hay as the cause of farmer's lung. *Proc. Soc. Exp. Biol. Med.* 113: 472.

Kokkarinen, J.I., H.O. Tukiainen, and E.O. Terho. 1992. Effect of corticosteroid treatment on the recovery of pulmonary function in farmer's lung. *Am. Rev. Respir. Dis.* 145: 3-5.

Kokkarinen, J.I., H.O. Tukiainen, and E.O. Terho. 1993. Recovery of pulmonary function in farmer's lung. *Am. Rev. Respir. Dis.* 147: 793-96.

Kolmodin-Hedman, B., and N. Stjernberg. 1986. Symptoms in farmers exposed to moldy wood chips. *Am. J. Ind. Med.* 10: 310.

Kotimaa, M.H., L. Oksanen, P. Koskela. 1991. Feeding and bedding materials as sources of microbia exposure on dairy farms. *Scand. J. Work Envir. Health* 17: 117-22.

Kristiansen, J.D., and A.X. Lahoz. 1991. Riding-school lung? Allergic Alveolitis in an 11-year-old girl. *Acta Paediatr. Scand.* 80: 386-88.

Kumar, A., N. Elms, and P.K. Viswanath. 1993. Monoclonal antibodies against farmer's lung antigens having specific binding to IgG antibodies. *Int. Arch. Allergy Immun.* 102: 67-71.

Kusaka, H., H. Ogasawara, M. Munakata, K. Tanimura, H. Ukita, N. Denzumi, Y. Homma, and Y. Kawakami. 1993. Two-year follow-up on the protective value of dust masks against farmer's lung disease. *Int. Med.* 32: 106-11.

Lacey, J. 1980. The microflora of grain dusts. In: *Occupational Pulmonary Disease: Focus on Grain Dust and Health.* J.A. Dosman JA and D.J. Cotton, eds. New York: Academic Press, pp. 417-440.

Lalancette, M., G. Carrier, M. Laviolette, S. Ferland, J. Rodrique, R. Begin, A. Cantin, and Y. Cormier. 1993. Farmer's lung. *Am. Rev. Respir. Dis.* 148: 216-21.

Larsson, K.A., A. Eklund, L-O. Hansson, B-M. Isaakson, and P.O. Malmberg. Swine dust causes intense airways inflammation in healthy subjects. *Am. J. Respir. Crit. Care Med.* 150: 973-77.

Larsson, K., A. Eklund, and P. Malmberg. 1992. Alterations in bronchoalveolar lavage fluid but not in lung function and bronchial responsiveness in swine confinement workers. *Chest* 101: 767-74.

Larsson, K., A. Eklund, P. Malmberg, L. Bjermer, R. Lundgren, and L. Belin. 1991. Hyaluronic acid (hyaluronan) in BAL fluid distinguishes farmers with allergic alveolitis from farmers with asymptomatic alveolitis. *Chest* 101: 109-14.

Laviolette, M., Y. Cormier, A. Loiseau, P. Soler, P. Leblanc, and A.J. Hance. 1991. Bronchoalveolar mast cells in normal farmers and subjects with farmer's lung. *Am. Rev. Respir. Dis.* 144: 855-60.

Leatherman, J.W., A.F. Michael, B.A. Schwartz, and R. Hoidal. Lung T cells in hypersensitivity pneumonitis. *Annals Int. Med.* 100: 390-92.

Leblanc, P., J. Belanger, M. Laviolette, and Y. Cormier. 1986. Relationship among antigen contact, alveolitis, and clinical status in farmer's lung disease. *Arch. Inter. Med.* 146: 153-57.

Lecours, R., M. Laviolette, and Y. Cormier. 1986. Bronchoalveolar lavage in pulmonary mycotoxicosis (organic dust toxic syndrome). *Thorax* 41: 924-26.

Leduc, D., P. Gris, P. Lheureux, P.A. Gevenois, P. De Vuyst, J.C. Yernault. 1992. Acute and long term respiratory damage following inhalation of ammonia. *Thorax* 47: 755-57.

Lefcoe, N.M., T.H. Wonnacott. 1974. Chronic respiratory disease in four occupational groups. *Arch. Envir. Health* 29: 143-46.

Leib, G.M.P., W.N. Davis, T. Brown, M. McQuiggan. 1958. Chronic pulmonary insufficiency secondary to silo-filler's disease. *Am. J. Med.* 24: 471-74.

Lenhart, S.W., and L.D. Reed. 1989. Respiratory protection for use against organic dusts. *Env. Assess. Health Prom.* 193-96.

Leskinen, L., and T. Klen. 1987. Storage mites in the work environment of farmers. Kuopio Regional Institute of Occupational Health, Kuopio, Finland. *Eur. J. Resp. Dis.* 71: 101-12.

Levin, P.J., L.J. Klaff, A.G. Rose, and A.D. Ferguson. 1979. Pulmonary effects of contact exposure to paraquat: A clinical and experimental study. *Thorax* 34: 150-60.

Lewis, D.M., P.A. Romeo, and S.A. Olenchock. 1986. Prevalence of IgE antibodies to grain and grain dust in grain elevator workers. *Env. Health Persp.* 66: 149-53.

Louhelainen, K., P. Vilhunen, J. Kangas, and E.O. Terho. 1987. Dust exposure in piggeries. *Eur. J. Respir. Dis.* 71: 80-90.

Lowrey, T., L.M. Schuman. 1956. Silo-filler's disease - a syndrome caused by nitrogen dioxide. *JAMA* 162: 153-60.

Lundholm, B.S., R. Rylander. 1980. Occupational symptoms among compost workers. *J. Occup. Med.* 22: 256-57.

Lynch, D.A., C.S. Rose, D. Way, and T.E. King. 1992. Hypersensitivity pneumonitis: Sensitivity of high-resolution CT in a population-based study. *AJR* 159: 469-72.

Madsen, D., L.E. Klock, F.J. Wenzel, J.L. Robbins, and C.D. Schmidt. 1976. The prevalence of farmer's lung in an agricultural population. *Am. Rev. Respir. Dis.* 113: 171-72.

Making swine buildings a safer place to work. 1990. From the proceedings of a symposium of the National Pork Producers Council in Davenport, Iowa.

Malmberg, P., and K. Larsson. 1993. Acute exposure to swine dust causes bronchial hyperresponsiveness in health subjects. *Eur. Respir. J.* 6: 400-04.

Malmberg, P., A. Rask-Andersen, U. Palmgren, S. Hoglund, B. Kolmodin-Hedman, and G. Stalenheim. 1985. Exposure to microorganisms, febrile and airway-obstructive symptoms, immune status and lung function of Swedish farmers. *Scand. J. Work Environ. Health* 11: 287-93.

Malmberg, P., A. Rask-Andersen, S. Hoglund, B. Kolmodin-Hedman, J.R. Guernsey. 1988. Incidence of organic dust toxic syndrome and allergic alveolitis in Swedish farmers. *Int. Arch. Allergy Appl. Immun.* 87: 47-54.

Marx, J.J., J. Guernsey, D.A. Emanuel, J.A. Merchant, D. Morgan, and M. Kryda. 1990. Cohort studies of immunologic lung disease among Wisconsin dairy farmers. *Am. J. Ind. Med.* 18: 263-68.

Marx, J.J., and R.L. Gray. 1982. Comparison of the enzyme-linked immunosorbent assay and double immunodiffusion test for the detection and quantitation of antibodies in farmer's lung disease. *J. Allergy Clin. Immun.* 70: 109-113.

Marx, J.J., J.T. Twiggs, B.J. Ault, J.A. Merchant, and E. Fernandez-Caldas. 1993. Inhaled aeroallergen and storage mite reactivity in a Wisconsin farmer nested case-control study. *Am. Rev. Respir. Dis.* 147: 354-58.

Marx, J.J., F.J. Wenzel, and R.C. Roberts, et al. 1973. Migration inhibition factors and farmer's lung antigens. *Clin. Res.* 21: 852-57.

Marx, J.J., M.P. Arden-Jones, M.W. Treuhaft, R.L. Gray, C.S. Motszko, and F.F. Hahn. 1981. The pathogenetic role of inhaled microbial material in pulmonary mycotoxicosis as demonstrated in an animal model. *Chest* 80: 76S-78S.

Matson, S.C., M.C. Swanson, C.E. Reed, and J.W. Yunginger. 1983. IgE and IgG-immune mechanisms do not mediate occupation-related respiratory or systemic symptoms in hog farmers. *J. Allergy Clin. Immun.* 72: 299-304.

May, J.J., and M.B. Schenker. 1996. Agriculture. In: *Occupational Environment Respiratory Disease*. St. Louis: Mosby, pp 617-636.

May, J.J., L. Stallones, D. Darrow, and D.S. Pratt. 1986a. Organic dust toxicity (pulmonary mycotoxicosis) associated with silo unloading. *Thorax* 41: 919-23.

May, J.J., L.H. Marvel, D.S. Pratt, and D.P. Coppolo. 1990. Organic dust toxic syndrome: A follow-up study. *Am. J. Ind. Med.* 17: 111-13.

May, J.J., D.S. Pratt, L. Stallones, P. Morey, S.A. Olenchock, I.W. Deep, and G.A. Bennett. 1986b. A study of silo unloading: The work environment and its physiologic effects. *Am. J. Ind. Med.* 10: 318.

Meneely, G.R., A.D. Renzetti, Jr., and J.D. Steele. 1962. Chronic bronchitis, asthma, and pulmonary emphysema. *Am. Rev. Resp. Dis.* 85: 762-68.

Merchant, J.A. and K.J. Donham. 1989. Health risks from animal confinement units. In: *Principles of Health and Safety in Agriculture*. J.A. Dosman and D.W. Cockcroft, eds. Boca Raton, FL: CRC Press, pp. 58-61.

Meyer, V.M., and D.S. Bundy. 1991. Farrowing building air quality survey. *Swine Research Report,* pp. 68-70.

Miadonna, A., A. Pesci, A. Tedeschi, G. Bertorelli, M. Arquati, and D. Olivieri. 1994. Mast cell and histamine involvement in farmer's lung disease. *Chest* 105: 1184-89.

Miller, A., J.C. Thornton, H.A. Anderson, and I.J. Selikoff. 1988 Clinical respiratory abnormalities in Michigan. *Chest* 94: 1187-94.

Mink, J.T., J.W. Gerrard, D.W. Cockcroft, D.J. Cotton, and J.A. Dosman. 1980. Increased bronchial reactivity to inhaled histamine in nonsmoking grain workers with normal lung function. *Chest* 77: 28-31.

Mönkäre, S., T. Haahtela, and L. Ikonen. 1981. Bronchial hyperactivity to inhaled histamine in patients with farmer's lung. *Lung* 159: 145-50.

Mönkäre, S., T. Haahtela. 1987. Farmer's lung-a 5-year follow-up of eighty-six patients. *Clin. Allergy* 17: 143-51.

Moore, V.L., J.N. Fink, J.J. Barboriak, et al. 1974. Immunologic events in pigeon breeder's disease. *J. Allergy Clin. Immun.* 53: 319-28.

Morbidity and Mortality Weekly Report. 1992. Fatalities attributed to entering manure waste pits-Minnesota, 1992. Silo-Filler's Disease. July 23.

Müller-Wening, D., M. Schmitt. 1990. Comparison of the effectiveness of two respirators in farmers suffering from farmer's lung. *Pneumologie* 44: 781-86.

Müller-Wening, D., H. Repp H. 1989. Investigation on the protective value of breathing masks in farmer's lung using an inhalation provocation test. *Chest* 95: 100-05.

Mulvey, J.J., H. Rothschild, A. Ciegler, Y. Fan, and R.L. Marier. 1983. The association of fungi with sugarcane production and lung cancer mortality in Louisiana. *J. Louis.State Med. Soc.* 135: 3-4.

NIOSH Update. 1993a. *NIOSH Warns farmers of deadly risk of grain suffocation.* April 28.

NIOSH Update. 1993b. *NIOSH Warns of deadly carbon monoxide hazard from using pressure washers indoors.* May 10.

NIOSH Update. 1993c. *NIOSH Warns: Manure pits continue to claim lives.* July 6.

NIOSH Health Hazard Evaluation Report. Report 83-195-1426.

O'Connell, E.J., J.A. Zora, D.N. Gillespie, and E.C. Rosenow. 1989. Childhood hypersensitivity pneumonitis (farmer's lung): Four cases in siblings with long-term follow-up. *J. Pediatrics* 995-997.

Ojanen, T. 1992. Class specific antibodies in serodiagnosis of farmer's lung. *Brit. J. Ind. Med.* 49: 332-36.

Ojanen, T., E.O. Terho, H. Tukiainen, and R.A. Mantyjarvi. 1990. Class-specific antibodies during follow up of patients with farmer's lung *Eur. Resp. J.* 3: 257-60.

Olenchock, S.A., D.M. Lewis, and J.C. Mull. 1986. Composition of extracts of airborne grain dusts: Lectins and lymphocyte mitogens. *Envir. Health Persp.* 119-123.

Olenchock, S.A., J.J. May, D.S. Pratt, and L.A. Piacitelli. 1990. Presence of endotoxins indifferent agricultural environments. *Am. J. Ind. Med.* 18: 279-84

Olenchock, S.A., J.C. Mull, P.C. Major. 1980. Extracts of airborne grain dusts activate alternative and classical complement pathways. *Ann. Allergy* 44: 23-28.

Olenchock, S.A., and R. Burrell. 1976. The role of precipitins and complement activation in the etiology of allergic lung disease. *J. Allergy Clin. Immun.* 58: 76-88.

O'Neill, D.H., and V.R. Phillips VR. 1992. A review of the control of odour nuisance from livestock buildings: Part 3, properties of the odorous substances which have been identified in livestock wastes or in the air around them. *J. Agri. Eng. Res.* 53: 23-50.

Orefice, U., P.L. Struzzo, F. Pignatelli, and P.L. Ferrazzano. 1991. Cellular modifications in the BAL-fluid exposed farmers to antigens of extrinsic allergic alveolitis. *Sarcoidosis* 8: 189-90.

Organic Dusts and Lung Disease. 1990. Special Issue in Proceedings of an International Workshop held in Skokloster, Sweden, October 24-27, 1988. Guest Editors: Ragnar Rylander and Yvonne Peterson. *Am. J. Ind. Med.* 17: 1-144.

Osbern, L.N., and R.O. Crapo. 1981. Dung lung: a report of toxic exposure to liquid manure. *Ann. Intern. Med.* 95: 312-14.

Pahwa, P., A. Senthilselvan, H.H. McDuffie, and J.A. Dosman. 1994. Longitudinal estimates of pulmonary function decline in grain workers. *Am. J. Respir. Crit. Care Med.* 150: 656-62.

Parnell, C.B., D.D. Jones, R.D. Rutherford, and K.J. Goforth. 1986. Physical properties of five grain dust types. *Envir. Health Pers.* 66: 183-88.

Parry, R.R., and E. Schlenker. 1994. Respiratory, atopic, and serological characterization of Hutterite farmers in South Dakota. Supplement to Agricultural Health and Safety: Workplace, Environment, Sustainability. Centre for Agricultural Medicine, University of Saskatchewan, Saskatoon, Saskatchewan. Canada, pp.1-6.

Pepys, J., and P.A. Jenkins. 1965. Precipitin (F.L.H.) test in farmer's lung. *Thorax* 20: 21-35.

Pepys, J., P.A. Jenkins. 1963. Farmer's Lung. Thermophilic actinomycetes as a source of "farmer's lung hay" antigen. *The Lancet* 607-11.

Pepys, J., R.W. Riddell, K.M. Citron, and Y.M. Clayton. 1962. Precipitins against extracts of hay and moulds in the serum of patients with farmer's lung, aspergillosis, asthma, and sarcoidosis. *Thorax* 17: 366-74.

Pesci, A., G. Bertorelli, P.P. Dall'Agleo, et al. 1990. Evidence in bronchoalveolar lavage for third type immuno reactions hypersensitivity pneumonitis. *Eur. Respir. J.* 3: 359-61.

Pesci, A., G. Bertorelli, M. Marvisi, D. Olivieri. 1991a. In patients with farmer's lung mast cells in BAL reflect mast cells in lung tissue. *Sarcoidosis* 8: 177-79.

Pesci, A., G. Bertorelli, and D. Olivieri. 1991b. Mast cells in bronchoalveolar lavage fluid and in transbronchial biopsy specimens of patients with farmer's lung disease. *Chest* 100: 1197-1202.

Pforte, A., U. Schild, G. Breyer, K. Haussinger, and H.W.L. Ziegler-Heitbrock. 1992. A role for IgE in extrinsic allergic alveolitis? *Clin. Investig.* 70: 277-82.

Pickrell, J.A., A.J. Heber, J.P. Murphy, M.M. May, D. Nolan, F.W. Oehme, D. Schoneweis, J.R. Gillespie, and S.C. Henry. 1992. Endotoxin, ammonia, and total and respirable dust in swine confinement buildings: the effect of recirculated air and respiratory protecive masks. Kansas State University Swine Day, Report of Progress 6676, Agricultural Experiment Station, Marc A. Johnson, Director. pp. 167-172.

Pratt, D.S., L. Stallones, D. Darrow, and J.J. May JJ. 1986. Acute respiratory illness associated with silo unloading. *Am. J. Ind. Med.* 10: 328.

Pratt, D.S., and J.J. May. 1984. Feed-associated respiratory illness in farmers. *Arch.Envir.Health* 39: 43-48.

Preller, L., D. Heederik, J.S.M. Boleij, F.J. Vogelzang, and M.J.M. Tielen. 1995. Lung function and chronic respiratory symptoms of pig farmers: focus on exposure to endotoxins and ammonia and use of disinfectants. *Occup. Envir. Med.* 52: 654-60.

Quinn, T.J., K.J. Donham, J.A. Merchant, and D.A. Schwarts. 1995. Peak flow as a measure of airway dysfunction in swine confinement operators. *Chest* 107: 1303-08.

Ramazzini, B.N. 1969. A treatise of the diseases of tradesmen. *JAMA* 210: 2391-92.

Ramirez-R, and A.R. Dowell. 1971. Silo-filler's disease: nitrogen dioxide-induced lung injury. *Annals of Int. Med.*, pp. 569-576.

Ramsdell, J.W., F.J. Nachtwey, and K.M. Moser. 1982. Bronchial hyperreactivity in chronic obstructive bronchitis. *Am. Rev. Respir. Dis.* 126: 829-32.

Rankin, J., M. Kobayashi, R.A. Barbee, et al. 1967. Pulmonary granulomatosis due to inhaled organic antigen. *Med. Clin. N. Amer.* 51: 459-82.

Rask-Andersen, A., P. Malmberg, and M. Lundholm. 1989a. Endotoxin levels in farming: absence of symptoms despite high exposure levels. *Brit. J. Ind. Med.* 46: 412-16.

Rask-Andersen, A. 1992. Inhalation fever: A proposed unifying term for febrile reactions to inhalation of noxious substances. *Brit. J. Ind. Med.* 49:40.

Rask-Andersen, A. 1989b. Organic dust toxic syndrome among farmers. *Brit. J. Ind. Med.* 46: 233-38.

Rask-Andersen, A. 1995. The organic dust toxic syndrome - a review. *Agricultural Health and Safety: Workplace, Environment, Sustainability.* Chapter 1.21:101.

Radtogi, S.K., B.N. Gupta, T. Husain, N. Mathur, and N. Garg. 1989. Study of respiratory impairment among pesticide sprayers in mango plantations. *Am. J. Ind. Med.* 16: 529-38.

Rastogi, S.K., B.N. Gupta, T. Husain, N. Mathur, and N. Garg. 1989. Study of Respiratory impairment among pesticide sprayers in mango plantations. *Am J Ind Med* 16: 529-38.

Rautalahti, M., E.O. Terho, I. Vohlonen, J. Nuutinen, K. Husman, and O. Korhonen. 1987. Effect of indoor feeding season for cattle on lung function of dairy farmers. *Eur. J. Respir. Dis.* 71: 188-96.

Reynolds, H.Y., J.D. Fulmer, J.A. Kazmierowski, W.C. Roberts, M.M. Frank, and R.G. Crystal RG. 1977. Analysis of cellular and protein content of broncho-alveolar lavage fluid from patients with idiopathic pulmonary fibrosis and chronic hypersensitivity pneumonitis. *J. Clin. Invest.* 59: 165-175.

Reynolds, H.Y. 1988. Hypersensitivity pneumonitis: correlation of cellular and immunologic changes with clinical phases of disease. *Lung* 166: 189-208.

Reynolds, H.Y. 1982a. Today's practice of cardiopulmonary medicine. Immunologic lung diseases (Part 1). *Chest* 81: 626-31.

Reynolds, H.Y. 1982b. Immunologic lung diseases (Part 2). *Chest* 81: 745-51.

Richerson, H.B., I.L. Bernstein, J.N. Fink, G.W. Hunninghake, H.S. Novey, C.E. Reed, J.E. Salvaggio, M.R. Schuyler, H.J. Schwartz, and D.J. Stechschulte. 1989. Guidelines for the clinical evaluation of hypersensitivity pneumonitis. *J Allergy Clin Immunol* 84:839-844.

Richerson HB. 1974. Varieties of acute immunologic damage to the rabbit lung. *Annals New York Academy of Sciences* 221:340-360.

Rijcken B, Schouten JP, Weiss ST, Speizer FE, Van der Lende, R. 1987. The relationship of nonspecific bronchial responsiveness to respiratory symptoms in a random population sample. *Am Rev Respir Dis* 136:62-68.

Rylander R. 1994. Symptoms and mechanisms--inflammation of the lung. Am J Ind Med 25: 19-23.

Rylander R, Bake B, Fischer JJ, Helander IM. 1989. Pulmonary function and symptoms after inhalation of endotoxin. Am *Rev Respir Dis* 140:981-986.

Rylander, R., K. Andersson, L. Belin, G. Berglund, R. Bergstrom R, Hanson L, Lundholm M, Mattsby I. 1976. Sewage worker's syndrome. *The Lancet* 478-479.

Sandström T, Bjermer L, Rylander L. 1994. Lipopolysaccharide (LPS) inhalation in health subjects causes bronchoalveolar neutrophilia, lymphocytosis, and fibronectin increase. *Am J Ind Med* 25:105-107.

Sastre, L., M.D. Ibanez, M. Lopez, and S. Lehrere SB. 1990. Respiratory and immunological reactions among Shiitake (Leninus edodes) mushroom workers. *Clin. Exp. Allergy* 20: 13-19.

Schaller, R.E., and P.M. Nicholson RM. 1980. The nature, behavior, and characterization of grain dust. In: Dosman JA, Cotton DJ, eds. *Occupational Pulmonary Disease: Focus on Grain Dust and Health*. New York: Academic Press, pp513-525.

Schenker M, Ferguson T, Gamsky T. 1991. Respiratory risks associated with agriculture. *Occupational Medicine: State of the Art Reviews*. Vol. 6, No. 3, July-September. Philadelphia, Hanley & Belfus, Inc., 415-429.

Schlenker, E.H., G.R. Lenardson, C. McClain. et al. 1989. The prevalence of respiratory symptoms among farmers and ranchers in Southeastern South Dakota. In: *Priniciples of Health and Safety in Agriculture*. J.A. Dosman and D.w. Cockcroft DW, eds. Boca Raton, FL: CRC Press, pp85-7.

Schlenker, E.H., and R.R. Parry. 1990. Utilization of masks by Hutterite farmers. U of South Dakota Press; 13-17.

Schullian, D.M. 1982. Farmer's lung disease: A new historical perspective from Iceland. Notes and Events: 440-443.

Schuyler M. 1995. Pathogenesis of hypersensitivity pneumonitis. Cotton and Other Organic Dusts Conference. Special Invited Papers in San Antonio, Texas, January 6-7.

Schwartz D.A., S.K. Landas, D.L. Lassise DL, Burmeister L.F., G.W. Hunninghake, and J.A. Merchant JA. 1992. Airway injury in swine confinement workers. *Ann. Int. Med.* 116:b630-35.

Schwartz, D.A., K.J. Donham, S.A. Olenchock; W.J. Popendorf, D.S. Van Fossen, L.F. Burmeister, and J.A. Merchant. 1995a. Determinants of longitudinal changes in spirometric function among swine confinement operators and farmers. *Am. J. Respir. Crit. Care Med.* 151:47-53.

Schwartz, D.A., P.S. Thorne, S.J. Yagla, L.F. Burmeister, S.A. Olenchock, J.L. Watt, T.J. Quinn TJ. 1995b. The role of endotoxin in grain dust-induced lung disease. *Am. J. Respir. Crit. Care Med* 152:603-8.

Seal, R.M.E., E.J. Hapke, G.O. Thomas, J.C. Meek, and M. Hayes. 1968. The pathology of the acute and chronic stages of farmer's lung. *Thorax* 23:469-89.

Semenzato, G., M. Chilosi, E. Ossi, L. Trentin, G. Pizzolo, A. Cipriani, C. Agostini, R. Zambello, G. Marcer, and G. Gasparotto. 1985. Bronchoalveolar lavage and lung histology. *Am. Rev. Respir. Dis.* 132: 4000-04.

Semenzato, G., L. Trentin, R. Zambello, C. Agostini, A. Cipriani, and G. Marcer. 1988. Different types of cytoxic lymphocytes recovered from the lungs of patients with hypersensitivity pneumonitis. *Am. Rev. Respir.* 137: 70-74.

Semenzato, G. 1991. Immunology of interstitial lung diseases: cellular events taking place in the lung of sarcoidosis, hypersensitivity pneumonitis and HIV infection. *Eur. Respir. J.* 4: 94-102.

Sennekamp, J., M. Rust, R. Kroidl, and M. Spyra. 1990. IgA-Antibodies in extrinsic allergic alveolitis are more specific than IgG-antibodies. *Pneumologie* 44: 522-23.

Senthilselvan, A., H.H. McDuffie, and J.A. Dosman. 1992. Association of asthma with use of pesticides. *Am. Rev. Respir. Dis.* 146: 884-87.

Senthilselvan, A., H.H. McDuffie, J.A. Dosman. 1994. Asthma in farmers: association with cholinesterase inhabiting insecticides. Supplement to Agricultural Health and Safety: Workplace, Environment, Sustainability 17-22.

Senthilselvan, A., J.A. Dosman. 1995. Risk factors for allergic rhinitis in farmers. *Agricultural Health and Safety: Workplace, Environment, Sustainability*. Chelsea, MI: Lewis Publishers, Chapter 1.15:75.

Sharma, O.P. 1991. Hypersensitivity Pneumonitis. *Disease of the Month*, p. 411-471.

Shelley, E., G. Dean G, D. Collins, et al. 1979. Farmer's lung: a study in north-west Ireland. *J. Irish Med. Assoc.* 72:261-64.

Sibille, Y., J.B. Martinot, P. Staquet, L. Delaunois, B. Chatelain, and D.L. Delacroix. 1988. Antiproteases are increased in bronchoalveolar lavage in interstitial lung disease. *Eur. Respir. J.* 1:498-504.

Silver, S.F., N.L Muller, R.R. Miller, and M.S. Lefcoe. 1989. Hypersensitivity pneumonitis: Evaluation with CT. *Radiology* 173: 441-45.

Snider, G.L. 1994. History and Physical Examination. *Textbook of Pulmonary Diseases*. Boston, MA: Little, Brown and Company, pp. 243-271.

Solal-Celigny, P.H., M. Laviolette, J. Hebert, and Y. Cormier. 1982. Immune reactions in the lungs of asymptomatic dairy farmers. *Am. Rev. Respir. Dis.* 126: 964-67.

Skea, D.L., D. McAvoy, and I. Broder. 1988. Grain dust contains a tannin-like material which fixes complement. *Envir. Res.* 46: 181-89.

Sorenson, W.G. 1990. Mycotoxins as potential occupational hazards. *Develop.Ind. Micr.* 31: 205-11.

Stanford, C.F., G. Hall, A. Chivers, D.P. Nicholls, and J. Evans. 1990. Farmer's lung in Northern Ireland. *Brit. J. Ind. Med.* 47: 314-16.

Tabona, M., M. Chan-Yeung, D. Enarson, L. MacLean, E. Dorken, and M. Schulzer. 1984. Host factors affecting longitudinal decline in lung spirometry among grain elevator workers. *Chest* 85: 782-86.

Takizawa, H., K. Ohta, K. Hirai, M. Yoshikata, T. Horiuchi, N. Kobayashi, J. Shiga, T. Miyamoto. 1982. Mast cells are important in the development of hypersensitivity pneumonitis. *J. Immun.* 143: 1982-88.

Terho, E.O. 1986a. Diagnostic criteria for farmer's lung disease. *Am. J. Ind. Med.* 10: 329.

Terho, E.O., R.A. Mantyjarvi, O.P. Heinonen, T.H. Ojanen, I. Vohlonen, and H. Tukiainen. 1985. Familial aggregation of IgG antibody response to antigens associated with farmer's lung. *Int. J. Epid.* 14: 589-93.

Terho, E.O., K. Husman, I. Vohlonen, O.P. Heinonen. 1987a. Atopy, smoking, and chronic bronchitis. *J. Epidem. Comm. Health* 41: 300-05.

Terho, E.O. 1986b. Diagnostic criteria for farmer's lung disease. *Am. J. Ind. Med.* 10: 329.

Terho, E.O., O.P. Heinonen, R.A. Mantyjarvi, and J. Vohlonen. 1984. Familial aggregation of symptoms of farmer's lung. *Scand. J. Work Envir. Health* 10: 57-58.

Terho, E.O., O.P. Heinonen, and S. Lammi S. 1986c. Incidence of clinically confirmed farmer's lung disease in Finland. *Am. J. Ind. Med.* 10: 330.

Terho, E.O., K. Husman, and I. Vohlonen. 1987b. Prevalence and incidence of chronic bronchitis and farmer's lung with respect to age, sex, atopy, and smoking. *Eur. J. Resp. Dis.* 71: 19-28.

Terho, E.O., K. Husman, T. Kauppinen. 1987c. Proceedings of the international symposium on work-related respiratory disorders among farmers. *Eur. J. Respir. Dis.* 71 S 154: 120-35.

Terho, E.O., S. Lammi, O.P. Heinonen. 1980. Seasonal variation in the incidence of farmer's lung. *Intl. J. Epidem.* 9: 219-20.

Terho, E.O., I. Vohlonen, K. Husman, M. Rautalahti, H. Tukiainen, and M. Viander. 1987d. Sensitization to storage mites and other work-related and common allergens among finnish dairy farmers. *Eur. J. Respir. Dis.* 71: 165-74.

Terho, E.O., K. Husman, I. Vohlonen. 1987e. Work-related respiratory diseases among Finnish farmers. *Eur. J. Respir. Dis.* 71: 1-15.

Teschler, H., A.B. Thompson, W.R. Pohl, N. Konietzko, S.I. Rennard, and U. Costabel. 1993. Bronchoalveolar lavage procollagen-III-peptide in recent onset hypersensitivity pneumonitis: correlation with extracellular matrix components. *Eur. Respir. J.* 6: 709-14.

Thedell, T.D., J.C. Mull, and S.A. Olenchock. 1980. A brief report of gram-negative bacterial endotoxin levels in airborne and settled dusts in animal confinement buildings. *Am. J. Ind. Med.* 1: 3-7.

Thorne, P.S., M.S. Kiekhaefer, P. Whitten, and K.J. Donham. 1992. Comparison of bioaerosol sampling methods in barns housing swine. *Appl. Envir.Microb.* 58: 2543-51.

Thorne, P.S., K.J. Donham, J. Dosman, P. Jagielo, J.A. Merchant, P.S. Thorne, and S. Von Essen. 1995. Report of the Occupational Health Group: Proceeding from "Understanding the Impacts of Large Scale Swine Production: An Interdisciplinary Scientific Workshop." June 29-30: 1-41.

Trentin, L., G. Marcer, M. Chilosi, M.C. Sci, R. Zambello, C. Agostini, M. Masciarelli, R. Bizzotto, C. Gemignani, A. Cipriani, G.D. Vittorio, and G. Semenzato. 1988. Longitudinal study of alveolitis in hypersensitivity pneumonitis patients: an immunologic evaluation. *J. Allergy Clin. Immun.* 82: 577-85.

Trentin, L., N. Migone, R. Zambello, P. Celle, F. Aina, C. Feruglio, P. Bulian, M. Masciarelli, C. Agostini, A. Cipriani, G. Marcer, R. Foa, G. Pizzolo, and G. Semenzato. 1990. Mechanisms accounting for lymphocytic alveolitis in hypersensitivity pneumonitis. *J. Immun.* 145: 2147-54.

Tse, K.S., N. Craven, R.M. Cherniack. 1980. Allergy to saprophytic fungi in grain workers. *Occup. Pulm. Dis.* pp. 335-46.

Tse, K.S., P. Warren, M. Janusz, D.S. McCarthy, R.M. Cherniack. 1973. Respiratory abnormalities in workers exposed to grain dust. *Arch. Envir. Health* 27: 74-77.

Tupi, K., I. Vohlonen, E.O. Terho, and K. Husman. Effects of respiratory morbidity on occupational activity among farmers. 206-11.

Turner, F, and P.J. Nicholls. 1995. The role of the epithelium in the airway: Relevance to teh response to inhaled noxious agents. Cotton and Other Organic Dusts Conference. Special Invited Papers in San Antonio, Texas, January 6-7.

The University of Iowa. 1989. Swine Confinement and Respiratory Health. Cooperative Extension Service, Iowa State University of Science and Technology and the United States Department of Agriculture Cooperating. December.

U. S. Department of Health and Human Services. 1994. Work-Related Lung Disease Surveillance Report.

Vergnenegre, A., X. D'arco, B. Melloni, M.T. Antonini, C. Courat, Dupont-Cuisinier, and F. Bonnaud. 1995. Work related distal airway obstruction in an agricultural population. *Occup. Envir. Med.* 52: 581-86.

Virolainen, R., K. Tupi, E.O. Terho, K. Husman, V. Notkola, and I. Vohlonen. 1987. Characteristics of farmers who have obtained personal dust respirators. *Eur. J. Respir. Dis.* 71: 199-205.

Vogelmeier, C., F. Krombach, S. Munzing, G. Konig, G. Mazur, T. Beinert, and G. Fruhmann. 1993. Activation of blood neutrophils in acute episodes of farmer's lung. *Am. Rev. Respir. Dis.* 148: 396-400.

Vohlonen, I., E.O. Terho, M. Horsmanheimo, O.P. Heinonen, and K. Husman. 1987a. Prevalence of chronic bronchitis in farmers according to smoking and atopic skin sensitization. *Eur. J. Respir. Dis.* 71: 175-80.

Vohlonen, I., K. Husman, E.O. Terho, and K. Tupi. 1987b. Prevalence of serum precipitins against microbes in mouldy hay, and of chronic bronchitis and farmer's lung with respect to farmers occupational health hazards. *Eur. J. Resp. Dis.* 71: 139-43.

Von Essen, S. 1993. Bronchitis in agricultural workers. *Seminars in Respiratory Medicine* 14: 60-69.

Von Essen, S.G., D.P. O'Neill, S.A. Olenchock, R.A. Robbins, and S.I. Rennard. 1995a. Grain dusts and grain plant components vary in their ability to recruit neutrophils. *J. Toxicol. Env. Health* 46: 425-41.

Von Essen, S.G., D.P. O'Neill, S. McGranaghan, S.A. Olenchock, and S.I. Rennard. 1995b. Neutrophilic respiratory tract inflammation and peripheral blood neutrophilia after grain sorghum dust extract challenge. *Chest* 108: 1425-33.

Von Essen, S.G., A.B. Thompson, R.A. Robbins, K.K. Jones, C.A. Dobry, and S.I. Rennard. 1990a. Lower respiratory tract inflammation in grain farmers. *Am. J. Ind. Med.* 17: 75-76.

Von Essen, S.G., R.A. Robbins, A.B. Thompson, R.F. Ertl, J. Linder, and S. Rennard. 1988. Mechanisms of neutrophil recruitment to the lung by grain dust exposure. *Am. Rev. Respir. Dis.* 138: 921-27.

Von Essen, S., R.A. Robbins, A.B. Thompson, and S.I. Rennard. 1990b. Organic dust toxic syndrome: An acute febrile reaction to organic dust exposure distinct from hypersensitivity pneumonitis. *Clin. Toxic.* 28 (4): 389-420.

Wang, Z., P. Malmberg, P. Larsson, B. Larsson, and K. Larsson. 1996. Time course of interleukin-6 and tumor necrosis factor-alpha increase in serum following inhalation of swine dust. *Am. J. Respir. Crit. Care Med.* 153: 147-52.

Warren, C.P.W., R.M. Cherniack, and K.S. Tse. 1974. Hypersensitivity reactions to grain dust. *J. Allergy Clin. Immun.* 53: 139.

Warren, C.P.W., V. Holford-Strevens, M. Cheang, and J. Manfreda. 1989. Acute disorders attributable to exposure to grain in farmers. In: *Principles of Health and Safety in Agriculture.* J.A. Dosman and D.W. Cockcroft, eds. Boca Raton, FL: CRC Press, pp. 91-93.

Warren, C.P.W. 1977. Extrinsic allergic alveolitis: a disease commoner in non-smokers. *Thorax* 32: 567-69.

Warren, C.P.W., K.S. Tse. 1972; 1995. Extrinsic allergic alveolitis owing to hypersensitivity to chickens - significance of sputum precipitins. *Am. Rev. Respir. Dis.* 672-77.

Warren, C.P.W., and J. Manfreda. 1980. Respiratory symptoms in grain farmers: a brief report. In: *Occupational Pulmonary Disease: Focus on Grain Dust and Health.* J.A. Dosman and D.J. Cotton, eds. New York: Academic Press, pp. 383-90.

Warren, C.P.W., V. Holford-Strevens, and R.N. Sinha. 1983. Sensitization in a grain handler to the storage mite lepidoglyphus destructor (schrank). *Annals of Allergy* 50: 30-33.

Wathes, C.M. 1987. Airborne microorganisms in pig and poultry houses. "Environmental Aspects of Respiratory Disease in Intensive Pig and Poultry Houses, Including the Implications for Human Health," pp. 57-71.

Watson, R.D. 1986. Prevention of dust exposure. *Am. J. Ind. Med.* 10: 229-43.

Weber, S., G. Kullman, E. Petsonk, W.G. Jones, S. Olenchock, W. Sorenson, J. Parker, R. Marcelo-Baciu, D. Frazer, and V. Castranova. 1993. Organic dust exposures from compost handling: Case presentation and respiratory exposure assessment. *Am. J. Ind. Med.* 24: 365-74.

Weill, H., M.M. Ziskind, R.C. Dickerson, and V.J. Derbes. 1964. Epidemic asthma in New Orleans. *JAMA* 190: 811-14.

Wenzel, F.J., D.A. Emanuel, and R.L. Gray. 1971. Immunofluorescent studies in patients with farmer's lung. *J. Allergy Clin. Immunol.* 48: 224-229.

Wenzel, F.J., R.L. Gray, R.C. Roberts, and D.A. Emanuel. 1974. Serologic studies in farmer's lung. *Am. Rev. Respir. Dis.* 109: 464-68.

Wick, G. 1992. Complete clarification of a case of farmer's lung. *Int. Arch. Allergy Immun.* 98: 89-92.

Wilhelmsson, J., I-L. Bryngelsson, and C-G. Ohlson. 1989. Respiratory symptoms among Swedish swine producers. *Am. J. Ind. Med.* 15: 311-18.

Williams, J.V. 1963. Inhalation and skin tests with extracts of hay and fungi in patients with farmer's lung. *Thorax* 18: 182-95.

Yamaguchi, E., N. Okazaki, A. Itoh, K. Furuya, S. Abe, and Y. Kawakami. 1991. Enhanced expression of CD2 antigen on lung T cells. *Am. Rev. Respir. Dis.* 143: 829-33.

Yoshida, K., and J. Maybank J. 1980. Physical and environmental characteristics of grain dust. In: *Occupational pulmonary Disease: Focus on Grain Dust and Health.* J.A. Dosman and D.J. Cotton, eds. New York: Academic Press, pp. 441-61.

Yoshizawa, Y., M. Tanoue, H. Yano, T. Sato, M. Ohtsuka, S. Hasegawa, and Y. Kimula. 1991. Sequential changes in lung injury induced by preformed immune complexes. *Clin. Immun. & Immunopath.* 61: 376-86.

Zejda, J.E., T.S. Hurst, C.S. Rhodes, E.M. Barber, H.H. McDuffie, and J.A. Dosman. 1993. Respiratory health of swine producers. Focus on young workers. *Chest* 103: 702-09.

Zejda, J.E., S. Gomez, T.S. Hurst, E.M. Barber, C. Rhodes, H.H. McDuffie, and J.A. Dosman. 1994. Respiratory health of swine producers working in livestock confinement buildings. Supplement to Agricultural Health and Safety: Workplace, Environment, Sustainability, pp. 7-15.

Zhou, C., T.S. Hurst, D.W. Cockcroft, and J.A. Dosman. 1991. Increased airways responsiveness in swine farmers. *Chest* 99: 941-44.

Zuskin, E., B. Kanceijak, E.N. Schachter, J. Mustajbegovic, S. Goswami, S. Maayani, Z. Marom, and N. Rienzi. 1991. Immunological and respiratory findings in swine farmers. *Envir. Res.* 56: 120-130.

23

HEALTH AND SAFETY OF MIGRANT AND SEASONAL FARMWORKERS AND THEIR FAMILIES

Judy Hayes Bernhardt, R.N., Ph.D., M.P.H.
Department of Community and Mental Health Nursing
East Carolina University

> The most vulnerable population in the hazardous field of agriculture is migrant and seasonal farmworkers. These workers have the most dangerous jobs, often have substandard housing and sanitation facilities, practice high risk behaviors, and have a high risk of communicable diseases. Work hazards extend to the home because residential quarters are usually in close proximity to fields and orchards. Their migratory lifestyles limit the availability of data on this population. They are the least protected population in terms of financial resources, education, access to medical care, unionization, workers compensation, and programs and strategies to improve health and safety.

INTRODUCTION

Migrant and seasonal farmworkers and their families are the most vulnerable population in the high risk occupation of agriculture. As with many other agricultural workers, there is often no real separation of workplace and residential environments. In addition to working at high risks jobs, they also live in high risk housing, in high risk environments, and practice high risk behaviors (Palerm 1992). High risk behaviors include acts related to the transience itself, lifestyle choices, and low socioeconomic conditions.

Sources used to generate health and occupational data generally exclude migrants and other seasonal workers or include them in the general category of agricultural workers or hired farm laborers. This severely limits the type and amount of information available on migrants as a distinctive group. Other issues creating problems in studying migrants as a category of farmworkers are: the shift of some migrant workers to other types of work; the practice of some workers to change from following the crops to becoming permanent residents in enclaves of seasonal workers; the difficulty in studying individuals over time as they migrate back and forth to their home base which is often in their native country; the illegal alien status of some migrant farmworkers; their underutilization of health care and public assistance services which serve as contact points for reporting systems; the difference in ethnic, cultural, and social backgrounds of migrant and seasonal farmworkers as compared with the farm owner/operator; and, the number of workers for whom farm work is not their only or primary source of income/occupation.

Because various agencies of the government, service organizations, and surveyors define "migrant" and "seasonal farmworker" in different ways, even basic information such as the size of these groups is unclear. The Office of Migrant Health estimates there are one million migrant workers and about three million migrants, dependents, and seasonal workers in the United States (Wilk 1986). This office defines a migrant worker as one whose primary employment, for at least the past two years, is seasonal agricultural work and who establishes a temporary home for the purpose of this employment. A seasonal farmworker is one who works cyclically but who does not move about enough to establish a temporary home. Very little data is available on seasonal workers separate and apart from migrant workers. They are relatively comparable in terms of occupational tasks performed, income levels, rural lifestyles, and their periods of cyclic employment. Despite the difference in residential stability, migrants and seasonal farmworkers are usually considered together, as they are in this chapter.

Migrant and seasonal farmworkers form a diverse group. Race, ethnicity, and degree of acculturation vary with the geographic itinerary or "stream". Locations of the three largest streams are the east coast, stretching from Florida through the Carolinas to the northern Atlantic states; the central area, spreading from Texas and parts of Arizona through the plains and up the Mississippi River Valley to the Midwest states; and the west coast, based in California and parts of Arizona and spreading north to Oregon and Washington. The west coast and central streams are about 98% Hispanic with some Southeast Asians and American Indians. The east coast stream is more ethically diverse with African and Mexican Americans, Mexicans, Puerto Ricans, Haitians, and other groups from Central America and the Caribbean islands. A large but unknown segment of these workers and their families are undocumented and in the United States illegally (Meister 1991).

WORK ENVIRONMENT

Employment Conditions and Regulations

Many farm owners/operators negotiate directly with migrant and seasonal farm workers to tend and/or harvest their crops. Some farm owners hire farm labor contractors or crew leaders who recruit, hire, employ, furnish, or transport migrant or seasonal farm workers. The migrant worker then is employed by the labor contractor. One purpose of the Migrant and Seasonal Agricultural Worker Protection Act (SPA) is to require farm labor contractors to register and to assure necessary protection of migrants and other seasonal farm workers, regardless of who employs them.

Most migrant workers are paid by the piece rate or perweight system so wages are the direct result of time and effort spent in the fields. Problems with crops due to weather, plant disease, or equipment failure means less income. Time to walk to the toilet, rest breaks, and time to eat means time away from work and lost wages. There is no paid sick leave. During peak harvest, especially long days in the field are expected with no additional pay incentive (Meister 1991).

The federal Fair Labor Standards Act provides regulations concerning minimum wages. Farmers covered under the Fair Labor Standards Act must pay hourly employees the minimum wage or pay employees who work on a piece-rate basis an amount at least equal to the minimum wage. Farm workers are not subject to the overtime provisions of the act. Farmers are covered by this law if certain conditions exist:

- They grow crops that directly or indirectly leave the state.
- The farm business has gross annual sales volume of $500,000 or more.
- They use more than 500 man-days of agricultural labor in any quarter of the preceding calendar year (about the equivalent of seven employees working full-time in a calendar quarter).

An employer generally must withhold income taxes, withhold and pay social security and Medicare taxes, and pay unemployment taxes on wages paid to an employee, but not on payments made to independent contractors. Few migrant and season farm workers receive unemployment compensation for the months they do not work. The out-of-work farm worker begins the process by completing a form listing every employer he has for worked for in any state. These employers are then checked to see if they qualify to pay unemployment compensation. Farm employers pay federal unemployment tax (FUTA) if, during the current or preceding year, they either paid cash wages of $20,000 or more to farm workers in any one calendar quarter or employed ten or more farm workers for some part of at least a day during each of twenty different calendar weeks. Thus only the larger agricultural establishments are required to pay unemployment tax and only those farm workers who are employed by large farm organizations usually qualify for unemployment benefits (Slesinger and Pfeffer 1992).

The Fair Labor Standards Act establishes minimum ages for covered employment in agriculture, except where a specific exemption applies. Covered employment includes employees whose occupations involve producing crops or livestock that will be involved in interstate commerce. While most farm workers are covered as nearly all farm commodities directly or indirectly cross state lines, specific exemptions exclude many of the younger workers. The minimum age for working in agricultural jobs is 14 years unless the work is declared hazardous by the Secretary of Labor or performed during school hours. Twelve and 13-year olds may be employed with parental consent or on a farm where the minor's parent is also employed. Minors less than 12 years of age may be employed with parental consent on farms where employees are exempt from federal minimum wage provisions. Children of any age may be employed by their parent or person standing in the place of their parent at any time in any occupation on a farm owned or operated by their parent or person standing in their place.

Like most other agricultural workers, very few migrants and seasonal farm workers are unionized. Probably less that 5% of agricultural workers belong to a union. Collective bargaining for farm workers is excluded from the provisions of the National Labor Relations Act which ensures workers' rights to such representation. Only Hawaii and California have granted farm workers this type of legislation. Even the United Farm Workers union has made little effort to recruit unskilled migrant workers (Slesinger and Pfeffer 1992).

Health, Safety, and Housing Regulations

The federal Occupational Safety and Health Act of 1970 assures safe and healthful working conditions for all working men and women, including agricultural workers. Some states chose to remain under federal OSHA regulations and other states are administering their own plans, which include regulations approved by OSHA as being at least as stringent as the federal law. One exemption under both federal and state OSHA plans is small businesses. While the exact number varies from state to state, farms and other businesses employing ten or less employees during the preceding 12 months are exempt from OSHA inspections. Members of the farmer's immediate family are not included in the count of employees to determine exemption. This means that farms employing about 85% of all migrant and seasonal farm workers are exempt from OSHA inspection.

For those covered, state and federal OSHA regulations applicable to agriculture include the following general areas:

- Standards for temporary labor camps.
- Standards for field sanitation.
- Safety requirements for farm equipment.
- Safety requirements for tractors, including roll-over protection devices and operation of slow moving vehicles on the highways.
- Information and protection for employees working with and exposed to pesticides, anhydrous ammonia, and other hazardous chemicals.
- OSHA regulations regarding posting of notices to inform employees of OSHA protections and obligations, any violations, and annual summaries of occupational injuries and illnesses; required maintenance of injury and illness records for 5 years; and, notification of OSHA office within 48 hours of any accident resulting in a fatality or hospitalization of 5 or more employees.
- Requirement for readily available first aid supplies; provision of a person trained to render first aid who is readily accessible at all times.

While provision of housing is not required, those who choose to be migrant housing operators are required to provide housing free from recognized hazards that cause, or are likely to cause, death or serious injury or serious physical harm. Table 23.1 lists some of the minimal requirements included in OSHA Regulation 1910.142 which covers temporary labor camps.

Several agencies may have authority to regulate migrant housing. Under federal law, the Wage and Hour Division of the U.S. Department of Labor has the right to inspect migrant labor camps before occupancy. In many states, migrant housing inspections are required by law and enforced by the state departments of labor. In some states, jurisdiction over migrant housing is retained by state OSHA and complaints regarding migrant housing are inspected by state OSHA inspectors. Ideally, authority and responsibility for migrant housing is part of a cooperative arrangement among the various governmental agencies. For example, the Migrant Housing Act of North Carolina enacted by the General Assembly established a single set of standards for all agricultural migrant housing and vested enforcement authority for the standards in the state Department of Labor. Beginning January 1, 1990 anyone who owns or operates a housing unit in North Carolina for one or more migrant workers must register with the department and have housing inspected before the migrant(s) moves in.

In addition to migrant housing regulations, OSHA field sanitation standards are required of an employer providing migrant housing to anyone or employing eleven or more workers. Sufficient amounts of potable drinking water must be provided in accessible locations to meet the needs of all workers engaged in hand-labor operations in the fields. One toilet facility and one hand washing facility must be provided for each twenty employees or fraction thereof who work in the field for 3 or more hours per day. These facilities must be located within a 1/4 mile walk of each laborer's work position or at the closest point of vehicular access to the field. Potable drinking water and toilet and handwashing facilities must be maintained in accordance with appropriate public health practice. Workers must be allowed reasonable opportunities during the day to drink the water and use the sanitation facilities. Unfortunately, infrequent inspections in the farming industry allow some farm owners/operators to ignore these minimum hygiene standards.

Worker's compensation is covered by individual state regulations; there is no federal worker's compensation legislation. Definitions of who qualifies as an employer, who meets the test for an employee, what injuries and illnesses are covered, in what amounts, and for how long, vary from state to state. For example, in North Carolina a farmer must regularly employ ten or more full-time nonseasonal workers before he is required to purchase worker's compensation insurance from a private carrier to cover employees if they sustain an injury or illness on the job. This insurance pays medical expenses, rehabilitation costs, lost wages, permanent disability, and death benefits. Such coverage generally bars an injured worker from bringing legal action against the farm employer. As most migrants and seasonal farm workers are employed on farms with ten or less paid employees, they are not usually covered by worker's compensation insurance.

The 1992 Worker Protection Standard consists of regulations designed to protect agricultural workers and pesticide handlers from injury or illness due to pesticides. Included are requirements for posting safety information and displaying facts about each pesticide application, pesticide training for handlers and farm workers, provision of personal

Table 23.1. OSHA Standard 1910.42. Temporary Labor Camps SubPart Number J--General Environmental Controls.

Site	- Adequately drained - Adequately sized to prevent overcrowding of necessary structures - Grounds surrounding shelter maintained in clean and sanitary conditions		
Shelter	- Constructed to provide protection from the elements - Heating, cooking and water heating equipment installed in accordance with local and state ordinances, codes and regulations - Adequate heating equipment provided for camps used during cold weather		
	Sleeping Rooms - At least 50 square feet of floor space provided for each occupant sleeping in a room - At least a seven foot ceiling provided - Beds, cots, or bunks provided and elevated at least one foot from the floor and at least three feet apart - Adequate storage facilities available for clothing and personal articles - Floors constructed of wood, asphalt, or concrete - Wooden floors tightly constructed, smooth, in good repair, and at least one foot above ground level	All Living Quarters - Windows provided with the total area not less than one-tenth of the floor area - At least half of each window constructed to be opened for ventilation - All exterior openings screened with 16-mesh material - External doors closed with self-closing devices - A minimum of 100 square feet per person provided when workers cook, live and sleep in the same room.	Cooking Facilities - Sanitary facilities provided for storing and preparing food - Stoves (one per ten people or one per two families) provided in enclosed and screened shelters when cooking facilities are used in common
Water Supply	- Adequate and convenient supply, approved by appropriate health authority, provided for drinking, cooking, bathing and laundry. - Outlets distributed so no shelter is more than 100 yards from a hydrant, if water is not piped to the shelters - Common drinking cup prohibited		
Toilet Facilities	- Adequate number provided for capacity of camp - Satisfactorily ventilated - Located in a room used only for toilet purposes - Privy located not closer than 100 feet to any living area - Located no more than 200 feet from sleeping rooms - When facilities are shared, as in multifamily shelters and barracks type facilities, provide separate toilet room for each sex - Continuously lighted by safe type of lighting - Adequate supply of toilet paper provided - Kept in sanitary condition, cleaned at least daily		
Laundry, Handwashing, and Bathing Facilities	- Handwashing basin provided per family shelter of six persons in a shared facility - Shower head provided for every 10 persons - Slop sink provided in each building used for handwashing, bathing and laundry - Adequate supply of hot and cold running water provided for bathing and laundry purposes - Every service building provided with equipment capable of maintaining temperature of at least 70 degrees F during cold weather - Facilities for drying clothes provided		
Lighting	- Where electric service available, each habitable room in camp provided with at least one ceiling-type fixture and one separate floor or wall-type outlet, provided at least 30 foot-candles 30 inches from the floor		
Refuse Disposal	- At least one fly-tight, rodent-tight, impervious garbage storage container, approved by the appropriate health authority, provided for each family shelter - Container located within 100 feet of each shelter on a wooden, metal or concrete stand - Containers emptied when full, but not less than twice weekly		
Kitchens, Dining Hall, and Feeding Facilities	- In camps with multiple family feeding operations permitted or provided, food handling facilities comply with requirements of the "Food Service Sanitation Ordinance and Code," USPHS		
Insect and Rodent Control	- Protective measures taken to prevent infestation by and harborage of animal or insect vectors or pests		
First Aid	- Adequate first aid facilities, approved by a health authority, maintained and made available in every camp for emergency treatment - person trained to administer first aid in charge of facility and readily accessible at all times		
Reporting Communicable Disease	- Camp superintendent reports to the local health officer the name and address of anyone known to have, or suspected of having a communicable disease and also any case of suspected food poisoning or an unusual prevalence of any illness in which fever, diarrhea, sore throat, vomiting, or jaundice is prominent symptom		

protective equipment required by the pesticide label, establishment of a decontamination site, and provision of emergency assistance if needed. The Environmental Protection Agency's (EPA) regulations require commercial applicators to keep records of restricted use pesticides; agricultural growers are generally excluded. Several states have enacted their own laws governing recordkeeping for pesticide usage (Meister 1991).

State hazard communication standards may be identical to or more stringent than the program of federal OSHA. Hazard communication standards have as their goal the reduction of injury and illness resulting from improper use and storage of chemicals in the workplace. Regulations regarding information standards are imposed on the employer, as well as the distributors and manufacturers. Information must be provided to the employer by way of product labels and Material Safety Data Sheets (MSDS). The employer must prepare a written program telling how his business will comply with the standard and the program must be made available to employees. Employers must provide information and training to all employees who may be exposed to hazardous chemicals (including pesticides).

Occupational Illnesses and Injuries

Occupational hazards of migrant and seasonal farmworkers include the risk of work-related injuries; cumulative trauma disorders from fast paced, repetitive manual labor; dermatitis from contact with plant materials for extended periods of time; pesticide-related illnesses; musculoskeletal and soft tissue problems from heavy lifting and carrying and difficult work positions; climate-related illnesses; respiratory conditions from antigens in dusts from plants and from organic and inorganic chemical exposure; reproductive problems; and spread of communicable disease (Mobed et al. 1992).

The risk of work-related injuries is high in the working environment of the migrant laborer. The ground may be uneven, wet, and slippery, leading to falls. The use of ladders is necessary to harvest some crops. Heat and humidity, long work days, infrequent rest periods, and poor posture caused by bending and stooping all lead to increased fatigue, which has been proven to increase the risk of accidents. Noise and vibration from farm equipment and the monotonous routine of the work also play a part in injury causation.

Studies of occupational injuries in farmers have generally made no distinction between farm owners and hired workers; very few studies of occupational injuries among migrant workers have been conducted. Minor injuries are usually treated superficially in the field using first aid supplies that may be furnished by a crew leader or farm owner. Study of a random sample of 287 migrant farm workers (70% Latino, 24% U.S. born black, and 6% Haitian) in eastern North Carolina found that 8.4% reported an occupational injury during the previous 3 years (Ciesielski et al. 1991). Eighty percent of these injuries were broken bones, sprains, and cuts. Motor vehicles and machinery caused about one-fifth of the injuries and most often resulted in lost work days. Of the injured workers who considered medical care necessary, 41% received delayed treatment (more than 24 hours later) and 24% never received any treatment. Problems with the crew leader and lack of transportation were cited as factors in delayed treatment and also in failure to keep follow-up appointments. Medical expenses were paid by the farmer or crew leader for 38% of the injured workers. Only 20% received any compensation for lost work time and none of this was from workers' compensation (which is not required for seasonal farmworkers in North Carolina). The incidence of injury was believed to be underreported in this study due to social and economic pressures. Some workers did not believe that the crew leader would provide transportation even though it was legally required. Fear of retribution prevented some workers from reporting their injuries and seeking treatment. Loss of wages was also cited as a reason for not seeking treatment or follow-up care.

Migrant and seasonal workers are exposed to many physical, chemical, and infectious agents that can cause or exacerbate dermatological disorders. Included are ultraviolet radiation, plant materials, fertilizers, pesticides, and zoonoses. Contact, irritant, and allergic dermatitis are found in this population. Fruit and vegetable growers depend almost entirely on the manual labor of seasonal workers. Fruit and vegetable crops are very labor intensive and require an average of 120 hours of manual labor per acre, compared with grain crops which average 3 hours per acre. The Bureau of Labor Statistics reports that crop production workers have over 5 times the incidence of occupational skin conditions reported for all private sector employees combined, 33.4 per 10,000 person years. The true incidence of skin conditions is probably much higher since only the most severe cases would lose work time to seek medical treatment and thus be reported (Schenker and McCurdy 1990).

Seasonal workers do strenuous manual labor in hot and humid environments. Because of the heat, the expense, and the bulkiness of the garments which may slow the worker down, protective clothing is seldom worn. Sunburned, cracked, scratched, scraped, and otherwise damaged skin is more susceptible to skin irritants. Inadequate facilities for handwashing, bathing, and laundering of work clothes prolong the contact. Risk to the individual worker is determined by the specific exposures encountered, the types of job activities performed, the use of personal protective equipment, personal hygiene, and individual susceptibility (Schenker and McCurdy 1990).

The skin is the most common route of pesticide exposure. Skin rashes and inflammation, corneal irritation, and systemic poisoning may result from acute pesticide poisoning. Chronic health problems related to pesticides are not well understood but are thought to include dermatitis, anxiety, headaches, sleep disturbance, memory and concentration problems, fatigue, as well as possibly blood disorders, reproductive problems, birth defects, cancer, and disturbances in liver and kidney functions (Wilk 1986).

Because illnesses due to pesticide exposures are often undocumented, the true extent of these problems is unknown. The migrant worker generally seeks health care only for those conditions that are acute and prevent him or her from working for an extended period of time. The health care provider who does not routinely treat agricultural workers may fail to consider pesticide exposure as an etiological factor; misdiagnosis may occur. Even when the care provider asks specifically about chemicals, the worker may have no or incomplete knowledge of his exposure. Different chemicals are used on different crops and many different chemicals may be used on one crop during the growing season (Meister 1991).

Seasonal farmworkers come in contact with pesticide residue in various ways: touching the foliage of plants with their bare skin; being inadvertently sprayed in a field by either ground or aerial means; reentering a field too soon after spraying or after an inadequately protective reentry interval; and living in camps where pesticide residue may drift from the field, contaminating their food, water, and living quarters (Wilk 1986). Workers may also be exposed by eating or smoking with pesticide contaminated hands; they may eat fruits or vegetables with pesticide residue from the fields in which they are working. Drinking water and water used for personal hygiene in the fields may be contaminated by residue.

Heavy demands of manual labor required of the migrant worker may lead to musculoskeletal problems such as joint irritation and degenerative diseases including arthritis of the hands, knees, and hips. Occupational risk factors include heavy lifting and carrying, and working prolonged periods in difficult postures involving bending, stooping, or reaching. The fast work pace, repetitive motions, and long work hours intensify musculoskeletal wear and trauma. Fatigued muscles are at increased risk of injury (Wilk 1986).

Migrant and seasonal workers are particularly concerned with thermal conditions causing work-related problems ranging from prickly heat rash, heat cramps, and heat exhaustion to heat stroke. As heat and humidity rise, fatigue increases and endurance decreases (Brown 1991). Accident frequency increases in hot weather. Seasonal workers generally work outdoors in the hottest part of the day, often long beyond an 8-hour day. Frequent rest periods result in lost wages and may be discouraged by crew leaders. Housing is seldom air conditioned or provided with ventilation fans. Workers may sleep poorly and become chronically fatigued. Their propensity toward living and working in unsanitary environments makes them more likely to suffer from diarrhea, nausea and vomiting, fever and dehydration. Risk of acute pesticide poisoning also increases with dehydration. Other risk factors for developing heat related disorders are poor nutrition, inadequate physical conditioning, inappropriate clothing, diabetes, alcoholism, and pregnancy.

Knowledge of the effects of migrant farm work on reproduction is limited; paternal exposure effects in particular are not generally known. Information related to pregnancy outcomes in migrants is almost nonexistent. Description of the effects of pesticides and other agricultural chemicals on the fetus is largely anecdotal. Dehydration, a common occurrence for seasonal workers, may concentrate fetal exposure to harmful chemicals (Meister 1991). The physiological stresses of pregnancy itself are likely to be exacerbated by the demands of long hours in hot fields and living in substandard and unsanitary housing conditions. Prenatal care is generally absent or inadequate, particularly in second and subsequent pregnancies.

SOCIOECONOMIC CONDITIONS

Housing Conditions

Although structural and sanitary conditions vary widely, migrant housing is often found to be substandard. Poverty, combined with the migratory lifestyle, leads to particularly poor living conditions. Regulation of housing and hygiene in migrant camps typically is not strictly enforced; inspections generally occur at infrequent intervals. Migrant workers are at greater risk of exposure to chemicals than other agricultural workers when their homes are very close to, or actually in, the fields; exposures continue during their nonworking hours.

The farm owner may provide barrack or dormitory-style housing for single male migrant workers. Tenant houses or mobile homes are often used as dwellings for small groups of workers or families. Sometimes the migrant may be left to obtain his own housing. Some rent apartments or motel rooms in nearby towns. Some live in their cars, trucks, or vans. Workers are sometimes found living in makeshift cardboard and plastic structures in or beside the fields.

Even housing that meets minimal OSHA requirements for temporary labor camps provides a challenge to the health and safety of migrant and seasonal workers and their families. Maintaining adequate personal hygiene and sanitary living conditions takes time and effort. Removing residual soil, plant material, and pesticides from the workers' skin and clothing may not always be accomplished, increasing the time in contact with these materials. Contaminants may also be transferred to family members in camp (Meister 1991).

Some migrant farmworkers maintain close ties to their native country but settle permanently with their families in rural areas and small towns located in the regions of the U.S. where seasonal farmwork is most plentiful. This influx of young families with small children is in contrast with most of rural America. These enclaves of former migratory workers still contain families with very low annual earnings, generally at or below the federal poverty level. Their private housing arrangements often are at the same substandard level they endured as migrant workers. They seldom settle in communities that have resources available to help them improve their poverty situation. These settled workers often end up competing with migrant workers for employment, thus generally keeping wages low, affordable housing scarce and resources to help those living in poverty depleted (Palerm 1992).

Child Laborers

Some migrant workers travel alone, some in crews, and some as family units. If the wife works in the field, child care is a problem. Sometimes a pregnant or older female will remain in camp and care for the small children of several families. At other times, young children accompany their parents and play in or near the fields.

Whether the family unit travels as a work group or the adults travel alone often depends on the type of farm activities to be done. It is generally helpful to have family members to harvest field crops when the pay is for piecework; children are not useful in canneries where the pay is an hourly rate.

Agriculture is the only industry in the U.S. that allows children under the age of 16 to work legally. Children as young as 10 years may harvest potatoes and strawberries due to specific exemptions granted by the Department of Labor. Children ages 12 and above may legally work, with parental permission, in any crop. Because of family poverty, children younger than the legal age are often found working in the fields. All of these children are subject to interruptions in their education and delays in social development (Martin et al. 1995).

HEALTH AND HEALTH CARE UTILIZATION

General Health Status

It is believed that migrant and seasonal farmworkers suffer from the same causes of morbidity and mortality as agricultural workers in general. Depending on the amount of time spent living and working in the United States, the incidence and prevalence of these conditions may more closely resemble those of the native/home country. However, even such crude data as death rates, median survival, and infant mortality rates cannot be calculated due to the lack of a reasonably precise denominator (Rust 1990).

Studies of migrant farmworkers in Wisconsin found in both 1978 and 1989 that about 13% of those surveyed said their health was excellent, about half reported good health and about one third described their health as only fair. These figures are very different from those for the U.S. population in general where 40% describe their health as excellent, half as good, and less than 10% as fair or poor. This same study found 27% of those interviewed had never had a physical examination when they were not sick, 26% had never had a dental exam and 43% had never had an eye examination (Slesinger and Ofstead 1993).

Common Health Problems

Much of the available data about migrant health problems comes from reports of the most common types of visits made to migrant and community health centers and hospitals in rural areas where migrants work or live during the off season. Follow-up after diagnosis is extremely difficult due to the migratory lifestyle. Even those seasonal workers who return to the same areas year after year often receive treatment for chronic conditions in their home base and only present for acute problems during the work season. Even if morbidity and mortality rates were available, it would be impossible to separate out those segments of risk attributable to race, ethnicity, cultural practices, lifestyle, occupational hazards, and the effects of poverty.

Health problems most frequently reported at the migrant health centers include dermatitis, injuries, respiratory problems, musculoskeletal ailments including back pain, eye problems, gastrointestinal problems, and diabetes (Wilk 1986). Treatment is usually sought for acute and emergency conditions rather than chronic and preventive services. Basic preventive health care is not a priority and even immunizations may be incomplete, particularly for adults. One study of migrant children in South Carolina found that only 35% of the students in a summer education program had completed their series of immunizations by age 10. This leaves children, already at risk because of their lifestyles, at increased risk for infectious disease (Lee et al. 1990).

Injuries, both occupational and nonoccupational, are an important problem for the male migrant worker. A study of migrants working in the west-central "Ridge" region of South Carolina found, in general, crews of single black males who lived on the farms where they were currently working, and were being paid by piece rate to work the peach crop (McDermott and Lee 1990). A record review found 47 visits to the Migrant Health Center by males for injuries, accounting for 12% of all visits by males. Of the 175 total visits made by migrant farmworkers to the local emergency room, 113 visits were made by black males, with 63% due to injury and 58% due to injury from violence. During the same time period, interviews were conducted with 116 male migrant workers. The primary characteristics associated with injury were: black (68%), single (65%), 41 to 54 years of age (46%), over 5 years as a migrant, primary family with six or more children, and heavy alcohol use. Personal violence accounted for 50% of the injuries found in this group with 83% of the episodes occurring in camps after dark with heavy drinking involved in a majority of the cases (McDermott and Lee 1990).

For migrant and seasonal workers and their families, it is difficult to separate health problems due to unsanitary working conditions from those associated with poverty level living conditions. Diseases related to poor sanitation include intestinal, skin, and respiratory conditions as well as general nutrition problems. These diseases are commonly spread by sharing eating and drinking utensils, drinking nonpotable water, and fecal contamination due to inadequate or nonexistent handwashing and toilet facilities. In Utah, clinic charts of migrant farmworkers without access to water and sanitation facilities in the fields were compared with clinic charts of low income patients in Salt Lake City who had available sanitary facilities for the incidence of sanitation and water-related symptoms and diseases (Arbab and Weidner 1986). Diarrhea was found to be twenty times more prevalent, nausea and vomiting 13 times more frequent, and gastroenteritis, abdominal or intestinal pain, and bloody stools 6 to 26 times more frequent among the migrants than the urban poor. Tuberculosis was 24 times more frequent and helminthic infestations 35 times more frequent among the migrants than the urban poor. Fevers of unknown origin occurred 120 times as frequently in the migrants. Urinary tract infections were found 3 times more often in migrants than in the urban poor, even though there were 15% fewer females in the migrant group. While the sample size was too small to allow any definitive statements, it is clear that migrant patients are treated much more often at clinics for symptoms or conditions related to poor sanitation, inadequate hygiene, and/or impure drinking water than the urban poor.

Intestinal parasites are also a problem for migrant workers and their families. In both a 1987 convenience sample of 265 and a 1988 random sample of 181 migrant farmworkers in eastern North Carolina, parasite prevalence was found to be high with more than 50% of the subjects testing positive (Ciesielski et al. 1992). Hookworm was the most prevalent pathogenic parasite with the highest rates found in those born in Central America and Haiti. Although many farmworkers remained asymptomatic, reported symptoms of nausea, appetite loss, abdominal pain, and low hematocrits (means were 39.4%) indicated morbidity due to parasite infections. Parasitic infection is also viewed as a sentinel health event, indicating that living and working conditions are such that these migrants are at risk for other enteric infectious diseases, which are transmitted in a similar fashion.

The majority of migrants working in the United States were born in Latin America, often in areas where cysticercosis, malaria, and Chagas' disease are endemic. In Mexico, mortality from cysticercosis exceeds that of diabetes. All recent outbreaks of malaria in the United States have occurred among farmworkers. In a random sample of migrants in eastern North Carolina, 138 Hispanic and Haitian migrant workers were found to have a seroprevalence of 10% for cysticercosis, 2% for Chagas' disease and 4.4% for plasmodia species, with one case of active malaria found (Ciesielski et al. 1993). While it was unclear what the clinical significance of seropositivity is, it does show that migrant farmworkers from Latin American countries are at risk of infection. Poor field sanitation and inadequate hygiene facilities in camp increase the risk of transmission. Health care providers treating workers from these countries should be aware of the possible presence of these infections.

Tuberculosis (TB) is also a risk factor as many migrants come from countries and socioeconomic backgrounds where the rates of TB are much higher than the general national rate. Crowded, poorly ventilated living quarters and crowded buses and trucks used for transportation contribute to the spread of this respiratory disease. Alcoholism and malnutrition may also be risk factors for some of these workers (Jacobson et al. 1987). While there is no specific information on migrants, the Center for Disease Control (CDC) estimates that the risk of TB among farm workers is 6 times higher than the general employed population (CDC 1992a), an estimate much lower than any published study. A random sample of 543 migrant farmworkers conducted in North Carolina in 1988 found the prevalence of active tuberculosis in Hispanics to be 0.5 and 3.6% in American-born blacks. This case rate for American-born blacks was more than 3,000 the national case rate of TB (Ciesielski et al. 1991b).

The rate of asymptomatic TB infections as demonstrated by positive skin tests is very high among migrant farmworkers and calls for ongoing and preventive activities in this population group. Rates vary considerably by country of origin. A 1987 study of migrant workers in the Delmarva Peninsula found 27% of American-born blacks and 55% of Haitian workers with positive skin tests (Jacobson et al. 1987). A study of migrant farmworkers in North Carolina in 1988 found rates ranging from 37% in Hispanics to 62% in American-born blacks to 76% in Haitians (Ciesielski et al. 1991b). These results were presented as indication that TB among farmworkers is an occupational problem, not one of national origin.

One reason for the high rates of positive tuberculin tests as well as infectious disease probably is the difficulty in enforcing the prophylaxtic treatment recommended. Getting patients to take isoniazid daily for a year is difficult in a stable population. With the migratory movements of these workers and the lack of ongoing health supervision, adequate compliance is almost impossible to achieve.

High Risk Behaviors

High risk behaviors are often adopted under the pressure of migration. Such behaviors put migrant and seasonal farmworkers and their families at increased risks of related health and social problems. These risks include problems of alcohol and drug abuse, violence, increased risk of sexually transmitted diseases (STD's) and HIV infection.

Living away from their families is often cited as the hardest part of the job for male workers traveling alone. Social isolation is a problem for migrant workers and also for the families who may travel with them. Few camps are large. Those migrants with cars and trucks mainly use them to transport workers to and from the fields and between jobs. There is often a language barrier. The migrant and his family may speak little English and some are illiterate in their native language. Living in crowded, sparsely furnished rooms with no privacy, lacking recreational opportunities, and some fearing detection of their undocumented legal status, may contribute to high rates of personal violence and resulting injuries.

One study in North Carolina of the children of migrant and seasonal farmworkers found that more than half had been victims of violence, 33% having witnessed it, and 13% having both witnessed violence and been a victim of it. These rates are said to be as high or higher than rates of exposure found in poverty stricken, high crime urban areas. Exposure to violence was found to be related to emotional problems in the children, behavioral problems, and weapon carrying behavior (Martin, Gordon, and Kupersmidt 1995). Few migrant children ever receive professional counseling for these problems as available rural mental health services for them are very limited.

A study by the President's Commission on Mental Health of 96 migrant health centers, 66 community mental health centers, and 33 projects of the National Institute of Alcohol Abuse and Alcoholism found that alcoholism was viewed by 60% of the center and project directors who responded as the most significant migrant health problem. Anxiety was cited by 50% and depression by 40% as significant problems (Chi and McClain 1992).

Drinking patterns vary widely among different cultural, racial, and ethnic groups. Blacks and more acculturated Hispanics may embrace the social norm of drinking "to party" rather than the traditionally held cultural norm of Hispanic men which is to drink to relive tension (Chi and McClain 1992). Hispanic males tend to gather after work and on weekends, usually outdoors, to drink, eat, and socialize with other men. Males also drink in cantinas and at dances. Some drink during work. Cultural restrictions generally do not permit women to drink except in small amounts at family gatherings. Women who go out drinking on their own are treated with less respect and viewed as less virtuous (Alaniz 1994).

Drinking serves as a major social activity for many male migrant farmworkers. Chi and McClain (1992) studied a culturally diverse group of 286 migrant farmworkers in Orange County, New York and found 58% were regular drinkers, 23% were occasional drinkers, and 18% nondrinkers. In general, both regular and occasional drinkers drank more frequently in the migrant camps than when they were in their home communities. Females were 3 1/2 times more likely to be nondrinkers than males. Currently married workers were far less likely to be drinkers than any other category. Those with family members living in camp were far less likely to be drinkers. About 90% of the Puerto Ricans and Jamaicans, 88% of the blacks, and 67% of the Mexicans claimed to be frequent or occasional drinkers, in contrast to 23% of the Haitians. More than the other cultural groups, the Haitians tended to duplicate the strong family and social support systems of their rural heritage which may have decreased their need for drinking as an escape mechanism.

Voluntary screening in 1992 of 310 predominately male migrants in Florida found high prevelences of syphilis, HIV infection, and tuberculosis (CDC 1992b). Eight percent had reactive serologic tests for syphilis; cases of primary, secondary, early latent, and late latent syphilis were diagnosed; less than one-fifth had previously been treated for the condition. There was a 5% HIV antibody seropositive rate, most with newly diagnosed infections. Tuberculin skin tests were read as positive for 44% of the workers including one case of active TB. Use of crack cocaine was associated with positive serological tests for syphilis. Intravenous drug use and homosexuality were rarely reported by any of these migrant workers. Forty-seven percent of those surveyed had never used a condom, taking a major risk with unprotected heterosexual intercourse.

A less well-known risk is that of self-injecting antibiotics and vitamins reported by 12% of the migrants in one sample and 20% in another study; the services of a "lay injectionist" may also be bought for a small fee. Sharing needles for these types of self-treatment may play a role in the spread of HIV in farmworkers. This practice is rooted in the Latin American cultures where injectable medications are freely available over the counter without a prescription. One reason self injection is practiced is because some people believe there is no need to seek health care for certain problems when you can get medicine without it. Another reason given is the difficulty encountered in accessing health care. Supplies are obtained from relatives through the mail, at flea markets, in Mexican stores, or on trips to their native countries. Antibiotics are used primarily for minor infections, fevers, colds, toothaches, and cuts. Vitamins are taken to keep healthy, to prepare for pregnancy, to treat certain conditions such as poor eyesight, headaches, and fatigue. The medication regimens provide no therapeutic value (McVea 1995).

Barriers to Health Care

Among the many problems migrant and seasonal farmworkers face is poor access to health care services. Potential barriers to health care utilization include lack of transportation, inaccessibility of care provider, no or limited health insurance, long waits for appointments, inconvenient hours of service, losing income while away from work, language and cultural differences, and undocumented legal status.

From the Hanes 1982-1984 survey, Estrada et al. (1990) found that about one-third of the adult Mexican American population encountered barriers during their most recent medical encounter and the barriers were severe enough to prevent them from obtaining the care they sought 75% of the time. Those experiencing the most problems in obtaining care were those with the lowest educational levels and the lowest incomes. Of this group, those speaking only Spanish, foreign born, and less acculturated were the ones most likely to obtain no care. The barrier most commonly identified was the cost of the health care. Another barrier identified involved lack of services available during nonworking hours which related to cost (Estrada et al. 1990).

Lack of knowledge and education concerning diagnostic and treatment procedures, as well as cultural barriers, may prevent migrant and seasonal farmworkers from using preventive services and seeking early detection of serious health conditions. One attitude that often slows Hispanic farmworkers from seeking health care is that of fatalism, that there is little the individual, or the health care provider, can do to prevent or cure diseases. If a person becomes ill or if a sick person is healed then it is often said to be "God's will". Feelings of shame, embarrassment, and fear of discomfort may also serve as barriers. Hispanic females in particular express discomfort and embarrassment with having a male health care provider, particularly when the visit involves gynecological and obstetrical questions and examinations. Hispanic males strive to display their masculinity and avoid situations that would be potentially embarrassing or might make them appear weak (Lantz et al. 1994).

Migrant health clinics were established in 1962 as part of the Community Health Centers under the Public Health Service as a part of a program to provide services to the poor and medically underserved population. In 1985, 122 federally funded migrant health centers in 300 rural areas provided health care to 460,000 migrants, seasonal farmworkers, and their families, thus reaching only 17% of the estimated population (Wilk 1986). These centers have been able to remain open thus far but budgets have not kept pace with inflation and most clinics today are able to offer fewer services than twenty years ago (Slesinger and Pfeffer 1992).

Poor access to basic health care services applies to many farmworkers not only during the seasons they migrate to do farmwork, but also during the times they are located in their home bases. Many are from isolated rural areas in the impoverished countries of Mexico, Central America, and the Caribbean islands. They have a lifetime of inadequate health services which is only intensified by their migratory lifestyle and/or their seasonal employment (Palerm 1992).

Few migrants have any form of health insurance. Many work in states where they are not protected by workers' compensation for on-the-job injuries and illnesses. Most are unaware of what, if any, public assistance is available to them. Few migrant families receive Medicaid, food stamps, or WIC (Women, Infants, and Children) benefits; for most the paperwork is insurmountable. To enroll means to overcome many obstacles. First the worker must know that such services are available. The determination of who qualifies in which states is a complex, frustrating process, particularly for someone whose command of the English language is absent or limited. Even when materials are available in Spanish, some migrants are unable to use them because they are illiterate in their own language. Waiting periods vary, and to qualify for some programs, the worker must rent a post office box to supply an acceptable permanent mailing address. Agencies are usually open during days and times that work in the field is required, which means lost wages. Agencies are generally understaffed and communication is often strained; sometimes there is prejudice against these workers, documented or not, from other countries (Palerm 1992).

Undocumented workers generally fear that any contact with the health care system or attempt to seek public assistance will cause them to be reported to the Immigration and Naturalization Service. Even those workers who are now enrolled in the amnesty program (Immigration Reform and

Control Act of 1986) usually avoid health care because of residual fear from their undocumented days and because the rules and regulations regarding the amnesty agreement are often ambiguous and unclear to them (Meister 1991).

When injuries and illnesses do occur, many of these farm laborers and their families rely on the traditional remedies and treatments of their native lands and the self-prescription of over-the-counter medicines available to them. Workers who become acutely ill are generally treated in emergency departments of hospitals in the areas where they are working or quickly transported back to their native lands by family or friends to avoid any attention from those in authority. Minor injuries generally go untreated by any health professional. Health promotion and preventative health measures are seldom practiced. This includes regular prenatal care and well-child visits as well as routine dental checkups.

SUMMARY

It is clear that migrant and seasonal farmworkers and their families constitute a very vulnerable population. While much remains unknown about their health status, it is obvious that risk factors from their hazardous work, transient lifestyle, frequently risky personal behaviors, and poverty level socioeconomic conditions are interrelated. Multiracial and multicultural differences and language barriers generally intensify their problems.

Other chapters in this book have addressed specific occupational hazards associated with agriculture and strategies for their prevention. For migrant and seasonal farmworkers the issues of field sanitation, housing, and health care access are as important as prevention of occupational hazards.

Easy access to potable drinking water, clean toilets, and adequate handwashing facilities constitute the absolute minimum requirements. Clean, safe housing that prevents crowding and provides proper ventilation must also be available. Facilities for bathing and clothes laundering and appliances for safe food storage and meal preparation are needed, as well. Existing laws and regulations pertaining to field and housing sanitation for migrant workers must be actively enforced. Where such regulations do not exist, they must be enacted. The high prevalence of tuberculosis and enteric diseases in this population demonstrate these needs. Reasonable access to affordable services for primary health care and prevention programs, urgent and emergency care, and continuity of care for chronic conditions must also be provided.

Despite rapidly increasing mechanization of many agricultural processes, there is no objective indicator that the need for seasonal farmworkers to tend and harvest the labor intensive fruit and vegetable crops is declining. For all these reasons, there is a serious, unrelenting need for strategies and programs that address not only their occupational safety hazards and health exposures but also their underlying social conditions. Current measures are clearly inadequate.

REFERENCES

Alaniz, M.L. 1994. Mexican farmworker women's perspectives on drinking in a migrant community. *Intl. J. Addictions* 29(9): 1173-88.

Arbab, M.D., and B.L. Weidner. 1986. Infectious diseases and field water supply and sanitation among migrant farmworkers. *Am. J. Public Health* 76(6): 694-95.

Brown, W.D. 1991. Heat and cold in farm workers. *Occup. Med.* 6(3): 371-89.

Center for Disease Control (CDC). 1992a. Prevention and control of tuberculosis in migrant farmworkers: recommendations from the Advisory Council. *MMWR* 41(RR-10).

Center for Disease Control (CDC). 1991b. HIV infection, syphilis, and tuberculosis screening among migrant farmworkers— Florida, 1992. *J. Am. Med. Assoc.* 268 (15): 1999-2000.

Chi, P.S.K., and J. McClain. 1992. Drinking, farm and camp life: a study of drinking behavior in migrant camps in New York State. *J. Rural Health* 8(1): 41-51.

Ciesielski, S., S.P. Hall, and M. Sweeney. 1991a. Occupational injuries among North Carolina migrant farmworkers. *Am. J. Public Health* 81(7): 926-27.

Ciesielski, S., J.R. Seed, D.H. Esposito, and N.H. Hunter. 1991b. The epidemiology of tuberculosis among North Carolina migrant farmworkers. *J. Am. Med. Assoc.* 265(13): 1715-19.

Ciesielski, S., J.R. Seed, J. Estrada, and E. Wrenn. 1993. The seroprevalence of cysticercosis, malaria, and trypanosoma cruzi among North Carolina migrant farmworkers. *Public Health Reports* 108(6): 736-40.

Ciesielksi, S., J.R. Seed, J.C. Oritz, and J. Metts. 1992. Intestinal parasites among North Carolina migrant farmworkers. *Am. J. Public Health* 82(9):1258-62.

Estrada, A.L., R.M. Trevino, and L.A. Ray. 1990. Health care utilization barriers among Mexican Americans: Evidence from HHANES 1982-84. *Am. J. Public Health* 80 (suppl.): 27-31.

Jacobson, M.L., M.A. Mercer, L.K. Miller, and T.W. Simpson. 1987. Tuberculosis risk among migrant farm workers on the Delmarva peninsula. *Am. J. Public Health* 77(1): 29-32.

Lantz, P.M., L. Dupuis, D. Reding, M. Krauska, and K. Lappe. 1994. Peer discussion of cancer among Hispanic migrant farm workers. *Public Health Reports* 109(4): 512-20.

Lee, C.V., S. McDermott, and C. Elliott. 1990. The delayed immunization of children of migrant farmworkers in South Carolina. *Public Health Reports* 105(3): 317-30.

McDermott, S., and C.V. Lee. 1990. Injury among male migrant farmworkers in South Carolina. *J. Comm. Health* 15(5): 297-305.

McVea, K.L.S.P. 1995. Self injection among North Carolina migrant farmworkers [MCN Clinical Supplement]. *Migr. Health Newsline* March/April: 2-4.

Martin, S.L., T.E. Gordon, and J.B. Kupersmidt. 1995. Survey of exposure to violence among the children of migrant and seasonal farmworkers. *Public Health Reports* 110(3): 268-76.

Meister, J.S. 1991. The health of migrant farmworkers. *Occup. Med.* 6(3): 503-18.

Mobed, K., E.B. Gold, and M.B. Schenker. 1992. Occupational health problems among migrant and seasonal farm workers. *West. J. Med.* 157(3): 367-73.

Palerm, J.V. 1992. A season in the life of a migrant farm worker in California. *West. J. Med.* 157(3): 362-66.

Rust, G.S. 1990. Health status of migrant farmworkers: A literature review and commentary. *Am. J. Public Health* 80(10): 1213-17.

Schenker, M.B., and S. McCurdy. 1990. Occupational health among migrant and seasonal farmworkers—The specific case of dermatitis. *Am. J. Ind. Med.* 18: 345-51.

Slesinger, D.P., and C. Ofstead. 1993. Economic and health needs of Wisconsin migrant farmworkers. *Migr. Health Newsline* 10(6): 3-4.

Slesinger, D.P. and M.J. Pfeffer. 1992. Migrant farm workers. In *Rural Poverty in America*. C. Duncan, ed. Westport, CT: Auburn House.

Wilk, V.A. 1986. *The Occupational Health of Migrant and Seasonal Farmworkers*. Supported by Farmworker Justice Fund, Inc. Kansas City, MO: National Rural Health Care Association.

24

SAFETY ON THE FARMSTEAD

Wilma S. Hammett, Ph.D.
Judieth E. Mock, Ed.D.
North Carolina Cooperative Extension Service
North Carolina State University

> Children and the elderly are the most vulnerable members of the farm family. They are more likely to the be the victims of accidents and unintentional injuries. It is vitally important that farm families know and understand their children's developmental capabilities and that they as parents develop appropriate responses to protect them. Falls are the leading cause of unintentional deaths for people 80 years old and above. By far, stairs and floors are the leading cause of hospitalization for persons over 65 as compared to other homes, home furnishings, and fixtures. This chapter offers information and guidelines to consider when choosing homes, home furnishings, and fixtures, disposing of household chemicals, child safety, and creating a safe environment on the farm for both youth and the elderly.

INTRODUCTION

The family farm is a multifaceted entity where work, play, and everyday living activities are combined. Often several generations of family members live and work together. However, today farming is a high-technology industry with heavy machinery, toxic chemicals, large animals, power tools, ponds, liquid manure storage facilities, grain processing areas, etc., that can be hazardous to adults and children.

Two groups of farm family members, children and elderly, need special precautions to make their farmstead environments safe. The physical limitations of the elderly and the inability of children to perceive danger produce added risks for them.

Farm ponds and firearms present special concerns since they represent both a means of recreation and a necessary component of the farm's operations. Regular, consistent training and appropriate examples set by adults are the best safety lessons for children. Although it is impossible to remove all the potential hazards on the farm and farmstead, awareness and education can help to limit the dangers.

CREATING SAFE FARMSTEAD ENVIRONMENTS FOR CHILDREN

In 1992, unintentional injuries were the leading cause of death for children and youth, 1 to 24 years of age, accounting for 40% of the 49,505 deaths. Drownings, fires, and burns were the leading causes of death after motor vehicle accidents for this age group. For infants under 1 year of age, unintentional injuries were the fourth leading cause of death after certain perinatal conditions such as congenital anomalies, and sudden infant death syndrome. Infants under age 1 have a higher death rate from unintentional injuries than any other age less than 16 (NCS 1995).

In 1994, more than 1,750,000 children under age 5 were treated in hospital emergency rooms for injuries associated with the home and products in and around the home. Home furnishings and fixtures, home structures and construction materials, personal use items, child nursery equipment and supplies, toys, housewares, space heating, cooking, and ventilating appliances were the leading consumer products associated with the injuries (CPSC 1994).

Farm children are exposed to potential dangers every day since they live and play on the worksite. Almost 24,000 farm children are injured each year while working or playing on the farm. Tractors and equipment, falls, and encounters with animals are the leading causes of injuries. It is vitally important that farm families know and understand their children's capabilities at the various stages of development and growth and that they as parents develop appropriate responses to protect their children.

Each child's development in physical capabilities, maturity and judgment is unique. While predicting how a child will react in a specific situation is next to impossible, there are some general characteristics based on age group development that provide some insight and understanding. As children grow, their ability to interact with their environment change rapidly (Abend and Hallman 1995).

Developmental Growth

0 to 2 years. The totally dependent infant at birth becomes an active explorer by age 2. At first the baby is tasting and touching everything within reach. Then, as it begins to crawl, its world expands to include the floor and everything on or near it. As mobility increases, the toddler starts to climb onto and into everything and his world expands to the tops of tables and counters and appliances. Skills are inconsistent and falls are common. Toddlers have no concept of "safe or unsafe" so parents must be aware at all times where their toddlers are and what they are doing.

3 to 5 Years. Water, interesting noises, and moving parts fascinate children 3 to 5 years old. They love to climb and can now see those high storage areas previously out of their range. They do not think logically and believe they *can* do anything they want to do. In other words, they do not understand risk.

6 to 8 Years. Children 6 to 8 years old can understand danger. However, they have short attention spans and can easily be distracted. They are curious and are trying to improve their physical skills. Also, they want to perform activities that may not be appropriate for their physical capabilities in an effort to appear competent and be accepted by peers.

9 to 11 Years. Children of this age begin to think logically, but not consistently as evidenced by the fact that sometimes they seem to understand cause and effect and, other times, they don't. They are "hands-on" oriented and want to be involved in family or group activities, yet they are striving to develop their own sense of self and to experience achievement. There are great variations in size during this developmental stage and parents may misleadingly think that taller children are more capable of performing adult tasks, when in fact that the reverse may be true.

12 years and Older. Early teens are most often clumsy and awkward. Their feet and hands grow more than the rest of their bodies, often resulting in lack of coordination. Adolescents are extremely interested in experimenting, resist supervision and authority, and think they are immortal. These traits enhance their accident potential as well as the fact that they are most likely to have chores on the farm that are necessary to the farm's daily operation (Abend and Hallman 1995).

The ability to understand danger and react safely is a learned behavior. Children need appropriate examples to follow as well as consistent training to develop good judgment. The parents' role should not only include supervision, but making and enforcing rules, education, and safe handling and storage practices that provide protection for everyone in the family as well as being a safety conscious role model (Abend and Hallman 1995).

Common hazards for children include falls, burns, drownings, poisoning, electrical shock, choking, strangulation, suffocation, and accidents associated with toys and nursery and home furnishings. Following are some safety guidelines for each of these hazards:

Falls

If a window must be open where a small child is sleeping or playing, open double hung windows from the top or only open others 2 inches or less. A screen over an open window will keep out bugs but will not be strong enough to keep an infant inside. Install locks to prevent sliding windows from opening wide enough for a child to fall out.

Keep furniture away from open windows as a child may climb on the furniture to reach the window. Children should not be allowed to climb ladders, either. A ladder should never be left in an area where a child is playing.

Gates can be used at the top and bottom of stairs to keep infants and toddlers away from stairs until they can climb them safely.

A baby should never be left alone in a highchair, even with a safety strap (CPSC #1004; CPSC #4241; AAFP 1995; Fontana 1973).

Burns

Children 3 years old or less are most susceptible to burn injuries. Grease and hot water are most often the causes of burn injuries to this age group. Scalds from water 140° F or above can occur in 2 seconds.

The kitchen is a dangerous area. Scalding hot water, hot foods, hot grease are all potential hazards. Some safety precautions to practice to prevent burn injuries from occurring include:

- Turn handles on pots and pans away so the handles face the side or rear of the stove.
- Turn the hot water temperature down to 105° F as tap water can seriously burn a small child.
- Store matches and lighters out of reach of small children.
- Use the back burner of the stove.

Other precautions include:

- Never leave a small child alone, even for a minute.
- Toys with heating elements are only recommended for children 8 years old and up.
- Before bathing an infant, test the temperature of the water with the elbow or the inside part of the arm. (Fontana 1973; CPSC #1004).
- Do not leave children unsupervised near fireplaces, barbecue grills, or any open flame. By law, children's sleepwear should be flame resistant. Look for the label to be sure. Improper washing in either chlorine bleach or fabric softener can reduce the flame resistant protection (Gunderson 1995).

Household batteries can cause chemical burns. Household batteries can overheat and rupture when the wrong battery or the wrong charger is used. Rechargeable batteries should be recharged in the proper size battery charger. Using a charger intended for larger batteries can cause the batteries to overheat and rupture. Safety guidelines for household batteries follow:

- Do not place regular household batteries not intended to be recharged in a battery charger as the batteries could overheat and rupture.
- Do not mix alkaline and carbon-zinc batteries together in the same appliance or toy as it may cause overheating and rupture.
- Always replace the entire set of batteries at once and use a complete set of new ones to prevent overheating and rupture.
- Do not put batteries in backwards. If a battery is reversed (positive end to positive end instead of positive end to negative end) it can overheat and rupture.
- Parents should warn young children not to take out batteries or install them. Parents, not children, should install batteries (CPSC 1994).

Drowning

A young child can drown in a very small amount of water. Cases of drownings have occurred in 2 inches of water. Children have drowned in bathtubs, basins, toilet bowls, and even 5-gallon buckets. Young children should be kept out of the bathroom unless supervised.

Empty 5-gallon buckets after using them and don't leave a bucket with even 2 inches of liquid in it around a child who is unsupervised because a young child can fall into the bucket and drown. A baby or toddler should never be left unattended in a bathtub supporting ring because that child can drown in the time it takes to answer the doorbell or telephone.

Since children are naturally attracted to water, farm families should fence in the pool area and farm ponds to keep small children out of the water without adult supervision (CPSC 1994; CPSC #5006; CPSC #1004).

Suffocation

Abandoned and open refrigerators are seen by children as great places to hide and play but there is only about 10 to 15 minutes of air for a child in a closed refrigerator. If a refrigerator is discarded and left on the farm property, remove the door completely.
It would be best to have it picked up by a used appliance dealer because many of the parts are recyclable.

To childproof an old refrigerator:

- Turn the unit with the door facing against a wall.
- Lock the door with a pad lock, or use a chain wrapped around the refrigerator and secured by a pad lock.

Plastic bags from dry cleaners, produce, grocery, or trash bags should be destroyed or stored on a high shelf if they are to be recycled. This will keep a small child from playing with them. Never use plastic coverings or handy bags as crib mattress covers or pillow covers as they can cling to a child's face and cause suffocation (Fontana 1973; CPSC #4241).

Do not leave infants 12 months or younger on adult beds with a regular mattress or a waterbed mattress. The infant can become trapped between the mattress and frame or wall.
An infant can also suffocate when he or she becomes wedged against an adult and the mattress and when it is placed in the prone position on a waterbed mattress. Infants should not be placed on soft bedding products such as pillows, sheepskins, quilts, comforters, and toys. When infants sleep in the prone position on these products, they are inhaling air with high levels of carbon dioxide (exhaled air) instead of air with high levels of oxygen. Also a mesh playpen or portable crib can be hazardous to an infant if the side is down since the mesh can form a loose pocket into which an infant can roll and thus suffocate (CPSC 1994).

Electrical Shock

Never use a radio, TV, heater, or any other electrical appliance near a tub, sink, or pool. Electric rollers and hairdryers should be disconnected when not in use to prevent electrical shock.

All electrical cords should be checked periodically and any worn or exposed wires repaired. All unused electrical outlets should have safety caps to prevent children from inserting metal objects into them.

Children should be taught proper use of electric toys and adult supervision of play with electric toys is recommended (Fontana 1973; CPSC 1994; CPSC #4241).

Choking/Strangulation

Children under 3 years old should not play with balloons. If they bite them or put the balloon in their mouths, they can choke or suffocate if the balloon pieces become lodged in the throat. Even common items like coins, pins, buttons, or small batteries can choke a child.

Strings and cords are often used in a child's environment, yet they can be hazardous to infants and children.

If a crib is placed against the wall, any wall decoration, ribbons, or streamers should be out of reach to prevent the infant's entanglement or strangulation. Crib toys strung across the crib or playpen should be removed when a child begins to push up on his hands or knees or is 5 months old. Even toys with long strings or cords can wrap around an infant or young child's neck causing strangulation. Toys with long strings, cords, loops, or ribbons should never be hung in cribs or playpens as a child may become entangled in them.

Strings and cords on children's clothing can be dangerous as well. Loose clothing or clothing with strings may get caught on playground equipment when the child is playing (CPSC #5094). Pacifiers or other items tied around a child's neck can also cause strangulation. (CPSC #5095).

Window covering cords are one of the home furnishings products most frequently associated with strangulation of children under age 5. Children under 18 months old get entangled when their crib is placed near window covering pull cords. Older children get entangled while climbing on furniture. "Break away" cords can be used on any 2-corded window treatments such as horizontal blinds, pleated, and cellular shades, Roman Shades, etc. Hardware for draw draperies and sheers can be permanently installed to the floor or wall so it does not become a hazard (CPSC #5030).

All toy chests with lids should have a safety latch to prevent the lid from falling unexpectedly on a child. A free-falling lid can seriously injure or kill a child (CPSC #4241).

Poisonings

Total poisoning deaths have risen 110% from 1984 to 1994 with a 94% increase in poisoning deaths from 1993 to 1994. In 1992, more than twice as many 1 to 2 year olds died of poisoning than any other children below 15 years of age. Drugs, medicaments, and biological agents were the leading ones involved while other solids and liquids such as petroleum products, other solvents, agricultural and horticultural chemicals, and pharmaceutical preparations were also frequently involved (NSC 1995).

To prevent poisonings:

- Vitamins or mineral supplements that contain iron should be in child-resistant containers. Iron medications contain ferrous sulfate, ferrous gluconate, or ferrous fumarate which are extremely hazardous to young children. Only a few pills can kill a child. Poison Control Centers indicate that iron supplements are responsible for 30% of the child poisoning deaths from medications.
- Avoid taking medicines in front of children.
- Two tablespoons of Pinesol or Drano can kill a small child. Remember young children will eat and drink almost anything so keep all hazardous products out of their reach. Store household cleaning products in high locked cabinets away from children, rather than under the kitchen sink.
- Keep all medicines in a locked cabinet out of reach of children.
- Never refer to medicine as candy.
- Keep all medications in child-resistant closures.
- Never transfer hazardous household products or medicines to a bottle without a child resistant closure.
- Be sure to reclose the child-resistant closure. Don't let any inconvenience in opening the product deter closing it.
- Do not allow a child to play with a closed container of hazardous household products or medications. A child resistant closure does not guarantee that it is impossible for the child to open it. A child resistant closure is not necessarily child proof (CPSC #4241; Fontana 1973; Wyant et al. 1982).

(For other guidelines on using and storing hazardous products refer to the hazardous household products section of this chapter.)

Plants in the home, yard and even the garden may contain a wide range of potent chemicals that can be harmful to a child. Philodendron and diffenbachia, two common houseplants, contain oxalate, a substance that irritates the mouth and throat causing swelling severe enough to impair breathing. Foxglove, rhododendron, laurel, and oleander contain cardiac glycosides, which affect the heart. Even using a rhododendron stick for a hot dog skewer can result in poisoning. Cherry, plum, and peach pits, hydrangea and cotoneaster can release cyanide in the small intestines. Nightshade, horse nettle, tomato, and potato leaves contain solanine which causes severe vomiting and is toxic to the stomach. Many yard shrubs and berries can cause nausea, vomiting, and diarrhea. However, most deaths from plant poisonings are the result of the ingestion of wild mushrooms. Generally, cut flowers are not poisonous but can be hazardous if they get lodged in a child's throat.

Below are some safety guidelines for parents:

- Learn to recognize toxic plants in the home, yard, and garden
- Keep toxic houseplants away from children. Discard dead leaves that fall from plants.
- Avoid using toxic flowering plants or shrubs, especially those with berries, in the landscape plan.

Table 24.1. List of poisonous plants.

The following plants contain a wide variety of poisons and symptoms may vary from a mild stomach ache, skin rash, and swelling of the mouth and throat to involvement of the heart, kidneys, or other organs. The Poison Control Center can give you more specific information on these or other plants that can be poisonous and may not be on this list.

Anemone	Holly Berries	Narcissus
Angel Trumpet Tree	Horsetail	Nightshade
Apricot Kernels	Hyacinth	Oleander
Arrowhead	Hydragea	Periwinkle
Avocado Leaves	Iris	Peyote
(Mescal)	Ivy (Boston, English, other)	Philodendron
Azaleas	Jack in the	Poison Hemlock
Betel Nut Palm	Pulpit	Poison Ivy
Bittersweet	Jequirity Bean	Poison Oak
Buckeye	Jerusalem Cherry	Poppy
Buttercups	Jessamine (Jasmine)	(California
Caladium	Jimson Weed	Poppy
Calla Lily	(Thorn Apple)	Excepted)
Castor Bean	Jonquil	Pokeweed
Cherries,	Lantana Camara	Potato Sprouts
Wild and	(Red Sage)	Primrose
Cultivated	Larkspur	Ranunculus
Crocus, Autumn	Laurels	Rhododendron
Daffodil	Lily of the Valley	Rhubarb Blade
Daphne	Lobelia	Rosary Pea
Delphinium	Marijuana	Star of
Devil's Ivy	Mayapple	Bethlehem
Dieffenbachia	Mistletoe	Sweet Pea
(Dumb Cane)	Moonseed	Tobacco
Elderberry	Monkshood	Tomato Vines
Elephant Ear	Morning Glory	Tulip
English Ivy	Mother-in-Law Plant	Water Hemlock
Four o'clock		Wisteria
Foxglove	Mushroom	Yew

Table 24.2. List of nonpoisonous plants.

Abelia	Coleus	Lipstick Plant
Absynnian	Corn Plant	Magnolia
Sword Lily	Crab Apples	Marigold
African Daisy	Creeping Charlie	Monkey Grass
African Palm	Creeping Jennie	Mother-in-Law
African Violet	(Moneywort,	Tongue
Airplane Plant	Lysima)	Norfolk Island
Aluminum Plant	Croton (House Pine	Peperomia
Aralia	Variety)	Petunia
Araucaria	Dahlia	Prayer Plant
Asparagus Fern	Daisies	Purple Passion
(Dermatitis)	Dandelion	Pyracantha
Aspidistra	Dogwood	Rose
(Cast Iron Plant)	Donkey Tail	Sansevieria
Aster	Dracaena	Schefflera
Baby's Tears	Easter Lily	Sensitive Plant
Bachelor buttons	Eugenia	Spider Plant
Bamboo	Eucalyptus	Swedish Ivy
Begonia	(caution)	Umbrella
Birds Nest Fern	Eugenia	Violets
Blood Leaf Plant	Gardenia	Wandering Jews
Boston Ferns	Grape Ivy	Weeping Fig
Bougainvillea	Hedge Apples	Weeping Willow
Cactus (Certain	Hens and Chicks	Wild Onion
Varieties)	Honeysuckle	Zebra Plant
California Holly	Hoya	
California Poppy	Jade Plant	
Camelia	Kalanchoe	
Christmas Cactus	Lily (Day, Easter)	

- Properly dispose of pruning clippings properly to keep them away from children.
- Do not pick any wild mushrooms. There are no "rules of thumb" to adequately identify safe varieties from poisonous varieties.
- Remove wild mushrooms from the yard as soon as they appear, which usually is after rainy spells in the fall and spring.
- Remember, even garden plants such as tomato and potatoes can have toxic parts. (Wiley 1988; Fontana 1973).

Hazardous Art Materials

In 1988, the Hazardous Art Materials Labeling Act was passed requiring all art materials be reviewed and appropriate warning labels added (effective November 1990). The act covers children's crayons, chalk, paint sets, modeling clay, coloring books, pencils, etc. Art materials that are nonhazardous and therefore safe for preschoolers and elementary school children to use will have a label that indicates the product conforms to ASTM D-4236. Hazardous materials may still be used in children's art materials since the act did not ban their sale or use. However, they are required to have a label that gives adequate directions and warnings for their safe use. It is the parent's responsibility when buying hazardous art materials that they be purchased for children who have enough maturity and will read and heed the directions for use and the warnings (CPSC 1994).

Structure and Furnishings

Furnishings and certain elements of the home structure itself can be potentially hazardous to a child. Garage doors installed prior to March 1982 do not have the automatic reverse feature. If small children live in the home, parents should replace the door opener with one that has an automatic reverse feature or disconnect the garage door opener and operate the garage door manually. Another alternative might be to relocate the wall switch as high as practical to prevent a child from operating the door (CPSC 1994).

Older sliding glass doors were made of regular plate glass and can break into large pieces which are dangerous. Newer sliding glass doors have tempered glass that shatters into small pieces. A film can be added to the older type glass to prevent injury from broken glass or decorative decals can be used so that children can see that the door is closed (Gunderson 1995).

Young children may climb on bookcases, shelves, chests of drawers, TV carts, stands or tables. Any of these pieces of furniture can tip over on a child. Angle braces or anchors attached to the furniture and to the wall in a child's room can prevent tipping. The anchors should be screwed to the wall studs. Televisions should be placed on low furniture to prevent accidents. Do not use tablecloths on lamp tables or dining tables as infants and toddlers can pull them off causing heavy objects to fall or hot liquids to spill (CPSC 1994).

When buying bunk beds, check all eight mattress support fin tabs and pockets for cracks in the metal or welds as they can cause the bed to collapse (CPSC 1994). Check the guardrail spacing on bunk beds. The space between the guardrail and the mattress should be small enough to prevent a child's head from becoming trapped between them (Selling Safety 1993).

When buying a used crib, be sure it meets the following guidelines from the Consumer Product Safety Commission:

- No missing, broken, loose, or poorly installed screws, brackets or other hardware on the crib or the mattress support.
- No more than 2 3/8 inches (or four fingers wide) between the slats to keep a baby's body from fitting through the slats.
- A firm, snug-fitting mattress to prevent baby from being trapped between the mattress and the side of the crib.
- No corner posts over 1/16" above the end panel (with the exception of canopy-style cribs) to prevent clothing from getting caught.
- No cutout areas on end panels which can entrap a baby's head.
- A mattress support that doesn't pull apart easily, to prevent entrapment between the mattress and the crib.
- No cracked or peeling paint, which may be a lead hazard.
- No splinters or rough edges (CPSC #5020; CPSC #5027; CPSC 1994).

Toys can be a potential hazard for children, too. Toys should be checked periodically for broken parts and potential hazards. Damaged toys are dangerous should be immediately repaired or discarded. Make sure that any small parts which might become lodged in a child's windpipe, ears, or nose do not become exposed. Select toys that are age appropriate (CPSC #4241).

Guides for Supervision

Supervision is important to the safety of every child who lives, works, and plays on the farm, no matter what age they are. Parents should know where children are at all times. If both parents are working, they should have adult supervision for children. Concentrating on work and supervising a child is risky for both the parent and the child. Some general guidelines include:

- Provide safety education for children. Train and familiarize children with a job's procedures and equipment when they are helping with chores. Explain the procedures, observe them as they work, and correct their mistakes using positive comments.
- Do not allow young children to play in areas with high levels of noise or dust. Exposure to these environments can lead to cumulative hearing loss and permanent damage to respiratory systems.
- Require older children to wear dust masks and hearing protection when they are working with loud machinery or in grain processing areas.
- Do not allow children to remove objects such as hay bales or containers that are too heavy for them.

- Work areas should be "off-limits" for playing. Tour the farm and tell children where they can and cannot play.
- Children and adults should always tell each other where they are and what they are doing.

Lastly, look around the farm for potential hazards and eliminate them. Parents must act responsibly and think safety at all times. Below are some potential hazards found on the farm.

- Remove junk piles. Young children may think these areas are great for play and climbing.
- Unplug power tools when not in use.
- Empty pails of water, waste oil, or other liquids that may attract toddlers. Drowning can occur in only a few inches of water.
- Don't leave heavy objects such as tractor tires leaning against a wall where children might pull them over on top of themselves. Store those large and heavy objects out of the way in a barn or shed with a lock.
- Ladders to structures like silos should be at least 7 feet off the ground to prevent children from climbing them.
- Fence in farm ponds and liquid manure storage facilities. Use fencing that children cannot climb or squeeze through.
- Make sure livestock gates and fences are sturdy and secure to prevent unwanted contact between children and animals (Abend and Hallman 1995).

SAFETY FOR THE ELDERLY

Accidental falls were the leading cause of injury-related hospital emergency room visits in 1992 accounting for over 7.7 million visits to the emergency room that year (NSC 1995). For younger groups, falls may mean bruises, sprains, etc. However, for older adults, whose bones are more brittle, hip fractures are often the result of nonfatal accidents from falls. According to Harvard Medical School's Health Letter, almost half of the persons over 65 who suffer a hip fracture from a fall cannot walk well after their injury and 1 out of 8 of them die from complications of that fall within 4 months (*How to Fall-Proof Your Home for the Elderly* 1988-1989). Falls are the leading cause of unintentional deaths for people 80 years old and above, accounting for over half of the unintentional injury deaths for that age group (NSC 1995).

In 1994, over one million people 65 and over were treated in hospital emergency rooms for injuries associated with the home and the use of products in the home, including floors, stairs, handrails, beds, chairs, rugs and carpets, tables, ladders, bathtubs and showers, hot tubs, spas, whirlpools, stools, electric fixtures, windows and doors—including glass and non-glass. Almost 400,000 adults 65 years old and over were treated in hospital emergency rooms for injuries associated with home furnishings and fixtures while over 600,000 were treated for injuries associated with home structure and construction materials. These hospital emergency room treatments cost over 5.8 billion dollars for this age group alone (CPSC 1994).

Home furnishings and fixtures and the home structure and construction materials were the two leading consumer product categories associated with accidental deaths of people 65 and older in 1994 (CPSC 1994).

In rural areas, older persons are more likely to live in their own homes longer rather than moving to a retirement community specifically designed to meet their limited physical abilities. When living independently is no longer an option, they are more apt to move in with one of their adult children since fewer retirement communities and intermediate care facilities exist in rural areas and small towns (Longino Jr. 1993).

Sensory and Physical Changes

An understanding of the sensory and physical changes that occur during the aging process is important. While the aging process affects individuals differently and at different rates, it is important to be aware of the changes and recognize them to make the home an adaptive and accommodating environment instead of a hazardous environment. As a person ages there is a gradual decline in physical strength, flexibility, dexterity, and endurance (Salmen 1991). The senses of vision, hearing, taste, smell, and touch also change with age. Such physical and sensory changes can result in reduced mobility, inaccurate perception of the living environment, and difficulty in accomplishing tasks, thus affecting an older person's ability to function safely in their own home (Hooyman and Kiyak 1988).

Several aspects of vision change with age including decreased visual acuity, farsightedness, changes in color perception, decreased sensitivity to light, and decreased ability to adapt to glare. As a person ages, the lens tissue thickens and hardens, making it less elastic while the muscles around the lens deteriorate. This results in reduced ability to change focus from near to far or visual acuity. Visual acuity begins to decline gradually between 45 to 50 years of age and increases in severity after age 65. The Center for Health Statistics indicates that one-half of all people have a visual acuity of 20/70 at age 65. Presbyopia or farsightedness is another age-related change and is usually remedied with bifocal or trifocal lenses for people in their 40s and 50s. However, even with bifocal lenses, many older people still experience difficulty in seeing small details such as numbers in a telephone directory or directions on medications. The lens also becomes more opaque with age allowing less light to go through it. Thus, the older person has poor vision in low light and their ability to adjust to changes in light level is reduced.

Differential hardening of the lens causes uneven light refraction and the older person becomes extremely sensitive to glare (Hooyman and Kiyak 1988). Reflecting surfaces in the home, including bathroom fixtures, high gloss furniture finishes, waxed wood and resilient flooring and "sun-drenched" windows, can be distracting to an older person. Often falls are caused by an older person's inappropriately shifting body weight when reacting to the glare on a high gloss waxed floor. Even fluorescent light flickers can be distracting to the aging person while many younger people may not even notice them at all (Hiatt 1980).

The loss of convergence of images formed in the two eyes results in the deterioration of depth and distance perception. This makes stairs more hazardous to the older person and can increase the risk of falls.

With age, the blind spot increases in size and retinal metabolism deteriorates causing narrower peripheral vision. The field of vision in some older adults reduces from 270 degrees to as narrow as 120 degrees.

The elevator muscles of the eye, which move the eyeball up and down in the socket, also weaken, resulting in reduced range of upward and downward gaze. Thus, the older person may not see objects or low furniture on the floor and objects above eye level.

As the lens of the eye yellows with age, its ability to see the full spectrum of colors is diminished. Thus, an older person may not be able to distinguish between shades of blue, and between blues, greens, and violets. This decline in color sensitivity affects individuals at different rates. It may be noticeable in some people around 70 years of age while others do not experience difficulty for another 10 to 15 years (Hooyman and Kiyak 1988).

Hearing loss or presbycusis is a common disability for older persons as ear bone conductivity and/or nerve sensitivity decrease with age. The ability to hear higher frequencies declines gradually and then extends to the lower frequencies interfering with everyday communications. Presbycusis can also amplify loud noises and create a ringing in the ears. At this stage these background noises interfere with the ability to hear normal conversations. This inability to hear high-pitched sounds such as bells, fire sirens, smoke detectors, etc., can create potential safety problems when audible emergency signals are not heard (Hiatt 1980; Salmen 1991).

Agility becomes more limited due to changes in elasticity of connective tissue and cartilage and musculature sluggishness. These changes affect the gait and the ability to turn the head from side to side and up and down. Other physical changes include the postural slump in which the head is held slightly in front of the body and the shoulders are rounded and rolled. Some chronic illnesses can cause dizziness, disorientation, and frailty, thus affecting the older person's ability to function in the home environment safely (Salmen 1991).

Changes in the central nervous system also occur with age. The central nervous system controls the kinesthetic mechanism and researchers have found that the older person may be 5 to 20 degrees off in estimating their position in space. As a result they tend to take slower, shuffling, more deliberate steps. These age-related changes, combined with slower reaction time, muscle weakness, and decreased visual acuity, increase the possibility of the older person falling and injuring himself or herself (Hooyman and Kiyak 1988). Stairs, furnishings with sharp corners, rugs, high gloss-floors, and rooms filled with furniture and accessories can become a hazard instead of a safe environment (Salmen 1991).

Age-related changes in touch can affect the older person's perception of pain, especially in the extremities. Thus, an older person is at risk of scalding because of reduced ability to detect high water temperatures.

Common sense precautions and a home safety checklist can prevent falls and other injuries of older persons living on the farmstead. To reduce the risk of falls by older adults, modifications can be made to the home using the following checklist as a guide.

Exterior Entrances

- All outside walks, steps, patios, and ramp surfaces should be firm and even with a non-slip surface (even when wet). Non-skid strips should be added if the surface is slippery when dry or wet.
- Doorways should have no or a low threshold. Beveled thresholds should be no higher than 1/2" for interior doors and 3/4" for exterior doors, or use the type of threshold at entrance doors which compresses under weight, or eliminate the threshold entirely.
- Install motion lights so that the older person does not have to fumble for a light switch or put porch and patio lights on a timer so they burn at night and go off at bedtime or early morning.
- Open risers on steps can be a hazard for the older person whose vision is not good. Close open risers with strips of wood.

- Install rails on both sides of exterior stairs and porches. Doormats can create a tripping hazard. If the doormat is over 1/2" in height, replace it with a low-pile type with a slip resistant-backing or secure it to the floor surface with tacks, staples, or double-sided carpet tape.

Ramps

- Exterior ramps should have a maximum slope of 1:20 (1" rise for every 20" of length.)
- Interior ramps should have a maximum slope of 1 to 12 (1" for every 12" of length.) A steeper incline is difficult for an older person in a wheelchair to maneuver and can cause the wheelchair to tip backwards.
- If the exterior surface of the ramp is wood, apply paint mixed with sand (1 pound of silica sand per gallon of paint) for an easy non-skid surface. Rolled roofing material tacked down is effective, also.
- "Broom-finished" concrete ramps also provide a good non-slip surface. The broom strokes should be perpendicular to the slope of the ramp.

Stairs and Floors

- Falls on stairs and floors are overwhelmingly the cause of injuries that most frequently send persons over age 65 to hospital emergency rooms.

- Add handrails on both sides of stairways for extra support to the elderly person. Be sure the handrail is continuous beyond the last step because the older person will need it as a support to get on and off the last step. Also, an older person may forget and think they are on the last step when the handrail stops (Figure 24.1)

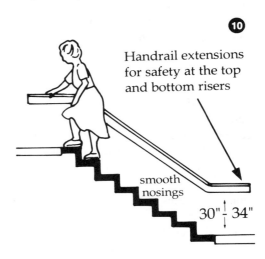

Figure 24.1. Handrails on both sides of the stairways provide additional safety for the older person.

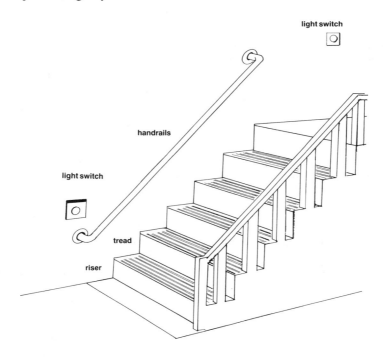

Figure 24.2. Stairs should be well-lighted with a switch at the top and bottom of the stairs.

- Mark any steps that are narrower, higher, or lower than any other steps by using contrasting tape, non-skid strips, or white paint, etc., to make them more visible. The same techniques can also be used to highlight bottom and top steps for contrasting visibility so the elderly person denotes the change in level. Painting the edges of outside steps white or a bright color will make them more visible at night.
- Keep all stairways and floors free of objects such as newspapers, children's toys, books, shoes, clothing, etc.
- Make sure stairs are well-lighted. There should be a light and switch at the top and bottom of the steps. (Figure 24.2) Add wall mounted fixtures that project light upward at the top and bottom of the stairs if a ceiling light fixture does not provide adequate lighting. Wall mounted fixtures with direct lighting on the steps can create a glare on the steps and cause falls.
- Don't use throw rugs at the bottom or top of the stairs.
- Handrails should be designed so that an older person can grip the rail between the thumb and fingers.
- Handrails should be mounted about 1 and 1/2 inches from the wall to allow space for the hand. Mount them so they support up to 150 pounds at any point. Wider spacing from the wall can be dangerous if a person slips and their elbow lodges between the wall and the rail.
- Wooden handrails should be sanded and finished to prevent splinters.
- Tread and riser designs of steps are important. Risers greater than 6 to 7 inches are difficult for the older person to climb. The tread should be wide enough for the foot to rest completely on it. If tread width is too narrow, it can be extended by installing a projecting edge or a nosing to the front of the treads. Any nosing which is added should be beveled to allow the toes to slide up and over it. Without the beveled piece on the riser, any nosing will create a tripping hazard to older people or people with leg paralysis.

Do not use carpet runners on stairs. An elderly person may not be able to distinguish between the tread and risers effectively, causing a tripping hazard. If the floors in the kitchen, bath, laundry area and entry are not slip-resistant when wet, add non-skid strips to them.
- Highly waxed resilient floors or glossy finishes can produce reflected glare when sunlight from a window shines onto it, giving a sense of instability to an elderly person. Add a window treatment such as blinds or sheers that will allow the light to enter, but reduce the glare.
- Add non-skid adhesive strips, flowers or dots to ceramic tile or other slippery floor surfaces in the bathroom or cover the present floor with indoor/outdoor carpeting or replace the present floor with a non-skid ceramic tile.
- Carpet with a low dense pile and level loop construction is best. Pile height should be no greater than 1/4 inch. Use thin padding underneath or no padding and glue the carpet directly to the subfloor.
- Avoid patterned carpet with strong color contrasts and sculptured carpet. Those "busy" designs can cause visual problems and thus an uncertain sense of balance.
- Avoid rugs if possible. If not, use a non-skid bath rug or mat or use non-skid backings under rugs. Be sure the edges of the rug stay in place and do not turn up, thus tripping an elderly person with a shuffling gait.
- Keep hallways and traffic areas in the home free of furniture, especially furniture 18 inches or lower and furniture with sharp corners.

Interior Thresholds

Interior thresholds should be no greater than 1/2 inch to prevent a tripping hazard for older people with walking problems. If the threshold is above 1/2 inch replace it with a lower threshold with beveled edges to prevent tripping or remove it completely.

Grab Bars

- Install grab bars near the toilet and in the bathtub area. They should be attached firmly to studs in the wall and be capable of supporting 250 pounds of pressure in either direction. Molly bolts, nails, or screws attached into sheet rock do not provide adequate support.
- Consider the needs and capabilities of the elderly person when selecting between wall-mounted sheltering arms or pivoting grab bars. Pivoting grab bars offer the best stability when mounted correctly. Sheltering arms provide the best support for getting up and down from the toilet. Remember grab bars designed for people who use crutches, canes, or walkers aren't always satisfactory for people in wheelchairs.
- Towel bars and shower curtain bars should be installed to the same specifications as grab bars since they may become potential grab bars for any person when needed.

- Locate a vertical grab bar on the wall at the foot of the tub and close to the entry side of the tub and a horizontal grab bar on the long wall of the tub.
- All grab bars should have a slip resistant finish.
- Grab bars should be 1 1/4 to 1 1/2 inches in diameter for easy grasping by the user.
- Mount grab bars 1 1/2 inches away from the wall to provide adequate space for the user's hand to wrap around the bar. (Drafts and Jumper Architects 1989; Kirby 1995; Salmen 1991; CPSC 1985).

Electrical

- Don't overload electrical outlets as this can cause a fire.
- Cords to lamps, telephone, appliances, TV, etc., should be placed out of the traffic area to prevent a tripping hazard.
- Do not place furniture on an electrical cord because the cord can be damaged and cause a fire or be a shock hazard.
- Do not attach cords to the wall or baseboard with nails or staples as they can damage the cord and cause a fire or a shock hazard. Use tape to attach them to the wall or baseboard.
- Locate space heaters away from traffic areas and flammable materials such as curtains, rugs, and furniture.
- If any electrical appliances are used in the bathroom, replace present circuits with Ground Fault Interrupted (GFI) ones to eliminate the possibility of electrical shock.
- If an extension cord is used, place it on the floor against a wall where an older person cannot trip over it.
- If an extension cord is needed, use one with sufficient amp or wattage rating for the appliance being used. Standard 18-gauge extension cords can carry 1,250 watts. Overloading an extension cord can cause a fire so do not attach a number of extension cords to each other.
- Check electrical cords on lamps, appliances, and equipment, etc., periodically and replace any that are damaged, frayed, or cracked.
- Periodically check outlets and switches. If they are warm or hot, it may indicate that an unsafe wiring condition exists. Unplug any cords to the outlet and do not use the switches. Call an electrician to check the wiring immediately.
- "Tucking in" electrical blankets or placing additional coverings on top of them can cause overheating and a possible fire.
- When replacing fuses use the correct size amperage rating. A larger size will allow excessive current to flow and overload the outlet and the wiring to the point of starting a fire. (Salmen 1991; CPSC 1985).

General Safety

- Install fire or smoke alarms near an elderly person's bedroom. If hearing the alarm is a problem, there are smoke alarms with a large blinking light. There are also smoke alarms that shake the bed.
- Install carbon monoxide alarms.
- Set the hot water heater temperature at 115 to 120° F or the low setting if an elderly person is living in the home.
- Use bathmats, decals, or non-slip strips in bathtubs.
- Use a night light or motion lights in the bathroom and hallway to provide light for a person getting up at night.
- Use luminous cover plates for light switches in the bedroom and bath areas to help in locating the switch at night.
- Buy a tub seat for a person who has poor leg strength or who tires easily to use for bathing.
- If an individual has trouble turning door knobs and switching on lamps, replace them. A lever type door handle will be easier to operate. Lamps can be connected to portable dimmer devices that operate by sliding a button.
- Do not place potholders, dish towels, or plastic utensils on or near the range where they may ignite. Fasten loose fitting long sleeves with pins or elastic bands while cooking to prevent them from catching fire.
- Provide a step stool with a handrail for added stability to reach heights. Furniture should not be used.
- Make sure all flammable products are stored away from ignition sources including gas water heaters. Incidents have been reported where gasoline stored 10 feet away from a gas water heater exploded (Kirby 1991; Salmen 1991; CPSC 1985).

Color and Furnishings

Furniture should have a firm cushion for support, arms that extend the full depth of the chair seat, and a seat that is not too deep or too low. If an elderly person needs chair arms for leverage when sitting or standing from a chair, use a chair with arms that extend the full depth of the chair. Chair arms that are too short can cause the elderly person to lean over too much and easily lose his or her balance.

- Rocking chairs and recliners may not be good choices for the elderly person who has trouble getting up and down without leverage support from chair arms. An elderly person can trip over the protruding rockers.
- Some recliners have mechanisms that are difficult for an elderly person with reduced arm strength to operate. Recliners with an electronic button are best. Tables in the bedroom of an elderly person should have rounded corners. Use adaptive devices available from children's stores.
- Low profile furniture such as coffee tables, stools, ottomans, etc. should be eliminated.
- Coffee tables should be at least 24 inches high, otherwise they will be out of the elderly person's line of sight. Also less bending will be required, thus reducing the chances of tripping or falling.
- Add colored strips of tape or colored decal strips across the front of the lavatory vanity to readily distinguish it.
- Add colored strips of tape or colored decal strips across the bathtub rim and base to distinguish the top edge and bottom edge of the tub from the surrounding areas. This is especially helpful when the walls and floor are the same color as the tub.
- Add colored strips of tape or colored decal strips across the shower curb to distinguish the edge of the curb from the walls and floor.
- Use window treatments such as adjustable blinds, sheets, shades, etc. to control glare on the floor and furnishings (Hammett 1992).

HAZARDOUS HOUSEHOLD PRODUCTS

A hazardous product has chemical, biological, or physical properties that make it potentially dangerous to humans or the environment. Many of the products are used for crop production, housework, gardening, home improvement, equipment maintenance, and even hobbies in and around the farmstead. In fact, it is estimated that the average home has 100 to 200 products used for cleaning, painting, and disinfecting.

These materials when not used, stored, or disposed of properly can contaminate drinking water, can ruin septic systems, can cause injuries, poisoning, and air pollution (Powell and Bradshaw 1990; Hammett 1990).

What Makes a Household Product Hazardous?

Household products are hazardous if they are:

- *Ignitable*—capable of being set on fire or of bursting into flame either spontaneously or from interaction with another substance or material. Regulations classify a material as ignitable if it is a liquid with a flash point less than 140° F. Charcoal lighter fluid, gasoline, kerosene, nail polish remover (acetone), and various oils ignite easily. What's the difference between a flammable or combustible liquid? Flammable liquids have a flashpoint of under 100°F. Common flammable liquids include gasoline, acetone, methyl, ethyl, and many alcohols. Combustible liquids have a flashpoint of 100° F or more. Combustible liquids include kerosene and motor oil. Sometimes flammable and combustible liquids do not need a flame or spark to make them start to burn. They will burn as soon as their ignition temperature is reached. Ignition temperature is the temperature at which those liquids will start to burn spontaneously. For example, oil soaked rags will start to burn anytime the ignition temperature is reached, and if there are common combustible materials like wood, paper, or cloth in the trash also, the fire can spread easily (Bureau of Business Practices 1993).
- *Corrosive*—capable of eating away materials or destroying living tissue when contact occurs (pH less than 2 or greater than 12.5). Battery acid and strong alkali products such as lye are examples.
- *Explosive and/or Reactive*—can cause an explosion or release poisonous fumes when exposed to air, water, or other chemicals. For example, acids when mixed with water can result in rapid heating and spattering.
- *Toxic*—poisonous, either immediately (acutely toxic) or over a long period of time (chronically toxic). Many pesticides, cleaning products, paints, photographic supplies, and art supplies are toxic (HHWP 1989).

How Do Consumers Know if a Product Is Hazardous?

The Federal Hazardous Substances Act of 1960 established labeling requirements for consumer products containing hazardous substances. If a household product has a hazardous substance, the front label must include a warning and a description of the hazard.

DANGER—substances which are extremely flammable, corrosive or highly toxic.
POISON—substances which are highly toxic.
WARNING, or CAUTION—substances which are moderately or slightly toxic.

A statement telling how to avoid the hazard must appear with safe use instructions on hazardous household products and pesticides. Examples might be:

KEEP OUT OF REACH OF CHILDREN   or
USE IN A WELL-VENTILATED AREA

These labels must also include the following information:

1. Brand name
2. Common and/or Chemical Name (example: sodium hypochlorite or bleach)
3. Amount of Contents (example: 16 oz.)
4. Signal Word—Danger, Poison, Warning or Caution.
5. Instruction for Safe Handling and Use (example: recommended amount to use).
6. Name and address of manufacturer, distributor, packer or seller.
7. Description of Hazard and Precautions (example: Irritant to skin and eyes, harmful if swallowed).
8. First Aid Instructions, when necessary or appropriate (example: If swallowed, feed milk).

As part of the hazard label, many manufacturers now provide a toll-free number on the container. Consumers and/or physicians can use that number to get more information about chemical ingredients used in the product. Consumers can request Material Safety Data Sheets (MSDS). These will provide more information regarding safety and health. Pesticides and some household products now have safe disposal recommendations on the label as well.

Pesticides are regulated by the Federal Insecticide, Fungicide and Rodenticide Act (FIFRA) which sets the minimum standards for label information. Thus, the category description of the "signal words" is different between pesticides and other household hazardous products.. On pesticides, DANGER or POISON means the product is highly toxic. A few drops to one teaspoon can kill a 150 pound adult. WARNING means the product is moderately toxic. One teaspoon to one ounce can kill a 150 pound adult. CAUTION means the product is slightly toxic. It would take over one ounce to kill a 150 pound adult.

Unlike pesticides, there is no distinction between CAUTION and WARNING on household products. These are the signal words found on most cleaning products. Many laundry and automatic dishwasher detergents, disinfectants, and all-purpose cleaners fall into this category.

DANGER indicates that greater precaution should be taken. This signal word most often is found on specialty products used for tough jobs such as oven cleaners or drain openers. DANGER will also be used on products that can ignite if exposed to open flame.

POISON rarely appears on household cleaning products. This is the strongest hazard indication and means that accidental exposure could cause severe medical effects. The term may be found on household lye and on some car care products like antifreeze (HHWP 1989; Niemeyer et al. 1992).

What the Labels Don't Tell

Label information is directed at "acute" or immediate effects only. There is no given information about "chronic" or long-term hazards of chemical products, such as cancer or birth defects.

There are other concerns about labels, as well. Some products contain ingredients that have not been officially recognized by the federal government as hazardous but still are cause for concern. "Inert" ingredients are chemicals added as "carriers" for the active ingredients in cleaners and pesticides. Only the percentage of inert ingredients are required on the label, not their identity and some inert ingredients are hazardous (Patrick 1992).

There are no standardized lists of chemical names. Many chemicals have numerous trade and/or scientific names and this makes it hard for consumers to compare products. Antidotes listed on the labels may be incomplete, out-of-date, or even dangerously wrong.

Today as a result of the environmental concerns of consumers, many products are using the term "non-toxic" on the label The use of the term "non-toxic" is for advertising only and is not a regulatory definition by the federal government. The product may be promoted as "non-toxic," yet still have a hazardous label with signal words on it.

It is very important that consumers know as much as possible about products before they use them so that they can adequately protect themselves (Hammett 1990).

Types of Household Hazardous Products

Most hazardous household products can be grouped into five major categories:

- *Automotive products* include motor oil, brake and transmission fluid, antifreeze and car batteries, gasoline, kerosene, diesel fuel, and car wax with solvent.
- *Household cleaners* include drain cleaners, oven cleaners, toilet cleaners, spot removers, silver polishes, furniture polishes, cleansers and powdered cleaners, window cleaners, bleach, liquid cleaners, and dyes.
- *Paints and solvents* include latex, oil-based, auto and model paint, paint stripper, primer, rust remover,

turpentine, varnish, wood preservative, mineral spirits, and glues.
- *Pesticides*.
- Miscellaneous.

Other hazardous products include: aerosol products, dry cell and disc or button batteries, hearing aid batteries, moth balls and flakes, shoe polish, photographic chemicals, smoke detectors and air fresheners and deodorizers (Hammer 1988).

Exposure to Hazardous Products

Hazardous substances may enter one's body in three ways—ingestion, inhalation, and absorption through the skin.

Toxins can be ingested by eating or drinking hazardous substances or contaminated food and water. Ingestion is a major cause of poisoning in children 6 years old and under. People using hazardous products should avoid eating, smoking, or drinking around the products. Toxins can also be inhaled. Gases, vapors, and sprays can be absorbed through the lungs and enter the blood. Good ventilation is essential when using hazardous products. Air conditioners alone do not provide sufficient ventilation since they recirculate air, even when set on "vent." Thus, they do not adequately remove contaminants. If a person can smell a toxic chemical, the ventilation is usually not sufficient (although some harmful chemicals have no odor). Special masks or respirators should be used to protect a person from inhaling toxic chemicals.

Toxins can be absorbed through the skin. Hazardous products containing irritants or corrosives will injure the skin and may be absorbed and some hazardous chemicals can even be absorbed without causing any overt damage to the skin. Gloves and/or protective clothing should be worn when working with corrosive products and the skin should be washed thoroughly with soap and water if the chemical is spilled on the skin.

The eyes also are vulnerable to injury. Many hazardous products can cause eye damage if splashed into the eye. Oven cleaners, drain cleaners, and paint thinners are just three examples. Goggles should be worn when working with these products. Regular eyeglasses do not provide enough protection and contact lenses (especially soft lenses) should not be worn at all. The lenses may absorb the vapors and then hold the irritant against the eye. Safety goggles should be worn, are inexpensive, and can be purchased at hardware, automotive supply, and farm equipment stores. If a chemical gets in the eye, the eye should be flushed immediately with water and medical attention sought (HHWP 1989).

Guidelines for Using and Storing

Selection can be the first step toward minimizing danger. Follow these guidelines:

- Read the label. Are the ingredients safe to use in and around the home?
- Make sure the product will do the job that needs to be done.
- Buy the least hazardous product for the job. Let the signal words (Poison, Danger, Warning, Caution) be a guide.
- Check the label to see if a product has several uses so that a different product won't be needed for each job.
- Avoid aerosol products as they may contain hazardous or toxic propellants, and the fine mist that they produce may be more easily inhaled. Pressurized cans can cause problems or explode when they are crushed, punctured, or burned.
- Know how to properly dispose of the container.
- Remember, the word "non-toxic" is for advertising only and does not mean the product meets any federal regulations for non-toxicity.
- Buy only as much of each product as is needed for a short period of time.
- Choose water-based paint, glue, shoe polish, and similar products rather than solvent-based products.

The following guidelines will help keep the home and environment safe when using hazardous products.

- Read the directions on the label and follow them. Twice as much doesn't mean twice the results.
- Use the product only for the tasks listed on the label.
- Wear protective equipment recommended by the manufacturer.
- Handle the product carefully to avoid spills and splashing. Close the lid as soon as the product is used to control vapors and reduce chances of spills. Secure lids tightly after use.
- Use products in well-ventilated areas to avoid inhaling fumes. Work outdoors if possible. When working indoors is the only option, open windows and use a fan to circulate the air toward the outside. Air conditioners alone do not provide sufficient ventilation because they recirculate air, even when set on "vent." Take plenty of fresh-air breaks. If dizziness, headaches, or nausea occur, take a break and go outside.
- Do not mix or use a pesticide near a well head.

- Avoid mixing hand dishwashing detergents with chlorine bleach. Some formulations contain ingredients such as ammonia that are incompatible with chlorine and mixing them can release hazardous gases.
- Do not eat, drink, or smoke while using hazardous products. Traces of hazardous chemicals can be carried from hand to mouth and smoking can start a fire if the product is flammable.
- Do not mix products unless directions indicate that it is safe to do so. This can cause explosive chemical reactions or generate poisonous compounds. Even different brands of the same product may contain incompatible ingredients.
- If pregnant, avoid chemical exposure as much as possible. Many toxic products have not been tested for their effects on unborn infants.
- Avoid wearing soft contact lenses when working with solvents and pesticides as they can absorb vapors, and hold the chemical near the eyes.
- Carefully and tightly seal products when finished. Escaping fumes can be harmful and spills can occur.
- Never mix swimming pool chemicals with other products. Swimming pool chemicals contain chlorine, calcium, hypochlorite, strong bases or strong acids. Mixing these chemicals with others can present a fire hazard, and may generate toxic gases.

Use Common Sense

The following guidelines are recommended for storing products safely in the home:

- Follow label directions for proper storage conditions.
- Leave the product in its original container with original label attached.
- Never store hazardous products in food or beverage containers.
- Never store hazardous products in the same cabinet or pantry with food.
- Make sure lids and caps are tightly sealed.
- Store hazardous products on high shelves or in locked cabinets out of reach of children and animals.
- Store incompatibles separately. Keep flammables away from corrosives. Bleach and ammonia can produce toxic fumes if combined. Other combinations such as dry swimming pool chlorine and organics such as brake fluid may produce an explosion or fire. Keep oxidizers away from heat sources and anything that will burn. When an oxidizer gets warm, it can release oxygen and provide the perfect environment for a fire. Chlorine and iodine are two common oxidizers.
- Store volatile products in a well-ventilated area, out of reach of children and pets.
- Keep containers dry to prevent corrosion. If a product container does begin to corrode, place it in a plastic or metal bucket with a lid. The new container should be a material that will not be affected by the leaking product. Clearly label the outside of the container with the contents and appropriate warnings. A second alternative would be to properly dispose of the product.
- Store rags used with flammable products (furniture stripper, paint remover, etc.) in a sealed, marked container.
- Keep flammable products away from heat, sparks, or sources of ignition.
- Know where flammable materials in the home are located and know how to extinguish them.
- Keep a working fire extinguisher in the home or have materials to control fires where you can get to them.
- Store gasoline only in a safety-approved container, away from all heat sources, flames, or sparks and in a well-ventilated area.
- Never store used car batteries near an area where children play. Batteries contain large quantities of sulfuric acid (corrosive) and substantial amounts of lead (highly toxic). Lead ingestion can cause permanent learning disabilities, poor motor coordination, stunted growth, and reduced intelligence. These and other developmental effects often are not apparent until children start school.
- Store oil, grease, and lubricants 400 feet from any wells located on the farm.
- Store pesticides in original containers, properly labeled, and in a locked cabinet or building out of the reach of children. To offer the greatest protection to drinking water, store pesticides at least 400 feet downslope from a well (Hammett 1990, 1991).

Disposal

In rural locations, some household and farm wastes may be disposed of onsite. Common disposal methods include burning or simply piling or burying trash in a ditch on the "back 40." Waste disposed of in an open dump, or even underground, can take many years to degrade or breakdown and represents a potential hazard.

It is estimated that over 90% of rural residents in the United States rely on groundwater as a source of drinking water. When household hazardous waste is poured in a ditch, dumped, or buried in a field, it can contaminate streams, rivers, lakes, poison plants and wildlife, and even move through the soil and contaminate groundwater.

Pouring household hazardous waste down the drain or toilet is not a safe option for rural residents. Large quantities of certain chemicals such as chlorine bleach can destroy the bacteria in the septic system that helps to break down waste. Without the bacteria to break down the waste, the septic system will not function properly, the drainfield can get clogged and the system will fail (Gover 1995).

"Vertical separation distance" describes the distance between the bottom of the septic system's drainfield and the water table. This distance allows the soil, which acts as a filter, to remove pathogens (disease-causing bacteria, viruses, or protozoa) in the effluent before it reaches the groundwater. Factors such as temperature, high groundwater tables, groundwater mounding, and soil type affect pathogen removal and the separation distance needed for removal. Absorption of viruses is effective when the soil pH is below 7.0 with 5.0 being the optimum level. When the soil pH is higher, absorption is not as effective. Some household chemicals poured down the drain or toilet can change the soil's critical pH level just below the drainfield. Other household chemicals, such as solvents, pass unchanged through the drainfield and into surrounding soil and groundwater. Many paint removers and aerosol paint products contain methylene chloride which can pass through a septic system without breaking down (Wats Q and A 1995).

Some rural residents may be on a sewage treatment system from small towns or suburbs of a city. Sometimes suburbs and small towns use "packaged" treatment plants and thus are less able to break down hazardous wastes. Undestroyed toxic chemicals can pass through the sewage treatment plant into the sewage sludge or wastewater. If the wastewater is discharged into a stream, the toxic chemicals may potentially end up in the drinking water downstream. If the sludge is contaminated, it cannot be recycled and used to fertilize crops for animal feed. It will have to be taken to a hazardous waste landfill for disposal (Gover 1995).

Open burning of dry combustibles in small amounts is appropriate where permitted by local ordinance. Dry combustibles include untreated and unpainted wood, paper, and cardboard. Burning outdoors in well-ventilated areas minimizes adverse health effects from smoke. Materials or products containing toxic or harmful substances—including empty pesticide bags—should not be burned. While burning may destroy some toxic substances, others will become concentrated in the smoke, ash, and sludge which results from burning. Burning in the same location repeatedly may cause the toxic substances to accumulate around the burn area.

Safe disposal of household hazardous products protects farm land, people, animals, and the water supply. It is important to read the labels and follow disposal recommendations carefully. Many household hazardous products such as paint, used oil, antifreeze, and automotive batteries are recyclable. For example, the lead in the battery can be recycled for use in new batteries and other products. The plastic battery casing is also recyclable. Batteries should be stored in a safe, dry place out of direct sunlight, out of reach of children and pets, and away from the well, since a car battery contains over 18 pounds of lead. Most places that sell batteries will take back used batteries. Some service stations and scrap metal dealers will take used batteries, as well. Many communities also have recycling centers which handle old automotive batteries.

Pesticides used in and around the home should be disposed of in a safe and proper manner. Never burn or reuse old pesticide containers. Always triple rinse containers, return the rinse water to the spray tank and apply following label instructions. If the plastic containers cannot be returned to the place of purchase, take the triple-rinsed containers to a licensed landfill. Many county extension centers offer collection programs to collect pesticide containers for recycling. Even triple-rinsed pesticide containers may still contain enough pesticide residue that they should not be used for any other purpose.

Remember, clean out the storage area regularly. Clean up any spills. Leaking containers, oily rags, and other combustible items not used should be disposed of properly.

Take household hazardous waste to a community waste collection site often designated on certain days of the year or to a household hazardous waste collection facility if there is one in the local community (Hervel and Powell 1993).

PESTICIDE RESIDUES AND THE FARMSTEAD

The volume of pesticides used in crop protection and management has increased dramatically in the past half century. The use of pesticides has contributed to the productivity and quality of the agricultural industry and the food supply. At the same time, however, pesticides are also associated with both acute and chronic exposure hazards. Concern with the chronic effects of human exposure and prevention of accidental exposure are receiving increasing attention due to the health hazards related to pesticide use.

Pesticides are classified into three levels of toxicity. The toxicity level is identified by signal words on the pesticide label. Pesticides labeled *Danger* are highly toxic; those labeled *Warning* are moderately toxic; while *Caution* labels identify a low level of toxicity. Routes of human exposure to pesticides include dermal absorption, inhalation, and

Table 24.3. Percent of pesticide residue on fabric after laundering under laboratory conditions.

Insecticides	% of Residue	Herbicides	% of Residue
Carbamate		Acetanilide	
carbaryl FL (Sevin)	10	alachlor ED (Lasso)	2
carbaryl WP (Sevin)	0		
Organophosphate		Carbamate	
methyl parathion L	28	triallate EC (Far-Go)	42
fonofox EC (Dyfonate)	14		
Pyrethroid		Dinitoaniline	
cypermethrin WP (Ammo,Cymbush)	17	trifluralin EC (Treflan)	55
cypermethrin EC (Ammo, Cymbush)	25		
cyfluthrin WP (Baythroid)	10	Triazine	
cyfluthrin EC (Baythroid)	30	atrazine FL (Aatrex)	21
deltamethrin EC (Decis)	48	atrazine WP (Aatrex)	11

FL-flowable liquid WP-wettable powder ED-emulsifiable concentrates (Source: Nelson, et al. 1992).

ingestion. The dermal route is the most common means of pesticide exposure and absorption with estimates as high as 97% (Leonas 1993). Dermal exposure can occur directly, or indirectly through transfer from contaminated clothing. Dermal exposure can be reduced appreciably through use of recommended laundry procedures to remove pesticide residues from contaminated clothing. Indirect exposure of farm family members through pesticide transfer in laundering can also be prevented through use of recommended laundry procedures (Nelson and Fleeker, 1988).

Clothing can act as a barrier, keeping pesticides away from the skin, however that same clothing can become a source of contamination if pesticide residues are not laundered out after each use. Research has shown that pesticides can be transferred from the outer layer to inner layers of clothing, increasing exposure risk to the wearer. Clothing soiled with highly toxic or concentrated pesticides require careful handling. Liquid, oil-based concentrates are difficult to remove from fabrics; therefore clothing soiled with highly toxic liquids should be discarded. In Table 24.3, Olson and Nelson (1993) provide an indication of the ease of removing certain pesticides based on the percentage of pesticide residue remaining after one laundering. As a general guideline, the more water-soluble the pesticide, the greater the ease of removal.

Guidelines for Handling and Storing Pesticide-Soiled Clothing

When handling and storing pesticide-soiled clothing, these precautions, based on findings from numerous studies, help protect all family members.

- Wear rubber gloves to handle pesticide-soiled clothing. Carefully wash the gloves in hot water after each use and store and use them only for this purpose.
- Designate a separate, covered container such as a plastic garbage can for storing pesticide-soiled clothing. Never place these garments with other clothing to be laundered.
- Launder pesticide-soiled clothing after each use (daily). The longer the pesticide remains in the clothing, the more difficult it is to remove.
- Launder pesticide-soiled clothing separately from family laundry. Pesticides can be transferred from one garment to another in the laundry process. The individual responsible for the laundry should be informed when pesticides have been used, and what the pesticide label recommended so that clothing can be properly laundered.
- Remove clothing immediately that has become saturated with highly toxic pesticides and discard in an appropriate manner to minimize the risk of exposing others. Do not attempt to launder (Mock and Jennings, 1991).

Laundering Recommendations for Pesticide-Soiled Clothing

The following laundering procedures are recommended for removal of pesticide residues from fabrics.

- Pre-rinsing clothing before washing will aid in the removal of pesticide particles from fabric. Pre-rinsing

is especially effective in removing particles when a wettable powder pesticide has been used. Pre-rinse in a safe area away from food or drinking water.
- Commercial pre-wash products help remove some pesticide residues. Follow directions on product label.
- Generally laundering in hot water removes more pesticide residue from clothing. Use the normal 12 to 14 minute wash cycle.
- Launder only one or two garments in a single load and use the full water level. This allows the water to thoroughly flush the fabric. Launder garments soiled by the same pesticides together.
- Use heavy duty liquid detergents for oil-based pesticide formulations. Using 1 1/2 times the recommended amount of detergent is also helpful in removing pesticide residues from heavily soiled clothing.
- When low toxicity pesticides are used, clothing may be effectively laundered in one machine washing. However, if the pesticide is highly toxic or concentrated, use repeated launderings.
- Small amounts of pesticide residue remain in the washing machine after the wash cycle is complete. To clean the machine and prevent residue transfer to subsequent laundry loads, run a complete cycle with an empty load and detergent.
- Line dry clothing to avoid contamination of the dryer.
- To clean chemical-resistant gloves, first wash gloves thoroughly before removing them. To remove gloves, peel glove off by holding the cuff, turning it wrongside out. Wash in a bucket or tub with warm water and heavy-duty detergent. Use a dowel to submerge gloves. Line dry. Check for rips and leaks after cleaning by using water to fill the gloves or by holding gloves up to the light. (Mock and Jennings, 1991, 1994).

FOOD SAFETY

Principles of food safety apply universally. The farmstead, however, is a setting in which foods may be produced, preserved, and consumed in one location, therefore, a discussion of food safety principles is appropriate. The following information is summarized from *Keeping Food Safe to Eat* (Lackey and Kolasa 1995).

Food-borne illness may be mild, or deadly. Individuals react differently but the very young, the elderly, and people with some diseases suffer more severely. Recent evidence suggests that food-borne illness also can lead to long-term health problems. Bacteria, molds, viruses, and mycotoxins are among the microorganisms that cause food-borne illness. Food-borne illness can be prevented by handling food so that it does not become contaminated with microorganisms and by making conditions unfavorable for any microorganisms that are present to grow.

Control Temperature to Keep Food Safe

Control microbial growth by controlling temperature. The temperatures between 41°F (5°C) and 140°F (60°C) are known as the danger zone. Microorganisms multiply quickly between these temperature points.

Heating to Keep Food Safe. When heated above 140°F, most microorganisms are killed. Microorganisms that produce a toxin are killed by heating but the toxin may not be destroyed. Therefore it is important that bacteria are not allowed to multiply or produce toxin in the first place, because heating does not always make food safe.

Heat food quickly so it does not stay in the danger zone any longer than necessary. Large quantities of a food need to be divided into several pots so it can be heated above 140°F as quickly as possible. Oven temperatures below 325°F are not recommended for cooking large meats such as roasts and turkey for the same reason. Table 24.4 outlines the temperature and visual appearance foods must reach for safety.

Cooling to Keep Food Safe. When food is refrigerated (held below 41°F), any microorganisms present stay alive but they do not multiply very quickly. Foods should be chilled as quickly as possible. Divide large portions of food into smaller portions in shallow pans where food is no more than 4 inches deep. Cut large pieces of meat into smaller portions to hasten cooling. Food should not be left at room temperature longer than 2 hours, or no longer than 1 hour if very warm. Many people believe that hot food should not be placed in the refrigerator, that it will harm the refrigerator and cause other foods in the refrigerator to spoil. This thinking dates back to the days of the "ice box" cooler. Hot foods do not stress the cooling power of the refrigerator.

Thawing Foods

Place foods to be thawed on a plate to catch any dripping liquid. Place thawing food on the lowest shelf in the refrigerator so it does not drip onto other foods. Foods may be thawed in a microwave oven, but thawing should be followed by immediate cooking.

Keep Surface Clean

Hand washing removes microorganisms that can be transferred to food. Keeping equipment clean does the same thing. Microorganisms can be transferred from raw meat to foods that won't be cooked, such as salad greens, by using dirty cutting boards or unwashed knives. Microorganisms move around the kitchen on drops of raw meat juices, dirty towels and sponges, and utensils. Keep kitchen towels and washcloths clean. Wipe up spills immediately. Use plenty of hot, soapy water to wash dishes. Air drying dishes keeps microorganisms from being spread by damp and dirty towels. Cutting boards and other utensils may be sanitized in a chlorine solution of 2 tablespoons chlorine bleach to 1 gallon warm water. Soak utensils in sanitizing solution for 1 to 2 minutes, remove, and air dry.

If In Doubt, Throw It Out

The cardinal rule of food safety is "If In Doubt, Throw It Out." Contaminated food may not look, smell, or taste bad. If there is any question that food may have been improperly handled while being prepared, cooked or stored, Do Not Eat It. The Food Safety Actions listed in Table 24.4 are important to the farmstead kitchen.

Table 24.4. Temperature and appearance foods must reach for safety.

Food Appearance	End-Point Cooking Temperature °F	Visual
Pork	160	Grey color inside
Poultry	180	Juices run clear
Ground beef, pork	160	Juices run clear
Ground poultry	170	Juices run clear
Beef	160	Medium doneness
Eggs		White is firm, yolk begins to thicken. Scrambled eggs are firm, no liquid egg visible
*Reheating leftovers**		
Solid foods like meat	165	
Liquids like soup		Reheat liquid foods to a rolling boil

Reheating should be done quickly. Foods reheated to 165° F should then be held above 140° F.

Source: Lackey and Kolasa 1995.

THE FARM POND

Farm ponds can be an important economic resource in farming operations, as a source of irrigation, water for livestock, fire protection, and recreation. These ponds also can be the source of potential safety hazards. They contribute to accidental drownings, with children under the age of four constituting the largest group of accidental drowning victims, and the overall majority of victims ranging in age from toddlers to young adults (Murphy, 1990). It is the responsibility of the farm operator to take preventive measures to make the farm pond as safe as possible. In most cases, these measures include fencing and posting. Liability may be increased when access to the pond is not restricted to deter uninvited persons. If the pond is used for recreation and swimming, signs warning of specific dangers and signs identifying safe areas for swimming can prevent accidents. All farm ponds used for swimming should have a rescue post with a life ring, rope and a 12 to 14 foot pole (Bean, 1991). Bean (1991) outlines farm pond safety precautions as follows:

- Teach children to swim.
- Mark safe swimming areas with posts or floats.
- Remove submerged rocks, stumps, broken bottles, and other hazards.
- Keep simple rescue devices near the pond.
- Post life saving warning signs.
- Have the pond tested for water quality/contamination.
- Do not permit any individual to swim alone.
- Do not permit small children near a farm pond without adult supervision.

Health hazards associated with farm ponds include potential contamination. Water exhibiting a cloudy appearance, a foul odor, or excess algae may be contaminated by fertilizer or pesticide runoff, or other pollutants and should not be used for swimming. Any ponds used for swimming should be analyzed to determine water quality by a qualified laboratory.

Other safety hazards include submerged rocks, broken glass, and miscellaneous debris often found in farm ponds.

FIREARM SAFETY

In 1994, firearms were the fifth most common type of home unintentional-injury death in the United States, with 900 deaths reported. The majority of unintentional-injury deaths occurred among the 5 to 24 year old age group (NSC 1995). Firearm death rates often are higher in rural than in urban areas, suggesting a relationship to higher rates of firearms ownership. For example, in North Carolina almost half of rural adolescent boys own a shotgun, rifle, or handgun (Sadowski et al. 1993). Parents in rural areas are approximately twice as likely as urban/suburban parents to own guns (Webster et al. 1993).

Education for achieving firearms injury control in rural areas may be contingent upon recognizing cultural differences between rural and urban residents. Differences in injury risk and in the effectiveness of control measures are dependent on availability of firearms, and on differences in risk perception.

> Certain products, such as all-terrain vehicles (ATVs) and firearms, are often considered by rural residents as tools for living just as automobiles are considered a 'necessity' for work and recreation by rural and urban residents alike....
>
> People who live in rural areas may be comfortable with driving ATVs and with having loaded firearms around the house. They may not perceive them as dangerous if used correctly, any more than urban residents consider the lawnmower or barbecue grill a threat." (Cole 1993, p. 510)

Increasing the population's perception of injury risks associated with firearms is a major component of injury control.

Perceptions of the potential for crime may also impact firearm ownership and use. Statistically, rural areas are still safer than urban living areas. However, rural crime rates have increased dramatically in the past two decades. Rural crime is primarily associated with property, and may be reduced with basic security measures (National Crime Prevention Council, n.d.).

The use of firearms for protection and/or recreation is a much discussed issue. If firearms are kept in or around the farmstead, these safety rules should be observed. Instruct all family members in care and safety. Store guns in a safe place away from children and be sure they are never loaded during storage. Store ammunition in a locked enclosure separate from the guns. Secure the guns with trigger locks that can be removed only by using a special wrench or key. Lock the wrench or key up separately.

SUMMARY

Children and the elderly are the most vulnerable members of the farm family. They are more likely to be the victim of accidents and unintentional injuries. It is vitally important that farm families know and understand their children's developmental capabilities and that they as parents develop appropriate responses to protect them. Parents should provide adult supervision, safety education, and be safety role models for their children.

Falls are the leading cause of unintentional deaths for people 80 years and above. The home and home furnishings and fixtures in the home were most associated with injuries of the elderly with falls on stairs and floors as overwhelmingly the cause that send persons over 65 to hospital emergency rooms. Understanding physical changes and making adaptations to accommodate the physical limitations can encourage and promote an active lifestyle for the older farm family generation.

Many of the products used for farming, housework, gardening, home improvement, and equipment maintenance are hazardous. Safety guidelines should be followed when using, storing, and disposing of these materials to protect family members, the farm land, the well water, and surface and ground water.

Prevention of accidental exposure to pesticides and pesticide residues is a major concern not only to the farm operator and/or farm worker, but to the entire farm family. Exposure can be reduced appreciably through use of recommended handling, storage and laundry procedures for pesticide soiled clothing.

Farm ponds are an economic resource in farming operations as well as a resource for recreation. It is the responsibility of the farm operator to make the farm pond as safe as possible. Contamination of farm ponds represents a potential health hazard as well. Water quality should be analyzed by a qualified laboratory.

The use of firearms for protection and recreation is a much debated issue. Increasing the farm population's perception of injury risks associated with firearms is a major component of injury control.

REFERENCES

Abend, E., and E. Hallman. 1995. *Safer Farm Environments for Children.* Ithaca, NY: Cornell Cooperative Extension, Cornell University.

American Academy of Family Physicians (AAFP). 1995. *Child Safety: Keeping Your Home Safe for Your Baby.* Kansas City, MO: American Academy of Family Physicians.

Bean, T.L. 1991. *Farm Pond Safety.* AEX-390. Columbus, Ohio: Ohio State University Extension.

Bureau of Business Practice. 1993. *Basic Chemistry Lesson: Working Safely with Hazardous Materials.* Waterford, CT: Bureau of Business Practice.

Cole, T.B. 1993. An injury control strategy for rural North Carolina. *N. Car. Med. J.* 54(10): 508-10.

Consumer Products Safety Commission.
  [1994?] *Safety Facts.* 2nd ed.
  [n.d.] *Poison Awareness: A Discussion Leader's Guide*
  [1995] *Protect Your Child.* Available from gopher://cpsc.gov:70/00/Whats_New/4241.txt.
  [1995] *Your Used Crib Could Be Deadly: Safety Alert.* Available from gopher://cpsc.gov:70/00/Whats_New/5020.txt.
  [1995] *Children Can Strangle in Window Covering Cords.* Available from gopher://cpsc.gov:70/00/Whats_New/5030.txt.
  [1995] *Some Crib Cornerposts May Be Dangerous: Safety Alert.* Available from gopher://cpsc.gov:70/00/Whats_New/5027.txt.
  [1995] *A Hidden Hazard in the Home: Infants and Toddlers Can Drown in 5-Gallon Buckets.* Available from gopher://cpsc.gov:70/00/Whats_New/5006.txt.
  [1994] *CPSC Annual Report.* Available from gopher://cpsc.gov:70/00/0093annl.txt.
  [1995] *Baby Safety Checklist.* Available from gopher://cpsc.gov:70/00/Whats_New/1004.txt.
  [1995] *Strings Can Strangle Children on Playground Equipment: Safety Alert.* Available from gopher://CPSC.gov:70/00/Whats_New/5094.txt.
  [1995] *Prevent Child Drownings in the Home.* Available from gopher://CPSC.gov:70/00/Whats_New/5013.txt.
  [n.d.] *Strings, Cords and Necklaces Can Strangle Infants: Safety Alert.* Available from gopher://CPSC.gov:70/00/Whats_New/5095.txt.
  [1985] *Safety for Older Consumers: Home Safety Checklist.*

Drafts and Jumper Architects. 1989. *Checklist-Residential Design for Retirement Living.* Columbia, SC: Drafts and Jumper Architects.

Fontana, V. 1973. *A Parent's Guide to Child Safety.* New York: Thomas Y. Crowell Company.

Gover, N. 1995. *Keeping Wastes Out of Drains, Toilets, Small Flows Prevents Pollution.* Morgantown, WV: National Small Flows Clearinghouse, West Virginia University, Spring 9:5.

Gundersen, B. [1995]. *Is Your Home A Safe Place for Kids?* Available from http://www.ci.la.ca.us/department/LAFD/homekid.htm.

Hammer, M. 1988. *Hazardous Household Substances: A Primer for Extension Professionals.* Gainesville, FL: Florida Cooperative Extension Service, Florida State University.

Hammett, W.S. 1992. *Furnishing A User-Friendly Home.* HE-391. Raleigh, NC: NC Cooperative Extension Service, North Carolina State University.

Hammett, W.S. 1991. *Reducing Hazardous Products in the Home.* HE-368-2. Raleigh, NC: North Carolina Cooperative Extension Service, North Carolina State University.

Hammett, W.S. . 1990. *Hazardous Household Products.* HE-368-1/WQWM-61. Raleigh, NC: North Carolina Cooperative Extension Service, North Carolina State University.

Hervel, K.L., and G.M. Powell. 1993. *Reducing the Risk of Groundwater Contamination by Improving Farm and Home Waste Management.* Manhattan, KS: Kansas Cooperative Extension Service, Kansas State University.

Hiatt, L.G. 1980. Architecture for the aged: design for living. *Inland Arch.* 24: 6-17.

Hooyman, N.R., and H.A. Kiyak; 1988. *Social Gerontology: A Multidisciplinary Perspective.* Needham Heights, MA: Allyn and Bacon, Inc.

Household Hazardous Waste Project (HHWP). 1989. *Guide to Hazardous Products Around the Home.* Springfield, MO: Household Hazardous Waste Project, Missouri State University.

How to fall-proof your home for the elderly. 1988-1989. *Family Safety and Health* (Winter): 15.

Kirby, S.D. 1991. Aging in place: a housing educational program. Unpublished manuscript. Stillwater, OK: Oklahoma Cooperative Extension Service, Oklahoma State University at Stillwater.

Lackey, C.J. and K.M. Kolasa. 1995. *Keeping Food Safe to Eat.* HE-451. Raleigh, NC: The North Carolina Cooperative Extension Service, North Carolina State University.

Leonas, K.K. 1993. Applicator exposure and pesticide deposition patterns: a review. In: *Consumer Environmental Issues: Safety, Health, Chemicals and Textiles in the Near Environment.* Extension Workshop Papers. Ames, IA: Iowa State University, University Extension.

Longino, F., Jr., and W.H. Haas III. Migration and the rural elderly. In: *Aging in Rural America.* C.N. Bull, ed. Newbury Park, CA: Sage Publications. pp. 17-29.

Mock, J. and H.T. Jennings. 1991. *Laundering Pesticide-Soiled Clothing.* HE-355. Raleigh, NC: The North Carolina Cooperative Extension Service, North Carolina State University.

Mock, J. and H.T. Jennings. 1994. *Personal Protective Equipment: Gloves.* HE-446. Raleigh, NC: The North Carolina Cooperative Extension Service, North Carolina State University.

Murphy, D.J. 1990. *Farm Pond Safety.* Fact Sheet Safety 27. University Park, PA: Pennsylvania Cooperative Extension Service.

National Crime Prevention Council. n.d. *Got a Minute to Talk About Rural Crime?*

National Safety Council (NSC) 1995. *Accident Facts*. Itasca, Illinois: National Safety Council.

Nelson, C.N. and J.R. Fleeker, eds. 1988. *Limiting Pesticide Exposure Through Textile Cleaning Procedures*. Fargo, ND: North Dakota State University at Fargo.

Nelson, C., J. Laughlin, C. Kim, K. Regakis, M. Rahell, and L Scholten. 1992. Laundering as decontamination of apparel fabrics, residues of pesticides from six chemical classes. *Arch. Envir. Contam. Toxic.* 23: 85-90. In: *Washing Clothing Worn While Applying Pesticides*, HE-FS-2312-A. 1993. W. Olsen and C. Nelson. Minneapolis, MN: Minnesota Extension Service, University of Minnesota.

Niemeyer, S., A. Ziebarth, L. Rottmann, M. Braunn, G. Hopp, G. Hall, D. Stevens, and J. Schwab, 1992. *Household Waste Management: For Your Health and Environment's Sake*. Lincoln, Nebraska: Nebraska Cooperative Extension Service, University Nebraska-Lincoln.

Olson, W., and C. Nelson. 1993. *Washing Clothing Worn While Applying Pesticides*. HE-FS-2312-A. Minneapolis, MN: Minnesota Extension Service, University of Minnesota.

Patrick, Esther. 1992. Product labeling for human safety. In: *Cleaning Products . . . In Our Homes, In Our Environment, Detergents. . .In-Depth.'92*, Columbus, Ohio: Seventh Symposium and Second Videoconference, April 9, p. 35.

Poisonous Plants. 1995. Maryland Poison Center. Available from http://www.pharmacy.ab.umd.edu/webhome/MPC/Plant.html/

Powell, G., and H. Bradshaw. 1990. Household Product Disposal Guide. Manhattan, KS: Kansas Cooperative Extension Service, Kansas State University.

Sadowski, L.S., R.B. Cairns, and J.A. Earp. 1993. Firearm Ownership Among Nonurban Adolescents, *Am J Dis Child* 143 (1989): 1410-3, In: An injury control strategy for rural North Carolina. T. B. Cole, ed. *N. Car. Med. J.* 54 (10): 508-10.

Salmen, J.P.S. 1991. *The Do-Able Renewable Home*. Washington, D.C.: American Association of Retired Persons,.

Selling safety. 1993. *Furniture World* 1, pp 20-22.

Wats, Q. 1995. *Small Flows*. Morgantown, WV: National Small Flows Clearinghouse, West Virginia University, Spring, 9:24.

Webster, D.W. et al. 1993. Parents' beliefs about preventing gun injuries to children, *Pediatrics* 89 (1992). In: T. B. Cole, ed. An injury control strategy for rural North Carolina,. *N. Car. Med. J.* 1 54, no.10 (1993): 508-10.

Wiley,. Scott. 1988. Plants can be potent. *For Kid's Sake Newsl.* 6(3): 2.

Wyant, A., K.R. Bloom, E. Bowman, and A. Mitchell. 1982. *Guide to Consumer Product Information*. New York: Bristol-Myers Company.

25

INDOOR ENVIRONMENTAL HAZARDS IN ANIMAL HOUSING

Robert W. Bottcher, Ph.D., and Roberto D. Munilla, M.E.
Department of Biological and Agricultural Engineering
North Carolina State University

> Modern production techniques for poultry and livestock often require the use of large indoor environmentally controlled housing facilities. While increasing production efficiency, many undesirable conditions may develop within these facilities that threaten the health and safety of the animals and the workers around them. Ventilation and environmental control are key elements of proper building design to minimize the effects of harmful gases, heat and humidity. Other hazards discussed include air pollutants such as dust, ammonia, hydrogen sulfide, carbon monoxide, and carbon dioxide. In addition, mechanical hazards, electrical hazards, noise and fire hazards must be analyzed. Management and control technologies are discussed for each type of hazard.

INTRODUCTION

During the past few decades, the stocking densities of poultry and livestock being raised in animal production buildings have increased tremendously. These increases have sought to improve the industry's economic efficiency by serving more animals with each building structure and set of housing subsystems. Environmental control and waste management systems have been developed to provide productive indoor conditions for the large animal populations housed in these facilities. In addition, improvements in veterinary medicine and pharmaceuticals have increasingly reduced the risk of losses to diseases. Unfortunately, as animal production operations have become more intense, they have also become more susceptible to sudden variations or extremes in the indoor environment. The operation's profitability, the animals' well-being, and the health and well-being of humans within and around the facilities, are all greatly affected by environmental hazards such as poor indoor air quality and excessive noise in production buildings. The purpose of this chapter is to explain the more prevalent hazards in animal housing, as well as the nature and operation of the systems that modify and control indoor environments in animal production buildings.

Although considerable attention has been focused on controlling indoor environments to improve animal productivity, intensive animal production operations can sometimes create undesirable conditions, such as elevated levels of noxious gases, dust, and microorganisms within the buildings and emissions of odors from them. Indoor air quality is often poor compared to human standards and microorganisms are typically plentiful on indoor surfaces (Wathes 1994). Prolonged exposure to high levels of ammonia, dust, or airborne microbes is known to cause a range of respiratory problems in animal production workers and is also suspected of adversely affecting animal performance (although these latter effects have sometimes been difficult to quantify). While the industry acknowledges the need to create wholesome production environments, efforts in this direction have to be weighed against the need to keep operating costs low. Economic considerations can often favor minimizing ventilation or controlling other parameters that impact air quality in animal production facilities. As a result, environmental conditions can sometimes become marginal.

Exposure to poor indoor air quality constitutes a health risk for facility workers and visitors. For the most part, current environmental control systems can provide acceptable indoor air quality if they are properly managed. In those instances where problems arise, proper maintenance of ventilation equipment and provision of good air mixing and fresh air exchange can often dramatically improve indoor air quality, particularly in cool weather. While research and development efforts to improve the control and quality of indoor air and reduce the associated health risks are continuing, for now the most effective means of managing

Figure 25.1. Broiler chicken house with sidewall curtains opened and fans along centerline.

Figure 25.2. Inside a tunnel-ventilated swine finishing house, facing the exhaust fans.

such risks often lie in the use of personal protective equipment, particularly respirators. In extreme cases, it may be necessary to remove affected individuals from the environments altogether, e.g., by changing their occupations.

VENTILATION AND ENVIRONMENTAL CONTROL

Provision of productive indoor environments is of course essential to profitable animal production. Ventilation and waste management systems have been developed to economically remove the large volumes of animal heat, moisture, and manure produced in the buildings. Generally speaking, agricultural buildings are considerably less expensive than residential, commercial, or industrial buildings, and economic constraints continually force animal producers to contain costs.

For example, broiler chicken buildings fully equipped with watering, feeding, ventilation, and lighting systems may cost from $5 to $7 per square foot of floor area, which is roughly an order of magnitude less expensive than residential buildings. Figures 25.1 and 25.2 show indoor views typical of poultry and swine buildings. Economic returns to animal producers must pay for the building systems and utilities, whether the animals are grown for integrated production companies or by independent producers. Consumers of animal products collectively exert considerable economic pressure to minimize costs, as can be surmised by observing grocery store customers' behavior when selecting meat and poultry items. The net result is that the current system of intensive animal production offers little opportunity to introduce expensive additions to building systems, unless governmental regulations or lawsuits force the issue.

Environmental control in animal buildings is geared toward ensuring productivity. Figure 25.3 depicts processes typically affecting the thermal environment and air quality in a poultry building in which birds are reared on a layer of "litter" composed of, say, wood shavings. The litter base provides a buffer to absorb poultry manure and control ammonia production. Producers remove the litter from time to time, and it is generally spread directly as a soil amendment and fertilizer or composted for later use. This litter, however, is also a source of dust and when the litter moisture becomes high, so does the level of ammonia. The birds themselves, the feed bins, and feed troughs—even the incoming air—are all sources of dust. In other poultry or livestock buildings, different approaches to manure management are employed; e.g., the manure drops from the animal pens or cages to shallow pits underneath them, where it is flushed frequently from the building to a "lagoon" where it is biologically degraded. The chosen method of manure management obviously affects the production and concentration of pollutant gases and dust in the building, but the basic processes shown in Figure 25.3 remain pertinent to intensive animal production environments.

Ventilation Rate

Ventilation systems are designed to move air through the buildings to remove certain undesirable constituents of the indoor air, especially heat (in warm weather), humidity (in warm and cool weather), and carbon dioxide (in all weather) (Albright 1990). The design procedure typically assumes

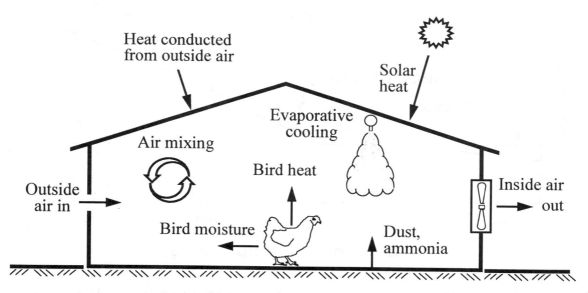

Figure 25.3. Processes affecting the thermal environment and air quality in poultry buildings.

(perhaps erroneously) that successful control of these constituents will also achieve adequate control of other aerial pollutants. A graph of design ventilation rates vs. outdoor air temperature, showing the relationship between the criteria for temperature, moisture, and carbon dioxide control, is shown in Figure 25.4. For any given outdoor temperature, a building's ventilation rate should equal or exceed the greater of the design ventilation rates required to adequately control the indoor temperature, moisture, or carbon dioxide levels. Thus, in very cold winter climates, such as those in Canada, carbon dioxide control may dictate the ventilation rate. In such cases, the outdoor air will be cold and dry and supplemental heating will be required, which (theoretically) eliminates the problem of moisture control. At moderately cold temperatures, moisture control dictates the ventilation rate, and at warmer temperatures, temperature control takes precedence (Figure 25.4). The values for the ventilation rate curves in Figure 25.4 will vary depending on the animal species and parameters such as solar heating and building insulation level. In some climates, such as those in the southern U.S., ventilation for carbon dioxide control is not considered since ambient air temperatures are never (or very rarely) low enough.

Since the ventilation system must work over a wide range of conditions, a method of adjusting ventilation rate to approximate the design values exemplified by Figure 25.4 is needed. *Staging* is an example of such control; at low temperatures a minimum ventilation rate is maintained by running one or two fans, and as temperature increases, additional ventilation fans are turned on in an incremental fashion.

Natural Ventilation

Ventilation of animal buildings is provided either by natural means (primarily by wind blowing through openings in the side walls) or mechanically (by fans blowing into or out of the buildings). With mechanical ventilation, the ventilation rate is reasonably predictable based on the fan specifications. However, natural wind ventilation can present problems if the wind dies down or the "wind environment" around the building is unfavorable (e.g., where surrounding trees or hills impede the wind). In general, obstacles to wind create a sheltering effect on the leeward side, which is the principle behind windbreaks. Put another way, wind travels some distance downstream of obstacles before "returning to the ground", as shown in Figure 25.5. Wind tunnel and full scale studies have determined this distance to be in the range of 4 to 10 obstacle heights, depending on the obstacle geometry and other factors (Evans 1957; Krishnan 1965). Hence, it can generally be assumed that placement of buildings less than four

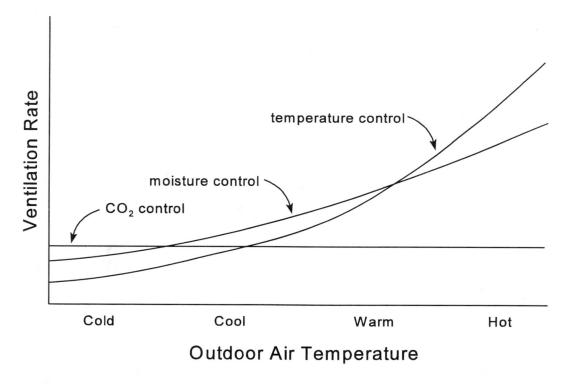

Figure 25.4. Ventilation graph showing ventilation rate design criteria for carbon dioxide, moisture, and temperature control. Modified from Albright (1990), Figure 6-2.

"obstacle heights" away from obstructions such as forested areas, cornfields, or other buildings will provide less wind ventilation than placement farther away (Figure 25.5).

Warm Weather Ventilation

In warm weather, the environmental control system is designed to control temperature; theoretically, this should also provide adequate control of humidity (Albright 1990). Heat enters the building interior through the walls and roof, and with the incoming ventilation airflow, additional heat is produced by the animals (See Figure 25.3). Heat leaves the building primarily in the outgoing air flow generated by the ventilation system. Thus the ventilation is really the only means to remove heat from the building. Evaporative cooling, provided by mists of water or within evaporative cooling pads at the air inlets, reduces air temperature but increases its humidity. Since most domesticated animals cannot lose heat through sweating, reducing indoor air temperatures and forcing airflow over the animals are imperative in hot weather. Evaporative cooling provides such temperature reduction, and is considerably less expensive than conventional air conditioning. It is a myth that evaporative cooling is ineffective in humid climates—it works well in animal buildings even in the *most humid* regions of the United States, since the periods requiring temperature reduction (in the middle of the day) are also when relative humidity is lowest and the potential for evaporative cooling is substantial. However, evaporative cooling does not remove heat; it merely "trades" temperature for humidity.

Thus, in warm weather, animal producers use *ventilation* to move large volumes of air through the building and thus remove heat and moisture from it. For example, a fan ventilation system for a typical broiler chicken building may be designed to move up to 150,000 cubic feet of airflow per minute (cfm) through the building, while a system for a swine finishing building may be designed for 90,000 cfm. Enormous ventilation rates, although essential for temperature control, can make it difficult to economically clean pollutants from the indoor air. Since air cleaning costs are directly related to the airflow rate to be cleaned, some cost estimates for livestock building systems have ranged from $200,000 to $600,000. This would greatly exceed the initial costs of typical buildings. Hence, many researchers are attempting to develop less expensive air cleaning systems for livestock buildings. It may be that requiring cleaner livestock building air will force adoption of air conditioning systems. While this would significantly reduce building ventilation rates, it would require cooling and dust and gas removal and thus be expensive in buildings with substantial animal mass.

Cool Weather Ventilation

Ventilation can remove moisture, dust, ammonia, and other gases produced in the building, provided they are mixed into air which leaves the building. Hence, effective air mixing is another necessary function of the environmental control system (Figure 25.3). Poor air quality is often a problem in animal buildings in cool weather because of inadequate ventilation as well as poor air mixing. Producers are notorious for minimizing ventilation in cool weather in order to conserve heat (especially when they are paying for heating fuel). Thus, indoor pollutant levels increase in concentration and the ventilation rate may not be sufficient to bring them down to acceptable levels. Even with

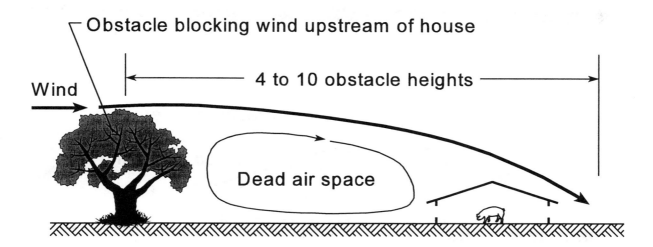

Figure 25.5. Wind travels a distance beyond an obstacle before returning to the ground. Wind ventilation of buildings in the lee of such obstacles may be poor.

ostensibly adequate ventilation, poor air mixing can render the ventilation ineffective. If air entering the building "short circuits" and leaves the building immediately, localized "dead air" spaces are formed, where high levels of pollutants can persist. Poor air mixing also contributes to temperature stratification in heated buildings, which can increase heat loss and waste energy. In order to provide good air mixing, the ventilation system's inlet air jets must have sufficient momentum or supplemental indoor mixing fans are needed.

To impart momentum to incoming air jets in mechanically (fan) ventilated buildings and enhance air mixing, the inlet openings are typically controlled to create a sufficient static pressure difference across them. In the U.S., static pressure is typically measured in inches of water column, and it represents the "pull" the fans (which are blowing air out of the house) are exerting on the air coming into the house. Adequate static pressure cannot be developed unless the house is tight, with all cracks and holes sealed up. At a high static pressure, incoming air has momentum and tends to move in one direction. If it can be made to shoot across the ceiling, it will warm up and slow down before circulating back down to the animals (Figure 25.6). The warmed air can then pick up substantially more moisture than the cold outside air. If the static pressure is low, however, incoming cold air will immediately fall to the floor and the animals (Figure 25.6).

This will not only chill the animals, but can create wet spots when the floor is chilled below the indoor dew point temperature; this, in turn, can increase ammonia production. A static pressure in the range of 0.03 to 0.08 inches of water is typically recommended. If the static pressure is any lower, the air will not mix as well, and if it is higher the fans may "work too hard" (i.e., their energy efficiency is worse, electricity is wasted, and they may not move enough airflow). Inlet controllers—devices that adjust inlet openings automatically to control the static pressure—can maintain this pressure range, but manually setting the inlet openings can also work if the grower will adjust them as needed. Baffled slot inlets near the eaves work very well. Cracking side wall curtains can also work, but it is important to avoid having the incoming, cold air shoot downward toward the animals.

Although maintaining adequate ventilation rates and inlet airflow control can theoretically provide good air mixing, in practice the use of indoor mixing fans is often necessary. Cracks and holes in curtains and side walls may prevent development of adequate static pressures, and some producers still rely entirely on natural, wind ventilation. Paddle fans, propeller fans, and plastic air distribution ducts (with fans blowing air through the duct and out holes along the duct) are widely used in livestock and poultry buildings

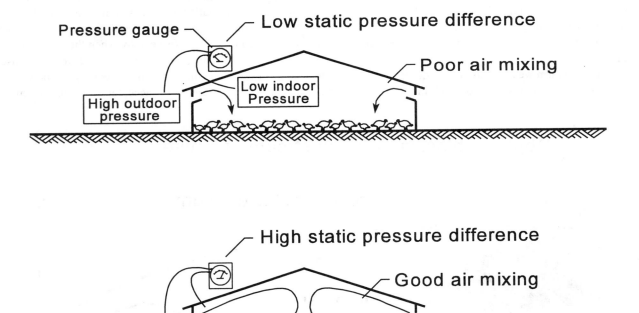

Figure 25.6. Without a high pressure difference to impart sufficient momentum to the inlet air jet, cold outside air falls to the floor upon entry into warm animal buildings.

for air distribution and mixing. In cool weather it is important to avoid forcing airflow from the fans directly onto the animals, which chills them, so the fans or distribution duct holes are typically oriented to blow horizontally or upward. Unfortunately, few if any reversible paddle fans suitable for use in the dusty, sometimes corrosive environment of poultry or livestock housing are commercially available in the U.S., so paddle fans are typically run at low speed, blowing downward in cool weather. However, Boon and Battams (1988) found that ceiling fans blowing upward were more effective in reducing temperature stratification than fans blowing downward, and did not cause problematic air velocities near the floor-reared chickens.

Provision of indoor air movement to entrain air pollutants in the building exhaust airflow, reduce temperature stratification, cool animals in hot weather, etc., may have detrimental effects on indoor air quality. Generation of dust, aerosols, and ammonia and resuspension of settled dust from indoor surfaces can be increased by forced air movement (Maghirang et al. 1993). However, the need to entrain pollutants into the exhaust airflow, and certainly to cool the animals in hot weather, have long been considered to be critical functions of the environmental control systems. Unless other economical means of cooling animals and removing pollutants are developed, indoor air quality control will apparently have to contend with such forced indoor air movement.

System Control

Ambient conditions vary diurnally, seasonally, with changing animal size, and due to weather; the ventilation, heating, and cooling systems must be controlled to accommodate such changes and still provide productive indoor environments. Thermostatic control, in which the fans, heaters, or cooling systems are controlled based solely on the indoor temperature level, is typical. Programmable controllers and relatively complex computer programs are also being increasingly employed. However, it is axiomatic that the more sophisticated control systems are also more subject to unexpected failures.

Since controller failures and power outages are relatively common occurrences on livestock farms, and lack of ventilation can quickly lead to lethal indoor conditions, fail-safe systems are highly recommended. These can include warning systems which broadcast an alarm or alert the producer by telephone, "curtain-drop" devices which open side-wall curtains in the event of a power failure, and standby electric generators and switching systems. Figure 25.7 shows a broiler house full of dead birds probably killed by a lethal combination of high temperature and humidity which occurred when the ventilation system failed during a hot summer night.

Such disastrous mortalities can readily occur due to a combination of high temperature and humidity, and may in some cases also occur due to high levels of ammonia and/or carbon dioxide. Since the buildings are typically not airtight, lack of oxygen per se is not as likely a cause of mortality, except when animals such as chickens pile up when attempting to migrate into a small area. The risk of such disasters has induced the animal industry to rely on relatively simple environmental control systems, and can be expected to weigh against the adoption of air cleaning and odor control equipment which impose airflow restrictions or utilize substantial electric power. Those interested in demonstrating or requiring air cleaning technologies for animal producers should understand the need to avoid, or be able to deal with the consequences of, disasters exemplified by Figure 25.7.

Instruments

Effective management of environments in animal production facilities requires the ability to measure and evaluate the various parameters which affect the quality of this environment. A set of measurement devices and a basic knowledge of the functional aspects of these devices, as well as the facility's environmental control systems, is needed.

As stated previously, good indoor air quality depends primarily on effective ventilation. In animal production housing, the ventilation system's effectiveness depends on the exhaust fans, cracks, holes, and other undesirable openings, the air inlets, and their control. The inlet air jets created by the negative static pressure will only develop if the fans are working near their design capacity with no major unplanned air leaks.

Simple, effective devices for observing air flow are smoke generators such as smoke sticks or smoke bombs. Theatrical smoke generators, beehive smokers, and other equipment have also been effective. Smoke generators can be used to study the effective throw of the air inlets, indoor patterns of air mixing, and to detect air leaks (Figure 25.8). The information obtained using smoke is often visually dramatic and can be grasped intuitively. Smoke devices are available in different colors and sizes; the volume and the duration of the smoke emission should be chosen based on the application. For example, large volumes of smoke should not be produced indoors with animals or people present. Table 25.1 lists several vendors of smoke emitters.

A more quantitative measurement of indoor air flow can be obtained using anemometers. These devices use either a rotating vane or a heated wire element and electronically

Figure 25.7. House of dead chickens killed by heat stress due to ventilation failure. Photograph courtesy of Dr. I. L. Berry, University of Arkansas Department of Agricultural Engineering.

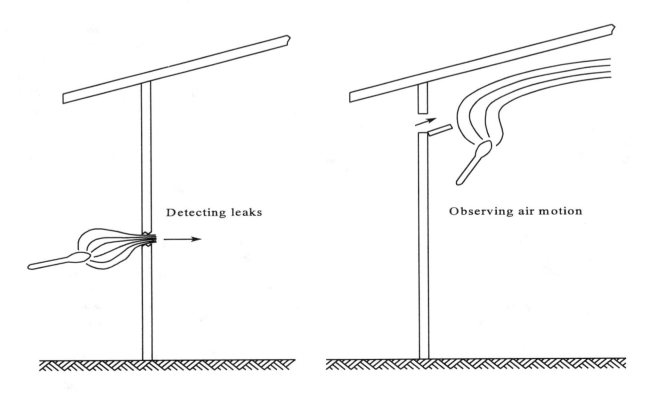

Figure 25.8. Smoke devices can reveal leaks, inlet air jets, and other flow patterns.

Table 25.1. Sources for ventilation monitoring equipment.

Vendor and Telephone No.	Smoke Emitters	Vane Anemometers	Hot-Wire Anemometers	Tachometers	Static Pressure Meters
Animal Environment Specialists Columbus, OH 1-800-969-0114	X	X	X		X
Cole Parmer Instruments, Inc. Chicago, IL 1-800-323-4340	X	X	X	X	X
Davis Instruments Baltimore, MD 1-800-368-2516	X	X	X	X	X
Dwyer Instruments Marietta, GA (404) 427-9406		X	X		X
E. Vernon Hill, Inc. Benicia, CA (707) 747-5577	X				
W.W. Grainger, Inc. many branches nationwide 1-800-225-5994			X	X	X
Mine Safety Appliances, Inc. Decatur, GA 1-800-MSA-2222	X				
Mitchell Instrument Annandale, NJ (619) 744-2690	X	X	X	X	X
Superior Signal Co. Spotswood, NJ (908) 251-0800	X				

Note: this is not an exhaustive list. The use of brand names does not imply endorsement of the products named or criticism of similar ones not named.

measure and display the air velocity across the sensing head (see Figure 25.9). Careful and patient use of anemometers will produce an accurate record of the air flow patterns in a house. They are useful in detecting variations in the desired flow patterns, dead spots, drafts, leaks, and air velocities caused by mixing fans or entering through inlets. Table 25.1 lists some sources of anemometers.

If a fan-ventilated house has ventilation problems, measurement of the static pressure will be in order. This static pressure is the pressure difference between the inside and outside air causing the airflow through the air inlets (Figure 25.6). Measurement requires the use of devices such as an inclined manometer or pressure gauge (see Figure 25.10). Effective ventilation in production housing requires negative static pressures of at least 0.04 inches of water.

These devices are simple to use: one port is connected via flexible tubing, through any convenient opening, to the outside air (if using the meter indoors) or the inside air (if using it outdoors). The ports on static pressure meters must be connected properly; for example, with a negative pressure building (fans blowing air out of the building), the inside static pressure is low and outside static pressure is high. Table 25.1 also lists some sources of static pressure meters.

If the measured static pressure falls below 0.04 inches of water and the house is airtight, there may be problems with the fans or the inlet vents. A tachometer is used to measure fan rotational speed. Both mechanical and stroboscopic tachometers are available; some sources are listed in Table 25.1. Mechanical tachometers require physical contact with the fan propeller shaft to measure the motor RPM, and

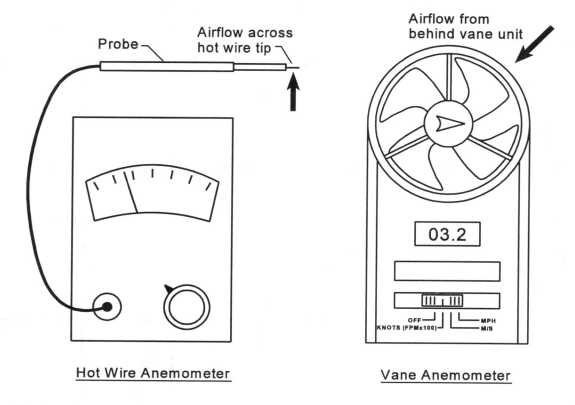

Figure 25.9. Anemometers are used to measure air velocity.

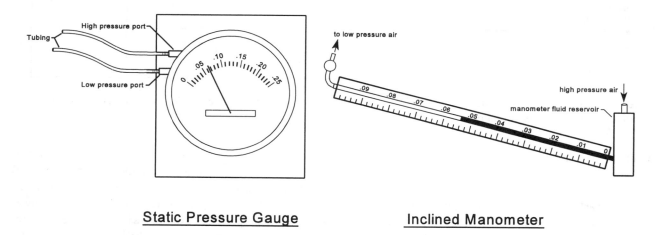

Figure 25.10. Devices for measuring pressure difference.

therefore can be dangerous to use. Stroboscopic tachometers use high intensity lights which are made to flash intermittently at a controllable rate; these devices rely on the optical illusion of "frozen action" which results when the fan RPM and the rate of flashing are synchronous. At this point the fan blade seems to stop and the fan speed in RPM is presumably equal to the tachometer flash rate. It is important to note that simply adjusting the flash rate until the fan blades appear to be stopped does not ensure an accurate reading because the same blade may not be in the same position at each flash (e.g., flashing the strobe at 3/4 or 1 to 1/4 times the correct rate with a 4-bladed fan). By stopping a unique rotating part such as an oil fitting or key shaft, or by placing a shiny sticker on a fan blade, the correct strobe flash rate can be obtained.

AIR POLLUTANTS IN ANIMAL HOUSING

Aerial pollutants commonly occurring within animal facilities include dust, bioaerosols, ammonia, and other noxious gases, endotoxins, and malodorous compounds. In many cases, these are not in sufficient concentration to pose acute health risks or may not even be measurable, but acute and chronic effects of exposure to animal house environments have been documented (De Boer and Morrison 1988; Donham 1990; Lenhart et al. 1990; Schlenker et al. 1987). Severe acute effects such as respiratory arrest and death have occurred due to exposure to high levels of hydrogen sulfide in manure pits (MMWR 1993) or when liquid manure is agitated under the floor of buildings (Donham 1990). The hazards associated with hydrogen sulfide release when livestock manure is agitated in confined spaces have prompted state and federal warnings.

The concentration of pollutants in animal buildings will vary both in space (due to incomplete air mixing, animal and manure locations, and other geometric factors), time (due to diurnal and seasonal variations, animal growth, etc.) and composition (due to animal type, size, and production system). This can make accurate measurement of contaminant levels and interpretation of such measurements difficult. Hence, determination of exposure levels for humans, and possibly for the animals, is best conducted using personal air sampling equipment. Other instruments for measuring air quality parameters are useful in identifying problems and evaluating control measures.

Dust and Other Aerosols

Aerosols in animal production housing generally consist of solid particles or "dust", liquid droplets such as mist from evaporative cooling nozzles, and microbial contaminants, with particle diameters ranging from 0.6 to 50 microns. Particles larger than 8 to 10 microns that are inhaled are removed by the natural cleaning mechanisms of the upper respiratory tract while smaller, respirable particles can be inhaled deeper into the lungs (EPA 1991; Leonard and Donham 1990). The 8-hour permissible exposure levels (PELs) set by OSHA for dust are 15 mg/m^3 for total dust (i.e., particles up to 10 microns in diameter), and 5 mg/m^3 for respirable dust (up to 2.5 microns) (EPA 1991). Published data suggest these levels are not usually exceeded in livestock buildings (Baucom et al. 1995; De Boer and Morrison 1988; Lenhart et al. 1990). However, Donham (1987) and Robertson (1993) have argued for reduction of exposure limits since exposures at levels below current limits have evidently caused health problems.

Dust found in animal housing is primarily composed of feed components and dried fecal material but can also contain dander (hair and skin cells), molds, pollen, grains, insect parts and mineral ash. Some constituents such as animal dander, feed, and feathers can cause allergic responses (Levy 1982; Virtanen, et al. 1990). Due to its effects on animals, humans, sensors, and equipment, dust is arguably the most pervasive and important air quality problem in poultry and livestock housing. Dust particles are also involved in the conveyance of odors (Hammond et al. 1981), microorganisms, and endotoxin (De Boer and Morrison 1988).

Endotoxin, viable bacteria, and other microbial contaminants are important constituents of animal building air pollution and are ventilated from the buildings. Homes et al. (1996) found that viable bacteria were carried up to 300 m downwind of a swine production facility. Endotoxins, which are toxic substances secreted by bacteria, occur in respirable particulate form and have been measured at levels above the recommended exposure limit of 10 ng/m^3 in poultry and swine facilities (Baucom et al. 1995; Lenhart et al. 1990).

Control of dust and the associated health risks has proven especially problematic due to large dust generation rates, the high ventilation rates, and the complex composition of the dust (Carpenter 1986; Dawson 1990; Maghirang et al. 1993). Several methods of removing dust or preventing dust generation in animal buildings have been proposed or are under evaluation, including wet, electrostatic, cyclonic, and dry dust filters, and oil sprays that bind up the dust particles and keep them out of suspension. While these methods have all been effective at filtering dust, future implementation will depend on their cost effectiveness. Sprinkling or spraying of vegetable oils has been notably successful in reducing dust levels in swine facilities (Takai et al. 1995). It is ironic that many of the more effective dust control methods currently being evaluated also pose nontrivial safety risks: 1) spraying of oils or fogging may produce respirable aerosols, 2) electrostatic dust collection or ionizing mist systems utilize large voltages and may pose electric shock risks, and 3) ozone gas, while effective in sanitizing and deodorizing facilities, is also toxic to humans and animals; the OSHA 8-hour PEL for ozone is 200 g/m^3 or 0.1 ppm by volume (EPA 1991). Regulatory and legal pressure to reduce emissions of odors, bioaerosols, or other pollutants may prompt adoption of dust control measures in the near future.

Ammonia

Ammonia is produced by bacterial action on urine and feces during decomposition. Almost all production systems store animal waste inside the buildings, at least temporarily.

It can drop to the floor where it mixes in with absorbent litter, or through gratings or slats into pits under the floor. Most of the ammonia present in animal confinement housing comes off of the floors and from the manure pits (Leonard and Donham 1990). Ammonia has a sharp pungent odor and is colorless, lighter than air, highly water soluble, and detectable at concentrations as low as 5 ppm. Ammonia levels in poultry buildings typically range from 0 to 50 ppm (Sneath et al. 1996; Wilhelm and Snyder 1996) although levels can exceed 100 ppm in poorly ventilated structures (De Boer and Morrison 1988). The OSHA PEL for ammonia is 27 ppm for a 15 minute exposure period (EPA 1991), although again, researchers have argued for lower limits for the benefit of workers (Donham 1987; Robertson 1993). Thus, ammonia levels in animal buildings can be sufficiently high to adversely and chronically affect human health.

Ammonia levels in animal buildings can be controlled by frequent removal of waste, management of indoor moisture, and adequate ventilation. Commercial litter and feed additives are also available to control ammonia levels; for example, alum has proven effective as a litter additive in some trials (Moore et al. 1995). However, ventilation is the principle method of controlling indoor ammonia. Increasing ventilation effectively dilutes ammonia concentration and also tends to dry floors and litter, which reduces the rate of ammonia release. One method of implementing ventilation control for this purpose in poultry buildings is based on time-averaging indoor humidity levels, which affect litter and floor moisture (Timmons and Gates 1987); i.e., prolonged humid conditions lead to increased litter moisture and ammonia. However, increasing the ventilation rate also increases the building's heat loss, which in turn requires increased fuel use or larger heater capacity during the heating season. Such costs are, of course, understood by the producers and are the main reason they limit ventilation in cool weather.

Hydrogen Sulfide

Hydrogen sulfide is an acutely toxic gas produced by the decomposition of animal manure and is often released into the air when liquid manure is agitated. It is colorless, denser than air, and at concentrations below 150 ppm has a smell characteristic of rotten eggs. However, its odor is not an indication of its concentration; above 6 ppm the odor increases slightly as concentration increases significantly (Barker et al. 1991). At levels above 150 ppm it can completely deaden the sense of smell. The OSHA and ACGIH exposure limits for hydrogen sulfide are 10 ppm for an 8 hour, 5 day exposure period (EPA 1991) and at levels above 50 ppm human evacuation is recommended (Barker et al. 1991). Most of the time, hydrogen sulfide in animal buildings occurs at levels of 5 ppm or less; Baucom et al. (1995) measured levels below 2 ppm in five swine buildings during mild weather. However, levels up to 1500 ppm may be reached when a swine manure pit is agitated, while levels above 500 ppm cause unconsciousness and death (Barker et al. 1991). Thus workers should wear a self contained respirator if potential exposure to hydrogen sulfide is expected (Derthick et al. 1994) and should enter such pits with a lifeline attached to them. The hazard created during manure agitation may also be controlled by ensuring adequate ventilation during manure pumping and by removing the manure only when people, and preferably animals as well, are absent from the buildings. Ventilating the manure pits separately from the animal space should also control gas levels (Driggers 1987), although in many cases pit ventilation systems have been omitted from building designs due to system costs.

Other Gases

Carbon dioxide, methane, and carbon monoxide are three other gases occurring in animal production indoor environments. They normally do not cause problems but a word of caution about each is in order.

Methane is a natural product of manure decomposition and it is nontoxic. At very high concentrations it can cause dizziness and even asphyxiation. However, the main safety concern is the flammability of methane; at concentrations over 50,000 ppm (5%) it can be explosive (Barker et al. 1991). Of course, this also makes methane valuable as an energy source and has prompted research to recover methane from animal manure, e.g., by anaerobic digestion. Because it is less dense than air it tends to rise and can accumulate in the upper levels of airtight buildings and enclosed manure pits (Barker et al. 1991). The NIOSH recommended exposure limit for methane is 1,000 ppm for an 8-hour work period (Barker et al. 1991). When some ventilation is provided, methane generally dissipates from animal buildings (De Boer and Morrison 1988).

Carbon dioxide is produced by manure decomposition and animal respiration. It is a nontoxic gas, denser than air and the earth's atmosphere contains about 300 ppm. At considerably higher concentrations it can cause asphyxiation by reducing available oxygen (Barker et al. 1991). Concentrations inside well-ventilated buildings can range from 1,000 ppm during summer to 10,000 ppm during winter. The OSHA 8-hour and 15-minute PEL's for carbon dioxide are 10,000 and 30,000 ppm, respectively, while the ACGIH recommended 8-hour TLV for carbon dioxide is 5,000 ppm (EPA 1991). As noted above carbon dioxide control is important in cold climates (Figure 25.4), but it usually does not pose a major health risk in animal housing.

Carbon monoxide is the product of the incomplete combustion of hydrocarbons. It is colorless, odorless, and has nearly the same density as air. In humans and animals it combines with the blood's hemoglobin and causes systemic intoxication (De Boer et al. 1988). In animal production operations, carbon monoxide hazards are primarily caused by operation of heaters or internal combustion engines without venting the combustion products outdoors (De Boer and Morrison 1988) or when combustion heaters malfunction (Barker et al. 1991). Winter is the most dangerous period because buildings are usually closed and ventilation rates are at their lowest (Barker et al. 1991). The OSHA and NIOSH recommended 8-hour TLV for carbon monoxide is 40 mg/m³ or 35 ppm (EPA 1991). Pregnant female workers should be aware that fetuses are much more sensitive to carbon monoxide than adults (Barker et al. 1991). Extreme precautions should be exercised when the potential for carbon monoxide generation exists; its ability to cause drowsiness and unconsciousness at relatively low levels make it a dangerous contaminant. Combustion heaters and engines should always be vented to the outside.

Odors

Unpleasant odors have long been associated with domestic animal production. Attempts to control the odors associated with livestock farming have been constrained by economic factors as well as the complex nature of odor generation, composition, perception, and measurement. Although many people have expressed the opinion that odors per se are not a health hazard, Schiffman et al. (1995) showed that odors can impact the moods of people living near swine operations, including reductions in vigor and increases in tension, depression, anger, fatigue, and confusion. The nuisance odor emissions from livestock farms are a complex mixture of several different compounds resulting from anaerobically decomposing manure, rotting feed and dead animals, and pesticide sprays (Miner and Barth 1988). Odors are emitted from the buildings, lagoons, and other manure storage facilities, and by land spreading of manure as fertilizer. Odorous compounds found in animal houses include amines, amides, alcohols, carbonyls, mercaptans, sulfides, fatty acids, phenols, cresols, indoles, and skatols (Schiffman et al. 1995). Since many of these compounds are conveyed in dust particles, dust filtering as well as bio-scrubbing and biofiltering have been effective at reducing odorous emissions from livestock buildings (Klarenbeek and van Harreveld 1996). In addition, recent experimental odor control approaches have included odor neutralizing and masking agents, oil or water sprays, and ozone; as noted above these can pose their own safety risks. Due to the wide variety of odor control products and manufacturer's claims, laboratory methods of evaluating their efficacy have recently been developed and several promising compounds have been identified (Bundy et al. 1996; Williams et al. 1996).

Air Quality Control and Management

Proper installation and operation of a well-designed ventilation system is a producer's best assurance of adequate indoor air quality. Dirty fans, fan guards, and inlet gratings, coupled with loose and cracked fan belt drives can cut fan efficiency in half. A good ventilation system will provide thorough air mixing, eliminate dead spaces having stagnant air, and move fresh air through the house. Air should enter through the planned ventilation inlets and not through other unplanned openings (cracks, holes, open doors, or windows). Ventilation vents should open only enough to provide high velocity jets which insure proper "throw" and thorough air mixing. During summer months, evaporative cooling using pad or misting systems is often needed to reduce the indoor air temperature. During winter months when ventilation rates are reduced to a minimum to conserve heat, supplemental mixing fans are needed. Indoor air that is very humid and smelly is usually an indicator of insufficient ventilation.

Prevention and early detection of toxic gas levels will reduce risks posed by toxic gases, so gas monitoring and prevention practices should be prioritized according to risk. Since carbon monoxide is an "odorless killer", installation of CO detectors near combustion heaters is advisable. The heaters should be vented to the outside if possible, cleaned and inspected thoroughly at the beginning of each heating season, and visually monitored on a daily basis while they are in use to ensure that they burn efficiently and produce minimal levels of carbon monoxide. Extreme caution must be exercised during manure removal since agitation of manure slurries will release hydrogen sulfide; the slightest hint of "rotten eggs" should be cause for concern. Since hydrogen sulfide can quickly inure the sense of smell as concentrations increase and become deadly, it must be considered an acutely toxic hazard.

Odor auditing procedures involving regular evaluations of the facility odor levels and their composition will assist in reducing odor problems. Producers who fastidiously clean up spilled manure and generally maintain clean standards can be expected to produce less odorous emissions. Where feasible, power washing of production housing is an effective way to cut down on dust generation. Air washers and other filtration and air cleaning systems may soon be more widely applied to animal buildings as regulatory and legal pressures force the issue.

Indoor dust control is likely to remain problematic for some time. During cold weather, the relative humidity of the heated indoor air can drop below 25% and dust generation will increase if supplemental moisture is not provided. Although adding humidity to the indoor air—say through misting—can reduce dust levels, it also results in a heating cost as the indoor air must be maintained at the target temperature. The use of feed additives such as oil, fat, and lecithin can help reduce dust emission from feed meals. Also, pelletized feed is generally less dusty than feed meals (Derthick et al. 1994). Installing feed covers and extending feed spouts will also help keep feed particles from being blown into the air. The effectiveness of indoor oil spraying for dust control will likely stimulate more experimentation with oil spraying in animal buildings.

Personal Protection/Avoidance

Problems due to respiratory exposure to airborne contaminants can obviously be avoided if the air quality is good enough. However, ventilation, air cleaning, and other sanitation measures have in many cases been incapable of completely removing contaminants or have been too expensive for producers to install or utilize. As a result, the most effective means of minimizing respiratory hazards in animal housing buildings is to consistently utilize NIOSH-approved respiratory protection appropriate for the situation. Requirement of respiratory protection legally necessitates a respiratory protection program to include evaluation of worker's ability to work with the respirator, regular training of personnel, routinely monitoring air quality, selection of appropriate NIOSH-approved respirators, respirator fit testing, medical evaluations, and maintenance, inspection, cleaning, and storage of respirators (Lenhart 1995; NIOSH 1994). Respirators used in animal production facilities include disposable dust/mist masks, reusable dust/mist masks, chemical cartridge respirators (which can include particulate matter prefilters), powered air-purifying respirators (which can provide eye protection as well), and self-contained respirators for dangerous gases such as hydrogen sulfide or carbon monoxide (Derthick et al. 1994; Prather 1992).

Besides the use of NIOSH-approved respiratory protection, limiting the duration of exposure of workers to house environments may prevent the development of respiratory problems. Certainly workers who develop a chronic cough or breathing problem should be evaluated by a physician. Medication, if necessary, may improve the symptoms so they can return to work. However, in rare instances the severity of the respiratory problems may require that a worker avoid further exposure to confinement environments (Bottcher et al. 1994; Derthick et al. 1994).

MECHANICAL HAZARDS

Several mechanical hazards typically exist in animal housing, such as fans, motors, winches, and feed augers. Unguarded fans in animal buildings are especially dangerous. If fans hanging inside a building are within reach, they must have guards or screens so people cannot touch any moving parts. Several state OSHA programs differ in their maximum allowable screen mesh sizes for guarding fans; some require 1/2 by 1/2-inch mesh screens which can rapidly clog with dust in many poultry or livestock buildings. Some state OSHA's have accepted larger screen sizes such as a 1 x 2 inch mesh screen at least 7 inches from all moving parts (Bottcher et al. 1994); many agricultural fans are now routinely shipped with such a screen covering the fan box. Fans sufficiently far above the floor (e.g., paddle fans placed 7 to 10 ft high) generally do not require guarding.

Workers operating winches for raising or lowering items such as fans, curtains, and feeder lines must be careful to avoid releasing the winch before the object is fully raised (and secured), or fully lowered. Accidentally striking a winch under tension can cause it to release; the sudden, violent spinning of the winch handle can then cause severe injury. Also, augers for conveying feed into the buildings must be properly guarded. Before any maintenance is performed on any equipment it must be disconnected electrically; it should be unplugged or switched off at the control and at the breaker box.

Steel cables are used extensively in animal housing; when worn or frayed they could produce gashes and puncture wounds, especially on the operator's hands. Regular use of a sturdy pair of work gloves can prevent these wounds. Production housing floors can be slippery and obstructed by equipment and railings; regular use of a good pair of work boots will help prevent falls and foot injuries.

ELECTRICAL HAZARDS

Electrical hazards in enclosed animal production buildings due to faulty electrical wiring or components include the risk of shock and the potential for fire, in addition to potential destruction of good equipment such as motors and pumps. Several environmental factors in animal facilities can be particularly destructive to the metal and plastic parts of conventional wiring materials (Hiatt and McFate 1993):

- Dust generated from the animals, bedding or litter, and feed which may also contain salts and corrosive minerals.
- Moisture originating from the respiration of the animals, evaporation from animal waste, and from washdown operations.
- Corrosive vapors such as ammonia and other gases emanating from animal litter and manure which accumulates within the building.

Wiring materials that are acceptable for "dry" applications may be unacceptable for use in animal production facilities because of these destructive factors (Collins et al. 1995; Hiatt and McFate 1993). Wiring practices that protect electrical cable and system components from abuse by livestock and rodents is required for use in animal buildings. Special attention must also be paid to avoid exposure to and damage from hazards such as tractors and feeding equipment..

The National Electrical Code® (NEC) is the standard for electrical work in the United States (Collins et al. 1995); Article 547 of the NEC, "Agricultural Buildings", provides specific requirements for the wiring of animal production facilities where excessive dust, moisture, and corrosive vapors may be present. Although many rural areas do not have an inspection program to verify that wiring meets the NEC (Hiatt and McFate 1993), the potential impact of a fire loss necessitates good wiring practices (Collins et al. 1995). Proper wiring practices include properly sized wiring, circuits, and power supply panels, use of moisture-resistant wiring such as type UF or non-metallic conduit (not NM or NM-B cable), moisture-proof boxes and fixtures with moisture- and dust-tight connections, lighting fixtures rated for use in wet environments or animal facilities, and both system and equipment grounding (Collins et al. 1995; Hiatt and McFate 1993).

Appropriate design and reliable installation of livestock facility electrical systems are crucial to using electricity efficiently, providing a safe environment for workers and animals, and minimizing the potential for fire loss. A majority of all farm losses due to fire are related to electrical system failure (Collins, et al. 1995).

NOISE

The human ear is made up of delicate structures that are irreversibly damaged by exposure to high levels of sound, and cumulative exposure to loud noise can result in noise-induced hearing loss (NIHL) or *sociocusis* (Kryter 1994). The first evidence of noise-induced hearing loss is difficulty hearing high-pitched sounds; noise levels may be too loud when the ears ring or buzz, speech or other noises are temporarily muffled, music tones become hard to differentiate, or following normal conversation becomes difficult (Gay 1990).

Sound levels are measured in decibels (dB); each 10 dB increase represents a 10-fold increase in sound level. Sound created by a soft whisper is about 30dB while 120 dB will cause pain. The Occupational Safety and Health Association limits noise exposure to an average of 90 dB over an 8-hour work shift (OSHA 1996). Table 25.2 shows the maximum permissible exposure levels for various durations. A hearing conservation program is required of employers when 8-hour time-weighted average sound levels exceed 85 dB (OSHA 1996).

Table 25.2. Permissible noise exposures.

Duration (hours per day)	Noise level (dB(A))
8	90
6	92
4	95
3	97
2	100
1.5	102
1.0	105
0.5	110
0.25	115

Farming has often been associated with hearing loss (Plakke and Dare 1992); although tractors and other equipment cause the most noise exposure, significant noise can arise in livestock housing from the animals as well as machinery such as fans, feeding equipment, cages, and feeder chains and augers. The combined intensity of these noises may reach 100 dB. In swine buildings 115 dB can be reached at feeding time and when hogs are moved or disturbed (Baucom et al. 1995; Gay 1990). Pazzona and Murgia (1993) measured a peak noise level of 106 dB and average level of 93.8 dB during mechanical milking of sheep, and estimated the risk of NIHL to be 12% for sheep farm workers in Sardinia.

The most effective way to prevent NIHL is to reduce indoor noise levels, e.g., with low noise fans, rubber fan mounts, or automated feeding systems that reduce animal produced noise by feeding all of the animals at once (Derthick et al. 1994). Since noise levels in animal housing often exceed recommended levels, a hearing protection or conservation program should be provided for those exposed to excessive noise levels (Baucom 1995; Gay 1990; OSHA 1996). When properly worn, disposable foam or reusable rubber earplugs, or hearing protector earmuffs, effectively reduce noise. A good fit is essential to provide adequate protection.

Any hearing conservation program should be integrated into the total safety program of a livestock rearing facility. Separate safety programs will often result in the worker weighing the relative importance of one program against another. Unlike other injuries, hearing loss occurs painlessly and slowly; many people do not realize how great an impact hearing loss will have on their lives until it is too late (Derthick et al. 1994; Gay 1990).

FIRE

Several factors facilitate livestock and building loss due to fire, such as poor management and maintenance, improper storage of combustibles, and unsafe electrical wiring (Arble and Murphy 1989). Because the building and fire codes governing livestock buildings tend to vary by county, the burden for insuring adequate safety standards typically falls to the producers. Assistance from fire officials or safety extension specialists should be requested as appropriate. The extent of damage caused by a fire often depends on the availability of adequate water supplies to extinguish it and the time it takes to alert firefighters and get them to the scene. Other causes of fire damage not related to poor fire prevention practices are lightning, arson, and accidents.

The Northeast Regional Agricultural Engineering Service (NRAES) has summarized fire control principles and methods for livestock buildings (Arble and Murphy 1989), including building design, early warning devices, fire suppression, and example system installations. Important principles include:

A practical, cost effective building design will minimize damage if a fire does occur; e.g., use of proper building materials and design can aid in fire self-termination and compartmentation of fires to reduce the extent of damage. Buildings should be constructed with fire retardant materials such as concrete or masonry wherever possible. Combustible building materials should be treated with a fire-retardant. Combustibles not frequently used in the building should be discarded; those frequently used should be stored in a fire retardant compartment and inventory kept to a minimum.

Only wiring materials and equipment meeting the requirements of Article 547 of the National Electrical Code, to include wire size and correct load, should be used. Electrical equipment should be installed according to manufacturer's specifications, used for specified purposes only, and adequately maintained or replaced. All electrical equipment such as fuses, junctions, and outlet boxes should be kept free of grease and dust.

Heating devices should be approved by the American Gas Association or UL (Underwriter's Laboratories). A reasonable distance should be maintained between combustibles and heat producing equipment such as heaters, stoves, furnaces, and welding and grinding equipment. A fire extinguisher should be present during hazardous activities such as welding or cutting.

Fire walls can be effective. Although livestock buildings are often long, single compartment structures, the use of compartmentation practices can delay the spread of fire within a building, increasing the time needed to respond and suppress a fire. Fire walls that extend from the floor to ceiling, constructed of fire-resistive materials and utilizing self-closing fire rated doors should be placed where there is a change in operation, where practical.

Provision for natural ventilation is important in removing smoke, gases, and heat when electrical power is out, so that firefighters can reach the base of a blaze and increase the chance of survival for livestock. Ventilation diverts toxic gases and smoke away from occupants, livestock, and firefighters. It also controls the spread of fire by directing the flow of air currents and releasing unburned combustion gases before they ignite. Ventilation is best achieved by utilizing skylights, roof hatches, emergency escape exits, and similar devices installed during construction.

Fires may be detected by early warning devices that can detect smoke or heat. Smoke and products of combustion can indicate the presence of fire but smoke detectors are susceptible to false alarms due to airborne elements present in livestock buildings, such as dirt, dust, dander, or high moisture. Smoke detectors should be serviced every 4 to 6 months to reduce false alarms. Industrial quality heat detectors work well in agricultural environments. Early fire detection systems may include automatic telephone dialers which contact farm houses, neighbors, or the fire department in the event of a fire. Early warning devices should be regularly cleaned, inspected, and tested by professionals.

Fire suppression systems reduce risk of loss. Manual fire suppression involves using extinguishers or the fire department while automatic fire extinguishing depends solely on sprinkler systems. A manual fire suppression system requires adequate numbers of properly placed fire extinguishers and training for the farm workers to use them. Ten pound ABC type extinguishers should be placed in all major buildings near exits, especially in buildings with extensive electrical equipment or stored flammable liquids.

Measures can be taken to assist the fire department. Farms should have a layout that provides space for fire trucks and ease of movement through buildings. On-site water should be available in addition to a municipal supply. Placement of buildings sufficiently far apart (e.g., 100 ft) can prevent flame spread.

CHILDREN IN BUILDINGS

Animal production facilities can seem like attractive playgrounds to small children, especially during inclement weather. Because of their complexity and potential for danger, no one should treat animal production facilities as play areas. Lack of experience and caution can make children especially suspectable to injury in agricultural environments. Young children visiting the facilities should always be under the close supervision of trained production personnel. Older children should only be allowed to work in these environments after adequate training and with parental supervision.

SAFETY SIGNS

Safety signs can describe and depict hazards utilizing the safety-alert symbol, a signal word such as DANGER or CAUTION, a symbol or pictorial panel, and a word message panel (ASAE 1994). Given the numerous indoor hazards in animal facilities, safety signs could be helpful to animal producers. Figure 25.11 shows such a safety sign developed for the North Carolina Poultry Federation in 1995, which they produced for poultry growers in black-on-yellow colors. This sign warns of respiratory hazards, unguarded equipment, winches, electrical hazards, and the danger of children playing in buildings. Such signs can also be readily produced in Spanish. Educational efforts by animal production industries as well as by veterinary, medical, industrial hygiene, engineering, and other professionals can positively impact producer awareness of health and safety hazards, and should be encouraged.

SUMMARY

Hazards in animal production buildings include those due to poor air quality as well as machinery, electrical, fire, and noise hazards. Indoor air quality is greatly affected by

RECOMMENDATIONS FOR SAFETY IN THE POULTRY HOUSE

RESPIRATORS GIVE REAL PROTECTION.
- wear tight-fitting NIOSH/MSHA-approved respirators as necessary.

GUARDS ON FANS AND MOTORS PREVENT ACCIDENTS.
- keep fingers, hands away from moving parts

WINCHES CAN BE RISKY - BE CAREFUL.
- gears, cable, spinning handles can cause injury

BE CAUTIOUS DURING ELECTRICAL WORK & STORMS.
- eliminate electrical hazards and minimize exposure to shock

SUPERVISE CHILDREN IN POULTRY HOUSES.
- poultry houses are not playgrounds

For more information refer to N.C. Cooperative Extension bulletin AG-481-3 on "Improving the Health and Safety of Poultry Facility Workers", the N.C. Dept. of Labor-OSHA Standards and Guidelines, and the National Electric Code.
PROVIDED BY THE NORTH CAROLINA POULTRY FEDERATION

Revised 15 June 1995

Figure 25.11. Safety sign produced for poultry workers by the North Carolina Poultry Federation.

ventilation, which is needed to remove animal heat and moisture as well as other contaminants from indoor air. Poor air quality is often due to inadequate ventilation in cool weather. Both chronic and acute hazards are caused by poor air quality. Toxic gases, dust, bioaerosols, and odors all affect human health and should be controlled; regulatory and legal pressures may force adoption of air cleaning methods. When control measures are insufficient, respiratory protection and complete respiratory protection programs are recommended. Mechanical hazards due to unguarded fans, motors, and winches can be controlled with proper guarding and careful management. Electrical hazards are often caused by inadequate or faulty wiring materials or techniques and are a leading cause of fire in livestock buildings. Fire hazards are also caused by poor management and improper storage of combustibles. Noise can sometimes reach hazardous levels in swine buildings at feeding time and in other situations with loud equipment; taking steps to reduce noise levels or using adequate hearing protection may be needed to avoid noise-induced hearing loss. Children in buildings merit special attention due to the hazards present; children should not be allowed in the buildings to play. Safety signs and other educational programs are needed to inform workers in animal buildings of these hazards.

REFERENCES

Albright, L.D. 1990. *Environment Control for Animals and Plants*. St. Joseph, MI: American Society of Agricultural Engineers.

Arble, W.C., and D.J. Murphy. 1989. *Fire Control in Livestock Buildings*. NRAES-39. Ithaca, NY: Northeast Regional Agricultural Engineering Service.

American Society of Agricultural Engineers (ASAE). 1994. Safety signs. ASAE Standard S441.1 In: *ASAE Standards 1994*. St. Joseph, MO: ASAE, pp. 274-90.

Barker, J., S. Curtis, O. Hogsett, and F. Humenik. 1991. Safety in swine production systems. *Pork Industry Handbook*. Raleigh, NC: North Carolina Cooperative Extension Service, PIH-104.

Baucom, D., P. Cowan, and L. Allen. 1995. *The Evaluation of Potential Health Hazards to Farmers Working in Swine Confinement in Martin County, North Carolina*. Occupational and Environmental Epidemiology Section, North Carolina Department of Environment, Health and Natural Resources.

Boon, C.R., and V.A. Battams. 1988. Air mixing fans in a broiler building—their use and efficiency. *J. Agr. Eng. Res.* 39: 137-47.

Bottcher, R.W., R.L. Langley, and R. McLymore. 1994. *Improving the Health and Safety of Poultry Facility Workers*. Raleigh, NC: North Carolina Cooperative Extension Service, AG-481-3.

Bundy, D.S., J. Zhu, S.J. Hoff, and Q. Liu. 1996. Development procedures and experiences in evaluating manure additives for odor control. *Proc. of the International Conference on Air Pollution from Agricultural Operations*. Ames, IA: Midwest Plan Service, pp. 271-73.

Carpenter, G.A. 1986. Dust in livestock buildings—review of some aspects. *J. Agric. Eng. Res.* 33: 227-41.

Collins, E.R., G.R. Bodman, and L.E. Stetson. 1995. Electrical wiring for swine buildings. *Pork Industry Handbook*. Raleigh, NC: Cooperative Extension Service.

Dawson, J.R. 1990. Minimizing dust in livestock buildings: possible alternatives to mechanical separation. *J. Agric. Eng. Res.* 47: 235-48.

De Boer, S., and W.D. Morrison. 1988. *The Effects of the Quality of the Environment in Livestock Buildings on the Productivity of Swine and Safety of Humans: A Literature Review*. Guelph, Ontario: University of Guelph, Department of Animal and Poultry Science.

Derthick, S., R.W. Bottcher, R.L. Langley, and R. McLymore. 1994. *Agricultural Health Promotion System: Swine Buildings and Worker Health*. Raleigh, NC: North Carolina Cooperative Extension Service, AG-481-4.

Donham, K.J. 1987. Human health and safety for workers in livestock housing. Latest developments in livestock housing, *Procedings of the International Commission of Agricultural Engineers (CIGR) Meeting, Urbana-Champaign, Illinois, June 22-26*, pp. 86-95.

Donham, K.J. 1990, Health effects from work in swine confinement buildings. *Am. J. Ind. Med.* 17: 17-25.

Driggers, L.B. 1987. *Ventilation of Swine Buildings Using the North Carolina Underslat Ventilation System*. Raleigh, NC: North Carolina Cooperative Extension Service, AG-132.

Environmenal Protection Agency (EPA). 1991. *Introduction to Indoor Air Quality—A Reference Manual*. U.S. Environmental Protection Agency, July.

Evans, B.H. 1957. *Natural Air Flow Around Buildings*. Texas Engineering Experiment Station, Research Report 59, March.

Gay, J.E. 1990. Plan on hearing tomorrow? Wear your protection today. *National Hog Farmer* (Spring) pp. 30-31.

Hammond, E.G., C. Fedler, and R.J. Smith. 1981. Analysis of particle-borne swine house odors. *Agri. Envir.* 6: 395-401.

Hiatt, R.S., and K.L. McFate. 1993. *Electrical Wiring for Livestock and Poultry Structures.* Columbia, MO: National Food and Energy Council.

Homes, M.J., A.J. Heber, C.C. Wu, L.K. Clark, R.H. Grant, N.J. Zimmerman, M.A. Hill, B.R. Strobel, M.W. Peugh, and D.D. Jones. 1996. Viability of aerosols produced from a swine facility. *Proc. of the International Conference on Air Pollution from Agricultural Operations.* Ames, IA: Midwest Plan Service, pp. 127-31.

Klarenbeek, J.V., and A.P. van Harreveld. 1996. On the regulations, measurement and abatement of odors emanating from livestock housing in the Netherlands. *Proc. of the International Conference on Air Pollution from Agricultural Operations.* Ames, IA: Midwest Plan Service, pp. 16-21.

Krishnan, P.V. 1965. Spacing of buildings for natural ventilation. *Trans. of the ASAE* 8: 208-209; 215.

Kryter, K.D. 1994. *The Handbook of Hearing and the Effects of Noise.* San Diego, CA: Academic Press.

Lenhart, S.W. 1995. Components of a model respiratory protection program for the poultry industry. *Proc. of the 1995 Poultry Supervisor's Short Course.* Raleigh, NC: North Carolina State University, pp. 8-15.

Lenhart, S.W., P. D. Morris, R. E. Akin, S. A. Olenchock, W. S. Service, and W. P. Boone. 1990. Organic dust, endotoxin, and ammonia exposures in the North Carolina poultry processing industry. *Appl. Occup. Environ. Hyg.* 5(9): 611-18.

Leonard S, and K. Donham. 1990. Dust and gases. *National Hog Farmer* (Spring), pp. 16-21.

Levy, D.A. 1982. Allergic response and agricultural activities. *Ann. Am. Conf. Gov. Ind. Hyg.* 2: 161-70.

Maghirang, R.G., L. L. Christianson, H. B. Manbeck, and G. L. Riskowski. 1993. *Control of Dust in Animal Buildings—A Review.* ASAE Paper No. 934549, St. Joseph, MI: ASAE.

Manninen, A., J. Kangas, M. Linnainmaa, and H. Savolainen. 1989. Ammonia in Finnish poultry houses: effects of litter on ammonia levels and their reduction by technical binding agents. *Am. Ind. Hyg. Assoc. J.* 50(4): 210-15.

Miner, J.R., and C.L. Barth. 1988. Controlling odors from swine buildings. *Pork Industry Handbook.* Raleigh, NC: N.C. Cooperative Extension Service, PIH-33.

MMWR. 1993. Fatalities attributed to entering manure waste pits—Minnesota, 1992. *MMWR* 42(17): 325-29.

Moore, P.A., T.C. Daniel, and D.M. Miller. 1995. Effect of chemical amendments on ammonia volatilization from poultry litter. *J. Env. Quality* 24(2): 293.

National Fire Protection Agency (NFPA). 1993. *National Electrical Code®.* Published by and a registered trademark of the National Fire Protection Association, Quincy, MA.

National Institute for Occupational Safety and Health (NIOSH). 1994. Request for assistance in preventing Organic Dust Toxic Syndrome. *NIOSH Alert,* April 1994. NIOSH Publ. No. 94-102. Cincinnati, OH: NIOSH.

Occupational Safety and Health Administration (OSHA). 1996. *Occupational Noise Exposure.* Standard No. 1910.95, SubPart G.

Pazzona, A., and L. Murgia. 1993. Estimation of noise-induced hearing impairment risk in sheep dairy farming. *J. Agr. Eng. Res.* 55: 107-12.

Plakke, B.L., and E. Dare. 1992. Occupational hearing loss in farmers. *Public Health Reports* 107(2): 188-92.

Prather, T.G. 1992. *Agricultural Respiratory Hazards.* Knoxville, TN: The University of Tennessee Agricultural Extension Service.

Robertson, J.F. 1993. Dust and ammonia concentrations in pig housing: the need to reduce maximum exposure limits. In: *Livestock Environment IV, Proc. of the Fourth International Livestock Environment Symposium,* ASAE Publication 03-93, St. Joseph, MI: ASAE, pp. 694-700.

Schiffman, S, E.A. Sattely Miller, M.S. Suggs, and B.G. Graham. 1995. The effect of environmental odors emanating from commercial swine operations on the mood of nearby residents. *Brain Res. Bull.* 37(4): 369-75.

Schlenker, E.H., Parry, R.R., and M.A. Hellickson. 1987. Respiratory characteristics of poultry laborers. In: *Latest Developments in Livestock Housing, Proc. of the International Commission of Agricultural Engineers (CIGR) Meeting,* Urbana-Champaign, Illinois, June 22-26, pp. 127-36.

Sneath, R.W., M.R. Holden, V.R. Phillips, R.P. White, and C.M. Wathes. 1996. An inventory of emissions of aerial pollutants from poultry buildings in the UK. *Proc. of the International Conference on Air Pollution from Agricultural Operations.* Ames, IA: Midwest Plan Service, p. 207-14.

Takai, H., F. Moller, M. Iversen, S.E. Jorsal, and V. Bille-Hansen. 1995. Dust control in pig houses by spraying rapeseed oil. *Trans. ASAE* 38(5): 1513-18.

Timmons, M.B., and R.S. Gates. 1987. Relative humidity as a ventilation control parameter in broiler housing. *Trans. ASAE* 30(4): 1111-15.

Virtanen, T., P. Kalliokoski, P. Vilhunen, A. Taivainen, and R. Mantyjarvi. 1990. Concentrations of specific dusts in swineries and the humoral response of swinery workers. *Allergy* 45: 354-62.

Wathes, C.M. 1994. Air and surface hygiene. Chapter 6 in *Livestock Housing*. C.M. Wathes, and D.R. Charles, eds. Wallingford, Oxon, UK: CAB International, pp.123-48.

Wilhelm, L.R., and S. Snyder. 1996. Field comparison of broiler house management systems in east Tennessee. *Proc. of the International Conference on Air Pollution from Agricultural Operations*. Ames, IA: Midwest Plan Service, p. 479.

Williams, C.M., and S.S. Schiffman. 1996. Effect of liquid swine manure additives on odor parameters. *Proc. of the International Conference on Air Pollution from Agricultural Operations*, Ames, IA: Midwest Plan Service, pp. 409-412.

26

INDUSTRIAL HYGIENE EVALUATIONS IN AGRICULTURE

William A. Popendorf, Ph.D., C.I.H.
Department of Biology, Utah State University

Stephen I. Reynolds, Ph.D., C.I.H.
Institute of Rural and Environmental Health, The University of Iowa

---

As with other industries, the anticipation, recognition, evaluation, and control of heath and safety hazards is the key to decreasing injuries and illnesses in agriculture. This chapter begins with a overview of exposure guidelines and limits. A discussion of non-quantitative and quantitative evaluation methods follows. Methods and equipment used to measure the concentration of various potentially hazardous substances in the farmer's working environment are described. Besides environmental measurements, a discussion of biological and medical monitoring is included. Additionally a detailed description of numerous hazardous gases, aerosols, dusts, microbial organisms, and natural toxins and the diseases associated with these agents is presented. The chapter provides useful tables listing hazards associated with different agricultural activities and sources where equipment used to monitor the agricultural environment can be purchased.

INTRODUCTION

Occupational diseases among farmers have been documented for nearly three hundred years (Ramazzini 1713). Reviews of modern morbidity and mortality studies suggest that respiratory hazards still represent the greatest health risk to farmers (Popendorf et al. 1985; Donham 1986; Popendorf and Donham 1991). Respiratory hazards in agricultural or any occupational setting can be classified as either gaseous or particulate (aerosol) in nature. The physical nature and differences between gases and aerosols affect the way they behave in the environment, how and where they affect the body, and how they can be sampled or evaluated.

Gaseous hazards include gases and vapors. Physically speaking, gases are chemicals that exist only in the gaseous state under normal conditions. Vapors are from chemicals that are less volatile than true gases and can coexist in equilibrium with their liquid counterparts under normal use conditions. Both gases and vapors can dissolve into and be slowly released from liquids such as water, a ubiquitous commodity on farms. Once airborne, both gases and vapors are molecular in nature and move much like air. Toxic gases and vapors will stay mixed in air and will only dissipate via molecular and turbulent diffusion or by being reabsorbed into some environmental surface.

Aerosols are either solid particles (like dusts and powders), mists made of liquid particles (like a fog), or fumes that form from thermal reactions (like smoke). No intrinsic property drives particles to become airborne by themselves; they require some mechanism like grinding, agitation, or combustion to become airborne. Once airborne, aerosol particles are much larger than air molecules. Aerosol particles tend to settle out of still air due to gravity or become separated from air flowing in tightly curved paths such as around obstructions in ducts, in the nose, etc. This simple difference between gases vapors and aerosols affects both their health hazards and the ways they can be sampled.

This chapter will explain some of the health hazards and basic methods used to collect samples of the most important respiratory hazards in agricultural settings as exemplified by the gaseous and aerosol hazards listed in Table 26.1. Table 26.1 lists the settings in which each respiratory hazard is likely to exist, broken out into subcategories of livestock type, phase of crop production, and tasks not unique to agriculture such as machinery operations or maintenance and construction on the farm. Finally, specific known sources of each of these respiratory hazards are listed on the right side of Table 26.1.

AGRICULTURAL OCCUPATIONAL DISEASE

Human Anatomy and Physiology

Before discussing these hazards, the structure and function of the respiratory system will be reviewed as they pertain to the differences between gaseous and particulate respiratory hazards. Air normally enters the respiratory system through the nose, specifically through the nostrils, although people at work often breath through their mouth. In either case, inertia causes most very large aerosol particles (diameters >100 μm) to separate from the air that flows into these orifices. The air then follows a tortuous curving path through the nose and back of the mouth. The velocity of the air is maximal in this region, causing nearly all particles >10 μm in diameter to impact onto airway walls due to centrifugal force. As the air passes into the trachea, bronchi, and progressively smaller branches of the bronchioles (the

Table 26.1. Overview of Agricultural Respiratory Hazards (adapted from Popendorf and Donham 1991).

	Livestock			Crops			Other		
	beef/dairy	poultry	hogs	planting	harvest	storage	machinery	construction	
Gases and Vapors									
Ammonia	x	x	x	x					Animal urine or fertilizer
Carbon dioxide	x	x	x			x	x		Animal respiration or plant decay
Carbon monoxide		a	a				x	x	Combustion sources
Disinfectant derived gases	x	x	x						e.g. formaldehyde, chlorine, cresylic acid
Fumigant Pesticides						x			Phosphine, methyl bromide, chloropicrin
Hydrogen sulfide	x		x						Manure gas with other mercaptans
Nitrogen dioxide					b	b	c		"Silo gas", arc welding, diesel exhaust
Oxygen deficiency		d	d			d			Confined space
Inorganic (Dust) Aerosols									
Soil (inorganic dirt)	e			x	x			x	Dirt, may contain silica or microbes
Welding fumes						f	f	f	Especially when welding galvanized steel
Organic Aerosols									
Livestock production buildings	x	x	x					g	"Enclosed animal production buildings"
Mycotoxins	x	x	x		x	x			Endotoxin, aflatoxin, fumonison, ochratoxin
Noninfectious microbes	x	x	x					x	Newcastle, Q fever, histoplasmosis
Pesticides	x	x	x	h	h				e.g. OPs, carbamates, herbicides
Veterinary biologicals	x	x	x						Antibiotic feed additives and vaccines
Crop specific organic aerosols									
Cotton dust					x	x			Mostly in processing and textiles
Grain dust	x	x				x			Both clean but particularly moldy grain
Mushrooms					x	x			Mold spores
Silo uncapping dust						x			Moldy silage causing FL or ODTS
Sugar cane dust					x	x			Causing bagassosis
Wood dust						x			Especially moldy wood chips causing ODTS

footnotes:
a) a possible hazard if gas fired heaters are used.
b) produced in silos from 1-14 days after harvest and may therefore overlap days for a multi-day harvest.
c) a particular hazard from diesel powered machinery.
d) a particular hazard in airtight silos or when livestock ventilation fails.
e) a possible hazard to beef feed lot operators.
f) a particular hazard from zinc-fume fever when welding galvanized steel in poorly ventilated conditions.
g) histoplasmosis is a particular hazard during renovation of poultry buildings.
h) although pesticides are airborne during both application and harvest, skin is the major route of exposure.

conducting portion of the lung), its velocity decreases and little additional separation occurs. On the other hand, water soluble gases and very small particles will be adsorbed into the moist tissues lining these passageways. Those gases that are acidic, basic, or corrosive will cause rapid irritation to these tissues. Also, small amounts of certain gases or/and particles can cause constrictions and inflammation in the conducting portions of airways within some people, creating symptoms known as bronchitis or asthma.

At the end of the air's journey are the alveoli, highly vascular tiny sac-like compartments where most of the gas exchange occurs by molecular diffusion. Because of the alveoli's small size (400 μm across), a large fraction of even the small particles between 2 and 10 μm diameter have enough time in the quiet air to fall by gravity onto the bottom of the sac. This sedimentation could create a large buildup of debris within the lung if that were the end of a particle's story. However, the body's defense mechanisms can either move most particles to a passageway where they can be removed in the mucus swept by the cilia to the throat, pass it through the wall of the alveoli and on through the lymph system, digest the particle at the site, or encase it with scar tissue. If a particle is not prevented from reaching the alveoli by one of the separation mechanisms or is not readily removed or digested, its presence can easily damage this sensitive deep lung tissue via various toxic effects.

Chemical Injury Versus Traumatic Injury

Because of the above mechanisms, the rate of disease progression, its severity, the symptom(s) of respiratory disease, and therefore its diagnosis all differ among exposed individuals. While there are many similarities in the practices of preventive health and safety, two characteristics of interest separate them: 1) the variability in the definition of a health injury, and 2) the potential for a protracted duration over which a health injury can take place.

Traditionally there was little question whether and when an occupational injury occurred. For instance, it is visually obvious whether a laceration either occurred or did not occur; after an amputation there is no question when a body part is missing. However, whether a chemical can or does cause an adverse respiratory health effect is often a matter of its definition. For instance, a study showing a 10% change in a measurable lung function such as Forced Expiratory Volume in 1 second, FEV_1, might be defined by some as an injury, by others as "statistically significant," and by still others as inconsequential. Other times, health injury may be defined as a certain adverse symptom such as irritation or a recurring cough, by a change in enzyme level, or by an x-ray reading. The degree to which a person's health is changed and the frequency with which such changes occur within a group defines the "threshold of injury." Such definitions are not consistent among chemicals or even among studies of any one chemical (ACGIH 1995b).

The second characteristic is the time of action. Fatalities are an end point common to both safety and health hazards. But in contrast to a fatal traumatic injury, fatalities from exposure to a single acute health hazard are believed to be less common than are fatalities resulting from "chronic" chemical exposures repeated over weeks, months, or even years. This protracted nature makes an association between chronic exposure and adverse health effect (whether it's disease or death) difficult to prove.

The net result has been ambiguity regarding the importance of a clinically measurable effect, equivocation regarding the work relatedness of an adverse outcome, and uncertainty or disagreement on appropriate values for allowable occupational exposure limits. Allowable chemical exposure limits that have been set are *not* a guarantee of no effect. They *must* be viewed only as our best collective opinion, at one point in time, above which a defined effect is deemed an unacceptably frequent outcome. Both the definition of an adverse health effect, the method by which the health effect is measured in individuals or in groups, and even the method by which exposure to the chemical is measured can change over time. Occupational health exposure limits should never be taken as indicative of conditions that will never produce a health injury.

Exposure Guidelines and Limits

In anticipation that a diverse audience may not be familiar with industrial hygiene, the definitions of a few terms and concepts ingrained into industrial hygiene folklore are provided. The first is the paradigm of recognition, evaluation, and control. Historically, this process began with the recognition of adverse health effects that already existed within a working population. The primary scientific tool of the recognition phase is epidemiology, although simple observation and medical surveillance can also play a more subjective part. Today, we can anticipate (and hopefully avoid) adverse health effects based on toxicology or experience in other work settings; thus, anticipation has recently been elevated to a status equal to and in fact preceding "recognition."

To evaluate the risk, hygienists have developed and rely upon a system of "performance based" exposure limits and guidelines. Prominent among these are Threshold Limit Values (TLVs) (ACGIH 1995b) and the OSHA regulatory equivalents called Permissible Exposure Limits (PELs) (OSHA 1993). The goal of performance standards is to prevent adverse health effects by keeping exposures and doses to acceptably low levels without specifying the method or "work practices" that must be used to achieve those levels. TLVs, as stated by its authors, "*are not* fine lines between safe and dangerous concentration(s) nor are they a relative index of toxicity. They should not be used by anyone untrained in the discipline of industrial hygiene" (ACGIH 1995b). Nonetheless, they are provided herein as guidance.

Most TLVs are 8-hour time-weighted average concentrations (TLV-TWA), "a concentration to which nearly all workers may be repeatedly exposed, day after day, without adverse effect" (ACGIH 1995b). For these chemicals, variations above and below the average over a workday are assumed to balance out. For certain other chemicals (including many agricultural gases), variations above the average are important and peak exposures are limited to either a short-term exposure limit (TLV-STEL) or a ceiling value (TLV-C). It is important that chemicals with a STEL and especially with a C limit should *not be evaluated with long-term sampling methods*. For these chemicals, conditions capable of causing harm can occur over much shorter intervals than 8 hours; for STELs, a nominal 15-minute interval can be assumed. The vast majority of PELs were either TLV-TWAs or TLV-Cs in 1969 (OSHA 1993); many TLVs have changed over the years (nearly always lowered), while legal constraints have kept most OSHA PELs from changing.

Two other quite different higher concentrations are sometimes useful to know. The first is the concentration considered immediately dangerous to life or health (IDLH). The IDLH represents the maximum concentration from which someone (without a respirator) could escape within 30 minutes without experiencing either any irreversible health damage or effects that could impair their ability to escape (e.g., severe eye irritation) (NIOSH 1990a). The IDLH is also not an exact value, and in fact was originally determined for the purpose of respirator selection. Sometimes a lethal concentration is known either from laboratory tests (such as the LD_{50}, a concentration that killed 50% of a test species) or from single events such as a human fatality accident investigation (usually referred to as an Ld_{low}) (NIOSH 1990b).

Finally, knowing the odor threshold is also useful in practice. It is important to know, for instance, if one can detect a given chemical before it is a hazard, meaning that the odor threshold must be lower than the exposure limit. Where these numbers are reversed, one could be overexposed without even sensing the chemical's odor. Sometimes odor thresholds are studied systematically by using a test panel; usually they are more experiential. It is almost universally true that the threshold of detection varies widely among individuals; some people may not detect a chemical until 10- or even 100-fold above the reported threshold (Ruth 1986; AIHA 1989). Thus, odor thresholds are useful for most of us, but they are not an assurance for all of us.

Performance standards require that users evaluate their exposure to determine compliance or the level of safety. The alternative is a work practice standard. The work practice approach is not common in industrial hygiene but is used for agricultural pesticides. In the process of registering pesticides, the EPA promulgates what amounts to a "use practice standard" in the form of label instructions that specify the ways the chemical can be safely and legally used. The implication is that if a user follows these instructions, exposure will be sufficiently low to prevent adverse health effects. This "work practice" standard for pesticides contrasts sharply with general industry standards where "to assure . . . safe and healthful working conditions," employers are expected to comply with "performance standards" expressed as PELs (PL 91-596).

One might ask, is a "work practice standard" or a "performance standard" approach better for agriculture? Performance standards would grant farmers their independent way to avoid airborne hazards, but the widespread application of these standards, as envisioned for OSHA, would require an additional administrative and support structure to conduct on-site monitoring on more than a million farms or "places of agricultural employment" nationwide. And many agricultural activities, working environments, and chemical exposure levels vary sufficiently by season, by day, and even by hour as to make such measurements moot. Which is not to say that measurements and even performance standards have no place in agriculture. For instance, ammonia in animal production facilities is amenable to the application of traditional environmental monitoring and health interpretations.

In principle, it is simpler to determine compliance with "work practice" standards, but they have their own limitations. Perhaps the most important is that in order to prevent overexposure, work practice standards have to account for the worst foreseeable circumstances; therefore, they have to be overly restrictive for the average workplace, which makes them conducive to low compliance. To make a work practice standard efficient, requires a thorough understanding and control of the conditions that could create an unacceptably high hazard. That level of understanding, let alone an adequate ability to control the workplace, is often lacking in agriculture. It remains a challenge for the future to define the conditions favoring the best form of standard or whether either form is adequate for agriculture. For the present, this difference can only influence the decision on whether to evaluate a particular agricultural setting.

Non-Quantitative Evaluation Methods

Exposure assessment usually requires measurements. However, for some purposes a relatively simple calculation from tabulated data yields a ratio that can be qualitatively useful either to compare the relative likelihoods of being overexposed to different chemicals or to approximate exposure before measuring. The Vapor Hazard Index is derived from the ratio of the vapor pressure of a chemical (which must usually be converted from mmHg to ppm) to its exposure limit such as its TWA-TLV, STEL, or C (either given in or converted to ppm). The vapor pressure is the concentration of vapors *immediately* above a liquid surface. In an acceptable workplace, this concentration at the source must be diluted at least to its exposure limit in the worker's breathing zone. The magnitude of the dilution needed from the liquid surface to an acceptable breathing zone comprises the Vapor Hazard Ratio (VHR) for each chemical. The Vapor Hazard Index (VHI) is simply the common logarithm of the Vapor Hazard Ratio. The VHI can be calculated according to Equation 26.1 for any other chemical for which a vapor pressure is known and an allowable exposure limit has been or can be set (Popendorf 1984).

$$VHI = \log(VHR) = \log \frac{\text{Vapor Pressure (mmHg)} \times 10^6}{760 \text{mmHg} \times TLV} \quad (26.1)$$

The amount of dilution actually present between a vapor source and a worker's breathing zone depends on the setting

Table 26.2. The Vapor Hazard (VHR) and Index (VHI) for selected chemicals. TLVs are for 1994-1995, and vapor pressures are at 25°C unless otherwise indicated. Indented compounds are industrial solvents and/or pesticide inert carriers shown for comparison.

Compound	TLV ppm		Vapor P mmHg	VHR ppm/ppm	VHI
benzene	.3	A1,skin	95	416,700	5.6 a
chloropicrin	.1		20 @20	263,160	5.4
hydrogen cyanide	~5	C,skin	620 @20	173,600	5.2
benzene (29CFR1910.1028)	1	PEL	95	125,000	5.1 b
carbon disulfide	10	skin	361	47,500	4.7
dichloropropene	1	skin	36	47,400	4.7
carbon tetrachloride	5	A2,skin	113	29,700	4.5
benzene	10	A2	95	12,500	4.1 b
n-hexane	50		151	3,975	3.6
perchloroethylene	25	A3	18	968	3.0
pentane	600		426 @20	934	3.0
toluene	50	skin	28	737	2.9 a
ethylene dibromide (EDB)	20	PEL only	11	724	2.8
mercury	.006		.00185	406	2.6
acetone	750		230	400	2.6
phosdrin (Mevinphos)	.01	skin	.003 @20	395	2.6
dichlorvos (DDVP)	.1	skin	.03	395	2.6
toluene	200	PEL	28	184	2.3 b
n-heptane	400		40 @20	131	2.1
isopropyl alcohol (IPA)	400		33 @20	110	2.0
o-xylene	100		6.6	87	1.9
m- and p-xylenes	100		3.6	48	1.7
diazinon (c 0.1 mg/m³)	.008	skin	2×10^{-4}	33	1.5
parathion (c 0.1 mg/m³)	.008	skin	6×10^{-5}	10	1.0
azinphos-methyl (Guthion)	.2	skin	.0004 @20	2.6	0.4
dibrom [Naled] (c 3 mg/m³)	.19	skin	3×10^{-4}	2.1	0.3
Lindane	.042	C	3×10^{-5} @20	0.9	0.0
Malathion (c 10 mg/m³)	0.74	skin	4×10^{-5}	.07	-1.1

Footnotes:

a   the higher of two vapor hazards shown in this table because the TLV used in this line either is lower than the current PEL or is a proposed lower TLV, or both (see b).

b   the lower of two vapor hazards shown in this table (see a); listed as "PEL" where its limit is higher and the VHR and VHI are lower than the TLV.

c   value converted to ppm but listed in TLV publication only as mg/m³ meaning it is considered only an aerosol.

Table 26.3. Example air sample data sheet.

Site Code _____ Site Name _____ Date __ / __ / __
Field Investigator (initials or name) __ __ __ __   mo da yr
Process name: _____
Worker name (optional) or Sample location _____

AIR MONTORING

Pump ID # _____   Flow Rate _____ mL/min   Colorimetric sample
Air pump data   (1)   (comment : reading)   Start time _____
  Start time _____ _____   End time _____
  Flow check(s) _____ _____   Sample duration _____
  Stop time _____ _____   Stain length _____
  Run duration _____ minutes = _____ L.
Sample identification for Lab: _____   Concentration _____

ENVIRONMENTAL INFORMATION

Air Temperature _____ Relative Humidity _____ % or Wet Bulb _____
Air Movement: Source of air _____ Velocity _____ fpm.
  Direction Relative to Operator from _____ to _____
  Direction Relative to Source _____
Protective Clothing and Equipment in Use: circle Y or N on each line
  Respirator: [N] [Y] if Y (**U**napproved) (**D**isposable) (**Q**uarter mask)
    (**H**alf mask) (**P**owered air)
  Eyes: [N] [Y] if Y (**S**pectacles) (**G**oggles) (**S**ideshields) (**F**ace shield)
  Gloves [N] [Y] if Y (**C**anvas) (**L**eather) (**P**VC) (**R**ubber) (**O**ther)
  Boots: [N] [Y] if Y (**R**ubber) (**L**eather) (**F**abric)

OBSERVATIONS

Description of Activities (for instance list operations being conducted):
  1._____
  2._____
  3._____
  4._____
  5._____
  6._____
Any visible dermal exposure? [Y] or [N] Deposition Location? _____
_____
Other Comments (Diagram on Back?): _____
_____

but is expected to be the same no matter what chemical is present. Thus, in the relative sense, the chemical whose vapor is least likely to exceed its TLV is the chemical with the lowest VHI. The potential for any vapor to reach or exceed its allowable exposure limit in any setting is indicated directly by the Vapor Hazard Ratio and its analogous Vapor Hazard Index as shown in Table 26.2. Chemicals with VHIs of zero or less *cannot* create a vapor hazard, even with no dilution. Vapors from chemicals with VHIs of 2 or less require very little dilution. Exposures to chemicals with VHIs of 2 to 4 are unlikely to exceed recommended limits in outdoor settings. The low VHI of many agricultural pesticides indicates they are generally more of an aerosol than a vapor hazard.

Quantitative Exposure Assessments

The quantitative evaluation of exposures can be broken down into four phases: planning, method calibration, sample collection and/or analysis, and data interpretation. Before selecting an appropriate sampling method and even before choosing a sampling strategy, meaningful and useful exposure assessment starts with a clearly defined objective (Leidel et al. 1977; Ness 1991; Lynch 1994). Three common objectives include 1) measuring the effectiveness of engineering controls or production practices, 2) estimating risk of human or animal disease, or 3) evaluating compliance with occupational exposure standards or guidelines.

Components of a sampling strategy include the number samples to be collected, their location (whether area or personal), and the sampling duration. The effectiveness of engineering controls such as ventilation, or production practices such as frequency of cleaning, can be evaluated by using area samples that are placed in fixed locations. Because of the significant effects of individual work habits, mobility, and varying tasks over time, area samples are not good measures of individual exposure. To estimate the risk of disease or compliance with exposure guidelines requires personal samples where devices are placed directly on individuals. The generic exposure assessment methods presented below are discussed later within selected agricultural settings along with some further comments on specific strategies.

As mentioned earlier, calibration is an important step in assuring that the sample collection and/or analysis steps are operating properly, in terms of the air volume collected or the response of the device to the amount of chemical it contained. Calibration methods for direct reading devices should be provided by the specific manufacturer. Flow rate calibration for sampling pumps will be discussed later in this chapter.

Exposure limits with which to compare results were discussed earlier. But the interpretation of quantitative exposure assessment results can go beyond comparisons to available exposure limits by integrating descriptive information recorded at the time of the measurement. Table 26.3 suggests a format for recording both the data necessary for quantifying results and some of the descriptive information useful to interpret these sampling results.

HAZARD EVALUATION METHODS

Table 26.4 presents a matrix of two basic conceptual options for selecting exposure assessment methods for airborne hazards. Additional selection factors to be

Table 26.4. Options for assessing airborne hazards (Ness 1991; NIOSH 1994; Lynch 1994; Eller 1994; ACGIH 1995a).

		QUANTIFICATION	
		Laboratory Analysis	Direct Reading
COLLECTION	Active	An air pump and collection media for the specific gas or vapor.	Colorimetric pull-tube and its corresponding pump or an adaptor available with some solid state instruments.
		An air pump and filter for aerosols.	An adaptor available with some solid state aerosol detection instruments.
	Passive	Diffusion badge or tube with collection media for the specific gas or vapor.	Colorimetric diffusion tube or a solid state gas or vapor detection instrument.
		Not applicable to aerosols.	Only one passive solid state aerosol detection instrument on the market.

discussed later include 1) the availability of valid methods for the specific contaminant to be measured, 2) the measurement sensitivity (the method's limit of detection), 3) the accuracy needed, 4) the difficulty or ease of the method, and 5) its cost. These concepts are presented in the context of airborne hazards but are applicable to all exposure assessments.

Assessing Airborne Hazards

Airborne concentrations may be determined either through sample collection on selective sampling media such as filters, sorbents, or liquid solutions, followed by analysis in a laboratory ("laboratory-based methods"), or through measurement using direct reading methods that provide immediate or at least quick results. Direct reading devices provide a distinct advantage when decisions need to be made quickly, and they eliminate the need for handling and shipping samples; however, they are not available for all the hazards of interest, and they are usually not as sensitive or accurate as laboratory methods. Validated methods have been developed for industrial hygiene sampling and analysis by the National Institute for Occupational Safety and Health (NIOSH), the Occupational Safety and Health Administration (OSHA), and several other organizations (NIOSH 1994). However, sample preparation and analytical techniques developed by the Environmental Protection Agency (EPA) and the Association of Analytical Chemists (AOAC) are, in many cases, for environmental samples and are not appropriate for occupational exposure assessment. Gaseous chemicals can often be collected either by actively drawing a known volume of air through a device that retains the gas or vapor of interest or by means of passive diffusion of the gas or vapor molecules into the collector or indicator. The former method is called "active sampling"; the latter, "passive sampling." With one exception, aerosols require an active sample collection and analysis method. Active sampling is conducted by using a vacuum pump to pull known quantities of air onto the sample collection media. For personal breathing zone sampling, a battery-powered pump, capable of pulling 0.5 to 4.0 liters of air per minute (Lpm) but small enough to be worn on a belt, is connected via tubing to a sample collection device attached to a person's collar. High volume pumps, capable of moving 20 to 30 Lpm of air, may be used for area samples but are much too big to wear and require AC electrical power. Some direct reading devices also use pumps to pull a known volume of air into their sampling chamber.

Passive sampling devices have been developed for both direct reading and laboratory analysis methods. Passive sampling relies upon diffusion of the chemical into the sampler. These devices do not require the use of a calibrated pump, and their small size often makes them much more acceptable for personal sampling. A fairly wide variety of passive colorimetric and electronic direct reading devices have been developed which rely on the rapid molecular diffusion of gases and vapors. Colorimetric devices utilize the length of a color change of granules contained within a glass tube to indicate concentration. Other direct reading devices depend upon the reaction of the contaminant of interest with a sensor to create an electronic signal. But because aerosol particles do not diffuse as readily, small and inexpensive direct reading devices for aerosol measurement are very limited.

Sampling methods that determine the cumulative airborne concentration over the whole period of time sampled are termed integrative methods. Alternatively, a method that measures the concentration at one point in time is termed a grab sample. With the development of miniature data logging systems, many electronic direct reading instruments can now provide both a history of short-term concentrations as well as integrated data for time-weighted average exposures.

When a method requiring laboratory analysis is selected, communication with the laboratory selected to analyze samples is an extremely important first step. The industrial hygienist or other professional collecting samples cannot treat the laboratory as a "black box" that will automatically generate valid results. Industrial hygiene laboratories specialize in analyzing samples from occupational environments. The laboratory should have a Quality Assurance program in place that addresses sample receiving, handling, and storage; calibration of analytical equipment and the use of standard reference materials; training of personnel; and reporting and documentation. The laboratory may participate in a Proficiency Analytical Testing (PAT) or other inter-laboratory quality assurance program or be certified by the American Industrial Hygiene Association (a list of AIHA certified laboratories is published quarterly in the *AIHA Journal*), the EPA, or other professional organization. At this time certification of laboratories performing analyses of environmental microorganisms is not available, but criteria are being developed by the AIHA.

Communication with the laboratory should begin before samples are collected. The laboratory can be helpful in selecting the appropriate method and sampling media. In addition, the amount of analyte required by the laboratory (i.e., the limit of detection of the method) needs to be known to set the sampling rate and duration. A chain of custody form should be used to document samples sent to the laboratory and the analysis requested. Clarifying expectations concerning reporting is important to ensure that results are clearly understood and interpretable.

Direct Reading Measurements

Direct reading methods do not involve a separate laboratory to which samples would have to be sent. They are therefore quicker, can be cheaper, but are usually less accurate, less sensitive, and may be affected by cross-reactions with other similar or sometimes even dissimilar materials.

Colorimetric Detector Tubes for Gases and Vapors. Detector tubes are also called "pull-tubes." Because only a small air sample is pulled through the tube for a few seconds to less than 2 minutes, it is sometimes also called a grab sample. Detector tubes provide a reading as soon as the air sample is collected. Detector tubes must be purchased from commercial vendors such as those listed in Table 26.5. Initially a user would need the following (usually sold as a kit):

- A pump. Either a piston-like pump or a 1-handed bellows pump. The pump "pulls" a measured amount of air through an attached detector tube, open at each end.
- Glass tubes filled with crystalline material that changes color in the presence of a gas or vapor. Tubes (sold by the box of 5 to 10 tubes) are made for specific chemicals and in some cases for different concentration ranges.
- An optional extension hose available from most manufacturers is likely to be needed to take a measurement remotely when it may be too dangerous to stand there for several minutes. Some examples of environments potentially immediately dangerous to life or health include a building where livestock are down, any enclosure over manure being agitated or pumped, a building that has had ventilation failure, and a freshly filled silo.

Calibration of detector tube pumps should include checking that the pump does not leak, testing the accuracy of its stroke volume, and perhaps checking that the sample collection interval is within a specified range (see the manufacturer's instructions).

To take a detector tube measurement:

Step 1. Select the appropriate tube for the gas to be measured.
Step 2. Use the tip breaker hole on the pump to break the tips off both ends of the glass tube.
Step 3. Insert the tube into the pump inlet with the arrow pointing toward the pump. *For cases where the air quality may present an immediate danger to life or health*, an extension hose should be used between the tube and pump to allow the person to sample remotely.
Step 4. Actuate the pump either by pulling out on the piston or syringe pump handle or by squeezing the bellows pump hand grip. Either of these actions is counted as one "stroke"; some tubes require multiple strokes.
Step 5. Wait until the stroke is completely collected (the indication of completeness will vary by pump design). It may take from about 0.2 to 2 minutes for the stroke to finish.
Step 6. Read the gas concentration at the "line" between the stained and unstained portions of the tube. If that line seems blurred, read at the midpoint between the farthest partial stain and a fully developed stain. Write the results on an appropriate annotated sampling form or record (see example in Table 26.3).
Step 7. Steps 4 and 5 may need to be repeated several times (multiple strokes), as directed in the manufacturer's directions to get a reading within the concentration range specified.

Detector tubes are especially advantageous to measure a gas concentration quickly, such as when they could be either dangerously high or high for a brief period. Examples of such conditions include when manure is agitated and pumped, potentially releasing a large pulse of hydrogen sulfide, or when most livestock in a building are found dead due to ventilation failure (which could either be from an oxygen depletion or carbon dioxide buildup) or from an internal combustion engine or malfunctioning heater (carbon monoxide). Detector tubes are much faster than diffusion tubes but slower than most electronic meters. A grab sample only represents conditions over the couple of minutes the sample is collected; when conditions vary, multiple tube readings scattered over perhaps several hours would need to be taken to estimate the average concentrations. The cost for an initial kit ranges from $200 to $375. Tubes cost from $3 to $5 each, depending upon the chemical and manufacturer. Although tubes and pumps made by different manufacturers are not interchangeable, tubes for many different gases, including all of those in Tables 26.1, 26.6, and 26.7, can be used in the same pump.

Colorimetric Diffusion Tubes. Diffusion tubes, also called "dosimeters," are a more convenient way than detector tubes to measure the *average* gas concentration over a longer period of time in which variations in concentrations are not important. Like detector tubes, diffusion tubes contain a color-indicating material and are available for a variety of gases and concentration ranges. Unlike detector tubes, they have no pump and are open at

Table 26.5. Equipment manufacturers applicable to airborne agricultural hazards.

Colorimetric Tubes and Pumps for Agricultural Gases in Table 6 and Organic Vapors

Name	Address	City State Zip	Phone
Matheson/Kitagawa	166 Keystone Dr.	Montgomeryville, PA 18936-9969	800-828-4313
MSA Instrument Div.	Box 427	Pittsburgh PA 15230	800-MSA-2222
National Draeger Inc.	Box 120	Pittsburgh, PA 15230	800-922-5518
Roxan/Kitagawa	5425 Lockhurst Dr.	Woodland Hills, CA 91367	800-228-5775
Sensidyne/Gastec	16333 Bay Vista Drive	Clearwater, FL 34620	800-451-9444

Electronic Direct Reading Gas Meters in Table 6

Name	Address	City State Zip	Phone
Biosystems Inc	PO Box 158	Rockfall, CT 06481	203-344-1079
CEA Instruments	16 Chestnut St.	Emerson, NJ 07630	201-967-5660
Enmet	PO Box 979	Ann Arbor, MI 48106	313-761-1270
Industrial Scientific	1001 Oakdale Rd.	Oakdale PA 15071-1500	800-338-3287
MSA Instrument Div.	Box 427	Pittsburgh PA 15230	800-MSA-2222
National Draeger Inc.	Box 120	Pittsburgh, PA 15230	800-922-5518
Neotronics of NA	PO Box 2100	FloweryBranch, GA 30542-2100	800-535-0606
Sensidyne	16333 Bay Vista Drive	Clearwater, FL 34620	800-451-9444

Electronic Direct Reading Vapor Meters

Name	Address	City State Zip	Phone	Comment
PE Photovac	25-B Jefryn Blvd W	Deer Park, NY 11729	516-254-4199	generic ppm
Foxboro	PO Box 500	EastBridgewater, MA 02333	800-321-0322	IR and GC
MSA Instrument Div.	Box 427	Pittsburgh PA 15230	800-MSA-2222	generic ppm
Bacharach	625 Alpha Dr.	Pittsburgh, PA 15238	800-736-4666	generic LEL

Aerosol Direct Reading Equipment applicable to agricultural settings.

Name	Address	City State Zip	Phone
MIE Inc.	1 Federal St. #2	Billerica, MA 01821-3500	508-663-7900

Generic Air Sample Collection Equipment

Name	Address	City State Zip	Phone
Air sampling pumps (for aerosols and gases/vapors)			
MSA Instrument Div.	Box 427	Pittsburgh PA 15230	800-MSA-2222
Sensidyne/Gilian	16333 Bay Vista Drive	Clearwater, FL 34620	800-451-9444
SKC INC.	334 Valley View Rd	Eighty Four, PA 15330	800-SKC-84PA
Spectrex	3580 Haven Av.	Redwood City, CA 94063	800-822-3940
Microbial Air Sampler (for microbial aerosols only)			
Graseby	500 Technology Ct.	Smyrna, GA 30082	800-241-6898

only one end. While diffusion tubes are convenient, they should *not* be used to evaluate potentially IDLH concentrations, since a diffusion tube takes several hours to react. Most of the vendors in Table 26.5 manufacture both detector and diffusion tubes.

Components of the Diffusion Tube sampling apparatus include:

- One or more TUBES containing a color-indicating material that reacts in the presence of a selected gas or vapor.
- (Optionally) a plastic TUBE HOLDER that clips to a variety of surfaces such as structures in the building (when taking an area sample) or a person's collar (when measuring personal exposure).

To take a Diffusion Tube measurement:

Step 1. Break open the tube at the score mark. Diffusion tubes open only at one end, indicated by a score mark or red dot. If the tube has a red dot, point the dot away from you while pulling the ends toward you.
Step 2. Insert the tube in the holder so the opened end is exposed to the air to be tested. If a holder is not available, use tape to secure the tube to a surface, making sure the open end of the tube is not covered.
Step 3. To collect an area sample, clip the holder to something in the room, usually at a height about 4.5 to 6 feet above the floor. To collect a personal sample, attach the holder to the person's collar or pocket lapel.
Step 4. Write down the time of day that you start the sample on a data sheet (see, for example Table 26.3).
Step 5. Leave the tube exposed to the air to be tested, usually for 3 to 10 hours.
Step 6. When you are ready to take a reading:
  a. Write down the time of day that you end the sample.
  b. Read and record the strain length as you would for the detector tube.
  c. Determine the number of hours the tube was exposed.
Step 7. Calculate the average gas concentration. A formula provided by the manufacturer is used to convert the stain length reading in ppm × hours to the average gas concentration (in parts per million, ppm, or percent by volume) over the time period that it was monitoring.

$$\text{Concentration} = \frac{\text{stain length}}{\text{hours exposed}} \quad (26.2)$$

Example:
A diffusion tube was broken open and hung at 7:00 a.m. It was removed at 2:00 p.m. The stain length was read as 80 ppm x hours.

$$\text{Concentration} = \frac{80 \text{ ppm x hours}}{7 \text{ hours}} = 11 \text{ ppm}$$

Startup costs for diffusion tube evaluations are reduced because no pump is required. Tube holders are about $8, much less than pull tube pumps. Diffusion tubes seem to range between $4 to $5 each, about the same as the most expensive pull tubes.

Electronic Direct Reading Measurements. Hand-held electronic direct reading instruments are available to evaluate each gas in Table 26.1. All of the vendors listed in Table 26.5 make a hand-held, battery-powered meter that responds to one or more of the gases of agricultural interest. Electronic meters continuously indicate the current concentration (minus a few seconds response lag time). Many meters have a warning tone that is set at or near the STEL, which may or may not be useful for agricultural settings. Most of the vendors' meters listed in Table 26.5 have a data logging capability, which means they store a record of both the short-term values (which could be compared with a TLV-STEL) and the long-term average (which can be compared with a TLV-TWA).

In comparison to colorimetric tubes, electronic meters respond more rapidly, are probably more accurate when new (although both can cross-react to other gases in these environments), and are more cost effective if used frequently. Thus, nitrogen dioxide and hydrogen sulfide meters are particularly useful because of the potential in farm settings for these gas concentrations to change rapidly either with distance or time. An ammonia meter would be similarly useful to assess fluctuating exposures during anhydrous ammonia applications but would not be advantageous in more stable livestock confinement buildings unless it were tied into an automated ventilation control system. Electronic gas meters with digital readout capability cost at least $450 to $800, depending upon the chemical and manufacturer; some of these can have data logging capability or have sensors that are interchangeable for various gases. Meters with only an alarm can cost only $300. Disadvantages of meters include their high initial cost relative to tubes, a limited sensor life of one to two years, and a requirement for battery power.

In comparison to the often acute hazards characteristic of gases and vapors, aerosols typically represent chronic respiratory hazards. Dust is the major cause of respiratory

Table 26.6. Anticipated health hazards from agricultural gases studied in other settings; the sequence of gases matches Table 26.1.

NH3

Odor detection threshold (pungent smell)	1-5 ppm
Irritating to eyes, nose, and trachea (wet body parts)	10-15 ppm
Adsorbed onto an aerosol (contributor to bronchitis and asthma)	unknown
TLV-TWA = recommended exposure limit	25 ppm
TLV-STEL = recommended exposure limit	35 ppm
PEL = 8-hour permissible exposure limit (OSHA)	50 ppm
IDLH (damage to the upper respiratory tract)	500 ppm
Lethal	~10,000 ppm

CO2

No odor detection threshold	none
TLV-TWA = recommended exposure limit = PEL	5,000 ppm
TLV-STEL = recommended 15-minute exposure limit	30,000 ppm
IDLH (drowsiness and headache)	50,000 ppm

CO

No odor detection threshold	none
TLV-TWA = recommended exposure limit	25 ppm
PEL = 8-hour permissible exposure limit (OSHA)	50 ppm
Headaches and nausea	8-hours at 100 ppm or 3-hours at 300 ppm
Induces spontaneous abortions in swine	100-150 ppm
Asphyxiation dependent upon duration of exposure	3 hrs at 500 ppm
IDLH	15 mins 1500 ppm

H2S

Odor detection threshold (rotten-egg smell)	.01-.1 ppm
Offensive odor	3-5 ppm
TLV-TWA = recommended exposure limit	10 ppm
TLV-STEL = recommended 15-minute exposure limit	15 ppm
PEL = 10 minute exposure limit (OSHA)	20 ppm
PEL = **ceiling** exposure limit (OSHA)	50 ppm
Olfactory paralysis (cannot be smelled)	50-100 ppm
Bronchitis (dry cough)	100-150 ppm
IDLH (pneumonitis and pulmonary edema)	300 ppm
Rapid respiratory arrest (death in 1-3 breaths)	1000-2000 ppm

NO2

Odor detection threshold	~0.1 ppm
TLV-TWA = recommended exposure limit	3 ppm
PEL = ceiling exposure limit (OSHA) = TLV-STEL	5 ppm
Increased airway and pulmonary resistance	5 ppm
IDLH (coughing, headache)	50 ppm
Nausea, vomiting, delayed pulmonary edema	100-500 ppm
Rapid pulmonary damage (death)	500-1500 ppm

O2 deficiency (from CH4 or N2 as simple asphyxiants at sea level)

Normal O_2 in clean air	21.6 %
O_2 at the LEL for CH_4 = 5.3%	19.8 %
OSHA limit requiring ventilation 1910.94(d)(9)	19.5 %
OSHA requires atmosphere supplying respirators 1910.134(g)(5)	16.0 %
headache or nausea	11 %
euphoria and loss of coordination	8 %
unconsciousness within a minute	4 %

Table 26.7. Anticipated health hazards from agricultural vapors studied in other settings.

Chlorine
- Odor detection threshold (irritant) — 0.1 ppm
- TLV-TWA = recommended exposure limit — 0.5 ppm
- TLV-STEL = recommended 15-minute exposure limit — 1 ppm
- PEL = **ceiling** exposure limit (OSHA) — 1 ppm
- IDLH — 25 ppm

Cresol
- Odor detection threshold (sweet or tarry smell) — .001 ppm
- TLV-TWA = recommended exposure limit = PEL — 5 ppm
- IDLH — 250 ppm

Gasoline and diesel fuel
- Odor detection threshold for diesel — ~1 ppm
- Odor detection threshold for gasoline — ~100 ppm
- TLV-TWA = recommended exposure limit — 300 ppm
- TLV-STEL = recommended 15-minute exposure limit — 500 ppm
- IDLH (headache, dizziness, coughing vomiting) — ~4000 ppm

Formaldehyde
- Odor detection threshold (pungent smell) — 0.1-1 ppm
- TLV-C = recommended ceiling exposure limit — 0.3 ppm
- PEL = 8-hour permissible exposure limit (OSHA) — 0.75 ppm
- PEL = 15-minute exposure limit (OSHA) — 2 ppm
- Irritating to eyes, nose, and trachea (wet body parts) — 2 ppm
- IDLH (damage to the upper respiratory tract) — 30 ppm
- Lethal — ~500 ppm

Insecticides
- (see Vapor Hazard Ratio table)

Phosphine
- Odor detection threshold (fishy smell) — 0.15 ppm
- TLV-TWA = recommended exposure limit = PEL (OSHA) — 0.3 ppm
- TLV-STEL = recommended 15-minute exposure limit — 1 ppm
- IDLH — 200 ppm

symptoms among pork producers, contributing to cough, phlegm, allergies, and permanent lung damage with long term exposure. First among the many limitations of direct reading aerosol meters on the market is their high cost. Other limitations include an indication only of mass or particle count concentration rather than differentiating between inorganic and organic dusts, among particulate components of organic dust, or the chemical toxins on some organic particles. And the readings that aerosol meters provide will differ as a function of the physical nature of the aerosol (its size and shape). We have found a small hand-held passive meter called the Mini-RAM (from MEI, Bellerica MA) to be a useful screening device, but, at a current cost of $3200, this instrument is only cost effective if many evaluations will be made of similar settings, such as screening enclosed livestock production buildings or grain handling facilities. For research purposes, we still rely primarily upon more traditional aerosol sampling pumps and filters.

SAMPLE COLLECTION AND ANALYSIS METHODS

Sample collection with laboratory analysis is often the preferred method of industrial hygiene sampling primarily because of the accuracy gained by avoiding cross-reactions with other similar or sometimes even dissimilar materials and the improved sensitivity, compared with direct reading methods.

Gas and Vapor Sampling. In comparison to aerosol sampling, an active sample for gases or vapors requires a lower flow sampling pump (usually a low flow pump capable of 10 to 500 mL/min). The flow rate of this pump must be calibrated by a manner analogous to that described below for aerosol sampling. Gas and vapor samples are typically collected on a solid sorbent rather than a filter. Common solid sorbents include charcoal for nonpolar organics, silica gel for polar organics, and various molecular sieves (Ness 1991; NIOSH 1994). Air is then drawn through the gas or vapor collector for a limited time period. The minimum sample collection time is determined by the laboratory method detection limit and the anticipated environmental concentration. The maximum time to collect a sample is also limited by the sample-holding capacity of the collector before it saturates and lets part of the contaminant pass on through with the air. Again instructions for an individual method, will be specific to the agent and method as suggested by such sources as the NIOSH Manual of Analytical Methods (NIOSH 1994).

Aerosol Sampling. In a gross sense, aerosol sampling is much like gas and vapor sampling; however, because particles of different sizes behave differently in the air and deposit at different locations within the respiratory track, several variations in aerosol sampling have been developed. The most common differences are between *respirable* aerosols and *inhalable* aerosols (the latter used to be called the *total* aerosol). An intermediate option, called a *thoracic* aerosol, is more specialized (ACGIH 1995b). Sampling for *respirable aerosols* removes the large particles that would be deposited in the upper and conducting portions of the respiratory system and only collects that which would reach the alveoli. The actual removal is a curve with 100% removal for particles 10 μm in diameter and 50% removal for 4 μm diameter particles (ACGIH 1995b). Since many respiratory effects from organic dust are believed to be on the airways, it is probably better (and certainly easier) to collect total rather than respirable aerosol samples of organic dusts.

Components of an aerosol sampling apparatus include the following equipment:

- A filter in a cassette filter holder. The 2-piece cassette is sealed in the laboratory and contains a support pad behind a filter.
- Tubing (to connect the cassette and the pump).
- An air sampling pump (see Table 26.5).
- If a respirable sample is desired, a cyclone pre-selector will be needed to remove the large particles and pass only those capable of reaching the alveoli, as described earlier.
- A pump battery charger (although not used at the field sampling site).

The sampling pump draws air through the cassette and tubing at a known rate that must be calibrated. As air passes through the cassette, dust accumulates on the filter inside, which is either weighed or analyzed by another means.

NIOSH standard methods for total dust (method 0500) and respirable dust (method 0600) can be employed to collect either area samples or personal breathing zone samples (NIOSH 1994). Total airborne dust may be collected on pre-weighed non-hygroscopic filters, in 37 mm plastic cassettes, using personal sampling pumps at flow rates of 1 to 2 liters per minute. Acceptable filters include PVC, Teflon, and polycarbonate (Ness 1991; NIOSH 1994). Neither cellulose acetate filters nor glass fiber filters are recommended if the filter will only be weighed; the former tends to absorb moisture and gain weight under humid conditions and the latter tends to loose friable glass fibers and weight during handling. On the other hand, both filter materials work well for selected chemical analyses (NIOSH 1994). Open faced filter cassettes have been found to under-sample in these environments (Mulhausen et al. 1987). It is recommended that sampling be conducted with a closed face filter cassette (air enters through the hole in the top of the cassette).

Respirable dust samples are collected in a similar fashion, except that the cassette is fitted onto a cyclone (NIOSH 1994). Two basic types of cyclones are available. The nylon device based on the ACGIH criteria for respirable dust has a collection efficiency of 50% for particles with an aerodynamic diameter of 4 microns, and must be calibrated at 1.7 liters per minute. The other is an aluminum device based on the criteria established by the British Medical Research Council, with 50% collection efficiency at 5.0 μm aerodynamic diameter particle size, and must be calibrated at a flow rate of 1.9 liters per minute (Ness 1991). Care must be taken to ensure that battery-powered pumps do not lose energy and decrease flow rate during the sampling period.

The IOM inhalable dust sampler is generally similar to a total aerosol sample but collected with a newer device that has a defined collection efficiency beginning at 50% for particles with an aerodynamic diameter of 100 microns, and increasing progressively for smaller particles (Ness 1991). IOM samplers should be calibrated at a flow rate of 2 liters per minute, and can be used for both area and personal samples. The IOM has a removable filter holder that is weighed along with the filter, and we have found that care must be exercised during handling and weighing to avoid losing the sample. While it has been suggested that the IOM may have more significance for human health than total dust samples, this has not been demonstrated, and there are currently no standards or recommended guidelines for comparison to IOM samples.

To take a total aerosol dust sample:

Step 1. Attach one end of the tubing to the air inlet of the pump.

Step 2. Attach the free end of the tubing to the open "back" side of a cassette (usually engraved with a bulls-eye pattern and space for a sample I.D. number). Make sure that the cassette is oriented so dust will collect on the front of the filter instead of the back of the support pad.

Step 3. Hang, clip, or tape the pump at the desired sampling location. Align the tubing so there are no kinks and the cassette does not face up.

a. For personal air samples, the pump is hung from the operator's belt (or an adjustable belt provided). Attach the cassette to the person's collar or to the front of the shoulder.

b. For area samples, the pump may be attached to any locally convenient fixture including stanchions, bars, wires, the back end of a hog crate, etc. If using adhering tape, slide it under the metal clip on the back of the pump rather than wrapping it completely around the front of the pump. Indoors, the cassette should be located 4 to 6 feet above the floor, out of animals' reach and should *not* be located directly in line with a fresh air inlet.

Step 4. Remove the red or blue cap from the front of the cassette.

Step 5. Turn the pump on. Write down the time of day the sampling starts (see for example Table 26.3).

Step 6. Check flow rate.

a. Allow the pump to run for approximately a minute to stabilize. While holding the rotameter vertical, adjust the flow (usually with a screwdriver) so the rotameter ball is centered at the desired setting. (Or use an external rotameter placed momentarily on the front of the cassette.)

b. Depending upon the type of pump used and your experience with them, it is recommended to check the flow rate once or twice during the sampling period; adjust as needed. A sampling period of 4 to 8 hours is recommended.

Step 7. Before turning the pump off at the end of the sampling period, check the flow rate and record the position of rotameter float on the form provided (e.g., Table 26.3).

Step 8. Turn the pump off. Record the time of day (and the number of minutes displayed on the timer, if it is equipped with an elapsed time window).

Step 9. Carefully remove the cassette from the tubing so as not to dislodge the mound of dust on the filter (holding it essentially face up helps to prevent sample loss).

Step 10. Cap both sides of the cassette with the red or blue plastic caps provided. If possible, store the cassette face up (with the back side down) for shipment to a laboratory for analysis by weight and for other components.

To determine the mass of dust collected, the filter should have been pre-weighed in a laboratory before it was used and weighed again after sampling. Other laboratory analyses may also be conducted on the sample to determine the chemical or microbial content of the sample. In either case, the concentration is determined by dividing the amount of contaminant collected and determined by analysis by the volume of air passed through the filter, and is usually expressed as mg of contaminant per cubic meter of air (mg/m^3 or smaller units as appropriate).

Air Sampling Pump Operations. The air sampling pump is an important element of most aerosol and some gas and vapor sampling methods. Most air sampling pumps will have the following features:

- Air inlet—usually oriented vertically on the pump and found either on its top or side; the free end of tubing will be connected to this inlet.
- On/Off switch—found either on the top of the pump or behind a small plate on the front of the pump.
- Rotameter—a tapered, transparent tube with a ball (called a "float"); lines or marks on or behind the tube indicate the air flow rate through the pump and where the ball should "float" when the pump is adjusted properly. Rotameters must be calibrated periodically. Not all pumps have an internal rotameter.
- Flow adjuster—usually a screw found near the on/off switch; the flow adjuster regulates the volume rate of air through the pump.
- Recharging port—found either on the back or on the side of the pump below the air inlet; the battery charger plugs into this port.

Some pumps will also have the following features:

- Timer—found on the front side of some pumps; shows the number of minutes the pump has run. The time display usually remains after the pump is shut off, but automatically resets the next time the pump is turned on.
- Stroke counter—found on many low flow pumps; after calibration of volume per stroke, the stroke count can be used in parallel with sample time to determine the volume of air sampled.
- Flow fault light—illuminates to indicate the flow was interrupted or did not remain constant; the elapsed time to flow fault will be saved on the timer.
- Battery light or other charged battery or low voltage indicator.

Most air sample pumps use a rechargeable Ni-Cad battery. A full charge should be adequate for at least 8 hours of use. If present, the battery charge indicator may indicate the charge status of the battery or at least that it was fully charged at the start of the sampling interval. When the pump needs recharging, most rechargers have two modes. The "normal" or "high" mode is intended to recharge the pump fully in no more than 16 hours. A fully discharged battery will normally take 16 to 20 hours to fully recharge. If the battery will be left on the charger for more than 24 hours, it should be charged on the "trickle" or "low" mode to avoid overcharging, which can damage the battery and cause it to "take" less charge than it should. The battery should be charged in a clean, dry environment at room temperature.

Flow Rate Calibration. Whether a rotameter is integral to a pump, or a rotameter is placed momentarily onto the front of the sampler, all rotameters must be calibrated periodically in reference to a known standard. Typical reference standards are either a buret soap bubble meter (Figure 26.1) or its electronic equivalent (Ness 1991).

Operation of a buret soap bubble meter consists of applying suction (usually from an air sampling pump) to the tapered end of a laboratory buret, creating a soap film bubble across the open end, and allowing the bubble to sweep up through a volume marked on the side of the buret. A 1-liter buret works well for aerosol sampling; a 100 or 200 mL buret works well for gas-vapor sampling. The volume flow rate can be determined by measuring the interval required for the bubble to pass between any two scale markings on the buret. Manual timing for between 10 and 30 seconds is recommended, with the flow rate calculated from Equation 26.3. For example, the bubble would travel from the 0 mL to the 1000 mL marks in 30 seconds at a flow rate of 2 L/min. The flow rate should be verified at least three times at the desired flow rate to estimate the precision of the calibration and flow setting procedure.

$$\text{Flow Rate (Lpm)} = \frac{\text{buret volume swept (mL)}}{\text{bubble time (sec)} \times 1000/60} \quad (26.3)$$

For pumps with rotameters, mark or record the position of the float at the desired flow rate in a manner accessible to the field user. The position of the float is indicated at its *widest* point, i.e., the midpoint of a spherical float. The person conducting the sampling in the field will check the pump flow rate and adjust as necessary by positioning the rotameter float to that mark both before starting to sample (step #6a) and during sampling (step #6b), and will check and record its position at the end of sampling (step #7).

Biological and Medical Monitoring

Biological and medical monitoring complement air sampling to provide a more complete indication of exposure. A comprehensive occupational health and safety program integrates exposure assessment information obtained by environmental sampling with data from biological and medical monitoring to form a complete picture of the relationship between the work environment and measurable health effects.

Biological Monitoring. Biological monitoring provides a quantitative measure of the absorbed or effective dose experienced by individuals exposed to hazardous agents. This dose represents an integration of exposures from inhalation, ingestion, and skin absorption and can reflect cumulative exposures away from work as well as at work. Biological samples can include samples of blood, urine, and breath, although hair, nails, teeth, and other tissues may also be sampled. The biological half-life of the contaminant in the human body and the pharmacokinetics of metabolism are important factors in biological sampling. These parameters will determine what should be sampled for and when the sample should be collected relative to exposure. For example, a compound such as benzene that has a very short biological half-life (several hours) should be sampled either during or immediately after exposure. Other issues that arise with the use of biological sampling include informed consent, medical-legal ethics, and privacy of information.

Blood provides an indirect but proportional measure of the analyte or metabolite concentration reaching internal organs and is commonly used to measure exposure to metals such as lead, carbon monoxide, and organophosphate cholinesterase inhibitors. While blood sampling is invasive and sometimes not well accepted by individuals, it is often less variable than urine samples (e.g., for metals) and can be less subject to contamination. The appropriate use of

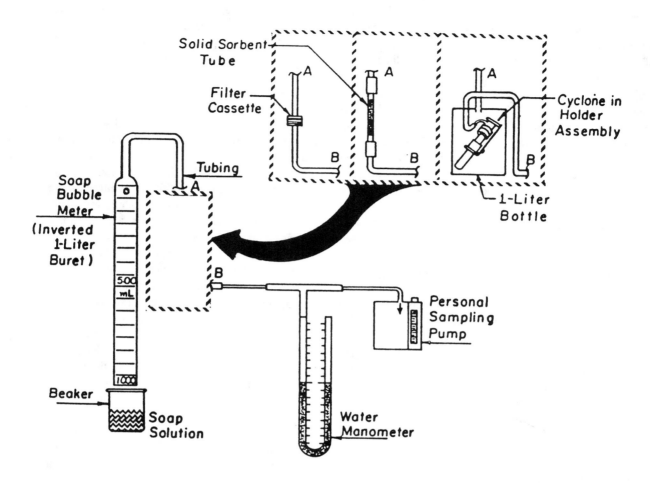

Figure 26.1. Calibration apparatus for either filter cassette, solid sorbent tube, or cyclone (NIOSH 1994).

preservatives such as heparin to prevent clotting and procedures to prevent potential exposure to bloodborne pathogens such as hepatitis or HIV are important issues to consider when collecting and shipping samples.

Urine samples may be collected for hydrophilic chemicals such as arsenic, heavy metals, and a variety of metabolites such as from parathion. Either a spot sample (midstream) or a 24-hour integrated sample may be collected. The balance of solids to water in urine is an important variable, and density, or better yet, creatinine content may be used to normalize the concentration of contaminants measured.

Exhaled breath can be collected using gas bags, sorbent materials, pipettes, or even respirator filters. Exhaled breath is less reproducible than blood or urine samples but may be better accepted since it is not invasive. It is most commonly used to measure volatile organics or a gas such as carbon monoxide, which are excreted through the lungs. End exhaled air (cf. total or mixed exhaled air) provides the best approximation of the actual concentrations in alveoli. The American Conference of Governmental Industrial Hygienists publishes a list of Biological Exposure Indices (BEIs) in their annual TLV booklet (ACGIH 1995b). These BEIs are intended for use in occupational settings and are based on the body burdens expected to be observed if workers are exposed at the TLV. For each of 39 compounds listed as of 1995 (only four of which have likely relevance to agriculture), the BEI specifies the analyte to be measured, the type of biological sample to be collected, the timing of when to collect the sample, and the recommended concentration. The NIOSH Manual of Analytical Methods also specifies a small number of biological sampling methods. The NIOSH Manual also presents data on the method's accuracy, range, limit of detection, interferences, and alternative methods (NIOSH 1994).

Medical Monitoring. Besides supporting biological monitoring, medical evaluations provide a means of monitoring health status and evaluating measurable clinical effects of exposure and uptake. Procedures such as pulmonary function testing or audiometry provide quantitative measures of physiologic function, and can be used with clinical tests for blood constituents, enzymes, and metabolic products indicative of liver, kidney, and other organ status. Baseline medical evaluations should be performed before individuals begin working in a new environment to provide information on their initial health status and to identify any potential concerns that could be exacerbated by the anticipated exposures. For example, an individual with asthma would not be a good candidate for work in an enclosed livestock environment with potential exposures to organic dust and ammonia. It is noteworthy that with the development of genetic screening tools, ethical issues have been raised concerning the identification of potentially sensitive individuals that might exclude them from the workplace or insurance coverage. A balance needs to be struck between identifying sensitive individuals and making reasonable accommodation to allow them to work in their chosen field. Periodic medical evaluations should be conducted to identify any changes compared with their baseline before they become significant health problems. The frequency of these evaluations may depend upon the specific exposures experienced and can vary from several months to several years. At the conclusion of employment, an exit medical examination should be conducted to provide documentation of final health status.

SETTINGS WITH GASEOUS RESPIRATORY HAZARDS

A wide range of gaseous respiratory hazards exists in agriculture: some in confined spaces, others in enclosed (semi-confined) spaces, and a few in open spaces. Many are not unique to agriculture; others are associated with commodities and/or processes unique to agriculture. The gases and vapors are listed in Table 26.1 in simple alphabetical order. This section will list settings in which they occur. Where these gaseous hazards occur in other industries, their health hazards are already well understood. What generally is not well understood are the factors controlling their generation on farms—factors that must yet be defined in order for farmers to anticipate and thereby avoid unusually hazardous conditions.

Anhydrous Ammonia Applications (NH_3)

Ammonia is a well known industrial hazard whose health effects are primarily upper respiratory and eye irritation (see Table 26.6). More serious respiratory health effects from ammonia are rare, probably because of its strong odor, allowing for early detection and maximizing the opportunity for avoidance where possible (Helmers et al. 1971; Holness et al. 1989; Swotinsky and Chase 1990). There are two primary sources of ammonia on farms: from animal urine and when it is used as a fertilizer.

Anhydrous ammonia is the most heavily used fertilizer in production agriculture. It is stored and sold in liquid form under pressure by farm supply firms scattered in agricultural areas. It is transported to the field in small portable nurse tanks, and applied by either the farmer or an employee of the farm supply firm. Because of this use pattern, farmers, farm supply employees, and local residents are all at risk of exposure. Anhydrous ammonia poses an additional hazard because it is highly hygroscopic, highly caustic, and extremely cold (-28°F under pressure). When this material contacts the skin, it will desiccate, penetrate, and freeze tissue. Anhydrous ammonia is particularly hazardous to the eyes, because almost any eye contact with this chemical can result in permanent blindness (Helmers et al. 1971). While the above acute effects typify injuries from accidental anhydrous spills, no data has been located characterizing routine airborne ammonia concentrations surrounding anhydrous ammonia applicators.

Ammonia is amenable to all categories of evaluation methods ranging from electronic meters for peak pulses, to detector tubes for intervals of several minutes, to data logging meters, diffusion tubes, and sample collection methods (NIOSH 1994). Peak concentrations of ammonia around applicators can be measured using direct-reading devices with a specific ammonia detector, or with direct-reading colorimetric detector tubes operated with a hand pump. These devices will provide immediate information that can be compared with the STEL and used to make decisions concerning the immediacy of the hazard. Direct-reading ammonia detectors have also been developed for personal monitoring and can be used to evaluate both acute and longer term exposure over the duration of the application process. Data logging capabilities of direct-reading ammonia detectors allow the evaluation of fluctuations in concentration over time, as well as the determination of the overall TWA exposure. While high levels of dust and other factors may interfere with the operation of direct reading sensors, this should not be a particular problem in outdoor applications of anhydrous ammonia. Colorimetric dosimeters, which are smaller and more convenient for personal sampling will not provide useful information concerning short term exposures unless ammonia concentrations are extremely high. Dosimeters are useful in determining the integrated TWA exposure over periods of time up to 8 or 10 hours (Manninen 1988).

Freshly Filled Silo Gas (NO₂)

A combination of carbon dioxide (CO_2), nitric oxide (NO), and nitrogen dioxide (NO_2) produced from fresh silage by the breakdown of carbohydrates and organic nitrates is commonly called "silo gas" (Hayhurst and Scott 1914; Ramirez and Dowell 1971; Scott and Hunt 1973; Horvath et al. 1978).

While the concentrations of each of these gases can be above their lethal level inside the silo, nitrogen dioxide (NO_2) is of greatest concern (Wang and Burris 1960). NO_2 is very damaging to the alveoli. Fatalities can be nearly instantaneous, but most reported hospitalizations suggest a pattern of respiratory distress characterized by the buildup of lung fluids delayed after exposure by several hours (Ramirez and Dowell 1971; Scott and Hunt 1973; Horvath et al. 1978). Concentrations of NO_2 as high as 4% have been measured 2 days after filling silos (Wang and Burris 1960), which is 800x its IDLH in comparison to 15x the IDLH for CO_2 (NIOSH 1990a). After harvest the process of NO_2 production in silos takes several hours to begin, peaks in 2 to 5 days, but can last for at least 2 weeks. It is less concentrated at the base of the silo chute but still potentially capable of causing chemical injury (Ramirez and Dowell 1971; Scott and Hunt 1973; Horvath et al. 1978). Nitrogen dioxide is also one of several toxic gases produced at lower concentrations by diesel engines and arc welders.

Because of the rapid respiratory damage that NO_2 can cause and the potential for high concentrations to be generated in silos, direct reading methods are required. Colorimetric detector tubes are adequate to evaluate conditions at the base of the chute. Electronic detectors are better suited for early entry into a silo, even after ventilation, because of the potential for rapid changes and the need for multiple measurements. To safely enter a silo, readings should be monitored beginning at the base, and continuing up the chute. Measurements should be taken before a door is fully opened; if NO_2 concentrations have built up within the silo, its high molecular weight will allow it virtually to pour down the quiescent air inside the chute. Odor detection for NO_2 is possible but may not be an adequate warning in a rapidly changing environment.

Gases Around Livestock

Carbon dioxide is a direct product of respiration from livestock and metabolizing microbes (molds and bacteria) and a product of combustion as from propane-fueled space heaters, wood stoves, or engines. While virtually undetectable by odor (see Table 26.6), carbon dioxide is not inert; it does affect the respiratory center of the brain, eventually causing asphyxiation. The concentration of CO_2 in swine barns exceeds the TLV less frequently than does ammonia. TWA measurements of CO_2 concentration can be viewed as an indicator of adequate general ventilation in relation to the numbers of animals within a building; diffusion detector tubes are quite adequate for this purpose. Data logging electronic instruments are available if one wanted to know the magnitude in diurnal variations in CO_2. However, in cases of ventilation failure in these buildings, CO_2 will build up and is likely to exceed the fatal concentration for livestock in a matter of a few hours. Grab sample measurements should always be taken (along with O_2 measurements) before entering a building where ventilation failure is suspected; the need for a rapid response dictates that pull-tubes or an electronic direct reading instrument be used in this case.

Ammonia also emanates from the breakdown of urea in animal urine, which creates a problem particularly when the livestock are raised indoors—a common and cost-effective practice in much of the U.S. (Carlile 1984; Donham and Popendorf 1985). Prior studies have shown that ammonia concentrations in poultry barns during fall and winter frequently exceed the 25 ppm exposure limit and sometimes even exceed the 35 ppm TLV-STEL (Popendorf et al. 1985; Donham and Popendorf 1985; Mulhausen et al. 1987; Manninen et al. 1989; Reynolds et al. 1994). When enclosed for the winter, the ammonia concentration in swine barns also exceeds the 25 ppm TLV-TWA most of the time (Donham and Popendorf 1985). While ammonia is severely irritating before it is toxic (Holness et al. 1989), there is still ample reason to assess TWA ammonia exposures in most livestock buildings.

Carbon dioxide, ammonia, and other gases inside enclosed livestock facilities usually do not pose an acute hazard. Therefore, exposure assessment should focus on evaluating TWA concentrations over the workshift. While direct reading instruments that can measure carbon dioxide and ammonia are available, we have observed problems with these sensors in enclosed livestock environments due primarily to their high dust and humidity levels. Nonetheless, some computer-based livestock building monitoring and ventilation control systems initially installed to respond to temperature, have been modified to include detectors for carbon dioxide or ammonia at fixed locations.

Colorimetric passive dosimeters have been used with success to measure both carbon dioxide and ammonia concentrations for extended time periods in enclosed livestock environments. A pump is not required, and either area sampling or personal sampling can be performed. The major limitations of dosimeters are their lack of accuracy (up to ±35%) and the potential for saturation if the combination of gas concentration and exposure time exceeds the dosimeter's limit. Colorimetric detector tubes can be useful in situations where particular tasks, such as the tilling of bedding in turkey barns or the "loadout" of poultry can generate short term, high concentrations of ammonia (Reynolds et al. 1994).

Midget impingers containing a liquid absorbing solution (dilute sulfuric acid) can also be used to sample ammonia in these environments (Ness 1991; NIOSH 1994). This method provides greater sensitivity and accuracy than dosimeters. It does require a pump (either battery-powered or from an AC source), which should be calibrated at a flow rate of no more than 1 Lpm to avoid loss of the absorbing solution. The sample is usually analyzed at a laboratory using an ion-specific electrode, spectrophotometry, or ion chromatography (NIOSH 1994; Manninen 1988). Midget impingers (even micro-impingers) are not suitable for personal sampling, since they are subject to spilling. Impingers are susceptible to freezing in cold weather.

Manure Pit Gases (H_2S)

Hydrogen sulfide (H_2S), sulfur-containing mercaptans, and organic acids of many sorts have been identified among the odorous gases emanating from manure typically stored and undergoing anaerobic decay in a pit, either under or adjacent to most hog and many dairy barns (Burnett 1969; Merkel, et al. 1969; Banwart and Brenner 1975).

Under normal conditions, hydrogen sulfide contributes to odor but is well below levels of health concern in these buildings (Donham and Popendorf 1985; Donham et al. 1977). However, sometimes when manure is agitated before pump-out to be returned to the fields as fertilizer, sufficient H_2S is rapidly released to create the fatal levels shown in Table 26.6 (Morese and Woodbury 1981; Donham et al. 1982; Hagley and South 1983; CDC 1989). One of its insidious first effects is to numb the senses of the nose, making further detection impossible. Multiple deaths are common if someone enters the pit soon after agitation and others follow in a futile attempt at rescue.

Again no predictive data exists to anticipate or control this hazard. Air sampling during pump-out should be preplanned; the use of a remote sampling tube is recommended to keep the investigator out of danger until a safe zone or condition has been established. Detector tubes are a little slow but otherwise quite adequate to evaluate this hazard. Small electronic H_2S detectors are also available that will emit an audible alarm when the TLV concentration is reached. Detector tubes are much too slow to be of any value for this hazard.

Combustion Driven Sprayers, Washers, Heaters (CO)

Carbon monoxide (CO) is generated not only from space heaters but at even greater rates from all internal combustion engines (cars, trucks, tractors, electric generators, high pressure sprayers). CO is a chemical asphyxiant, preventing the blood from carrying oxygen to the brain and the rest of the body. CO is also virtually undetectable by odor (see Table 26.6). As a chronic hazard at lower concentrations it can cause headache and miscarriage among pregnant livestock or humans. Higher concentrations will cause unconsciousness and death in a predictable pattern related to time of exposure. Measurements of CO are warranted only if a combustion source is present.

Either integrated or grab sampling could be appropriate, depending upon the nature of the concern, based either upon observed symptoms or upon suspected sources. Measurements using long term diffusion tubes could be evaluated against the TLV-TWA for CO if low level symptoms consistent with headaches, nausea, or spontaneous abortion of livestock are observed or as a periodic check of emissions from regularly used heaters. The more rapidly responding detector tubes could also be used in these cases (and should be used if more acute symptoms are the concern), but they will only reflect the conditions at that moment. Concentrations would increase if the source emission rate were to increase or the ventilation rate were to decrease.

The greater concern about carbon monoxide is usually acute poisoning from high concentrations. A variety of passive colorimetric and electronic carbon monoxide dosimeters have been developed and are readily available to consumers. They can be used as either area or personal samplers. Electronic instruments are usually capable of emitting a warning signal when carbon monoxide concentrations exceed exposure guidelines. Because all electronic direct reading chemical sensors have a limited shelf life, periodic calibration with a standard carbon monoxide gas is critical to assure its continued performance. Data logging capabilities are also available for several of these instruments.

Confined Space (O_2 Deficiency)

Oxygen can be depleted anytime a biological process (animal or microbial) is consuming oxygen from within an enclosure with virtually no air exchange with the outside atmosphere. The most obvious example would be the hazard of asphyxiation from oxygen depletion in an airtight silo on a dairy farm. The normal concentration of oxygen is 21%. A concentration less than 19% should be ventilated. Below 16%, OSHA recommends that a supplied air respirator be used. While oxygen depletion is a potential problem in any confined space, on the farm the same processes that consume oxygen will generally create a greater hazard by also releasing one or more of the toxic gases noted above. Even the generation of toxicologically inert methane gas from the anaerobic decay of manure creates a danger from combustion before it displaces sufficient air to cause asphyxiation or even a noticeable shortness of breath. Since all conditions on farms conducive and preclusive to oxygen depletion have not been fully explored, and prior experience is not a reliable predictor of all future conditions, it is recommended that a test for oxygen content precede any other test for toxic gases in a confined or enclosed space where biologic activity (either livestock or microbes) could have been present. Detector tubes and electronic meters are equally effective. It is important to note that oxidizing gases such as chlorine or ozone will be detected by oxygen sensors and can result in what looks like a normal oxygen atmosphere, when in reality oxygen is deficient.

Commercial instruments called "confined space monitors" are now available with two to four sensors in one device in combinations such as oxygen, combustible gas, carbon monoxide, and/or hydrogen sulfide. Confined space monitors normally include alarms that will warn when parameters approach dangerous levels. Before entry of the confined space, remote monitoring can be performed by use of an extension tube or sometimes by lowering or placing the confined space monitor into the area, then retrieving and reading it. The sensors should be calibrated periodically using standard gas mixtures. Methane or hexane is normally used to calibrate the combustible sensor, depending upon the environment.

The combustible gas detector provides a reading in percent of the Lower Explosive Limit (LEL) or lower flammable limit of the gas mixture. Detection of combustible gases at 10 to 20% of the LEL is usually cause for concern, especially since concentrations can increase quickly. If oxygen levels are low, the combustible sensor will not act properly and concentrations of combustible gases will be underestimated. Another problem with combustible gas sensors is that they can become saturated at high concentrations and stop responding. If the person sampling cannot observe the readout rise to its peak and drop, there is a danger of interpreting a zero as a safe condition. Most instruments now include data logging and a peak hold capability.

Concentrations of gases in a confined space may vary drastically, depending on location, due to stratification and lack of air movement. For example oxygen levels may be 20% at the top entrance, but only 16% 6 feet down the ladder. Before entry into a confined space, measurements should be taken at several elevations and distances from the entrance. Sampling should continue for the duration of the entry. While monitoring is an important component of confined space entry, entry should never be performed without appropriate ventilation, respiratory protection, other personal protection devices, and a backup person capable of either performing a rescue or calling for assistance.

Other Potential Gases and Vapors

In comparison to the above gases, a small number of agricultural chemicals are used which can present serious vapor hazards, as listed in Table 26.7. On the other hand, the volatility of the chemicals near the bottom of this table is so low, they will be mentioned later when talking about agricultural chemical aerosols.

A variety of disinfectants are used in livestock operations, especially dairy farms and many large hog buildings. Disinfectants are usually applied as a high pressure spray of active biocides including chlorine and quaternary ammonia compounds, organic iodines, cresol-based compounds (cresylic acid), and formaldehyde emitters. While the spray is a dilute aqueous aerosol that is not considered a significant hazard (and is covered by EPA's work practice standard as described on each label), the vapors generated by mixing some of these cresol or formaldehyde emitting products (particularly with bleach) can be quite irritating. If measurements are to be taken of these gases, it is important to record *all* the compounds included within the treatment solution. Detector tubes (pull tubes) are available for formaldehyde and chlorine gas. Electronic meters are also available for these compounds but are not likely to find frequent usage on individual farms.

Fumigants (highly volatile insecticides) such as phosphine (usually aluminum phosphide or Phostoxin®) and a decreasing range of volatile organics (like carbon disulfide or ethylene dichloride) are used in produce storage areas. Fumigants usually present a clear airborne hazard that must be avoided according to the manufacturer's label instructions. Evaluation of some fumigants can be accomplished by detector tubes; others require either more expensive electronic direct reading vapor meters (like those listed in Table 26.5) or sample collection and analysis.

The remaining majority of pesticides have very low vapor pressures; that is, air concentrations of most insecticides and herbicides can barely if ever reach their exposure limits as a vapor, even in an enclosed environment.

Table 26.8. Guidelines mostly developed from aerosols studied in other settings.

	TOTAL mg/m³	RESPIRABLE mg/m³
Quartz (crystalline free silica) TLV	n.a	0.1
Cotton dust TLV	0.2 a	
Swine dust recommended exposure limit b	2.4	n.a
Grain dust TLV	4	n.a.
Nuisance dust TLV (PNOC)	10	5

footnote a) cotton dust as sampled by a vertical elutriator which collects particles in a diameter range between that of a total and respirable sample.
b) Donham et al. 1989.

However, application spray can linger in indoor settings such as greenhouses at higher concentrations than when used in the field (Stamper et al. 1988; Liesivuori et al. 1988). All of these less volatile chemicals fall into the "Other Potential Organic Aerosols" category described later in this chapter.

AEROSOL RESPIRATORY HAZARDS

Inorganic Dusts

Soil and dust are ubiquitous in agriculture. Soil-derived dust is mostly mineral in nature and can contain many respiratory hazards. The major soil minerals are silicates, for which a TLV was set many years ago at 10 mg/m³ based on the nuisance of dust accumulation in the nose or mouth and limitations of the lung to remove large masses. However, there is some evidence that pneumoconiosis can develop among farm workers exposed to common soil silicates (Sherman et al. 1979). Moreover, chronic exposure to airborne quartz is well known to produce a debilitating form of pneumoconiosis called "silicosis." Quartz exposure depends upon its concentration in the dust but is recommended not to exceed 0.1 mg/m³(ACGIH 1995b). Quartz in some agricultural soils ranges from 1 to 10% (Popendorf et al. 1982). And asbestos can also be a component of some western U.S. soils (Popendorf and Wenk 1983).

Inorganic soil-derived aerosols can be generated from both mechanized and manual farming operations as a function of activity and soil moisture (Popendorf et al. 1982; Popendorf and Wenk 1983). Field implements can generate dust plumes ranging from 100 to 200 mg/m³ (Casterton 1982). Exposures to drivers range from 10 to 20 mg/m³ in open tractors to less than 2.5 mg/m³ in an enclosed cab (Casterton 1982). When manual fruit harvesters in arid climates such as central California, disturb foliage on which dust has built up for several months, total aerosol concentrations of 15 to 30 mg/m³ can be generated (Popendorf et al. 1982). Such conditions exceed the health exposure guidelines in Table 26.8. Clearly, the potential for classical pneumoconiosis from long-term exposure to inorganic dust during both mechanized and manual agricultural operations in dry-to-arid climates is sufficient to justify aerosol exposure sample collection and monitoring using standard methods (Ness 1991; NIOSH 1994).

An additional category of inorganic aerosols are welding fumes. Welding and gas cutting will vaporize whatever metal is being used. This vapor will immediately condense into very small aerosol particles called "fumes." Of particular interest on farms is construction welding on galvanized steel. Fresh zinc fumes formed when welding galvanized steel can cause a short-lived form of fever and body soreness (muscle aches, headaches, and chills) called "zinc fume fever." (McCord 1960) This is normally only a problem when welding in poorly ventilated conditions. Farmers and field researchers should be aware of this potential and could sample for it using an appropriate filter to be analyzed in the laboratory for zinc (NIOSH 1994).

Organic Dusts

In moist climates, organic dust is a pervasive and increasingly recognized agricultural respiratory hazard (Petro et al. 1978; Donham et al. 1984; Jones et al. 1984; Rylander et al. 1986; Mulhausen et al. 1987; Holness et al.

Table 26.9. Major groups of field-use agricultural pesticides.

Major Group	Common commercial names
Insecticides	
Organophosphates	Counter, Parathion, Guthion, Lorsban, Rabon
Carbamates	Temik, Furidan, Lannate, Baygon, Sevin
Organochlorines	Thiodan, dieldrin, lindane, chlordane
Herbicides	
Phenoxy-aliphatic acids	2,4-D, Weed-B-Gone, 2,4,5-T, Trioxone
Bipyridyls	Paraquat, Diquat
Triazines	Atrazine, Bladex, AAtrex
Other/miscellaneous	
Thiocarbamates (fungicides)	Thiram, AAtack, Nabam, Maneb, Zineb
Arsenicals (herbicides)	Paris Green, cacodylic acid
Acetanilides (herbicides)	Alachlor, Lasso, Ramrod
Dicarboximides (fungicides)	Difolitan, Captan
Dinitrotoluidine (herbicides)	Amex, Prowl, Treflan

Table 26.10. Agricultural aerosol categorical constituents and examples.

Inorganic minerals	dirt; primarily silicate minerals.
Feed (grains)	corn, soy beans, oats, etc.
Animal dander	hair, skin, etc.
Insect parts	especially mites.
Fecal material	a source of endotoxin-containing gram negative bacteria.
Bacteria	esp. gram-negative bacteria such as enterobacter and pseudomonads.
Thermophylic actinomyces	spore forming bacteria.
Mold spores	e.g. Aspergillus, Penicillium, Rhizomucor.
Microbial toxins	e.g. endotoxin, aflatoxin, trichothecenes.

1987; Donham et al. 1989; Rylander and Peterson 1990; Cormier et al. 1991; Reynolds et al. 1993; Reynolds et al. 1994; Donham et al. 1994). Common agricultural respiratory diseases caused by organic dust aerosols such as atopic asthma, occupational asthma, and bronchitis are chronic in their nature and not specific to agriculture (Petro et al. 1978; Donham et al. 1984; Rylander et al. 1986). Agricultural organic dust aerosols may be from a specific crop such as corn or sugar cane, or a mixture of materials such as in livestock buildings. Examples of components in organic dust are listed in Table 26.10. In comparison to inorganic dirt, it is important to know that one or more components of organic dust are often biologically active, either by being toxic to lung tissue directly or by triggering people's immune system. Unfortunately, identifying which of the constituents within organic dusts are responsible for a given health condition is usually quite difficult. Differential methods of analysis for microbes are described later in this chapter.

A common source of organic dust is livestock raised in enclosed buildings. For instance, total dust levels in poultry buildings were found to range broadly from 1 to 10 mg/m^3 (prominantly higher in the winter) and up to 25 to 30 mg/m^3 during "loadout" (when the birds are gathered and taken to market) (Jones et al. 1984; Mulhausen et al. 1987; Reynolds et al. 1993; Reynolds et al. 1994). About 20% of this dust was in the respirable size range (<10 µm). Dust levels in hog buildings were similar to those in poultry, ranging from 2.5 to 15 mg/m^3 (Banwart and Brenner 1975; Donham et al. 1986; Holness et al. 1987). And organic dust is also a concern in dairy barns (Cormier et al. 1991). These concentrations commonly exceed recommended guidelines of 2.4 mg/m^3 for exposure to swine dust (Donham et al. 1989, 1994).

Respirable dust usually makes up only a small percentage of the total dust in these environments, and it may be necessary to sample for at least 6 to 8 hours to get a measurable respirable sample during the summer when dust levels are typically lowest. In contrast, during the winter, a full shift total dust sample may easily become overloaded. The use of sequential multiple filters (e.g., one in the morning and one in the afternoon) for the same person or area can be used to calculate a time weighted average exposure for the full shift, and will provide more specific information about exposures during the shorter time periods sampled. Total and respirable dust samples can be interpreted by comparison to recommended standards for these environments, as seen in Table 26.8.

The Mini-RAM direct reading aerosol meter has a data logging capability that can be used to provide a record of changes in concentrations over time. This can be especially useful in identifying specific tasks or situations that contribute to elevated dust concentrations. Other particle counting devices are available that can be used to evaluate particle size distributions. These devices do need to be calibrated periodically and zeroed before each use; unfortunately, most are not calibrated for the type of dust found in enclosed livestock environments and, therefore, comparison of results to exposure guidelines should be done with caution.

Microbes

Microbes comprise an important fraction of organic aerosols that present both acute and chronic risks. Examples of acute risks include settings with moldy produce like silo uncapping that may produce toxic responses discussed later, fungi in soil habituated by chickens or wildfowl that may produce histoplasmosis or blastomycosis (Donham 1985), and arid soils of the southwest that can contain arthrospores capable of producing coccidioidomycosis (Johnson 1981). Agricultural workers may be exposed to a variety of chronic risks from microorganisms including bacteria, fungi, viruses, or mites during their routine work, especially inside enclosed livestock buildings, during planting or harvesting, while chopping hay or silage, and during loading and unloading of grain storage. Exposures to bacteria, fungi, vegetable dusts, animal danders, and insects can create direct inflammation of lung tissues or allergic responses resulting in chronic bronchitis, asthma, hypersensitivity pneumonitis, and acute febrile illness. Bacteria and fungi also produce chemical components that can result in illness when inhaled or ingested.

Airborne microorganisms, called bioaerosols, are most commonly measured using microbial cascade impactors (Andersen Microbial Samplers - AMS) containing selective agar media, or All-Glass Impingers (AGI-30) (Ness 1991; Reynolds et al. 1994). The AMS provides data on the size distribution of microorganisms, but can be easily overwhelmed by the high concentrations found in agricultural environments. The AGI-30 method offers advantages in that serial dilutions can be made from the liquid sampling media, thus extending the upper limits of the method. Both methods require laboratory culturing of the sampling media followed by counting and identification of the microorganisms recovered. Both methods require an electrical power source, and therefore, are not suitable for personal sampling. Alternative methods are being developed that will measure total culturable plus nonculturable microorganisms by collection on filters followed by examination with fluorescent microscopes and DNA probes. These methods may also eventually be used for the evaluation of filters collected as personal exposures.

Zoonoses are infectious diseases common to animals and man. Pathogenic microorganisms may cause disease when inhaled, ingested, or by direct skin contact. Examples of infections associated with farming include anthrax—which can appear as severe hemorrhagic pneumonia (often fatal) or as a passing dermatitis; aspergillosis—a lung infection caused by inhalation of fungal spores from certain Aspergillus species; and dermatitis from Trichophyton species of fungi. At least 24 of the over 150 such diseases known worldwide are occupational hazards for agricultural workers in North America (Acha and Szyfres 1980; Donham 1985). Because the route of most zoonotic infections is thought not to be via aerosols, sampling for airborne zoonotic microbes is possible but rare (Donham 1975). The more common diagnostic or evaluation method is serological monitoring of animals or people (Acha and Szyfres 1980; Schnurrenberger and Hubbert 1981; Donham 1985).

Natural Toxins

Natural toxins are low vapor pressure organic molecules that can be hazardous only as aerosols or by ingestion. In fact, because natural toxins are produced by many bacteria and molds, their route of exposure is tied to the organic aerosols containing the microbes to which they are bound.

Perhaps the natural toxin of greatest agricultural interest is endotoxin from Gram-negative bacteria that are strongly linked to respiratory disease from cotton (Rylander 1987; Jacobs 1989), from livestock (Thelin et al. 1984; Rylander and Peterson 1990; Hagmar et al. 1990), and from moldy hay and grains (Donham et al. 1986; Rylander and Peterson 1990). Air samples for endotoxins may be collected on a variety of 25 mm or 37 mm filter media using sampling pumps at 1 to 2 Lpm. Glass fiber, polycarbonate, cellulose acetate, Teflon, and PVC have all been used, although glass fiber (without binder) may be most suitable for agricultural environments (Reynolds and Milton 1993; Reynolds et al. 1994). Samples are extracted from the filters in the laboratory with pyrogen-free water, and most commonly analyzed for endotoxin with a biochemical assay that uses an enzyme from horseshoe crabs (Limulus Amoebocyte Lysate) (Olenchock 1988; Reynolds and Milton 1993). Work continues on standardizing and validating these methods.

Toxins produced by molds are referred to generically as mycotoxins. Mycotoxins can cause a variety of toxic effects including cancer. Each mycotoxin is an organic chemical produced by one or more fungi. Examples include aflatoxin, vomitoxin, ochratoxin, zearalenone, and a family of trichothecenes (Rodricks et al. 1977). Nearly all known mycotoxins can be analyzed by either Gas Chromatography (GC) or High Pressure Liquid Chromatography (HPLC), perhaps enhanced by Mass Spectroscopy (MS). Airborne samples are usually collected on glass fiber filters using personal or area sampling pumps. Bulk samples of material suspected of harboring fungal contamination may also be collected to help identify the source and to permit analysis of constituents of interest that may otherwise be below the limit of detection in the air samples.

Prominent among the mycotoxins is aflatoxin, a highly carcinogenic chemical produced by the mold Aspergillus flavus commonly associated with corn, peanuts, and cottonseed. Although the most publicized route for human and animal exposure to aflatoxins is dietary, evidence from two epidemiologic studies of agricultural produce handlers support the possibility that aflatoxins in airborne dust particles may also contribute to elevated cancer rates to the respiratory and digestive systems (Hayes et al. 1984; Alavanja et al. 1987). Aflatoxin becomes airborne in grain dust. Farmers can be exposed to airborne grain dust throughout the year from harvest and grain storage (at perhaps 1 mg/m^3) to animal feeding in confined buildings (4 mg/m^3) and bin clean-out operations (40 mg/m^3). Calculations of annual airborne doses to Midwest livestock farmers extrapolated by multiplying typical airborne dust levels (Donham et al. 1986) by the concentration of aflatoxin found in grain dust on farms in studies by Sorenson et al (1981), Burg and Shotwell (1984), and Zennie (1984), show that one-third of their annual dose may come from a few hours spent cleaning out moldy corn from the bottom of a grain bin, while about two-thirds could come from chronic exposures while doing their daily chores. Aflatoxin appears to be so toxic that a dose capable of producing cancer could result from repeated low level exposures at concentrations that are usually below the detection limits of current analytical methods. Thus, further evaluations will be hampered until a method is available to detect low-levels of aflatoxin, normally collected using routine aerosol sampling. Meanwhile, the evidence is sufficiently strong to justify recommending the routine use of respiratory protection when working around moldy produce to prevent not only the acute symptoms of ODTS but also to diminish the chronic risks of cancer.

Grain Bin Cleanout and Silo Unloading

Compared with the repeated low level exposures of most agricultural aerosols described above, certain less common tasks can result in high levels of organic dust aerosol exposures that can create two ailments rather unique to agriculture: organic dust toxic syndrome (ODTS) and hypersensitivity pneumonitis (HP).

ODTS is an acute influenza-like illness with headache, muscle aches and pains, fever, and malaise. ODTS is believed to be a direct effect of exposure to very high concentrations of endotoxin and/or other toxic microbe-related components in agricultural dusts (Rylander et al. 1986; Rask-Andersen 1989; Rylander and Peterson 1990). The onset of these symptoms is delayed, following exposure, by 3 to 6 hours, and the duration of these symptoms usually last 24 to 72 hours.

The agricultural workers' version of hypersensitivity pneumonitis is called Farmers Lung (FL). Although its clinical symptoms are similar to ODTS, FL is thought to be a specific, delayed hypersensitivity reaction (sensitization) to certain thermophilic bacteria and fungal spores found, for instance, in moldy hay. The symptoms of Farmers Lung are generally more severe in the acute stage than ODTS and may lead to chronic lung scarring and interstitial fibrosis while ODTS does not (Arden-Jones 1982; Terho 1986). The major differential is a blood test for precipitins (indicators of the body's immune system), which are only present in FL, along with observable infiltrates in chest x-rays and possible restrictive pulmonary function.

The exposure threshold for either FL or ODTS is unknown. Thus, evaluation efforts are largely a research topic. Because aerosols that produce these symptoms are sufficiently concentrated to be highly visible dust clouds (Popendorf et al. 1985), the criterion for adopting personal respiratory protection is within the power of the farmer to discern. Respiratory protection is justified not only to prevent ODTS and FL, but also to avoid the chronic hazard from other mycotoxins discussed in the previous section.

Other Potential Organic Aerosols

Pesticides are often viewed as potential vapor hazards, but because of their low vapor pressure most present primarily an aerosol hazard (see Table 26.2). Most pesticides are formulated as solids (such as granules and wettable powders) or liquids (in an organic or water base); fumigants may be gases or highly volatile liquids. Pesticides can present a hazard to applicators, (Nigg and Stamper 1983; Devine et al. 1986; Popendorf 1988) to harvesters re-entering a sprayed field, (Popendorf and Leffingwell 1982; Nigg et al. 1984), and to rural residents via air (Maybank et al. 1978), water (Fairchild 1987), and even food (NRC 1987) contamination. Toxicologically, the major field-use pesticides can be broken down into six major chemical groups, as shown in Table 26.9. As a group, all of these chemicals have a very low vapor pressure; that is they are not very volatile (see those listed in Table 26.2). They are generally suspended in water, applied as large-diameter aerosols, and, even then, present a small respiratory hazard to applicators in comparison to their exposure hazard via the dermal route (Nigg and Stamper 1983; Devine et al. 1986; Popendorf 1988). And the primary route of their post-application hazard to harvesters continues to be dermal.

Except for paraquat, the primary effect of pesticides is not to the respiratory system. The most well known toxic effect of insecticides is the acute cholinesterase inhibition caused by organophosphates and carbamates. In one study from 1971 to 1973, acute pesticide poisoning accounted for 10% of all hospital admittances of farmers and agricultural workers in Colorado, Iowa, and South Carolina (Savage 1976). Pesticides as a group are also believed by some to represent a chronic health hazard, especially their potential association with cancer. While mortality studies have revealed that farmers have lower overall cancer rates than the general population, they have also revealed statistically significantly elevated specific cancers such as leukemia, non-Hodgkins lymphoma, multiple myeloma, and lip, prostate, and skin cancers. Among the reasons that no clear etiology for these cancers has been established are the many other suspect cancer causative agents including zoonotic viruses, mycotoxins, diesel exhaust, and ultraviolet radiation (sunlight) and the lack of accurate estimates of pesticide exposure (Schenker and McCurdy 1986).

The evaluation of potential air, skin, and food routes of pesticide exposure requires sample collection and costly analysis and is therefore not recommended herein. As previously stated, pesticide usage and resulting exposure is addressed by EPA work practice standards that do not *require* any evaluation by the user. For those wanting to measure exposure, the airborne methods combining the particulate and vapor sampling techniques previously described are less important than either biological or dermal monitoring (EPA 1987; Ness 1994). The most common dermal monitoring method involves the extraction and analysis of the chemical from either whole body clothing, or a gauze or other cellulose patch with an impervious backing attached to the skin at strategic locations (Popendorf and Leffingwell 1982; EPA 1987). Using whole body suits may be advantageous when exposure may not be spatially uniform, the patch technique provides better assurance of sample retention without contamination. All these techniques require prior method validation for sample recovery.

Popendorf and Leffingwell (1982) found a quantitative relationship between the dermal deposition of organophosphate cholinesterase inhibitors measured using the gauze dosimeter technique (derm mg/kg) and the percent change in red-blood cell acetylcholinesterase (% AChE), which may be useful in this context:

$$\% \text{ AChE} = 100\{1 - e^{(-7(\text{derm mg/kg})/\text{LD50})}\} \qquad (26.4)$$

Many farmers, especially those with livestock, are exposed to veterinary chemicals (Popendorf et al. 1985). They are grouped into two broad categories of "biologicals" and "antibiotics." Veterinary "biologicals" include vaccines, bacterins, and toxoids from living products to enhance the immunity of an animal to a specific infectious disease or diseases. While accidental injection is a broad hazard, the only biological presenting an airborne hazard is Newcastle disease vaccine, which is nebulized inside poultry buildings as a live product (Keeney and Hunter 1950; Popendorf et al. 1985). Antibiotics are products derived (or synthesized) from living organisms, mainly mold species of the genus Streptomyces. Because of their use as feed additives, exposure to penicillin, tetracycline, sulfamethazine, erythromycin, and virginiamycin may occur via feed dust aerosols. Air sampling for veterinary chemicals, although not common, would be via an aerosol technique described above, with the sample analyzed for the labeled product.

SUMMARY

It should be clear that farmers and workers on farms could be exposed to a wide range of respiratory hazards. Methods exist to evaluate exposures to most of the known or suspected hazards, although other natural hazards on farms may still be unknown and new man-made hazards may still be created.

The effects of gaseous hazards are well known. Several direct reading methods are available for farmers to evaluate gaseous exposures in near-real time. But, the conditions controlling their concentrations are neither well known nor uniform. Perhaps the complacency of customarily low concentrations breeds indifference toward "work practice" instructions that can appear overly cautious. The random occurrence of high hazards and the resulting health injuries are often a surprise to the victim. The acutely toxic nature of most agricultural gases gives a distinct advantage to direct reading instruments over collected samples that must be analyzed in a laboratory.

On the other hand, the effects of aerosol hazards, particularly organic dust aerosols, are not well known. Some acute effects like ODTS are understood, but the agent and long-terms risks are still in question. Based on projections from other industries, widespread chronic lung damage among farmers without respiratory protection habitually exposed to organic dust will come as no surprise to occupational health researchers. The chronic nature of most aerosol exposures allows the use of aerosol sample collection techniques, followed by laboratory analysis for individual aerosol components like endotoxin, other mycotoxins, pesticides, or quartz. One rather expensive direct reading aerosol method is available which can be used in either chronic or acute exposure settings, although exposure limit guidelines are only available for the former.

Future needs include the development of a better infrastructure to inform farmers of their own hazards and to make equipment more easily available to them, the personal and financial incentives to evaluate their air quality, and the technical knowledge to take efficacious and economical corrective action to improve their airborne environment. However, both measurements of and protection from these hazards are currently possible by informed and interested individuals.

REFERENCES

American Conference of Governmental Industrial Hygienists (ACGIH). 1995a. *Air Sampling Instruments.* 8th edition. Cincinnati, OH: ACGIH.

American Conference of Governmental Industrial Hygienists (ACGIH). 1995b. *Threshold Limit Values and Biological Exposure Indices.* Cincinnati, OH: ACGIH.

Acha, P.N., and B. Szyfres. 1980. *Zoonoses and Communicable Diseases Common to Man and Animals.* Scientific Publication No. 354. Washington, DC: Pan American Health Organization.

American Industrial Hygiene Association (AIHA). 1989. *Odor Thresholds for Chemicals with Established Occupational Health Standards.* Akron, OH: AIHA.

Alavanja, M.C., H. Malker, and R.B. Hayes. 1987. Occupational cancer risk associated with the storage and bulk handling of agricultural foodstuff. *J. Tox. Envir. Health* 22: 247-54.

Arden-Jones, M. 1982. Farmer's lung: An overview and prospects. *Ann. Am. Conf. Gov. Ind. Hyg.* 2: 172-82.

Banwart, W.C., and J.M. Brenner. 1975. Identification of sulfur gases evolved from animal manures. *J. Envir. Qual.* 4(3): 363-66.

Burg, R.W., and O.L. Shotwell. 1984. Aflatoxin levels in airborne dust generated from contaminated corn during harvest and at an elevator in 1980. *J. Assoc. Off. Anal. Chem.* 67: 309-12.

Burnett, W.E. 1969. Air pollution from animal wastes. Determination of malodors by gas chromatographic and organoleptic techniques. *Envir. Sci. Tech.* 3(8): 744-49.

Carlile, F.S. 1984. Ammonia in poultry houses: a literature review. *World Poultry Sci. J.* 40(E): 99-113.

Casterton, R.H. 1982. Enclosed Environments on agricultural tractors. *Ann. Am. Conf. Gov. Ind. Hyg.* 2: 121-27.

Centers for Disease Control (CDC). 1989. Fatalities attributed to methane asphyxia in manure waste pits--Ohio, Michigan, 1989. *MMWR* 38(33): 583-86.

Cormier Y., L. Boulet, G. Bedard, and G. Tremblay. 1991. Respiratory health of workers exposed to swine confinement buildings and dairy barns. *Scand. J. Work Env. Health* 17: 269-75.

Devine, J.M., G.B. Kinoshita, R.P. Peterson, and G.L. Picard. 1986. Farm worker exposure to terbufos [phosphorodithioic acid] during planting operations of corn. *Arch. Envir. Contam. Toxic.* 15: 113-19.

Donham, K.J. 1975. Infectious diseases common to animals and man of occupational significance to agricultural workers. In: *Proc. of Conference on Agricultural Health and Safety.* New York: New York Society for Occupational and Environmental Health, pp. 160-75.

Donham, K.J., M. Rubino, T.D. Thedell, and J. Kammermeyer. 1977. Potential health hazards to agricultural workers in swine confinement buildings. *J. Occup. Med.* 19(6): 383-87.

Donham, K.J., L.W. Knapp, R. Monson, and K. Gustafson. 1982. Acute toxic exposure to gases from liquid manure. *J. Occup. Med.* 24(2): 142-45.

Donham, K.J., D.C. Zavala, and J.A. Merchant. 1984. Respiratory symptoms and lung function among workers in swine confinement buildings: a cross-sectional epidemiological study. *Arch. Env. Health* 39: 96-101.

Donham, K.J., and W. Popendorf. 1985. Ambient levels of selected gases inside swine confinement buildings. *Am. Ind. Hyg. Assoc. J.* 46: 658-61.

Donham, K.J. 1985. Zoonotic diseases of occupational significance in agriculture: A review. *Intl. J. Zoonoses* 12: 163-91.

Donham, K.J., L.J. Scallon, W. Popendorf, M.W. Truehaft, and R.C. Roberts. 1986. Characterization of dusts collected from swine confinement buildings. *Am. Ind. Hyg. Assoc. J.* 47(7): 404-10.

Donham, K.J. 1986. Hazardous agents in agricultural dusts and methods of evaluation. *Am. J. Ind. Med.* 10: 205-20.

Donham, K.J., P. Hagland, Y. Peterson, R. Rylander, and L. Belin. 1989. Environmental and health studies of farm workers in Swedish swine confinement buildings. *Brit. J. Ind. Med.* 46: 31-37.

Donham, K.J., S.J. Reynolds, P. Whitten, J.A. Merchant, L. Burmeister, and W. Popendorf. 1994. Respiratory dysfunction associated with enclosed swine facilities: dose-response of pulmonary function to environmental exposures. *Am. J. Ind. Med.* 27: 405-18.

Eller, P.M. 1994. Measurement of worker exposure. In: *Patty's Industrial Hygiene and Toxicology Vol. III, Theory and Rationale of Industrial Hygiene Practice.* 3rd edition. New York: Wiley-Interscience.

Environmental Protection Agency (EPA). 1987. *Pesticide Assessment Guidelines. Subdivision U. Applicator Exposure Monitoring.* (NTIS Order No. PB 87-133286). Washington DC: Office of Pesticide Programs, U.S. EPA.

Fairchild, D. 1987. *Ground Water Quality and Agricultural Practices*, Chelsea, MI: Lewis Publishers, Inc.

Hagley, S.R., and D.L. South. 1983. Fatal inhalation of liquid manure gas. *Med. J. Australia* 2: 459-60.

Hagmar, L., A. Schutz, T. Hallberg, and A. Sjoholm. 1990. Health effects of exposures to endotoxins and organic dust in poultry slaughterhouse workers. *Int. Arch. Occup. Envir. Health* 62: 159-64.

Hayes, R.B., J.P. Van Nieuwenhuize, J.W. Raatgever, and F.J.W. Ten Kate. 1984. Aflatoxin exposures in the industrial setting: an epidemiological study of mortality. *Food Chem. Toxic.* 22(1): 39-43.

Hayhurst, E.R., and E. Scott. 1914. Four cases of sudden death in a silo. *JAMA* 63: 1570-72.

Helmers, S., F.H. Top, and L.W. Knapp. 1971. Ammonia injuries in agriculture. *J. Iowa Med. Soc.* 61(5): 271-80.

Holness, D.L., E.L. O'Blenis, A. Sass-Kortsak, C. Pilger, and J.R. Nethercott. 1987. Respiratory effects and dust exposures in hog confinement farming. *Am. J. of Ind. Med.* 11: 571-80.

Holness, D.L., J.T. Purdham, and J.R. Nethercott. 1989. Acute and chronic respiratory effects of occupational exposure to ammonia. *Am. Ind. Hyg. Assoc. J.* 50(12): 646-50.

Horvath, E.D., G.A. do Pico, R.A. Barbee, and H.A. Dickie. 1978. Nitrogen dioxide-induced pulmonary disease. *J. Occup. Med.* 20: 103-10.

Jacobs, R.J. 1989. Airborne endotoxins: an association with occupational lung disease. *Appl. Ind. Hyg.* 4(2): 50-56.

Johnson, W.M. 1981. Occupational factors in coccidioidomycosis. *J. Occup. Med.* 23(5): 367-74.

Jones, W., K. Morring, S.A. Olenchock, T. Williams, and J. Hickey. 1984. Environmental study of poultry confinement buildings. *Am. Ind. Hyg. Assoc. J.* 45(11): 760-66.

Keeney, A.H., and M.C. Hunter. 1950. Human infection with Newcastle virus of fouls. *Arch. Opthalmology* 44: 573-80.

Leidel N.A., K.A. Busch, and J.R. Lynch. 1977. *Occupational Exposure Sampling Strategy Manual.* DHEW (NIOSH) Publication No. 77-173. Cincinnati, OH: NIOSH.

Liesivuori, J., S. Liukkonen, and P. Pirhonen. 1988. Reentry intervals after pesticide application in greenhouses. *Scand. J. Work Envir. Health*, 14 (Supp. 1): 35-36.

Lynch, J.R. 1994. Measurement of worker exposure. In: *Patty's Industrial Hygiene and Toxicology Vol. III, Theory and Rationale of Industrial Hygiene Practice.* 3rd edition. New York: Wiley-Interscience, pp. 27-80.

Manninen, A., J. Kangas, M. Linnainmaa, and H. Savolainen. 1989. Ammonia in Finnish poultry houses: effects of litter on ammonia levels and their reduction by technical binding agents. *Am. Ind. Hyg. Assoc. J.* 50(4): 210-15.

Maybank, J., K. Yoshida, and R. Grover. 1978. Spray drift from agricultural pesticide applications. *J. Air Pollu. Control Assoc.* 28: 1009-14.

McCord, C.P. 1960. Metal fume fever as an immunlogical disease. *Ind. Med. Surg.* 29: 101-07.

Merkel, J.A., T.E. Hazen, and J.R. Miner. 1969. Identification of gases in a confinement swine building environment. *Trans. ASAE* 12: 310-15.

Morese, D.L., and M.A. Woodbury. 1981. Death caused by fermenting manure. *JAMA* 245(1): 63-64.

Mulhausen, J.R., C.E. McJilton, P.T. Redig, and K.A. Janni. 1987. Aspergillus and other human respiratory disease agents in turkey confinement houses. *Am. Indr. Hyg. Assoc. J.* 48(11): 894-99.

National Institute for Occupational Safety and Health (NIOSH). 1990b. *Registry of Toxic Effects of Chemical Substances*. U.S. DHHS (NIOSH) Publ. No. 90-117. Washington, DC: U.S. Government Printing Office.

National Institute for Occupational Safety and Health (NIOSH). 1990a. *Pocket Guide to Chemical Hazards*. U.S. DHHS (NIOSH) Publ. No. 90-117. Washington, DC: U.S. Government Printing Office.

National Institute for Occupational Safety and Health (NIOSH). 1994. *Manual of Analytic Methods*, 4th edition. P.M. Eller and M.E. Cassinelli, eds. U.S. DHHS (NIOSH) Publ. No. 90-113, Washington, DC: U.S. Government Printing Office.

National Research Council (NRC). 1987. *Regulating Pesticides in Food*. Washington, DC: National Academy Press.

Ness, S.A. 1991. *Air Monitoring for Toxic Exposures*. New York: Van Nostrand Reinhold.

Ness, S.A. 1994. *Surface and Dermal Monitoring for Toxic Exposures*. New York: Van Nostrand Reinhold.

Nigg, H.N. and J.H. Stamper. 1983. Exposure of spray applicators and mixer-loaders to chlorobenzilate miticide in Florida citrus groves. *Arch. Envir. Contam. Tox.* 12:477-482.

Nigg, H.N., J.H. Stamper, and R.M. Queen. 1984. The development and use of a universal model to predict tree crop harvester pesticide exposure. *Am. Ind. Hyg. Assoc. J.* 45(3): 182-86.

Olenchock, S.A. 1988. Quantitation of airborne endotoxin levels in various occupational environments. *Scan. J. Work Envir. Health* 14:72-73.

Occupational Safety and Health Adminstration (OSHA). 1989. *Occupational Safety and Health Standards for General Industry*. 29 CFR 1910. Washington, DC: Commerce Clearing House.

Petro, W., K.C. Bergmann, R. Heinze, E. Muller, H. Wuthe, and J. Vogel. 1978. Long-term occupational inhalation of organic dust-effect on pulmonary function. *Int. Arch. Occ. Health* 42: 119-27.

Popendorf, W., A. Pryor, and H.R. Wenl. 1982. Mineral dust in manual harvest operations. *Ann. Am. Conf. Gov. Ind. Hyg.* 2: 101-15.

Popendorf, W. and J.T. Leffingwell. 1982. Regulating OP pesticide residues for farmworker protection. *Residue Reviews* 82: 125-201.

Popendorf, W., and H.R. Wenk. 1983. Chrysotile asbestos in a vehicular recreation area: a case study. In: *Environmental Effects of Off-Road Vehicles - Impacts and Management in Arid Regions*. H.G. Wilshire and R.H. Webb, eds. New York: Springer-Verlag, pp. 375-96.

Popendorf, W. 1984. Vapor pressure and solvent vapor hazards. *Am. Ind. Hyg. Assoc. J.* 45(10): 719-26.

Popendorf, W., K.J. Donham, D.N. Easton, and J. Silk. 1985. A synopsis of agricultural respiratory hazards. *Am. Ind. Hyg. Assoc. J.* 46(3): 154-61.

Popendorf, W. 1988. Mechanisms of clothing exposure and dermal dosing during spray application. In: *Performance of Protective Clothing: Second Symposium*. S.Z. Mansdorf, R. Sager, and A.P. Nielsen, eds. Philadelphia: American Society for Testing and Materials, pp. 611-24.

Popendorf, W., and K.J. Donham. 1991. Agricultural hygiene. In: *Patty's Industrial Hygiene*. 4th edition. New York: J. Wiley & Son, pp. 575-608.

Public Law 91-596. Occupational Safety and Health Act of 1970. Washington, DC.

Ramazzini, B. 1713. *Diseases of Workers*. Transl. by W.C. Wright, 1993. Thunder Bay, Ontario, Canada: OH&S Press.

Ramirez, R.J., and A.R. Dowell. 1971. Silo filler's disease: nitrogen dioxide-induced lung injury. Long-term follow-up and review of the literature. *Ann. Intern. Med.* 74: 569-76.

Rask-Andersen, A. 1989. Organic dust toxic syndrome among farmers. *Brit. J. Ind. Med.*, 46: 233-38.

Reynolds, S.J., D. Parker, D. Vesley, D. Smith, and R. Woellner. 1993. Cross-sectional epidemiological study of respiratory disease in turkey farmers. *Am. J. Ind. Med.* 24: 713-22.

Reynolds, S.J. and D.K. Milton. 1993. Comparison of methods for analysis of airborne endotoxin. *Appl. Occup. Envir. Hyg.* 8(9): 761-67.

Reynolds, S.J., D. Parker, Vesley, K. Janni, C. McJilton. 1994. Occupational exposure to organic dusts and gases in the turkey growing industry. *Appl. Occup. Envir. Hyg.* 9(7): 493-502.

Rodricks, J.V., C.W. Hesseltine, and M.A. Mehlman. 1977. *Mycotoxins in Human and Animal Health*. Park Forest South, IL: Pathotox Publications.

Ruth, J.H. 1986. Odor thresholds and irritation levels of several chemical substances: a review. *Am. Ind. Hyg. Assoc. J.* 47(3): A142-A155.

Rylander, R., Y. Peterson, and K.J. Donham, eds. 1986. Health effects of organic dusts in the farm environment. Proc. of an International Workshop held in Skokolster Sweden, April 23-25, 1985. *Am. J. Ind. Med.* 10: 193-340.

Rylander, R. 1987. Role of endotoxins in the pathogenesis of respiratory disorders. *Eur. J. Respir. Dis.* 71(154): 136-44.

Rylander, R. and Y. Peterson, eds. 1990. Proceedings of an International Workshop held in Skokolster Sweden, October 24-27, 1988. *Am. J. Ind. Med.* 17: 1-147.

Savage, E.P. 1976. Acute pesticide poisonings. *Pesticide Residue Hazards to Farm Workers, Proceedings of a Workshop Held February 1976.* HEW (NIOSH) Publ. No. 76-191, pp. 63-65.

Schenker, M., and S. McCurdy. 1986. Pesticides, viruses, and sunlight in the etiology of cancer among agricultural workers. In: *Cancer Prevention: Strategies in the Workplace.* C.E. Becker and M.J. Coye, eds. Washington, DC: Hemisphere Publishing Corp., pp. 29-37.

Schnurrenberger, P.R. and W.T. Hubbert. 1981. *An Outline of Zoonoses.* Ames, Iowa: Iowa State University Press.

Scott, E.G., and W.B. Hunt. 1973. Silo-filler's disease. *Chest* 63: 701-06.

Sherman, R.P., M.L. Barman, and J.L. Abrahams. 1979. Silicate pneumoconiosis of farm workers. *Lab. Invest.* 40(5): 576-82.

Sorensen, W.G., J.P. Simpson, M.J. Peach, T.D. Thedell, and S.A. Olenchock. 1981. Aflatoxin in respirable corn dust particles. *J. Tox. Envir. Health.* 7: 669-72.

Stamper, J.H., H.N. Nigg, W.D. Mahon, A.P. Nielsen, and M.D. Royer. 1988. Pesticide exposure to greenhouse foggers. *Chemosphere* 17(5): 1007-23.

Swotinsky, R.B., and K.H. Chase. 1990. Health effects of exposure to ammonia: scant information. *Am. J. Ind. Med.* 17: 515-21.

Manninen, A. 1988. Analysis of airborne ammonia: comparison of field methods. *Ann. Occup. Hyg.* 32(3): 399-404.

Terho, E.O. 1986. Diagnostic criteria for farmer's lung disease. *Am. J. Ind. Med.* 10: 329.

Thelin, A., O. Tegler, and R. Rylander. 1984. Lung reaction during poultry handling related to dust and bacterial endotoxin levels. *Eur. J. Respir. Dis.* 65: 266-71.

Vehicular recreation area: a case study. In: *Environmental Effects of Off-Road Veichles--Impacts and Management in Arid Regions.* H.G. Wilshire and R.H. Webb, eds. New York: Springer-Verlag, pp. 375-96.

Wang, L.C., and R.H. Burris. 1960. Mass spectrometric study of nitrogenous gases produced by silage. *J. Agric. Food Chem.* 8(3): 239-42.

Zennie, T.M. 1984. Identification of aflatoxin B1 in grain elevator dust in central Illinois. *J. Tox. Envir. Health* 13:589-93.

27

OCCUPATIONAL SAFETY AND HEALTH REGULATIONS IN AGRICULTURE

Regina C. Luginbuhl, M.S.
Director, Agricultural Safety and Health Division
North Carolina Department of Labor

> Occupational safety and health standards were created in response to high numbers of workplace injuries. Agriculture is among those industries with the highest injury and fatality rates. This chapter outlines the occupational safety and health regulations that apply to agriculture and indicates how they are implemented. Topics include recordkeeping, tempory labor camps, field sanitation, chemical hazards, and farm machinery. Applications of the OSHA standards are discussed. Case studies are presented to illustrate the danger associated with noncompliance as well as the advantages that can be gained by the producers if the guidelines and regulations are properly implemented.

INTRODUCTION

Every day, sounds of alarm clocks buzzing, coffee perking, refrigerator doors slamming, and door keys jangling reverberate all across the United States as 127 million men and women rush to work. And every day, some families will wait in vain for a family member who won't return. How safe are workplaces in the United States? As each of us go off to work in our modern "mine," who is the canary?

For many years, the lack of accurate data on the extent and patterns of occupational injuries was a major impediment to understanding and controlling workplace injuries.

In 1912, in an attempt to improve this lack of understanding, the U.S. Bureau of Labor Statistics (BLS) began measuring workplace safety. This bureau was created as the statistical branch of the U.S. Department of Labor, and is charged with compiling accurate statistics on work injuries and illnesses.

The Occupational Safety and Health standards, regulations affecting workplace conditions, were created in response to workplace injuries. These regulations were preceded by various state initiatives in an attempt to slow workplace injuries and fatalities.

FROM THE FARM TO THE FACTORY

The first industrial inspection programs began in the late 19th century. At that time, men and women were leaving agriculture for small towns and cities, where industrial jobs in textile mills and factories were burgeoning. Recent immigrants to the United States, seeking economic security and religious or political freedoms, were desperate for employment and often overlooked the safety hazards inherent in their new found work sites.

Accidents could not be overlooked, and Massachusetts passed the first worker safety law in 1877, requiring guards on spinning machinery in textile mills (NCIPC 1989; Hamilton 1925).

From 1927 to 1969, several federal safety rules were established for coal miners, construction workers, and maritime workers, industries where accident rates were particularly high.

In 1970, Congress examined figures revealing that job-related accidents accounted for more than 14,000 worker deaths each year, and nearly 2 1/2 million worker injuries annually. Legislation to safeguard U.S. workers was signed by President Richard M. Nixon on December 29, 1970 (Ford 1991).

The Williams-Steiger Act, better known as the Occupational Safety and Health Act of 1970, was passed to "assure so far as possible every working man and woman in the Nation safe and healthful working conditions."

STATE OSH PLANNING

Under the OSH Act of 1970, states could assume responsibility for enforcing workplace safety and health regulations that are "at least as effective" as federal OSHA.

Like Massachusetts, several individual states had regulations addressing worker safety as early as the 19th century. Michigan adopted laws relating labor, employees, and safety soon after statehood was granted in 1837, and began a factory inspection program in 1893. California began operating a safety enforcement program in 1913 (OSPHA 1995).

Currently, 23 states and 2 territories operate state OSHA enforcement programs provided for under Section 18 of the Occupational Safety and Health Act of 1970.

CURRENT SAFETY AND HEALTH RISKS

Statistics indicate that 6,271 people died on the job in 1993. Men, the self-employed, and workers aged 55 and older appear to have a higher risk of workplace fatalities relative to their share of employment. The fatality rate was highest in mining and agriculture, each with 26 fatalities per 100,000 workers. The service industries had one of the lowest rates, at 2 fatalities per 100,000 workers. Workers in farming-related occupations accounted for 10% of all fatal work injuries, but represent only 3% of the Nation's employment total (Toscano 1995).

Ten states combined accounted for about half of the fatal job-related injuries: California, Florida, Georgia, Illinois, Louisiana, New York, North Carolina, Pennsylvania, Tennessee, and Texas. In contrast, half of the remaining states had fewer than 100 fatalities each. Consider the industry mix when reviewing this information. The states with a large agricultural economy, and a small service economy, will have higher rates, based on the dominant, high-hazard industry. (See Figures 27.1 and 27.2.)

As farm work has become increasingly mechanized, farm accidents have increased. The graph below illustrates that agriculture is the only major industry in the United States that has not experienced a declining fatality rate.

Hazards faced by agricultural workers include tractor rollovers and runovers (Ehlers 1993), injuries and amputations from machinery such as augers and power takeoff devices (PTOs) (Stoskopf and Venn 1985), electrocutions, and pesticide poisoning (NCIPC 1989).

WHAT IS BEING DONE?

The agricultural industry and its proponents have been successful in avoiding environmental safety, health, and labor regulations. As a result, few OSH standards apply to agricultural operations. In addition, this subset of regulations applies to a small number of farms, those that have more than 10 full-time workers, excluding family members. Those OSH standards that do apply to agricultural operations are listed below.

29 CFR Part 1910

(a) The following standards in the Code of Federal Regulations (CFR), Title 29, Part 1910 apply to agricultural operations:

1) Recordkeeping requirements
2) Temporary labor camps—1910.142;
3) Storage and handling of anhydrous ammonia—1910.111(a) and (b);
4) Pulpwood logging—1910.226;
5) Slow-moving vehicles—1910.145;
6) Hazard communication—1910.1200.

(b) Except to the extent specified in paragraph (a) of this section, the standards contained in Subparts B through T and Subpart Z of Part 1910 of title 29 do not apply to agricultural operations.

The intent of the exemption (b) noted above "is to avoid placing an undue financial burden on small farms and to comply with an annual amendment to the House appropriations bill, which prohibits the Department of Labor from regulating farms with 10 or fewer workers" (GAO/HRD 1992).

29 CFR Part 1928

1928.51—Rollover protective structures (ROPS) for tractors used in agricultural operations.
1928.52—Protective frames for wheeltype agricultural tractors—test procedures and performance requirements. 1928.53—Protective enclosures for wheeltype agricultural tractors—test procedures and performance requirements.
1928.57—Guarding of farm field equipment, farmstead equipment, and cotton gins.
1928.110—Field Sanitation

The following sections will now explore each of the above-mentioned standards.

RECORDKEEPING REQUIREMENTS

Under the Williams-Steiger Occupational Safety and Health Act, the agricultural employer or representative must maintain a record of work-related deaths, injuries, or illnesses of employees. The employer or employer representative must make a report to the nearest OSHA office within 8 hours if there are work-related fatalities. An annual summary of occupational injuries and illnesses must be posted in the workplace.

Figure 27.1. States and territories currently operating state OSHA programs.

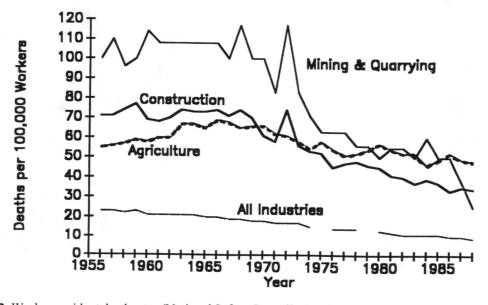

Figure 27.2. Worker accident death rates (National Safety Council, *Accident Facts*).

The agricultural employer must provide a person who is trained to render first aid when an infirmary, hospital, or clinic is not near the workplace and must make first aid supplies readily available.

A farming operation is exempt from federal OSHA standards if the grower employed 10 or fewer workers at all times during the preceding 12 months, unless the grower operates a migrant labor camp. Members of the grower's immediate family are not considered when making this determination. States with their own OSH enforcement program may have differing requirements.

REGULATIONS PROTECTING AGRICULTURAL WORKERS

Temporary Labor Camps, 29 CFR 1910.142

Section 1910.142. Sets out standards governing *any* labor camp that is to be used to house workers on a temporary basis. These sites are usually constructed based on the owner's or operator's convenience, and are geographically close to the actual workplace. As such, the standards have been used in connection with oil well drilling, construction, and, for the purposes of this chapter, with housing sites for migrant farmworkers who do work of a temporary agricultural nature.

The 1910.142 standards cover the health and safety standards at a housing site, such as living space, bathroom facilities, and sanitation requirements for the grounds and kitchen areas.

State-plan states, such as Michigan and North Carolina, may have more encompassing requirements than the federal standards (N.C. Migrant Housing Act; Labor Rules 333.12421, Michigan). There, a full inspection of a single dwelling may take as long as an hour and a half and concern a "myriad of details" (The Independent, Fuquay-Varina 1992).

The federal standards also specifically include the following:

- An adequate water supply...capable of delivering 35 gallons per person per day ...;
- An adequate supply of hot and cold running water;
- One toilet for 15 occupants;

Figure 27.3. Photo collage of news clippings on housing hazards.

- One showerhead for 10 people;
- Fifty square feet of space per occupant in quarters used for sleeping.

Working Without

Substandard housing conditions may lead to accidents, and substandard conditions in farmworker housing are no exception. Headlines in local newspapers document farmworker fatalities from household fires, and local hospitals treat many farmworker emergencies resulting from food contaminated by inadequate refrigeration, drinking water polluted with fecal coliform bacteria, and respiratory diseases spread through inadequate living space or improper ventilation.

These problems affect the lives and the quality of life of the farmworker residents, but also the character of the communities in which they reside, if only temporarily.

Go by the Rules

Has there ever been a newspaper headline heralding the absence of an accident? What does a labor camp look like that is in compliance with the regulations? What does it take to not only provide housing but be successful?

The Edward Lindsay Pope family in Clinton, North Carolina, farms 30 acres of peppers, 20 of cucumbers, and 25 of tobacco. They also raise hogs.

Mrs. Pope spoke one day in October, 1990, about her experiences with the Fernando Reyes family who had worked with them for the previous 2 years picking cucumbers and priming tobacco.

When asked how she provides a place that is set up for (farmworker) families, Mrs. Pope responded, "It's very important for us to have a place where they can cook. It's very important that we furnish a washing machine.

"I've been there. I'm there every day in the field and I know. You just pack those clothes down for a week and they rot. I think it's because I have been a part of farming, and I know how important it is to have something to eat, and I know how important it is for somebody to have clean clothes."

When asked about complying with the housing standards, she added, "First thing we're going to do is go by the rules. Whatever we have to do or are supposed to do. Not try to beat the system" (NCDOL 1991).

Benefits of safe, healthy migrant farmworker housing include:

- Farmworkers that return to work for a grower because the housing is more than the minimum required;
- An accident-free workplace;
- Community resources (hospital care, fire departments) that are wisely expended.

Alternatives

In the recent past, alternative housing options to labor camps owned or operated by growers have been increasing. In California's Monterey County, farmworkers formed a coalition aimed at teaching farm laborers and other low-income people how to improve their housing conditions. Farmworkers who were unsatisfied with their living conditions were taught ways to become advocates for their own farm labor camps or apartment complexes, and were given training in ways that they could purchase their living space (*Mercury News* 1992).

In another California location, the Villa de San Joaquin provides clusters of 2-story townhomes, each with its own backyard, and a common play area. It is being built in Stockton, where approximately 6,000 seasonal workers come to work but have encountered problems finding suitable shelter (*The Record* 1995).

Financed by a housing grant from Rural Economic Community Development (RECD), formerly the Farmers Home Administration, Casa Vista, a 24-unit dormitory for 48 single men, opened in June, 1991 in Longmont, Colorado. Requests for the housing came from local and state officials, church groups, and nonprofit social service agencies. Said Rob Culligan, director of Christian Community Services in Alamosa, "Building this housing is to everyone's benefit." (*Colorado Post* 1991).

Recently, health clinics have become a catalyst for better housing. Health workers are constantly reminded of the impact of poor housing conditions on health when they see respiratory infections, gastrointestinal distress, and influenza in high percentages among farmworkers. Clinics can be involved in the housing issue in a variety of ways, ranging from offering referrals, performing simple rehabilitation of existing housing, or managing farmworker housing units. Some projects include self-help housing, cooperative barracks, apartments, and private individual homes (*SHE News* 1995).

Field Sanitation, 29 CFR 1928.110

"Agricultural employers shall provide, without cost to the employees, drinking water, handwashing facilities, and toilets. Employees must be notified and be given access to these facilities" (29 CFR 1928.110).

On the face of it, this standard may look deceptively noncontroversial. Who would not agree that people need cool water to drink? But this standard was promulgated only after a 14-year battle, which included a lawsuit brought by farmworkers against the United States Department of Labor, OSHA (Wilk 1987).

Field Sanitation standards had been in effect for industrial employers since 1971, but agriculture had been exempt.

This standard took effect on May 30, 1987. At that time, the standard was expected to cover 471,600 hand laborers, of whom approximately 118,000 were interstate migrant workers (*Current Report* 1987).

Farming operations that employ 11 or more workers on any given day in the year preceding an OSHA inspection are covered under the agency's field sanitation standard, according to a compliance directive issued June 22, 1992.

The states were given 6 months to adopt the federal standard. In states where a standard was not yet adopted, interim federal enforcement assistance was provided in response to complaints.

Several states already had field sanitation requirements that were more rigorous than the federal standard. Of the 21 states with state field sanitation standards, four (Alaska, Arizona, Oregon, and Washington) required compliance on farms that employ 10 or fewer workers (GAO/HRD 1992).

Since that time, states with their own OSH enforcement program have continued to extend the benefits of field sanitation, to include more than "11 or more" mandated by federal standard. In 1989, the legislature of North Carolina passed a bill stating that anyone owning or operating migrant farmworker housing must provide field sanitation, in addition to those who employ 11 or more workers on any given day. In Michigan, the federal standard was extended in 1991 so that all agricultural employers must provide, at no cost to the agricultural employee, potable water in locations that are readily accessible to all agricultural employees.

Working Without

The absence of field sanitation has been the source of much hand-wringing, philosophical debate, and illness. No one will deny that there is a relationship between health and access to potable drinking water facilities, excreta disposal, and hand washing. This relationship was the basis for the United Nations Drinking Water Supply and Sanitation Decade, 1981-1990.

Though many Third World studies propose a causal link between health status and sanitary facilities, recent studies can be found that document the same relationship and inherent problems here in the United States.

Several studies have been conducted to investigate sources of common fecal-oral disease indicators such as diarrhea, nausea and vomiting, and gastroenteritis. It has been documented that the number of migrant farmworkers in Utah who used several Salt Lake County clinics for gastrointestinal diseases was significantly higher than that of the neighboring urban poor who had the same access to the clinics (Arbab 1986).

Table 27.1 illustrates that the migrant farmworker population suffered greatly elevated rates of illness when compared with the control group. The rate of diarrhea per 1,000 clinic users was approximately 20 times higher in the migrant population; gastroenteritis was 26 times higher; fevers of unknown origin, 120 times higher.

Table 27.1. Frequencies and rates per 1,000 clinic users.

Symptom	Migrant Farmworkers Number	Rate	Control Group Number	Rate
Diarrhea	143	153	72	8
Nausea, Vomiting	48	51	36	4
Abdominal or Intestinal Pain	62	66	89	10
Nonspecific Gastritis Gastroenteritis	24	26	—	—
Bloody Stools	8	9	9	1
Fever	31	33	3	3

* January 1981 to May 1984 (Arbab)
Source: *American Journal of Public Health* 76(16).

This study was conducted in response to a request for testimony at an OSHA hearing regarding fecal-related disease among migrant farmworkers in Utah, where, at that time, no sanitation standard existed. Migrant labor is used in Utah orchards or row crops. Seventy-seven percent of the symptoms listed in the chart above occurred in the field work season of April through October. No seasonal distinction occurred for symptoms among the control group.

A survey conducted in 1989 in North Carolina (Sweeny 1990) indicated that 32% of North Carolina's farmworkers surveyed lacked sanitary drinking water, 64% lacked handwashing facilities, and 94% lacked adequate toilet facilities. Only 4% had access to the requirements of the standard.

A 1993 report cites poor quality drinking water supplies for migrant farmworkers, stating that in many work sites "water supplies are inadequate, and (the farmworkers) must drink irrigation water which may be contaminated with pesticides or infectious agents (Smith 1993).

Consumer interest has had an impact on field sanitation standards as well. Public health officials expressed concern that improper field sanitation, particularly the lack of toilets in the field, will lead to contaminated produce. The news media reflects this concern in articles such as the following.

> "...For those middle class persons...here is some disquieting information: A continued lack of sanitation facilities will mean that millions of field laborers will perpetuate the practice of relieving themselves wherever they can, turning fruit and vegetable farms into open toilets. . . That is worth considering the next time you approach a restaurant salad bar or visit the fresh produce section of the neighborhood supermarket" (Walters 1985).

Given the tough conditions and low pay of tobacco harvesting, Dr. Daniel A. Sumner, speaking at the "Outlook for Tobacco" seminar at the U.S. Department of Agriculture, said, "It is in the economic interest of most growers to encourage more and better work by their laborers and to retain their better workers by providing drinking water and other facilities in the fields that satisfy workers' needs" (Sumner 1989).

Proper field sanitation benefits all of us, those who labor in the fields, and those who consume the fruits and vegetables grown there.

Go by the Rules

In essence, the provisions are as follows:

- Drinking water
    potable
    single-use cups or a fountain
    suitably cool
    sufficient quantities

- Toilet facilities
    one toilet for every 20 employees
    adequate supply of toilet paper
    waste container
    properly maintained
    located within 1/4 mile walk of each
        hand laborer's place in the field

- Handwashing facilities
    soap
    disposable towels
    waste container

The sanitation requirements that are covered under the field sanitation standards are basic to all who want to preserve and maintain good health. Some of the recommendations would seem to be common sense health precautions that apply to all who work outdoors in hot weather. The precautions are:

- Drink water frequently and especially on hot days;
- Urinate as frequently as necessary;
- Wash hands both before and after using the toilet;
- Wash hands before eating and smoking.

If field sanitation is freely provided, facilities are clean and well-maintained, and employees informed of the availability, these recommendations become possible.

CHEMICAL HAZARDS

In our struggle to feed many more mouths than the small, family farms fed in the past, we rely on many more chemicals. As a result, people working in agriculture are exposed to a number of chemical hazards. These hazards include contact with chemical fertilizers and pesticides.

Storage and Handling of Anhydrous Ammonia 29 CFR 1910.111(a) and (b)

Anhydrous ammonia is a commonly used fertilizer that presents atypical hazards. Anhydrous ammonia is liquid when stored under pressure in the applicator tank. It becomes a gas when released in the soil. Accidental release into the air can cause very serious injuries. This chemical freezes the skin upon contact, and, unless treated rapidly, it continues to dehydrate and destroy cells. The process is particularly active with soft tissue such as eyes, the mouth, and the respiratory tract since it leaches water from the tissue (Popendorf 1991; Roberts 1992).

OSH standards concern the design, construction, location, installation, and operation of anhydrous ammonia systems including refrigerated ammonia storage systems.

Hazard Communications 29 CFR 1910.1200

In 1983, the Occupational Safety and Health Administration issued a rule called "Hazard Communication." At that time, this rule applied only to employers in the manufacturing sector of industry. In 1987, the scope of the rule was expanded to include employers in the nonmanufacturing sector. The basic goal of the standard is to ensure that employers and employees know about chemical hazards in the workplace and how to protect themselves.

The Environmental Protection Agency has overriding authority in regulating pesticides, but has delegated to the states much of the responsibility for implementing pesticide laws and regulations. The designated "lead" agency may vary from state to state. The state Department of Agriculture is most typically the lead agency statewide, with the Department of the Environment in second place (Mather 1994).

The largest single user of pesticides in the U.S. is agriculture. In 1985, agriculture accounted for 77% of nationwide usage of 1.08 billion pounds (not including wood preservatives); and 78% of expenditures of $4.6 billion (EPA 1986).

In 1991, those percentages remained steady. Agriculture accounted for three-fourths of the total pesticide usage by volume, according to the EPA. Industry, business, and government represented another 18% of the total pesticide use, while home and garden use accounted for the remaining 6% (Mather 1994).

Both farmers and farmworkers are exposed to pesticide exposure through many mediums. Acute pesticide poisoning is one of the most common chemical-related health problems among agricultural workers (Ehlers 1993).

Farm families are increasingly concerned about their water supply, as water may become contaminated through spills or improper storage of pesticides, fertilizers, or other chemicals. An estimated 10.5 million rural households use private well water (*NC Farm Bureau News* 1991).

Farmworker families may suffer additional hazards due to housing location. Dr. Richard Andres, working in Maryland's Eastern Shore, states, "The most frustrating aspect of all of this (illness) is the ... constant exposure to herbicides, pesticides and insecticides. Many (farmworkers) live in camps that are located in or near the fields which are sprayed with chemicals during various stages of the harvest (of tomatoes, green peppers, cucumbers, squash)" (Dahlberg 1992). The Pesticide Farm Safety Center (PFSC), a panel of medical experts and farmworker advocates, found that "In agriculture, $0.30 is spent annually on employee health protection compared to $4.34 spent on all other industrial workers. . . Especially with the use of pesticides, the agriculture workplace must be regulated as any other place of employment" (NC Right-to-Know 1991).

DANGEROUS GASES

Of the approximately 2.3 million livestock farms in the United States, an unknown number contain manure pits or tanks. Inside the pit, the manure undergoes anaerobic digestive fermentation and forms fertilizer. In this process, four potentially dangerous gases may be generated:

- Methane
- Hydrogen sulfide
- Carbon dioxide
- Ammonia

On July 26, 1989, five farm workers died after consecutively entering a manure pit on their dairy farm. The victims were a 65-year-old dairy farmer, his two sons aged 37 and 28, a 15-year-old grandson, and a 63-year-old nephew. The younger son entered the pit to repair the machinery. The others entered the pit one at a time, in a rescue attempt, and each individual fell victim to methane asphyxiation (NIOSH Alert, 90-103).

No Occupational Safety and Health Administration standard exists for work in and around manure pits.

Go by the Rules

To come into compliance with the Storage and Handling of Anhydrous Ammonia 29 CFR 1910.111(a) and (b), a grower needs to do the following:

- Know the standard;
- Carry water on all anhydrous ammonia tanks for emergency purposes;
- Do not attempt to remove clothing that has been saturated with ammonia because it may be frozen to the skin;
- Allow skin burns to remain open to the air (NRAES-12, First on the scene).

To come into compliance with 29 CFR 1910.1200, the Hazard Communication standard, an employer needs to do the following:

- Read the standard.
- List the hazardous chemicals in the workplace.
- Obtain material safety data sheets for all chemical substances.
- Make sure that all containers are labeled.
- Use personal protective equipment.
- Develop and implement a written hazard communication program.

In all states, the Occupational Safety and Health Hazard Communications Standard covers all industries including farming. It requires that information on hazardous chemicals be communicated to employees through labels, material safety data sheets (MSDS), and training programs. A written hazard communication program and recordkeeping are required. Compliance by nonmanufacturing sectors is now required.

In Washington state, three agencies have joined forces to provide a coordinated effort to improve conditions for Washington farmworkers in terms of pesticide exposure. The state departments of Agriculture, Labor and Industries, and Health work in three related categories: prevention; communication; and education, training, and feedback (Brookreson 1994).

FARM MACHINERY

Farm machinery has led to greater production but has also led to an increase in farm accidents. Machines have been invented to harvest almost every kind of fruit and vegetable produced in this country. The cotton harvester (see Figure 27.4) does the work of 100 man days in one day. Tomatoes, peaches, and sweet potatoes can now be picked and graded by machine. But as our work pace accelerates, caution and care often decline. Are we in too much of a hurry to turn off the machine before we try to fix it? Do we work long hours without a break?

The leading cause of reported work-related death for U.S. farmers involves machinery (NIOSH 1992). Agricultural machinery is also a leading cause of nonfatal injuries. The most recent data indicate that those working on farms are often struck by falling trees, caught in farm machinery, and engulfed in collapsing grain or other storage facilities. The accident rate for agricultural services was particularly high, with 94 workers killed per 100,000 workers employed. Workers included in this industrial group perform activities such as crop planting, dusting, and harvesting; veterinary and other livestock services; and landscape and other horticultural services. The primary causes of fatal injuries on farms were transportation-related incidents, such as tractor rollover and falls from tractors, and in various contacts with running machinery, falling objects, or collapsing grain and other materials (Toscano 1995).

A particular problem with injuries of this sort is that many go unreported. OSHA reporting requirements are limited to farms with 11 or more employees, but few farms are this large. If an injury is reported, data categories may vary from state to state.

The majority of Occupational Safety and Health standards pertaining to agriculture concern farm machinery. They are as follows:

   29 CFR 1910.266: safety standards for pulpwood logging
   29 CFR 1910.145: slow moving vehicle emblem
   29 CFR 1928.51: rollover protective equipment
   29 CFR 1928.51: safety for agricultural equipment
   29 CFR 1928.57: guarding farm field equipment

   29 CFR 1928.21: safety and health standards for agriculture; agricultural machine guards for farm field equipment, farmstead equipment, cotton gins.

In 1990, Congress appropriated $11 million to study agricultural hazards and to initiate reductions in the health and safety problems in agriculture. The National Institute for Occupational Health and Safety (NIOSH) created five agricultural funding initiatives to collect data on farm hazards and then initiate reductions in these hazards. Results from the North Carolina study are shown in Table 27.2.

Working Without

Tractors are involved in more farm accidents than any other piece of machinery (Schafer and Kotrlik 1986). Tractor overturns account for more farm work deaths then any other farm accident type.

Go by the Rules

In general, certain rules apply to the safe operation of farm equipment. These general rules are as follows:

- Know the standard;
- Dress for the job;
- Take frequent breaks;
- Keep equipment in good condition.

In dealing with machinery:

- Use machine guarding;
- Disengage equipment before attempting repairs.

When working with chain saws:

- Never cut alone;
- Keep anti-kickback devices in working order;
- Cut on the uphill side of a downed tree to prevent rollover.

When driving a tractor:

- Use a rollover protective structure (ROPS);
- Never mount or dismount a moving tractor;
- Do not allow extra riders.

When driving tractors on state roads and highways:

- Obey highway regulations;
- Attach slow-moving emblem.

Figure 27.5. Tractor accidents are the leading cause of farm injury.

CHILD LABOR

Farm work is notably unregulated when it comes to young workers. Most states exempt agriculture from child labor laws altogether. And on the federal level, various exemptions in the labor laws allow farm children to work at much younger ages than in other industries.

For example, 16- and 17-year-olds can do hazardous work in agriculture, while the age for similar work in other industries is 18. Ten- and 11-year-olds may be employed if the farmer gets a waiver from the Department of Labor, simply by proving that *not* employing 10- and 11-year-olds would cause severe economic hardship to the farm (Nixon 1995).

Children may work as agricultural workers in nonhazardous jobs at the age of 12 with the written consent of their parents or together with their parents on a farm. At 14, children may work full-time in nonhazardous jobs outside of school. At 16, they can perform hazardous jobs and don't have to be in school.

There are no comprehensive data on the number of children killed annually in farm accidents. The United Farm Workers Union and studies on migrant children estimate that 800,000 children work in agriculture. According to the Wall Street Journal, 23,500 are injured and another 300 die on the farm each year.

Child labor is necessary, in addition to the labor of adult migratory workers, in spite of the increase in the capability of farm machinery. This is because almost every fruit found in the kitchens of the diet-conscious is still picked by hand; every bunch of grapes, every avocado, apple, peach, plum, and strawberry. Lettuce is the only major acreage vegetable

Figure 27.4. Harvesting cotton mechanically.

When working with grain:

- Make grain bins off limits for children;
- Never enter a bin of flowing grain;
- Have a plan for safe entry, and exit when you need to enter the bin;
- Install ladders inside the bins.

But until reporting is required and standardized, injuries and fatalities will continue unchecked and unknown.

Table 27.1. 1994 North Carolina farm injuries/illnesses. Type of case report (N.C. Farm Injury Project).

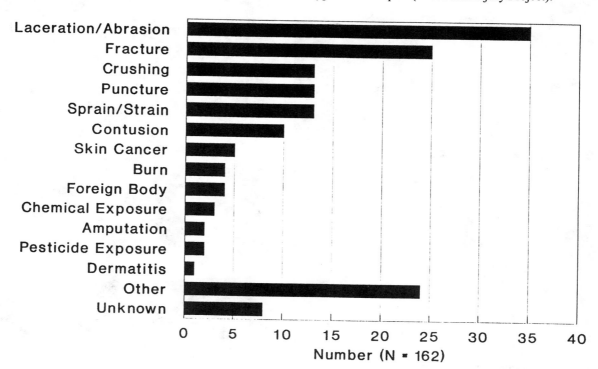

that is completely hand-harvested, but on smaller farms, tomatoes, cucumbers, sweet potatoes, and white potatoes may be hand-harvested or hand-graded. It is also because United States residents are accustomed to a cheap and plentiful food supply.

Go by the Rules

The secretary of labor has found and declared that the following occupations in agriculture are hazardous for minors under the age of 16. Excluding certain exemptions, no minor under 16 may be employed at any time in a set of specific occupations. Some of these occupations are listed below:

- Operating a tractor of over 20 horsepower;
- Operating a trencher, earth-moving equipment, forklift, potato combine, or power-driven circular, band, or chain saw;
- Operating or assisting in operating a corn picker, cotton picker, grain-combine, hay mower, forage harvester, hay baler, potato digger, feed grinder, crop dryer, mobile pea viner, forage blower, auger conveyer, self-unloading wagon, power-post hole digger, power-post driver, or nonwalking rotary tiller;
- Working in a pen or stall occupied by a bull, boar, stud horse, sow with pigs, or cow with newborn calf.

Figure 27.6. Packing cherry tomatoes.

Figure 27.7. Grading cherry tomatoes.

SUMMARY

It was data collection, analysis, and public outcry that brought workplace regulations into being. If data are missing or inaccurate, accidents will continue because nothing will be done to acknowledge or prevent them.

The safety regulations needed by farmworkers could be useful to farm operators as well. But because Congress, in an attempt to shield small-scale farmers, continues to exempt agricultural employers who have 10 or fewer workers, many farm-related injuries and illnesses continue to go unreported. When accidents go unreported, unsafe operating procedures or equipment go unreported as well.

Agricultural labor issues are integral forces in our economy. While prices of the chemical fertilizers, seeds, and machinery are fixed, and the grower can't guarantee a firm, fixed price for the agricultural commodity produced, labor costs may be the only way to cut corners.

"Cheap labor benefits agriculture in the short run... But it also helps to blind farmers to the technological changes they will have to make in order to compete with foreign producers, who have access to even cheaper labor. The most effective way to improve the lives of farm workers is simply to enforce the existing labor and immigration laws" (Martin 1983).

This chapter has discussed in detail the OSHA regulations that are applicable to agriculture. They are few. The chapter has also explored the exemptions to these regulations, specifically the exemption of "10 or fewer workers".

In 1994, the state of Washington became the first to remove the exemption of "10 or fewer workers." The effects of this rulemaking took effect on January 16, 1996, and the agricultural standards apply to all farm operations, regardless of size. This rule has been hotly contested by agribusiness groups.

The exemption from safety and health regulations is paid for in kind, with heavy injuries of both farmers and farmworkers, adults, and children. Agricultural regulations exist to protect these workers from illness and injury.

REFERENCES

29 CFR 1928.110. 1987. *Field Sanitation Standard*, (52 FR 16050-16096).

Arbab, D., and B. Weidner. 1986. Infectious diseases and field water supply and sanitation among migrant farm workers, *Am. J. Public Health.* 76(6).

Aspelin, A. 1992. *Pesticide Industry Sales and Usage: 1990 and 1991 Market Estimates*, Washington, DC.: U.S. EPA, Office of Pesticide Programs, H-7503W: 1-11.

Brookreson, W.E., S. Cant, and C. Sageser. 1994. *Coordinating Farmworker Protection Efforts of Washington State Department of Agriculture, Washington State Department of Health, and The Department of Labor and Industries: An Action Plan.* February 3.

Caring crosses state lines--teenagers aid in self-help construction. 1995. *SHE News*. Visalia, CA: Self-help enterprises 15(2).

The Colorado Post, 1991. Boulder County project opens door to solutions. 4 August.

Dahlberg, V.I. 1992. Sin Descanso. *Mas*, March-April.

Department of Public Health, Bureau of Environmental and Occupational Health. 1978. Section 12421 of Act No. 368 of the Public Acts of 1978, as amended, S333.12421 of the Michigan Compiled Laws. Agricultural Labor Camp Rules.

Ehlers, J.K., C. Connon, C.L. Themann, J.R. Myers, and T. Ballard. 1993. Health and safety hazards associated with farming. *Am. Assoc. Am. Health Nurses* 41: 9.

Ford, R.D., 1991. Twenty years of change at OSHA. *Job Safety and Health Quarterly*, Spring, pp. 20-32.

Grassroots Worker Protection, The Case for State Occupational Safety and Health Programs. 1995. A report from OSHPA, the Occupational Safety and Health State Plan Association, September: 1-6.

Hamilton, A. 1925. *Industrial Poisons in the United States.* New York: The Macmillan Company.

Hamilton, A., 1943. *Exploring the Dangerous Trades: The Autobiography of Alice Hamilton*. Boston: An Atlantic Monthly Press Book; Little, Brown and Company.

Hayes, S. 1992. Dreaded inspection proved painless. Fuquay-Varina, NC: *The Independent*, September 9.

Martin, P.L. 1983. Labor-intensive agriculture; the $18 billion U.S. fruit and vegetable industry is increasingly reliant on illegal-immigrant labor. *Sci. Amer.* 249, pp. 54-6.

Mather, T. 1994a. Searching for hens' teeth: information scarce on pesticide usage. *N. Car. Insight* 15(2-3): 20-31.

Mather, T. 1994b. How North Carolina stacks up against other states in the regulation of pesticides. *N. Car. Insight*, 15(2-3): 61-95.

Mercury News, Morning Edition. 1992. Coalition to help farm laborers improve their housing conditions. San Jose, CA. 25 February.

National Committee for Injury Prevention and Control. 1989. Occupational injuries. (U.S.) injury prevention: meeting the challenge. *Am. J. Prev. Med.*, 5: 177-89.

Nixon, R. 1995. Working in harm's way. *Southern Exposure* XXIII: 3/4.

North Carolina Department of Labor. 1993. *Introduction to Migrant Housing Inspections in North Carolina*, pp. 1-9.

North Carolina Department of Labor. 1991. One Family's Success Story. *Migrant Housing Bulletin*, May.

North Carolina Farm Bureau News. 1991. Testing well water for contamination. April 23.

NRAES-12. 1989. First on the scene. Ithaca, NY: Northeast Regional Agricultural Engineering Service, Cooperative Extension.

Occupational Safety and Health Reporter. 1987. *Current Report.* May 6, p. 1348. text p. 1363.

Planners OK housing for farm workers. 1995. Stockton, CA: *The Stockton Record*, 29 February.

Popendorf, W., and K. J. Donham. 1991. Agricultural hygiene. In: *Patty's Industrial Hygiene and Toxicology.* 4th edition. G.D. Clayton and F.E. Clayton, eds. New York: Wiley, pp. 721-61.

RCAP/MESA Newsletter. 1995. Creating a tool box for housing. Leesburg, VA: Rural Community Assistance Program, Vol. 2, Issue 4, Summer.

Roberts, D.L. 1992. *A Guide to Farm Safety and Health*. Division of Occupational Safety and Health, North Carolina Department of Labor. NC-OSHA #10.

Schafer, S.R., and J.W. Kotrlik. 1986. Factors affecting farm safety practices. *J. Safety Res.* 17(3): 123-27.

Smith, R.B. 1993. Perils in the fields: agricultural industry profile. *Occup. Health Safety Mag.*, May.

Stoskopf, C.H. and J. Venn. 1985. Farm accidents and injuries: a review and ideas for prevention. *J. Envir. Health* 47(5): 250-52.

Sumner, D.A. 1989. Professor discusses immigrant tobacco workers', conditions. Raleigh, NC: *Raleigh News and Observer*, 3 December.

Sweeny, M. and S. Ciesielski. 1990. Where work is hazardous to your health: a survey of occupational injuries and field sanitation among North Carolina farmworkers. In: *Migrant Health Newsline*.

Texas 'crop sheets' get boost; panel wants symbols in field. 1992. *Right-To-Know News*, 6: 14. May 22, pp. 7-8.

Toscano, G. and J. Windau. 1995. The changing character of fatal work injuries. In: *Fatal Workplace Injuries in 1993; A Collection of Data and Analysis*. Bureau of Labor Statistics, U.S. Department of Labor, Robert B. Reich, Secretary, Katharine G. Abraham, Commissioner Report 891.

U.S. General Accounting Office, Human Resources Division. 1992. *Hired Farmworkers Health and Well-Being at Risk*, GAO/HRD-92-46.

U.S. Department of Health and Human Services. 1990. *NIOSH Alert: Request for Assistance in Preventing Deaths of Farm Workers in Manure Pits*. DHSS (NIOSH) 90-103.

Walters, R. 1985. *The Washington Times*, 30 April.

Wilk, V.A. 1987. A field sanitation standard at last: the 14-year struggle. *Farmworker Justice News*, Spring.

EPA WORKER PROTECTION STANDARD FOR AGRICULTURAL PESTICIDES

Kay Glenn Harris, B.S.
Worker Protection Specialist
North Carolina Department of Agriculture

In the 1990s, the Environmental Protection Agency (EPA) significantly amended the scope and protection of the existing Worker Protection Standard for Agricultural Pesticides, which focused on the occupational health and safety related concerns of farm workers. This chapter addresses the employees affected by these regulations and identifies the tasks, such as mixing, loading, and cleaning application equipment, or applying pesticides. Worker Protection Standards are addressed in detail to help the reader determine necessary requirements for compliance with EPA rules and regulations.

INTRODUCTION

In 1992, the Environmental Protection Agency (EPA) significantly amended the scope and protection of the existing Worker Protection Standard for Agricultural Pesticides, which had focused on the occupational health and safety-related concerns of farm workers. The amendments expand coverage to include the working conditions of agricultural workers in four specific areas of agricultural production: on farms, forests, greenhouses, and nurseries. The employees affected by this regulation are identified by the tasks they perform. A *worker* performs hand labor tasks associated with the cultivation and harvesting of agricultural commodities. This includes such tasks as watering, weeding, or harvesting. A *handler* performs tasks involving direct contact with pesticides. These tasks include, mixing, loading, cleaning application equipment, or applying pesticides.

The intent of the revised rule is to protect workers by preventing or minimizing the potential risk of pesticide exposure by providing basic pesticide safety information, including the pesticide hazard information and the means to mitigate exposures.

The regulation places responsibility on the *agricultural employer* for ensuring that the employees who perform worker or handler tasks are provided at least the minimum of the standard requirements.

The information provided in the following chapter should serve as a guide for compliance to the Worker Protection Standard. It presents the *maximum* in requirements and does not address the exemptions or exceptions that may allow alternatives, which involve different requirements.

Implementation of the Worker Protection Standard regulation varies from state to state. For compliance with your state's requirements, contact the agency responsible for pesticide law enforcement or the regional EPA office for assistance.

THE AGRICULTURAL EMPLOYER'S RESPONSIBILITIES

The agricultural employer must identify *all* employees affected by the Worker Protection Standard (WPS). Therefore, clear definitions of the tasks performed on the establishment must be made. The WPS defines *a worker* as any person, including a self-employed person, who is employed for any type of compensation and who performs activities relating to the production of agricultural plants on an agricultural establishment. A *handler* is defined as any person, including a self-employed person, who is employed for any type of compensation by the agricultural establishment or commercial pesticide handling establishment and who performs tasks including mixing, loading, transferring, or applying pesticides, disposing of pesticides or pesticide containers, handling opened containers of pesticides, or acting as a flagger. The agricultural employer must provide both workers and handlers with the information found in the following five areas. All information *must be* provided in a language the workers understand.

Information Displayed at Central Location

Information displayed at a central location including:

- EPA WPS safety poster.
- Information about the nearest medical facility (name, address, telephone number).
- Information about each pesticide application (EPA registration number, active ingredient(s), description and location of the treated site, restricted entry interval (REI), time and date of the application).
- Location of the posted material.
- Any changes or updates to any of the posted materials
- Information posted should be legible.

Pesticide Safety Training. Workers and handlers must be trained at least once every 5 years. The minimum requirements include:

- Training conducted by a certified pesticide applicator. The training should be presented in an oral and \ or audiovisual format, using easily understood terms and in a language understood.
- EPA worker training manual for workers and handlers and any other EPA approved training materials.
- Use of any other EPA approved training materials.

Decontamination Supplies. Decontamination supplies must be reasonably accessible and within 1/4 of workers and handlers and equipped with the materials listed below:

For Workers

- Soap and single-use towels in sufficient quantities to meet the needs of the workers.
- Water for emergency eye flushing or whole body rinsing. The water must be safe for drinking and of a quality and temperature that will not injure the eyes or skin upon contact. Use of water stored in tanks used for mixing pesticides is prohibited.
- Decontamination supplies for workers are not allowed in areas being treated or under restricted entry intervals.

For Handlers

The same supplies as workers including:

- A coverall at all task (mixing, loading) locations including the area(s) designated for removing personal protective equipment.
- Enough water for routine washing, for emergency eye flushing, or washing the entire body.

Decontamination supplies must be containerized when handlers are required to be in a treated area(s).

Employer Information Exchange. Employer information exchange must be made between the agricultural employer and the commercial applicator to insure that the following information is acknowledged.

- Time, date, location and description of the area to be treated, the products' restricted entry interval requirements, the EPA registration number, and active ingredient(s).
- The type of warnings required for workers and handlers—oral and\or treated area posting notifications.
- Any additional safety precaution prescribed by labeling for workers and other persons.
- Information describing the specific locations of *all* areas on the agricultural establishment, which have been treated or under a REI and any restrictions for entering those areas.

Emergency Assistance. Emergency assistance is required to be provided when a handler or worker has been poisoned or injured by pesticides. Specific requirements are as follows:

- Transportation must be readily available.
- The victim and the medical personnel must be immediately provided with:
  - information about the victims' exposure and how the pesticide was being used.
  - the product name, EPA registration number, active ingredient(s), and all of the first aid and medical information from the product's label.

The additional employer duties for worker and handler address other restrictions and limitations during pesticide applications or entry prior to the expiration of a REI.

RESTRICTIONS DURING APPLICATIONS FOR WORKERS

- Early entry into fields during application is restricted to appropriately trained and equipped handlers.
- In a nursery establishment, workers must remain 100 feet away from the area being treated.
- Workers are not allowed in greenhouses during a pesticide application.

- Contact by workers with anything treated with a pesticide, and for which a REI applies, is prohibited.
- If required by labeling, post and / or orally inform workers of the treated areas and of the REI.
- Post all greenhouse applications.
- All entrances of the treated areas must be posted so they can be readily seen, including entrances from labor camps.
- The EPA WPS designed poster (14"X 16"), must be posted no sooner than 24 hours before the scheduled application, and removed before workers enter and within 3 days after the expiration of the REI.
- Inform workers before each application of the location and description of the treated area, and of the REI. Workers employed after the scheduled application must receive the same warnings.

HANDLER APPLICATION RESTRICTIONS

- Voice and sight contact must be made every 2 hours with anyone handling a pesticide with skull and crossbones on the label.
- Handlers are not allowed to make applications in a manner that would allow contact with anyone other than a trained, personal protective equipment (PPE) equipped handler.
- Fumigant-related application tasks performed in a greenhouse must be monitored (visually or voice) by a trained handler equipped with the appropriate PPE.

SPECIFIC INSTRUCTIONS FOR HANDLERS

- Inform all handlers of the tasks to be performed, and about information concerning the specifics of the pesticide and application equipment for safe application.
- All labeling must be made accessible at all times.
- Commercial handlers must be informed of all treated areas and any entry restrictions before any application is made on the establishment.

Application Equipment Safety

- All equipment must be inspected and/or repaired before each use.
- Only trained and properly equipped handlers are allowed to clean, inspect, or adjust equipment containing pesticides or residues.

Requirements for Personal Protective Equipment Care, Replacement, Storage, and Disposal

- Label-specified PPE must be clean, inspected before each day's use, repaired or replaced as needed.
- Respirators must fit correctly, dust/mist filters and vapor-removing cartridges/canister types should be replaced as prescribed by the manufactures or when damaged, or when breathing becomes difficult.
- PPE should be washed and stored separately from other clothing and laundry.
- If PPE can be reused, clean according to the manufacture's instructions, or as described by the pesticide label, or wash separately in detergent and hot water.
- PPE contaminated with undiluted pesticides baring DANGER or WARNING as a signal word must be discarded and disposed of in accordance with federal, state, and local regulations.

For complete guidance on the requirements, exceptions, and exemptions of the Worker Protection Standard, contact your state lead agency, or your Regional EPA Office for a copy of the *How To Comply: What Employers Need To Know* manual.

SUMMARY OF PROVISIONS: GENERAL SCOPE AND APPLICABILITY

Pesticide uses included in the scope are those involved in the production on agricultural plants on farms, and in forests, nurseries, and greenhouses.

Exceptions:

- Government-sponsored public pest control;
- Livestock uses;
- Habitations, gardens, lawns;
- Pasture/rangeland, rights-of-way and structures;
- Vertebrate pests;
- Attractants/repellents in traps;
- Post-harvest;
- Research uses of unregistered pesticides.

Exemptions: Owner and immediate family exempt from generic provisions, principally training, notification, decontamination, and emergency assistance. They must comply with pesticide-specific requirements, such as personal protective equipment (PPE) and restricted-entry intervals (REI's).

Responsibility

Employers have the responsibility to make sure the protections of this standard are provided to agricultural workers and pesticide handlers.

Employers may not prevent or discourage any agricultural worker or pesticide handler from complying with the standard and may not take retaliatory action against handlers or workers who attempt to comply.

Key Definitions

Agricultural employer:

- Hires or contracts for the services of agricultural workers OR
- Owns or is responsible for the management and condition of an agricultural establishment that uses such workers.

Commercial pesticide handling establishment is any establishment other than an agricultural establishment (farm, forest, nursery, or greenhouse) that:

- Employs handlers to apply pesticides on agricultural establishments OR
- Employs persons to perform tasks as crop advisors on agricultural establishments.

Handler:

- Mixes, loads, transfers, or applies pesticides;
- Disposes of pesticides or unrinsed containers;
- Handles opened containers;
- Flags;
- Cleans, adjusts, handles or repairs contaminated equipment;
- Assists with application;
- Enters enclosed area after use of airborne pesticide before PEL or ventilation criteria are met;
- Enters area treated with soil fumigant to adjust or remove tarps;
- Performs tasks as a crop advisor during application or an REI.
- Worker performs tasks (other than handler tasks) related to the production of agricultural plants on an agricultural establishment.

PROTECTIONS FOR ALL EMPLOYEES

Employers of pesticide handlers and agricultural workers must make sure the following protections are provided to workers/handlers in their employ.

Centrally Located Information

If workers or handlers are employed on an agricultural establishment, the employer must establish a central location to:

- Display a poster containing WPS-specified information;
- List the location of the nearest emergency medical facility;
- Post information about each pesticide application on the establishment, including:
  --location and description of treated area;
  --product name, EPA registration number and active ingredient(s);
  --time and date of the application; and
  --restricted entry interval for the pesticide.

The employer must also:

- Keep information about applications posted until at least 30 days after the REI expires; and
- Inform workers/handlers where the poster is located and allow them access.

Emergency Assistance

In the case of a suspected pesticide poisoning, the employer must make available prompt transportation to an appropriate medical facility.

The employer must also provide the worker or handler or the treating medical personnel with information from the pesticide labeling and information about how the suspected exposure occurred.

Decontamination

Employers must provide workers with decontamination supplies while the workers are performing permitted activities in a treated area where an REI is in effect or performing any activities in a treated area where an REI has expired within the past 30 days.

Employers must provide pesticide handlers with decontamination supplies while handlers are performing handling activities.

Supplies for washing pesticides from the skin and eyes must be provided within 1/4 mile of all workers/handlers, including:

- Enough water for washing (Water must be of a quality and temperature that will not cause illness or injury when it contacts the skin or eyes or if it is swallowed);

- Enough soap and single-use towels; and
- Clean coverall (at handler sites).
- Eyeflush water must be made immediately available to handlers and early-entry workers if they are required to wear protective eyewear.

Information Exchange

An agricultural employer must be informed when a pesticide is to be applied on the agricultural establishment and be provided the information needed to be posted at the central location by a commercial handler:

- Whether both oral warnings and treated area posted are required; and
- Any other protection requirements on the label for workers or other people.

A commercial handler employer must be informed by the agricultural employer of entry restrictions for and the location/description of any areas on the agricultural establishments that the commercial handler may be in (or within 1/4 mile of) which may be treated with a pesticide or be under an REI while the commercial handler is on the establishment. The commercial handler employer must provide this information to the commercial handler.

PROTECTIONS FOR WORKERS

Application Restrictions

The employer must keep workers other than trained and protected pesticide handlers out of an area being treated.

- Under some application conditions, employers must keep nursery or greenhouse workers out of locations that are near an area being treated.

Entry Restrictions

If contact with pesticides is possible, the employer must keep workers from entering a treated area until the REI is over.

Exceptions:

- Non-hand labor tasks can take place up to 8 hours/worker/day;
- Tasks can take place, if necessary, due to a declared agricultural emergency; and
- Additional exceptions can be requested of EPA.

Employers must protect early-entry workers by making sure of the following:

- No entry for the first 4 hours following the end of the application and until any label-specified inhalation exposure level or the WPS ventilation criteria have been met;
- Workers are informed about health effects and safety information from pesticide labeling;
- PPE provided, cleaned, and maintained *for* the worker;
- Worker wears and uses PPE correctly;
- Workers are instructed how to put on, use, and remove the PPE and about the importance of washing thoroughly after removing PPE;
- Workers provided a clean place to put on and take off PPE and to store personal clothing;
- Action taken, if necessary, to prevent heat-related illness while wearing PPE;
- Soap, towels, and water provided when PPE is removed, and
- Make sure no contaminated PPE is worn home or taken home.

Training for Workers

Unless already a certified applicator or a trained handler, each early-entry worker must be trained before performing permitted tasks in a treated area which remains under an REI.

Unless already a certified applicator or trained handler, workers must receive basic pesticide safety training and the complete WPS training before their sixth day of entry into treated areas on an agricultural establishment within 30 days of the REI expiration.

The training must include written or audiovisual materials and be presented in a manner the worker can understand, using nontechnical terms.

The trainer must be a certified applicator, a trainer of certified applicators, someone who has completed an approved train-the-trainer course, or a trained handler.

The training program must contain the general pesticide safety information specified in the WPS.

Notice of Applications

On farms, nurseries, and forests, each worker who might enter a treated area or walk within 1/4 mile of a treated area during application or an REI must be warned orally or by posting warning signs at the treated area.

In greenhouses, each worker who might enter a greenhouse during an application or an REI must be warned by posted warning signs at entrances to treated areas.

Some pesticides will have a statement on the product labeling requiring both posting of warning signs and oral warnings to workers.

The posted warning sign must:

- Include the words: "Pesticides/Pesticidas—Danger/Peligro—Keep Out/No Entre";
- Contain the WPS warning sign symbol (a stern face and an upraised hand);
- Meet size and color requirements; and
- Be visible at all usual entrances to the treated area.

The oral warning must:

- Give location and description of treated area;
- State the time during which entry is restricted; and
- Instruct workers not to enter the treated area until the REI is over.

PROTECTIONS FOR HANDLERS

Application Restrictions

The employer and the handler must make sure that no pesticide is applied so as to contact, either directly or through drift, any person other than a trained and protected handler.

The employer must make sure that any handler who is handling a pesticide with a skull and crossbones symbol on the label is monitored visually or by voice contact at least every 2 hours.

The employer must make sure that any handler who is handling a fumigant in a greenhouse maintains continuous visual or voice contact with another handler.

Personal Protective Equipment

When personal protective equipment is required by the product labeling for the handling activity, the employer must:

- Provide the PPE to each pesticide handler;
- Clean and maintain the PPE correctly;
- Make sure that each handler wears and uses the PPE correctly;
- Provide each handler a clean place to put on and take off PPE and to store personal clothing;
- Take action, if necessary, to prevent heat-related illness while PPE is being worn;
- Provide soap, towels, and water to each handler at the end of the handling activity when PPE is removed; and

- Not allow any handler to wear home or take home PPE worn for handling activities.

Knowledge of Pesticide Labeling

The employer must make sure that each handler has either read the pesticide labeling or been informed of the information on the labeling.

The pesticide product labeling must be accessible to the handler during the handling activity.

Safe Operation of Equipment

The employer must make sure that each handler is instructed in the safe operation of handling equipment.

The employer must make sure that all handling equipment is inspected and in good operating condition before each use.

Training for Handlers

Unless already a certified applicator or trained to use restricted-use pesticides, handlers must be trained before performing handler tasks.

The training must include written or audiovisual materials and be presented in a manner the handler can understand.

The trainer must be a certified applicator, a trainer of certified applicators, or someone who has completed an approved train-the-trainer course.

The training program must contain the general pesticide safety and correct handling practice information specified in the WPS.

Cleaning and Maintaining PPE

The employer must make sure that anyone cleaning PPE is informed:

- That the PPE may have pesticides on it;
- Of the potentially harmful effects of pesticides; and
- Of the correct ways to handle and clean PPE.

The employer must make sure that:

- PPE is inspected and repaired before each use;
- PPE is cleaned according to manufacturers' instructions or in detergent and hot water;
- PPE that cannot be cleaned is disposed of;
- Clothing drenched with concentrates of Danger or Warning pesticide are disposed of;

- PPE is kept, washed, and stored separately from personal clothing;
- Clean PPE is dried appropriately; and
- Respirator filters, cartridges, and canisters are replaced as often as required.

PPE Substitutions and Exceptions

Substitutions and exceptions to PPE are permitted when engineering controls are used.

Pilots in open cockpits are exempted from any chemical-resistant footwear requirement; a helmet may be substituted for chemical-resistant headgear, and a visor may be substituted for protective eyewear.

Pilots in closed cockpits are exempted from all PPE requirements; long-sleeved shirt, long pants, shoes and socks are required.

Handlers using closed systems for mixing and loading are exempted from all PPE except chemical-resistant gloves and apron; long-sleeved shirt, long pants, shoes, and socks are required. If the closed system is pressurized, protective eyewear is also required.

Handlers using enclosed cabs are exempted from all PPE except for any respirator requirement; long-sleeved shirt, long pants, shoes, and socks are required. Respirators are waived if the enclosed cab offers respiratory protection equal or greater to the type of respirator specified.

Handlers or early entry workers working with plants with sharp thorns may wear leather gloves over chemical-resistant glove liners.

Handlers or early entry workers working in rough terrain may wear leather boots instead of chemical-resistant footwear.

REQUESTS FOR EXCEPTION TO REI

Affected parties may request that EPA grant an exception to the prohibition of routine hand labor tasks during an REI.

Exception request must include specified information about the need, nature, feasibility, and basis for the request.

EPA will issue a notice that a request has been received and allow at least 30 days for public comment.

If no decision is issued by 9 months after close of comment period, applicants may presume the request is denied, unless the deadline is extended due to the complexity of the request or the number of requests being reviewed.

If a request is granted, employers must provide the early-entry workers with the protections required by the WPS for other early entries.

EPA may withdraw exceptions to REI's if information indicates that the health risks to workers are unacceptable or if it is no longer needed. Affected parties may request a hearing when an exception is withdrawn.

WORKER PROTECTION STANDARD QUESTIONS AND ANSWERS

What Is the Revised Worker Protection Standard (WPS)?

The Environmental Protection Agency revised the Worker Protection Standards to reduce the risk of illness or injury resulting from pesticide handlers' and agricultural workers' occupational exposures to pesticides used in the production of agricultural plants on farms, nurseries, greenhouses, and forests.

How Does the Revised Standard Differ from the Older Standard?

The revised standard increases protection for agricultural workers and pesticide handlers who are often the most highly exposed employees. This standard does not permit workers to perform hand labor tasks in treated areas during re-entry intervals without adequate safety/protective clothing and equipment. The revised WPS not only addresses the agricultural worker (nurseries, greenhouses, and forest operations), but also the commercial pesticide handling establishment, which employs handlers to apply pesticides on agricultural establishments *or* employs persons to perform tasks as crop advisors on agricultural establishments.

The WPS expands requirements for:

- Use of personal protective equipment (PPE),
- Pesticide application warnings (posted, oral)
- Restricted-entry intervals (REI)
- Decontamination supplies
- Emergency assistance (transportation and pesticide product information)
- Pesticide safety training

Whom Does the Worker Protection Standard Affect?

Persons primarily responsible for compliance are growers or employers involved in the production of agricultural plants, on farms, and in forests, nurseries, and greenhouses. Also included are commercial pesticide establishments, which employ handlers to apply pesticides or persons to perform tasks as crop advisors on agricultural establishments (Table 28.1.).

Table 28.1. Compliance at a glance.

	Workers	Handlers
Central Information	X	X
Safety Training	X	X
Decontamination Sites	X	X
Employer Info Exchange	X	X
Emergency Assistance	X	X
Post\Oral Notification (see label for both)	X	X
Applications/ Monitors		X
Entry During REI		X
PPE		X

The agricultural worker and the pesticide handler are primary beneficiaries of the revised rule and are defined as follows:

Handler: performs any of the following tasks:

- Mixes, loads, transfers, or applies pesticides;
- Disposes of pesticides or unrinsed containers;
- Handles opened containers;
- Flags;
- Cleans, adjusts, handles, or repairs contaminated equipment;
- Assists with application;
- Enters enclosed area after use of airborne pesticide before ventilation criteria are met;
- Enters area treated with soil fumigant to adjust or remove tarps;
- Performs tasks as a crop advisor during application or a REI.

Worker: performs tasks (other than handler tasks) related to the production of agricultural plants on an agricultural establishment.

Substantial labeling revisions will be made to approximately 8,000 to 10,000 agricultural pesticides to reflect the WPS requirements.

Are There Any Exceptions to This Standard?

Yes. The following are allowable exemptions:

- Government-sponsored public pest control;
- Livestock uses;
- Habitations, gardens, lawns, etc.;
- Pasture/rangeland, right-of-way and structures;
- Vertebrate pests;
- Attractants/repellents in traps;
- Post-harvest uses;
- Research uses in unregistered pesticides.

The owner and immediate family are exempt from generic provisions, principally training, notification, decontamination, and emergency assistance. However, the owner and family must comply with pesticide-specific requirements such as personal protective equipment (PPE) and restricted-entry intervals (REI's).

How Will Agricultural Employees Be Informed About Pesticide Hazards?

The revised WPS ensures that this information is provided by:

- Requiring notification either orally, posted, or in a language or manner that the worker can understand;
- Requiring safety training for all workers and handlers;
- Informing and providing pesticide handlers and early entry workers with pesticide label information;
- Requiring that a list of the most recently applied pesticides be kept on the establishment.

Certain provisions of the Worker Protection Standard have been amended. For information about the exceptions, exemptions and policy changes to the WPS, contact your regional office of the Environmental Protection Agency.

The following list provides the agricultural employer with WPS compliance requirements for workers and handlers at a glance.

For additional information on the Worker Protection Standard, contact the EPA regional office in your area.

Region 1 (MA, CT, RI, NH, VT, ME)
U.S. Environmental Protection Agency Region 1
Pesticides an Toxic Substances Branch (APT)
1 Congress Street
Boston, MA 02203
(617) 565-3800

Region 2 (NY, NJ, PR, VI)
U.S. Environmental Protection Agency, Region 2
Pesticides and Toxic Substances Branch (MS-105)
2890 Woodridge Ave., Building #10
Edison, NJ 08837-3679
(908) 321-6765

Region 3 (PA, MD, VA, WV, DE)
U.S. Environmental Protection Agency, Region 3
Toxics and Pesticides Branch (3AT-30)
841 Chestnut Building
Philadelphia, PA 19107
(215) 597-8598

Region 4 (GA, NC, SC, AL, MS, KY, FL, TN)
U.S. Environmental Protection Agency, Region 4
Pesticide and Toxic Substances Branch
12th Floor, AFC
100 Alabama Street, SW
Atlanta, GA 30303-3104
(404) 562-9012

Region 5 (IL, MI, MN, IN, OH, WI)
U.S. Environmental Protection Agency, Region 5
Pesticides and Toxic Substances Branch (5SPT)
77 W. Jackson Blvd.
Chicago, IL 60604
(312) 886-6018

Region 6 (TX, OK, AR, LA, NM)
U.S. Environmental Protection Agency, Region 6
Pesticides and Toxics Branch (6T-P)
1445 Ross Ave.
Dallas, TX 75202-2733
(214) 665-7235

Region 7 (MO, KS, IA, NB)
U.S. Environmental Protection Agency, Region 7
Toxics and Pesticides Branch (TOPE)
726 Minnesota Ave.
Kansas City, KS 66101
(913) 551-7020

Region 8 (CO, MT, ND, SD, UT, WY)
U.S. Environmental Protection Agency, Region 8
Toxic Substances Branch (8ART-TS)
One Denver Place, Suite 500
999 18th Street
Denver, CO 80202-2405
(303) 293-1440

Region 9 (CA, NV, AZ, HI, GU)
U.S. Environmental Protection Agency, Region 9
Pesticides and Toxics Branch (A-4)
75 Hawthorne Street
San Francisco, CA 94105
(415) 744-1090

Region 10 (WA, OR, ID, AK)
U.S. Environmental Protection Agency, Region 10
Pesticides and Toxic Substances Branch (AT-083)
1200 Sixth Ave.
Seattle, WA 98191
(206) 553-1198

SUMMARY

A very large segment of the population is employed or indirectly involved with agricultural work. It is therefore imperative that the occupational hazard associated with one of the hazards, pesticide exposure, be addressed in a manner that can minimize the risks.

The provisions of the Worker Protection Standard are geared toward the conditions of two types of employees on agricultural establishments: those who handle pesticides and those who perform tasks related to the cultivation and harvesting of plants on farms, forests, nurseries, and greenhouses. These provisions are intended to: 1) eliminate or reduce exposure to pesticides through proper use and maintenance of personal protective equipment and pesticide application equipment; 2) mitigate exposures that occur by providing decontamination supplies and emergency assistance in the event of pesticide poisoning or injury; and 3) inform employees about the hazards of pesticides by providing complete information and safety training about the pesticides and the risks or hazards associated with exposure.

The current changes made to the Worker Protection Standard were done in response to comments received by the U.S. Environmental Protection Agency in an effort to improve the implementation of the Standard nationwide. If you have any questions, concerns, or comments about the implementation of the WPS in your state, contact the state lead agency or your county extension service for assistance.

REFERENCES

40 CFR, Protection of Environment, Parts 170 Worker Protection Standard, as amended, 1992, pp. 222-247.

Environmental Protection Agency (EPA). 1993. *The Worker Protection Standard for Agricultural Pesticides--How To Comply*, pp. 1-79.

Environmental Protection Agency (EPA). 1993. *Protect Yourself from Pesticides--Guide for Pesticide Handlers*.

Environmental Protection Agency (EPA). 1993. *Protect Yourself from Pesticides--Guide for Agricultural Workers*.

North Carolina Pesticide Law of 1971, as amended. Regulation 2 NCAC 9L .1800. Worker Protection Standard.

SECTION 3:

SAFETY AND HEALTH IN FORESTRY

29

EPIDEMIOLOGY OF FORESTRY INJURIES AND ILLNESSES

R. Lane Tippens, M.D., M.P.H.
Ricky Langley, M.D., M.P.H.
Duke University School of Medicine

> The forestry industry is involved with the planting and harvesting of trees for human usage. Logging is the most hazardous occupation in the forestry industry. It is estimated that 1 out of 5 loggers is injured at work yearly and about 2 out of 1,000 die each year from a job related injury. As described in this chapter, deaths occur from a variety of causes: falling objects, motor vehicle incidents, machinery incidents, and being caught in, under or between objects. Besides the acute traumatic events that are too common, chronic injuries are frequently noted in forestry workers. Repetitive strain injuries such as back strain and vibration-induced white finger are common. Other hazards discussed include noise, exhaust, environmental infections, allergic and contact dermatitis from trees, pesticide exposures, and respiratory disorders from wood dust. The use of personal protective equipment and a brief discussion on various methods to prevent injuries is presented.

INTRODUCTION

The forestry industry involves the science of cultivating, maintaining, and ultimately, the harvesting and transporting of forest products. There are two main components: The cultivation and maintenance, referred to as silviculture, and logging, the harvesting of the commodity. The more significant hazards are associated with the latter activity. Logging consists of felling trees, removing the limbs and branches (limbing), and cutting or splitting the trees into manageable logs (bucking). Trees and logs are then moved to central locations (landings) by one of several methods (skidding or forwarding). When terrain is not ideal, logs may be transported by steel cables attached to a winching apparatus (cable yarder) via a system of cables, blocks, pulleys, and carriages (cable yarding). Logs are then partially suspended and dragged over the ground (high-lead yarding) or hoisted into the air and conveyed on overhead cables (skyline yarding) to the landing site. At the landing, logs are mechanically loaded onto vehicles for transport to sawmills or pulpmills.

Traditional forest work requires demanding physical labor, and while this has been lessened somewhat by modern mechanization, the machinery used in harvesting wood products have their own intrinsic hazards and may create sources of monotony and social isolation.

Generally, logging can be characterized as an extremely arduous activity that entails the handling of materials of immense mass, using potentially dangerous tools, out-of-doors often in remote and isolated environments, often with unstable or rough, uneven terrain, vulnerable to climatic extremes and endemic disease.

INJURY ANALYSIS

The NIOSH-sponsored National Traumatic Occupational Fatality (NTOF) surveillance system utilizes death certificate data to yield an average annual rate of fatality per 100,000 workers. During 1980 to 1989 nearly 6,400 total workplace deaths were recorded (NIOSH 1993). Forestry workers were grouped with agriculture and fishery occupations, and had an annual death rate of 20.3/100,000, an incidence that was only exceeded by mining, construction, and transportation. Logging is the most hazardous activity within the forestry industry and 1,492 deaths occurred to loggers during this time period. The nonmanagerial employment in the logging industry during this time period averaged approximately 87,500 workers. When considered exclusively, the annual fatality rate among loggers was found to be 164/100,000, nearly 23 times that for all other U.S. workers (NIOSH 1994).

The Bureau of Labor Statistics (BLS) found 158 fatalities in the logging industry in 1992 (BLS 1994). This amounts to a

2 in 1,000 risk of death per year. NIOSH estimates that there are 16,500 compensable injuries each year, which amounts to an incidence rate of 1 in every 5 loggers. The U.S. Department of Agriculture states that the logging industry accident rate has pushed workers' compensation insurance to 40% of payroll cost (Federal Register 1994).

The NTOF data files found the largest number of fatalities were among white males working in the combined occupation group of logging occupations/general laborers (Myers and Frosbroke 1994). The majority of logging deaths (65.5%) occur to workers classified as heavy logging occupations (e.g., fellers, limbers, buckers, choker setters) followed by truck drivers (11%), general laborers (7.5%), and material moving machine operators (5.5%).

The fatality rate also varied by region of the country with the Central and East regions having the highest rates and the Northeast and Lake states the lowest rates.

Myers and Fosbrake analyzed NTOF data for the years 1980 through 1988 and reported on their study of 1,278 logging fatalities (Myers and Frosbroke 1994). The majority of deaths (87%) were due to being struck by a falling object, motor vehicle incident, machinery incident, and being caught in, under, or between objects.

Of 636 deaths related to falling objects, 436 were related to felled trees, 102 to falling logs, 77 due to limbs and snags, and 26 involved other falling objects. Loading or unloading logging trucks were involved in 51 of the 102 deaths due to falling logs.

Deaths from motor vehicles were usually due to logging trucks (65.6%) and were primarily traffic related (83.1%). Of 185 machinery deaths, equipment rollovers were responsible for 76 deaths, being run over caused 29 deaths, being caught between pieces of equipment caused 19 deaths, and cable breaks were responsible for 14 deaths.

Of 114 deaths due to being caught in, under, or between objects, rolling logs were responsible for 68 deaths and trees caused 15 deaths.

Environmental hazards, falls, and electrocutions were also responsible for a moderate number of logging deaths yearly.

The Bureau of Labor Statistics estimated in 1991 an injury incidence rate among loggers of 15.6 cases per 100 full time workers compared to 7.9 per 100 for the combined private sector (Federal Register 1994). The 1991 lost-workday incidence rate for logging was 9.9 per 100 full time workers compared to 3.7 per 100 for the combined private sector. The lost-workday rate, an indicator of severity of injury, was 274.8 lost workdays per 100 full time workers, compared to 79.8 per 100 for the private sector. The BLS reveals that there were 8130 nonfatal injuries to loggers per 100,000 in 1993 (BLS 1995).

ACUTE TRAUMA

Injuries in forestry have also attracted the attention of epidemiologists internationally. A Finnish study, which compared census records with death certificates, found that the mortality for active forestry workers was the second highest of all occupations over a 15-year period, encompassing 10 million person years (Notkola et al. 1993; Green 1993). A New Zealand study, which investigated logging specifically, revealed a fatal injury rate of 203 per 100,000 per year among loggers (Marshall et al. 1994). During the lifetime of Swiss forestry workers, 1 in 3 can expect some permanent disability attributable to their work, and 1 in 17, death. Recent data from the Swedish Bureau of Statistics revealed a relative incidence of forestry accidents to be 120 per million hours worked. By comparison, agricultural workers experienced 50 accidents per million hours worked during the same period (Hansson 1987). In Canada, the forestry industry had the highest rate of lost time injury (161 per 1,000 workers) and the highest fatality rate (1.04 per 1,000) of all industries in 1986 (Salisbury et al. 1991).

A study of deaths (n=87) among fellers and buckers in British Columbia from 1981 to 1987 sheds some light on the factors which contribute to adverse incidents (Salisbury et al. 1991). Falling trees were responsible for 86.2% of the cases and rolling logs for 9.2%. Chain saws caused two deaths, one person was run over by a skidder, and one person was killed in a work-related MVA. Head injury was responsible for 48.8% of the fatalities, followed by multiple injuries (22.1%), chest crush injuries (11.6%), asphyxia (9.3%), and exsanguination (6.9%). An environmental factor was involved in 55.2% of the incidents, while work practice errors were involved in 58.2%, inadequate supervision in 39%, and personal factors such as financial problems or impending retirement were felt to be contributory to the incident in 5.7% of the cases.

Environmental factors deemed important in this study included chain saw kickback (2.3%), lean of the tree (3.4%), the terrain (8%), weather (4.6%), previously fallen trees or windfalls (2.3%), hangups (13.8%), and excessive snow depth (2.3%).

Work practices which were considered in error included snags (43%), domino falling (10.1%), falling trees into standing timber (5.1%), working too close (1.3%), or "Dutchman" (7.6%). The Dutchman procedure involves an incomplete cut used to affect the direction of tree fall. Domino falling is an unsafe practice using one tree to cause several other trees to fall in succession.

In New Zealand from 1975 to 1988, 81 work-related fatalities among loggers and ten fatalities among silviculture workers occurred yielding fatality rates of 2.03 and 0.15 per 1,000 workers per year, respectively. This compares to an overall rate among the New Zealand workforce of 0.07 deaths per 1,000 workers per year (Marshall et al. 1994). The hospitalization rate among loggers was 38.93 and among silviculture workers was 9.58 per 1,000 workers per year during this time period. The leading cause of death among loggers was due to falling trees accounting for over half of the fatal injuries. Among silviculture workers, falling trees accounted for a third of their deaths. The three commonest types of contact resulting in hospitalization among loggers were chainsaw injuries (33%), falling trees (25%), and rolling logs (8%). Among silviculture workers, chainsaw injuries (23%), falling trees (23%), and slips, falls, and trips (20%) were the most common.

In the United States, the main causes of acute injury are chainsaw injuries, falling wood, rolling logs, and trips and falls (BLS 1984). The body parts most frequently involved are the lower extremities (knee, ankle, and foot). The leading cause of death is from contact with falling trees, branches, or limbs, and these accidents accounted for 60% of logging deaths (NIOSH 1994). Indeed, the logging industry leads all other sectors in incidence of injuries caused by being struck by an object. The Bureau of Labor Statistics found 342 cases of this type of injury per 10,000 workers in 1993.

Other acute mechanical injuries which frequently occur include foreign body eye injuries, and abrasions to the eye and to exposed skin (Chapel 1992).

Accidents during transportation contribute to the total injury rate in this industry. Logging roads are typically primitive, may course through rugged terrain, and vehicular mishaps are not rare. Also, workers, especially those not living in a logging camp, may commute significant distances. A more unusual form of vehicular trauma has been associated with the use of helicopters for timber harvesting in remote Alaska. Over a recent 18-month period, 19 injuries, 9 of them fatal, resulted from helicopter crashes during logging operations (CDC 1994).

After studying regional differences in forestry practices and fatality rates in the U.S., possible factors influencing the risk of logging injuries have been evaluated (Myers 1994). Potentially higher fatality risks include hardwood harvesting, sawtimber harvesting, selective harvesting, natural stands, cyclical wood markets, steep terrain, small business operation, unmanaged forest stands, and logging safety unregulated. Potentially lower risks were associated with softwood harvesting, pulpwood harvesting, clear-cut harvesting, plantation stands, mechanized harvesting, steady wood markets, level terrain, corporate operation, managed forest stands, and regulated logging safety.

In 1984, the Bureau of Labor Statistics published a detailed analysis of injuries among 1086 logging employees in the U.S. (BLS 1984). A summary of some of the findings is presented in the following tables (See Tables 29.1, 29.2, 29.3).

Table 29.1. Description of work site and activity performed at time of injury.

Worksite	Percent
Location of Injury	
Cutting site	53
Landing	20
Between cutting and landing site	18
Employer built road	3
Highway	4
Terrain Where Injury Occurred	
Flat ground	44
Medium slope	36
Steep slope	19
Ground Cover at Injury Site	
Little or no brush	35
Moderate brush	37
Heavy brush	26
Swampy, marshy, boggy	3
Activity at Time of Injury	
Felling trees	23
Limbing	15
Choker setting or hookup	14
Bucking	12
Tractor or cable skidding	9
Chasing	5
Loading/unloading	5
Rigging	4
Servicing equipment	4
Hauling logs to mill	1
Other	8

Source: BLS 1984

CHRONIC INJURY

Repetitive strain injury is the most common subacute or chronic malady experienced by forestry employees. Saw operators may work for a considerable proportion of time holding awkward static postures which may contribute to musculoskeletal stress.

A typical tree-planter will travel 2.4 km carrying 16.8 kg of equipment and materials in order to plant an average of 1250 seedlings daily. Fifty percent of the planters in this study reported a work-related accident during their planting career. The lower extremities, the skin, the eyes, and the wrist were most frequently injured (Giguere et al. 1993). There is evidence that the work rate and rest time of a tree planter are not well ordered within the daily and seasonal working routine, and may compromise worker well-being in the silviculture industry (Trites 1993).

Table 29.2. Cause of logging incident (n=1086).

Etiology	Number	Percent
Hit by limb, tree, log	259	24
Falling wood	127	
Rolling logs	37	
Logs rigged for yarding	30	
Other	65	
Slip, trip, or fall	258	24
Falls from elevation	105	
Falls to same level	117	
Injured by chain saw	222	20
(See Table 29.3)		
Muscular strain	85	8
Hit by cable, hook, chain	60	6
Foreign body eye injury	55	5
Mobile equipment accident	33	3
Skidder	9	
Log truck	17	
Mobile equipment n.e.c.	2	
Ground surface	1	
Other	4	
Other	114	10

Source: BLS 1984

Table 29.3. Cause of chain saw injuries.

Etiology	Percent*
Chain saw kicked back	64
Fell on saw	13
Did not have tight grip on saw	7
Hand slipped into chain	6
Wrong cutting method	3
Chain on saw broke	3
Using wrong size saw	1
Saw ran after shutoff	1
Improper saw maintenance	<1
Other	18

*For chain saw percentage in Table 29.3, because more than one response is possible, the sum of the percentages may not equal 100. Percentages are calculated by dividing each response by the total number who answered the question.

Source: BLS 1984

An appreciable number of chain saw operators develop vibration-induced white finger disease (VWFD) (Hutton et al. 1993). This primarily vasomotor disorder is also known as "occupational Raynaud's syndrome", "hand-arm vibration syndrome", and "traumatic vasospastic disease". According to NIOSH, 20 to 50% of employees in felling operations may be affected by hand-arm vibration syndrome. In Japan, 9.6% of overall forestry workers reported VWFD but the rate was 21% in those with over 30 years exposure (Mirbod et al. 1992). In Finland, the prevalence was 40% in the early 1970s and 78% reported nocturnal hand numbness (Koskimies et al. 1992). Without question, this disease is associated with frequent exposure to vibration, predominantly in the 30 to 500-Hz range.

Symptoms progress from numbness precipitated by exposure to vibration, which then persists for increasing duration of time as exposures continue. Next, finger blanching may follow, and finally may come loss of some degree of motor function. In some cases, there have been associated nerve conduction abnormalities, which may confuse diagnosis with carpal tunnel syndrome. The greater the vibration acceleration in the saws, the greater the likelihood of VWFD. After modifications were made to chainsaws to decrease the weighted vibration acceleration from 14 to 2 m/s^2 and the saws made lighter, the prevalence of VWFD among Finnish loggers decreased to 5% and numbness to 28% and complaints of hand weakness decreased from 19% to 9% (Koskimies et al. 1992). Other risk factors seem to be exposures in excess of 2 hours a day, and work in colder climates (Nagase et al. 1992).

PHYSICAL WORKLOAD

Silviculture and logging can be exceptionally arduous occupations. Due to the high physical demands of the job, forest workers have a high aerobic capacity (Kurumatani et al. 1992). In a study of loggers cutting logs with a chainsaw, the heart rate increased as well as hormonal levels of cortisol, adrenaline, noradrenaline, ACTH, and dopamine. It is postulated that vibration may have an effect on the whole body, including the hypothalamus and the limbic lobe of the brain (Matoba 1985).

In an occupation where physical workload may approach 4,000 kcal/day energy expenditure, the fatigue threshold may be easily exceeded, increasing the likelihood of musculoskeletal injuries, accidents, and heat exhaustion. Additionally chronic exposure to noise and vibration can fatigue body structures, which may lead to permanent physiologic damage (Crutchfield and Spanks 1991). Upper limb disorders including upper limb pain, muscle-tendon syndromes and carpal tunnel syndrome have been noted to occur frequently in chainsaw forestry operators. These upper extremity disorders increased as vibration exposure increased (Bovenzi 1991; Chernial 1994).

NOISE AND EXHAUST

Chainsaws and heavy machinery such as skidders and complex harvesting machines may generate noise often in excess of 110 dB. Hearing loss has been reported among forestry workers in many countries (International Labor Organization 1983). Studies have shown that loggers with VWFD had a 10dB greater noise-induced permanent threshold shift than in loggers without VWFD. It is proposed that vibration may produce vasoconstriction in both cochlear and digital blood vessels as a result of sympathetic nervous system activity (Pyykko et al. 1981).

Loggers may experience discomfort from chainsaw exhaust. In one study, 34% of loggers often experienced discomfort and another 59% occasionally experienced discomfort from the exhaust. The symptoms included cough and irritation of the nose and throat (45%), eye irritation (27%), headache (27%), nausea (17%), and eczema (12%). Loggers may be exposed to hydrocarbons, benzene, formaldehyde, and carbon monoxide in exhaust from chain saws (Nilsson 1987). The long term effects from this exposure are unknown.

ENVIRONMENTAL HAZARDS

Work in remote forested areas entails exposure to a number of natural hazards. Biting insects are the most frequently encountered hazard. These encounters usually represent a mere nuisance, but nationally more deaths are attributable to insect stings than all other animal encounters (Langley 1994). Severe systemic anaphylactoid reactions are not unusual after stings by members of the order *Hymenoptera*, including fire ants, wasps, and bees. Severe reactions as well may occur following stings from scorpions belonging to the order *Scorpionida*. Hymenoptera stings were significantly more frequent and the percentage of systemic reactions was much greater in forestry workers as compared to controls in one study (Shimizu et al. 1995). Anaphylaxis has also been witnessed after bites from deerflies, blackfies, and horseflies. The roster of other biting or stinging arthropods encountered in the forest includes ticks, mosquitoes, biting midges, blister beetles of the order *Coleoptera*, stinging caterpillars of the order *Lepidoptera*, and spiders (Frazier 1987). Stings of some insects, such as fire ants and stinging caterpillars, result in lesions that may be unique to these animals. The medical practitioner providing care for these workers should learn to recognize all lesion patterns that may be pathognomonic to species indigenous to local forestry sites. The role for insects as disease vectors is detailed elsewhere.

Snake bites are a recognized occupational hazard in the forestry industry. In the continental United States, most venomous snakes belong to the *Crotalidae* (pit viper) family, characterized by mostly hemopathic venom. Members include several species of rattlesnakes, copperheads, and cottonmouth water moccasin, and representatives can be found in all but a few states. The coral snake, native to the deep south region of the country, is the lone representative of the *Elapidae* family, and its venom is characterized as being rich in neurotoxins. So-called harmless snakes may deliver a bite wound prone to infection, particularly with microorganisms unique to reptiles, namely various *Salmonella* species. The Gila Monster lizard (*Heloderma suspectum*), which can produce a neurotoxin, can deliver a bite resulting in edema and much pain, but this animal is usually only found in the desert southwest. A number of "harmless" snakes such as the *Thamnophis elegans vagrans* (garter) can produce amounts of low toxicity venom which can cause localized symptoms (Gomez et al. 1994). Since most snakes are poikilothermic and undergo hibernation, snake bites generally occur during warmer months. Numerous venomous snakes live in foreign countries and many fatalities occur yearly from snakebites, especially in tropical regions.

Year-round work in the open air necessarily implies exposure to all extremes of weather. Hazards include both hypo- and hyperthermia, and heat exhaustion. The latter problems may be magnified during summer months by the need to wear heavy personal protective equipment (PPE).

Other climatic and environmental concerns include lightning injuries, wind hazards from severe storms, and forest fires.

INFECTIOUS DISEASES

Due to their exposure to a variety of animals and plants that may harbor infectious organisms, it is not surprising that loggers are at an increased risk of developing certain infectious diseases (Moll et al. 1994). A discussion of some of the infection hazards faced by forestry workers follows.

Cutaneous sporotrichosis has been documented among silvaculture workers, associated with the handling of tree seedlings packed in sphagnum moss (Coles et al. 1992).

Lyme disease is clearly more likely in the forestry worker population because of their risk of tick bites from the *Ixodes* genus. The seroprevalence of antibodies to *B. burgdorferi* has been found to be significantly higher among forestry workers compared to controls (Kuiper et al. 1993; Moll et al. 1994; Nakama et al. 1994), and almost 19% of outdoor workers in New Jersey have tested positive for Lyme antibodies (Schwartz et al. 1993).

Several studies of Lyme disease have been done in foreign countries among forestry workers. In the Netherlands the seroprevalance of Lyme disease ranges from 20 to 28% (Kuiper et al. 1993). In Italy, the seroprevalance ranges from 3 to 19% depending on the location (Nuti et al. 1993; Cristofolini et al. 1993). The prevalance of Lyme disease is also reported to be higher in forestry workers in Japan (Nakama et al. 1994). Other *Borrelia* species represent a potential risk for foresters, particularly those who may be lodging in primitive conditions.

Other tick-borne diseases have been reported in forestry workers. Numerous cases, including a 1973 Arizona outbreak of tick-borne Relapsing Fever, carried by argasid ticks of the *Ornithodros* genus, demonstrate that outdoor workers in endemic areas may be at some risk for these illnesses (CDC 1973).

Other diseases associated with arthropod vectors that are of concern to outdoor workers include Rocky Mountain Spotted Fever, due to *R. rickettsii*. RMSF is especially problematic for forestry workers in the southeastern United States, with most cases occurring between April and October. Transmission, which results from the bite of the *Dermacentor* and *Amblyomma* tick, is unlikely unless the tick has remained attached for an extended number of hours (Sexton et al. 1992).

Another tick-borne illness, Colorado Tick fever, may often be subclinical with 15% of perennial campers being seropositive for antibodies. The responsible virus is in the *Reoviridae* family, and is most prevalent in the mountains of the western U.S., where it is transmitted by the *Dermacentor andersoni* tick.

Tularemia should be in the differential for unexplained febrile illness of any forestry worker. The primary mode of transmission of *F. tularensis* to humans is not by exposure to wild rabbit flesh according to the traditional wisdom, but via the bites of deer flies and ticks belonging to the genus *Dermacentor* or *Amblyomma*. Peak occurrence in the U.S. is during summer months in the midwestern and southern states (Kligman et al. 1991; Taylor 1991).

The same *Ixodes* genus of tick that is responsible for transmission of Lyme disease may also serve as the vector of the considerably rarer disease Babesiosis. The protozoa of the genus *Babesia* is the infecting agent of this malaria-like illness, and the bulk of these cases have been identified in the northeastern U.S. (Ryan 1987). Another tick-borne disease, Ehrlichiosis, caused by organisms of the rickettsial genus *Ehrlichia*, resembles RMSF in symptomatology. The specific species of tick vector for this organism has yet to be determined, although there are indications that *Amblyomma americanum* is a likely suspect (Anderson 1993).

Tick-borne encephalitis seroprevalance studies in Eastern France have detected antibodies in 8% of forestry workers (Collard et al. 1993) and a 1% prevalence rate was found in Northen Italian foresters (Cristofolini et al. 1993)

Forestry employees may be at a higher risk for infection with the arbovirus group, which consists of over 450 arthropod-borne viruses (Sanford 1991). In the United States, St. Louis encephalitis, and Eastern and Western equine encephalitis are the most common of these mosquito-borne infections. Approximately 2,000 cases are reported annually in the U.S. and often occur as summer outbreaks. Japanese encephlatis is a concern in Asian countries.

Leishmania mexicana, a causative agent of mucocutaneous leishmaniasis, transmitted by the sand fly, *Phlebotomus longipes*, has been associated with forest clearing in Latin America where it is known as chiclero's ulcer. In an evaluation of mucocutaneous leishmaniasis in a rural area of Peru, 60% of the cases occurred among wood laborers or farmers (Heimgartner and Heimgartner 1976). Rare cases have also been reported in south-central Texas (Baum and Barens 1994).

The risk of contracting tetanus is greater among workers in any occupation susceptible to open wounds. A strict anaerobe, *C. tetani* may flourish in the reduced environment of the grossly contaminated, complex wounds often associated with logging trauma. A Finish study, examining Tetanus, found that agricultural and forestry accidents resulted in 26% of the cases of Tetanus over a 16-year period in that country (Luisto and Seppalinen 1992).

During the recent decade, epizootics of rabies, rhabdovirus of the genus *Lyssavirus*, have occurred in several regions of the United States (Krebs et al. 1993) Human cases are rare, and canine rabies is largely controlled, but wild animals represent a significant reservoir, posing an occupational hazard to workers in rural areas. In third world countries, rabies is frequently present in dogs as well as bats and other mammals. Thousands of human infections occur yearly in third world countries.

The Hanta viruses of the family *Bunyaviridae* are endemic among many North American rodent populations. Transmission is typically via contact with rodent excreta or by inhalation of contaminated dust particles. Although this disease has not yet been associated with forestry work in the United States, theoretically any outdoor worker may be at risk wherever the local animal reservoir population has become dense. Studies of Hantaan virus and Puumala virus seroprevalance found antibodies in 7% of forestry workers in northeast Italy and 15.9% in northern Sweden (Ahlm et al. 1994; Nuti et al. 1993).

Forestry work may also increase one's risk of developing malaria (Oemijati 1992; Camargo et al. 1994). Leptospiral antibodies have been found in 10 to 12% of farmers and foresters in the Italian Alpines (Nuti et al. 1993).

The preceding is a fairly comprehensive list of infectious diseases that may be associated with forestry operations. Logging activities outside of the United States, particularly in tropical zones, could entail exposure potential to an extensive array of exotic diseases unique to locale. The occupational health professional providing care for foresters anticipating work outside of the U.S. should contact the CDC for the International Travel Directory for disease risk and prevention information by region, and for disease outbreak bulletins. Vaccines are available to prevent some of these diseases.

OCCUPATIONAL DERMATITIS

Occupational dermatitis is common in the forestry industry. Irritation of the skin and eyes, allergic contact dermatitis, contact urticaria, pigmentary changes, and possibly photosensitivity may occur from contact with wood (Mitchell 1980).

Among forest workers in Oregon insured by the State Accident Insurance Fund (SAIF) in 1976 to 1977, dermatitis was responsible for 27.7% of their injuries and illnesses compared to 1.7% of all other Oregon SAIF combined injury-illness claims (Storrs 1982).

During 1960 through 1965 occupational dermatitis claims in Oregon constituted 51% to 61% of all occupational disease claims. In 1961, 26% of the occupational skin disease claims were from the forest products industry and 25% of these were attributed to contact with wood or trees. In 1964, 24% of dermatitis claims were due to the forest products industry (Suskind 1967).

In a study of skin disease in California in 1979, the Agriculture industry accounted for 13% of all cases reported. The case rate for occupational skin disease was 6.1 per 1,000 employees compared to 1.6 per 1,000 for all industries combined. Poison oak, shrubs, plants, and trees caused 59.8% of the cases (Mathias 1989).

A study of forestry workers in Indonesia found that 38 of 45 workers (84%) living in a jungle camp suffered from skin disease (Siregar 1975). Thirty percent of the workers had contact dermatitis to wood, 8% dermatitis from insect and leech bites, 8% had contact dermatitis from petrol and oils, and 26% had a mycotic or bacterial skin infection that was attributed to a lack of hygeine facilities in the jungle.

In a study of of wood machinery workers in Belgium, 3% of the workers in contact with tropical wood had dermatitis or mucosal signs (Oleffe et al. 1975).

Contact dermatitis resulting from skin exposure to plants of the *Toxicodendron* species—poison ivy, poison oak, and poison sumac—represent the most frequent dermatologic complaint among forestry workers. Not only the leaves, but the woody vines which typically wind about trees, contain the antigenic oleoresins. The dermatitis-producing resins of these species can remain allergenic for up to a week when on clothing or equipment, and may be even more stable in cool dry environments. Dried, withered leaves and dormant woody parts of the plants retain some antigenicity, and this is important to remember when wood cutting during cooler months. Air-borne particles of the plant can disseminate the antigen, and this may present a hazard during chainsawing and burning. Dermatitis resulting from air-borne exposure may produce a characteristic diffuse lesion pattern often involving the face, as contrasted with the classical linear arrangement normally associated with direct contact with these plants.

Over 50 species of trees are known to cause dermatitis. Most contact dermatitis cases are irritation reactions through local mechanical trauma as from wet sawdust, splinters, and thorns (Adams 1969). Irritant chemicals are often found in the sap or latex, or in the bark of trees and may cause a skin rash. Sensitizing chemicals are generally found in the heartwood and may also cause erythema and blistering of the skin.

True wood allergies occur less frequently than irritant reactions but may be seen with many exotic woods. Some pine species may cause dermatitis due to allergenic compounds in sap and buds such as coniferin, coniferylic alcohol and vanillin. Allergic contact dematitis may occur to beech, sugar maple, balsam poplar, elm wood, and oak shavings due to the presence of 2,6-dimethoxbenzoquinone (Hausen 1981). A variety of phenols, quinones, saponins, stilbenes, terpenes, and furocoumarins are found in exotic wood and can cause both irritant and allergic contact dermatitis (Mitchell 1980).

More common is sensitivity to d-usnic acid found in lichens on tree bark, and allergy to a group of chemicals known as sesquiterpene lactones (STL) (Storrs 1982). STL is found in thousands of species of plants, mostly weeds, but is also concentrated in the liverwort genus *Frullania*, which grows on the bark of trees in moist forested areas. This may pose a problem for sensitive workers, particularly in the northwest.

Wet work, which constitutes a great proportion of work in the Pacific northwest, results in an increased risk of irritant dermatitis, but this may also represent a risk wherever there is arduous work in a humid environment.

A list of trees that may cause dermatitis is presented in Table 29.4. (NIOSH 1987).

Table 29.4. Trees causing dermatitis.

Brazilian Rosewood	Makore
Cocobolo	Yew
Honduras Rosewood	African black walnut
Nigerian Satinwood	Elm
Agba	Teak
Mesquite	Peroba rose
African mahogany	Alder
African irko	Peroba de Campos
Fir	Lapacho
Eastern red cedar	Gaboon
Incense cedar	Topolite
Pine	Antilles greenheart
Spruce	Brazilian walnut
Douglas Fir	Louro
Northern white cedar	Partridge wood
Western red cedar	Cocus wood
Hemlock	Tagayasan
East Indian satinwood	Kingswood
West Indian satinwood	Costa Rico rosewood
Olon	East Indian rosewood
Moah	African blackwood

Source: NIOSH 1987.

PESTICIDE EXPOSURE

The greatest insecticide and herbicide exposures would be expected among tree nursery workers. Although there has been documentation of a mild degree of pesticide absorption in this worker population, in general, the magnitude of exposure does not appear to approach that of the agricultural occupations (Lavy 1993; Robinson 1993). In a study of a helicopter crew applying 2,4-dichlorophenoxyacetic acid for brush control in forestry, urine concentrations of 2,4-D were determined. The calculated absorbed dose was less than the no observed adverse effect level (NOAEL) (Knopp and Galss 1991).

A study of airborne herbicide residue after a prescribed fire on field sites treated 30 to 169 days prior with a variety of herbicides was unable to detect herbicide residues in the smoke samples. Previous modelling assessments and laboratory experiments had shown the risk of airborne herbicide residues to workers was insignificant (Mcmahon and Buch 1992).

In a study of forestry workers mixing and applying a triclopyr herbicide, biomonitoring was performed to determine if any herbicide was being absorbed (Middendorf et al. 1994). Six of 21 volunteers in the study had low levels of the herbicide found in their urine. All levels were significantly below the no observed effect level.

However, in a study of 45 workers involved in dipping conifer seedlings in lindane, transporting and planting the seedlings over a 4-month period, their was evidence of absorption of the chemical in all workers at the end of the fourth month (Drummond et al. 1988). Additionally, two of the workers developed illness and had elevated levels of lindane in their blood.

Therefore it is important that all workers who apply pesticides should be thoroughly trained in procedures for mixing and application of pesticides. All equipment should be maintained in top operating conditions, and personal protective equipment worn during mixing and application. Additionally, workers should use good personal hygiene such as thorough handwashing after working with pesticides (Middendorf et al. 1994).

OCCUPATIONAL RESPIRATORY DISEASE

Although work-related respiratory problems have been identified among woodworkers, and sawyers, particularly associated with western red cedar exposures, occupational respiratory disease has generally been linked to workers in sawmills and furniture factories instead of forestry work. However, even cutting of live trees has been associated with respiratory disease (Dykewicz et al. 1988). Respiratory diseases reported among sawmill workers include extrinsic allergic alveolitis (Halpin et al. 1994; Belin 1980), organic dust toxic syndrome (Weber et al. 1993), asthma (Cote et al. 1991; Vedal et al. 1988), non-asthmatic chronic airflow obstruction (Carosso et al. 1987), chronic bronchitis, mucus membrane irritation syndrome (Enarson and Chin-Yueng 1990; Goldsmith and Shy 1988) and nasal cancer (Wills 1982; Nylander and Dement 1993; Gallagher et al. 1985; Kawachi et al. 1989).

The most common allergic respiratory effect due to wood dust is asthma. (NIOSH 1987) About 5% of workers exposed to Western red cedar develop asthma. Plicatic acid has been identified as the cause of the asthma. (Cartier 1986) A list of some woods that have been associated with allergic respiratory disease is provided. (NIOSH 1987)

Depending on the type of wood and amount of exposure to wood dust, it is possible for loggers also to develop these respiratory diseases. An increased rate of nasal cancer has been noted in loggers (Kawachi et al. 1989, Gallagher et al. 1985).

Table 27.5. Wood dust reported to cause allergic repiratory effects.

African maple	California redwood
Spruce	African zebrawood
Fir	Teak
Pine	Okume
Eastern white cedar	Quillaja bark
Boxwood	Sambe
Kejaat	Tanganyika aningre
Congo hardwood	Central American walnut
Oak	
Abiruana	Western red cedar
Mahogany	Iroko
Cedar of Lebanon	Cocobolo

Source: NIOSH 1987

PSYCHO-SOCIAL ISSUES

Forestry workers may be employed in relatively remote locations, removed from family and other primary support groups, for extended periods of time. In many regions of the country, where logging operations are performed mostly on federal land, the mere existence of logging jobs is subject to an ever-shifting political climate, engendering an atmosphere of uncertainty and job insecurity among these workers. Complex harvesting machines used in large scale timbering operations may increase the mental workload, while at the same time creating monotony. As expected, alcohol abuse and depression are not uncommon among these workers. A Finnish study investigating cause-specific mortality in the forestry industry found a relatively high suicide rate among these workers that persisted after other variables had been adjusted (Notkola et al. 1993). A study reported that forestry workers as compared to constuction workers had higher scores of external locus of control on an Accident Locus of Control Test (Salminen and Klen 1994). Subjects in both groups who had high scores on external locus of control took more risks than subjects with higher scores on internal locus of control.

PROTECTIVE EQUIPMENT, SAFE OPERATIONS, AND PREVENTIVE HEALTH

The appropriate use of personal protective equipment can decrease the exposure to hazardous agents. The use of proper clothing such as hard hats and hearing protection can not only prevent acute injuries but also the development of chronic illness/injury, e.g., noise-induced hearing loss.

Safety equipment on vehicles can also prevent injuries from occurring. Seat belts have been shown to prevent many accidents from occurring if worn while operating the machinery. However, safety equipment is not enough. Workers must receive proper training and field supervision on safe work practices and on the operation of machines and vehicles.

A summary of the 1984 BLS survey on the use of safety equipment in the logging industry is summarized in Table 29.6 (BLS 1984).

Table 29.6. Safety equipment worn or incorporated on the equipment.

Type of Safety Equipment	Percent Utilized
Equipment worn	
Calk or cork-soled boots	62
Dust masks	2
Hearing protectors	25
Glasses or goggles	17
Gloves	75
Hard hat	87
Leg protectors	29
Steel-toed boots	28
Other worn equipment	2
Not wearing safety equipment	4
Equipment on the vehicle	
Falling object protective structure	59
Rollover protective structure	53
Seat belt	63
Other equipment device	8
Not aware of safety device	10

Source: BLS 1984

The 1994 OSHA Standard for Logging Operations requires proper PPE, including face, foot, and hearing protection; avoidance of loose clothing; and availability of extensive on-site first aid kits (*Federal Register* 1994). Also standards are issued concerning machine stability and slope limitations; discharge of hydraulic and pneumatic storage devices on forestry machines; machine breaking specifications; protective structures to prevent machine rollover and injury from cutting machines; and vehicle and machine maintenance and inspection. Also, the rule requires that when sawing, backcuts be above the horizontal line of the undercut. In addition, cutting with the return side or tip of saw should be avoided. Saw operators should maintain a 2 meters distance from other workers (U.S. Dept. Of Labor

1995). In addition, in order to reduce the number of injuries related to falling objects, NIOSH has recommended that workers ensure that loose bark, broken limbs, or other damaged sections of a tree are removed before tree is felled (NIOSH 1995). Other recommendations include ensuring that work areas are assigned so that no tree can fall into an adjacent, occupied work area; removal of trees that become lodged against other trees; use of safety belts when loggers are operating heavy machinery or vehicles; and development of a comprehensive, enforceable safety program. Workers not involved in an operation should keep a distance of at least two tree lengths away from a felling.

For helicopter logging, the Alaska Department of Labor has urged employers engaged in this activity to follow certain precautions. Specific recommendations include clear identification of the logging areas; equipping vehicles with an obstructed rear view with an adequate motion warning alarm, or stationing a designated signalman; providing ground workers with adequate fluorescent garments; and the carrying out of in-depth routine safety inspections (COC 1994).

The incidence of VWFD in chain saw operators should be reduced by minimizing individual worker exposure to vibrating hand-held tools. Design improvements in newer tools have resulted in relative decreases in both vibration acceleration and tool weight, and this should lessen the hazard. NIOSH guidelines include the use of vibration dampening hand grips and improved tool design, adequate maintenance for vibratory hand tools, improved work practices and medical surveillance, and keeping hands warm with gloves. Administrative controls should include limiting exposures to no more than two hours a day. Smoking reduction should be encouraged, as abstinence should lessen frequency of vasomotor attacks (NIOSH 1983).

Workers involved in chainsaw and other loud machinery operation qualify for hearing conservation programs. Strict adherence to hearing protector use must be emphasized.

During warmer months efforts should be made to reduce heat stress. The Amreican Conference of Governmental Industrial Hygienists (ACGIH) offers a reasonable guideline to follow using a heat stress threshold limit value (TLV) to derive a work/rest regimen, based on ambient wet globe bulb temperature, workload category, and type of clothing worn (ACGIH 1992). Potable liquids must be readily available at all times, and workers taught that regular fluid intake is a mandatory part of their job. Likewise, workers must be protected from overexposure to cold. In addition to adequate warm clothing, portable shelters should be made accessible for weather extremes.

Because of the arduous nature of these occupations, work tasks should have maximal ergonomic organization, and mechanization should be exploited wherever possible.

Worker education should be provided covering topics such as lifting kinetics, proper equipment use, liberal fluid intake, and prompt reporting of injuries. Strength and agility testing are not considered to be medical procedures under present Americans with Disabilities Act (ADA) law, and should be considered before a job offer in these occupations.

The incidence of arthropod-borne disease can be minimized by encouraging outdoor workers to use insect repellant compounds such as diethyltoluamide, especially about the cuffs of trousers and on shoes. The avoidance of brightly colored clothing discourages ticks and at the same time lessens the probability of *Hymenoptera* stings. Workers should be reminded to check themselves after each shift for the presence of ticks, and these should be removed promptly. The medical professional should be ever vigilant for the signs and symptoms of arthropod-borne illness among forestry workers, as prognosis is dependent on early treatment in most of these diseases.

The risk of Sporotrichosis infection in silvaculture could be minimized by strict adherence to use of thick work gloves and arm protection and the use of packing material other than sphagnum moss when transporting seedlings (Coles et al. 1992).

The tetanus immuno-status for all forestry employees should be current.

The care provider should always remain cognizant of potential psychological pathology among forestry workers. Routine surveillance should include vigilance for the signs or symptoms of alcohol abuse, and depression. Any pronouncements from the worker indicative of work-related anxieties or resentments should be dealt with compassionately and with deliberation. Outreach counseling, such as offered within employee assistance programs (EAP), should be available and accessible, and participation should be encouraged.

Operating a company safely has numerous benefits. Low accident rates are correlated with higher levels of financial stability, solvency, and management efficiency (Smith et al. 1978). Fewer accidents decrease the cost of workers' compensation insurance as well as avoiding pain and suffering in the injured worker. An injured worker frequently has a loss in wages, and the morale of fellow workers is usually affected when a co-worker is injured. Indirect cost from an accident may include lost production, equipment repair cost, and cost of training a new replacement worker.

A study of 26 "safety successful" logging companies identified characteristics that made their operations safer (Reisinger et al. 1994). They had effective people managers who were able to keep crew turnover to a minimum and maintain a stable crew, hired primarily experienced wood workers, had highly mechanized operations, promoted teamwork, insisted on mandatory use of personal protective equipment, and demonstrated a strong management commitment to safety.

SUMMARY

Forestry work has historically been a hazardous occupation. Logging can be characterized as an extremely ardous activity, entailing the handling of materials of great mass, using dangerous tools, often in remote, isolated environments. Risk of injury or death continues to be elevated, as evidenced by recent NIOSH data showing the annual fatality rate among loggers to be 23 times that of other U.S. workers.

Acute trauma remains the problem most in need of intervention. There is greater risk for injury and death when the cutting involves hardwoods, in unmanaged natural stands, and on uneven terrain. Workers should be made aware that tree snags are associated with a large proportion of fatalities. Certain practices such as "domino falling" should be discouraged. Adherence to the 1994 OSHA Standard for Logging Operations, outlined in this chapter, should greatly reduce the incidence of acute logging injuries.

Physicians and safety professionals who deal with forestry workers need also to be aware of the other hazards associated with this industry. Musculoskeletal injuries, and hypo- and hyperthermia, are possibilities in any heavy outdoor work. More peculiar to forestry are specific issues such as vibration induced white finger disease (VWFD), and loud noise exposure in chain saw operators. A higher incidence of respiratory disease may also be found among loggers. Work in remote forested areas entails exposure to a number of environmental hazards. Insect and reptile bites, *toxicodendron* dermatitis, and arthropod-borne infectious disease, are all likely to be encountered in the forestry worker population. Tree nursery workers and others who may mix or apply pesticides may be at a greater risk for insecticide and herbicide exposure, especially if sound work practices and personal protective equipment are neglected. Finally, anyone concerned about the well-being of the forestry worker should not overlook the complex psycho-social issues seen in an employee faced with a demanding workload, often in remote locations, removed from family, and in an atmosphere of uncertain job security.

Worker and management education is of paramount importance. Companies should be encouraged to follow the example of the twenty-six "safety successful" companies mentioned in this chapter. The ideal company should stress good communication, promote teamwork, use highly mechanized operations, good vehicle maintenance and inspection, and insist on mandatory use of personal protective equipment. Management must demonstrate a strong commitment to safety.

REFERENCES

Adams, R.M. 1969. *Occupational Contact Dermatitis*. Philadelphia and Toronto: J.B. Lippincott Co.

Ahlm, C., M. Linderholm, P. Juto, B. Stegmayr, and B. Settergren. 1994. Prevalence of serum IgG antibodies to Puumala virus (haemorrhagic fever with renal syndrome) in northern Sweden. *Epidem. Infection.* 113: 129-36.

American Conference of Governmental Industrial Hygienists (ACGIH). 1992-1993. *Threshold Limit Values for Chemical Substances and Physical Agents and Biological Exposure Indices*. Cincinnati: ACGIH, pp. 91-98.

Anderson, B.E., K.G. Sims, J.G. Olson, J.E. Childs, J.F. Piesman, C.M. Happ, G.O. Maupin, and B.J. Johnson. 1993. Amblyomma americanum: a potential vector of human ehrlichiosis. *Am. J..Tropical Med. Hyg.* 49: 239-44.

Baum, K.F., and R.L. Barens. 1994. Successful treatment of cutaneous leishmaniasis with allopurinol after failure of treatment with ketoconazole. *Clinic. Inf. Dis.* 18: 813-15.

Belin, L. 1980. Clinical and immunological data on "wood trimmer's disease" in Sweden. *Eur. J. Resp. Dis.* 61: 169-76.

Bovenzi, M., A. Zadini, A. Franzinelli, and F. Borgogni. 1991. Occupational musculoskeletal disorders in the neck and upper limbs of forestry workers exposed to hand-arm vibration. *Ergon.* 34: 547-62.

Bureau of Labor Statistics (BLS). 1984. *Work Injury Report (WIR) Injuries in the Logging Industry*, Bulletin 2203.

Bureau of Labor Statistics (BLS). 1994. *1992 Census of Fatal Occupational Injuries.*

Bureau of Labor Statistics. 1995. *Work injuries and illnesses by selected characteristics, 1993 (*Summary Text Only). Filename BLS95.142.

Camargo L.M., M.U. Ferreira, H. Krieger, E.P. DeCamargo, and L.P. DaSilva. 1994. Unstable hypoendemic malaria in Rondonia (western Amazon region, Brazil): epidemic outbreaks and work-associated incidence in an agro-industrial rural settlement. *Am. J. Tropical Med. Hyg.* 51: 16-25.

Carosso, A., C. Ruffino, and M. Bugiani. 1987. Respiratory diseases in wood workers. *Brit. J. Ind. Med.* 44: 53-56.

Cartier, A., H. Chan, J.L. Malo, L. Pineau, K.S. Tse, and M. Chan-Yeung. 1986. Occupational asthma caused by eastern white cedar (Thuja occidentalis) with demonstration that plicatic acid is present in this wood dust and is the causal agent. *J. Allergy Clin. Immun.* 77: 639-45.

Centers for Disease Control (CDC). 1973. Relapsing fever. *MMWR* 22: 242-46.

Centers for Disease Control (CDC). 1994. Risk for traumatic injuries from helicopter crashes during logging operations—southeastern Alaska, January 1992-June 1993. *MMWR* 43: 472-75.

Chapel, T.A. Skin and soft tissue emergencies. 1992. In: *Emergency Medicine.* J.E. Tintinalli, R,L. Krome, E. Ruiz, eds. New York: McGraw-Hill.

Cherniak, M. 1994. Upper extremity disorders. In: *Textbook of Occupational and Environmental Medicine.* L. Rosenstock, M.R. Cullen, eds. Philadelphia: W.B. Saunders Co., pp. 379-80.

Coles, F.B., A. Schuchat, J.R. Hibbs, S.F. Kondracki, I.F. Salkin, D.M. Dixon, H.G. Chang, R.A. Duncan, N.J. Hurd, and D.L. Morse. 1992. A multistate outbreak of sporotrichosis associated with sphagnum moss. *Am. J. Epid.* 136: 475-87.

Collard, M., J.P. Gut, D. Christmann, E. Hirsch, G. Nastorg, F. Sellal, and X. Haller. 1993. Tick-borne encephalitis in Alsace. *Revue Neur.* 149: 198-201.

Cote, J., H. Chan, G. Brochu, M. Chan-Yeung. 1991. Occupational asthma caused by exposure to neurospora in a plywood factory worker. *Brit. J. Ind. Med.* 48: 279-82.

Cristofolini, A., D. Bassetti, and G. Schallenberg. 1993. Zoonoses transmitted by ticks in forest workers (tick-borne encephalitis and Lyme borreliosis): preliminary results. *Medicina del Lavoro* 84: 394-402.

Crutchfield, C.D., and S.T. Sparks. 1991. Effects of noise and vibration on farm workers. In: *Occup. Medicine: State of the Art Reviews.* D.H. Cordes and D.F. Rea, eds. Philadelphia: Hanley & Belfus, Inc, pp 355-69.

Drummond, L., E.M. Gillanders, H.K. Wilson. 1988. Plasma y-hexachlorocyclohexane concentrations in forestry workers exposed to lindane. *Brit. J. Ind. Med.* 45: 493-97.

Dykewicz, M.S., P. Laufer, R. Patterson, M. Roberts, and H.M. Sommers. 1988. Woodman's disease: hypersensitivity pneumonitis from cutting live trees. *J. Allergy Clin. Immun.* 81: 455-60.

Enarson, D.A., and M. Chan-Yeung. 1990. Characterization of health effects of wood dust exposures. *Am. J. Ind. Med.* 17:33-38.

Federal Register. 1994. 29 CFR. Parts 1910 and 1928. Logging Operations; Final Rule. 59: 51672-51748.

Frazier, C.A. 1987. *Insect Allergy.* St. Louis: W.H.Green.

Gallagher, R.P., W.J. Threlfall, P.R. Band, and J.J. Spinelli. 1985. Cancer mortality experience of woodworkers, loggers, fishermen, farmers, and miners in British Columbia. *Nat. Cancer Inst. Monographs.* 69: 163-67.

Giguere, D., R. Belanger, J.M. Gautier, and C. Larue. 1993. Ergonomic aspects of tree-planting using 'multipot' technology. *Ergonomics.* 35: 963-72.

Goldsmith, D.F., C.M. Shy CM. 1988. Respiratory health effects from occupational exposure to wood dusts. *Scand. J. Work Envir. Health* 14: 1-15.

Gomez, H.F., M. Davis, S. Phillips, P. McKinney, J. Brent. 1994. Human envenomation from a wandering garter snake. *Annals Emer. Med.* 23: 1119-22.

Green L.M. 1993. Mortality in forestry and construction workers in Finland. *J. Epid. Comm. Health* 47: 508-09.

Halpin, D.M., B.J. Graneek, M. Turner-Warwick, and A.J. Newman Taylor. 1994. Extrinsic allergic alveolitis and asthma in a sawmill worker: case report and review of the literature. *Occup. Envir. Med.* 51: 160-64.

Hansson, R., E. Broberg, A. Johansson, U. Jorner, R. Selander, and B. Karlberg-Nilsson. 1989. *Study of Accidents in Farming and Forestry in 1987.* Stockholm, Sweden: Lantbrukshalsan AB (The Swedish Farmers' Safety and Preventive Health Association.)

Hausen, B.M. 1981. *Woods Injurious to Human Health.* Hawthorne, New York: Walter DeGruyter, Inc.

Heimgartner, E., and V. De Heimagartner. 1976. Experience with endemic dermatological diseases in the Peruvian wilderness: mucocutaneous leishmaniasis and Brazilian foliaceous pemphigus. *Midicina Cutanes Ibero-Latino-Americana.* 4: 1-6.

Hutton, S.G., N. Paris, R. Brubaker. 1993. The vibration characteristics of chainsaws and their influence on vibration white finger disease. *Ergonomics.* 36: 911-26.

International Labor Organization. 1983. Forestry industry. *Encyclopedia of Occupational Safety and Health.* 3rd edition. Geneva, Switzerland: ILO Publications, pp. 908-12.

Kawachi, I., N. Pearce, and J. Fraser. 1989. A New Zealand cancer registry-based study of cancer in woodworkers. *Cancer* 64: 2609-13.

Kligman, E.W., W.F. Peate, and D.H. Cordes. 1991. Occupational infections in farm workers. *Occup. Med.: State of the Art Reviews* 6: 429-43.

Knopp, D., and S. Galss. 1991. Biological monitoring of 2,4-dichlorophenoxyacetic acid-exposed workers in agriculture and forestry. *Intl. Arch. Occup. Envir. Health.* 63: 329-33.

Koskimies, K., I. Pyykko, J. Starck, R. Inaba. 1992. Vibration syndrome among Finnish workers between 1972 and 1990. *Intl. Arch. Occup. Envir. Health.* 64: 251-56.

Krebs, J.W., T.W. Strine, and J.E. Childs. 1993. Rabies surveillance in the United States during 1992. *J. Am. Vet. Med. Assoc.* 203: 1718-31.

Kuiper, H., A.P. van Dam, A.W. Moll van Charante, N.P. Nauta, and J. Dankert. 1993. One year follow-up study to assess the prevalence and incidence of Lyme borreliosis among Dutch forestry workers. *Eur. J. Clin. Microb. Inf. Dis.* 12: 413-18.

Kurumatani, N., B. Yamaguchi, M. Dejima, Y. Enomoto, and T. Moriyama. 1992. Aerobic capacity of forestry workers and physical demands of forestry operations. *Eur. J. Appl. Phys. Occup. Phys.* 64: 546-51.

Langley, R.L. 1994. Fatal animal attacks in North Carolina over an 18-year period. *Am. J. Forensic Med. Path.* 15: 160-67.

Lavy, T.L., J.D. Mattice, J.H. Massey, and B.W. Skulman. 1993. Measurements of year-long exposure to tree nursery workers using multiple pesticides. *Arch. Envir. Contam. Toxic.* 24: 123-44.

Luisto, M., and A.M. Seppalainen. 1992. Tetanus caused by occupational accidents. *Scand. J. Work Envir. Health.* 18: 323-26.

Marshall, S.W., I. Kawachi, P.C. Cryer, D. Wright, C. Slappendel, and I. Laird. 1994. The epidemiology of forestry work-related injuries in New Zealand, 1975-88: fatalities and hospitalizations. *N. Zeal. Med. J.* 107: 434-37.

Mathias, C.G.T. 1989. Epidemiology of occupational skin disease in agriculture. In: *Principles of Health and Safety in Agriculture.* J.A. Dosman, D.W. Cockcroft, eds. Boca Raton, FL: CRC Press, pp. 285-87.

Matoba, T., M. Chiba, and T. Sakurai. 1985. Body reactions during chain saw work. *Brit. J. Ind. Med.* 42: 667-71.

McMahon, C.K., and P.B. Buch. 1992. Forest worker exposure to airborne herbicide residues in smoke from prescribed fires in the southern United States. *Am. Ind. Hyg. Assoc. J.* 53: 265-72.

Middendorf, P., C. Timchalk, B. Kropscott, and D. Rick. 1994. Forest worker exposure to Garton TM 4 Herbicide. *Appl. Occup. Envir. Hyg.* 9: 589-94.

Mirbod, S.M., H. Yoshida, C. Nagata, R. Inaba, Y. Komura, and H. Iwata. 1992. Hand-arm vibration syndrome and its prevalence in the present status of forestry enterprises in Japan. *Intl. Arch. Occup. Envir. Health.* 64: 93-99.

Mitchell, J.N.S. 1980. Woods. In: *Contact Dermatitis.* E. Cronin, ed. Edinburgh: Churchill Livingstone, pp. 548-74.

Moll van Charante, A.W., J. Groen, A.D. Osterhaus. 1994. Risk of infections transmitted by arthropods and rodents in forestry workers. *Eur. J. Epidem.* 10: 349-51.

Myers, J.R., and D.E. Fosbroke. 1994. Logging fatalities in the United States by region, cause of death, and other factors--1980 through 1988. *J. Safety Res.* 25: 97-105.

Nagase, H., H. Nakamura, S. Nohara, K. Miura, and T. Ohnishi. 1992. Mutivariate analysis on the relationship between subjective symptoms and risk factors for the development of symptoms including working conditons, life habits and physical status in forestry workers using chain saw. *Sangyo Igaku - Japanese J. Ind. Health.* 34: 551-59.

Nakama, H., K. Muramatsu, K. Uchikama, and T. Yamagishi. 1994. Possibility of Lyme disease as an occupational disease--seroepidemiological study of regional residents and forestry workers. *Asia-Pacific J. Public Health.* 7: 214-17.

Nilsson, C., R. Lindahl, and A. Norstrom. 1987. Occupational exposure to chain saw exhausts in logging operations. *Am. Ind. Hyg. Assoc. J.* 48: 99-105.

National Institute for Occupational Safety and Health (NIOSH). 1983. Vibration syndrome. *Current Intelligence Bulletin 30.* Washington, DC, pp. 83-110.

National Institute for Occupational Safety and Health (NIOSH). 1993. *Fatal Injuries to Workers in the United States, 1980-1989: A Decade of Surveillance; National Profile.* Cincinnati, OH: DHHS (NIOSH) Publication No.. 93-108.

National Institute for Occupational Safety and Health (NIOSH). 1994. *NIOSH Alert: Request for Assistance in Preventing Injuries and Deaths of Loggers.* Cincinnati, OH: DHHS (NIOSH) Publications No. 95-101.

National Institute for Occupational Safety and Health (NIOSH). 1987. *Health Effects of Exposure to Wood Dust: A Summary of the Literature.* Cincinnati, OH: NIOSH Publications.

Notkola, V.J., P. Martikainen, and P.I. Leino. 1993. Time trends in mortality in forestry and construction workers in Finland 1970-1985 and impact of adjustment for socioeconomic variables. *J. Epidem. Comm. Health.* 47: 186-91.

Nuti, M., D. Amaddeo, M. Crovatto, A. Ghionni, D. Polato, E. Lillini, E. Pitzus, and G.F. Santini. 1993. Infections in an Alpine environment: antibodies to hantaviruses, leptospira, rickettsiae, and Borrelia burgdorferi in defined Italian populations. *Amer. J. Tropical Med. Hyg.* 48: 20-25.

Nylander, L.A., and J.M. Dement. 1993. Carcinogenic effects of wood dust: review and discussion. *Amer. J. Ind. Med.* 24: 619-47.

Oemijati, S. 1992. Risk behavior in malaria transmission in Indonesia. *S.E. Asian J. Tropical Med. Public Health.* 23 Suppl: 47-50.

Oleffe, J.A., J. Sporcq, and P. Hublet. 1975. Epidemiological study of the wood industry in Belgium. *Contact Dermatitis.* 5: 315-6.

Pyykko, I., J. Starck, M. Farkkila, M. Hoikkala, O. Korhonen, and M. Nurminen. 1981. Hand-arm vibration in the aetiology of hearing loss in lumberjacks. *Brit. J. Ind. Med.* 38: 281-89.

Reisinger, T.W., R.G. Sluss, and R.M. Shaffer. 1994. Managerial and operational characteristics of "safety successful" logging contractors. *Forest Products J.* 44: 72-77.

Robinson, D.G., D.G. Trites, and E.W. Banister. 1993. Physiological effects of work stress and pesticide exposure in tree planting by British Columbia silvaculture workers. *Ergonomics.* 36: 951-61.

Ryan C.P. 1987. Selected arthropod-borne diseases: plague, Lyme disease, and babesiosis. In: *The Veterinary Clinics of North America: Small Animal Practice.* J.R. August and A.S. Loar, eds. 17: 179-94. Philadelphia: W. B. Saunders Co.

Salisbury, D.A., R. Brubaker, C. Hertzman, and G.R. Loeb. 1991. Fatalities among British Columbia fallers and buckers, 1981-1987. *Canadian J. Public Health* 82: 32-37.

Salminen, S., and T. Klen. 1994. Accident locus of control and risk taking among forestry and construction workers. *Perceptual & Motor Skills.* 78: 852-54.

Sanford, J.P. 1991. Arbovirus infections. *Harrison's Principles of Internal Medicine*, 12th Ed. J.D. Wilson et al, eds. New York: McGraw-Hill, Inc., pp. 725-43.

Schwartz, B.S., M.D. Goldstein, and J.E. Childs. 1993. Antibodies to Borrelia burgdorferi and tick salivary proteins in New Jersey outdoor workers. *Amer. J. Public Health* 83: 1746-48.

Sexton, D.J., and H.P. Willet. Rickettsiae. 1992. In: *Zinsser Microbiology.* 20th Edition. W.K. Joklik et al, eds. Norwalk, CT: Appleton & Lange, pp. 700-18.

Shimizu, T., T. Hori, K. Tokuyama, A. Morikawa, and T. Kuroume. 1995. Clinical and immunologic surveys of Hymenoptera hypersensitivity in Japanese forestry workers. *Annals Allergy, Asthma, & Immun.* 74: 495-500.

Siregar, R.S. 1975. Occupational dermatoses among foresters. *Contact Dermat.* 1: 33-37.

Smith, M.J., H.H. Cohen, A. Cohen, and R.J. Cleveland. 1978. Characteristics of successful safety programs. *J. Safety Res.* 10: 5-15.

Storrs, F.J. 1982. Dermatitis in the forest products industry. In: *Occupational and Industrial Dermatology.* H.I. Maibach and G.A. Gellin, eds. Chicago: Year Book Medical Publishers, Inc., pp. 323-31.

Suskind, R.R. 1967. Dermatitis in the forest product industries. *Arch. Envir. Health* 15: 322-26.

Taylor, J.P, G.R. Istre, T.C. McChesney, F.T. Satalowich, R.L. Parker, and L.M. McFarland. 1991. Epidemiologic characteristics of human tularemia in the southwest-central states, 1981-1987. *Amer. J. Epidem.* 133: 1032-38.

Trites, D.G., D.G. Robinson, and E.W. Banister. 1993. Cardiovascular and muscular strain during a tree planting season among British Columbia silvaculture workers. *Ergonomics* 36: 935-49.

U.S. Department of Labor, Occupational Safety and Health Administration. 1995. OSHA partial stay of enforcement of final rule governing logging operations. 60 *FR* 7447.

Vedal, S., D.A. Enarson, H. Chan, J. Ochnio, K.S. Tse, and M. Chan-Yeung. 1988. A longitudinal study of the occurrence of bronchial hyperresponsiveness in western red cedar workers. *Amer. Rev. Resp. Dis.* 137: 651-55.

Weber, S., G. Kullman, E. Petsonk, W.G. Jones, S. Olenchock, W. Sorenson, J. Parker, R. Marcelo-Baciu, D. Frazer, and V. Castranova. 1993. Organic dust exposures from compost handling: case presentation and respiratory exposure assessment. *Amer. J. Ind. Med.* 24: 365-74.

Wills, J.H. 1982. Nasal carcinoma in woodworkers: a review. *J. Occup. Med.* 24: 526-30.

30

TIMBER HARVESTING SAFETY

Thomas L. Bean, Ed.D. and Linda K. Isaacs, M.S.
Department of Food, Agriculture and Biological Engineering
The Ohio State University

> This chapter discusses the high risks associated within the logging industry and the importance of developing effective safety protocols. The unique characteristics of the typical timber harvesting operations, the most common logging-related injuries and illnesses, and human factors which may contribute to injuries and illnesses are addressed. The authors present eight components of an effective formal logging safety program that will help any logging operation be more safety conscious.

INTRODUCTION

The purpose of this chapter is to discuss several approaches to preventing logging-related accidents. It is critical that the reader understand the specific hazardous working conditions loggers face every day on the job. First, the unique characteristics of the typical timber harvesting operation will be discussed, including how the crew works together as a four-step system to prepare the tree from stump to loadout. Second, the most common logging-related injuries and illnesses will be discussed for each specific job. The factors (e.g., human, equipment and environmental) which may have contributed to the injury or illness occurring are also described. Finally, the authors will present the eight components of an effective formal logging safety program.

WHO WORKS IN THE LOGGING INDUSTRY?

Logging often attracts a unique kind of worker. Because of the adverse working conditions and strenuous physical labor, these workers have a special love for outdoor work, work well alone and prefer the seasonal nature of the job. It is a common belief that logging requires little special skills or education, consequently many inexperienced workers are attracted to the crews.

The availability of timber harvesting work is often dependent upon worksite conditions, the influx of workers in the area and the market demand for timber. The result is that the experience and skill level of the employee can vary considerably.

There is a high turnover rate among loggers. Hiring logging workers is not a problem, retaining them over a season is indeed a feat. Although several factors may explain this high dropout rate, the most common are the inherent seasonality of the work and labor migration to other industries. During the peak season, skilled workers are in high demand. As a result experienced employees may be lured away from an employer. Logging work is often perceived as a low-level temporary job, leading to little promotion or career satisfaction. Unfortunately, continuously hiring and training new employees can be costly and reduce productivity. Not only are hiring costs high, but a new, untrained employee may be at a higher risk of injury or illness. In addition, other crew members may be at a higher risk of injury since there is a short transition period where everyone must become accustom to working with the new man or woman.

THE TIMBER HARVESTING SYSTEM: MAN AND MACHINE WORKING TOGETHER

Harvesting lumber has been referred to as a four component system, involving a crew of workers using specialized machinery or equipment—timber cutting, primary transportation, loading and secondary transportation. The ultimate goal of the harvesting crew is to work together to prepare timber or chips for easy transport in an efficient and safe manner. The crew works as one entity, passing the tree down through a series of steps. It is critical that the safety conscious employer, manager and employee fully understand the sequence of these operations, including: the specific tasks performed by each crew member, the type of hazardous equipment utilized and how each crew member's unsafe work habits may put him or his co-workers in danger. By taking the extra time to visualize how the workers work as a team, the owner or foreman can easily identify hazards and take the appropriate preventative measures. As an example, the sequence of steps involved in a typical truck system of hauling random-length timber is depicted below.

Timber Cutting

Prior to moving the harvesting crew into the area the owner or foreman should inspect the site, build access roads and measure/mark whole trees for harvest. Unlike other manufacturing industries, the owner of a timber harvesting operation may have little control over the working conditions his crew must work under. Special attention must be made to inspect the area for hazardous environmental conditions, marking or removing hazardous washouts, loose debris & timber, leaning trees or steep terrain which could cause landslides, roll-overs or side overturns. In deciding how the crew will progress from tree to tree, every effort should be made to allow safe access to the timber, including an escape route. Often leaning trees or loose limbs, called widow-makers, must be removed or stabilized before the workers can be moved into the area.

Trees are cut down either by a worker (called a feller, faller or sawyer) using a power bar saw or mechanical means using a tree shear or feller buncher.

Safe removal begins with preparation. Sets of workers should be evenly and safely spaced so that their activity does not endanger other workers. Assess the situation and prepare the correct tools for the job; for example, ensure that there is plenty of fuel for saws. Plan escape routes. Snags are dead or dying trees, that are still standing. Because they have started the decaying process, they are particularly dangerous. Regardless of appearance, consider the snag to be rotten. By following appropriate safety precautions, snags can be removed safely. Improperly removing a snag can cause injury and death. To alleviate some of the danger, snags should be harvested before live trees.

Many operators use a skidder's piling blade or fairlead to push hazardous snag from the work area. In some cases, a snag can be pulled by a line extended from a rubber-tired skidder. Do not remove a snag in high winds or if the tree is holding heavy, wet snow. Removal techniques vary and should be based on the size, height, and condition of the snag. Before cutting snags with a chain saw, check for loose limbs, bark or other materials that could present a problem. Be alert for tops that could break and fall due to saw vibrations. Remove any remaining bark with an axe. Look at the base for splits, rot, cat-faces, and cracks. Check the surrounding area and remove objects from the base of the tree and larger work area. Establish the lean of the snag by using a plumb-bob or axe. Remember, the snag is rotten. As a result the snag could crumble and fall during the facing procedure. Vibrations can be problematic. Heavy blows from tools send shock waves through the tree and dislodge loose materials. If the tree is wedged over, the vibrations of pounding could cause the tree to fall backwards or break off. Keep an eye on the top of the snag while sawing the undercut to gauge tree movement. If the snag falls on its own, it could fall in any direction. If rotten, the snag may immediately sit down on the bar when faced.

There is no practical way to remove high, dead limbs from a tree but a feller can take some steps to make the work area safer. Bump the tree with a skidder or bulldozer with overhead protection to knock the limbs free. Avoid having trees with dead crowns or limbs fall into one another. Identify a safe area during felling.

Felling the Standard Tree. Trained fellers and buckers guarantee the crews' safety while on the job and should adapt felling equipment to the environmental conditions. Never work in heavy winds, fog or electric storms. Selecting and using proper tools is the first step towards safe felling practices. Tool selection should be geared towards the tree structure and condition and the surrounding area. A single-bitted axe and wedges are essential safety tools. The wedges help guide the tree and their use promotes safe felling. An axe should have a good square pounding surface and be in good repair.

Felling takes a small portion of the time needed to harvest a tree however, it can be the most dangerous time. Chain saws make the process easier and the ease can cause the feller to forget the dangers presented by large trees. Plan how to cut down the tree before the actual felling begins. Consider such things as terrain, wind direction, lean, lay and soundness of the tree. Check the lay or bed of the tree. Look for dead tree tops, snags, and widow makers. Before felling begins, remove undergrowth and tree limbs up to breast height. Walk the intended lay to check for objects that might damage the tree or endanger the feller. A qualified feller knows the hazards and plans an escape route. As tree lift begins, be alert to the direction that the tree will fall. Establish the direction of the fall before harvesting. If the tree is felled in the right direction, gravity will provide valuable free transportation. As the tree is falling set the saw down and exit along the escape route. Watch for kickbacks and falling materials and whiplash branches. Be sure to stay clear of the butt. If possible, stand behind a sound, shielding tree, far back from the planned direction of fall.

There are two methods to felling, directed and random. Directed felling is the preferred way to operate, which allows for a concentration of timber in a planned spot. It is an effective means to facilitate the bunching process.

Never fell a tree that is considered too dangerous. If it can be safety removed, clear the brush and small trees away from the base. An appropriate felling notch assists the direction of the fall. Two felling cuts are recommended, the conventional cut and the open face cut. Cut the felling notch on the side of the tree in the planned direction of the fall.

Making good cuts is the pride of a feller. Slanted falling cuts can cause a tree to slide off a stump during felling. An undercut that is too small or shallow causes the undercut to close too quickly, or will allow excessive stump pull or buttsplits. Do not cut off rolling roots. They act as support limbs to the stump.

Staaf and Wilsten (1984) recommend the following felling safety rules:

- Use the correct equipment.
- Nearest co-worker should be at least two tree lengths away.
- Do not go under jammed or stuck trees.
- Clear undergrowth before felling.
- Make correct guide cut.
- Leave a breaking crest.
- Keep thumb under the front handle bar.
- Keep close contact with the chain saw.
- Work with knees bent and feet apart.
- Make felling cut above guiding cut to prevent tree from sliding backwards.
- Do not walk with a moving saw.

An experienced feller always carries the power saw by the handle. If it is carried over the shoulder it can cut the hands and neck.

Felling Problem Trees. Only fell trees with excessive lean or against the natural lean after receiving proper instruction from trained instructors. When felling a tree with a lean, establish the direction of the lean and plan an escape route. Make an open-faced notch width approximately 80% of the tree diameter. Bore directly into the tree behind the junction of the notch, leaving sufficient holding wood to guide the tree. Depending on the severity of the lean side, notching may be required to prevent pulling out stump roots. From the bore cut, continue to remove the remaining holding wood while placing a safety wedge under the side lean on the backside of the cut. Make the final cut from the backside of the tree by placing a horizontal cut below the bore cut, overlapping the bore cut with the horizontal backcut. If a line is used to fell a leaner, to prevent falling prematurely when pulling, do not put too much tension on the line before cutting is completed.

Felling on a slope can cause the tree to jump back and could be fatal in some situations. Trees felled down a slope can slide uncontrollably. Trees felled across a slope are difficult to collect. Slope felling must start at the top of the hill and work down at a 45 degree angle to the incline.

Hang ups are the result of trees getting tangled. Cutting the supporting tree to dislodge the hang up is dangerous and should not be attempted, however a feller can fall a tree into the hang up.

Mark and lower lodged trees prior to working within two tree lengths of the work area. Vibrations from the work area can cause the lodged tree to fall. Marking the tree gives the cutter an idea of where the safe work areas lie. Be careful if you handspike, pull, or cut short butt sections off the trunk of a lodged tree but never climb it. Use safe felling practices to control the direction of fall and enforce a mandatory procedure for handling hung trees to prevent fatalities and injury. Do not cut down a lodged tree or the supporting tree.

Bucking. Chain saw bucking is done to cut whole trees into shorter lengths for easier skidding or loading. The segments may be called logs, bolts, or if only the top is removed, tree-length logs.

Bucking specifications vary according to logging operations but by using a tape measure every log can be cut as specified. Buckers must assess the work environment to ensure that there is no danger from debris, falling or rolling timber, and plan a safe escape route.

Before cutting the log, ensure that the lay of the timber is such that it can be bucked safely. Anticipate and plan for hazards such as pivot points, natural skids, and soundness of the log. Discard brush, debris, snag and other unstable trees from the work area. Be aware of other logs and keep a safe distance away. Touching another log with the tip of the bar can cause a kickback. Ensure that the saw has plenty of gasoline. If the worker stops to refuel, the log may fall.

Limbing and Topping. Do not stand on the downhill side of the log being cut. To prevent a kickback, do not stand directly behind the saw. Block logs when bucking on sloped ground, working from the uphill side.

Choose appropriate cuts and log lengths. Only buck a tree at double length if it is dangerous to do otherwise. When a log is not completely bucked, cut or mark a cross on the end. Notify those who might be endangered.

Using a chain saw, limbs and branches are cut off where the tree fell or at "cold decks" (where the skidder operator drags whole trees into piles near the loader for limbing). Delimb smaller branches before bucking a tree. Branches that obstruct work or intensify the cutting situation should be removed before branches that cross cut the main branches. Plan the operation and use proper cutting techniques to avoid pinching and splitting of the branches. Mechanical delimbing is also common in softwood logging, whole trees are forced into a "limbing gate" where the branches and limbs are sheared off. Topping is really a form of bucking except that the worker climbs (or a feller buncher is utilized) to a certain height in the tree, then cuts only what is called the "merch topU," the smallest utilizable top. Topping by a worker is rarely done today.

Primary Transportation

Skidding, Forwarding and Yarding. After the cutting is done, the logs must be transported to a landing for loading, this is termed primary transportation. A landing may be either a cleared area where logs are temporarily stacked or simply a portion of the shoulder on the access road. Depending upon the terrain and type of operation, several different methods are used to transport logs to the landing.

"Skidding" involves dragging whole trees or logs of any length from the stump to the landing area. The skidder operator must go to the area where the trees have been cut down, pick up a load, carry (or forward) it to the landing area for bucking or loading, then repeat the process. On even and stable ground, this may be done by machine (i.e., rubber-tired skidders, crawler tractors or farm tractor). Often, if the terrain is wet or steep, timber may be skidded by using a rigging of wire rope or cable and powered winches (called yarding). Helicopters have been used successfully on special sites to accomplish this task.

Operators and choke setters are responsible for inspecting equipment at least daily. After the choke setter sets the choke, the setter should move out of danger and into the view of the operator. The operator and setter should watch for falling trees. Hook and unhook chokers from the uphill side or end of the log, unless the log is blocked to prevent rolling or swinging. Chokers should be position near the end of the log or the tree length.

While skidding, the first rule of safety is to keep all bystanders out of the way of harm. Secondly, it is important to discuss and agree on hand communication signals. Thirdly, move the machinery smoothly because the wrong speed, weather or obstacles can cause a rollover or a tree top to break off. Do not bunch trees in a larger bundle than the skidder can pull.

Position the skidder so that the winch cable can be spooled out in a fairly straight line. Winch the load in as straight a line as possible. When more than one log is being skidded, attach the further one to the end of the main cable first. Pull the load up and under the fairlead before traveling. Drive with the dozer blade in the highest position to avoid obstructions that might flip or damage the machine. Drop the load "on the fly" to prevent getting stuck in soft ground. To dislodge from a stuck position, pull the winch control lever down to the free spool position without stopping the skidder. Stop the skidder on solid ground and reload the winch.

Use the correct gear when climbing or descending a grade so that the engine is not overloaded or over-speed. Lower the load to the ground when traveling downhill to add drag to the brake. Drop the load parallel to the pile at the landing. Abrupt turns on steep terrain can cause rollovers. Stay to the inside of the road on side hills to prevent the turn of logs from rolling off the outer edge. If a load does slide over the edge and is followed by the machine, put the winch in free spool and dump the load.

Never work directly below a skidder parked on a grade. Park in a safe position with the park break on, the dozer blade lowered and the transmission in neutral before working behind the skidder. Set the skidder wheels against a rock, tree, or other secure object before dismounting.

The Loading Zone. It is important to prepare the loading area to maximize safety. Clear debris and note obstacles at the work site. The best site for loading is a level area. Check soil firmness and loader response by making a test lift.

Warn people away from the loading zone. Only the loading and unloading personnel shall be in the area. Other workers should be clear of suspended loads and within sight of the loader operator. Never allow riders on the equipment. This ensures that everyone will be out of the way before backing up or swinging the load. Skidder operators should stay clear of loading equipment when dropping a load or removing chokers. When loading or unloading always lower the boom and grapple into a cradle or place them firmly on the ground before leaving the cab. Do not allow anyone underneath the raised boom or its load.

There must be adequate clearance between a vehicle and the pile or deck of forest product. Allow three feet between the counterweight and nearby objects. Balance the load to minimize shifting. Place logs and bolts in a secure orderly manner. Do not put tongs on any logs until delimbers are clear. Never attempt to delimb logs carried by loading equipment. Do not unhook chokers until all lines are slack.

Loading

This component involves using a loader to move whole trees or logs from the landing area onto a truck for transport. Special care must be taken in stacking the heavy trees on the truck bed. Never overload a truck or stack logs above the tops of the standards. Anyone in this hazardous area (e.g., truck driver, foreman or visitors) may be at a high risk of being crushed or run-over by vehicles or falling logs.

Secondary Transportation

The final component of the harvesting system is secondary transportation where the lumber is hauled on public roads or highways to the point of delivery, storage areas, further processing or another mode of transportation (i.e., truck, rail or barge). Truck drivers are at a high risk of highway accidents.

HOW ARE LOGGERS INJURED OR KILLED?

Working either alone or in close proximity with other crew members, loggers run the risk of injury from a variety of hazards. Peters (1990) states that felling trees with a chain saw is the most dangerous activity in logging. Being struck by snags, broken limbs, tops or butt rebound can result in massive trauma and asphyxiation. Falling trees has long been recognized as the major cause of fatalities.

Traumatic head injury has been identified as a leading problem among injured loggers (Myers and Mainwaring 1992). The authors attribute this to the loggers inability to react, possibly due to obstructed vision or that the hazard was outside of the victim's visual domain. Even skidder operators, who must climb out of their protective cabs to attach choker lines, run the risk of being hit by timber or other debris. Logging is inherently dangerous because it involves extensive cutting and heavy lifting of unstable materials. In an analysis of Worker's Compensation claims in Louisiana (Hoop et al. 1993), cuts and lacerations to the thighs, legs, knees and feet accounted for the largest percentage (29%) of the claims.

Power saws accounted for nearly all cases involving hand tools. Sprains and strains due to overexertion (i.e., lifting, pulling and pushing) and falls from vehicles are also common injuries.

Although little is known about the prevalence of occupational illness among loggers, working with the loud, vibrating power saws and heavy equipment can increase his or her risk of hearing loss, cumulative trauma disorders and eye injury. Often the loggers must work long hours in extremely cold or hot temperatures, resulting in hypothermia, frost bite or heat stroke. In addition, nonavailability of clean restroom facilities and potable water may increase the likelihood of gastrointestinal and kidney damage.

Roll-overs and side turnovers are typical skidder accidents. Because of the rough and unstable terrain, skidder operators may also be injured due to side overturns or rear roll-overs on steep grades. Additionally, the heavy impact of a falling tree or earth-moving equipment can trigger a landslide, resulting in massive trauma and possible asphyxiation of the skidder operator and crew members in the area.

HAZARDS INVOLVED IN LOGGING

Working safely on a logging crew requires special teamwork skills, talents and safety attitude. Rather than focus on one task, the logging worker must be able to perform several tasks at once while operating dangerous heavy equipment or machinery in rough terrain or adverse weather conditions. A small mistake in this risky job may not only seriously hurt or kill the worker, but also endanger the lives of those working around him and result in costly property damage. Often the worker must not only be aware of his or her own safety while felling or moving unpredictable timber in the rough terrain, but also must watch out for the safety of others. Because the logging crew work as a team, accident prevention is everyone's responsibility—including the owner, foreman, and worker. One careless worker can endanger the whole crew.

Several different approaches or models have been used to explain why logging-related accidents occur. A complex combination of different factors may contribute to the accident. Rummer (1993) utilized a modified Haddon Matrix to identify appropriate countermeasures to prevent or minimize loss. The contributing factors were divided into four main classifications—Human, Equipment, Environment and Socio-economic. Table 30.1 illustrates the use of a Haddon Matrix in explaining how a worker was killed in a chain saw kickback accident.

It is readily apparent that the deadly combination of the inexperienced logger's unsafe behaviors, adverse working conditions and hazardous equipment contribute to the accident (event) occurring. Tragically, these same factors can also greatly reduce the loggers chances of surviving the injury while help is coming. Each of the four main classifications of contributing factors are explained in detail below.

Human Factors

The worker's physical, mental, and skill level may increase his or her risk for injury. The physical and mental fitness of the operator often contributes to the likelihood of an accident occurring. "Age, experience, health, predisposition to risk, and training all influence an accident" (Peters 1990). Fatigue can be a major catalyst for an accident. Certain tasks, such as skidding or limbing, may become repetitive or monotonous causing the operator to be "hypnotized" or mentally fatigued. In addition, the vibration and jerky movement of the equipment (e.g., power saw) requires that the worker contract his or her muscles over long periods of time increasing fatigue. The result is that operator fatigue plays a significant role in felling injuries and deaths.

Table 30.1. A modified Haddon matrix describing a chain saw kickback accident.

	Human	Equipment	Environment	Socio-Economic
Pre-event	Untrained, less than 2 years experience	Chain saw w/ brake Standard Chain No protective chaps worn.	Heavy brush, Hardwood Stand	Working alone, under quota for day
Precursor/ Behavior	Fails to sharpen chain, worked through breaks			
Event	Cuts Leg	Saw kicks back	Hidden limb catches saw	
Post-event	Cries for help; Bleeds to death		Victim lying in depression in heavy brush, cold and wet conditions accelerate shock.	No one in range, no formal emergency system in place.

No matter what time or weather constraints you are working under, never ignore fatigue.

Mental fatigue can affect reaction time and judgement. Unfortunately, by the time you become aware of being tired, your ability to operate safely and react quickly to hazards have already been seriously compromised. It is critical that you establish a realistic work pace, allowing for frequent breaks (i.e., at least every 2 hours) during the work day. Harvesting timber is a physically and mentally challenging job. Just as a feller may "gas up" his power saw, so should he "refuel" his body throughout the day to fight fatigue. Safety experts suggest drinking at least one 8 oz. glass of fluids (e.g., water) every hour, increasing to two glasses per hour if the weather is very hot or humid. Never operate equipment or machinery while under the influence of alcohol or drugs. Even certain nonprescription drugs (i.e., antihistamines and cold medicines) may cause drowsiness and seriously impair the operators ability to react quickly.

A five year study of logging injuries in Louisiana (Hoop et al. 1993), indicates that most logging accidents happen to inexperienced workers (82% of claimants had less than three years training). Although on the job training is one of the most common training methods, safety cannot be learned along the way. Prior to starting on the job, the new employee should complete a formal, new employee, safety orientation.

Years of experience does not guarantee safe work habits. Regrettably, years of logging experience does not necessarily guarantee safe work habits. Often an experienced worker who has escaped serious injury becomes overconfident, modifying or becoming lackadaisical in following proper safety procedures (e.g., neglecting to sharpen the saw blade). Frequent, standardized training of all workers, both inexperienced and experienced, is critical to insure that everyone is following proper safety procedures and rules. Integrating a high standard of safety into all aspects of the hiring, training and supervision of worker's reflects the company's commitments to reducing losses due to injury and illness and encourages everyone to take responsibility for his or her own safety and the safety of the other crew members.

The success of the logging operation is dependent upon the number of logs the crew can harvest per day. Subsequently, many fellers are paid piece rate or on a minimum production quota. This compensation system may increase their risk of injury by pressuring loggers to work too quickly, encouraging them to alter or abandon safe work practices and personal protective equipment. The safety-conscious manager must take special consideration in assessing and scheduling the work so that safety of all employees is never compromised. Additionally, those workers who work safely and actively participate in recognizing and reducing hazards at the worksite should be formally recognize and rewarded (either monetary or nonmonetary).

Environmental Conditions

There are several environmental conditions which greatly increase the logger's risk of injury and illness. Felling trees in standing timber and/or heavy brush increases the risk of hangups, snags and widowmakers. The feller must be aware of what will be the consequences of felling a tree in a certain direction, insuring that other crew members and equipment are out of the way and that the impact of the fall will not trigger additional fall or landslide movement. Visibility and communication between crew members is critical! Adverse weather conditions can seriously impede the logger's ability to recognize potential hazards, plan an escape route and communicate with other crew members. Electrical storms, high wind, and any other weather condition that could injure workers mandate that all work stop. A brisk wind makes directed and controlled tree felling difficult, work should be suspended under these conditions. Fog can also create a dangerous work environment, since fellers cannot easily see broken off tree tops, hanging limbs or snags. Similarly, heavy snow can cover ground debris and trees, causing limbs to fall, impeding safe escape or rescue, and/or obstructing a feller's view. Written procedures and verbal instructions concerning these situations should address each of these weather hazards, describing exactly what the loggers should do and not do under these conditions.

During severe weather conditions, all work should be stopped and the workers moved to safety. Often hard-working fellers and buckers are isolated and unaware of impending weather changes. During these potentially hazardous conditions workers should be monitored closely and kept abreast of any weather changes, either visually or by walkie-talkie. A comprehensive written procedure will provide all employees the necessary direction and guidance for fellers and foremen to work (or stop work) during hazardous weather conditions.

Fellers must be aware of what other obstructions are in the fall path, i.e., electric power lines, equipment and structures. The OSHA standard recommends working at least two tree lengths away from utility lines. Once the area has been scheduled for felling the foreman or owner should contact the power company so that the line can be de-energized. A written (and verbal) standardized checklist of hazards and specific procedures should be distributed to all employees. If a tree falls on a line, assume it is electrified until the power company says otherwise.

Unfortunately, in the case of an accident where a logger makes contact with the line and is electrocuted, the first reaction of his co-workers is to go help him. The result is an accident where the possible rescuers become the next victims. Whenever the machinery contacts power lines, the crew should move to a safe distance from the powerline and request assistance from the power company. Should a loader become entangled in the powerline, the operator may consider carefully jumping clear, so that the body is not touching the loader and ground at the same time. However, it may be safer to stay in the operator's seat until help arrives.

Equipment Factors

A preliminary walk-around ensures the operator that equipment and work area are properly prepared.

Loggers utilize a variety of hand power tools and heavy equipment in harvesting timber. Safe equipment operation in the forest doesn't begin when you turn the key. Rather, it starts with a four step procedure by the operator:

1. Operator has successfully completed standardized safe operating training.
2. Equipment is properly maintained and serviced.
3. Equipment is equipped with the proper safety devices.
4. The work area has been inspected and all hazards (e.g., washouts, steep grades, loose ground debris, etc.) have been removed, marked, or noted.

Formal, standardized training to operate equipment should include specific details describing how to safely operate and maintain the piece of machinery. Although the operator's manual is a valuable source of information, it should not be the sole means of training. After intensive one-on-one training with the foreman or an employee trainer in an isolated area concerning the new hazards unique to the operator, the operator should be required to: describe his or her responsibilities; identify the major hazards unique to that particular type of equipment (and what safety procedures/devices reduce that risk); show the foreman or trainer how to safely maintain and operate the equipment.

Each operator should be provided a standardized maintenance and service checklist for his or her particular piece of equipment. It is especially critical that the worker take the initiative and responsibility to report any down equipment or maintenance needs to the foreman or maintenance crew as soon as possible. Similarly, the foreman (and maintenance crew) must react as quickly as possible to the request for repair. The ideal situation is that the worker feels that he "owns" the piece of equipment, taking an active role in ensuring that the equipment is running smoothly and safely. Prompt correction of an unsafe work condition increases employee morale and productivity. All equipment and hand tools should also be inspected either by the foreman or a qualified employee at the beginning of the work day.

Each tractor, skidder, yarder, log stacker, log loader and mechanical felling device (e.g., tree shearer and feller-buncher) must be equipped with a falling object protective structure (FOPS) and/or roll-over protective structure (ROPS). The protective canopy protects the operator from injury due to falling trees, limbs, saplings or branches which might enter the compartment area and from snapping winch lines or other objects. A seatbelt should always be worn, to keep the worker in the zone of protection and increase his or her chance of survival.

Strict enforcement and daily inspection of seatbelts by all employees should be a top priority. In order to avoid falls, mounting steps and handholds should be provided and kept clean throughout the day.

Before entering the work area, the operator should dismount and walk over the terrain. Whenever possible remove or mark loose ground debris, soft earth and settling on hills, stumps, overhanging rocks, rock slides and near dead trees. Slopes over 40 degrees are considered hazardous and should be marked. This preliminary step by the operator is critical! By traveling over the area on foot, he or she can get a better feel for the stability of the ground and hidden hazards. It is suggested that, especially when the operator is inexperienced, the foreman should accompany him or her in the inspection guiding them in recognizing hazards and planning how to safely operate around it. The operator should also note where the rest of the crew members are working and how their work area may overlap. These hazardous areas where man, machine and falling timber may interact should be identified and explained by the foreman during the daily pre-shift meeting. Everyone on the crew should be fully aware of what the other members are doing throughout the shift. It is the foreman's responsibility to keep everyone informed of any changes in work scheduling or location throughout the day. Surrounded by loud heavy machinery and heavy brush, the unaware crew member can be struck by falling timber or equipment.

There are special types of equipment used in skidding and loading timber, each having their own unique hazards.

Tow Lines. Tow lines are utilized to skid the whole trees from the stump to the "cold deck" or loading area. Under very high tension, if the wire rope should break the skidder operator (and any observers) may be seriously injured by the backlash. Anyone in close proximity is at risk of serious injury from tow line recoil.

Both crew members and management should stay well away from the tow line (1-2 lengths) to avoid being struck by the recoil. Operators may also be injured while hooking up the lines or chokers, either by: falling while climbing over timber; catching fingers, arms, feet and toes between the line and tree; or being struck by falling debris, snags or timber. Wire rope failure is a result of kinking which usually occurs when it is improperly unwound from the shipping reels. Leather gloves should be worn when handling wire ropes to protect the hands from cuts, contusions and amputations. Wire rope tends to fray after extensive use, inspect the ropes before initial installation and at least once a week. To maintain the life of wire rope, lubricate the wire regularly to protect it against corrosion and excessive wear.

Winch. Similarly to the tow lines, the winch is also under extreme tension. After hitching the winch to the load, have everyone stand clear (at a distance at least equal to the length of rope being used). It is suggested that if the cable is under extreme tension and a break is possible, lay a blanket or knapsack over the cable to retard its recoil if it does break.

Chains. Consult the manufacturer when selecting chains. Inspect chains for cracks, corrosion, pits and deformed, stretched, weak, or gorged links. Immediately replace damaged or deformed chains. Never splice a broken or damaged chain, replace it. When hooking a chain, put the hook completely over the chain to prevent the chain from slipping and the hook from bending. Place the hook as far from the load as possible with the hook opening away from the object so that the pull is on the back of the hook. Stand the chain's length away from a hitched or hooked load.

Chipper. The chipper is one of the most hazardous pieces of equipment utilized by logging operators. Safety begins before starting the chipper. Special care should be taken to isolate the chipper away from other traffic, trimming trees away from the area and clearly marking the area as a danger zone. The operator is responsible for the safety of those around the equipment. In feeding the chipper, many operators (and unauthorized observers) become overconfident and hypnotized by the fast speed the trees are drawn into the infeed chute. Tragically, the operator's hand or foot can easily be drawn in before he or she has time to react. Only allow trained professionals (including management) to work in the area of the loader and chipper conveyor. Even these workers must stay clear of the working radius of the chipper, unless the operator indicates otherwise. Pull-in injuries are the most common type of chipper accident.

The chipper operator communicates by using horn signals; a series of repeated short blasts is recognized as the signal for danger.

Power Saws. The chain saw is the primary cause of amputations, cuts, lacerations and contusions at the logging site. Although a much more powerful version of the typical hardware store chain saw, many fellers do not follow proper safety procedures or tend to improvise when operating the power saw. Frequent inspection and maintenance is the most effective preventative measure for power saw accidents.

The three most common operational hazards involved in operating a power saw are: kickback, pushback, and pull-in. Kickback is the most common hazard, caused by the saw chain suddenly stopping due to contact with a solid object, resulting in the operator losing control of the saw. Tip contact in some cases may cause a lightening fast reverse reaction, kicking the guide bar up and back towards the operator. All saws should have a "chain break" feature that provides some protection.

To reduce the risk of being injured by kickback, do not stand in line with or behind the saw during felling. Do not stand behind the saw bar unless a wedge is in the tree.

Picking the proper chain is another important safety factor. The way the chain is filed and the kind of chain used can make the saw easier (or more difficult) to handle. Chain trouble can be demonstrated as vibrations, kickbacks, inaccurate sawing or feller fatigue (Dent 1974). The chain should be checked often, properly maintained and sharpened frequently. Inspect the saw regularly to ensure that handles and guards are snugly in place, controls function properly, and that the muffler and brakes are operable. Adjust the saw so that the chain will not be driven after the throttle is released. The continuous throttle control system shuts off power when throttle pressure is released. Follow the manufacturers' instruction for operation and adjustments.

Before turning on the power, ensure that the saw has enough fuel to complete the job. Only refuel when the saw is off and not moving. Do not smoke while refuel and stay at least twenty feet away from any potential ignition source. Wipe spilled fuel off the saw. Start the saw at least ten feet from the fueling area. Firmly support the saw on the ground or by gripping the rear handle of the saw between the thighs and holding the upper handle with the left hand. Safe starting procedure varies so consult the operators' manual for the model used.

Do not work alone. Before starting a cut, clear the surrounding area of brush and debris. While cutting, keep your weight balanced and feet firmly planted.

Maintain a stance that ensures a tight grip. During operation, firmly hold the saw with the thumb and fingers encircling the handles. Do not cut overhead or lean to make a cut and NEVER cut towards the body. Avoid contact with the cutting chain and muffler when carrying the saw. Always carry the saw at the side with the bar pointed forward when traveling uphill and backwards when traveling downhill. Use a bar guard during transport.

Escape Route

Good logging practices begin with a safe escape route. Plan and clear a complete escape route from the stump to the end of where the tree should land. This usually is a distance of about twenty feet (Dent 1988). Choose an escape route on the contour, at 45 degrees to the planned direction of the fall, and at least 20 feet long. Head down the escape path as soon as the tree is committed to fall. Shut off the saw before retreating. If the saw is hung up leave it behind.

Regardless of how carefully the lay is chosen, the weather, the lean or misjudgments can still present danger. Revaluate the escape route periodically to ensure it provides a path to safety.

DEVELOPING A FORMAL SAFETY PROGRAM

An accident is an unplanned event that causes near-miss, injury or property loss. The primary objective of any safety program should be accident prevention. It is only by reducing accidents that the incidence of serious personal injuries can be reduced. The other effect's are: lost production, property damage, medical expenses and Worker' Compensation claims. Estimating the cost of an accident is very difficult, since both direct and indirect costs must be considered. It is estimated that the direct costs (e.g., replacement or repair of equipment & machinery, wage and hospitalization benefits, increase in Worker's Compensation insurance premiums, etc.) account for only 20% of the total cost of an accident. The remainder of the costs are indirect and are more difficult to quantify (e.g., time spent on investigating accidents and near-misses, preparing paperwork, reduction in employee morale and productivity, etc.). Ultimately, more than 60% of the total cost of an accident is absorbed by the employer. Public administration (30%) and the individual (less than 10%) also suffer economic loss due to logging accidents (Klen 1989).

Accident prevention is very difficult in a logging operation since the very nature of felling and moving tons of timber in isolated rough terrain is inherently hazardous and often difficult to avoid. There must be a strong commitment from everyone working on the crew to work safely all the time. Working safely is accepted as the only way to do the job. The most effective approach to reducing accident-related losses is to implement a formal logging safety program. Acting as a foundation for the development of this safety philosophy among all employees, both management and workers work together toward a common goal - reducing both injury and property damage at the job site.

An effective accident prevention program is composed of eight important components:

Component 1: Active participation of both management and employees

An effective accident prevention program requires total commitments from the owner down to the newest worker. Visible and active participation by management in every phase of the program (e.g., attending weekly crew safety meetings) communicates to crew members that safety is a top priority. It also lends support to the foreman or supervisor who has direct contact with the loggers on a daily basis.

Component 2: Frequent, standardized safety training

An unsafe behavior (e.g., failing to sharpen a saw) acts as a catalyst for the accident and injury to occur. One approach is to hire only skilled, experienced loggers who have the proper safety attitude and who are familiar with the unique characteristics of the jobsite. Considering the high turnover rate in logging, this is not realistic. Instead, management must foster the safety attitude and develop a certain level of skill through standardized training.

The content of the training should include:

- Safe performance of assigned work tasks.
- Safe use, operation and maintenance of tools, machines and vehicles the employee operates, including emphasis on understanding and following the manufacturers operating and maintenance instructions, storage and precautions.
- Recognition, prevention and control of other safety and health hazards in the logging industry.
- Procedures, practices and requirements of the employer's work site.
- The requirements and responsibilities of both the employer and employees under the OSHA standards.

Component 3: Written (and verbal) formal safety procedures and rules.

Everyone on the jobsite is working toward a common goal - reducing accidents. This goal should be broken down into clear, measurable objectives (e.g., conducting weekly safety training meetings). By clearly describing and assigning specific responsibilities to management and employees he or she is aware of what is expected and may feel ownership in improving the worksite conditions.

Component 4: Maintenance of safe working conditions through daily frequent employee and management inspection and maintenance of the tools, equipment, and worksite.

Frequent, documented inspections is a valuable preventative measure. Working as a team, both management and crew members are responsible for identifying unsafe working conditions. A formal checklist should be provided for each piece of equipment, tool and work area. While the operator is responsible for daily inspections and servicing checks, the foreman may conduct a more comprehensive inspection of the whole worksite. The results of the inspections (and their corrective action) should be shared with the entire crew at the weekly safety meeting, not only to make everyone aware of possible dangers but also to reinforce the crew members potential in improving working conditions.

Component 5: Strict and immediate enforcement of formal safety procedures and rules.

One valuable outcome of the effective formal logging safety program is the positive attitude and expectation of all employees, involving a commitment that the safe way is the only way to do the job. Even though the crew may be under a stringent timber quota or inclement weather, the formal safety procedures and rules must be followed.

Component 6: Formal emergency and medical assistance plan for on-site rescue.

One possible factor which may contribute to the high fatality rate in the logging industries is that the crews are often working alone, isolated and far from medical assistance. With that in mind, the safety-conscious employer must have in place a formal emergency plan to provide on-site medical assistance and/or transport the injured worker to the hospital or clinic. So that valuable minutes are not wasted, every employee and foreman should be aware of the proper procedures and basic first aid. Employees working alone or in obscured visibility should be periodically checked and given the means to signal for help (e.g., signal horn or walkie-talkie).

Component 7: Formal injury, illness, and near-miss investigation, involving analysis and reporting to both management and employees.

Similar to the formal inspection, investigating past injuries, illness, and near-misses helps you pinpoint potential dangers. Sharing the investigation results with both management and crew not only reminds everyone to work safely, but may also generate some suggestions for improving the working conditions. Once again, this promotes the crew member becoming more involved in his or her own safety.

Component 8: Provide proper personal protective equipment and first aid kits.

Every employee should be properly dressed for the job, this would include:

- Durable, close-fitting pants and long sleeved shirt
- Steel-toed shoes with slip-resistant soles (water-resistant or water repellent)
- Hearing protection (either ear muffs or ear plugs) if working within 50 feet of chain saws
- ANSI approved hard hat
- Puncture-resistant gloves
- Eye and face protection (e.g., goggles, face shield or screen visor)
- Special protective equipment (e.g., ballistic nylon chaps, respirator).

Since all personal protective equipment must be inspected every day by the operator, he or she must be properly trained in how to properly fit and check for wear and tear. Defective or damaged personal protective equipment must be immediately repaired or replaced. Any equipment provided by the worker must meet the minimum standards outline in the formal safety procedures and rules. A fully-equipped first aid kit should be provided at each site where trees are being cut, at each active landing and on each employee transport vehicle. The number, and content of each kit should reflect the degree of isolation, the number of employees, and the hazards reasonably anticipated at the work site. These kits must be restocked when needed and approved/reviewed annually by a health care provider.

SUMMARY

We have discussed the special safety issues that must be considered to reduce the incidence of logging-related accidents. Logging was identified as a "high-risk" occupation, where massive trauma and head injuries are common due to struck-by or chain saw accidents. Often multiple factors were involved in contributing to the injury or illness. In addition, because the logger often works with other crew members, his unsafe behavior may endanger their lives as well. Using the cooperative spirit of the crew, it is suggested that employers implement a formal logging safety program where both management and crew members take an active role in identifying and correcting unsafe work conditions.

REFERENCES

American Pulpwood Association. 1988. *Timber Harvesting.* Danville, IL: The Interstate Publishing Company.

Dent, D.D. 1974. *Professional Timber Falling: A Procedural Approach.* Portland, Oregon: Ryder Printing Co.

Hoop, C.F. J.C. Pine, and B.D. Marx. 1993. *Major Logging Injuries in Louisiana: Nature and Trends.* Presented at 1993 American Society of Agricultural Engineers Winter Meeting, Paper No. 510, St. Joseph, MI: ASAE.

Klen, T. 1989. Costs of occupational accidents in forestry. *J. Safety Res.* 20, pp. 31-40.

Lown, J., T.G. Prather, and P. Peters. 1993. *Multimedia Training for Loggers, Arborists, and Woodcutters.* Presented at 1993 American Society of Agricultural Engineers Winter Meeting, Paper No. 937523, St. Joseph, MI: ASAE.

Myers, J. and D. Fosbroke. 1994. Logging fatalities in the United States by region, cause of death, and other factors-1980 through 1988. *J. Safety Res.* 25(2): 97-105.

Myers, M.L. and J. Manwaring. 1992. *Preventing Logging Fatalities in Alaska.* Presented at 1992 American Society of Agricultural Engineers Winter Meeting, Paper No. 927508. St. Joseph, MI: ASAE.

Peters, P.A. 1990. *Logging Fatalities and Injuries Due to Felling Trees.* Presented at 1990 American Society of Agricultural Engineers Winter Meeting, Paper No. 937536. St. Joseph, MI: ASAE.

Rummer, B. 1993. *Engineering Solutions for Logging Safety.* Presented at 1993 American Society of Agricultural Engineers Winter Meeting, Paper No. 937526, St. Joseph, MI: ASAE.

Staaf, K.A.G., and N.A. Wilsten. 1984. *Tree Harvesting Techniques.* Dordrecht, The Netherlands: Martinus Nijhoff Publishers.

31

WILDLAND FIRES AND FIREFIGHTING

Ricky L. Langley, M.D., M.P.H.
Duke University School of Medicine

Wildland fires destroy thousands of acres of timber and grassland, kill thousands of animals, and injure scores of humans each year. Of the 80,000 U.S. Firefighters whose job is to control wildland fires, an average of 16 deaths occur annually as a result of firefighting activities. This chapter describes factors associated with injuries among firefighters, including direct thermal injury, smoke inhalation, heat stress, elevated levels of carbon monoxide and other chemicals, low oxygen levels, and equipment hazards, falling trees, and dangerous terrain. Also discussed are illnesses which may be attributed to wildland firefighting activities such as musculoskeletal injuries, hearing loss, mental stress, communicable diseases, respiratory dysfunction, cardiovascular and zoonotic diseases. Finally, information on fire behavior, wildland fire suppression including the ten standard orders of the USDA Forest Service, and fire prevention is provided.

INTRODUCTION

Fires are one of the most spectacular as well as dangerous events that occur in the world. While many fires are started naturally, such as from a lightening strike in a dry forest, unfortunately many are intentionally or unintentionally started by man. In the United States, over two million structural and nonstructural fires occur each year. An estimated 4,000 to 4,300 Americans die due to fires each year. Millions of acres of land and forest are destroyed annually, frequently as a result of carelessness among campers or due to the illegal activities of an arsonist. The economic loss due to fires in the U.S. was estimated at 8.2 billion dollars in 1994 (NSC 1995).

EPIDEMIOLOGY

Fires were responsible for 1.4% of all fatal occupational injuries during 1994. Most fatal fires are house fires due primarily to smoking products (NSC 1995). However, it is estimated that 0.2% of nonfatal agricultural injuries causing lost work time are due to fires. As more people move into rural areas, the chance of encountering wildland fires increases. Between 1871 and 1947, more than 2,000 people were killed in wildland fires (Davis and Mutch 1995). Because of increased distance from fire stations, people in rural areas need to be prepared to respond to wildland fires.

In the United States, about 80,000 firefighters are involved with firefighting activities on 70,000 woodland fires that burn an average of 2 million acres each year (Ward et al. 1989). Wildland firefighters are exposed to many hazards. During the years 1981 to 1990, 1,221 firefighters died in the line of duty (NFPA 1991). Of their deaths, 162 were a result of wildland fires. Almost three-fourths (117) of the deaths occurred during fire suppression activities and 45 occurred when responding to or returning from such fires. During this century, some of the deadliest blazes for firefighters include St. Joe, Idaho in 1910 with 72 killed, Los Angeles in 1933 with 25 killed, Shoshone National Forest, Wyoming in 1937 with 15 deaths, Mendocino National Forest, California with 15 deaths in 1953, and Glenwood Springs, Colorado with 14 killed in 1994. An analysis of the causes of death among the wildland firefighters killed between 1981 and 1990 is listed in Table 31.1.

The circumstances surrounding the deaths included stress (35 deaths), caught or trapped by fire progress (26), exposure to or contact with an object (17), smoke exposure (6), exposure to electricity (6), hot weather (3), struck by lightening (1), struck by falling object (1). An additional 20 were killed in firefighting apparatus, including 16 in aircraft crashes. Twelve of the 35 deaths from stress were due to physical overexertion at the fire scene (NFPA 1991).

An average of 16 deaths (range of 6 to 22 deaths) occurred annually to wildland firefighters during 1981 through 1990. Regional distribution showed that 64 of the wildland firefighter deaths occurred in the Western United

States, 58 in the South, 21 in the North Central U.S. and 19 in the Northeast.

Table 31.1: Causes of deaths associated with wildland fires (NFPA 1991).

Etiology	Number
1. Heart attack	34
2. Internal trauma	26
3. Burns	15
4. Asphyxiation	11
5. Electric shock	8
6. Crushing	8
7. Heatstroke	3
8. Amputation	2
9. Stroke	2
10. Bleeding	1
11. Drowning	1
12. Fracture	1
13. Aneurysm	1
Total	114

FACTORS ASSOCIATED WITH INJURIES AND ILLNESSES AMONG FIREFIGHTERS

Injuries occurring among individuals involved in wildland fire suppression activities are due to the following factors: direct thermal injury, smoke inhalation, heat stress, low oxygen levels, elevated levels of carbon monoxide, carbon dioxide and other chemicals, particulate matter, the equipment, falling trees, and dangerous terrain (Davis and Mutch 1995).

Flames are responsible for burns. Heat may produce thermal injuries of the respiratory tract. Singed facial hair, burns of the nose, mouth, and stridor or dysphonia are clues of respiratory injury. Smoke inhalation may transport toxic components to the lungs that may lead to pulmonary edema (Bizovi and Leikin 1995). Smoke is also an irritant to the eyes and may make it difficult to keep the eyes open. This may impede the ability to safely escape.

Heat stress problems frequently occur in firefighters. Insulative clothing, strenuous physical work, air temperature, humidity, radiant heat, and poor air movement all contribute to elevating the body temperature leading to heat exhaustion or heat stroke (Sharkey 1979).

Elevations of carbon monoxide and carbon dioxide occur in most wildland fires. However, most studies report moderate elevations, and the risk to man appears to be related to prolonged exposure which can cause headache, impaired judgement, lethargy, and psychomotor difficulties (Harrison et al. 1995). Additionally, angina may be induced in individuals with coronary artery disease (Proctor and Hughes 1978). Additional contaminants that wildland firefighters may be exposed to include nitrogen oxides, sulfur dioxide, aldehydes, polyaromatic hydrocarbons, semivolatile and volatile organic compounds, particulate matter, fire retardants, and herbicides (Harrison et al. 1995; Materna et al. 1992). While wildland firefighters are not likely to experience the extreme acute exposures that structural firefighters may encounter while working in enclosed spaces, they may spend several days or weeks working in smoke with shifts of 12 or more hours. Additional off shift exposures may occur when fire base-camps are located in areas that may fill with smoke. The use of self-contained breathing apparatus is often not feasible and respiratory protective equipment in wildland firefighters may only consist of bandannas tied over the nose and mouth (Sutton et al. 1990).

Wildland firefighters are also at risk for many of the hazards faced by their urban counterparts. These risks include musculoskeletal injuries, mental stress, communicable disease exposures, hearing loss, respiratory dysfunction, and cardiovascular disease. There is also concern that toxic exposures may affect the reproductive system and may cause cancer.

Studies of pulmonary function among wildland firefighters have demonstrated declines in airflow rates across the fire season (Harrison et al. 1995). Transient increases in airway responses as measured by methacholine challenge have also been reported (Sheppard et al. 1986). There is concern that chronic obstructive lung disease may develop (Scannell and Balmes 1995).

Both chemical and nonchemical exposures may be reproductive hazards. Numerous chemicals detected in the fire environment have been noted to cause reproductive toxicity in animals. Examples include acrolein, benzene, carbon dioxide, carbon monoxide, formaldehyde, hydrogen chloride, hydrogen cyanide, and sulfur dioxide, to name a few (McDiarmid and Agnew 1995). Carbon monoxide has been found to adversely affect the human fetus, and pregnant firefighters may need to seek alternate duty during their pregnancy if exposure to high levels of carbon monoxide is likely (McDiarmid and Agnew 1995). Nonchemical exposures that have adverse human reproductive effects include hyperthermia, physical activity, noise, and psychological stress (Agnew et al. 1991). While firefighters are exposed to all these factors, unfortunately, few studies have evaluated the reproductive health of firefighters.

It is not uncommon for urban firefighters to provide medical care to victims of fires or other accidents. It is likely then that wildland firefighters may also be called upon to provide these services as well as treat injured co-workers.

Thus, firefighters are potentially exposed to many infectious agents. Probably of most concern are the blood-borne pathogens including Hepatitis B, Hepatitis C, and Human Immunodeficiency Virus (HIV) (Weaver and Arndt 1995). Firefighters who provide medical treatment should be instructed in universal precautions and offered the Hepatitis B vaccine.

Wildland firefighters are also at risk of contracting a zoonotic infection (Weaver and Arndt 1995). Lyme disease, Rocky Mountain Spotted Fever and Erhlichiosis are infectious agents transmitted by tick bites. Mosquitoes can transmit several infectious agents, including malaria and encephalitis viruses. Wildland firefighters should inspect themselves daily for ticks and consider wearing insect repellent. Depending on the geographic area, certain vaccines or chemoprophylactic medications may be warranted. For example, yellow fever vaccine or malaria chemoprophylaxis may be recommended for firefighters in third world countries. All firefighters should ensure their tetanus immunization status is up to date.

Firefighters have been found to have elevated rates of hearing loss (Tubbs 1995). Noise exposures include sirens, air horns, diesel engines, and chain saws. Often noise exposures, some exceeding 120 decibels, occurring over short time periods during emergency response are not uncommon.

Chemicals in the fire environment may also have a toxic effect on the auditory system and contribute to hearing loss among firefighters (Morata et al. 1994). A hearing conservation program is recommended as a way to help prevent hearing loss among firefighters. Evaluation and possibly repositioning of sirens and air horns may reduce noise exposure. Appropriate use of hearing protection devices should be worn during noisy activities.

Firefighters are exposed to various chemicals contained in fire smoke and building debris, many of which are carcinogenic or mutagenic (Golden et al. 1995). Examples of mutagens/carcinogens include acrolein, benzene, formaldehyde, sulfur dioxide, asbestos, diesel exhaust, and polycyclic aromatic hydrocarbons (Golden et al. 1995: McDiarmid and Agnew 1995). Working as a firefighter increases the risk of developing leukemia, non-Hodgkin's lymphoma, multiple myeloma, brain, urinary, and possibly intestinal, prostate, and skin cancers (Golden et al. 1995). Surprisingly, most studies do not report an increase in lung cancers among nonsmoking firefighters (Scannell 1995). Periodic medical examinations may be useful in detecting early signs of cancer in firefighters.

Cardiovascular disease is frequently noted as a health risk among firefighters. Firefighting duties require strenuous activities which often mandate the firefighters to work at near maximal heart rates for long periods of time (Melius 1995).

High heat load increases the physiologic demands on the heart. Additionally, exposure to certain chemicals, especially carbon monoxide and cyanide, may be cardiotoxic (Proctor and Hughes 1978). Several mortality studies have found an increase in heart disease mortality among firefighters (Melius 1995). Preventive measures to reduce the risk of cardiovascular stress include proper rest breaks, proper fluid replacement, controlling exposure to carbon monoxide and other contaminants through the use of appropriate respiratory protective equipment, appropriate medical screening for cardiovascular disease risk factors and maintenance of good physical fitness (Melius 1995). For wildland firefighters, the use of self contained breathing apparatus is often impractical. However, they should try to ensure their camps are a safe distance from smoke to reduce exposure to carbon monoxide (Harrison et al. 1995).

Due to the strenuous nature of firefighting, it is not surprising that musculoskeletal injuries account for almost half of on the job injuries among the one million firefighters in the United States (IAFF 1994). Most of these injuries consist of sprains, strains, and myalgias that usually involve the back (Matticks et al. 1992). Musculoskeletal injuries are the primary cause of employee absenteeism and disability (Reichelt and Conrad 1995). Several factors have been identified that may be related to the likelihood of becoming injured. Personal, workplace, and external environmental factors are listed in Table 31.2.

Formal instructions on lifting properly, handling equipment, and maintaining physical fitness may decrease the number of injuries that occur.

Other frequent injuries reported among wildland firefighters include cuts, scratches, fractures, and eye injuries due to falling trees or limbs, rolling logs or rocks, and dust particles. Exposure to poison oak and poison ivy, stinging insects, and poisonous snakes may also occur (Davis and Mutch 1995).

FIRE BEHAVIOR

The behavior of a wildland fire is influenced by several factors. Heat, oxygen, and fuel in the proper combination are required for ignition and combustion to occur. Heat energy is transferred by conduction, convection, radiation, and spotting (Wenger 1984). Spotting refers to the spread of fire by airborne embers.

Climate, fuel and the topography of the land all influence the behavior of a fire (Wenger 1984). Climate influences the vegetation growth in the area and is modified by local topography. Six broad climates are recognized in the United States—arid, semi-arid, subhumid climates with deficient summer rainfalls, subhumid climate with adequate precipitation in all seasons, humid, and wet climates.

Table 31.2. Factors related to musculoskeletal injury.

Personal

Commitment to job and team	Fatigue levels
Skill and knowledge level	Stress levels
Lifestyle practices	Body strength
Age and gender	Experience
Preexisting heredity conditions	Physical fitness level

Workplace

Job task	Equipment
Staffing level	Safety training
Selection, promotion, and discipline	Health culture
	Medical services

External Environment

Emergency situations	Unpredictability
Number of runs	Rescue/EMS situations
Weather	Heat/cold
Structural conditions	Water

Adapted from Conrad et al. 1994.

Fifteen fire climate zones have been defined for North America based on temperature and precipitation patterns, latitudinal differences, and physiographic provinces (Schroeder and Buck 1970). The fire climate zones are as follows:

1) Interior Alaska and the Yukon
2) North Pacific Coast
3) South Pacific Coast
4) Great Basin
5) Northern Rocky Mountains
6) Southern Rocky Mountains
7) Southwest
8) Great Plains
9) Central and Northwest Canada
10) Subarctic and Tundra
11) Great Lakes
12) Central States
13) North Atlantic
14) Southern States
15) Mexican Central Plateau

Most wildland fires tend to be a mixture of various kinds and types of fuels. The more fuel burning, the hotter the fire. Additionally, the size and arrangement of fuel and the amount and type of leaves and dead vegetation all influence the behavior of the fire (Davis and Mutch 1995).

Topographic conditions primarily affect the direction and rate of spread of a fire (Wenger 1984). The steepness of the slope affects the rate of spread of the fire while the elevation of the land affects the length of the fire season and fuel availability.

There are factors which are warning signs for hotter, faster burning conditions as shown in Table 31.3.

Table 31.3. Warning signs for hot, fast burning conditions in wildland fires.

Fuel	Weather	Topography
More fuel	Faster winds	Steeper slopes
Drier fuel	Unstable atmosphere	South and south west-facing slopes
Dead fuel	Down draft winds	Gaps or saddles
Flashy fuel	Higher temps	Chimneys and narrow canyons
Aerial fuel	Drought conditions Low humidity	

Modified from Davis and Mutch 1995.

In an analysis of 125 wildland fires that resulted in numerous deaths and injuries, the incidents were precipitated by the following situations (Smith et al. 1981). In 29.6% of cases the fire was running up slope; sudden wind shift occurred in 20.8% of cases; rapid rate of fire spread occurred in 13.6% of fires; spot fires in 9.6%; fire running downslope in 6.4%; concentrated fuel flare up in 4.8%; downdrafts and gusts in 4.0%; aircraft wake turbulence in 0.8%; equipment failure in 0.8%; and other miscellaneous events including medical illness or injury in 9.6% of wildland fires.

SAFETY AND WILDLAND FIRE SUPPRESSION

As previously noted, wildland fire suppression is a very hazardous occupation. Fire suppression involves discovering, attacking, extinguishing, or keeping the fire within predetermined borders. To suppress a fire, the fuel must be removed or the temperature of fuels reduced or oxygen excluded.

Numerous injuries and fatalities could be prevented if a few principals are followed (Wenger 1984). The firefighter should receive adequate training in use of firefighting tools, fire suppression techniques, and fire behavior. Experience and attitude are important in safely fighting fires. Maintenance of physical fitness and periodic medical examinations help assess the condition of the firefighter and may identify early sign of illness due to firefighting activities such as changes in pulmonary function. The equipment must be properly maintained to assure it will function when most needed. Firefighters must use personal protection gear and clothing and the issue of a fire shelter is recommended. Fire resistant clothing may mitigate the effects of exposure to heat. A person designated as a safety officer that monitors all aspects of firefighter health and safety, participates in planning sessions, initiates corrective actions concerning unsafe practices, and eliminates or modifies safety hazards is recommended. All firefighters should receive training in first aid.

The USDA Forest Service developed standard firefighting orders to provide for basic safety in wildfire management. The ten standard orders follow:

1. Keep informed on fire weather conditions and forecasts.
2. Know what your fire is doing at all times—observe personally, use scouts.
3. Base all actions on current and expected behavior of fire.
4. Have escape routes for everyone and make them known.
5. Post a lookout when there is possible danger.
6. Be alert, keep calm, think clearly, act decisively
7. Maintain prompt communications with your crew, your boss, and adjoining forces
8. Give clear instructions and be sure they are understood
9. Maintain control of your crew at all times
10. Fight fire aggressively but provide for safety first

Additionally, there are several situations that indicate potential danger. Following is a list of these "Watch Out" situations.

1. You are moving downhill toward a fire.
2. You are fighting fire on a hillside where rolling material can ignite fuel below you.
3. You notice the wind begins to blow, increase, or change direction.
4. You feel the weather getting hotter and drier.
5. You are on a line in heavy cover with unburned fuel between you and the fire.
6. You are away from burned area where terrain and/or cover makes travel difficult and slow.
7. You are in country you have not seen in the daylight.
8. You are in an area where you are unfamiliar with local factors influencing fire behavior.
9. You are attempting a frontal attack on a fire with mechanized equipment.
10. You are getting frequent spot fires over your line.
11. You cannot see the main fire and you are not in communication with anyone who can.
12. You have been given an assignment or instructions not clear to you.
13. You feel drowsy and feel like resting near the fire line.
14. You have not scouted or sized up the fire.
15. You have not identified safety zones and escape routes.
16. You are uninformed on strategy, tactics, and hazards.
17. You have not established communication link with crew members or supervisor.
18. You have constructed a line without a safe anchor point.

FIRE PREVENTION

The objective of fire prevention is to decrease the number of human caused fires (Wenger 1984). Prevention of fires requires the following information: where fires occur, when fires occur, what causes fires, who causes fires, how fires start, and why fires occur.

Actions to prevent fires include education of campers, hunters, and outdoor workers to act in a fire safe manner and to reduce fire hazards or risks. Law enforcement is necessary to reduce violations of fire laws. Examples of violations include arson, failure to have a burning permit, railroad engine spark-arrester code violation, failure to extinguish a fire before leaving a campsite, failure to provide adequate clearance around a home or industrial operation, or open burning during a restricted time.

A large number of fires can be prevented by the proper application of engineering technologies (Wenger 1984). This involves the elimination or modification of sources of ignition and ignitable material. Mechanical or electrical equipment used in woodlands is often a source of uncontrolled fires. Additionally, broadcast burning, railroads, rubbish dumps, harvesting operations, electrical power transmission, construction operations, and recreational areas are all potential sources for uncontrolled fires.

Proper land management can decrease the risk of fires in rural residential regions (Wenger 1984). It is important to have comprehensive plans to develop hazard zoning areas and fire safety ordinances. Rural residents should work with government officials to reduce the amount of flammable material in critical areas

Periodic equipment inspection, proper engineering and land management practices, including manipulation, modification, and reduction of flammable vegetation will reduce the number of human caused fires.

SUMMARY

Wildland fires cause many deaths to rural residents, firefighters, and wildlife each year. Besides the heat exhaustion and burns, of which most people are aware, numerous other acute injuries frequently occur. Musculoskeletal strains and sprains, eye irritation and injuries, arthropod and snake bites and cardiovascular stress may lead to disabling illnesses and injuries. There is concern that exposure to the many combustion products in fires may have long term health effects such as cancer or lead to chronic pulmonary problems. More research is needed to develop lightweight, reusable, inexpensive, and comfortable personal protective equipment, especially respirators, for wildland firefighters. Training must be thorough and lectures and demonstrations on health and safety topics should be given utmost priority by firefighting agencies.

Additionally, individuals moving to rural areas need to be made aware of the hazards of improper burning activities. Improper use or insufficient knowledge or training in the use of equipment such as chainsaws and generators often leads to injuries as well as starting fires. Forest Service personnel, cooperative extension agents, 4-H clubs, and rural volunteer firefighters should provide educational information to these rural residents. Especially in times of prolonged drought conditions, rural residents need to keep abreast of fire advisories and know the best escape route in case of a wildland fire in their vicinity.

REFERENCES

Agnew, J., M.A. McDiarmid, P.S.J. Lees, and R. Duffy. 1991. Reproductive hazards of fire fighting 1. Non-chemical hazards. *Am. J. Ind. Med.* 19: 433-45.

Barrows, J. 1951. *Fire Behavior in Northern Rocky Mountain Forests.* USDA Forest Service, Northern Rocky Mountain Forest and Range Exp. Station, Paper No 29.

Bizovi, K.E., J.D. Leikin. 1995. Smoke inhalation among firefighters. In: *Firefighters' Safety and Health.* P. Orris, J. Melius, and R. Duffy, eds.. *Occup. Med.: State of the Art Reviews.* Philadelphia, PA: Hanley and Belfus, Inc. October-December, 10: 721-33.

Conrad, K., G. Balch, P. Reichelt, et al. 1994. Musculoskeletal injuries in the fire services. Views from a focus group study. *AAOHN J* 42: 572-81.

Davis, K., and R. Mutch R. 1995. Wildland fires: dangers and survival. In: *Wilderness Medicine: Management of Wilderness and Environmental Emergencies.* 3rd edition. P. Auerbach, ed. St Louis: Mosby-Year Book, Inc, pp. 213-42.

Golden, A., S. Markowitz, and P. Landrigan. 1995. The risk of cancer in firefighters. In: *Firefighters' Safety and Health.* P. Orris, J. Melius, and R. Duffy, eds.. *Occup. Med.: State of the Art Reviews.* Philadelphia, PA: Hanley and Belfus, Inc. October-December, 10: 803-20.

Harrison, R., B. Materna, and N. Rothman. 1995. Respiratory health hazards and lung function in wildland firefighters. In: *Firefighters' Safety and Health.* P. Orris, J. Melius, and R. Duffy, eds.. *Occup. Med.: State of the Art Reviews.* Philadelphia, PA: Hanley and Belfus, Inc. October-December, 10: 857-70.

International Association of Fire Fighters (IAFF). 1994. *International Association of Fire Fighters: 1993 Death and Injury Survey.* Washington, DC: IAFF.

Materna, B.L., J.R. Jones, and P.M. Sutton et al. 1992. Occupational exposures in California wildland firefighting. *Am. Ind. Hyg. Assoc. J.* 53: 69-76.

Matticks, C.A., J.J. Westwater, H.N. Himel, et al. 1992. Health risks to fire fighters. *J. Burn Care Rehabil.* 13: 223-35.

McDiarmid, M, and J. Agnew. 1995. Reproductive hazards and firefighters. In: *Firefighters' Safety and Health.* P. Orris, J. Melius, and R. Duffy, eds.. *Occup. Med.: State of the Art Reviews.* Philadelphia, PA: Hanley and Belfus, Inc. October-December, 10: 829-41.

Melius, J. 1995. Cardiovascular disease among firefighters. In: *Firefighters' Safety and Health.* P. Orris, J. Melius, and R. Duffy, eds.. *Occup. Med.: State of the Art Reviews.* Philadelphia, PA: Hanley and Belfus, Inc. October-December, 10: 821-27.

Morato, T., D. Dunn, and W. Sieber. 1994. Occupational exposure to noise and ototoxic organic solvents. *Arch. Envir. Health* 49: 359-64.

National Fire Protection Association (NFPA). 1991. *Wildland Fire Fatalities (1980-1991).* Quincy MA: NFPA.

National Safety Council (NSC). 1995. *Accident Facts, 1995 edition.* Itasca, IL: NSC.

Proctor, N.H., and J. Hughes. 1978. *Chemical Hazards of the Workplace.* Philadelphia, PA: J.B. Lippincott Company.

Reichelt, P., and K. Conrad. 1995. Musculoskeletal injury: ergonomics and physical fitness in firefighters. In: *Firefighters' Safety and Health.* P. Orris, J. Melius, and R. Duffy, eds.. *Occup. Med.: State of the Art Reviews.* Philadelphia, PA: Hanley and Belfus, Inc. October-December, 10: 735-46.

Scannell, C., and J. Balmes. 1995. Pulmonary effects of firefighting. In: *Firefighters' Safety and Health.* P. Orris, J. Melius, and R. Duffy, eds.. *Occup. Med.: State of the Art Reviews.* Philadelphia, PA: Hanley and Belfus, Inc. October-December, 10: 789-801.

Schroeder, J., and C. Buck. 1970. *Fire Weather Agricultural Handbook 360*, USDA Forest Service.

Sharkey, B.J. 1979. *Heat Stress.* Missoula, MT: USDA Forest Service, Missoula Equipment Development Center.

Sheppard, D., S. Distefano, L. Morse, and C. Becker. 1986. Acute effects of routine firefighting and lung function. *Am. J. Ind. Med.* 9: 333-40.

Smith, A., et al. 1981. *Report of U.S.-Canadian Task Force Study of Fatal and Near-Fatal Fire Accidents.* National Wildlife Coordinating Group (unpublished report).

Sutton, P., J. Castorina, and R. Harrison. 1990. *Carbon Monoxide Exposure in Wildland Firefighters.* Berkeley, CA: California Department of Health Services, Field Investigation FI-87-008.

Tubbs, R. 1995. Noise and hearing loss in firefighting. In: *Firefighters' Safety and Health.* P. Orris, J. Melius, and R. Duffy, eds.. *Occup. Med.: State of the Art Reviews.* Philadelphia, PA: Hanley and Belfus, Inc. October-December, 10: 843-56.

Ward, D., N. Rothman, and P. Strickland. 1989. *The Effects of Firesmoke on Firefighters: A Comprehensive Study Plan.* Missoula, MT: United States Department of Agriculture Forest Service Intermountain Research Station.

Weaver, V.M. and S.D. Arndt. 1995. Communicable disease and firefighters. In: *Firefighters' Safety and Health.* P. Orris, J. Melius, and R. Duffy, eds.. *Occup. Med.: State of the Art Reviews.* Philadelphia, PA: Hanley and Belfus, Inc. October-Deember 10: 747-62.

Wenger, K. 1984. *Forestry Handbook.* 2nd edition. New York: John Wiley and Sons, pp. 189-252.

Wilson, C. 1977. Fatal and near-fatal forest fires, the common denominators. *Int. Fire Chief* 43: 9.

32

EPIDEMIOLOGY AND PREVENTION OF HELICOPTER LOGGING INJURIES

George A. Conway, M.D., M.P.H.
Jan C. Manwaring, B.S.
U.S. Public Health Service, Centers For Disease Control and Prevention (CDC),
National Institute For Occupational Safety and Health (NIOSH), Division of Safety Research
Alaska Field Station

> As environmental concerns regarding how humans may adversely affect the wilderness increase, new technological methods must be developed to protect these areas while allowing for selective logging. In the forestry industry, new methods for harvesting timber include the use of helicopters, especially in inaccessible territory. Helicopters have been successfully utilized in transporting timber while helping prevent erosion secondary to decreased road construction in these areas. Unfortunately, numerous helicopter crashes and fatalities have occurred. This chapter will describe several of these crashes. Additionally, recommendations on the following topics will be discussed in order to decrease helicopter crashes: equipment, maintenance, human factors, training, management, oversight, interagency/company cooperation, and the environment.

INTRODUCTION

To many logging companies, the helicopter represents a viable option for yarding and transporting timber recently felled in areas that are otherwise inaccessible and/or unfeasible for conventional logging (because of rugged terrain, steep mountain slopes, increasing environmental restrictions, and rising costs) (Proctor 1994; Stehle; Georgia Forest Commission 1986). Because of their unique capabilities, the use of helicopters in hauling logs and recently felled trees ("helicopter logging," "helicopter long-line logging," or "heli-logging") has steadily increased in the logging industry. Unfortunately, helicopter logging in some areas, such as southeast Alaska, has been an extremely high-risk operation, resulting in helicopter crashes with severe traumatic injury and death to pilots and loggers. A series of serious crashes in Alaska during 1992 and 1993 brought these operations to our attention. We believe that much can be learned from the Alaskan experience with this technology. We present the recent experience with these operations in Alaska, followed by a summary of the larger U.S. experience with this rapidly expanding industry, and recommendations for prevention of injuries.

Figure 32.1. Helicopter logging operation, Southeast Alaska (Jan Manwaring, NIOSH, Alaska).

ALASKA INVESTIGATIVE FINDINGS

The National Transportation Safety Board (NTSB) investigated six helicopter crashes related to transport of logs by cable (long-line) that occurred in southeastern Alaska during January 1992 to June 1993, and which resulted in nine worker fatalities (five loggers and four pilots) and ten worker injuries (five loggers and five pilots) (Table 32.1). The following summarizes case investigations of these incidents:

Incident 1. On February 23, 1992, a helicopter crashed while transporting nine loggers. The copilot and five loggers were fatally injured; the pilot and four loggers were seriously injured. The NTSB investigation revealed that a long-line attached to the belly of the helicopter became entangled in the tail rotor during a landing approach, causing an in-flight separation of the tail section with subsequent crash (NTSB 1993a, 1993b). Passenger flights with long-line and external attachments are illegal and violate industry safety standards (FAA 1992).

Incident 2. On March 6, 1992, a helicopter crashed while preparing to pick up a load of logs with a long-line while in a 200-foot hover. The pilot and copilot were seriously injured. According to the pilot and copilot, the engine failed, and the pilot immediately released the external log load and attempted autorotation (NTSB 1993a). NTSB investigation revealed a hole in the side of the rear section of the engine case, which had occurred when the engine failed (NTSB 1993a, 1993c). Further NTSB investigation revealed fatigue

Table 32.1. Helicopter logging incidents, Alaska, 1992-1993.

Date	# Killed	# Injured	Type of Helicopter	Logging Company
2/23/92	6 (co-pilot and 5 loggers)	5 (pilot and 4 loggers)	Manufacturer A, type A Single-engine	Company A
3/6/92	0	2 (pilot and co-pilot)	Manufacturer A, type A Single-engine	Company A
11/10/92	0	0	Manufacturer A, type B Single-engine	Company A
2/19/93	2 (pilot and co-pilot)	0	Manufacturer A, type A Single-engine	Company B
5/2/93	1 (solo pilot)	1 (ground crew logger)	Manufacturer A, type C Single-engine	Company B
5/8/93	0	2 (pilot and co-pilot)	Manufacturer A, type A Single engine	Company B

failure of the compressor assembly impeller, and inadequate quality control by the manufacturer. Inadequate routine maintenance by the operator was also cited in this incident.

Incident 3. On November 10, 1992, a helicopter crashed while attempting to land at a logging site, sustaining substantial damage. The solo pilot was not injured. NTSB investigation revealed that the helicopter's long-line had snagged on a tree stump during the landing. Further investigation revealed that the company had no documented training program (NTSB 1993a, 1993d). Thorough training in long-line lift-load techniques might have averted this occurrence.

Incident 4. On February 19, 1993, a helicopter crashed from a 200-foot hover after transporting two logs to a log drop area (Figure 32.2). The pilot and copilot were fatally injured. NTSB investigation revealed an in-flight metal fatigue failure of a flight-control piston rod. Evidence indicated that log loads routinely carried by the helicopter exceeded the aircraft's weight and balance limitations. Laboratory examination of the flight-control hydraulic system revealed a degree of binding and wearing not consistent with normal wear (NTSB 1993a, 1993c).

Incident 5. On May 2, 1993, a helicopter crashed during an attempted emergency landing after using a long-line to lift a log to an altitude of 1,200 feet above ground level followed by rapid descent to a 75-foot hover (Figure 32.3). The solo pilot was killed, and a logger on the ground was injured. NTSB investigation revealed an in-flight separation of the tail rotor and tail rotor gear box from the helicopter. Investigative evidence indicated that log loads routinely carried by the aircraft exceeded its weight and balance limitations. Additionally, according to NTSB, on the day of the crash the company "... was reportedly using a procedure that would have heavily loaded the helicopter drive train, e.g., autorotating with a heavy external load from a point near the logging site to a drop point at a lower altitude where a full power recovery to a hover was executed before dropping the external load" (NTSB 1993a). Further, records associated with the helicopter gear box showed that it had been purchased (by the company) as surplus from the U.S. Army, which had removed it from service in 1986 because of "excessive wear" (NTSB 1993a, 1993f).

Incident 6. On May 8, 1993, a helicopter crashed after attempting to lift a log from a logging site with a long-line (Figure 32.4). The pilot and copilot sustained minor injuries, and the aircraft was substantially damaged. NTSB investigation revealed that company maintenance personnel had recently installed the engine and that the engine failed because machine nuts had come loose from the engine or its housing and became caught in the engine. The helicopter crashed as the pilot attempted autorotation. Investigative evidence indicated that log load weights for flights over the preceding 2 weeks had substantially exceeded the maximum authorized gross weight of the helicopter (NTSB 1993a, 1993g).

Figure 32.2. Feb. 19, 1993: Crash from 200 foot hover. Dora Bay, Alaska (NTSB, Anchorage, Alaska).

Figure 32.3. May 2, 1993: Mechanical failure associated with overloading machine and use of work surplus parts (Bell 204B. Copper Harbor, Alaska).

Figure 32.4. May 8, 1993: Crash after attempted autorotation after mechanical failure while logging (NTSB, Anchorage, Alaska).

Epidemiologic Analysis

Statewide occupational injury surveillance in Alaska through a federal-state collaboration was established in mid-1991, with 1992 being the first full year of comprehensive population-based occupational fatality surveillance for Alaska.

During the time these incidents occurred, an estimated 25 helicopters in Alaska were capable of conducting long-line logging operations. A survey of local jurisdication agencies (FAA and Alaska Department of Labor 1993) revealed that in early 1993 there were approximately 20 single-engine models from one manufacturer in service, 5 other multi-engined aircraft, and approximately 50 helicopter pilots employed in heli-logging operations in southeastern Alaska. Using these denominators, the events reported here were equivalent to an annual crash rate of 16% (6 crashes/25 helicopters/18 months), 0.24 deaths per long-line helicopter in service per year (9 deaths/25 helicopters/18 months), and an annual fatality rate for long-line logging helicopter pilots of approximately 5,000 deaths per 100,000 pilots, or 5% (4 pilot deaths/50 pilots/18 months) (CDC 1994). In comparison, during 1980 to 1989, the U.S. fatality rate for all industries was 7.0 per 100,000 workers per year; Alaska had the highest overall occupational fatality rate of any state (34.8 per 100,000 per year) for the same period (NIOSH 1993).

According to NTSB investigations, all six crashes involved "...improper operational and/or maintenance practices that reflected a lack of FAA surveillance of logging operations (routine regulatory inspections of long-line helicopter logging) at remote sites in southeast Alaska" (NTSB 1993a). NTSB further stated that, "The inadequate surveillance allowed unsafe operations and maintenance practices to continue until fatal accidents caused those practices to be detected" (NTSB 1993a). In one-half of these incidents (numbers 4, 5, and 6 above) investigative evidence also indicated that log loads routinely exceeded weight and balance limits for the aircraft.

All of these severe incidents occurred among helicopters operated by two companies using single-engine aircraft (Table 32.1). To enable a more thoughtful approach to this analysis, proven and putative risk factors for these events have been arranged in a time-phase or Haddon's matrix (Table 32.2). These events are often the result of the interaction of many different factors.

Table 32.2. Features of Alaska helicopter logging injury events (after Haddon).

	Host/Human	Agent/Vehicle	Environment

Pre-event/ Pre-injury	Pilot
	   Training,
	   Experience,
	   Fatigue,
	   Stress,
	   Rx, illegal
	     drugs
	   Alcohol
	Ground crew
	   Training
	   Experience
		Helicopter design
		   Lift, durability;
		   Maintenance & repairs
		   Engines & controls
		   Ergonomics
		   Unstable work platform;
		   Surplus/improvised
		     equipment
			Terrain
			Weather
			Landing zones
			Oversight
			   FAA (CFR pt 133)
			   industry

Event/Injury	Pilot
	   Reaction to
	     emergency situation
	     (i.e. autorotation),
	   Task overload
	Ground crew
	   Reacting, avoiding
		Helicopter
		   Autorotation
		   performance:
		   deformation on
		     impact;
		   Fires & explosions
			Terrain
			Weather

Post-event	Types of injury, severity		Little assistance available EMS not available

OVERVIEW OF HAZARDS OF HELICOPTERS AND SLING-LOAD LOGGING OPERATIONS

Helicopters are very complex machines with an inherent requirement for constant vigilance and input from the pilot during flight, and extraordinary maintenance requirements between flights. In contrast to conventional fixed-wing aircraft, helicopters can take off and land vertically, but are not self-trimming (i.e., able to maintain stable or level flight when control surfaces are in a neutral position), and cannot successfully move or hover without constant input to the controls by the pilot. The aerodynamics of these machines are fundamentally unforgiving, as they do not glide, and when the engine stops, free fall commences immediately, and can only be arrested by successfully restarting the engine or by autorotation maneuvers. Autorotation allows a helicopter to make an unpowered descent by maximizing on the windmilling effect and orientation of the main rotor. Forward airspeed and altitude can be converted to rotor energy to reduce the rate of descent. However, successful autorotation depends on helicopter airspeed and altitude when the maneuver is attempted (see Figure 32.5, Height-Velocity Curve—often referred to as the Dead Man's Curve—for Autorotation) (Roland and Detwiler 1967). Most helicopter logging operations are conducted at an altitude of less than 500 feet while at a hover or very slow airspeed, which is dangerously within the height-velocity curve for single-engine helicopters, as illustrated. This chart also displays the location of the six previously mentioned crashes, all dangerously inside this curve.

Even if successful autorotation were possible, suitable emergency landing sites in logging areas are rare. In short, such complex operations under these extreme and demanding circumstances, combined with frequent overloading of equipment (whether inadvertent or intentional), greatly increase the likelihood of both human error and machine failure (USC 1992).

Helicopter flight is also comparatively dangerous: according to NTSB 1990 data, the rate of fatal crashes for unscheduled flights in helicopters is 14.5 per million hours flown, 18-fold that for fixed-wing aircraft, 0.82 per million hours flown (Bertoldo 1996). Helicopter pilots have been well-documented to be an especially high-risk group for fatal occupational injuries (Conroy et al. 1992).

Helicopter logging operations place heavy demands on helicopter machinery and associated equipment. A typical logging helicopter carries an approximately 200-foot cable or long-line, which is attached by a hook to the belly of the helicopter (Figure 32.1). A second hook is attached to the free end of the long-line, where a choker cable (a cable apparatus designed to cinch or "choke" around suspended logs) is connected to haul from 1 to 4 logs per load (a load may weigh from 6,000 to 10,000 pounds); the hook is opened and closed electronically by a hand control located in the helicopter cockpit.

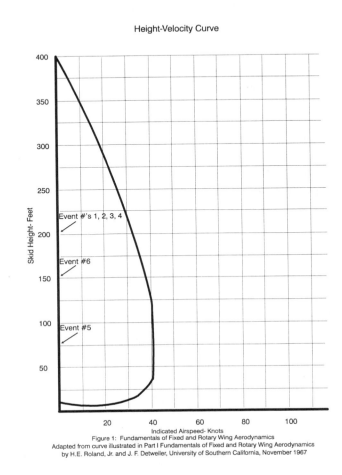

Figure 32.5. Fundamentals of fixed rotary wing aerodynamics.

A helicopter used in logging operations may complete up to 250 to 320 load/lift cycles, or turns, each day; each turn takes 1 to 3 minutes to complete. These highly repetitive lift/transport/drop turns are frequently conducted at or beyond maximum aircraft capacity in remote areas where rugged forest terrain, extremely steep mountain slopes (as great as 70 degrees), and adverse weather conditions prevail. Under these conditions, and when combined with poor equipment maintenance, helicopter flight components and equipment have been known to fail with tragic regularity (CDC 1994; TSB 1991).

The majority of experienced pilots we have interviewed reported having much greater confidence in multi-engine helicopters. One pilot with 20 years experience (and over

18,000 flight hours) in helicopter logging stated his concerns about heli-logging with single-engine helicopters and the need for redundant systems: "Imagine accelerating your automobile engine from idle, to full throttle 320 times a day, 6 days a week, 52 weeks a year. Engines do quit, clutches fail, inputs fail, fuel pumps stop pumping, accessory drive shafts break, power plants do stop running—most often when the pilot least expects it to happen. However, if he is flying a helicopter with two engines, things are a whole lot easier to handle. These helicopters have: (in addition to two engines), two servo systems, two electrical systems, two fuel systems, and two pilots!" (Lindamood 1996).

Another human factors concern is the tremendous potential for task overload in solo pilots. The experienced heli-logging pilot quoted above also stated, "The small engine aircraft usually have only one pilot who flies all day long. He must divide his attention between the engine temps, pressures, power gauges, warning lights, fuel quantity, and weight cell, as well as watching his load to keep it clear of obstacles and ground personnel and his rotor blades out of the trees. While concentrating on all this, he very often lifts off too much weight and over grosses the aircraft before he realizes it."(Lindamood 1996). Under such complex operations, it is not uncommon for a helicopter long-line to suddenly snag on a tree, log, stump, or forest debris with occasional disastrous outcome.

From a human factors standpoint there are other concerns. Helicopter logging, as well as other external load helicopter operations (which are regulated under the Code of Federal Regulations [CFR], Part 133) include (non-regulated) flight crew duty periods which can exceed 10 hours per day for 10 consecutive days, and flying much of the time under conditions that are unfavorable for successful autorotation of single engine helicopters in the event of engine failure or loss of power. This practice may lead to fatigue of sufficient magnitude to be hazardous. Recent analyses of NTSB data for fixed-wing aircraft show that pilots involved in repeat crash incidents ("accident-prone" pilots) were twice as likely to have crashed in Alaska and have flown for more hours in the previous 3 months (Baker et al. 1995). Decrements in aircrew function due to fatigue from overlong and repeat missions were also documented during the Desert Shield Operation (Bisson et al. 1993).

Table 32.3. Alaska helicopter logging injury recommended countermeasures (from Alaska Interagency Working Group for the Prevention of Occupational Injuries, July 1993).

	Host/Human	Agent/Vehicle	Environment
Pre-event/ Pre-injury	Increased training for pilots and ground crew Improved work/ rest cycles	Maintenance per manufacturer's recommendations, Impact (g)- resistant seats NTSB- to prohibit surplus equipment	Improved interagency communication, Increased FAA oversight
Event/Injury	Practical training in autorotation		Emergency (backup) landing zones
Post-event			

Ground operations also pose unique hazards in helicopter logging. Fallers and buckers must be especially cautious of downwash (air moved at high velocity by the helicopter's propeller[s], which may also knock limbs and debris onto ground crew). The long-line cable is made of steel, and must be grounded prior to being handled, because of its high static electrical (shock) potential. Loads can be accidentally released in transit, and crush those below. Ground crewmen have also been killed by walking or falling into a moving prop or tail rotor, as has been well documented in military settings (Crowley and Geyer 1993). Rigorous attention to communications, procedures, and protective equipment can mitigate these risks (Workers Compensation Board of BC 1990). Also, many loggers and ground crew are transported to their worksites by helicopters, with the attendant risk of such transport (see incident 1, above).

Lastly, major attention in private-sector helicopter design to adequate occupant restraint, crash attenuation, and fire prevention has been relatively recent (Vyrnwy-Jones 1985). Modest modifications in occupant restraint, such as headrests and chest harnesses, as well as G-absorbing or crash-attenuating seats and well-tested fire attenuation systems, when combined, could prevent up to an estimated 95% of all helicopter fatalities, and likely a substantial proportion of those associated with heli-logging (Bertoldo 1996; Krebs et al.; Glatz; Springate et al. 1989).

Hazard Reduction and Injury Prevention: The Interagency Response in Alaska

In response to the six Alaska logging helicopter crashes of 1992 and 1993, we convened a meeting in Anchorage on July 8, 1993 to discuss approaches for reducing the number of such crashes and ameliorating the outcome of crash injuries. The meeting was attended by representatives from the Alaska Interagency Working Group for the Prevention of Occupational Injuries (consisting of the Alaska Department of Health and Social Services, Alaska Department of Labor, FAA, NTSB, OSHA, U.S. Coast Guard, the U.S. Forest Service, and NIOSH). The working group noted that there were no formalized training programs or standards of performance required by the FAA for helicopter long-line logging operations. Furthermore, crash investigation teams had previously observed that operating standards did not comply with manufacturers' recommendations.

Based on these and other findings, the following eight recommendations were made by the working group (CDC 1994; Alaska Dept. of Health and Human Services 1993a, 1993b).

- All helicopter logging pilots and ground crews should receive specific training in long-line logging operations.
- Companies should follow all manufacturers' recommendations for more frequent helicopter maintenance (because of intensity and use) and for limits on maximum allowable loads.
- Companies should establish and observe appropriate limits on helicopter crew flight time and duty periods.
- Companies should consider the additional safety factor of using multi-engine helicopters for long-line logging.
- Specific industry-wide operating standards and procedures should be developed.
- Companies should provide training in on-site emergency medical care for helicopter logging crews at all work locations.
- State, regional, and local agencies involved in emergency medical services education should make low-cost emergency medical training available to persons likely to work in a helicopter logging environment.
- All flights over water should include appropriate survival equipment for all crew, who should wear personal flotation devices at all times during flights over water.

When these preventive interventions are superimposed on the risk time-phase matrix for these events (Table 32.3), it becomes clear that the emphasis chosen by the Working Group was on pre-event factors.

One other major concern was discussed during the July 1993 Interagency Group meeting: according to CFR, Part 133, regardless of where helicopter logging operations are conducted, the jurisdictional responsibility for inspection currently resides with the FAA office nearest the main or registered corporate office for the helicopter logging company, no matter how distant the FAA office may be from the actual helicopter logging site. In the six Alaska cases in this report, these FAA offices were in Salt Lake City, Utah, and Riverside, California (this necessitates travel of great distances to conduct helicopter logging inspections, and therefore remote operations, such as encountered in Southeastern Alaska may not be inspected for long periods). According to the NTSB, in the six Alaska heli-logging accidents, one operator had not received an on-site FAA inspection for 2 years, and the other operator had never received an on-site FAA inspection (NTSB 1993a). The NTSB has therefore recommended that operational and maintenance oversight responsibilities for remote heli-logging sites be assigned to the nearest FAA office (NTSB 1993a).

During the summer of 1993, the FAA and the Alaska Department of Labor increased their inspectional oversight of flight and ground operations at helicopter logging sites. Also, during this same period of time, two of the helicopter logging companies with the most operating problems, and who each accounted for three of these serious crashes, closed down their Alaska operations. Fortunately, there are other helicopter logging companies in southeast Alaska with outstanding safety records, that had already implemented the safety recommendations made by the Interagency Working Group. As a result, there have not been any logging helicopter crashes or injuries in Alaska through 1995, as shown in Figure 32.6.

U.S. HELI-LOGGING EXPERIENCE

NTSB data from 1980 through 1994 (compiled from information contained in the NTSB's Accident Briefs of CFR Part 133 helicopter external load operations in the U.S.) was analyzed for common factors for all known heli-logging accidents and crashes investigated by the NTSB. There were 220 such accidents known, of which 214 (or approximately 97%) were considered crashes. It should be understood that crashes and accidents are not interchangeable words for the purposes of this study. Crashes are a subset of accidents that involve impact of the helicopter body or its parts, resulting in serious injury or fatality to humans, and/or damage or destruction of the aircraft.

Perhaps the most notable findings involving NTSB data are those that reveal the most common probable causes and other common factors contributing to these helicopter crashes, even though denominator data from the helicopter logging industry were not available for determining rates. Each of these NTSB "Accident Briefs" was abstracted for 62 variables and entered them into a computer database (using Paradox software).

Of the 220 external load accidents, 64 (29%) occurred during helicopter logging operations and represent the largest portion of helicopter external load accidents by type of operation (Figure 32.7). These 64 accidents resulted in 24 fatalities, of which 17 were pilots, 5 were loggers being transported as passengers in 1 crash event, and 2 were ground crew involved in 2 separate accidents.

When we examined the distribution of all U.S. heli-logging accidents by month (Figure 32.8), we saw that the majority of accidents occur during warmer weather, which is when most heli-logging operations are conducted. Although denominator data for both the number of pilots and helicopters involved in heli-logging is lacking, it has been suggested that the main reason for the increased number of accidents during the spring and summer months has been due to the increased amount of heli-logging operations during that time of year, not that these operations are necessarily inherently more hazardous during the spring and summer.

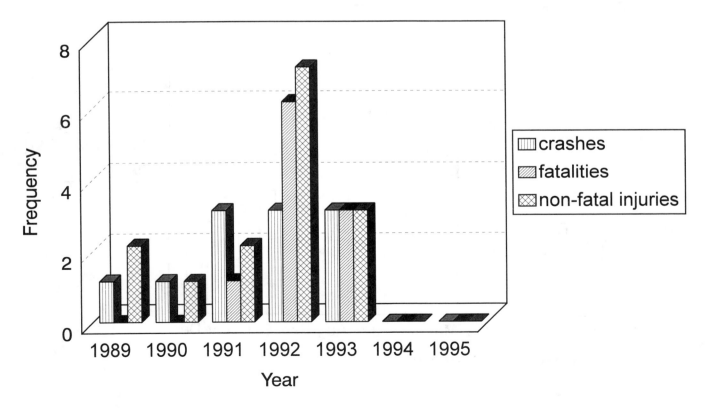

Figure 32.6. Crashes, fatalities, and non-fatal injuries in Alaska helicopter logging operations, 1989-1995.

538 / SAFETY AND HEALTH IN AGRICULTURE, FORESTRY, AND FISHERIES

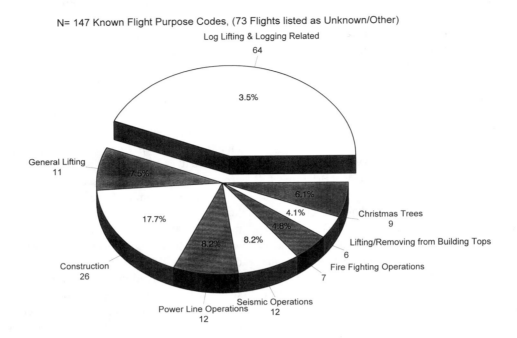

Figure 32.7. External helicopter crashes by purpose of use catagories, all U.S., 1980-1994 (NTSB 1995).

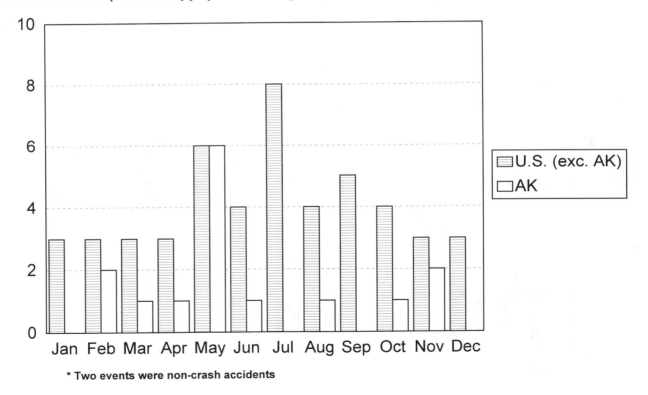

Figure 32.8. Logging helicopter crashes by month, 1980-1994, U.S. and Alaska, U.S. N=49, Alaska N=15 (NTSB 1995).

Alaska leads the U.S. in the number of heli-logging accidents, followed by Washington, California, and Oregon (Table 32.4). Of the 64 total U.S. long-line logging helicopter crashes from 1980 to 1994, Alaska had 15 (23%) of the crashes.

There are differences in external load helicopter accidents by operational phases in which the accidents occurred (Figure 32.9). When non logging-related helicopter accidents take place, they most often occur during the landing phase (22%), with the hovering phase coming in second (19%). When we look only at long-line logging helicopters and further sort for geographic variations between Alaska and the rest of the U.S., the differences become even more striking. Helicopter accidents in the rest of the United States are evenly divided between maneuvering, hovering, and landing phases, with descent and other phases coming in smaller portions behind these categories. In Alaska, however, over half occur in the hovering phase. These differences between Alaska's operational phase statistics and those for the rest of the country raise many as-yet unresolved questions: Do these data suggest anything about the lack of suitable emergency landing sites? Could this be due to the rough, and often steep terrain of a logging environment, as compared to all other external load operations, and heli-logging in states other than Alaska where suitable emergency landing sites may be perhaps more plentiful and accessible? If so, then what can be done in the Alaska heli-logging environment to improve the accessibility and suitability of emergency landing sites?

Table 32.4. Known long-line logging helicopter crashes by state (all U.S., 1980-1994, N=64**).

State	Number	Percent*
Alaska	15	23
Washington	11	17
California	9	14
Oregon	8	13
Idaho	6	9
Alabama	4	6
South Carolina	4	6
Georgia	2	3
Florida	2	3
Montana	2	3
Lousiana	1	2

*Percentages rounded off
** Two events were non-crash accidents.

Approximately 91% (59) of the heli-logging accidents were in single-engine helicopters, and 9% in twin-engine helicopters (Figure 32.10). Unfortunately, there are not adequate data available for number of helicopters of single or dual engine design currently in use in helicopter logging operations, nor are there available tabulations by number of engines.

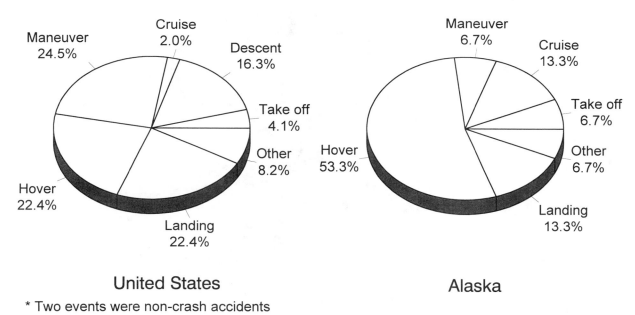

Figure 32.9. Logging helicopter crashes by operation phase, 49* United States (N=49), and Alaska (N=15), 1980-1994 (NTSB 1995).

Logging helicopter crashes within the lower 49 states (minus Alaska) were caused primarily by part failures (41%), with flight crew error a distant second (27%) (Figure 32.11). In Alaska, however, the data are strikingly different: maintenance error and flight crew error contribute 40% each, with parts failures contributing the remaining 20%. What does this suggest regarding helicopter maintenance operations in Alaska heli-logging? Several NTSB investigations of logging helicopter crashes in Alaska indicated inadequate maintenance facilities contributing to the probable causes of those accidents. Our office is still in the initial phases of data analysis of Part 133 helicopter long-line-related crashes. Continued research will provide a better idea of what safety issues are involved in this form of logging transportation. For the near future, research efforts will be focusing on obtaining denominator data for a better assessment of helicopter logging risks.

Because of the previous Alaskan experience, a rising concern for heli-logging safety nationwide, and a projected increase in heli-logging due to environmental restrictions and economic factors, the Alaska Interagency Working Group for the Prevention of Occupational Injuries and NIOSH sponsored the first Helicopter Logging Safety Workshop in Ketchikan, Alaska, on March 1 and 2, 1995. The objectives of the workshop were to: describe and analyze the risks of helicopter logging; share new aerologging technology; foster safety research in aerologging operations and technology; review current regulations governing helicopter logging; consider helicopter logging safety training opportunities and options; and draft consensus safety recommendations for helicopter logging.

The 65 workshop participants, representing 12 helicopter logging companies, 4 helicopter manufacturers, 4 industry associations, 5 federal agencies, 2 state agencies, 6 logging companies, 1 university, and a representative from the Helicopter Association of Canada, used a consensus-building group process to determine possible root causes, countermeasures, and action plans. Workshop participants drafted the following safety recommendations for injury prevention in heli-logging (Klatt et al. 1996).

Equipment

- The use of multi-engine helicopters is recommended for aerologging.
- The design, weight & balance, and operating limitations established by the manufacturer must not be exceeded.
- Aerologging equipment and components should be certified by the FAA, and overhauled in accordance with the manufacturers documentation or manuals.

Maintenance

- The aerologging industry should establish standards for sound maintenance procedures.

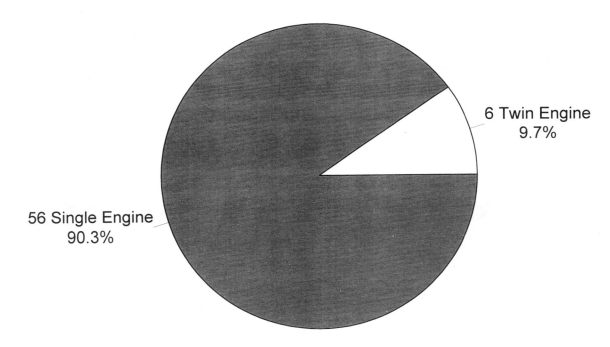

Figure 32.10. Long-line logging helicopter accidents (N=62*) by number of engines, all U.S., 1980-1994 (NTSB 1995).

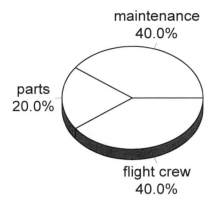
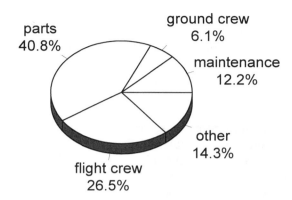

Figure 32.11. Logging helicopter crashes by primary probable cause (NTSB 1995).

- Adequate facilities should be available for the level of maintenance to be accomplished.
- An FAA-approved maintenance program should be established.
- Only FAA-approved parts should be used.
- All flight critical components should have accurate historical records.
- All maintenance work should be inspected prior to sign-off by certificated authority.

Human Factors

- The use of a qualified second pilot is recommended for aerologging.
- Companies should develop and publish standards for maximum flight and duty time.
- Companies should establish and enforce standards and methods to monitor unsafe attitudes and unsafe types of competition.
- The use of drugs and alcohol in aerologging should be prohibited, and aerologging camps should be dry.
- There should be random drug and alcohol testing, and mandatory testing in the event of a mishap.
- The FAA should not be permitted to sanction by way of irrevocable certificate action, those individuals entering voluntary drug and alcohol rehabilitation programs.
- It was also recommended that NIOSH conduct or sponsor a study of cockpit environment design for improvement of comfort and safety, and chronic injury reduction.

Training

- Helicopter model-specific and flight-specific training should be provided for aerologging operations.
- Flight and ground crew coordination training should be provided for all aerologging crews.
- Companies should provide maintenance training in specific helicopter models, special inspections, and documentation of maintenance operations.
- Companies should provide recurrent documented training for flight crews and mechanics.

Management

- An aerologging association should be established to serve as a forum and spokesman for the aerologging industry.
- Companies should be encouraged to develop a strong safety culture within upper level management.
- Mid-level managers should be trained on the concepts and responsibilities of developing a strong safety management culture.
- Employees should be encouraged to report safety violations without fear of punishment.
- Companies should specifically designate a safety manager, with a specific job description.
- The safety manager should receive formal training on a continuous basis.
- Companies should establish an employee/management safety committee.

- All employees should participate in the management of safety.
- Company officials and employees should be made aware of the cost benefits of an accident free operation.
- Companies should establish job/task termination safety rules.

Oversight

- The group strongly recommended that FAA promptly enforce all known rule violations.
- Staff of all local FAA, Flight Standards District Offices (FSDO'S) should be trained in all pertinent aspects of aerologging operations.
- Companies should be required to give prior notification to the local FAA, FSDO'S concerning any proposed helicopter logging operations in their service area.

Interagency/Company Cooperation

- Establish a helicopter logging association and encourage membership.
- Companies should establish communication between each other when conducting aerologging operations in close proximity.
- Companies conducting aerologging in the same areas should establish joint EMS and emergency action plans.

- Companies and agencies should develop and disseminate a contractor's safety check list.
- Companies and agencies should assist each other in writing and disseminating incident and accident reports.
- Companies and agencies should develop and disseminate Standard Operations Procedures manuals.

Environment

- Companies should provide improved and continual training concerning environmental hazards for all helicopter logging crews.
- Companies should establish improved communication and educate U.S. Forest Service, state agencies, and environmental group personnel concerning the necessity of more adequate helicopter emergency landing zones, and concerning the potentially hazardous combination of danger trees and rotor downwash.

These recommendations are also summarized in matrical format (Table 32.5).

SUMMARY

Helicopter logging is an expanding industry in the United States and abroad. Helicopter pilots and ground crews involved in long-line logging operations face an extremely

Table 32.5. Alaska helicopter logging injury countermeasures - proposed at March 1995 workshop in Ketchikan, AK.

	Host/Human	Agent/Vehicle	Environment
Pre-event/ Pre-injury	Qualified second pilot, Flight/duty time limits, Drug/alcohol/ testing, Availability of alcohol/drug rehabilitation	Multi-engine only, Dual drive train, Improved controls, Improved crash worthiness Limit to certified parts with valid FAA history	Industry SOP for maintenance, safety culture & management, Heli-logging association, Educate FAA FSDO, Improve communications
Event/Injury	Qualified second pilot	Crash-resistant fuel tanks, Controlled deformation	
Post-event		EPIRBs	Improve EMS availability, CPR/first aid training for crews

high risk for severe traumatic injuries resulting from helicopter crashes. Inadequate equipment, improper operational and/or maintenance practices, and the lack of adequate inspectional surveillance of helicopter long-line logging operations in Alaska have been frequently cited as the factors most strongly associated with the risk of crashes. The risks for fatal and serious injuries in this industry should and can be reduced by scrupulous attention to the needs of pilots, crew, and equipment. To minimize these extreme hazards, pilots and crew need more rest and better training; helicopters and equipment need more frequent and intensive maintenance; and operators must adhere not only to existing regulations, but also to manufacturer recommendations for load, lift cycle, and other appropriate applications. The Alaska experience has shown that helicopter logging can be extremely hazardous. However, careful attention to identifying and minimizing the risks and hazards can make it safer.

Acknowledgments

The authors wish to thank: Rick Kelly for his assistance with graphic materials for this manuscript; Linda Ashley for her diligent assistance with all phases of preparation of this document; Timothy Pizatella and Larry Garrett for their editorial suggestions; Doug Herlihy, formerly of the National Transportation and Safety Board, for sharing his technical insights and investigative photographs; and Diana Hudson for her assistance in editing the text and preparation of graphics.

REFERENCES

Alaska Department of Health & Social Services. 1993. *Alaska EMS Goals Document.* Juneau, AK: Alaska Department of Health & Social Services.

Alaska Department of Health and Social Services, Division of Public Health, Section of Epidemiology. 1993. Helicopter logging: Alaska's most dangerous occupation. *State of Alaska Epidem. Bull.*, Bulletin #32.

Baker, S.P., G. Li, M.W. Lamb, and M. Warner. 1995. Pilots involved in multiple crashes: "accident proneness" revisited. *Aviation, Space, Envir. Med.* 66: 6-10.

Bertoldo, R. 1996. Operational and aircrew factors in helicopter application. *Proc. of the Helicopter Logging Safety Workshop, March 1-2, 1995, Ketchikan, Alaska.* M. Klatt, D. Hudson, and G.A. Conway. eds. Anchorage, AK: Alaska Interagency Working Group for the Prevention of Occupational Injuries, pp. 76-81.

Bisson, R.U., T. J. Lyons, and C. Hatsel. 1993. Aircrew fatigue during Desert Shield C-5 transport operations. *Aviation, Space, Envir. Med.* 64: 848-53.

Centers for Disease Control and Prevention (CDC). 1994. Risk for traumatic injuries from helicopter crashes during logging operations—southeastern Alaska, January 1992-June 1993. *MMWR*, July 8.

Conroy, C., J.C. Russell, W.E. Crouse, T.R. Bender, and J.A. Holl. 1992. Fatal occupational injury related to helicopters U.S. 1980-1985. *Aviat. Space Envir. Med.* 63: 67-71.

Crowley, J.S., and S.L. Geyer. 1993. Helicopter rotor blade injury: a persistent safety hazard in the U.S. Army. *Aviation, Space, Envir. Med.* 64: 854-58.

Federal Aviation Administration. 1992. *Code of Federal Regulations, Volume 14.* Part 133—rotorcraft external load operations. Washington, DC: U.S. Government Printing Office.

Georgia Forestry Commission. 1986. Harvest by helicopter—a new way of logging. *Georgia Forestry,* September.

Glatz , J.D. 1992. *Energy Attenuation for Crashworthy Seating Systems: Past, Present, and Possible Future Development.* Warminster, PA: Naval Air Development Center.

Klatt, M., D. Hudson, and G.A. Conway, eds. 1996. *Proc. of the Helicopter Logging Safety Workshop, March 1-2, 1995, Ketchikan, Alaska.* Anchorage, AK: Interagency Working Group for the Prevention of Occupational Injuries.

Krebs, M.B., G. Li, and S.P. Baker. 1995. Factors related to pilot survival in helicopter commuter and air taxi crashes. *Aviation, Space, Envir. Med.* 66: 99-103.

Lindamood, M. 1996. Helicopter logging from a pilot's perspective. *Proc. of the helicopter logging safety workshop, March 1-2, 1995, Ketchikan, Alaska.* M. Klatt, D. Hudson, and G. A. Conway, eds. Anchorage, AK: Alaska Interagency Working Group for the Prevention of Occupational Injuries, pp. 76-81.

National Institute for Occupational Safety and Health (NIOSH). 1993. *Fatal Injuries to Workers in the United States, 1980-1989: A Decade of Surveillance.* Center for Disease Control and Prevention, U.S. Dept. of Health and Human Services.

National Transportation Safety Board (NTSB). 1993a. NTSB safety recommendation A-93-78 through -80. Washington, DC: National Transportation Safety Board, June 17.

National Transportation Safety Board (NTSB). 1993b. *Final Report, Aviation,* NTSB accident/incident number: ANC92FA040. Washington, DC: NTSB.

National Transportation Safety Board (NTSB). 1993c. *Final Report, Aviation,* NTSB accident/incident number: ANC92LA044. Washington, DC: NTSB.

National Transportation Safety Board (NTSB). 1993d. *Final Report, Aviation,* NTSB accident/incident number: ANC93LA015. Washington, DC: NTSB.

National Transportation Safety Board (NTSB). 1993e. *Final Report, Aviation*, NTSB accident/incident number: ANC93FA033. Washington, DC: NTSB

National Transportation Safety Board (NTSB). 1993f. *Final Report, Aviation*, NTSB accident/incident number: ANC93FA056, Government Printing Office, Washington, DC: NTSB.

National Transportation Safety Board (NTSB). 1993g. *Final Report, Aviation*, NTSB accident/incident number: ANC93FA061, Washington, DC: NTSB.

Proctor, P. 1994. Ecological benefits boost heli-logging. *Aviation Week & Space Techn.* 140: 65, May 9.

Roland, H.E. Jr., J. F. Detwiler. 1967. *Fundamentals of Fixed and Rotary Wing Aerodynamics, Part I.* Los Angeles: University of Southern California.

Springate, C.S., R. R. McMeekin, and C. J. Ruehle CJ. 1989. Fire deaths in aircraft without the crashworthy fuel system. *Aviation, Space, Envir. Med.* 60 (10, Suppl.): B35-8.

Stehle, T.C. Helicopter logging of valuable furniture timber from natural rain forest in the Southern Cape. *South Afr. Forestry J.* #155: 51-53.

Transportation Safety Board of Canada (TSB). 1991. Aviation occurrence report: hydra management atd. Aerospatiale 332c Super Puma (Helicopter) C-GQRL, Quatam River, British Columbia, 03 October 1987. TSB.

University of Southern California, Institute of Safety and Systems Management. 1992. *Aircraft Accident Investigation Manual.* Los Angeles: USC.

Vyrnwy-Jones, P. 1985. A review of Army Air Corps helicopter accidents, 1971-1982. *Aviation, Space, Envir. Med.* 56: 403-09.

Workers Compensation Board of British Columbia. 1990. *Helicopter Operations in the Forest Industry, A Manual of Standard Practices.* Richmond, British Columbia.

33

LOGGING AND THE LAW: OSHA RULES AND REGULATIONS FOR FORESTRY WORKERS

George E. Jeter, A.B.
North Carolina Department of Labor

> In 1992, more than 150 loggers died from work related injuries. Logging appears consistently in the Bureau of Labor Statistics (BLS) ranking of the 10 most deadly American occupations. The 1995 OSHA Logging Operations Standard (LOS) is the United States government's response to this problem. This chapter focuses on OSHA and its structure and examines in detail the 1995 Logging Operations Standards that apply to the logging industry. It also documents safety procedures to follow while engaged in logging operations to prevent injury and/or deaths.

INTRODUCTION

How important are safe logging practices to federal Occupational Safety and Health Act (OSHA) administrators? A good indication is the fact that the new OSHA Logging Operations Standard was signed into law in early 1995, during a time when the agency was under intense political pressures to adopt a more business friendly stance. In recent years, many states have been bolstering their own public safety efforts concerning logging safety. Logging industry groups and associations have also become leaders in the growing push for timber worker safety.

There is one major reason for this vigorous safety drive coming from OSHA and others: logging kills and maims far too many people. The United States Bureau of Labor Statistics (BLS) estimates that 1 in 5 loggers are injured on the job every year in America. In 1992, more than 150 loggers died from work-related injuries. Logging appears consistently in the BLS ranking of the ten most deadly American occupations.

The 1995 OSHA Logging Operations Standard is the United States government's response to this problem. Every logging company in this country is supposed to comply with this law. Keeping that thought in mind, this chapter will first take a brief look at OSHA and its structure and then examine in detail the 1995 Logging Operations Standard (which appears as 29 CFR 1910.266 in the OSHA General Industry Standards).

CREATION OF OSHA STANDARDS

The now replaced OSHA Pulpwood Logging Standard was created shortly after the Occupational Safety and Health Act itself, in the early 1970s. Technically, this standard only covered pulpwood harvesting, and its language and terminology reflected the concerns of that time. Another reality was that many people in the logging industry, particularly small logging operators, often had little knowledge or understanding of the standard, or sometimes even OSHA itself.

Today, logging employers and employees typically know more about OSHA, and the new logging standard is receiving more publicity than the old one did. However, to really understand how this standard—or any other standard—works, it is important to understand the whole OSHA framework.

OSHA laws are called *standards* and these standards are divided into general industrial classifications. They include general industry, construction, agriculture, and maritime standards. Logging operations fall into the general industry category, under 29 CFR Part 1910.

Every employer or manager should have a copy of the OSHA standards book that applies to his or her business. For example, many logging company managers may be diligently studying the OSHA Logging Operations Standard. However, are they aware of the OSHA Bloodborne Pathogens Standard, which could easily apply in a medical emergency, or with the many other standards covering their office and vehicle storage areas?

All these topics cannot be discussed at one time, but it is important for employers to familiarize themselves with these occupational safety and health laws. Federal OSHA, many state government agencies, and private publishers offer OSHA standards books to the public, usually at a very reasonable price. Having a general industry standards book, with a clear index, is one of the best $35 investments an employer will ever make.

OSHA ENFORCEMENT

There is a federal government act, administered by the U.S. Department of Labor (USDOL), and it does contain laws governing the safety and health conditions of you and/or your employees. This leads to a big question that often concerns employers: how are OSHA laws enforced?

OSHA is enforced in every state either through the USDOL directly or an OSHA designee—usually that state's department of labor, commerce, or industry. Presently, USDOL enforces OSHA laws directly in about half the states and U.S. territories and provides funding and oversight to selected designees in the other half. Every employer should find out exactly what agency administers OSHA in his or her area. This is important because there may be additional local or state occupational safety and health laws enforced in that location.

The government tries to bring workplaces into OSHA compliance through two main avenues. One effort is through a variety of educational and consultation programs designed to teach employers and employees good occupational safety and health practices. These programs include speakers' bureaus, free on-site voluntary visits, educational publications, and similar efforts. Education and consultation are generally offered to employers and employees without fear of surprise inspections or fines. A wide array of such programs and materials is available—another good reason to learn more about what is offered from the local OSHA office.

OSHA's compliance inspection program is the classic way to bring worksites into occupational safety and health compliance. A particular employer's site may be targeted for inspection through random selection, an employee complaint, or perhaps because of its past accident history. It is true that OSHA fines can be large at times. A detailed penalty structure exists indicating how to match monetary penalties to violations, depending upon their likelihood to cause injury, employer's knowledge of the hazard, and so forth. Basically, an OSHA citation can cost an employer from $0 to $70,000 per violation. Criminal legal action can be pursued in rare, extreme cases.

Citations can cost employers in other ways. Employees or former employees may use OSHA violations as evidence to pursue their own legal actions. Adverse publicity from being cited can cost an employer business and create customer ill will. Good employees may quit if they come to believe a worksite is unsafe. For all these reasons, it makes smart business sense to follow OSHA standards.

Standard Operating Procedures: Looking at the Logging Operations Standard

The OSHA Logging Operations Standard became effective on February 9, 1995. It replaced the old Pulpwood Logging Standard. Several parts of the 1995 standard were "stayed" during the year as OSHA and various logging industry representatives hammered out some of the new language. Eventually, satisfactory compromises were reached in most cases, and the full standard went into effect. This background is important because the new logging standard needs to be given credit as a very clearly written, performance-oriented set of rules. These rules incorporate many ideas from logging employers and spell out safety and health responsibilities in clear English.

Readers possessing a copy of the 1995 OSHA Logging Operations Standard may want to follow along in it throughout the rest of this chapter. The standard's section numbers will appear in parenthesis as major points are discussed. Please remember that this is a new standard at the time this book is being printed. Changes may be made to it at a later date.

The standard begins with a table of contents and a listing of the three appendixes that apply. The first significant statement (1910.266(b)(1-3)) immediately follows. This scope and application section is important because OSHA now states for the record that this law covers all types of logging. Remember, the old standard was centered around pulpwood harvesting only. OSHA also uses a catchall statement to cover dangerous conditions not addressed in the standard itself: "Hazards and working conditions not specifically addressed by this section are covered by other applicable sections of Part 1910."

The standard then defines a list of logging terms (1910.266(c)) that it will use throughout the text. Most of these words and phrases are standard terms that people familiar with the logging industry will recognize easily. The terminology also shows OSHA's continuing drive toward sensible, specific instructions. There are no more "widow maker" definitions in the new standard.

There are a few important terms, however, that are worth a closer look. *Ballistic nylon* is a key phrase concerning the standard's view of personal protective equipment (PPE). Ballistic nylon is a fabric "of high tensile properties designed to provide protection from lacerations." This material is, in OSHA's view, the state-of-the-art protective material for leggings. All chain saw operators must have leg protection that is either made from ballistic nylon or shows itself to be as protective as that material.

The word *machine* is a new term in the standard (the old logging standard often referred to *equipment*) that is featured heavily throughout the logging rules. It is defined as "a piece of stationary or mobile equipment having a self-contained power plant, that is operated off-road and used for the movement of material." This term refers to tractors, skidders, front-end loaders, feller-bunchers, and similar heavy machinery.

The term *danger tree* is a clear, general-purpose term referring to any standing tree that poses a distinct hazard to employees. Conditions causing a tree to be defined as a danger tree include damage or deterioration to the tree's root structure, trunk, or limbs. The tree's degree of lean, or direction of lean, can also make it a danger tree.

GENERAL REQUIREMENTS: FROM PPE TO EXPLOSIVES

Once past the introduction and definitions, we come to perhaps the most important part of the Logging Operations Standard, the General Requirements (1910.266(d)) section. This section contains the rules for personal protective equipment (PPE), first-aid kits, seat belts, fire extinguishers, environmental conditions, work areas, signaling, overhead electrical lines, flammable liquids, and explosives.

Employers, employees, and OSHA inspectors will probably spend more time dealing with General Requirement issues than any other part of the standard in the coming years. Let's start at the Personal Protective Equipment (1910.266(d)(1)(i-vii)) paragraph, which contains this key requirement: "The employer shall assure that personal protective equipment, including any equipment provided by an employee, is maintained in a serviceable condition."

Basically, *serviceable condition* means that the equipment in question must function as it was intended at the time it was manufactured. A cracked hard hat, ripped leggings, and other similar worn-out items would not be deemed serviceable. Also, note the phrase about employee-provided equipment. An employee may have a favorite pair of safety shoes (or the employer may require employees to provide their own safety shoes) but if that pair of shoes fails to protect the employee due to wear, deterioration, or another similar problem, it is the employer, not the employee, who will be held accountable for the problem.

The General Requirements section further requires employers to provide cotton gloves, or equivalent hand protection, to any employee who handles wire rope on the job. Ballistic nylon leggings or chaps must be provided to employees who operate chain saws. Such leg protection must cover the thigh and leg all the way to an employee's boots. (OSHA makes one exception to this leg protection requirement: if an employer can demonstrate that wearing such leggings would create a greater hazard due to climbing, etc., then they do not have to be provided in such cases.) Employers must provide their loggers with appropriate head protection, hearing protection, and eye and face protection. Typically, this can be handled by providing a logger's model hard hat that has ear muff and face screen attachments.

Remember, all these PPE devices must be provided free of charge to the employees by their employers. Once provided, it remains an employer's responsibility to see they are used and fixed or replaced when worn or damaged.

Safety Footwear

When it comes to PPE, nothing raises more questions than the debate about who pays for safety footwear. Let's answer that question after a brief look at what OSHA requires logging employees to have on their feet. The foot protection paragraph for logging (1910.266(d)(1)(v)) is very specific. "The employer shall assure that each employee shall wear foot protection," it states, equal to heavy-duty logging boots that are waterproof or water resistant and cover the ankle.

These boots ("shoes" would be an incorrect term for loggers given the ankle coverage requirement) also must provide support to the ankle and some degree of protection against chain saw penetration. Calk soled or other slip-resistant type boots are acceptable in cases where an employer can demonstrate they meet these requirements and there is a need for them based on the terrain, timber type, or weather conditions.

Now, there are two ways to answer the question about who pays for the boots. OSHA states in Subpart I of its General Industry Standards (this is a good example of why it pays to have all the standards) that "it is the responsibility of the employer to provide, at no cost to the employee, all personal protective equipment which the employee does not wear off the jobsite for use off the job."

This statement means that an employee can legally request that the employer buy his or her safety boots, as long as the employee only wears them during working hours. Many employers can see advantages to this arrangement as well, since they can know precisely how the boots are being used at all times. Of course, an employer can agree to buy all needed safety boots and still let the employees wear them home, if so inclined. If the employer does not want to shoulder the entire cost of the boots or if the loggers want to use them during personal time, then management and workers can negotiate an agreement regarding a boot allowance. In any case, the boots used must meet the standard's requirement and be in serviceable condition at all times.

Here's the bottom line to remember about personal protective equipment and OSHA: whenever a piece of PPE fails to protect an employee due to wear, poor sizing, damage, or any other similar problem, the employer will be held responsible.

First Aid: A New Level Of Responsibility

The old Pulpwood Logging Standard mentioned first aid only briefly, mostly to stress the need for logging crews to carry a snakebite kit. The new Logging Operations Standard calls on loggers to provide a much higher first-aid response when an injury occurs.

OSHA now gives a detailed description of the minimally required first-aid kit (1910.266 Appendix A) that must be available at every worksite where timber is being felled, at each active landing area, and on each company transport vehicle. These kits are mandated to contain 15 specific items at a minimum: gauze pads (4" X 4"), 2 large gauze pads (8" X 8" or larger), a box of adhesive bandages, one package of gauze roller bandage (2" or wider), 2 triangular bandages, wound cleaning agent (such as sealed moistened towelettes), scissors, a blanket, tweezers, adhesive tape, latex gloves, resuscitation equipment (such as a resuscitation bag, airway, or pocket mask), 2 elastic wraps, a splint, and directions for requesting emergency help.

These first-aid kits must be inspected and approved by a health care provider annually. Obviously a kit that contained the 15 items described above would be fairly large, especially if other items were added (such as a snakebite kit) to suit the given site conditions. However, those items are mandated under the new standard. (Note: You should check your local area office regarding first-aid kit requirements. The North Carolina Department of Labor, the OSHA designee agency in that state, for example, was negotiating with federal OSHA to only mandate that kits be physician approved in its jurisdiction when this book was published.)

OSHA goes on to explain, in detail, what type of first-aid training (1910.266(i)(7)) is now required for loggers as well. All logging company employees (including supervisors) now must be trained in first aid and CPR. An employee must take first-aid training at least once every 3 years, or as required to meet that state's requirements. CPR training must be administered to the employee at least once a year, or as required to meet that state's requirements. All such training must be administered by a designated person. The employer must make sure that the material is presented in a way all the employees understand. Finally, employers must keep written certifications of this training showing when an employee received his or her first-aid and CPR training last and who did the training.

First-aid and CPR training (1910.266 Appendix B) has to cover the following: the definition of first aid, legal issues including Good Samaritan laws, basic anatomy, respiratory arrest, cardiac arrest, hemorrhage, lacerations and abrasions, amputations, musculoskeletal injuries, shock, eye injuries, burns, loss of consciousness, exposure, paralysis, poisoning, loss of mental functions and artificial ventilation, drug overdose, CPR, using dressings and slings, treating sprains and fractures, handling and transporting an injured person, and treating bites, stings, and exposure to poisonous plants and animals.

Obviously, OSHA has put a new, highly structured mandate on the nation's logging community when it comes to first aid and CPR. The agency clearly believes that any logger should be ready, able, and equipped to render immediate assistance when an accident occurs in the woods.

Seat Belt Requirements

The OSHA Logging Operations Standards has new, detailed rules (1910.266(d)(3)) for using seat belts. Any piece of equipment featuring a roll-over (ROPS) or falling object (FOPS) protective structure or overhead guards must have a seat belt. Again, just as with the PPE rule, this requirement even includes vehicles and equipment provided by the employee. OSHA continues to hold the employer responsible for what happens on his or her worksite.

Just to cover the details, the new safety belt rule goes on to state that the employee operating the piece of equipment must use the provided safety belt by "securely and tightly" fastening himself or herself into it. Belts cannot be removed from equipment (except for replacement or repair) and must be maintained in serviceable condition.

Each vehicle and piece of heavy machinery must have a portable fire extinguisher as well. These extinguishers are required to meet the necessary conditions listed in the OSHA Fire Protection Standards (1910, Subpart L).

Environmental Conditions and Work Areas

Logging by definition takes place outdoors on difficult terrain. OSHA attempts to address this aspect of the job in its Environmental Conditions (1910.266(d)(5)) and Work Areas (1910.266(d)(6)) sections. Regarding Environmental Conditions, the standard simply requires work to stop and employees to move to a safe place whenever "electrical storms, high winds, heavy rain or snow, extreme cold, dense fog, fires, mudslides, and darkness" or other unspecified conditions make continued work too hazardous.

Regarding Work Areas, OSHA requires loggers to stay spaced so that "the duties of each employee shall be organized so the actions of one employee will not create a hazard for any other employee." This means loggers should be assigned to places in the work area where they cannot be hit by a tree being felled by another employee or work crew. OSHA considers two tree lengths to be the distance that should be kept between adjacent work zones. That distance should be expanded when a felled tree could roll or slide due to sloping.

Logging employees must stay within visual and/or audible contact with other employees working in adjacent areas. OSHA defines *audible contact* to mean distinct, directed noise such as whistles, horns, or radios. Generalized engine noise from a chain saw or a machine does not count as audible contact. When signals are used to give direction or instruction, only a designated person should give them, except in emergencies. The employer must be able to account for everyone who started work at the site when the shift ends.

Power Lines, Flammable Liquids, and Explosives

Any logging or tree cutting work done near overhead electrical power lines must be done so that the loggers do not violate OSHA 1910 Subpart S, Electrical Work. In general, this means loggers must not work close to overhead lines until they are deenergized and grounded (1910.333(c)(3)). If a felled tree or branch makes contact with a power line, the appropriate power company is to be contacted immediately. Every logging employee must stay clear of that area until the power company says that the line(s) in question are safe and there is no electrical hazard.

Flammable and Combustible Liquids (1910.266(d)(9)) have to be treated as Hazardous Materials as described in OSHA 1910 Subpart H. This means following some basic safety measures including not transporting them in the driver compartment or in any passenger occupied area of a vehicle, and shutting off any vehicle or machine during fueling. (An exception: diesel-powered machines and vehicles can be refueled while at idle, as long as they are going to continue operating and safe fueling procedures are followed.) Some flammable liquids, such as chain-saw and diesel fuel, can be used to start slash pile fires. The employer must ensure that whenever fuel is used this way, no hazard is created.

Explosives and blasting agents (1910.266(d)(10)) also must be handled as described in Subpart H. Two key requirements include limiting access to blasting agents and explosives so that only a designated person can handle them and never transporting them in the driver's compartment or passenger occupied area of any vehicle.

Chain Saws and HPPTs

Now that we've examined all the General Requirements in the OSHA Logging Operations Standard, we need to focus on the rules for equipment and machinery. Let's start with the Hand and Portable Powered Tools (HPPT) section, called 1910.266(e) in OSHA language. Chain saws are considered to be HPPT.

First, the standard requires that all HPPTs be kept in serviceable condition. The employer is responsible for ensuring that any tool given to an employee has been inspected for damage or wear at the beginning of a work shift. This inspection should include checking handles and guards for splinters, sharp edges and fit, checking tool controls, chain saw chains and mufflers, chain brakes and shielding devices, the heads of shock and impact-driven tools for mushrooming, cutting edges, and all other safety features.

OSHA requires employers to make sure that HPPTs are only used for their designed purposes. Other HTTP requirements include removing a tool immediately when the head of any shock, impact-driven, or driving device begins to chip. Dull cutting edges have to be sharpened in accordance with the tool manufacturers instructions. Concerning tool storage, tools need to be stored in a provided location when not in use. Racks, boxes, holsters, or other arrangements should be used when tools are transported so they cannot become a hazard in a moving vehicle.

When it comes to the Chain Saws (1910.266(e)(2)) section of the HPPT rules, the American Pulpwood Association (APA) has done an excellent analysis of the new expanded OSHA requirements. The APA has condensed these requirements down to the 15 key points:

- Chain saws placed into service after the effective date of these OSHA regulations shall be equipped with a chain brake which must be in place and function properly.
- Chain saws placed into service before the effective date shall be equipped with a fully functional protective device that minimizes chain saw kickback.
- No chain saw kickback reduction device shall be removed or otherwise disabled.
- Chain saw chains must be kept in proper adjustment and chain saw mufflers must be operational and in place.
- Each gasoline-powered chain saw shall be equipped with a continuous throttle system which stops the

running chain when pressure on the throttle is released.
- Each chain saw shall be operated and adjusted in accordance with the manufacturer's instructions.
- Chain saws shall be refueled at least 10 feet away from any open flame or other source of ignition.
- Any chain saw will be started at least 10 feet away from any fueling area.
- Chain saws will be started on the ground or where otherwise firmly supported (no drop starting allowed).
- Chain brakes shall be engaged when the saw is started.
- The chain saw shall be held with both hands during operation unless the employer can demonstrate that a greater danger is posed by keeping both hands on the saw during that particular situation.
- The saw operator shall clear away brush or other potential obstacles that might interfere with cutting or using a retreat path prior to felling a tree.
- Cutting directly overhead is prohibited.
- Chain saw must be carried in a manner that will prevent operator contact with the chain and muffler and may not be carried further than 50 feet without shutting the saw off, or engaging the chain brakes.
- Chain saw must be shut off or at idle before the operator starts a retreat after felling a tree.

The preceding points were published in the APA's *A Field Guide for Complying with the New OSHA Safety Requirements* (American Pulpwood Association, January 1995). The APA has many excellent publications. Its address is given at the end of this chapter in the Additional Resources listing.

Machines

The Machines section (1910.266(f)) of the 1995 OSHA Logging Operations Standard requires the serious attention of any logging employer. A *machine* is any piece of off-road equipment having a self-contained power plant that is used to move materials. It is much longer and more detailed than the old Equipment section it replaces. One key point to remember: when OSHA discusses machinery, its perspective is not primarily concerned with machine life or the quickest way to do the job. OSHA sees heavy logging equipment as powerful, dangerous machines that can kill or maim a human being in the blink of an eye.

Once again, the section begins with a statement that says the employer is responsible for seeing that any machine operated by his or her employees, including machines brought by an employee to the site, shall be kept in serviceable condition.

Each machine must be inspected before a work shift begins. Damaged or unserviceable machines cannot be used during that shift until repaired. The operating and maintenance instructions must be available to the operator. Employees shall comply with all such instructions.

Keeping those instructions in mind, let's look at the Machine Operation requirements (1910.266(f)(2)) themselves. First, a machine can only be operated by a designated person. The employer (or site manager) should know exactly who is operating machinery and that they are qualified to do so. Next, stationary logging machines and their components have to be anchored or somehow stabilized to prevent movement during use. The rated capacity of any machine cannot be exceeded.

When a machine has to be operated on sloping terrain, first check for instructions, directions, and limitations found in the manufacturer's operating guide regarding sloping. Machines must be operated in compliance with these manufacturers' instructions.

Operators must check their machines' paths before starting or moving them to ensure that no other employee is in the way. Once a machine is being operated, remember that it must keep a safe distance between it and other employees and/or machines. No employees but the operator are allowed to ride on a machine unless there are seat belts, or protection equal to the operator's, provided for them. No employee is ever allowed by OSHA to ride on a load. Finally, a machine must be operated from the operator's station or as indicated by the manufacturer.

When it is time to shut down a machine, the machine's brake locks or parking brakes have to be applied first (and before the operator ever leaves the cab for any reason). The machine's transmission must be placed in park. Each moving element (blades, buckets, shears, etc.) needs to be grounded. Once the machine's engine is shut down, pressure or stored energy in the machine's hydraulic or pneumatic devices needs to be discharged if it is possible such a hazardous part could move while the machine is shut down. If another vehicle is going to be used to transport the machine, then that vehicle's rated capacity cannot be exceeded. Machines that are transported by other vehicles are required to be "loaded, secured, and unloaded" using safe practices that do not create any hazards to employees.

Protective Structures

The OSHA Logging Operations Standard Protective Structures section (1910.266(f)(3)) covers ROPS, FOPS, safety glass, overhead guards, and similar protective structures on machines. There is a key general instruction for this section: every tractor, skidder, swing yarder, log loader, and mechanical felling device (tree shears, feller bunchers, etc.) placed into service for the first time after February 9, 1995, must have a FOPS and/or ROPS. Logging machines that began service after that time but have had their FOPS/ROPS taken off, must be reequipped with them. OSHA makes one exception to the rule. This requirement does not apply to machines that make 360 degree rotations (such as knuckleboom loaders and swing loaders).

In general, ROPS must be installed and maintained (in machines built after August 1, 1986) as described in recommendations made for logging machine ROPS by the Society of Automotive Engineers (SAE J1040, April 1988). FOPS must be installed and maintained as described in the SAE recommendations for logging machine FOPS (SAE J231, January 1981). See the Additional Resources listing at the end of this chapter for the SAE's address.

OSHA then goes on to flesh out some minimal requirements for the cab areas of logging machines. Protective structures have to be big enough not to impede the operator's normal movements. The overhead covering of each cab must be a solid material and it must extend over the entire canopy. The entire cab, including entrances, must be completely enclosed with a mesh or other acceptable material to prevent objects from entering the cab.

Each cab's upper portion must be fully enclosed with "mesh material with openings no greater than 2 inches" or by "other materials which the employer demonstrates provides equivalent protection and visibility." These enclosures should be designed to allow for "maximum visibility" according to the standard.

When a transparent material is used to enclose a cab's upper portion, the material must be safety glass or another material that provides the same level of protection and visibility. Transparent materials must be kept clean. Transparent materials that have cracked, broken, been scratched, or otherwise pose a hazard to the operator have to be replaced.

Sapling and branch deflectors need to be installed in front of each cab. These deflectors must be installed so that they do not block the operator's visibility or impede access to the cab. (The cab entrance itself should extend at least 52 inches from the cab's floor.) In addition, OSHA states that machines used near cable yarding operations must be equipped with roofs or sheds strong enough to provide protection from breaking lines.

Forklift trucks used in logging operations must be equipped with an overhead guard that meets the American Society of Mechanical Engineers (ASME) specifications listed in its *Safety Standard for Rough Terrain Forklift Trucks* (B56.6-1992) publication. (See the Additional Resources listing at the end of this chapter for the ASME's address.)

Logging machine guarding (1910.266(f)(8)) is required to protect employees whenever they could be exposed to such hazards as moving elements from shafts, pulleys, conveyor belts, and gears. Guarding is required on all debarking, limb cutting, and chipping machines to protect employees from flying wood chips, chunks, logs, and limbs. All machine guarding must be used whenever the machine in question is being operated.

Machine Access, Exhaust Systems, and Brakes

The Machine Access section of the new OSHA logging standard is fairly simple. It dictates that machine access systems all meet the specifications of the ASE's *Recommended Practice for Access Systems for Off-Road Machines* (SAE J185, June 1988). This basically means that hazard-free machine access must be provided on machines where any employee must climb onto the machine to enter the cab. The standard further states that machine cabs must have at least two means of egress. All walking and working surfaces, and work areas, on or in a machine also must have slip-resistant surfaces to ensure safe footing.

Exhaust Systems (1910.266(f)(6)) are covered by a few straightforward paragraphs in the standard. The key points are:

- Exhaust pipes must direct expelled gases away from the operator.
- Exhaust pipes must be mounted or guarded on the machine so as to protect the employee from accidental contact.
- The exhaust pipes must be equipped with spark arresters. (Engines equipped with turbochargers do not require arresters.)
- Each machine must feature a factory-installed (or equivalent) muffler that is in place whenever the machine is being operated.

Brakes (1910.266(f)(7)) is a new, but very short, section in the OSHA standard. The first rule here requires employers to make sure that the brakes on any machine are sufficient to hold its rated load capacity on the slopes where it is being operated. A second rule requires machines to be equipped with a secondary braking system, such as an emergency or parking brake, which can effectively stop and maintain the machine's parking performance, regardless of the machine's direction of travel or whether the engine is running. Machines put into initial service after September 8, 1995 must have three braking systems: service, secondary, and parking brakes.

Vehicles

OSHA rules for logging vehicles (cars, buses, trucks, and semi-trailers that are used to carry employees or machines on business-related activities) are spelled out in another new Logging Operations Standard section simply titled Vehicles (1910.266(g)). Vehicle requirements were less clear in the old standard.

The section begins with OSHA's standard requirement that any vehicle (including one provided by the employee) used to transport an employee off a public road, or to perform any logging operation, be kept in "serviceable condition." Any vehicle owned, rented, or leased by the employer must be inspected before initial use each work shift. Hazardous defects must be repaired before use or the vehicle must be replaced.

The employer must make sure that the operating and maintenance instructions are available in each vehicle. Vehicle operators and maintenance workers must comply with these instructions. It is also the employer's duty to ensure that only employees with valid driver licenses (for the class of vehicle in question) operate vehicles.

Vehicles must be equipped with mounting steps and handholds whenever necessary to prevent an employee from being injured while entering or leaving the vehicle. Every seat in the vehicle must be securely fastened to it. Finally, OSHA states that several key parts of the Machines section also apply to logging vehicles. These requirements include the parts discussing rules for rated capacity, ensuring no employees are in the travel path, operating machines (vehicles) at a safe distance from other employees, the shut down procedure, loading and securing in a safe manner, and the rules regarding brakes.

Tree Felling

The OSHA Logging Operations Standard's Tree Felling section (1910.266(h)) is basically the heart of this standard. OSHA gives this section its own detailed General Requirements paragraph to emphasize that point.

These general requirements (remember, the standard itself also has a General Requirements section) state that trees must not be felled in a manner that could create a hazard from, "but not limited to," striking a rope, power line, or machine. It requires immediate supervisors to be consulted whenever unfamiliar or unusually hazardous conditions are present. A supervisor must give his or her approval to continue cutting under such conditions.

The introduction to tree felling also states that no yarding machine can be operated within two tree lengths of trees being manually felled. No employee is to approach within two tree lengths of another employee who is felling trees until that employee has acknowledged it is safe to do so (unless the employer demonstrates that multiple employees are needed to manually fell a particular tree). No employee is to approach within two tree lengths of a mechanical felling operation until the machine operator acknowledges it is safe to do so.

OSHA states in this opening to Tree Harvesting that every danger tree is to be felled, removed, or avoided. When such a tree is to be felled, the least hazardous means must be used, such as using heavy machines. When such a tree is to be left standing in the work zone, then it must be marked and no work can be conducted within two tree lengths of it (unless the employer can demonstrate that working within a shorter distance is not hazardous). Each danger tree must be carefully checked for loose bark, broken branches, and other damage before being felled or removed. Such hazards must be removed or held in place before the tree is felled or removed.

When the trees to be felled are on a slope where it is reasonable to believe trees or logs could roll or slide, work is to be done uphill of, or on the same level as, previously felled trees. The final general requirement for tree felling is that "domino felling" is prohibited. Trees are to be felled one at a time.

Manual Tree Felling

The Manual Felling part (1910.266(h)(2)) of the Tree Harvesting section makes good companion reading to the HPPT section on chain saws we examined earlier. This part makes it clear that OSHA expects a logging employee (feller) to actually go through several steps before the saw ever touches the tree. First, the feller must pick a cleared retreat path that extends diagonally (unless it can be shown that an alternative angle is less hazardous) from the expected felling line. The feller is required to evaluate conditions that could make felling the tree hazardous. Such conditions include, but are not limited to, snow and ice accumulation, the wind, the lean of the tree, dead branches, and the location of surrounding trees. In the case of snow and/or ice, if the accumulation on a tree could create a hazard to the employee, it must be removed before felling commences, or the area around the tree must be avoided. When a spring pole or other tree under stress is going to be cut, no employee except the feller can be within two tree lengths of the area until the stress is released.

Each tree being manually felled must be undercut, unless the employer can show that not making an undercut will not create a hazardous condition. The undercut must be of a size that the tree does not split and that it falls in the intended direction. Each tree being manually felled by the Humboldt cutting method must be backcut, again unless the employer can demonstrate that not making a backcut will pose no hazard. The backcut should allow for sufficient hinge wood to guide the tree and prevent it from prematurely slipping or twisting off the stump. The backcut must be above the level of the horizontal cut of the undercut. (An exception: backcuts may be at or below the horizontal cut in tree pulling operations.)

Bucking, Limbing, and Chipping

The Bucking and Limbing section (1910.266(h)(3)) of the OSHA logging standard stipulates that bucking and limbing work done on a slope, where it is reasonable to think a tree could roll or slide, must done on the uphill side of each tree. When dealing with windblown trees, precautions must be taken before bucking or limbing begins. These precautions include, but are not limited to, chocking or moving the tree to a stable location. The goal is to keep the root wad, tree butt, or logs from striking an employee.

The Chipping section (1910.266(h)(4)) is concerned with chipping operations taking place in logging and forestry work "in-woods locations." The standard requires workers to keep chipper access covers and doors closed until the drum or disc comes to a complete stop. Chipper infeed and discharge ports must be guarded to prevent employee contact with the disc, knives, or blower blades. Whenever service or maintenance is performed, the chipper must be shut down and then locked out in accordance with the OSHA Lockout/Tagout Standard (1910.147). Detached trailer chippers need to be chocked during use on any slope where it is reasonable to assume that rolling or sliding could occur.

Yarding

Yarding (1910.266(h)(5)) gets far more attention in the new logging standard than was given to this activity under the old law. This activity was generally termed *skidding* and *prehauling* in the old standard. The first yarding rule OSHA gives is that "no log shall be moved until each employee is in the clear."

No yarding line is to be moved until the machine operator has clearly received and understood the signal to do so. If the yarder operator has any doubt about the signal, he or she must repeat it as understood and wait for a confirming signal before moving any line. Each machine must be positioned so that during winching, both the machine and the winch stay within their design limits. The yarding machine or vehicle, including the load, must be operated with a safe clearance from all obstacles. No load can exceed the rated capacity of the pallet, trailer, or other carrier.

Each choker must be hooked and unhooked from the uphill side or end of the log. (Exception: when the employer can show it is unfeasible to do so in a particular situation, the log can be securely chocked to prevent sliding or rolling, and then hooked/unhooked from the downhill side or end.) Each choker needs to be positioned near the end of the log or tree section.

Towed equipment such as (but not limited to) skid pans, pallets, arches, and trailers must be attached to their machine or vehicle so that a 90-degree turn can be safely made. Finally, each yarded tree needs to be placed in an orderly fashion so that it does not create a hazard. It must be stable before bucking and limbing work commences.

Loading and Unloading

The Loading and Unloading section (1910.266(h)(6)) is distinct from the Vehicles section we covered earlier. Keeping those earlier rules in mind, let's examine OSHA's regulations for log transport vehicles. First, a transport vehicle must be positioned so that it provides a safe working clearance between the vehicle and the deck of trees or logs. Only the loading/unloading machine operator and other essential personnel can be in the work area during this process. No transport vehicle driver is to remain in the cab if logs will be carried over it during the procedure. (Exception: if the employer can demonstrate why there is a necessary reason for the driver/operator to stay in the cab. There should be specific driver/operator protection in these cases, such as reinforced cabs.)

Each log must be placed on the transport vehicle in an "orderly manner" and secured tightly. (The whole load should be positioned to prevent slippage or loss during transport.) Each stake and chock used to trip loads must be constructed so that the tripping mechanism is opposite the side where the load is released.

Every tie down must be left in place over the load's peak log to secure the logs until the unloading lines (or other equivalent protection) are in place. Each tie down can be released only from one side except in the following cases: when a tie down is released by remote control, or when the employee making the release is protected by racks or stanchions capable of withstanding the force of the logs.

The transport vehicle operator must be sure each tie down is tight before transporting the load. While transporting the load, the operator must stop and inspect (and tighten if needed) the tie downs if there is reason to think that the load has shifted or the tie downs have loosened.

Training

This brings us to the last topic in the OSHA Logging Operations Standard—Training. The Training section

(1910.266(i)) states from the first sentence that all training must be provided free of charge to employees (including supervisors). OSHA gives several specific times when training must be provided to employees. They include:

- Initial training, where required, as soon as possible (and no later than the effective date of the OSHA requirements) for each current and new employee.
- Prior to each employee's initial assignment.
- Whenever an employee is assigned new work tasks, tools, equipment, machines, or vehicles.
- Whenever an employee demonstrates unsafe work performance.

All training must include at least these elements: performing work tasks safely; operating and maintaining tools, machines, and vehicles used by the employee safely with an emphasis on understanding and using the manufacturers' instructions; recognizing hazards associated with the employee's specific work assignments; recognizing and preventing safety and health hazards in the logging industry; knowing the employer's specific policies and requirements; and knowing the requirements of the OSHA Logging Operations Standard.

Remember, that the employer is responsible for ensuring that each employee, new or veteran, can safely perform all assigned duties and operate all the tools and equipment associated with those tasks. Each employee being trained needs to work closely with a designated person until the trainee demonstrates that he or she is performing the tasks safely.

As we discussed in the First Aid section, all training must be conducted by a designated person. The employer needs to make sure that the material is presented in a way all the employees understand. The employer must keep written certificates of all training showing when each employee received his or her training and who did the training.

Finally, every employee must attend a safety and health meeting as necessary (at least once a month), to be sponsored by the employer. These meetings can held for groups or conducted individually. They can be part of other staff meetings.

SUMMARY

We have now taken a thorough look at the 1995 OSHA Logging Operations Standard. Please remember that more information and assistance are available from many sources, including your local OSHA office. This standard sets some tough rules. But they are clearly written and deal with the biggest, specific safety problems encountered in the logging and forestry industry. By following this standard, American loggers can continue to earn a good living, and return home safe and whole at the end of the day.

SOURCES OF ADDITIONAL INFORMATION

American National Standards Institute, 11 West 42nd Street, New York, New York 10036.

American Pulpwood Association, Inc., 600 Jefferson Plaza, Suite 350 Rockville, Maryland 20852. The APA offers a wide range of useful materials related to forestry work, timber harvesting, and the pulpwood industry.

American Society of Mechanical Engineers, United Engineering Center, 345 East 47th Street, New York, New York 10017-2392.

OSHA Publications Office, U.S. Department of Labor, 200 Constitution Avenue, N.W., Room N3101, Washington, D.C. 20210.

Society of Automotive Engineers, 400 Commonwealth Drive, Warrendale, Pennsylvania 15096.

SECTION 4:

SAFETY AND HEALTH IN FISHERIES AND AQUACULTURE

34

EPIDEMIOLOGY OF FATAL INJURY IN THE U.S. COMMERCIAL FISHING INDUSTRY

Richard D. Kennedy, M.S.
Jennifer M. Lincoln, B.S.
National Institute for Occupational Safety and Health

From 1990 through 1994, workers in the U.S. commercial fishing industry experienced an average of 84 fatalities and thousands of injuries each year. In fact, the commercial fishing industry has a fatality rate 31 times higher than the average U.S. industry fatality rate. Fishing aboard unstable vessels, lack of training in cold water survival techniques, and failure to use personal flotation devices have been identified as risk factors for fatal injuries. This chapter reviews the epidemiology of commercial fishing fatalities including the causes of death, location of death, types of vessels involved, specific fishery associated with the fatal injury, and the strengths and weaknesses of agency databases in recording commercial fishing fatalities.

INTRODUCTION

The U.S. commercial fishing industry is a multi-billion dollar business, employing tens of thousands of persons, primarily along the coastal regions of the nation. Commercial fishing has long been identified as one of the highest-risk occupations since the first fishing fleets harvested cod and flounder off New England shores over 200 hundred years ago (Morris 1979). During the decade of the 1980s, estimates place the average annual number of U.S. commercial fishing fatalities between 64 (NIOSH 1994) and 108 (NRC 1991). These fatalities equate to one of the highest occupational fatality rates among all U.S. workers (NIOSH 1995a). Likewise, nonfatal injuries among fishermen are a substantial source of medical costs and losses to local fishing economies and communities. The magnitude and nature of nonfatal injury in the commercial fishing industry are often recorded only partially, or not at all. Six years (1982 through 1987) of Coast Guard injury data detail nearly 14,000 injuries nationwide (NRC 1991, p. 5).

Recent national fatality data from the U.S. Department of Labor's (DOL) annual Census of Fatal Occupational Injury (CFOI) for the 3-year period 1992 and 1994 (BLS 1995a) and from the USCG for the 5-year period 1990 through 1994 (U.S. Coast Guard 1995a, 1995b) showed a small yearly increase between 1990 and 1993, followed by a sharp decrease in the number of 1994 fatal events in the U.S. commercial fishing industry. Utilizing the latest (1993) national employment estimates, the occupational fatality rate for commercial fishermen is 155 per 100,000 workers—the highest among all occupations (Toscano and Windau 1995). Additionally, workers in specific fishing industry segments such as harvesting of crab in Alaskan waters, still experience injury rates a hundred times that of the average U.S. worker (Schnitzer et al. 1995; Knapp and Ronan 1991; CDC 1993).

In addition to a thorough description of the known risk factors and the epidemiology of fatal injuries among workers in the U.S. commercial fishing industry, chapter topics include sections on 1) recent historical background; 2) a description of fishermen, fishing vessels, and fisheries; and 3) injury data repositories.

RECENT BACKGROUND

To improve the overall safety of U.S. commercial fishing vessels, Congress passed the U.S. Commercial Fishing Industry Vessel Safety Act of 1988 (CFIVSA). The provisions of the CFIVSA charged the Coast Guard "to issue a series of specific regulations for U.S. documented or state-numbered uninspected fishing, fish processing, and fish tender vessels" (46 CFR Part 28: 100-500). The following CFIVSA guidelines were intended to improve the overall safety of commercial fishing industry vessels by:

1. Having the Department of Transportation, through the Coast Guard, assume the leadership role and coordinate the national effort to improve safety within the commercial fishing industry.

2. Establishing, in stages, an integrated safety program.
3. Enhancing education and training of vessel captains and crew.
4. Upgrading data gathering, recordkeeping, and database management facilities.

By addressing known risk factors, such as vessel instability or the lack of proper lifesaving equipment, safety would now be a priority. Requirements for compliance with specific CFIVSA regulations may be based upon the following six factors: type and length of vessel, geographical location(s) of operations, seasonal conditions, number of persons on board, whether documented or state registered vessel, and date vessel built or converted (46 CFR Part 28: 115). For example, since November 15, 1991, each vessel must carry Coast Guard-approved throwable flotation devices. If the vessel's length is 65 feet or more, then three orange Ring Life Buoys of at least 24-inch size with 90 feet of line are required; if vessel length equals 26 or more feet, but less than 65 feet, then one orange Ring Life Buoy of at least 24-inch size with 60 feet of line is required (46 CFR Part 28: 270). The CFIVSA was passed into law effective September 15, 1991; however, certain regulations, such as the fleet-wide implementation of survival craft, were phased in over the 4-year period, 1991 to 1994.

To address the human-factors component of safe operation of fishing vessels, the CFIVSA called for specific instruction, drills, and safety orientation of commercial fishers (46 CFR PArt 28: 270). "Individuals serving aboard a commercial fishing vessel must be familiar with their assigned duties and the proper responses to a number of contingencies including (but not limited to) abandoning the vessel, fighting a fire in different locations on board the vessel, and recovering an individual from the water" (NRC 1991, v).

Dividends from shipboard drills, training, and the use of survival equipment called for by the CFIVSA started almost immediately (Morris 1979).[1] USCG commercial fishing safety activities include licensing of vessels, voluntary vessel inspections, training and education, and search and rescue operations.

In response to congressional "concern over unabated losses of commercial fishing vessels and fishermen, and to fishermen's concerns over the rising costs of insurance" (NRC 1991, v-vi) experienced in the early 1980s, the CFIVSA also mandated that the Secretary of Transportation conduct a *Fishing Industry Vessel Inspection Study*. The Secretary requested that the National Research Council (NRC) of the National Academies of Sciences and Engineering "conduct a comprehensive assessment of vessel and personnel safety problems and develop a full range of safety management alternatives, including vessel inspection" (NRC 1991, vi). The NRC conducted "a broad-based study of safety problems in the U.S. commercial fishing industry, including identification and characterization of safety problems" (NRC 1991, xvi). The results of the study, *Fishing Vessel Safety: Blueprint for a National Program,* was published in 1991.

The NRC publication reviewed the commercial fishing safety record for the period 1982 through 1987, the relationships between human factors and vessel casualties, and the state of safety and survival equipment, and generated a list of conclusions and recommendations. The review provided an additional source of statistical tabulations of fishing industry vessel casualties. It offered the following candid summarization of the state of U.S. commercial fishing by the end of the 1980s: "Other leading fishing nations—such as New Zealand, Japan, Norway, and Great Britain—use formal measures to improve professional competence among fishermen and the material condition of their vessels. In contrast, the U.S. fishing industry and government have pursued voluntary, piecemeal safety measures that lack cohesive leadership or coordination and are constrained by limited resources. While improvements to safety have been experienced on a vessel-by-vessel, person-by-person basis industry wide, voluntary measures have not achieved measurable results."

The Public Health Service

The National Institute for Occupational Safety and Health (NIOSH), as one of the U.S. Centers for Disease Control and Prevention (CDC), is the agency responsible for conducting research and making recommendations for the prevention of work-related illnesses and injuries. In mid-1991, NIOSH established an office in Alaska, primarily to investigate and develop preventive strategies for occupational injuries in three key industries: aviation, logging, and commercial fishing (Helmkamp et al. 1993). Data for the period 1980 to 1989 (NIOSH 1995), along with other independent research (Schnitzer et al. 1995; Knapp and Ronan 1991), showed that Alaskan fishermen were dying at a rate approximately 10 times the state average (35/100,000 workers per year) and 50 times that of the average U.S. worker (7/100,000 workers per year) (NIOSH 1995). In response to a continuing loss of commercial fishermen,

---

[1] Benefits of the implementation of Electronic Position Indicating Radio Beacons (EPIRB) and immersions suits was starkly demonstrated when the fishing vessel *Majestic* went down in the 36-degree waters of the Bering Sea on September 22, 1992. EPIRB signals, automatically transmitted by the *Majestic* giving her geographic position, were received by satellite and relayed to the Operations Center at Coast Guard Station Kodiak, which instructed the USCG cutter *Rush* to launch a search and rescue helicopter. Five crew members were rescued after five hours in the water in their immersion suits. Without an EPIRB to alert the USCG of vessel sinking, and without immersion suits to protect her crew from hypothermia, the *Majestic* and her crew of five probably would have vanished without a trace.

particularly in Alaskan waters, the National Fishing Industry Safety and Health (FISH) Workshop was convened in Anchorage, Alaska, in October, 1992 (NIOSH 1994).

FISHERMEN, FISHING VESSELS, AND FISHERIES

The Fisherman

We will refer to the men and women who work as commercial fishermen as fishers. The NRC points out that "some are first-generation fishers; some come from generations of fishers, often with one or more family-owned fishing vessels... as a general rule, fishers receive no guaranteed wage, no overtime pay, and few fringe benefits—they accept only the assurance of hard work, long hours, and a high-risk workplace" (NRC 1991, p. 23). Their compensation is working on one of the few remaining frontiers in exchange for a share of the profits at the end of the fishing expedition.

"A fisher may be a fishing vessel owner who serves as captain of his own vessel (owner/operator), a person employed by the owner to operate the vessel (... captain), or a crewman" (deckhand) (NRC 1991, p. 23). Most U.S. commercial fishers are involved in small-scale operations, are regarded as self-employed and generally work for only a portion of the year; many hold other part-time jobs when not fishing, doing regular vessel maintenance or participating in business-related activities such as attending trade shows, receiving educational instruction, or doing routine paperwork (Dzugan 1995).

A commercial fishing license (NIOSH 1994)[2] is required for persons participating in the U.S. commercial fishing industry. An exception to this is the family-owned and operated vessel where adolescents often participate as crew members. Individual U.S. states set the license requirements and establish payment of fees for the privilege of conducting harvesting operations within their waters. Usually there are no formal training or experience prerequisites for obtaining a commercial fishing license. On a national level, commercial fishers have no certification program, nor is there any statutory authority mandating the USCG to require licensing of commercial fishers working aboard vessels less that 200-gross tons (Commercial Fishing Industry Vessel Safety Act 1988, Executive Summary). However, captains and certain other operators of vessels greater than 200-gross tons require an USCG-issued license with training endorsement stating they can operate a ship of this size (Commercial Fishing Industry Vessel Safety Act 1988, Chapter 106).

Fishing Vessels

"Commercial fishing vessels are self-propelled or wind-driven platforms" used to harvest marine resources for a profit (NRC 1991, p. 22) They may also be considered a "workplace, a means of transportation to and from the fishing grounds, an itinerant domicile for overnight or extended trips, and in some cases, an industrial plant for processing" (NRC 1991, p. 22) their catch or the harvest from other smaller catcher vessels. Fishing vessels in the U.S. commercial fishing industry range from under 25 feet in length to the huge 600-foot-plus catcher-processor factory trawlers of the North Pacific. The majority of the fleet are small fishing vessels (around 40 feet) with nearly 99% of all vessels being less than 79 feet in length (U.S. Coast Guard 1995c). Fishing vessels incorporate mechanical riggings (also called gear-type) to harvest their catch. Some common varieties of fishing apparatus "include various types of nets, trolling gear, trawls, hooks, dredges, rakes, and traps" (NRC 191, p. 42) or pots.

United States commercial fishing vessels of at least 5 net tons are required to be documented by the USCG, and the Certification of Documentation must be renewed annually (46 CFR, Chapter 33). Information recorded on the Certificate of Documentation includes vessel and owner name, a full description of the vessel (e.g., vessel length, width, depth, tonnage, year built, hull material, place and year built), any entitlements, restrictions or operational endorsements, and the USCG official number (33 CFR, Part 173).

A vessel engaged in commercial fishing must have a fisheries endorsement—commonly referred to as a license. This endorsement entitles the vessel to fish and land its catch in the United States. Licensing of vessels for the purpose of harvesting marine resources within state waters is mainly the domain of individual state governments, often through departments of fish and game (NRC 1991, p.159).[3]

The Fisheries

Management of marine resources is accomplished through a number of state and federal agencies by partitioning all harvesting activities into well-defined units called fisheries. Fisheries are both species-specific, such as lobster, halibut, or dungeness crab, and geographic-specific, such as the North Atlantic, Gulf of Mexico, or Bering Sea areas. The vast majority of fisheries limit harvest to a fixed, pre-announced time window lasting from minutes to months. Some fisheries such as Alaska salmon are closed-entry, meaning special permits are required of vessel owners and

---

[2] There are two main catagories of licenses--the captain's license and the deckhand or species-specific harvesting license issued to the vessel crew.

[3] Certain fisheries such as groundfish (i.e., pollock or black cod) also require vessel licensing permits managed by the Federal Government's National Marine Fisheries Service.

captains in order to fish. Readers interested in obtaining more detail on the conservation and management of all fishery resources within the U.S. 200-mile Exclusive Economic Zone are encouraged to consult the Magnuson Fishery Conservation and Management Act (MFCMA), Public Law 94-265 as amended (Magnuson Act).

The ability to classify commercial fishing operations by fishery has important implications in conducting epidemiological research (Kennedy et al. 1994). Stratification by fishery may control for the possible confounding effects of the four key variables common to all commercial fishing: time, location, equipment, and management (Table 34.1). A fishery provides a natural partition for meaningful analysis, interpretation of findings, and development, implementation, and evaluation of intervention strategies.

Table 34.1. Factors with descriptions and examples common to all commercial fisheries.

Factor	Description	Example
Time	Harvest activities occur during the same portion of the calendar year	Spawning salmon runs occur during the summer months
Location	Fishing grounds (regions) are well established	The Grand Banks fishing grounds
Equipment	Catch-specific harvesting gear are mechanically equal	Pots (cages) are used to harvest Maine lobster
Management	Vessels and fishers operate under identical rules and regulations	The International Halibut Commission governs all U.S. commercial halibut fishing

POPULATION-AT-RISK

Accurate estimates of the workforce in the U.S. commercial fishing industry have always been difficult to obtain (NRC 1991, p. 38). Unlike most other U.S. industries, the seafood industry's employment and payroll have not been provided on a regular basis through standard economic data systems and reports. "The U.S. Department of Labor captures data on the vast majority of the U.S. economy (including seafood processing) through a system of quarterly, nonagricultural, civilian, employer reports which are required by law for reporting unemployment insurance contributions" (McDowell et al. 1989). Two industry sectors are not captured with this system: seafood harvesting (commercial fishing) and the military. The seafood-harvesting sector is classified as an agricultural industry (OMB 1987) and the method of pay most often used (crew shares) does not fit the normal reporting system (McDowell et al. 1989). A major consequence of this is a lack of workforce estimates and the inability to calculate meaningful injury and fatality rates in one of the nation's most hazardous industries.

Fatality rates published by the CFOI program are calculated using employment estimates from the Current Population Survey (CPS) (Toscano 1994). The CPS is a household survey conducted for the Department of Labor that generates monthly employment estimates (Toscano 1994). The CFOI fatality rates "relate the total number of workplace deaths ... to the annual average number of workers facing that risk [and] measurements are developmental and do not reflect the movement of persons into and out of the labor force, the length of their work week or work year, or the effect of multiple jobholders" (Toscano 1994).

Fishery management guidelines require that fishers tailor their work day, week, or even month around the various fishery openings. Fishing vessel captains and crew often spend several days or weeks preparing for a fishery opening that may last only 20 minutes (i.e., the Prince William Sound Herring Fishery). These intense, derby-style fisheries are management schemes resulting from the biological nature of the species and the consequence of too many vessels competing for a limited number of fish (Knapp 1994, p. 81; NRC 1991, p. 28). Likewise, some fishery openings are regulated by a total harvest quota system: fishers fish as hard as they can and catch as much as possible until a certain, predetermined catch-weight limit is reached. During these intense periods of fishing, fishers may work in excess of 18 hours per day for up to a week or more (NRC 1991, p. 249). These examples illustrate that standard 9-to-5, 40-hour weeks are unusual among fishers. Comparing the attributes of the standard "work week" of a fisher to that of the vast majority of other professions is therefore unrealistic.

The seasonality and lure of high wages in commercial fishing also attract a substantial percentage of migrant workers. For example, nearly one-third of workers in the Alaskan commercial fishing industry are residents of other states that come to Alaska (primarily during the summer months) to fish commercially and then return home (Focht 1986). As mentioned earlier, fishers often hold other full-time jobs when their regional fisheries are closed. For example, the spiny Florida lobster fishery is closed for just over one-third year (April 1 to August 5); vessel skippers and crew either regear their vessels for another fishery or seek work at a second occupation.

Difficulties such as those stated above have severely limited enumeration of the number of workers in the commercial fishing industry. While the numbers of commercial fishing licenses and permits (vessels and individuals) are widely available, there appears to be no direct correlation between the number of licenses and average annualized employment (NRC 1991, p. 20), nor are there industry-accepted estimates of average crew size (NRC 1991, p. 41). Only in the state of Alaska, where commercial fishing is one of the major private industries, have significant collaborative efforts been made to successfully generate employment counts (Rodgers et al. 1980; Thomas 1987; McDowell et al. 1989).

CAPTURING FATAL AND NONFATAL INJURIES

Nationwide

At present, several agencies collect nationwide data on fatal and nonfatal injuries in the U.S. commercial fishing industry; however, "the only source of detailed information available on fishing vessel casualties" resides with the USCG (NRC 1991, p. 188). The Coast Guard's Marine Investigations Module (MINMOD) was implemented in 1992 as part of an updated and enhanced Marine Safety Information System (MSIS) (U.S. Coast Guard 1995d). The MINMOD captures vessel damage and personnel casualty data, in addition to information related to pollution incidents.

MINMOD's unique data base design enables users to view a marine incident as a chain of events—record-linking to other components of the MSIS as appropriate (Tansey 1995). The USCG MINMOD data base began January 1, 1992, replacing the Main Casualty (CASMAIN) data base. Both USCG data bases are coded summaries of incidents reported on USCG Marine Casualty Reports (Form 2692). USCG data can provide sufficient information on the general nature of the fatal event such as date, time, and location of the incident, environmental conditions, routine vessel information, and basic demographic information of the decedent. Missing from the USCG MINMOD and CASMAIN data bases are a limited number of fatal injury events that occur on or near shore, often involving one or two fishers using small, state-numbered boats (e.g., a skiff)

Table 34.2. Strengths and limitations of national systems for recording fatal events in the U.S. commercial fishing industry.

Agency	Database	Strengths	Limitations
Coast Guard	CASMAIN (1982-91) MINMOD (1992-present)	1. Centralized data base 2. High sensitivity for vessel-related incidents 3. Updated several times per year 4. Record-linked to other modules 5. Narrative information - may mention fishery	1. May miss non-vessel related fatalities 2. Only closed cases are included 3. Codings not con-ducive to epidemiology research 4. Quality of record keeping varies from unit to unit
Department of Labor, Bureau of Labor Statistics	Census of Fatal Occupational Injuries Program (1992-Present) *	1. Centralized data base 2. Uses multiple data sources 3. Highly sensitive 4. Narrative information-- may mention fishery	1. Quality of recordkeeping and interpretation of case definition may vary from state to state
National Institute for Occupational Safety and Health	National Traumatic Occupational Fatalities surveillance system (1980-Present)	1. Centralized data base 2. Legal document 3. Narrative information-- may mention fishery	1. Depends upon the injury-at-work box correctly marked 2. Quality of recordkeeping varies from state to state 3. Coded for usual occupation and industry

* For an excellent discussion on the CFOI program, please see *An Evaluation of the Census of Fatal Occupational Injuries as a System of Surveillance* by Connie Austin, U.S. Department of Labor, Bureau of Labor Statistics, Report 891, pp. 51-54.

(Schnitzer et al. 1995; Kennedy et al. 1995; Davis et al. 1994) or dockside incidents where the jurisdictional agency most often resides with the local department of public safety.

As previously mentioned, the U.S. Department of Labor, Bureau of Labor Statistics, provides information about all known work-related fatal injuries through the Census of Fatal Occupational Injuries (CFOI) program. The goal of the CFOI program is to collect information on all fatal occupational injuries. To accomplish this, the CFOI uses multiple data sources such as death certificates, workers' compensation reports and claims, and other federal and state administrative records. "Work relationship is verified for each fatality by using at least two independent source documents" (BLS 1995b, iii). Included in the CFOI data base are occupation and industry of the decedent, along with other information about the fatally injured worker and the events leading to the fatality (BLS 1994, pp. 5-7). Nationwide fatality statistics for the CFOI program first appeared for the year 1992.

Death certificates provide an alternate, albeit limited, source of data on commercial fishing fatalities (Stout and Bell 1991; Kraus et al. 1995). The National Traumatic Occupational Fatalities (NTOF) surveillance system developed by NIOSH is a census of fatal occupational injuries. Death certificates collected from the 50 states, New York City, and the District of Columbia must satisfy three criteria: 1) decedent is 16 years of age or older; 2) the "injury at work" box is checked on the death certificate; and 3) external cause of death (E Code), E800-E999 (NIOSH 1993). "Analysis of occupational injury death by demographic, employment, and injury characteristics facilitates effective use of resources aimed at preventing injuries in the workplace" (NIOSH 1993).

A summary of the strengths and limitation of each of the above data collection and analysis systems can be found in Table 34.2.

FATALITY STATISTICS

The following statistics were compiled in July 1995. The sources of these fatality data were both the USG CASMAIN (years 1990 and 1991) and MINMOD (years 1992 through 1994) data bases. Although no concerted effort was made to record-match these fatalities with incidents or deaths from other state or national fatality data bases such as the nationwide CFOI program, the distribution by year of USCG fatalities for 19192 through 1994 (246) is similar to those recorded by CFOI for the same 3-year period (238). The number of fatalities presented here, however, may vary from those tabulated by the USCG due to case definition.

During the 5-year period, 1990 to 1994, at least 421 U.S. commercial fishers died while working on the job. Of these, 268 (63.7%) were confirmed where the body was recovered. Of the remaining 153 cases, 150 were presumed deaths where a body was not recovered; and in 3 cases it was unclear whether the victim's body was recovered or not (Figure 34.1)

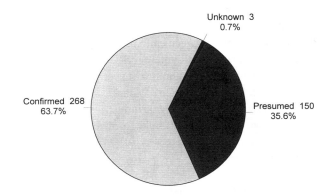

* Confirmed fatality equates to a recovered body; a presumed fatality is where no body is recovered.

Figure 34.1. Number of confirmed and presumed fatalities, U.S. commercial fishing industry, 1990-1994, N=421.

In no other occupation is there a larger proportion of presumed deaths to confirmed deaths.

The overwhelming cause of death among workers in the U.S. commercial fishing industry was drowning (Figure 34.2). Drowning, drowning due to hypothermia, or presumed drowning was responsible for 330 (78.4%) fatalities. No single, definable cause could be identified as the number two factor accountable for fatal injury, due to a large number of

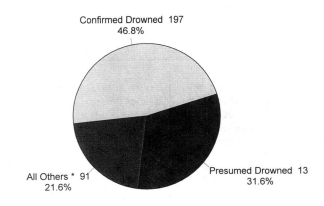

* Includes 72 fatalities coded as other (22), not-elsewhere classified (18), and unknown (32).

Figure 34.2. Distribution and number of fatalities by cause of death, U.S. commercial fishing industry, 1990-1994, N=421.

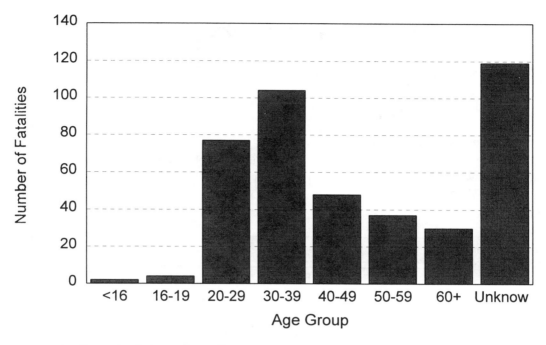

Figure 34.3. Number of fatalities by age group, U.S. commercial fishing industry, 1990-1994, N=421.

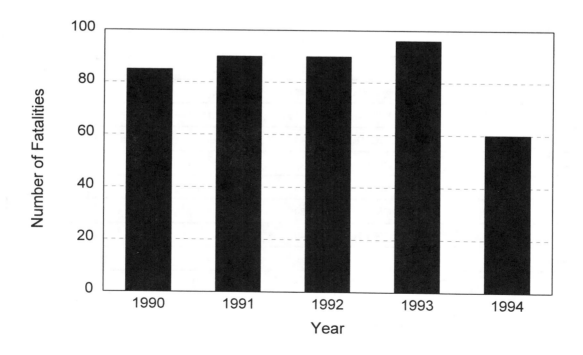

Figure 34.4. Number of fatalities by year, U.S. commercial fishing industry, 1990-1994, N=421.

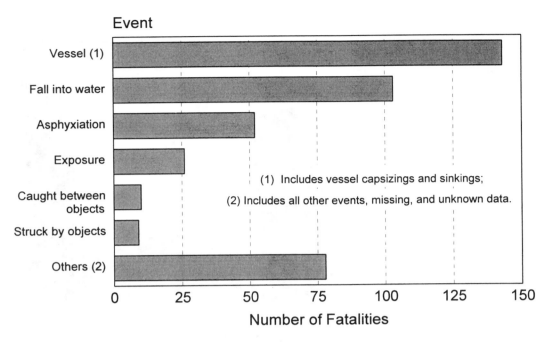

Figure 34.5. Number of fatalities by event type, U.S. commercial fishing industry, 1990-1994, N=421.

Table 34.3. Number of fatalities by water body, U.S. commercial fishing industry, 1990-1994, N=421.

Water Body	Zone	n	Percent * Water body	n	Percent All Waters
Atlantic	*All*			118	28.0
	ICW (1)	8	6.8		
	Coastal	27	22.9		
	3 to 200 mi	83	70.3		
Gulf of Mexico	*All*			73	17.4
	ICW (1)	18	24.7		
	Coastal	11	15.1		
	3 to 200 mi	44	60.3		
Pacific	*All*			189	44.9
	Bering Sea	60	31.8		
	Coastal (2)	80	42.3		
	3 to 200 mi (3)	49	26.0		
Others	*All*			41	9.7
	Navigable Waters (4)	23			
	Others	11			
	Unknowns	7			
Totals		421		421	100.0

(1) ICW = Intercoastal waterways; (2) Includes Intercoastal waterways of SE Alaska and the Aleutians; (3) Includes the Gulf of Alaska and Hawaiian waters; (4) Includes navigable rivers and lakes.
* Column percents may not sum to 100% due to rounding.

SOURCE: United States Coast Guard [1995]

missing victims, unknown causes, not-elsewhere-classified, or other coding classifications.

The mean age at death for the 302 fatalities where age data were available was 39.4 years. The distribution of age at death for decedents by age group appears in Figure 34.3. The gender of the victim was not available, but estimates based on the distribution by gender from other sources (BLS 1995a; Schnitzer et al. 1995; NIOSH 1995b) suggests that approximately 10 (2%) of all U.S. commercial fishing fatalities occurring between 1990 and 1994 would have been women.

For the 5-year period, 1990-1994, the number of fatalities in the U.S. commercial fishing industry averaged just over 84 per year. The distribution of fatalities climbed slowly between 1990 and 1993, but dropped substantially in 1994 (Figure 34.4).

Figure 34.5 indicates the type of events that most often claim the life of a fisher. Just over one-third (34.2%) of all fatalities were due to known or presumed vessel capsizings and/or sinkings. Nearly one-out-of-four (24.5%) of all fatalities were due to falls into water.

The USCG data identified marked regional variation in fatalities among U.S. fishers. The largest number, 189, (44.9%) of all fatalities took place in the waters of the Pacific (Table 34.3). Waters of the Atlantic claimed 118 lives (28.0%), while 73 (17.4%) fishers died in waters of the Gulf of Mexico. One hundred forty-one (33.5%) of all fishers died in waters along the coast or within lakes or rivers.

Data on vessel length were available for fatal events involving 368 fishers. Events involving circumstances other than those occurring onboard commercial fishing vessels or where vessels' data were unknown or missing, accounted for 53 lives.

For those vessel-related fatalities where vessel length was available, 146—or 4 out of 10 fishers—died on commercial fishing vessels in the 40- to 64- foot class. An approximately equal number of fishers died on vessels in the 26- to 39-foot (67 fatalities), the 65- to 78-foot (75 fatalities) and the 79- to 199-foot (56 fatalities) class (Figure 34.6).

Fatalities due to vessel capsizings and sinkings accounted for 129 deaths. The largest proportion (86.0%) of fatalities associated with vessel capsizing and/or sinkings occurred for vessels in the 40- to 199-foot range. During the 1990s, there were no vessel capsizings and/or sinkings of U.S. commercial fishing industry vessels of 200 feet or greater. Additionally, type of event data indicate that fishers on vessels of 200 feet or greater encounter more fatal injuries due to falling overboard events as compared to being fatally injured in a vessel-related casualty incident. For the 5-year period 1990 to 1994, nearly 8 out of every 10 fisher fatalities (78.4%) drowned or were presumed to have drowned. Confirmed drownings, meaning the body was recovered, accounted for 60% of this total. The distribution of confirmed versus presumed drownings (body not recovered) varied considerably by geographical location (Figure 34.7).

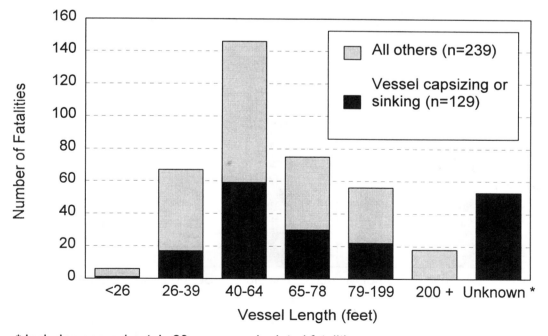

* Includes approximately 20 non-vessel related fatalities.

Figure 34.6. Number of fatalities by vessel length and event status, U.S. commercial fishing industry, 1990-1994, N=421.

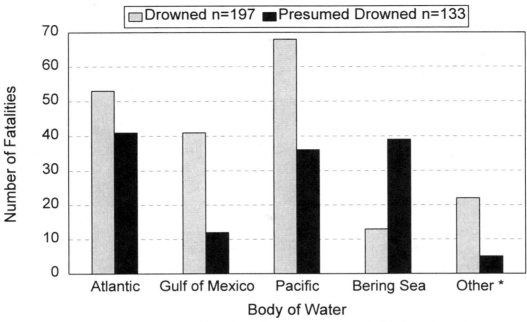

Figure 34.7. Distribution of presumed versus confirmed drownings by body of water, U.S. commercial fishing industry, 1990-1994, N=421.

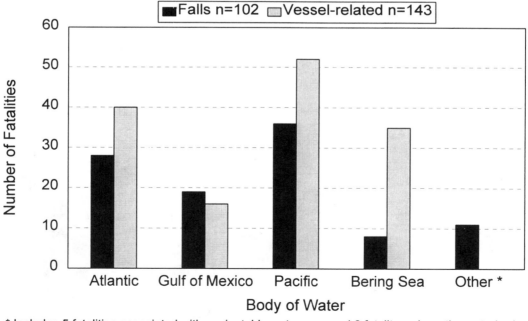

Figure 34.8. Number of falls and vessel-related fatalities by body of water, U.S. commercial fishing industry, 1990-1994, N=421.

The greatest number of confirmed drownings (68) occurred in the waters of the Pacific. Confirmed drowning fatalities ranked second (53) for the Atlantic and third (41) for the Gulf of Mexico. The largest number of presumed drownings (41) occurred in the Atlantic, followed closely (39) by the Bering Sea and the Pacific (36). Proportionally, confirmed drownings versus presumed drownings were highest in the waters of the Gulf of Mexico (3.4:1), a 10-fold reverse being true for drownings occurring in the Bering Sea (1:3.0).

The two leading circumstances of death among fishers were falls into water (also referred to as man overboard, n=102) and vessel-related casualties such as vessel capsizings and/or sinkings (n=143). The distribution of these two factors was contrasted against the region (body of water) of occurrence for 245 fatalities (Figure 34.8). Distribution of circumstance of event was similar between the Pacific and Atlantic regions, with the former experiencing approximately 30% additional fatalities (88 versus 68, respectively). The distribution of circumstance type for the Gulf of Mexico region was nearly equal for falls into water (19 events) and vessel-related fatalities (16 events). Four times as many fishers (35 fatalities) lost their lives in vessel-related events compared to falls into water (8 fatalities) for the region of the Bering Sea.

FATALITY RATES

The effort to calculate fatality rates in the U.S. commercial fishing industry is hampered by the lack of reliable workforce estimates on the number of commercial fishers. The latest estimate of employment for fishers in the U.S. commercial fishing industry was produced by the Department of Labor, Bureau of Labor Statistics (BLS), using average employment estimates from its 1993 Current Population Survey (Toscano and Windau 1995).

The 1993 occupational fatality rate for U.S. commercial fishers, based on the BLS numerator (90) and the BLS denominator (58,000), was estimated at 155 per 100,000 workers (Toscano and Windau 1995). Utilizing USCG fatality counts (96) as the numerator with the BLS denominator yields a slightly higher annual fatality rate of 166 per 100,000 workers. Citing either rate, commercial fishers have the highest occupational fatality rate in the nation—at least 31 times the average U.S. worker fatality rate (5 per 100,000) (Figure 34.9).

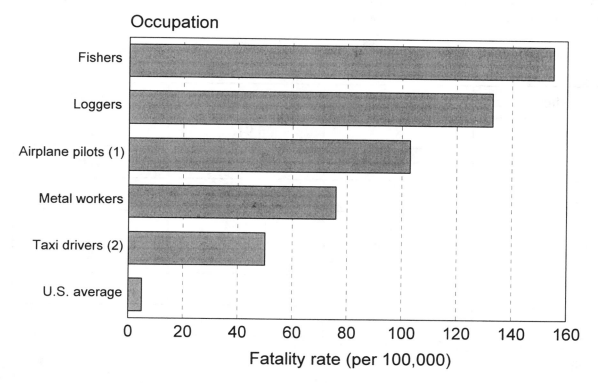

(1) Includes airplane navigators; (2) Includes chauffeurs.

Figure 34.9. Fatality rates (ranked) for selected high-risk occupations, U.S. industries, 1993.

FISHERY-SPECIFIC FATALITY RATES

The high-risk groups within the commercial fishing industry may be identified by examining fatality rates by fishery. This requires workforce estimates that are produced by similar algorithm, to standardize time-at-risk. Estimates of workforce in the five major Alaskan fisheries have been periodically made for the past 25 years (Rodgers et al. 1980; Thomas 1987; McDowell et al. 1989; Kennedy 1994). An understanding of how fatality rates vary among Alaskan fisheries is important since that particular state has the highest proportion of fishers among all workers of any state and, more importantly, has the distinction of having the most fisher fatalities of any state.

The latest published Alaska fishery-specific fatality rates are based on data for the calendar years of 1991 and 1992 (CDC 1993). Fatality frequency counts and workforce estimates presented here may vary from those that appeared in the published report. This is due to refinements in workforce estimates that were done in 1994 (Kennedy 1994) and follow-up information on the individual fatality events, (NIOSH 1995b) enabling a more precise classification of fishery.

Of the 67 fatalities recorded, 63 were identified as participating in one of the five major Alaska fisheries (groundfish, halibut, herring, salmon, and shellfish). Two fishers died while harvesting sea cucumbers (CDC 1993a) and information on the type of fishery was not available for another two fishers. "Both the number of fatalities (32) and the average annual fatality rate (681/100,000) were the highest for the shellfish fishery (Table 34.4). Half (16) of these fatalities resulted from the disappearance and presumed sinking of three vessels in the Bering Sea in three separate incidents; all three vessels were fishing for king crab in the vicinity of the Pribilof Islands during winter months (February 1991, November 1991, and January 1992)" (CDC 1993b).

Additional Risk Factors

"Workers at greatest risk for fishing-related fatal injury are those who operate aboard unstable (i.e., easily capsized) vessels and those who have insufficient training in shipboard safety, especially regarding cold-water survival techniques and the use of lifesaving equipment such as the personal flotation device" (CDC 1993b).

For the 5-year period 1990 to 1994, 129 out of 421 (30.6%) fatalities can be attributed to vessel capsizings and/or sinkings (U.S. Coast Guard 1995a, 1995b). "Capsizings occur when vessels are made to operate in environmental

Table 34.4. Fishery-specific fatality rates, Alaska commercial fishing industry, 1991-1992.

Fishery	Fatalities	Workforce (1)	Rate (2)
Groundfish	8	5,916	135
Halibut	9	2,114	426
Herring	0	1,680	0
Salmon	14	15,864	88
Shellfish	32	4,702	681
Misc./Unknown	4		
Total	67	30,276	221

(1) Workforce estimates are for the 2-year period, 1991 and 1992.
(2) Rate equates to numerator divided by denominator times 100,000.

conditions (e.g., wind, sea, or ice) or in ways (e.g., improper loading) that exceed their righting capability. Often, sudden overwhelming events with high fatality potential occur because of insufficient warning time to abandon ship. Persons can be trapped inside the hull, entangled in the rigging, or thrown into the water without personal survival equipment" (NRC 1991, p. 77). Fatality data thus far presented include incidents involving fishing vessels that are believed to have capsized and sunk leaving no survivors, or in some cases no evidence of mishap. "The single most preventable problem related to drowning in the fishing industry is vessel instability" (Myers et al. 1994, p. 9; NIOSH 1995a).[4]

Environmental conditions may also contribute to the severity of work-related incidents. The USCG has classified all waters in the North Pacific and Atlantic (both above 32 degrees north latitude) as "cold" waters (<60 F [<15.6 C]) (U.S. Coast Guard); in these waters, hypothermia can lead to death by drowning within minutes of immersion. Because "immersion suits provide a measure of thermal resistance to temperatures to allow enough time for rescuers to reach individuals in the water ... they are required equipment for all vessels" (46 CFR Part 28:110) in these environments.

SUMMARY

Fatality statistics presented here have confirmed that commercial fishing remains the highest-risk occupation in the United States. Additionally, a substantial number of commercial fishing vessels are lost during fishing operations and in transit to and from the fishing grounds. The findings of the present study closely agree with those cited by the NRC study committee (NRC 1991, 38-65). Although fatality counts have decreased from an annual average of 108 for the 6-year period 1982 to 1987, to 84 fatalities for the 5-year period 1990 to 1994, the proportions of fishers lost to vessel capsizings and/or sinkings and those fatally injured from falls remain basically unchanged. Death due to drowning is still overwhelmingly the most common cause of death.

These data provide a national prospective for other studies on the nature of specific fatal injury, causes, or risk factors in the U.S. commercial fishing industry. Exploration of risk factors and their interactions and relationships with fatal injury in the U.S., state, or regional fisheries can contribute insight and may ultimately lead to the development of new interventions for prevention of fatal injury in American's riskiest workplace.

---

[4] Vessel instability was cited as a major factor in three of the most celebrated fishing vessel sinkings in the North Pacific, the *Aloutian Enterprise* (9 deaths), *Western Sea* (6 deaths), and *Lasseigne* (deaths). Investigations on each were done and extensive reports of these and other prominent fishing vessel sinkings and are available from the National Transportation Safety Board and the U.S. Coast Guard.

REFERENCES

Alaska Occupational Injury Surveillance System data base. 1995. Anchorage, AK: National Institute for Occupational Safety and Health, Division of Safety Research, 1990-1994.

Bureau of Labor Statistics (BLS). 1994. *Fatal Workplace Injuries in 1993: A Collection of Data and Analysis*. U.S. Department of Labor Report 870, April, pp. 5-7.

Bureau of Labor Statistic (BLS). 1995a. *Census of Fatal Occupational Injury for 1991-1994: Data Diskette*. Department of Labor, August.

Bureau of Labor Statistics (BLS). 1995b. *Fatal Workplace Injuries in 1993: A Collection of Data and Analysis*. U.S. Department of Labor Report 891, June, p. iii.

Center for Disease Control (CDC). 1993. Commercial fishing fatalities—Alaska, 1991-1992, *MMWR* 42.

Commercial Fishing Industry Vessel Safety Act of 1988, Public Law 100-424.

Davis, L.K., D.R. Brooks, L.E. Taylor, and C. Schwartz. 1994. Data sources for fatality surveillance in commercial fishing: Massachusetts, 1987-1991, In: *Fatal Occupational Injuries in 1992: A Collection of Data and Analysis*. U.S. Department of Labor report 870, May 1994. pp. 42-48.

Dzugan, J. Executive Director, Alaska Marine Safety Education Association, Personal Communication, Sitka, AK, September 1995.

Focht, R. 1986. *Employment and Gross Earnings in Alaska's Commercial Fisheries: Estimates for all Participants and Residents of Alaska, Washington, Oregon, and California, 1983-1984*. Juneau, AK: Commercial Fisheries Entry Commission, Report No. 86-88.

Helmkamp, J.C., R.D. Kennedy, D.E. Fosbroke, and M.L. Myers. 1993. Occupational fatalities in the fishing logging and air transport industries in Alaska, 1991; *Scan. J. Work Envir. Health* 18 Suppl 2: 55-57.

Kennedy, R.D. 1994. *Final Report: Occupational Injury Rates in the Alaska Commercial Fishing Industry*. Morgantown, WV: Unpublished document, NIOSH, March 9.

Kennedy, R.D., M.E. Veazie, G.A. Conway, and H. Amandus. 1994. Fishing deaths in Alaska vary by fishery. Letter to the editor, *Amer. J. Public Health*, 84(3): 496.

Kennedy, R.D., J.C. Helmkamp, and T.M. Ungs. 1995. The epidemiologic utility of the MINMOD data base. Morgantown, WV: Unpublished abstract, July.

Knapp, G., and N. Ronan. 1991. *Fatality Rates in the Alaska Commercial Fishing Industry*. Anchorage, AK: University of Alaska, Institute of Social and Economic Research; Alaska Sea Grant College Program, Publication No. AK-SG-90-03.

Knapp G.P. 1994. Safety implication of derby fisheries. In: *Proc. of the National Fishing Industry Safety and Health Workshop*. U.S. Department of Health and Human Services, Public Health Service, CDC, NIOSH, p. 81.

Kraus, J.F., C. Peek, T. Silberman, and C. Anderson. The accuracy of death certificates in identifying work-related fatal injuries. *Amer. J. Epidem.* 141(10): 973-79.

McDowell E., J. Calvin, N. Gilbertsen. 1989. *The Alaska Seafood Industry Study: A Technical Report*, prepared for the Alaska Seafood Industry Study Commission, Anchorage, AK.

Morris, M. 1979. *Our Maritime Heritage: Marine Developments and Their Impact on American Life*. University Press of America, pp. 1-17.

Myers, M.L., M.L. Klatt, and G.A. Conway. 1994. Executive summary. In: *Proc. of the National Fishing Industry Safety and Health Workshop*. U.S. Department of Health and Human Services, Public Health Service, CDC, NIOSH, p. 9.

National Institute for Occupational Safety and Health (NIOSH). 1995. The National Traumatic Occupational Fatalities Database. Morgantown, WV: Division of Safety Research.

National Institute for Occupational Safety and Health (NIOSH). 1994a. *Proc. of the National Fishing Industry Safety and Health Workshop*. U.S. Department of Health and Human Services, Public Health Service, CDC, NIOSH Publ. 94-109, p. iii.

National Institute for Occupational Safety and Health (NIOSH). 1994b. *Preventing Drownings in Commercial Fishermen*. NIOSH Publication No. 94-107: U.S. Department of Health and Human Services, Public Health Service, CDC, April, p.3.

National Institute for Occupational Safety and Health (NIOSH). 1993a. *Fatal Injuries to Workers in the United States, 1980-1989: A Decade of Surveillance—National Profile*. U.S. Department of Health and Human Services, Public Health Service, CDC, August, p. 1.

National Institute for Occupational Safety and Health (NIOSH). 1993b. *Fatal Injuries to Workers in the United States, 1980-1989: A Decade of Surveillance—National Profile*. U.S. Department of Health and Human Services, Public Health Service, CDC, August, p. 1.

National Research Council. 1991. The commercial fishing safety record. In: *Fishing Vessel Safety: Blueprint for a National Program*. Washington, DC: National Academy Press.

National Transportation Safety Board (NTSB). Department of Public Affairs, NTSB Headquarters, Washington, DC.

Rodgers, G.W., R.F. Listowski, and D. Mayer. 1980. *Measuring the Socioeconomic Impacts of Alaska's Fisheries*. Anchorage, AK: University of Alaska, Inst. Social and Economic Research.

Schnitzer, P.G., D. Landen, and J. Russell. Occupational injury deaths in Alaska's fishing industry, 1980-1988. *Amer. J. Public Health* 85(5): 685-88.

Standard Industrial Classification Manual. 1987. Executive Office of the Present, Office of Management and Budget, p. 7.

Stout, N., and C. Bell. Effectiveness of source documents for identifying fatal occupational injuries: a synthesis of studies. *Amer. J. Public Health*, 81(6): 725-28.

Tansey, T.A. 1995. USCG Commander, Marine Investigations Division. 1995. Personal communication. Washington, DC.

Thomas, K. 1987. *Alaska Seafood Employment 1977-1984*. Juneau, Alaska: Alaska Department of Labor.

Toscano G, and Windau J. 1994. *Fatal Workplace Injuries in 1992: A Collection of Data and Analysis*. Department of Labor, Bureau of Labor Statistics, Report 870, April, p. 9.

Toscano, G., and J. Windau. 1995. *Bureau of Labor Statistics, Fatal Workplace Injuries in 1993: A Collection of Data and Analysis*. U.S. Department of Labor, Report 891, June, p. 9.

U.S. Coast Guard. 1995a. United States Coast Guard, Marine System Information System (MSIS), Washington DC: U.S. Department of Transportation.

U.S. Coast Guard. 1995b. *United States Coast Guard Main Casualty (CASMAIN) database, 1982-1991*. Washington D.C.: U.S. Department of Transportation, Coast Guard, Marine System Information System.

U.S. Coast Guard. 1995c. *United States Coast Guard Marine Information Module (MINMOD) database, 1992-1994*. Washington D.C.: U.S. Department of Transportation, Coast Guard, Marine System Information System.

U.S. Coast Guard. 1995d. United States Coast Guard, Marine System Information System, Marine Safety Evaluation Branch, G-MMI-3, Washington, DC.: U.S. Department of Transportation.

U.S. Coast Guard. U.S. Department of Transportation. 1994. 33 CFR, Part 173 Vessel Numbering and Casualty and Accident Reporting, p. 645.

U.S. Coast Guard. *Marine Casualty Reports*, Marine Casualty Branch, USCG Headquarters, Washington DC.

35

SAFETY AT SEA

Dewayne Hollin, B.B.A., M.B.A.
Sea Grant College Program
Texas A&M University

Fishing vessel operators and crews face a host of hazards unique to their industry. Collisions with other vessels, flooding, groundings, floundering, capsizing, and marine organisms are but a few of these hazards. In addition, common hazards similar to those found in other industries, such as mechanical components, noise, fires and explosions, and hazardous fumes, are present as well. In this chapter, guidelines to establishing and maintaining an effective safety program on fishing vessels are discussed. Safety policy, personnel issues, and safety training are identified as key elements of the safety program. Hazard checklists are included to facilitate evaluation and assessment of individual vessels. Procedures to document and investigate accidents are explored. Survival techniques for fishing vessels are outlined in terms of emergency equipment, food and water, and other skills.

INTRODUCTION

Commercial fishing is a hazardous occupation, according to recent U.S. Coast Guard fishing vessel casualty statistics, and the industry trend shows no signs of improvement. Based on an analysis of the annual death rate per 100,000 fishermen, the U.S. commercial fishing industry has an annual death rate seven times greater than the national average for all types of U.S. industry groups. Statistically, it is the most dangerous American industry based on this data.

Analysis of casualty investigations indicate human error as the major contributing factor in most fishing vessel accidents. The human element most commonly attributed to these casualties is a lack of technical knowledge in vessel operations not related to actually catching fish. Poor watchkeeping practices, navigational errors, rules of the road violations and a lack of understanding of vessel stability are common causes of vessel accidents. Another cause is material failure resulting in vessel flooding and fire or explosion. A greater awareness of the need to maintain the vessel and its equipment may have averted some of these casualties.

There is no means of removing the dangers of fishing nor can human error be entirely eliminated, but we can reduce the chances of accidents by altering the behavior of the vessel captain and crew toward a greater safety awareness. While each member of the crew is responsible for his own safety, the captain is ultimately responsible for the safe operation of the vessel and its equipment, and for the safety of everyone aboard. The vessel is only as safe as the people who own and operate it. Better knowledge about the vessel, its equipment, and the fishing operation can provide less risk to all concerned. It is the purpose of this chapter to provide the knowledge commercial fishermen need to be safer at sea.

SAVE LIVES, SAVE LOSSES

As the number of fishing vessels in the Gulf of Mexico increases, and a significant number of vessel accidents and casualties continues to occur, it is imperative that each vessel in the Gulf of Mexico implement a sound safety program. Currently the Gulf of Mexico is home to about one-fourth of all fishing vessels of 5 net tons or more in the U.S. fishing fleet.

Between 1982 and 1987, 269 vessels and 125 human lives were lost in the Gulf of Mexico; losses of over 100 fishing vessels and 25 lives were recorded off Texas alone. Many of these losses might have been avoided if safety programs had been established by vessel owners and operators, and followed by captains and crews.

This chapter is designed to provide fishing vessel owners and operators with guidelines to improve their individual safety programs. It includes information about the prevention of accidents, safety inspections, safety rules for crew members, accident procedures, and safety training. It offers sample forms to be used in collecting safety-related data and includes a section on how to survive at sea after abandoning ship.

Because of the diversity of fishing operations within the industry, it is recognized that no ready-made safety package can fill the needs of all operators. While this chapter's guidelines form a sound basis for organizing an effective safety program, other programs may meet fishing industry and individual needs just as effectively and the suggestions and goals listed here can be used as a supplement to other procedures and safety programs in the marine work environment.

While this chapter provides basic guidelines from which an individual program suitable for a particular fishing operation can be designed, individual programs should be coordinated with applicable federal, state, and local safety requirements. Before adopting any of the sample forms in this chapter, vessel owners and operators should consult a legal advisor about legal aspects associated with the use of such forms.

If one boat or life is saved because of the information in this chapter, the effort to compile and publish it will be worthwhile.

SAFETY: A COMMON PROBLEM

The U.S. Coast Guard published the following fishing vessel casualty statistics for Gulf of Mexico operations during 1982-1987:

Operational Collisions. 284 collisions occurred in the Gulf; 60 percent occurred during daylight hours under clear, calm conditions. The highest number of incidents occurred during June, July, and August, with the largest primary causes being personnel fault (48%) and rules-of-the road violations (45%). Total losses were due mainly to collision with a submerged object.

Groundings. 150 groundings occurred in the Gulf with 23 total losses. Most groundings (58%) occurred in calm seas during clear to partly cloudy weather. The largest single cause was personnel fault (62%) followed by navigational error (44%).

Explosions and Fires. 117 occurred in the Gulf; 63 resulted in total losses. The largest single cause listed was equipment failure (39%), followed by electrical equipment malfunctions (38%).

Flooding, Foundering and Capsizing. 282 occurrences in the Gulf resulting in 97 losses—the highest number of vessels lost. The largest single cause was listed as "unknown" (24%), followed by personnel fault (18%) and weather and equipment failure (15% for each).

Material Failure. 192 reports in the Gulf included 24 resulting in total losses. The largest single cause was failure of on-board equipment (75%). Structural failure accounted for 75% of the total losses and 70% of the deaths; however, this category contained only 19% of the total incidents.

Overall, shrimping vessels recorded the highest number of total losses with flooding, foundering and capsizing, and structural failure being the most common causes of loss. Shrimp boats accumulated the greatest number of fatalities.

TOWARD A SAFER VESSEL

Why Have a Company Safety Policy?

A policy statement from the owner or manager of a fishing vessel operation establishes authority, responsibility, accountability and endorsement of the company's safety program. The company policy should be publicized and promoted to crewmen to both reduce human suffering resulting from accidents and lower insurance costs and indirect costs of accidents. A safety policy is important because:

- A safety policy makes it easier to enforce safe practices and conditions aboard vessels.
- It makes it easier for supervisors to implement company policy.
- It makes it easier for crewmen to follow safety rules and instructions.
- It makes it easier to justify preventive maintenance for equipment and to select proper equipment at the time of purchase.

A company safety policy declaration should include these ideas:

- The safety of crewmen, the public and of company operations are most important.
- Safety takes precedence over expediency or shortcuts.
- Wherever possible, attempts will be made to reduce the chance of an accident.
- The company will comply with all governmental safety rules and regulations.
- Cooperation and active support for the safety program will be expected from each crewman.

How to Identify and Hire Safe Personnel

It is extremely difficult to identify an accident-prone job applicant, but every effort should be made to hire individuals suitable for maritime employment, who are mentally and physically capable of performing their duties without jeopardizing their personal safety or the safety of fellow crew members and the vessel on which they are employed. An effective safety program starts with a careful review of each application for employment. Elimination of those applicants who appear to be physically or mentally unsuitable for the job, or whose records indicate a history of accidents, can reduce the potential for accidents.

The following procedures are recommended for evaluating job applicants:

Use a "new crewman checklist" as part of the hiring process; keep it with the crewman's personnel file. Include in it all information gathered from the following procedures.

- Have all applicants complete a standard application-for-employment form, designed specifically for your company.
- Carefully review the application form and conduct a personal interview to appraise the applicant's suitability for employment. This will help determine mental ability, past job experience, and appropriate physical condition to perform the job.
- Check all injury records and past performance records on the application. Accident repeaters can be eliminated by checking past performance records, which are maintained by industry trade associations.
- Contact the applicant's employers of the previous 5 years by telephone or letter to find out about the applicant's work habits, type of work, job skills, injuries, claims, settlements, and reasons for leaving.
- Require applicants to take a physical exam performed by a doctor familiar with requirements for fishing vessel operations. A medical authorization form (see Figure 35.10) may be useful in gaining background medical information.
- Assign the new crewman to a vessel according to his experience. The less experienced crewman should be assigned to a vessel whose captain is qualified to supervise and train the new crewman.
- After a suitable period of time, have the crewman's supervisor evaluate him for his attitude and capability, compatibility with crew members, work habits, safety consciousness, and safety awareness and practices.

Safety Training and Orientation

Fishing operations often have difficulty offering safety orientation and training. New personnel are hired and assigned to a vessel with very little information about the operator, the crewman's duties, safety procedures or benefits. A common practice in the fishing industry has been to ignore any orientation or training and simply tell a new crewman to board a fishing vessel at a designated point. No concerted effort has been made industry-wide to establish procedures for orientation and assignment of new personnel.

New crewmen, however, should be given a thorough orientation—particularly in the areas of safety and specific duties and responsibilities.

Topics to be covered in an orientation include:

- Where and how to join the vessel.
- How long the fishing trip will last.
- Sleeping and eating arrangements.
- Type of clothing and shoes to wear.
- Additional equipment required.
- How to get along with the other crew members.
- Safety precautions to follow.
- Recognition of possible safety hazards aboard the vessel.
- Emergency procedures and basic first aid.
- Emergency communication.

The safety orientation should begin with a tour of the vessel and an explanation about the use of safety equipment aboard. Include a demonstration whenever possible. Point out to the new crewman the more common safety hazards aboard the vessel; recognition of these may help him avoid an accident. All crew members should be aware of basic emergency procedures and first aid in case something happens to the captain and other crew members. Familiarize each new crewman with emergency communication procedures and communication equipment.

The new crewman should read all rules and regulations that are important to his safety while he is aboard the vessel and should sign a statement saying that he has read and understands these rules and regulations. Date the statement and keep it on file.

To be sure crewmen follow the guidelines of the safety program, the owner/operator needs to establish these basic procedures:

- Plan all work to minimize the possibility of personal injury, property damage and loss of productive time.
- Maintain a system for prompt detection and correction of faulty equipment and unsafe procedures and practices.
- Enforce the use of personal protective equipment, safety equipment and mechanical guards.
- Establish an effective program of vessel and equipment inspection and maintenance.
- Establish an educational program to promote cooperation and interest by investigating all accidents to determine cause and then taking corrective action; posting and enforcing minimum safety requirements; and promoting the safety program through safety meetings, safety incentive programs and other safety materials.

Each new crewman should be given a set of general safety rules and any other safety materials when he is hired. Duties and responsibilities of the captain and crew members should be outlined in clear, easy-to-understand terms with the captain having major responsibilities to enforce the company safety program and safety rules while the vessel is at sea.

Making the Work Environment Safe

The duty of every vessel owner is to furnish a seaworthy vessel for his crew, keeping the work environment as free as possible from recognized and avoidable hazards that could cause physical harm. Many accidents can be avoided by close observation of work practices, by attention to design and details of the equipment, by safety education and by promotion of safety awareness among crew members both for their own safety and the safety of others aboard.

Accident prevention efforts should include:

Inspection of Equipment. Frequent and regular inspection of marine equipment should be performed with corrective measures for unsatisfactory equipment. Persons responsible for vessel safety inspections should look for hazardous conditions that already exist as well as conditions that are potentially hazardous. (See the checklists in this chapter for examples of safety inspection forms.) Common sense, job knowledge and awareness are most important in making these inspections. Potential hazards can be identified only by an inspector who is aware and knowledgeable about the type of vessel being inspected and the operation of the equipment on that vessel. The vessel's crew represents another good source of information on the safety hazards that exist or may develop on board. Existing or potentially unsafe conditions, particularly those conditions identified by the crew, should always be corrected at the earliest opportunity.

Safety Education Program. A safety education program should be directed to vessel personnel pointing out the mutual safety benefits for crew and owner. A sincere interest in the welfare of the crew by the vessel owner or operator is of major importance in gaining support for a safety education program. Safety posters pointing out potential hazards aboard ship or listing safety rules or tips can be placed in prominent locations.

Safety Incentive Program. Some form of safety incentive program with bonuses, prizes, or awards should be given to vessel captains and crew members who have not had a "lost time" accident. These incentive programs should be well publicized to crew members.

Safety Training Program. Safety training, conducted on a regular basis, should include emergency drills, safety meetings and safety equipment demonstrations. All crew members should participate in all drills, meetings, and vessel inspections. Knowing how to handle emergency situations can often be the difference between losing and saving a life. A printed list of emergency procedures should be prominently displayed on the vessel, and each crew member should be made aware of these procedures. Specific emergency procedures should be developed for fires aboard the vessel, abandoning ship, man-overboard, use of emergency lighting and power, adverse weather conditions, rules-of-the-road, galley safety, and general first aid and medical treatment.

Safety Recommendations. Safety recommendations should be made periodically. Recommendations are developed by vessel inspectors, marine surveyors, insurance claims people and others checking the vessel for safety equipment and general condition. These recommendations often include more detail than required, though frequently they present just enough information to prevent an accident or serious injury.

How Safe Is Your Equipment?

Recommendations for Gulf of Mexico shrimping vessels:

- Carry on board a minimum of one 30-inch diameter U.S. Coast Guard approved life-ring buoy with 75 feet of 3/8-inch diameter line attached.
- Carry on board one U.S. Coast Guard approved life-raft float certified to accommodate all persons on board.
- Install one, two, or three portable fire extinguishers, each with a minimum U.S. classification of 16-B:C in engine room entrance, galley-mess area, pilot house, or crew berth area.
- Carry on board six day-night type U.S. Coast Guard approved distress flares.
- Install locking devices on packing gland nuts to prevent possible loosening.
- Install a marine radiotelephone near the helmsman position.
- Carry on board one extra anchor and anchor rope of suitable shape and size to properly anchor vessel. Secure bitter end to vessel.
- Install flex line between fuel feed line and fuel supply at engine.
- Install shutoff valve at galley range.
- Renew butane feed line from supply tanks to galley range.
- Keep watertight doors closed and secured while not in use.
- Install an audio bilge alarm system to indicate excessive amount of bilge water.
- Secure batteries in acid resistant container.
- Secure wiring with fireproof clips.

- Protect all electric circuits with breaker switches or fuses.
- Periodically have qualified serviceman check fire extinguishers.
- Exercise care in preparing surfaces for painting.
- Upgrade protective coatings.
- Install guards on exposed machinery.
- Maintain navigation lights in accordance with U.S. Coast Guard regulations.
- Carry a minimum of three crewmen including captain.
- Carry on board, in addition to the two engine-driven bilge pumps, one portable pump with a minimum 2-inch diameter. Locate in protective container, preferably on top of deck house and secure. Pump should have 20 feet of 2-inch suction hose with strainer attached and 10 feet of 2-inch discharge hose.
- Have engine room visually checked by responsible crew personnel every 4 hours while at sea.
- Install bilge piping system independent of seacock suction.
- Have shutoff valve for sea water suction fitted at through-hull fitting and connected to independent bilge pump with discharge outlet at least 2 inches above vessel's loaded water line.
- Have all crew personnel wear U.S. Coast Guard approved work vests while on weather deck.
- Carry bulkhead-mounted, battery-powered emergency lights on board. Locate one in deck house and one in engine room.
- Clean bilges of oil deposits.
- Clean propulsion engine of oil deposits.
- Insulate exhaust pipe at bulkhead.
- Construct tail shaft of bronze or stainless steel in lieu of cold roll steel.
- Install anchor line hawse pipe on foredeck. Store anchor rope below deck.
- Disconnect master switch on electric circuit while vessel is unattended.
- Keep all weather deck hatches, doors, and port lights secured while at sea.
- Carry on board one U.S. Coast Guard approved life preserver for each person on board.
- Install deck-mounted cable guides near winches in order to properly guide cable on winch drums.
- Install guardrails all about the deck winch to prevent possible personal injuries to the operator.

Fishing the Safe Way

General safety rules for the fishing industry are difficult to present in a form that will capture the attention of the crew and inspire compliance. Printed and posted safety rules, tips or guidelines are as much a part of a safety program as accident reporting or a safety policy.

The safety tips presented here were compiled in cooperation with the Texas fishing industry, the U.S. Coast Guard, the Marine Advisory Service at Texas A&M University, the Texas Shrimp Association and other fishing industry groups.

Safety Tips for Crew Members

- No alcoholic beverages or drugs. If you are taking prescribed medication, inform the captain and show him the prescription.
- Never smoke in your bunk, in any confined area or while fueling at oil docks.
- Take your time as you move about the vessel. Always use handrails.
- Never run up or down steps and never run on deck. Avoid "horseplay", wrestling, running, or jumping aboard the vessel.
- Always wear life vests or life jackets during rough seas.
- Learn how to lift, load, and unload cargo. Bend your knees, not your back.
- Never stand under stressed rigging. Do not walk on or straddle rope. Never stand in the loop of a line.
- When tying up, always place your hands over the line but never beneath the line. Keep fingers from in-between lines and solid objects. Make certain adequate slack is present in anchor line before making wrap on anchor cleat.
- Always face the ladder when climbing up or down. Keep stairs and ladders clear of tripping hazards.
- Know the location of all fire extinguishers and how to operate them. Advise operator when extinguishers have been used or discharged.
- Never walk barefoot aboard the vessel; always wear appropriate boots or deck shoes.
- Become familiar with emergency procedures and all alarms and whistles.
- Do not lean over edge of vessel to grab a line. Use a boat hook.
- Beware at all times of slick decks, open hatches, loose or swinging rigging, and loose lines or gear on deck.
- Keep all watertight doors and hatches closed when underway and in open waters. Replace and re-secure all manhole covers, hatch covers and deck plates prior to getting underway.
- Do not obstruct passageways with gear or cargo.
- Learn the location of the vessel's first-aid kit and use it when needed. Report all injuries, scratches, cuts, burns, sprains, etc., to the captain at once, no matter how minor.
- Do not remove guardrails or other safety guards from around winch, power takeoff and chain or belt-driven equipment.

- Do not jump from vessel to dock before vessel has come to a complete stop. Use rubber tire as a step.
- Use hand tools properly. Keep tools clean. Check their condition before use and do not carry sharp tools in pockets. Use the right tool for the job. Ground all portable electric tools.
- Do not discharge oil or oily waste into water. Violators are subject to stiff penalties.
- When repairing, checking, oiling, cleaning, or adjusting equipment, be sure equipment is turned off and that the switch will not be turned on by other crew members.
- Avoid loose clothing and loose fitting rain gear near winches and chain, belt- or gear-driven equipment.
- In the galley be a good housekeeper. Be careful with knives, keep pot handles turned away from front of stove, keep cabinet doors and drawers closed, and clean up all spills immediately.
- Wear safety glasses when performing chipping, grinding, or other work which produces flying particles.
- When pulling in the net, look out for catfish, hardheads, stingrays, sharks, jellyfish, and other marine organisms that may cause injury. Report any injuries at once for proper medical attention.
- When fishing at night be sure to have proper light array and be cautious of platforms and other fishing vessels in the area. Use radar if it is available and have the wheel manned at all times.
- Do not use hands or feet directly on towing wire to guide cable on winch drums.
- On vessels with sliding pelican hooks, take care not to be underneath when lowering.
- Wear safety hard hats when working around or under the power block.
- Do not swim off the boat.
- If defective or damaged equipment is noted, or if any hazardous or dangerous condition is discovered, notify the captain immediately. Keep safety in mind at all times. Know the proper way to perform your job.
- Always keep holds clear of trash and greasy rags, and wipe up all oil or fuel spills.
- Do not go into the fish or shrimp hold until proper ventilation is assured and no dangerous or poisonous gases are present. Do not go into the hold without notifying a fellow crew member. Follow manufacturer's recommendations on use of shrimp dip or other chemicals. Do not stand on bitts or elevated air passages to throw a line on or off a cleat, dolphin, bitt or piling.

Checklists to Spot Potential Hazards

A good vessel safety-inspection form provides an effective method of checking for hazardous and potentially hazardous conditions on a fishing vessel. Blank copies should be provided to personnel, whose responsibilities include safety, to help them carry out periodic inspections. Each form should be signed, dated and submitted to the vessel owner. Common sense, job knowledge, and awareness are important tools in these inspections, made by either vessel- or shore-based personnel.

Each checklist is designed primarily as a guide for conducting a thorough inspection. Existing hazardous conditions can be determined easily by checking items on the list, but the less apparent, potentially hazardous conditions will probably require the trained eye of an inspector who is familiar with and knowledgeable about the type of vessel and the nature of its operations.

Several examples of checklists are provided, but the best checklist is the one you design specifically for your vessel. There are six basic areas to check on any vessel—safety equipment and alarms, deck, engine room, living quarters, galley and wheelhouse. Inspect each of these areas *before* designing a vessel safety inspection form for your boat.

Figure 35.2 is a vessel inspection report form which can be used for almost any type of boat. It is very general, but it covers most items to be checked.

We believe the health and safety of crewmen should receive first consideration in our activities. The protection of our most important asset, our people, is of major concern to us all. At production facilities and aboard our vessels the personal safety of our crewmen takes precedence over operating productivity, whenever necessary.

It is our desire to eliminate accidents that can result in painful loss or suffering to crewmen or damage to company property. The "cost" of any accident is always too high because it can mean possible loss of earnings and discomfort to crewmen. It can have an unfavorable impact on company profits, which in turn can also adversely affect crewmen's jobs.

Our goal regarding safety first can take place only if the frequency of accidents is kept to a minimum. Each supervisor is charged with seeing that operations under his supervision are performed safely and efficiently. A high degree of safety consciousness among crewmen and wholehearted cooperation with safety procedures, regulations, and standards are necessary for the common good of all concerned.

President's signature

Figure 35.1. Sample safety policy statement.

Figure 35.3 is a vessel inspection safety checklist for a large fishing vessel. It is fairly specific as to the items and areas to be checked.

Figure 35.4 is a vessel inspection form to be completed by the captain. Although it is very general in some areas, the form is specific in those areas with which the vessel captain should be most familiar.

Figure 35.5 and 35.6 are safety equipment checklists to be used for an overall inspection of any type vessel. These lists help the inspector determine which vital items are not aboard the vessel.

Figure 35.7 is a checklist similar to one used by the Massachusetts Inshore Draggermen's Association (MIDA) to qualify for special insurance discounts. At the request of the owner, members of the MIDA Safety Committee inspect each vessel to qualify the vessel for an insurance renewal discount or dividend. The list is reasonably easy to use and covers most of the safety equipment found on fishing vessels.

Figure 35.8 is a motor vehicle inspection form to check automobiles or trucks which transport crews or equipment while in port.

Figure 35.9 is an example of a pre-employment medical report suggested previously in the section on how to identify and hire safe personnel.

Vessel _____

1. **VESSEL IN GENERAL**   GOOD FAIR POOR
   a. Hull
   b. Docks
   c. Bulwarks
   d. Ladders and stairways
   e. Handrails
   f. Watertight doors
   g. Vents, plugs, covers
   h. Storage of flammables
   i. Other

2. **ENGINE ROOM**   GOOD FAIR POOR
   a. Guards (pulleys, v-belt, etc.)
   b. Relief valves (tanks, hot water heater, etc.)
   c. Deck plates
   d. Insulation
   e. Bilges
   f. Leaks (fuel, oil, etc.)
   g. Other

3. **ELECTRICAL/LIGHTING**   GOOD FAIR POOR
   a. Lighting
   b. Protective globes/cages
   c. Small appliances/tools
   d. Emergency lighting
   e. Searchlight(s)
   f. Running lights
   g. Position lights
   h. Navigation equipment
   i. Other

4. **LIVING QUARTERS**   GOOD FAIR POOR
   a. Galley in general
   b. Stove
   c. Refrigerator
   d. Freezer
   e. Sleeping quarters
   f. Heads
   g. Other

5. **SAFETY EQUIPMENT AND ALARMS**   GOOD FAIR POOR
   a. Life preservers — How many? ___
   b. Work vests — How many? ___
   c. Life rings
   d. Life raft or float
   e. Life ring lights
   f. Fire extinguishers — How many? ___
   g. Fire hose
   h. Fog bell
   i. Whistle
   j. Flares
   k. Radios
   l. First-aid kit
   m. Bilge pump
   n. Other

REMARKS/SUGGESTIONS: (Explain any item marked FAIR or POOR.)

_____
_____
_____
_____

Captain's Signature _____

Date _____

Figure 35.2. Vessel Inspection Report Form.

Vessel _____ Date _____ Inspected by _____

DECK	OK	NOT OK	ACTION/REMARKS
Davits (check wirerope, hooks, hydraulic system)			
Deck free of oil/grease			
Fire hose stations			
Life rings with line			
Mooring lines			
Hatch covers			
Flammable liquid storage			
ENGINE ROOM			
Machine guarding (belts, chain drives, shafting)			
Tools secured - condition			
Deck/ladders free of oil/grease			
Flammable liquid/solvent storage			
Fire extinguishers			
Electric wiring			
Ammonia gas masks			
Bilges clean			
Metal receptacle for oily rag disposal			
Watertight doors (check dogs and gaskets)			
Steering gear, cables			
Parts stowage - secure			
Eye protection (provided/worn)			
Hearing protection (provided/worn, signs posted)			
Fixed CO_2 system			
FISH HOLD			
Adequate lighting			
Adequate ladders			
Adequate stanchions and pen boards			
Hatch covers			
Routing of shaft			
Access to pipes and lines			
Adequate ventilation			
Non-skid coatings			

Figure 35.3. Vessel Inspection Safety Equipment Checklist.

Vessel_____ Date_____ 19____		Voyage_____ Time_____ AM/PM__	
ITEM	SAFE/OK	NOTE DEFECTS, COMMENTS	DATE CORRECTED
ENGINES & EQUIPMENT			
BILGE PUMPS & BILGE CONDITION			
BATTERIES — WATER/CHARGE			
ELECTRICAL SYSTEMS			
STUFFING BOXES			
FUEL AND SYSTEMS			
RADIO FREQUENCIES			
WORKING			
DISTRESS			
RADAR AND AUTOMATIC PILOT			
RUNNING AND OTHER LIGHTS			
CHARTS AND INSTRUMENTS			
COMPASS AND DEVIATION CARD			
MEDICAL SUPPLIES			
EMERGENCY SIGNALS AND FLARES			
ALL ALARMS			
FIRE EQUIPMENT			
SAFETY EQUIPMENT			
LIFE VESTS			
LIFE RAFT			
HULL CONDITION			
TRAWLING EQUIPMENT			

Vessel inspected as above is properly equipped
and considered ☐ Safe ☐ Unsafe for voyage

Signature _____ Title _____

Figure 35.4. Captain's Report Form.

_____ 1. Ring buoy/life raft/distress flares

_____ 2. Fire extinguishers

    ___a. Engine room

    ___b. Galley area

    ___c. Pilot house

    ___d. Crew berth

_____ 3. Radio telephone near helmsman position

_____ 4. Additional anchor

_____ 5. Shutoff valve at galley range

_____ 6. Renew butane feed line from supply tanks to galley range

_____ 7. Bilge alarm system

_____ 8. Batteries in acid resistant container

_____ 9. Wiring: fireproof chips

_____10. Breaker switches or fuses

_____11. Guards on protective covers on exposed machinery

_____12. Navigation lights: U.S. Coast Guard regulations

_____13. Portable bilge pump

_____14. 4-hour intervals on engine room check

_____15. Bilge piping system independent of seacock suction

_____16. Battery-powered emergency lights: deck house and engine room

_____17. Clean oil deposits: bilge propulsion engine

_____18. Insulate exhaust pipe at bulkhead

_____19. Bronze or stainless steel tailshaft

_____20. Cable guides (deck mounted) near winches

_____21. Guardrails around deck winch

Vessel _____ Inspector _____

Figure 35.5. Vessel Inspection Safety Equipment Checklist.

Vessel _____ Date _____ Location _____

NAVIGATION AIDS	OPERATIVE	NEEDS REPAIR	CONDITION	COMMENTS
Alarm system				
Bell (anchor, fog)				
Compass light				
Engine emergency shutdown				
Foghorn				
Navigation lighting				
Emergency lighting				
General lighting				
Pump system-bilge				
Pump system-ballast				
Pump system-fire				
Radar				
Radio				
Winches (anchor)				
Anchor line				
Rope condition				
Lights				
Chains				
Exhaust system working				
Check hatches and moldings				
Check handrails				
First-aid kits				
Check engine for leaks				

Figure 35.6. Vessel Inspection Form.

Figure 35.6 (*continued*)

	OPERATIVE	NEEDS REPAIR	CONDITION	COMMENTS
Check hull for leaks				
Check for rust spots				
Check window glass				
Adequate tie-up lines				
Combustible materials				
Windshield wipers				
General housekeeping				
Check captain's log				
Lids on all containers				
Check ladders				
Check electrical 3 prong				
Check fuel tank leaks				
Potholders on stove				
Check captain's chair				
Air conditioning and heating				
Batteries (lights)				
Navigation charts				
Compass				

	YES	NO	ACTION

DECK

1. Non-skid decks
   Comments:

2. Proper handrails (in way of deck house doors)
   Comments:

3. Ladders checked and secured
   Comments:

4. Safety straps on overhead blocks installed
   Comments:

5. All bottled gas to be strapped with metal belting
   Comments:

MACHINERY

1. Engine manifold cooling and exhaust properly installed
   Comments:

2. No soft plastic lines or hoses below deck where subject to pressure
   Comments:

3. Screening or filtering on bilge suction lines adequate
   Comments:

4. Alarm signals for oil pressure and oil temperature in working order
   Comments:

5. Check rigging for wear of pins, blocks and lines
   Comments:

6. Check mast and boom — safety chains used where possible
   Comments:

7. All portable equipment to have adequate tie-downs
   Comments:

8. Anchor should be proper size
   Comments:

9. Reel drive and belt guards where necessary
   Comments:

10. Guardrails on winch
    Comments:

Figure 35.7. Insurance Safety Incentive Program Checklist.

Figure 35.7 (*continued*)

Inspected by: _____
Safety group member

SAFETY EQUIPMENT
1. Fire and smoke alarms in engine
   room and galley   _____ _____
   Comments:

2. Bilge alarm   _____ _____
   Comments:

3. All safety equipment to be checked
   on a scheduled basis   _____ _____
   Comments:

4. Distress flares and beacons (if not
   included in life raft)   _____ _____
   Comments:

5. Adequate first-aid equipment   _____ _____
   Comments:

6. Emergency rations (if not included
   in life raft)   _____ _____
   Comments:

7. Automatic life raft   _____ _____
   Comments:

8. Safety glass in front windows   _____ _____
   Comments:

9. VHF radio with CH 16 capability   _____ _____
   Comments:

FIRE PROTECTION
1. Extinguishers handy in all areas of need   _____ _____
   Comments:

2. Extinguishers checked every 12 months
   or after use   _____ _____
   Comments:

3. Extinguishers NOT TO BE USED AS
   COAT HANGERS, etc.   _____ _____
   Comments:

4. Galley stove, properly insulated with automatic
   shutoff on fuel lines. No pilots
   without thermocouple shutoff valve   _____ _____
   Comments:

ELECTRICAL EQUIPMENT
1. All wiring should be permanent, with
   overload protection   _____ _____
   Comments:

2. Should have main battery disconnect   _____ _____
   Comments:

Signatures: _____   _____
Owner or captain   Vessel name

Vehicle _____ License No. _____
Inspection Expires _____ Inspected by _____
Mileage _____ Date _____
Return From Crew Change For F/V _____

Tires	Good	Fair	Poor		
LF	☐	☐	☐	Engine	Smooth ☐ Rough ☐ Burns Oil ☐
LR	☐	☐	☐	Transmission	Smooth ☐ Slips ☐ Leaks Oil ☐
RR	☐	☐	☐	Brakes	Good ☐ Pull to ___ side ☐
				Front end	
RF	☐	☐	☐	Steering	OK ☐ Play ☐ Pulls to ___ side ☐
Spare	☐	☐	☐	Other	

Windshield or glass damage _____

No glass damage ☐

ITEM	WORKING ORDER	ITEM	DAMAGED
Headlights	Yes ☐ No ☐	Top	Yes ☐ No ☐
Tail lights	Yes ☐ No ☐	Hood	Yes ☐ No ☐
Brake lights	Yes ☐ No ☐	Grill and front bumper	Yes ☐ No ☐
Turn signals	Yes ☐ No ☐	LF Fender	Yes ☐ No ☐
Emergency flashers	Yes ☐ No ☐	LH Doors	Yes ☐ No ☐
Horn	Yes ☐ No ☐	LR Fender	Yes ☐ No ☐
Windshield wipers	Yes ☐ No ☐	Trunk/tailgate and rear bumper	Yes ☐ No ☐
Windshield washers	Yes ☐ No ☐	RR Fender	Yes ☐ No ☐
		RH Doors	Yes ☐ No ☐
Speedometer	Yes ☐ No ☐	RF Fender	Yes ☐ No ☐
Jack and lug wrench	Yes ☐ No ☐	Interior	Yes ☐ No ☐

REMARKS:

Figure 35.8. Vehicle Inspection Report.

Dr. _____

Please furnish (company name and address)
_____
_____

the following report regarding my condition.

Signed _____
(To be detached and retained by reporting physician if desired)

MEDICAL REPORT

PATIENT	Name _____ Age ___ Address _____ Occupation _____ Employed by _____
HISTORY OF CONDITION	Date of Accident _____ 19 __ History as described by patient _____ _____ _____ _____ Date of your first treatment _____ 19 __
X-RAY	Date taken _____ 19 __ Where taken _____ Findings _____
DIAGNOSIS (Describe and locate character and extent of injury)	_____ _____ _____ _____ _____
CONTRIBUTING FACTORS	In your opinion, is disability solely a result of above described accident? _____ _____ _____
PROGNOSIS (Your estimate)	Total Occupational Disability ____ weeks __ days. Ended __ 19 __ Partial Occupational Disability ____ weeks __ days. Ended __ 19 __ (50 percent or more) Probable period of measurable discomfort __ weeks __ days. __ Ended __ 19 __ Estimated amount of your bill _____ If patient has been discharged give _____ and _____ Date of discharge Amount of your bill Estimated cost of medical treatment other than your own _____

Signed _____

Date _____ 19 ___   Address _____

Figure 35.9. Medical Report Form.

REPORTING AND INVESTIGATING ACCIDENTS

An accident reporting and investigation system is essential in any safety program. Not only is it often necessary to record and report employee accidents to process insurance claims and satisfy state and federal industrial safety regulations, but an investigation is also vital to determine and remove the fundamental and contributing causes of the accident to avoid recurrence. Accumulated investigation records represent historical data that can be readily analyzed to identify trends and problem areas, and provide clues to future accident prevention.

Since accident prevention depends heavily on information gathered from accident reports, it is important that personal injuries and vessel damages be reported accurately and promptly. Each new crewman should know the procedures and requirements for reporting injuries to the vessel captain.

Accident Reporting Procedures

Notify the shore side base immediately. The more serious the accident (or illness), the more important it is to report without delay. *Do not discuss fault* over your marine radio. But do include in your verbal report 1) date, exact time, and location of accident or injury; 2) name of the injured or ill person and his job aboard the vessel; 3) nature and severity of injuries or illness; and 4) description of the accident.

- If the injury is caused by cable or rope or some similar piece of equipment, save as much of the cable, rope, or equipment as possible for evidence. Turn this over to the vessel owner as soon as possible. Photographs of the scene of the accident are helpful.

- Complete a Report of Personal Injury (or illness) on each injury or illness no matter how minor. (Examples of report forms mentioned are included at the end of this section.) The forms should be completed accurately, in as much detail as possible. Space for the captain's signature should appear on the form to encourage closer supervision of the crew and continued interest in maintaining safe working conditions.

- Require each witness to an accident to complete a Personal Injury Witness Report in his own words. Additional statements from crew members who did not actually witness the accident or injury might be helpful.

- Have the injured or ill person complete a Personal Injury Report as soon as possible. The captain should make sure the report form is complete, accurate and detailed; however, he should not instruct the injured person as to what answers to give on the report.

- Enter the accident in the vessel daily log giving date, time, location and name of the injured or ill person.

- Turn over all report forms to the vessel owner or operator as soon as possible along with a cover letter explaining any inconsistencies in the reports.

- Upon receipt of these reports, the responsible personnel should review the information to satisfactorily answer these questions: 1) What caused the accident? 2) Why did the cause exist? 3) Is the accident likely to occur again? 4) What actions can be taken to eliminate the cause?

- Hazards that can result in further injuries should be corrected immediately, and items of lesser importance should be corrected at the earliest opportunity.

Loss Reporting Procedures for Physical Damages

Every case of loss or damage to vessels or to property of others should be reported promptly to the dispatcher by telephone, if possible. If not possible, use your radio, but remember, *do not discuss fault* over the marine radio. If the loss is likely to exceed $1,500, use the fastest means of reporting. The more serious the loss, the more important it is to report it without delay.

Report: 1) the date, exact time, and location at the time of loss; 2) the name and address of the owner of the other vessel or property damaged or involved (if available, give the official number and type of other vessel involved); 3) name of the captain (and names of crew, if available) of the other vessel; 4) a brief, accurate description of the casualty or loss; and 5) a description of the damage to all vessels and/or property involved along with your best estimate of the amount of damage to each.

Enter the casualty or loss in the vessel's daily 109. Prepare the vessel accident report, an official written, complete report to the vessel owner. If the vessel will be away for more than 1 week, mail the form to the owner's office.

Dispatcher Procedures Following Accident. As a minimum, the following information should be recorded in the dispatcher's log:

- The date, exact time, and location of the accident or injury.

- The name (firm) of vessel(s) involved.

- The name(s) of other vessel(s) involved. Include name(s) of owner(s) or operator(s) of other vessel(s) involved.

- The name(s) of injured person(s) and their job aboard the vessel.

- Nature and severity of injuries.

- Description of damage to vessel(s).

- Brief description of the accident.

Requirements for Reporting Marine Accidents

The owner, agent, master, or person in charge of a fishing vessel involved in a marine casualty shall give notice as soon as possible to the nearest Coast Guard marine safety or marine inspection office whenever the casualty involves any of the following:

- All accidental groundings and any intentional grounding which also meets any of the other reporting criteria or creates a hazard to navigation, the environment, or the safety of the vessel.

- Loss of main propulsion or primary steering, or any associated component or control system, the loss of which causes a reduction of the maneuvering capabilities of the vessel. Loss means that systems, component parts, subsystems, or control systems do not perform the specified or required function.

- An occurrence materially and adversely affecting the vessel's seaworthiness or fitness for service or route, including but not limited to fire, flooding, or failure or damage to fixed fire extinguishing systems, lifesaving equipment, auxiliary power generating equipment, or bilge pumping systems.

- Loss of life.

- Injury causing a person to remain incapacitated for a period in excess of 72 hours.

- An occurrence not meeting any of the above criteria but resulting in damage to property in excess of $25,000. Damage includes the cost necessary to restore the property to the service condition which existed prior to the casualty, including the cost of salvage, gas freeing, and drydock. It does not include such items as demurrage.

Fishing vessels which are not documented, but have state or Coast Guard numbers are required to report to the authority that issued the numbers whenever the casualty results in any of the following:

- Loss of life.

- Injury requiring medical treatment beyond first aid.

- Damage to the vessel and other property totals more than $200 or a complete loss of a vessel.

- A person disappears from the vessel under circumstances that indicate death or injury.

A report required by this section must be made within 48 hours of the occurrence if a person is injured and requires medical treatment beyond first aid or disappears from a vessel and within 10 days of the ocurrence of death if an earlier report is not required by this paragraph.

Investigation of Marine Accidents and Claims

Vessel owners or operators should use the following checklist in the investigation of an accident or claim:

- Investigate every accident as soon as possible. Report any claim to your attorney or insurance underwriter promptly for an early investigation; cooperate with the marine underwriter and investigator assigned to the claim.

- Statements from vessel operations personnel should be taken as quickly as possible. Always put them at ease during interviews. Get correct names, addresses, and phone numbers from all witnesses and other involved personnel.

- Be sure vessel operations personnel know the purpose of the investigation (to ensure that the same thing does not happen again.) Do not try to place the blame for the accident on anyone.

- Encourage witnesses to offer their thoughts on the accident and on how it might have been prevented.

- Do not begin the investigation with fixed opinions on the accident. Be objective in your investigation.

- Do not downplay the seriousness of any accident. An accident that results in a minor injury could, under different circumstances, result in a casualty.

- Look for unsafe conditions as well as unsafe acts.

- Photographs of the accident made for documentation will not only aid the investigation, but if properly dated and signed, can be used as evidence if a court case results.

- Summarize the findings of the investigation and make these findings known to those people interviewed. Thank those people who helped with the investigation.

- Publicize the results of an accident investigation to ensure that a similar accident will not happen again.

Analyzing Your Safety Program

For a safety program to be meaningful, it is necessary that it contribute measurable results. Comparison of accidents and injury records before and after program implementation does not always tell the whole story. Assess the types of accidents and injuries occurring. Evaluate all accidents reported and safety hazards corrected following an accident and injury analysis. Compile and analyze simple records of all accidents which occur in a company.

By analyzing injury records and frequency rates, injury causes, and peak accident periods, a general program can be plotted. This allows definite conclusions about measures to take for a maximum-effect safety program. The results should be circulated and discussed openly at all levels.

Forms to Document Accidents

The samples of accident report forms that follow offer specific sets of questions to be answered after an injury or accident. It is important to use the forms as soon as possible after an incident, so that information is accurate and detailed. Each crew member should know how to report an injury or accident. Each form should be signed, dated, and submitted to the vessel owner. Blank copies of the forms should be kept by the vessel captain at all times.

These forms not only supply information needed for insurance claims and government reports, but also provide a valuable record for vessel owners and operators. Over a period of time, the forms can be analyzed to determine the most common causes of accidents and injuries so that the company's safety policies and training sessions can be updated.

The first five forms, Figures 35.10 through 35.14, are examples of forms to be filled out in case of an injury. The reports will help give an overview of the incident and help piece together what happened and why. There are two examples of forms to be completed by the injured party and the captain. A vessel owner may want to use these forms as a guide in developing a set of forms for a particular vessel.

Figure 35.15, a non-injury statement, can be used at the end of a trip or during a crew change. Coercion should not be used to get a crew member to sign the form.

Figure 35.16, a report of physical damage, can be used by the captain to record an accident and describe the details of the accident. This becomes a record of the accident and should be kept on file for reference during the investigation of the accident.

To be Completed by Injured Party

The following detailed information is requested by owners:

Name of Vessel _____

Owner, Operator _____

Injured's Full Name _____ Social Security No. _____

Home Address _____

Age _____ Position _____ Married/Single _____

Name of Nearest Relative _____ Address _____

_____ Number of Dependents _____

Length of Employment _____ Earnings Per Month _____

Date Joined Vessel _____ Date Left Vessel _____

Date and Time of Injury or Illness _____ To Whom First Rerported _____

Location of Vessel at Time of Accident or Illness _____

Injured's Statement of How the Accident or Illness Occurred and Cause Thereof, If Known

_____

_____

_____

_____

_____
(Injured's Signature)

Nature of Injuries or Illness _____

_____

Had Crew Member Been Drinking Intoxicants? _____ Name and Address of Former Employer _____

_____

What Was Done for Man After the Accident or Illness? _____

Name and Address of Doctor and Hospital if Any _____

Condition of Injured When Leaving Vessel _____

Remarks of Captain in Charge of Vessel at Time of Accident _____

_____

Figure 35.10. Report of Personal Injury or Illness.

TO BE COMPLETED BY INJURED CREWMAN
THIS REPORT IS IMPORTANT; PLEASE ANSWER IN COMPLETE DETAIL

YOUR NAME _____

ADDRESS _____

YOUR OCCUPATION _____ HOW MUCH EXPERIENCE IN THIS OCCUPATION? _____

ON WHICH VESSEL DID ACCIDENT HAPPEN? _____

WHERE ON VESSEL DID ACCIDENT HAPPEN? _____

WHEN DID ACCIDENT HAPPEN? DATE _____ HOUR _____ AM/PM

WHAT PART OF YOUR BODY WAS INJURED? _____

DID YOU LOSE CONSCIOUSNESS? YES _____ NO _____ IF SO, FOR HOW LONG? _____

WHAT WERE YOU DOING WHEN ACCIDENT HAPPENED? _____

_____

DID ANYONE ELSE SEE THE ACCIDENT? YES _____ NO _____ IF SO, WHO? _____

WHAT WERE THEY DOING WHEN THE ACCIDENT HAPPENED? _____

_____

WAS FIRST AID NECESSARY? YES _____ NO _____ IF SO, WHO PROVIDED IT? _____

_____

NAME AND ADDRESS OF DOCTOR WHO TREATED YOU, IF TREATED _____

_____

HAVE YOU EVER HAD THE SAME OR SIMILAR INJURY BEFORE? _____

EXPLAIN _____

SIGNED, INJURED CREWMAN _____

DATE _____

CAPTAIN _____

DATE _____

TO WHOM IT MAY CONCERN:

YOU ARE HERBY AUTHORIZED TO RELEASE TO THE BEARER HEREOF ALL HOSPITAL RECORDS AND MEDICAL INFORMATION CONCERNING MY PHYSICAL CONDITION

SIGNED _____ DATE _____

Figure 35.11. Personal Injury Report Form.

Report to be submitted in all cases of injury

Crewman's Full Name _____

Position Held _____ Social Security _____

Vessel _____

How Long In This Position? _____

Date & Time of Injury _____, 19 _____ at _____ AM/FM

To Whom Reported? _____ Date Reported _____

Where Did Injury Take Place?   Vessel _____ Purse Boat _____ Ashore _____

Exact Location _____

Was Crewman on Duty at Time of Injury?   Yes _____ No _____

Nature of Injury (Please describe in detail. Use back of page if necessary.)

_____

_____

How was Injury Treated? _____

_____

Was Crewman Placed Ashore?   Yes _____ No _____ If yes, How and When _____

_____

Has Crewman Returned to Work?   Yes _____ No _____

Weather Conditions at Time of Injury:

    Wind Direction _____ Wind Velocity _____ MPH

    Sea State _____ Wave Height _____ FT

    Rainy or Dry _____ Location of Vessel _____

What Happened? _____

_____

Why Did It Happen? _____

_____

What Action Have You Taken to Prevent A Similar Occurrence? _____

_____

What Action Do You Recommend To Prevent A Similar Occurrence? _____

_____

Captain's Signature _____ Date _____

Figure 35.12. Captain's Report of Personal Injury Form.

TO BE COMPLETED BY CAPTAIN OF VESSEL

Name of Injured _____ Occupation _____

F/V _____ Enroute From _____ To _____

Date of Report _____ Date of Accident _____ Hour _____

Exact Location of Vessel _____

State What Crewman Was Doing When Accident Occurred; Give Exact Location on Boat; Draw Sketch on Back of This Report _____

_____

_____

Weather Conditions _____

Gear, Equipment, Tools or Machinery Involved, If Any _____

_____

Name of Immediate Supervisor _____ Position _____

When was Immediate Supervisor Made Aware of Injury? _____ A.M. ____ P.M. ____

Description of Injury _____

_____

Was First Aid Given? _____ By Whom _____

Did Injured Go Ashore for Treatment? _____ Where _____

Did Injured Return to Duty After Receiving Treatment? _____ If Not, Why _____

_____

What Verbal Statement Did Injured Make as to Cause of Accident and To Whom? _____

_____

List Name, Rating and Address of All Witnesses _____

_____

_____

Captain's signature _____ Date _____

Figure 35.13. Personal Injury Report Form for Captains.

To be Completed by Witness to Injury

NAME OF INJURED CREWMAN _____

OCCUPATION _____ DATE _____

VESSEL INVOLVED _____

EXACT LOCATION ON VESSEL ACCIDENT OCCURRED _____
_____

DESCRIBE IN DETAIL WHAT INJURED WAS DOING AT TIME OF INJURY _____
_____
_____
_____
_____

WHAT WERE YOU DOING AT TIME OF ACCIDENT? _____
_____
_____

HOW FAR WERE YOU FROM THE INJURED CREWMAN? _____
_____

GIVE IDENTITY OF ANY OTHER WITNESSES _____
_____
_____
_____

SIGNATURE _____ DATE _____

POSITION _____

ADDRESS _____
_____

HOME TELEPHONE _____

Figure 35.14. Personal Injury Report Form for Witnesses.

VESSEL: _____		DATE: _____
I certify that I have not had an injury during this trip:		
SIGNATURE	POSITION	CREW CHANGE From _____ To _____
MASTER'S COMMENTS		

INSTRUCTIONS TO MASTER

1) The above information must be completed immediately upon completion of crew change.
2) It should be attached to the vessel log and turned in along with the log.
3) If any crew member reports an injury, a personal injury report must be completed immediately and turned into the general manager or designated owner's representative.

Figure 35.15. Non-Injury Statement Form.

Name of Vessel _____

Owner/Operator _____

Date and Time of Incident _____ Location _____

Wind _____ Seas _____ Visibility _____

F/V Enroute From _____ To _____

Name of other Vessel(s) or Property Involved _____

Owner of other Vessel(s) or Property _____

Name of Captain of other Vessel _____

Describe the Incident _____

_____

_____

_____

_____

_____

_____

Describe Damage to all Vessel(s) and/or Property _____

_____

_____

_____

Names of Personnel Injured (also fill out Report of Personal Injury or Illness form)

_____

_____

_____

_____

_____
Captain's Signature

Use Reverse Side for Additional Information or Diagrams

Figure 35.16. Report of Physical Damage Form.

SURVIVAL AT SEA*

Most of this section has been taken directly from a University of Alaska Sea Grant Publication entitled, "Safety Notes for the North Pacific Fisherman," Marine Advisory Bulletin No. 3, March 1975.

No vessel owner or operator deliberately jeopardizes his own or his boat's safety, and every experienced vessel captain considers himself competent in operating his vessel. However, accidents can happen anywhere, at any time, to even the most experienced mariner. So it is important to know how to deal with an emergency situation, e.g. how to survive if the vessel is heavily damaged or lost, or what to do if a crew member is injured or ill.

This section deals with emergency procedures and survival techniques and equipment.

Emergency Procedures

Every case of sickness or injury has its own special problems but these general rules apply to almost every serious situation.

- Get the victim to a doctor. Use your radio to get medical advice from the Coast Guard.
- Don't push the "panic button"—keep calm.
- Don't worry about using manufactured first-aid materials—do the best you can with what's handy.
- Keep the victim lying down with his head on the same level as his body until you know the extent of his injury; don't move him more than needed.
- Look for signs of serious bleeding, difficulty in breathing, shock, burns, and broken bones. Be sure you find all injuries. While looking for (or treating) injuries you may have to remove some of his clothing. Be as gentle as you can you may have to cut the clothing off.
- Don't remove more clothing than necessary because it may increase the chance of shock. Be sure there is no tight clothing around his neck, chest, stomach, legs, or ankles.
- Treat the injuries in order of importance: 1) bleeding, 2) breathing, 3) shock. Bleeding and "no breathing" are the two biggest worries. Shock will usually follow all serious injuries. Remember these injuries must be treated immediately—delay can cause death.
- Don't move a person with a broken bone until the broken area is stabilized by a splint.
- An injured person sometimes will vomit. If this happens and the person is lying down, turn his head to the side to prevent him from choking.
- If the victim is unconscious, vomiting, or has been injured in the chest or stomach, do not try to give him liquids.
- Keep the person warm.
- Do not move the injured person unless you have to.

More information on how to deal with specific first-aid procedures can be obtained from a first-aid manual. The Marine Advisory Service at Texas A&M University has an abbreviated, easy-to-read chart entitled "Emergency First Aid" which can be ordered by any vessel owner or operator. Written in both English and Spanish, it serves as a ready reference in the event of an emergency aboard a vessel.

Marine Organisms

Injuries can occur not only from ship-related activities, but also from numerous marine organisms in the Gulf which can inflict painful and serious wounds.

Sharks, even small ones, with their razor-sharp teeth, are capable of serious injury, and "dead" sharks have reflex actions which can close their mouths on hapless hands or feet.

Smaller organisms, including jellyfish, rays, sea catfish, and other spiny-rayed fishes, may pose problems for fishermen sorting the catch or bringing nets aboard. A strong ammonia solution should be applied to areas stung by jellyfish tentacles. Puncture wounds inflicted by the dorsal spines of sea catfish and spiny-rayed fishes should be soaked, for at least 2 hours, in a bucket of hot water with a liberal amount of Clorox, Purex, or other disinfectant added. The slime on these spines, that protects the fish, causes rapid infection in human wounds.

While the dorsal spine of a sea catfish can pierce the bottom of a tennis shoe, rays can leave the sheath from their stinger embedded in a man deeply enough to require surgical removal. A wound caused by a ray should be treated by a physician as soon as possible.

Not all injuries are as serious as these, but any wound should be cleansed with an antibacterial agent and covered with an antiseptic.

Fumes

Even after you've sorted out hazardous marine organisms, your catch is still a source of possible injury. The bacteria associated with fish and shrimp degradation, *pseudomonas putrefaciens,* can become concentrated in the bilge of the ice hold where it thrives in organic rich waters of the drip from your catch. As the bacteria proliferates in the water, substantial quantities of oxygen are consumed, and when the level of oxygen reaches anaerobic conditions, the bacteria begin to generate hydrogen sulfide.

Hydrogen sulfide is highly irritating to the body mucosa—eye irritation results from even low concentrations, while irritation to the upper respiratory tract results from slightly higher concentrations. In low concentrations, hydrogen sulfide provides a warning odor suggestive of rotten eggs. At larger concentrations, however, the gas desensitizes the sense of smell. Thus an individual could become exposed to a lethal dose without any forewarning. Exposure to high concentrations acts as a systemic poison on the nervous system and causes respiratory arrest, coma and death. In its toxic affect on humans, hydrogen sulfide is as lethal as hydrogen cyanide.

To prevent toxic accumulations of hydrogen sulfide, fish holds should be vented prior to entry and the ice hold bilge kept as dry as possible.

Burns

Minor burns should be cooled with tap water to remove heat from the area. Apply commercial topical dressings to assist in healing. Serious burns require prompt and special attention. *Do not attempt any treatment of the burn itself.* The burned area should be covered with sterile, lint-free material and soaked with sterile water if possible (if sterile water is not available, any potable water may be used). Treat for shock immediately and get medical attention as soon as possible. Give the person as much fluid as possible to counteract the loss of body fluids. Any person burned over one-tenth or more of his body will usually go into serious shock and the result of shock will be the key to whether he lives—so treat for shock.

Crushed Appendages

An all too common injury, this usually occurs when someone puts his hands or feet where they don't belong. Unless the appendage is severely traumatized, the only first aid available is cold compresses and elevation of the injured area. If within 2 or 3 days the effected area has lost its feeling, a physician should be contacted.

Fires

The best way to fight a fire is to remove every possible cause of fire you can. Remember a fire needs three things: something that will burn, air (oxygen), and enough heat to start it. Carbon dioxide (CO_2) extinguishers have stamped into them how much they should weigh when full—a fish scale will tell you when one needs a recharge. Dry chemical extinguishers are excellent for all types of fire. They have gauges or their tops unscrew to show the brass seal. (A broken seal means the extinguisher is no good.)

If you have a "Class A" fire (paper, clothes, wood, bedding, etc.) your best move may be to simply try to dispose of the burning material overboard. (Mattresses afire, for instance, are extremely difficult to extinguish.) If it is a "Class C" or electrical fire—cut off the current. (You may have to secure your engine and pull your battery cables.) If it is a "Class B," or fuel fire, it will probably occur in the bilges where spilled or dripped fuel collects, and where fumes (which are heavier than air) collect.

Depending on its location you may or may not be able to secure your fuel line to the engine (a good early move). For any type of fire either carbon dioxide (CO_2) or dry chemical is good extinguisher. Aim your nozzle at the base (source) of the flames, and rapidly sweep the nozzle back and forth—"pushing the flames into a corner." *Don't* aim the nozzle steadily into the center of the fire. Remember, both types of extinguishers go empty very quickly. Also, remember a fire can "reflash" even after it seems to be completely out. Embers may be smoldering and fuel in the bilges may be hot enough to start burning again. Electrical shorts can restart fires.

A fire inside a boat can produce deadly carbon monoxide gases in addition to heavy smoke. Once the fire is completely out, thoroughly ventilate the area.

A firefighting guide entitled, "Marine Fires: Preventing Them, Fighting Them" is available from Oregon State University as published by the Oregon State Sea Grant Program. Table 35.1 provides basic information about various firefighting systems as reproduced from that publication.

SURVIVAL TECHNIQUES AND EQUIPMENT

Abandoning Ship

When all attempts to save the vessel have been exhausted and the crew is in jeopardy, you may decide to abandon the vessel. You will normally have a few minutes to "get organized" before having to abandon. First, launch your rubber raft or skiff downwind. If you have a raft, push the plunger on the release mechanism to free the grips. Pick the raft up by the finger slots at each end and toss it overboard, after being sure the cord on it is secured to your boat (it's a good idea to keep it secured always). Unless you have shortened the operating cord for some reason, you will have to pull about 100 feet of cord before being able to give the sharp tug needed to inflate the raft. Inflation takes about 30 seconds, and the cord acts as a painter. Meanwhile be sure that your life jacket is securely fastened. If additional food, liquids, blankets, or clothing can be put in the raft or

Table 35.1. Basic firefighting systems and how to use them.

System	Advantages	Disadvantages	Types of fire to use it on	Where to use it
WATER (bucket, hose, hand pump)	Cools the fire.	Heavy. Can cause drastic change in vessel stability — and possibly capsize your vessel. Not to be used near engine air intakes or on electronic, electric, galley (grease), and oil-heating stove fires.	Wood, mattresses, paper.	Topside.
WATER (under pressure and applied to fire through a nozzle that emits a spray)	Cools the fire. Spray allows firefighters access to burning area.	Difficult to use on grease or oil fires.	Wood, paper, rags. To cool oil fire.	Berthing spaces. Engine-room spaces.
CARBON DIOXIDE (CO_2) (extinguisher)	Compact. Easy to use. Easily transportable throughout the boat. Engines and electronic gear easy to clean after being sprayed with CO_2.	Does not cool the fire, setting up the possibility of a reflash. Requires space to be secured (no drafts or ventilation). Blanket must sometimes be left in place for extended periods. *Cuts down oxygen;* firefighters will collapse if they inhale too much CO_2. Not good for topside fires.	Electrical, electronic, engine room fires; oil and grease (be careful not to spread the fire by blowing hot oil or grease all around with CO_2).	Electronic spaces, engine room, galley.
HALON (FREON) (extinguisher)	Same as CO_2 but less hazard to firefighters after breathing. Heavier than CO_2 — somewhat more useful than CO_2 topside and in drafty spaces.	Expensive to recharge. Possibility of reflash if blanket is not left in place long enough.	Electrical, electronic, engine room, oil and grease.	Electrical and electronic gear, engine room.
CARBON DIOXIDE (CO_2) HALON (FREON) (built-in system)	Fast extinguishing for inaccessible areas; can be triggered remotely — if properly installed; allows firefighters to keep away from fire.	More expensive; must be installed by a professional; only covers space where installed; needs periodic check to insure that no modification to boat has made system inoperable.	Electric, electronic, engine room, oil and grease.	Engine room; confined space where electronic gear is concentrated.
DRY CHEMICAL	Compact. Easy to use. Easily transportable.	Bottle should be vertical for most effective discharge. After-fire cleanup of chemical powder is time-consuming and difficult. Electronic gear is almost impossible to clean after exposure to dry chemical.	Engine, oil and grease, electrical.	Galley (unless the galley is close to electronic gear); engine room.
PYRENE	DO NOT USE.	When put in contact with fire, phosgene gas is emitted — *toxic to those in the vicinity.*	DO NOT USE.	DO NOT USE.
CAN OF BAKING SODA	Cheap, effective, easy to clean up.	Difficult to apply to fire without sustaining hand burns.	Grease fires on galley stove tops.	Galley.

skiff, do so. Don't be overly choosy—there isn't that much time. (Even canned turnips taste good on the third day adrift.) Be sure you have your knife with you.

A life raft is a wonderful piece of equipment. But like any other piece of equipment, it can be no better than its operator. The life raft will serve its purpose—which is to save your life in case of a shipwreck—only if you, the operator, handle it properly.

A copy of the instruction manual supplied with the life raft should be available on board the vessel, and should be studied—not only read, but studied—not only by the skipper but by every man in the crew. Because the vessel's life raft is in its capsule, ready to go in case of need, it is not available for drills. Life raft boarding/behavior demonstrations (by dealers, inspectors, Coast Guard, etc.) should be attended whenever possible by all potential users, skippers and crew alike. Not that such a demonstration is as valuable as a drill. But it is the best substitute available.

A raft canopy—which automatically raises—is strong enough to jump on to board. When jumping, be aware of the canopy fixtures: protect your face with your arms; throw the trunk of your body onto the canopy. To minimize the height of the jump, watch the waves, and hit the canopy as the raft surges up on a sea. When all are aboard, ride out the full length of the painter. Unless you are riding badly (e.g., shipping water), keep secured to the painter. The painter holding you to your boat is designed to break at a point much less than would pull you under. There is no danger of getting "sucked under," though you should be far enough out to ensure against being struck by masts, booms, or trolling poles should the boat roll over on its way down. There are a couple of good reasons for keeping attached to your boat. First—many vessels have been abandoned, only to stay afloat due to some inexplicable quirk, and later be saved. Second—if help is on its way, you'll be closer for a little longer to your broadcast position, and as long as your boat still floats (even if lying on its beam) you'll be easier to spot from air and sea, both visually and on radar.

When and if you are drifting "free" in the raft, check to be sure that your "sea anchor" (drogue) is not fouled. It will stream automatically on many rafts. In addition, most rafts have a spare packed aboard. When it is doing its job, you'll drift less than 1 knot in a 40-knot wind.

More often then not, a skipper has a certain "feeling" that abandoning may be necessary long before the word is given. Why not "swallow your pride" and launch your raft early—and hopefully unnecessarily. Fill it with provisions: blankets, clothes, water, food, etc. Should the launching prove unneeded, the raft can be towed to port for repacking.

Life rafts of identical manufacture and quality, and approved by the U.S. Coast Guard, may be quite diverse in "lifesaving quality." Rafts may be purchased supplied with different equipment, such as "ocean pack" and "limited pack." The limited pack is just what the name implies—limited. *The ocean pack is strongly recommended.* A life raft so equipped will include flares, smoke signals, air/water pump, patching kit, canned food and water, can opener, knife, sea anchor (drogue), rainwater catch rainwater storage bags, first-aid kit, flashlight, signal mirror, heaving line, and repair kit. Lights atop and inside the canopy operate automatically for 12 or more hours (enough time to get organized). When you must leave your vessel, pull the raft up as close as possible and hold it while everyone gets in. Try to stay as dry as you can; being dry will save you from discomfort later and may enhance your chances of survival.

If your boat is small you may not be willing to pay $1,000 or so for a raft. This doesn't mean you should not have some form of escape—how about that skiff you've got aboard? You can adapt it. Plan ahead. Fill the space under the thwarts (seats)— and anywhere else that's practical—with Styrofoam blocks (secured, of course). Prepare an "Abandon Ship Kit" with items as follows:

- One knife and one small hatchet
- Three cans of water per man and one can opener
- One 6 ft. x 6 ft. polyethylene sheet per man (thin)
- Two smoke flares and two night flares
- One ball of cotton twine
- One box of wooden safety matches

Place these items in a dry, sturdy, plastic bag and seal it. This small package can be either permanently stowed in the boat or kept handy nearby. The polyethylene sheets are cheap and compact. They will provide some cover from the wind and rain while afloat.

Water

Your body is about 70 percent water. Maintaining the "water balance" of your body is the prime requirement for survival. Thirst is not always due to water need. The sensation of thirst can be created by sugar and salt—and even by sweetened beverages. So when water is scant, avoid such food and drink. If you have no water supply, *don't eat.* Digestion of food will drain needed water from your body—you can probably last up to over a month without any food, but not much more than a week without water. Every bit of body water you conserve increases the length of your survival. Thus you should eat minimum quantities of food and avoid perspiring. A person stranded adrift should take no water the first day, and 1 pint a day each day thereafter until rescue or water depletion. If you are without water completely you will probably get delirious in about 4 days. If one of your members becomes delirious it may take physical force to keep him aboard.

Thirst may be reduced by chewing on gum—or practically anything. However, this relief does not reduce the body's need for water. Seawater has a salt content of 3 1/2 percent. That is equivalent to a full teaspoon of salt in a 6-ounce glass of water. The salt content of seawater is three times greater than that of human blood. Drinking seawater will exaggerate thirst, promote water loss through the kidneys and intestines, and shorten your survival time.

Under the conditions of lack of water, urine is too concentrated to be drunk. Its toxic waste products will add to the agony of thirst, contribute to dehydration, and lead to excessive body heat of 105 degrees and over. Drinking urine will cut down your survival time.

Alcohol will promote water loss through the skin and kidneys. Drinking alcohol under the conditions of lack of water is suicidal.

When no food is eaten during water deprivation, energy must be obtained from the body's own fats and proteins. In the process of turning the body's fats and proteins into energy, water is manufactured, and this body-made water helps maintain kidney activity. By not eating when you don't have enough water available, you will actually prolong your life.

Catch a fish. If it is large, carve a cup-shaped hole in the meat. If it is small, cut up the meat and squeeze it through a cloth. The water which collects in the hole, or is squeezed, is totally salt-free and will quench your thirst even if it doesn't taste very good.

Food

In gathering your abandon-ship rations make every possible effort to obtain "double-duty" food (foods with high water content plus nutritional value).

Water contents of vegetables: beets, 87 percent; sweet potatoes, 68 percent; radishes, 93 percent; carrots, 88 percent; potatoes, 77 percent; tomatoes, 94 percent; turnips, 90 percent; onions, 87 percent; lettuce, 94 percent; cabbage, 92 percent; spinach, 92 percent; squash, 95 percent.

While water content should be the first consideration for selecting abandon-ship rations, it is advisable to add (if possible) some compact food of higher calorie (or energy) value. (Figures given in calories per pound.) Jam, 1300; coconut (dried), 2600; condensed milk, 1500; walnuts, 3300; cheese, 1800; butter, 3500; chocolate candy, 2500.

Liquor contains considerable energy value but tends to dehydrate the body.

Cold Weather Foods. Sugar and fats are rapidly absorbed by the tissues, and supply a quick source of heat and energy. Foods rich in sugar and fats should be rationed during cold weather. Butter, cheese, dried milk, nuts, coconut, chocolate and cocoa drinks are rich in fats. Chocolate candy is an excellent "cold-weather food," in both sugar and fats.

Seaweed as Food. Laver, Irish moss, and agar are used as human food; but unless you have a plentiful supply of water, you are advised not to obtain nourishment from seaweed. Not only are seaweeds tough and salty, but they also absorb large quantities of water, leaving you with an intolerable thirst.

Staying Warm

This is not too great a problem in a canopied raft where the temperature can be kept at about 70 degrees during freezing weather. In addition to wearing warm clothing if at all possible, there are several other ways to stay warm. Most rafts have seat tubes and floors which must be inflated by a hand pump packed aboard the raft. These provide good insulation from the cold, damp, raft bottom. The doors of all rafts are easily closed. If sufficient warmth cannot be attained while the doors are partly open to provide adequate ventilation, then button them up tight. When the "stuffiness" becomes unbearable, open *both* doors simultaneously for a moment. Wind will whisk the bad air out in a moment. . .then button her tight again and warm up. All survivors should huddle together as much as possible, and take particular care of their hands and feet. Placing your bare hands under your armpits is a good way to keep them warm. Place your bare feet under another's thigh for maximum warmth.

Grease. Grease spread over the body will *not* protect you from the cold. Grease will make your clothes less protective by filling the air pockets, which are the real source of warmth. In a high wind, grease (of any type) on the face will prove helpful, but it is useless in still cold. A small amount of oil may be applied to the feet, and rubbed in until the surface is dry. A large quantity will prove harmful.

Clothing. Warmth from your clothing is supplied by air pockets between the fibers and air layers between the layers of clothing. These air layers prevent too much heat from getting out and too much cold from getting in. Clothing should not be too tight. Particularly avoid tight shoes. Loosen the shoestrings. Wear two or more pairs of socks—woolen over cotton.

Wool and Fur. Woolens and furs offer excellent protection against the cold because nonconducting air pockets are formed in the meshes, and these air pockets provide insulation between the skin and the outer temperature.

Heat

During most of the season in the Gulf of Mexico, the heat and not the cold is the environmental factor with which you must be concerned. Preventing sunburn is much easier than treating it. Many people are severely burned because they fail to realize that the effects of sunburn are not felt until several hours after exposure. If you wait until your skin turns pink or feels hot before you cover it, it will already be too late. Second and third degree burns can result from exposure to the sun just as from exposure to an open flame, although it does take somewhat longer.

Two other related illnesses are sunstroke and heat exhaustion. Sunstroke is the result of direct exposure to the sun. It may affect you suddenly, but is usually preceded by dizziness, nausea, and headache. Heat exhaustion is the result of long exposure to heat when the temperature and humidity are high. It may occur without exposure to the sun.

After addressing heat injuries, it may seem somewhat strange to talk about hypothermia, which is usually associated with cold water. But be aware that you can suffer from hypothermia even in tropical waters. Your body will begin to cool whenever you are immersed in water cooler than 92° F. The warmest ocean waters that can be expected anytime of year is 84° F.

Sickness

These are ailments frequently experienced by survivors:

Vitamin Deficiencies: If your gums bleed and have a bruised appearance, you have a touch of scurvy, caused by a lack of vitamin C, which is found in fresh fruits and vegetables.

The inability to see well in a dim light indicates night blindness, the result of vitamin A deficiency. Vitamin A is found in carrots, peaches, cod liver oil, butter, eggs, milk and beef liver.

The peeling and scaling of your skin, knee jerks and other indications of nervousness, which may or may not be accompanied by headaches, back pain and depression, show that you are a victim of pellagra. Pellagra is associated with a deficiency in niacin and protein. These symptoms are mentioned so you will not worry about having a dangerous disease; because scurvy, night blindness, and pellagra are diseases caused by a bad diet. After you reach land, proper food will cause these symptoms to disappear. Forget them.

Staying Cheerful. Do not undervalue morale. Your state of mind can give you courage and confidence. The lack of morale has proved fatal far more often than the lack of water. Morale is the total of little things. Don't underestimate these trivialities.

Extra Food. If food beyond the usual rations has been salvaged, save it for periods of gloom.

Water. Your canned water may taste "flat" —but it is not bad and is still water! Remember that a man who has a fever or is a diabetic requires more water than others. Allow him an extra ration.

Liquor. Liquor should be rationed out during squally weather, and to men on watch, during stormy nights. Remember the effect of dehydration, though, and save the "party" for after the rescue.

Bodily Functions. With short rations bowel action is often stopped. This is normal and no cause for alarm.

Bedding. Spread life preservers on the bottom for those who are turning in.

Impartiality. From Captain Bligh we quote a method that was time-honored even in 1789: "I divided it (a noddy, about the size of a pigeon), with its entrails, into 18 portions, and by a well-known method of the sea, of 'who shall have this', it was distributed with the allowance of bread and water for dinner...One person turns his back on the object that is to be divided: another then points separately to the portions, at each of them asking aloud, 'Who shall have this?' to which the first answers by naming somebody. This impartial method of division gives every man an equal chance of the best share."

Recreation. Do not overlook recreation. A portable radio, tuned to dance music and news for the crew will help morale.

Plastic cards will eliminate long stretches of daytime boredom, since they are not affected by salt air, water, heat or humidity.

Suicide. The suicidal impulse is no stranger among survivors. The victim of hysteria cannot help himself. The man who suddenly starts over the side, saying, "I'm going down to the corner for a glass of beer," is suffering from hallucination, disassociation of time and place, and it is your duty to restrain him.

Your principal hope after a disaster has struck is to get someone to pick you up. As mentioned above, however, you must take all action as if you did not expect rescue for weeks. This is necessary because although most rescues are made within hours of the disaster, some stretch on for many days. So *plan ahead.* Don't waste anything—food, water, physical strength—nor signaling gear.

Rescue

All signaling gear is good. "Radar" flares are visible on radar 250 miles and more away—well worth the investment. Flares (either hand-held or "Very" lights) can be seen at night as far as 20 to 24 miles, and smoke as far as 16 miles (with little wind) by day. Dye will persist over half an hour (in moderate seas) and is easily seen from the air (at up to 4 miles distance). Since a skiff, raft, or even a fishing vessel is difficult to see from sea or air at any substantial distance, it is essential that you properly use signals. You must never assume you have been sighted, but continue your efforts until certain. An aircraft will always clearly show when he sees you— perhaps by "buzzing." Remember that an aircraft may have to leave you due to weather, fuel, or darkness. Sit tight and save further signals for his return—he'll be back.

Hand-held flares must be held out over the water on the downwind side at an angle to prevent hot "drippings" from burning either you or the skiff or raft. Burning skiffs and rafts make bright signals— but don't float well.

Many signaling devices work from both ends (don't trigger both ends at the same time, of course). Be sure to "douse" the end you have used in water and save the signal at the other end—if it is that type.

Signaling mirrors are effective devices, (visible about 8 miles) but not as good as flares or smoke. Don't forget that even your shirt afire on the end of a paddle is better than nothing.

Searchers use carefully computed patterns. If you can detect the type of pattern being used, you should save your signaling device for when the searcher is closest. Searches will usually be made in either "parallel sweeps" or "expanding squares" (Figures 35.2 and 35.3).

Figure 35.2. Parallel sweeps.

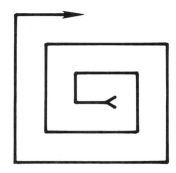

Figure 35.3. Expanding square.

Raft Navigation

This may seem an impossible subject—but you can control your movements. First of all, be sure your "sea anchor" (drogue) is working properly. This will reduce your drift rate tremendously, and thus make you easier to find. For instance with a 10-knot wind you will drift 7 miles a day with a drogue as opposed to 20 miles a day without it.

However, if you are reasonably close to shore, and the wind is onshore, it may be wiser to let the raft drift. You can steer it with your paddle up to about 45 degrees off the wind without too much difficulty or loss of speed. You can control your direction downwind up to 40 degrees off the wind depending on where you secure the "sea anchor" to the raft. Unless you are specifically trying to "sail" to a particular point, you should attempt only to sail directly downwind, and have your "sea-anchor" streamed. This will keep you in the middle of the search patterns being computed ashore.

You can't really "navigate" a raft, but you can have some control over where you're going and how fast you'll get there.

Man Overboard

Action taken aboard when a man is overboard is somewhat dependent on circumstances. If you actually see the man go in the water, immediately toss him a life ring, life jacket, or anything else that will float. You'll probably get the boat around to pick him up before he needs such things—but who knows? Bring the boat around and pick him up on your leeward side. If the water is cold and your freeboard high, he will probably need help— possibly someone (with a lifeline attached to his life jacket) to go into the water to help him. Vessels of high freeboard should have a plan for picking up someone who has fallen into the water—it is an all too common and frequent accident. All crew members should know the procedures.

Searching. If a man fell over without being immediately noticed, call the Coast Guard as soon as you discover him missing, and retrace your course to the farthest distance back he could possibly be. Then come back again about 1/4 mile off to the side—preferably on the side to which you would expect him to drift. Figure 35.4 is a useful pattern to use in searching.

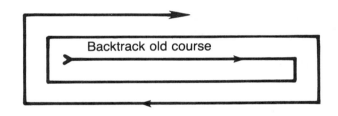

Figure 35.4. Backtracking.

Treatment. Once the man is aboard, check to see if he needs artificial respiration. This should be followed by treatment for shock, regardless of how "healthy" he seems. Immediately strip off all his clothes and wrap him in plenty of towels and blankets. Don't force liquor on a man—it doesn't really warm him, but it can be a good morale factor. However, excessive liquor when his body temperature has been reduced by immersion in cold water can be extremely hazardous.

OTHER SAFETY PUBLICATIONS AND PROCEDURES

Other survival techniques, procedures and equipment are covered by various U.S. Coast Guard publications. Rescue procedures including helicopter evacuation rules for vessel inspections, rules of the road, vessel light signals and shapes and other regulations can be obtained through the U.S. Coast Guard Safety Office nearest you. Also, the Coast Guard Auxiliary will examine safety items aboard your vessel and issue you a decal for the boat if it passes. Some commercial fishing industry associations also have established safety inspection procedures for member's vessels.

SUMMARY

It is the duty of every crew member to become familiar with basic procedures and systems that affect his or her safety, the safety of other crew members, and the safety of the vessel. Crewmen should follow the vessel's rules at all times, and promptly report injuries, accidents, and safety defects to the captain. Accidents generally happen when someone fails to take appropriate action in dangerous circumstances. Good seamanship is common-sense awareness and thinking ahead, which means preparation and forethought are crucial.

The purpose of this guide is to provide a general orientation to fishing vessels and their operation; identify measures to maintain safe operation of the vessel and safety awareness of the crew; recognize potential safety hazards; establish procedures for reporting and investigating accidents; plan vessel emergency procedures and survival techniques and check out equipment; and provide suggested reporting forms, documents, and checklists to make accident reporting easier and recognizing safety hazards faster.

It is each crewman's duty to ask as many questions as possible about basic safety and vessel operations before he or she leaves the dock. The more each crewman knows about safety and seamanship, the better the odds that the vessel will be a success story rather than a casualty statistic.

REFERENCES

Committee on Fishing Vessel Safety. 1991. *Fishing Vessel Safety, Blueprint for a National Program.* Washington, DC: Marine Board, National Research Council.

Hollin, D. and E.D. Middleton, Jr. 1989. *Gulf Coast Fishing Vessel Safety Manual.* College Station, TX: Sea Grant College Program, Texas A&M University.

Hollin, D. and E.D. Middleton, Jr. 1989. *Safety at Sea, A Guide for Fishing Vessel Owners and Operators.* College Station, TX: Sea Grant College Program, Texas A&M University.

Myers, M., and M. Klatt. 1994. *Proceedings of the National Fishing Industry Safety and Health Workshop.* Anchorage, AK: National Institute for Occupational Safety and Health (NIOSH).

36

DROWNING AND COLD WATER SURVIVAL

Thomas B. Faulkner, M.D., M.H.A.
Rollins School of Public Health, Emory University

> This chapter will focus on the hazards at sea. The events involved in drowning will be discussed, briefly explaining what happens to a victim submerged underwater. More importantly, cold water immersions, and how these injuries can pre-dispose the healthiest of individuals to eventual drowning will be presented. While treatment procedures will be described for each of these topics, the major focus will be on prevention and survival from these maritime accidents.

DROWNING

Within the maritime-based occupations, drowning and immersion-related injuries account for a higher toll of lost life than any other form of accident. The sea is unforgiving, and because of this, a man overboard is in significant peril unless he is removed rapidly from this treacherous environment.

Drowning actually involves a spectrum of injuries and death related to submersion under water. Literally, drowning is the death that results from prolonged immersion under water due to lack of oxygen, or asphyxia. However, any temporary survival after a drowning is called near drowning, and death that follows this recovery, for whatever reason, is called secondary drowning. Regardless of the definitions, drowning continues to be a leading cause of death in the world.

By World Health Organization (WHO) estimates, five to six individuals per 100,000 will die each year from drowning—roughly 140,000 deaths annually (NIOSH 1994). Nine thousand of these deaths occur in the United States, out of a total of 80,000 submersion incidents. This makes drowning second only to motor vehicle accidents as a leading cause of accidental death in children. These numbers are staggering; yet, since many cases are not reported, they probably represent a conservative measure.

In the commercial fishing industry, these rates are much higher. The United States Coast Guard (USCG) estimated that between 1982 and 1987, the annual occupational fatality rate for United States commercial fishermen on all coasts was 47 deaths per 100,000 workers, with drowning being the major cause. Additionally, the National Institute for Occupational Safety and Health (NIOSH 1994) measured the fatality rate among commercial fishermen in Alaska. Between 1991 and 1993, the fatality rate of this population was 195 per 100,000, with 91% of these deaths being attributed to drowning.

Several risk factors have been associated with drowning events. Obviously, children are at higher risk if they do not know how to swim, but numerous adults are similarly untrained. Among adults, the unquestionable major contributor to submersion mishaps is drug use, especially alcohol (USCG 1975). Studies in Australia and the United States have shown that a majority of otherwise healthy victims of drowning had elevated levels of alcohol in their system at autopsy.

By impairing judgment, alcohol directly contributes to the individual's risk of death by drowning. Impaired judgment leads to reckless behavior, dangerous situations, and accidents. When in the water, the drunken individual could become easily confused and disoriented, all but sealing his or her fate. The link between drugs and alcohol and drowning events can not be emphasized enough.

To better understand the event of drowning, and how to treat and prevent it, a discussion of how immersion effects the body must be discussed. When submerged, the victim will initially hold his or her breath, panic, and begin to violently thrash about. Despite this breath holding, a hyperventilation reflex will ensue, and the victim will actually swallow large volumes of water. This water intake may result in vomiting and aspiration of vomit in addition to water. The subsequent decrease in oxygen in the blood stream, or hypoxia, causes unconsciousness. Thus, when retrieved from the water, the victim will likely be unconscious, and have his or her lungs and stomach filled will water. In this situation, delivering oxygen back to the brain will be a difficult, if not impossible task, but one that still should be attempted if time is on your side.

Management of the drowning victim is directed toward returning oxygen to the brain and other tissues, and supporting circulation. First, the victim must be rescued from the water. This effort must be carefully weighed against the risk to rescuers, as there are thousands of stories of rescuers actually drowning or seriously injuring themselves in an attempt to help a drowning victim. With this in mind, evaluate the situation and environmental risks thoroughly before letting anyone put themselves at great peril. Coordinate efforts, and use tethers or flotation devices to assist the rescuer. More information on victim retrieval is described in the section on rescue at sea.

In the event that the victim is still submerged, retrieval should be left only to those with the proper equipment (wet or dry suits, SCUBA gear, lights) and training. Perhaps the most difficult decision will be to know when a rescue is too dangerous, and victim retrieval should be stopped. Although difficult, knowing when things are too dangerous can prevent any further needless fatalities.

If the victim is recovered in a timely manner, all efforts should be made at resuscitation until the rescuer is told to stop by qualified medical personnel, or is exhausted. As will be discussed in the section on immersion, although the victim may appear dead, there is still the possibility of recovery from a drowning. Thus the ABC's of cardiopulmonary resuscitation (CPR) should be utilized, with mouth-to-mouth ventilation and chest compressions. While it is beyond the scope of this chapter, the risk of drowning, and the need for rapid rescue and support of the drowning victim should prompt ship captains in commercial fishing to obtain CPR training for themselves and their crew.

Communication is paramount in the rescue of the drowning victim. Rescuers should be organized and aware of what each is doing. This also allows for accounting of each of the rescuers. As soon as possible, contact should be made with the nearest Coast Guard facility or rescue agency. Inform them of your situation, including the number of victims, their time submerged, and your location. Assign a member of the crew to maintain continuous contact with these services to keep them informed of any changes. Early and appropriate contact with these rescue facilities can hasten assistance to you, and increase the chances of successful resuscitation of the victims.

We have focused primarily on the unconscious victim of drowning. However, what about the conscious victim? While the victim is alert and breathing unassisted, this does not mean he or she is not still at risk. Remember, the victim may have swallowed, which impairs the ability to deliver oxygen effectively to the tissues, especially the brain. Therefore, these individuals should be monitored closely for any change in their status, and taken to the closest hospital for complete evaluation.

COLD WATER IMMERSION

A more common event, and one that can lead to drowning, is injury due to cold water immersion. As an example of its lethality, consider the case of R.M.S. Titanic, the "unsinkable" passenger liner that sank after hitting an iceberg with the loss of 1,489 passengers and crew. By most accounts, a majority of the victims of this tragedy entered the 32° Fahrenheit water alive (USCG 1975). But with the arrival of rescue ships less than 2 hours later, only those in the lifeboats survived. It was not a violent death by drowning that took these lives, but rather the gradual incapacitation brought on by immersion in the cold waters of the North Atlantic.

This is still a significant cause of death today, with crews or individuals being swept overboard, and though recovered from the water within minutes, dying from the effects the cold water took on their systems. To understand the insidious nature of cold water immersion, we need to understand the body's normal methods of maintaining body temperature, and how they apply in a cold water environment.

Body temperature is kept constant at the center, or core, of our body by changes in the cardiovascular system. Specifically, the 98.6° F temperature we enjoy is maintained by altering the bloodflow to our skin and limbs. When in a hot environment, the blood flow to these areas will increase, allowing heat to be released from body surfaces. In colder situations, this blood flow is limited, to protect against heat loss. All of these changes are driven to keep the body's core (i.e., heart, lungs, and brain) at a constant temperature. Muscles may also begin to shiver in an effort to generate heat for the body.

Heat loss occurs by various methods. Conduction of heat from the body to any surrounding matter such as water or metal is common. Especially when in the water, body heat is rapidly conducted to the surrounding water, at a rate 25 times faster than loss to air alone. When there is a passing current of air or fluid, convection can also drain body heat. This windchill effect is very dangerous and actually lowers the true outside temperatures. Table 36.1 shows the changes in temperature, given various wind speeds and their effect on exposed skin.

The convection effect of windchill can be especially dangerous to the individual pulled from the water into the cold air. If not sheltered from this danger, the victim is at greater risk of hypothermia.

Evaporation of moisture from the skin also acts to cool the body. While this is quite beneficial when in a warm environment, in colder climes it adds to rapid heat loss. For this reason, someone removed from the water should be dried with towels as quickly as possible to prevent this form of heat loss.

Table 36.1. Wind chill temperatures and potential effects on exposed skin.

Estimated wind speed (mph)	Actual temperature reading (° F)											
	50	40	30	20	10	0	-10	-20	-30	-40	-50	-60
	Equivalent chill temperature (° F)											
calm	50	40	30	20	10	0	-10	-20	-30	-40	-50	-60
5	48	37	27	16	6	-5	-15	-26	-36	-47	-57	-68
10	40	28	16	4	-9	-24	-33	-46	-58	-70	-83	-95
15	36	22	9	-5	-18	-32	-45	-58	-72	-85	-99	-112
20	32	18	4	-10	-25	-39	-53	-67	-82	-96	-110	-121
25	30	16	0	-15	-29	-44	-59	-74	-88	-104	-118	-133
30	28	13	-2	-18	-33	-48	-63	-79	-94	-109	-125	-140
35	27	11	-4	-20	-35	-51	-67	-82	-98	-113	-129	-145
40	26	10	-6	-21	-37	-55	-69	-85	-100	-116	-132	-148
(Wind speeds over 40 mph have little additional effect.)	LITTLE DANGER. In less than 1 hour with dry skin. Maximum danger of false sense of security.				INCREASING DANGER. Danger from freezing of exposed flesh within 1 minute.				GREAT DANGER. Flesh may freeze within 30 seconds.			

Finally, the body also releases heat in a fashion similar to a radiator. This heat of radiation does not cause as rapid a loss of heat as the other forms mentioned, and can actually be used to keep the individual or group of victims warm. As will be explained shortly, by preventing rapid heat loss, and controlling the rate and method of heat loss, the chances for survival in cold water or outdoor environments are increased.

SIGNS AND SYMPTOMS OF HEAT LOSS (HYPOTHERMIA)

If core body temperature becomes too cold, as with cold water immersion, then the individual becomes hypothermic. Hypothermia essentially acts to short-circuit the body and can result in shock. With shock, the body's ability to function and regulate itself is gradually damaged and lost. As this progresses, life is slowly taken from the victim, and makes them susceptible to drowning.

Figure 36.1 and Table 36.2 provide comparative data demonstrating the effects hypothermia has on the individual as core body temperature is lowered.

Both indicate that as core body temperature decreases, physical and mental processes gradually diminish, and death can occur if the individual is not rescued in time.

Several factors can influence the rate at which hypothermia and these symptoms can overwhelm the body. The first concerns water temperature. Figure 36.2 and Table 36.3 show estimated survival times for unprotected individuals in water of various temperatures.

Simply put, the colder the water, the shorter the survival time for the unprotected individual.

Because of this, it is important to dress for protection against these cold temperatures in an effort to improve survival time. Dressing for survival means covering up as much of the body as possible to prevent heat loss. This includes a wool cap for the head as well as gloves or mittens over the hands. Dressing in layers of clothing is also important, using wool and polypropylene materials to cover one another. Perhaps the best protection is an immersion suit, or dry suit, that keeps the body away from direct contact with the water. Even with the protection of the immersion suit, the use of layered clothing is still important to protect against the loss of body heat from conduction and convection. Figure 36.3 shows samples of layered clothing to consider when preparing for work at sea.

Though it may be obvious, everyone should be wearing a personal flotation device (PFD) while working on the water. Keeping afloat without a PFD will be a rapidly exhausting task especially in layered and bulky clothing. Many PFDs available are easy to work with and can be worn over or under work clothing such as those shown in Figure 36.3.

The use of a PFD will enhance survival time by preventing the energy consuming efforts of swimming,

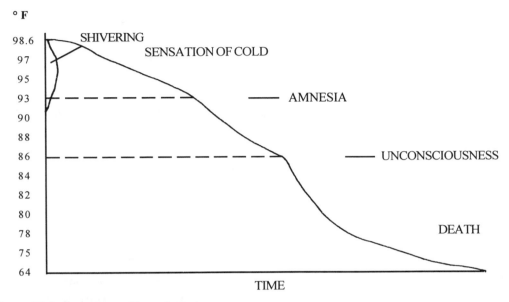

Figure 36.1. Symptoms of hypothermia.

Table 36.2. Symptoms of hypothermia.

°F	°C	
99.6	37	Cold sensations; Skin vasoconstriction; Increased muscle tension; Increased oxygen consumption
97	36	Sporadic shivering suppressed by voluntary movements; Gross shivering spells; Further increases in oxygen consumption; Uncontrollable shivering
96	35	Voluntary tolerance limit in laboratory experiments; Mental confusion; Impairment of rational thought; Drowning possible; Decreased will to struggle
93	34	Loss of memory -- speech impairment; Sensory function impaired; Motor performance impaired
91	33	Hallucinations, delusions, clouding of consciousness; In shipwrecks and survival experience, 50% do not survive
90	32	Heart rhythm irregularities; Motor performance grossly impaired
88	31	Shivering stopped
86	30	Loss of consciousness; No response to pain
80	27	Death

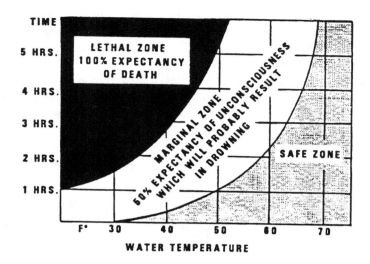

Figure 36.2. Survival rate zones for unprotected individuals.

Table 36.3. Estimated survival times (in hours) for different water temperatures.

Water Temperature (°F)	Survival Time (hours)
over 70	Indefinite (depending of fatigue)
60-70	less than 12
50-60	less than 6
40-50	less than 3
35-40	less than 1 1/2
less than 35	less than 3/4

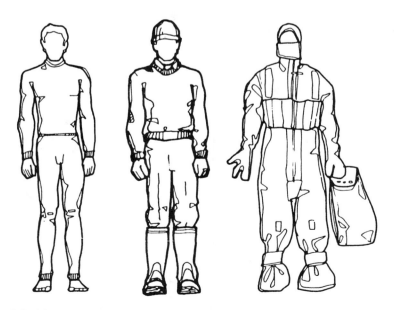

Figure 36.3. Layering of clothing for work on the water (adapted from Hollin and Middleton 1986).

treading water, or drown-proofing, which will be discussed shortly. Figure 36.4 demonstrates the effects of flotation devices and activity on survival time.

As shown in Table 36.3, survival times are much longer if the individual has a method of flotation available. However, while treading water may slightly prolong survival in the individual without a PFD, swimming with a PFD on actually lowers survival in cold water.

The H.E.L.P. (or Heat Escape Lessening Position) and Huddle positions are methods where the individual or group of survivors can minimize heat loss and prolong survival. With the H.E.L.P. technique, the arms and legs are drawn up and crossed close to the body in an effort to minimize heat loss from convection and conduction (Figure 36.5). With the Huddle, radiation of body heat is shared among the individuals who hold each other close together (Figure 36.6).

Figure 36.4. Types of personal flotation devices (PFD) (adapted from Hollin and Middleton 1986).

Table 36.4. Survival times in relation to use of flotation equipment and physical activity.

Situation	Predicted Survival Time (hrs)
No Flotation	
Survival Floating	1.5
Treading Water	2.0
With Flotation	
Swimming	2.0
Holding Still	2.7
H.E.L.P.	4.0
Huddle	4.0

Figure 36.5. The H.E.L.P. position for individual heat loss prevention (adapted from Hollin and Middleton 1986).

Figure 36.6. The Huddle position for groups or survivors (adapted from Hollin and Middleton 1986).

This method is also excellent for close communication and morale boosting under such stressful circumstances. For each of these positions, a PFD is mandatory to get the most benefit from them. Without a PFD, individuals is required to put more energy into keeping their heads above water, and will become fatigued faster. Similarly, other survivors in the Huddle position will have to assist those without a flotation device, and put themselves at higher risk. If done correctly, the H.E.L.P. and Huddle positions can increase survival time by up to 50%.

Additional equipment to carry includes signaling devices to aid in rescue and contact with others in the water. Simple equipment like a whistle, a signal mirror, flares or a waterproof flashlight can mean the difference between rescue and being lost at sea (Figure 36.7). Make sure that the equipment is in good repair and that batteries are fresh and protected from the water.

DROWN-PROOFING

Pending the arrival of help, the individual must stay afloat often for many hours until rescue. In warmer water (above 72° F), when flotation equipment is not available or working adequately, drown-proofing provides a method of staying afloat until helps arrives.

Position

As shown in Figure 36.8, drown-proofing involves alternating between two positions of the body. The resting position, X-1a, has the individual bend slightly forward at the waist with the head down and face submerged in the water. The individual should rest with the head, back, and arms afloat on the surface, and the arms raised above the head. The legs should be allowed to relax and dangle below the body.

The breathing position, X-1b, requires periodic raising of the head and face out of the water by pivoting at the neck. During this movement, a slight bending back at the waist may assist in raising the head, but should not be so extreme as to raise the shoulders out of the water, and thus cause the legs to sink. Similarly, the arms may scull back and forth at the surface to provide support for the head. The legs should kick only if necessary to offset negative buoyancy, and only when attempting to breath.

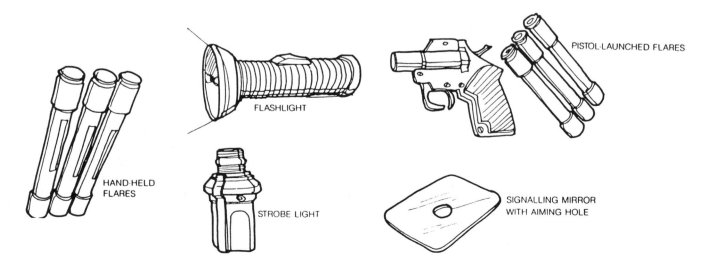

Figure 36.7. Types of signal devices that the individual can carry (adapted from Hollin and Middleton 1986).

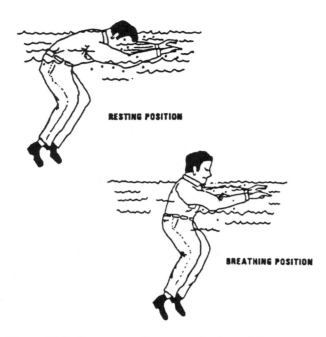

Figure 36.8. Drown proofing proper body position (Source: U.S. Navy).

others. Keep in mind that treading water, while not as physically demanding as swimming, does require more muscle movement and can lead to fatigue faster than drown-proofing. Therefore, use treading sparingly to conserve energy.

Position

Unlike drown-proofing, treading water requires almost continuous movement of the arms and legs, but also dictates that these movements be controlled and coordinated. The trunk, neck, and head should be leaning slightly forward in a vertical, sitting position (Figure 36.9) that is almost identical to the breathing position in drown-proofing. The head should be tilted slightly back as needed, so that the mouth is just above the water.

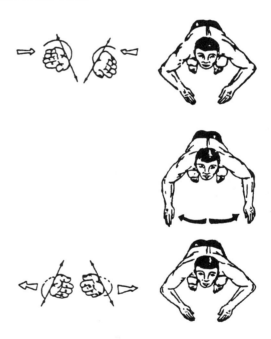

Figure 36.9. Proper technique for treading water (Source: U.S. Navy)

A full cycle should begin in the resting position. The arms should scull outward as the head and face are raised out of the water. With the face clear, a complete air exchange is performed. The head is then lowered back into the water and the arms, legs, and back are returned to the resting position. It is often helpful to count to yourself in an effort to relax and control the alternation between the two positions.

Once the face clears the water, exhale through the mouth and nose and inhale through the mouth in a controlled manner. It is important to try to maintain a controlled and regular breathing pattern when drown-proofing. With the goal to stay afloat while minimizing energy requirements and avoiding exhaustion, keep movements slow, deliberate, and coordinated. Keeping close to the surface, relaxing, and using a controlled breathing pattern that minimizes movements are the keys to effective drown-proofing.

Even in rough seas, drown-proofing allows the individual to stay afloat for many hours, and to periodically look for rescuers or flotation equipment. Understanding of this basic survival maneuver is extremely important to those who travel and work on the water.

TREADING WATER

In the event that rescue is imminent, drown-proofing may not be necessary, and the need to keep the head above the surface is important. In this situation, treading water will help keep the victim floating without excessive energy movements, and allow him or her to look for help and talk to

The arms and hands should scull back and forth below, but parallel to the surface of the water to provide support. Bending at the elbows, the arms scull inward at a 45-degree angle with thumbs up, then the thumbs rotate down to scull outward, keeping movements equal and controlled (Figure 36.9).

The legs are very important in treading water, and far more active then in drown-proofing. Like the arms, the legs kick to help keep the head above the surface. A modified frog kick is suggested that provides the needed support with the smallest amount of energy expenditure. Figure 36.10 shows the three steps of this kick.

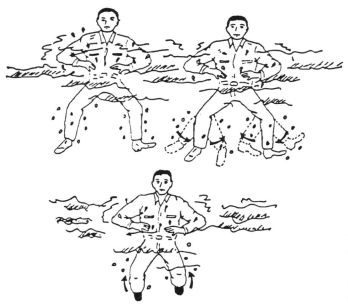

Figure 36.10. Proper treading water leg and arm action (Source: U.S. Navy)

1) The knees are raised and separated to shoulder width apart, with heels bent slightly in toward the body.
2) The feet are kicked out beyond the knees, and the toes are pointed up.
3) Keeping the knees stationary, the lower legs push down on the soles as the toes are pointed downward.

Coordination of the arms and legs is best when you kick as you are sculling inward.

Breathing while treading water is not as limited as when drown-proofing. Breathing should be continuous, avoiding breath holding and shallow breathing. Similarly, exhale through mouth and nose, and inhale through the mouth to get a complete air exchange, and prevent exhaustion and hyperventilation.

As indicated before, treading water allows one to look for help or flotation devices, with minimal energy expenditures. However, it should not be considered for periods where rescue is more than 5 minutes away. Conserving energy and avoiding exhaustion are fundamental aspects of drown-proofing and treading water. If help is not readily available, revert to drown-proofing to stay afloat and save strength.

RESCUE DEVICES AND PROCEDURES

Rescue from a water environment may be in the form of another sea vessel or an aircraft. Although "help has arrived," getting a survivor from the water into the rescue vessel has its own share of hazards. Common rescue methods and devices are presented here to familiarize those that work at sea with what they may expect in the event of an accident.

Surface Rescue

If conditions permit, a small rescue craft or motor whale boat may be launched to pick up survivors (Figure 36.11).

The craft will approach the survivor and attempt to keep him close to the bow. Equipment, such as a flotation device or rescue hook, may be used to assist the victim. A rescue swimmer may also be used to assist in getting the victim on board. It is important that one individual, ideally the boat driver, be identified as the rescue leader. Follow the leader's directions, and let the rescuers do the work of positioning and retrieving. Attempting to swim to the vessel may actually hinder the rescue, endanger all involved, and exhaust the victim.

If conditions do not favor deployment of a small boat, a surface ship may lower a rescue hook or embarkation net or ladder alongside the victim (Figure 36.12). Rough seas make

Figure 36.11. Small boat rescue (Source: U.S. Navy).

rescue very difficult, and the need for rescuer coordination and communication is vital to its success.

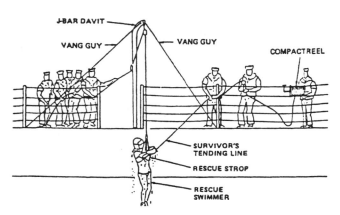

Figure 36.12. Surface ship rescue (Source: U.S. Navy).

Again, the rescued party needs to follow the directions of the rescuers, and avoid trying to free themselves unassisted.

Rescue by Air

Assisted—Often, Search and Rescue (SAR) helicopters are the first on the seen after an accident at sea. Usually, SAR helicopters carry rescue swimmers that will enter the water and assist the victims in getting on board the aircraft. These rescue swimmers usually will want to get a brief report on the medical condition of the victims in the water. Those with the worst injuries will be retrieved first. All survivors should follow the directions of the rescue swimmer without question, and continue to perform water survival procedures until time for hoisting to the rescue aircraft.

In most circumstances, a harness will be lowered by the helicopter, and the survivor will be secured in it by the rescue swimmer. The survivor and rescue swimmer may then be lifted to the aircraft together, and then the rescue swimmer will be lowered to pick up the next survivor (Figure 36.13).

Unassisted—In the event that there is not a rescue swimmer, or that he is not deployed due to weather conditions, it is important that those in the water coordinate the rescue with whatever device is lowered. Ideally, the same order of retrieval should be maintained as if there were a rescue swimmer; that is, the most injured or exhausted should be raised first.

As the device is lowered, make sure everyone stays clear of it until it has settled in the water. This is important to prevent anyone from being hit by the device, as well as to allow the static electricity to be safely discharged. Two common types of rescue devices used by the U.S. Coast Guard are shown in Figures 36.14, 36.15, 36.16.

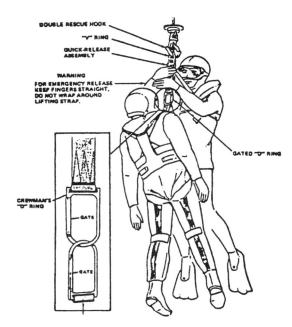

Figure 36.13. Harness lift with Rescue swimmer assist (Source: U.S. Navy).

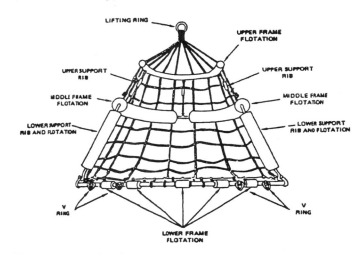

Figure 36.14. Flotation rescue net (open front) (Source: U.S. Navy).

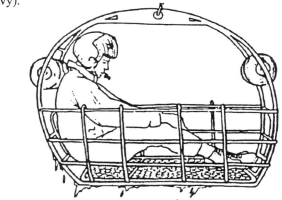

Figure 36.15. U.S. Coast Guard rescue basket (older version).

Figure 36.16. U.S. Coast Guard rescue basket (newer version) (Source: U.S. Navy).

The rescue net is large enough to lift more than one survivor, however putting more that one survivor in at a time can increase the chances of accident and injuries. Once in the water, the net should be positioned so that a survivor enters the open side. Once in, the survivor should turn around with his or her back against the netting and legs and arms within the net as it is raised.

The rescue basket is similar to the rescue net but is limited to one survivor at a time. The survivor should enter it over either side, and sit down with hands holding on to the side rails. Again, arms and legs should be kept safely inside the basket.

Prior to being hoisted to the aircraft it is important to make sure that the survivor is safely in the device, and no one else is tangled to the device or the hoist cable. With this assured, the survivor should give the thumbs up to the hoist operator. On the way up, movement should be minimized, as this may lead to dangerous swinging of the device. Once at the helicopter door, the survivor should remain in the hoisting position until the device is secured to the aircraft, and follow the crewman's directions to enter the helicopter. If other survivors are being hoisted, the survivor should stay away from the aircraft hatch, and not assist the crewman unless told to do so.

SUMMARY

Working on the water, whether an inland lake or the open seas, is a dangerous occupational setting for even the most experienced seaman. Drowning continues to take an alarming toll of fisherman and sailors despite regulations and education. Because of this, it is the responsibility of those who work on the water to protect themselves from these dangers and be ready for any emergency or accident that may arise.

Boat captains and supervisors need to ensure that their crewmen know the basic survival skills and the prevention methods mentioned above. Coordination and communication with rescue agencies will also help prevent needless loss of life from the harsh elements of work on the water.

REFERENCES

Hollin, D., E.D. Middleton, Jr., eds. 1989. *Gulf Coast Fishing Vessel Safety Manual*. National Council of Fishing Vessel Safety and Insurance.

United States Naval Aviation Schools Command. 1992. *Naval Aviation Water Survival Training Program*. AFCM John H. Uptegrove, ed.. May.

U. S. Department of Health and Human Services. 1994. Request for assistance in preventing drownings in commercial fisherman. In: *ALERT*, April. Public Health Service, Centers For Disease Control and Prevention, National Institute for Occupational Safety and Health.

United States Coast Guard, Department of Transportation. 1975. *A Pocket Guide to Cold Water Survival*. CG 473.

37

DIVING HAZARDS

Edward D. Thalmann, M.D.
Captain, Medical Corps, U.S. Navy (retired)
Occupational Medicine, Anesthesiology, and Divers Alert Network
Duke University Medical Center

> While diving is usually considered a recreational hobby, divers commercially harvest millions of dollars of abalones, oysters, and other shellfish yearly. Divers also repair boat hulls, work on aquaculture equipment and help construct piers and levees. Without proper training and equipment, diving can be a hazardous activity. This chapter discusses several aspects of diving including the following topics: physics of diving, underwater diving apparatus and tools, gases used in underwater diving, and decompression tables. Information on diseases and hazards associated with diving including decompression illness, barotrauma, gas toxicity, infectious hazards, near drowning, and hazardous marine life is presented. Finally, information on physical requirements for divers, regulations for commercial diving activities, and a list of resources is provided.

INTRODUCTION

Within the fields of agriculture, forestry, and fisheries, diving may be done in support of a variety of activities or in order to accomplish certain underwater tasks. Fisheries provide the most obvious example of diving employment. Diving is used to construct or maintain pens, barriers, levees, piers, etc., repair boat hulls, or in the actual capture or harvesting of seafood. Within agriculture and forestry, diving may be employed for the construction or maintenance of underwater structures, clearing of conduits for passage of water, or removal of obstacles in rivers or dams, as a few examples. Diving is a specialized field where individuals are totally dependent on life support equipment in the form of underwater breathing apparatus (UBA) and thermal protection garments, as well as a host of other equipment necessary to allow them to remain at their underwater worksite and accomplish their job. Failure of any one of these pieces of equipment can result in a routine task quickly turning into an emergency where loss of life may occur. For this reason, individuals responsible for the health and safety of divers must be thoroughly familiar with all types of diving equipment, their operation, limitations, and failure modes, and how this equipment will impact the diver's ability to do work as well as how it might interfere or limit emergency procedures.

Diving may be done in waters as shallow as a few feet or to depths of 1,000 feet or more in the open ocean. It may be done from the shoreline, from a pier, from a small boat or large ocean-going vessel, or in a large land based experimental chamber. The breathing gas may be air—various mixtures of nitrogen and oxygen—or helium and oxygen, or in some specialized cases 100% oxygen. For the purposes of this chapter, I will concentrate on the type of diving most likely to be encountered in agriculture, forestry, or fisheries, namely dives done to depths of 220 feet or less while breathing air. I will also mention the use of various nitrogen/oxygen mixtures that are now coming into widespread use. Dives done deeper than 220 feet using helium oxygen require specialized knowledge and training and are outside the scope of this chapter.

Within the United States there are four main groups recognized by government regulatory agencies that are involved in diving: the military, commercial, recreational, and scientific. Up until the last 20 years or so the military had the most experience in diving and they were the first to establish a comprehensive set of rules and regulations which still are used as a benchmark (USN 1993a, 1993b). In addition, through its various research and technical evaluation facilities, the U.S. Navy has funded almost all diving research up to the present. Almost all of this research is in the public domain and has been used extensively by the other three communities to develop procedures unique to their requirements. The Occupational Safety and Health Administration (OSHA) and the U.S. Coast Guard are the federal regulatory agencies involved with oversight of commercial diving operations. Basically, any diver who receives a fee for his diving activity is considered a commercial diver, with a few notable exceptions. These exceptions are diving instructors, search and rescue divers, those involved in experimental diving, and those whose diving is exclusively scientific in nature.

The first part of the chapter is an introduction to the diving environment. Diving physics will be covered because it forms the basis for understanding many of the safety issues and risks associated with diving. The influence of the diving environment on ergonomics will be discussed because it impacts a diver's ability to perform his job and the management of some types of injuries. A discussion of diving practice follows to orient the reader with diving equipment and common procedures, such as decompression tables. A thorough understanding of the operation and function of diving hardware is essential to provide effective medical coverage and health monitoring. The same can be said of diving procedures. Also, there will be a discussion of the role of the various regulatory agencies in diving operations and what their requirements are. Both the recognition and treatment of diving related diseases are covered although emphasis is placed on the former. For details of treatment the reader will be directed to the appropriate references. The final section deals with the resources available in the areas of diving medicine and physiology. There is an extensive network in place that can put one in touch with experts in the field who can provide information on any aspect of diving. In addition there are several excellent books that can provide a ready reference on any area of diving.

Where section headings are referenced, this indicates places for more in depth information on the subject, not necessarily the detailed contents of the section.

THE DIVING ENVIRONMENT

Divers work underwater and that environment imposes significant restrictions on their ability to perform work. The first and most obvious problem is that they must be supplied with a source of breathing gas. Next, they must cope with the increased density of the aqueous environment that surrounds them. This environment will impede motion and cause significant alterations in cardiovascular and respiratory function. In addition, water has a heat capacity by volume some 2,500 times that of air at one atmosphere and a thermal conductivity some 26 times greater. These properties impose a significant thermal load on the diver, requiring that he or she wear specialized protective equipment in most cases. Next, as the diver proceeds to increased depths, the density of breathing gas will increase, imposing an additional load on the respiratory system. Both the equipment and the environment will place restrictions on sensory input; visibility is usually significantly reduced and visual fields are usually restricted by a helmet or mask. Sound travels faster underwater than in air so it can be heard at further distances but may be difficult to localize. Finally, the increased ambient pressure itself may have a direct impact on the diver, separate from its role in increasing breathing gas density.

Immersion

Once a diver enters the water, the effect of gravity on the circulatory system is eliminated. The normal pooling of blood in the legs is eliminated and there is a shift of blood into the thoracic cavity, resulting in increased blood return to the right heart. The increased stroke volume allows maintenance of cardiac output at lower heart rates. In addition, the right atrium becomes distended resulting in secretion of atrial natriuretic peptide (ANP) which causes increased urine output (Lanphier and Camporesi 1993; Lundgren 1996). This immersion diuresis may result in dehydration during long periods of submergence, especially in cold water even when wearing thermal protection (Deuster et al. 1989, 1990).

Depending on the diver's position in the water, the thoracic cavity may be at a higher or lower pressure than the diver's mouth. (The pressure gradient of water is 76 mmHg per meter of depth). If the diver is submerged and breathing through a long tube extending to the surface, a snorkel, he or she will have to generate a considerable negative pressure during inhalation to overcome the hydrostatic gradient. This makes breathing through a snorkel longer than a foot in length impractical.

The breathing apparatus that supplies the diver with gas will do so at a pressure which may differ from the pressure surrounding the thorax. Open circuit SCUBA regulators supply gas at essentially the same pressure as the mouth, and in a horizontal position little or no effort is required to inhale. If the diver becomes vertical, some effort is required since the thorax is at a higher pressure than the mouth. Some types of diving apparatus use breathing bags, which act as a reservoir of gas during respiration. Placement of these breathing bags can result in the application of a constant positive or negative pressure to the lungs—a static lung load. Optimal placement occurs when there is a slight positive load; applying negative loads can result in significant dyspnea during exercise (Thalmann et al. 1979; Hickey et al. 1987).

Pressure

Pressure units in diving are presented in several ways. The International System of Units (SI) unit most commonly used is the kilopascal (kPa), defined as 1000 newtons/m². In operational diving, the units usually used are depths, measured in either feet or meters. Freshwater has an approximate density of 1 gm/l over the temperature range usually encountered in diving. Therefore, for every 10 meters of depth the pressure will increase by 98 kPa, or about 0.97 atmospheres (atm). This is equivalent to a 1 atm increase in pressure for every 10.3 meters or 34 feet of depth increase. The density of seawater is related to its salinity and varies throughout the world. In the United States, seawater density was defined by the U.S. Navy to be 64.0 lbs/ft³ at 4 °C giving it a specific gravity of 1.0248—the value used to calibrate depth gauges in feet of seawater (fsw). This gives a 1 atm increase in pressure for each 33.1 foot increase. In Europe, it was decided to define the density of seawater so that a depth of 10 meters of seawater (msw) was exactly 1 bar or 100 kPa, giving it a specific gravity of 1.01972 at 4°C. This difference in specific gravity means that a simple feet to meters conversion between depth gauge readings will give a small error. A depth of 100 feet is 30.48 meters but a depth gauge reading of 100 fsw is equivalent in pressure to a metric depth gauge reading of 30.63 msw. The maximum error over a depth range of 200 fsw is about a foot, making the difference of little practical importance until much greater depths are encountered. The various pressure units encountered in diving are given in Table 37.1.

Hydrostatic pressure is the force exerted by the water on any submerged body and is equal in all directions. Also, since water is essentially incompressible at pressures compatible with human life, its density is independent of depth. Absolute pressure is the pressure due to depth plus the ambient atmospheric pressure. Gauge depth is the difference between atmospheric pressure and the pressure being measured. At the surface, the absolute pressure is one atmosphere while the gauge pressure would read 0. If a diver is at 100 fsw then the gauge pressure is about 3 atm, but the absolute pressure would be about 4 atm.

Buoyancy refers to the force resulting from the displacement of water by an immersed object. If an object floats, it is positively buoyant; if sinks, it is negatively buoyant. In seawater, a material with a volume of 1 ft³ weighing exactly 64 lbs will have a buoyancy of 0 lbs, meaning that no additional force is required to move it up or down in the water column. If it weighs only 60 lbs, it will be 4 lbs positive, meaning that it can support and additional 4 lbs of weight before sinking. If it weighs 70 lbs, then a upward force of at least 6 lbs will be required to move it upward through the water column. Buoyancy is important in diving. If a diver is to do useful work, he or she must carry enough weight to remain in one position when exerting force on underwater objects. By the same token, a free swimming diver must not carry so much weight that he is constantly trying to counteract sinking. Buoyancy compensation is discussed in a later section.

Gas Laws

The behavior of gases in diving can be described by the ideal gas law:

$$P \cdot V = n \cdot R \cdot T \qquad (37.1)$$

where P denotes the pressure, V the volume, n the number of gas molecules, R a constant of proportionality called the gas constant, and T the absolute temperature (°K). The value of the constant R depends on the units of the other variables. A

Table 37.1. Pressure units conversion.

1 atm = 1.013247 bar	1 atm = 33.08 fsw [a]	1 atm = 10.13 msw [b]
1 atm = 101.3247 kPa	1 bar = 32.646 fsw	1 bar = 10.00 msw
1 atm = 14.6959 psi	1 fsw = 3.063 kPa	1 msw = 10.000 kPa
1 atm = 760.00 torr	1 fsw = 22.98 torr	1 msw = 1.450 psi
1 bar = 100.000 kPa	1 psi = 2.251 fsw	1 msw = 75.01 torr
1 bar = 10⁵ Pa		
1 bar = 14.50377 psi	a: All fsw conversions assume seawater density of 1.02480 at 4 C.	b: All msw conversions assume seawater density of 1.01972 at 4 C.
1 bar = 750.064 torr		
1 mPa = 10.000 bar		
1 psi = 6,894.76 Pa		
1 psi = 51.7151 torr		
1 torr = 133.322 Pa		

good way to approximate R is to remember that 1 mole of gas (an amount in grams equal to its molecular weight) occupies a volume of 22.4 liters at a pressure of 1 atm and a temperature of 0° C (273° K). Using these units R is approximately 0.0821 atm l/mole deg. However, for most applications, the ideal gas law is not used as it appears in Equation 37.1, but rather as a ratio between two conditions where either temperature or pressure are held constant.

Because gas is compressible, the conditions under which volume is measured must be specified. Usually gas volumes are expressed in volume STPD (Standard Temperature and Pressure and Dry). This means the volume occupied by the gas at 0° C, at 1 atm, where no water vapor is present. The other condition that may be encountered is volume ATPS (Actual Temperature and Pressure Saturated with water vapor). This condition is used when referring to a diver's lung volumes. For example, the temperature would be body temperature (37° C), the pressure equal to ambient, and the gas completely saturated with water vapor.

Boyle's Law. If temperature is held constant, the following relationship holds:

$$P_1 \cdot V_1 = P_2 \cdot V_2 \qquad (37.2)$$

Simply stated, if the pressure is doubled, the volume halves; or if the volume is halved, then the pressure is doubled. So if a syringe is filled with 10 cc of gas and capped so none will escape, decreasing the syringe volume to 5 cc will raise the pressure to 2 atm. Conversely, if one takes the capped syringe to a pressure of 2 atm by descending to a depth of 33 fsw, the volume with halve to 5 cc. At 3 atm it will have decreased to 1/3 or 3.33 cc at 4 atm to 1/4 or 2.5 cc, and so on. Boyle's law is especially important in understanding the mechanism of pulmonary barotrauma during ascent. If a diver is at 66 fsw and is breathing from a regulator he will be able to fill his lungs completely with air. If he then begins ascent but holds his breath, the gas will attempt to expand. Since the diver is holding his breath this expansion will be impeded and at some point the pressure inside the lung relative to the ambient pressure will increase because the volume is held constant and the lungs will rupture.

The other common application of Boyle's law is in estimating breathing gas duration. Air cylinders carried by a diver have a fixed volume of gas when charged, usually 100 ft³ STPD (about 2832 l). As a rule of thumb, a diver doing moderate work will ventilate his or her lungs with about 1 ft³ of gas per minute ATPS (about 28 l/min) giving the diver a maximum duration of 100 min breathing from the cylinder (in this case the correction for body temperature and water vapor can be ignored). At a depth of 33 fsw, the pressure is doubled so the gas in the cylinder will occupy only half the volume it would at 1 atm but the volume of the diver's lungs is the same. Moving the 1 ft³ of gas each minute at 2 atm (ATPS) is equivalent, therefore, to moving 2 ft³ min STPD, and the cylinder will last half as long or about 50 min. At 66 fsw the pressure is tripled and the gas will last a third as long, and so on. As will be seen later this dependence of breathing gas duration on depth is one of the disadvantages of using certain types of underwater breathing apparatus.

Charles' Law. If the volume of gas is held constant, such as in a closed compressed gas cylinder, the following relationship holds:

$$P_1 / P_2 = T_1 / T_2 \qquad (37.3)$$

This means that if the absolute temperature is doubled the pressure will double. When a compressed gas cylinder is filled it is usually charged to its full pressure, commonly 2500 psi (170 atm) at room temperature, say 68 °F (20 °C). If it sits out in the sun and is heated to 140 °F (60 °C) the pressure will increase. It will not, however, double because the temperature in equation 37.3 must be expressed as absolute temperature or °K. Thus the temperature ratio would be [(60+273)/(20+273) =] 1.14 and the tank pressure would increase to 2850 psi. This increase is within the safety design limit of the cylinder. If the temperature where much higher, say from being caught in a fire, the cylinder pressure could increase to the bursting point.

Dalton's Law. If there is only a single gas in a mixture, then the pressure will describe the amount of gas in a specified volume at a specified temperature. If there is more than a single gas in the mixture then it is the partial pressure of the gas that is of interest. Dalton's Law states that the total pressure of a gas is the sum of its partial pressures; that is, the pressure that would be exerted by each gas individually if it were the only one present in that volume. Suppose we start with a cylinder of air at 1 atm then increase the pressure to 2 atm by adding only nitrogen. Air consists of 20.95% O_2, 78.08% N_2 and 0.07% other gas, the majority of which is Argon. However, for most purposes, air can be thought of as consisting of 79% N_2 and 21% O_2. The partial pressure of a gas is found by multiplying its fraction by the total pressure, so at 1 atm the cylinder has 0.21 atm of O_2 and 0.79 atm of N_2. Since the pressure was increased 1 atm by adding only N_2, the final partial pressure of the two gases is 0.21 atm O_2 and 1.79 atm N_2. Referring to the ideal gas law above, a partial pressure of 1 atm means that the number of moles of gas per liter of actual volume is approximately 1/22.4 or 0.0446 moles per liter. Remembering that 1 mole of gas contains $6.022 \cdot 10^{23}$ molecules, the molecular concentration is $2.69 \cdot 10^{22}$ molecules per liter. The fraction of gas in the cylinder is the partial pressure divided by the total pressure. In this case, the O_2 fraction is $(0.21/2 =) 0.105$ and the N_2 fraction is $(1.79/2 =) 0.895$.

Partial pressure may be expressed in any pressure units. In the above example, the total pressure of 2 atm is equivalent to 1,520 mmHg, so the final mixture has an O_2 partial pressure of 160 mmHg and an N_2 partial pressure of 1360 mmHg. Another unit of partial pressure commonly seen in diving is *Surface Equivalent Value* (sev). This refers to the fraction of gas which would have to be breathed at 1 atm to have the same partial pressure as the gas breathed at depth and is usually used only to express partial pressures less than 1 atm. In the above example, the gas containing 10.5% O_2 at 2 atm would have a partial pressure of 0.21 atm, the same as a 21.0% gas at 1 atm. So at 2 atm a gas containing 10.5% O_2 can be said to have an oxygen partial pressure of 21.0% sev.

Henry's Law. Henry's law relates the amount of gas dissolved in a liquid to its partial pressure at a particular temperature is given by the formula:

$$C = \sigma_{gas,liquid} \cdot P_{gas} \tag{37.4}$$

C is the concentration of the gas in the liquid, P_{gas} is the partial pressure of the gas and $\sigma_{gas,liquid}$ the solubility of the gas in the particular liquid in question. To compute the actual concentration, values for $\sigma_{gas,liquid}$ are obtained from standard handbooks. However, for many purposes the exact concentration does not have to be known—one must simply recognize that the amount of gas in solution is proportional to its partial pressure. For inert gases such as helium or nitrogen the value of $\sigma_{gas,liquid}$ is essentially constant over the pressure ranges of interest in diving, so Equation 37.4 describes a linear relationship. This is not the case for all gases. The solubility of oxygen and carbon dioxide in blood is a function of their partial pressure and the relationship between partial pressure and concentration is not linear. While solubility is a function of temperature, when considering the effects inside the body it is assumed that all tissues are at normal body temperature. Gas solubility is important in understanding the events behind decompression sickness.

Light and Vision

Even crystal clear water absorbs light such that the intensity is reduced by 75% at about 16 feet, and to about 3% of ambient levels by 130 feet. Small solid particles suspended in the water will cause turbidity, causing light to scatter and further reduce visibility, even if a light source is carried by the diver. In many instances, visibility is reduced to the point where divers must work by touch.

As light passes through water the color quality also changes. Blue light is scattered preferentially and almost all red color is gone below a depth of 30 feet. This is important to consider if underwater objects are color coded. Suggestions for using colors that are distinguishable underwater are found in Adolphson and Berghage (1974).

Another factor affecting underwater vision has to do with light refraction. In order to focus, the human eye must have an air phase in contact with the cornea. The diving mask or helmet provides this but when the light passes from the water into this air phase it is refracted, just as it is when passing form the air phase into the cornea. This additional refraction causes a slight magnification The air-water refractive index is such that an object 12 feet away underwater will appear to be about 9 feet away. This fact is important to remember when a diver is called upon to estimate sizes or distance while submerged. The only reliable way to predict distances is by direct measurement unless the diver is very experienced in compensating for this magnification.

Sound

Underwater Noise. The speed of sound underwater is about four times that of air allowing sounds to travel much longer distances underwater than in air. However, this increased speed decreases the time delay between the ears making the localization of sound underwater more difficult. If divers are wearing a wet suit hood or a helmet, sound levels will be significantly attenuated and localization may be all but impossible.

Diving equipment such as compressors and high pressure air escaping from pipes, valves, or hoses can be a significant source of noise and hearing loss is not uncommon among divers (Molvaer and Albreksten 1990). Ear protection is the best defense from this type of noise. Another source arises from gas passing into helmets (Curley and Knafelc 1987) and chambers. Modern diving helmets and chambers are designed to minimize this problem but older equipment may have significant noise levels (Adolphson and Berghage 1974). Hearing protection provides relief in chambers but may not be suitable for wearing in some types of helmets. Currently, the U.S. Navy tries to keep helmet noise levels within OSHA standards but calibrating equipment and measuring sound levels at increased ambient pressure or underwater poses problems which require specialized expertise (Curley and Knafelc 1987).

While there are OSHA sound level exposure limits for air, there are no such limits for submerged divers or for divers wearing helmets where the gas density is increased. The effect of underwater noise on the ear depends on whether the ear canal is filled with water and whether there is any type of covering over the canal. A water-filled canal creates a large impedance mismatch between the middle and outer ear across the tympanic membrane resulting in an effective attenuation of sound. Having the ear covered with a neoprene wetsuit hood will further attenuate sound. These considerations make setting limits on the noise produced by underwater tools difficult and so far no agreed upon limits exist.

Sonar can be a source of underwater noise. Low intensity devices such as fish finders or depth finders do not present a problem. Sonar found on military ships and submarines can produce dangerously high pressure waves and diving should not be done in areas where these devices are active. Divers using hand held sonar devices with power levels no greater than 100 watts will be adequately protected by a quarter-inch neoprene wet suit hood. (USN 1993a).

Because a comprehensive set of noise exposure limits and design guidelines for minimizing noise do not exist for diving equipment, a comprehensive hearing conservation program with yearly (or more frequently as indicated) audiograms is essential. If there is evidence of hearing damage, an investigation should be undertaken to find the sources of the sound, the diver avoiding all significant noise exposure in the meantime.

Underwater Explosions. An underwater explosion creates a series of hydraulic shock waves with an initial wave being followed by waves of diminishing intensity. The effect of an explosion depends on its size, the depth of the water (the deeper the explosion the more attenuated the shock waves), and the distance from the explosion. In addition, a fully submerged diver is more prone to injury than one who is partially submerged. The target organs for explosion damage are the gas filled organs and spaces, the sinuses, middle ear, lungs, and intestines (Kindwall 1994). Divers should leave the water before any underwater blasting takes place (OSHA 1991; USN 1993a).

Heat Transfer

Water has a high thermal conductivity and heat capacity compared to air. Thermal conductivity refers to the rate at which heat is conducted through a medium and the heat capacity refers to the amount of heat required to raise the temperature of a substance by 1 °C. Because of the high thermal conductivity and heat capacity, heat generated by a diver in the water will cause essentially no change in the water temperature next to his skin, resulting in "clamping" of his skin temperature at water temperature. Water at 34 to 35 °C is thermoneutral, that is a resting individual will generate just enough heat to compensate for loss through the skin so that the core temperature will remain constant. Water colder than thermoneutral will cause a drop in core temperature and if higher than thermoneutral the core temperature will rise.

Thermal protection in cold water is governed by the heat flow equation:

$$H = TC \cdot (dT/dX) \qquad (37.5)$$

where H is the heat flux, TC is the thermal conductivity (1/TC is the insulative value) and dT/dX the temperature gradient across the insulative layer of thickness dX. This equation shows that for a given temperature difference (say between the skin and the water), the heat flux can be decreased by increasing the thickness of the insulating layer or by reducing thermal conductivity. The insulative value of a thermal garment is usually expressed as the ratio dX/TC. The smallest value for TC is obtained with a vacuum but while it works in containers with rigid walls (e.g., Thermos® bottles) it is impractical in flexible thermal garments. Gases have a very low thermal conductivity and thermal garments use materials which will place a gas-filled space, usually air, around the diver. The thickness of this gas-filled space is chosen to provide adequate insulation with the smallest possible bulk. Additional improvement can be gained by using a gas with a lower thermal conductivity than air, such as argon, to fill the matrix. Helium has a much higher thermal conductivity than air and would greatly degrade insulative characteristics.

DIVING PRACTICE

Underwater Breathing Apparatus

Underwater breathing apparatus (UBA) supplies breathing gas to the diver and may be one of several designs. All types of UBA place some additional respiratory load on the diver either in the form of resistance or static lung loading. The latter has been discussed above under *Effects of Immersion*. Resistance results from the valving, orifices, and hoses through which the gas must pass and in a poorly designed system can pose significant limitations on a divers work capacity. Most modern UBAs are designed to standards which minimize their impact on diver work capacity (Bevan 1993).

Surface Supplied. The UBA used by divers can be either surface-supplied or self-contained. Surface-supplied UBAs are usually a helmet or full face mask with air being supplied from a compressor or pressure cylinders on the surface through a long hose or umbilical. This hose is attached to a strength member, or tether, which functions as a safety line and contains communication wires, allowing the diver to talk to personnel at the surface. The diver may breathe through a demand valve, similar in function to that of the open circuit SCUBA regulator discussed below, or the helmet may be ventilated. In ventilated helmets, a constant flow of gas is maintained through the helmet sufficient to wash out CO_2. The helmet functions like a well mixed closed space and the CO_2 partial pressure within the helmet is given by the ratio of CO_2 production and helmet ventilation rate by the formula:

$$PCO_2 (mmHg) = 816 \left| \frac{VCO_2(STPD)}{VHelmet(ATPS)} \right| \quad (37.6)$$

The constant of proportionality (816) is a conversion factor between STPD and ATPS volume at 20°C. A diver exercising at a moderate work rate having a CO_2 production of 1.6 l/min and a helmet ventilation rate of 113 l/min (4 ft³/min) would have a helmet CO_2 level of just over 11 mmHg—close to the maximum inspired value of 0.015 atm allowed by the U.S. Navy. OSHA regulations require that the minimum helmet ventilation be 4.5 ACFM at any depth or that the helmet be capable of keeping the inspired CO_2 level below 0.02 atm at a CO_2 production of 1.6 l/min. Note that the helmet volume is measured at depth so if the diver were at 66 fsw (3 atm) the compressor output at the surface would have to be 330 l/min or about 12 ft³/min. Helmet hypoventilation because of inadequate compressor capacity or because the diver does not adjust his flow to a high enough level during periods of work can lead to hypercapnia within a space of a few minutes.

The gas supply for a surface supplied UBA is usually substantial giving the diver essentially an unlimited breathing gas source. Also helmets and full face masks do not require mouthpieces so that the diver can breathe even if unconscious. Helmets provide head protection and are attached in such a way that accidental removal is nearly impossible, thus providing the maximum degree of protection. A full face mask provides less protection and some types can be torn from the head accidentally. They are usually used where the bulk of a helmet would encumber the dive, such as working in very tight confined areas.

SCUBA. Self-Contained Underwater Breathing Apparatus (SCUBA) are those units where the gas supply is carried in a pressure cylinder on the diver's back. These divers are free to swim about as they wish and are generally unencumbered. However, the amount of breathing gas is limited to what they can carry. Also, SCUBA divers may or may not have communication to surface personnel and if they have no lifeline to the surface their status or location may not be known.

The biggest disadvantage of SCUBA over surface supplied types of UBA is the duration of the breathing gas. The most common type of SCUBA is the familiar demand type open circuit demand regulator. When the diver creates a slight negative pressure by beginning inhalation, the regulator supplies gas until exhalation begins at which point flow stops. All of the diver's exhaled gas is dumped into the water through an exhaust valve. Since no gas is reused, the duration of the gas supply will depend only on the diver's minute ventilation. The example given in the section on *Boyle's Law* illustrates how this duration varies with depth. Typically, open circuit SCUBA will provide durations of only an hour or so, but they may be much shorter if heavy work is undertaken. For jobs requiring longer stays underwater the open circuit SCUBA diver must either return to the surface frequently or have extra breathing gas cylinders available at the worksite.

There are two other types of SCUBA that will prolong the breathing gas duration to varying degrees (Bevan 1993; Egstrom 1993; USN 1993a, 1993b). The most efficient design is the fully closed circuit apparatus in which oxygen is added by an electronic circuit at exactly the rate it is consumed and so long as the depth is constant no gas is lost from the breathing loop. A less complex but more inefficient design is the semi-closed SCUBA where a high O_2 gas is injected at a constant rate using pneumatic controls. Since the gas injection rate exceeds oxygen consumption, some gas is constantly vented from the breathing loop. While the closed circuit design can control the inspired O_2 to a constant partial pressure independent of depth or level of exertion, the semi-closed apparatus will have an inspired PO_2 dependent on both depth and level of exertion. Both of these types of SCUBA use breathing bags and CO_2 absorbent canisters. The operating manual for each of these type of UBA will give details on their operation and how to choose appropriate breathing gases. In addition, special training is required in order to dive safely with these types of apparatus.

Heavy work or underwater construction is usually accomplished with a surface supplied UBA which will provide a stable platform, the large amounts of air required for heavy work, and two-way communication. SCUBA diving is usually reserved for lighter work where large volumes of air are not required and where unencumbered mobility is essential. Performing heavy work safely, using SCUBA, requires extensive experience. The logistics involved in having sufficient breathing gas readily available may be considerable, especially when using open circuit systems. SCUBA diving involving heavy work is usually done only in situations where the presence of a surface umbilical would not allow the job to be accomplished, and should not be undertaken lightly.

Breath Hold Diving (Lin and Hong 1996). With the advent of SCUBA, breath hold diving has decreased dramatically as a means of performing underwater tasks although in some areas it is the only method allowed when spear fishing for sport. In this type of diving the diver may enter the water and immediately begin his or her descent. More likely the diver will be swimming face down wearing a diving mask and breathing through a snorkel. Once the diver sees an objective, he or she will inspire and execute the dive, returning to the surface in a few minutes. Breath hold diving with the intent of setting depth or time records is hazardous and not dealt with here.

Once the diver inspires and closes his glottis the only gas available to him is that in his lung. Once he begins his descent the increased ambient pressure will increase the partial pressure of all his alveolar gases. The increase in oxygen partial pressure will increase the rate of diffusion into the blood meeting metabolic needs. However, the increase in CO_2 partial pressure will reverse the normal gradient in the lung, causing the arterial CO_2 to rise and eventually exceed venous. It is the rising arterial CO_2 that eventually causes the sensation of air hunger (the so-called break point), telling the diver to return to the surface. During ascent, the partial pressure of the alveolar gases drop and at surfacing the alveolar PO_2 may have fallen to almost mixed venous values indicating a critical state of hypoxia. If a diver hyperventilates at the surface before executing his breath hold he will begin with a lower than normal arterial CO_2 partial pressure allowing additional time at a depth before the break point is reached. However, during this time the alveolar oxygen fraction has fallen to lower values than during a normal breath hold. During ascent, as the oxygen partial pressure falls, a situation could be reached where the arterial PO_2 falls low enough to cause unconsciousness.

Breath hold diving causes little excess nitrogen to accumulate in body tissue and therefore should not be associated with decompression sickness. However, a syndrome of vertigo, nausea, partial or complete paralysis, unconsciousness and even death has been reported in some pearl divers who make 40 to 60 dives during a typical 6-hour day, each lasting about 2 minutes. These symptoms are thought by some to be due to N_2 retention.

1 Atm Suits. While in-water diving has been done to depths of 1000 fsw or more, the life support requirements and the decompression time are considerable. In addition, compressing a diver to great depths may take hours or days in order to avoid problems with compression pains (see below) and the High Pressure Nervous Syndrome (Bennett and Rostain 1993). These problems can be circumvented with 1 atm diving suits. Basically, these are small one-man pressure vessels which keep the occupant at 1 atm, much like the occupants of a submerged submarine. There is no physiological constraint on either ascent or descent rates and the maximum depth is a function of the hardware design. The disadvantages are the large bulk, diminished dexterity from using manipulators instead of hands, and the restricted range of motion and visibility.

Other Diving Gear

Masks/Helmets. A diving mask or helmet is necessary for underwater vision. Helmets and full face masks provide some protection for the head and face and also supply breathing gas to the diver. If a diver is using a SCUBA regulator with a mouthpiece, the mask will usually cover the nose and face. As the diver descends, having the nose in the mask allows easy equilibration of the mask's internal pressure as the diver descends preventing a condition called mask squeeze. Goggles are almost never used in modern day diving operations because there is no easy way to equilibrate the pressure in the goggles as descent occurs.

The type of mask used is usually an individual preference. Most will significantly reduce the diver's visual field and may introduce distortions, especially in the peripheral vision area (Adolphson and Berghage 1974). Mask fogging is a persistent problem, and there are many agents designed to prevent this. However, irritating residues may be left on the faceplate, resulting in eye or skin irritation.

Fins and Boots. A free-swimming diver will usually wear fins to propel him through the water. These are individual preference items. Fins do not, however, provide a stable platform from which a diver can do heavy work. In these cases boots are usually worn. They are usually weighted so the diver will always remain in an upright position when moving through the water column. When wearing boots or shoes, divers must enter the water close to the work site since movement through the water by walking, once on the bottom, can be slow and difficult.

Buoyancy Compensation. Unless weighted, divers will tend to float. This is prevented by using weights placed around the waist or in pockets on the diving dress. Most divers will weight themselves so they are negatively buoyant, giving them some stability when laying or standing on the bottom. In order to compensate for this negative buoyancy a flotation device such as a variable volume dry suit or buoyancy compensator is used. These devices are inflated when a condition of positive or neutral buoyancy is needed or when ascent to the surface is required. When completely deflated, the diver achieves his maximum negative buoyancy. Surface-supplied divers wearing boots who need a stable platform to accomplished work may wear enough weights to be 20 lbs or more negative. SCUBA divers will usually only weight themselves a few pounds negative.

The buoyancy compensator should be able to lift the fully weighted diver when fully inflated. However, the weights themselves are usually worn so that they can be quickly jettisoned in an emergency.

Clothing. Divers' clothing may range from only swim trunks, to a fully enclosed abrasion resistant garment. The type of garment worn depends on the job site, the water temperature and what contaminants might be found in the water. At a minimum, enough clothing should be worn to prevent abrasions or foot injury. If the water is contaminated with either potentially infectious material (such as near a raw sewage discharge) or toxic chemicals (such as from a spill) then special garments must be worn (NOAA 1991). Basically these garments cover all areas of the body and uncontaminated water is constantly pumped through them at a slight positive pressure to keep any potentially hazardous material out. Helmets are almost always used to prevent any exposure to the head or face. Avoiding potential exposure to toxic or infectious hazards requires meticulous planning. Immunization against infectious threats should be done when possible. Decontamination of diving gear can usually be accomplished using agents such as Betadine® (povidone-iodine) or Zephiran® (benzalkonium chloride) solutions. The U.S. Navy has determined that these solutions are compatible with most materials used in diving equipment. Other agents should be used with caution and only after compatibility with the materials to be decontaminated has been demonstrated. In some cases, consideration should be given to complete disposal of equipment in the appropriate manner.

Diving in toxic waste utilizes the same type of diving dress as used for infectious hazards but now consideration must be given to the potential breakdown of the dress from the contaminants (NOAA 1991). In many cases, the diving dress will have absorbed so much material that it has to be properly disposed of after diving.

Thermal Protection. Most thermal problems in diving arise from cold water. There are two basic ways of providing protection to cold water divers, passive and active systems.

Passive systems simply place an insulative layer around the diver to minimize the amount of heat lost to the water. The most familiar of these is the wet suit, so called because no attempt is made to keep water from entering the suit. Wet suit material is usually neoprene that has been filled with air cells making it a foam. These cells are closed so that the air cannot escape, and it is the air in the cells that supplies the insulation. While these suits are not waterproof, they must be carefully tailored to conform to the contours of the body (much like a body stocking) so that water will not "flush through" the suit. Wet suits come in various thicknesses in order to provide protection suitable to the water temperature. In warmer water, thin (1/8") material provides comfort and abrasion protection. In colder water, 1/4" or thicker material is used, and additional protection is achieved by adding layers such as an additional jacket. Wet suits are usually adequate protection in water as cold as 60 °F (15° C). If the water is colder, exposure times must be reduced. Also, since the air cells will collapse as the pressure increases, the insulative value will decrease as depth increases. No specific guidelines on exposure limits for wet suits currently exist but exposures as long as 2 3/4 hours at 10 °C have been done with less than a 1° C drop in core temperature (Wolf et al. 1985). Experienced cold water divers are usually able to judge the amount and type of thermal protection needed for a particular dive. Even so, one should always be on the lookout for possible hypothermia during or following cold water operations since diver comfort is not always a reliable index of thermal protection adequacy.

The variable volume dry suit can provide adequate thermal protection in near freezing water for duration's in excess of 8 hours (Sterba 1993). With these garments, the diver dons an undergarment looking much like a coverall, which contains a material providing a loft, so that an area of trapped air surrounds the skin. In this regard these undergarments are equivalent to Arctic exposure clothing. Since it is the air that provides the insulation, the material must be kept dry; for this purpose, the diver wears a waterproof outer garment. Seals at the neck and wrist and waterproof zippers ensure no water enters the suit. According to Boyle's Law, unless additional gas is added, the air space around the diver will collapse with descent. In order to prevent this, a source of compressed gas is used to fill the suit. The amount of gas can be varied to provide different degrees of buoyancy, thus the "variable volume" descriptor. Using these suits safely requires training and experience. On the one hand gas volume must be sufficient to maintain adequate insulation, but it must not be so much that the diver begins to ascend uncontrollably.

As noted above, as a diver enters the water, mechanisms come into play that will increase urine output. If a diver urinates in a wet suit, thermal protection is not compromised, but this is not so for dry suits. During long-duration dives, dry-suited divers may find it impossible not to urinate, and, if this occurs, the area where the undergarment becomes wet will loose much of its insulative properties, which can result in hypothermia. For male divers, some dry suit systems avoid this problem by having a catheter attached to a penile sheath that directs urine out of the suit through a one-way valve. This system is not without its own peculiar hazards since air trapped in the catheter may compress during descent, resulting in a hematoma on the head of the penis. This is avoided by ensuring that the catheter is filled with fluid by urinating before beginning descent. Similar systems for female divers have not yet been developed.

Hands and feet present special problems. If fins are worn, the insulation around the foot is compressed, increasing heat loss. Special fins designed to avoid this problem have been developed. Gloves that allow a reasonable degree of manual dexterity may not provide adequate hand insulation, resulting in cold and painful fingers. These conditions are what usually limits duration during dry suit dives. While frostbite cannot occur in the water, a condition called nonfreezing cold injury may result, causing tissue damage and long-term sensitivity to further cold exposures (Francis and Golden 1985).

Active heating is usually accomplished by pumping hot water to a diver from the surface through an umbilical. The maximum water temperature to which the skin should be exposed is 110° F (43° C) to prevent the possibility of burns from long-term exposures. Since the water reaches the diver through a long umbilical, considerable cooling may take place so that the temperature at the hot water supply end may be considerably more than 110° F. The amount of cooling is dependent on the rate of flow to the diver (more flow, less cooling) and safety devices or procedures must be in place to prevent an inadvertent slug of this very hot water from reaching the diver. Accidents where this occurred have resulted in severe burns.

Diving in warm water is usually considered a luxury, but, in some climates, such as the Red Sea, the water may be hot enough to cause rises in core temperatures. Diving near hot water discharges poses similar problems. However, studies of this problem have shown that hyperthermia is in fact unlikely in water temperatures as hot as 95 °F (35 °C,) even with heavy exercise (Doubt et al. 1990; Hyde et al. 1990). However, using a variation of the hot water suit, where chilled water is pumped down to the diver, will provide comfort and protection, especially in water exceeding 95 °F. Another strategy for prolonging exposure time in warm water is to have the diver wear a vest containing ice (Holmer 1989).

Underwater Tools

Power tools used underwater are operated by compressed air, hydraulics, or electricity. Tools are never powered up until they are needed, and all electrical circuits must be protected by ground fault interrupters (ADC 1994; OSHA 1991). Underwater arc welding is very common and equipment designed for underwater use is designed to eliminate electrical shock hazards. However, degradation of dental amalgams has been seen in some divers engaging in underwater arc welding, so frequent dental examinations may be warranted (Ortendahl 1987; Ortendahl et al. 1985).

Air-powered tools and those using high pressure water jets, such as devices for cleaning surface debris from underwater structures, can generate considerable noise. At present, there are no specific noise standards applicable to underwater tools but diving helmets and wet suit hoods do afford a degree of protection. Injury from single-day exposures is unlikely but chronic exposure may cause hearing problems. A well established hearing conservation program is required for early detection of hearing damage and to prevent permanent injury.

Breathing Gases

The breathing gas most commonly used in diving is air that, for practical purposes, contains 79% N_2 and 21% O_2. When using air, depth is limited by three factors, nitrogen narcosis, oxygen toxicity, and increased breathing gas density. OSHA limits air diving to 190 fsw but dives of 30 minutes or less may be done to 200 fsw. While deeper dives on air have been done, the divers ability to do useful work is severely limited. The depth limitation due to narcosis can be eliminated by replacing the nitrogen with non-narcotic helium and the oxygen toxicity problem eliminated by keeping the oxygen partial pressure below toxic limits. Helium/oxygen mixtures are generally used when diving below 130 fsw and are almost always used at depths below 200 fsw.

Oxygen may be added to air to make enriched oxygen/nitrogen mixtures, known as NITROX. Since decompression sickness is mainly a function of the nitrogen content of the breathing gas, these mixes will have less of a decompression obligation at a given depth than when breathing only air. Their disadvantage is that the higher oxygen fraction imposes a shallower maximum depth compared to air, because of oxygen toxicity. NITROX mixes are very advantageous at shallow depths where oxygen toxicity will not be a problem. They may greatly extend the time the diver is allowed at depth before decompression stops become necessary.

Helium-nitrogen-oxygen mixtures and hydrogen-oxygen mixtures have been used in diving under very specialized circumstances. These gas mixes are unlikely to be encountered in routine diving operations.

Decompression Procedures

According to Henry's Law, once a diver enters the water and begins to descend, the increased partial pressure of the breathing gas constituents will cause more gas to dissolve in the tissues (Vann and Thalmann 1993; Thalmann 1996). After

24 hours or so, the tissues will be in equilibrium with the gas dissolved in the blood, a condition known as saturation. However, air divers rarely spend that much time in the water; times are more likely to be a few hours at most. As the divers begin to ascend, the ambient pressure surrounding the body tissues is reduced and, if the ascent is faster than gas can be eliminated, a point will be reached where the total dissolved gas pressure in body tissue exceeds the ambient pressure, a condition known as supersaturation. If this supersaturation becomes excessive, the dissolved tissue gas will come out of solution, forming gas bubbles and resulting in a condition known as decompression sickness (DCS). DCS is an occupational disease of diving and the clinical and long-term health aspects will be covered later. In order to safely reach the surface without getting DCS, divers will follow a decompression schedule. These are protocols specifying how shallow divers may come before stopping to take a decompression stop, how long to stay there until they are allowed to ascend to the next shallower stop, and then how long to stay at each shallower stop until they reach the surface. Each entry in the table is the amount of time which must be spent at that particular stop in order to allow enough time for gas elimination so that bubbles will not form. These decompression schedules are based on mathematical models and are computed for a specific depth and time combination (Vann and Thalmann 1993), assuming that the divers have spent the full time at the specified depth. If the divers have been at several depths, they will have to use the schedule for the maximum depth attained, assuming they have spent the entire time at that depth, which makes for some degree of inefficiency.

A technique sometimes employed to minimize the amount of time a diver must spend in the water to decompress is Surface Decompression or Sur-D (USN 1993a). Special decompression tables have been developed for this procedure in which divers spend only a portion of the total decompression in the water, at which point they are rapidly brought to the surface and recompressed to a predetermined depth in a dry recompression chamber where they complete their decompression. By minimizing in-water time, problems with hypothermia or strong currents are minimized and, once divers are safely in a chamber aboard a diving vessel, it may get under way while the decompression continues.

A decompression table is a set of decompression schedules for many depth and time combinations. There are many sets of decompression tables in use throughout the world. Up until recently the most widely used tables were the U.S. Navy Standard Air Tables (USN 1993a). The British Royal Navy has its own set of decompression tables and the Canadian Forces uses the DCIEM tables, a version of which has been adapted for sports diving (Edmonds et al. 1992).

Other sports diving tables include the BSAC, French Navy 90 (Edmonds et al. 1992) and PADI (DSAT 1994; Rogers 1988) tables. Commercial diving companies also have their own decompression tables, which may or may not be available to the public.

Diving at altitude results in a reduced ambient pressure at the surface. The U.S. Navy air decompression table can be used at altitudes up to 2300 ft (700 m) without modification. Above this, one can either modify the Navy Tables (Bell and Borgwardt 1976; Egi and Brubakk 1995; NOAA 1991) or use tables specifically designed for altitude diving (Boni et al. 1976). Flying after diving, or driving over a high mountain pass after a dive subjects the diver to a reduced ambient pressure that could cause DCS, which would not have otherwise occurred. Current USN guidelines suggest waiting a minimum of 2 hours after a no-decompression dive and 12 hours after a decompression dive. Recent research suggests that a minimum of 12 hours after any dive (longer for multiple repetitive decompression dives) may be wise (Vann et al. 1996).

Modern technology has allowed the computer algorithms used to compute decompression tables to be programmed into a computer small enough to be carried on a diver's wrist. These devices constantly monitor depth and compute a decompression table specific to the exact depth/time profile for that dive. This generally requires less decompression time than when using tables, since tables require that divers assume they have been at the maximum depth of the dive for the entire time. These dive computers have gained wide acceptance among the sport diving community and have the theoretical advantage of eliminating unnecessary decompression. Some of the decompression schedules computed by these devices are significantly shorter than those using currently available tables, and the safety of these has not been verified.

A discussion of the advantages and disadvantages of these various tables and dive computers is beyond the scope of this chapter. What is important is that divers be following some recognized type of decompression procedure. Ideally, statistics should be available on the incidence of DCS experienced for each table used, but this is not always the case. For instance, for the USN Standard Air Table the overall incidence is low, but certain schedules have high incidences (Berghage and Durman 1980). OSHA (1991) requires there be some procedure in place to investigate incidents of DCS to determine its cause. In many cases, the decompression schedule will not have been followed, but DCS can occur even if schedules are followed meticulously. This is because any dive has a risk of DCS occurring, decompression tables are designed only to keep this risk low, they cannot eliminate it (Vann and Thalmann 1993). Keeping records of DCS incidence over hundreds or thousands of man dives is the only way to verify the risk, and if it appears excessive, some corrective action should be taken, such as modifying the decompression schedules in use or switching to another table.

Dive Profiles

If a diver has been at depth for only a short period of time, he or she may not have taken up enough nitrogen into the tissues to require decompression stops and can ascend directly to the surface. These types of dives are known as no-decompression dives and the maximum amount of time that may be spent at a particular depth before decompression stops are required is called the no-decompression time or no-decompression limit. No-decompression diving is generally regarded as being safer than decompression diving because the diver can always ascend directly to the surface in an emergency without having to worry about missing decompression stops. Decompression diving, on the other hand requires that a diver take decompression stops when ascending to the surface and, in an emergency, some of these stops may be missed, putting the diver at increased risk of DCS. If the amount of missed decompression time is large, death may follow rapidly after reaching the surface. Procedures for dealing with missed decompression have been developed (USN 1993a; NOAA 1991).

In many diving operations, the job is at a specific depth, so the diver descends to that depth, does the job, then ascends following the appropriate decompression schedule. This is known as a single or bounce dive. Bounce dives may be decompression or no-decompression dives. Divers who spend time at several depths during a dive have done a multi-level dive. An example might be working on some pilings where cross braces must first be repaired at 30 fsw taking 30 minutes, and then another at 60 fsw taking 20 minutes. When using decompression tables for multilevel dives, the schedule is chosen as if all time was spent at the deepest depth, in this case the diver would use the 60 fsw for 5-minute schedule. A decompression computer, on the other hand, would compute a schedule for the exact profile as above.

Divers will occasionally complete a dive profile, leave the water for a period of time (called a surface interval), and then perform another dive sometime later. This is repetitive diving, and procedures have been worked out to account for the effects of previous dives on subsequent dives when determining decompression requirements (USN 1993a). In some areas, such as in fish farming and sea urchin harvesting, divers may make many dives to a relatively shallow depth, interspersed with short (less than 10 to 20 minutes) surface intervals throughout the day. This type of diving is known as "yo-yo" diving and symptoms ascribed to DCS have been reported, even though the total time of all the dives has never exceeded the no-decompression limit for the deepest depth (Douglas 1995; Wilcock et al. 1992). Whether "yo-yo" diving will require specialized decompression procedures (Parker et al. 1994) awaits the gathering of more data to define the scope of the problem.

Diving several hours a day, day after day, is called multi-day diving and problems have been reported in divers who have done this for several days or more (AAUS 1991). In the sport diving community, a 1-day break after every 3 days of diving has been recommended but in some commercial settings this advice may not be followed.

Deep diving where decompression times may become quite long require additional equipment such as open diving bells or more sophisticated closed diving bells, called personnel transfer capsules (Bevan 1993; NOAA 1991). This type of diving requires specialized training and support equipment and is outside the scope of this chapter, and unlikely to be encountered in agriculture, forestry, or commercial fisheries.

DIVING-RELATED DISEASES

Diseases peculiar to diving result from the direct physical effects of pressure (barotrauma, compression pains), pharmacological properties of gases (narcosis, oxygen toxicity), the physical effects of gases (DCS, arterial gas embolism, osteonecrosis), and the aquatic environment (thermal problems, near drowning, aquatic animal and plant life). What follows is a brief overview to orient those unfamiliar with the field. The references should be consulted for more details, especially with regard to diagnosis and treatment.

Decompression Illness (DCI)

DCI is used to describe those diseases that result from too rapid a reduction in ambient pressure, decompression sickness (DCS), cerebral arterial gas embolism (CAGE), and chokes. These diseases are not confined to divers but may also be seen in compressed air workers and aviators. Aviators have their tissues saturated at 1 atm and may undergo sudden reductions in ambient pressure during rapid ascents to altitude.

Decompression Sickness (DCS). DCS is thought to arise when gas dissolved in body tissue comes out of solution to form gas bubbles. While this can happen, in principal, in any tissue, the skin, joints, inner ear and central nervous system account for the vast majority of symptoms. If caught early, symptoms of DCS can be alleviated by reversing the process, that is putting the diver in a chamber and increasing the pressure to redissolve the gas phase, so called recompression treatment. If treatment is delayed too long, other mechanisms of tissue injury, such as hypoxia and inflammation, may develop decreasing the efficacy of recompression and prolonging recovery (Francis and Gorman 1993).

DCS is usually divided into musculoskeletal and central nervous system disease. Musculoskeletal DCS may involve the skin, producing rashes; the lymphatic system, producing local edema; or result in joint or muscle pain. These conditions are usually not life threatening but treatment will relieve the discomfort. Potential long term consequences may include osteonecrosis. Also, if these symptoms are severe, it is thought they may be a harbinger of more serious symptoms, and early treatment will abort a more serious pathological process.

DCS of the central nervous system (CNS) resulting from diving affects the spinal cord more than the cerebrum (this is not true of DCS in aviators). Common symptoms are paresthesias, muscle weakness or paralysis, and bladder dysfunction. Injuries may arise at several random locations within the spinal cord, sometimes producing scattered symptoms at several unrelated levels. When the disease involves the cerebrum almost any presentation may occur. Disorientation, confusion, decreased mental acuity, amnesia, and loss of consciousness are just a few of the possibilities.

The inner ear may be involved in DCS by itself or along with other CNS symptoms. It is usually associated with ascents from great depth or switching of breathing gases from helium/oxygen to air during decompression. It is manifested by the sudden onset of vertigo, nausea, and nystagmus. It is rare during diving at depths less than 220 fsw but can occur and may be indistinguishable from a perilymph fistula resulting from barotrauma. In these cases recompression is recommended, along with suitable precautions to prevent exacerbation of a perilymph fistula, if it is in fact present (USN 1993a).

Cerebral Arterial Gas Embolism (CAGE). If a diver ascends too quickly, with his glottis closed, the intra-alveolar pressure may increase to the point where the lung parenchyma, tears allowing gas to enter the pulmonary capillaries forming bubbles. This situation usually occurs in emergency ascents but may also occur if an obstruction such as scar tissue or mucus prevents even a small area of the lung from ventilating. Once the bubbles enter the blood stream their size increases, according to Boyle's Law, as the pressure is reduced during ascent and as gas rapidly diffuses from the supersaturated tissue into the bubble. This bubble will enter the left heart and as with any embolus, preferentially end up in the brain. The usual presentation is loss of consciousness or other profound cerebral symptoms occurring within minutes of reaching the surface. Severe residual injury may result if prompt recompression is not instituted.

Chokes. If a diver ascends to the surface, missing significant amounts of decompression, then large numbers of bubbles may form in the venous circulation and become trapped in the pulmonary capillaries, causing pulmonary congestion resulting in a disease called chokes. The result of chokes ranges from mild dyspnea to severe congestion with pulmonary edema and hypoxia, resulting in death.

Compression Pains. During descent, some divers may experience joint pains similar to those of DCS. The occurrence of these compression pains is usually related to descent rate, slowing this will lessen the chance of occurrence or severity. There appears to be no injury from these pains, but they may last throughout the dive and be confused with decompression pains. A careful history will differentiate between the two (USN 1993a).

Treatment of DCI

The first symptom of DCI is denial and, unless symptoms are severe, a diver may not seek immediate treatment. From a medical standpoint, the highest hurdle is recognition. Once DCI is included in the differential diagnosis, resources are available worldwide that can put individuals in contact with diving medical experts 24 hours a day.

The treatment for any of the forms of DCI is recompression in a treatment chamber. The U.S. Navy recompression procedures are the most widely used and have the best documented efficacy (Thalmann 1990). Royal Navy procedures are close to USN procedures and some commercial dive companies have their own procedures (Edmonds et al. 1992). Whenever a diving operation takes place, the location of the nearest treatment chamber should be obtained and a plan to rapidly evacuate stricken personnel developed. The U.S. Navy specifies 3 degrees of urgency as a guide for determining how rapidly transport should be done. *Emergent* divers are severely ill, have major CNS system involvement or are deteriorating. Rapid transport by air, if available, is indicated. Divers falling in the *urgent* category will have either only pain or mild CNS symptoms that are stable. Here rapid ground transport is appropriate without breaking any speed limits. Divers falling in the *timely* category have had symptoms that have been stable for several hours and are evident only with a detailed examination. These individuals can wait until a thorough medical examination determines if transport to a recompression facility is needed.

It is not always possible to obtain complete resolution of symptoms on the first treatment and, in some cases, complete resolution may not be possible. If residual symptoms remain, follow up recompression treatments are indicated until it is clear they are no longer effective. A rule of thumb in this regard is that if there is no significant improvement on two consecutive treatments then further treatments are probably not warranted (USN 1993a).

Barotrauma

Barotrauma is caused by expansion or compression of gas within closed spaces. The target organs are usually the sinuses, middle ear, and lungs, although the GI tract and teeth may be affected in some cases. Normally there are no closed off air spaces in the body but disease or injury can plug a sinus, block the Eustachian tube, or block a terminal air passage creating a closed gas filled space.

Pulmonary Barotrauma. Lung barotrauma usually occurs during ascent after having breathed from a source of compressed gas at depth. Lung squeeze during descent is possible but even in breath-hold divers, who descend to depths in excess of 100 fsw, it is extremely rare. The most serious form of pulmonary barotrauma, CAGE has already been discussed. If the lung parenchyma tears and gas leaks only into the intrathoracic space then a pneumothorax or pneumomediastinum may occur. These conditions are not usually life threatening and are usually treated by breathing 100% oxygen at the surface. If the pneumothorax is large, or if it continues to increase in size with breathing (a tension pneumothorax) then insertion of a chest tube is indicated. A pneumothorax that occurs during recompression treatments may complicate treatment by expanding during ascents. Unless recognized and treated, serious injury may result. Pneumothorax or pneumomediastinum may impact a divers future diving, especially if they occurred without any evidence of injury (see below).

Ear and Sinus Squeezes (Farmer 1990). Paranasal sinus barotrauma may occur during descent or ascent. The affected individual will usually have congestion of some sort or may have an anatomical abnormality that prevents venting. If the occlusion occurs during descent, sharp stabbing needle like pains will be felt, which may become unbearable. Since the ambient hydrostatic pressure is transmitted to the lining of the sinus by the blood, the blood vessels will enlarge and eventually rupture, filling the sinus with blood. When this happens, the pain may be relieved but a bloody discharge form the nose will be noted on surfacing. If a diver has difficulty equalizing during descent but manages to get to depth, the sinus may become blocked during ascent, resulting in a reverse squeeze.

Middle ear barotrauma has the same basic mechanism as sinus barotrauma except that the tympanic membrane will rupture before any bony structures. Rupture of the tympanic membrane will suddenly relieve the intense pain of distention but the sudden movement of the ossicles may send a pressure wave through the inner ear fluid, causing a tear in the round or oval window. This should be suspected and ruled out in all cases of tympanic membrane rupture.

Many divers will not be able to descend without some active form of equalizing the pressure in the middle ear or the sinuses (so called clearing). This is usually accomplished by a Valsalva or other related maneuver, which raises the pressure in the oropharynx forcing air into sinuses or the middle ear. If this clearing cannot be accomplished easily diving should be avoided since an over enthusiastic Valsalva maneuver may cause damage in and of itself. Decongestants will relieve mild congestion and allow diving but, in the case of severe congestion, diving should be avoided until the problem resolves.

Other Squeezes. Gas spaces under dental fillings may cause pain during descent, and a filling replacement may be required to avoid problems. Failure to keep dry suits properly inflated will cause whole body squeeze as the suit collapses around the diver, which may result in serious injury. This is unlikely to occur unless the diver has an uncontrollable descent, such as being overweighted and falling into the water or letting go of a descent line, where he cannot inflate his suit. In wet suits, occasionally a gas pocket will form between the suit and the skin. As descent continues, the gas volume decreases and the gas pocket pressure may become less than the ambient hydrostatic pressure, drawing the adjacent skin into it. The result is a hematoma, which may not be noticed until after the dive.

Gastrointestinal Rupture. During descent, divers may swallow gas that will later expand during ascent. This usually results only in uncomfortable abdominal distention, which is relieved by either belching or passing the gas rectally. However, cases of rupture of the GI tract following diving have been reported, but this condition is quite rare (Molenat et al. 1995).

Vertigo. Vertigo is a sensation of spinning, which can range from being an annoyance to a disability. Caloric vertigo occurs when one ear canal suddenly fills with cold water while the other remains dry, a not uncommon occurrence while wearing a wet suit hood. The effect is transient and will resolve in a few moments as the water

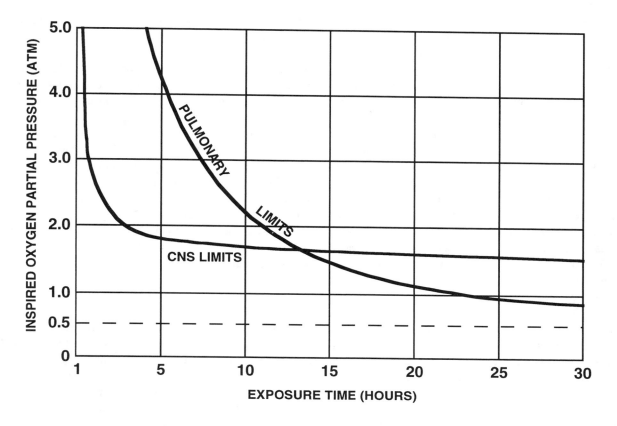

Figure 37.1. Pulmonary and CNS limits for inspired oxygen partial pressure.

warms or the other ear canal fills with water. Alternobaric vertigo occurs when there are unequal pressures in the middle ear, usually the result of being able to clear one ear but not the other. It is usually a transient phenomenon, lasting only a few seconds, as both ears eventually become fully equalized. Vertigo lasting more than a few minutes could be due to a perilymph fistula or cochlear damage and deserves a thorough workup by a specialist.

Effects of Gases

Narcosis. Nitrogen is narcotic at increased partial pressures and its effects can be measured experimentally as shallow as 66 fsw (Bennett 1993). Nitrogen narcosis using air is the reason air diving below 220 fsw is not recommended. Helium is not narcotic, making it the inert gas of choice for deep diving.

High levels of CO_2 may occur from a decrease in minute ventilation due to increased gas density and UBA flow restrictions, or may occur when the CO_2 absorbent in closed or semi-closed UBAs becomes spent. Carbon dioxide is narcotic at high partial pressures but it also causes extreme dyspnea, which will usually cause dive termination before this occurs. However, when breathing high O_2 NITROX mixtures, this dyspnea may be blunted and the narcotic effect of CO_2 will cause the diver to ignore these warning signs of hypercapnia to the point that unconsciousness may occur (Gray et al. 1991; USN 1993a).

Oxygen Toxicity (Clark 1993; Harabin 1996). Oxygen at high partial pressure is toxic, the target organs in diving being the lung and the CNS. It is a partial pressure time phenomenon and the curves are different for the two organs, Figure 37.1. Pulmonary oxygen toxicity occurs from long exposure at PO_2 levels which may have no effect on the CNS. Lung oxygen toxicity is usually encountered during repeated daily diving breathing 100% O2 or during recompression treatments involving large amounts of 100% O_2 breathing. It usually begins with a substernal burning and may progress to pain on inhalation. The symptoms resemble a flu-like upper respiratory tract infection. Decreases in vital capacity and inspiratory capacity may result but these will usually return to normal with cessation of oxygen breathing. Occasionally interrupting 100% O_2 breathing with short periods of air breathing (air breaks), can significantly delay the onset of oxygen toxicity symptoms.

At higher partial pressures, CNS symptoms may occur well before pulmonary symptoms. The most serious CNS symptom is the oxygen convulsion, which is usually a grand mal type of seizure. Oxygen convulsions are of prime concern in diving operations. Besides convulsions, other symptoms of CNS oxygen toxicity include nausea, vertigo, tinnitus, muscle twitching, and scotomata. These symptoms do not form a hierarchy and some may be present for a considerable amount of time without a seizure ever occurring. However, a seizure may occur without any other preceding symptom. Also, there may be so little time between initial symptom onset and the convulsion that the diver can take no preventive action. CNS symptoms are very unlikely (although still possible) at PO_2 levels less than 1.3 atm. Above this level, the time to onset of symptoms decreases as the PO_2 increases, although there is a very large individual variation (Donald 1992). Oxygen seizures in and of themselves have no residual effects (Curley and Robin 1987); the danger to the diver is severe injury or drowning as a result of the convulsion.

Oxygen depth/time limits have been developed for NITROX diving, which if followed should make convulsions unlikely (NOAA 1991). However, any time oxygen partial pressures above 1.0 atm are breathed, a plan should be in place to deal with the possibility of oxygen convulsions.

Hypoxia and Hypercarbia. Hypoxia may occur because of interruption of the breathing gas supply or sending a breathing gas with insufficient oxygen to the diver. In closed and semi-closed UBAs, it may occur from failure to add sufficient oxygen to the breathing loop. Interruption of breathing gas supply is immediately obvious to the diver. If gas cannot be restored immediately then the dive should be aborted. Supplying the wrong gas to the diver usually occurs from errors in the lineup of gas supply valving. The first symptom is usually an unresponsive diver. If this occurs, corrective action to get the proper O_2 breathing mix to the diver must be taken immediately to prevent permanent damage. In closed circuit UBAs and in some semiclosed UBAs oxygen monitors will warn the diver if the O_2 falls below acceptable limits. If the situation is not immediately correctable, the dive should be aborted.

CO_2 buildup causing hypercarbia may occur from the failure of the CO_2 absorbent canister in closed or semi-closed circuit rebreathers, from reduction in minute ventilation due to excessive breathing resistance, or inadequate ventilation of ventilated helmets. Symptoms include dyspnea and hyperventilation but during exercise these may be ignored by the diver and the first sign may be unconsciousness. Even if the diver looses consciousness, hypercapnia will rarely cause permanent damage so long as adequate oxygen is available. In some cases, CO_2 buildup may cause severe headaches shortly after leaving the water. If this occurs then the possibility of hypercapnia should be considered and the reason identified.

Breathing Gas Contamination. If breathing gas is compressed from the atmosphere then contamination may occur from ambient pollutants or from carbon monoxide buildup from improperly maintained compressors. Adhering to good maintenance procedures and periodic analysis of compressor output will minimize the chance of this occurring. However, if there is any unusual odor from the breathing gas or if divers complain of symptoms such as headache, nausea, or dizziness, or if he becomes unconscious from breathing compressed gas then contamination should be suspected. The suspected cylinder or source should be isolated and gas samples analyzed for possible contaminants.

Infections

Skin infections may occur from wearing wet suits or dry suits that have not been properly cleaned after a dive. A good soap and water washing and thorough drying are usually sufficient to prevent this from happening. Other types of skin infections may result from contact with various forms of aquatic life or chemical contaminants (Fischer 1978).

Diving masks and regulators should also be thoroughly washed and dried after each dive. For most applications, soap and water is sufficient but Betadine® or Zephiran® solutions may be used for increased protection. The U.S. Navy has found these compatible with most diving gear and to provide adequate protection. Other disinfectants should be used with caution until their compatibility with diving gear has been established.

Otitis externa, swimmer's ear, was a common and sometimes debilitating problem at one time in saturation diving and will occasionally become a problem in surface-oriented diving during prolonged operations. The U.S. Navy has developed a simple and effect prophylactic procedure using otic Domeboro® solution, which has been very effective at controlling otitis externa (Thalmann 1974; USN 1993a).

Dysbaric Osteonecrosis

The lesions of dysbaric osteonecrosis are indistinguishable from those due to other causes, and the diagnosis is made by the association of diving activity with the disease. The femur, tibia and upper arms are the most common sites affected. Juxta-articular lesions have the most potential for causing permanent joint injury and, if present, will preclude further diving activity. Shaft lesions have not been associated with disability and would not affect diving activity. Periodic long bone x-rays will usually spot the problem before serious consequences result, and protocols for this have been developed (Kindwall 1994; Walder 1990).

Near Drowning

Near drowning in divers is treated the same as in swimmers with the exception that the possibility of DCS or CAGE complicating treatment must be considered. The depth and time of the dive, amount of decompression missed, and the rate of ascent of the diver must all be considered. If either of these diseases is present, recovery may be hampered unless recompression treatment is instituted. If the recompression chamber has full ACLS capability, including defibrillation, CPR can be carried out in combination with recompression treatment. In cases where defibrillation cannot be carried out in the chamber then compression should not take place until cardiac rhythm has been restored. If defibrillation is required during treatment, bringing the individual to the surface for a short period should be considered (USN 1993a).

Sea and Aquatic Life Injuries

Injury from sea life or aquatic life are very real problems. Diagnosis and treatment of these injuries have been well covered by Edmonds (1989). Divers who are involved in harvesting or fish farming must be thoroughly familiar with the possible aquatic life injuries that may be encountered before entering the water. In other cases consultation with locals familiar with the aquatic environment may provide invaluable information and suggest ways to avoid possible injury. Wearing of protective clothing, especially gloves, will provide sufficient protection in most cases.

Other Conditions

Dyspnea is a sensation of air hunger. It may occur while diving with or without exertion. Hypoxia and hypercarbia are possible causes, as is a UBA with a high breathing resistance. However, it may occur for no apparent reason during periods of exertion and when it does there should be an immediate reduction in the level of exertion. If it does not subside with rest, or returns with exertion, the dive should be aborted until the cause is found. Another cause of dyspnea, albeit rare, is pulmonary edema (see below).

Transient unilateral facial paralysis has been reported following dives and is usually transient. It is probably due to inadequate middle ear equilibration during ascent (Becker 1983; Farmer 1990; Molvaer and Eidsvik 1987).

As noted above, immersion diuresis may result in dehydration during extended dives. This may result in transient postural hypotension immediately after exiting the water from peripheral venous pooling. Another cause occurs in mildly hypothermic divers who rewarm themselves in a hot shower. The immersion dehydration combined with the sudden elimination of peripheral vasoconstriction from the warm water on the skin may cause peripheral pooling and fainting. Having these divers seated while showering will eliminate this phenomenon.

Sea sickness medications may impact diving because many produce drowsiness. One agent that provided good prophylaxis with minimal drowsiness was transdermally administered scopolomine (Transderm Scop®). This agent was investigated during long air dives and found to produce few serious side effects (Schwartz and Curley 1985; Williams et al. 1988). Since some persons have idiosyncratic reactions to this drug the patch should be applied at least 24 hours before diving commences in first time users. If during this period no serious side effects are noted then use during diving operations can be done with reasonable safety.

Pulmonary edema was originally described in otherwise healthy SCUBA divers swimming in water 12° C or colder by Wilmshurst et al. (1989). Symptoms included dyspnea at depth without exertion, expectoration of froth, hemoptysis, and syncope. The proposed mechanism was an abnormal vascular reactivity to cold. Pons et al. (1995) reported on four cases in otherwise healthy individuals where symptoms occurred in water as warm as 20.6 °C. They were unable to find the abnormal vascular responses found by Wilmshurst. In addition, a survey by Pons of 1,250 sport divers found only one additional case, attesting to its rarity. More recently, cases of pulmonary edema have been described in 27° C water in one individual (Hampson and Dunford 1996). The mechanism for this rare occurrence remains elusive.

HEALTH MONITORING

The descriptions of the various hazards that may result from diving reflect the wide variety of physiological mechanisms involved, not the frequency of symptoms. The most common injuries from diving are barotrauma of the middle ear and paranasal sinuses. The other conditions are comparatively rare. The yearly incidence of DCS, as reported by the U.S. Navy, is less than 0.1% and in some commercial companies only a few cases occur each year (AAUS 1992). This does not call for neglect of the possibility of long-term consequences from diving, indeed there have been several workshops looking at the possibility of long-term neurological sequelae in commercial divers (Elliott and Moon 1993). The key to preventing possible long-term consequences is to target health monitoring at the areas most likely to cause problems. These areas include the sinuses, chest, and lungs, which are the main target organs for barotrauma. The ears are a target for noise injury, barotrauma, or DCS. Long bones are involved in dysbaric osteonecrosis and the central nervous system is involved in decompression illness. Also the fingers and toes may suffer damage from non-freezing cold injury.

Physical Requirements

Diving is a strenuous profession and the first criteria that must be applied is cardiovascular and musculoskeletal fitness. In this regard commercial divers should be held to the same fitness standards as firemen or heavy construction workers. However, the diving environment occurs in a unique environment, imposing some special considerations. There are no legislated or universally agreed upon conditions that would be disqualifying for diving. The Consensus Standards for Commercial Diving Operations published by the Association of Diving Contractors (ADC 1994) provides a voluntary set of guidelines for the conduct of the physical examination and disqualifying conditions. According to the ADC the following are disqualifying from diving:

- History of seizure disorder other than early childhood febrile convulsions.
- Cystic or cavitary disease of the lungs, significant obstructive or restrictive lung disease, or recurrent pneumothorax.
- Chronic inability to equalize sinus and middle ear pressure.
- Significant cardiac abnormalities or dysrhythmias.
- Chronic alcoholism, drug abuse, or history of psychosis.
- Significant hemoglobinopathies.
- Significant malignancies.
- Grossly impaired hearing.
- Significant osteonecrosis.
- Chronic conditions requiring continuous control by medication.
- Pregnancy.
- Deformities of the skull that would prevent the individual from wearing required equipment.
- Dental or jaw deformities that would interfere with the proper use of a mouthpiece.
- Symptomatic cervical ribs, congenital brachial cleft fistulas, any chronic draining fistula of the head or neck, and persistent and chronic spastic contraction of the muscles of the head and neck.

Table 37.2. Medical tests for diving.

Test	Initial	Periodic (At least every 2 years, annually after age 35)	Comments
Chest X-ray	x	x	14" x 17" minimum.
Bone & Joint X-rays	x		Initially and as medically indicated
12 lead EKG	x		Annually after age 35 and as medically indicated.
Stress EKG			Only as medically indicated.
Pulmonary Function	x	x	FVC and FEV1.
Audiogram	x	x	Pure tone audiogram. Bone conduction as medically indicated.
EEG			Only as medically indicated.
Visual Acuity	x		Reexamination as medically indicated.
Color Blindness	x		
Hemoglobin, Hematocrit, White Blood Count	x	x	
Routine Urinalysis	x	x	

- Any acute disease of the ear, nose, throat, or Eustachian tubes, chronic otitis media or serious otitis externa, perforation of the tympanic membrane, any significant nasal or pharyngeal obstruction, and speech impediments due to organic defects.
- Chronic obstructive or restrictive pulmonary disease.
- Diabetes mellitus, whether insulin dependent or treated by oral hypoglycemics or diet.

Bilateral vision corrected to 20/40 is required. If there is a hearing loss of 35 dB or more up to frequencies of 3000 Hz, and 50 dB or more at 3000 to 6000 Hz, referral to a specialist familiar with diving is required. Specific attention is paid to hearing loss in the speech range (500 to 2000 Hz). Unilateral hearing loss or labyrinthine dysfunction are considered disqualifying by some, because if the remaining organ is injured, severe disability could result.

Pulmonary function testing with specific attention to FVC and FEV1 is recommended and FEV1/FVC ratios less than 75% should be investigated. In many cases, these individuals will have no evidence of active pulmonary disease and may just have larger than normal vital capacities. Pulmonary function testing should be used as an adjunct to the overall evaluation of an underlying disease of the respiratory system and should never be the only criteria for determining fitness to dive.

In the U.S. Navy, divers who have suffered a spontaneous pneumothorax are disqualified from further diving because of the increased probability of a later pneumothorax in such individuals. If the pneumothorax was due to trauma, however, diving may resume after a suitable period for healing (USN 1993a). This relates to the finding that injured lung parenchyma is more resistant to rupture after healing than normal lung tissue (Calder 1985).

As mentioned earlier, lesions of osteonecrosis confined to the long bone shaft are not disqualifying unless there is concern that it may progress to involve the juxta-articular region with continued diving.

There are many medical conditions for which no hard and fast rule regarding fitness to dive can be made (Elliott 1994). In many cases there is simply not enough experience with the condition to adequately assess its impact on diving. These must be handled on a case-by-case basis, with referral to a specialist familiar with diving, along with considerations of the possible physiological mechanisms, the disposition of the involved individual, and the ability to closely monitor the effects of the condition on diving activity without putting anyone at undue risk (Linaweaver and Bove 1990).

Physical Examinations and Recordkeeping

Initially, a complete history and physical examination should be done. The ADC Consensus Standards recommend a periodic examination at least every 2 years and annually after age 35. Some state agencies (e.g., CALOSHA) require annual physical examinations. Table 37.2 lists the medical tests recommended by the ADC. NOAA has similar recommendations (NOAA 1991). These recommendations are a minimum requirement; additional tests should be performed as medically indicated.

There are no specific requirements on how bone and joint surveys should be conducted but the protocol established by the British Medical Research Council Decompression Sickness Registry has been recommended in the past (Walder 1990). After the initial survey there are no specific time requirements for conducting the survey, and periodic examination is probably unnecessary in divers who have never exceeded a depth of 30 msw (100 fsw). In divers at risk for osteonecrosis, a periodic examination every 3 to 5 years and upon leaving diving would be adequate.

In addition to periodic physical examinations, additional examinations should be done after a diving-related injury or illness of more than 72-hours duration, any episode of unconsciousness, or known DCS with audio-vestibular or central nervous system dysfunction (OSHA 1991; ADC 1994). State agencies may have additional requirements.

Returning to diving after injury or illness requires judgment and must be handled on a case-by-case basis. The U.S. Navy has regulations regarding returning to diving after treatment for DCI, where all symptoms where completely relieved (USN 1993a). Times to return to normal diving activity for musculoskeletal symptoms with no neurological involvement range from 48 hours to 7 days, depending on response to treatment. Neurological symptoms consisting only of mild subjective sensory changes, which respond rapidly to treatment, require a 14-day wait, and more severe DCS symptoms or symptoms of CAGE require a 4-weeks wait. Very severe symptoms which are completely resolved require a 3-month wait. After the waiting period the diver must still obtain the recommendation of a physician experienced in diving medicine. Commercial diving companies may have their own criteria for the waiting period but they all require that the diver be released to full activity by a physician.

Even in cases where complete relief of all symptoms is obtained, some circumstances require that careful consideration be given to returning a diver to full diving status. If a diver has a history of frequent DCS on dives where his fellow divers experience little or no DCS he may be uniquely susceptible to DCS. Some of these individuals may go on to develop severe symptoms on otherwise innocuous dives. Even if they fully recover from DCS they should restricted from further diving. Divers who have residual symptoms of DCS after treatment or who require extended periods of treatment to resolve neurological symptoms should not continue diving. Divers who suffer repeated bouts of barotrauma should likewise discontinue diving unless the problem can be found and corrected. Divers suffering trauma, who have undergone surgery, or who have recovered from a medical disease should be restricted from diving until they can meet all physical requirements. Some conditions may require consultation with an expert in diving medicine.

A complete medical record of all examinations and physicians notes should be maintained for a minimum of 5 years from the date of the last hyperbaric exposure (ADC 1994; OSHA 1991).

Breathing Gas Purity and Analysis

Specifications for diver's breathing gas purity and for monitoring of gas compressor output are given by the U.S. Navy (1993a 1993b) and NOAA (1991). Generally, gas suppliers will have documentation attesting to the levels of trace contaminants in bottled gas but periodic independent analyses are prudent. Compressor outputs are usually analyzed semi-annually for purity. Gases from an untested source should never be used unless there is and emergency and no alternative is available. In these cases, the diver should breathe the air at the surface for several minutes before diving (NOAA 1991).

Whenever breathing gases are mixed by combining oxygen with any other gas, the oxygen content of the gas should be verified after the gas is removed from the charging station.

REGULATIONS

Commercial diving operations are regulated at both the federal and state level. At the federal level, OSHA has regulatory authority over all diving that is land based or from vessels within the United States waters not subject to Coast Guard inspection. The U.S Coast Guard has jurisdiction over not only regulated vessels but also over artificial islands or installations on the Outer Continental Shelf (e.g., oil rig platforms) (USCG 1989). Coast Guard regulations follow OSHA regulations for the most part. Certain states, such as California, may have their own regulations.

Any diver who gets paid for diving is considered a commercial diver. OSHA specifically exempts the following types of commercial diving from its regulations:

- Open-circuit compressed-air SCUBA within the no-decompression limits, performed solely for instructional purposes.
- Search and rescue or diving related to public safety by or under the control of a government agency.
- Diving done for research, development, or related purposes involving human subjects.
- Scientific diving under the direction and control of a diving program that has either a diving safety manual or a diving control (safety) control board.

The Coast Guard regulations give specific exception from its regulations to all but the first category listed above (USCG 1989).

Commercial diving companies in the United States will generally fall under either the OSHA or Coast Guard regulations. One requirement of these regulations is that the employer must develop and maintain a safe practices manual. Large commercial companies will generally develop their own manuals but the ADC Consensus Standards for Commercial Diving Operations (ADC 1994) is designed to meet all of the requirements of a safe practices manual. Commercial divers may use the ADC Consensus Standards as is or modify them for their own use, all they need do is to supply their own decompression tables since the standard does not cover this. The U.S Navy Diving Manual (USN 1993a ,1993b) or the NOAA Diving Manual (NOAA 1991) may also form the basis for developing a safe practices manual.

In the U.S., there are no legislated requirements for physical examination content or frequency for civilian divers at the federal level, although the ADC Consensus Standards are considered a reasonable minimum. Some states do have specific requirements. In Europe many countries have specific requirements concerning physical qualifications and for examining physicians (Elliott 1995).

In some types of commercial operations, semi-professional divers are employed. These are usually sport divers who periodically work for a fee, or commercial divers doing additional part time work. Their level of qualification may range from expert to almost none and they may follow no particular set of rules or regulations when diving. Abalone and sea urchin divers (Butler 1995) and part-time underwater repairmen who do small boat repair fall into this category. Although these divers have no standards per se, by the letter of the law they are commercial divers and fall under OSHA regulations. Recently, OSHA has required some commercial operations using semi-professional divers to treat them as commercial divers and to require that they meet minimum training standards (Lewis 1994).

MEDICAL COVERAGE OF DIVING OPERATIONS

In order to adequately cover a diving operation, medical personnel should be familiar with all aspects of the planned operation as well as all the equipment that will be used. Decompression procedures, emergency procedures, and evacuation procedures should all be agreed upon in advance. In addition, the location and method of access to a recompression facility should be determined as well as who will be medically responsible for conducting treatments.

OSHA regulations require that there be a designated person-in-charge at the dive location in charge of all aspects of the diving operation; this individual is usually called the Dive Supervisor. While not required by OSHA, a medically responsible individual should also be identified who would be responsible for determining fitness to dive, manage the treatment of medical conditions that may arise and direct any recompression treatments that may take place. The medically responsible individual should be a health professional who has received in depth training in the recognition and treatment of diving-related illness. If the on-scene individual is not a physician, a physician trained in diving medicine should be available for consultation by radio or telephone.

The size of the dive team depends on the type of operation to be carried out. The minimum dive team should consist of the Dive Supervisor, the diver, the standby diver and a dive tender. The standby diver is available anytime a diver is in the water and his job is to assist the diver should an emergency arise. The tender remains at the surface and is responsible for knowing the diver's location and status at all times. In some diving operations, an additional person at the dive site may function as a life support technician whose job is to ensure that the proper breathing gas and other life support functions are being supplied to the diver. On complex diving operations, a standby tender and a timekeeper/recorder may also be needed. Diving operations involving only a single diver without a standby diver or where there are divers in the water with no surface support personnel are hazardous and should be avoided.

OSHA requires that all dive team personnel be trained in cardiopulmonary resuscitation and first aid. In addition, they should have training in the use of equipment and tools used to perform their assigned tasks and have had training in diving physics and physiology.

OSHA specifies that a dual lock multiplace recompression treatment chamber be within 5 minutes of the dive site for any air dive in excess of 100 fsw or outside of the no-decompression limits and for any diving using a breathing gas other than air. In addition, a first aid kit approved by a physician must be at the dive site. Suggested contents for such kits have been published (USN 1993a; NOAA 1991) but they should be tailored to the specific type of diving operation. There should always be a sufficient supply of 100% O_2 delivered by a mask capable of continuously supplying >90% inspired oxygen fractions for at least the amount of time it will take to transport the diver to a recompression facility. Immediate availability of 100% O_2 generally improves treatment outcomes even if there is a delay to treatment (Moon et al. 1995).

The environmental conditions of the dive should be clearly spelled out, including the expected dive profiles, water temperature, sea state or currents, hazards associated with the work site, and marine life hazards. Once these are identified, a careful inventory of diver training, equipment, and emergency procedures should be conducted to ensure all anticipated events can be handled.

If marine life hazards at the proposed dive site are not known, local individuals or local government sources may be helpful. It is especially important to identify potential poisonous hazards for which anti-venom should be available. Edmonds (1989) is a good source of information on the subject.

A careful survey of the dive site should be done to identify any areas where injury or entrapment may occur. In spite of this, accidents may still happen resulting in injury that may be complicated by being trapped underwater or by decompression requirements. If the individual is trapped, the first concern is ensuring that adequate life support, especially breathing gas, is available until he can be extracted. Once the diver is extracted, the rapidity with which he is brought to the surface depends on his risk from not having his injury treated compared to his risk of injury from DCI. If a chamber is on site the diver can usually be brought to the surface and then immediately recompressed in the chamber where his injuries can be treated. If there will be a considerable period of time between surfacing and reaching a chamber the problem becomes more complex and there are no specific guidelines. A physician who is experienced in diving medicine will need to be consulted.

AVAILABLE RESOURCES

There are a variety of good publications, many of which have been referred to here, which can be consulted for more in depth information on diving. Some are listed in Table 37.3. In addition, the USN and NOAA diving manuals not only give a good basic background in diving physics and medicine but address may operational and emergency issues. The ADC Consensus Standards are designed to meet the requirements for a safety manual as specified by OSHA and Coast Guard Regulations.

Table 37.3. Diving resources in print.

Textbooks

Dangerous Marine Creatures. Edmonds, C. 1989. Reed Books Pty. Ltd. 2 Aquatic Drive, Frenches Forest, NSW 2086, Australia.

Diving and Subaquatic Medicine, Third edition. Edmonds, C., C. Lowry, and J. Pennefather. 1992. Butterworth-Heinemann Ltd, Boston.

Diving Medicine, Second edition. (Bove, A.A., and J.C. Davis eds.), 1990. W.B. Saunders, Philadelphia.

The Physiology and Medicine of Diving, Fourth edition. (Bennett, P.B. and D.E. Elliott, eds.),1993. W.B. Saunders, Philadelphia.

Journals

Aviation, Space and Environmental Medicine. Aerospace Medical Association, Alexandria, VA.

Journal of Applied Physiology. The American Physiological Society, Bethesda, MD..

Undersea and Hyperbaric Medicine (formerly Undersea Biomedical Research). Undersea and Hyperbaric Medical Society, Rockville, MD.

The textbooks provide more in depth information on diving medicine and are indispensable for the health professional. The four textbooks listed would provide a more than adequate reference library for almost any diving medicine problem. While papers addressing diving may be found in almost any journal, the three listed are where the majority of papers are published. Additionally, a great volume of diving literature is contained in government reports from facilities such as the Navy Experimental Diving Unit (Panama City, FL), and the Naval Medical Research Institute (Bethesda, MD) in the U.S., the Defense and Civil Institute of Environmental Medicine (North York, Ontario) in Canada, and the Defense Research Establishment (Alverstoke, Hants) and the Institute of Naval Medicine (Gosport, Hants) in England.

Consultations

There are several resources available that will provide access to experts in diving medicine. The Divers Alert Network (DAN) at Duke University in Durham North Carolina, provides a 24-hour emergency hotline for diving accidents and has access to physicians who are expert in the management of diving-related diseases, (919) 684-8111. Nonemergency information can be obtained during normal business hours by calling (919) 684-2948. DAN can provide the names of physicians throughout the world who are trained in diving medicine. DAN also provides many education and training courses available to the public.

The U.S. Navy has two facilities that maintain a 24-hour watch for emergencies. The Navy Experimental Diving Unit, (904) 230-3100, and the Naval Medical Research Institute, (301) 295-1839. While these facilities are concerned primarily with military diving, they can assist civilians in *bone fide* emergencies where no civilian assistance is available. The Diving Division of the Defense and Civil Institute of Environmental Medicine, (416) 635-2000, in Canada, and the Institute of Naval Medicine in England ([44] 1705 722-351) provide similar services.

The Undersea and Hyperbaric Medical Society (UHMS) in Rockville, Maryland, (301) 942-2980, can provide the names of its worldwide physician members who are available for consultations during normal business hours. They also have an extensive list of diving medicine-related publications, can provide a literature search service, and can direct individuals to educational resources.

At present, there is no agency in the United States or Canada that provides certification of physicians in diving medicine. Within the U.S. Navy ,there is a medical specialty in Undersea Medicine, which includes formal training in diving medicine. This requires attendance at a 6-month course that is generally considered to provide the best training in the field currently available. Unfortunately, it is not generally open to civilians. There are a number of courses in diving medicine, ranging from very basic to advanced, and the times and locations of these courses can be obtained through DAN or the UHMS. Physician qualification in the field is currently judged by where they received their training and the depth and breadth of their experience. Military training and experience are currently held in high regard, as is a wide variety of experience in the commercial sector.

In Europe, many countries require special licenses for physicians who perform diving physicals (Elliott 1995).

Certification of technical personnel and nurses is currently provided by the National Board of Diving and Hyperbaric Medical Technology in New Orleans, Louisiana, (504) 366-8871.

SUMMARY

Diving may be encountered in agriculture, forestry, and fisheries in support of a variety of underwater tasks, including construction, maintenance, and repair of underwater structures and the capture or harvesting of seafood. In these arenas the most likely type of diving which will be encountered will be at depths 100 fsw and shallower with the diver breathing air from a surface supplied underwater breathing apparatus or SCUBA.

Medical management of diving operations requires a knowledge of the physics and ergonomics of the diving environment. Especially important is knowledge of breathing apparatus function and how it will affect the diver's ability to work. In addition, a knowledge of the ergonomic restrictions of diving equipment will assist in evaluating work limitations and identifying tasks which may be especially hazardous. Medical conditions most likely to result directly from diving include barotrauma, hypothermia, decompression illness, hypercarbia, and oxygen toxicity. Recognition and treatment of these disorders again requires an in-depth knowledge of the diving environment.

Health monitoring should concentrate on periodic evaluation of the function of the target organs usually involved in diving diseases. These include the ear, lung, central nervous system, and the bones and joints. Determining whether an individual is physically qualified to participate in diving operations may not always be straightforward and in many instances must be handled on a case-by-case basis. Guidelines for physical qualification standards and establishing operational safety standards have been published for commercial divers by the Association of Diving Contractors and are available to anyone who needs them. Diving is regulated at the federal level by OSHA and the Coast Guard and also at the state level in many cases. A well-established network of diving expertise is available through such organizations as the Divers Alert Network and the Undersea and Hyperbalic Medical Society. This network can be easily accessed to assist in diagnosis and treatment of diving accidents, to identify training and education opportunities, and to assist in identifying sources of information relevant to diving operations and medicine.

REFERENCES

American Academy of Underwater Sciences (AAUS). 1991. *Proceedings of the American Academy of Underwater Sciences Repetitive Diving Workshop.* M.L. Lang and R.D. Vann, eds. Costa Mesa, CA: AAUS, Publication AAUSDSP-RDW-02-92.

Association of Diving Contractors (ADC). 1994. ADC Consensus Standards for Commercial Diving Operations, Change 1, 1994. Houston, TX: Association of Diving Contractors, Inc.

Adolphson, J.A., and T.E. Berghage. 1974. *Perception and Performance Under Water.* John Wiley and Sons, New York.

Becker, G.D. 1983. Recurrent alternobaric facial paralysis resulting from SCUBA diving. *Laryngoscope* 93(5): 596-98.

Bell, R.L, and R.E. Borgwardt. 1976. The theory of high-altitude corrections to the U.S. Navy standard decompression tables. The Cross corrections. *Undersea Biomed. Res.* 3(1): 1-23.

Bennett, P.B. 1993. Inert gas narcosis. In: *The Physiology and Medicine of Diving.* Fourth edition. P.B. Bennett, P.B. and D.E. Elliott, eds. Philadelphia: W.B. Saunders, pp. 170-93.

Bennett, P.B., and J.C. Rostain. 1993. The high pressure nervous syndrome. In: *The Physiology and Medicine of Diving.* Fourth edition. P.B. Bennett, P.B. and D.E. Elliott, eds. Philadelphia: W.B. Saunders, pp. 194-237.

Berghage, T.E. and D. Durman. 1980. *U.S. Navy Air Decompression Schedule Risk Analysis.* Bethesda MD: Naval Medical Research Institute, Report NMRI 80-1.

Bevan, J. 1993. Commercial diving equipment and procedures. In: *The Physiology and Medicine of Diving.* Fourth edition. P.B. Bennett, P.B. and D.E. Elliott, eds. Philadelphia: W.B. Saunders, pp. 33-52.

Boni, M.R., R. Schibili, P. Nussberger, and A.A. Buehlmann. 1976. Diving at diminished atmospheric pressure. *Undersea Biomed. Res.* 3(3): 189-204.

Butler, W.P. 1995. Maines urchin diver: a survey of diving experience, medical problems, and diving-related symptoms. *Undersea Hyperbaric Med.* 22(3): 307-13.

Calder, I.M. 1985. Autopsy and experimental observations on factors leading to barotrauma in man. *Undersea Biomed. Res.* 12(2): 165-82

Clark, J.M. 1993. Oxygen toxicity. In: *The Physiology and Medicine of Diving.* Fourth edition. P.B. Bennett, P.B. and D.E. Elliott, eds. Philadelphia: W.B. Saunders, 121-69.

Curley, M.D., and M.E. Knafelc. 1987. Evaluation of noise within the MK 12 SSDS helmet and its effect on divers hearing. *Undersea Biomed. Res.* 14(3): 187-204

Curley, M.D., and G.J. Robin. 1987. Effects of acute hyperbaric CNS oxygen toxicity on divre neurophysiological functioning. *Undersea Biomed. Res.* 14(2) (Supplement) Abstract #83.

Deuster, P.A., D.J. Smith, B.L. Smoak, L.C. Montgomery, S. Singh, and T.J. Doubt. 1989. Prolonged whole body cold water immersion: fluid and ion shifts. *J. Appl. Physiol.* 66(1): 34-41.

Deuster, P.A., D.J. Smith, B.L. Smoak, L.C. Montgomery, and T.J. Doubt. 1990. *COLDEX-86: Fluid and Electrolyte Changes During Prolonged Cold Water Immersion.* Bethesda MD: Naval Medical Research Institute, Report NMRI 90-133.

Donald, K. 1992. *Oxygen and the Diver.* Flagstaff, AZ: Best Publishing.

Doubt, T.J., J.R. Roberts, N.A.S. Taylor, R.P. Weinberg, and N.E. Holmes. 1990. *Pyridostigmine and Warm Water Diving Protocol 90-05: IV. Physical Performance.* Bethesda MD: Naval Medical Research Institute, Report NMRI 90-98.

Douglas, J.D.M. 1995. Salmon farming: occupational health in a new rural industry. *Occup. Med.* 45(2): 89-92.

Diving Sciences and Technology Corp. (DSAT). 1994. Recreational Dive Planner, Diving Sciences and Technology Corp. Product number 60054. Santa Anna, CA: International PADI, Inc.

Edmonds, C. 1989. *Dangerous Marine Creatures.* Frenches Forest, Australia: Reed Books Pty. Ltd.

Edmonds, C., C. Lowry, and J. Pennefather. 1992. *Diving and Subaquatic Medicine.* Third edition. Boston: Butterworth-Heinmann Ltd.

Egi, S.M. and A.O. Brubaak. 1995. Diving at altitude: a review of decompression strategies. *Undersea Hyperbaric Med.* 22(3): 281-300

Egstrom, G.H. 1993. SCUBA-Diving procedures and equipment. In: *The Physiology and Medicine of Diving.* Fourth edition. P.B. Bennett, and D.E. Elliott, eds. Philadelphia: W.B. Saunders, pp. 19-32.

Elliott, D.H. 1995. Medical Assessment of Fitness to Dive. Biomedical Seminars, Surrey England. U.S. distribution by Best Publishing, Flagstaff, AZ.

Elliott, D.H. and R.E. Moon. 1993. Long-term health effects of diving. In: *The Physiology and Medicine of Diving.* Fourth edition. P.B. Bennett, and D.E. Elliott, eds. Philadelphia: W.B. Saunders, pp. 583-604.

Farmer, J.C., Jr. 1990. Ear and sinus problems in diving. In: *Diving Medicine.* Second edition. A.A. Bove and J.C. Davis, eds. Philadelphia: W.B. Saunders, pp. 200-22.

Fischer A.A. 1978. *Atlas of Aquatic Dermatology.* New York: Grune and Stratton.

Francis T.J., and F.S. Golden. 1985. Non-freezing cold injury: the pathogenesis. *J. Royal Naval Med. Serv.* 71(1): 3-8.

Francis, T.J.R. and D.F. Gorman. 1993. Pathogenesis of the Decompression Disorders. In: *The Physiology and Medicine of Diving.* Fourth edition. P.B. Bennett, and D.E. Elliott, eds. Philadelphia: W.B. Saunders, pp. 454-80.

Gray, C.G., E.D. Thalmann, and R. Syklawer. 1981. *United States Coast Guard Emergency Underwater Escape Rebreather Evaluation.* Panama City FL: Navy Experimental Diving Unit, NEDU Report 2-81

Hampson, N.B. and R.G. Dunford. 1996. Pulmonary edema of SCUBA divers may occur in "cold" or "warm" water. *Undersea Hyperbaric Med.* 23 (supplement) Abstract #50.

Harabin, A. 1996. Gas physiology in diving: oxygen toxicity. In: *Handbook of Physiology, Section 4: Environmental Physiology Volume II.* M.J. Fregly and C.M. Batteis, eds. The American Physiological Society. New York: Oxford University Press, pp. 1005-08.

Hickey, D.D., W.T. Norfleet, A.J. Pasche, and C.E.G. Lundgren. 1987. Respiratory function in the upright, working diver at 6.8 ata (190 fsw). *Undersea Biomed. Res.* 14: 241-62.

Holmer, I. 1989. Body cooling with ice for warm-water diving operations. *Undersea Biomed. Res.* 16(6): 471-79.

Hyde, D., R.P. Weinberg, D.M. Stevens, and T.J. Doubt. 1990. *Pyridostigmine and Warm Water Diving Protocol 90-05: II Thermal Balance.* Bethesda MD: Naval Medical Research Institute, Report NMRI 90-96.

Kindwall, E.P. 1994. Medical aspects of commercial diving and compressed air work. In: *Occupational Medicine.* Third edition. C. Zenz, C., O.B. Dickerson, and E.P. Horvath, eds. pp. 343-83.

Lanphier, E.H. and E.M. Camporesi. 1993. Respiration and exertion. In: *The Physiology and Medicine of Diving.* Fourth edition. P.B. Bennett and D.E. Elliott, eds.) Philadelphia: W.B. Saunders, pp. 77-120.

Lewis, R. 1994. Fatalities associated with harvesting of sea urchins—Maine, 1993. *MMWR* 43(13): 235; 241-42.

Lin, Y-C., and S.K. Hong. 1996. Hybrebaria: breath-hold diving. In: *Handbook of Physiology, Section 4: Environmental Physiology Volume II.* M.J. Fregly and C.M. Batteis, eds. The American Physiological Society. New York: Oxford University Press, pp. 979-995.

Linaweaver, P.G., and A.A. Bove. 1990. Medical evaluation for diving. In: *Diving Medicine.* Second edition. A.A. Bove and J.C. Davis, eds. Philadelphia: W.B. Saunders, pp. 302-10.

Lundgren, C.E.G. 1996. Gas physiology in diving. Breathing under water: ventilatory needs. In: *Handbook of Physiology, Section 4: Environmental Physiology Volume II.* M.J. Fregly and C.M. Batteis, eds. The American Physiological Society. New York: Oxford University Press, pp. 999-1005.

Molvaer, O.I. and G. Albreksten. 1990. Hearing deterioration in professional divers: an epidemiological study. *Undersea Biomed. Res.* 17(3): 231-46.

Molvaer, O.I. and S. Eidsvik. 1987. Facial baroparesis: a review. *Undersea Biomed. Res.* 14(3): 277-95.

Moon, R.E., D. Uguccioni, J.A. Dovenbarger, G. deL Dear, G.Y. Mebane, B.W. Stopl, and P.B. Bennett. 1995. Surface oxygen for decompression illness. *Undersea Hyperbaric Med.* 22 (Supplement): Abstract #51.

NOAA. 1991. *NOAA Diving Manual, Oct. 1991*. U.S. Department of Commerce, Washington, DC: Superintendent of Documents.

Occupational Safety and Health Administration (OSHA). 1991. Code of Federal Regulations. 29 CFR, Chapter XVII, 1910, Subpart T- Commercial Diving Operations.

Ortendahl, T. 1987. Oral changes in divers working with electrical welding/cutting underwater. *Swed. Dent. J. Suppl.* 43: 1-53.

Ortendahl, T.W., G. Dahlen, H.O. Rockert. 1985. Evaluation of oral problems in divers performing electrical welding and cutting under water. *Undersea Biomed. Res.* 12(1): 69-76.

Parker, E.C., S.S. Survanshi, E.D. Thalmann, and P.K. Weathersby. 1994. Analysis of the risk of decompression sickness due to yo-yo diving using the USN probabilistic decompression model. *Undersea Hyperbaric Med.* 21 (Supplement): Abstract #21.

Pons, M., D. Blickenstorfer, E. Oechslin, G. Hold, P. Greminger, U.K. Franzeck, and E.W. Russi. 1995. Pulmonary oedema in healthy persons during scuba-diving and swimming. *Eur. Respir. J.* 8: 762-67.

Rogers, R.E. 1988. Renovating haldane. *Undersea J.* (3):16-18.

Schwartz, H.J.C., and M.D. Curley. 1985. Transderm scopolamine in the hyperbaric environment. *Undersea Biomed. Res.* 12(1) (supplement) Abstract #52.

Sterba, J.A. 1993. Thermal problems: prevention and treatment. In: *The Physiology and Medicine of Diving*. Fourth edition. P.B. Bennett and D.E. Elliott, eds. Philadelphia: W.B. Saunders, pp. 301-41.

Thalmann, E.D. 1974. *A Prophylactic Program for the Prevention of Otitis Externa in Saturation Divers*. Panama City FL: Navy Experimental Diving Unit, NEDU Report 10-74.

Thalmann, E.D. 1990. Principles of U.S. Navy recompression treatments for decompression sickness. In: *Treatment of Decompression Sickness. Forty-fifth Workshop of the Undersea and Hyperbaric Medical Society*. R.E. Moon and P.S. Sheffield, eds. Kensington, MD: Undersea and Hyperbaric Medical Society.

Thalmann, E.D. 1996. Gas physiology in diving: decompression. In: *Handbook of Physiology, Section 4: Environmental Physiology Volume II*. M.J. Fregly and C. M. Batteis, eds. The American Physiological Society. New York: Oxford University Press, pp. 1012-15.

Thalmann, E.D., D.K. Sponholtz, and C.E.G. Lundgren. 1979. Effects of immersion and static lung loading on submerged exercise at depth. *Undersea Biomed. Res.* 60: 259-90.

U.S. Coast Guard (USCG). 1989. Code of Federal Regulations. 46 CFR, Chapter 1, 197, Subpart B-Commercial Diving Operations.

U.S. Navy (USN). 1993a. US Navy Diving Manual Volume 1, Revision 3. (Air Diving). Naval Sea Systems Command, Washington, DC: NAVSEA 0994-LP-001-9010. Superintendent of Documents.

U.S. Navy (USN). 1993b. US Navy Diving Manual Volume 2, Revision 3. (Mixed-Gas Diving). Naval Sea Systems Command, Washington, DC: NAVSEA 0994-LP-001-9020. Superintendent of Documents.

Vann, R.D and E.D. Thalmann. 1993. Decompression physiology and practice. In: *The Physiology and Medicine of Diving*. Fourth edition. P.B. Bennett and D.E. Elliott, eds. Philadelphia: W.B. Saunders, pp. 376-432.

Vann, R.D., W.A. Gerth, P.J. Denoble, C.R. Sitzes, and L.R. Smith. 1996. A comparison of recent flying after diving experiments with published flying after diving guidelines. *Undersea and Hyperbaric Med.* 23(supplement): Abstract #49.

Walder, D.N. 1990. Aseptic necrosis of the bone. In: *Diving Medicine*. Second edition. A.A. Bove and J.C. Davis, eds. Philadelphia: W.B. Saunders, pp. 192-199.

Wilcock, S.E., S. Cattanach, P.M. Duff, and T.G. Shields. 1992. The incidence of decompression sickness arising from diving at fish farms. In: *Proc. of the Joint Meeting on Diving and Hyperbaric Medicine, Basel Switzerland, 15-19 Sept 1992*. European Undersea and Biomedical Society, *Stiftung fur Hyperbarmedizin*, Basel. J. Schmutz and J. Wendling, eds. Basel, Switzerland: Foundation for Hyperbaric Medicine.

Williams T.H., A.R. Wilkinson, F.M. Davis, and C.M.A. Frampton. 1988. Effects of transcutaneous scopolamine and depth on diver performance. *Undersea Biomed. Res.* 15(2): 89-98.

Wilmshurst, P.T., M. Nuri, A. Crowther, and M.M. Webb-Peploe. 1989. Cold-induced pulmonary oedema in SCUBA divers and swimmers and subsequent development of hypertension. *Lancet* 1(8629): 62-65.

Wolff, A.H., S.R.K. Coleshaw, C.G. Newstead, and W.R. Keatinge. 1985. Heat exchanges in wet suits. *J. Appl. Physiol.* 58(3): 770-77.

***NMRI and NEDU Reports can be obtained through the National Technical Information Service. 5285 Port Royal Road, Springfield VA 22161.

38

MOTION SICKNESS

David P. Thomson, M.S., M.D.
Air Medical Services, East Carolina University School of Medicine

> Motion sickness is a cluster of unpleasant symptoms occurring during or immediately after travel that can be problematic in many occupations including commercial fishing. The symptoms of motion sickness fall into two syndromes, the sopite syndrome and the nausea syndrome. The nausea syndrome consists of stomach discomfort, nausea, sweating, and pallor. The sopite syndrome consists of yawning, drowsiness, and a disinterest in activity. These two syndromes may occur independently or together. Motion sickness is seen with all forms of travel except walking on land. A discussion is given of current theories of the pathophysiology of motion sickness and the pharmacology of drugs used to treat the disorder.

INTRODUCTION

Few maladies, except perhaps the common cold, are as ubiquitous as motion sickness. Since mankind began using animals and boats for transportation, motion sickness has been a scourge. The attractiveness of sail as a form of transportation has undoubtedly been diminished by this problem. Indeed, the word *nausea* is derived from the Greek word for boat (Reason 1975, p.2). Since the time of Hippocrates, physicians have been attempting to combat this malady. The effect of motion sickness on commerce is unknown, but it has had a profound effect on military operations. Cicero said he would prefer death to seasickness, and Admiral Lord Nelson reportedly suffered from seasickness throughout his naval career (Reason 1975, p.4).

Motion sickness is a cluster of unpleasant symptoms occurring during or immediately after travel. These symptoms have been clustered into two groups, known as the sopite syndrome and the nausea syndrome. The nausea syndrome, which consists of stomach discomfort, nausea, sweating, pallor, and vomiting, is what most people think of as motion sickness. The sopite syndrome consists of yawning, drowsiness, and a disinterest in activity (Wright et al. 1995). The two syndromes may occur independently or together. Motion sickness is seen with nearly all forms of travel, walking on land being the lone exception.

The nineteenth and twentieth centuries have seen an explosion of transportation. Some form of motion sickness has been associated with land, sea, and air transportation. Carsickness, seasickness, and airsickness are well known among travelers. The advent of long duration spaceflight brought with it yet another variety of motion sickness: space motion sickness. Mal de debarquement, a sensation of continued motion after leaving the vehicle or ship, is another form of motion sickness.

Despite the ubiquitous nature of this problem, little research has focused on how it affects the fishing industry. A search of the literature, as well as an Internet search, revealed no research related to the effects of seasickness on fishing. Probably because of funding, research into seasickness has focused on naval crews, with some recent research involving oil rig crews. Although fishermen are thought of as a hearty group, it seems unlikely that they would all escape such a common malady.

The toll in economic damage and human suffering is staggering. In a series of articles published in early 1992, Landolt outlines the problem associated with just one type of boat, the totally enclosed survival craft. These boats are used by oil rig crews as escape devices for capsizing oil rigs (Landolt et al. 1992a, 1992b). He reports that about 90% of the personnel using these boats experience seasickness, regardless of the wave height. At least one death has been reported where seasickness was a factor, although details are sketchy. He makes a plea for increased research to treat this problem, stating that rescue of these personnel could require days, and current treatments are clearly inadequate.

Motion sickness resembles the common cold in one other way: despite its widespread occurrence, no clear cut etiology has been discovered, and no definitive treatment has been found. This chapter will provide the reader with a background in the anatomy and physiology of motion sickness, some theories on the causes and the current treatments available for the problem.

ANATOMY AND PHYSIOLOGY

The nervous system is provided with a constant stream of information that must be integrated to control motion and balance. This information is received from the eyes, vestibular system, and proprioceptors. Although all of these systems must work together, much of the research into motion sickness is centered on the vestibular system, located in the inner ear.

The vestibular system consists of the bony labyrinth, located in the petrous portion of the temporal bone, which contains the membranous labyrinth (see Figure 38.1).

The bony labyrinth consists of the vestibule, three semicircular canals, and the cochlea. Within it is a clear fluid, perilymph, and the membranous labyrinth. The vestibule forms the center of this structure, with the semicircular canals and cochlea projecting from it. Anterior to the vestibule is the cochlea, which functions primarily as an organ of hearing. The superior and posterior semicircular canals are vertically oriented, with the superior canal being positioned transversely, while the posterior canal is located in a more anteroposterior plane. The lateral semicircular canal is positioned in the horizontal plane. The point at which each of the canals attaches to the vestibule is called the ampulla (Warwick and Williams 1973, p. 1147). The membranous labyrinth is located within the bony labyrinth, and consists, in part, of a cochlear duct as well as semicircular ducts and ampullae corresponding to each of the canals. The utricle and saccule, contained within the vestibule, constitute the remaining portion of the membranous labyrinth. Nerves arise from the three ampullae, as well as from the saccule and utricle, eventually coming together to form the vestibular nerve (Clemente 1987, fig. 803).

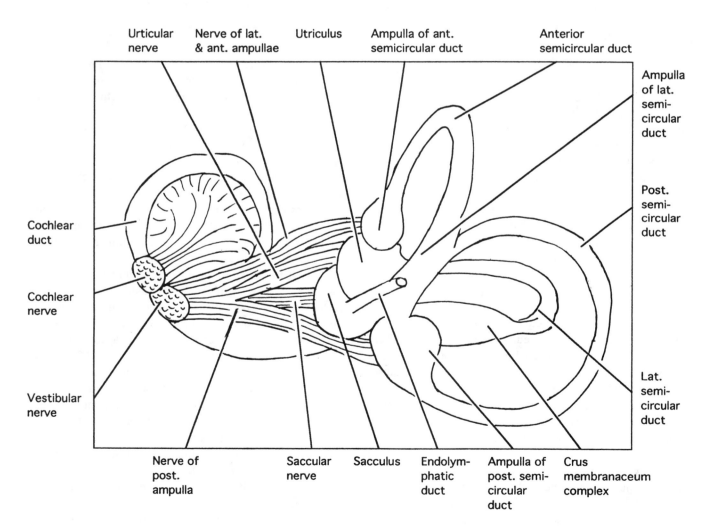

Figure 38.1. The right membrous labyrinth (adapted from CD Clemente, *Anatomy: A Regional Atlas of the Human Body*, 3rd Edition, Baltimore: Urban & Schwartzenberg 1987.)

Within the ampullae of the semicircular ducts, hair cells detect motion of the endolymph. Similar cells are located within the saccule and utricle, but these interact with the otolithic membrane, a specialized surface containing calcite crystals called otoliths. These hair cells deliver information to their respective portions of the vestibular nerve (Warwick and Williams 1973, p. 1152).

The semicircular ducts provide the organism with information regarding dynamic motion of the head in space. The otoliths are affected by the pull of gravity, indicating the position of the head relative to gravity (Reason and Brand 1975, p.83-85).

Along with the vestibular apparatus, people receive position information from the proprioceptive nerves within the joints and muscles. The eyes also function to provide information about the location of the person in space. Projections from the vestibular nerves send information to the ocular nerves via the medial longitudinal fasciculus, creating the vestibuloocular reflex. Projections from the vestibular nerves to the cerebellum act as modulators of this reflex. The vestibular nerves interact with the proprioceptive nerves in the vestibulospinal pathway, providing information about posture. Vestibular information is brought to the conscious level by projections to the cerebral cortex (Daroff 1994, p.96-97).

The three position-sensing systems: the eyes, vestibular system, and proprioceptive system, function together to help the organism locate itself in space. In the case of the vestibular system, patients who are born without this system may never know of its loss, unless one of the other systems fails. Disruption of any of these systems in the adult, however, results in derangement, which is only partially compensated for by the other systems.

ETIOLOGY

Although motion sickness is a common malady, we understand little about its causes. Several theories have been advanced. Nineteenth century theories included suggestions that there was an increase or decrease in blood flow to the head. Another hypothesis was that motion of the visceral organs produced the nauseogenic stimulus. This hypothesis and other similar theories were dominant until early in this century, because they addressed the site of discomfort: the stomach. Late in the nineteenth century physicians became aware of the role of the vestibular system, the eyes, and the proprioceptors in the detection of motion (Reason and Brand 1975, p. 10-16). This realization eventually developed into the currently dominant theory: the sensory conflict theory.

Sensory Conflict Theory

The sensory conflict theory proposes that motion sickness derives from conflicts between the visual system, vestibular system, and the proprioceptive system. Reason and Brand, in their classic work, describe several types of conflicts that may result in motion sickness. Their theory is that when the three motion-detecting systems provide inputs that the individual feels are incompatible, motion sickness results. One example of this would be the person who is reading in a moving vehicle or ship. The vestibular and proprioceptive systems appreciate the motion of the ship, but the eyes are fixed on an immobile target. Another example would occur when one sits near the screen in a large theater where the film is depicting motion. The eyes perceive motion, but the vestibular and proprioceptive systems know that the person is seated and immobile. Both of these are examples of "visual-inertial" conflicts.

They also describe "canal-otolith" conflicts in which the semicircular canals and the otoliths provide conflicting signals. An example is space motion sickness, in which the otoliths receive no gravitational pull but the semicircular canals continue to perceive motion. Closer to earth, positional alcoholic nystagmus, with its associated room spinning and nausea, is thought to result from alcohol affecting the specific gravity of the cupula faster than it affects the endolymph. This relative change causes the cupula to become buoyant, and therefore responsive to the pull of gravity. Since the body assumes that the semicircular canals give input regarding angular accelerations, the gravitational pull is assumed to be angular, and a sensation of room spinning results. At the same time, however, the otoliths and proprioceptors are at rest, with the resulting conflict producing motion sickness (Reason and Brand 1975, p. 123-126). Support for this theory is demonstrated by the lack of motion sickness susceptibility among subjects without an intact vestibular system (Cheung et al. 1991).

Visceral Gravireceptors

Although Reason's conflict theory is currently most accepted, other theories have been proposed. In a combination of the old and the new, von Gierke has suggested that a conflict occurs between the otolith system and visceral gravireceptors. These gravireceptors, whose existence has been postulated but not demonstrated anatomically, consist of the abdominal organs as they undergo stretching and compression. For purposes of this theory, the abdominal viscera are considered to be one fluid-filled sac. When individuals are subjected to cephalocaudad oscillations at certain frequencies, the abdominal viscera resonate at a different frequency than that of the otoliths, producing a conflict. How the nervous system acquires this vibratory information from the viscera is at present unclear. The extremely nauseogenic nature of cephalocaudal vibration, when compared to the benign nature of anteroposterior vibration, is felt to be evidence for this theory. With both of these stimuli, the otoliths will travel in a similar path, but the viscera will travel further during cephalocaudad vibrations. It is postulated that the difference must be due to the different perception of the viscera (von Gierke and Parker 1994). This theory has only recently been developed, and clearly further is needed before this model can be fully accepted.

Oculocardiac Reflex

An alternate hypothesis is based on the oculocardiac reflex, a phenomenon produced when traction is placed on the oculomotor muscles during strabismus surgery. Ebenholtz postulates that the vagal symptoms of motion sickness, such as nausea and bradycardia, are produced because of the stretch of eye muscles during the high frequency nystagmus produced by vestibular stimulation.

The author proposed to prove this theory by attempting to block motion sickness by inducing retrobulbar anesthesia (Ebenholtz et al. 1994). To date, this theory remains untested, but an experiment by a Pennsylvania State University group provides some support. Using an optokinetic drum stimulus, this group compared the degree of motion sickness induced with and without vision restricting devices. When the subject's visual field was reduced to approximately 15°, significantly less motion sickness was noted (Stern et al. 1990).

Fluid Shift Theory

Fluid shifts have been implicated as an element in space motion sickness, a unique form of motion sickness. The microgravity (near weightless) environment of orbiting spacecraft, such as the Space Shuttle and Mir Space Station causes body fluids to move from the peripheral tissues, especially in the lower limbs, into the central circulatory system. This "central volume expansion" is thought to make some people more susceptible to motion sickness. This has been offered as an explanation of why some astronauts and cosmonauts who are not susceptible to motion sickness on earth become violently ill in space (Simanonok and Charles 1994).

Heredity

Susceptibility to motion sickness is thought to have some basis in heredity. Using a rotating optokinetic drum to provoke motion sickness, Stern and his associates at Pennsylvania State University found that subjects of Chinese descent were considerably more likely to become motion sick than either African-American or European-American subjects (Stern et al. 1993). Hamid has reported that subjects with a history of motion sickness reported at least one member of their family who also had motion sickness, also alluding to a genetic connection (Hamid 1991).

Neurologic Connections

Others have suggested connections with various neurologic phenomena. The nausea that is frequently associated with narcotic administration suggests that endorphins, the endogenous opiates, may play a role in motion sickness. In one study, naloxone, an opiate antagonist, was found to increase the susceptibility to motion sickness (Allen et al. 1986). Other noxious stimuli, such as pain, also increase endorphins, so it would seem logical that there would be an interaction with motion sickness, but further work is needed to refine this theory.

The similarities in electroencephalographic recordings of motion sickness and partial complex seizures led Chelen to suggest that motion sickness may be a form of seizure activity (Chelen 1993a). In this study, electroencephalographic recordings were performed while subjects were made motion sick in a rotating (Coriolis) chair. The recordings demonstrated increased brain waves in areas and patterns similar to some types of partial complex seizures. The National Aeronautics and Space Administration has taken advantage of this by using phenytoin to counteract seasickness in booster recovery ship crews (Woodard 1993). Personnel assigned to ships recovering the Space Shuttle boosters were given phenytoin in amounts sufficient to produce therapeutic blood levels. A placebo control was used. Subjects were tested both with a rotating chair stimulus and under actual conditions at sea in a 41-foot vessel. Additionally, since divers would be given this medication, subjects were tested in a hyperbaric chamber. The study indicated that phenytoin was effective in preventing motion sickness under both experimental and operational conditions. No impairment in the capabilities of either the shipboard sailors or the divers was noted.

Heightened sympathetic nervous system tone, as measured by the rate of salivary secretion during motion stimulus, has been implicated in making subjects more susceptible to motion sickness (Gordon et al. 1992). The authors admit, though, that this work is preliminary, and further study is indicated. Heightened vagal tone (parasympathetic tone) has also been suggested as being protective against motion sickness (Uijtdehaage 1993). Although commonly thought of as an antagonist of parasympathetic tone, scopolamine appears to stimulate central parasympathetic centers, such as the vagal nucleus, while the anticholinergic effects are primarily peripheral. Studies of this interaction are in their infancy, with more data being needed to fully understand the effects of motion sickness on the various components of both the central and peripheral nervous systems.

Personality

Personality traits appear to influence motion sickness. Subjects were evaluated with the Eysenck Personality Questionnaire, a 90-question test that provides ratings of extroversion, neuroticism, psychoticism, and lying. A positive correlation was found between psychoticism and lack of susceptibility to seasickness. The authors are uncertain as to the relevance of this particular trait, but their work clearly indicates how higher functions, such as personality, affect this problem (Gordon et al. 1994).

Aerobic Fitness

Two authors have noted a correlation between increasing aerobic fitness and motion sickness. Banta tested subjects for their aerobic capacity, initially rating their aerobic capacity as high, medium, or low. When the high aerobic capacity group was compared with the low aerobic capacity group in their ability to tolerate a motion challenge, it was noted that the high aerobic capacity group had a significantly lower tolerance (Banta et al. 1987). The second study evaluated susceptibility to motion sickness before and after an aerobic training period of eight weeks. They noted a marked increase in motion sickness following this aerobic training (Cheung et al. 1990). Both of these authors suggest a role for increased vagal tone as well as hormonal changes. Neither study, however, provides a definitive answer in this regard.

Mal de Debarquement

Mal de debarquement is a phenomenon where the person continues to perceive motion even after that motion has stopped. The patient continues to experience the sensation of a pitching ship after return to dry land. This malady might be thought of as an "inverse" motion sickness, however there is no associated nausea, vomiting, or headache (Murphy 1993). Three-quarters of naval crew members studied by Gordon experienced mal de debarquement (Gordon et al. 1992). Essentially no research has been done on mal de debarquement, which is a common problem that may help explain motion sickness.

Etiology: Summary

To summarize, the etiology of motion sickness remains controversial. Both increased parasympathetic tone and increased sympathetic tone have been implicated; both have been suggested as being protective. Several sites within the nervous system and the sensory organs clearly are involved, but how they interact to produce motion sickness remains a mystery. The dominant theory—the sensory conflict theory—connects the trigger to the symptoms, but does not explain the pathways by which this connection is made. Fortunately, empiric research into treatments has not been slowed by this lack of scientific evidence.

TREATMENT

Because no clear-cut etiology has been determined, a variety of treatment modalities have been tried. Many of these are pharmacological treatments, including drugs from many different classes. Nonpharmacologic means have been tried as well. Some of these employ psychological strategies, while others use nontraditional and Far Eastern medical technologies.

Psychological and Alternative Medicine Treatments. Treatment of motion sickness using behavioral methods dates to the notion that sailors needed to acquire their "sea legs." This was an unknowing application of extinction, a basic principle of behavior modification. Today this principle is used primarily with military aircrew. Bagshaw has reported that 20 to 30% of pilot trainees develop motion sickness, but that this frequency diminishes as the trainee becomes used to his new environment (Bagshaw and Stott 1985).

Some individuals never completely adapt, and for them specialized programs have been developed. Golding and Stott, working for the Royal Air Force, reported a system in which air crew were methodically desensitized to motion sickness. Using a Coriolis chair, subjects were subjected to nauseogenic stimuli, the intensity of which was rated by the subject on a 1 (no symptoms) to 4 (moderate nausea) scale. The speed of the chair was gradually increased until the subject reached the level appropriate for their group. Subjects in the low intensity group were stopped when they reached a level of 2, while the high intensity group was kept in the chair until their symptoms reached a level of 4. The sessions were conducted twice daily for one week, at which time the subject was tested to see how much stimulus was needed to provoke a symptom level of 4. Subjects were crossed over into the opposite group during the next week. No differences in tolerance were noted between the groups, suggesting that it was not the intensity, but more likely the frequency of the challenge, which increased the subject's tolerance (Golding and Stett 1995).

Similar programs are being developed to improve adaptation to space motion sickness, a nearly universal phenomenon among Shuttle and Mir crews (Harm and Parker 1994). These systems are new, and data regarding their effectiveness is not yet available.

All of these programs suffer from one problem: desensitization to one type of motion sickness may not generalize to other types of motion sickness. Thus the pilot who overcomes airsickness may still become violently ill when placed aboard a ship. The frequency of space motion sickness among space crews, who are generally chosen because of their aviation experience, is a striking example of this lack of generalizability. Many space crew members who are resistant to motion sickness during earth-based training become ill shortly after exposure to microgravity. Some slight increase in tolerance to all stimuli may be possible, however, if the desensitizing stimulus is of sufficient magnitude (Dobie and May 1990). Thornton writes that Space Shuttle crew members, once exposed to the significant stress of space motion sickness, rarely experience motion sickness on return to earth (Thornton et al. 1987).

Several programs use relaxation and confidence building techniques along with nauseogenic challenges to promote motion sickness resistance. Dobie notes that incapacitation due to motion sickness may be in part due to how the individual perceives the challenge and reacts to it (Dobie et al. 1987). His program, developed at the U.S. Naval Biodynamics Laboratory, uses cognitive and behavioral therapy as an adjunct to desensitization training. Subjects with a history of motion sickness susceptibility were given confidence-building counseling, where they were reassured that their motion sickness was a normal response. Following this counseling, subjects were given desensitization training, where they were exposed to a rotating optokinetic drum. As they learned to relax, longer and longer runs with the drum were tolerated. Subjects were tested with the rotating optokinetic drum before and after treatment and asked to stop the experiment should they become uncomfortable. Subjects who had undergone this training performed better than a control group, which received no similar training. Their results suggest that this two pronged approach of behavioral and cognitive therapy combined with desensitization works better than desensitization alone. The Canadian armed forces have developed a program that uses biofeedback relaxation therapy prior to desensitization training. Subjects are trained in the use of a biofeedback device and relaxation tapes. They report a 77% success rate with this system (Banks et al. 1992). The U.S. Air Force has used a similar program. Fliers who experience chronic severe motion sickness are treated with biofeedback, relaxation therapy, and brief psychotherapy, in addition to the desensitization stimulus. They claim an 85% success rate with this technique (Jones et al. 1985).

Aids to Habituation

Research suggests that any aid that could increase the ability of the subject to tolerate exposure might speed the habituation process. It also has an impact on the practical operational problem of whether giving medications to combat motion sickness improves performance or delays adaptation. Several studies have looked at this. In a 1986 study, researchers noted an acceleration of habituation when subjects received scopolamine/d-amphetamine combination (Wood et al. 1986). They postulated that the improved habituation was due to greater stimulation being tolerated. They did, however, note a rebound sensitivity when the medication was stopped but the stimulus continued. Lackner and Graybiel gave subjects promethazine injections before exposing them to motion in a slow rotation room. When compared to placebo, there appeared to be no difference in the ability to adapt to this stimulus (Lackner and Graybiel 1994).

At about the same time, Wood took another look at this problem, but with different results. He and his colleagues found that individuals treated with scopolamine and exposed to a rotating chair stimulus did not adapt as fast as subjects given placebo (Wood et al. 1994). Because there were differences in the medications used, it is difficult to compare these studies, but the suggestion is that there may be medications which suppress the symptoms of motion sickness without delaying habituation. Specific medications, combinations, and dosages still need to be worked out.

Acupuncture

Acupuncture has been used by the Chinese for thousands of years to treat nausea and other gastrointestinal complaints. This suggests that acupuncture and the related technique, acupressure, might be effective in the treatment of motion sickness. Commercially available accupressure bands have even been advertised as treatment for travelers susceptible to motion sickness. Two studies (Warwick-Evans et al. 1991; Bruce et al. 1990) looked at the efficacy of these bands. Both of these studies used a rotating chair with head motion as the nauseogenic stimulus.

Warwick-Evans was unable to demonstrate any improvement over a placebo band. Bruce found that scopolamine provided a significantly better treatment when compared with the band.

In contrast to these studies, Hu looked at the effect that manual accupressure at the P6 point had on nausea. The P6 point is a site about 3 cm proximal to the distal palmar crease, between the flexor carpi radialis and palmaris longus tendons. It is commonly used by acupuncturists to treat gastrointestinal symptoms. The stimulus for this experiment was a rotating optokinetic drum consisting of alternating black and white vertical stripes on the inside of a 50 cm diameter container. The subject sat on a stool within the drum, which was spun, producing a visual stimulus (Hu et al. 1995). His group found that manual accupressure produced a statistically significant improvement in nausea scores when compared to control and sham groups. Objective signs, such as tachygastria, also decreased. It is worth noting, though, that some improvement was also seen with subjects who were treated with accupressure at a point other than P6. Hu also used electrical acustimulation at point P6, and in this experiment was able to show a significant improvement over sham and control populations (Hu et al. 1992). Once again, though, the sham group had some improvement over control, although not as much as the acustimulation group.

It would appear difficult to separate the treatment from the halo effect produced by the researcher's attention to the subject. Taken together, these studies suggest that acupressure and acustimulation in and of themselves may not produce any benefit, but larger trials are clearly indicated.

Pharmacologic Therapies

Early Trials. Throughout the years, many drugs have been used for the treatment of motion sickness. Probably the first true systematic studies of drugs for the treatment of motion sickness occurred during World War II. Holling, McArdle, and Trotter surveyed several classes of drugs by giving large groups of soldiers various medications and compounds. The subjects were then placed on small ships, such as minesweepers, in heavy seas. They returned to port when 40% of the crew had become sick, or at 6 hours. Medications included several proprietary remedies, atropine, methedrine, barbiturates, and scopolamine (hyoscine). They noted that scopolamine produced dry mouth, and were concerned about the possibility of drowsiness and diminished sweating, but because of wartime limitations were unable to pursue other substances (Holling et al. 1944). Smith reviewed the literature and came to the conclusion that scopolamine is the drug of choice for the treatment of airsickness (Smith 1948).

In 1949, dimenhydrenate was discovered, and eventually marketed as Dramamine® (Chinn 1956). The success of this development effort led the U.S. armed forces to investigate several medications. Chinn led this effort, in which soldiers on regular trans-Atlantic troop transports were used as subjects. His group reported that meclizine, cyclizine, promethazine, and scopolamine produced the best results against motion sickness. They noted, however, that scopolamine (1 mg initial dose, then 0.5 mg daily) produced significant side effects, including dizziness, drowsiness, dry mouth, and blurred vision, when used for several days.

Scopolamine

Scopolamine, the prototypical motion sickness medication, is still used today, despite its side effects. The primary side effect that users complain of is sedation. Various doses and routes of administration have been tried in an effort to minimize sedation. Holling's researchers recommended 0.6 to 1.2 mg doses, administered every 6 hours. They also suggested that 0.3 mg doses may be used for maintenance during longer voyages, this reduction resulting in reduced side effects (Holling et al. 1944). A 1966 review of the literature reported that the 0.6 mg oral dose was the most common dose, and that it resulted in approximately 60% effectiveness (Wood et al. 1966). They also noted that dry mouth, blurred vision, and sleep disturbances have been associated with its use.

Combinations of scopolamine with stimulants have been used, as well. In a 1968 paper, Wood and Graybiel noted that scopolamine with amphetamine (1.2 mg scopolamine with 20 mg dextroamphetamine) was superior to any single drug therapy tested, and had minimal side effects (Wood and Graybiel 1968).

The effort to minimize side effects and make administration more convenient led to the development of transdermal scopolamine. The Israeli Navy evaluated the use of transdermal scopolamine in clinical trials at sea. Their results demonstrated a significant protective effect in the group using transdermal scopolamine. Both treatment and placebo groups experienced sleepiness, and the transdermal scopolamine subjects reported headache, dry mouth, and blurred vision more frequently than the control subjects. This difference did not, however, reach statistical significance (Attias et al. 1987). To assess the operational significance of the side effects, Gordon, also with the Israeli Navy, administered professional naval skills tests and cognitive tests to subjects given transdermal scopolamine and placebo. Although subjects in the test state reported higher rates of dry mouth and fatigue, there were no significant differences on any of the objective tests (Gordon et al. 1986).

Table 38.1. Motion sickness medications.*

Medication	Class	Route	Dosage	Efficacy	Comments / Side Effects
Cinnarizine	calcium channel blocker	po	50 mg	Effective	Investigational - Probably useful in mild motion sickness
Dimenhydrinate	antihistaminergic	po	50-100 mg / 6 hrs	Effective	Available without prescription; may cause drowsiness
Dimenhydrinate	antihistaminergic	im	50 mg / 6 hrs; children 1mg/kg / 6 hrs IM only	Effective	IV must be diluted in 10 cc saline; may cause drowsiness
Doxepin	antidepressant / antihistamine	po	10-100 mg	Effective	Very sedating
Flunarizine	calcium channel blocker	po	Investigational	Possibly effective	Investigational - Not avaliable in the U.S.
Ginger root	herbal	po	one capsule	Not demonstrated	A "natural" remedy, with many followers, but not proven by scientific trials; no known side effects
Lidocaine	anesthetic	IV	100 mg	Not effective	May cause seizures in overdose
Meclizine	antihistaminergic	po	12.5 - 50 mg daily	Possibly effective	May be available without prescription; some preparations are chewable; Somewhat sedating
Metoclopramide	antidopaminergic	po	10 mg	Not effective	Increases gastrointestinal motility - may be a useful adjunct
Metoclopramide	antidopaminergic	IV	10 mg	Not effective	See above
Phenytoin	antiepileptic	po	1 gram	Probably effective	Must be administered well in advance of the challenge; Some patients complain of sedation; ataxia may occur with loading dose, but usually resolves
Phenytoin	antiepileptic	IV	2 gram	Probably effective	See above
Promethazine	antihistamine / anticholinergic	IM/IV	12.5 - 50 mg / 6 hours	Very effective	May be used after the onset of symptoms; may produce significant sedation Risk of dystonic reaction
Promethazine	antihistamine / anticholinergic	po, pr	25 mg / 12 hours		Suppositories may be used after the onset of nausea. Also see above.
Scopolamine	anticholinergic	td	1.5 mg/3 days	Very effective	Must be placed on patient at least 6 hours before stimulus; not currently available; Side effects include dizziness, headache, difficulty concentrating. Reports of mild psychosis with chronic use.
Scopolamine	anticholinergic	sq	0.6 mg, initially; 0.3 mg/ 6 hours; children:0.006 mg/kg	Very effective	May have significant side effects, such as blurred vision and amnesia, especially if used for prolonged periods.
Terfenadine	peripheral antihistamine	po	300 mg	Effective	Not sedating. Only one study thusfar. Should not be taken with macrolide

* Check with your physician before taking any of the medications listed in this chapter.

In another trial, under operational conditions, a Dutch group evaluated transdermal scopolamine against placebo during a 7-day cruise on a frigate. These trials were conducted in moderate to heavy seas, with waves of 1 to 6 meters and winds of from 5 to 9 on the Beaufort scale. They noted a reduction in nausea in the transdermal scopolamine group during the first two days. After about the first 2 to 3 days, there was no difference between the groups, which the authors attribute to adaptation by the placebo group. Side effects included salivation, headache, dizziness, difficulty concentrating, diplopia, palpitations, and skin irritation. They concluded that the patches were effective for treating seasickness, as long as they were in place at least 4 hours before embarkation (van Marion et al. 1985). Using microcomputer-based testing, Kennedy noted decreasing performance with increasing doses of scopolamine. He states that oral doses over 1 mg are likely to impact on performance (Kennedy 1990).

To evaluate the effects of these drugs on performance, Wood used a "pursuit meter." This consisted of a pattern on a computer screen, which the subject was to trace using a joystick. The subject was scored on the number of errors made while attempting to trace each of the several patterns which could be displayed. He noted that scopolamine at 0.25, 0.5, and 0.6 mg, marezine 50 mg, meclizine 50 mg, and dimenhydrenate 50 mg produced no decrements on the pursuit test when compared with placebo. He also found that scopolamine 1mg, combined with d-amphetamine 5 mg and promethazine with d-amphetamine 10 mg, did not cause any performance decrement (Wood et al. 1985). In another study, though, he found promethazine alone produced performance decrements equal to approximately one to two drinks of an alcoholic beverage (Wood et al. 1994). Similar results have been noted by the U.S. Federal Aviation Administration's Aviation Psychology Laboratory (Schroeder et al. 1985). Dahl points out that the transdermal route has the advantage of being useful even when nauseated, but that the kinetics of transdermal scopolamine suggest that, to be effective, it needs to be applied several hours before the motion challenge (Dahl et al. 1984). Other researchers have noted that at least 8 hours must elapse between application and the motion challenge (Levy and Rapaport 1985). At the present time Ciba Self-Medication, the manufacturer of the transdermal scopolamine patches is experiencing difficulties with the manufacturing process and anticipates production will resume no sooner than 1997 (Ciba Self-Medication Consumer Affairs Hotline, Personal Communication, 23 January 1996).

Because of the delay associated with transdermal scopolamine, and the problems associated with taking oral scopolamine when nauseated, NASA investigators have looked at other routes of administration. One of these is a buccal form of scopolamine. This preparation has the advantage that it is absorbed through the oral mucosa and it appears to reach therapeutic blood levels in about 20 minutes (Norfleet et al. 1992). At the present time, the author is not aware of any commercially available buccal preparation. Injectable medications are another alternative in settings where trained personnel are available. Graybiel and Lackner used injectable scopolamine (0.5 mg IM) and promethazine (50 mg IM) and found both to be effective in preventing sickness induced by parabolic flight (Graybiel and Lackner 1987).

Stimulants

Since the major complaints with scopolamine use have been related to sedation, combinations of scopolamine with stimulants have been tried. When used in combination with ephedrine, no differences in efficacy between scopolamine alone, scopolamine with ephedrine, or placebo were noted (Tokola et al. 1984). The volunteers were asked to provide a subjective rating of the effects of the treatment on their ability to perform their duties. In this measure, scopolamine alone was rated best, while the placebo and scopolamine with ephedrine groups were equal. There were no differences in the side effects between the treatments. The authors note that they used only 28 subjects in an open sea trial and 30 subjects in a rotating chair trial, and that they may not have had sufficient power to demonstrate differences. Using a combination of 5 or 10 mg dextroamphetamine with 1 mg scopolamine, Wood and his colleagues were able to reduce the incidence of side effects when compared to scopolamine alone (Wood et al. 1985).

Amphetamine combinations may also have effects against the Sopite Syndrome—the lethargic feeling, gastric immotility, and brain wave slowing that often accompany motion sickness. In a 1990 study, Wood recommended the scopolamine/dexedrine combination as the preparation most effective against all the symptoms of motion sickness (Wood et al. 1990).

The use of stimulants to counteract the side effects of scopolamine has also been noted to increase the efficacy of the preparation over scopolamine alone. This suggests that sympathomimetics may themselves be effective against motion sickness. Kohl and his colleagues investigated this phenomenon, publishing their results in two articles (Kohl et al. 1986, 1987c). In a placebo controlled study they administered 5 different sympathomimetic medications, 20 mg methamphetamine, 25 mg phenmetrazine, 37.5 mg phentermine, 20 mg methylphenidate, and 75 mg pemoline, to subjects who were then subjected to a rotating chair motion challenge. All of the medications produced statistically significant improvement over placebo. His second article attempted to explain the neuropharmacology of this phenomenon, suggesting that enhanced dopaminergic transmission in the basal ganglia is the basis for the improvement seen with these drugs.

Dimenhydrinate

Dimenhydrinate is another drug with a long history in the treatment of motion sickness. Some authors rate its effectiveness as equal to that of transdermal scopolamine (Bennett 1995, pg. 470). It is available over the counter in both liquid and tablet forms, making it the most convenient and most popular of the motion sickness medications. It is also available in a parenteral form for IM and IV use. Unfortunately, little recent research has been performed on this drug. In the only recent study of this drug, Muth was able to demonstrate a statistically significant improvement in symptoms when dimenhydrinate is compared with placebo (Muth et al. 1995).

Ginger Root

Several groups have looked at the use of ginger root for the treatment of motion sickness. This use is based on the historical use of ginger (Zingiber officinale) to quiet gastrointestinal distress. In a study comparing ginger root powder, dimenhydrenate, and placebo, Mowrey was able to demonstrate improved resistance to motion sickness in subjects given ginger root, as compared to the other substances (Mowrey and Clayson 1982).

Grøntved and his colleagues conducted a laboratory trial of ginger root, followed by an operational clinical trial. These produced mixed results. Their first study used a caloric stimulus, and evaluated vertigo and nystagmus. Subjects given ginger root had improved tolerance to this stimulus, when compared to placebo (Grøntved and Hentzer 1986). When he took this drug to trial on the open sea, the trend was a decrease in symptoms with ginger root, but the difference did not reach clinical significance (Grøntved et al. 1988). Using a rotating chair stimulus, Stewart could find no improvement over placebo when subjects were given either powdered ginger or fresh ginger (Stewart et al. 1991).

Lecithin

Another attempt to find a "natural" treatment actually discovered a potentiator of motion sickness. Phosphatidylcholine, better known as lecithin, was tested as a treatment in this experiment. Subjects receiving an acute dose of lecithin were more susceptible to sickness, and this seemed to diminish with chronic use. The authors postulate that the stimulation of the cholinergic system by choline explains this phenomenon (Kohl et al. 1985). It appears that we will need to look for other substances to find a "natural" cure for motion sickness.

Phenothiazines

Phenothiazines are frequently prescribed for nausea related to infectious or toxic causes. Although all phenothiazines, including prochlorperazine and thiethylperazine are effective for the treatment of nausea, only promethazine appears to have protective effects against motion sickness (Wood 1979). The problem with all of these drugs, like so many of the anti-motion sickness medications, is that they produce sedation in most cases. Wood suggests this is a problem only for the ship's or aircraft's crew, but that it might actually be beneficial to the passengers. The incidence of space motion sickness in the Space Shuttle program has led NASA researchers to look closely at promethazine, despite its apparent drawbacks. Although not a rigidly controlled study, they noted that 75% of crew members who experienced space motion sickness achieved relief with an intramuscular dose of promethazine (Davis et al. 1993). Unlike earth based studies, the incidence of sedation among shuttle crews treated with promethazine appears minimal (Bagian and Ward 1994). The reasons for this difference have not yet been elucidated.

Calcium Antagonists

Flunarizine, a calcium channel blocker, has been shown effective in treating patients with peripheral vertigo. Lee investigated its use in motion sickness (Lee et al. 1986). He noted that subjects exposed to a spinning chair stimulus experienced less nystagmus when given flunarizine, compared to prochlorperazine and placebo. Also of importance is his finding that subjective drowsiness was lessened with the use of flunarizine when compared with placebo.

Cinnarizine, another calcium channel blocker, has also been investigated for treatment of motion sickness. It is thought to act as a sedative of the labyrinth by blocking calcium ion influx to the vestibular sensory cells. A combined trial of cinnarizine in both the laboratory and clinical settings was performed by a group from the Israel Naval Medical Institute. Vestibulo-ocular reflex gain, the ratio of the maximal eye velocity during the slow phase of nystagmus to the head rotation velocity in a rotating chair, was measured. When 50 mg of cinnarizine was used, a significant improvement in vestibulo-ocular reflex gain was seen in the laboratory. In a trial on ships in rough seas, 69% of the subjects improved with cinnarizine, versus 31% with placebo, a significant difference (Shupak et al. 1994). When compared with scopolamine, cinnarizine was felt to be superior during light seas, owing to its lower incidence of side effects. However, scopolamine was the drug of choice for severe motion sickness, where side effects were less important than efficacy (Pingree and Pethybridge 1994).

Phenytoin

The association with seizure activity noted earlier led several groups to look at phenytoin, a common anti-epileptic medication, for the treatment of motion sickness. Phenytoin has a good safety profile, with most of the complications associated with its use being related to chronic therapy. When therapeutic levels were achieved, Knox found that there was a significant improvement in the ability of subjects to tolerate both rotating chair and small boat challenges (Knox et al. 1994). Unfortunately, this treatment suffers from the same problem that transdermal scopolamine has: the subject must receive the medicine well in advance of the challenge. Additionally, achievement of therapeutic levels requires that the patient be loaded with the drug.

One study attempted to overcome the latter problem by giving a single low dose of phenytoin 4 hours prior to the motion challenge. This showed that the phenytoin reduced the incidence of tachygastria and allowed a significantly greater number of subjects to complete the entire test. The difference in subjective symptom scores was not, however, significant (Stern et al. 1994). Although subjects reported some subjective side effects, no change in sensory, cognitive, or performance capabilities was seen when subjects were given phenytoin in a standard dose for seizures (approximately 1 gram) (Chelen 1993). When used as a prophylactic during sea operations, levels of 9 g/L produced protection against motion sickness (Woodard 1993). No problems were noted among surface crews or divers who took this medication while working in space shuttle booster recovery operations.

Peripherally Acting Antihistamines

Kohl has investigated two peripherally acting H1 antihistamines, astemizole and terfenadine, with mixed results. In a 1987 study, he tested subjects following a week of astemizole administration at 30 mg daily. He could find no improvement over placebo using a rotating chair stimulus test (Kohl 1987b). Using terfenadine, however, he was able to demonstrate a significant improvement after giving one 300 mg dose (Kohl et al. 1991). The inability of terfenadine to cross the blood-brain barrier suggests that it acts at a yet undetermined peripheral site, perhaps directly on the vestibular apparatus. Nicholson recommends terfenadine as the H1 antihistamine of choice for allergic reactions among aircrew, because neither sedation or other side effects are seen at doses as large as four times the recommended dose (Nicholson 1985). Further studies are needed before making similar recommendations regarding motion sickness.

Meclizine

Meclizine is an H1 antihistamine commonly used in the treatment of vertigo. It should, therefore, be a natural choice for the treatment of motion sickness. Dahl looked at this medication and found that after the first 30 minutes it performed better than placebo, but not as well as transdermal scopolamine. During the first 30 minutes, subjects actually reported worse motion sickness when given meclizine as compared to the placebo (Dahl et al. 1984).

Metoclopramide

Not all anti-emetic agents are useful in the treatment of motion sickness. Using both parabolic flight and a rotating chair as stimuli, subjects were treated with metoclopramide (Kohl 1987a). No improvement over placebo was noted, although this agent is frequently used in vomiting due to cancer chemotherapy. Of interest, though, is that metoclopramide works to increase gastric emptying during motion sickness (Wood et al. 1990). This suggests it may be a useful adjunct, although the endpoints of most studies (nausea, short of vomiting) may preclude demonstrating any statistical benefit.

Doxepin

Doxepin, commonly used as an antidepressant, has been investigated for treatment of motion sickness. Although doxepin is classically thought of as an inhibitor of norephinephrine and serotonin reuptake, it is also a potent antihistamine. Kohl took advantage of this, using both a chronic (25 day) and acute (3 day) dosing schedule. When he compared this to a scopolamine-amphetamine combination drug, he found equivalent efficacy. Sedation did, however, produce problems for many of the subjects while taking doxepin (Kohl et al. 1993).

Local Anesthetics

Lidocaine, a commonly used anesthetic, and tocanide, an orally administered derivative of lidocaine, were investigated by Pyykkö and his colleagues (Pyykkö 1985). Both of these medications have been used in treating vertigo and nausea. They hypothesized that these drugs would have an effect on the vestibular end organs and supratentorial brain structures, resulting in decreased motion sickness. They were, unfortunately, unable to demonstrate any benefit for motion sickness.

Experimental Approaches

There have been some novel approaches to the pharmacological treatment of motion sickness. Kohl and MacDonald describe the use of dexamethasone to suppress ACTH release that occurs in response to the motion stressor (Kohl et al. 1991). They also reviewed the data regarding the use of a vasopressin antagonist. Because of the expense of these medications, they noted that Kappa opioid receptors can be used to inhibit vasopressin, and therefore pain medications such as nalbuphine and butorphine, which are agonists at the kappa receptor, may be useful in the treatment of motion sickness. Clinical trials of this theory have yet to be performed.

Pharmacological Treatments: Conclusion

Scopolamine and promethazine appear to be the best medications for treating motion sickness. Scopolamine is commonly available only in the transdermal preparation, and therefore requires a significant lead time. Its side effects profile is tolerable for all but the most sensitive of jobs. Promethazine may be administered orally, intravenously, intramuscularly, or rectally. It may produce significant sedation, but its rapid action makes it ideal for short-term treatment of patients with severe symptoms. Augmentation of either of these medications with dextroamphetamine improves both the efficacy of the medication and the side effects profile. The addictive nature of amphetamines and restrictions regarding their use may limit their utility for civilian applications. Dimenhydrinate, owing to its over-the-counter availability, should be considered for mild to moderate motion sickness. Other medications appear promising, but more research is needed before recommending their routine use.

SUMMARY

Motion sickness is a common, yet poorly understood phenomenon. The sensory conflict theory is the leading explanation for the problem, but provides little insight into what the specific sensory inputs are that cause motion sickness. Avoidance of conflicts is the best solution, but may be unavoidable, given the individual's duties. The effects of motion sickness on the fishing industry are unknown, and research in this area is sorely needed. Current treatments are less than ideal, either requiring a significant lead time or producing problematic side effects. Given sufficient time, most people can adapt; for some, other treatment is required. Further research into the causes and treatments is clearly needed.

REFERENCES

Allen, M.E., C. McKay, D.M. Eaves, and D. Hamilton. 1986. Naloxone enhances motion sickness: endorphins implicated. *Aviat. Space Envir. Med.* 57: 647-53.

Attias, J., C. Gordon, J. Ribak, O. Binah, and A. Rolnick. 1987. Efficacy of transdermal scopolamine against seasickness: a 3-day study at sea. *Aviat. Space Envir. Med.* 58: 60-62.

Bagian, J.P., Ward, D.F. 1994. A retrospective study of promethazine and its faliure to produce the expected incidence of sedation during space flight. *J. Clin. Pharmacol.* 34: 649-51.

Bagshaw, M., and J.R.R. Stott. 1985. The densitization of chronically motion sick aircrew in the Royal Air Force. *Aviat. Space Envir. Med.* 56: 1144-51.

Banks, R.D., D.A. Salisbury, P.J. Ceresia. 1992. The Canadian Forces Airsickness Rehabilitation Program, 1981-1991. *Aviat. Space Envir. Med.* 63(12): 1098-1101.

Banta, G.R., W.C. Ridley, J. McHugh, J.D. Grissett, and F.E. Guedry. 1987. Aerobic fitness and susceptibility to motion sickness. *Aviat. Space Envir. Med.* 58: 105-8.

Bennett, D.R., ed. 1995. *AMA Drug Evaluations*. Chicago: American Medical Association.

Bruce, D.G., J.F. Golding, N. Hockenhull, and R.J. Pethybridge. 1990. Accupressure and motion sickness. *Aviat. Space Envir. Med.* 61: 361-65.

Chelen, W.E., M. Kabrisky, and S.K. Rogers. 1993a. Spectral analysis of the electroencephalographic response to motion sickness. *Aviat. Space Envir. Med.* 64(1): 24-9.

Chelen, W., N. Ahmed, M. Kabrisky, and S. Rogers. 1993b. Computerized task battery assessment of cognitive and performance effects of acute phenytoin motion sickness therapy. *Aviat. Space Envir. Med.* 64 (3 Pt 1): 201-05.

Cheung, B.S., I.P. Howard, K.E. Money. 1991. Visually-induced sickness in normal and bilaterally labyrinthine-defective subjects. *Aviat. Space Envir. Med.* 62(6): 527-31.

Cheung, B.S.K., K.E. Money, and I. Jacobs. 1990. Motion sickness susceptibility and aerobic fitness: a longitudinal study. *Aviat. Space Envir. Med.* 61: 201-4.

Chinn, H.I. and the Army, Navy, Air Force Motion Sickness Team. 1956. Evaluation of drugs for protection against motion sickness aboard transport ships. *JAMA* 160: 755-60.

Clemente, C.D. 1995. *Anatomy: A Regional Atlas of the Human Body*. Baltimore: Urban & Schwarzenberg.

Dahl, E. D. Offer-Ohlsen, P.E. Lillevold, and L. Sandvik. 1984. Transdermal scopolamine, oral meclizine, and placebo in motion sickness. *Clin. Pharmacol. Ther.* 36: 116-20.

Daroff, R.B. 1994. Dizziness and vertigo. In: *Harrison's principles of internal medicine*. 13th edition. K.J. Isselbacher, E. Braunwald, J.D. Wilson, et al., eds. New York: McGraw-Hill, Inc.

Davis, J.R., R.T. Jennings, B.G. Beck, and J.P. Bagian. 1993. Treatment efficacy of intramuscular promethazine for space motion sickness. *Aviat. Space Envir. Med.* 64: 230-33.

Dobie, T.G., May, J.G. 1990. The generalization of tolerance to motion environments. *Aviat. Space Envir. Med.* 61: 707-711.

Dobie, T.G., J.G. May, W.D. Fischer, S.T. Elder, and K.A. Kubitz. 1987. A comparison of two methods of training resistance to visually-induced motion sickness. *Aviat. Space Envir. Med.* 58(9, Suppl.): A34-41.

Ebenholtz, S.M., M.M. Cohen, and B.J. Linder. 1994. The possible role of nystagmus in motion sickness: a hypothesis. *Aviat. Space Envir. Med.* 65: 1032-35.

Golding, J.F., J.R.R. Stott, 1995. Effect of sickness severity on habituation to repeated motion challenges in aircrew referred for airsickness treatment. *Aviat. Space Environ. Med.* 66: 625-30.

Gordon, C.R., Y. Jackman, H. Ben-Aryeh, I. Doweck, O. Spitzer, R. Szargel, A. Shupak. 1992. Salivary secretion and seasickness susceptibility. *Aviat. Space Environ. Med.* 63(5): 356-59.

Gordon, C., O. Binah, J. Attias, and A. Rolnick. 1986. Transdermal scopolamine: Human performance and side effects. *Aviat. Space Environ. Med.* 57: 236-40.

Gordon, C.R., O. Spitzer, A. Shupak, and I. Doweck. 1992. Survey of mal de debarquement. *Brit. Med. J.* 304(6826): 544.

Gordon, C.R., H. Ben-Aryeh, O. Spitzer, I. Doweck, A. Gonen, Y. Melamed, and A. Shupak. 1994. Seasickness susceptibility, personality factors, and salivation. *Aviat. Space Envir. Med.* 65: 610-14.

Graybiel, A., J.R. Lackner. 1987. Treatment of severe motion sickness with antimotion sickness drug injections. *Aviat. Space Envir. Med.* 58: 773-76.

Grøntved, A., T. Brask, J. Kambskard, and E. Hentzer. 1988. Ginger root against seasickness. A controlled trial on the open sea. *Acta Otolaryngol* (Stockholm) 105: 45-49.

Grøntved, A., and E. Hentzer. 1986. Vertigo-reducing effect of ginger root: A controlled clinical study. *ORL* 48: 282-86.

Hamid, M.A. 1991. Vestibular and postural findings in the motion sickness syndrome. *Otolaryngol. Head Neck Surg.* 104(1): 135-36.

Harm, D.L., D.E. Parker. 1994. Perflight adaptation training for spatial orientation and space motion sickness. *J. Clin. Pharm.* 34: 618-27.

Holling, H.E., I. McArdle, and W.R. Trotter. 1944. Prevention of seasickness by drugs. *Lancet* Jan 22, pp. 127-29.

Hu, S., R.M. Stern, K.L. Koch. 1992. Electrical acustimulation relieves vection-induced motion sickness. *Gastroenterology* 02: 1854-58.

Hu, S., R. Stritzel, A. Chandler, and R.M. Stern. 1995. P6 acupressure reduces symptoms of vection-induced motion sickness. *Aviat. Space Envir. Med.* 66: 631-34.

Jones, D.R., R.A. Levy, L. Gardner, R.W. Marsh, J.C. Patterson. 1985. Self-control of psychophysiologic response to motion stress: using biofeedback to treat airsickness. *Aviat. Space Envir. Med.* 56: 1152-57.

Kennedy, R.S., R.C. Odenheimer, D.R. Baltzley, W.P. Dunlap, and C.D. Wood. 1990. Differential effects of scopolamine and amphatamine on microcomputer - based performance tests. *Aviat. Space Envir. Med.* 61: 615-21.

Knox, G.W., D. Woodard, W. Chelen, R. Ferguson, and L. Johnson. 1994. Phenytoin for motion sickness: clinical evaluation. *Laryngoscope* 104.

Kohl, R.L., D.S. Calkins, and R.E. Robinson. 1991. Control of nausea and autonomic dysfunction with terfenadine, a peripherally acting antihistamine. *Aviat. Space Envir. Med.* 62(5): 392-96.

Kohl, R.L., G.R. Sandoz, M.F. Reschke, D.S. Calkins, and E. Richelson. 1993. Facilitation of adaptation and acute tolerance to stressful sensory input by doxepin and scopolamine plus amphetamine. *J. Clin. Pharmac.* 33(11):1092-1103.

Kohl, R.L. 1987a. Failure of metoclopramide to control emesis or nausea due to stressful angular or linear acceleration. *Aviat. Space Envir. Med.* 58: 125-31.

Kohl, R.L., and S. MacDonald. 1991. New pharmacologic approaches to the prevention of space/motion sickness. *J. Clin. Pharmac.* 31(10): 934-46.

Kohl, R.L., D.S. Calkins, and A.J. Mandell. 1986. Arousal and stability: the effects of five new sympathomimetic drugs suggest a new principle for the prevention of space motion sickness. *Aviat. Space Envir. Med.* 57: 137-43.

Kohl, R.L., J.L. Homick, N. Cintron, and D.S. Calkins. 1987b. Lack of effects of astemizole on vestibular ocular reflex, motion sickness, and cognitive performance in man. *Aviat. Space Envir. Med.* 58: 1171-74.

Kohl, R.L., and M.R. Lewis. 1987c. Mechanisms underlying the antimotion sickness effects of psychostimulants. *Aviat. Space Envir. Med.* 58: 1215-18.

Kohl, R.L., P. Ryan, and J.L. Homick. 1985. Efficacy of phosphatidylcholine in the modulation of motion sickness susceptibility. *Aviat. Space Envir. Med.* 56: 125-28.

Lackner, J.R. and A. Graybiel. 1994. Use of promethazine to hasten adaptation to provocative motion. *J. Clin. Pharmac.* 34: 644-48.

Landolt, J.P., I.M. Light, M.G. Greenen, and C. Monaco. 1992a. Seasickness in totally-enclosed motor-propelled survival craft: five offshore oil rig disasters. *Aviat. Space Envir. Med.* 63(2): 138-44.

Landolt, J.P., and C. Monaco. 1992b. Seasickness in totally-enclosed motor-propelled survival craft: remedial measures. *Aviat. Space Envir. Med.* 63(3): 219-25.

Lee, J.A., L.A. Watson, and G. Boothby. 1986. Calcium antagonists in the prevention of motion sickness. *Aviat. Space Envir. Med.* 57: 45-48.

Levy, G.D., and M.H. Rapaport. 1985. Transderm scopolamine efficacy related to time of application prior to the onset of motion. *Aviat. Space Envir. Med.* 56: 591-3.

Mowrey, D.B. and D.E. Clayson. 1982. Motion sickness, ginger, and psychophysics. *Lancet,* March 20.

Murphy, T.P. 1993. Mal de debarquement syndrome: a forgotten entity? *Otolaryngol. Head Neck Surg.* 109(1): 10-13.

Muth, E.R., M. Jokerst, R.M. Stern, K.L. Koch. 1995. Effects of dimenhydrinate on gastric tachyarrhythmia and symptoms of vection-induced motion sickness. *Aviat. Space Envir. Med.* 66: 1041-45.

Nicholson, A.N. 1985. Central effects of H1 and H2 antihistamines. *Aviat. Space Envir. Med.* 56: 293-98.

Norfleet, W.T., J.J. Degioanni, D.S. Calkins, M.F. Reschke, M.W. Bungo, F.A. Kutyna, and J.L. Homick. 1992. Treatment of motion sickness in parabolic flight with buccal scopolamine. *Aviat. Space Envir. Med.* 63(1): 46-51.

Pingree, B.J., and R.J. Pethybridge. 1994. A comparison of the efficacy of cinnarizine with scopolamine in the treatment of seasickness. *Aviat. Space Envir. Med.* 65(7): 597-605.

Reason, J.T., J.J. Brand. 1975. *Motion Sickness.* New York: Academic Press.

Schroeder, D.J., W.E. Collins, and G.W. Elam. 1985. Effects of some motion sickness suppressants on static and dynamic tracking performance. *Aviat. Space Envir. Med.* 56: 344-50.

Shupak, A., I. Doweck, C.R. Gordon, and O. Spitzer. 1994. Cinnarizine in the prophylaxis of seasickness: laboratory vestibular evaluation and sea study. *Clin. Pharmac. Ther.* 55(6): 670-80.

Simanonok, K.E., and J.B. Charles. 1994. Space sickness and fluid shifts: A hypothesis. *J. Clin. Pharmac.* 34: 652-63.

Smith, P.K. 1948. Treatment of airsickness with drugs. *Am. J. Med.* 4: 649-52.

Stern, R.M., S. Hu, R.B. Anderson, H.W. Leibowitz, and K.L. Koch. 1990. The effects of fixation and restricted visual field on vection-induced motion sickness. *Aviat. Space Envir. Med.* 61: 712-15.

Stern, R.M., S. Hu, R. LeBlanc, and K.L. Koch. 1993. Chinese hyper-susceptibility to vection-induced motion sickness. *Aviat. Space Envir. Med.* 64(9 Pt 1): 827-30.

Stern, R.M., S.H. Uijtdehaage, E.R. Muth, and K.L. Koch. 1994. Effects of phenytoin on vection-induced motion sickness and gastric myoelectric activity. *Aviat. Space Envir. Med.* 65(6): 518-21.

Stewart, J.J., M.J. Wood, C.D. Wood, and M.E. Mims. 1991. Effects of ginger on motion sickness susceptibility and gastric function. *Pharmacology* 42(2): 111-20.

Thornton, W.E., T.P. Moore, S.L. Pool, and J. Vanderploeg. 1987. Clinical characterizatiion and etiology of space motion sickness. *Aviat. Space Envir. Med.* 58 (9, Suppl.): A1-8.

Tokola, O., L.A. Laitinen, J. Aho, G. Gothoni, and H. Vapaatalo. 1984. Drug treatment of motion sickness: scopolamine alone and combined with ephedrine in real and simulated situations. *Aviat. Space Envir. Med.* 55: 636-641.

Uijtdehaage, S.H., R.M. Stern, and K.L. Koch. 1993. Effects of scopolamine on autonomic profiles underlying motion sickness susceptibility. *Aviat. Space Envir. Med.* 64(1): 1-8.

van Marion, W.F., M.C.M. Bongaerts, J.C. Christiaanse, H.G. Hofkamp, and W. van Ouwerkerk. 1985. Influence of transdermal scopolamine on motion sickness during 7 days' exposure to heavy seas. *Clin. Pharm. Ther.* 38(3): 301-05.

von Gierke, H.E., D.E. Parker. 1994. Differences in otolith and abdominal viscera graviceptor dynamics: implications for motion sickness and perceived body position. *Aviat. Space Envir. Med.* 65(8): 747-51.

Warwick, R., and P.L. Williams, eds. 1973. *Gray's Anatomy*, 35th ed. Philadelphia: W.B. Saunders.

Warwick-Evans, L.A., I.J. Masters, S.B. Redstone. 1991. A double-blind placebo controlled evaluation of acupressure in the treatment of motion sickness. *Aviat. Space Envir. Med.* 62(8): 776-78.

Wood, C.D., and A. Graybiel. 1968. Evaluation of sixteen anti-motion sickness drugs under controlled laboratory conditions. *Aerospace Med.* December.

Wood, C.D., J.J. Stewart, M.J. Wood, F.A. Struve, J.J. Straumanis, M.E. Mims, and G.Y. Patrick. 1994. Habituation and motion sickness. *J. Clin. Pharmac.* 34: 628-34.

Wood, C.D. 1979. Antimotion sickness and antiemetic drugs. *Drugs* 17: 471-79.

Wood, C.D., R.E. Kennedy, A. Graybiel., R. Trumbull, and R.J. Wherry. 1966. Clinical effectiventss of antimotion-sickness drugs: computer review of the literature. *JAMA* 198: 133-36.

Wood, C.D., J.E. Manno, B.R. Manno, R.C. Odenheimer, and L.E. Bairnsfather. 1987. The effect of antimotion sickness drugs on habituation to motion. *Aviat. Space Envir. Med.* 57: 539-42.

Wood, C.D., J.E. Manno, B.R. Manno, H.M. Redetzki, M.J. Wood, and M.E. Mims. 1985. Evaluation of antimotion sickness drug side effects on performance. *Aviat. Space Envir. Med.* 56: 310-16.

Wood, C.D., J.E. Manno, B.R. Manno, H.M. Redetzki, M. Wood, W.A. Vekovius. 1984. Side effects of antimotion sickness drugs. *Aviat. Space Envir. Med.* 55: 113-16.

Wood, C.D., J.J. Stewart, M.J. Wood, J.E. Manno, B.R. Manno, and M.E. Mims. 1990. Therapeutic effects of antimotion sickness medications on the secondary symptoms of motion sickness. *Aviat. Space Envir. Med.* 61: 157-61.

Woodard, D., G. Knox, K.J. Myers, W. Chelen, and B. Ferguson. 1993a. Phenytoin as a countermeasure for motion sickness in NASA maritime operations. *Aviat. Space Envir. Med.* 64(5): 363-66.

Wright, M.S., C.L. Bose, and A.D. Stiles. 1995. The incidence and effects of motion sickness among medical attendants during transport. *J. Emerg. Med.* 13(1): 15-20.

39

AQUACULTURAL HAZARDS

Robert M. Durborow, Ph.D.
Cooperative Extension Program
Kentucky State University

Fish is considered to have many positive health benefits in humans and the demand for fish is expected to increase by more than 70% in the next 35 years. As the consumption of fish has increased, aquaculture has become a rapidly growing industry. This chapter reviews the hazards associated with feral and farm-raised aquatic organisms such as zoonotic infections, diseases from ingestion of toxins in fish and shellfish, sewage in ponds, hydrogen sulfide gas on pond bottoms, adverse reactions from shrimp preservatives, and the dangers of handling antibiotics and vaccines used in aquaculture. Many of the zoonotic conditions are avoided by consuming cultured fish rather than those caught from the wild, and by properly cooking fish. Safety hazards around fish production facilities include electrical shock when in the culture water, drownings, musculoskeletal strains from feeding and cleaning activities, tractor rollovers, and PTO-related injuries. Accidents occurring during fish processing include spine punctures, band saw lacerations and amputations.

INTRODUCTION

Aquaculture is the farming of aquatic animals or plants. Valued at $660 million in 1988, it is the fastest growing sector of American agriculture compared to the $192 million value of the industry in 1980 (FAO 1990). On a worldwide level, aquaculture reached 10,068,443 tons of aquatic animal production and 3,139,473 tons of aquatic plant production in 1987 for a total value of approximately $19 billion (Nash and Kensler 1990).

In the United States, per capita fish consumption rose 24% between 1980 and 1990, and cultured fish comprised 5 to 10% of the consumed fish (Evans 1992). Fish is the main source of animal protein in many countries and is considered a healthy food by western countries (Higgins and Kolbye 1984). Global demand for fish is expected to increase 70% in the next 35 years. Because commercial fishery stocks are approaching maximum sustainable yield, and are being overfished in some cases (NOAA 1988), the needed supply of fish will have to come from an aquacultural source. Aquaculture production will have to increase 7-fold by the year 2025 to meet this demand (Rominger 1993). As the industry grows nationwide and worldwide, human health and safety concerns will grow in attention and importance. Health concerns affecting people range from contracting bacterial infections by consuming or handling fish to inhaling antibiotics while preparing medicated feed. Obvious safety threats include drowning, hypothermia, and tractors rolling over. The following discussion covers the details of health and safety concerns of aquaculture.

HEALTH ISSUES

The main health issue surrounding the consumption of cultured fish is its *positive* effect on man's diet as a healthy protein source. Fish not only contains less fat than other protein sources, but has a higher proportion of its fatty acids in the mono-unsaturated or polyunsaturated form, which are associated with reducing the occurrence of human cardiovascular disease (Browne 1990). As with any animal-derived food, however, there are some risks involving human health when contamination of the food animal occurs. This section deals primarily with these risks. Reviews of zoonoses caused by pathogens being transmitted from fish to people have been written by Murrell (1995), Nemetz and Shotts (1993), Inglis, et al. (1993), Sindermann (1990a, 1990b), Eastaugh and Shepherd (1989), and Shotts (1980, 1987). The incidence of fish-related diseases of people in the United States is low, and often involves no more than mild gastroenteritis if a contaminated fish is consumed, or localized infection of the skin if pathogens come into contact with open wounds or are injected into the person (Nemetz and Shotts 1993). Cases of human infection from consuming contaminated fish are more common than skin infections, and in most cases shellfish, rather than finfish, are the cause (Sindermann 1990b). Of all foodborne disease outbreaks in the U.S., however, only about 10% are attributed to seafood; which amounts to 1 illness per 250,000 servings (and only 1 illness per 5,000,000 servings of finfish) (Otwell 1993). This compares to 1 illness per 25,000 servings of chicken. Based on these figures, Otwell (1993) estimates that 1 illness out of 10,000,000 servings of farm-raised catfish may occur (although there are presently no CDC reported outbreaks from eating cultured catfish). Cases may go unreported, because the symptoms would usually involve only mild gastrointestinal problems.

Incidences of parasite infections from seafood, in general, are much more prevalent now than 15 years ago, most likely due to the increased consumption of raw or undercooked fish products such as sushi and sashimi. Mortalities sometimes occur from infections originating from a fish source, but the patient must usually be immunocompromised, as in the case of acquired immunodeficiency syndrome (AIDS).

Quality Assurance

Aquaculture provides its own quality control to assure the return of the customer. Pond-raised fish in the southeastern United States are brought to processing plants alive and stored in vats flushed with clean, aerated water. When it comes time to process them, they are electrocuted, beheaded, eviscerated, skinned, and put into an ice bath before they are filleted or put on ice as a whole dressed product. The fillets are usually individually quick frozen in carbon dioxide freezer tunnels and are glazed with a protective layer of ice (Figures 39.1 and 39.2). The whole process, from the time the fish is alive until the filet is frozen, takes about 20 minutes. Catfish are also flavor-checked before harvesting and before processing to assure that no off-flavors are present. Channel catfish processing plants in the southeastern United States usually voluntarily pay a U.S. Department of Commerce food health inspector to certify their products as satisfactory. This procedure almost eliminates chances of contamination of the processed farm-raised fish. Cultured rainbow trout processing plants also have strict quality assurance (Flick, in press).

Mishandling of fish at the retail level may lead to high bacterial concentrations on aquacultured products. To help prevent these problems, temperature requirements for holding seafood have changed from < 45° F and > 140° F (required for most other foods) to < 40° F and > 145° F according to Cindy Holden (1993), Quality Assurance Director of Shoney's, Inc.

The catfish industry also has a quality assurance program at the producer level that advises the farmers to construct ponds on pesticide-free soil, use clean water sources, feed fresh uncontaminated food, and avoid the presence of residues in the fish flesh by observing prescribed withdrawal times for antibiotics and any

Figure 39.1. At the processing plant, whole dressed farm-raised channel catfish are put into ice baths to stop growth of any bacteria that may be in the flesh.

Figure 39.2. Farm-raised catfish fillets are individually quick frozen (IQF) and glazed with a layer of ice in carbon dioxide tunnels.

chemicals used to treat pond problems (Brunson 1993). This program is part of incorporating HACCP (Hazard Analysis Critical Control Point) principles into aquaculture. The quality and safety of cultured trout are also safeguarded by a quality assurance program administered by the U.S. Trout Farmers Association (USTFA 1994).

Commercially raised catfish are grown in ponds in static water that is "purified" by biological processes occurring naturally in the ponds (e.g., soil and water bacteria metabolizing nitrogenous wastes). Rainbow trout, on the other hand, are usually farmed in long, rectangular concrete tanks called raceways with fresh, cold spring or mountain stream water constantly flowing through and flushing out any fish waste products. In states such as Washington, salmon are reared in net pens suspended in coastal ocean waters. Very few fish are cultured in closed-system tanks using recirculated water; those that are, tend to become infected more readily with bacteria contagious to man. The pathogens most likely are harbored in the biofilters used to metabolize nitrogenous fish wastes in these closed systems. *Mycobacterium* spp. and *Vibrio vulnificus* are two bacteria that have been found in these systems. They are infective to both fish and man. Because aquaculture pathogens contagious to man are not usually found in culture facilities other than in these recirculating systems, and because fish production in the recirculating systems constitutes such a small percentage of national and world production (probably less than 1%), the overall threat of zoonoses originating from aquaculture production facilities is extremely small.

Bacteria and Parasites

The following bacteria, which affect the health of both fish and man, will be discussed in this section:

- *Leptospira* sp.
- *Edwardsiella tarda*
- *Yersinia ruckeri*
- *Vibrio* spp.
- *Aeromonas hydrophila*
- *Plesiomonas shigelloides*
- *Clostridium perfringens* and *C. botulinum*
- *Erysipelothrix rhusiopathiae*
- *Mycobacterium fortuitum, M. chelonei* and *M. marinum*
- Bacteria from processing contamination such as *Salmonella* spp., *Shigella* spp., *Staphylococcus aureus, Pseudomonas* spp., and *Streptococcus iniae*
- Bacteria whose action causes scombroid poisoning and other microorganisms that produce toxins

The following parasites will also be covered in this section:
- The nematodes *Anasakis simplex, Pseudoterranova decipiens*, and *Eustrongylides* spp.
- The cestode *Diphyllobothrium latum*
- The trematodes *Clonorchis sinensis, Opisthorchis* spp., *Heterophyes heterophyes* and *Nanophyetus salmincola*
- The protozoa *Cryptosporidium* sp. and *Giardia* sp.

Bacteria

Mycobacterium spp. Mycobacterium spp. are acid-fast bacteria; the genus includes the bacterium that causes tuberculosis (*M. tuberculosis*). *M. fortuitum, M. chelonei*, and *M. marinum* are capable of infecting both fish and people; the later two grow at 30° C but not at 37° C, so they are usually able to infect only the extremities in man. They also may be missed during culturing in hospitals because incubator temperatures are usually maintained at only 37° C. *M. fortuitum* grows at both 30° C and 37° C. Growth on TSA blood agar plates is very slow. Signs of the bacteria's presence become apparent after about 5 days of incubation at 25° C when slight, greening or alpha hemolysis appears. *Mycobacterium* spp. typically cause cutaneous lesions on the hands and arms; skin granulomas caused by *M. marinum*

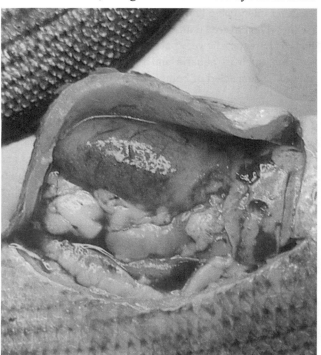

Figure 39.3. This enlarged, granular spleen at the bottom of the fish's open body cavity is more than 20 times larger than a normal spleen. The condition was caused by *Mycobacterium marinum* infecting this hybid striped bass which was raised in a closed, recirculating tank culture system.

were first reported by Linell and Norden (1954). Similar lesions caused by *M. fortuitum* were reported on the hands of individuals cleaning contaminated aquaria (Shotts 1987). Treatment of these lesions can be with topical antibiotic ointment, surgery to remove the lesion, or simply waiting several weeks for the infection to resolve itself. Surgeries to remove lesions caused by *M. marinum* were performed on several infected workers at a recirculating fish culture system in Kentucky in 1991.

M. marinum occasionally spreads to lymph nodes and rarely to tissues adjacent to the original lesion except when occurring in persons infected with the human immunodeficiency virus (HIV). HIV-positive individuals may develop arthritis, osteomyelitis, and tenosynovitis caused by the spreading of the bacteria. Respiratory disease may also be manifested in these immunocompromised patients (CDC 1987a). People infected with HIV are often advised not to clean aquaria due to the threat of *Mycobacterium* spp. infections. Cases of this bacteria were found in angelfish from an aquarium in May 1995 (Durborow 1996), in fathead minnows held in aquaria as bioassay test animals (Durborow 1993), and in hybrid striped bass from a large-scale commercial recirculating system in Louisville, Kentucky in 1991.

Fish infected with *Mycobacterium* spp. typically exhibit extreme splenomegaly with a granular spleen and a very pale (white) liver covered with petechial hemorrhaging. Gills also have a characteristic very pale appearance. Fish from closed systems, such as aquaria and recirculating culture tanks, should be suspected of being carriers of this bacteria.

***Nocardia* spp.** *Nocardia* spp. produce similar granulomatous lesions as *Mycobacterium* spp., are acid-fast, and also cause disease in fish and man (Shotts 1980, 1987). These two genera are sometimes confused for these reasons, and because *Mycobacterium* spp. are more notorious, they may be credited many times with causing diseases when *Nocardia* sp. is the actual pathogen.

***Clostridium* spp.** *Clostridium* spp. occur naturally in the intestine of fish and rarely cause disease in the fish. *Clostridium perfringens* and *C. botulinum* have caused disease in humans, usually resulting from the consumption of uneviscerated contaminated fish (CDC 1987b; Bean and Griffin 1990; Telzak et al. 1990). Practices of processors in the aquaculture industry practically eliminate the chances of this bacteria contaminating the processed fish. The fish are gutted and cleaned in the first stages of processing. The *Clostridium* spp. toxins are heat-labile, so routinely cooking fish (avoiding raw or undercooked fish) should also help to avoid this infection. A small percentage of farm-raised fish are sold whole and undressed either directly to the consumer or more typically through fee-fishing ponds (pay lakes). The consumer then should be advised to properly clean and cook his fish before consumption.

Symptoms of *C. perfringens* often involve gastroenteritis, which typically resolves itself after 24 hours. *Clostridium botulinum*, on the other hand, causes weakness, headaches, respiratory distress, vision disorders, vomiting, abdominal pain, and diarrhea. Neurotoxins produced during cases of botulism cause neurological problems within 1 to 6 days of ingestion. Clinical signs include difficulty in swallowing and speaking, ptosis, and general muscle dystonia. Partial paralysis (especially respiratory paralysis) may lead to death, usually within the first 10 days of food ingestion. Death results in 15 to 33% of botulism cases. If the patient survives, paralysis may persist for months (Bryan 1986).

***Erysipelothrix* spp.** *Erysipelothrix rhusiopathiae* (formerly *E. insidiosa*) is passively transmitted to man from the mucus on the skin or spines of fish, causing erysipeloid (Sheard and Dicks 1949). It is known as a disease of fish filleters and is common among persons such as butchers, veterinarians, and fish health workers who come into frequent contact with animals, particularly fish. It has the potential of occurring in aquaculture processing plants. The bacteria cause no disease in the carrier fish (Wood 1975); clinical manifestations in man are reviewed by Reboli and Farrar (1989).

Erysipelothrix rhusiopathiae causes three different clinical syndromes in man: a skin infection, usually on the fingers or hands (referred to as erysipeloid or "fish rose"); a diffuse cutaneous form, which spreads from the skin to adjacent tissues; and a septicemic form. The author has found that consistent use of hand protection (e.g., latex gloves) helps to prevent various skin conditions caused by frequent handling of sick fish specimens in a diagnostic laboratory. The septicemia condition in man is extremely rare and has been reviewed (Gorby and Peacock 1988): Only 49 cases were reported in the United States between 1912 and 1988; it caused endocarditis in 44 of these patients; only 11 of the cases were caused by fish; and 19 of the 49 patients died. The bacteria itself survives pickling, smoking, and salting, but does not survive moist heat at 55° C for 15 minutes (Gorby and Peacock 1988).

***Vibrio* spp.** *Vibrio* spp. are commonly isolated from bodies of water, especially in marine and estuarine environments. They are the most frequently encountered infectious bacteria of marine fish, the most notorious being *V. anguillarum*, which does not infect humans (Frerichs and Roberts 1989). *Vibrio alginolyticus, Vibrio cholerae* 0 group 1, *V. cholerae* non-01, *V. parahemolyticus*, and *V. vulnificus* have been credited with causing diseases in man (Eastaugh and Shepherd 1989; Colwell and Grimes 1984).

Infection usually occurs when raw or undercooked fish (usually shellfish) are ingested. *Vibrio alginolyticus* causes skin infections, stump ulcer, conjunctivitis, and enteritis in humans (Blake et al. 1980; Hiratsuka et al. 1980). *Vibrio cholerae* 0 group 1 causes severe vomiting and diarrhea, while *V. cholerae* non-01 produces less severe symptoms (Morris et al. 1981; Lowry et al. 1989). *Vibrio parahemolyticus*, which is probably the most common *Vibrio* species recovered from freshwater aquaculture disease diagnostic laboratory cases, often causes mild enteritis in man, but may cause abdominal pain, vomiting, watery diarrhea, fever, chills and headache (Venugopal and Karunasagar 1988). Lowry et al. (1986) reported a *V. hollisae* infection caused by the consumption of contaminated catfish, but it was not farm raised.

Vibrio vulnificus infections in aquacultured fish are very rare. A case occurred in 1991 in which *Tilapia* sp. were raised in a closed tank system at Louisville, Kentucky using recirculating water. The system's ozone water-disinfecting apparatus had been dysfunctional for several weeks before the disease outbreak. Other cases are discussed by Johnston et al. (1985), Klontz et al. (1988), Hoffman et al. (1988), Vartian and Septimus (1990), and Nelson (1996). A septicemic syndrome in man (usually from ingesting raw oysters) can result in chills, fever, blister formation, ecchymotic hemorrhages, pain in the lower extremities, changes in mental status, and sometimes vomiting and diarrhea (Blake et al. 1980). The mortality rate is approximately 50%, despite prompt diagnosis and treatment (Morris 1988). Even human wound infections caused by *V. vulnificus* that are diagnosed and treated promptly result in 25 to 50% mortality (Blake et al. 1980). Clinical signs of the wound infections include blisters, cellulitis, edema, purpura, hemorrhagic bullae, necrotic eschar at the bullous lesion sites, and ulcers on the extremities (Hoffman et al. 1988).

Plesiomonas shigelloides. *Plesiomonas shigelloides* is reported as a fish pathogen in the catfish farming industry (Durborow et al. 1991) (Figure 39.4) and in the trout production industry (Cruz et al. 1986). It infects man when uncooked fish or shellfish are ingested, and causes gastroenteritis, usually involving fever, abdominal pain, vomiting, and diarrhea (Brenden et al. 1988). Risks such as foreign travel or seafood consumption increase the chances of contracting the infection, but it is not commonly seen in otherwise healthy individuals; 70% of patients with *P. shigelloides* have a serious preexisting health problem such as cirrhosis or cancer (Nemetz and Shotts 1993).

An extraintestinal form of *P. shigelloides* infections in people is rarely found (Kennedy et al. 1990). Meningitis, sepsis, cellulitis, and pancreatic abscess have been reported.

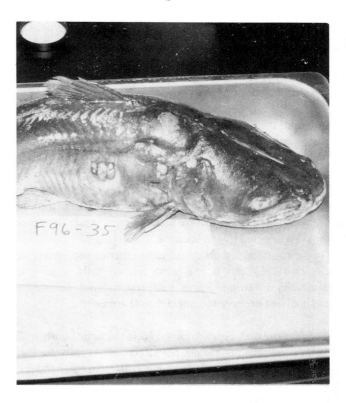

Figure 39.4. A *Plesiomonas shigelloides* bacterial infection caused open lesions on this channel catfish.

***Aeromonas* spp.** The motile *Aeromonas* complex bacteria include *Aeromonas hydrophila*, *A. sobria*, *A. caviae*, *A. schubertii*, and *A. veronii*. They are capable of infecting cultured and non-cultured fish and are commonly found in aquatic environments (Shotts and Teska 1989). These aeromonads are capable of infecting humans, causing gastroenteritis or localized wound infections. *Aeromonas salmonicida*, on the other hand, is not associated with infections in man; it causes furunculosis in trout and salmon.

Aeromonas spp. infections in people are very rare, occurring in less than one person in a million each year (CDC 1990). Aeromonad gastroenteritis is usually caused by ingesting contaminated water and involves diarrhea of various degrees of seriousness (George et al. 1985). Wound infections of *Aeromonas* spp. are caused by lacerations or punctures of the skin, and may progress to cellulitis, deep muscle necrosis, or septicemia. These infections usually occur in persons with a compromised immune system such as AIDS patients (Flynn and Knepp 1987), but have been reported from immunocompetent hosts (Heckerling et al. 1983; Karam et al. 1983).

Edwardsiella tarda. *Edwardsiella tarda* causes edwardsiellosis, also called emphysematous putrefactive disease in warmwater fish, and although it is not a common problem in farm-raised fish, neither is it rare. It can produce a gas-filled malodorous area of decomposing muscle, which starts as a fistula deep in the muscle (Lewis and Plumb 1979). It occurs mostly when water temperatures are above 30° C (86° F). Infections of *E. tarda* in man have been reviewed by Wilson et al. (1989). Receiving a wound from infected fish or ingesting them is the source of human infections. Diarrhea (sometimes with blood), fever, nausea, vomiting, colonic ulceration, pseudomembranes, terminal ileum nodularity, osteomyelitis, septicemia, and localized infections are the clinical manifestations of *E. tarda*, and the high mortality rate of affected individuals (44%) is mostly attributed to the high number of patients who have preexisting illnesses at the time they are infected by the bacteria.

Edwardsiella ictaluri causes enteric septicemia in catfish (ESC) and is the most economically significant disease in the United States' catfish production industry. It has not been observed to infect man (Farmer et al. 1985).

Yersinia ruckeri. *Yersinia ruckeri* is in the same family as *E. tarda* (Enterobacteriaceae), but affects only coldwater fish, causing enteric redmouth disease in cultured trout. Farmer et al. (1985) reported a case of human infection with this bacteria, which was isolated from the patient's bile. No clinical signs were reported.

Leptospira spp. Leptospirosis occurred in four English fish farmers in 1981 (Robertson et al. 1981; Gill et al. 1985a; 1985b). Three of the farmers had influenza-like symptoms and two had antibody titers against *Leptospira* sp. One fish farmer had jaundice and high leptospiral antibody titers. The source of the *Leptospira icterohaemorrhagica* infections was a mystery at first, but was later attributed to being transmitted via rat urine present in the feed storage areas or via the water of the fish farms. Storing feed in rat-proof storage buildings, using rat poison, or keeping pet cats in the feed storage area are a few ways to combat the presence of rats. (Figures 39.6 and 39.7). Keeping pond levees mowed and maintaining a well-groomed fish farm also make the pond habitat less inviting to rats.

Bacterial contamination after harvesting or processing. Bacteria such as *Salmonella* spp., *Shigella* spp., and *Staphylococcus aureus* may be introduced by those handling fish or from contaminated objects (fomites) coming into contact with the fish (Eyles 1986). At the processing plant, heat process failure and cross contamination of cooked fish with raw fish are potential sources of bacterial contamination.

It is obviously very important to maintain sanitation in fish processing plants. The author has observed that

Figure 39.5. The white "cataract" eye condition contrasts with the dark body of this rainbow trout infected with *Yersinia ruckeri*, the causative bacterium for enteric redmouth disease.

Figure 39.6. Bags of feed are stored on pallets in an area that should be able to exclude rats and mice.

Figure 39.7. Cats can be effective for controlling mice and rat populations.

commercially produced catfish and trout are processed in plants that usually lack any kind of bad or fishy odor. These aquaculture processing plants voluntarily pay USDC inspectors to approve the cleanliness of their processed products. These strict, self-imposed health standards of the aquaculture industry are essential for maintaining satisfied customers and thus staying in business.

Fish can be an ideal substrate upon which introduced bacteria may multiply to a level dangerous to consumers. *Pseudomonas* spp., for example, is able to actively grow even at temperatures near 0° C, producing by-products such as hydrogen sulfide (Connell 1990). Fortunately, however, fish often deteriorate and spoil due to natural breakdown in the flesh before bacteria can reach high concentrations, and the bad smell and appearance of the fish convinces the potential consumer not to ingest the fish, thereby avoiding bacterial food poisoning (Inglis et al. 1993).

Recently, in the summer of 1996, several cases of septicemia and skin infections of *Streptococcus iniae* were reported in Chinese citizens residing in Toronto, Canada. One patient developed meningitis and transient arthritis (Holden 1996). All affected individuals had recently handled tilapia that were raised in closed, recirculating systems in the United States, and they either had open wounds or were stuck by the fish spines during the cleaning process.

Antibiotics were successfully used to treat the infections, but, in one case, an elderly patient died after the infection, although other causes for mortality were also suspected.

Toxins produced by bacteria and other microorganisms. Toxins on or in fish resulting from bacterial action (e.g., scombroid poisoning) or from toxins ingested by the fish (e.g., ciguatera poisoning) cause problems in fish caught in the wild, and have the potential of occurring in aquaculture, especially in fish cultured in marine waters. Mycotoxins and a neurotoxin in the feed consumed by fish have also been investigated for human health concerns.

Scombroid poisoning. Scombroid poisoning results from ingestion of fish containing significant concentrations of histamine, which causes symptoms similar to an allergic reaction. These signs include a burning sensation in the mouth, itching, hives and skin rash, flushing, headache, palpitations, and gastrointestinal disturbance including nausea, vomiting, and diarrhea (Taylor et al. 1989); the condition usually resolves within a day. Scombroid poisoning may often be misdiagnosed as food allergy because of its similar symptoms.

Fish in the family Scombroidae (tuna, mackerel, bonito, and skipjack) are most frequently linked to scombroid poisoning, while bluefish, mahi-mahi, herring, and sardines have also been implicated (Taylor et al. 1989). It has been reported in fresh, canned, dried, and salted fish (Main and Yates 1990). The high histamine levels result from bacterial degradation of the amino acid histidine in the flesh of fish (often dark-fleshed fish). *Proteus margonii*, *Klebsiella* spp., *Vibrio* spp., and *Clostridium* spp. residing in the fish muscle convert histidine to histamine, which can build up to toxic concentrations within 3 to 4 hours if fish are not refrigerated (Pan 1985; Kow-Tong and Malison 1987). Some researchers are of the opinion, however, that histamine is not the substance in the fish causing the poisoning, but rather a potentiator causing histamine activity in the patient (Taylor et al. 1984). They argue that histamine itself, when ingested directly, degrades and does not produce signs of food poisoning. High histamine levels in the contaminated fish may only indicate the simultaneous presence of this potentiator.

Treatment of scombroid poisoning is with antihistamines, and prevention is best achieved by refrigerating fish during all stages from harvest to consumption (Bartholomew et al. 1987). Processing practices in the aquaculture industry render the occurrence of this kind of food poisoning very unlikely. Fish are electrically stunned while alive and then are beheaded, gutted, skinned and immediately put into an ice bath. The time lapsed from death by beheading until they are chilled in the ice bath is usually less than 5 minutes. The whole dressed fish are removed from ice only long enough to process further (filleting, cutting into steaks, etc.), and are individually quick frozen in carbon dioxide tunnels or stored on ice within about 20 minutes after beheading. Time would not allow for any buildup of bacteria or bacterial metabolites on the fish. In addition, the fish from which scombroid poisoning has been reported are not currently being cultured, so it has only a potential of occurring in an aquacultural environment.

Ciguatera poisoning. Carnivorous fish including grouper, snapper, kingfish, and barracuda (none of which are cultured fish) can accumulate significant concentrations of ciguatera toxin in their flesh (Halstead 1988). They do this by eating herbivorous fish that have eaten plants covered with the dinoflagellate *Gambierdiscus toxicus* (which actually produces the ciguatoxin). Persons consuming the carnivorous fish that have ingested the ciguatera toxin become sick. The fish themselves may also exhibit signs of neurotoxicity.

Clinical signs in humans involve gastroenteritis including nausea, vomiting, and diarrhea within 5 to 6 hours of eating contaminated fish (Frenette et al. 1988). Conditions usually resolve on their own in 1 or 2 days. Ciguatera poisoning also involves neurologic signs such as abnormal sensations of the mouth and extremities, most notably extreme cold sensitivity of the mouth and the extremities, itching, muscle and joint pain, headaches and weakness. Severe toxicity may lead to bradycardia, shock, and coma, which may be a result of cerebral edema (Palafox et al. 1988). Intensive medical monitoring and treatment may be necessary in cases of severe intoxication.

Ciguatera poisoning was suspected to have originated from a farm-raised salmon (possibly imported from Chile) in one case in January 1992 (Ebesu et al. 1994). The affected woman became seriously ill one and a half hours after eating the fish. She experienced nausea, diarrhea, abdominal pain, hyperventilation, increased heart rate, tooth pain, a metallic taste in her mouth, numbness in her extremities, a freezing sensation, and the inability to distinguish hot from cold temperatures. She was also unable to eat solid foods for more than 3 months. More than a year after ingesting the fish, she was still experiencing severe headaches, numbness in her arms, and a heavy sensation in her chest.

Neurotoxin in fish feed. The neurotoxin domoic acid is produced by diatoms in the genus *Pseudonitzschia* (Shumway 1989; Bates et al. 1991; Buck et al. 1992; Garrison et al. 1992; Horner and Postel 1993) and accumulates in fish such as anchovies and herring, which are used as ingredients in fish feed (White 1977, 1981). The toxin also concentrates in mussels, clams, and crabs (Wekell et al. 1994), and can kill pelicans and cormorants that feed on these animals (Work et al. 1993). In one case, cultured mussels from eastern Canada containing domoic acid were responsible for 3 human deaths, 6 permanent hospitalizations, and over 100 cases of permanent short-term memory loss (Todd 1993).

Regarding finfish culture, Hardy et al. (1995) found that domoic acid does remain in fish feed made with anchovies containing domoic acid, but that it does not accumulate in the tissues of trout consuming the contaminated feed. There is, therefore, no danger posed to people consuming the farm-raised trout.

Mycotoxins in fish feed. The possibility of the mycotoxin aflatoxin B_1 (AFB_1), which sometimes occurs in fish feed, accumulating in the fish and consequently posing a health hazard to humans consuming the fish was investigated by Plakas et al. (1991). Rainbow trout are so sensitive to mycotoxins that they would die before having a chance to be eaten by man. And even channel catfish, which are more tolerant of mycotoxins, would not be able to accumulate enough toxin to be harmful to people: aflatoxin's half life was found to be only 4 hours and the fish's muscle retained the least amount of aflatoxin of any tissue analyzed (Plakas et al. 1991). Withholding any suspected feed from affected fish for 24 hours before harvesting would help to further ensure that no significant aflatoxin levels would be present in processed fish.

Pesticides in wild-caught fish. Wild fish caught in many river systems in the United States and from some parts of the ocean contain toxic pesticide residues. Fillet samples of fish from the Missouri River near Nebraska contained DDT, PCB, dieldrin, heptachlor, trifluralin, and chlordane (Christiansen et al. 1990 to 1991). Farm-raised fish provide the assurance of not having these kind of pollutants in their flesh if the producer makes sure that toxic chemicals are not present in the pond soils or water source.

Parasites

Anasakiasis. People develop anasakiasis by ingesting fish containing the third stage larvae of the nematodes *Anasakis simplex* or *Pseudoterranova* (*Phocanema*) *decipiens*. Marine mammals are the normal definitive host of these round worms, and humans assume this role when they consume raw or undercooked infected fish such as salmon, herring, and Pacific cod (Kliks 1986). The involvement of salmon, a major aquaculture species, in this infection indicates at least a potential for this condition to arise from an aquaculture environment. Deardorff and Kent (1989), however, point out that no parasites of public health significance occur in net-pen reared salmon, fed commercially prepared diets.

Of the anasakiasis cases reported in the 30-year span between 1958 and 1987, 70% occurred in the most recent 8 years (McKerrow et al. 1988). This, of course, can be attributed to the recent popularity of consuming raw fish. The condition is common in the United States in the western states and around the Great Lakes (Schmidt and Roberts 1981). Cases are much more abundant in Scandinavia and Russia. Japan has an estimated 1,000 new cases of anisakiasis each year (Oshima and Kliks 1987).

Anasakis simplex can penetrate the wall of the stomach or intestine and cause abdominal pain, fever, nausea, vomiting, or diarrhea (Cross 1994). Infections are sometimes misdiagnosed as gastric ulcers or appendicitis (Nemetz and Shotts 1993). Patients may cough up the worm and can sometimes feel the parasite in their throat, causing "tingling throat syndrome" (McKerrow et al. 1988). *Pseudoterranova decipiens* rarely penetrates the stomach or intestinal wall of man and causes milder symptoms or none at all. Anasakiasis can be prevented by heating fish to at least 60° C (140° F) for 5 minutes, freezing at -20° C (-4° F) for 60 hours (Wittner et al. 1989), or blast freezing at -35° C for 15 hours (Murrell 1995). In addition, preserving the fish with salt in 20 to 30% brine for 10 days kills the *Anasakis* parasite.

Eustrongylidiasis. Eustrongylidiasis is caused by humans ingesting fish meat infected with the nematode *Eustrongylides* sp. Three Maryland fishermen became infected after eating live bait minnows (Centers for Disease Control 1982). After experiencing severe abdominal pain 24 hours after eating the minnows, followed by progressive abdominal cramping pain, appendectomy surgery was performed on one of the fishermen. During the surgery, two *Eustrongylides* sp. nematodes in the fourth stage were discovered; one in the abdominal cavity and one in the process of penetrating the intestinal cecum. Although it has never become a problem, one farm-raised channel catfish was found to be infected with *Eustrongylides* sp. at a Mississippi processing plant in the late 1980s. No other reports of this or similar parasites have been made. Because cultured channel catfish are typically cooked before consumption (there is no custom of eating them raw or undercooked), they present practically no health risk to the consumer.

***Philometra* sp.** A case of an adult nematode migrating into a person's open hand wound was reported by Deardorff et al. (1986). The patient had filleted a jack fish he had caught from the wild; 3 hours later he noticed that the distal end of the thin red worm was protruding 1.1 cm out of his wound. His doctor removed the nematode which had penetrated 2.5 cm into the patient's flesh. The parasite was identified as *Philometra* sp.

Diphyllobothrium latum. The broad fish tapeworm or broad tapeworm of man, *Diphyllobothrium latum*, infects man when he ingests a raw or undercooked fish containing the larva (plerocercoid or "grub") of this cestode. The normal definitive hosts are bears, dogs, cats, coons, and other mammals able to consume freshwater fish. Diagnosis is by finding ova in the patient's feces (Nemetz and Shotts 1993).

The condition is prevalent in the United States among people along the Great Lakes; there, the practice of eating cold-smoked fish (which does not kill the plerocercoid) can lead to the establishment and growth of the adult tapeworm in the intestine. It is also commonly found in Russia and Scandinavian countries, especially Finland (Cross 1994). Adult *D. latum* can reach lengths of 50 to 60 feet in humans and may live up to 10 years (Murrell 1995). The potential exists for its occurrence in aquacultured fish. Symptoms are similar to those caused by other tapeworms, such as abdominal discomfort, diarrhea, nausea, and weakness (Schmidt and Roberts 1981). Pernicious anemia may develop in patients due to a vitamin B_{12} deficiency caused, in part, by excessive absorption of B_{12} by the tapeworm and, in part, by the patient's normal inefficiency in absorbing the vitamin.

Infection can be prevented by heating all parts of the fish to at least 56° C (133° F) for 5 minutes in order to kill the plerocercoid. Freezing at -18° C (0° F) for 24 hours or -10° C (14° F) for 72 hours is also effective in killing the grub (Wittner et al. 1989). The frequency of *D. latum* in the wild can be controlled by properly treating sewage from lakeside dwellings and ships, and by deworming domestic pets (Murrell 1995).

Clonorchis sinensis. The fishborne trematode or liverfluke *Clonorchis sinensis* is very prevalent in Southeast Asia. An estimated seven million people are infected worldwide (Rim et al. 1994). Symptoms include indigestion, epigastric discomfort, and diarrhea; acute pancreatitis may result when the worm inhabits the pancreatic duct (Cross 1994). Chronic infection of this trematode may lead to cholangiocarcinoma.

Man serves as the definitive host, and eggs from the adult *C. sinensis* exit the host in the feces. Snails (serving as intermediate hosts) ingest the eggs that develop inside the snail. The larval trematode eventually leaves the snail as a motile larval cercaria that finds a fish to parasitize. The cercaria penetrates the fish (usually a cyprinid such as a carp, minnow, or goldfish) and encysts in the muscle. It becomes infective in 3 to 4 weeks (Murrell 1995). Properly cooking, salting, pickling, or smoking fish before consumption can prevent infection, and the use of raw animal manure or untreated human sewage in aquaculture ponds (a common practice in many Asian countries) should be avoided.

***Opisthorchis* spp.** Another trematode infecting humans, *Opisthorchis* spp., produces symptoms similar to *Clonorchis sinensis* but is considered a greater health threat because it more frequently causes cholangiocarcinoma. The parasite is, in fact, a major cause of death in rural Northeast Thailand (Haswell-Elkins et al. 1992); an estimated 7 million people are infected in Thailand. *Opisthorchis viverrini* occurs chiefly in Southeast Asia, and *O. felineus* is found primarily in Eastern Europe (Cross 1994).

Opisthorchis is transmitted similar to *Clonorchis* and is controlled in the same way. The life cycle also involves snails, fish, and man.

***Heterophyes heterophyes*.** Human infections of the digenetic trematode *Heterophyes heterophyes* are not endemic to the United States but are found in Egypt, Greece, Israel, China, Japan, Korea, Taiwan, and the Philippines. A case in the United States was reported from imported sushi served at an oriental restaurant (Adams et al. 1986). It has the potential of occurring in fish cultured in the listed countries.

H. heterophyes infections cause week-long diarrhea and slight eosinophilia (Adams et al. 1986). They can be prevented by avoiding raw fish from the above-listed countries or by properly cooking fish before consumption.

***Nanophyetus salmincola*.** *Nanophyetus salmincola* is another digenetic trematode of fish which can infect man when infected raw salmon are ingested (Fritsche et al. 1989) or when the worm is accidentally transferred from hand to mouth while working with infected salmon (Harrell and Deardorff 1990). Abdominal pain, diarrhea, nausea, vomiting, weight loss, and fatigue are common symptoms, but many cases are asymptomatic (Eastburn et al. 1987). It is deeply imbedded in the wall of the small intestine and is notorious for infecting natives of eastern Siberia (Schmidt and Roberts 1981). *N. salmincola* is becoming more common in the Pacific Northwest of the United States. The two most common natural mammalian hosts for the adult stage of this fluke are raccoons and skunks. The necessity of having snails, fish, and mammals as hosts to complete the complex life cycle of this parasite makes it unlikely for it to occur in an aquaculture environment.

Nanophyetus salmincola is involved in causing salmon poisoning in dogs that eat raw salmon infected with this trematode. The trematode carries the rickettsia *Neorickettsia helminthoeca,* which is the organism that actually kills the dog if the condition is untreated.

Cultured Prawn for Trematode Control. The digenetic trematode, *Schistosoma* spp., causes schistosomiasis (or bilharzia), a serious disease in tropical regions around the world. The snail intermediate host *Biomphalaria glabrata* is essential to the survival and spread of the *Schistosoma* parasite. Potential control of this snail with cultured freshwater prawn *Macrobrachium rosenbergii* has been explored (Roberts and Kurts 1990). The prawn prefer *B. glabrata* as a food and aggressively consume them.

***Giardia lamblia*.** The flagellated protozoan intestinal parasite *Giardia lamblia* was reported in canned salmon due to fecal contamination (Osterholm et al. 1983). It may, therefore, have some potential relevance to cultured salmon, although *Giardia* infections are mostly associated with hikers and campers consuming contaminated mountain stream water without first treating it. Symptoms include yellowish to white diarrhea (due to blockage of the bile duct), lethargy, intense hunger followed by a strong feeling of satiation after eating very little, and nausea often associated with certain smells. Malabsorption of fats and fat soluble vitamins may occur. Extreme sanitation is necessary to prevent the spread to family members.

SAFETY ISSUES

Safety issues are becoming officially recognized in aquaculture. For example, a forum on aquaculture products safety was held at Auburn University, Alabama, in 1993. Rainbow trout facilities with 25 or more employees are now inspected by OSHA for things such as maintaining adequate first aid kits in the work place. OSHA also oversees worker

Figure 39.8. OSHA safety guidelines must be followed during the installation of netting that prevents bird depredation on hatchery trout.

safety in the trout industry in Idaho when bird netting and the metal frames supporting the bird netting are installed at trout hatcheries for preventing bird depredation of the trout (personal communication, Gary Fornshell 1995: Aquaculture Specialist, University of Idaho Cooperative Extension Service, 246 3rd Avenue East, Twin Falls, Idaho 83301).

Sewage in Ponds. Night soil (human excrement used as fertilizer) and human wastewater are used in Southeast Asia and Mexico to fertilize aquaculture ponds. An analysis was made on health risks to fish farm workers and people consuming raw fish from ponds fertilized with either treated or untreated night soil in China (Ling et al. 1993). Pond workers and the general public consuming animals, plants, or water from ponds using untreated night soil had high incidences of ascariasis, trichuriasis, clonorchiasis, fasciolopsiasis, schistosomiasis, *Salmonella,* and hepatitis B. When the night soil was treated in an anaerobic digester for 10 days and then used as pond fertilizer, parasite eggs were decreased by 99.8%, fecal coliforms were reduced 10-fold, *Salmonella* was eliminated, and general water quality improved, including chemical and biological oxygen demands and dissolved oxygen levels. The growth rate of fish doubled in ponds receiving treated night soil compared to ponds with untreated night soil. In a study in Mexico, risks of *Ascaris* infection and diarrhea decreased when 1989 World Health Organization guidelines for restricted irrigation were followed for wastewater used in fish production ponds (Blumenthal et al. 1991).

In another tropic environment, fish from Nigerian aquaculture ponds carried several enteric bacteria that are capable of infecting man (Ogbondemtnu 1993). They included *Escherichia coli, Citrobacter, Enterobacter, Proteus, Salmonella, Serratia,* and *Klebsiella*. The warmer year-round climate in the tropics and the custom of using human sewage to help fertilize tropical aquaculture ponds create an environment different than that encountered in the temperate or subtropical settings in the U.S. Therefore, the risk of dangerous levels of enteric bacteria accumulating in domestic aquaculture products is minimal.

Hydrogen Sulfide Gas. Hydrogen sulfide gas (H_2S) is produced naturally on pond bottoms during anaerobic reactions in the mud and mud-water interface. H_2S is highly toxic to cultured fish. Channel catfish fry die when exposed to concentrations as low as 0.005 ppm (Tucker 1991). Recently, four adult men fell into a flatfish culture pond, and three of them drowned. Autopsies showed that the men had been exposed to nearly lethal levels of hydrogen sulfide gas which led to their drownings (Kimura et al. 1994).

Shrimp Preservatives. Sodium metabisulfite is routinely used in the shrimping industry as a preservative; it is referred to as "shrimp dip." Two crewmen on a shrimp trawler were found dead in the ship's hold due to asphyxia. They died while applying dry sodium metabisulfite (Atkinson et al. 1993). The chemical is known to affect sulfite-sensitive persons, causing bronchospasm, oculonasal symptoms, and urticaria/angioedema. In this case, however, it appeared that sulfur dioxide (SO_2) was produced from a chemical reaction of the sodium metabisulfite with acid and water. SO_2 gas reacts with respiratory tissue forming sulfurous acid, inducing a pulmonary reaction, and causing hypoxemia (Atkinson et al. 1993).

All individuals working with shrimp should be educated on the potential hazards involved with handling sodium metabisulfite. The culture of marine peneid shrimp continues to grow as an industry—the potential for the occurrence of this kind of accident will increase in aquaculture as the industry increases in size.

Sodium bisulfite is typically applied to marine shrimp to prevent melanosis (black spot). The potential exists for sensitive persons to have allergic reactions to sulfite residuals; however, if the FDA limit of 100 mg/l sulfite in edible portions of shrimp is followed, no allergy problems should occur (Otwell 1993). Most aquacultured shrimp do not require these preservatives, because they are usually sold rapidly after harvest, and they have less tendency to develop melanosis.

Anesthetic. The anesthetic, 2-phenoxyethanol, used by three women to anesthetize small salmon, caused headache and symptoms of intoxication in all three (Morton 1990). They soon developed diminished sensation and strength in their hands and fingers with the symptoms being worse in their preferred hand, but no persistent neuropathy developed. However, after 1 to 2 years of exposure, all three women developed focal cognitive impairments, and one woman had problems with hearing (Morton 1990). Skin and eye protection should be used as safeguards against anesthetics as well as other chemicals that may be used around fish. Problems with this anesthetic in aquaculture are unlikely to occur because it is not labeled for use on food fish. Only chemicals approved for use by FDA should be used in aquaculture. Consumers of fish are not the only people benefiting from chemical-use guidelines and laws; fish farm and fisheries workers also benefit by using only approved chemicals.

Antibiotic Safety. Allergies or hypersensitivities to antibiotics are safety risks faced by fish pathologists, researchers, fish producers, and feed mill operators when administering them or incorporating them into feed. Localized skin reactions, respiratory problems, or generalized anaphylactic shock may result after a repeat exposure to the drug (Inglis et al. 1993). Contact of the antibiotic powder with the skin of the handler is a common route of sensitivity development (Schulz 1982). Gloves, long sleeves, and face masks or other respiratory protection are recommended for use around antibiotic powders.

Misuse or overuse of antibiotics such as administering the drug at less-than-adequate doses or for fewer than the prescribed number of days, or excessive, unnecessary use of the drug often lead to the development of resistant bacterial strains. This is a potentially dangerous situation for people if the fish pathogen being treated is also pathogenic to man (Young 1994). Antibiotic-resistant strains of the human pathogens *Salmonella* spp., *Aeromonas hydrophila*, and *Plesiomonas shigelloides* were isolated from a fish farm in Southeast Asia (Twiddy 1995). The bacteria showed varying degrees of resistance to nalidixic acid, oxolinic acid, chloramphenicol, neomycin, oxytetracycline, tetracycline, furazolidone and sulphamethoxazole combined with trimethoprim. Fish in this part of the world are often raised in waters using human sewage for fertilizer, so this study cannot accurately be applied to the United States aquaculture industry. It does, however, give our domestic industry a warning to avoid situations that could lead to hazards such as these.

Withdrawal times imposed by FDA for the antibiotics oxytetracycline, and sulfadimethoxine + ormetoprim provide adequate time for residues to reach safe levels in edible portions of catfish and salmonid fish (Plakas 1993). The 21-day withdrawal after oxytetracycline (Terramycin®) is fed to catfish and salmonids allows enough time for drug concentrations to drop below 0.1 mg/l in the flesh (Plakas et al. 1988). The 3-days withdrawal time for sulfadimethoxine + ormetoprim (Romet-30®) in catfish (Plakas et al. 1990) and the 6-week withdrawal in salmonids (Droy et al. 1990) permitted drug levels to drop below 0.1 mg/l in both species. FDA, however, does not examine residues in imported seafoods (Schnick 1990; Ahmed 1991).

Vaccine Safety. In aquaculture, when fish are vaccinated they are usually immersed in a bacterin and are sometimes given a booster later on in the feed. In the salmon industry, individual salmon are injected with vaccines by handheld innoculating guns; the value of salmon is high enough to justify this expensive procedure. They are immunized against diseases such as vibriosis and furunculosis. During the injection procedure, the workers occasionally inject themselves accidentally in the hand or fingers holding the fish. Inflammatory reactions occur at the injection site, while less often a fever, muscle ache, and general malaise are reported (Leira and Baalsrüd 1992). Endotoxins in the vaccine may be the cause of these conditions. A few serious anaphylactic reactions have resulted, which involved coma, dyspnea (difficulty in breathing), and confusion; all those affected recovered after hospitalization. Needle guards have been developed to help prevent these self-injection accidents (Figure 39.9).

Injuries While Processing Fish and Manufacturing Feed. In aquaculture processing plants, injuries occur from workers interacting with fish and machinery. An obvious risk involves puncture wounds from fish spines; channel catfish have very pronounced, serrated pectoral and dorsal fin spines (Figure 39.10). The spines can lock into an erect position (Grizzle and Rogers 1976), and the spine and covering epithelial layer are associated with a toxin that produces a stinging sensation in persons who are punctured (Birkhead 1967).

In cultured rainbow trout processing plants, government regulations now require workers to periodically rotate from positions at risk for carpal tunnel syndrome (personal communication, Gary Fornshell 1995).

Fish bile has caused superficial corneal erosions in fish industry workers in Norway (Christoffersen and Olsen 1993). Immediate rinsing of the eyes affected by the bile greatly reduces the chance of permanent eye damage. In one case, delayed attention to the accident led to serious corneal

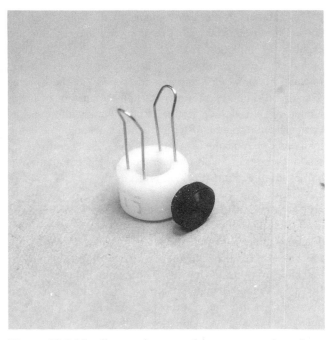

Figure 39.9. Needle guards are used to protect workers from accidental self-injection during the vaccination of young salmon.

opacity. Aquaculture processing plant employees and fish health workers should wear eye protection when handling fish, and an eye-wash station should be located in the work area.

The band saw used to behead catfish is a source of some serious cut injuries, including severed fingers, despite the metal-link gloves worn by band saw operators. Some highly skilled operators can exceed beheading 60 fish per minute. One problem arises when the fish are not completely stunned by the electrical shock as they enter the processing plant; a quick jerking movement by the fish passing through the band saw may sometimes pull the worker's hand toward the saw (Figure 39.11). Other potential hazards also exist such as cuts from the skinning machine or fillet knives, and typical problems that are encountered while working with or repairing heavy equipment such as being struck by a moving fork lift or by a falling equipment part during repair.

Similar equipment-associated accidents may occur in aquaculture feed mills; hard hats and hearing protection are required for workers' safety. There is also the risk of respiring the dust of feedstuffs as well as antibiotic powders that are incorporated into medicated feed. Protective dust masks should be worn in those situations.

Electrical Accidents. Electrically powered equipment is commonly used around water in aquaculture. In commercial channel catfish culture ponds, floating electric aerators are positioned in the water near the pond bank. The electricity usually travels from the power line to a meter and "disconnect" box with circuit breakers, and from there to circuit breaker boxes at each aerator. The power can be manually switched off at either box, and the circuit breakers will trip and interrupt the electrical flow if the circuit becomes overloaded. Correct size fuses must be used in order for the circuit breakers to trip when necessary.

All wires around the pond levees are buried underground and are encased in metal conduit and/or PVC pipe. The electrical cord connecting the breaker box with the floating aerator is coated with a heavy plastic, waterproof covering and is encased inside metal conduit to prevent the plastic from being cut by equipment or aquatic animals (Figure 39.12). Many fish farms use qualified electricians to do the electrical work around water. All pond workers generally are required to leave the pond before a newly installed aerator is tested, and electrical power switches are turned off at the power boxes when workers are in the water. One electrocution occurred in the early 1980s in west-central Mississippi, when a catfish farm manager was trying to dislodge an aerator's electrical wire that had become entangled underneath his truck. The aerator had recently been installed and the cable had not yet been buried in the

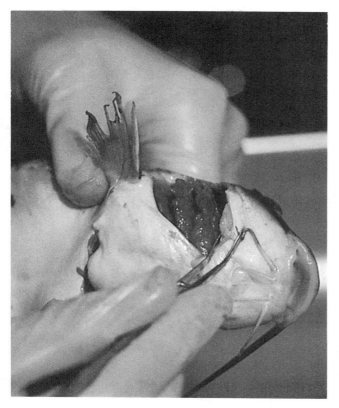

Figure 39.10. Puncture wounds from channel catfish spines are a potential safety hazard.

Figure 39.11. Open blade band saws used to behead fish pose a safety risk to certain processing plant workers.

Figure 39.12. Floating paddlewheel aerator electrical wiring must be protected from mechanical or aquatic animal damage.

Figure 39.13. Movement of fish loading equipment near overhead power lines during harvest has much potential danger.

levee. No known electrically related accident around aerators has occurred in west-central Mississippi since that time.

Movement of fish loading equipment near overhead power lines during harvest has much potential danger. Long booms attached to a front-end loader or "cherry-picker" truck are used to lift a basket loaded with fish from the pond to a hauling truck that sits relatively high on top of the levee. When the movement comes close to power lines, the farm manager often has the local power company turn off power to the whole line during the harvest (Figure 39.13) (Brent Johnson, personal communication 1995: Aqua Farms, Inc., Route 2 Box 620, Greenville, Mississippi 38701).

Other Safety Hazards in Aquaculture. Tractors are very powerful and perform much vital heavy-duty work on all farms, including fish farms. This power, however, must be highly respected to avoid mishaps such as rolling over. Levee slopes are often built to be on a safe incline, but erosion may cause sudden drops or dips in the path of the tractor, leading to a roll-over. Roll bars should be installed routinely on tractors used in all kinds of farming. OSHA has mandated rollover protection systems (ROPS) on most farm tractors (Figure 39.14).

Figure 39.14. Roll bars on tractors help to protect the aquaculturist from being crushed if the tractor turns over.

Figure 39.15. A tractor's power take-off (PTO) torque capacity can cause serious injury if a person's clothing gets caught.

Figure 39.16. Injury can be caused by the high-torque-capacity axle that holds the paddles on an electric-powered hatching trough.

The power take-off (PTO) of a tractor is frequently used to power pond aerators; the PTO torque capacity can cause serious injury to a man if some of his clothing gets caught in it (Figure 39.15). Protective shields should cover all PTO connections and the moving parts of equipment powered by the PTO.

Another piece of aquaculture equipment with powerful torque is the axle that holds the paddles on an electric-powered hatching trough (Figure 39.16). If a fish farm worker is caught in this device, he would typically be too far from the on/off switch to turn off the power. Catfish hatching troughs, recently constructed near Indianola, Mississippi, have wires running the length of the troughs that can be pulled to shut off the power of the turning paddles.

Muscle strains can be avoided if precautionary measures are taken such as keeping ponds free of weeds to avert bogging-down of seine nets during harvest. Tractors or 4-wheelers may also be used to aid in pulling the seine. These kinds of problems are less prevalent in the trout growing industry because bar graders are slid down the raceway to crowd the fish, and then they are lifted with a fish pump into the hauling tanks. Tenosynovitis conditions have been reported from using a handheld scoop to broadcast feed into ponds (Douglas 1995).

Figure 39.17. Trout hatchery workers routinely walk on top of narrow concrete raceway walls which poses safety risks from falls.

Foam-padded vests have increased fish farm safety in several ways: they serve as floatation devices that prevent drowning, they help to insulate workers for protection against hypothermia, and they provide padding, which prevents chest and rib injuries (Douglas 1995). They appear to be more effective than life jackets that utilize CO_2 cylinders. In one case study, 10 of 15 life jackets failed to inflate when the subjects fell into the water (Douglas 1995). Drowning and falling-related injuries appear to be a risk at many trout growing facilities when workers routinely walk on top of the concrete walls (often less than a foot wide) dividing raceways (Figure 39.17).

Drowning is also a risk among SCUBA divers who clear dead and dying fish from salmon net pens in the Pacific Northwest; entanglement and entrapment in the netting has caused at least one drowning accident. These divers use multiple descents and ascents when diving from cage to cage collecting dead fish. This is referred to as "yo-yo diving," and no decompression tables are available for that style of diving. Decompression (dysbaric) illness cases have resulted (Douglas 1995).

Ice-related injuries are a threat on fish production ponds in the winter. Numerous accidents may occur such as breaking through thin ice and drowning or receiving bruises and/or broken bones from slipping on the ice. This could occur if producers are working with their aerators in an attempt to keep them running to maintain zones of ice-free water. Some pond owners also seine underneath the ice to harvest fish during the winter when fish are needed to supply markets.

Two other miscellaneous safety hazards are getting sunburn from excess ultraviolet radiation (working in an environment surrounded by water makes this occurrence more likely), and being exposed to high noise levels around equipment, resulting in possible ear damage. Sunscreen and ear protection should be used.

SUMMARY

Aquaculture, the fastest growing agricultural business in the United States, needs to continue focusing on health and safety potentials in its farm-raised products. Much effort has been put into quality assurance of fish products as evidenced by the Aquaculture Products Safety Forum held in Auburn, Alabama; manuals sponsored by Catfish Farmers of America and U.S. Trout Farmers Association dealing with HACCP guidelines to assure food safety; and the stricter temperature-range guidelines that many restaurants voluntarily impose on themselves during preparation of fish.

Precautions must be taken while handling fish before and during processing. Bacteria (such as *Streptococcus*) and/or parasites associated with fish can be introduced into the handler by way of an open wound or by accidental injection of the pathogen from a fish spine. Before eating fish, all consumers should take proper steps to ensure the killing of bacteria and parasites (such as nematodes or cestodes) on or in the fish. Heating or freezing fish to the required temperature for the necessary length of time will result in a product safe from bacterial and/or parasitic contamination. Care must also be taken to prevent cross-contamination of the finished product with raw product or contaminated surface areas.

Areas of safety concern include apparent water-related hazards such as drowning and electrical shock, as well as dangers in processing facilities such as open band saws. Usual dangers from farm equipment such as the tractor and tractor PTO exist, making it necessary to practice strict safety procedures around such machinery. More bizarre safety hazards include hydrogen sulfide gas coming from a pond bottom, causing unconsciousness in fish pond workers, and hypoxemia of shrimpers exposed to the shrimp preservation sodium metabisulfite. Even antibiotics and vaccines, which are designed to alleviate fish health problems, can themselves be hazardous to unprotected people handling them.

No profit or time advantages are worth taking health or safety risks. Injuries to owners or workers during production or processing reduce their quality of life and even reduce profits if workers must take time off work to recover from an accident. Unsafe shortcuts may save money initially, but real economic losses will be realized over time with the loss of customers. Aquacultural producers, processors, and consumers all stand to profit (health-wise and economically) by following precautionary procedures in health and safety issues.

REFERENCES

Adams, K., D. Jungkind, E. Bergquist, and C. Wirts. 1986. Intestinal fluke infection as a result of eating sushi. *Am. J. Clin. Pathol.* 86: 688-89.

Ahmed, F.E., ed. 1991. *Seafood Safety*. Committee on Evaluation of the Safety of Fishery Products, Food and Nutrition Board, Institute of Medicine. Washington, DC: National Academy Press, p. 25.

Atkinson, D.A., T.C. Sim, J.A. Grant. 1993. Sodium metabisulfite and SO_2 release: an under-recognized hazard among shrimp fishermen. *Ann. Allergy* 71: 563-66.

Bartholemew, B.A., P.R. Berry, J.C. Rodhouse, R.J. Gilbert, and C.K. Murray. 1987. Scombro-toxic fish poisoning in Britain: features of over 250 suspected incidents from 1976-1986. *Epidem. Infect.* 99: 775-82.

Bates, S.S., A.S.W. DeFreitas, J.E. Milley, R. Pocklington, M.A. Quilliam, J.C. Smith, and J. Worms. 1991. Controls on domoic acid production by the diatom *Nitzschia pungens* f. multiseries in culture: nutrients and irradiance. *Can. J. Fish. Aquat. Sci.* 48: 1136-44.

Bean, N. and P. Griffin. 1990. Foodborne disease outbreaks in the United States, 1973-1987: Pathogens, vehicles, and trends. *J. Food Protect.* 53: 804-17.

Birkhead, W.S. 1967. The comparative toxicity of stings of the Ictalurid catfish genera *Ictalurus* and *Schilbeodes*. *Comp. Biochem. Physiol.* 22: 101-11.

Blake, P.A., R.E. Weaver, and D.G. Hollis. 1980. Diseases of humans (other than cholera) caused by vibrios. *Annu. Rev. Microbiol.* 34: 341-67.

Blumenthal, U.J., B. Abisudjak, E. Cifuentes, S. Bennett, and G. Ruiz-Palacios. 1991. Recent epidemiological studies to test microbiological quality guidelines for wastewater use in agriculture and aquaculture. *Public Health Reviews* 19 (1-4): 237-42.

Brenden, R., M. Miller, and J. Janda. 1988. Clinical disease spectrum and pathogenic factors associated with *Plesiomonas shigelloides* infections in humans. *Rev. Infect. Dis.* 10: 303-16.

Browne, S. de la R. 1990. Food safety concerns and aquaculture. In: *Aquaculture International Congress Proceedings.* September 4-7, 1990, Vancouver BC., Canada, pp. 210-12.

Brunson, M., ed. 1993. *Catfish Quality Assurance*. Catfish Farmers of America and Mississippi State University Cooperative Extension Service.

Bryan, F.L. 1986. Seafood-transmitted infections and intoxications in recent years. In: *Seafood Quality Determination*. D.E. Kramer, D.E. and J. Liston, eds. Amsterdam: Elsevier Science, pp. 319-37.

Buck, K.R., L. Uttal-Cooke, C.H. Pilskaln, D.L. Roelke, M.C. Villac, G.A. Fryxell, L. Cifuentes, and F.P. Chaves. 1992. Autoecology of the diatom *Pseudonitzschia australis* Frenguelli, a domoic acid producer, from Monterey Bay, California. *Mar. Ecol. Prog. Ser.* 84: 293-302.

Centers for Disease Control (CDC). 1982. Intestinal perforation caused by larval *Eustrongylides*. *MMWR* 31: 383-84.

Centers for Disease Control (CDC). 1987a. Diagnosis and management of mycobacterial infection and disease in persons with human immunodeficiency virus infection. *Ann. Intern. Med.* 106: 254-56.

Centers for Disease Control (CDC). 1987b. Botulism associated with commercially distributed kapchunka-New York City. *MMWR* 34: 546-47.

Centers for Disease Control (CDC). 1990. *Aeromonas* wound infections associated with outdoor activities-California. *MMWR* 39: 334-41.

Christiansen, C.C., L.W. Hesse, and B. Littell. 1990-1991. Contamination of the channel catfish *Ictalurus punctatus* by organochlorine pesticides and polychlorinated biphenyls in the Missouri River. *Trans. Neb. Acad. Sci.* 18: 93-98.

Christoffersen, T., and E.G. Olsen. 1993. Injury to the cornea due to fish bile. *Scand. J. Work Envir. Health* 19: 358-59.

Colwell, R.R. and D.J. Grimes. 1984. *Vibrio* diseases of marine fish populations. *Helgol. Wiss. Meeresunters* 37: 265-287.

Connell, J.J. 1990. Control of Fish Quality. 3rd edtion. Oxford: Fishing News Books, Blackwell Scientific Publications, Ltd.

Cross, J.H. 1994. Fish and invertebrate-borne helminths. In: *Foodborne Disease Handbook*, Vol. 2. Y.H. Hui, J.R. Gorham, K.D. Murrell, D.O. Cliver, eds. New York: Marcel Dekker, pp. 279-329.

Cruz, J.M., A. Saraiva, J.C. Eiras, R. Branco, and J.C. Sousa. 1986. An outbreak of *Plesiomonas shigelloides* in farmed rainbow trout, *Salmo gairdneri* Richardson, in Portugal. *Bull. Eur. Assoc. Fish Path.* 6: 20-22.

Deardorff, T.L. and M.L. Kent. 1989. Prevalence of larval *Anisakis simplex* in pen-reared and wild-caught salmon (Salmonidae) from Puget Sound, Washington. *J. Wildlife Dis.* 25: 416-19.

Deardorff, T., R. Overstreet, M. Okihiro, and R. Tam. 1986. Piscine adult nematode invading an open lesion in a human hand. *Am. J. Trop. Med. Hyg.* 35: 827-30.

Douglas, J.D.M. 1995. Salmon farming: occupational health in a new rural industry. *Occup. Med.* 45(2): 89-92.

Droy, B.F., M.S. Goodrich, J.J. Lech, and K.M. Kleinow. 1990. Bioavailability, disposition and pharmacokinetics of ^{14}C-ormetoprim in rainbow trout (*Salmo gairdneri*). *Xenobiotica* 20: 147-57.

Durborow, R.M. 1993. 1992 fish disease cases at Kentucky State University fish disease diagnostic laboratory. *Kent. Fish Farming Newsl.* 6(2): 6.

Durborow, R.M. 1996. 1995 fish disease cases at Kentucky State University fish diseases diagnostic laboratory. *Kent. Fish Farming Newsl.* 9(1):3.

Durborow, R.M., P.W. Taylor, M.D. Crosby, and T.D. Santucci. 1991. Fish mortality in the Mississippi catfish farming industry in 1988: Causes and treatments. *J. Wildlife Dis.* 27(1): 144-47.

Eastaugh, J. and S. Shepherd. 1989. Infectious and toxic syndromes from fish and shellfish consumption: A review. *Arch. Inter. Med.* 149: 1735.

Eastburn, R.L., T.R. Fritsche, C.A. Terhune, Jr. 1987. Human intestinal infection with *Nanophyetus salmincola* from salmonid fishes. *Amer. J. Tropical Med. Hyg.* 36: 586-91.

Ebesu, J.S.M., H. Nagai, and Y. Hokama. 1994. The first reported case of human ciguatera possibly due to a farm-cultured salmon. *Toxicon* 32: 1282-86.

Evans, M. 1992. Aquaculture laying groundwork for future growth. Glynn, P.B., ed. *Farmline*, USDA Economic Research Service, 13(5): 12-14.

Eyles, M.J. 1986. Microbiological hazards associated with fishery products. *CSIRO Food Research Quarterly* 46 (1): 8-16.

FAO. 1990. Aquaculture production (1985-1988). FAO Fisheries Circular No. 815, Revision 2, FAO, Rome, Italy.

Farmer, J.J., B.R. Davis, and F.C. Hickman-Brenner. 1985. Biochemical identification of new species and biogroups of Enterobacteriaceae isolated from clinical specimens. *J. Clin. Microbiol.* 21: 46-76.

Flick, G.J. In Press. *Total Quality Assurance and Hazard Analysis Critical Control Point Manual for the Trout Processing Industry*. Virginia Sea Grant College Program Publication No. VSG-95-09.

Flynn, T., and I. Knepp. 1987. Seafood shucking as an etiology for *Aeromonas hydrophila* infection. *Arch. Intern. Med.* 147: 1816-17.

Frenette, C., J. MacLean, and T. Gyorkos. 1988. A large common-source outbreak of ciguatera fish poisoning. *J. Infect. Dis.* 158: 1128-31.

Frerichs, G. and R. Roberts. 1989. The bacteriology of teleosts. In: *Fish Pathology*. R. Roberts, ed. Philadelphia: Bailliere Tindall, p. 189.

Fritsche, T., R. Eastburn, L. Wiggins, and C. Terhune. 1989. Praziquantel for treatment of human *Nanophyetus salmincola* (*Troglotrema salmincola*) infection. *J. Infect. Dis.* 160: 896-99.

Garrison, D.L., S.M. Conrad, P.P. Eilers, and E.M. Waldron. 1992. Confirmation of domoic acid production by *Pseudonitzschia australis* (Bacillariophyceae) cultures. *J. Phycol.* 28:604-07.

George, W., M. Nakata, J. Thompson, and M. White. 1985. *Aeromonas*-related diarrhea in adults. *Arch. Intern. Med.* 145: 2207-11.

Gill, O., J. Coghlan, and I. Calder. 1985a. The risk of leptospirosis in United Kingdom fish farm workers: Results from a 1981 serological survey. *J. Hyg.* 94: 81-86.

Gill, N., S.A. Waitkins, and I.M. Calder. 1985b. Further update on leptospirosis: continuing risk in fish farmers. *Brit. Med. J. Clin. Res. Ed.* 290 (6486): 1988.

Gorby, G. and J. Peacock. 1988. *Erysipelothrix rhusiopahtiae* endocarditis: Microbiologic, epidemiologic, and clinical features of an occupational disease. *Rev. Infect. Dis.* 10: 317-325.

Grizzle, J.M. and W.A. Rogers. 1976. *Anatomy and Histology of Channel Catfish*. Opelika, AL: Auburn University Agricultural Experiment Station. Craftmaster Printers, Inc.

Halstead, B. 1988. *Poisonous and Venomous Marine Animals of the World*. Princeton, NJ: Darwin Press.

Hardy, R.W., T.M. Scott, C.L. Hatfield, H.J. Barnett, E.J. Gauglitz, Jr., J.C. Wekell, and M.W. Eklund. 1995. Domoic acid in rainbow trout (*Oncorhynchus mykiss*) feeds. *Aquaculture* 131: 253-60.

Harrell, L., and T. Deardorff. 1990. Human nanophytiasis: Transmission by handling naturally infected coho salmon (*Onchorhynchus kisutch*). *J. Infect. Dis.* 161: 146-48.

Haswell-Elkins, M.R., P. Sithithaworn and D. Elkins. 1992. *Opisthorchis viverrini* and cholangiocarcinoma in Northeast Thailand. *Parasitol. Today* 8: 86-89.

Heckerling, P., T. Stine, J. Pottage, S. Levin, and S. Harris. 1983. *Aeromonas hydrophila* myonecrosis and gas gangrene in a nonimmunocompromised host. *Arch. Intern. Med.* 143: 2005-07.

Higgins, B.C. and A.C. Kolbye. 1984. Risks and benefits of seafood. In: *Seafood Toxins*. E.P. Regelis, ed. American Chemical Society Symposium Series No. 262, pp. 59-67.

Hiratsuka, M., Y. Saitoh, and N. Yamane. 1980. Isolation of *Vibrio alginolyticus* from a patient with acute enterocolitis. *Tohoku J. Exp. Med.* 132: 469-72.

Hoffman, T., B. Nelson, R. Darouiche, and T. Rosen. 1988. *Vibrio vulnificus* septicemia. *Arch. Intern. Med.* 148: 1825-27.

Holden, Cindy. 1993. Aquaculture product safety: a view from the food service industry. In: *Aquaculture Products Safety Forum Proceedings*. B.E. Perkins, ed. Southern Regional Aquaculture Center, Auburn, Alabama, pp. 48-52.

Holden, Constance, ed. 1996. Fish fans beware. *Science* 273: 1049-51.

Horner, A.R. and J.R. Postel. 1993. Toxic diatoms in western Washington waters (U.S. west coast). *Hydrobiologia* 269/270: 197-205.

Inglis, V., R.H. Richards, K.N. Woodward. 1993. Public health aspects of bacterial infections of fish. In: *Bacterial Diseases of Fish.* V. Inglis, R.J. Roberts, and N.R. Bromage, eds. New York: Halsted Press: John Wiley and Sons, Inc., pp. 284-302.

Johnston, J., S. Becker, and L. McFarland. 1985. *Vibrio vulnificus*: Man and the sea. *JAMA* 253: 2850-53.

Karam, G., A. Ackley, and W. Dismukes. 1983. Posttraumatic *Aeromonas hydrophila* osteomyelitis. *Arch. Intern. Med.* 143: 2073-74.

Kennedy, C., M. Goetz, and B. Mathisen. 1990. Postoperative pancreatic abscess due to *Plesiomonas shigelloides*. *Rev. Infect. Dis.* 12: 813-16.

Kimura, K., M. Hasegawa, K. Matsubara, C. Maseda, M. Kagawa, S. Takahashi, and K.T. Tanabe. 1994. A fatal disaster case based on exposure to hydrogen sulfide-an estimation of the hydrogen sulfide concentration at the scene. *Forens. Sci. Intl.* 66 (2): 111-16.

Kliks, M. 1986. Human anasakiasis: An update. *JAMA* 255: 2605.

Klontz, K., S. Lieb, M. Schreiber, H. Janowski, L. Baldy and R. Gunn. 1988. Syndromes of *Vibrio vulnificus* infections: Clinical and epidemiologic features in Florida cases, 1981-1987. *Ann. Intern. Med.* 109: 318.

Kow-Tong, C., and M. Malison. 1987. Outbreak of scombroid fish poisoning, Taiwan. *Am. J. Public Health* 77: 1335-36.

Leira, H.L. and K.J. Baalsrüd. 1992. Self-injection of fish vaccine can cause anaphylaxis. *Scan. J. Work Envir. Health* 18: 410.

Lewis, D.H. and J.A. Plumb. 1979. Bacterial diseases. In: *Principal Diseases of Farm-Raised Catfish*. J.A. Plumb, ed. Southern Cooperative Series No. 225, pp. 15-24.

Linell, F. and A. Norden. 1954. A new acid-fast bacillus occurring in swimming pools and capable of producing skin lesions in humans. *Acta Tuberc. Scand.*, Suppl. 33: 1-85.

Ling, B., T. Den, Z. Lu, L. Min, Z. Wang, and A. Yuan. 1993. Use of night soil in agriculture and fish farming. *World Health Forum* 14: 67-70.

Lowry, P., L. McFarland and H. Threefoot. 1986. *Vibrio hollisae* septicemia after consumption of catfish. *J. Inf. Dis.* 154: 730-31.

Lowry, P., L. McFarland, B. Peltier, N. Roberts, H. Bradford, J. Herndon, D. Stroup, J. Mathison, P. Blake and R. Gunn. 1989. *Vibrio* gastroenteritis in Louisiana: A prospective study among attendees of a scientific congress in New Orleans. *J. Inf. Dis.* 160: 978-84.

Main, R., and B.D. Yates. 1990. Two cases of scombrotoxin fish poisoning. In: *Surveillance Programme for Foodborne Infections and Intoxications.* P.W. Collier, ed. Scotland, Report 9.

McKerrow, J., J. Sakanari, and T. Deardorff. 1988. Anasakiasis: Revenge of the sushi parasite. *N. Engl. J. Med.* 319: 1228-29.

Morris, J. 1988. *Vibrio vulnificus*-A new monster of the deep? *Ann. Intern. Med.* 109: 261-62.

Morris, J., R. Wilson, B. Davis, I. Wachsmuth, C. Riddle, H. Wathen, R. Pollard and P. Blake. 1981. Non-0 group 1 *Vibrio cholerae* gastroenteritis in the United States: Clinical, epidemiologic, and laboratory characteristics of sporadic cases. *Ann. Intern. Med.* 94: 656-58.

Morton, W.F. 1990. Occupational phenoxyethanol neurotoxicity: a report of three cases. *J. Occup. Med.* 32 (1): 42-45.

Murrell, K.D. 1995. Foodborne parasites. *Intl. Envir. Health Res.* 5(1): 63-85.

Nash, C.E. and C.B. Kensler. 1990. A global overview of aquaculture production in 1987. *World Aquaculture*, 21(2): 104.

Nelson, S.S., ed. 1996. Crab bite fatal to Alabama woman. Janesville, LA: *The Aquaculture News*, pg. 2.

Nemetz, T.G. and E.B. Shotts, Jr. 1993. Zoonotic diseases. In: *Fish Medicine.* M.K. Stoskopf, ed. Philadelphia: W.B. Saunders Company, pp. 214-20.

NOAA. 1988. *Aquaculture and Capture Fisheries: Impacts in U.S. Seafood Markets.* Washington, DC: National Marine Fisheries Service, U.S. Dept. Of Commerce.

Ogbondemtnu, F.S. 1993. The occurrence and distribution of enteric bacteria in fish and water of tropical aquaculture ponds in Nigeria. *J. Aquac. Tropics*, 8 (1): 61-66.

Oshima, T. and Kliks, M. 1987. Effects of marine mammal parasites on human health. *Int. J. Parasitol.* 17: 415-21.

Osterholm, M., J. Forgang, T. Ristinen, A. Dean, J. Washburn, J. Godes, R. Rude, and J. McCullough. 1983. An outbreak of foodborne giardiasis. *N. Engl. J. Med.* 304: 24-28.

Otwell, S.W. 1993. Overview of aquaculture product safety. In: *Aquaculture Products Safety Forum Proceedings.* B.E. Perkins, B.E., ed. Auburn, AL: Southern Regional Aquaculture Center, pp. 53-71.

Palafox, N., L. Jain, A. Pinano, T. Gulick, R. Williams, and I. Schatz. 1988. Successful treatment of ciguatera fish poisoning with intravenous mannitol. *JAMA* 259: 2740-42.

Pan, B.S. 1985. Histamine formation in the canning process. In: *Histamine in Marine Products.* B.S. Pan and D. James, eds. FAO Fisheries Technical Paper No. 252, pp. 41-44.

Plakas, S.M., R.M. McPhearson, and A.M. Guarino. 1988. Disposition and bioavailability of ^{3}H-tetracycline in the channel catfish (*Ictalurus punctatus*). *Xenobiotica* 18: 83-93.

Plakas, S.M., R.W. Dickey, M.G. Barron, and A.M. Guarino. 1990. Tissue distribution and renal excretion of ormetoprim after intravascular and oral administration in the channel catfish (*Ictalurus punctatus*). *Can. J. Fish. Aquat. Sci.* 47: 766-71.

Plakas, S.M., P.M. Loveland, F. S. Bailey, V.S. Blazer, and F.L. Wilson. 1991. Tissue disposition and excretion of ^{14}C-labelled aflatoxin B_1 after oral administration in channel catfish. *Food Chem. Toxicol.* 29: 805-08.

Plakas, S.M. 1993. Method development for drugs and other chemicals in aquaculture products. The role of tissue distribution and metabolism studies. In: *Aquaculture Products Safety Forum Proceedings.* B.E. Perkins, B.E., ed. Auburn, AL: Southern Regional Aquaculture Center, pp. 30-47.

Reboli, A. and E. Farrar. 1989. *Erysipelothrix rhusiopathiae*: An occupational pathogen. *Clin. Microbiol. Rev.* 2: 354-59.

Rim, H.J., H.F. Faraq, S. Sornmani and J.H. Cross. 1994. Foodborne trematodes: Ignored or emerging. *Parasitol. Today* 10: 207-09.

Roberts, J.K. and A.M. Kurts. 1990. Predation and control of laboratory populations of the snail *Biomphalaria glabrata* by the freshwater prawn *Macrobrachium rosenbergii*. *Ann. Tropical Med. Paras.* 84 (4): 401-12.

Robertson, M.H., I.R. Clarke, J.D. Coghlan, and O.N. Gill. 1981. Leptospirosis in trout farmers. *Lancet* 2 (8247): 626-27.

Rominger, R. 1993. Talking points for the Joint Subcommittee on Aquaculture, Washington, DC. Address by Richard Rominger, Deputy Secretary of Agriculture on June 30, 1993.

Schmidt, G.D., and L.S. Roberts. 1981. *Foundations of Parasitology.* St. Louis: The C.V. Mosby Company, St.Louis.

Schnick, R.A. 1990. Record of meeting with FDA on the new seafood fish definition proposed by the Fish and Wildlife Service. Memorandum, National Fisheries Research Center, LaCrosse, WI.

Schulz, K.H. 1982. Allergy to chemicals: problems and perspectives. In: *Allergy and Hypersensitivity to Chemicals.* Copenhagen: World Health Organization (WHO).

Sheard, K. and H.G. Dicks. 1949. Skin lesions among fishermen at Houtman's Abrolhos, Western Australia, with an account of erysipeloid of Rosenbach. *Med. J. Aust.* 2: 352-54.

Shotts, E.B., Jr. 1980. Bacteria associated with fish and their relative importance. In: *CRC Handbook Series in Zoonoses.* Section A: Bacterial, Rickettsial and Mycotic Diseases. Steele, J., ed. Boca Raton, FL: CRC Press, pp. 517-25.

Shotts, E.B., Jr. 1987. Bacterial diseases of fish associated with human health. *Vet. Clin. N. Am. Small Anim. Pract.*, 17 (1): 241-47.

Shotts, E.B., Jr., and J.J. Teska. 1989. Bacterial pathogens of aquatic vertebrates. In: *Methods for the Microbiological Examination of Fish and Shellfish.* B. Austin, and D. Austin, eds. New York: John Wiley and Sons, pp. 164-86.

Schulz, K.H. 1982. Allergy to chemicals: problems and perspectives. In: *Allergy and Hypersensitivity to Chemicals.* Copenhagen: World Health Organization (WHO).

Shumway, S.E. 1989. Toxic algae, a threat to shellfish farming. *World Aquaculture.* 20 (3): 65-74.

Sindermann, C.J. 1990a. *Principal Diseases of Marine Fish and Shellfish.* Vol. 1. 2nd edition. San Diego, CA: Academic Press, Inc.

Sindermann, C.J. 1990b. *Principal Diseases of Marine Fish and Shellfish.* Vol. 2. 2nd edition. San Diego, CA: Academic Press, Inc.

Taylor, S.L., J.Y. Hui, and D.E. Lyons. 1984. Toxicology of scombroid poisoning. In: *Seafood Toxins.* E.P. Ragelis, ed. Washington, DC: American Chemical Society, pp. 417-30.

Taylor, S.L., J.E. Stratton, and J.A. Nordlee. 1989. Histamine poisoning (scombroid fish poisoning): an allergy-like intoxication. *Clin. Toxic.* 27 (4 and 5): 225-40.

Telzak, E., E. Bell, D. Kautter, L. Crowell, L. Budnick, D. Morse, and S. Schultz. 1990. An international outbreak of type E botulism due to uneviscerated fish. *J. Infect. Dis.* 161: 340-42.

Todd, E.C.D. 1993. Domoic acid and amnesiac shellfish poisoning- a review. *J. Food Protect.* 56: 69-83.

Tucker, C.S. 1991. *Water Quantity and Quality Requirements for Channel Catfish Hatcheries.* Southern Regional Aquaculture Center Publication No. 461.

Twiddy, D.R. 1995. Antibiotic-resistant human pathogens in integrated fish farms. *Asian Food J.* 10 (1): 22-29.

United States Trout Farmers Association (USTFA). 1994. *Trout Producer Quality Assurance Program.* R. MacMillan, D. Campbell, R. Cooper, G. Fornshell, D. Garling, J. Hinshaw, J. McCraren, P. Mamer, A. Morton, J. Parsons, and D. Ramsey, eds. Harpers Ferry, WV: USTFA.

Vartian, C., and E. Septimus. 1990. Osteomyelitis caused by *Vibrio vulnificus. J. Infect. Dis.* 161: 363.

Venugopal, M.N. and I. Karunasagar. 1988. Effect of chemical preservatives on *Vibrio parahemolyticus* in fish. In: *The First Indian Fisheries Forum.* M.M. Joseph, ed. Asian Fisheries Society, Indian Branch, Mangalore, pp. 407-08.

Wekell, J.C., E.J. Gauglitz, Jr., H.J. Barnett, C.L. Hatfield, and M. Eklund. 1994. The occurrence of domoic acid in razor clams *(Siliqua patula),* Dungeness crab *(Cancer magister),* and anchovies *(Engraulis mordax). J. Shellfish Res.* 13 (2): 587-93.

White, A.W. 1977. Dinoflagellate toxins as probable cause of an Atlantic herring *(Clupea harengues harengus)* kill, and steropods as apparent vector. *J. Fish. Res. Can.* 34: 2421-24.

White, A.W. 1981. Marine zooplankton can accumulate and retain dinoflagellate toxins and cause fish kills. *Limnol. Oceanogr.* 26 (1): 103-09.

Wilson, J., R. Waterer, J. Wofford, and S. Chapman. 1989. Serious infections with *Edwardsiella tarda*: A case report and review of the literature. *Arch. Intern. Med.* 149: 208-10.

Wittner, M., J. Turner, G. Jacquette, L. Ash, M. Salbo, and H. Tanowitz. 1989. *Eustrongylidiasis*-a parasitic infection acquired by eating sushi. *N. Engl. J. Med.* 320: 1124-26.

Wood, R. 1975. Erysipelothrix infection. In: *Diseases Transmitted from Animals to Man.* W. Hubbert, W. McCullough, and P. Schnurrenberger, eds. Springfield, IL: Charles C. Thomas, pp. 271-281.

Work, T.M., B. Barr, A.M. Beale, L. Fritz, M.A. Quilliam and J.L.C. Wright. 1993. Epidemiology of domoic acid poisoning in brown pelicans *(Pelecanus occidentalis)* and Brandt's cormorants *(Phalacrocorax penicillatus)* in California. *J. Zool. Wildl. Med.* 24: 54-62.

Young, H.K. 1994. Do nonclinical uses of antibiotics make a difference? *Inf. Control Hosp. Epidem.* 15 (7): 484-87.

40

SKIN DISEASES IN FISHERMEN

William A. Burke, M.D.
Associate Professor and Section Head, Dermatology
East Carolina University School of Medicine

A high proportion of occupationally-related health problems in fishermen and watermen involve the skin. Cutaneous problems in fishermen include those derived from exposure to 1) high levels of ultraviolet radiation; 2) environmental extremes; 3) traumatic injuries; 4) various skin irritants and allergens; 5) various environmental infectious agents; and 6) a variety of dangerous and/or toxic marine fauna and flora. There currently is a paucity of well-established data on health-related problems in this group of workers. Increased industrialization and commercialization of the occupations of fishing and aquaculture will likely stimulate interest in improving the health and safety of this workforce. Funding of future research studies will likely have a positive impact on our knowledge of work-related skin problems in watermen.

INTRODUCTION

As with the occupation of agriculture, and for similar reasons, there is little data available on cutaneous diseases in fishermen and watermen. There are many occupational similarities between farmers and fishermen. Fishing boats have traditionally been owned by individual fishermen, and the few onboard mates and other workers are often part-time or family members. This generally excludes these workers from workman's compensation or Occupational Safety and Health Administration (OSHA) laws. Indeed, OSHA regulatory activities have been directed primarily toward nonfishing industrial activities aboard the larger fish processing vessels, and the Coast Guard remains the principal federal agency with safety compliance programs affecting fishermen (NRC 1991).

Fierce, prideful independence characterizes the commercial fisherman, and many operate out of their pockets on a cash basis (NC Dept. of Natural Resources and Community Dev. 1979). This leads to an occupational group of stoic individuals who are hard-working yet distrustful of institutions including medicine, and who rarely seek or can afford medical care, especially for perceived minor cutaneous ailments. While the fisherman is similar to the farmer, in that he works outdoors and is exposed to high levels of ultraviolet radiation and extremes of the elements, his exposure to infectious agents as well as fauna and flora, which can potentially affect the skin, are markedly different.

In a study (Scrimgeour 1994). of the prawn trawling industry in Australia, skin infections and dermatitis accounted for approximately one-fourth of health problems in this group of watermen. "Traumatic injuries" (lacerations, sprains, fractures) was the only category with a higher percentage (30.9%), and thus a significant proportion of occupationally related health problems in fishermen involve the skin.

SKIN DISEASES IN FISHERMEN

Exposure to Ultraviolet Light

Because of their high occupational exposure to ultraviolet (UV) radiation, fishermen are at high risk for a variety of acute and chronic cutaneous diseases. These diseases are similar to sun-related disorders in other outdoor occupations and have been previously discussed in the chapter on skin diseases in farmers. Physicians must remember the potential for phototoxic and photoallergic medications when prescribing pharmacologic agents for use in fishermen because potentially serious eruptions may occur (Figure 40.1).

In a study (Vitasa et al. 1990) of Maryland watermen, risk of developing actinic keratoses (AK's), squamous cell carcinomas (SCC's), and basal cell carcinomas (BCC's) was increased in fishermen who were older, had childhood freckling, and blue eyes (Skin types I, II). The risk of development of AK's was 1.5 times higher in those watermen who exceeded the median of cumulative UVB exposure. Those in the upper quartile of cumulative exposure had a 2.5 times increased risk for development of SCC. Interestingly,

BCC was not correlated with cumulative UVB exposure, and the BCC:SCC ratio was extremely low at 1:1 (compared to 4:1 in the general population). This suggests that BCC prevalence is already at saturation in this population with extreme UVB exposure, or that other factors might perhaps be playing a role.

In another study (Nicolini et al. 1989) of 566 fishing workers in Chile, 43% had actinic cheilitis, and presence of this problem also correlated with low skin type (I to II). This precancerous condition can progress to SCC (Figure 40.2), and fishing has also been shown to be an occupational risk factor for cancer of the lip (Spitzer et al. 1975).

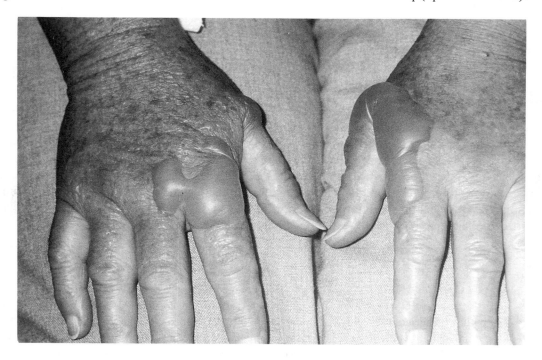

Figure 40.1. A blistering phototoxic drug eruption which occurred after a day's fishing while on a quinolone antibiotic.

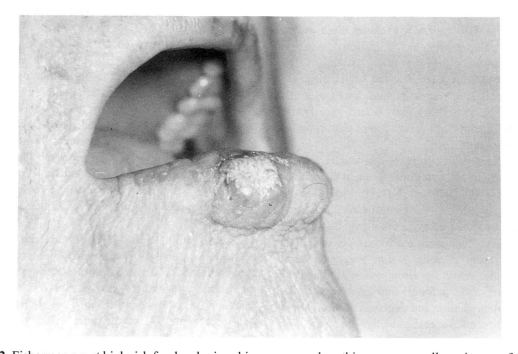

Figure 40.2. Fishermen are at high risk for developing skin cancers such as this squamous cell carcinoma of the lower lip.

Exposure to Environmental Extremes

Like other outdoor workers, fishermen can be exposed to extremes of heat and cold, and these have been previously discussed in the chapter on skin diseases in farmers. One unique syndrome due to non-freezing cold injury in boat rowers has been termed "pulling boat hands" (Toback et al. 1985). This superficial blistering eruption on the dorsal hands is presumably due to a vasospastic phenomenon coupled with the mechanical trauma of rowing.

Immersion foot can also be seen in watermen, especially after prolonged cold water immersion following a shipwreck, and this has been termed "lifeboat foot." Fishermen commonly wear rubber boots, and if cold water should get inside for extended periods due to leakage, immersion foot might also occur.

Warm water immersion foot can also incapacitate a seaman. In this condition, the stratum corneum of the soles becomes hyperhydrated and results in thickened, wrinkled macerated skin, which is painful with weightbearing. Persons with thick, callused soles are mostly at risk, and nondraining footwear such as rubber boots may aggravate the condition. Drying the feet for 6 to 8 hours (overnight) helps to prevent this condition, and rest and drying for 2 to 3 days is generally required to treat the condition once it occurs (Adnot and Lewis 1994).

Chronic exposure of feet to wetness, leading to maceration, can also lead to skin infections due to dermatophytes (tinea pedis) or corynebacteria (pitted keratolysis—Figure 40.3). Antifungal/antibiotic medications can help clear these mild but sometimes incapacitating skin infections, but maintenance of a drier environment for the feet is key to preventing recurrence.

Skin Conditions Due To Trauma

While major traumatic injuries in fishermen are more studied, minor traumatic injuries such as cuts, punctures, or lacerations are common and can be due to sharp objects such as fish hooks, knives, fish/crustacean spines, mollusk shells, fish bites or contact with a variety of other sharp objects. Fishing lines can cut bare skin (Figure 40.4), and ropes can cause abrasions ("rope burn"). Denticles on the skin of elasmobranch fish (sharks, skates, rays) can cause abrasions and skin tears (Mandojana and Sims 1987). Most minor traumatic injuries to the skin of fishermen occur on the fingers, hands, or upper extremities (NRC 1991). Disruption of the normal epidermal barrier function in even superficial wounds can lead to secondary infection.

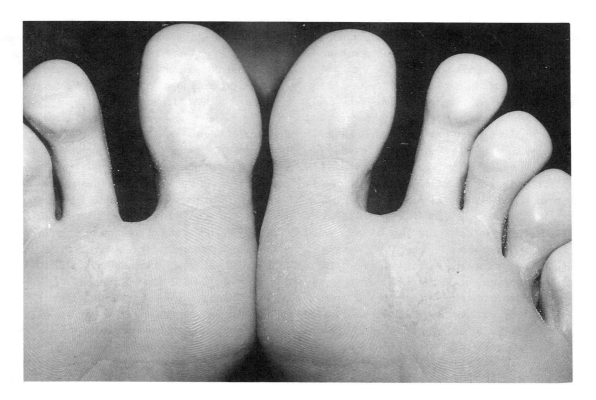

Figure 40.3. Pitted keratolysis is a superficial bacterial infection of the stratum corneum and can be seen in watermen whose feet are chronically moist.

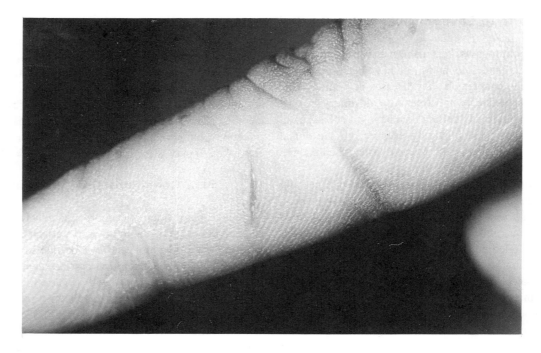

Figure 40.4. Minor trauma to the skin such as this fish line cut is common in fishermen and can lead to secondary infection.

Contact Dermatitis in Fishermen

The skin of fishermen can come in contact with a variety of cleansing agents used in boat upkeep, which can lead to an irritant dermatitis. Rubber gloves and boots are commonly worn and can be the cause of an allergic contact dermatitis in sensitized persons (Ross 1969).

Hand dermatitis due to fish handling can be seen in atopic individuals and is due to sensitive skin being subjected to multiple wet-dry cycles leading to chapping and fissuring. Irritating skin toxins (crinotoxins) may be found in fish slime, especially in soapfish, trunkfish, toadfish, pufferfish, and clingfish (Mandojana and Sims 1987). Scombroid dermatitis (or "tuna dermatitis") is due to irritating chemicals in the skin of scombroid fish (tuna, albacore, bonito, mackerel, and jacks) (Fisher 1978). Contact dermatitis and contact urticaria can also be seen when handling other raw fish or shrimp (Fisher 1990; Kavli et al. 1985). Patch testing or skin-prick testing is often positive in these individuals indicating an allergic mechanism is usually responsible. "Red feed dermatitis" is most commonly seen in fishermen who clean mackerel that have ingested a red-orange crustacean (*Calanus* sp.), and skin contact can lead to erythema, swelling and ulceration (Fisher 1978; Edmonds 1989).

Contact with certain algae, dinoflagellate "red tide" organisms, bryozoans, and a variety of other sea inhabitants can also cause a eczematous dermatitis. These will be discussed later in this chapter.

Cutaneous Infections in Fishermen

As with terrestrially acquired skin infections, the majority of skin infections in fishermen, or otherwise derived from an aquatic environment, are primarily due to *Staphylococci* and *Streptococci*. However, there are more unusual infections that need to be considered when an infection presents in mariners that have an occupational exposure to water and fish.

Erysipeloid is an erysipelas-like illness that most commonly begins on the hands (often over a bony prominence) following a traumatic inoculation injury derived from an aquatic environment. A small erythematous papule usually begins in 1 to 3 days and further spreads into a well-demarcated area of bright-red cellulitis. Unlike erysipelas, the infected individual generally feels well and often lacks constitutional signs and symptoms such as fever, malaise, and regional adenopathy. In most cases, the disease is self-limited and fishermen may not seek medical assistance. However, the disease can occasionally disseminate with endocarditis as a frequent complication (Barrett et al. 1983).

The causative organism, *Erysipelothrix rhusiopathiae*, is a Gram-positive pleomorphic rod which is distributed widely in nature. Fishermen and seafood industry workers refer to erysipeloid colloquially as "fish poison," "crab poison," "shrimp-picker's disease," and, in other parts of the world, "seal finger" or "blubber finger." The organism also causes swine erysipelas and can also be seen in farmers, veterinarians, and abattoir workers.

Treatment of erysipeloid is similar to erysipelas with penicillin, ampicillin, and first-generation cephalosporin antibiotics being mainstays of therapy. Unlike most other Gram-positive bacteria, *Erysipelothrix rhusiopathiae* is commonly resistant to vancomycin.

Mycobacterium marinum is an acid-fast atypical mycobacterium (Runyon Group I) that is commonly encountered by fishermen (Figure 40.5) and tends to be more insidious in its onset than erysipeloid. A pink-red papule grows over several weeks at the site of inoculation, which is usually over a bony prominence. The primary lesion continues to enlarge over months into a verrucous granulomatous lesion that may exhibit sporotrichoid spread along lymphatic channels (Raz et al. 1984). Since *M. marinum* does not grow well at 37°C, the organism does not disseminate throughout the body, but it can locally invade bone, tendons, bursa, joints, and the carpal tunnel (Wagner et al. 1981; Harth et al. 1994; Clark et al. 1990).

Diagnosis is usually made on a clinical basis and by obtaining a biopsy specimen of the lesion. Histologically,

Table 40.1. Vibrios which may cause infection in man.

Skin/soft tissue	Ear
V. alginolyticus	V. alginolyticus
V. cholerae 01	V. cholerae non-01
V. cholerae non-01	V. mimicus
V. damsela	V. parahemolyticus
V. hollisae	
V. parahaemolyticus	
V. vulnificus	

Sepsis	Gastrointestinal
V. alginolyticus	V. cholerae 01
V. cholerea non-01	V. cholerea non-01
V. fluvialis	V. fluvialis
V. hollisae	V. furnissii
V. mimicus	V. hollisae
V. parahaemolyticus	V. mimicus
V. vulnificus	V. parahaemolyticus
	V. vulnificus

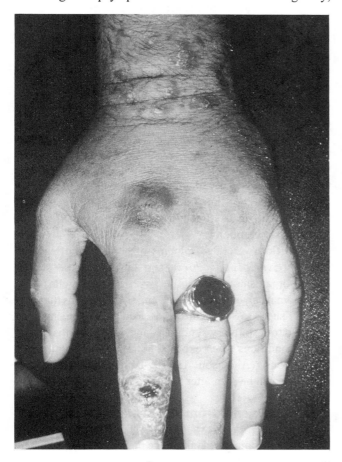

Figure 40.5. An ulcerated and sporotrichoid skin infection due to *Mycobacterium marinum* occurred in this commercial fisherman after traumatizing his finger while cleaning out the bilge in his boat.

granulomatous inflammation is commonly seen, but organisms are sparse and may not be seen readily on acid-fast staining. Culture from tissue usually confirms the diagnosis, but it may take weeks for the bright yellow (photochromogen) colonies to grow. Since the organism does not tolerate higher temperatures, it is important to ensure that the microbiology laboratory is incubating the cultures at 30 to 33°C (Edelstein 1994).

Tentative diagnosis of *M. marinum* is usually suggested by the occupational or recreational history (Kullayanijaya et al. 1993) and may be seen in fishermen, fishmongers, seafood processors, water sports enthusiasts, and aquaria owners. Treatment is usually with minocycline in adults or sulfamethoxazole-trimethoprim in children, although tetracycline, doxycycline, clarithromycin, or rifampin with ethambutol have also been recommended. Again, the organism does not tolerate heat well, and hot water soaks or a heating pad (for 30 to 60 minutes twice daily) should be used as an adjunct to systemic antibiotic therapy.

Aeromonas hydrophila is a well-known fish pathogen that is found primarily in fresh and brackish water and can cause serious infection in humans. This facultative Gram-negative rod can be acquired via the gastrointestinal tract where it may cause gastroenteritis, or via traumatic inoculation leading to a wound infection (Davis et al. 1978). Myonecrosis, fasciitis and gas gangrene have been reported in serious wound infections (Heckerling et al. 1983; Vukmir 1992).

This bacterium has multiple virulence factors and can disseminate throughout the body, leading to septic shock. Risk factors for life-threatening infections include liver disease (alcoholism, hemochromatosis, etc.) and immunosuppressed patients (especially those with lymphoproliferative disorders). Treatment of milder cases is with an oral fluoroquinolone (ciprofloxacin or norfloxacin) or sulfamethoxazole-trimethoprim. For more serious disseminated infection, an intravenous third-generation cephalosporin and an aminoglycoside or imipenin/cilastatin is recommended.

The most serious infections encountered by fishermen are due to the halophilic vibrios, which are found primarily in brackish and salt water. These Gram-negative motile organisms are some of the most common bacteria found in sea water, and those at risk for these infections include fishermen, seafood processors, and water sports enthusiasts. Infection can also be acquired via the gastrointestinal tract (especially after eating raw or undercooked seafood—most commonly oysters) (Wickbolt and Sanders 1983). The various vibrios-causing infections in man are listed in Table 40.1 (Limpert and Peacock 1988; Howard and Bennett 1993; Levine and Griffin 1993; Morris and Black 1985; Klontz 1990; Perez-Tirse et al. 1993; Bonner et al. 1983; Shandera et al. 1983; Lopes et al. 1993).

The most virulent of the vibrios is *Vibrio vulnificus*. When acquired via the gastrointestinal tract in a primary septicemia, the affected individual usually becomes ill within 24 hours after eating raw or undercooked seafood. Commonly, the clinical presentation is septic shock. This is also true of a wound infection (leading to a secondary septicemia) caused by this organism, and myonecrosis, fasciitis, and gas gangrene are not uncommon. *Vibrio vulnificus* wound infections are most commonly reported in commercial/recreational fishermen, shrimpers, crabbers, shellfishers, persons with open wounds exposed to sea water, and traumatic injuries such as crab bites (Newell 1984; Musher 1981; Johnston et al. 1985; Tyring and Lee 1986; Kaye 1990; Burke and Jones 1987).

Risk factors for dissemination and septicemia are similar to those for *Aeromonas hydrophila*: liver disease, immunosuppressed patients, and those with lymphoproliferative disorders. The Food and Drug Administration currently recommends that persons with certain risk factors should *not* eat raw or undercooked seafood as disseminated infection due to *Vibrio vulnificus* can have a 40 to 60% fatality rate even with treatment! Risk factors listed in this warning include: liver disease (alcoholism, hemochromatosis, viral hepatitis, thalassemia major), immunosuppressed persons (AIDS, persons on long-term corticosteroid medications, cancer chemotherapy patients), diabetics, and patients with gastrointestinal disorders including therapeutically induced or naturally low gastric acid (FDA 1993).

Early and mild cases of *Vibrio* infection in adults are usually treated with oral doxycycline. Serious infections are treated with intravenous doxycycline and ceftazidime or chloramphenicol. The quinolones and sulfamethoxazole-trimethoprim have been used successfully in some of the vibrioses.

It must be again emphasized that most skin infections in fishermen are due to the more usual skin pathogens such as *Staphylococcus* or *Streptococcus* species. However, it is important to consider these other environmental pathogens since treatment is often different, and in some cases, the infection can progress rapidly to a fatal outcome. Other, even rarer infectious agents have been implicated in aquatic-derived skin infections (Auerbach 1987; Czahor 1992) and only the more common and serious ones have been discussed here.

Skin Problems Due to Cyanobacteria and Algae

Lyngbya majuscula, a blue-green alga (or cyanobacterium, as some taxonomists prefer) secretes lyngbyatoxins, which are irritating to skin and mucous membranes (Grauer and Arnold 1961). Mats of this algae are often broken apart during storms and may clog nets. Exposure in fishermen may lead to an irritant dermatitis, which most commonly begins on the hands. Conjunctivitis and respiratory symptoms may occur with windy conditions where the organism can be present in the "salt spray." Organisms allowed to remain in contact with the skin for an extended period of time under occlusive clothing, such as bathing suits, leads to an irritant dermatitis known as "seabather's eruption."

"Red tides" are algal blooms of certain dinoflagellate species (*Gymnodinium breve* and others). These single-celled organisms are motile with two flagella (one longitudinal and one transverse), and secrete toxins that may lead to fish kills (See Figure 40.6). Red tide organisms may also lead to respiratory symptoms and conjunctivitis during windy conditions when exposed to "salt spray" containing their toxins (Anderson 1994). Irritant dermatitis and ulceration can be seen after handling contaminated fish.

Skin Problems from Zooplankton (Sea Lice)

If microscopic zooplankton are present in large numbers in water, they can cause an itchy sensation when filtered by bathing suits, often termed "sea lice." Some species of zooplankton, reportedly from crab larvae, contain sharp spines that can lead to an irritant dermatitis similar to

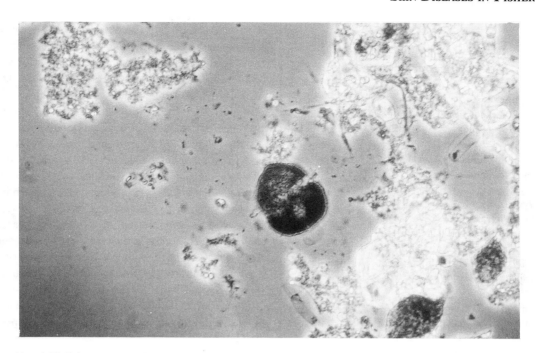

Figure 40.6. "Red tide" due to toxin-producing microscopic dinoflagellates (such as the *Gymnodinium sp.* pictured here) can lead to an irritant skin rash or, when present in "salt spray," to conjunctivitis or respiratory symptoms.

"fiberglass dermatitis" (Burnett and Cargo 1979). Cymothoids are biting zooplankton that are found in marine and estuarine areas and can attack humans leading to multiple painful or pruritic punctate erythematous papules (Best and Sablan 1964). The rash, termed cymthoidism, is named for the suborder Cymthoidea, yet are more likely due to a wide variety of zooplankton in a variety of taxonomic groups.

Stinging "sea lice" eruptions on exposed surfaces can be seen with jellyfish larvae (e.g., sea thimble or thimble jellyfish) (Tomchick et al. 1993). Broken jellyfish tentacles in areas of heavy surf may also mimic "sea lice" skin reactions.

Skin Problems Due to Coelenterates (Cnidarians)

The phylum Coelenterata (or Cnidaria) contains approximately 9000 species and is divided into three classes: the Scyphozoa, the Hydrozoa, and the Anthozoa (Table 40.2). True jellyfish are in the Scyphozoa and have a motile "bell" with trailing tentacles, which, in many species, can sting man. The sting is due to modified cells called nematocysts (Figure 40.7), which fire coiled "spring-loaded" venom tubes attached to sharp harpoon-like needles when triggered. The venom tube is attached to a venom sac that contracts after nematocyst discharge thereby delivering venom through the epidermis and into the dermis of the victim. Numerous nematocysts are grouped into "batteries," which may be visible as small "dots" along jellyfish, man-o-war, anemone, or hydroid tentacles.

Fishermen and other mariners are usually quite familiar with the different varieties of jellyfish. When jellyfish occur in large numbers (jellyfish "blooms"), fishing with nets becomes impossible, and a net engorged with jellies can easily burst. Drag nets full of stinging jellies hoisted above the deck can fall on deckhands in a "stinging rain" of jellyfish parts, and shrimpers and other watermen have termed this "hot jelly" (Rudlow and Rudlow 1991). In China, Korea and Japan, there are about a dozen species that support a major fishery. Jellyfish are a delicacy costing up to $12 apiece and are considered by some to be a cure for hypertension and bronchitis (Rudlow and Rudlow 1991).

True jellyfish (Class Scyphozoa) vary in size—in bell diameter from a few millimeters to over 2 meters, and tentacles can be as long as 36 meters (Halstead 1988). Although all jellyfish have nematocysts, those in some species are incapable of penetrating human epidermis and are considered harmless.

The box jellyfish (*Chironex fleckeri*) is considered one of the most venomous animals on earth and can kill a human in from only seconds to a few minutes (Halstead 1988; Hamner and Doubilet 1994). Found commonly in the coastal waters of northeastern Australia, it is feared by all on or in the water, and this jelly has caused the death of at least 72 people over the last century (Auerbach and Halstead 1989).

Table 40.2. The phylum Coelenterata (Cnidaria): classes and families which contain creatures potentially harmful to man.

1. **Class Scyphozoa**
   a. Family Carybdeidae, Chirodropidae (sea wasp, box jellyfish, jimble)
   b. Families Catostylidae, Cyaneidae, Lychnorhizoidae, Nausithoidae, Pelagiidae, Rhizostomatidae, Ulmaridae (other true jellyfish)

2. **Class Hydrozoa**
   a. Family Physaliidae (Portuguese man-o-war)
   b. Family Milleporidae (fire coral)
   c. Families Coynidae, Plumulariidae (hydroids)
   d. Family Olindiadidae (stinging medusa, orange-striped jellyfish)

3. **Class Anthozoa**
   a. Family Acroporidae (true corals)
   b. Families Actiniidae, Actinodendronidae, Actinodiscidae, Activiidae, Aliciidae, Hormathiidae, Sagartiidae (sea anemones)

Other members of the box jellyfish/sea wasp (Cubomedusae), which are relatively dangerous stingers, include the genera *Chiropsalmus* and *Chirodropus* (Halstead 1988). The jimble (*Carybdea rastoni*) and fire jelly (*Tamoya haplonema*) produce less dangerous but very painful stings (Auerbach and Halstead 1989). The Cubomedusae are named for the box-like shape at the base of the bell from which four tentacles or groups of tentacles extend.

The Irukandji jellyfish (*Carukia barnesi*) is also in the Cubomedusae and is found in Indopacific waters including those surrounding Australia, Indonesia, and Fiji (Edmonds 1989). The name Irukandji refers to an aboriginal tribe living near Cairns, Australia where the Irukandji syndrome was originally described. Following an initial sting, the skin becomes inflamed and swollen. Generalized symptoms usually follow within 2 hours and include severe abdominal and back pain, myalgias, nausea, vomiting, diaphoresis, dyspnea, malaise and parasthesias. The illness usually lasts several hours to 2 days (Edmonds 1989; Halstead 1988).

The stinging or sea nettle (*Crysaora quinquecirrha*—Figure 40.8) is a common jellyfish, occurring in large numbers in the summer months in the estuaries, sounds, and inlets of the eastern coast of the United States. While not especially dangerous, the sting is quite painful. The bell and oral arms may very from white to pink-red to brownish-red in color. Adults generally have 24 or more tentacles and 4 larger oral arms dangling from the 4 to 8 inch bell (Ruppert and Fox 1988).

Similar in appearance to the sea nettle is the oceanic or mauve jelly (*Pelagia noctiluca*) which is most frequently seen in the open ocean, but may be carried inshore by currents (Calder and Pridgen 1989). This smaller, bioluminescent jelly has a 3-inch pink-brown warty surfaced bell and 8 marginal stinging tentacles.

The lion's mane jelly (*Cyanea capillata*) is a cold water stinging jellyfish and is the largest jelly known. The pink to pink-yellow "inverted dinner plate" shaped bell can be as large as 7 feet in diameter and the numerous tentacles descend from eight clusters reaching as long as 98 feet

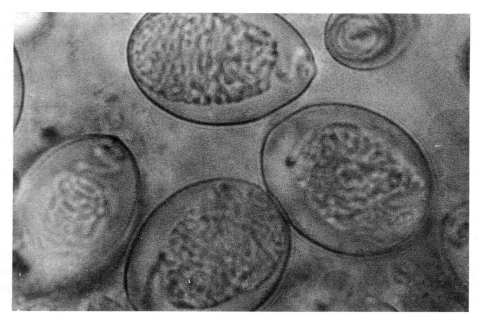

Figure 40.7. "Spring-loaded" stinging cells, or nematocyts, of the sea nettle, *Chrysaora quinquicirrha*.

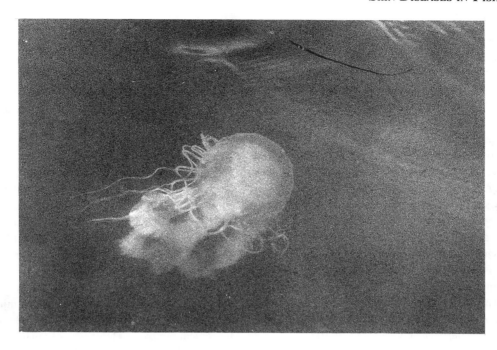

Figure 40.8. The sea nettle (*Chysaora quinquecirrha*) with multiple tentacles which contain multiple batteries of stinging nematocysts.

(Ruppert and Fox 1988). The largest examples are found in Arctic and Antarctic areas, but smaller specimens (5 to 6 inches in diameter) may be seen as far south as Florida (Calder and Pridgen 1989). Dust from desiccated lion's mane jellies caught in fishermen's nets has been reported to cause respiratory symptoms and conjunctivitis (Edmonds 1989). The bottom jellyfish (*Cassiopea* sp.) is extremely abundant in mangrove salt flats and lagoons, and rests inverted, using its motile bell to keep itself on the bottom. Symbiotic algae on the tentacles are exposed to the sun and provide this stinging jelly with food.

Many jellyfish cause only minimal stings or no sting at all. The moon jelly (*Aurelia aurita*) has a flattened translucent bell with a "four-leafed clover"-shaped gonadal structure centrally. Although the bell diameter may exceed 15 inches in diameter, most specimens are in the 6 to 8 inch range. The very short tentacles around the rim can occasionally cause a very mild sting and smaller specimens are safely handled.

The cannonball or cabbagehead jellyfish, also commonly called the jellyball (*Stomolophus meleagris*), is the most detrimental jellyfish to commercial fishing in the Southeast United States (from Chesapeake Bay to Florida) because of huge "blooms" during the summer months, which clog nets and limit trawl time (Rudlow and Rudlow 1991; Ruppert and Fox 1988; Calder and Pridgen 1989; Anderson and Renshaw 1989). Although very abundant, this jelly is considered harmless and can be safely handled. The jellyball is easily recognized by its 8 to 10 inch hemispherical bell with a maroon-brown band around the base. It has large mouthparts below the bell but no tentacles. Because of the abundance of these jellies, University of Georgia Marine Extension Agent Service scientists have suggested exporting the jellyball to oriental edible jellyfish markets. Simple adaptations of trawl nets could convert this nuisance jelly into a sought-after marketable product (International Oceanic Foundation 1995).

The mushroom jelly (*Rhopilema verrilli*) are often mistaken for the smaller jellyball. The 10 to 12 inch umbrella is translucent to yellow-tan and lacks the brownish band of the jellyball. In addition, below the mouthparts are multiple finger-like projections. This jelly is also considered harmless to human skin during most of its life cycle, but can be a nuisance to fishermen during blooms.

Some other jellies known to sting man include: the marbled jelly (*Lychnorhiza* sp.), "stinging alga" (*Nausithoë punctata*), stinging cauliflower (*Drymonema dalmatinum*), blue-tinted jellyfish (*Phyllorhiza punctata*), sea thimble (*Linuche unguiculata*), and the Australian brown blubber jelly (*Catostylus mosaicus*) (Halstead 1988; Humann 1982). Commonly feared but completely harmless are the unrelated comb jellies (Phylum Ctenophora). These small, transparent, free-floating, oval to pear-shaped, iridescent, and bioluminescent jellies move through the water via cilia located in longitudinal bands.

The Portuguese man-o-war (*Physalia physalis*) is often included in discussions about true jellyfish since the clinical presentation and treatment are similar; however, this coelenterate is in the class Hydrozoa. Unlike true jellies, there is no bell and motility is dependent on a wind-driven sail atop its blue float (Figure 40.9). This pneumatophore can be as large as a football with tentacles as long as 30 meters in length, although most specimens are smaller. The tentacles may each contain up to 750,000 nematocysts and the man-o-war can deliver a dangerous sting; several fatalities have been reported (Auerbach 1984).

Commonly seen in the waters off Florida, the Bahamas, and the Caribbean Islands, large numbers of man-o-war organisms are carried via the Gulf Stream and can be driven ashore in "beach invasions" by windy storms as far north as Cape Hatteras. Because of the severe sting, which can blister and scar, fishermen are usually well aware of this dangerous sea creature. The smaller Pacific man-o-war or bluebottle (*Physalia utriculus*) has a single tentacle which can be as long as 15 meters (Auerbach and Halstead 1989).

Jellyfish and man-o-war injuries are similar in presentation but vary considerably in severity. Mild stings usually exhibit painful linear urticarial or papular lesions, and often it is possible to see where the individual batteries of nematocysts have envenomated the victim. More serious envenomation such as with the box jelly and man-o-war can lead to purpura, vesiculation, hemorrhagic bullae and necrosis, which may heal with scarring. Generalized constitutional symptoms are common with the more serious stings. Fatal jellyfish stings are usually due to the box jelly or the Portuguese man-o-war and can be due to severe envenomation, anaphylaxis, or panic in open water, leading to drowning.

Delayed reactions following jellyfish envenomation typically occur 1 to 2 weeks following the initial sting, and often occur along the same distribution. Unlike the acute reaction, which is painful, the delayed reaction is pruritic. A delayed reaction at the site of a previous sting may also occur within days of a subsequent sting. Delayed reactions usually resolve within 2 weeks to 2 months (Reed et al. 1984; Ohtaki et al. 1986).

Treatment of jellyfish and man-o-war stings is similar. If available, vinegar (or 5% acetic acid) should be poured over adherent tentacles to inactivate unfired nematocysts (Turner et al. 1980; Schwartz and Smale 1984; Hartwick et al. 1980; Auerbach 1991; Burnett et al. 1983). For stings from the sea nettle or lion's mane jelly, sprinkling the injured area with baking soda is preferable (Burnett et al. 1983; Burnett 1992). Mechanical removal of tentacles is then done using a towel, gloves, or a knife blade. The skin should then be flushed with ambient sea water but *not* with fresh water as changes in osmotic pressure will cause any remaining nematocysts to fire. Many other "remedies" have been anecdotally advocated including applications of alcohol, urine, household ammonia, household bleach, salt, talcum powder, sugar, gasoline, kerosene, or simply rubbing the area with sand. These, however, are not recommended and may actually be

Figure 40.9. The Portuguese man-o-war (*Physalia physalia*) can produce a severe sting and fatalities have occurred following massive, severe stings.

harmful. Use of such commercial products as Accent® meat tenderizer and Stingose®, is controversial.

Acute jellyfish reactions generally last from several hours to several days and may be treated symptomatically with antihistamines, analgesics, cool compresses, or menthol-containing lotions. Ice or cold packs have also been recommended for symptomatic relief (Exton et al. 1989). For serious envenomations, especially with box jelly or man-o-war stings, the victim should be treated with first-aid measures and then rushed to the nearest medical facility. An antivenom is available in Australia for *Chironex fleckeri* stings. Although topical or systemic corticosteroids are not recommended for acute stings, they can be very effective in suppressing delayed jellyfish reactions.

Hydroids are feathery plant-like hydrozoan animals that contain nematocysts and attach themselves to ocean bottoms, reefs, docks, and pilings. A short-lived urticarial reaction follows an immediate "splash of grease" sensation on exposed skin. Hydroids are extremely common, and fishermen may come into contact with them while working around docks, boat hulls, fish and crab traps, or cleaning out nets. Symptomatic care is at times needed, but most cases are mild enough to not require any treatment.

Fire coral (*Millepora* sp.) is a pale-yellow to yellow-brown to orange-red coral-like hydrozoan that contains nematocysts on short tentacles extending from surface gastropores. The skin reaction is similar to a mild jellyfish sting and symptoms last approximately 1 to 2 hours. Although uncomfortable, the sting is not dangerous and is treated symptomatically. As with jellyfish stings, delayed reactions may occur and will usually respond to topical or oral corticosteroid medications.

True corals and sea anemones are in the class Anthozoa. Sharp coral exoskeletons made of calcium carbonate can cause abrasions and lacerations in watermen who frequent reefs. Coral injuries are often slow to heal and can lead to secondary infection. If there is concern about coral fragments remaining in a wound, the calcium carbonate is radiopaque on radiographs. Treatment of coral injuries consists of cleansing the wound and removing all foreign material, using x-rays as necessary for visualization. Application of a topical antibiotic is also recommended as is attention to tetanus prophylaxis (Edmonds 1989).

Red soft corals (*Dendronephthytia nipponica*) can cause an allergic dermatitis as well as conjunctivitis, rhinitis, and asthma in fishermen who come in contact with these organisms in contaminated drag nets. This problem was initially described in spiny lobster fishermen in Japan (Onizuka et al. 1990).

Sea anemones are sessile marine coelenterates, with numerous nematocyst-containing tentacles (Figure 40.10), and some species are capable of stinging man. Known in the fishing industry as "sponge fisherman's disease" or "sponge diver's dermatitis," the skin reaction is actually due to small sea anemones, which grow on sponges. Initially, there is itching and a burning pain at the sting site, followed by erythema and edema. Vesiculation, necrosis, and ulceration may follow in more severe cases. At times,

Figure 40.10. Sea anemones are sessile coelenterates with tentacles which may sting in some species.

generalized symptoms such as fever, chills, malaise, nausea, vomiting, and headache may occur (Halstead 1988). A fatality due to fulminant hepatic necrosis and failure was reported following an anemone (*Condylactis* sp.) sting (Garcia et al. 1994). Treatment is similar to that used for jellyfish stings.

Dermatitis Due to Bryozoans
(Phylum Ectoprocta)

Bryozoans are sessile, benthic plant-like animals commonly referred to as "sea moss" or "sea-chervils." In some oceanic areas, these animals are extremely common on the sea floor and frequently contaminate fish trawls and drag nets. The dermatitis often seen in North Sea fishermen is commonly termed "Dogger Bank itch" (Newhouse 1966). An allergic contact dermatitis as well as photosensitivity have been reported due to the bryozoans *Alcyonidium hirsutum, A. gelatinosum,* and *Electra pilosa* (Halstead 1988; Jeanmougin et al. 1987; Leroy et al. 1988).

The pruritic eczematous eruption usually begins on the extensor hands, forearms and antecubital fossae. The face and especially periorbital areas often becomes edematous and eczematous. In severe cases, the rash may generalize and become disabling to a fisherman. In a study in the Baie de Seine (France), 13 of 120 trawlermen had positive patch tests to *Alcyonidium gelatinosum* (Audebert and Lamboureau 1978). In the North Sea areas, as many as 7% of fishermen are sensitized, and in 1939, the Danish Workmen's Compensation Act included skin disorders related to *Alcyonidium* (Halstead 1988).

Treatment consists of corticosteroid medications as well as oral antihistamines. Discontinuation of exposure to the offending organism is the only cure (Halstead 1988).

Skin Diseases Due to Sponges
(Phylum Porifera)

Sponges are sessile animals that can usually be handled safely, but may be host to numerous other creatures (e.g., echinoderms, anemones, crustaceans, marine annelids, etc.) that can at times harm man. The unique porous and elastic properties of sponges are due to their labyrinthine structure as well as a fibrous protein known as spongin. Some species of sponges contain calcium carbonate or siliceous spicules (Figure 40.11) that can penetrate the epidermis and lead to a pruritic eruption similar to fiberglass dermatitis. Some sponges also contain toxins that can penetrate the epidermal barrier in areas traumatized by the spicules. Toxic sponges include the touch-me-not or fire sponge (*Tedania ignis*), the poison bun sponge (*Neofibularia nolitangere*), the green sponge (*Haliclona viridis*), the Australian stinging sponge (*Neofibularia mordens*) and the red sponge (*Microciona prolifera*). "Red moss" or "sponge poisoning" is due to contact with the red sponge and is common in oyster fishermen in the northeastern United States (Fisher 1978; Halstead 1988).

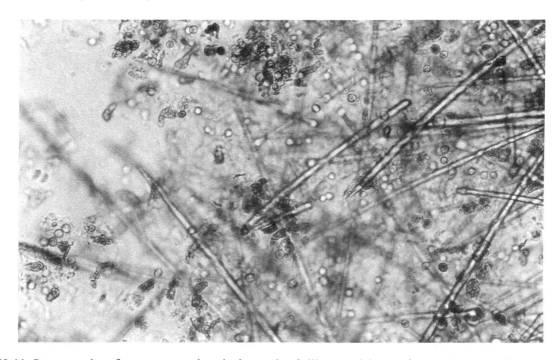

Figure 40.11. Some species of sponges contain spicules made of silica or calcium carbonate. In non-toxic species, these may cause pruritis, but in toxic species the spicules help to deliver toxin to the victim.

Skin reactions due to toxic sponges usually present with a stinging or itching sensation beginning within minutes to hours after contact (Burnett et al. 1987). Erythema and edema with pain and paraesthesias follow. The reaction may persist for several weeks to several months and desquamation often occurs. In severe cases, there may be vesiculation or bulla formation. Eye contact can lead to iritis. As mentioned previously, "sponge fisherman's disease" or "sponge diver's dermatitis" is actually due to a symbiotic anemone.

Sponge spicule dermatitis can be treated by drying the skin and then using adhesive tape to remove adherent spicules. Treatment of dermatitis due to a toxic sponge envenomation is largely symptomatic as antihistamines and corticosteroids have little effect (Burnett et al. 1987).

Skin Diseases Due to Echinoderms
(Phylum Echinodermata)

Echinoderms that can at times cause skin problems are the sea cucumbers, sea stars (starfish) and sea urchins. Sea cucumbers (Figure 40.12) are aptly named marine animals that are common on the sea floor. Worldwide, sea cucumbers support a major fishery with an estimated 27,000 tons harvested in 1983. Most such fishing and processing operations are small, however, and harvesting takes place by wading, free-diving and dredging from small boats. Hong Kong and Singapore are major markets, with France the largest European market (Conand and Sloan 1989).

The sea cucumber can cause a mild irritant dermatitis when it rubs against bare skin, such as when children playfully throw these creatures at each other. This is presumably due to anchor-like spicules on the skin of the sea cucumber (Zahl 1961). Children also commonly play with these animals as "squirt guns." The "squirt" contains a digestive enzyme, holothurin, which can cause an irritant dermatitis, or with eye contact, severe conjunctivitis or even blindness (Halstead 1988). Most cases are mild and treated symptomatically.

Sea stars or starfish are free-living, star-shaped echinoderms that are generally safe to handle. The only fishery of this sea creature is located in Denmark where the 150 to 4000 tons/year catch is used to supplement domestic animal feed stocks (Conand and Sloan 1989). The crown-of-thorns starfish (*Acanthaster planci*) is the only venomous member of the sea stars, and this reef predator is found throughout much of the Indian and Pacific Oceans as well as the Red Sea (Edmonds 1989; Birkeland 1989). This spiny starfish grows to a diameter of 60 cm and usually has 13 to 16 arms containing numerous 3 to 6 cm magnesium calcite spines. The spines are covered with an integumental sheath that produces a toxin. The color of this creature varies widely from red to orange to yellow to green (Edmonds 1989). Envenomation via puncture leads to an extremely painful wound with erythema, and edema, often with nausea and vomiting. Contact with the slime covering this (and other) starfish may cause a contact dermatitis.

Figure 40.12. The sea cucumber has skin that can irritate as well as a digestive enzyme-containing squirt which can lead to irritation of skin as well as conjuctivitis and, rarely, blindness.

Figure 40.13. Sea urchins contain calcium carbonate spines which can penetrate skin and are sometimes venomous.

Treatment of crown-of-thorns stings involves removal of the spines. Although they can break during extraction, they are usually easier to extract than are sea urchin spines. The spines may be visualized on x-ray if necessary. Hot water (110 to 115° F) soaks are useful in reducing pain (Edmonds 1989). Topical (or systemic if needed) antibiotics should be used and tetanus prophylaxis is indicated. Contact dermatitis from sea stars can be treated symptomatically or with a low-potency topical corticosteroid.

Sea urchins (Figure 40.13) are bottom-dwelling, slowly mobile, mostly nocturnal spiny creatures that for generations have been well known to be occupational hazards to pearl divers and fishermen. The roe of some species is edible, and is considered a delicacy by some and an aphrodisiac by others (Edmonds 1989). Major fisheries exist in Japan, Chile, France, Canada, Korea, Peru, Mexico, Barbados, and the United States, and during the early 1980s, world landings of sea urchins averaged almost 50,000 tons annually (Edmonds 1989; Conand and Sloan 1989). Mortality in the sea urchin fisheries is primarily from drowning (CDC 1993) and ingestion of the ovaries of some species during their reproductive season can cause poisoning with symptoms of gastroenteritis (Halstead 1988).

Cutaneous injury due to sea urchins is primarily due to the calcium carbonate spines which are of two types: the longer primary spines, and the shorter secondary (e.g., oral, aboral) spines. Although most sea urchins have solid or hollow primary spines that can penetrate the skin of man, some are considered more dangerous—especially those in the families Diadematidae and Echinothuridae (Halstead 1988). The most dangerous primary spines are seen in the black-spined sea urchin (*Diadema* sp.). This urchin has long (up to 30 cm), sharp, hollow, fragile and brittle spines that can easily penetrate human skin causing immediate severe pain, which often appears out of proportion to the visualized injury and is likely due to a toxin produced by the cells of the integumental sheath of the spine. The spines usually break off in the wounded skin and the wound area may be discolored a violet-black due to pigments in the sheath. The immediate pain usually increases in intensity and can be accompanied by erythema, edema, parasthesias, partial motor paralysis, syncope, tachycardia, hypertension, or, very rarely, respiratory distress and death (Edmonds 1989; Halstead 1988; Auerbach and Halstead 1989; Burnett et al. 1986). Primary oral and secondary aboral spines of some echinothurids have venom sacs on their tips (Halstead 1988). With these spine injuries, only minimal penetration is needed for envenomation to occur.

Pedicellariae are specialized spines that have unique pincers which can grasp human skin with a three-pronged fang and, in some species, can envenomate via venom sacs surrounding the pincer tips. Some pedicellariae are used primarily for grasping while others may be quite toxic. Advanced globiferous pedicellariae seen in the family Toxopneustidae are considered the most advanced for envenomation (Halstead 1988; Endean 1961). In this family, the open pedicellariae appear as minute (3 to 4 mm) purple and white "flowers" scattered amongst the primary spines, and when handled, the "flowers" close, grasping the skin, thereby envenomating the victim with a neurotoxin. Severe

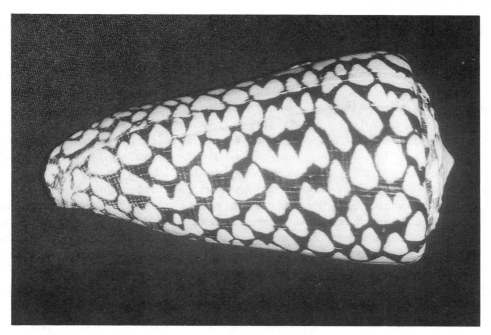

Figure 40.14. Cone shells are prized by collectors for their beautiful shells. When live specimens are handled, they can deliver a potent sting and fatalities have been reported.

pain, muscle paralysis (especially facial) and respiratory distress often occur.

Delayed reactions to sea urchin stings have also been reported. Inflammatory papular and vesicular eruptions may occur approximately ten days following an acute sting and may last for several months (Burke et al. 1986; Asada et al. 1990). These reactions are likely Type IV delayed hypersensitivity reactions due to antigenic proteins in sea urchin spines.

Granulomatous delayed reactions are due to embedded inorganic substances ($CaCO_3$, $MgCO_3$, phosphates, $CaSO_4$ and SiO_2) found in the spine (Rocha and Fraga 1962; Kinmont 1965) and these reactions can at times be debilitating and chronic (O'Neal et al. 1964). Spines embedded near or into tendons, bursae, joints, or neurovascular structures can be especially problematic—both in acute and delayed reactions (Fisher 1978; Auerbach and Halstead 1989; Coombs and Mutimer 1993).

Treatment of sea urchin stings first involves ignoring many of the anecdotal remedies including applications of urine, mud, alcohol, ammonia, and mechanical pounding of the area. Attached pedicellariae should be immediately removed as these spines continue to envenomate even after they have partially broken off. Primary spines should also be removed, although they are extremely brittle and crumble easily during attempts at extraction.

Pain control can usually be achieved by immersing the involved area for 60 to 90 minutes in hot (110 to 115°F) water mixed 1:1 with vinegar. The acetic acid is presumably to help dissolve calcium carbonate spines near the surface, while the heat serves to inactivate the heat-labile toxin. If there is concern about embedded spines, they can be visualized on x-ray and surgical consultation obtained as needed. The wound should be watched closely for secondary infection and attention to tetanus prophylaxis is indicated. Type IV delayed hypersensitivity reactions should be treated with topical, intralesional or systemic corticosteroid medications. Intralesional corticosteroids are most commonly used with sea urchin granulomas when needed although gradual resolution over many months usually occurs.

Skin Problems Related to Mollusks
(Phylum Mollusca)

Marine mollusks are well known to commercial fishermen and watermen and there are many important fisheries in this group of animals: oysters, clams, abalone, conchs/whelks, scallops, squid, cuttlefish and the octopus. Many mollusks contain ornamental shells, and there is an ornamental shell trade fishery, which primarily supplies the demands of shell collectors and shell crafters. There is also a commercial shell trade fishery for mollusks harvested for pearls and mother-of-pearl (Wells 1988).

Shells of mollusks are composed of calcium carbonate and can have sharp edges (especially when broken) that can lead to lacerations in watermen. Some mollusk shells are enormous such as the giant clam, *Tridacna gigas*, which grows to nearly 5 feet in length and can weigh up to 500 pounds. Found in the South Pacific, there are many fables of divers being trapped by this "man-eater". These are likely untrue as this creature rarely closes its shell completely. A smaller species, *Tridacna derasa*, can grow to 2 feet in length and uses strong adductor muscles to close its shell rapidly and completely, which can cause crushing injuries or even amputation of digits (Wexler 1994).

Cone snails (Figure 40.14) are mollusks, which can cause toxic envenomation. Prized by collectors for their beautiful shells, envenomation occurs while handling live specimens. The venom is a neurotoxin, which has curare-like properties (Pearce 1990; Brown and Shepherd 1992). The toxin is synthesized in a tubular duct and stored in a muscular venom bulb. Hollow-barbed, harpoon-like radular teeth are stored in a sheath and deliver the venom when forcefully ejected via the proboscis (Halstead 1988; Miner 1961). Although all of the approximately 400 species of cone shells are venomous, only some are large enough to be dangerous to man and these occur primarily in the Indian and Pacific Ocean, although *Conus textile* may be found in Florida and the Caribbean (Edmonds 1989).

The initial cone snail "sting" is painful, and local numbness and tingling usually occur and may spread to the entire body. Muscular paralysis may follow and serious envenomations can lead to respiratory depression, cardiac arrest, and death. A fatality rate of 15 to 25% has been reported (Mandojana and Simes 1987; Edmonds 1989; Halstead 1988). Since no antivenom is available, treatment is symptomatic and supportive. Hot water (110 to 115° F) immersion of the site of envenomation for 30 to 60 minutes is worthwhile as is attention to tetanus prophyloxis protocols.

Sea stories of giant "man-eating" squid and octopuses (cephalopoda) have been around since man began sailing the seas. Giant squid (*Architeuthis harveyi*) can be up to 5 meters in length with 12-meter tentacles. Sea fables have reported these creatures attacking seagoing craft, and human attacks have actually been documented on rare occasions (Edmonds 1989). At the center of the eight short tentacles is a mouth with a parrot-like beak which, with larger squid can cause serious injury or loss of fingers (Miner 1961). As squid are a delicacy in many parts of the world and are an excellent fish bait, there exists several fisheries to supply the world's market demand (DeLoguna 1988; Murata 1988).

Octopuses are generally reclusive creatures and rarely cause problems in man. However, the giant Pacific octopus (*Octopus dofleini*) can grow to more than 100 kg and 9 meters in diameter, and there exist anecdotal accounts of divers being attacked by the curious creature (Edmonds 1989; Sanders 1993; Bavendam 1995). Like the squid, it also has sharp parrot-beak-like mouthparts which can bite (LaGorce 1961).

While most octopuses are nonvenomous, the bite of the blue-ringed octopus (*Hapalochlaena lunulata* and *H. maculosa*) can be deadly. These small (10 to 100 grams; 2 to 20 cm), tan to brownish creatures contain bright blue rings, making identification easy. Found primarily in tidal pools and shallow reef areas of Australia and the Indo-Pacific region, they are unlikely to be seen by most fishermen. However, because these marine animals contain the very deadly venom maculotoxin, it is important for fishermen to be knowledgeable about this creature. The venom is found in its salivary glands and is injected into the skin when bitten by the parrot-like beak when the octopus is handled or allowed to crawl on the skin. Maculotoxin is an extremely potent neurotoxin that contains tetrodotoxin and, with envenomation, death often occurs within minutes due to peripheral as well as phrenic nerve blockade resulting in muscular paralysis and respiratory failure. Treatment is supportive as no antivenom is available (Edmonds 1989; Auerbach and Halstead 1989; Williamson 1987).

Nudibranchs (sea slugs) are unshelled, slug-like marine mollusks, and some of these creatures can sting man. Several species (e.g., *Learchis poica, Dondice occidentalis* and *Glaucus* spp.) feed on jellyfish, hydroids, or the Portuguese man-o-war. Undischarged and undigested nematocysts are transported via the digestive tract to tiny finger-like projections (cerata) on the sea-slug's back. Cnidosacs are small capsules containing these undischarged nematocysts at the tips of the cerata, and provide the nudibranch with a stinging defense (Ruppert and Fox 1988; Humann 1992). Treatment of nudibranch stings is symptomatic.

Skin Problems from Marine Worms
(Phyla Platyhelminthes, Annelida)

Cercarial dermatitis ("swimmer's itch," "clam-digger's itch") is a pruritic erythematous papular eruption due to schistosome-like animal blood flukes (phylum Platyhelminthes) which can burrow into the skin of man. It can be acquired in virtually any aquatic area worldwide from ditches/canals to ponds/lakes to rivers/streams to estuaries or ocean bays harbors. Cercariae that can penetrate the skin of man as an accidental host generally parasitize waterfowl or rodents, with various snails as intermediate hosts (Baird and Wear 1987).

Cercarial dermatitis in the United States is merely a nuisance as the eruption and discomfort is self-limited, and there are no long-term sequelae. Vigorous toweling immediately following water exposure may reduce the number of cercariae penetrating the skin. Antihistamines and antipruritics may provide symptomatic relief.

One of the more commonly used fishing baits is the annelid bloodworm, *Glycera dibranchiata*, which may be

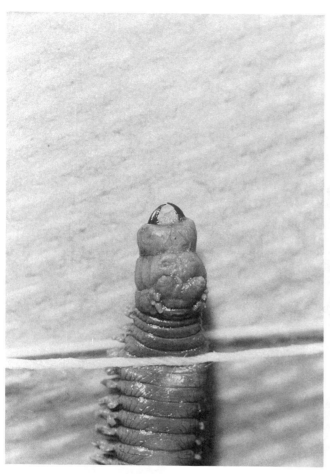

Figure 40.15. A marine annelid, the clamworm is used as a fishing bait and can bite with chitinous jaws in its eversible probiscus.

purchased alive in bait shops. At the tip of the long muscular pharynx of this polychete, are four poison glands that open onto four chitinous fangs. While many fishermen have been bitten by this worm, it is not a severe bite and symptomatic care is all that is required (Ruppert and Fox 1988).

Clamworms or northworms (*Nereis* sp.) are extremely common estuarine worms and are also used as bait by fishermen. This polychete has an eversible pharynx with a pair of sharp chitinous jaws (Figure 40.15), which are used defensively as well as for feeding (Ruppert and Fox 1988; Ballantine 1983). Contact urticaria, angioedema, conjunctivitis and dyspnea has also been reported in a fisherman, as well as chronic fingertip eczema in persons handling *Nereis diversicolor* (Camarasa and Serra-Baldrich 1993). The larger (up to 1.5 m) biting reef worms, *Eunice aphroditois* and *Onuphis teres*, are found on Australian reefs and are capable of more severe bites (Edmonds 1989; Halstead 1988). Another biting worm is the ringneck worm, *Eunice antennata* (Ruppert and Fox 1988). A related marine annelid,

the rockworm (*Marphysa sanguinea*) is a commonly used fishing bait, and angioedema, asthma and rhinitis have been reported in fishermen handling this bait (Valero et al. 1989).

Certain marine annelids contain numerous bristle-like setae that can sting. Stinging species of bristleworms/fireworms include *Chloeia flava*, *Chloeia viridis*, *Hermodice carunculata*, and *Eurythoë complanata*. These brightly colored worms afflict their sting via penetration of the setae into the skin thereby releasing the venom. Fishermen are especially at risk when catching fish that feed on these worms (Edmonds 1989) or when the worms are caught on a baited hook (Mandojana and Sims 1987). Treatment of bristleworm stings is initially directed at removal of the bristles from dried skin using adhesive tape or forceps. Hot water soaks, or application of vinegar, alcohol, or dilute ammonia to the involved area have been anecdotally advocated to help alleviate the pain (Mandojana and Sims 1987; Edmonds 1989; Halstead 1988).

Skin Problems Due to Arthropods
(Class Crustacea)

As previously discussed, a wide variety of small to microscopic marine arthropods and arthropod larvae (zooplankton) can cause a pruritic skin condition commonly termed "sea lice." In addition, urticaria (generalized as well as contact), angioedema, and even anaphylaxis have been reported after ingestion (or skin contact) of certain crustacea—especially shrimp (Hjorth and Reed-Peterson 1976; Hoffman et al. 1981; Lehrer 1990). Crustaceans also have sharp exoskeletal spines, which can cause puncture or laceration injuries that may lead to secondary infection (Burnett 1989). Pinching injuries due to claws of lobsters, crayfish and crabs often lead to lacerations or crushing injuries that can also lead to secondary infection (Figure 40.16). Large muscular claws, as seen in the American/Northern lobster (*Homarus americanus*) and the Florida stone crab (*Menippe mercenaria*), are especially dangerous as they may easily crush a finger (Ruppert and Fox 1988).

Much feared but actually quite harmless is the horseshoe crab (*Limulus polyphemus*), which is occasionally used by fishermen as eel bait (especially female crabs with eggs) (Barash and Bavendam 1993). Not related at all to true crabs, these ancient creatures are chelicerates, more related to spiders, scorpions and ticks. Their weak pincers on the legs are relatively harmless and the spiny spike like tail which, although fabled to be poisonous, is not. Known as a telson, this "tail" is used primarily in burrowing and in flipping the creature when accidentally inverted (Heston 1987). In the past, Indians used this spike for a fish spear, yet it is quite harmless unless one is extremely careless.

Figure 40.16. Crabs can cause traumatic pinching injuries which may lead to secondary infection.

Figure 40.17. Sharks can cause severe bite injuries to fishermen when handled improperly.

Nonvenomous Traumatic Injuries Due to Fish
(Phylum Chordata)

Fish bites are common injuries in fishermen. These can range from mild injuries from smaller toothed fishes such as flounder or mackerel, to more serious injuries from larger fish such as sharks and barracuda. Sharks (Figure 40.17) in particular are quite dangerous due to their large teeth, large mouths and extremely strong jaw musculature, giving a biting force estimated at up to 18 tons per square inch (Auerbach and Halstead 1982).

While the wealth of information about shark bites is about in-water shark attacks, most bites in fishermen occur out of water while handling sharks, such as when attempting to remove a shark from a net or removing a hook from a live shark. Thinking they have control of the fish, inexperienced fishermen may hold up a live shark by the tail to display it to others. This can lead to a bite on the leg, trunk, or even arm as the shark's cartilaginous vertebral column allows for incredible flexibility.

Shark attacks, although rare, can occur when fishermen are in the water. This risk is obviously higher in fisheries requiring diving for the catch, such as in the abalone industry (Klimley 1994). Although at least 32 of the 300 to 350 species of shark have been reported to attack man, (Auerbach and Halstead 1982) the more dangerous sharks include the great white (*Carcharodon carcharias*), mako (*Isurus* spp.), tiger (*Galeocerdo cuvieri*), bull (*Carcharinus leucas*), grey reef (*Carcharinus amblyrhynchos*), blacktip reef (*Carcharinus melanopterus*), oceanic whitetip (*Carcharhinus longimanus*), spinner (*Carcharhinus brevipinna*), blacktip or spotted (*Carcharhinus limbatus*), reef whitetip (*Triaenodon obesus*), blue shark (*Prionace glauca*), lemon (*Negaprion* spp.) and hammerhead (*Sphyrna* spp.) sharks (Edmonds 1989; Brown and Shepherd 1992; Howard and Burgess 1993; Halstead 1980; Curtsinger 1995). Shark bites can be fatal since the numerous sharp teeth, coupled with massive muscular jaws, can cause major damage to muscle, fascia, tendons, ligaments, bone, nerves, blood vessels, organs, and body cavities.

The great barracuda (*Sphyraena barracuda*) can also cause serious bites in fishermen when carelessly handled. They often continue to bite viciously when brought aboard a boat before being subdued. In contrast to the arcuate bite of a shark, barracuda often leave straight or V-shaped lacerations (Auerbach and Halstead 1982).

Bluefish (*Pomatomus saltatrix*) also have sharp teeth and the larger "chopper" sized blue can cause a serious bite during careless handling (Lange 1988). Large grouper, sea bass, snappers, mackerel, and wahoo can also inflict serious bite injuries when improperly handled (Hoffman et al. 1992).

Moray eels (family Muraenidae—Figure 40.18) are often caught on hook and line during reef or shelf fishing or may be caught as bycatch in trawl nets or fish traps. The strong muscular jaws contain numerous recurved teeth, which can inflict a serous injury. Since the bite is most frequently on the hand, tendon, nerve, and blood vessel damage is often seen. Closed-space infections occur relatively frequently due to the depth of the puncture wounds. Once a fisherman is bitten, the moray eel will tenaciously grip its victim, and decapitation or jaw disarticulation may be necessary to dislodge the fish (Erickson et al. 1992). Rather than attempting to unhook a writhing moray caught on a hook and line, experienced fishermen will generally cut the line rather than risk the eel wrapping around a forearm while biting viciously.

Some fish, such as the toadfish (family Batrachoididae), are relatively small, yet have very muscular jaws as well as stinging spines. Known in some areas as an "oystercracker," this fish can cause crushing injuries and amputation of digits.

The Atlantic wolffish or wolf eel (*Anarhichas lupus*) has large teeth and powerful jaws, which can cause severe bite injuries to fishermen (Amos and Amos 1988). These fish,

Figure 40.18. Moray eels have sharp recurved teeth which can inflict a serious bite when carelessly handled.

which can grow to 1 1/2 meters in length and 1.75 kg in weight, have large canine teeth and molars to feed on molluscs, crustacea, and echinoderms and can lead to a penetrating or crushing injury to a careless fisherman. These fish are popular food fish in Europe, and there exists a major fishery in the Gulf of Maine and Georges Bank where about 1 1/2 million pounds per year are caught (McClane 1978). Ratfish (*Chimaera monstrosa* or *Hydrolagus colliei*) can also sting and are feared by fishermen for their formidable bite (Edmonds 1989). Parrotfish (family Scaridae) and puffers (family Tetraodontidae) have specialized parrot-beak-like jaws, which can amputate fingertips with careless handling.

Figure 40.19. Sawfish have extremely sharp teeth on their bills and should not be boated or handled until completely subdued.

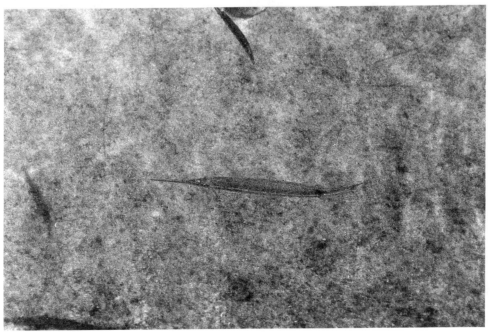

Figure 40.20. Needlefish are capable of leaping from the water at high speed when frightened and can cause serious penetrating injuries.

Billfish (family Istiophoridae) are highly prized by fishermen, with commercial fishermen primarily seeking swordfish for culinary markets while recreational fishermen also seek marlin and sailfish. The "bill" can cause laceration as well as penetration injuries when the fish is brought aboard prior to being completely subdued. Injuries also occur while attempting to hold the fish for a "trophy picture." Deaths have been reported (Edmonds 1989) not only by severe traumatic injuries, but also by drowning—when a person is pulled overboard after wrapping a leader/line around the wrist in order to get a better grip in landing the fish.

Sawfish can grow to 6m in length and contain a bill/snout that contains numerous sharp projections (Figure 40.19). This bill is used to kill smaller fish for food by "slashing" the bill back and forth through schooling fish. Man is usually injured while trying to handle or bring aboard a live hooked fish or attempting to extract one from a net (Edmonds 1989).

Needlefish (or garfish) (Figure 40.20) have sharp "beaks" which can cause penetrating injuries when the 0.5 to 1.5m fish leaps from the water at high speed when "spooked." Traveling in schools, it is often frightened by lights used by fishermen while fishing at night. Various penetrating injuries have been reported: into soft tissue leading to infection, into the abdominal cavity leading to peritonitis and death, into the neck and spinal cord causing a hemisection, into the thorax leading to pneumothorax, and into the eyes/orbit leading to blindness or brain penetration and death. Although the surface injury may appear deceptively small, deep penetration leading to severe internal injuries may occur, and the beak can break off deeply within the wound. Most fishermen injured by needlefish were in low freeboard boats at night. Japanese lobstermen who dive at night have also been injured (Barss 1985, 1982; McCabe et al. 1978).

Fish spines (Figure 40.21) are mostly nonvenomous but can still cause penetrating injuries while handling, sorting, or removing fish from a hook. Spines are used defensively, and some fish, such as the porcupine fish, have numerous spines. Sharp gill plates can cause laceration injuries when handling such fish as croaker (*Micropogonias undulatus*), black sea bass (*Centropristis striata*), and snook (*Centropomus undecimalis*) (Amos and Amos 1985). Abrasion injuries can be caused by denticles on the skin of sharks and rays (Mandojana and Sims 1987).

Treatment of nonvenomous fish injuries involves exploration, cleansing, debridement, and close observation for secondary infection. Attention to tetanus prophylaxis guidelines is recommended. Major fish bite or fish bill injuries may result in extensive blood loss and first aid measures are directed at stopping hemorrhage by pressure or, if required, a tourniquet followed by rapid delivery to a medical center. If there is concern about retention of a foreign body such as a fish tooth or beak, x-ray examination is helpful. Surgical consultation may be necessary in these cases or with large wounds or those with tendon/nerve involvement.

Figure 40.21. Fish spines may be venomous or non-venomous, yet all can cause deep puncture injuries which can lead to secondary infection.

Venomous Fish Injuries (Phylum Chordata)

While most fish spines are nonvenomous, some are attached to a venom apparatus and can cause a painful and, at times, fatal injury to fishermen. Venomous fish stings in fishermen usually occur while attempting to remove the fish from hook and line, from a net in which the fish has become entangled or trapped, or while fish sorting without protective gloves.

Venomous sharks include the horned sharks or hornshark (*Heterodontus francisci*) and spiny dogfish (*Squalus acanthias*). Horned sharks are found in kelp beds on the California coast and have a sharp venomous spine located just anterior to each of its two dorsal fins. Other *Heterodontus* species are found in other parts of the world and are commonly known as the bullhead and Port Jackson shark (Edmonds 1989).

The spiny dogfish is found worldwide and has similar spines anterior to its two dorsal fins. Spine injuries usually occur on the hand while handling, or after a sudden lunge by the shark—which can drive a spine into an unwary fisherman's leg (Halstead 1988). Envenomation consists of sharp, immediate, intense pain which lasts for hours, and which is followed by erythema, swelling and occasionally numbness (Edmonds 1989; Halstead 1988).

Ratfish or chimaeras (*Chimaera monstrosa* or *Hydrolagus colliei*) are sometimes encountered by trawler and deep sea fishermen as they are primarily found in deep, cool water worldwide. As mentioned earlier, they can inflict a serious bite injury, and fishermen actually fear the bite more than the sting (Halstead 1988). The single sharp venomous spine is located anterior to the first of two dorsal fins. When a cutaneous puncture injury occurs, there is an immediate sharp burning pain followed by numbness and swelling (Magerog et al. 1991).

Stingrays (Figure 40.22) are in the order Rajiformes (skates and rays) and are the single most important group of venomous fish, with 1,500 to 2,000 human injuries per year (Halstead 1988; Brown and Shepherd 1992). There are six different families that contain rays which can sting man. While the numerous stingrays vary in their appearance, they all have a venomous stinger located on or near the tail. The calcium carbonate spine is covered by an epidermal integumental sheath which contains the venom. Most injuries to humans occur on the feet or lower extremities when a person steps (while wading or on deck) on a stingray, thereby trapping it. The tail thrusts the spine into the victim's foot, ankle, or calf, damaging the sheath and releasing venom into the wound. Recurved teeth hold the sharp spine in place while grooves on the spine allow for deeper penetration of the venom (Figure 40.23). Fishermen can also be stung on the upper extremities while holding a live stingray aloft on hook and line. While smaller stingrays typically give a penetrating puncture injury, large stingrays can cause severe slashing lacerations, which usually occur on the calf.

Figure 40.22. There are numerous species of stingrays and all have a barbed poisonous spine located on or near the tail. This species is an Atlantic stingray (*Dasyatis sabina*).

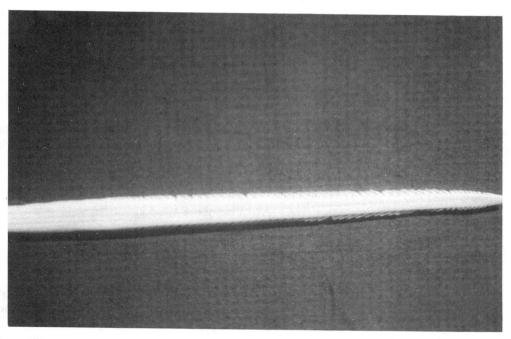

Figure 40.23. A stingray spine has recurved teeth making extraction difficult. Grooves on the surface of the calcium carbonate spine allow for venom flow into the wound.

Figure 40.24. Toadfish have dorsal and opercular spines which are venomous, and a bite injury may lead to crushing injuries or distal digital amputation.

Immediate sharp throbbing pain—often excruciating and out of proportion to the injury seen—is the initial presentation. Generalized symptoms often follow including nausea, vomiting, tachycardia, diaphoresis, diarrhea, myalgias, hypotension, weakness, and paralysis. Death may rarely occur (Halstead and Vinci 1987; Kizer 1990; Bel 1991).

Catfish are found in both fresh and salt water and usually have a single venomous dorsal spine at the anterior portion of the dorsal fin, as well as bilateral spines at the anterior portion of the pectoral fins. Found worldwide, there are over 1,000 species of catfish (Brown and Shepherd 1992). Puncture injuries to the hands and forearm are most commonly seen in fishermen. Immediately following envenomation, there is an intense burning and throbbing sensation with pallor surrounding the puncture site. Erythema and edema follow, and the wound may sometimes become necrotic and ulcerate (Halstead 1987; Burnett et al. 1985; Zeman 1989).

Weeverfish (*Trachinus* spp.) are primarily found in the eastern Atlantic Ocean from southern Norway to northern Africa, and the Baltic, Mediterranean, and Black Seas. As the most venomous and dangerous fish of temperate zones (Halstead 1988), they are commonly feared by fishermen. These small fish (not known to exceed 46cm [Halstead 1988]) bury themselves in sand or mud in coastal areas. Weevers have 5 to 7 venomous dorsal spines as well as bilateral venomous opercular spines. Spine injuries lead to puncture wounds with immediate stabbing pain which becomes excruciating over the next 30 minutes such that the victim often screams in agony, becoming delirious. The pain may last 1 to 2 days, and the sting site usually becomes erythematous and may ulcerate taking months to heal. Systemic symptoms can include fever, chills, diaphoresis, nausea, vomiting, syncope, loss of speech, hypotension, cardiac arrhythmias and, rarely, death (Edmonds 1989; Halstead 1988; Auerbach and Halstead 1989; Brown and Shepherd 1992). Fishermen are usually stung while wading or attempting to remove weevers from a net. The spines may even penetrate boots. Weeverfish are fished commercially in Denmark and other local European markets because of their tasty flesh; however, the venomous nature of these fish has likely limited development of this fishery (Halstead 1988).

Like the weevers, toadfish (Family Batrachoididae—Figure 40.24) also have venomous dorsal and opercular spines. They are found worldwide in coastal waters and, as mentioned earlier, can cause a severe bite injury as well as sting. The sting is extremely painful and has been described as similar to a scorpion sting. Initial symptoms of pain are soon followed by erythema, calor, and edema (Halstead 1988, 1987).

Stargazers (family Uranoscopidae) are small (40 cm or less), venomous fish found in temperate and tropical coastal areas of the Atlantic, Pacific, and Indian Oceans, as well as the Mediterranean Sea. They have a flattened head and spend most of their lives buried in mud or sand with only their upturned eyes and mouth exposed. The shoulder area (cleithral) spines are venomous and are usually encountered when the fish is stepped on while wading, or when a fisherman removes a stargazer from a net or hook. Fatalities due to stargazer stings have been reported in the Mediterranean area (Edmonds 1989; Halstead 1988; Halstead and Vinci 1987). Not only can the stargazer deliver a venomous sting, these fish are also capable of discharging an electrical shock, as will be discussed later in this chapter.

Scorpionfish (Family Scorpaenidae) include a wide variety of more than 300 species of venomous fish (Edmonds 1989; Doubilet 1987), and are second only to stingrays in annual incidence of fish stings in humans, with approximately

Figure 40.25. The lionfish (*Pterois* sp.) is prized by aquaria enthusiasts and can sting with its dorsal, anal, and pelvic spines.

300 stings per year (Auerbach and Halstead 1989; Brown and Shepherd 1992). In the clear water of coral reefs, scorpionfish are often brightly colored, while in murky waters with mud, sand, or rocky bottoms, they are often drab colored, unsightly creatures. Lionfish (turkeyfish, zebrafish, firefish, feather fingers or butterfly cod—*Pterois* spp., *Brachirus* spp. and *Dendrochinus* spp.) are brightly colored fish with long colorful spines. Members of the genus *Pterois* are the most colorful (Figure 40.25) and there exists a market for these fish for sale to saltwater aquaria enthusiasts. There are 13 dorsal spines, 3 anal spines and 2 pelvic spines which are venomous and can result in a painful sting (Trestrail and Al-Mahashneh 1989). Although dangerous appearing, the long pectoral spines are nonvenomous (Halstead 1988).

Other scorpionfish (*Apistus* spp., *Centropogon* spp., *Erosa* spp., *Gymnapistes* spp., *Heliocolenus* spp., *Inimicus* spp., *Minous* spp., *Notesthes* spp., *Scorpaena* spp., *Scorpaenopsis* spp., *Sebastes* spp., *Sebasticus* spp., and *Sebastodes* spp.) are most commonly found in coastal waters as bottom dwellers. Some bury themselves in mud or sand, while others hide in crevices or plant beds, while still others are free-swimming. The dorsal, anal, and pelvic spines are venomous, while the opercular spines do not histologically demonstrate venomous glandular tissue. Common names for these scorpionfish include: bullrout, waspfish, cobbler, rock cod, devilfish, demon stinger, bearded ghoul, fortesque, sculpin, sea scorpion, rockfish and stingfish (Edmonds 1989; Halstead 1988). These fish are readily caught on hook and line and, in some areas of the world, the flesh is eaten.

The envenomation syndrome of the lionfish and scorpionfish groups is similar, and recovery is often within several hours (Ell and Yates 1989). Following the sting, there is immediate pain followed by erythema, edema, pallor, and central cyanosis. As the pain radiates and increases in intensity, the sting site may become numb. Motor difficulties and temporary paralysis may ensue. Systemic effects may include headache, anxiety, nausea, vomiting, pallor, diaphoresis, tremors, generalized skin rash, paralysis, syncope, tachycardia and seizures. Death is rare but reported (Halstead 1988; Auerbach and Halstead 1989; Brown and Shepherd 1992).

The most venomous fish, the stonefish (*Synanceja* spp.), is also in the Scorpaenidae. These fearless fish inhabit shallow water in the eastern Indian and southwest Pacific Oceans, and are greatly feared by fishermen in these areas. Most envenomations occur with careless handling or while wading barefoot and stepping on the fish. The brown, unattractive stonefish is well-camouflaged and may not be seen by the victim. A sting results in immediate excruciating pain, radiating proximally over minutes and which may lead to edema and extremity paralysis. Cyanosis of the sting site with surrounding pallor and erythema usually occurs with accompanying sharp, piercing tenderness. If the victim survives, the site may abscess and ulcerate over days and become necrotic and gangrenous. Systemic symptoms occur within minutes to hours and include headache, nausea, vomiting, diarrhea, malaise, diaphoresis, tachycardia, cardiac arrhythmias, hypothermia, delirium, loss of consciousness,

Figure 40.26. The blue tang (*Acanthurus coeruleus*) is a surgeonfish with a retractable blade-like appendage on the caudal penduncle.

seizures, and respiratory distress (Edmonds 1989; Halstead 1988; Auerbach and Halstead 1989; Lehmann and Hardy 1993). Death may occur within one hour and is usually within 6 to 8 hours of envenomation (Auerbach and Halstead 1989). If the victim survives, the agonizing pain may last for days with gradual reduction of extremity edema, and residual pain, numbness and paralysis resolving over weeks to months.

Surgeonfish (including doctorfish and tangs) include a variety of brightly colored to tan-brown marine reef fishes that have blade-like spines on the caudal peduncle. In some species, such as the true surgeonfishes, doctorfish, and blue tang (*Acanthurus* spp.—Figure 40.26), the spines are muscle-powered and sometimes venomous and are fully retractable into a sheath. In other species, such as the yellow tang (*Zebrasoma flavescens*), the spine exits the groove as the tail moves in the opposite direction. In yet other species, such as the unicornfish (*Naso* spp.), there are fixed spines on either side of the tail. Fishermen are usually harmed while handling these fish when removing them from nets or hooks (Edmonds 1989; Halstead 1988).

Rabbitfish (*Siganus* spp.) resemble surgeonfish and can also cause venomous stings in fishermen. Found in the Indo-Pacific Oceans and Red Sea, their 13 dorsal, 7 anal, and 4 pelvic spines have associated venom glands. The sting is similar to scorpionfish stings (Edmonds 1989; Halstead 1988; Halstead and Vinci 1987).

Leatherbacks (*Scomberoides sanctipetri*) are in the family Carangidae (jacks and pompano) and were initially reported by fishermen to have venomous anal spines. Found in Indo-Pacific waters, other members of this family are found in the Atlantic and may prove to have similar venomous spines (Halstead 1988).

Old wife (*Enoplosus armatus*) has knife-like venomous dorsal spines, and is found in sheltered Australasian bays and harbors (Edmonds 1989). Other fish reported as possibly venomous include some species of moray eels, sea robins, flying gurnards, butterfly fish, squirrel fish, scats, and goosefish (Auerbach and Halstead 1989).

Treatment of the various venomous fish stings is similar—immediate first aid directed at immersion of the affected area in hot (110 to 115° F) water for 30 to 90 minutes or until pain is relieved (Auerbach and Halstead 1989; Brown and Shepherd 1992). This helps to inactivate the heat labile toxins. Exploration for any spine or integumental sheath remnants is indicated, and the spines may be visualized on radiographic examination if needed. The wound should then be irrigated with sterile warm water. Infiltration of the area with 1 to 2% lidocaine may help with pain control (Auerbach and Halstead 1989). Hyoscine butylbromide or emetine HCl injected into the sting site are traditional yet controversial treatments for stonefish envenomation (Edmonds 1989; Auerbach and Halstead 1989). With serious envenomations, first aid measures may include cardiopulmonary resuscitation, and rapid delivery to a medical facility is in order. An equine-derived stonefish antivenom is available in Australia. Tetanus prophylaxis guidelines should be followed with all fish sting injuries. As many venomous fish injuries are of the puncture variety, close observation for secondary infection is needed, yet regular use of prophylactic antibiotics is controversial.

Electric Fish (Phylum Chordata)

Certain fish are capable of harming fishermen by delivering an electrical shock. The electrical charge is produced by acetylcholine-rich cells known as electrocytes. While each individual electrocyte produces only about one ten-thousandth of a volt, large numbers of these cells are triggered to fire simultaneously, producing an electric field of significant voltage (Rudloe and Rudloe 1993; Wu 1984). An uneducated and unwary fisherman may be surprised when handling these fish while sorting catch from a net, or when handling metal wire leader material while catching one of these fish on hook and line.

Electric rays (Family Torpedinidae) are the most common electric fish encountered by modern fishermen. In ancient times, Theophratus, a Greek pupil and successor of Aristotle, observed that the shock of torpedo rays could be transmitted to fishermen through metal fishing spears. Plotinus, a Roman philosopher, observed that fishermen could feel the torpedo ray's numbing shock either via direct contact or through a reed (Wu 1984).

Torpedo rays are commonly called numb rays, numbfish, or wishfish (Edmonds 1989). Seasoned fishermen may jokingly tell an inexperienced deckhand when asked about a torpedo ray: "That's a wishfish; make a wish, close your eyes, and rub it between your eyes and the wish will come true" (Rudloe and Rudloe 1993). A gullible deckhand is in for a shocking experience.

In all, there are about 40 species of electric rays (Rudloe and Rudloe 1993). The kidney-shaped electric organs can produce electric discharges from 8 to 220 volts in a series of short bursts passing from the negative ventral side of the ray to the electrically positive dorsal side (Edmunds 1989). The Atlantic torpedo ray (*Torpedo nobiliana*) is found throughout North Atlantic temperate regions and may grow to 5 to 6 feet and up to 200 pounds (although 30 pounds is more common). The largest specimens have been measured to discharge from 170 to 220 volts (McClane 1978; International Oceanic Foundation 1993).

Stargazers (Family Uranoscopidae) are the only other marine fish that have the ability to shock. Also capable of stinging, these fish produce a maximum shock of about 50 volts. Shrimpers often refer to stargazers as "440's" while electric rays are "220's," yet the rays are capable of delivering the more powerful shock (Rudloe and Rudloe 1993).

In the freshwater Amazon basin, electric eels have been measured to discharge up to 650 volts (Rudloe and Rudloe 1993). Although high-voltage formidable shocks can stun a fisherman, most electrical fish shocks are of low amperage and, although uncomfortable, are not otherwise serious and do not require specific therapy, except in unusual instances.

Sea Snakes (Phylum Chordata)

With few exceptions, sea snakes are found in the marine environment and spend their entire life in water. Perhaps the most venomous vertebrate, it is also the most numerous of all snakes and is found throughout the Pacific and Indian Oceans. For unknown reasons, sea snakes do not inhabit the Atlantic Ocean, Mediterranean Sea, or Red Sea (MacLeish and Cropp 1972; Tu and Folde 1987). While some species are aggressive (e.g., *Enhydrina schistosa*, *Hydrophis cyanocinctus*), most are considered docile, and yet even these species can become very dangerous and aggressive at times such as during breeding season (Halstead 1988). The yellow-bellied sea snake (*Pelamis platurus*) has the widest distribution and is even seen in the eastern Pacific from the Baja Peninsula to Peru. Although some species are pelagic, most prefer coastal waters and are especially abundant in river mouths (Halstead 1988).

Sea snakes are commonly encountered by Asian fishermen after the snakes become entrapped in trawl nets (Edmonds 1989). A single net haul may capture as many as 100 snakes in areas where they are abundant. Occasionally, sea snakes are caught on hook and line. Fishermen are usually bitten while handling nets, sorting catch, removing them from hook and line, or when stepping on them while wading or on deck (Halstead 1988). There actually exists an Asian fishery for sea snakes as their flesh is eaten in the Orient. Sea snake skin is also used for making western-style cowboy boots and belts. Parts of sea snakes are used for medicinal purposes and as an aphrodisiac (Tu and Fulde 1987).

Following a sea snake bite, the injured area is usually asymptomatic, and no inflammation or edema is present. Because the fangs (usually 1 but up to 3 pair) are relatively short and set posteriorly in a comparatively small mouth, envenomation may not occur with many bites. Although only 20% of persons bitten suffer systemic symptoms, half of those that do will die if untreated (Tu and Fulde 1987).

The venom is primarily a very potent neurotoxin and myotoxin, although hemotoxins, cytotoxins, cardiotoxins, and vasoactive compounds are also present (Brown and Shepherd 1992; Tu and Fulde 1987). When envenomation occurs, symptoms usually develop within several minutes to 6 to 8 hours depending on the amount of venom injected, and a true medical emergency exists (Brown and Shepherd 1992). Initial symptoms include malaise, euphoria, anxiety, myalgias, dysphonia, dysphagia, diplopia, ptosis, nausea, and vomiting. An ascending motor paralysis, eventually involving respiratory musculature is the usual cause of death. Necrosis of skeletal muscle leads to myoglobinuria within several hours, which can help with diagnosis and lead to acute renal failure in severe cases (Halstead 1988).

First aid treatment of sea snake bites is directed at immediate suction (without incision) of the wound. Use of a lymphatic/venous tourniquet is controversial. Rapid movement of the victim to a medical facility is mandatory as sea snake antivenom as well as supportive measures are often necessary and may be lifesaving (Tu and Fulde 1987; Patterson and Swallow 1991). As with all animal bite injuries, attention to tetanus prophylaxis protocols is indicated.

HAZARD PREVENTION

Fishing is an outdoor occupation and protection from carcinogenic wavelengths of light is a very important part of health maintenance in mariners. As sun damage is cumulative, use of proper sunscreens at an early age is recommended. A "waterproof" non-PABA sunscreen with a sun protection factor of at least 15 is recommended for fair-skinned fishermen. Physical sunscreens such as broad-brimmed hats and long-sleeved shirts are also useful in protection from solar radiation. Lip balms containing sunblocking agents are useful in protection of the lips, both from extremes of the environment as well as from excessive ultraviolet exposure, which can lead to actinic cheilitis and lip cancer.

Protective clothing such as rain gear, waterproof gloves, and rubber boots are important in protecting watermen during foul weather, and in some fisheries bib-coverall boots are helpful in protection. Practical protective clothing that is affordable, convenient to wear, and offers flotation lifesaving protection is not widely available in the United States. In Canada, regulations requiring safety standards for fishermen worksuits were expected to be in place by the late 1990s (NRC 1991).

Attention to maintenance of dry feet is important in prevention of immersion foot. Wearing of protective gloves can prevent many of the traumatic injuries in fishermen, many of which occur on the hand and upper extremity. Protective gloves can also prevent repeated wet-dry cycles, as well as contact with irritants or allergens, all of which may lead to recurrent hand eczema. A decrease in traumatic injuries by wearing protective gloves will also help to reduce the incidence of skin infections in watermen.

Protective clothing is also important in prevention of stings due to coelenterates and sponges as most nematocysts and spicules cannot penetrate even relatively thin products. Gloves and protective clothing will also protect from most stings of nudibranchs and bristleworms as well as from bites of marine annelids.

When wading, fishermen should "shuffle" their feet to prevent stepping on such sea creatures as stingrays, weeverfish, toadfish, scorpionfish, stonefish, and sea snakes. Thin rubber-soled footwear are not sufficiently protective against stings from many of these vertebrates.

Even when wearing heavy gloves while handling, some sea creatures are capable of causing traumatic penetrating and bite injuries, as well as envenomation syndromes. Education of watermen is important to prevent injury while handling some of these creatures. Tongs to help sort out potentially dangerous bycatch may be useful, but use of metal tongs can transmit an electric shock when handling electric fish; use of insulating materials/gloves is necessary in these instances.

SUMMARY

It is generally not the experienced or learned fisherman that is at high risk for skin problems related to the occupation of fishing. Rather it is the inexperienced "apprentice" mariner who is at high risk. While a regulatory approach toward mandatory safety requirements through such agencies as the U.S. Coast Guard, National Marine Fisheries Service, and the Occupational Safety and Health Administration may be necessary in some areas, education of watermen is likely to be a more reasonable and cost-effective approach in obtaining a reduction of skin problems in this occupational group. There is a notable paucity of research data related to the occupation of fishing, and funding of studies in this area is long overdue. Promotion of safety and exchange of research ideas through national/regional workshops as well as continuing education activities and training programs through local universities and community colleges are worthwhile pursuits. Dissemination of safety data and education of information to fishermen via mailings, newsletters, magazines, journals, manuals, and videotapes utilizing financial grant support from state and federal programs is likely to have the most immediate and long-term impact on occupational safety in watermen.

REFERENCES

Adnot, J., and C.W. Lewis. 1994. Immersion foot syndromes. In: *Military Dermatology. Textbook of Military Medicine Part III. Disease and the Environment*, W.D. James, ed. Washington DC.: The Surgeon General at TMM Publications, pp. 55-68.

Amos, W.H., and S.H. Amos. 1985. *Atlantic and Gulf Coasts*. New York: Alfred A Knopf Inc.

Anderson, D.L., and L. Renshaw. 1989. Jellies. *S. Car. Wildlife* 36: 32-37.

Anderson, D.M. 1994. Red tides. *Scientific American* 271: 62-68.

Asada, M., J. Komura, H. Hosokawa, T. Akaeda, and Y. Asada. 1990. A case of delayed hypersensitivity reaction following a sea urchin sting. *Dermatologica* 180: 99-101.

Audebert, C., and P. Lamoureau. 1978. Professional eczema of trawlermen by contact with Bryozoires in the "Baie de Seine" (First French cases 1975-1977). *Ann. Dermat. Venereol.* 105: 187-92.

Auerbach, P.S. 1991. Marine envenomations. *New Engl. J. Med.* 325: 486-93.

Auerbach, P.S. 1984. Hazardous marine animals. *Emerg. Med. Clin. N. Amer.* 2: 531-44.

Auerbach, P.S., and B. Halstead. 1982. Marine hazards: attacks and envenomations. *J. Emerg. Nursing* 8: 115-22.

Auerbach, P.S., B.W. Halstead. 1989. Hazardous aquatic life. In: *Management of Wilderness and Environmental Emergencies*, P.S. Auerbach and E.C. Gechr, eds. St Louis: C.V. Mosby Co., pp. 933-1028.

Auerbach, P.S. 1987. Natural microbiologic hazards of the aquatic environment. *Clinics in Dermatol.* 5: 52-61.

Baird, J.K., and D.J. Wear. 1987. Cercarial dermatitis: the swimmer's itch. *Clinics in Dermatol.* 5: 88-91.

Ballantine, T. 1983. *Tideland Treasure*. Hilton Head, SC: Deerfield Publishing.

Barash, L., F. Bavendam. 1993. Mass appeal. *Nat. Wildlife* 31: 14-19.

Barrett, J.H., S.A. Estes, J.A. Wirman, J.A. Morris, R.E. Morris, and J.L. Staneck. 1983. Erysipeloid. *J. Am. Acad. Dermatol.* 9: 116-23.

Barss, P.G. 1985. Penetrating injuries caused by needle-fish in Oceania. *Med. J. Austral.* 143: 617-22.

Barss, P.G. 1982. Injuries caused by garfish in Papua New guinea. *Brit. Med. J. Clin. Res.* 284(6309): 77-79.

Bavendam, F. 1995. Memories of your embrace. *Intl. Wildlife* 25: 31-37.

Bel, M. 1991. Fish stings and other marine envenomations. *W. Virg. Med. J.* 87: 301-03.

Best, W.C, and R.G. Sablan. 1964. Cymothoidism (sea louse dermatitis). *Arch. Dermat.* 90: 177-80.

Birkeland, C. 1989. The Faustian traits of the crown-of-thorns starfish. *Amer. Scientist* 77: 154-63.

Bonner, J.R, A.S. Coker, C.R. Berryman, and H.M. Pollock. 1983. Spectrum of *Vibrio* infections in a Gulf Coast community. *Ann. Intern. Med.* 99: 464-69.

Brown, C.K., and S.M. Shepherd. 1992. Marine trauma, envenomations, and intoxications. *Emerg. Med. Clin. N. Amer.* 10: 385-408.

Burke, W.A., J.R. Steinbaugh, and E.J. O'Keefe. 1986. Delayed hypersensitivity following a sea urchin sting. *Int. J. Dermat.l* 25: 649-50.

Burke, W.A., and B.E. Jones. 1987. Cutaneous infections of the coast. *N.C. Med. J.* 48: 421-24.

Burnett, J.W. 1992. Human injuries following jellyfish stings. *Md. Med. J.* 41: 509-13.

Burnett, J.W., H. Rubinstein, and G.J. Calton. 1983. First aid for jellyfish envenomation. *S. Med. J.* 76: 870-72.

Burnett, J.W. 1989. Crustaceans: crabs, crayfish, lobsters and shrimp. *Cutis* 43: 208.

Burnett, J.W., G.J. Calton, and R.J. Morgan. 1985. Catfish poisoning. *Cutis* 41: 208.

Burnett, J.W., and D.G. Cargo. 1979. Cutaneous irritation induced by crab larvae. *J. Am. Acad. Dermat.* 1: 42-43.

Burnett, J.W., G.J. Calton, and R.J. Morgan. 1987. Dermatitis due to stinging sponges. *Cutis* 39: 476.

Burnett, J.W., G.J. Calton, and R.J. Morgan. 1986. Venomous sea urchins. *Cutis* 38: 151.

Calder, D.R., and M.C. Pridgen. 1989. *Guide to Common Jellyfishes of South Carolina*. Charleston: South Carolina Sea Grant Consortium.

Camarasa, J.G., and E. Serra-Baldrich. 1993. Contact urticaria from a worm (*Nereis diversicolor*). *Contact Derm.* 28: 248-49.

Center for Disease Control (CDC). 1993. Fatalities associated with harvesting of sea urchins—Maine, 1993. *J. Am. Med. Assoc.* 271: 1477.

Clark, R.B., H. Spector, D.M. Friedman, K.J. Oldrati, C.L. Young, and S.C. Nelson. 1990. Osteomyelitis and synovitis produced by *Mycobacterium marinum* in a fisherman. *J. Clin. Microbiol.* 28: 2570-72.

Conand, C., and N.A. Sloan. 1989. World fisheries for echinoderms. In: *Marine Invertebrate Fisheries: Their Assessment and Management.* J. F. Caddy, ed.. New York: John Wiley and Sons, pp. 647-663.

Coombs, C.J., and K. L. Mutimer. 1993. Echinoidea tenosynovitis. *Austral. N. Zeal. J. Surg.* 63: 309-11.

Curtsinger, B. 1995. Close encounters with the gray reef shark. *Nat. Geogr.* 187: 45-67.

Czachor, J.S. 1992. Unusual aspects of bacterial water-borne illnesses. *Am. Fam. Phys.* 46: 797-804.

Davis, W.A, J.G. Kane, and V.F. Garagusi. 1978. Human *Aeromonas* infections: a review of the literature and a case report of endocarditis. *Medicine* 57: 267-77.

DeLoguna, J.B. 1988. Managing an international multispecies fishery: The Saharan trawl for cephalopods. In: *Marine Invertebrate Fisheries: Their Assessment and Management.* J. F. Caddy, ed. New York: John Wiley and Sons, pp. 591-612.

Doubilet, D. 1987. Scorpionfish: danger in disguise. *Nat. Geogr.* 172: 634-43.

Edelstein, H. 1994. *Mycobacterium marinum* skin infections: report of 31 cases and review of the literature. *Arch. Int. Med.* 154: 1359-64.

Edmonds, C. 1989. *Dangerous Marine Creatures.* Frenchs Forest, New South Wales: Reed Books Pty., Ltd.

Ell, S.R., and D. Yates. 1989. Marinefish stings. *Arch Emerg. Med.* 6: 59-62.

Endean, R. 1961. The venomous sea urchin *Toxopneustes pileolus.* *Med. J. Austral.* 1: 320.

Erickson, T., T.L. Vanden Hoek, A. Kuritza, and J.B. Leiken. 1992. The emergency management of moray eel bites. *Ann. Emerg. Med.* 21: 212-16.

Exton, D. R, P.J. Fenner, and J.A. Williamson. 1989. Cold packs: effective topical analgesia in the treatment of painful stings by *Physalia* and other jellyfish. *Med. J. Austral.* 151: 625-26.

Fisher, A.A. 1990. Contact urticaria due to occupational exposures. In: *Occupational Skin Disease*. R.M. Adams, ed. Philadelphia: W.B. Saunders Co., pp. 113-26.

Fisher, A.A. 1978. *Atlas of Aquatic Dermatology*. New York: Grune and Straton.

Food and Drug Administration (FDA). 1993. To prevent *Vibrio* infections, high-risk patients should avoid eating raw molluscan shellfish. *FDA Med. Bull.* (March); 23:6.

Garcia, P.J., M.H. Schein, and J.W. Burnett. 1994. Fulminant hepatic failure from a sea anemone sting. *Ann. Intern. Med.* 120: 665-66.

Grauer, F.H., and H.L. Arnold. 1961. Seaweed dermatitis. *Arch. Dermat.* 84: 62-74.

Halstead, B.W. 1980. *Dangerous Marine Animals*. Centerville, MD: Cornell Maritime Press.

Halstead, B.W., and J.M. Vinci. 1987. Venomous fish stings (Ichthyoacanthotoxicoses). *Clin. Dermatol* 5: 29-35.

Halstead, B.W. 1988. *Poisonous and Venomous Marine Animals of the World*. Princeton NJ: Darwin Press, Inc.

Hamner, W.M., and D. Doubilet. 1994. Australia's box jellyfish: a killer down under. *Nat. Geogr.* 186: 116-30.

Harth, M., E.D. Ralph, and R. Faraawi. 1994. Septic arthritis due to *Mycobacterium marinum*. *J. Rheumatol.* 21: 957-60.

Hartwick, R., V. Callanan, and J. Williamson. 1980. Disarming the box-jellyfish. *Med. J. Austral.* 1: 15-20.

Heckerling, P.S., T.M. Stine, J.C. Pottage, S. Levin, and A.A. Harris. 1983. *Aeromonas hydrophila* myonecrosis and gas gangrene in a nonimmunocompromised host. *Arch. Intern. Med.* 143: 2005-07.

Heston, T.B. 1987. Living history: the horseshoe crab. *Sea Frontiers* 33: 195-99.

Hjorth, N., and J. Roed-Peterson. 1976. Occupational protein dermatitis in food handlers. *Contact Dermatitis* 2: 28.

Hoffman, D.R., E.D. Day, Jr, and J.S. Miller. 1981. The major heat stable allergen of shrimp. *Ann. Allergy* 47: 17-22.

Hoffman, J., G.R. Hack, and B. Clark. 1992. The man did fine, but what about the wahoo? *JAMA* 267: 2039.

Howard, R.J., and N.T. Bennett. 1993. Infections caused by halophilic marine *Vibrio* bacteria. *Ann. Surg.* 217: 525-31.

Howard, R.J., and G.H. Burgess. 1993. Surgical hazards posed by marine and freshwater animals in Florida. *Am. J. Surg.* 166: 563-67.

Humann, P. 1992. *Reef Creature Identification*. Jacksonville, FL: New World Publications.

International Oceanic Foundation. 1993. Sea secrets. *Sea Frontiers* 39: 11.

International Oceanic Foundation. 1995. Sea secrets. *Sea Frontiers* 41: 22-24.

Jeanmougin, M., F. Lemarchand-Venencie, X.D. Hoang, J.L. D'Hondt, and J. Civatte. 1987. Occupational eczema with photosensitivity caused by contact with Bryozoa. *Ann. Dermatol. Venereol.* 114: 353-57.

Johnston, J.M., S.F. Becker, and L.M. McFarland. 1985. *Vibrio vulnificus*: man and the sea. *JAMA* 253: 2850-53.

Kavli, G., I.T. Gram, D. Moseng, and G. Orpen. 1985. Occupational dermatitis in shrimp peelers. *Contact Derm.* 13: 69-71.

Kaye, J.J. 1990. *Vibrio vulnificus* infections in the hand: report of three patients. *J. Bone Joint Surg.* 72-A: 283-85.

Kinmont, P.D.C. 1965. Sea urchin sarcoidal granuloma. *Brit. J. Dermatol.* 77: 335-43.

Kizer, K.W. 1990. When a stingray strikes. *The Physician and Sportsmedicine* 18: 93-109.

Klimley, A.P. 1994. The predatory behavior of the white shark. *Amer. Sci.* 82: 122-33.

Klontz, K.C. 1990. Fatalities associated with *Vibrio parahaemolyticus* and *Vibrio cholerae* non-01 infections in Florida (1981-1988). *S. Med. J.* 80: 500-02.

Kullavanijaya, P., S. Sirimachan, and P. Bhuddhavudhikrai. 1993. *Mycobacterium marinum* cutaneous infections acquired from occupations and hobbies. *Int. J. Dermatol.* 32: 504-07.

LaGorce, J.O. 1961. Devilfishing in the Gulf Stream. In: *The Book of Fishes*. J.O. LaGorce, ed. Washington, DC: National Geographic Society, pp. 189-96.

Lange, W.R. 1988. The perils of bluefish: handle with care! *Md. Med. J.* 37: 475-77.

Lehmann, D.F., J.C. Hardy. 1993. Stonefish envenomation. *N. Engl. J. Med.* 329: 510-11.

Lehrer, S.B. 1990. Hypersensitivity reactions in seafood workers. *Allergy Proc.* 11: 67-68.

Leroy, D., A. Dompmartin, P. Lauret, M.C. Boullie, and C. Audebert. 1988. Allergic contact dermatitis to Bryozoa and photosensitivity. *Photo-dermatology* 5: 227-29.

Levine, W.C., P.M. Griffin, and the Gulf Coast *Vibrio* Working Group. 1993. *Vibrio* infections on the Gulf Coast: results of first year of regional surveillance. *J. Infect. Dis.* 167: 479-83.

Limpert, G.H., and J.E. Peacock. 1988. Soft tissue infections due to non-cholera vibrios. *Am. Acad. Fam. Physicians* 37:193-98.

Lopes, C.M., E.M. Rabadão, C. Ventura, S. daCunha, R. Côrte-Real, and A.A. Melico-Silvestre. 1993. A case of *Vibrio alginolyticus* bacteremia and probable sphenoiditis following a dive in the sea. *Clin. Infect. Dis.* 17: 299-300.

MacLeish, K., and B. Cropp. 1972. Diving with sea snakes. *Nat. Geogr.* 141: 564-78.

Magerog, N., and A. Baerheim. 1991. Ratfish (*Chimaera monstrosa*) sting. *Tidsskrift for Den Norske Laegeforening* 111: 2102-03.

Mandojana, R.M., and J.K. Sims. 1987. Miscellaneous dermatoses associated with the aquatic environment. *Clin. Dermat.* 5: 134-45.

McCabe, M.J., W.M. Hammond, B.W. Halstead, and T.H. Newton. 1978. A fatal brain injury caused by a needlefish. *Neuroradiology* 15: 137-39.

McClane, A.J. 1978. *Field Guide to Saltwater Fishes of North America*. New York: Henry Holt and Co., Inc.

Miner, R.W. 1961. Sea creatures of our Atlantic shores. In: *The Book of Fishes*. J.O. La Gorce, ed. Washington, DC: National Geographic Society, pp. 285-291.

Morris, J.G., and R.E. Black. 1985. Cholera and other vibrioses in the United States. *New Engl. J. Med.* 312: 343-50.

Murata, M. 1988. Population assessment, management and fishery forecasting for the Japanese common squid, *Todarodes pacificus*. In: *Marine invertebrate fisheries: Their assessment and management*. J.F. Caddy, ed. New York: John Wiley and Sons, pp. 613-36.

Musher, D.M. 1981. "Cellulitis" with sepsis in a fisherman. *Hosp. Prac.* 16: 124-28.

National Research Council (U.S.) Marine Board. Committee on Fishing Vessel Safety. 1991. *Fishing Vessel Safety: Blueprint for a National Program*. Washington DC: National Academy Press.

Newell, G.C. 1984. Hemorrhagic bullae after a day's saltwater fishing. *Hosp. Prac.* 19: 236-37.

Newhouse, M.L. 1966. Dogger Bank itch among Lowestoff trawlermen. *Proc. Royal Soc. Med.* 59: 1119-20.

Nicolini, .S, C. Ascorra, C. Guzman, and A.V. Latife. 1989. Actinic cheilitis in Quinta fishing workers: prevalence and associated histopathologic aspects. *Odontologia Chilena* 37: 169-74.

North Carolina Department of Natural Resources and Community Development/North Carolina Department of Commerce. 1979. *Feasibility Study for Seafood Industrial Parks for the Central and Southern Coastal Areas of North Carolina*. Wilmington, NC: Henry von Oesen and Associates, Inc., pp. 5-1, 5-2 (unpublished).

Ohtaki, N., A. Satoh, H. Azuma, and T. Nakajima. 1986. Delayed flare-up reactions caused by jellyfish. *Dermatologica* 172: 98-103.

Onizuka, R, K. Inoue, and H. Kamiya. 1990. Red soft coral-induced allergic symptoms observed in spiny lobster fishermen. *Arerugi--Jap. J. Allerg.* 39: 339-47.

O'Neal, R.L, B.W. Halstead, and L.D. Howard, Jr. 1964. Injury to human tissues from sea urchin spines. *Cal. Med.* 101: 199-202.

Patterson, L., and S. Swallow. 1991. Sea snake envenomation. *Med. J. Austral.* 155: 850.

Pearce, J.B. 1990. Snail toxins. *Science* 250(4978): 191-92.

Perez-Tirse, J., J.F. Levine, and M. Mecca. 1993. *Vibrio damsela*: a cause of fulminant septicemia. *Arch. Intern. Med.* 153: 1838-40.

Raz, I., M. Katz, H. Aram, and H. Haas. 1984. Sporotrichoid *Mycobacterium marinum* infection. *Int. J. Dermat.* 23: 554-55.

Reed, K.M., B.R. Bronstein, and H.P. Baden. 1984. Delayed and persistent cutaneous reactions to coelenterates. *J. Am. Acad. Dermat.* 10: 462-65.

Rocha, G., and S. Fraga. 1962. Sea urchin granuloma of the skin. *Arch. Dermat.* 85: 146-48.

Ross, J.B. 1969. Rubber boot dermatitis in Newfoundland: a survey of 30 patients. *Canad. Med. Assoc. J.* 100: 13-19.

Rudloe, A., and J. Rudloe. 1993. Electric warfare: the fish that kill with thunderbolts. *Smithsonian* 24: 94-106.

Rudlow, J., and A. Rudlow. 1991. Jellyfish do more with less than anything. *Smithsonian* 21: 100-11.

Ruppert, E.E., and R.S. Fox. 1988. *Seashore Animals of the Southeast*. Columbia, SC: University of South Carolina Press.

Sanders, D. 1993. In the realm of a giant. *Canad. Geogr.* 113: 34-41.

Schwartz, P.A., and M.J. Smale. 1984. Treatment of bluebottle stings at the beach. *S. Afr. Med. J.* 66: 45-46.

Scrimgeour, E.M. 1994. Prawn trawling in the Gulf of Carpentaria, northern Australia. *Med. J. Austral.* 160: 253-55.

Shandera, W.X., J.M. Johnston, B.R. Davis, and P.A. Blake. 1983. Disease from infection with *Vibrio mimicus*, a newly recognized *Vibrio* species. *Ann. Intern. Med.* 99: 169-71.

Spitzer, W.O., G.B. Hill, L.W. Chambers, B.E. Helliwell, and H.B. Murphy. 1975. The occupation of fishing as a risk factor for cancer of the lip. *N. Engl. J. Med.* 293: 419-24.

Toback, A.C., R. Korson, and P.A. Krusinski. 1985. Pulling boat hands: a unique dermatosis from coastal New England. *J. Am. Acad. Dermat.* 12: 649-55.

Tomchick, R.S., M.T. Russell, A.M. Szmant, and N.A. Black. 1993. Clinical perspective on seabather's eruption, also known as "sea lice." *JAMA* 269: 1669-72.

Trestrail, J., and Q.M. Al-Mahasneh. 1989. Lionfish sting experiences of an inland poison center: a retrospective study of 23 cases. *Vet. Hum. Toxicol.* 31: 173-75.

Tu, A.T., and G. Fulde. 1987. Sea snake bites. *Clin. Dermat.* 5:118-26.

Turner, B., P. Sullivan, and J. Pennefather. 1980. Disarming the bluebottle. *Med. J. Austral.* 2: 394-95.

Tyring, S.K., and P.C. Lee. 1986. Hemorrhagic bullae associated with *Vibrio vulnificus* septicemia. *Arch. Dermat.* 122: 818-20.

Valero, A., J. Huguet, J. Sanosa, A. Malet, and P.A. Garcia-Calderon. 1989. Dermato-respiratory allergy induced by a marine worm (*Marphysa sanguinea*) used as fishing bait. *Ann. Allergy* 62: 514-17.

Vitasa, B.C., H.R. Taylor, P.T. Strickland, F.S. Rosenthal, S. West, H. Abbey, S.K. Ng, B. Munoz, and E.A. Emmett. 1990. Association of nonmelanoma skin cancer and actinic keratosis with cumulative solar ultraviolet exposure in Maryland watermen. *Cancer* 65: 2811-17.

Vukmir, R.B. 1992. *Aeromonas hydrophila*: myofascial necrosis and sepsis. *Intensive Care Med.* 18: 172-74.

Wagner, R.F., A.B. Tawil, A.J. Colletta, L.C. Hurst, and L.D. Yecies. 1981. *Mycobacterium marinum* tenosynovitis in a Long Island fisherman. *N.Y. State J. Med.* 81: 1091-94.

Wells, S.M. 1988. Impacts of the precious shell harvest and trade: conservation of rare or fragile resources. In: *Marine Invertebrate Fisheries: Their Assessment and Management*. J.F. Caddy, ed. New York: John Wiley & Sons, pp. 443-54.

Wexler, M. 1994. The art of growing giants. *Nat. Wildlife* 32: 20-27.

Wickboldt, L.G., and C.V. Sanders. 1983. *Vibrio vulnificus* infection: case report and update since 1970. *J. Am. Acad Dermat.* 9: 243-51.

Williamson, J.A.H. 1987. The blue-ringed octopus bite and envenomation syndrome. *Clin. Dermat.* 5: 127-33.

Wu, C.H. 1984. Electric fish and the discovery of animal electricity. *Amer. Sci.* 72: 598-607.

Zahl, P.A. 1961. Man-o-war fleet attacks Bimini. In *The Book of Fishes*. J.O. La Gorce. Washington DC: National Geographic Society, pp. 163-88.

Zeman, M.G. 1989. Catfish stings: a report of three cases. *Ann. Emerg. Med.* 18: 211-13.

LEGAL CONCERNS IN THE FISHING INDUSTRY

Dennis W. Nixon, J.D., M.M.A., A.B.
University of Rhode Island

> The occupational health of the commercial fisherman is inextricably bound to the health of the industry itself. The commercial fishing industry is one of the oldest in our nation's history. Vessel safety issues, downsizing, recreational fishing expansion and changes in insurance programs for commercial fishermen are a few of the many problems facing the fishing industry. The National Research Council's recommendation for a program of education, licensing of operators, and vessel inspection has been met with mixed results, ranging from resistence by the industry to any new regulatory intrusion to the lack of funds for the Coast Guard's implementation of new programs. The only prospect for improvement is in industry-supported initiatives where fishermen see the potential for long-term benefits.

INTRODUCTION

The United States fishing industry is in trouble. The deadly combination of over fishing, habitat loss, and marine pollution has created an operational crisis for much of the nation's fishing fleet. In August of 1995, the Clinton Administration announced that it was committing $25 million toward buying out New England fishermen. This purchase was the first significant sum of money offered to simultaneously relieve the pressure on the local stocks of cod, haddock, and flounder and relieve the vessel owners of their economic loss. The $25 million is part of a $53 million disaster assistance program initiated by the Commerce Department in response to declining stocks nationwide. Even if the program is successful in reducing the total fishing effort, scientists are unsure if or when stocks will recover to their historic levels.

The operational difficulties of the fishing industry have a direct impact on occupational safety and health issues faced by commercial fishermen. Historically, fishermen do not get hurt as often in profitable fisheries, and if they do, they return to work much more quickly. The reasons are simple. With fewer fish to catch, vessels must work longer and harder to earn the same income. Fatigue of both personnel and equipment becomes a major hazard. With a reduced cash flow, expensive maintenance and repairs are often postponed. An already dangerous working platform becomes even more hazardous. In earlier examples where declining fisheries made the future look less than promising, some crew members have decided to "cash out" and either stage accidents or exaggerate minor injuries.

This chapter will address the occupational hazards of the commercial fishing industry, the regulatory framework under which it operates, and the legal remedies available to crew members when they are injured. It will conclude with recommendations to improve both the operational safety of the industry and the methods by which injured or disabled fishermen are compensated.

OCCUPATIONAL HAZARDS

For those who have not been to sea, it is difficult to describe the dangers involved in the commercial fishing industry. Depending on a vessel's size and location, it may be at sea for as short as a day or as long as several months. Except for transit time out to the fishing grounds, vessels typically work around the clock, in all kinds of sea conditions, with heavy equipment powered by hydraulics. Decks are routinely washed by seawater, covered with a moving cargo of fish. Fish slime, leaking hydraulic fluid, and ice can increase the hazard of deck work. Fatigue sets in quickly; small crews are pushed to the breaking point. Since fishermen are paid on the basis of what they catch, vessels are now sailing with smaller complements of crew to increase the share for those who remain. The crew aboard large factory processing vessels off Alaska routinely work 16-hour days for months at a time.

Unlike most vessels, which are carefully loaded in port to enhance the vessel's stability, fishing vessels leave port with only fuel and ice and perhaps traps on deck. During the course of the trip, fuel and ice are consumed, but the vessel is loaded with as many fish as it can find and carry. If the fishery is depleted, a vessel's stability is usually adequate for a safe return to port. But if fishing is still plentiful, such as in the Northwest, a vessel may be loaded down to the point that stability becomes a serious problem if it encounters heavy seas on its return to port. According to the U.S. Coast Guard's CASMAIN data base, over 60% of the approximately 100 fatalities per year in the fishing industry are the result of a vessel's capsize and foundering.

In addition to the high fatality rate, it can be no surprise that injury rates are high as well. In the early to mid 1980s, the U.S. fishing industry was dramatically expanding to take advantage of the 200-mile fisheries zone established in 1977. By eliminating foreign competition on fisheries grounds close to home, the industry enjoyed several "boom" years and the rapid expansion meant many more inexperienced personnel going to sea on any vessel that could take them offshore. The result was a significant increase in both vessel and crew casualties. By 1984, it became very difficult for owners to obtain insurance; without insurance, the banks holding the mortgages would not let the vessels leave the dock. Congressional hearings were held, and by 1985 a crisis was declared. Several members of Congress urged the Department of Commerce, the parent agency of the National Marine Fisheries Service, to find the resources for an analysis of the problem and recommendations. That funding was made available through the National Council of Fishing Vessel Safety and Insurance and this author, with graduate student Fred Fairfield, was asked to conduct the study.

The results of the investigation were announced in late 1985 and presented at Senate and House Committee Hearings. Since there is no national data base for fishing vessel injuries, data had to be collected from the closed files of marine insurance companies from each region of the country. A nationwide, proportional, systematic, stratified sample of insurance claims files was collected, with the number of cases selected from each region based on that region's contribution to the total national catch value each year. The database provided information on the characteristics, cause, and duration of disabilities resulting from commercial fishermen's injuries documented in nearly 450 cases over the 5-year period from 1980 to 1984, as well as the insurance compensation paid to the victims and their attorneys.

Table 41.1. Relative frequency and cost of injuries.

Injury Site	Frequency		Costs	
	#	%	Total Value	%
Back	94	22.1%	$3,446,035	26.2%
Fingers	56	13.1%	$1,957,200	14.9%
Knee/Lower Leg	39	9.2%	$802,780	6.1%
Hand/Wrist	37	8.7%	$393,250	3.0%
Foot/Ankle	31	7.3%	$1,981,890	15.1%
Internal	22	5.2%	$715,850	5.4%
Elbow/Forearm	22	5.2%	$206,440	1.6%
Head	18	4.2%	$152,250	1.2%
Ribs	17	4.0%	$1,079,000	8.2%
Shoulder/Upper Arm	16	3.8%	$109,300	0.8%
Hip/Upper Leg	15	3.5%	$45,800	0.3%
Fatalities	13	3.1%	$769,800	5.8%
Face/Mouth	12	2.8%	$82,000	0.6%
Pelvis/Groin	11	2.6%	$667,800	5.1%
Eyes	8	1.9%	$71,500	0.5%
Neck	7	1.6%	$70,300	0.5%
Toes	4	0.9%	$158,900	1.2%
Paralysis	3	0.7%	$452,000	3.4%
Ears	1	0.2%	$4,700	0.0%
TOTAL			$13,166,795	

Injuries recorded within the database were characterized by the body parts affected, the nature of the injury, as well as the severity and duration of the resulting disability. Table 41.1 lists the injury sites showing frequency as a percent of the total cases sampled, and the percentage of the total settlement costs represented by each site.

In an effort to gain insight into systematic safety problems within the fishing fleet and identify the causes of the most common and costly injuries, those sites which constituted the top 50% of all cases sampled in frequency and settlement cost were reviewed in detail. Specific cause categories were identified as follows: *Gear handling* injuries involved the operation and manipulation of the vessel's harvesting equipment, as well as power plant components. *Slip and fall* injuries involved those due to loss of footing aboard the vessel or ashore. *Catch handling* injuries are those suffered while sorting, packing, or processing the catch. *Gear failure* covered those instances where equipment failure lead to injury.

Back injuries were the most common source of problems for commercial fishermen, followed by finger injuries, injuries to the knee and lower leg, hand and wrist injuries, and then foot and ankle problems. Together, these accounted for over 60% of all injuries, and 65% of all settlement costs. Table 41.2 summarizes the primary causes of the injuries described below.

Sprains, bruises, pinched nerves and pulled muscles provided 70% of back injuries, with another 29% involving fractures or surgery. Thirty-four percent of all finger injuries involved accidental amputation, 29% were fractured or crushed, with lacerations and infections accounting for another 25%. Most injuries to the knee and lower leg were dislocations and bruises (64%) with another 31% involving fractures, crushing incidents, or requiring surgery. Fractures and sprains or bruises accounted for 43 and 30% of hand and wrist injuries, respectively, with lacerations and infections resulting in another 23%. Fifty-nine percent of foot and ankle injuries involved crushing or fracture accidents primarily due to falling equipment, while 25% were dislocations or bruises and nearly 10% involved burns.

Permanent injuries, including all disabilities greater than one year in duration and fatalities, accounted for 15% of all injuries in the study. However, these accounted for 60% of the total settlement costs. These were primarily crush and fracture cases including knee and back injuries requiring surgery (41%), followed by major sprains and dislocations (15%), with head injuries and amputations accounting for 8% each. Fatalities were primarily due to drowning or hypothermia (62%), with crushing and fractures involved in another 15%. Gear handling and slip and fall accidents contributed almost equally to those injuries, at 27 and 26%, respectively. Gear failures were involved in only 10% of

Table 41.2. Leading causes of injuries.

Injury Site	Cause	%
Back	Sprains, pinched nerves, pulled muscles	70%
	Fractures, surgery	29%
Fingers	Accidental amputation	34%
	Fractured or crushed	29%
	Lacerations and infections	25%
	Other	12%
Knee/Lower Leg	Dislocations and bruises	64%
	Fractures, crushing accidents, surgery	31%
	Other	5%
Hand/Wrist	Fractures	40%
	Sprains or bruises	30%
	Lacerations and infections	23%
	Other	7%
Foot/Ankle	Crushing or fracture injuries	59%
	Dislocations and bruises	25%
	Burns	10%
	Other	6%

these cases. Boarding, man-overboard, and vessel losses combined for only 16% of all permanent injuries, but accounted for 58% of all fatal accidents. Gear handling, and "other" causes were each responsible for 17% of the fatal accidents with the remaining 8% due to slip and fall events.

A more recent claims analysis was conducted by the Point Club, a self-insurance group founded by commercial fishermen in Southern New England during the peak of the 1980s fishing vessel insurance "crisis." Proposed members of the group must be nominated by another member in good standing. Their vessels must pass rigid inspection criteria developed by the club safety committee, and all crew members must submit resumes of their fishing experience. Several members that failed to maintain the high standards expected were asked to leave the group. This unpublished study provides an interesting comparison to the earlier work because the population of vessels is much smaller (approximately 75) and represents one of the most safety-conscious fleets in the country. The study examined 114 claims over the period from 1990 to 1994. Perhaps because of the emphasis placed on preventing both lifting injuries and slip and falls, back injuries were only 12% of the total in number and 8% of the value. Surprisingly, there was an increase in the number and cost of hand injuries: 36% of the total in number and 37% in value. Almost all of the hand injuries can be attributed to gear handling accidents. One possible reason for the disparity is that in the national sample, the severity of the back injury was often in dispute, with the insurance company suspecting fraud. With a strong effort at self-policing, it is likely that the Point Club has not experienced the same level of fraudulent claims seen by the industry as a whole.

REGULATORY FRAMEWORK

Although the actual catching of fish is highly regulated by the various coastal states and the federal government, the occupational safety and health of the people who do the catching is far less regulated than their prey. The National Research Council estimated in 1991 that there were approximately 31,000 federally documented fishing vessels and some 80,000 state numbered vessels. Extrapolating from those numbers, they estimated that there are in excess of 230,000 people who consider themselves commercial fishermen. The majority of the fleet is composed of small vessels: 99% are less than 79 feet long, and 80% are less than 40 feet. These small boat fishermen treasure their independent life. In many cases, one of the compelling reasons to go to sea was to escape the "regulatory framework" that represents working ashore. As a result, any form of additional regulation is typically viewed with outright hostility. Compared to most workplaces ashore, the occupational safety and health of fishermen is viewed not as a regulatory issue, but one in which common sense and good seamanship should take precedence.

That said, there are a number of roles the federal government plays in vessel and crew safety. The Congress, through its oversight function, can hold hearings, amend laws, and see that important functions are funded. During the mid-1980s crisis discussed above, Congressional hearings provided an important focal point for a national effort to improve vessel safety on a voluntary basis. Although the Congress failed to change the seriously flawed way in which injured fishermen are compensated, it did pass the Commercial Fishing Industry Vessel Safety Act in 1986, which expanded requirements for safety and survival equipment.

The U.S. Coast Guard has the primary role in vessel safety and maritime law enforcement. However, virtually all fishing vessels are considered "uninspected vessels" and their basic design and construction are *not* regulated by the Coast Guard. Essentially, because of the vast inspection burden a national program would entail, we have historically relied upon the self-preservation instincts of those who go to sea (and their insurance companies) to ensure the safety of the fishing fleet. Unfortunately, that reliance does not take into account the very human characteristics of greed and stupidity which occasionally surface in the fleet. When the Coast Guard does board a vessel, its inspection for safety equipment is largely directed at finding the equipment used in a worst-case scenario: immersion suits, flotation devices, and life rafts. It is not directed at the safety of the crew during its normal fishing operations.

Many have argued that OSHA, the Occupational Safety and Health Administration, should step in to improve the safety of the crew's working environment. However, the Congress has seen fit to leave crew safety issues with the Coast Guard. OSHA has only been involved with the nonfishing industrial activities aboard the relatively few large fish processing vessels operating in the North Pacific.

The National Marine Fisheries Service restricts its activities to fisheries management and research. Until recently, several of their fishery management plans had a negative impact on vessel safety. In the Pacific halibut fishery, the entire season was only 4 days in duration. From the opening gun, fishermen worked around the clock until they collapsed from fatigue. Vessels were grossly overloaded to take advantage of the brief season, and a few sank each year on the way back to port. A quota system was recently established to spread the season over a longer period, and the problems associated with a brief 4-day season were eliminated.

LEGAL REMEDIES FOR INJURED FISHERMEN

Unlike land-based workers, commercial fishermen are not covered by state workmen's compensation statutes or the federal Longshoremen and Harbor Worker's Compensation Act. Instead, their compensation is found in the ancient and arcane field of admiralty law. The principal differences between state compensation statutes and admiralty law are the nature of the fault system and the amount that can be recovered. Workmen's compensation statutes operate on a no-fault basis; if injured on the job, the employer must pay for those costs and a percentage (usually a tax-exempt 66%) of the worker's average weekly income during the term of the disability. Although admiralty provides basic medical coverage and a small daily allowance under the no-fault provisions of the doctrine of maintenance and cure, the more important remedies of the Jones Act and the unseaworthiness doctrine both assume a fault-based system, with a trial in federal district court to determine liability and a wide range of damages beyond lost wages.

The system is confusing, time-consuming, and unnecessarily costly. As early as 1957, a federally funded study of the method for compensating the injured found that

> ...it disregards completely the financial, economic, and operational characteristics of the industry. Furthermore, the system in itself is unjust because it is wasteful and slow and it fosters misunderstanding and bitterness between employer and employees. Moreover, it encourages the use of dishonest methods by both parties because court awards often are not in proportion to the employee's injury or need (Danforth and Theodore 1957).

However, despite serious efforts in the 1970s and 1980s to simplify and improve the system, strong lobbying efforts from maritime injury attorneys prevented all reform and the liability-based system remains intact.

How does the system work? Three distinct remedies are involved, known collectively as the "blessed trinity" by personal injury lawyers: maintenance and cure, the Jones Act, and the unseaworthiness doctrine. A vessel's Protection and Indemnity (P&I) insurance policy is responsible for paying any award.

Maintenance and Cure

Medical care was provided for injured seamen over a century before land-based workers through the admiralty law doctrine of maintenance and cure. It is defined very simply as the legal obligation of the vessel owner to maintain and cure a seaman injured or taken ill while in the service of the vessel. The roots of the remedy are obscure, but it was apparently well-established in 1823 when the famed admiralty Justice Joseph Story described why the life of a seaman was in need of what was then a unique remedy in law:

> Seamen are by the peculiarity of their lives liable to sudden sickness from change of climate, exposure to perils, and exhausting labor. They are generally poor and friendless, and acquire habits of gross indulgence, carelessness, and improvidence (Harden v. Gordon 1823).

Justice Story went on to explain why this earliest form of workmen's compensation was important from a national policy perspective:

> Beyond this is the great public policy of preserving this important class of citizens for the commercial service and maritime defense of the nation. (And) it encourages seamen to engage in perilous voyages with more promptitude and at lower wages. (Harden v. Gordon 1823)

Thus it was not any great sense of benevolence that created this remedy; it was born from the practical necessity of obtaining and keeping a good crew at a reasonable wage.

Today, maintenance and cure is still the injured seaman's first line of attack. It is a no-fault remedy, which provides limited—but still important—benefits. The vessel owner must provide "cure" to the maximum extent practicable; i.e., if a seaman's leg is amputated, the vessel owner would be responsible for medical care until the seaman is fitted with a prosthesis, at which point his responsibility ends, even if the seaman is no longer able to return to sea. "Maintenance" is the daily stipend the injured seaman receives while recovering; today, that amount is typically $30 to $40 per day.

A threshold question, disputed for many years, was whether a fisherman, paid under a share system, should even be considered a seaman for the purpose of qualifying for maintenance and cure. Since many fishing trips were made in coastal waters and were only a few days in duration, many courts questioned the use of the remedy for fishermen. The question was considered in the cases of Vitco v. Joncich (1955) and Sterling v. New England Fish Co. (1976). Vitco and Sterling were both ships's cooks; Vitco suffered a series of heart attacks aboard a California tuna boat, while Sterling injured his knee in disembarking from a purse seiner in Alaska. Both men were found to be seamen under admiralty law, and today the issue is seldom disputed.

Although one might assume that the phrase "in the service of the vessel" would limit recovery to injuries received while actually aboard the vessel, the courts have taken a broader view and have extended the remedy to incidents ashore if they relate to a fisherman's duties. Thus, in the case of the Betsy Ross (Luksich v. Misetich 1944), a fisherman was injured in a warehouse while getting a net for his vessel. The California Industrial Accident Commission argued that it was a workmen's compensation case and that it had exclusive jurisdiction. The court found, however, that his actions were consistent with maritime employment and the fisherman was allowed to recover maintenance and cure.

If maintenance and cure is allowed, there is the often disputed question of when it should be terminated, particularly with chronic back and shoulder injuries. In Luksich v. Misetich (1944), the court considered the appropriate period for a fisherman suffering from a dislocated shoulder, and held that recovery should not be extended beyond the time when the maximum degree of improvement in his condition was reached. They found no authority for the proposition that a longer period was justified, or that a seaman permanently injured in the course of his employment should receive maintenance for life. In Baun v. Hudson (1952), a fisherman sought maintenance for the period of disability resulting from the aggravation of an old back injury while working aboard the defendant's fishing vessel. The court held that although the injury was apparently permanent, maintenance would be allowed only for the period of time after the voyage when an improvement in his condition may reasonably be expected from nursing, care, and medical treatment.

A fisherman is required to disclose to a prospective employer the existence of a disease which is likely to incapacitate him. However, if he is unaware of the disease, he will typically be allowed to recover. In Dragich v. Strika (1962), the plaintiff seeking maintenance and cure was a tuna fisherman who was unable to continue working because of the progressively debilitating effects of Parkinson's Disease. In a review of previous cases on the issue, the court found that relief should be granted unless it could be shown that the fisherman had knowingly or fraudulently concealed the illness from the vessel owner. In this case, he was unaware that he was suffering from the illness when he joined the vessel and was therefore entitled to receive maintenance and cure.

The Jones Act

Congress recognized the limitations of the maintenance and cure remedy when it passed the Jones Act of 1920. By that time, land-based workers enjoyed a much more comprehensive level of protection under state workmen's compensation laws. This statute provided that

> Any seaman who shall suffer personal injury in the course of his employment may, at his election, maintain an action for damages at law, with the right of trial by jury (46 U.S.C. 688).

To prevail in a Jones Act case, the injured fisherman must be able to prove the following: 1) the plaintiff was a seaman and a crew member of the vessel involved; 2) the defendant was the owner of the vessel at the time of the accident; 3) the plaintiff was injured while in the service of the vessel; 4) negligence on the part of the defendant was a proximate cause of the plaintiff's injuries; and 5) the extent of the plaintiff's damages. The cases applying the Jones Act can be broken down into two basic categories: first, where it is alleged that the defendant vessel owner failed to provide a safe place to work; and second, when it is alleged that the injury occurred as the result of the negligence of a fellow servant—the captain or another crew member.

Several cases will illustrate the types of conditions that courts have found unsafe. In Esta v. Persohn (1950), the plaintiff fisherman suffered a serious foot injury when one of the vessel's trawl doors fell on his foot. The court found that there was negligence on the part of the vessel owner for failing to equip the vessel with a lifting line to move the door safely. Winch accidents are a frequent cause of injury. In Justillian v. Versaggi (1954), a crew member aboard a shrimp boat lost several of his fingers while attempting to keep a cable from jumping off the winch drum. The court held that the owner of the boat was negligent in a) failing to provide the fisherman with a safe place to work, b) in providing a cable of such length or equipment of such character that the cable would "jump," and c) in that the captain, knowing that the fisherman would attempt to push the cable back, failed to instruct him as to the proper method of attempting such a task.

Negligence of fellow crew members in the operation of winches has also caused its share of injuries. In Hudgins v. Gregory (1955), the plaintiff suffered a hand injury when a fellow crew member operating the winch mistakenly took up cable after being signaled by the plaintiff to back off. The defendant vessel owner was held liable. In Nolan v. General Seafoods Corp. (1940), the mate selected a line for hoisting the net, which was too small in diameter. It parted under load, injuring the plaintiff fisherman. The court found that the mate was careless in selecting the wrong size line and found the vessel owner liable for the injuries caused.

The greatest number of Jones Act cases occurred between 1920 and 1960, when the Supreme Court gave new vitality to the doctrine of unseaworthiness and in effect used the power of "judicial legislation" to create a remedy that was more desirable than the Jones Act. The continuing vitality of the Jones Act remedy is based upon the fact that, by statute, the plaintiff is entitled to a trial by jury. Both maintenance and cure and unseaworthiness are considered remedies afforded by general maritime law, and thus are tried only before a judge without a jury. Since a jury is very likely to be more sympathetic in the awarding of damages than a judge, the injured seaman will virtually always allege negligence under the Jones Act in his complaint. Through a legal doctrine known as "pendent jurisdiction," the jury is then able to decide not only the Jones Act negligence issue, but also the maintenance and cure and unseaworthiness issues as well.

Unseaworthiness

The most important remedy for an injured seaman today is the doctrine of unseaworthiness, which enables an injured fisherman to recover against his vessel if his injury was caused by an unseaworthy condition of the vessel, its equipment, or crew. This is true whether or not the unseaworthy condition was caused by the negligence of the vessel owner, which is the standard required under the Jones Act.

The landmark case that established unseaworthiness as the prime remedy for injured fisherman was Mitchell v. Trawler Racer (1960). The plaintiff had been unloading fish at a pier in Boston. Upon completion, he stepped on the vessel's rail to reach the pier ladder, slipped on some "slimy substance," fell, and injured his back. His claim against the vessel was that it was unseaworthy for the purpose of disembarkation. The issue presented to the Supreme Court was whether or not this condition of so-called "transitory unseaworthiness" should be judged according to the standards of common law negligence. That standard would have required that the owner had some knowledge of the defect and that he failed to take corrective action. The court held, however, that liability for unseaworthiness is independent of negligence or notice, and granted the plaintiff recovery for his injuries.

The most common type of unseaworthiness case involves injury to a fisherman as the result of the failure of a piece of equipment aboard the vessel. In Texas Menhaden Co. v. Johnson (1964), the plaintiff was operating a winch and boom, tightening the purse line on a menhaden purse seiner. The boom buckled suddenly, and the fisherman's hand was drawn into the winch. The court held that the failure of a piece of equipment under proper and expected use is a sufficient predicate for a finding of unseaworthiness. In Solet v. M/V Capt. Dufrene (1969), a weld failed on a padeye when a net was being hoisted aboard. The cables, block, and shackle fell, striking Solet. The court held that the weld was defective and the vessel unseaworthy, since the failure occurred when the gear was being used for its intended purpose.

Failure to have certain equipment aboard was the unseaworthy condition in Stevens v. Seacoast Co. (1969). A young and inexperienced fisherman was injured aboard an oyster dredger when he accidentally grabbed a chain for support which was being used to pull in the dredge. His hand was pulled through a block and severely injured. Although the court stated that the new crew member should have been more carefully supervised, they did not find the vessel unseaworthy for that reason. The court based its unseaworthiness determination on the lack of a radio to call for assistance after the injury occurred and the lack of a first aid kit with sedatives to lessen the pain and suffering felt by the plaintiff.

Despite the relative ease of demonstrating an unseaworthy condition, occasionally courts do find that a vessel is "reasonably suited for her intended purpose" and deny recovery to an injured fisherman—especially if the injury is caused by his own recklessness. In Little v. Green (1970), plaintiff Little was acting as a rigman aboard the defendant's shrimp trawler. He was operating winches to bring in the two nets when the cable attached to one net overrode and Little tried to correct the problem by kicking it. His leg was caught in the cable, drawn into the winch, and seriously injured. Acting without the Captain's knowledge, Little had chained together the vertical levers that controlled the power to the two winches. The effect of this chain device was to keep the power to the winches locked in the "on" position without holding the levers. Consequently, when Little's foot was caught, the winch power could not be stopped by simply releasing his hold on the levers. The court held that his injury was attributable to his own contrivance, not the machinery or procedures of the vessel, and denied his unseaworthiness claim.

When either a Jones Act or unseaworthiness claim is successful, it typically brings an award much higher than a workmen's compensation system would provide. The reason is that a full range of legal "damages" are available, in addition to the wages lost for the term of disability. When issues like pain and suffering, loss of society, and even post traumatic stress disorder are factored in, the damage awards can be quite large. However, as much as 50% of those awards ultimately flow to the attorneys representing the injured seamen, which likely explains their strong opposition to any effort to simplify the process and bring fishermen into a workmen's compensation type system. Why would a fisherman favor such a change? Although court awards bring in big dollars, it still is a fault-based system in which the insurance companies can afford very good attorneys to defend against all claims. In the tough adversarial system that still prevails, many valid claims go un- or undercompensated.

SUMMARY

The commercial fishing industry is one of the oldest in our nation's history. Unfortunately, poor resource management decisions have crippled the industry's future growth. It seems likely that, like many other industries, commercial fishing will undergo significant "downsizing" in the years ahead.

Vessel safety issues have faded into the background as the very existence of commercial fishing is called into question in many states that place a higher value on recreational fishing. Florida, for example, recently banned the commercial use of nets in all state waters. Although the Point Club, the safety-conscious self-insurance program for commercial fishermen in New England, remains quite successful, the parent organization from which it spawned recently declared bankruptcy because of the decline in landings. The closure of Georges Bank in New England has cast a pall over the future of the region's fishing industry.

The occupational health of the commercial fisherman is inextricably bound to the health of the industry itself. Although reform is long overdue for the method in which we compensate injured fishermen, it is not likely to occur soon, with such pressing survival issues being debated around the country. There is no question that commercial fishing could be made a much safer industry; the National Research Council recently recommended a program of education, licensing of operators, and vessel inspection towards that goal. However, the struggling industry continues to resist any new regulatory intrusion. The Coast Guard has stated that it did does not have the funds to implement regulatory programs. The only prospect for improvement is in industry-supported initiatives where fishermen see the potential for a long-term future.

REFERENCES

Allen, D., and D.W. Nixon. 1986. *Study on the Use of Fishery Management Regulations and Techniques to Improve the Safety of Commercial Fishing Operations*, T.J. Lassen and K. Van Olst, eds. Washington, DC: National Council of Fishing Vessel Safety and Insurance, pp.58-86.

Danforth, W.C., and C.A. Theodore. 1957. *Hull Insurance and Protection and Indemnity Insurance of Commercial Fishing Vessels.* U.S. Fish and Wildlife Service, Special Scientific Report—Fisheries Nos. 241 and 241 Supplement. Washington, DC: Department of the Interior.

Davis, C.M. 1994. *Maritime Law Deskbook.* Seattle: Compass Publishing Co. National Research Council. 1991.

Fishing Vessel Safety: Blueprint for a National Program. Washington, DC: National Academy Press.

Nixon, D.W. 1984. *A Commercial Fisherman's Guide to Marine Insurance and Law. Marine Bulletin 50.* Kingston, RI: The University of Rhode Island.

Nixon, D.W. 1985. Statement of Dennis W. Nixon, Coordinator, Marine Affairs Program, University of Rhode Island, Kingston. Pp.46-60 in *Fishing Vessel Safety and Insurance*, S.Hrg. 99-268, U.S. Congress, Senate. Washington, DC: Government Printing Office.

Nixon, D.W. 1986. Recent developments in U.S. commercial fishing vessel safety, insurance, and law. *J. Maritime Law Comm.* 17(3): 359-87.

Nixon, D.W., and F.M. Fairfield. 1986. Fishermen's personal injuries: characterization, compensation, and solutions. Chapter 13 in *East Coast Fisheries Law and Policy*, Bubier and Rieser, eds. Portland, ME: Marine Law Institute.

Nixon, D.W., R. Moran and C. Philbrick, eds. 1987. *Self Insurance Programs for the Commercial Fishing Industry.* Workshop Summary Report for Alaska Sea Grant College Program, University of Alaska. Marine Advisory Bulletin No. 31.

Schoenbaum, T.J. 1994. *Admiralty and Maritime Law.* St. Paul, MN: West Publishing Co.

Cases Cited:

Baun *v.* Hudson, 108 F.Supp. 523 (D. Alaska 1952).
Dragich *v.* Strika, 309 F.2d 161 (5th Cir. 1962).
Esta *v.* Persohn, 44 S.2d 202 (C.A. La. 1950).
Harden *v.* Gordon, 11 F.Cas. 480, 482 (No. 6047) (C.C.D. Me. 1823).
Hudgins *v.* Gregory, (219 F.2d 255 (4th Cir. 1955).
Justillian *v.* Versaggi, 169 F.Supp. 71 (S.D. Tex. 1954).
Little *v.* Green, 428 F.2d 1061 (5th Cir. 1970).
Luksich *v.* Misetich, 140 F.2d 812 (9th Cir. 1944).
Mitchell *v.* Trawler Racer, 362 U.S. 329, 80 S.Ct. 926 (1960).
Nolan *v.* General Seafoods Corp., 112 F.2d 515 (1st Cir. 1940).
Solet *v.* M/V Capt. Dufrene, 303 F.Supp. 980 (E.D. La. 1969).
Sterling *v.* New England Fish Co., 410 F.Supp. 164 (W.D. Wash. 1976).
Stevens *v.* Seacoast Co., 414 F.2d 1031 (5th Cir. 1969).
Texas Menhaden Co. *v.* Johnson, 332 F.2d 527 (5th Cir. 1964).
Vitco *v.* Joncich, 130 F.Supp. 945 (S.C. Cal. 1955).

42

FEDERAL REGULATION OF FISHING VESSELS

William J. Meggs, M.D., Ph.D., F.A.C.E.P.
East Carolina University School of Medicine

> The United States government regulates safety in the commercial fishing industry by safety standards, certification of fishing vessels, licensing requirements for fishing vessels, and inspections of vessels to insure compliance. The United States Coast Guard is the federal agency empowered to enforce regulations of fishing vessels. Requirements for commercial fishing industry vessels include maintaining life saving equipment and emergency position indicating radio beacons. Stricter requirements apply to larger vessels and those with more than 16 people on board. Federal statue requires the compilation of statistics concerning marine casualties to be collected from insurers of fishing vessels.

INTRODUCTION

The United States government regulates safety in the commercial fishing industry by safety standards, certification of fishing vessels, licensing requirements for fishing vessels, and inspections of vessels to insure compliance. The Commercial Fishing Industry Vessel Safety Act of 1988 (public law 100-424, September 9, 1988) and the Aleutian Trade Act of 1990 provides legal authority for regulation. The United States Coast Guard is the federal agency empowered to enforce regulations of fishing vessels.

REQUIREMENTS FOR COMMERCIAL FISHING INDUSTRY VESSELS

Life Saving Equipment

Each individual on a commercial fishing vessel must be provided with at least one immersion suit, exposure suit, or wearable personal flotation device that is stowed so that it is readily accessible by that individual, in proximity to the individual's work station and sleeping quarters. If the work station and sleeping quarters are far apart, two such devices must be provided.

A ring life buoy that is throwable is required on all commercial fishing vessels. At least one ring life buoy must be equipped with a line. The length of the required line depends on the vessel length, with vessels less than 65 feet requiring a 60-foot line, and vessels greater than 65 feet requiring a 90-foot line. Standards and types of throwable flotation devices required depends on vessel length and can be found in the federal regulations.

Each vessel must have a sufficient number of survival craft to accommodate the total number of individuals on board the vessel. Standards for survival craft depend on the number of individuals on board, the maximum distance that the vessel will operate from the coast line, water temperature, and whether the vessel operates on open seas, lakes, bays, or sounds. Regulations cover the stowage of life craft and required survival craft equipment. Markings are required on the survival craft and the wearable personal flotation devices.

The code of regulations requires that survival equipment be maintained in a state of operational readiness. The code further provides standards for maintenance and inspection. Escape routes must be maintained without obstruction. Each vessel must be equipped with distress signals, with different requirements for vessels operating on coastal waters and the ocean, depending on the distance from the coastline. Required equipment includes parachute flares, hand held flares, smoke signals, electric distress lights, and day visual distress signals such as distress flags.

Emergency Position Indicating Radio Beacon (EPIRB)

An EPIRB is a small battery-powered emergency radio transmitting device that can be activated manually or automatically in cases of emergency. These devices transmit signals that alert the coast guard of distress, and allow the distressed vessel to be located. All commercial vessels operating on the high seas, defined as greater than 3 miles from shore, are required to carry EPIRBs. The first emergency locator transmitting devices, which transmitted signals at 121.5 MHz (civilian) and 243 MHz (military), were designed for aircraft in the early 1970s. These signals were detected by aircraft and satellites. The EPIRB was an adaptation of this technology to ships and boats. A limitation of the early system was the dependence on overhead aircraft or satellites.

The COSPAS/SARSAT satellite system was developed to provide continuous monitoring for emergencies. These satellites detect and relay EPIRB signals to a ground system consisting of 30 stations called Local User Terminals (LUTs). LUTs then notify local control centers, which then notify local rescue units. This system had a 95% rate of false alarms, and a high rate of lost signals. Hence the 406 EPIRB system was developed, named for its operating frequency of 406 MHz.

The 406 system has greater accuracy than the 121.5 MHz system, with a 1 to 3 mile accuracy versus a 5 to 10 mile accuracy for older systems. There is worldwide coverage with reduced response times. Each 406 MHz EPRIB broadcasts a unique identification signal, allowing rapid identification. False alarms are reduced because the 406 MHz frequency is dedicated to distress calls. Honing signals and strobe lights add to the effectiveness of this system. The 406 system is being phased in on all U.S. vessels required to carry an EPIRB. The types of EPIRBs in use are given in Table 42.1, and Table 42.2 gives important safety information to prevent EPIRB malfunction in an emergency.

EPIRBs must be registered with the National Oceanic and Atmospheric Administration (NOAA). Information on the vessel type, communications equipment, radio call sign, registration number, home port, and berthing areas are kept on file. The vessel owner and alternative contacts in case of disaster are kept on file. Registration is free of charge through NOAA, at:

NOAA/NEDSDIS
SARSAT Operations Division, E/'SP3
Federal Office Building 4
Washington, DC 20233
301-763-4680

The fine for gross negligence or false activation secondary to hoax is up to $10,000. Failure to register can have dire consequences in case of disaster at sea, but only 70% of 406 EPIRBs are registered.

Table 42. 1. Types of EPIRBs.

Category 1
406.025 MHz
dedicated satellite distress frequency
monitored by COSPAS/SARSAT (4 polar orbiting satellites)
homing signal (121.5 MHZ) and strobe light
automatic free float and manual operation
phased in on all U.S. commercial vessels operating >3 miles from shore

Category 2
same as Category 1 except no free float operation

Class A
121.5 and 243 MHZ
monitored by overhead aircraft (200 mile range)
COSPAR/SARSAT detection if in range
automatic free float and manual operation
being phased out in favor of category 1
required since 1975 on inspected vessels with exceptions

Class B
same as Class A except no free float operation
being phased out in favor of category 1 and 2

Class C
VHF-FM frequencies
alerting signal on VHF channel 16
homing signal on VHF channel 15
manual activated only
used in Great Lakes and coastal areas
banned after February 1, 1999
being phased out in favor of category 1 and 2

Class S
same as Class B but for survival craft use only
voluntary

Table 42.2. Precautions for using 406 MHZ EPIRBs.

Registration with NOAA is necessary
Must be correctly mounted
Free floating devices must be free of obstructions
Regular servicing of battery and hydrostatic releases
Monthly testing
Maintain device in "armed" position, never in "off" position

Fishing vessels are required to post an injury placard in a highly visible location accessible to the crew, measuring at least 5 inches by 7 inches (127 mm by 178 mm). The required wording of the injury placard is given in Table 42.3.

Table 42.3. Wording of the injury placard required on commercial fishing vessels.

Notice
Report All Injuries

United States law, 46 United States Code 10603, requires each seaman on a fishing vessel, fish processing vessel, or fish tender vessel to notify the master or individual in charge of the vessel or other agent or the employer regarding any illness, disability, or injury suffered by the seaman when in service to the vessel not later than 7 days after the date on which the illness, disability, or injury arose.

Adherence to additional regulations is required of vessels that have larger numbers of individuals on board or carry certain chemicals. Requirements for these larger vessels include maintaining fireman's outfits and self-contained breathing apparatuses. Specific first aid equipment and training is required. Exposed hazards must be equipped with guards. Specific navigational equipment is required, including marine charts of areas of operation, tide and current tables, and other publications. Operable magnetic steering compasses, anchors and radar reflectors, and general alarm systems are required. Communication equipment, high water alarms, bilge pumps, electronic position fixing devices, emergency instructions, emergency drills, and safety orientation are required and must meet certain standards. Specific requirements for any vessel can be determined from the Code of Federal Regulations.

After September 15, 1991, vessels with more than 16 individuals on board were subjected to stricter requirements. These requirements include fire detection systems, galley hood and other fire protection equipment, regulations applicable to fuel systems and ventilation of enclosed engine and fuel tank spaces, and regulations concerning electrical systems. An emergency power supply must be provided to operate navigation lights, steering systems, bilge pumps, fire protection and detection systems, communication systems, general alarm system, and emergency lighting. Further restrictions are placed on vessels with more than 49 individuals on board, including means of escape, embarkation stations, radar and depth sounding devices, hydraulic equipment, deck rails, lifelines, storm rails, and hand grabs. There are stability requirements, requirements of watertight and weather tight integrity, icing, and freeing ports that must be met by vessels greater than 79 feet in length.

Casualty Statistics

The Commercial Fishing Vessel Safety Act of 1988 requires the compilation of statistics concerning marine casualties. This data is to be compiled from insurers of fishing vessels, fish processing vessels, and fish tender vessels. Insurers are required to submit periodic reports of marine casualties.

SUMMARY

The regulations for fishery safety are not applied uniformly. Vessels may be exempted from some requirement based on vessel size, crew size, and the area of operation. There are voluntary standards for uninspected fishing vessels that cover vessel design, construction, and equipment (U.S. Coast Guard 1986). These standards were first published in August of 1986 and distributed to trade and fishermen associations. Safety and survival equipment, overview on selected equipment performance, and guidelines on maintenance are provided.

This chapter is based on the U. S. Coast Guard publications cited in the references. These documents should be consulted for definitive information. In addition, the Coast Guard has district fishing vessel safety coordinators that can be contacted for further information on fishing vessel safety and regulations. Addresses and telephone numbers of the safety coordinators are given below.

Coast Guard District Fishing Vessel Safety Coordinators

Commander (mvs)
First Coast Guard District
408 Atlantic Avenue
Boston, MA 02210-2209
617-223-8444

Commander (mvs)
Second Coast Guard District
1222 Spruce Street
St. Louis, MO 63103-2832
314-539-2655

Commander (mvs)
Fifth Coast Guard District
431 Crawford Street
Portsmouth, VA 23704-5004
804-398-6414

Commander (mvs)
Seventh Coast Guard District
909 S. E. First Avenue
Miami, FL 33131-3050
305-536-5651

Commander (mvs)
Eighth Coast Guard District
501 Magazine Street
New Orleans, LA 70130-3396
504-589-6271

Commander (mvs)
Ninth Coast Guard District
1240 East 9th Street
Cleveland, OH 44199-2060
216-522-3994

Commander (mvs)
Eleventh Coast Guard District
400 Oceangate
Long Beach, CA 90822-5399
213-499-5330

Commander (mvs)
Thirteenth Coast Guard District
915 Second Avenue
Seattle, WA 98174-1067
206-553-1711

Commander (mvs)
Fourteenth Coast Guard District
300 Ala Moana Boulevard
Honolulu, HI 98650-4982
808-541-2114

Commander (mvs)
Seventeenth Coast Guard District
P.O. Box 25517
Juneau, AK 99802-5517
907-463-2212

REFERENCES

Office of the Federal Register. National Archives and Records Administration. 1994. Requirements for commercial fishing industry vessels. *CFR* Title 46, Part 28, pp. 247-84.

U.S. Coast Guard Fishing Vessel and Offshore Safety Branch. U.S. Coast Guard. Department of Transportation. EPIRBs.

U.S. Coast Guard. 1986. *Voluntary Standards for U. S. Uninspected Commercial Fishing Vessels*. NVIC 5-86.

U.S. Coast Guard. 1991. *Federal Requirements for Commercial Fishing Industry Vessels*.

U.S. Congress. 1988. *Commercial Fishing Industry Vessel Safety Act of 1988*. Public Law 100-424. 100th Congress.

APPENDICES

Appendix A

AGRICULTURAL SAFETY CHECKLIST

AGRICULTURAL CHEMICAL STORAGE

- Are passageways clean and clear of slippery substances?
- Is the building sound, weather tight, and well-ventilated?
- Is the storage place used for chemicals only and never for human or animal food or medicine?
- Are chemicals stored in original containers and not in food or drink containers?
- Are emergency phone numbers available?
- Are multipurpose fire extinguishers available and checked regularly?
- Do you have proper respirators, goggles, gloves, and clothing for mixing?
- Is there a means of secondary containment for leftover chemical solutions?
- Is equipment available for containment of a spill?
- Are all chemicals stored in a room or building that can be locked to keep out children and livestock?
- Is the entrance to the chemical storage area posted to warn others of the hazards inside and to inform firemen of the contents in the event of a fire?
- Is the chemical mixing area located outside or in an open, well-ventilated part of the building?
- Does chemical mixing occur in the same place always, or in the field?
- Is proper first aid equipment available, including a large amount of clean water?
- Are all empty chemical containers disposed of properly?
- Is the storage area reserved for chemicals only?
- Are **"No Smoking"** signs displayed in and around buildings where chemicals are stored?

ANHYDROUS AMMONIA SAFETY

Supply/Nurse tank Checklist

- Are the fittings and valves clean and free from rust and wear?
- Is the kingpin in good condition and well lubricated?
- Is the high-pressure hose secure, with no cracks or signs of wear? Replace the hose if you can see the braided layer.
- Is the hose free of cuts, soft spots, blistering, kinking, flattening, or indications that it may have been stretched? Is there slippage at any coupling connector? Check for leaks, kinks, and bulges, especially near the couplings.
- Is the tank free of rust? Is the paint in good condition?
- Does the tank have a slow-moving vehicle (S.M.V.) emblem mounted on the rear of the tank that is in good condition?
- Are the tires in good condition and properly inflated?
- Are both ends of the hose secured to prevent damage to the hose and connections during transport?
- Is the emergency water supply full? Is the grabhose clean and free of any slimy internal buildup that would impede or contaminate water flow?

Transport Checklist

Before you tow the tank home, make sure

- Towing connections are in good condition, correctly adjusted and firmly secured.
- Hitches are secure and equipped with locking pins.
- Safety chains are securely connected.

Application Equipment Checklist

- Are applicator knives in good condition and not plugged with soil or other material?
- Is the shut-off rope the right length and in good condition?
- Are you using locking hitch pins?
- Is the nurse tank secured with a safety chain?
- Is all low-pressure tubing securely clamped and free of pinches, nicks, weak spots and leaks? *Folding and unfolding the applicator wings can pinch hoses so always inspect them before using. Are hoses clear and flexible? Hoses that are weathered, cracked, discolored or brittle need to be replaced. Check all hoses, connections and tires daily.*

- Is the regulator working properly?
- Is the applicator tool bar equipped with a breakaway coupler that is in good physical condition?
- Is the emergency 5-gallon water tank filled with fresh water? *Always empty and refill the water tank at least once a day to ensure an adequate supply of fresh water.*
- Do you have a second 5-gallon water tank filled with fresh water on the tractor? *A ruptured hose and the failure of a safeguard may prevent you from reaching the water tank on the nurse tank. A 5-gallon water tank on the tractor may be your only available source of water.*
- Do you have a small squeeze bottle of fresh water in your pocket that will be readily accessible if ammonia gets in your eyes?

Safety Tips

- Always use gloves and goggles approved for handling anhydrous ammonia when transferring anhydrous ammonia or when checking for worn hoses and plugged applicator knives.
- An approved respirator and goggles should be available in the truck or tractor bringing the nurse/supply tank to the site and on the tractor used to apply ammonia. If wind conditions are right, a leak could result in a cloud of ammonia that may cover the equipment—the only safe way out of this cloud is by wearing a half-face respirator and goggles or a full-face respirator. Both respirators and goggles must be approved for anhydrous ammonia applications.
- The first treatment for any exposure to anhydrous ammonia exposure is to flush the area with clean water for at least 15 minutes. Always seek medical attention after any exposure to anhydrous ammonia.

Labels, Marking & Safety Signs: Legal Requirements

- Nurse tanks must be labeled "Anhydrous Ammonia" in 4-inch letters on contrasting background, on the sides and rear of the tank.
- Federal DOT regulations require the words "Inhalation Hazard" in association with the anhydrous ammonia label, in 3-inch lettering be placed on both sides of the tank.
- A "Nonflammable Gas" placard with the numbers **1005** (identifying it as anhydrous ammonia) must be located on both sides and both ends of the tank.
- A "Slow Moving Vehicle" (S.M.V.) sign must be prominently displayed on the rear of the tank with the bottom of the sign at least 2 and not more than 6 feet from the ground.
- The valves must be appropriately labeled by color or legend as vapor (Safety Yellow) or liquid (Omaha Orange).
- The letters of the legend must be at least 2 inches high on contrasting background and within 12 inches of the valves.

Additional ASAE safety standards require the following:

- An operator's manual. It is extremely important that this manual be with the equipment and available for reference at all times.
- A safety type hitch pin with a standard safety chain attachment for the nurse tank wagons or running gear.
- In addition to the mandatory S.M.V. sign, appropriate lighting for travel on roadways must include at least one red tail lamp and two amber flashing warning lamps. These may be on the towing vehicle or the tank wagon, but they must be visible from the rear. Additionally, the tank wagon must have at least two red reflectors visible to the rear.
- Turn signals, flashing warning lights, and a red brake light are recommended on all anhydrous ammonia tank wagons that are towed on public roadways. To accommodate these lights, a standard 7-terminal breakaway connector plug should be used on the tank wagon.
- At least one safety sign should be located in between the control valves and the 5-gallon water supply.

BUILDINGS

- Are tools and supplies stored neatly?
- Do doors open easily and smoothly?
- Can doors be locked?
- Do buildings have adequate lighting protection?
- Is electrical service ample to handle power tools, heaters, welder, etc.?
- Are keys removed from equipment to prevent children or intruders from starting them?
- Is machinery parked so there is ample walking space between units?
- Is there proper support for detached equipment and mower blades?
- Is proper ventilation present before starting or running an engine inside?
- Are buildings free of unnecessary accumulation of trash, litter, junk, and other debris that could fuel a fire, cause falls, or get in the way?

- Are buildings well lighted?
- Are above-ground fuel storage tanks at least 40 feet from any building?
- Are above-ground fuel storage tanks child-proofed?
- Is the area near fuel storage tanks free of weeds and other easily combustible material?
- Are "NO SMOKING" signs displayed near fuel storage or refueling areas?
- Does each major farm building have a fully charged ABC-type fire extinguisher readily available?
- Does each major farm area have a well-maintained first-aid kit for use in the event of personal injury?
- Are important emergency phone numbers clearly written and posted near each telephone?
- Does all electrical wiring insulation appear to be in good condition?
- Are the floors of the buildings free of badly broken and raised concrete, slippery spots, small curbs, or other spots that could cause falls?

ELECTRICITY

- Inspect every electrical power tool and power cord before each use.
- Never lift electrical tools or equipment by the power cord.
- Do not use aluminum or metal ladders for electrical work.
- Each employee must know the location of electrical power disconnections.
- Determine the location of buried power lines before digging.
- Only authorized and qualified personnel are allowed to make electrical connections or repair electrical equipment or wiring.
- Extension cords used with portable electric tools and equipment are of the 3-wire type.
- All exposed wiring and cords with frayed insulation should be promptly repaired.
- Low clearance overhead power lines should be clearly posted.
- Are electrical outlets shielded from water?
- Are portable power tools either double insulated, the 3-wire grounded type, or equipped with circuit interruption protection?
- Are all power tools UL-approved?
- Are all stationary power tools properly grounded?
- Are all electrically operated tools grounded?
- Are switch boxes locked out to keep children from starting machinery?

GRAIN AND SILAGE STORAGE STRUCTURES

- Are entrances to grain and silage storage areas closed to keep out children?
- Are drivelines and working parts of filling and unloading machinery properly shielded?
- Is hearing protection worn?
- Do grain bins have both outside and inside permanent ladders?
- Are workers warned to stay out of bins that could be crusted or where grain flow could start?
- Can power be locked out with an identification tag so the unloading mechanism cannot be started when someone is in the bin?
- Are workers alerted to the dangers of silo gas and oxygen deficiency?
- Are structures properly ventilated before entry?
- Are employees aware of the need to use lifelines when entering storage facilities?
- Is the fuel supply protected from animals or outside abuse that could cause leaks?
- Is electrical equipment protected from animals or outside abuse that could cause faults?
- Do structures have adequate lightning protection?

HAZARDOUS COMMUNICATION CHECKLIST

- List all of the hazardous chemicals or products in the workplace.
- Establish a file for information on hazardous chemicals or products.
- Obtain an MSDS for each hazardous chemical or product in use or stored on site.
- Develop a system to ensure the labeling of all incoming hazardous chemicals or products.
- Review each MSDS for completeness.
- Ensure that MSDSs are readily available.
- Write a Hazard Communication Plan.
- Develop a method for communicating hazards to employees and to others.
- Inform employees of protective measures for handling hazardous chemicals or products used in the workplace.
- Alert employees to other appropriate forms of warning.

LOCKOUT/TAGOUT CHECKLIST

- Audit all machinery and equipment for types and magnitudes of energy and potential hazards.

- Identify and document all machinery and equipment for which a lockout/tagout procedure must be developed.
- Identify and document, by name and by job title, all affected employees.
- Identify and document, by name and by job title, all employees authorized to perform lockout/tagout procedures.
- Describe and document types and locations of energy-isolating devices for all machinery and equipment identified above as needing a lockout/tagout procedure.
- Describe and document the types of energy involved and the methods to be used to dissipate or restrain the energy for all machinery and equipment identified as needing a lockout/tagout procedure.
- Describe and document the method established to isolate the energy (lock or tag) and any additional safety measures to be taken.
- Develop a written lockout/tagout program.

LIVESTOCK

- Understand the livestock you are working with. (Newly purchased livestock are more of a hazard than settled-in livestock.)
- Avoid loud noises and quick movements around animals.
- Never prod an animal when it has no place to go.
- Move slowly and deliberately around animals.
- Touching animals gently can be more effective than shoving or bumping them.
- Special facilities should be provided for breeding stock.
- Be especially careful around newborn animals as parents are very protective.
- Male animals should be considered potentially dangerous at all times.
- Always have an escape route.
- Exercise extra care around strange animals.
- Rope livestock only as a last resort.
- If you open a gate, you close it.
- Promptly remove all dead livestock.
- Promptly request veterinary care for sick or injured livestock.
- Use extra care around animal vaccines and medications.
- Properly dispose of medical devices and equipment used on livestock.
- Use mechanical sorting equipment whenever possible.
- Remove all wire and other obstacles from feeding areas.
- Are children forbidden to play in or around barns, pens, and feedlots?
- Do locks operate?
- Are walkways roughened to prevent slips and falls?
- Are pens and loading chutes adequate for the job, sturdy, and well repaired?
- Are gates and lot fences adequate for the job, sturdy, and well repaired?
- Are feeding, grinding, and other equipment properly shielded and all shields in place?
- Are loading troughs on augers, elevators, and conveyors covered with grating?
- Are vents clear and fans operative in confinement housing?
- Are there at least two exits from each area used for working livestock?
- Are walking surfaces in work areas and passageways clean and free of manure?
- Are ramps and steps protected from rain or spilled water that could freeze?
- Are stairs and permanent ladders in good condition?
- Are portable heaters in good condition, vented, and placed so they can't tip into combustible materials?
- Are heaters checked often for defects?
- Is the manure pit or tank thoroughly ventilated before anyone enters for maintenance?
- Are safety harnesses used when entering pits and tanks?
- Are lagoons fenced and posted with warning signs?
- Are animals properly restrained?
- Are buildings adequately lighted?
- Is there adequate lighting protection?
- Are the correct size and type fire extinguishers available and checked regularly?
- Are emergency phone numbers on hand and up-to-date?

MACHINERY

- Are all warning labels and SMV signs in place?
- Are keys removed from equipment to prevent children or intruders from starting them?
- Are machinery guards in place?
- Are cotter pins, keys, nuts and bolts secured?
- Are exhaust systems properly maintained on machinery?
- Is proper ventilation present before starting or running an engine inside?
- Are hydraulics in good condition and leak free?
- Do machinery operators avoid wearing torn or ragged clothing?

- Are wheels set in the widest practical position to increase stability?
- Do lights on machinery function properly, and is lighting adequate for night operation?
- Is the power always turned off before adjusting, servicing, or unclogging machinery?
- Is there a program of regular safety inspection for all machinery and equipment?
- All power can be locked out for maintenance and repair.
- All machinery must be locked out while repairs are being made.
- Proper safety supports must always be in place when working under machinery.
- Always check behind the vehicle before backing up.
- Stop engine, disconnect the power source, and wait for all machine movement to stop before servicing, adjusting, cleaning, or unclogging the equipment, except where the machine must be running to be properly serviced or maintained, in which case the employer shall instruct employees as to all steps and procedures that are necessary to safely service or maintain the equipment.
- Vehicle safety (seat) belts must be utilized.
- All equipment left unattended at night must have appropriate reflectors.
- A safety tire cage must be used when inflating tires.
- Do not get between towed equipment and the towing machine until both have completely stopped.
- Make sure everyone is clear of machinery before starting the engine, engaging power, or operating the machine. Lock out electrical power before performing maintenance or service on farmstead equipment.
- Vehicle steps must be covered with a nonslip material.
- Passengers are not allowed to ride unless a passenger seat has been installed.
- Machinery will not be repaired, maintained, cleaned, or fueled while the engine is running.
- All towed equipment or machinery must use a safety chain.
- All of the machine's attachments must be lowered to the ground before servicing or maintenance.
- Check all machinery prior to use to ensure it is in a safe operating condition.
- Is the auger inlet shielded to prevent contact with the auger?
- Is the winch cable free of corrosion, wear, or damage that would reduce its strength?
- Are there any power lines near those areas where portable augers/elevators are located or generally used?
- Is each combine equipped with a 10-pound (minimum) ABC-type fire extinguisher?
- Are ladders and steps in good repair and free of mud and grease?

MANAGING ANIMAL WASTE

- Develop and follow animal waste and nutrient management plans.
- Confine animals and their wastes to protected areas.
- Include milk house waste in your waste management plan.
- Use filter strips to treat milk house waste.
- Intercept runoff from land upslope of the barnyard.
- Use diversions and waterways.
- Use subsurface drains to manage seepage areas.
- Direct barnyard runoff away from streams and other bodies of water.
- Direct barnyard runoff toward grass filter strips, pasture fields, and croplands.
- Use gutters, downspouts, underground outlets, and diversions to keep clean water out of baryards and waste storage structures.
- Store manure to allow flexibility in time of application.
- Use manure as replacement for commercial fertilizer.
- Spread manure on crops that need nutrients.
- Avoid spreading manure on frozen ground.
- Avoid spreading manure near streams, sinkholes, and wells.
- Calibrate manure spreaders to prevent over-fertilization.
- Use conservation practices to reduce runoff and erosion on land receiving manure.
- Use manure as a component of integrated crop management.
- Use tests to determine the nutrient value of manure.
- Store manure in stacking sheds to reduce nutrient losses.

PESTICIDE HANDLING

- Pesticides should be inaccessible to everyone but the user.
- Read the label.
- Wear proper clothing and protective equipment.
- Store pesticides according to instructions on label, in original containers in a safe, dry, locked, well ventilated area.
- Safe and proper storage is necessary to protect children, livestock, the environment, and the pesticide quality.
- Dispose of containers according to label instructions.

- Follow state and local standards when disposing of containers.
- Always shower with soap and water after handling pesticides.
- Words on the label such as *Caution, Warning, Danger,* and *Poison* tell you the level of hazard (toxicity).
- Inform your family of your spraying plans: chemicals to be used; where you will be; time schedule.
- Plan ahead: Calibrate equipment; get proper mixing and safety equipment ready; read pesticide instructions.
- Safety guidelines for mixing chemicals include adequate ventilation; stay upwind; mix only the quantity needed; use proper equipment; no eating, drinking, or smoking until hands and face have been washed with soap and water. Always read and follow label directions.
- Do the mixing and filling at least 100 feet from your water source. It is best to fill the sprayer in the field if possible.
- Filling hose should never touch liquid in tank. Observe filling to prevent overflow.
- Take every precaution to prevent contamination of yourself and surroundings.
- Take immediate steps to remove contaminated clothing and wash contaminated skin.
- You and your family members should read the section of the label dealing with treatment in case of chemical exposure.
- Know symptoms of pesticide poisoning before you begin, and take immediate action if you start to develop symptoms. Do not delay treatment if symptoms of chemical exposure occur. Take the chemical label with you when going for medical treatment
- Calibrate your sprayer before applying pesticides.
- Overheating can cause pesticide containers to leak or burst.
- Wash the spray tank in the field if possible. You can haul a supply tank of water to the field to do this.
- Apply wash and rinse solutions to the field or another field according to the label. Don't just dump them!
- Pesticide containers must be disposed of properly.
- Handle contaminated clothing with rubber gloves and wash it separately from the rest of the family's laundry. Launder pesticide-contaminated clothing according to recommended guidelines. Wash all personal protective equipment with heavy-duty detergent and water.
- Contaminated clothing usually must be washed three times with heavy-duty detergent. Each time, the machine should be set for the highest water level and the hottest water temperature.
- Line-dry clothes, so you don't risk contaminating your dryer.
- Never apply pesticides where rainfall or irrigation runoff may cause pollution.
- Check label instructions to see how long you should wait before reentry or harvest following pesticide application. Follow label instructions for posting sprayed fields with re-entry warnings.
- Know the rules for moving pesticides from one place to another. Rules for transporting pesticides anywhere include:
    - Pesticides must be in their original packages.
    - Vehicle must have correct warning sign.
    - In case of an accident, if someone is killed, seriously injured, or the damage is more than $50,000, contact the Department of Transportation immediately.
    - Report spillage on a public road to a law enforcement officer immediately.
- Store pesticides in original containers in a safe, dry, locked, and well ventilated place.

Appendix B

FORESTRY SAFETY CHECKLIST

GENERAL OPERATION

- All employees properly trained in the safest way to perform their job(s).
- All employees wearing hardhats when exposed to overhead hazards.
- Adequate footwear being worn.
- Hearing protection worn where required.
- Protective clothing worn when working with chain saws.
- Safety eyeglasses with side shields being worn.
- Machine guards and/or protective shields, barricades, safety devices not to be removed.
- Operators ensure that all guards and shields are in place and in proper working condition.
- Horseplay and running not permitted on the premises including all work areas and parking lots.
- Seat belts required to be used in all vehicles and machinery equipped with rollover cab protection.
- Check with supervisor prior to operating unfamiliar machines.
- Any unsafe condition noted must be reported to your supervisor.
- When mobile equipment is not in operation or parked, then blades, buckets, cutting heads, etc., must be lowered to ground level.
- Employees working on the ground shall always observe for overhead hazards.
- Truck drivers comply with all State and Federal laws, statutes, and regulations relating to highway safety.
- Employee will be trained in, and required to use proper lifting techniques and body mechanics.

LOADER OPERATION

- Operator does not swing boom or loads over workers.
- Loader has protective cab guarding.
- Boom lowered to ground when not working.
- Logs placed properly on trucks (height and arrangement).
- Truck drivers stay adequate distance from loading area.
- Trucks equipped with adequate metal standards and cab racks.
- Deckmen stay adequate distance from loader.
- Operators never load log trucks more than half the height of the diameter of the outer logs over stationary standards.
- Round load in the middle so as to secure and balance the load.
- Employee near log truck aware of and positions himself to avoid being struck by material falling from truck.

FELLING OPERATION

- All lodged or hung trees pulled down as soon as possible.
- Dead trees and snags felled when possible.
- Overhead hazards checked before felling.
- Windy conditions avoided when felling.
- All workers a safe distance from tree being felled.
- Manual fellers have clear path of retreat.
- Manual fellers undercut properly and fell in correct direction.
- Always plan the direction of fall of any tree felled.
- Never cut a standing tree completely through.

BUCKING AND LIMBING OPERATION

- Area clear of equipment movement.
- Equipment operators have clear view of approach.
- Direction of log movement determined before cutting.
- Using tip of chain saw avoided.
- Spring poles cut and removed with caution.
- Control of chain saw maintained.
- Employees avoid standing between logs that may roll while being bucked.
- Do not cut limbs that are supporting the log.
- If on a hill, stand on the upside of the hill.
- Never cut limbs above chest height.
- Stand on the opposite side of the tree, if possible, using it as a barrier between you and the limb you are cutting.
- Stand at a 45-degree angle to the limb you are cutting to prevent the saw from striking your leg if it slips. Do not face the limb squarely.

SKIDDING OPERATION

- Ground workers in safe locations.
- Passengers prohibited.
- Adequate guarding of cab sides, front, top.
- Equipment operated at safe speed.
- Cable ends properly maintained.

LOG TRUCK CHECKLIST

- Tires and lugs checked.
- Wipers working.
- Front and rear lights visible and working.
- Air pressure checked.
- Gauges, defroster, low air warning working.
- Check for any signs of fluid leaks.
- Emergency equipment in truck and operational.
- Trailer lines connected and out of the way.
- Standards in good shape.
- Springs checked.
- Binders tight, load balanced, load light.
- Tandem tires checked.
- Mud flaps in good repair.
- Brake linkage working.
- No flat tires.
- Pin set-5th wheel.
- Engine fluid levels, fan, and belts checked.

CHAIN SAWS

- Know your saw and how to operate it before you use it.
- Clear your work area.
- Check your chain's condition and sharpness.
- Fuel the saw in a safe place.
- Check the conditions in which you are cutting.
- Plan your escape route.
- Assess your own physical condition.
- Wear sturdy, snug-fitting clothing that gives you complete freedom of movement.
- Wear sturdy boots with nonslip soles, protective toes and high tops.
- Wear heavy duty, nonslip gloves.
- Wear a nonfogging, vented face screen or safety goggles.
- Wear an approved safety hard hat.
- Ear muffs or ear plugs are essential.
- Wear chaps made for use when using chain saws.
- Always hold the saw firmly with both hands and keep the left arm as stable as possible.
- Use a saw equipped with a chain brake, anti-kickback chain, or other anti-kickback devices.
- Watch for twigs that can snag the chain—the area should be free from any obstructions.
- Cut with the lower part of the saw blade—not with the tip or nose.
- Maintain a high saw speed when entering, cutting, and leaving the wood cut.
- Keep the chain sharp.
- Do not cut above mid chest height—the saw is too close to your face.
- Do not cut brush or shrubbery.
- Use only replacement bars specified by the manufacturer.
- Never fuel a saw while the engine is running.
- Never fuel a hot saw.
- Never smoke or have any type of flame while you are around a saw or during fueling.
- Never start the saw where you fueled the engine.
- Keep leaves and dry materials away from the hot muffler.
- Before starting the engine, make sure the chain is not contacting anything.
- Do not let the saw rest on your leg or knee while you start the engine.
- Do not drop start the chain saw. Always maintain control by standing securely, holding the saw firmly, and taking your hand off the trigger between cuts.
- Do not work when you are fatigued.
- Keep the handles dry and clean and free from the oil and fuel mixture.
- Whenever you are cutting, be sure your body is clear of the natural path the saw will follow when the cut goes through.
- Never straddle the log to make a cut.
- Always shut off the engine before setting the chain saw down—even when you are retreating from a falling tree.
- Make sure the saw is off and the chain has stopped before making any adjustments or repairs.
- Do not run the saw indoors.

Appendix C

COMMERCIAL FISHING VESSEL CHECKLIST

REQUIREMENTS FOR ALL VESSELS

- Backfire flame control.
- Ventilation.
- PFD/immersion suit.
- Ring life buoys.
- Survival craft.
- Stowage of survival craft.
- Lifesaving equipment markings.
- Maintenance/inspection of lifesaving equipment.
- Distress signals.
- Emergency position indicating radio beacons (EPIRBS).
- Fire extinguishing equipment.
- Injury placard.
- Waste management plan for ocean-going vessels >40 feet.
- Marine sanitation devices.
- Inland navigation rules on board inland waters only.
- Rules of the road (sound producing devices: bell, lights, and shapes).
- Pollution placard for vessels >26 feet/pollution prevention.
- Garbage placard for vessels >26 feet.
- FCC SSL.
- Load line certificate.
- Numbering.
- Fireman's outfit and SCBA.
- Register your 406 EPIRB with NOAA.
- First aid equipment and training.
- Guards for exposed hazards.
- Navigational information.
- Compasses.
- Anchors and radar reflectors.
- General alarm system.
- Communication equipment.
- High water alarms.
- Bilge pumps, piping and dewatering system.
- Electronic position fixing devices.
- Emergency instructions.
- Instructions, drills, and safety orientation.

Index

A

Absolute pressure, 619
ACGIH, 282, 504
Acupuncture, 648-649
Acute bronchitis, 367
Acute contact dermatitis, 326
Acute fatigue, 21
Aerodynamics, fixed rotary wing, 534
Aeromonas, 663
Aeromonas hydrophila, 685-686
Aerosols, 429, 460-464
    sampling and analyzing methods, 452-453
Aflatoxin, 463
Agent Orange, 144, 327
Agricultural industry (*see also* Animals; Animal safety; Farmers; Farming)
    accident statistics
        fatal, 215-218
        nonfatal, 218-220
    ANSI standards pertaining to, 202-205
    cancer and, 222-225 (*see also* Cancer)
    checklist for machinery ergonomics, 38-40
    chemical hazards of, 249-265 (*see also* Chemical hazards; Fertilizers; Herbicides; Insecticides; Pesticides; Rodenticides)
    CTDs among workers in, 240-241
    diseases and illnesses in, 4-6 (*see also* Diseases; Injuries and illnesses)
    employer responsibilities, 483-484
    EPA standards for pesticides, 483-492
    hazard and hazardous nature of, 9, 172-173, 220-222 (*see also* Hazards)
    hygiene evaluations, 439-468
    machinery and equipment design, 18-19
    mental health issues, 153-154, 158-162 (*see also* Mental health)
    migrant workers, 385-396
    poisoning in, 4
    respiratory diseases related to, 353-383 (*see also* Respiratory diseases)
    systems standards, 41
    worker disabilities, 171-172
        goal prioritizing, 174-175
        mobility aids for, 173
        needs of workers with, 173-176
        prevention and safety, 175-176
        resources for, 176-178
        task difficulty rating, 173-174
Agricultural machinery, 233-247
    ASAE standards, 245-246
    hazards, 234-243
        categories of, 235
        chemical, 241, 258-259
        crush point, 237
        cumulative trauma, 240-241
        energy, 238
        entanglement, 236
        evaluating, 242-243
        falling, 239
        pinch point, 235-236
        prevention techniques, 243-246
        rollover, 242
        run over, 241-242
        sharp edge, 236-237
        speed, 239
        thrown object, 237-238
        visibility, 237
        work environment, 239-240
    operating instructions, 246
    persons at risk using, 234
    regulations, 477-478
    safety concepts, 233-234
    safety hierarchy, 243-244
    warning signs for, 244
Agriculture (*see also* Farmers; Farming)
    electrical safety in, 77-88
    envenomization hazards, 89-125 (*see also* Arthropods)
    ergonomics in (*see* Ergonomics)
    grain (*see* Grain industry)
    hazards in, 6-7 (*see also* Hazards)
    noise and hearing loss in, 59-65
    occupational vibration exposure in, 53-58 (*see also* Vibration)
    overview of safety/health in, 3-7
    safety and health regulations, 469-482
    zoonotic hazards, 127-141
AIHA, 446
Air pollutants, 428-432 (*see also* Respiratory hazards)
    aerosols, 429, 460-464
    air sample data sheet, 444
    ammonia, 371, 429-430, 456, 475
    dust, 429, 460-462
    exposure guidelines, 441-442
    gases (*see* Toxic gases)
    hazards
        assessing airborne, 446
        environment for gaseous respiratory, 456-460
        evaluation methods, 445-451
        from agricultural gases, 450
        from agricultural vapors, 451
    in animal housing, 428-432
    measuring, 447-451
        equipment manufacturers for, 448
        with colorimetric detector tubes, 447
        with colorimetric diffusion tubes, 447, 449
        using electronic direct reading measurements, 449, 451
    permissible exposure limit (PEL), 441-442
    pesticides (*see* Pesticides)
    quantitative exposure assessments, 445
    sampling/analyzing methods, 452-456
        aerosol sampling, 452-453
        air sampling pump operations, 453-454

biological monitoring, 454-455
flow rate calibration, 454
gas sampling, 452
medical monitoring, 455-456
vapor sampling, 452
threshold limit value (TLV), 441-442
vapor hazard index (VHI), 443
vapor hazard ratio (VHR), 443
Air supplying respirator, 187
Alcohol
cancer and, 147
consumption among fishermen, 166-167
consumption among migrant workers, 393
Algae, 686
Allergic contact dermatitis, 327
Allergic contact phytodermatitis, 329-333
Allergic reactions, 121
Allomone, 121
American College of Occupational Medicine (ACOM), 62
American Conference of Governmental Industrial Hygienists (ACGIH), 202, 282, 504
American dog tick, 102
American Industrial Hygiene Association (AIHA), 446
American National Standards Institute (ANSI), 202-205
American Society of Agricultural Engineers (ASAE), 245-246
Americans with Disabilities Act (ADA), 504
Ammonia, 429-430, 456
anhydrous, 371
regulations for storing/handling, 475
Anaphylactic reactions, 121
Anasakiasis, 666-667
Andres, Richard, 476
Anhydrous ammonia, 371
regulations for storing/handling, 475
Animal housing, 419-438
air pollutants, 428-432
aerosols, 429
ammonia, 429-430
dust, 429
carbon dioxide, 430
carbon monoxide, 430-431
hydrogen sulfide, 430
methane gas, 430
odors, 431
air quality control and management, 431-432
children in, 435
electrical hazards, 432-433
fires in, 434
mechanical hazards, 432
noise in, 433-434
respiratory protection, 432
safety signs for, 435
ventilation, 421-428
cool weather, 423-425
instruments for monitoring, 425-428
natural, 422-423
rate of, 421-422
system control, 425
warm weather, 423
Animals
bites from, 339-341
flight zone, 294-295
hazards
animal feed, 260
confinement, 259
manure pits, 369-370, 458
Animal safety
beef cattle (see Beef cattle)
horses (See Horses)
injury control, 291-292
Anorexia nervosa, 156
ANP, 618
ANSI, 202-205
Anthrax, 342
Anthropometry, 15-16
Antibiotics, 260, 669-670
Ants, 93, 115-116
harvester, 116
red imported fire ant (RIFA), 92, 115-116, 334-335
velvet, 119-120
venom from, 93
Anxiety disorders, 154-155
AOAC, 446
Aprons, 190
Aquacultural hazards, 597, 659-679 (see also Skin diseases)
anesthetic for salmon, 669
anthropods, 697-698
antibiotic safety, 669-670
Atlantic wolffish, 700
bacteria (see Bacteria)
barracuda, 699
billfish, 701
black sea bass, 701
blue fish, 699
blue tang, 705
bryozoans, 692
catfish, 660-661, 704
Coelenterates, 687-692
cone snails, 696
consumption, 659-660
croaker fish, 701
cyanobacteria and algae, 686
drowning, 674
echinoderms, 693-695
electric fish, 706-707
electric rays, 706-707
electrical accidents, 671-672
fire coral, 691
fish bites
nonvenomous, 698-702
venomous, 702-706
fish spines, 701
Florida stone crab, 697
horseshoe crab, 697
hydrogen sulfide, 669
hydroids, 691
injuries
ice-related, 674
while manufacturing fish feed, 670-671
while processing fish, 670-671
jellyfish (see Jellyfish)
leatherbacks, 706
lionfish, 704, 705
lobsters, 697
man-o-wars, 690
marine worms, 696-697
mollusks, 695-696
moray eels, 699
muscle strains, 673
mycotoxins in fish feed, 666

needlefish, 701
neurotoxins in fish feed, 666
nudibranches, 696
octopuses, 696
old wife fish, 706
parasites (see Parasites)
pedicelliriae, 694-695
pesticides in wild-caught fish, 666
puncture wounds from channel catfish, 671
quality assurance of marine products, 660-661
rabbitfish, 706
ratfish, 702
red soft coral, 691
safety issues, 668-674
sawfish, 700, 701
scorpionfish, 705
sea anemones, 691-692
sea cucumbers, 693
sea snakes, 707
sea urchins, 694, 695
sewage in ponds, 669
sharks (see Sharks)
shrimp preservatives, 669
snook, 701
spiny dogfish, 702
sponges, 692-693
squid, 696
starfish, 693-694
stargazers, 704-705
stingrays, 702-703
stonefish, 705-706
surgeonfish, 706
toadfish, 699, 700, 703
torpedo rays, 706
tractors on fish farms, 672-673
vaccine safety, 670
weeverfish, 704
wolf eel, 700
zooplankton, 686-687
Arachnida (see Arthropods; Spiders)
Arthropods, 89-125, 333-339, 346, 697-698
ants (see Ants)
assassin bugs, 112-113, 335
bed bugs, 111-112
bees, 334-335
beetles, 336
blister beetles, 114-115, 336, 499
blood feeding, 94
brown recluse spider, 98-99
bumblebees, 119, 334
butterflies, 336
caterpillars (see Caterpillars)
centipedes, 96, 339
chiggers, 100-101
cicada killers, 119
conenose bugs, 112-113, 335
controlling, 120-121
deer flies, 335
description, 90-91
diseases transmitted by bites from, 129
envenomization
epidemiology of, 92
Hymenoptera stings, 92
fleas, 108-109, 337
flies (see Flies)
glossary of terms for, 121-122
gnats, 105, 335
groups of
arachnida, 90
insecta, 90
growth and development, 91
complete metamorphosis, 91
gradual metamorphosis, 91
modified metamorphosis, 91
hematophagous, 89
honey bees, 93, 118-119, 334
hornets, 93, 116-118
horseflies, 335
Hymenoptera, 115, 120, 499
identification based on
injuries from, 94-95
time of day bitten, 96
lice, 109-111
millipedes, 339
mites (see Mites)
mosquitoes, 89, 106-108, 335, 499
moths, 336
mud daubers, 119
red imported fire ant (RIFA), 92, 115-116, 334-335
scorpions, 96-97, 339, 499
spiders (see Spiders)
sweat bees, 119
ticks (see Ticks)
venoms, 92-94
ant, 93
honey bees, 93
hornets, 93
Hymenoptera, 93
spider, 93
urticatious caterpillar, 93-94
wasps, 93, 116-118, 334-335
waterbugs, 335
waterscropions, 335
wheel bugs, 112-113, 335
yellowjackets, 116-118
ASAE, 245-246
Assassin bugs, 112-113, 335
Association of Analytical Chemists (AOAC), 446
Association of Programs for Rural Independent Living (APRIL), 177
Asthma, 363-366, 502
allergy testing, 365
causative agents, 364
clinical findings, 364-365
epidemiology, 363-364
pulmonary function testing, 365
treatment, 365-366
Atlantic wolffish, 700
Atrial natriuretic peptide (ANP), 618
Audiometry, 62-63
Australian brown blubber jellyfish, 689

B

Back belts, industrial, 22
Back injuries, 22, 715
industrial back belts for, 22
lifting guidelines (see Lifting guidelines)
Back pain, vibration and, 55
Bacteria, 127-128, 661-666
Aeromonas, 663

Clostridium, 662
contamination after harvesting/processing, 664-665
Cyanobacteria, 686
Edwardsiella tarda, 663-664
Erysipelothrix rhusiopathiae, 662
Leptospira, 664
Mycobacterium, 661-662
mycotoxins in fish feed, 666
neurotoxins in fish feed, 666
Nocardia, 662
Plesiomonas shigelloides, 663
toxins produced by, 665-666
Vibrio, 662-663
Yersinia ruckeri, 664
BAL, 362
Barotrauma, 4, 630-631
  middle ear, 630
  paranasal sinus, 630
  pulmonary, 630
Barracuda, 699
Basal cell epithelioma carcinoma (BCE), 324
Bass, black sea, 701
BCE, 324-325
Bed bugs, 111-112
Beef cattle, 291-313
  handling, 294-303
    changing direction, 298
    emptying pen and sorting at gate, 301-303
    flight zone, 294-295, 296
    moving cattle backward, 297-298
    moving cattle forward, 297
    moving cattle into pen, 301
    moving herd with one handler, 298-299
    moving herd with two handlers, 300
    moving stragglers back into group, 300-301
    point of balance, 295, 296
    practices, 294
    tips on, 295-296
  handling facilities
    adapting existing structures, 309, 311
    components, 303-312
    crowding pen, 304-305
    designing, 303
    equipment selection, 305
    for 25 to 50 head, 312
    headgate, 303
    holding chute, 303-304
    holding pen, 305
    installing, 303
    intermediate design, 311
    layout, 309
    loading chute, 305
    preventing balking, 306-308
    scales, 305
    tips on, 308-309
    tobacco barns as, 309
    working chute, 304
  injury control, 291-292
  psychology, 292-294
    herd instinct, 293-294
    maternal instinct, 294
    sense of hearing, 293
    sense of sight, 292-293
    sense of smell, 293
Bees, 334-335

Beetles, 336
  blister, 114-115, 336, 499
Bicycle helmets, 194
Billfish, 701
Bioaerosols, 462
Biological hazards, 6
Bipyridyl herbicides, 256
Birth defects, 225
Biting flies (see Flies)
Black flies, 103-104
Black-legged tick, 102
Black sea bass, 701
Blacktip reef shark, 699
Black widow spider, 97-98, 338-339
Blastomycosis, 345
Blister beetles, 114-115, 336, 499
BLS, 9, 215, 497
Bluefish, 699
Blue shark, 699
Blue tang, 705
Blue-tinted jellyfish, 689
Body lice, 109-111
Body protection equipment (see Personal protective equipment)
Box jellyfish, 687-688
Boyle's law, 620
Brain cancer, 145
Breathing gas
  contamination, 632
  purity and analysis, 636
Bronchitis, 366-368
  acute, 367
  clinical features, 367
  epidemiology, 366
  etiologic agents, 366-367
  prognosis, 368
  pulmonary function testing, 367-368
  treatment, 368
Bronchoscopy with bronchoalveolar lavage (BAL), 362
Brown dog tick, 102
Brown recluse spider, 98-99, 339
Brucellosis, 342
Bryozoans, 692
Bubonic plague, 342
Bugs (see Arthropods)
Bulimia nervosa, 156
Bullhead shark, 702
Bull shark, 699
Bumblebees, 119, 334
Bump caps, 193-194
Buoyancy, 619, 624-625
Bureau of Labor Statistics (BLS), 9, 215, 497
Burns, 398-399, 522, 598
Butterflies, 336

C

Cabbagehead jellyfish, 689
Cable mite, 100
CAGE, 629
Canadian Farmers with Disabilities, 177
Cancer, 4, 143-151
  alcohol intake and, 147
  among women in agriculture, 224-225
  anatomic sites of among farmers, 145
  basal cell epithelioma carcinoma (BCE), 324-325

brain, 145
causes of, 143-144
chemical carcinogens and, 144
epidemiology of, 222-225
genetic factors, 146-147
hazard education approach to, 147
Hodgkin's lymphoma, 5, 145
in farmers, 144-145
in fishermen, 145-146
in forestry workers, 145
infectious agents, 146
leukemia, 145
lip, 145
malignant melanoma (MM), 324-325
melanoma, 5
multiple myeloma, 145
nasal, 502
non-Hodgkin's lymphoma, 145
nutritional factors, 147
occupational factors linked to, 144-146
pesticides and, 223-224
prostate, 145
radiation exposure, 146
rate of in United States, 143
reducing the risk of, 146
regulatory issues and, 146-147
scrotal, 4
skin, 145, 324, 681-683
social factors, 147
soft tissue sarcoma, 5, 145
squamous cell carcinoma (SCC), 324-325
stomach, 145
testicular, 145
Cannonball jellyfish, 689
Carbamate insecticides, 253-254
Carbon dioxide, 430, 522
Carbon monoxide, 370, 430-431, 458, 522
Cardiovascular disease, 523
Carpal tunnel syndrome (CTS), 10, 30, 241
Caterpillars, 93-94, 113-114, 336, 499
    hackberry leaf slug, 113
    puss, 113
    saddleback, 114
    slug, 113
    venom from, 93
Catfish, 661, 704
Cattle, beef (see Beef cattle)
Catworms, 345
Census of Fatal Occupational Injuries (CFOI), 215
Centipedes, 96, 339
Cercarial dermatitis, 696
Cerebral arterial gas embolism (CAGE), 629
CFIVSA, 557-558, 559
CFOI, 215
CFR, 470
Chain saws, 517
    injuries, 498
    kickback accidents, 514
    OSHA standards for, 549-550
    safety chaps, 190
Charles' law, 620
Chemical cartridge respirators, 186
Chemical hazards, 6, 241-242, 249-265
    disinfectants, 260
    fertilizers, 5, 258

firefighting and, 523
fungicides, 256-257
gases (see Toxic gases)
herbicides (see Herbicides)
insecticides (see Insecticides)
mycotoxins, 260-261
of animal confinement, 259
of animal feed, 260
of crops, 261
of farm equipment, 258-259
pesticides (see Pesticides)
preventing injuries by, 261
regulations for, 475-477
rodenticides, 257
silo gas, 259-260
Chiggers, 100-101
    bite prevention tips, 120-121
Chilblains, 69, 326
Child labor, 391, 478-479
Children
    in animal housing, 435
    safety guidelines for, 397-403
Chlamydia, 128
Chlorinated aromatic fungicides, 257
Chlorinated hydrocarbon insecticides, 254-255
Chlorophenoxy herbicides, 255-256
Chokes, 629
Choking, 400
Chromomycosis, 345
Chronic fatigue, 10, 21
Chronic motor neuron disease, 6
Cicada killers, 119
Ciguatera poisoning, 666
Circuit breakers, 82
CISDs, 164-165
Clonorchis sinensis, 667
Clostridium, 662
Coccidioidomycosis, 345
Code of Federal Regulations (CFR), 470
Cold stress, 67-69
    injuries
        chilblains, 69, 326
        frostbite, 69, 326
        frost nip, 69, 326
        hypothermia, 26, 69, 121, 607-611
        vascular abnormalities, 69
    potential occupational exposures, 68
    protecting against, 69
Cold water immersion, 606-613
    drown-proofing techniques, 611-612
    hypothermia, 607-611
    rescue devices and procedures, 613-615
    treading water, 612-613
Colorado Tick fever, 500
Colorimetric detector tubes, 447
Colorimetric diffusion tubes, 447, 449
Comb jellyfish, 690
Commercial Fishing Industry Vessel Safety Act of 1988 (CFIVSA), 557-558, 559, 723
Composting, 5
Compression pains, 629
Conductors, 77-78
Cone snails, 696
Conenose bugs, 112-113, 335
Conjunctivitis, 277

Contact dermatitis, 4, 501, 684
  acute, 326
  allergic, 327
  irritant, 326
Contact phytodermatitis
  allergic contact, 329-333
  irritant, 329
Copperhead snakes, 340
Coral
  fire, 691
  red soft, 691
Coral snakes, 340
COSPAS/SARSAT satellite system, 722
Court cases
  Baun v. Hudson, 718
  Dragich v. Strika, 718
  Esta v. Persohn, 718
  Hudgins v. Gregory, 718
  Little v. Green, 719
  Luksich v. Misetich, 718
  Mitchell v. Trawler Racer, 719
  Nolan v. General Seafoods Corp., 718
  Solet v. M/V Capt. Dufrene, 719
  Sterling v. New England Fish Co., 717
  Stevens v. Seacoast Co., 719
  Texas Menhaden Co. v. Johnson, 719
  Vitco v. Joncich, 717
Cows (see Beef cattle)
Crabs, 697
Creeping eruption, 345-346
Critical incident stress debriefings (CISDs), 164-165
Croaker fish, 701
CTS, 10
Cubital tunnel syndrome, 30
Cumulative trauma disorders (CTDs), 10-15
  among agricultural workers, 240-241
  by type of job performed, 11
  carpal tunnel syndrome (CTS), 10, 30, 241
  cubital tunnel syndrome, 30
  De Quervain's disease, 10
  epicondylitis, 10
  fatigue (see Fatigue)
  ganglionic cyst, 30, 241
  guyon canal syndrome, 30
  hand-arm vibration syndrome, 10, 30 (see also Hand-arm vibration)
  lateral epicondylitis, 30
  lower back, 241
  medial epicondylitis, 30
  pronator teres syndrome, 30
  radial tunnel syndrome, 30
  risk factors, 10-16
    awkward postures, 12-13
    cold temperature, 14-15
    forceful exertions, 13
    localized contact stresses, 14
    nonoccupational, 12
    occupational, 12
    repetition, 12
    vibration, 13-14
  OSHA checklist for, 45-52
  rotator cuff tendinitis, 30
  stenosing tenosynovitis, 30
  stenosing tenosynovitis crepitans, 30
  tendinitis, 10, 30, 241
  tenosynovitis, 10
  thoracic outlet syndrome, 30
  trigger finger, 10
  upper extremity, 10
Cysts, ganglionic, 30, 241

D

Dalton's law, 620-621
DAN, 638
DBT, 71
DCI, 628-630
DCS, 627, 628-629
DDT, 5-6, 254
De Quervain's disease, 10
Decibel (dB), 60
Decompression illness (DCI), 628-630
Decompression sickness (DCS), 627
Deer flies, 105-106
Dengue fever, 108, 121
Dermatitis, 121, 277, 501
  allergic, 327, 329-333, 346
  cercarial, 696
  contact, 4, 501, 684
    acute, 326
    allergic, 327
    allergic contact phytodermatitis, 329-333
    irritant, 326
    irritant contact phytodermatitis, 329
  due to bryozoans, 692
  epidemiology of, 225-226
  irritant, 326, 329
  occupational, 4
  pesticides and, 327
  phytophotodermatitis, 333
  ragweed, 332
  related to
    chemical exposure, 326-327
    plant exposure, 327-333
    tree exposure, 502
Diquat, 256
Disabilities, 171-179
  agricultural industry and, 172-173
  farmers having, 171-172
  from fishing injuries, 715-716
  mobility aids for, 173
  needs of agricultural workers with, 173-176
  resources for persons with, 176-178
Diseases (see also Injuries and illnesses; Viruses)
  cancer (see Cancer)
  cardiovascular, 523
  chronic motor neuron, 6
  conjunctivitis, 277
  dermatitis (see Dermatitis)
  farmer's lung disease (see Farmer's lung disease)
  infectious, 499-501
  Lyme, 4, 94, 103, 121, 335, 343, 499-500
  milker's nodule, 343
  orf, 343, 344
  Parkinson's disease, 225
  related to diving (see Diving)
  respiratory (see Respiratory diseases)
  rhinitis, 277
  silo filler's, 5
  silo unloader's syndrome, 5

skin (*see* Skin diseases)
sporotrichosis, 4
transmitted by
    bites from insects, 129
    ingestion, 130
    inhalation, 129
    skin contact, 129
    skin penetration, 129
tuberculosis, 226, 392-393
vibration-induced white finger disease (VWFD), 498, 504, 505
zoonotic (*see* Zoonotic diseases)
Disinfectants, 459
Disorders
    cumulative trauma (*see* Cumulative trauma disorders)
    definition, 10
    mental health (*see* Mental Health)
Divers Alert Network (DAN), 638
Diving breathing gas, 636
    decompression procedures, 626-628
    diseases related to
        barotrauma, 630-631
        breathing gas contamination, 632
        cerebral arterial gas embolism (CAGE), 629
        chokes, 629
        compression pains, 629
        decompression illness (DCI), 628-630
        decompression sickness (DCS), 627, 628-629
        dysbaric osteonecrosis, 632-633
        dyspnea, 633
        effects of gases, 631-632
        gastrointestinal rupture, 630
        hypercarbia, 632
        hypoxia, 632
        narcosis, 631
        oxygen toxicity, 631-632
        pulmonary edema, 633
        transient unilateral facial paralysis, 633
        vertigo, 630-631
    environment, 618-622
    equipment
        1 Atm suits, 624
        breathing gas, 626
        buoyancy compensation, 624-625
        clothing, 625
        fins/boots, 624
        masks/helmets, 624
        thermal protection, 625-626
        underwater breathing apparatus (*see* Underwater breathing apparatus)
        underwater tools, 626
    gas laws, 619-621
        Boyle's law, 620
        Charles' law, 620
        Dalton's law, 620-621
        Henry's law, 621, 626
    hazards, 617-641 (*see also* Aquacultural hazards)
    health monitoring, 633-634
    heat transfer, 622
    immersion, 618
    injuries from sea life, 633 (*see also* Aquacultural hazards)
    light and vision, 621
    medical coverage of operations, 637-638
    near drowning of divers, 633
    physical examinations, 635-636
    physical requirements for, 634-635
    pressure, 618-169
        absolute, 619
        hydrostatic, 619
        pressure units conversion, 619
    profiles, 628
    regulations, 636-637
    resources, 637, 638
    skin infections related to, 632
    underwater explosions, 622
    underwater noise, 621-622
Diving Division of the Defense and Civil Institute of Environmental Medicine, 638
Drown-proofing, 611-612
Drowning, 399, 605-606, 674
Dry bulb temperature (DBT), 71
Dust, 182, 429 (*see also* Air pollutants; Respiratory hazards)
    inorganic, 460
    organic, 460-462
Dysbaric osteonecrosis, 632-633
Dyspnea, 633

E

Ear, anatomy of, 61-62 (*see also* Hearing; Noise-induced hearing loss)
Eating disorders, 156
Echinoderms, 693-695
Edema, pulmonary, 633
Edwardsiella tarda, 663-664
Eels, moray, 699
Ehrlichiosis, 102, 342, 500
Elderly, safety guidelines for, 403-408
Electric fish, 706-707
Electric rays, 706-707
Electric shock
    calling the electric company, 86
    first-aid for, 85, 86
    lightning, 85-86
Electrical hazards, 6
    in animal housing, 432-433
    removing, 80-82
    while processing fish, 671-672
Electrical safety, 77-88
    changing fuses, 82
    children and, 399-400
    circuit breakers, 82
    employee work spaces, 78
    extension cords, 81-82
    for elderly persons, 407
    grain bin regulations, 79
    ground-fault circuit interrupter (GFCI), 80, 81
    grounding, 82-83
    grounding adapters, 81
    in corrosive and damp environments, 84
    in dusty environments, 84
    installing/maintaining on farms, 83-84
    OSHA requirements for, 77-78
    portable elevators and, 84
    power lines, 79, 80, 549
    protection around electrical systems, 82
    requirements of systems of 600 volts or less, 78
    standby generators and, 84
    switch lockouts, 83
    training employees, 80
Electrocution hazards, 78-80
Elevators, portable, 84

Emergency position indicating radio beacon (EPIRB), 721-723
Encephalitis, 108, 121, 500
Endemic typhus, 342
Energy hazards, 238
Envenomization, 121
   epidemiology of, 92
   hazards, 89-125 (*see also* Aquacultural hazards; Arthropods)
Environment Protection Agency (EPA), 146, 389
Environmental hazards, 4, 6
   forestry, 499
Environmental Protection Agency (EPA), 7, 146, 202, 446, 476
   pesticide standards, 483-492
      compliance requirements, 491
      employee protection, 486-487
      employer responsibilities, 483-484
      exception requests, 489
      governmental provisions, 485-486
      handler application restrictions, 485
      handler instructions, 485
      handler protection, 488-489
      worker application restrictions, 484-485
      worker protection, 487-488
Epicondylitis, 10
   medial, 30
EPIRB, 721-723
Epstein-Barr virus (EBV), 146
Equipment, protection (*see* Personal protective equipment)
Ergonomics
   adult body dimensions, 16
   anthropometry, 15-16
   assessing, 10-16
   checklists
      for agricultural machinery, 38-40
      for performance enhancement, 32-37
   cumulative trauma disorders (CTDs) and (*see* Cumulative trauma disorders)
   definition, 9
   discomfort survey, 31
   industrial back belts, 22
   manual materials handling (MMH), 22, 26, 42-43
   meatpacking guidelines, 27
   NIOSH lifting guidelines, 22-26
   objectives of, 28
   OSHA risk factor checklist, 45-52
   physical work and, 20-21
   program development, 27-28
   recommended work heights, 17
   science of, 9
   temperature and (*see* Temperature)
   tool design, 16-22
      agricultural machinery/equipment, 18-19
      controls/displays, 19-20
   workplace design, 15-16
      for average individuals, 16
      for entire population, 16
      for extreme individuals, 16
      for range of adjustment, 16
Erysipelothrix rhusiopathiae, 662
Erysipheloid, 342, 684-685
Eustrongylidiasis, 667
Explosions, 572
   underwater, 622
Explosives, 549
Extension cords, 81-82

F

Fair Labor Standards Act (FLSA), 245-246, 386
Farmer's lung disease (FLD), 277, 358-363, 463
   clinical features, 360-361
   epidemiology, 358
   etiologic agents, 358
   findings from bronchoscopy with bronchoalveolar lavage, 362
   histopathology, 361-362
   immunopathogenesis of farmer's lung, 358-360
   prognosis, 362-363
   radiographic findings, 361
   treatment, 362-363
Farmers (*see also* Agricultural industry; Agriculture; Farming)
   cancer in, 144-145 (*see also* Cancer)
   dairy, 221-222
   grain producers and handlers, 222
   injuries and illnesses of, 3, 4 (*see also* Injuries and illnesses)
   mushroom workers, 222
   poultry, 222
   skin diseases in, 321-352 (*see also* Skin diseases)
   swine producers, 221
   with disabilities, 171-172
      goal prioritizing, 174-175
      mobility aids for, 173
      needs of, 173-176
      task difficulty rating, 173-174
      prevention and safety, 175-176
      resources for, 176-178
Farming (*see also* Agricultural industry; Agriculture; Farmers)
   chemical hazards of, 249-265 (*see also* Chemical hazards; Fertilizers; Fungicides; Herbicides; Insecticides; Pesticides; Rodenticides)
   electrocution hazards, 78-80
   grain bin regulations, 79
   hygiene evaluations, 439-468
   installing/maintaining electrical systems, 83-84
   mental health issues, 153-154, 158-162
   migrant workers, 385-396
   portable elevators, 84
   respiratory hazards associated with, 184-185 (*see also* Respiratory hazards)
   standby generators, 84
   training seasonal workers, 80
Farmstead safety, 397-418
   accident awareness on, 315
   barn fires, 316
   firearms and, 415-416
   food and, 414-415
   for children, 397-403
      art materials, 401-402
      burns, 398-399
      choking/strangulation, 400
      developmental growth, 398
      drowning, 399
      electrical shock, 399-400
      falls, 398
      furnishings, 402
      guidelines for supervising, 402-403
      poisonings, 400-401
      suffocation, 399
   for elderly, 403-408
      electrical concerns, 407
      exterior entrances, 404-405
      floors, 405-406

furnishings, 407-408
general considerations, 407
grab bars, 406-407
interior thresholds, 406
physical changes in, 403-404
ramps, 405
sensory changes in, 403-404
stairs, 405-406
hazardous household products, 408-412
pesticide residues and, 412-414
ponds and, 415
Fatigue
acute, 21
chronic, 10, 21
local, 21
mental, 514 (*see also* Mental health)
muscle, 10
Federal Aviation Administration (FAA), 536
Federal Food, Drug and Cosmetic Act (FFDCA), 257
Federal Hazardous Substances Act of 1960, 408
Federal Insecticide, Fungicide and Rodenticide Act (FIFRA), 7, 257
Fertilizers, 5, 144, 258, 261
FFDCA, 257
Field sanitation, 473-475
FIFRA, 7, 257
Firearm safety, 415-416
Fire coral, 691
Fires and firefighting, 521-527
  at sea, 598
  classes of fires, 598
  climate zones, 524
  deaths associated with, 522
  epidemiology, 521-522
  fire behavior, 524
  fire prevention, 525-526
  forest, 4
  in animal housing, 434
  injuries and illnesses, 522-524
  on fishing vessels, 572
  safety standards, 525
  suppressing, 525
  warning signs for fast-burning fires, 524
First-aid kit, 548
Fish
  hazards with (*see* Aquacultural hazards)
  health issues regarding consumption, 659-660
  quality assurance of, 660-661
Fisheries
  ANSI standards pertaining to, 202-205
  definition/description, 559-560
  electrical safety in, 77-88
  ergonomics in (*see* Ergonomics)
  factors of commercial, 560
  hazards in, 6-7
  management guidelines, 560
  noise and hearing loss in, 59-65
  occupational vibration exposure in, 53-58 (*see also* Vibration)
  overview of safety/health in, 3-7
  zoonotic hazards, 127-141
Fishermen
  cancer in, 145-146 (*see also* Cancer)
  definition/description, 559
  hiring, 572-573
  injuries and illnesses of, 4, 7 (*see also* Aquacultural hazards)
  legal remedies for injured, 717-720
  migrant workers, 560
  sea sickness and (*see* Motion sickness)
  skin diseases in, 681-712 (*see also* Skin diseases)
  substance abuse among, 166-167
Fishing industry
  CFIVSA regulations, 557-558
  environment, 166
  hazards
    aquacultural (*see* Aquacultural hazards)
    diving, 617-641
    preventing, 707-708
  injuries
    epidemiology of fatal, 557-570
    fatality statistics, 4, 562-567
    nationwide, 561-562
    fatality rates, 567-569
    risk factors, 568-569
  legal concerns, 713-720
  mental health issues, 153-154, 165-167 (*see also* Mental health)
  public health and, 558
  safety at sea, 571-604
Fishing spider, 339
Fishing vessels
  accidents
    claims, 588
    dispatcher procedures, 587-588
    documentation forms, 589-596
    captain's report of personal injuries, 592
      non-injury statement, 595
      personal injury report, 591, 593, 594
      report of personal injury or illness, 590
      report of physical damage, 596
    investigations, 588
    reporting procedures, 587-588
    reporting requirements, 588
  capsizing, 572
  casualty statistics, 723
  Coast Guard safety coordinators, 723-724
  definition/description, 559
  explosions, 572
  fires, 572
  flooding, 572
  floundering, 572
  groundings, 572
  hazards on, 713-716
  injury placard on, 723
  material failure, 572
  operational collisions, 571-572
  regulations, 716, 721-724
    emergency position indicating radio beacon (EPIRB), 721-723
    life saving equipment, 721
  safety, 571-604
    company policy, 572
    education program, 574
    equipment inspections, 574
    equipment recommendations, 574-575
    hiring safe personnel, 572-573
    incentive program, 574
    program analysis, 588-589
    recommendations, 574
    tips for crew members, 575-576
    training program, 573, 574
    work environment, 573-574
  safety checklists, 577-586
    captain's report, 579

insurance safety incentive program, 583-584
    medical report, 586
    vehicle inspection report, 585-586
    vessel inspection report, 577
    vessel inspection safety equipment, 578, 580
    vessel inspection sheet, 581-582
  survival techniques (*see* Survival techniques)
  unseaworthiness declaration, 719-720
Flammable liquids, 549
FLD (*see* Farmer's lung disease (FLD))
Fleas, 108-109, 337
Flies, 103-105, 335, 499
  biting, 89, 103-104
  black, 103-104
  deer flies, 105-106, 335
  horseflies, 105-106, 335
  sand, 105
  stable, 104-105
Flooring, 405-406
Florida stone crab, 697
FLSA, 245-246
Fluid shift theory, 646
Food and Drug Administration (FDA), 146
Food safety, 414-415
  discarding questionnable food, 415
  keeping surfaces clean, 415
  temperature control and bacteria, 414
  thawing foods, 414
Foot protection, 189-190
  rubber boots, 190
  steel-toed shoes, 189-190
Forest fires, 4
Forestry (*see also* Timber harvesting)
  ANSI standards pertaining to, 202-205
  electrical safety in, 77-88
  envenomization hazards, 89-125 (*see also* Arthropods)
  environmental hazards, 499
  epidemiology of injuries and illnesses, 497-508
  ergonomics in (*see* Ergonomics)
  fires and firefighting, 521-527
  hazards in, 6-7
  mental health issues, 153-154, 162-165 (*see also* Mental health)
  noise and hearing loss in, 59-65
  occupational vibration exposure in, 53-58 (*see also* Vibration)
  overview of safety/health in, 3-7
  zoonotic hazards, 127-141
Forestry workers
  cancer in, 145 (*see also* Cancer)
  dermatitis in, 501 (*see also* Dermatitis)
  exhaust exposure, 499
  exposure to pesticides, 502
  hazards involving, 505
  illnesses of, 4-6
  infectious diseases in, 499-501
  injuries of, 4, 7
  logger accidents, 513
  noise exposure, 499
  physical workload of, 498-499
  psycho-social issues, 503
  respiratory diseases in, 502
  safety equipment, 503-505
Frostbite, 69, 326
Frostnip, 69, 326
Fumigants, 279, 370-371, 459
Fungi, 128

Fungicides, 256-257
  cancer and, 223-224
  chlorinated aromatic, 257
  implicated in allergic contact dermatitis, 327
  organometallic compounds, 256-257
  thiazoles, 257
  thiocarbamate, 257
Fuses, changing, 82

G

Ganglionic cyst, 30, 241
Gas laws
  Boyle's law, 620
  Charles' law, 620
  Dalton's law, 620-621
  Henry's law, 621, 626
Gas laws, 619-621
Gas masks, 186
Gases
  breathing gas, 626, 632, 636
  oxygen depletion in confined spaces, 458-459
  toxic (*see* Toxic gases)
Gastrointestinal rupture, 630
Generators, standby, 84
GFCI, 80, 81
Giardia lamblia, 668
Gila monster, 340
Glanders, 342-343
Gloves, 192-193
  caring for, 193
  cut-resistant, 193
  fabric, 193
  leather, 193
  rubber, 193
  types of, 192
Gnats, 335
  biting, 105
Grab bars, 406-407
Grain fever, 277
Grain industry, 273-290
  health hazards, 273
    dust exposure, 275
    fatality rates, 276
    fumigants, 279
    grain suffocation, 371
    mycotoxins, 279-280
    nonpulmonary health disorders, 277
    pulmonary health disorders, 276-277
    silo gas, 278
  injury statistics, 275-276
  NIOSH checklists, 285-290
  rules and regulations, 282-283
  safe work practices, 280, 282
  safety hazards,
    explosive properties of grain dust, 281
    fire/explosion, 280
    machinery, 280
    silo loading/unloading, 280
    surface area, 280, 281
Great white shark, 699
Green tobacco sickness, 226, 261
Grey reef shark, 699
Ground-fault circuit interrupter (GFCI), 80, 81
Grounding, 82-83

Grounding adapters, 81
Guns, safety issues, 415-416
Guyon canal syndrome, 30

H

Hackberry leaf slug, 113
Hamilton, Dr. Alice, 55, 56
Hammerhead shark, 699
Hand-arm vibration (HAV), 10, 30, 53, 55-56
    controlling, 56-57
        engineering controls, 57
        medical surveillance, 57
        worker education, 57
    preventing exposure to, 56
    protective equipment for, 57
    treatment for, 57
    work practices, 57
Handicapped Farmers Program, 177
Hands, protecting, 192-193
Hanta viruses, 500
Harvester ants, 116
HAV (*see* Hand-arm vibration)
Hazardous Art Materials Labeling Act, 401-402
Hazards, 6-7
    agricultural, 172-173
    agricultural machinery (*see* Agricultural machinery)
    aquacultural (*see* Aquacultural hazards)
    assessing, 206-211
    biological, 6
    chemical (*see* Chemical hazards)
    controlling
        administrative methods, 208, 211
        engineering methods, 207-208
        using personal protective equipment, 208
    diving, 617-641 (*see also* Aquacultural hazards)
    electrical, 6, 77, 432-433 (*see also* Electrical safety)
    electrocution, 78-80
    envenomization, 89-125 (*see also* Arthropods)
    environmental, 4, 6, 499
    grain industry (*see* Grain industry)
    health, 207
    helicopter, 533-537
    household products, 408-412
    identifying, 205-206
    indoor environmental in animal housing, 419-438
    mechanical, 6, 432
    on fishing vessels, 713-716
    reporting, 206
    reproductive, 522
    respiratory (*see* Respiratory hazards)
    safety, 207
    timber harvesting, 513-517
    zoonotic, 127-141
Hearing
    audiometry, 62-63
    measuring acuity of, 62
    normal, 61-62
Hearing Conservation Amendment of 1983, 63
Hearing loss
    noise induced (*see* Noise-induced hearing loss)
    preventing, 64-65
Heat cramps, 73
Heat edema, 73
Heat exhaustion, 73

Heat index, 74
Heat stress, 67, 70-74
    at sea, 601
    categorization by temperature, 73
    climate evaluation methods, 71-72
    controlling, 44
    firefighters and, 522
    heat exposure limits, 71
    illnesses, 72-73
        heat cramps, 73
        heat edema, 73
        heat exhaustion, 73
        heat stroke, 73
        heat syncope, 73
        intertrigo, 73
        prickly heat, 73
    illnesses, 72-73
    metabolic heat production, 71
    metabolic heat rates, 70
    occupational exposures to, 72
    preventing, 27, 73-74
    prevention checklist, 75
    storage of heat in body, 70
Heat stroke, 73
Heat syncope, 73
Helicopter logging, 529-544
    aerodynamics, 534
    hazards, 533-537
        ground operations, 535-536
        human factors, 535, 541
        reducing, 536-537
    incidents involving, 530-532
    injury epidemiology, 532-533
    interagency/company cooperation, 542
    maintenance, 540-541
    NTSB data on crashes, 537-540
    safety management, 541-542
    training, 541
Helmets
    bicycle, 194
    for horseback riding, 318
    motorcycle, 194
    safety, 194
Hematophagous anthropods, 89
Henry's law, 621, 626
Herbicides, 144, 255-256
    bipyridyl, 256
    cancer and, 223-224
    chlorophenoxy, 255-256
    diquat, 256
    implicated in allergic contact dermatitis, 327
    paraquat, 256, 371, 464
    urea substituted, 256
Heterophyes, 668
High pressure nervous syndrome, 624
HIV, 146
Hodgkin's lymphoma, 5, 145
Honey bees, 118-119, 334
    venom from, 93
Hookworms, 345, 392
Hornets, 116-118
    venom from, 93
Hornshark, 702
Horse flies, 105-106
Horseback riding

at night, 318
clothing to wear for, 318
falling off horse, 318-319
helmets for, 318
on trails, 318
prevalence of accidents, 315
Horseflies, 335
Horses, 315-310
approaching, 317
choosing, 316-317
dismounting, 318
handling, 317
hauling, 319
leading, 317
mounting, 318
saddling, 317-318
safety considerations
for barn fires, 316
for stables, 316
tying the halter, 317
Horseshoe crab, 697
House dust mite, 100
Household products, 408-412
determining if hazardous, 408-409
disposing, 411-412
exposure to hazardous, 410
hazardous components, 408
labels on, 409
storing, 410-411
types of hazardous, 409-410
Household safety (see Farmstead safety)
House spider, 339
Human immunodeficiency virus (HIV), 146
Human lice, 109-111
Hydrogen sulfide, 430, 458, 598, 669
Hydroids, 691
Hydrostatic pressure, 619
Hygiene, industrial, 439-468 (see also Air pollutants; Respiratory hazards)
Hymenoptera, 115, 499 (see also Arthropods)
controlling, 120
stings, 92
Hypercarbia, 632
Hypersensitivity pneumonitis, 5, 277, 358-363
clinical features, 360-361
epidemiology, 358
etiologic agents, 358
findings from bronchoscopy with bronchoalveolar lavage, 362
histopathology, 361-362
immunopathogenesis of farmer's lung, 358-360
prognosis, 362-363
radiographic findings, 361
treatment, 362-363
Hypochondriasis, 155
Hypothermia, 69, 121, 607-611
preventing, 26
Hypoxia, 632

I

Illnesses (see Injuries and illnesses)
Industrial hygiene, in agriculture, 439-468 (see also Air pollutants; Respiratory hazards), 439
Infections
cutaneous, 684-686
skin (see Skin infections)
Infectious diseases, 499-501 (see also Diseases; Skin infections)
Injuries and illnesses, 3 (see also Diseases; Viruses)
among firefighters, 522-524
associated with cold and hot environments, 67-76
back, 22
barotrauma, 630-631
by bugs and insects (see Arthropods)
by marine animals (see Aquacultural hazards)
carbamate insecticide poisoning, 253-254
catch handling, 715
cerebral arterial gas embolism (CAGE), 629
chain saw, 498, 514
chemical vs. traumatic, 441
ciguatera poisoning, 666
cold stress (see Cold stress)
cumulative trauma disorders (CTDs) (see Cumulative trauma disorders)
decompression illness (DCI), 628-630
decompression sickness (DCS), 627, 628-629
dengue fever, 108, 121
disabilities, 715-716
dysbaric osteonecrosis, 632-633
dyspnea, 633
ehrlichiosis, 102
encephalitis, 108, 121
epidemiology of, 3-4
agricultural, 215-232
fishing industry, 557-570
forestry, 497-508
fatal traumatic, 215-218
fatality statistics in fishing industry, 562-567
from plant hairs, 328-329
from thorns, 328
gastrointestinal rupture, 630
gear handling, 715
grain dust exposure, 275
grain fever, 277
green tobacco sickness, 226, 261
heat stress (see Heat stress)
high pressure nervous syndrome, 624
hypercarbia, 632
hypoxia, 632
ice-related, 674
in farmers, 3, 4-6
in fishermen, 4-6, 7
in forestry workers, 4, 7
logging, 498, 513
malaria, 108
migrant workers and, 389-390
motion sickness (see Motion sickness)
musculoskeletal, 523
narcosis, 631
nonfatal traumatic, 218-220
on fishing vessels, 714
organic dust toxic syndrome (ODTS), 277
organophosphate insecticide poisoning, 251-253
oxygen toxicity, 631-632
pesticide poisoning, 220
prevention, 7
pulmonary edema, 633
pyrethroid poisoning, 255
recordkeeping logs, 211
respiratory (see Respiratory diseases; Respiratory hazards)
Rocky Mountain Spotted Fever, 103, 122, 341

scombroid poisoning, 665
scurvy, 601
silo filler's syndrome (SFS), 277
skin infections (*see* Skin infections)
slipping and falling, 715
statistics, 470
swimmer's ear, 632
tick paralysis, 122
transient unilateral facial paralysis, 633
vertigo, 630-631
worker accident death rates, 471
Insecticides, 5, 250-255
    cancer and, 223-224
    carbamate, 253-254
    chlorinated hydrocarbon, 254-255
    implicated in allergic contact dermatitis, 327
    organochlorine, 254
    organophosphate, 250-253
        acute poisoning from, 251
        biomarkers of exposure to, 252-253
        chronic neurological sequela after poisoning, 252
        chronic organophosphate exposure, 252
        intermediate syndrome after poisoning, 252
        preventing poisoning from, 253
        treatment for poisoning, 251-252
    pyrethroid, 255
Insects (*see* Arthropods)
Institute of Naval Medicine in England, 638
Intertrigo, 73
Intestinal parasites, 392
Irritant contact dermatitis, 326
Irukandji jelly fish, 688

J

Jellyfish, 687-697
    acute reactions to, 691
    Australian brown blubber, 689
    blue-tinted, 689
    box, 687-688
    cabbagehead, 689
    cannonball, 689
    comb, 690
    envenomation of, 690
    Irukandji, 688
    lion's mane, 689
    marbled, 689
    mauve, 688
    moon, 689
    mushroom, 689
    sea nettle, 688
    sea thimble, 689
    sizes, 687
    stinging alga, 689
    stinging cauliflower, 689
    treatment for stings, 690-691
Jones Act of 1920, 718-719
Jumping spider, 339

L

Labor camps, temporary, 472-473
Lateral epicondylitis, 30
Leatherbacks, 706
Legal issues, 713-720 (*see also* Court cases)
    fishermen and, 717-720
    fishing industry, 713-720
Legislation
    Americans with Disabilities Act (ADA), 504
    Commercial Fishing Industry Vessel Safety Act of 1988 (CFIVSA), 557-558, 559, 723
    Fair Labor Standards Act (FLSA), 245-246, 386
    Federal Food, Drug and Cosmetic Act (FFDCA), 257
    Federal Hazardous Substances Act of 1960, 408
    Federal Insecticide, Fungicide and Rodenticide Act (FIFRA), 7, 257
    Hazardous Art Materials Labeling Act, 401-402
    Hearing Conservation Amendment of 1983, 63
    Jones Act of 1920, 718-719
    Longshoremen and Harbor Worker's Compensation Act, 717
    Magnuson Fishery Conservation and Management Act (MFCMA), 559
    Migrant and Seasonal Agricultural Worker Protection Act, 386
    National Labor Relations Act, 386
    Occupational Safety and Health Act of 1970, 63, 386, 469
Lemon shark, 699
Leptospira, 664
Leukemia, 145
Lice, 109-111
Life saving equipment, 721
Lifting guidelines, 22-26
    asymmetric multiplier (AM), 22
    coupling multiplier (CM), 23
    distance multiplier (DM), 22
    equation for, 23-24
    frequency multiplier (FM), 23
    horizontal multiplier (HM), 22
    recommended weight limit (RWL), 22
    vertical multiplier (VM), 22
Lightning, 85-86
Lionfish, 704, 705
Lion's mane jellyfish, 689
Lip cancer, 145
Lobster, 697
Local fatigue, 21
Loggers (*see* Forestry workers)
Logging incidents, 498
Logging industry, types of people involved in, 509 (*see also* Forestry; Timber harvesting)
Logging Operations Standard, 545-554
    brakes, 551
    bucking, 553
    chain saws, 549-550
    chipping, 553
    environmental conditions, 548-549
    exhaust systems, 551
    explosives, 549
    first aid, 548
    flammable liquids, 549
    general requirements section, 547
    hand and portable powered tools (HPPT), 549-550
    limbing, 553
    loading and unloading, 553
    machine access, 551
    machines, 550
    power lines, 549
    protective structures, 550-551
    safety footwear, 547-548
    seat belt requirements, 548
    training, 554
    tree felling, 552-553

vehicles, 552
work areas, 549
yarding, 553
Lone star tick, 102
Longshoremen and Harbor Worker's Compensation Act, 717
Loxoscelism, 121
Lumberers (*see* Forestry workers)
Lyme disease, 4, 94, 103, 121, 335, 343, 499-500
Lymphadenitis, 121
Lymphoma (*see also* Cancer)
  Hodgkin's, 5, 145
  non-Hodgkin's, 145

M

Madura foot, 345
Magnuson Fishery Conservation and Management Act (MFCMA), 559
Mako shark, 699
Malaria, 108, 500
Malignant melanoma (MM), 324
Man-o-war
  Pacific, 690
  Portuguese, 690
Management, safety, 198
Manual materials handling (MMH), 22
  employer checklist for, 26
  equipment selection guidelines, 42-43
Marbled jellyfish, 689
Marine organisms, 597 (*see also* Aquacultural hazards)
Marine worms, 696-697
Mauve jelly fish, 688
Meatpacking guidelines, 27
Mechanical filter respirators, 186
Mechanical hazards, 6
  in animal housing, 432
Medial epicondylitis, 30
Melanoma, 5 (*see also* Cancer)
Mental health, 153-170
  agricultural industry and, 153-154, 158-162
  disorders, 154-158
    adjustment, 157
    anxiety, 154-155, 157
    eating, 156, 157
    mood, 154, 157
    pain, 155-156
    personality, 156-158
    psychotic, 154, 157
    post-traumatic stress disorder (PTSD), 164
    schizophrenia, 157
    somatoform, 155-156, 157
    substance-related, 157, 158, 166-167
  farming and, 153-154, 158-162
  fishing industry and, 153-154, 165-167
  forestry and, 153-154, 162-165
  interventions, 167, 169
  symptom checklist, 168
Metamorphosis, 91, 121
Methane gas, 430
MFCMA, 559
Microbes, 462
Microorganisms, 414
Middle ear barotrauma, 630
Migrant and Seasonal Agricultural Worker Protection Act, 386
Migrant workers, 385-396
  child laborers, 391
  commercial fishing, 560
  employment conditions, 386
  employment regulations, 386
  health and safety regulations, 386-389
  health care barriers, 394-395
  health problems among, 391-393
  health status, 391
  high risk behaviors of, 393-394
  housing regulations, 386-389, 390-391
  occupational illnesses and injuries, 389-390
  OSHA standards, 388
  socioeconomic conditions, 390-391
  worker's compensation insurance, 387
Milker's nodule, 343
Millipedes, 339
Minnesota Farming Health Survey, 218
Mites, 99-100, 337
  cable, 100
  house dust, 100
  northern fowl, 100
  paper, 100
  scabies, 99-100
  straw itch, 100
MM, 324
MMH (*see* Manual material handling)
Mollusks, 695-696
Mood disorders, 154
Moon jellyfish, 689
Moray eels, 699
Mosquitoes, 89, 106-108, 335, 499
  bite prevention tips, 121
Moths, 336
Motion sickness, 643-657
  aids to habituation, 648
  anatomy and physiology, 644-645
  etiology, 645-647
    aerobic fitness, 647
    fluid shift theory, 646
    heredity, 646
    mal de debarquement, 647
    neurologic connections, 646-647
    oculocardiac reflex, 646
    personality traits, 647
    sensory conflict theory, 645
    visceral gravireceptors, 645-646
  treatment, 647-654
    acupuncture for, 648-649
    pharmacologic, 649-654
      antihistamines, 653
      calcium antagonists, 652-653
      dimenhydrinate, 652
      doxepin, 653
      experimental approaches, 654
      ginger root, 652
      lecithin, 652
      lidocaine, 653
      list of medications, 650
      local anesthetics, 653
      meclizine, 653
      metoclopramide, 653
      phenothiazines, 652
      phenytoin, 653
      scopolamine, 651
      stimulants, 651-652

tocanide, 653
Motorcycle helmets, 194
Mud daubers, 119
Multiple myeloma cancer, 145
Muscle fatigue, 10
Musculoskeletal disorders, 226-227
Musculoskeletal injuries, 523
Mushroom jellyfish, 689
Mycetoma, 345
Mycobacterium, 661-662
Mycobacterium marinum, 685
Mycotoxins, 279-280, 371, 463
   in fish feed, 666
Myeloma cancer, multiple, 145
Myiasis, 90, 122, 335

N

Nanophyetus salmincola, 668
Narcosis, 631
Nasal cancer, 502
National AgrAbility Project, 176, 177
National Board of Diving and Hyperbaric Medical Technology, 638
National Electric Code (NEC), 83
National Electric Safety Code (NESC), 79
National Fire Protection Association (NFPA), 202
National Institute for Occupational Safety and Health (NIOSH), 215, 446, 558, 605
   lifting guidelines, 22-26
      asymmetric multiplier (AM), 22
      coupling multiplier (CM), 23
      distance multiplier (DM), 22
      equation for, 23-24
      frequency multiplier (FM), 23
      horizontal multiplier (HM), 22
      recommended weight limit (RWL), 22
      vertical multiplier (VM), 22
National Labor Relations Act, 386
National Marine Fisheries Service, 716
National Migrant Resource Program Inc., 177
National Oceanic and Atmospheric Administration (NOAA), 722
National Safety Council (NSC), 172, 215
National Transportation Safety Board (NTSB), 530, 537
National Traumatic Occupational Facility (NTOF), 215, 497
Nausea, 643
Naval Medical Research Institute, 638
Navy Experimental Diving Unit, 638
NEC, 83
Needlefish, 701
NESC, 79
Neurotoxins, in fish feed, 666
NIHL (see Noise-induced hearing loss)
NIOSH (see National Institute of Occupational Safety and Health (NIOSH))
NITROX, 626
NOAA, 722
Nocardia, 662
Noise
   in animal housing, 433-434
   permissible exposure limit (PEL), 63
   reducing exposure to, 64
   underwater, 621-622
Noise-induced hearing loss (NIHL), 59-65
   definition, 59, 62
   epidemiology of, 226
   in animal housing, 433
   physical principles of, 59-61
   preventing, 64-65
   regulatory issues, 63-64
   sound
      frequency of, 59-60
      intensity of, 60
      measuring, 60
   time of exposure, 61
Non-Hodgkin's lymphoma, 145
Northern fowl mite, 100
NSC, 172, 215
NTOF, 215, 497
NTSB, 530, 537
Nudibranches, 696

O

Occupational dermatitis, 4
Occupational Safety and Health Act of 1970, 63, 386, 469
Occupational Safety and Health Administration (OSHA), 7, 27, 146, 202, 245, 282, 387, 446, 716
   law enforcement, 546-547
   Logging Operations Standard, 545-554
      brakes, 551
      bucking, 553
      chain saws, 549-550
      chipping, 553
      environmental conditions, 548-549
      exhaust systems, 551
      explosives, 549
      first aid, 548
      flammable liquids, 549
      general requirements section, 547
      hand and portable powered tools (HPPT), 549-550
      limbing, 553
      loading and unloading, 553
      machine access, 551
      machines, 550
      power lines, 549
      protective structures, 550-551
      safety footwear, 547-548
      seat belt requirements, 548
      training, 554
      tree felling, 552-553
      vehicles, 552
      work areas, 549
      yarding, 553
   states and territories operating under state programs, 471
Occupational vibration exposure (see Vibration; Vibration exposure)
Oceanic whitetip shark, 699
Octopuses, 696
Oculocardiac reflex, 646
ODTS (see Organic dust toxic syndrome)
Old wife fish, 706
Orf, 343, 344
Organic dust toxic syndrome (ODTS), 277, 353, 355-358, 463
   clinical features, 357-358
   definition, 355-356
   epidemiology, 356
   etiologic agents, 356
   prognosis, 358
   treatment, 358
Organochlorine insecticides, 254
Organometallic compounds, 256-257

Organophosphate insecticides, 250-253
　acute poisoning from, 251
　biomarkers of exposure to, 252-253
　chronic neurological sequela after poisoning, 252
　chronic organophosphate exposure, 252
　intermediate syndrome after poisoning, 252
　preventing poisoning from, 253
　treatment for poisoning, 251-252
OSHA (*see* Occupational Safety and Health Administration)
Oxygen, 626
　depletion of in confined spaces, 458-459
Oxygen toxicity, 631-632

P

Pacific man-o-war, 690
Pain disorders, 155-156
Paper mite, 100
Paper wasps, 116-118
PAPR, 186
Paralysis, transient unilateral facial, 633
Paranasal sinus barotrauma, 630
Paraquat, 256, 371, 464
Parasites, 128, 666-668
　Anasakiasis, 666-667
　Clonorchis sinensis, 667
　cultured prawn for trematode control, 668
　Eustrongylidiasis, 667
　Giardia lamblia, 668
　Heterophyes, 668
　intestinal, 392
　Nanophyetus salmincola, 668
　Philometra, 667
　Schistosoma, 668
Parkinson's disease, 225
Pedicelliriae, 694-695
PEL, 441-442
Permissible exposure limit (PEL), 441-442
Personal flotation device (PFD), 609, 610
Personal protective equipment (PPE), 181-195
　body coverings, 190-191
　　chain saw safety chaps, 190
　　chemical resistant coveralls, 191
　　leather aprons, 190
　　rubber aprons, 190
　cleaning and maintenance, 489
　EPA requirements, 485
　foot protection, 189-190, 547-548
　　rubber boots, 190
　　steel-toed shoes, 189-190
　for forestry workers, 503-505
　for hand-arm vibration (HAV), 57
　for operating agricultural machinery, 245
　hand protection, 192-193
　head protection, 193-194
　　bicycle helmets, 194
　　bump caps, 193-194
　　motorcycle helmets, 194
　　safety helmets, 194
　recommendations, 209-210
　respiratory (*see* Respirators)
　substitutions and exceptions to, 489
　using to control hazards, 208
Personality disorders, 156-158 (*see also* Mental health)
Pesticide Farm Safety Center (PFSC), 476

Pesticides, 5, 144, 249-250, 464
　cancelled, 250
　cancer and, 223-224
　carbamate, 5
　DDT, 5-6, 254
　EPA standards for, 483-492
　　compliance requirements, 491
　　employee protection, 486-487
　　exception requests, 489
　　governmental provisions, 485-486
　　handler application restrictions, 485
　　handler instructions, 485
　　handler protection, 488-489
　　protective equipment requirements, 485
　　safety training, 484
　　worker application restrictions, 484-485
　　worker protection, 487-488
　exposure to, 502
　Federal regulation of, 257-258
　handling clothes soiled with, 413
　implicated in allergic contact dermatitis, 327
　in wild-caught fish, 666
　known to cause allergies, 250
　laundering clothes soiled with, 413-414
　paraquat, 256, 371, 464
　poisoning from, 220
　preventing injuries by, 261
　pyrethroid, 5
　pyrethrum, 5
　residues around farmstead, 412-414
　restricted, 250
　suspended, 250
PFD, 609, 610
PFSC, 476
Philometra, 667
Physical activity, 20-21
Physical examinations, for diving, 635-636
Phytodermatitis
　allergic contact, 329-333
　irritant contact, 329
Phytophotodermatitis, 333
Pit vipers, 340
Plague, 342
Plants
　allergic contact phytodermatitis from, 329-333
　allergic reactions to, 331-333
　irritant contact phytodermatitis from, 329
　mechanical injury by, 328
　nonpoisonous, 401
　pharmacologic injury by, 328-329
　poisonous, 330-331, 401
Plesiomonas shigelloides, 663
Pneumonitis, hypersensitivity (*see* Hypersensitivity pneumonitis)
Poison ivy, 330
Poison oak, 330
Poison sumac, 330
Poisonings, child, 400-401
Poisonous plants, 401
Pollutants, air (*see* Air pollutants)
Ponds, farm, 415
Port Jackson shark, 702
Portuguese man-o-war, 690
Post-traumatic stress disorder (PTSD), 164
Posture, awkward, 12-13
Pott, Sir Percivall, 4

Power lines, 549
    determining equipment clearances, 80
    overhead, 79
Powered air-purifying respirator (PAPR), 186
PPE (*see* Personal protective equipment)
Prickly heat, 73
Programs, safety and health (*see* Safety and health programs)
Pronator teres syndrome, 30
Prostate cancer, 145
Protection equipment (*see* Personal protective equipment)
Psychotic disorders, 154, 157
PTSD, 164
Public lice, 109-111
Pulmonary barotrauma, 630
Pulmonary edema, 633
Pulmonary health disorders, 276-277
Pyrethroid insecticides, 255

Q

Q-fever, 342

R

Rabbit fever, 342
Rabbitfish, 706
Rabies, 341, 346, 500
Radial tunnel syndrome, 30
Radiation, exposure to, 146
Radio beacons, emergency position indicating radio beacon (EPIRB), 721-723
Radon, 267
Ragweed dermatitis, 332
Ramazzini, Bernardino, 9
Ramps, 405
Ratfish, 702
Rattlesnakes, 340
Reason's conflict theory, 645
Recommended weight limit (RWL), 22
Recordkeeping
    employee medical and exposure records, 211
    fishing vessel documentation, 589-596
    injury and illness logs, 211
    requirements for health and safety, 470-472
    training records, 211
Red imported fire ant (RIFA), 92, 115-116, 334-335
Red soft coral, 691
Reef whitetip shark, 699
Regulations (*see also* Legislation; Standards)
    agricultural safety and health, 469-482
    CFIVSA, 557-558
    CFR Title 29
        Part 1910, 470
        Part 1910.111, 475, 476
        Part 1910.142, 472-473
        Part 1910.145, 477
        Part 1910.266, 477
        Part 1910.1200, 475-476
        Part 1928, 470
        Part 1928.21, 477
        Part 1928.51, 477
        Part 1928.57, 477
        Part 1928.110, 473-475
    chemical hazards, 475-477
    child labor, 391, 478-479

Coast Guard, 636
    diving, 636-637
    EPA (*see* Environmental Protection Agency)
    farm machinery, 477-478
    Federal, 202
    field sanitation, 473-475
    fishing vessel, 716, 721-724
    grain bin electrical safety, 79
    grain industry, 282-283
    hazard communications, 475-476
    life saving equipment, 721
    migrant workers
        employment, 386
        health and safety, 386-389
        housing, 386-389, 390-391
    OSHA (*see* Occupational Safety and Health Administration)
    pesticides, 257-258
    recordkeeping requirements, 470-472
    storing/handling anhydrous ammonia, 475
    temporary labor camps, 472-473
    timber harvesting (*see* Logging Operations Standard)
Repetition, 12
Reproductive hazards, 522
Reptiles, 339-341, 346
    copperhead snakes, 340
    coral snakes, 340
    Gila monster, 340, 499
    pit vipers, 340
    rattlesnakes, 340
    snake bites, 499
    water moccasin snake, 340
Respirators, 186-189
    air supplying, 187
    chemical cartridge, 186
    color codes for, 187
    contact lenses and, 188
    gas masks, 186
    health considerations of the user, 187-188
    life-span, 189
    maintenance, 189
    mechanical filter, 186
    package safety markings, 187
    powered air-purifying respirator (PAPR), 186
    selecting, 188
    self-contained breathing apparatus (SCBA), 187
    testing the fit of, 188
    testing the pressure of, 188
    using, 188
Respiratory diseases, 4, 5, 353-383
    airway, 363-368
    asthma (*see* Asthma)
    bronchitis (*see* Bronchitis)
    farmer's lung disease (*see* Farmer's lung disease)
    hypersensitivity pneumonitis (*see* Farmer's lung disease)
    nasal cancer, 502
    organic dust exposure, 355-363
    organic dust toxic syndrome (*see* Organic dust toxic syndrome)
    related to agricultural industry, 353-383
    rhinitis (*see* Rhinitis)
    toxic gas exposures, 368-371
Respiratory hazards, 182-186, 220-222 (*see also* Air pollutants)
    aerosols, 429, 456, 460-464
    ammonia, 371, 429-430
    anhydrous ammonia applications, 456, 475
    asbestos-containing dust, 182

associated with animal confinement, 259
associated with farming, 184-185
carbon monoxide, 370, 430-431, 458, 522
disinfectants, 459
dust, 182, 429, 460-462
epidemiology of, 220-222
   for dairy producers, 221-222
   for grain producers and handlers, 222
   for mushroom workers, 222
   for poultry producers, 222
   for swine producers, 221
from combustion driven sprayers, 458
fumes, 182
fumigants, 279, 370-371, 459
gaseous, 182-183, 456-460
grain dust exposure, 275
identifying contaminants, 183
inorganic dusts, 460
manure pit gas, 369-370, 458
measuring contaminants, 183
microbes, 462
mist, 182
mycotoxins, 279-280, 371, 463
natural toxins, 462-463
organic dusts, 460-462
oxygen-deficient atmospheres, 183
oxygen depletion in confined spaces, 458-459
particulates, 182
reducing, 183, 186
   administrative controls, 186
   engineering controls, 183, 186
   using respirators, 186
silo gas, 368-369, 457
vapors, 183
Respiratory system, 440-441
Rhinitis, 277, 363-366
allergy testing, 365
causative agents, 364
clinical findings, 364-365
epidemiology, 363-364
pulmonary function testing, 365
treatment, 365-366
Rickettsia, 128
RIFA, 92, 115-116, 334-335
Rocky Mountain Spotted Fever, 103, 122, 341, 500
Rodenticides, 257
Rotator cuff tendinitis, 30
Running spider, 339
Rural Institute on Disabilities, 178
RWL, 22

S

Saddleback caterpillar, 114
Safety
electrical (*see* Electrical safety)
firearm, 415-416
firefighting, 525
fishing vessels at sea, 571-604
food, 414-415
on farmsteads (*see* Farmstead safety)
timber harvesting, 509-520
water, 415
Safety and health programs, 197-211
committees, 201-202

company philosophy, 197
design reviews, 206
for fishing vessels, 574
functions, 205-206
hazard assessment, 206-211
hazard control, 207-211
   administrative methods, 208, 211
   engineering, 207-208
   using personal protective equipment, 208
hazard identification, 205-206
hazard reporting, 206
health hazards, 207
importance of, 197
management responsibilities, 197-198
   accident causes traced back to, 199-201
   controlling, 198
   directing, 198
   organizing, 198
   planning, 198
   staffing, 198
policies and procedures, 198-202
recordkeeping
   injury and illness logs, 211
   employee medical and exposure records, 211
   training records, 211
regulations, 202
   EPA, 202
   Federal, 202
   OSHA, 202
safety hazards, 207
staffing, 198, 201
standards, 202-205
   ANSI, 202-205
   consensus, 202
   NFPA, 202
surveys, 205-206, 205
timber harvesting, 517-519
training, 211
Safety helmets, 194
Safety management, 198
Sand flies, 105
Sanitation, field, 473-475
Sarcoma, soft tissue, 5, 145
Satellite systems, COSPAS/SARSAT, 722
Sawfish, 700, 701
Scabies mite, 99-100
SCBA, 187
SCC, 324-325
Schistosoma, 668
Schizophrenia, 157
Scombroid poisoning, 665
Scorpionfish, 705
Scorpions, 96-97, 339, 499
Scrotal cancer, 4
SCUBA, 623
Scurvy, 601
Sea anemones, 691-692
Sea cucumbers, 693
Sea nettles, 688
Sea sickness (*see* Motion sickness)
Sea snakes, 707
Sea thimble jellyfish, 689
Sea urchins, 694, 695
Seasonal workers (*see* Migrant workers)
Self-contained breathing apparatus (SCBA), 187

Self-contained underwater breathing apparatus (SCUBA), 623
Sensory conflict theory, 645
SFS, 277
Sharks, 597, 698-699
    blacktip reef, 699
    blue, 699
    bull, 699
    bullhead, 702
    great white, 699
    grey reef, 699
    hammerhead, 699
    hornshark, 702
    lemon, 699
    mako, 699
    oceanic whitetip, 699
    Port Jackson, 702
    reef whitetip, 699
    spinner, 699
    spotted, 699
    tiger, 699
Signature spider, 339
Silo filler's syndrome (SFS), 5, 277
Silo gas, 278, 457
    effects of exposure to, 368-369
Silo unloader's syndrome, 5
Silos, chemical hazards of, 259-260
Skin cancer, 145, 681-683
Skin diseases, 6, 321-352, 681-712 (see also Skin infections)
    aeromonas hydrophila, 685-686
    animal bites (see Animals)
    cancer, 324 (see also Cancer)
    basal cell epithelioma carcinoma (BCE), 324-325
        malignant melanoma (MM), 324-325
        squamous cell carcinoma (SCC), 324-325
    cutaneous infections, 684-686
    dermatitis (see Dermatitis)
    due to echinoderms, 693-695
    due to environmental extremes, 325-326, 683 (see also Cold stress; Heat stress)
    due to sponges, 692-693
    due to trauma in fishing operations, 683
    due to ultraviolet light (sun), 321-325, 681-683
    erysipeloid, 684-685
    farmers and, 321-352
    hazard prevention from, 346-347
    immersion foot, 683
    in fishermen, 681-712
    insect bites and stings (see Arthropods)
    medications that cause photoallergic reactions, 323
    medications that cause phototoxic reactions, 323
    mycobacterium marinum, 685
    reptile bites (see Reptiles)
    trenchfoot, 326
    vibrio infection, 686
Skin infections, 341-347, 665
    animal-related, 341-344
    anthrax, 342
    blastomycosis, 345
    brucellosis, 342
    bubonic plague, 342
    catworms, 345
    chromomycosis, 345
    coccidioidomycosis, 345
    creeping eruption, 345-346
    ehrlichiosis, 342, 500

endemic typhus, 342
    erysipheloid, 342
    fungal-related, 344-347
    glanders, 342-343
    madura foot, 345
    mycetoma, 345
    plague, 342
    Q-fever, 342
    rabbit fever, 342
    related to diving, 632
    sporotrichosis, 345, 499
    staphylococci, 341
    streptococci, 341
    tularemia, 342, 500
Skin types, 322
Slugs (see Caterpillars)
Snakes, sea snakes, 707 (see also Reptiles)
Snook, 701
Soft tissue sarcoma, 5, 145
Somatoform disorders, 155-156
Sonar, 622
Sound
    frequency of, 59-60
    intensity of, 60
    measuring, 60
Spiders, 97-98, 338-339
    bite prevention tips, 120
    black widow, 97-98, 338-339
    brown recluse, 98-99, 339
    fishing, 339
    house, 339
    jumping, 339
    running, 339
    signature, 339
    tarantulas, 339
    venom from, 93
    wolf, 339
Spinner shark, 699
Spiny dogfish, 702
Sponges, toxic, 692-693
Sporotrichosis, 4, 345, 499
Spotted shark, 699
Squamous cell carcinoma (SCC), 324
Squid, 696
Stable flies, 104-105
Stairs, 405-406
Standards (see also Regulations)
    ANSI, 202-205
    EPA, 202
    EPA pesticide, 483-492
        compliance requirements, 491
        employee protection, 486-487
        employer responsibilities, 483-484
        exception requests, 489
        governmental provisions, 485-486
        handler application restrictions, 485
        handler instructions, 485
        handler protection, 488-489
        worker application restrictions, 484-485
        worker protection, 487-488
    OSHA's logging operations (see Logging Operations Standard)
    worker protection standard (WPS), 483-484, 489-490
Staphylococci, 684
Staphylococcus, 686
Starfish, 693-694

Stargazers, 704-705
Stenosing tenosynovitis, 30
Stenosing tenosynovitis crepitans, 30
Stinging alga jellyfish, 689
Stinging cauliflower jellyfish, 689
Stingrays, 702-703
Stings, Hymenoptera, 92 (*see also* Arthropods; Envenomization)
Stomach cancer, 145
Stonefish, 705-706
Story, Joseph, 717
Strangulation, 400
Straw itch mite, 100
Streptococci, 684
Streptococcus, 665, 686
Stress (*see* Mental health)
Suffocation, 399
    grain, 371
Suicide, 602
Sun exposure, 321-325, 681-683
Sunscreen, 346
Surgeonfish, 706
Survival techniques at sea, 597-603
    abandoning ship, 598-600
    air rescues, 614-615
    burns, 598
    cold water immersion, 606-613
    crushed appendages, 598
    drown-proofing, 611-612
    emergency procedures, 597
    fires, 598
    food, 600-601
    fumes, 597-598
    heat stress, 601
    man overboard, 603
    marine organisms, 597
    raft navigation, 602-603
    recreation, 602
    rescue signals, 602
    sickness, 601-602
    staying warm, 601
    suicide, 602
    surface rescues, 613-614
    treading water, 612-613
    water, 600
Sweat bees, 119
Swimmer's ear, 632

T

Tarantulas, 339
Temperature
    cold stress (*see* Cold stress)
    cold water immersion, 606-613
    CTDs and, 14-15
    dry bulb temperature (DBT), 71
    excessive exposure to the sun, 321-325, 681-683 (*see also* Heat stress)
    heat assessment ISO index, 72
    heat index, 74
    heat stress (*see* Heat stress)
    injuries associated with extreme environments, 67-76
    internal, 26
    thermal extremes in, 26-27
    thermal stress, 67
    wet bulb globe temperature (WBGT), 71-72
    wind chill, 68, 607
Tendinitis, 10, 30, 241
    rotator cuff, 30
Tenosynovitis, 10
    stenosing, 30
Testicular cancer, 145
Tests and testing
    audiometric, 62-63
    for air pollutants, 446-451
Tetanus, 500
Therapy (*see* Treatment)
Thermal stress, 67
Thiazoles fungicides, 257
Thiocarbamate fungicides, 257
Thoracic outlet syndrome, 30
Threshold limit value (TLV), 441-442
Tick paralysis, 122
Ticks, 89, 94, 101-103, 337-338, 341, 499, 500
    American dog, 102
    bite prevention tips, 120-121
    black-legged, 102
    brown dog, 102
    lone star, 102
Tiger shark, 699
Timber harvesting, 509-520
    components, 509-513
    cutting timber, 510
    equipment,
        chain saws, 517
        chains, 516-517
        chippers, 517
        tow lines, 516
        winches, 516
    escape routes, 517
    felling trees, 510-512
        bucking, 511
        difficult, 511
        limbing and topping, 511-512
        methods, 510-511
        safety rules, 511
    hazards, 513-517
        environmental conditions, 515
        equipment factors, 515-517
        human factors, 514-515
    helicopter operations (see Helicopter logging)
    loading zone, 512-513
    regulations *(see* Logging Operations Standard)
    safety program, 517-519
    transporting, 512-513
TLV, 441-442
Toadfish, 699, 700, 703
Tools
    design of, 16-22
        agricultural machinery/equipment, 18-19
        controls/displays, 19-20
    double-insulated, 80
    electrical safety and, 80-81
Torpedo rays, 706
Toxic gases
    ammonia, 429-430, 456
    anhydrous ammonia, 371, 475
    around livestock, 457-458
    carbon dioxide, 430, 522
    carbon monoxide, 370, 430-431, 458, 522
    disinfectants, 459

diving and, 631-632
exposure to, 368-371
from manure pits, 369-370
fumigants, 370-371, 459
health hazards from, 450
hydrogen sulfide, 430, 458, 598, 669
manure pit, 369-370, 458
measuring, 447
methane gas, 430
mycotoxins, 371, 463
paraquat poisoning, 371
sampling and analyzing methods, 452
silo gas, 457
Training, safety and health programs, 211
Transient unilateral facial paralysis, 633
Trauma, definition, 10
Treatment
    for allergic contact dermatitis, 332-333
    for asthma, 365-366
    for bites/stings
        assassin bugs, 113
        bed bugs, 112
        blisher beetles, 115
        black flies, 104
        black widow spider, 98, 339
        brown recluse spider, 99, 339
        caterpillars, 114
        centipede, 96
        chiggers, 101
        conenose bugs, 113
        deer flies, 106
        fishes, 597, 701-702, 706
        fleas, 109
        gnats, 105
        harvester ants, 116
        honey bees, 119
        hornets, 118
        horseflies, 106
        Hymenoptera stings, 335
        jellyfish, 690-691
        lice, 110-111
        mosquitoes, 108
        red imported fire ants, 116
        scorpions, 97
        sea snake bites, 707
        sea urchin stings, 695
        snake bites, 340
        stable flies, 105
        ticks, 103
        velvet ants, 120
        wasps, 118
        wheel bugs, 113
        yellowjackets, 118
    for bronchitis, 368
    for carbamate insecticide poisoning, 254
    for decompression illness (DCI), 629-630
    for electric shock, 85, 86
    for farmer's lung disease (FLD), 362-363
    for hand-arm vibration (HAV), 57
    for irritant dermatitis, 327
    for malignant melanoma (MM), 325
    for motion sickness, 647-654
    for organic dust toxic syndrome (ODTS), 358
    for organophosphate insecticide poisoning, 251-252
    for rhinitis, 365-366
    for scombroid poisoning, 665
    for zoonotic diseases, 132
Trenchfoot, 326
Trigger finger, 10
Tuberculosis, 226, 392-393
Tularemia, 342, 500

U

U.S. Coast Guard, 716
    safety coordinators, 723-724
UBA (see Underwater breathing apparatus)
Undersea and Hyperbaric Medical Society (UHMS), 638
Underwater breathing apparatus (UBA), 617, 622-624
    breath hold diving, 623-624
    self-contained, 623
    surface supplied, 622-623
United States Department of Agriculture (USDA), 146
Urea substituted herbicides, 256
USDA, 146

V

Vaccines
    for fish, 670
    for zoonotic diseases, 130
Vapor hazard index (VHI), 443
Vapor hazard ratio (VHR), 443
Vascular abnormalities, 69
Ventilation, 421-428
    cool weather, 423-425
    in animal housing, 421-428
    instruments for monitoring, 425-428
    natural, 422-423
    rate of, 421-422
    system control, 425
    warm weather, 423
Vertigo, 630-631
VHI, 443
VHR, 443
Vibration, 53-58
    back pain and, 55
        CTDs caused by, 13-14
        effects of, 241
        hand-arm vibration (HAV), 10, 30, 53, 55-56
    controlling, 56-57
        preventing exposure to, 56
        protective equipment for, 57
        treatment for, 57
        work practices, 57
    measuring, 54
    physical principles of, 53-54
        acceleration, 53
        displacement, 53
        frequency, 53
        resonance, 54
        velocity, 53
    standards, 54
    whole-body vibration (WBV), 53, 54-55
Vibration exposure, definition, 53
Vibration-induced white finger disease (VWFD), 498, 504, 505
Vibrio, 662-663
Viruses, 128
    Epstein-Barr virus (EBV), 146
    human immunodeficiency virus (HIV), 146

Vitamins, 601-602
VWFD, 498, 504, 505

W

Wasps, 116-118, 334-335
  venom from, 93
Water
  cold water immersion, 606-613
  treading at sea, 612-613
Water contaminants, 267
  inorganic elements, 267
  microbial pathogens, 267
  organics, 267
  radioactive elements, 267
Water moccasin snakes, 340
Water safety, farm ponds, 415
Water supplies, 267-272
  lead coming into, 271
  wells (see Wells)
Water table, 269
Waterbug, 335
Waterscorpions, 335
WBGT, 71-72
WBV, 53
WBV, 54-55
Weeverfish, 704
Wells, 268-272
  age of, 270
  backflow prevention, 270-271
  casing and cap of, 269
  condition of, 268
  depth of, 269
  drilled vs. dug, 270
  grouting around, 269-270
  location, 268
  pollution and, 268
  protection from ground surface, 270
  soil surrounding, 269
  testing, 271-272
  unused, 271
  water table, 269
Wet bulb globe temperature (WBGT), 71-72

Wheel bugs, 112-113, 335
WHO, 605
Whole-body vibration (WBV), 53, 54-55
Wind chill, 68, 607
Wolf eel, 700
Wolf spider, 339
Worker protection standard (WPS), 483-484, 489-490
Worker's compensation, 717
World Health Organization (WHO), 605
Worms, marine, 696-697

Y

Yellowjackets, 116-118
Yersinia ruckeri, 664

Z

Zoonotic diseases, 6, 122, 462
  diagnosing, 130, 132
  epidemiology of, 226
  etiologic agents, 127-128
    bacteria, 127-128
    chlamydia, 128
    fungi, 128
    parasites, 128
    rickettsia, 128
    viruses, 128
  firefighters and, 523
  preventing/controlling, 128-130
    cleaning/disinfecting, 130, 131
    ingestion, 130
    inhalation, 129
    insect bites, 129
    skin contact, 128-129
    skin penetration, 129
    vaccines for, 130
  safety checklist for avoiding, 132
  table of occupational infections, 134-141
  transmission and infectivity, 127-128
  treatment, 132
Zoonotic hazards, 127-141
Zooplankton, 686-687

Powerful New Books from Government Institutes

Environmental

Property Rights: Understanding Government Takings and Environmental Regulation

This new book explains what exactly constitutes a governement taking, what the laws and regulations are, state issues, litigation disputes, adequate compensation and more. Landowners, developers, as well as activists in environmental and property rights movements, will find this book to be the most comprehensive source available on property rights law.
Hardcover, 350 pages, 1997, ISBN: 0-86587-554-5 *$79*

Wetlands Mitigation: Mitigation Banking and Other Strategies for Development and Compliance

Written in plain English, not legalese, this practical reference provides explanations of the relevant regulations, and guidance on mitigation requirements and the permitting process. the book also includes helpful tables and checklists, numerous "real life" mitigation case studies, and several useful appendices.
Hardcover, Index, 272 pages, 1997, ISBN: 0-86587-534-0 *$75*

Environmental Engineering and Science: An Introduction

The new **Environmental Engineering and Science: An Introduction** covers the basics environmental professionals need to comply with regulations, manage resources and technology, and control pollution. Case studies, illustrations, exercises and solved numerical problems make this book an excellent teaching tool and desk reference.
Softcover, Index, 400 pages, April 1997, ISBN: 0-86587-548-0 *$79*

Internet & Telecommunications

Internet and the Law: Legal Fundamentals for the Internet User

This book provides you with an understanding of the legal landscape within which you operate. The basic principles pertaining to laws of copyright, trademark, trade secret, patent, libel/defamation and related issues as well as the basic principles of licensing are explained. This book outlines steps you can take to avoid or minimize your chances of unknowingly engaging in unlawful activity. In addition, this book helps assure that your rights as an Internet user will be protected.
Hardcover, 249 pages, 1996, ISBN:0-86587-506-5 *$75*

Telecommunications Act Handbook: A Complete Reference for Business

The Telecommunications Act of 1996 will dramatically change the telecommunications industry. This new Comprehensive handbook provides a practical, non-legalese explanation of the challenges, risks and benefits confronting this new telecommunications age. It provides business executives and non-legal scholars with an understanding of the new framework that will govern telecommunication activities for years to come. It also summarizes the history and the development of each sector of the telecommunications industry, and then clearly explains how the new legislation will change each industry sector.
Hardcover, Index, 640 pages, 1996, ISBN: 0-86587-545-6 *$89*

Environmental Guide to the Internet, 2nd Edition

Find the environmental information you need quickly and easily on the Internet. From environmental engineering to hazardous waste compliance issues, you'll have no problem finding it with this guide. Also includes 255 World Wide Web Sites.
Softcover, 235 pages, 1996, ISBN:0-86587-517-0 *$55*

Visit our website for the latest news updates on books, courses, electronic products and other training materials! http://www.govinst.com

Safety and Health

Cumulative Trauma Disorders: A Practical Guide to Prevention and Control
Cumulative Trauma Disorders are the number one worker's compensation claim injury in the United States. This new book provides practical, cost-effective approaches to preventing CTD's. Using numerous case studies, diagrams, illustrations, and checklists, the author discusses the various issues involved with CTDs and what measure can be used to prevent their onset..
Hardcover, Index, 272 pages, 1997, ISBN: 0-86587-553-7B *$59*

Making Sense of OSHA Compliance
At last—a simple, easy-to-understand guide to how OSHA works and the fundamentals of complying with its regulations. Filled with real-world examples, this manual is a practical guide for you to use in applying OSHA policies and laws.
Hardcover, Index, 250 pages, March 1997, ISBN: 0-86587-535-9 *$59*

International References

ISO 14000: Understanding the Environmental Standards
ISO 14000: changes the way many companies approach environmental management. Smart companies will use the ISO-registered label as an internationally-recognized stamp of approval. ISO 14000 will help you gain entry into new international markets and attain an edge over your competitors.
Softcover, 226 pages, 1996, ISBN:0-86587-510-3 *$69*

ISO 14001: An Executive Report
This concise, in-depty report provides you with an understanding of the potential importance of the new ISO 14001: Environmental Management System Specification. An easy-to-use flow chart guides you through the Compliance Audit Process, and a useful parallel comparision of ISO 14001 with other environmental standards helps you select the best standard for your company.
Softcover, 124 pages, 1996, ISBN:0-86587-551-0 *$55*

Environmental Law Set

Environmental Regulatory Glossary, 6th Edition
This glossary defines and standardizes more than 4,000 terms, abbreviations, and acronyms, all compiled directly from the environmental statutes and regulations. You must understand what these words mean in the context of environmental laws in order to comply with the regulations.
Hardcover, 544 pages, 1993, ISBN: 0-86587-353-4 *$72*

Environmental Statutes, 1997 Edition
The complete text of each statute as currently amended is included, with a detailed Table of Contents for your quick reference.
Softcover, 1200 pages, March 1997, ISBN: 0-86587-562-9 *$69*

Environmental Law Handbook, 14th Edition
Comprehensive, straightforward, practical, and authoritative—these words are used to describe this best-selling book. Written by 14 highly-regarded environmental attorneys, the **Environmental Law Handbook** enjoys a reputation as one of the mos basic and useful texts in the environmental field.
Hardcover, Index, 550 pages, March 1997, ISBN: 0-86587-560-X *$79*

To obtain a catalog of our books, write or call:
4 Research Place Rockville, MD 20850 tel. (301) 921-2355 fax (301) 921-0373